INTERSECTIONS OF PARTICLE AND NUCLEAR PHYSICS

INTERSECTIONS OF PARTICLE AND NUCLEAR PHYSICS

7th Conference
CIPANP2000

Quebec City, Canada   22–28 May 2000

EDITORS
Zohreh Parsa
William J. Marciano
Brookhaven National Laboratory

Melville, New York, 2000
AIP CONFERENCE PROCEEDINGS ■ VOLUME 549

Editors:

Zohreh Parsa
William J. Marciano

Brookhaven National Laboratory
Physics Department
Building 510A
Upton, NY 11973-5000
USA

E-mail: parsa@bnl.gov
       marciano@bnl.gov

The articles on pp. 211–229, 249–254, 471–474, 547–548, 781–789, 831–840, and 976–978 were authored by U. S. Government employees and are not covered by the below mentioned copyright.

Authorization to photocopy items for internal or personal use, beyond the free copying permitted under the 1978 U.S. Copyright Law (see statement below), is granted by the American Institute of Physics for users registered with the Copyright Clearance Center (CCC) Transactional Reporting Service, provided that the base fee of $17.00 per copy is paid directly to CCC, 222 Rosewood Drive, Danvers, MA 01923. For those organizations that have been granted a photocopy license by CCC, a separate system of payment has been arranged. The fee code for users of the Transactional Reporting Service is: 1-56396-978-5/00/$17.00.

© 2000 American Institute of Physics

Individual readers of this volume and nonprofit libraries, acting for them, are permitted to make fair use of the material in it, such as copying an article for use in teaching or research. Permission is granted to quote from this volume in scientific work with the customary acknowledgment of the source. To reprint a figure, table, or other excerpt requires the consent of one of the original authors and notification to AIP. Republication or systematic or multiple reproduction of any material in this volume is permitted only under license from AIP. Address inquiries to Office of Rights and Permissions, Suite 1NO1, 2 Huntington Quadrangle, Melville, N.Y. 11747-4502; phone: 516-576-2268; fax: 516-576-2450; e-mail: rights@aip.org.

L.C. Catalog Card No. 00-110449
ISBN 1-56396-978-5
ISSN 0094-243X
Printed in the United States of America

CONTENTS

Preface xix
Organizing Committee and Sponsors xxi

Precision Electroweak Measurements 1
    H. Przysiezniak
CP-Violation and Physics Beyond the Standard Model 16
    R. N. Mohapatra
Progress in Heavy-Quark Physics from BaBar, Belle, and CLEO 29
    J. D. Richman
Structure Functions at High Q^2 49
    A. T. Doyle
Polarized Structure Functions and the Spin Structure of the Nucleon ... 62
    M. G. Vincter
The Proton and Neutron Form Factors 74
    G. G. Petratos
Exotic Hadrons 86
    J. Napolitano
The Phases of QCD in Heavy Ion Collisions and Compact Stars 95
    K. Rajagopal
Effective Field Theory in Nuclear Physics 120
    M. J. Savage
Parity Violation and Nucleon Strangeness 138
    P. A. Souder
Three Generations of Quarks and Leptons: Who Ordered That? 150
    D. Bryman
Neutrino Oscillation Experiments 164
    K. Nakamura
Detection of Dark Matter 177
    M. C. P. Issac
New Particle Searches at LEP 200++ 189
    R. A. McPherson
What's the Matter? Physics Opportunities and Future Facilities 201
    M. H. Shaevitz
Intersections 2000: What's New in Hadron Physics 211
    J. D. Bjorken

QCD SPECTROSCOPY AND DYNAMICS

**A Study of the $\eta\eta\pi^-$ System Produced in the Reaction
$\pi^- p \rightarrow p \pi^+ \pi^- \pi^- 4\gamma$ at 18 GeV/c** 233
    P. Eugenio and G. McNicoll for the Brookhaven E852 Collaboration
Partial Wave Analysis Results from JETSET 237
    R. T. Jones
Light Scalar Mesons 241
    D. Black, A. H. Fariborz, and J. Schechter

Scalar Meson Decay Constants and the Nature of the $a_0(980)$ 245
    K. Maltman

Pseudovector Mesons, Hybrids and Glueballs 249
    L. Burakovsky and P. R. Page

η Electroproduction in the $s_{11}(1535)$ Region with CLAS 255
    J. A. Mueller for the CLAS Collaboration

CLAS Electro-Omega Production 259
    V. Burkert, A. Coleman, H. Funsten, F. Klein, A. Larabee,
    and B. Mecking for the CLAS Collaboration

Neutral Hyperon and H-Dibaryon Results from KTeV 263
    N. Solomey

Fishing for Narrow Dibaryons in $pd \rightarrow pX$ Reaction 267
    L. V. Fil'kov, V. L. Kashevarov, E. S. Konobeevski,
    M. V. Mordovskoy, S. I. Potashev, V. A. Simonov,
    V. M. Skorkin, and S. V. Zuev

**Radial Excitations of Low-lying Baryons and the Structure
of the Z^+ Penta-Quark** 271
    H. Weigel

Mesons and Hybrids in a Relativistic Many Body Theory 275
    S. R. Cotanch and F. J. Llanes-Estrada

Description of Glueballs in Effective Meson Lagrangians 279
    M. K. Volkov and V. L. Yudichev

Scalar and QCD String Confinement 283
    T. J. Allen, M. G. Olsson, and S. Veseli

Strong Isospin Breaking in CP-even and CP-odd $K \rightarrow \pi\pi$ Decays 287
    C. E. Wolfe and K. Maltman

Meson Spectra-Power Law Potential in the Dirac Equation 291
    L. K. Sharma and J. O. Fiase

Measurement of the Pion Light-Cone Wave Function 295
    D. Ashery

**The Transverse Quark Distribution and Proton Electromagnetic
Form Factors in Skew Distribution Formalism** 302
    J. P. Ralston, P. Jain, and R. V. Buniy

**A New Measurement of the Energy Dependence of Nuclear
Transparency for Large Momentum Transfer $^{12}C(p,2p)$ Scattering** 306
    A. Leksanov, J. Alster, G. Asryan, Y. Averichev, D. Barton,
    V. Baturin, N. Bukhtojarova, A. Carroll, A. Schetkovsky,
    S. Heppelmann, T. Kawabata, A. Malki, Y. Makdisi,
    E. Minina, I. Navon, H. Nicholson, A. Ogawa, Y. Panebratsev,
    E. Piasetzky, S. Shimanskiy, A. Tang, J. W. Watson,
    H. Yoshida, and D. Zhalov

**Longitudinal Momentum Fraction X_L for Two High P_t Protons
in pp$\rightarrow$ppX Reaction** 310
    D. Zhalov, S. Heppelmann, J. Alster, G. Asryan, Y. Averichev,
    D. Barton, V. Baturin, N. Bukhtoyarova, A. Carroll, T. Kawabata,
    A. Leksanov, Y. Makdisi, A. Malki, E. Minina, I. Navon, H. Nicholson,
    A. Ogawa, Y. Panebratsev, E. Piasetzky, A. Schetkovsky, S. Shimanskiy,
    A. Tang, J. W. Watson, and H. Yoshida

**Dynamics of Glueball and $q\bar{q}$ Production in the Central Region
of pp Collisions** 314
    A. Sobol for the WA102 Collaboration
**Exotic Meson Production in the Reaction $\pi^-p \to \eta'\pi^-p$
at 18 GeV/c** 318
    N. M. Cason for the E852 Collaboration
**Search for the Tensor Glueball Candidate $\xi(2230)$
in an Antiproton-Proton Formation Experiment** 322
    W. Roethel for the CB Collaboration
4π Decays of Scalar Mesons 326
    C. A. Meyer for the Crystal Barrel Collaboration
Studies of Coulomb Gauge QCD 330
    A. P. Szczepaniak and E. S. Swanson

RELATIVISTIC HEAVY IONS

Pb-Pb at 158 GeV Results from a Theorist's View 337
    S. Jeon
Searching for QGP: The J/ψ Probe in the NA50/CERN Experiment 341
    M. C. Abreu, B. Alessandro, C. Alexa, R. Arnaldi, M. Atayan,
    C. Baglin, A. Baldit, M. Bedjidian, S. Beolè, V. Boldea,
    P. Bordalo, A. Bussière, L. Capelli, L. Casagrande, J. Castor,
    T. Chambon, B. Chaurand, I. Chevrot, B. Cheynis, E. Chiavassa,
    C. Cicalò, T. Claudino, M. P. Comets, N. Constans, S. Constantinescu,
    N. De Marco, A. De Falco, G. Dellacasa, A. Devaux, S. Dita, O. Drapier,
    L. Ducroux, B. Espagnon, J. Fargeix, P. Force, M. Gallio, Y. K. Gavrilov,
    C. Gerschel, P. Giubellino, M. B. Golubeva, M. Gonin, A. A. Grigorian,
    J. Y. Grossiord, F. F. Guber, A. Guichard, H. Gulkanyan, R. Hakobyan,
    R. Haroutunian, M. Idzik, D. Jouan, T. L. Karavitcheva, L. Kluberg,
    A. B. Kurepin, Y. Le Bornec, C. Lourenço, P. Macciotta, M. Mac Cormick,
    A. Marzari-Chiesa, M. Masera, A. Masoni, S. Mehrabyan, M. Monteno,
    A. Musso, P. Petiau, A. Piccotti, J. R. Pizzi, F. Prino, G. Puddu,
    C. Quintans, S. Ramos, L. Ramello, P. Rato Mendes, L. Riccati,
    A. Romana, I. Ropotar, P. Saturnini, E. Scomparin, S. Serci,
    R. Shahoyan, S. Silva, M. Sitta, C. Soave, P. Sonderegger,
    X. Tarrago, N. S. Topilskaya, G. L. Usai, E. Vercellin,
    L. Villatte, and N. Willis
$J/\psi,\psi'$ and Drell-Yan Nuclear Dependence in 800 GeV/c p$-$A Collisions 346
    M. A. Vasiliev, M. E. Beddo, C. N. Brown, T. A. Carey, T. H. Chang,
    W. E. Cooper, C. A. Gagliardi, G. T. Garvey, D. F. Geesaman,
    E. A. Hawker, X. C. He, L. D. Isenhower, D. M. Kaplan, S. B. Kaufman,
    D. D. Koetke, W. M. Lee, M. J. Leitch, P. L. McGaughey, J. M. Moss,
    B. A. Mueller, V. Papavassiliou, J. C. Peng, G. Petitt, P. E. Reimer,
    M. E. Sadler, W. E. Sondheim, P. W. Stankus, R. S. Towell,
    R. E. Tribble, J. C. Webb, J. L. Willis, and G. R. Young
An Overview of AGS: Physics Results 350
    J. G. Lajoie

Flow and HBT Measurements in the E895 Experiment 355
    M. A. Lisa for the E895 Collaboration

**Collective Evolution of Hot QCD Matter from the QGP
to Freeze-Out** 359
    A. Dumitru and S. A. Bass

**Strange and Multi-strange Baryon Production at SPS
as a Probe of QGP Formation** 363
    F. Antinori, H. Bakke, W. Beusch, I. J. Bloodworth,
    R. Caliandro, N. Carrer, D. Di Bari, S. Di Liberto, D. Elia,
    D. Evans, K. Fanebust, R. A. Fini, J. Ftáčnik, B. Ghidini,
    G. Grella, H. Helstrup, A. K. Holme, D. Huss, A. Jacholkowski,
    G. T. Jones, J. B. Kinson, K. Knudson, I. Králik, V. Lenti,
    R. Lietava, R. A. Loconsole, G. Løvhøiden, V. Manzari,
    M. A. Mazzoni, F. Meddi, A. Michalon, M. E. Michalon-Mentzer,
    M. Morando, P. I. Norman, B. Pastirčák, E. Quercigh, G. Romano,
    K. Šafařík, L. Šándor, G. Segato, P. Staroba, M. Thompson,
    T. F. Thorsteinsen, G. D. Torrieri, T. S. Tveter, J. Urbán,
    O. Villalobos Baillie, T. Virgili, M. F. Votruba, and P. Závada

Meson Mixing and Dilepton Production in Heavy Ion Collisions 369
    A. K. Dutt-Mazumder, C. Gale, and O. Teodorescu

**Measurements of Hadronic Observables with the Solenoidal
Tracker at RHIC** 374
    R. Bellwied for the STAR Collaboration

Physics via Lepton Channels at RHIC/PHENIX 378
    K. Shigaki for the PHENIX Collaboration

Results from pQCD for A+A Collisions at RHIC & LHC Energies 384
    K. Tuominen

From Crêpes to Pancakes in the MV Model 388
    G. Mahlon

QCD AND NUCLEAR STRUCTURE

**Measurement of the Neutron Electric Form Factor G_E^n
in $\vec{D}(\vec{e},e'n)p$ Quasi-Elastic Scattering at $Q^2=0.5$ $(GeV/c)^2$** 395
    M. Zeier for the E93-026 Jlab Collaboration

**Methods for the Nonperturbative Approximation of Form Factors
and Scattering Amplitudes** 399
    J. R. Hiller

**Rare Pion Double Radiative Capture Reactions
on Hydrogen and Deuterium** 403
    M. D. Hasinoff, D. S. Armstrong, J. Clark, T. P. Gorringe,
    M. Kovash, S. Tripathi, D. H. Wright, and P. A. Zołnierczuk

Ortho-Para Transition in Muonic Molecular Hydrogen 407
    J. H. D. Clark, D. S. Armstrong, T. P. Gorringe, M. D. Hasinoff,
    P. M. King, T. J. Stocki, S. Tripathi, D. H. Wright,
    and P. A. Zołnierczuk

Creation and Decay of η-Mesic Nuclei 411
    G. A. Sokol, T. A. Aibergenov, A. V. Koltsov, A. V. Kravtsov,
    Y. I. Krutov, A. I. L'vov, L. N. Pavlyuchenko, V. P. Parlyuchenko,
    and S. S. Sidorin

η and η' Mesons in the Improved Ladder Bethe-Salpeter Approach 417
    M. Takizawa, K. Naito, Y. Nemoto, and M. Oka

Subthreshold K^+-Production Studies with ANKE at COSY-Jülich 421
    S. Barsov, U. Bechstedt, G. Borchert, W. Borgs, M. Büscher,
    M. Debowski, W. Erven, R. Eßer, P. Fedorets, D. Gotta,
    M. Hartmann, H. Junghans, A. Kacharava, B. Kamys, F. Klehr,
    H. R. Koch, V. I. Komarov, V. P. Koptev, P. Kulessa, A. Kulikov,
    V. Kurbatov, G. Macharashvili, R. Maier, S. Merzliakov,
    S. Mikirtychiants, H. Müller, A. Mussgiller, M. Nioradze,
    H. Ohm, A. Petrus, D. Prasuhn, K. Pysz, F. Rathmann,
    B. Rimarzig, Z. Rudy, R. Schleichert, C. Schneider,
    H. Schneider O. W. B. Schult, H. Seyfarth, K. Sistemich,
    H. J. Stein, H. Ströher, and I. Zychor for the ANKE Collaboration

Antikaon Production and Medium Effects in Proton—
Nucleus Reactions at Subthreshold Beam Energies 425
    E. Y. Paryev

Measurement of Proton Polarization in the $d(\vec{y},\vec{p})n$ Reaction 430
    S. Strauch for the JLab Hall-A Collaboration

Single π^0 Electroproduction from CLAS Data at Jefferson Lab 434
    K. Joo for the CLAS Collaboration

Jlab Measurements of the Deuteron Electric and Magnetic Form Factors 438
    G. G. Petratos

Precise Electro-Pion Production Experiments in the Delta Region 442
    A. M. Bernstein

Backward Emitted High-Energy Neutrons in Hard Reactions
of p and π^+ on Carbon 447
    A. Malki, E. Piasetzky, J. Alster, G. Asryan, Y. Averichev,
    D. Barton, V. Baturin, N. Bukhtoyarova, A. Carroll,
    S. Heppelmann, T. Kawabata, A. Leksanov, Y. Makdisi,
    E. Minina, I. Navon, H. Nicholson, A. Ogawa,
    Y. Panebratsev, A. Schetkovsky, S. Shimanskiy, A. Tang,
    J. W. Watson, H. Yoshida, and D. Zhalov

n-p Short-Range Correlations from (p, 2p+n) Measurements 451
    A. Tang, J. W. Watson, J. Alster, G. Arsyan, Y. Averichev,
    D. Barton, V. Baturin, N. Bukhtoyarova, A. Carroll,
    S. Heppelmann, T. Kawabata, A. Leksanov, Y. Makdisi,
    A. Malki, E. Minina, I. Navon, H. Nicholson, A. Ogawa,
    Y. Panebratsev, E. Piasetzky, A. Schetkovsky, S. Shimanskiy,
    H. Yoshida, and D. Zhalov

Progress in Perturbative Color Transparency 455
    J. P. Ralston, P. Jain, B. Kundu, and J. Samuelsson

Testing the Spin-Dependence of the In-Medium Nucleon-Nucleon
Interaction Using Polarization Observables in $(\vec{p},\vec{p}')$ Scattering
at Intermediate Energies 459
    F. Sammarruca and E. J. Stephenson

Flavor Asymmetry of the Nucleon Sea: Unambiguous Role of Goldstone Bosons 463
    A. W. Thomas, W. Melnitchouk, and F. M. Steffens

Hadron Structure Functions in the Bosonized Nambu-Jona-Lasinio Model 467
    H. Weigel and L. Gamberg

Pseudospin Symmetry in Nuclei, Spin Symmetry in Hadrons 471
    P. R. Page, T. Goldman, and J. N. Ginocchio

Extension of the Delta-Hole Approach to Higher Baryon-Hole Excitations in Nuclei 475
    V. V. Balashov, A. V. Bibikov, V. K. Dolinov, and M. M. Kaskulov

Structure and Production of Lambda Baryons 479
    J. T. Londergan, C. Boros, and A. W. Thomas

Relativistic Calculations for Incoherent η-Photoproduction 483
    I. R. Blokland and H. S. Sherif

Phase Transitions in Finite Density Baryonic Matter 487
    O. Schwindt and N. R. Walet

The Density Dependence of Charge Symmetry Breaking 491
    C. J. Horowitz and J. Piekarewicz

Elastic Form Factors and Charge Densities of Medium and Heavy Nuclei and the Proton Occupancies of Shell States 495
    I. S. Gulkarov and B. P. Nigam

LEPTON-HADRON AND HADRON-HADRON SCATTERING

Structure Functions at Very High Q^2 from HERA 501
    C. M. Cormack for the H1 and ZEUS Collaborations

New Measurements of Nucleon Structure Functions from the CCFR/NuTeV Collaboration 509
    A. Bodek, U. K. Yang, T. Adams, A. Alton, C. G. Arroyo,
    S. Avvakumov, L. de Barbaro, P. de Barbaro, A. O. Bazarko,
    R. H. Bernstein, T. Bolton, J. Brau, D. Buchholz, H. Budd,
    L. Bugel, J. Conrad, R. B. Drucker, B. T. Fleming, J. A. Formaggio,
    R. Frey, J. Goldman, M. Goncharov, D. A. Harris, R. A. Johnson,
    J. H. Kim, B. J. King, T. Kinnel, S. Koutsoliotas, M. J. Lamm,
    W. Marsh, D. Mason, K. S. McFarland, C. McNulty, S. R. Mishra,
    D. Naples, P. Nienaber, A. Romosan, W. K. Sakumoto, H. Schellman,
    F. J. Sciulli, W. G. Seligman, M. H. Shaevitz, W. H. Smith,
    P. Spentzouris, E. G. Stern, M. Vakili, A. Vaitaitis, V. Wu, J. Yu,
    G. P. Zeller, and E. D. Zimmerman

Deeply Virtual Compton Scattering and Skewed Parton Distribution 514
    Z. Chen

Study of Neutron Spin Structure Functions at Low Q^2 with Polarized ^{3}He 520
    S. Choi for the Jefferson Lab E94-010 Collaboration

Parton Distributions from 1+1 QCD 524
    V. John, G. S. Krishnaswami, and S. G. Rajeev

Photon and Pion Production at High p_T 527
    M. Zieliński

**Measurement of Drell-Yan Cross Sections and Flavor Asymmetry
in the Nucleon Sea** 532
    J. C. Webb, T. C. Awes, M. E. Beddo, M. L. Brooks,
    C. N. Brown, J. D. Bush, T. A. Carey, T. H. Chang,
    W. E. Cooper, C. A. Gagliardi, G. T. Garvey,
    D. F. Geesaman, E. A. Hawker, X. C. He, L. D. Isenhower,
    D. M. Kaplan, S. B. Kaufman, P. N. Kirk, D. D. Koetke,
    G. Kyle, D. M. Lee, W. M. Lee, M. J. Leitch, N. Makins,
    P. L. McGaughey, J. M. Moss, B. A. Mueller, P. M. Nord,
    V. Papavassiliou, B. K. Park, J. C. Peng, G. Petitt, P. E. Reimer,
    M. E. Sadler, J. Selden, W. E. Sondheim, P. W. Stankus,
    T. N. Thompson, R. S. Towell, R. E. Tribble, M. A. Vasiliev,
    Y. C. Wang, Z. F. Wang, J. L. Willis, D. K. Wise, and G. R. Young

Challenges in the Global QCD Analysis of Parton Structure of Nucleons 536
    W.-K. Tung

Diffractive Physics at the Tevatron 542
    K. Hatakeyama for the CDF and DØ Collaborations

A Ring Imaging Čerenkov for HERMES 547
    D. De Schepper for the HERMES Collaboration

Overview of Exotic Strange Quark Matter Search Experiments 549
    J. L. Nagle

**Calculations of the Quark Masses and Smallest Eigenvalues
of Wilson-Dirac Operator in Domain-Wall Formalism** 554
    V. Gadiyak, X. Ji, and C. Jung

Hyperon Electroproduction and Strangeness Physics 558
    S. L. Mintz

**Precision Measurement of the Neutron Magnetic Form Factor
from $^3\vec{He}(\vec{e},e')$** 562
    D. Dutta for the Jefferson Lab Hall A, E95-001 Collaboration

HEAVY QUARK AND HEAVY LEPTON PHYSICS

D-Meson Dalitz Fit from Focus 569
    S. Malvezzi for the Focus Collaboration

Extracting m_s from Flavor Breaking in Hadronic τ Decays 575
    K. Maltman and J. Kambor

Electroweak Couplings of the τ from LEP 579
    T. Paul

Recent Charmed Baryon Results from FOCUS 583
    E. W. Vaandering for the FOCUS Collaboration

Recent CLEO Results on Charmed Baryon Spectroscopy 588
    J. Yelton for the CLEO Collaboration

Review of Top Quark Physics 593
    U. Heintz

Bound States in NRQCD and NRQED and the Renormalization Group 598
    I. W. Stewart
Extracting $|V_{ub}|$ from Cut Spectra 604
    I. Z. Rothstein
Search for New Physics in Rare B Decays 609
    H. Landsman and Y. Rozen
$b \to s l^+ l^-$ Decays in and Beyond the Standard Model 614
    G. Hiller
Radiative B Decays and $B \to \tau \nu$ at CLEO 619
    H. Schwarthoff
Neutral Meson Decays to Four Leptons in KTeV 624
    E. Halkiadakis
B Meson Decays to Charmonia at CLEO 628
    A. Ershov for the CLEO Collaboration
Lifetimes and Inclusive Decay Rates of c and b Hadrons 632
    M. B. Voloshin

SPIN PHYSICS

Measurement of Polarized Quark Distributions of the Nucleon at HERMES 639
    M. Beckmann for the HERMES Collaboration
Large Flavor Asymmetry of Polarized Antiquarks: Prediction and Possible Tests in Semi-inclusive DIS and Polarized DY 643
    P. Schweitzer, B. Dressler, K. Goeke, M. V. Polyakov, and C. Weiss
A Precise Measurement of the g_2 Structure Function of the Proton and Deuteron 647
    D. E. McNulty for the E155x Collaboration
The Q^2-Dependence of the Generalized GDH Integral for the Proton 651
    B. Seitz for the HERMES Collaboration
First Double Polarization Measurements towards the Test of the Gerasimov-Drell-Hearn Sum Rule 655
    W. Meyer for the GDH- and A2-Collaborations
Transversity Quark Distribution Function in the Large N_c-Limit 659
    P. Schweitzer, D. Urbano, K. Goeke, P. V. Pobylitsa, M. V. Polyakov, and C. Weiss
Spin-Flipping a Stored Polarized Proton Beam with an rf Dipole 662
    B. B. Blinov, Y. S. Derbenev, T. Kageya, D. Y. Kantsyrev, A. D. Krisch, V. S. Morozov, D. W. Sivers, V. K. Wong, V. A. Anferov, P. Schwandt, and B. von Przewoski
The HERMES Internal Polarized Deuterium Gas Target 666
    M. Henoch for the HERMES Collaboration

Analyzing Power in CNI-Region at AGS (Experiment E950) 670
    I. G. Alekseev, M. Bai, B. Bassalleck, G. Bunce, A. Deshpande,
    J. Doskow, S. Eilerts, D. E. Fields, Y. Goto, K. Imai,
    M. Ishihara, V. P. Kanavets, K. Kurita, H. Huang, V. Hughes,
    K. Kwaitowski, B. Lewis, B. Lozowski, Y. Makdisi,
    H. O. Meyer, B. V. Morozov, M. Nakamura, B. V. Przewoski,
    T. Rinckel, T. Roser, A. Rusek, N. Saito, B. Smith,
    D. N. Svirida, M. Syphers, A. Taketani, T. L. Thomas,
    J. Tojo, D. Underwood, D. Wolfe, K. Yamamoto, and L. Zhu

Polarized Atomic Hydrogen Beam Studies in the Michigan Ultra-cold Jet 674
    R. S. Raymond, B. B. Blinov, N. S. Borisov, J. Cheng,
    A. M. Davidenko, V. V. Fimushkin, S. E. Gladycheva,
    V. N. Grishin, T. Kageya, D. Y. Kantsyrev, D. Kleppner,
    A. D. Krisch, V. G. Luppov, V. S. Morozov,
    J. R. Murray, J. J. Neumann, and B. Yankama

Spin Asymmetries in Polarized Hadron Collisions and the Polarized Gluon Density 676
    W. Vogelsang

The COMPASS Experiment at CERN 681
    A. Bravar for the COMPASS Collaboration

Sensitivities of the Gluon Polarization Measurement at PHENIX 685
    Y. Goto for the PHENIX Collaboration

Spin Physics Program at STAR 689
    W. W. Jacobs for the STAR Collaboration

Polarized Fragmentation Functions at RHIC 693
    D. de Florian

Measurement of Spin Observables in Exclusive $\bar{p}p \to \bar{\Lambda}\Lambda$ Production 697
    K. Paschke for the PS185 Collaboration

Transversely Polarized Λ Production 701
    D. Boer

Exploring the Spin of the Gauge Sector with Non-Perturbative Coordinates 705
    J. P. Ralston and R. V. Buniy

Measurement of g_1^p in the low-x and low-Q^2 Region at HERMES 709
    R. Kaiser for the HERMES Collaboration

Single-Spin/Azimuthal Asymmetries in Semi-Inclusive and Exclusive DIS 712
    C. A. Miller

Physics Prospects for a Polarized Electron-Ion Collider 718
    J. T. Londergan

NUCLEAR AND PARTICLE ASTROPHYSICS

Status of the Borexino Solar Neutrino Experiment 725
    J. C. Maneira for the Borexino Collaboration

Charge Conjugation Violation in Supernovae and the Neutron Shortage for R-Process Nucleosynthesis 729
    C. J. Horowitz and G. Li

The GZK Bound and Strong Neutrino-Nucleon Interactions above 10^{19} eV: a Progress Report 733
    J. P. Ralston, P. Jain, D. W. McKay, and S. Panda

Recent Progress in Understanding Nucleosynthesis via Rapid Neutron Capture 737
    Y.-Z. Qian

The rp-Process at X-Ray Burst Conditions 744
    M. Wiescher and H. Schatz

NEUTRINOS

Recent Results from CHORUS and NOMAD Experiments 753
    S. Ricciardi

Recent Results from the KARMEN2 758
    C. Oehler for the KARMEN Collaboration

Recent Results Addressing the KARMEN Timing Anomaly 762
    E. D. Zimmerman

Atmospheric Neutrino Results from Super-Kamiokande Experiment 767
    Y. Obayashi for the Super-Kamiokande Collaboration

The First Results of K2K Long-Baseline Neutrino Oscillation Experiment 772
    T. Ishida for the K2K Collaboration

Upcoming Neutrino Oscillation Experiments at Fermilab 777
    C. James

Neutrino Factories—Physics Potentials 781
    Z. Parsa

Matter Effects on Neutrino Oscillations 790
    I. Mocioiu and R. Shrock

Results from the Palo Verde Neutrino Experiment 795
    J. Wolf for the Palo Verde Collaboration

Active-Sterile Neutrino Transformation and r-Process Nucleosynthesis 800
    G. C. McLaughlin

The Sudbury Neutrino Observatory 805
    F. Duncan for the SNO Collaboration

Electroweak Physics at NuTeV 809
    G. P. Zeller for the NuTeV Collaboration

The MUNU Experiment 814
    A. Tadsen for the MUNU Collaboration

Status of the NEMO 3 Experiment for the Study of Neutrinoless Double-β Decay 819
    C. Augier for the NEMO Collaboration

Observation of High Energy Atmospheric Neutrinos with AMANDA 823
A. Karle, E. Andres, P. Askebjer, X. Bai, G. Barouch,
S. W. Barwick, R. C. Bay, K.-H. Becker, L. Bergström,
D. Bertrand, D. Bierenbaum, A. Biron, J. Booth, O. Botner,
A. Bouchta, M. M. Boyce, S. Carius, A. Chen, D. Chirkin,
J. Conrad, J. Cooley, C. G. S. Costa, D. F. Cowen, J. Dailing,
E. Dalberg, T. DeYoung, P. Desiati, J.-P. Dewulf, P. Doksus,
J. Edsjö, P. Ekström, B. Erlandsson, T. Feser, M. Gaug,
A. Goldschmidt, A. Goobar, L. Gray, H. Haase, A. Hallgren,
F. Halzen, K. Hanson, R. Hardtke, Y. D. He, M. Hellwig,
H. Heukenkamp, G. C. Hill, P. O. Hulth, S. Hundertmark,
J. Jacobsen, V. Kandhadai, J. Kim, B. Koci, L. Köpke,
M. Kowalski, H. Leich, M. Leuthold, P. Lindahl, I. Liubarsky,
P. Loaiza, D. M. Lowder, J. Ludvig, J. Madsen, P. Marciniewski,
H. S. Matis, A. Mihalyi, T. Mikolajski, T. C. Miller, Y. Minaeva,
P. Miocinovic, P. C. Mock, R. Morse, T. Neunhöffer, F. M. Newcomer,
P. Niessen, D. R. Nygren, H. Ogelman, C. Pérez de los Heros, R. Porrata,
P. B. Price, K. Rawlins, C. Reed, W. Rhode, A. Richards, S. Richter,
J. Rodriguez Martino, P. Romenesko, D. Ross, H. Rubinstein,
H.-G. Sander, T. Scheider, T. Schmidt, D. Schneider, E. Schneider,
R. Schwarz, A. Silvestri, M. Solarz, G. Spiczak, C. Spiering,
N. Starinsky, D. Steele, P. Steffen, R. G. Stokstad, O. Streicher,
Q. Sun, I. Taboada, L. Thollander, T. Thon, S. Tilav, N. Usechak,
M. Vander Donckt, C. Walck, C. Weinheimer, C. H. Wiebusch,
R. Wischnewski, K. Woschnagg, W. Wu, G. Yodh, and S. Young

ACCELERATORS, FACILITIES AND DETECTORS

Muon Sources 831
Z. Parsa
HERA-B Status and First Results 841
L. Garrido for the HERA-B Collaboration
Performance of the *BABAR* Detector 844
U. Langenegger for the *BABAR* Collaboration
Status of the Belle Experiment 848
E. Prebys for the Belle Collaboration
KLOE at DAϕNE 852
M. Adinolfi, A. Aloisio, F. Ambrosino, A. Andryakov, A. Antonelli,
M. Antonelli, F. Anulli, C. Bacci, A. Bankamp, G. Barbiellini,
G. Bencivenni, S. Bertolucci, C. Bini, C. Bloise, V. Bocci, F. Bossi,
P. Branchini, S. A. Bulychjov, G. Cabibbo, A. Calcaterra, R. Caloi,
P. Campana, G. Capon, G. Carboni, A. Cardini, M. Casarsa, G. Cataldi,
F. Ceradini, F. Cervelli, F. Cevenini, G. Chiefari, P. Ciambrone, S. Conetti,
S. Conticelli, E. De Lucia, G. De Robertis, R. De Sangro, P. De Simone,
G. De Zorzi, S. Dell'Agnello, A. Denig, A. Di Domenico, S. Di Falco,
A. Doria, E. Drago, V. Elia, O. Erriquez, A. Farilla, G. Felici, A. Ferrari,
M. L. Ferrer, G. Finocchiaro, C. Forti, A. Franceschi, P. Franzini, M. L. Gao,
C. Gatti, P. Gauzzi, S. Giovannella, V. Golovatyuk, E. Gorini, F. Grancagnolo,
W. Grandegger, E. Graziani, P. Guarnaccia, U. v. Hagel, H. G. Han, S. W. Han,

X. Huang, M. Incagli, L. Ingrosso, Y. Y. Jiang, W. Kim, W. Kluge,
V. Kulikov, F. Lacava, G. Lanfranchi, J. Lee-Franzini, T. Lomtadze,
C. Luisi, C. S. Mao, M. Martemianov, M. Matsyuk, W. Mei, L. Merola,
R. Messi, S. Miscetti, A. Moalem, S. Moccia, M. Moulson, S. Mueller,
F. Murtas, M. Napolitano, A. Nedosekin, M. Panareo, L. Pacciani,
P. Pagès, M. Palutan, L. Paoluzi, E. Pasqualucci, L. Passalacqua,
M. Passaseo, A. Passeri, V. Patera, E. Petrolo, G. Petrucci,
D. Picca, G. Pirozzi, C. Pistillo, M. Pollack, L. Pontecorvo,
M. Primavera, F. Ruggieri, P. Santangelo, E. Santovetti, G. Saracino,
R. D. Schamberger, C. Schwick, B. Sciascia, A. Sciubba, F. Scuri,
I. Sfiligoi, J. Shan, T. Spadaro, S. Spagnolo, E. Spiriti, C. Stanescu,
G. L. Tong, L. Tortora, E. Valente, P. Valente, B. Valeriani, G. Veneziano,
Y. Wu, Y. G. Xie, P. P. Zhao, and Y. Zhou

Future Kaon Programs at BNL, FNAL 858
S. H. Kettell

A Polarized Electron-Nucleon Scattering Experiment at TESLA 864
R. Kaiser for the TESLA-N Study Group

Development of the ATHENA Vertex Detector 867
P. Riedler

The ATLAS Liquid Argon Electromagnetic Calorimeter 872
P. Pralavorio for the ATLAS Liquid Argon Group

**Status and Physics Prospects of the Joint Project of KEK
and JAERI in Japan** 875
T. Nagae

The Hall D Detector at Jefferson Lab 879
C. A. Meyer for the Hall D Collaboration

Physics Prospects with an Intense Neutrino Experiment 882
N. Solomey

An Ultracold Neutron Facility at PSI 888
M. Daum for the UCN Collaboration

Recent Highlights at the MIT-Bates Linear Accelerator Center 890
K. D. Jacobs

Status of the BLAST Project 893
R. Alarcon and the BLAST Collaboration

TESTS OF FUNDAMENTAL SYMMETRIES

Recent Nucleon Decay Results from Super-Kamiokande 897
M. Earl for the Super-Kamiokande Collaboration

On the Particle Oscillations in a Medium 901
V. I. Nazaruk

Tests of CPT with CPLEAR 905
M. Mikuž for the CPLEAR Collaboration

**SLAC E158: An Experiment to Measure Parity Violation
in Moller Scattering** 910
M. Woods for the E158 Collaboration

Parity Violating Electron Scattering on the Proton and Deuteron at Backward Angles 913
    T. M. Ito for the SAMPLE Collaboration

Recent Results on the Muon Anomalous Magnetic Moment from BNL E821 917
    C. J. G. Onderwater, H. N. Brown, G. Bunce, R. M. Carey,
    P. Cushman, G. T. Danby, P. T. Debevec, H. Deng, W. Deninger,
    S. K. Dhawan, V. P. Druzhinin, L. Duong, W. Earle, E. Efstathiadis,
    F. J. M. Farley, G. V. Fedotovich, S. Giron, F. E. Gray,
    M. Grosse-Perdekamp, A. Grossmann, U. Haeberlen, E. S. Hazen,
    D. W. Hertzog, V. W. Hughes, K. Jungmann, D. Kawall,
    B. I. Khazin, J. Kindem, F. Krienen, I. Kronkvist, R. Larsen,
    Y. Y. Lee, I. Logashenko, R. McNabb, W. Meng, J. Mi, J. P. Miller,
    W. M. Morse, Y. Orlov, C. Özben, C. Pai, J. Paley, C. C. Polly,
    J. Pretz, R. Prigl, G. zu Putlitz, S. I. Redin, O. Rind, B. L. Roberts,
    N. M. Ryskulov, R. Sanders, S. Sedykh, Y. Semertzidis, Y. M. Shatunov,
    E. Solodov, A. Steinmetz, L. R. Sulak, C. Timmermans, A. Trofimov,
    D. Urner, D. Warburton, D. Winn, A. Yamamoto, and D. Zimmerman

A Precision Measurement of the Michel Parameter ξ'' in Polarized Muon Decay 920
    R. Prieels, P. Van Hove, J. Deutsch, J. Govaerts, P. Knowles,
    R. Medve, A. Ninane, J. Egger, F. Foroughi, X. Morelle, L. Simons,
    N. Danneberg, W. Fetscher, M. Hadri, C. Hilbes, K. Kirch, J. Lang,
    O. Naviliat, and J. Sromicki

New Results on Strange Form Factors of the Proton 923
    R. Holmes for the HAPPEX Collaboration

Parity Violating Measurements of Neutron Densities 927
    C. J. Horowitz

Search for Right-Handed Currents in the β^+-decay of ^{118}Sb 931
    B. Vereecke, D. Beck, M. Beck, B. Delauré, T. Phalet,
    P. Schuurmans, N. Severijns, S. Versyck, J. Deutsch, and R. Prieels

A Novel Approach for Measuring the Beta-Neutrino Angular Correlation in Nuclear Beta Decay 934
    M. Beck, F. Ames, D. Beck, B. Delauré, J. Deutsch, G. Bollen,
    O. Forstner, T. Phalet, W. Quint, P. Schmidt, P. Schuurmans,
    N. Severijns, B. Vereecke, S. Versyck, and the Eurotraps Collaboration

Muon-Electron Conversion in Nuclei 938
    A. Czarnecki, W. J. Marciano, and K. Melnikov

Muon Conversion Experiments—Current and Future 942
    A. I. Mincer for the MECO Collaboration

Searches for Lepton Flavor Violation in τ and B^0 Decays 946
    I. Narsky

Measurement of Direct CP Violation in K Meson Decays at KTeV 949
    A. Glazov

New Measurement of Direct CP Violation Parameter $Re(\epsilon'/\epsilon)$ by the NA48 Experiment at CERN 953
    I. Wingerter-Seez for the NA48 Collaboration

First Direct Observation of Time-Reversal Violation 957
> A. Angelopoulos, A. Apostolakis, E. Aslanides, G. Backenstoss,
> P. Bargassa, O. Behnke, A. Benelli, V. Bertin, F. Blanc, P. Bloch,
> P. Carlson, M. Carroll, E. Cawley, M. B. Chertok, M. Danielsson,
> M. Dejardin, J. Derre, A. Ealet, C. Eleftheriadis, W. Fetscher,
> M. Fidecaro, A. Filipčič, D. Francis, J. Fry, E. Gabathuler, R. Gamet,
> H.-J. Gerber, A. Go, A. Haselden, P. J. Hayman, F. Henry-Couannier,
> R. W. Hollander, K. Jon-And, P.-R. Kettle, P. Kokkas, R. Kreuger,
> R. Le Gac, F. Leimgruber, I. Mandić, N. Manthos, G. Marel,
> M. Mikuž, J. Miller, F. Montanet, A. Muller, T. Nakada, B. Pagels,
> I. Papadopoulos, P. Pavlopoulos, G. Polivka, R. Rickenbach,
> B. L. Roberts, T. Ruf, M. Schäfer, L. A. Schaller, T. Schietinger,
> A. Schopper, L. Tauscher, C. Thibault, F. Touchard, C. Touramanis,
> C. W. E. Van Eijk, S. Vlachos, P. Weber, O. Wigger, M. Wolter,
> D. Zavrtanik, and D. Zimmerman

Measurement of $B(K^+ \to \pi^+ \nu\bar{\nu})$ 961
> S. H. Kettell for the E787 Collaboration

Measurement of the Transverse Polarization of Positrons from the Decay of Polarized Muons 966
> N. Danneberg

A Test of Time Reversal Invariance in Stopped K^+ Decay 969
> M. D. Hasinoff for the KEK-E246 Collaboration

Search for T-Violation in Reactions with Slow Neutrons 972
> E. Korobkina and G. Danilyan

Determination of γ in $B^- \to D^0 K^-$ and Related Processes 976
> D. Atwood, I. Dunietz, and A. Soni

Electric Dipole Moments of Neutron and Heavy Atoms and Constraints on CP SUSY Phases 979
> M. Pospelov

Physics with Polarized Cold Neutrons at the Spallation Source SINQ 983
> K. Bodek, P. Böni, N. Danneberg, W. Fetscher, W. Haeberli,
> C. Hilbes, S. Kistryn, J. Lang, M. Lüthy, M. Markiewicz,
> A. Pusenkov, A. Schebetov, A. Serebrov, and J. Sromicki

Search for $D^0 - \overline{D^0}$ Mixing and CP Violation at CLEO 986
> A. Smith

Conference Schedule 991
List of Participants 993
Photos 1001
Author Index 1009

Preface

The seventh Conference on the Intersection of Particles and Nuclear Physics (CIPANP2000) was held from May 22 to May 28, 2000, in Quebec City, Canada. The purpose of this meeting, as with the six previous conferences in this series, was to bring together particle and nuclear physicists in a pleasant setting where they could hear up-to-date scientific reports and discuss areas of research which overlap both their disciplines.

The need for such an interdisciplinary conference was recognized by Alan D. Krisch and Malcolm H. MacFarlane, founding fathers of the CIPANP series. Its relevance has steadily grown as the areas of overlap between particle and nuclear physics have increased. In addition, the success of the standard model has provided a common underpinning for both disciplines as well as similar fundamental goals. Indeed, Quantum Chromodynamics (QCD) has proven to be "the" theory of strong interactions. As such, it forms the basis for nuclear physics as well as high energy hadronic interactions. QCD is a perfect theory. It is parameter free. Similarly, all known electroweak phenomena can be described by the $SU(2)_L \times U(1)_Y$ sector of the standard model. It has been tested at the incredible $\pm 0.1\%$ in reactions ranging from nuclear beta decay through Z-pole studies. Nevertheless, important outstanding questions remain. Why are there three generations of fermions? What is the true origin of electroweak symmetry breaking, mass generation and CP violation? Why is parity violated? These and other important issues provide valuable input for ongoing discussions and in making decisions regarding the future direction of the field.

The conference on Intersections between Particles and Nuclear Physics started with opening remarks by Stanley Kowalski and ended with a closing presentation by James Bjorken. We would like to thank them, and all the authors for providing their contributions, given in the following chapters (1-11), arranged starting with overviews (plenary) followed by (parallel) contributions arranged by topics respectively.

On behalf of the participants we would like to thank the Conference Chair, organizing committee, session coordinators / chairs, staff, sponsors, and others who contributed to the success of the conference. Also thanks to those who provided photos, some of which were used in these proceedings. Special thanks to the City of Quebec for their hospitality and for providing the photo used inside the cover of these proceedings.

Zohreh Parsa
Brookhaven National Laboratory

Organizing Committee

S. Kowalski, Chair (MIT), W. Marciano, Vice-Chair (BNL),
J. Appel (Fermilab), A. Astbury (TRIUMF), E. Berger (ANL),
K. Berkelman (Cornell), T. Bowles (LANL), J. Cameron (Indiana),
W. Haxton (Washington), K. Kilian (Julich), Y. Kuno (KEK),
V. Luth (Stanford), K. de Jager (JLab), C. Papanicolas (Athens),
F. Plasil (ORNL), H. Schellman (Northwestern), H.C. Walter (PSI),
Anne B. MacInnis, Secretary (MIT-Bates).

Sponsors

The following laboratories and agencies provided funds to support the CIPANP2000 Conference. Some also provided personnel to support many activities connected with organizing the meeting:

BNL,
FNAL,
IUCF,
JLAB,
LANL,
MIT-Bates,
NSF, and
TRIUMF.

Precision Electroweak Measurements

H. Przysiezniak

CERN, EP Division
1211 Geneva, Switzerland

Abstract. This talk describes some of the precision electroweak measurements from around the world, namely those related to the Z and W bosons, the top quark mass, $\sin^2 \theta_W$ at NuTeV, and three other fundamental measurements: $\alpha^{-1}(m_Z^2)$, $(g-2)_\mu$ at the E821 BNL experiment as well as the atomic parity violation (APV) measurement for the Cesium atom. These and other measurements are set in the context of the Standard Model (SM) and of the electroweak fit predictions. Future prospects for forthcoming experiments are briefly discussed.

INTRODUCTION

The motivations to perform precision electroweak measurements today are as strong as ever. Today's generation of experiments now have data of such precision that the electroweak measurements are probing the quantum corrections to the SM, otherwise known as radiative corrections. These are tested by a wide variety of measurements ranging from the muon magnetic moment to precision measurements at the Z pole and above in e^+e^- collisions, as well as precision measurements at hadron colliders. This talk is not an exhaustive survey but rather a biased summary of recent and new results, in particular: those which have an influence on the indirect determination of the Higgs mass, and those which are devised to be extremely stringent tests of the SM.

The measurements that are described here and which enter the first category are: the Z line shape and branching ratio measurements as well as the Z peak asymmetries from which $\sin^2 \theta_{\text{eff}}^\ell$ is extracted, the W mass, the top quark mass, $\sin^2 \theta_W = 1 - m_W^2/m_Z^2$ at NuTeV and $\alpha^{-1}(m_Z^2)$. The relation between the electroweak quantities are affected by radiative corrections. The most precisely known quantities being $\alpha(m_Z^2)$, G_F and m_Z^2, the W mass is related to them as follows

$$m_W^2 = \frac{\pi \alpha(m_Z^2)}{\sqrt{2} G_F (1 - m_W^2/m_Z^2)(1 - \Delta r^{\text{ew}})}$$

and the effective weak mixing angle, $\sin^2 \theta_{\text{eff}}^\ell = (1/4)(1 - g_V^\ell/g_A^\ell)$, by the relation

$$\sin^2\theta^\ell_{\text{eff}}\cos^2\theta^\ell_{\text{eff}} = \frac{\pi\alpha(m_Z^2)}{\sqrt{2}G_F m_Z^2(1+\epsilon_1)(1-\epsilon_3/\cos^2\theta_W)}, \tag{1}$$

where ϵ_1, ϵ_3 and $\Delta r^{\text{ew}} = f(\epsilon_1,\epsilon_2,\epsilon_3)$ are the radiative corrections. They are functions of m_{top}^2 and of $\log(m_{\text{Higgs}}/m_Z)$. The W mass and effective weak mixing angle measurements help to constrain the yet unobserved Higgs mass. Still, it can be deduced from the above expressions that reducing the error on $\sin^2\theta^\ell_{\text{eff}}$ will constrain m_{Higgs} even more particularly if the error on $\alpha(m_Z^2)$ is reduced simultaneously. The same relationship exists between the W mass and the top quark mass: reducing the experimental errors simultaneously will help to constrain the Higgs mass better.

The measurements which enter the second category are: the Z measurements from which Universality tests are performed, fermion pair production and asymmetries from the LEPII e^+e^- collider above the Z pole, W production and decays, gauge boson self-interactions and atomic parity violation. These measurements stringently test family Universality, Universality of weak neutral and charged current couplings, symmetry breaking and radiative corrections.

Z BEST OF BOTH WORLDS

The measurements described in the following sections were made at e^+e^- colliders: at SLC using the SLD detector, and at LEP using the ALEPH, DELPHI, L3 and OPAL detectors, at the Z pole and above in the case of LEPII. Approximately 4×10^6 Zs were accumulated per LEP experiment, while 555k Zs were taken at SLD using a polarized electron beam. The measurements help to constrain the Higgs mass and serve as stringent tests of the SM. The Z electroweak observables (at the Z peak, after QED corrections) are

- Line Shape: m_Z, Γ_Z, $\sigma_h^0 = 12\pi\Gamma_{ee}\Gamma_{\text{had}}/(m_Z^2\Gamma_Z^2)$
- Branching Ratios: $R_\ell = \Gamma_{\text{had}}/\Gamma_\ell$, $R_b = \Gamma_{b\bar{b}}/\Gamma_{\text{had}}$ and $R_c = \Gamma_{c\bar{c}}/\Gamma_{\text{had}}$
- Unpolarized FB Asymmetries for $f=\ell,b,c$: $A_{FB}^f = \frac{\sigma_F^f - \sigma_B^f}{\sigma_F^f + \sigma_B^f} = 0.75 A_e A_f$
- Polarization of τ leptons:

$$\mathcal{P}_\tau(\cos\theta) = \frac{\sigma_R - \sigma_L}{\sigma_R + \sigma_L} = -\frac{A_\tau(1+\cos^2\theta) + 2A_e\cos\theta}{1+\cos^2\theta + 2A_\tau A_e\cos\theta}$$

- Left-Right Asymmetry:

$$A_{LR}^m = \frac{\sigma^f(-|\mathcal{P}_e|) - \sigma^f(+|\mathcal{P}_e|)}{\sigma^f(-|\mathcal{P}_e|) + \sigma^f(+|\mathcal{P}_e|)} = \mathcal{P}_e A_{LR}^0 = \mathcal{P}_e A_e \quad f\neq e$$

- Left-Right FB Asymmetries for $f=\ell,b,c,s$:

$$A_{FB}^{\text{pol}} = \frac{\sigma_F^f(-|\mathcal{P}_e|) - \sigma_B^f(-|\mathcal{P}_e|) - \sigma_F^f(+|\mathcal{P}_e|) + \sigma_B^f(+|\mathcal{P}_e|)}{\sigma_F^f(-|\mathcal{P}_e|) + \sigma_B^f(-|\mathcal{P}_e|) + \sigma_F^f(+|\mathcal{P}_e|) + \sigma_B^f(+|\mathcal{P}_e|)} = 0.75\mathcal{P}_e A_f$$

Z Resonance Parameters at LEP

The Z resonance parameters (m_Z, Γ_Z, σ_h^0 and R_ℓ) are measured at LEP using $q\bar{q}$ and $\ell^+\ell^-$ event samples collected during scans of the Z peak. Since the lepton asymmetries A_{FB}^ℓ are sensitive functions of $\sqrt{s}$, they are also extracted from a simultaneous fit to the $q\bar{q}$ and $\ell^+\ell^-$ lineshape data. The SM values are used for the $Z\gamma$ and γ cross sections and the radiatively corrected lineshape functions are fit to data.

Combining the results from the four LEP experiments and for all years, assuming lepton Universality [1]: $m_Z = 91.1871 \pm .0021$ (.0017syst) (SM: 91.18692) GeV, $\Gamma_Z = 2.4944 \pm .0024$ (.0013syst) (SM: 2.49589) GeV, $\sigma_h^0 = 41.544 \pm .037$ (.035syst) (SM: 41.4804) nb, $R_\ell = 20.768 \pm .024$ (.017syst) (SM: 20.7394), $A_{FB}^{0,\ell} = .01701 \pm .00095$ (.00060syst) (SM: .016342).

The dominant sources of systematic uncertainties are the normalization of the cross sections, the knowledge of the center-of-mass energy E_{cm} and the QED radiative corrections to the line shape function.

Left-Right Asymmetry at SLD

This is a powerful, almost systematic free measurement made using a polarized electron beam ($\mathcal{P}_e$=22% in 1992; 63% in 1993; 73 to 78% from 1994 to 1997). The helicity of the polarized electrons is changed pulse to pulse. A feedback system keeps the left and right-handed electron currents equal to 10^{-4}. The left and right-handed luminosities are equal to very good approximation. The asymmetry is defined as $A_{LR}^0 = (1/\mathcal{P}_e)[N_Z(L) - N_Z(R)]/[N_Z(L) + N_Z(R)] = A_e$ where $N_Z(L,R)$ are the number of hadronically decaying Zs counted with left and right-handed electron beams. A_e is a function of $\sin^2\theta^\ell$: $A_e = [2(1 - 4\sin^2\theta_{eff}^\ell)]/[1 + (1 - 4\sin^2\theta_{eff}^\ell)^2]$, such that the A_{LR} measurement is translated into an effective weak mixing angle measurement [1]

$$A_{LR} = .15108 \pm .00218 \rightarrow \sin^2\theta_{eff}^\ell = .23101 \pm .00028 \text{ (.00018syst) (SM : .23145)}.$$

The dominant sources of systematic uncertainties are the electron polarization and the center-of-mass energy. As a cross check, the polarization of the positron beam was measured, and it was found to be consistent with zero: $\mathcal{P}_{e^+} = -.02 \pm .07\%$.

Z Summed Up

The lineshape parameters are used to extract the partial decay widths [1]: not assuming lepton Universality $\Gamma_{ee} = 83.90 \pm .12$ MeV, $\Gamma_{\mu\mu} = 83.96 \pm .18$ MeV, $\Gamma_{\tau\tau} = 84.05 \pm .22$ MeV; assuming lepton Universality $\Gamma_{\ell\ell} = 83.96 \pm .09$ MeV, $\Gamma_{had} = 1743.9 \pm 2.0$ MeV, $\Gamma_{invisible} = 498.8 \pm 1.5$ MeV. Taking the measured value for $\Gamma_{invisible}/\Gamma_{\ell\ell} = 5.941 \pm .016$ and dividing it by the SM expectation for $\Gamma_{\nu\nu}/\Gamma_{\ell\ell} =$

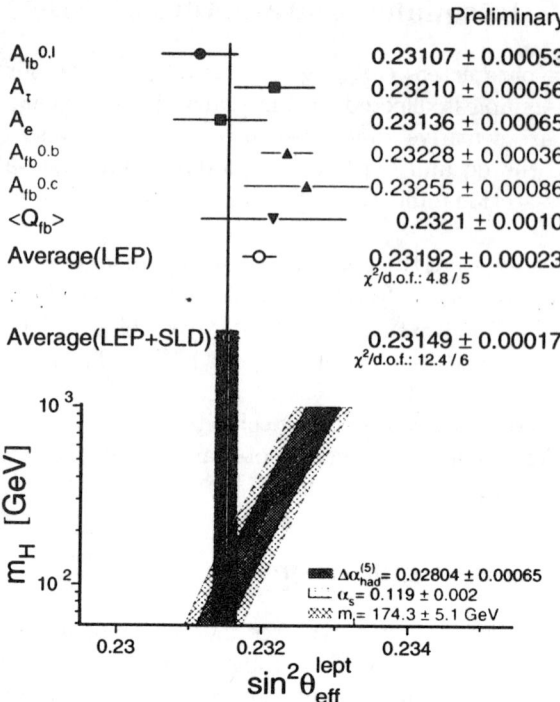

FIGURE 1. Summary of the LEP and SLD asymmetry measurements which enter in the determination of the world average effective weak mixing angle.

$1.9912 \pm .0012$, the number of neutrino families is extracted: $N_\nu = 2.9835 \pm .0083$. Converted into a 95% C.L. upper limit on an additional invisible width assuming $N_\nu = 3$ gives $\Delta\Gamma_{\text{invisible}} < 2.0$ MeV.

From the leptonic widths $\Gamma_{\ell\ell} = (1 + 3\alpha/4\pi)(G_F m_Z^3/24\pi\sqrt{2})[1 + (g_V^\ell/g_A^\ell)^2](1+\epsilon_1)$ and asymmetries $A_\ell = 2g_V^\ell g_A^\ell/(g_V^{\ell\,2} + g_A^{\ell\,2})$, the vector and axial vector coupling ratios are determined [1]: $g_A^\mu/g_A^e = 1.0001 \pm .0014$, $g_V^\mu/g_V^e = 0.981 \pm .082$, $g_A^\tau/g_A^e = 1.0019 \pm .0015$ and $g_V^\tau/g_V^e = 0.964 \pm .032$, consistent with Universality of the leptonic weak neutral couplings.

The effective weak mixing angle $\sin^2\theta_{\text{eff}}^\ell$ is extracted from the SLD and LEP Z-pole leptonic asymmetries. The asymmetries entering the world average (LEP+SLD) effective weak mixing angle determination are presented in Figure 1 [1]. The LEP quark asymmetry measurement $A_{fb}^{0,b}$ and the SLD leptonic asymmetry A_ℓ differ by $\sim 3\sigma$. This discrepency remains unresolved: possibly a statistical fluctuation, or an unkown systematic effect, or new physics. Nonetheless. $\sin^2\theta_{\text{eff}}^\ell$ remains the strongest constraint on the Higgs mass today.

Fermion Pair Production and Asymmetries at LEPII

Fermion pairs are produced through the radiative return diagram, when an initial state photon is emitted and the effective center-of-mass energy $\sqrt{s'}$ is approximately equal to the Z mass, and through the non-radiative diagram, when $\sqrt{s'/s} > 0.85$. The cross section and asymmetries are determined for these non-radiative events. From these measurements, limits on a wide range of physics scenarios can be set.

For example, the cross sections and asymmetries for $\mu^+\mu^-$ and $\tau^+\tau^-$ final states can be used to set limits on contact interactions between leptons expected to occur in the presence of composite fermions. These interactions are parametrised by an effective Lagrangian which is added to the SM one: $\mathcal{L}_{\text{eff}} = [g^2\eta/(1+\delta)\Lambda^2]\sum_{i,j=L,R}\eta_{ij}[\bar{e}_i\gamma_\mu e_i][\bar{f}_j\gamma^\mu f_j]$, where $g^2/4\pi = 1$ by convention, $\eta = \pm$ defines a constructive or destructive interference with the SM, $\delta = 1(0)$ for $f = e(f \neq e)$, $\eta_{ij} = \pm 1, 0$ is the helicity coupling between initial and final state, $e_{L,R}$, $f_{L,R}$ are the left and right-handed spinors, and Λ is the scale of the contact interactions.

For all LEP results combined, the excluded values of Λ for models leading to large deviations in $\mu^+\mu^-$ and $\tau^+\tau^-$ final states are [2]: $\Lambda_{AA}^{(+,-)} < (17.6, 13.9)$, $\Lambda_{VV}^{(+,-)} < (20.4, 17.2)$, $\Lambda_{RR}^{(+,-)} < (12.3, 9.7)$ and $\Lambda_{LL}^{(+,-)} < (12.8, 10.2)$ TeV, where $\eta_{LL} = 1, 1, 0, 1$, $\eta_{RR} = 1, 1, 1, 0$, $\eta_{LR} = -1, 1, 0, 0$ and $\eta_{RL} = -1, 1, 0, 0$ for the AA, VV, RR and LL models respectively.

W BOSONS ON TOUR

W boson related measurements were performed by the UA1 and UA2 experiments at the SPS collider ($p\bar{p}$, $E_{cm} = 630$ GeV, $\mathcal{L}_{\text{int}} \sim 12\text{pb}^{-1}$/expt.), at the Tevatron by the CDF and D0 experiments ($p\bar{p}$, $E_{cm} = 1.8$ TeV, $\mathcal{L}_{\text{int}} \sim 120\text{pb}^{-1}$/expt.), and at LEPII ($e^+e^-$, $E_{cm} = 161 \to 209$ GeV, $\mathcal{L}_{\text{int}} > 500\text{pb}^{-1}$/expt.). The W electroweak observables described here are: the W production and decays, the gauge boson self interactions and the W mass and width. These observables help to constrain the Higgs mass or serve as stringent tests of the SM.

W Pair Production and Decays at LEPII

W pairs are produced via the t-channel neutrino exchange and via the s-channel $\gamma - Z$ interference. The cross section is measured by the four LEP experiments for all decay channels: the fully hadronic channel (4q; BR~46%), the semileptonic channel ($\ell\nu q\bar{q}$; BR~43%), and the fully leptonic channel ($\ell\nu\ell\nu$; BR~11%). The combined measurements are shown in Figure 2 as a function of the center-of-mass energy [3]. The curves correspond to three calculations of the cross section: the two lower curve calculations (top:YSFWW3; bottom:RacoonWW) contain non leading $\mathcal{O}(\alpha)$ terms, whereas the upper curve calculation (Gentle) does not. The χ^2/dof for data versus Gentle is 11.6/6, whereas that of data versus Racoon is 5.6/6.

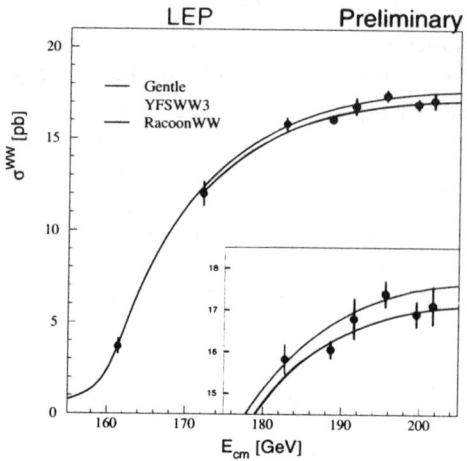

FIGURE 2. LEPII WW cross section as a function of the center-of-mass energy. The points are the data, the curves are the calculations from Gentle (top curve), YSFWW3 (middle) RacoonWW(bottom).

The branching ratio measurements from LEPII and the Tevatron BR(W → $e\nu$) measurement are consistent with Universality of the leptonic weak charged current [3,4]: BR(W → $e\nu$) = 10.63 ± 0.20%, BR(W → $e\nu$)(Tevatron) = 10.43 ± 0.25%, BR(W → $\mu\nu$) = 10.56 ± 0.19%, BR(W → $\tau\nu$) = 11.02 ± 0.26%. The result from LEPII assuming lepton Universality is given by: BR(W → $\ell\nu$) = 10.71 ± 0.10%, BR(W → $q\bar{q}$) = 67.85 ± 0.33%.

Using the measured hadronic branching ratio, the following relation

$$\frac{BR_{W \to qq}}{1 - BR_{W \to qq}} = (|V_{ud}|^2 + |V_{cd}|^2 + |V_{us}|^2 + |V_{cs}|^2 + |V_{ub}|^2 + |V_{cb}|^2)(1 + \frac{\alpha_s(m_W^2)}{\pi})$$

and the known values of the other CKM matrix elements, one finds $|V_{cs}|$ = .993 ± .016 [3]. This improves by a factor of ten the precision of the PDG value, which is also obtained without requiring Unitarity of the CKM matrix.

Gauge Boson Self Interactions

Any deviation from the SM prediction for gauge boson self-interactions is a true indication of non-SM physics and could indicate W compositeness. Triply charged (TGC), neutral or quartic gauge couplings have been investigated. Only the TGCs are discussed here. The general Lagrangian for TGCs contains 14 parameters. By requiring C, P and CP invariance and U(1)$_{em}$ invariance (q_W = e), 5 free parameters are left. Measurements made at LEP at the Z pole set bounds on the couplings. Finally, by requiring the additional $SU(2)_L \times U(1)_Y$ invariance, three couplings are left: κ_γ, g_1^Z and λ_γ where the SM predictions are 1, 1, 0. g_1^Z is the coupling strength of the W to the Z. κ_γ and λ_γ define the magnetic moment and the electric quadrupole moment of the W^+: $\mu_{W^+} = (e/2m_W)(1 + \kappa_\gamma + \lambda_\gamma)$ and $Q_{W^+} = -(e/m_W^2)(\kappa_\gamma - \lambda_\gamma)$.

At Tevatron

Limits on TGCs are set by looking at the $p_T^{\ell\nu}$ distribution of WW, WZ $\to (e,\mu) + \nu +$ jets events (λ, κ), the cross section of WZ $\to (e,\mu) + \nu +$ ee events (λ, g_1), the p_T^ℓ distribution of WW $\to$ dileptons events (λ, κ), and the E_T^γ distribution of W$\gamma \to (e,\mu) + \nu + \gamma$ events (λ, κ). Any excess of events is an indication for anomalous TGCs. Since the cross section with non-SM couplings increases with s, to avoid Unitarity violation, the anomalous couplings are expressed as form factors with a scale Λ e.g. $\lambda_V(s) = \lambda_V/(1 + s/\Lambda^2)^2$. Under the assumption that the WWγ couplings equal the WWZ couplings, for $\Lambda = 2.0$ TeV, the one dimensional 95% C.L. limits are $\Delta\kappa_\gamma = (-0.25, 0.39)$ and $\lambda_\gamma = (-0.18, 0.19)$ [5], where Δ indicates the deviation with respect to the SM prediction. Assuming the WWγ couplings are at the SM value, for $\Lambda = 2.0$ TeV, the one dimensional 95% C.L. limit is $\Delta g_1^Z = (-0.37, 0.57)$ [5].

At LEPII

Limits on TGCs are set through measurements of the cross section and W production and decay angles for WW events (κ, g_1, λ), and through measurements of the cross section, energy and θ of the lepton, jets or γ, for single W or single γ events (κ, λ). The largest sensitivity comes from the measurement of the W production and decay angles for WW events decaying semileptonically. The one dimensional 95% C.L. limits for the four LEP experiments combined are $\Delta\kappa_\gamma = (-0.09, 0.15)$, $\Delta g_1^Z = (-0.071, 0.024)$ and $\lambda_\gamma = (-0.066, 0.035)$ [6].

W Mass and Width

At Tevatron

Single Ws are produced in $q\bar{q}$ annihilations of the proton and anti-proton valence quarks. The W hadronic decays are lost in the QCD di-jet background and thus only leptonic decays are used for the measurement. The longitudinal information is lost at large η due to high background and lack of instrumentation, such that transverse quantities are used to determine the W mass. Two complementary variables are the transverse mass $m_T = \sqrt{2 p_T^\ell \not{E}_T (1 - \cos\phi_{\vec{\not{E}_T} - \vec{p_T^\ell}})}$ and the transverse momentum of the lepton p_T^ℓ. m_T and p_T^ℓ are respectively, to first order independent and linearly dependent, of p_T^W. They are less-more sensitive to the W production process, and more-less sensitive to detector resolutions. A Monte Carlo (MC) simulation of the W production, decay and detector response is used to generate the m_T and p_T^ℓ distributions as a function of m_W^{true}. Today's Tevatron W mass is $m_W = 80.450 \pm .063$ (.040stat; .049syst) GeV (CDF: $80.433 \pm .079$; D0: $80.474 \pm .093$) [4]. The W width is extracted from a fit to the high end of the m_T distribution by CDF $\Gamma_W = 2.06 \pm .13$ GeV [4].

The main systematics originate from the uncertainty on the energy response and resolution, and on the p_T and $\not{E}_T$ MC modelling. Since Z events are needed to set the energy-momentum scales and the W p_T and $\not{E}_T$ models, the systematic uncertainties are dominated by the limited Z statistics.

At LEPII

The W mass can be extracted via three different methods at LEPII. The first is a measurement of the W pair production cross section at the threshold center-of-mass energy (161 GeV). At this energy, the cross section is most sensitive to the W mass.

In the second method, the lepton energy spectrum of the fully leptonic or semileptonic events is measured: its endpoints are a function of the W mass $E_\pm^\ell = (\sqrt{s}/4)(1 \pm \sqrt{1 - 4m_W^2/s})$.

Finally, the direct reconstruction is the third and most powerful method. Only the fully leptonic events are not used here because they are underconstrained and the invariant mass of the Ws cannot be unambiguously determined. The center-of-mass energy is known very precisely and acts as a very strong constraint. The invariant mass is determined using a five or two constraint kinematic fit, for 4q or semileptonic events respectively. In addition, the 4q events require a jet pairing algorithm to determine which jets correspond to which W. Bad pairings bring about combinatorial background and contain little or no information on the W mass. A convolution or binned likelihood method is used to extract the W mass from the invariant mass distribution. Combining the direct reconstruction results from the four LEP experiments gives $m_W = 80.401 \pm .048$ (.027stat; .040syst) GeV (Aleph: $80.449 \pm .065$; Delphi: $80.308 \pm .090$; L3: $80.353 \pm .088$; Opal: $80.446 \pm .065$) [3]. The W width and mass are extracted simultaneously in a two dimensional fit and the LEP combined result for the width is $\Gamma_W = 2.19 \pm .15$ GeV [3]. The main systematics originate from the uncertainties on the LEP center-of-mass energy (17 MeV), the final state interactions (18 MeV), and the fragmentation (28 MeV).

Combining all results from all methods at LEPII with the Tevatron and UA2 results, the world average is given by [7]

$$m_W^{WA} = 80.419 \pm .038 \text{ GeV}.$$

TOP MASS AT TEVATRON

The top quark was discovered at Tevatron in 1995. Its mass is measured with a precision of 3% and makes it the most precisely known quark mass. The top mass measurement strongly constrains the Higgs mass, since the radiative corrections are functions of both of these parameters.

Top quark pairs are produced through $q\bar{q}$ annihilation into a gluon which splits into a $t\bar{t}$ pair. The gluon is not a colour singlet, such that colour strings are allowed to form between the top and the anti-top. This colour crosstalk is difficult to model and is one of the main sources of systematic uncertainties. The top (anti-top) then decays almost exclusively into a b (anti-b) quark and a W boson, and the W decays either hadronically or leptonically.

The combined CDF and D0 error on the top mass for the all hadronic channel, when both Ws decay hadronically (BR~44%, Signal/Noise~1/4), is approximately 8.5 ($7stat$) GeV. The invariant mass of the top decay products is determined in a three constraint kinematic fit. The di-lepton channel (BR~5%, S/N~4) has approximately the same precision and statistical weight. The events are under-constrained such that no invariant mass can be calculated. In this case, the decay dynamics of the data is compared with the MC expectation e.g. using the p_ν distribution. The top mass determined from the lepton+jet channel (BR~30%, S/N~1) has an error of ~ 5.5 ($4stat$) GeV, corresponding to a statistical weight of ~80% in the average. A two constraint kinematic fit is used to determine the invariant mass of the top quark decay products. The Tevatron top mass, for all channels and for CDF and D0 combined, is [4]

$$m_{top} = 174.3 \pm 5.1 \ (3.2stat; 4.0syst) \ \text{GeV}$$

with CDF: 176.0 ± 6.5 GeV and D0: 172.1 ± 7.1 GeV. The main systematics originate from the uncertainties on the jet energy scale, and on the MC modelling of the QCD effects.

$\sin^2 \theta_W$ AT NUTEV

The weak mixing angle has been measured by CCFR and more recently by NuTeV, at Fermilab. Early determinations of $\sin^2 \theta_W$ gave the SM's successful prediction of m_W and m_Z. More precise measurements gave the first useful limits on the top quark mass. The most recent results are the most precise to date and help to constrain the Higgs mass.

The Lewellyn Smith relation $R^\nu = NC/CC = f(\sin^2 \theta_W, \chi)$ can be used to extract $\sin^2 \theta_W$. NC (CC) is the number of neutral (charged) current events, and χ is the effect of the sea quark scattering e.g. the charm quark mass. Unfortunately, χ is the largest source of systematic uncertainties. NuTeV rather uses the Paschos Wolfenstein relation $R^- = (NC^\nu - NC^{\bar{\nu}})/(CC^\nu - CC^{\bar{\nu}}) = (R^\nu - rR^{\bar{\nu}})/(1-r) = 1/2 - \sin^2 \theta_W$ where $r = CC^{\bar{\nu}}/CC^\nu$. Using this relation, almost all sensitivity to

χ cancels out but separate ν and $\bar{\nu}$ beams are needed. These are supplied by the FNAL sign selected quadrupole train which selects mesons of the appropriate sign. Contamination of the beams remains small: $<1/1000$ ($<1/500$) of $\bar{\nu}_\mu$ (ν_μ) in the ν_μ ($\bar{\nu}_\mu$) beam; 1.3% (1.1%) of ν_e in the ν_μ ($\bar{\nu}_\mu$) beam.

The NC and CC events are characterized by their event length: CCs produce long event length muons, whereas NCs produce short hadronic showers. For ν and $\bar{\nu}$ separately, NuTeV measures $R_{\nu(\bar{\nu})}^{meas}$ = [# short evts in $\nu_\mu(\bar{\nu}_\mu)$ mode]/[# long evts in $\nu_\mu(\bar{\nu}_\mu)$ mode]. R_{MC}^ν and $R_{MC}^{\bar{\nu}}$ are both functions of $\sin^2\theta_W$ and χ. The relation $\tilde{R}^- = R^\nu - xR^{\bar{\nu}}$ is minimized with respect to χ, giving $x = .5136$. $\sin^2\theta_W$ is then varied until $\tilde{R}_{MC} = \tilde{R}_{data}$. For $m_{top} = 175$ GeV and $m_{Higgs} = 150$ GeV [8]

$$\sin^2\theta_W = 1 - m_W^2/m_Z^2 = .2253 \pm .0021 \;(.0018\text{stat}; .0010\text{syst})\; (\text{SM}: .2227)$$

with residual dependence on m_{top} and m_{Higgs}: $\delta\sin^2\theta_W = -.00435[(m_t/175)^2 - 1] + .00048\log(m_H/150)$ with m_t and m_H in GeV. This translates into a value of the W mass: $m_W = 80.26 \pm .11$ GeV.

EXTRA BEAUTIFUL MEASUREMENTS

A precise determination of $\alpha(m_Z^2)$ becomes extremely important in the context of the electroweak fits. From Equation 1, one can deduce that the size of $\Delta\alpha(m_Z^2)$ limits the precision on $\log m_H$ via the radiative corrections ϵ_1 and ϵ_3.

A precise measurement of the muon magnetic moment is also of great importance. The relation between the μ magnetic moment and the spin is given by $\vec{\mu} = g(e/m)\vec{s}$. Radiative corrections bring deviations from $g = 2$, defined as $a_\mu = (g-2)/2$, but these could also originate from non-SM effects.

Thirdly, the Cesium atomic parity violation measurement is discussed.

In the case of $\alpha^{-1}(m_Z^2)$

The electromagnetic constant at center-of-mass energy $\sqrt{s}$ can be written as [9] $\alpha(s) = \alpha(0)/[1 - \Delta\alpha_{lept}(s) - \Delta\alpha_{had}(s)]$. The leptonic contribution is precisely calculated to three loop order: $\Delta\alpha_{lept}(m_Z^2) = 314.97686 \times 10^{-4}$. The hadronic vacuum polarization term has the largest uncertainty and is determined via a dispersion integral: $\Delta\alpha_{had}(m_Z^2) = -[\alpha(0)m_Z^2/3\pi] \cdot Re \int_{4m_\pi^2}^{\infty} ds\{R(s)/[s(s-m_Z^2) - i\epsilon]\}$, where $R(s) = \sigma_{ee \to had}/\sigma_{ee \to \mu^+\mu^-}$. $R(s)$ is measured from $e^+e^- \to$ hadron data for $\sqrt{s} < 40$ GeV, and is evaluated using perturbative QCD (PQCD) for $\sqrt{s} > 40$ GeV, giving $\Delta\alpha_{had} = (280 \pm 7) \times 10^{-4}$. Summing these results, one obtains $\alpha^{-1}(m_Z^2) = 128.902 \pm .090$, which in today's electroweak fits translates into $\Delta\log m_H = \pm .30$.

However, it is possible to reduce the error on $\Delta\alpha_{had}$ by taking more data at $\sqrt{s} = 0.3 - 5.0$ GeV, by using PQCD down to $\sqrt{s} = 1.8$ GeV and by applying theory constraints from QCD sum rules [10]. These steps should bring the error down to $\Delta \alpha^{-1} = \pm .02$, which would translate into $\Delta \log m_H = \pm .20$.

$e^+e^- \to$ hadron data are presently being taken by various experiments, in particular by BESII in China ($2 < \sqrt{s} < 5$ GeV), and by CMD-2 and SND in Novosibirsk ($0.6 < \sqrt{s} < 1.4$ GeV). The first run of BESII in 1998 has brought the error on $R(s)$ from 15-20% down to 7% [11]. Combining the Novosibirsk and BESII data, a 1999 preliminary evaluation has given $\Delta \alpha^{-1}(m_Z^2) \simeq .035$. An update will be presented this summer (2000).

In the case of $(g-2)_\mu$ at E821 BNL

Both experimentally and theoretically, $a_\mu = (g-2)/2$ is known with a precision of $\sim 10^{-9}$. Theoretically, it is senstitive to large energy scales and to very high order radiative corrections. Its precision is limited by second order loop effects from the hadronic vacuum polarization. Experimentally, it is extremely sensitive to new physics.

The theoretical expression can be written as $a_\mu = a_\mu^{\text{qed}} + a_\mu^{\text{weak}} + a_\mu^{\text{had}}$. As in the case of α^{-1}, the hadronic contribution has the largest uncertainty. Today's theoretical calculation gives $a_\mu(SM) = (116\ 591\ 594.7 \pm 70) \times 10^{-11}$ [12]. Experimentally, a_μ is being measured at BNL by the E821 $g-2$ experiment. A 3.1 GeV π beam from the Alternating Gradient Synchrotron is used. The E821 goal is to achieve a precision of $\pm 40 \times 10^{-11}$. The present day world average on a_μ is $(116\ 592\ 050 \pm 450) \times 10^{-11}$.

The polarized muons from the decay $\pi \to \mu$ move in a uniform $\vec{B}$ field which is $\perp$ to the muon spin $\vec{s}_\mu$ and to the orbit plane. A quadrupole electric field $\vec{E}$ is used for vertical focusing. The spin precession ω_s minus the cyclotron frequency ω_c is given by $\vec{\omega}_a = -(e/m)\{a_\mu \vec{B} - [a_\mu - 1/(\gamma^2-1)]\vec{\beta} \times \vec{E}\}$. The second term in the brace cancels out for muons with the *magic* $\gamma_\mu = 29.3 \to p_\mu = 3.01$ GeV. This is exactly the energy chosen for the experiment. The decay time spectrum of the positrons from the decay $\mu^+ \to e^+ \nu_e \bar{\nu}_\mu$ is given by $N_0 e^{-t/\gamma \tau}[1 + A(E)\cos(\omega_a t + \phi(E))]$. One counts the number of positrons versus time and fits the above relation to extract ω_a.

Any deviation from the SM prediction could be interpreted as new physics e.g. muon sub structure, W compositeness, SuperSymmetry.

Atomic Parity Violation in Cesium

The atom is not a purely electromagnetic system. Parity violation can occur inside the atom. In order to detect this violation, the atoms in a gas are first given a preferential handedness e.g.right. Some suitable property is measured. The handedness is then reversed e.g.left, and the property is measured again. If the results of the two measurements differ, then parity is violated. The left-right asymmetry is expressed as $A_{LR} \propto Z^3/m_Z^2$, where Z in the numerator is the number of protons in the atom. The asymmetry is very small due to the large Z mass in the denominator. Cesium (Cs) 55 has been chosen because it has a reliable atomic structure calculation, it has one electron in its outer shell, and the remaining 54 electrons are tightly bound around the nucleus. Cs is the simplest of the heavy atoms, and the heaviest of the simplest.

In the Boulder experiment [13], a Cs beam passes through a region of perpendicular electric, magnetic and laser fields. The highly forbidden 6S→7S transition occurs in the Cs as the weak Parity Non Conserving (PNC) transition with a probability of 10^{-11}. An electric field provokes a Stark induced transition which has a probability 10^5 times larger than the PNC transition and which can interfere with it. In order to have non zero interference between the two, the 6S→7S transition is excited with an elliptically polarized laser field. The handedness of the region is changed by reversing each of the field directions. The parity violation is apparent as a small modulation in the 6S→7S excitation rate synchronous with all of these reversals. The modulation is related to the weak charge Q_W. The Boulder result for the weak charge of Cs is [14]

$$Q_W \, Cs^{133}_{55} = -72.06 \pm .28_{expt} \pm .34_{th}$$

The SM prediction is given by the relation [15] $Q_W^{SM} = -72.72 \pm .13 - 102\epsilon_3^{rad} + \delta_N Q_W$ where $\delta_N Q_W$ indicates new physics, and ϵ_3^{rad} are the radiative corrections which are evaluated to be $(4.9 \pm 1) \cdot 10^{-3}$ from high energy data results. Using this SM prediction and the experimental result, one finds $Q_W^{expt} - Q_W^{SM} = 1.18 \pm .46$ or $0.28 \leq \delta_N Q_W \leq 2.08$ at 95% C.L. , which corresponds to a 2.6σ discrepency with the SM. This result can be interpreted in the context of contact interations, as a measurement of Λ_{LL} and Λ_{RR}: $12.1 \leq \Lambda^+_{LL,RR} \leq 32.9$ TeV. The LEPII fermion pair cross section and asymmetry measurements exclude the regions of $\Lambda^+_{RR} < 12.3$ TeV and $\Lambda^+_{LL} < 12.8$ TeV.

HOW FIT IS THE STANDARD MODEL

Z Universality tests, fermion pairs at LEPII, W production and decays, triple gauge couplings, a_μ and atomic parity violation measurements are all consistent with the SM. All are stringent tests of the SM and help to set limits on a wide range of physics scenarios.

Amongst other measurements (see Figure 3), $\sin^2\theta_{eff}^\ell$, m_W, m_{top}, $\sin^2\theta_W$ and $\alpha^{-1}(m_Z^2)$ enter in the overall electroweak fit from which the Higgs mass is extracted. The χ^2/dof of the fit is 23/15, $m_{Higgs} = 67^{+60}_{-33}$ GeV and $m_{Higgs} < 188$ GeV at 95% C.L. . The direct searches at LEPII give $m_{Higgs} > 107.7$ GeV at 95% C.L. [16].

The weak mixing angle ($\sin^2\theta_{eff}^\ell = .23149 \pm .00017$) is the strongest constraint on the Higgs mass today [1]. m_W would be in the race if its error were ~ 25 MeV. The W mass measurement confirms the existence of the weak radiative corrections to $\sim 7\sigma$. Reducing the error on $\alpha^{-1}(m_Z^2)$ will make the $\sin^2\theta_{eff}^\ell$ measurement an even stronger constraint on the Higgs mass. The same can be said about the top

Moriond 2000

	Measurement	Pull	Pull -3 -2 -1 0 1 2 3
m_Z [GeV]	91.1871 ± 0.0021	.07	
Γ_Z [GeV]	2.4944 ± 0.0024	-.62	
σ^0_{hadr} [nb]	41.544 ± 0.037	1.72	
R_e	20.768 ± 0.024	1.19	
$A^{0,e}_{fb}$	0.01701 ± 0.00095	.70	
A_τ	0.1483 ± 0.0051	.13	
A_e	0.1425 ± 0.0044	-1.16	
$\sin^2\theta^{lept}_{eff}$	0.2321 ± 0.0010	.65	
R_b	0.21642 ± 0.00073	.85	
R_c	0.1674 ± 0.0038	-1.27	
$A^{0,b}_{fb}$	0.0988 ± 0.0020	-2.34	
$A^{0,c}_{fb}$	0.0692 ± 0.0037	-1.29	
A_b	0.911 ± 0.025	-.95	
A_c	0.630 ± 0.026	-1.47	
$\sin^2\theta^{lept}_{eff}$	0.23096 ± 0.00026	-1.87	
$\sin^2\theta_W$	0.2255 ± 0.0021	1.17	
m_W [GeV]	80.448 ± 0.062	.86	
m_t [GeV]	174.3 ± 5.1	.11	
$\Delta\alpha^{(5)}_{had}(m_Z)$	0.02804 ± 0.00065	-.20	

FIGURE 3. List of measurements entering the spring 2000 electroweak fit and pulls with respect to the SM prediction.

mass error with respect to the W mass measurement as a constraint on the Higgs mass.

FUTURE PROSPECTS

By the end of LEPII, each experiment will have accumulated well over 500 pb^{-1}. A realistic goal for the W mass measurement is to attain a world average error of 25 MeV, and a factor of two improvement on the TGCs. The goal of the E821 $g-2$ experiment is to reach an error on a_μ of 40×10^{-11}. New measurements of $R(s)$ for $\sqrt{s} = 0.3$ to 5.0 GeV at BESII and in Novosibirsk will help to reduce the error on α^{-1}, such that it will most likely reach $\Delta\alpha^{-1} \sim .02$.

In the near future (2001), Tevatron will start RUNII. During the four years of data taking, more than 14 fb^{-1} per experiment are expected at $\sqrt{s} = 2$ TeV. The top mass error will come down to 2-3 GeV per experiment, the overall Tevatron W mass error will be 20-40 MeV. Approximately 20 fb^{-1} will be needed for a 3σ

discovery of a SM Higgs of $m_{\text{Higgs}} < 180 \; GeV$ [17].

In the far future, each LHC (ATLAS and CMS) experiment will accumulate about 300 fb^{-1} over the ten years of running at $\sqrt{s} = 14$ TeV. The LHC combined experiments top mass error will be of the order of 2 GeV, the W mass error will be ≤ 25 MeV per experiment. Approximately 30 fb^{-1} will be needed for a 5σ discovery of a SM Higgs of $m_{\text{Higgs}} < 1000 \; GeV$ [18].

CONCLUSION

The SM is in good shape. Nothing really anomalous has been observed. A beautiful future exists for stringent tests e.g. fermion pair production, TGCs, a_μ and atomic parity violation. The same can be said about precision measurements constraining the Higgs mass e.g. at LEPII, Tevatron and LHC. There are many good reasons to be optimistic about the future of precision electroweak measurements !!

ACKNOWLEDGEMENTS

I would like to thank the LEPEWWG, D.Abbaneo, T.Barklow, A.Blondel, T.Diehl, P.Dornan, A.Höcker, K.Grupen, M.Lancaster, B.Pietrzyk, G.Quast, L.Roberts, D.Schlatter, Z.Zhao and many more... An extra special thanks to the CIPANP2000 organizers.

REFERENCES

1. The LEP Electroweak Working Group, *A Combination of Preliminary Electroweak Measurements and Constraints on the Standard Model*, CERN-EP-2000-016.
2. The LEP Electroweak Working Group f$\bar{\text{f}}$ Subgroup, *Combination of the LEP 2 $f\bar{f}$ Results*, LEP2FF/00-01.
3. The LEP WW Working Group, *LEP W-Pair Cross Section and W Mass and Width for Winter 2000 Conferences*, LEPEWWG/WW/00-01.
4. Lancaster, M., "Electroweak Measurements from Hadron Machines,", plenary talk presented at *XIX International Symposium on Lepton and Photon Interactions at High Energies*, Stanford, August 1999, hep-ex/9912031.
5. The D0 Collaboration, *Phys.Rev.* **D60**, 072002 (1999).
6. The LEP TGC Working Group, *A Combination of Preliminary Results on Triple Gauge Boson Couplings Measured by the LEP Experiments*, LEPEWWG/TGC/2000-01.
7. See http://lepewwg.web.cern.ch/LEPEWWG/plots/winter2000/m00_plot_mw.eps.
8. Zeller, G.P., representing the NuTeV Collaboration, "A Measurement of $\sin^2 \theta_W$ in νN Scattering from NuTeV,", talk presented at *American Physical Society Division of Particles and Fields*, University of California, Los Angeles, January 1999, hep-ex/9906024.

9. Davier, M., "Evaluation of $\alpha(m_Z^2)$ and $(g-2)_\mu$,", talk presented at *Tau'98*, Santander, September 98, hep-ph/9812370.
10. Davier, M., and Höcker,A., *Phys. Lett* **B 435**, 427-440 (1998).
11. Zhao, Z.G., "R-values in Low Energy e^+e^- Annihilation,", plenary talk presented at *XIX International Symposium on Lepton and Photon Interactions at High Energies*, Stanford, August 1999, hep-ex/0002025.
12. Roberts, L.B., "Status of the Muon (g-2) Experiment,", plenary talk presented at *XIX International Symposium on Lepton and Photon Interactions at High Energies*, Stanford, August 1999, hep-ex/0002005.
13. Wood, C.S.,Bennett, S.C.,Cho, D.,Masterson, B.P., Roberts, J.L.,Tanner, C.E., and Wieman, C.E., *Science* **275**, 1759-1763 (1997).
14. Bennett, S.C., and Wieman, C.E., *Phys. ReV. Lett* **82**, 2484-2487 (1999).
15. Dominici, D., "Bounds on New Physics from Parity Violation in Atomic Cesium,", talk presented at *International Europhysics Conference on High-Energy Physics: EPS-HEP '99*, Tampere, July 1999, hep-ph/9909290.
16. Straessner, A., "Measurement of the Mass of the W Boson at LEP and Determination of Electroweak Parameters,", talk presented at *XXXVth Rencontres de Moriond Electroweak Interactions and Unified Theories*, Moriond, March 2000.
17. Roco, M.T., and Grannis, P., "Higgs Boson Discovery Prospects at the Tevatron,", talk presented at *13th Topical Conference on Hadron Collider Physics*, Mumbai India, January 1999, FERMILAB-Conf-99/118-E.
18. Gianotti, F., "Precision Physics at LHC,", in *Barcelona 1998, Radiative Corrections*, edited by J. Solà, World Scientific, Barcelona, 1998, pp.270-288.

CP-Violation and Physics Beyond the Standard Model

R. N. Mohapatra*

Department of Physics, University of Maryland, College Park, MD-20742

Abstract. This article is meant to be a brief review of the present status of CP violation in the context of the standard model and beyond.

INTRODUCTION

Since the discovery of CP violation in 1964, understanding its origin has been a major focus of research in theoretical physics. On the experimental front, the field has moved far beyond the original discovery of the phenomenon in $K \to 2\pi$ decay and met many of its challenges by providing useful information on a wide variety of CP violating processes, even though many of the results appear only in the form of upper limits rather than a positive signal. On the theoretical side, the advent of gauge theories and the discovery of the three generations of fermions (and perhaps no more) has led to a unique setup for discussing CP violating phenomena in the context of the so-called Cabibbo-Kobayashi-Maskawa (CKM) model. It is however widely recognized that this model cannot be the final theory of CP violation since it leaves many questions unanswered and that there must be CP violating phenomena beyond the standard model. In this brief review, I start with a quick overview of the experimental situation that includes the recent measurement of ϵ'/ϵ, which has infused a new round of frantic theoretical activities to unravel its meaning. I, then, discuss the CKM model and note its predictions and deficiencies. The latter prompt us to look beyond the standard model for improvements.

Out of the myriads of scenarios of CP violation beyond the standard model, I pick only two, which seem to have strong independent theoretical motivations: (i) the left-right model of CP violation motivated by aesthetic considerations of quark-lepton symmetry, asymptotic parity conservation and neutrino mass discoveries and (ii) the syupersymmetric model (MSSM) which is strongly motivated by the attempts to understand the stability of the weak scale as well as the origin of the electroweak symmetry breaking. It turns out that the crisp understanding of CP violation in the context of the standard model goes into complete disarray once the theory is made supersymmetric. Thus if supersymmetry is the right theory of particle interactions at the TeV scale, one must go beyong MSSM. Here again we discover that a high scale left-right symmetry comes to rescue. We briefly mention this theory which combines the two ideas just described in a way that eliminates some of the vexing CP problems of the otherwise highly desirable MSSM, while keeping its good features in tact and further providing an understanding of neutrino masses.

BRIEF SUMMARY OF THE EXPERIMENTAL SITUATION

CP violation was first discovered in the decay of $K_L^0 \to 2\pi$ in the year 1964. In order to set the notation, we provide a brief sketch the CP violating parameters that describe this process. Defining $K_L^0 = (K_2 + \epsilon K_1)/(\sqrt{1 + |\epsilon|^2})$ and writing the ratio $\frac{K_L \to \pi_i \pi_j}{K_S \to \pi_i \pi_j} = \eta_{ij}$, we can write down [1]

$$\eta_{+-} = \epsilon + \epsilon' \qquad (1)$$
$$\eta_{00} = \epsilon - 2\epsilon'.$$

The parameter ϵ which describes the mass mixing between the CP even (K_1^0) and CP odd (K_2^0) states was measured in the famous Fitch-Cronin-Christenson-Turlay experiment in 1964 and its present value is $\epsilon = (2.285 \pm 0.019) \times 10^{-3}$ [2]. The parameter ϵ' was recently measured by the CERN NA48 [3] and Fermilab KTeV [4] and has led to very important consequences for the nature of CP violation. The present world average for this is given by:

$$\frac{\epsilon'}{\epsilon} = (21.2 \pm 4.6) \times 10^{-4} \qquad (2)$$

An immediate implication of this measurement is the demise of the superweak model [5] where all CP violation arises from an effective CP odd $\Delta S = 2$ operator and where $\epsilon' = 0$.

There are other evidences for CP violation in the K-system if one assumes the validity of CPT theorem. This comes from the recent measurement at CPLEAR of the time asymmetric observable

$$A = \frac{P(K^0 \to \bar{K}^0, t) - P(\bar{K}^0 \to K^0, t)}{P(K^0 \to \bar{K}^0, t) + P(\bar{K}^0 \to K^0, t)} \qquad (3)$$

The value of A is measured to be $(6.6 \pm 1.3 \pm 1.0) \times 10^{-3}$. As was noted by Kabir [7], this is a test of T violation. However, if one assumes CPT conservation, this also is a signal of CP violation and can be predicted from the observations of the ϵ parameter. One then gets, $A = (6.63 \pm 0.06) \times 10^{-3}$ in good agreement with observations.

The KTeV collaboration has also observed a large T-odd effect in the angular distribution between the e^+e^- and $\pi^+\pi^-$ decay planes in the K_L^0 decay mode $K_L^0 \to \pi^+\pi^- e^+e^-$. The magnitude of the observed asymmetry is $(13.6 \pm 2.5 \pm 1.2)\%$, in agreement with theoretical predictions. A similar number has also been found by the NA48 collaboration.

The only other possibly positive indication of CP violation is in the heavy quark system, in the decay of $B_d \to J/\psi K_S^0$ which predicts the parameter $sin2\beta = 0.79^{+0.41}_{-0.44}$. This question will be settled in the coming years as the data from BaBaR and BELLE are analyzed.

There are many other searches for CP violating effects in different systems. One of the most important among them is the search for the electric dipole moment (edm) of various elementary particles such as neutron, electron and possibly the muon. The present upper limit on the neutron edm is $d_n^e \leq 6.6 \times 10^{-26}$ ecm. That of the elctron is derived from the non-observation of the edm for different atoms. Since the atom is a complex system consisting of not only electrons but also protons and neutrons, a bit of an explanation is called for. In general, the edm of the atom has three contributions [8]:

$$d_A^e = R d_e^e + d_N^e + C_{eN} \tag{4}$$

It has been shown that in the case of paramagnetic atoms, the R can be anywhere from 100 (for Cs) to 600 (for Tl) whereas in the case of diamagnetic atoms such as mercury (Hg), R is a very small number. This happy theoretical accident has led to experiments both in Cs and Tl [2] yielding a very stringent limit on $d_e^e \leq 6 \times 10^{-27}$ ecm. Coming to higher generation leptons such as the muon, the present limit comes from the CERN (g-2) experiment and is given by $d_\mu^e \leq 10^{-18}$ ecm. There is a proposal to improve this limit by six orders of magnitude using the Brookhaven accelerator.

Other searches for CP violating effects have been carried out in the domain of nuclear physics [8]. Here there are two types of effects: the time reversal violating parity conserving (TVPC) and time reversal violating parity violating (TVPV). The first one is parameterized in terms of CP violating ρNN coupling $g_{\rho NN}^{(-)}$ and the second one by the corresponding πNN coupling $g_{\pi NN}^{(-)}$. The most stringent upper limits on the parameter $g_{\pi NN}^{(-)}$ comes from the limits on the atomic dipole moments of diamagnetic atoms such as Hg and TlF and is $g_{\pi NN}^{(-)} \leq 1.8 \times 10^{-11}$. The limits on the CP violating ρNN coupling derived from TVPC type processes are not very stringent. They are derived from studies of detailed balance in nuclear reactions as well as experiments involving transmission of polarized neutrons through ^{165}Ho target. The best limit is: $g_{\rho NN}^{(-)} \leq 2.5$.

CP VIOLATION IN GAUGE THEORIES

In the context of local field theories, constraint of the hermiticity of the Lagrangian implies that CP violating interactions would require the existence of complex couplings in the theory. In the framework of spontaneously broken gauge theories, the gauge interaction part prior to spontaneous breaking is always CP conserving since the gauge coupling constant is always a real number. This is guaranteed by group theory. CP violating complex coupling must therefore be supplied by other parts of the Lagrangian i.e. the fermion Yukawa coupling and/or the Higgs self interactions. These complex couplings in the final Lagrangian after spontaneous symmetry breaking can arise from two sources: (i) from intrinsically complex couplings (e.g. Yukawa couplings) in the Lagrangian and/or (ii) complex vacuum expectation values of the Higgs fields. The former will be called in this review as "intrinsic" CP violating models (ICPV) and the latter, "spontaneous CP breaking" (SCPV) models. The latter class of models generally have fewer parameters than the former.

With the above criterion for CP violation in mind, let us turn to CP violation in gauge models. The theoretical challenge is two-fold: (i) how does one introduce a complex coupling into the theory since fermions and Higgs bosons being complex fields, they can be redefined to absorb phases in the coupling constants ? and (ii) how does one understand naturally the smallness of the CP violating effects compared to the usual weak interaction effects ?

As regards, the first question, it was realized early on [9] in the development of gauge theories that, in the two generation pure lefthanded standard model, it is impossible to introduce a physical phase into any of the couplings due to the reason stated above. This observation led to the suggestion [9] that one needs to introduce right handed currents to describe CP violating effects. These arguments were sharpened in the celebrated paper of Kobayashi and Maskawa [10] who made the visionary proposal that an alternative way to introduce CP violation into gauge theories is to include three left handed generations. They further showed that in this model there is only one CP violating phase in the weak charged current interaction. This was before the discovery of the charmed particles and was therefore a bold proposal. As we know now, not only the charm particles but also the fermions corresponding to the third generation were discovered in subsequent years giving a major boost to the KM model of CP violation.

The year after the KM model was published, a two generation left-right symmetric model was published [11], where the idea is that since the two generation pure left-handed models having no CP violation, the CP violating effects all emerge in the LR model from the right handed currents and are therefore suppressed naturally by a factor $\left(\frac{M_{W_L}}{M_{W_R}}\right)^2$. Since the success of V-A theory of weak interaction tells us that the right handed currents are suppressed at low energies (i.e. $M_{W_R} \gg M_{W_L}$), no further assumption is needed to understand the suppression of the CP violating effects. The interesting point is that the precision measurements in muon decay as well as subsequent searches for the right handed current effects have so far led to bounds on W_R mass anywhere from 600 GeV to 1.8 TeV [2], which are in the range to give enough suppression for CP violating amplitudes. The discovery of the neutrino mass have given strong indication that the left-right symmetric theories may eventually be realized in nature, although the value of the scale depends on additional assumptions. Therefore, even after 25 years, the suggestion that CP violation may have something to do with left-right symmetry of nature remains alive and an active area of investigation.

Around the same time, the ideas of supersymmetry were being developed and towards late seventies, it was realized that they may provide answers to two of the vexing questions of the standard model i.e. why is the weak scale not destabilized by gravity and how does the electroweak symmetry breaking arise in nature. It was also noted shortly afterwards that the gauge couplings of the standard model may unify better in the presence of supersymmetry than without it. It was therefore appropriate to look at the profile of CP violation in the context of the so called minimal supersymmetric standard model (MSSM). It was realized that CP violation is a major unattractive aspect of the MSSM. One therefore needs to go beyond MSSM to solve the CP problems while at the same time keeping all the nice features of the MSSM valid.

In this talk, I will briefly touch on all these models and discuss their experimental tests. There are of course many other models of CP violations e.g. multi-Higgs models, heavy iso-singlet quark models etc. The scope of this talk is too limited to do justice to all these models. We therefore refer to an excellent recent book, where many of these topics and more have been discussed [12].

KOBAYASHI-MASKAWA MODEL OF CP VIOLATION

In the KM model of CP violation, there is only one CP violating phase which is a part of the mixing matrix that characterizes weak charged current interactionas follows:

$$\mathcal{L}_{wk} = \frac{g}{2\sqrt{2}}[\bar{\mathcal{P}}\gamma^\mu(1-\gamma_5)U\mathcal{N}]W_{L,\mu} + h.c. \tag{5}$$

The U matrix has three real angles and one phase. A particularly simple parameterization of this matrix was suggested by Wolfenstein, in which we have:

$$U = \begin{pmatrix} 1 & \lambda & \lambda^3 A(\rho+i\eta) \\ -\lambda & 1-\lambda^2/2 & \lambda^2 \\ \lambda^3 A(1-\rho-i\eta) & -\lambda^2 & 1 \end{pmatrix} \tag{6}$$

where detailed analysis of low energy weak data implies that $\lambda \simeq 0.22$, $A \simeq 0.8$. The parameters ρ and η take values in an area in the $\rho-\eta$ plane with a mid point somewhere around $\rho \simeq 0.2$ and $\eta \simeq 0.4$.

Kaon CP violation in the CKM model

We can now see very clearly how the suppression of the CP violating effects arise. Since the ϵ parameter measures the matrix element of a CP violating $\Delta S = 2$

operator and CP violating phase, η is only in the third generation mixing, the $\Delta S = 2$ operator must go through the the t-quark and therefore can be estimated to be of order $\sim \frac{G_F^2}{\pi^2}\lambda^{10}\eta m_t^2$. On the other hand, the $\Delta S = 2$ CP conserving operator that is responsible for the $K_L - K_S$ mass difference can go through the c-quark intermediate state and has a magnitude $\sim \frac{G_F^2}{\pi^2}\lambda^2 m_c^2$; one can then easily see from this very crude calculation that $\epsilon \equiv \frac{ImM_{12}}{ReM_{12}} \sim \lambda^8(m_t/m_c)^2 A\eta \approx 10^{-2}$ which is about a factor of 4 away from its observed value. Thus in this model the suppression of the CP violating effects arise in a natural manner.

The calculation of the ϵ' parameter in the standard model is more complicated and involves eleven operators denoted by Q_i ($i = 1,...11$). All these are $\Delta S = 1$ operators; all of them contribute to nonleptonic weak processes. Of these, the ones that involve the phase parameter η, contribute to the ϵ' observable via the ImA_0 term in Eq. (). We refer to the recent reviews for detailed discussions [13] of the various operators and the status of the evaluation of their matrix element between K and $\pi\pi$ states. Generally, the techniques used for the purpose can be broadly described as lattice methods, $1/N_c$ expansion methods and those that use chiral perturbation theory as a low energy model of mesons and baryons. Since the calculation involves the strong coupling regime of QCD, all these methods have uncertainties and precise numerical prediction for ϵ' is far from being at hand, although the general order of magnitude may be correct. Nonetheless to get an idea about the general order of magnitude of ϵ'/ϵ, we consider the strength of a socalled penguin graph with matrix element given by the vacuum saturation approximation. A rough estimate is

$$\frac{\epsilon'}{\epsilon} \approx \frac{1}{22\sqrt{2}} \frac{G_F \frac{\alpha_s}{4\pi} \lambda^5}{3 \times 10^{-7}(2.3 \times 10^{-3}} \ln\left(\frac{m_t^2}{m_c^2}\right) \frac{4m_K^4(f_K - f_\pi)}{m_s^2} \tag{7}$$

In the above expression, we have kept all dimensional terms in units of GeV and have used the vacuum insertion approximation for the four quark matrix elements and the $K^0 \to 2\pi$ amplitude $A_0 \approx 3 \times 10^{-7}$ GeV. Putting in all the numbers, it is easy to see that one gets an $\epsilon'/\epsilon \simeq$ few $\times 10^{-4}$. This again shows that the smallness of ϵ'/ϵ is understood in the CKM model as a consequence of the small flavor mixing effect.

In order to compare the prediction of the KM model with the KTeV and NA48 experiments, one needs more careful calculations. This question has been addressed in many papers [13] and a range of values between $4-20 \times 10^{-4}$ have been obtained. The paper by Jamin [13] gives the value $\epsilon'/\epsilon = (7.0^{+6}_{-3.5}) \times 10^{-4}$. The central value is perhaps a bit on the low side but not quite something that would be compelling evidence for physics beyond the standard model.

Predictions of the CKM model

Since all parameters of the model are fixed by observed weak processsses, rates for many other CP violating processes can be predicted in the CKM model. The most interesting ones seem to be the electric dipole moment of the neutron, bottom meson related processes and rare K-decays. Experiments are either under way or in planning stage to explore these processes to higher levels of precision. Let us comment briefly on these processes:

Electric dipole moments

Due to the unitarity of the mixing matrix, which is the only source of CP violation in the model, one loop order diagrams for edm of neutron, electron as well as muons vanish and one has go to the level of two loops and beyond to get the first nonvanishing contributions. For this reason, the edm of particles in this model are very highly suppressed. To give some examples, for the case of the neutron, the dominant diagram turns out to be the one involving strange particles in the loop [14]. The magnitude of this contribution is of order

$$d_n^e \simeq G_F^2 \frac{\alpha_s}{64\pi^3} \lambda^6 A^2 \eta \frac{m_t^2}{m_c^2} m_N^3 \tag{8}$$

This is roughly of order 10^{-30} ecm, which beyond the reach of present and planned experiments. Similarly for the elctron, $d_e^e \sim 10^{-41}$ ecm, whereas for the muon it is at the level of 10^{-38} ecm. Any evidence for edms above these values will be evidence for CP violating interactions beyond the standard model. One note of caution, however, is that the neutron edm of a magnitude larger than the above prediction could owe its origin to the QCD θ parameter. This is a very important aspect of CP violation beyond the standard model that will be beyond the scope this summary.

CP violation in B-systems

A very important test of the CKM model is in the domain of B-systems, where definite and observable CP violating effects are predicted. This is a vast topic, which is beyond the scope of this summary. We therefore refer to recent reviews on the subject [15]. One of the major ways to look for tests of CKM model in the B-system is to use the unitarity of the CKM matrix especially the bd element:

$$V_{ub}^* V_{ud} + V_{cb}^* V_{cd} + V_{tb}^* V_{td} = 0 \tag{9}$$

Dividing the equation by the real term in the sum i.e. $V_{bc}^* V_{cd}$, we get an equation for a triangle one of whose sides is along the x-axia and has length one. Denoting the base angles by β and γ, the triangle relation gives the vertex α to be $\alpha = 2\pi - \beta - \gamma$. The power of this triangle is that all the angles under certain assumptions can be related to observable physical processes. For instance, the process $B_d \to J/\psi K_S^0$ measures $sin2\beta$. This process has been searched for in the CDF experiment at Fermilab and also at ALEPH and OPAL collaborations. The CDF collaboration has published a number $sin2\beta = 0.79^{+0.41}_{-0.44}$. There are two dedicated experiments, the BABAR collaboration at SLAC and the BELLE collaboration at KEK, which are taking data that will soon provide a measurement of this parameter as well as others in the B-system. The standard model prediction for this number is $sin2\beta = 0.71 \pm 0.14$.

Similarly, the angle α can be measured in the process $B_d \to \pi\pi$. The CKM model prediction for this is $sin2\alpha = 0.06^{+0.35}_{-0.41}$. Similarly, the angle γ can be measured in the process $B_s \to K_S\rho$, the CKM prediction for this being $\gamma = 67° \pm 12°$.

WHY GO BEYOND THE STANDARD MODEL FOR CP VIOLATION?

Even though there is no compelling laboratory indication for CP violation beyond the standard model (with or without CKM phase), there are other indications which strongly suggest that the CKM model is not adequate for the purpose of understanding (i) the origin of matter in the universe and (ii) the solution to the strong CP problem that owes its origin to the periodic vacuum structure of QCD. Coupled with other obvious reasons such as the recent evidences for the neutrino mass (which most likely implies a left-right symmetric structure of weak interactions at high energies) and gauge hierarchy problem (which many think could be solved by the inclusion of supersymmetry above the TeV scale), one is naturally led to consider the possibility that there has to be new CP violating interactions beyond the standard model, even if the CKM description of the laboratory data is adequate.

This kind of reasoning has led to the suggestion of many interesting CP violating scenarios beyond the standard model. One can broadly classify them into two categories: the models of the first category are motivated by independent need to extend the standard model such as the existence of neutrino mass and supersymmetry. For instance, left-right symmetric extensions of the standard model are very well motivated by the seesaw mechanism to understand the neutrino masses. They in turn provide independent ways to understand the CP violation phenomena. Similar remarks apply to supersymmetry. The second category of models arise, if it turns out that CP violation has its origin in the vacuum structure rather than in the couplings in the basic theory. This is SCPV class of models mentined earlier. If one assumes SCPV, then the standard model is purely CP conserving and one must necessarily look beyond the SM for understanding the known CP breaking phenomena. The models that incorporate SCPV naturally are the multi-Higgs models [16] as well as the left-right models. One can also construct models with vector like singlet quarks [17]. In this summary, I will only present the left-right and supersymmetric models as two typical examples of testable models of CP violation beyond the SM.

LEFT-RIGHT SYMMETRIC MODELS OF CP VIOLATION

After the realization in 1972 that two generation standard model cannot support CP violating interaction, a left-right symmetric model two generation model was proposed in 1974. The reason for this is that in the presence of the extended left-right gauge structure, one can have non trivial phases in the weak currents even for one generation of quarks. In fact the general formula for phase counting for the left-right models with N_g generations is given by: $N = \frac{(N_g-1)(N_g-2)}{2} + \frac{N_g(N_g+1)}{2}$. This for two generations leads to three independent phases and for three generations to seven phases. It was observed in [11] that due to the fact that there are no phases in the left-handed weak currents, all CP violating observables in the kaon system receive a natural suppression factor $\frac{M_{W_L}^2}{M_{W_R}^2}$ regardless of how big the phases were. Thus inspite of the presence of more phases (which individually could be of order one), the suppression of CP violation was explained as a consequence of the suppression of the V+A currents in weak interactions. This provided a new way to understand the smallness of CP violation and made these models interesting.

Soon after the three generations of quarks became popular in the late 70's, it became clear that the left-handed KM phase could provide the dominant CP violating effect as in the CKM model thereby reducing the significance of the right handed current effects in CP phenomenology. However, it was noted in [18] that if CP is spontaneously broken, then there is only one phase in the theory and all the phases that appear in the weak mixing matrices are expressible in terms of this single phase. (to be called SCPV phase). The theory then becomes a single phase theory and becomes more predictive. It turns out [18,19] the weak mixing phases are related to the basic SCPV phase via the quark masses and one can show [19] that for the three generation case, the KM phase is at most 10-20%, whereas fitting all CP data in the KM model requires the phase to be of order one. This result is extremely interesting since it implies that in the large M_{W_R} limit one does not obtain enough CP violating effects. This has two implications: (i) the prediction of Kaon CP violation is again suppressed by the factor $\frac{M_{W_L}^2}{M_{W_R}^2}$, which then ties the smallness of CP and P-violaint effects together as in ref. [11]; (ii) it predicts smaller observable effects in B-systems. This version of the left-right models is therefore testable in the ongoing B-physics experiments. The model also predicts CP violating effects in beta decay [20].

Let us now provide the basic ideas of the left-right models and the profile of CP violation in these models. The model is based on the gauge group $SU(2)_L \times SU(2)_R \times U(1)_{B-L}$ with left-right symmetric fermion assignments as follows (see Table I):

The charged weak current Lagrangian in this model can be written as

$$\mathcal{L}_{wk} = \frac{g}{2\sqrt{2}} \left(\bar{P}\gamma^\mu(1-\gamma_5)U_L N W_{L,\mu} + \bar{P}\gamma^\mu(1+\gamma_5)U_R N W_{R,\mu} \right) + h.c. \qquad (10)$$

TABLE 1. Assignment of the fermion and Higgs fields to the representation of the left-right symmetry group

Fields	$SU(2)_L \times SU(2)_R \times U(1)_{B-L}$ representation
Q_L	$(2,1,+\frac{1}{3})$
Q_R	$(1,2,+\frac{1}{3})$
L_L	$(2,1,-1)$
L_R	$(1,2,+1)$
ϕ	$(2,2,0)$
Δ_L	$(3,1,+2)$
Δ_R	$(1,3,+2)$

Parity invariance of the theory imposes constraints on the left and right mixing matrices. For the case of intrinsic CP violation in the couplings, one has $U_L = U_R$. On the other hand for the case of SCPV, one gets $U_L = U_R^*$ [21]. Thus even though apriori, it may appear that there are many new parameters in this model, parity reduces the number of real angles to be same as in the standard model. As mentioned above, in the second case, the number of phases also reduce to only one as in the KM model since spontaneous CP violation is triggered by $<\phi> = e^{i\alpha} \begin{pmatrix} \kappa & 0 \\ 0 & \kappa' \end{pmatrix}$, which is the only CP violating phase in the theory. All the phases in $U_{L,R}$ thus become expressible in terms of α.

The model has several distinctive predictions from the standard model: (i) a lower value of the CP asymmetry $a_{CP}(B_d \to J/\psi K_S^0)$, typically of order 0.1 or so as against the KM model prediction of ≈ 0.75; (ii) similarly, the CP asymmetry in $B_s \to J/\psi K_S$ is expected to be of order $-0.1 - 0.4$ which an order of magnitude larger than the prediction of the standard model [19]; (iii) finally, the predictions for the edms of neutron, electron and muon are higher in this model e.g. one expects the $d_\mu^e \simeq 10^{-27}$ ecm and d_n^e also of the same order, both of which are considerably higher than the predictions of the standard model.

SUPERSYMMETRY AND CP VIOLATION

There are now a number of theoretical reasons that lead people to believe that supersymmetry with a effective breaking scale around a TeV or less is likely to be a part of the new physics beyond the standard model. They are: (i) stabilization of the Higgs mass in the TeV range (and not at the high new physics scale at the planck or similar large scale) due to natural cancellation of infinities caused by boson-fermion symmetry; (ii) radiative generation of electroweak symmetry breaking starting at high scale with a positive mass-square for the Higgs boson; (iii)

TABLE 2. The particle content of the supersymmetric standard model. For matter and Higgs fields, we have shown the left-chiral fields only. The right-chiral fields will have a conjugate representation under the gauge group.

Superfield	gauge transformation
Quarks Q	$(3, 2, \frac{1}{3})$
Antiquarks u^c	$(3^*, 1, -\frac{4}{3})$
Antiquarks d^c	$(3^*, 1, \frac{2}{3})$
Leptons L	$(1, 2, -1)$
Antileptons e^c	$(1, 1, +2)$
Higgs Boson $\mathbf{H_u}$	$(1, 2, +1)$
Higgs Boson $\mathbf{H_d}$	$(1, 2, -1)$
Color Gauge Fields G_a	$(8, 1, 0)$
Weak Gauge Fields $W^\pm, Z, \gamma$	$(1, 3+1, 0)$

better convergence of the gauge couplings at high scale, which could eventually point towards a grand unification of forces and (iv) existence of a stable supersymmetric particle that can play the role of the cold dark matter of the universe. Finally, a great plus for simple supersymmetric models is their prediction of a light Higgs (somewhere between 130 to 150 GeV), which can be hunted down in the current generation of experiments [22].

The simplest supersymmetric extension of the standard model is the so called minimal supersymmetric standard model (MSSM) with the particle content given in terms of superfields in table II.

Note that each field in the table has a scalar and a fermion part to it. Therefore MSSM has double the number of particles compared to the standard model plus the additional Higgs field and its fermionic partner. The Lagrangian for the theory has a supersymmetric part plus a part that breaks supersymmetry. In the supersymmetric part of the Lagrangian, there is only one phase like in the KM model but in the susy breaking part, there are 35 extra phases. Unlike the left-right model, there are no simple way to reduce the number of the phases. Thus CP violation is an extremely complex phenomenon in the MSSM. As such it leads to large uncontrolled CP violating effects in familiar processes like ϵ, ϵ' and the edm of the neutron.

The simplest example is the one loop contribution to the neutron edm [23], which arises via the exchange of the gluino which is the superpartner of the gluon. This leads to a value of $d_n^e \simeq \frac{e\alpha_s m_d}{4\pi M_{\tilde{q}}^2} sin\delta \approx 10^{-22} sin\delta$ ecm for $M_{\tilde{q}} \approx 100$ GeV. Thus one needs to tune the phase δ to the level of 10^{-4} to be compatible with present upper limits.

Similar large values appear for the ϵ and ϵ' unless one fine tunes parameters by large amounts [24]. This is particularly unsatisfactory since supersymmetry was introduced in the first place to avoid the fine tuning problems associated with the Higgs boson mass.

A simple extension of the MSSM that cures this problem is the left-right symmetric extension of the MSSM with high scale LR symmetry as suggested by the present data on neutrino masses and the seesaw way to understand them. That these models due to constraints of parity symmetry reduce the number of phases considerably has been shown in a series of recent papers [25,26]. The symmetry reason for the suppression of the edm is that edm is a parity odd operator; therefore in the aymptotic parity symmetric limit, it vanishes; therefore at low energies when parity is a broken symmetry of weak interactions, there appear additional suppressions. This model also provides a novel explanation of the ϵ' as being a purely supersymmetric effect. There are many tests of the model: a constrained super particle spectrum, an enhanced edm of the muon (10^{-23} ecm level) [27], neutron (10^{-27} ecm) etc. Its predictions in the B-sector are dependent on the superparticle spectrum and therefore B-physics will help to considerably restric the parameter space of the model, making it testable in the collider experiments. This model can also be tested in hyperon decays [28].

CONCLUSION

In summary, the standard model comes very close to explaining all known laboratory observations related to CP violation; yet there are compelling theoretical as well as cosmological reasons to believe that there must be CP violatiyong interactions beyond the standard model. We discuss only two scenarios out of the many possible ones in the literature due to lack of space and time. They are interesting from the point of view of their testability in the next generation of experiments. Important experiments besides the B-physics experiments already under way, are better searches for the edms of neutron and muon, searches for CP violating effects in rare kaon decays such as transverse polarization of muon in $K_{\ell 3}$ decay, $K_L \rightarrow \pi^0 e^+ e^-$ decay etc.

I would like to thank Peter Herczeg for many useful discussions on the topics covered in this review. This work is supported by the National Science Foundation grant no. PHY-9802551. I would like to dedicate this talk to the memory of Nimai Mukhopadhyay, who made many important contributions in the area of intersection between particle and nuclear physics.

REFERENCES

1. For a review and notation, see B. Winstein and L. Wolfenstein, Rev. Mod. Phys. **65**, 1113 (1993); also *CP Violation*, ed. C. Jarlskog, World Scientific (1988).
2. Particle Data Group, C. Caso et al. Euro. Phys. J. **C3**, 1 (1998).
3. V. Fanti et al. (NA48), hep-ex/9909022.
4. A. Alavi-Harati et al. Phys. Rev. Lett. **83**, 22 (1999).

5. L. Wolfenstein, Phys. Rev. lett. **13**, 562 (1964).
6. A. Angelopoulos et al. Phys. Lett. **B444**, 43 (1998).
7. P. K. Kabir, Phys. Rev. **D2**, 510 (1970).
8. For an excellent review of atomic and nuclear CP violating effects, see P. Herczeg, in *Symmetries and Fundamental Interactions in Nuclei*, ed. by Wick Haxton and Ernest M. Henley (World Scientific, Singapore, 1995); M. Musolf, Storrs preprint (1999).
9. R. N. Mohapatra, Phys. Rev. **D6**, 2023 (1972).
10. M. Kobayashi and T. Maskawa, Prog. Theor. Phys. **49**, 652 (1973).
11. R. N. Mohapatra and J. C. Pati, Phys. Rev. **D11**, 566 (1975).
12. G. Branco, L. Lavoura and J. Silva, *CP Violation*, (Oxford publishing, 1999).
13. For recent reviews, see A. Buras, hep-ph/9806471; M. Jamin, hep-ph/9911390; M. Ciuchini et al. hep-ph/0006056.
14. V. M. Khatsimovsky, I. B. Khriplovich and A. Zhitnisky, Z. Phys. **C36**, 455 (1987).
15. J. Rosner, hep-ph/0005258; I. Bigi, *Fundamental particles and interactions*, ed. R. Panvini and T. Weiler (AIP publications, 1997); page 1.
16. S. Weinberg, Phys. Rev. lett. **37**, 657 (1976); G. Branco, Phys. Rev. Lett. **44**, 504 (1978).
17. K. S. Babu and R. N. Mohapatra, Phys. Rev. Lett. **62**, 1079 (1989).
18. D. Chang, Nucl. Phys. **214B**, 435 (1983).
19. G. Barenboim, J. Bernabeu and M. Raidal, Phys. Rev. Lett. **80**, 4625 (1998); G. Barenboim, Phys. Lett. **B443**, 317 (1998); G. Barenboim et al., Phys. Rev. **D60**, 016003 (1999); P. Ball, J. M. Frere and P. Matias, Nucl. Phys. **B562**, 3 (2000); P. Ball and R. Fleisher, Phys. Lett. **B475**, 111 (2000).
20. P. Herczeg, Phys. Rev. **D28**, 200 (1983).
21. For a review, see R. N. Mohapatra, in *CP Violation*, ed. C. Jarlskog (World Scientific, 1988).
22. For a review of MSSM, see H. Haber and G. Kane, Phys. Rep. **117**, 75 (1985).
23. J. Ellis, S. Farrara and D. Nanopoulos, Phys. Lett. **B114**, 231 (1982); J. Polchinski and M. Wise, Phys. Lett. **125B**, 393 (1983); J. Gerard, W. Grimus, D. V. nanopoulos and A. Raychoudhuri, Nucl. Phys. **B253**, 93 (1985); P. Nath, Phys. Rev. Lett. **66**, 2565 (1991); Y. Kizhukuri and N. Oshimo, Phys. Rev. **D 45**, 1806 (1992); R. Garisto and J. Wells, Phys. Rev. **D 55**, 611 (1997).
24. For a review and references, see A. Masiero, hep-ph/9709242.
25. R. N. Mohapatra and A. Rasin, Phys. Rev. **D 54**, 5835 (1996).
26. K. S. Babu, B. Dutta and R. N. Mohapatra, Phys. Rev. **D61**, 019701R (2000).
27. K. S. Babu, B. Dutta and R. N. Mohapatra, hep-ph/0006329.
28. X. G. Ge, H. Murayama, S. Pakvasa and G. Valencia, hep-ph/9909562.

Progress in Heavy-Quark Physics from BaBar, Belle, and CLEO

Jeffrey D. Richman

Department of Physics
University of California
Santa Barbara, CA 93106

Abstract. I present an overview of progress in heavy-quark physics, focusing on results from e^+e^- colliding-beam experiments. The central goal of these experiments is to discover and precisely measure CP asymmetries in a variety of processes in the B-meson system. Together with related quantities, such as the magnitudes of Cabibbo-Kobayashi-Maskawa elements, these measurements will overconstrain the Standard-Model framework for CP violation, either confirming this framework or providing evidence for new physics. The heavy-quark physics program also includes many other fascinating processes that do not bear directly on CP violation but which nevertheless are of great interest. Such processes as electroweak penguin decays, rare hadronic B decays, and $D^0 - \bar{D}^0$ mixing have sensitivity to new physics or help to sharpen our understanding of physics within the Standard Model. I also discuss the new B-factory accelerators, whose remarkable success should ensure rapid progress in this field. My presentation is intended to highlight some of the most interesting aspects of the subject for nonexperts.

INTRODUCTION

I am delighted to speak to this audience of nuclear and particle physicists on recent progress in heavy-quark physics. My talk focuses on the program of the three e^+e^- colliding-beam experiments—BaBar, Belle, and CLEO—all of which operate at a center-of-mass energy equal to the mass of the $\Upsilon(4S)$ resonance: $\sqrt{s} = 10.58$ GeV. The overwhelmingly dominant decays are $\Upsilon(4S) \to B^0\bar{B}^0$ and $\Upsilon(4S) \to B^+B^-$, which provide a clean and kinematically constrained means to produce B mesons. Charm mesons are also produced at this energy, both as daughter particles of B mesons and from the continuum process $e^+e^- \to c\bar{c}$.

The CLEO experiment, which runs at the Cornell Electron Storage Ring (CESR), has a mature and ongoing heavy-quark physics program, while BaBar (at SLAC) and Belle (at KEK) are just beginning theirs. These new experiments operate at e^+e^- storage rings that have unequal beam energies, so that the resulting $B\bar{B}$ system is boosted relative to the detector. As we will see, this feature enables BaBar and Belle to measure time-dependent CP asymmetries in the neutral B meson system. First results from BaBar and Belle will be presented later this summer, so at this point I can only present a status report on the rapid progress being made toward the CP asymmetry measurements. Most of the results I will present are from CLEO, although I will also describe some important measurements from the LEP experiments at CERN, which use the process $e^+e^- \to Z \to b\bar{b}$ to study a variety of b-hadrons. In a separate talk [1], Barry Wicklund will describe results on heavy-quark physics from the Fermilab hadron-collider experiments. Other related plenary-session talks are by Mohapatra [2] on CP violation and El-Khadra [3] on lattice QCD calculations.

The title—"Conference on the Intersections of Particle and Nuclear Physics"—raises the question of whether the fields of heavy-quark and nuclear physics do, in fact, intersect! One example can be written directly in the form of an equation,

$$|V_{ud}|^2 + |V_{us}|^2 + |V_{ub}|^2 = 1, \tag{1}$$

which ties together measurements of V_{ud} and V_{ub}. The value of $|V_{ud}|$ is extracted from superallowed β decays of nuclei, while $|V_{ub}|$ is obtained from semileptonic decays of the B meson. More generally, both nuclear and heavy-quark physics have long focussed on the issues of discrete symmetries and their violation; today, searches for CP and T violation by physicists in both fields provide complementary approaches to these profound questions.

I will begin with an overview of the goals and ideas of heavy-quark physics and then describe the experiments and accelerators operating at the $\Upsilon(4S)$. I then discuss progress in four main areas: semileptonic decays and measurements of the Cabibbo-Kobayashi-Maskawa (CKM) matrix elements V_{cb} and V_{ub}; rare B decays, including electroweak and hadronic penguins; mixing in the $D^0\bar{D}^0$ system; and searches for CP asymmetries in B decays. First results on CP asymmetries from BaBar and Belle will be available later this summer, so I will simply present a status report on these experiments. (In the interests of full disclosure I should say that I am a member of the BaBar collaboration and a former member of the CLEO collaboration!)

GOALS AND IDEAS OF HEAVY-QUARK PHYSICS

The most compelling question in heavy-quark physics is whether the source of CP violation lies within the Standard Model, in physics beyond it, or in some combination of the two. The Standard Model can accomodate precisely one CP-violating (or "weak") phase, which resides in the three-generation CKM quark mixing matrix

$$V = \begin{pmatrix} V_{ud} & V_{us} & V_{ub} \\ V_{cd} & V_{cs} & V_{cb} \\ V_{td} & V_{ts} & V_{tb} \end{pmatrix} = \begin{pmatrix} 1 - \tfrac{1}{2}\lambda^2 & \lambda & A\lambda^3(\rho - i\eta) \\ -\lambda & 1 - \tfrac{1}{2}\lambda^2 & A\lambda^2 \\ A\lambda^3(1 - \rho - i\eta) & -A\lambda^2 & 1 \end{pmatrix} + \mathcal{O}(\lambda^4). \tag{2}$$

Here we have used the Wolfenstein parametrization, which shows explicitly that there are only four independent real parameters needed to specify the matrix. The expansion parameter is $\lambda = \sin\theta_C \approx 0.22$. The weak phase in the CKM matrix can manifest itself in many different ways, leading to CP asymmetries both small and large.

CP violation was discovered in 1964 in the neutral kaon system by Christensen, Cronin, Fitch, and Turlay [4], who observed the mode $K_L^0 \to \pi^+\pi^-$ with a branching fraction of around 10^{-3}. Jim Cronin made some prescient remarks in his Nobel-prize lecture [5]:

> ...the effect [of CP violation] is telling us that at some tiny level there is a fundamental asymmetry between matter and antimater, and it is telling us that at some tiny level interactions will show an asymmetry under the reversal of time. We know that improvements in detector technology and quality of accelerators will permit even more sensitive experiments in coming decades. We are hopeful, then, that at some epoch, perhaps distant, this cryptic message from nature will be deciphered.

In the B system, the Standard Model predictions for CP asymmetries are not at the 10^{-3} level, but in the range 0.1 to 1. This doesn't mean that they are easy to measure, because they are associated with decay processes that typically have small branching fractions, in the range 10^{-4} to 10^{-5}. Nevertheless, the outstanding feature of the B meson system is that the observed CP asymmetries may well be quite large.

Let's examine the conditions [6,7] required to produce a CP asymmetry in a decay process. First, we need a decay involving (at least) two amplitudes, A_1 and A_2, one of which has a CP-violating phase ϕ_2, which we explicitly display as $A_2 = a_2 \exp(i\phi_2)$. The total amplitude is given by

$$A = A_1 + a_2 e^{i\phi_2}. \tag{3}$$

Initially, we assume that A_1 and A_2 have no relative CP-conserving phase; A_1 and a_2 are therefore relatively real. Because overall phases do not affect the rate, we can take A_1 and a_2 to be real. The amplitude for the CP-conjugate (antiparticle) process is

$$\bar{A} = A_1 + a_2 e^{-i\phi_2}, \tag{4}$$

up to an overall phase [7]. It is easy to show that $|A| = |\bar{A}|$. Thus, even though a CP-violating phase ϕ_2 is present, and the rates for both processes are affected by the value of ϕ_2, without a relative CP-conserving phase there is no CP asymmetry.

We next introduce a CP-conserving phase (δ_2) in addition to the CP-violating phase. The total amplitudes for the original and CP-conjugate processes (up to an overall phase) are

$$A = A_1 + a_2 e^{i(\phi_2+\delta_2)} \qquad \bar{A} = A_1 + a_2 e^{i(-\phi_2+\delta_2)}. \tag{5}$$

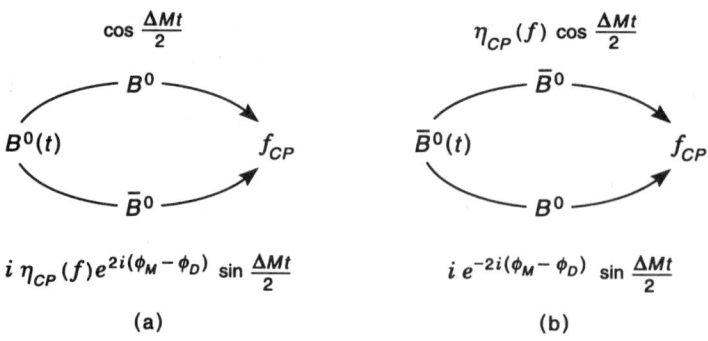

FIGURE 1. The interference between $B^0\bar{B}^0$ mixing and decay can be used to study CP asymmetries. The meson, a initially a B^0 or $\bar{B}^0$, is tagged using correlations with other particles produced at the same time. Due to B^0-$\bar{B}^0$ oscillations, the initial state evolves (becoming $B^0(t)$ or $\bar{B}^0(t)$) and develops both B^0 and $\bar{B}^0$ components. The decays of these states to a common final state (f_{CP}) interfere, and the time-dependent rates $\Gamma(B^0(t) \to f_{CP})$ and $\Gamma(\bar{B}^0(t) \to f_{CP})$ differ due to the CP-violating phase in the interference term.

In this case, one finds that $|A| \neq |\bar{A}|$, so that the process will display a CP asymmetry. We therefore need to study decays in which there is interference involving *both* the weak phase of interest and a well understood CP-conserving phase. Unfortunately, it is not yet possible to calculate phases due to strong final-state interactions. Even though so-called "direct CP violation" (from the interference between two direct decay amplitudes) can be quite interesting, it is therefore not the best source of information on weak phases.

However, a well understood CP-conserving phase arises from particle-antiparticle oscillations, or mixing. Figure 1 shows how we can use the interference between mixing and decay to study CP violation in B^0 decay. This method leads to *time-dependent CP asymmetries*. Mixing is especially interesting because it is sensitive to virtual intermediate states that can introduce weak phases other than those expected in the Standard Model. We note that the CP violation "in mixing" observed in the $K^0\bar{K}^0$ system arises from the interference between mixing amplitudes through real and virtual intermediate states. In the $B^0\bar{B}^0$ system, the contribution to mixing from real intermediate states is very small, and the CP asymmetry from mixing alone is expected to be difficult to measure.

The Standard Model framework for CP violation is specific and predictive, and it is possible in principle to infer the level of CP asymmetries from measurements of *CP-conserving* observables. In particular, the unitarity of the CKM matrix leads to six relations of the form

$$V_{ud}V_{ub}^* + V_{cd}V_{cb}^* + V_{td}V_{tb}^* = 1, \qquad (6)$$

where in this case we have taken the inner product between the first and third column of the matrix. This equation can be represented graphically as the so-called unitarity triangle, whose area is a direct measure of CP violation in the Standard Model. The B system is special in that each of the three terms in Eq. 6 is of order λ^3, leading to a triangle that is not as flat as those from the other five relations, and it means that the CP asymmetries are expected to be larger. The triangle can be specified by measurements of the lengths of the sides alone (non-CP-violating observables) or by measurements of angles between the sides (CP violating observables). (The overall orientation of the triangle is non-physical and depends on the convention used for the quark phases.) Thus, a second major goal of heavy-quark physics is to measure the magnitudes of CKM elements, partly to pin down all the parameters of the Standard Model, but also to achieve a comprehensive understanding of CP violation and its connection to the CKM matrix. Measurements of semileptonic decays, $B^0\bar{B}^0$ and $B_s\bar{B}_s^0$ oscillation rates, and the electroweak penguin processes $b \to tW^- \to s(d)$ provide information on $|V_{cb}|$, $|V_{ub}|$, $|V_{td}|$, and $|V_{ts}|$.

CP asymmetries are measured in hadronic processes, where the necessary interference terms are present in the Standard Model. Because the *phases* of CKM matrix elements are extracted from asymmetries rather than decay rates, the hadronic matrix elements can cancel, removing the dependence on theory. In contrast, to extract the *magnitudes* of CKM elements, we measure decay rates and rely on theoretical calculations that predict each rate up to an unknown overall constant, which is the CKM factor. For example, the branching fraction for a semileptonic decay of a meson M with quark content $Q\bar{q}$ can be expressed in the form

$$\mathcal{B}(M_{Q\bar{q}} \to X_{q'\bar{q}} \ell \bar{\nu}) = \gamma_{\text{theory}} |V_{q'Q}|^2 \tau_M. \tag{7}$$

Experimentalists measure the branching fraction and lifetime τ_M, while theorists must predict γ_{theory}. Note that the relative error on $|V_{q'Q}|$ is half that on the branching fraction due to the square root; this helps to make poor results look better! The processes used to extract the magnitudes of CKM elements are sufficiently simple that it is possible to make theoretical predictions for the decay rates with a precision typically in the 10% to 50% range in the rate, or in the 5% to 25% range for the magnitude of the CKM element.

The need to calculate the factor γ_{theory} and other related quantities has led to an intensive effort, both theoretical and experimental, to better understand the dynamics of processes involving heavy quarks. This represents a third major area of research in heavy-quark physics. There have been long-running programs of measuring and calculating form factors for semileptonic decays, decay constants for leptonic decays and mixing, and the rates for electroweak penguins. Sophisticated methods such as Heavy Quark Effective Theory (HQET) [8] and lattice QCD [3] are being used to constrain or predict various hadronic parameters. The relatively large masses of the bottom and charm quarks (compared to Λ_{QCD}) have been key to achieving advances in these calculations.

Although it has not been possible to carry out comparable calculations of the decay rates for hadronic final states, there has nevertheless been a recent effort to extract useful CKM information from hadronic decay rates. In this approach,

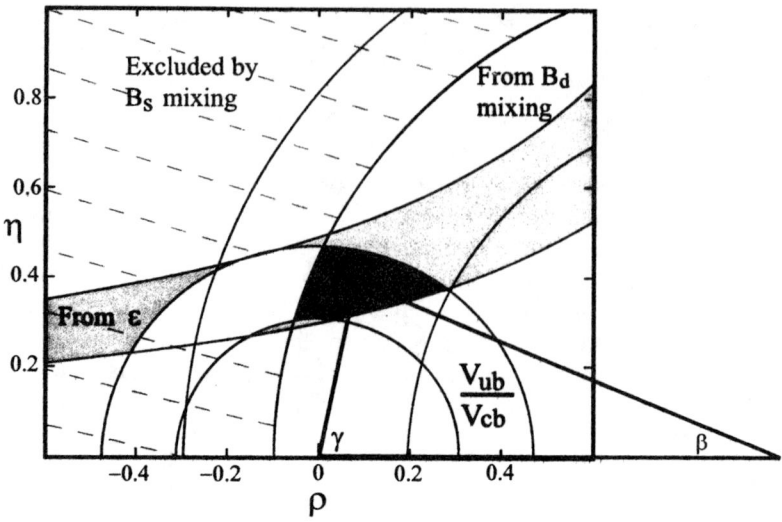

FIGURE 2. Constraints on the ρ-η plane from different types of measurements. The lengths of the sides are determined from rates for semileptonic decays, $B^0\bar{B}^0$ mixing, and $B_s\bar{B}_s$ mixing. Measurements from kaon physics can also supply important constraints, as in the quantity ϵ from CP violation in the $K^0\bar{K}^0$ system. Current investigations of CP asymmetries will determine the angles of the triangle α, β, and γ. When the constraints are combined statistically, the sharp corners are rounded off. (Figure used with permission of S. Stone.)

measurements are used to constrain the unknown hadronic parameters as much as possible.

Finally, heavy-quark decays involving loops are a fourth major area of investigation. Such processes, which include the electroweak penguins already mentioned, can be affected by new heavy particles that circulate in the loops. I will review some of the progress made in these studies, including the important case of $D^0\bar{D}^0$ mixing. This process is expected to be highly suppressed in the Standard Model due to GIM cancellations, and it provides a particularly sensitive way to search for the effects of new physics.

EXPERIMENTS AND ACCELERATORS AT THE $\Upsilon(4S)$

The success of the B factories running at the $\Upsilon(4S)$ resonance has been a remarkable achievement by the accelerator-physics community. These are the highest luminosity colliding-beam machines in the world! There are many complex problems that have been solved in the design of such machines, which involve extremely high-current, multi-bunch operation. The PEP-II ring, for example, is designed to store 1658 particle bunches in each beam; the issues of multibunch instabilities and the avoidance of parasitic crossings (bunch crossings not at the interaction

TABLE 1. Parameters of accelerators operating at the $\Upsilon(4S)$. The integrated luminosities of BaBar and Belle are given as of August 2000. For all experiments, the integrated luminosity includes both on-$\Upsilon(4S)$ and off-resonance (continuum) running.

Machine	CESR	PEP-II	KEKB
Detector	CLEO	BABAR	Belle
Date 1st events recorded	1979	May '99	June '99
Circumference (km)	0.768	2.199	3.016
Number of Rings	1	2	2
E_{e^+} (GeV)	5.3	3.1	3.5
E_{e^-} (GeV)	5.3	9.0	8.0
$\beta\gamma$	0.06 (B)	0.56 ($\Upsilon(4S)$)	0.42 ($\Upsilon(4S)$)
Num. of beam bunches (design)	$9 \times 4 \to 9 \times 6$	1658	2700
I_{beam} (A) (design)	$0.26 \to 0.5$	2.16 (e^+), 1.0 (e^-)	2.6 (e^+), 1.1 (e^-)
Δs_{bunch} (m) (design)	4.2	1.26	1.2
Δt_{bunch} (ns) (design)	14	4.2	4.0
Beam crossing angle (mrad)	± 2.1	0 (magnetic sep.)	± 11
Achieved luminosity cm^{-2}s^{-1}	8.3×10^{32}	$\sim 2 \times 10^{33}$	$\sim 2 \times 10^{33}$
Best daily integ. lum. (pb^{-1})	140	90	40
Integrated luminosity (fb^{-1})	13	14.8	6.8

point) are extremely important. These machines must store very high beam currents, while at the same time maintaining tolerable background conditions for the detectors, and there are stringent requirements on the vacuum quality.

Table 1 summarizes some of the main features of CESR, KEKB, and PEP-II. CESR has operated for many years and has produced a long series of results in B physics going back to the first observation of exclusive B meson decays. The electron and positron beams are at equal energy, so the $\Upsilon(4S)$ is produced at rest in the lab frame, and the daughter B mesons from $\Upsilon(4S) \to B\bar{B}$ have very low velocity: $\beta\gamma \approx 0.06$. Each beam (electrons or positrons) consists of 9 bunch trains with 4 bunches each, for a total of 36 bunches. To avoid parasitic collisions, the beams are put into special "pretzel" orbits that weave around each other with transverse displacements of up to 2 cm from the centerline of the accelerator vacuum chamber.

PEP-II and KEKB are only a year old, but already they have reached unprecedented beam currents and luminosities. Because these machines store beams at unequal energies, they require separate rings for electrons and positrons, except for a complicated region around the interaction point where the beams intersect. As a benefit, parasitic beam crossings are an issue only in the interaction region itself. Successive arriving beam bunches are 4.2 ns (1.26 m) apart (with 1658 bunches/beam) in PEP-II, and to prevent an outgoing bunch from interacting with the next incoming bunch, the beams are deflected with dipole magnets starting about 20 cm from the interaction point. In KEKB a different approach is taken, where the beams enter and exit with a ± 11 mrad crossing angle.

PEP-II has achieved a peak luminosity $\mathcal{L} = 2.3 \times 10^{33}$ cm^{-2}s^{-1} with total beam currents of $I_{e^-} = 0.75$ A and $I_{e^+} = 1.3$ A, and 606 bunches/beam. KEKB has achieved nearly the same luminosity with 1146 bunches per beam. At $\mathcal{L} = 3 \times 10^{33}$ cm^{-2}s^{-1}, BaBar and Belle will each accumulate about 30 million $\Upsilon \to B\bar{B}$ events in 10^7 s. All three machines—CESR, KEKB, and PEP-II—should reach even higher luminosities in the future.

SIMPLE PROCESSES: SEMILEPTONIC DECAYS AND MAGNITUDES OF CKM ELEMENTS

Semileptonic B decays are sufficiently simple to allow us to extract the magnitudes of CKM elements from decay rates. Strong interaction effects are, of course, still present, but they are isolated to a single hadronic current that can be rigorously parametrized by form factors. The form factors are Lorentz-invariant functions of q^2, the square of the mass of the virtual W.

Because semileptonic decays produce three-body final states (at least), there is a range of allowed kinematic decay configurations, whose population in the Dalitz plot of q^2 vs. E_ℓ (lepton energy) provides detailed information on the decay dynamics. For any semileptonic decay $B \to X \ell^- \bar{\nu}$, q^2 characterizes the recoil velocity of the hadronic daughter system X, since

$$w \equiv \gamma_X = \frac{E_X}{m_X} = \frac{M_B^2 + m_X^2 - q^2}{2 M_B m_X}. \tag{8}$$

At maximum q^2 ($q^2_{\max} = (M_B - m_X)^2$), $w = 1$, and the daughter hadron has zero recoil velocity relative to the B.

There are two main classes of semileptonic B decays, the dominant $b \to c \ell^- \bar{\nu}$ modes, which can be used to extract $|V_{cb}|$, and the much rarer and more difficult to study $b \to u \ell^- \bar{\nu}$ decays, which are used to determine $|V_{ub}|$. There is a large difference in how well we understand these two types of decays. The main $b \to c \ell^- \bar{\nu}$ modes, $B \to D^* \ell^- \bar{\nu}$ and $B \to D \ell^- \bar{\nu}$, have been studied in some detail; together they represent a substantial fraction (about two-thirds) of the inclusive semileptonic rate ($B(B \to Xe\nu) = (10.5 \pm 0.8)\%$) [9]. Form factors for these processes have been measured [11–13] by studying the kinematic distributions and their correlations, and their mild variation within the small kinematically accessible range of w (about 1 to 1.6) is consistent with theoretical expectations [8]. In contrast, the allowed range in w is typically much larger in $b \to u \ell^- \bar{\nu}$ transitions, so the recoil velocity of the daughter hadron can be quite large and the variation in the form factors much greater than for $b \to c \ell^- \bar{\nu}$ decays.

There has been a large effort, both experimental and theoretical, to extract $|V_{cb}|$ from the $B \to D^* \ell^- \bar{\nu}$ decay rate at zero recoil. At $w = 1$, a $B \to D^{(*)}$ transition produces very little disturbance to the hadronic wave function, since one heavy quark (b) is simply replaced by another (c), with little momentum transfer to the spectator quark. In the heavy-quark symmetry limit, $m_b, m_c \gg \Lambda_{\text{QCD}}$, the meson

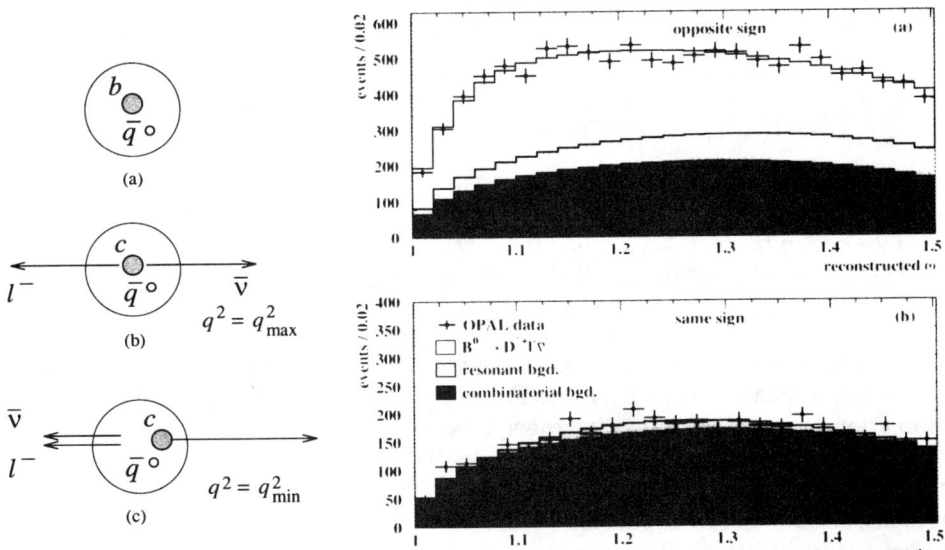

FIGURE 3. $B \to D^*\ell^-\bar{\nu}$ and the measurement of $|V_{cb}|$. The figures at the left show (a) a B meson before decay, (b) semileptonic decay configuration at high q^2 (low recoil velocity of daughter hadron), and (c) semileptonic decay configuration at low q^2 (high recoil velocity of daughter hadron). The goal of measurements is to determine $|V_{cb}|$ from the decay rate at high q^2, where the meson wave function is least disturbed and theoretical calculations are most reliable. The figures at right show OPAL data for $B \to D^{*+}\ell^-\bar{\nu}$ using a partial reconstruction technique for the D^{*+}. This technique gives a larger signal yield (and higher backgrounds) than the usual full reconstruction. The signal is seen in the opposite-sign lepton-D^* charge correlations (a) at right, while a test of the understanding of backgrounds is provided by the same-sign charge correlation data, shown in (b) at right. The rate is extrapolated to $\omega = 1$, where ω is the same as the quantity w discussed in the text.

is completely undisturbed by the decay. The decay rate can therefore be predicted with a relatively small uncertainty. (The decay $B \to D\ell^-\bar{\nu}$ has a smaller branching fraction, more background, and is forbidden by angular-momentum conservation at the zero-recoil point.) The differential decay rate for $B \to D^*\ell^-\bar{\nu}$ can be written

$$\frac{d\Gamma(B \to D^{*+}\ell^-\bar{\nu})}{dw} = \frac{G_F^2 M_B^5}{48\pi^3} K(w)|V_{cb}|^2 \mathcal{F}(w)^2, \tag{9}$$

where $w = E_{D^*}/m_{D^*}$) and $K(w)$ is a known kinematic function. The theoretical uncertainties are contained in $\mathcal{F}(w)$, which is a linear combination of the three form factors that govern the decay. Numerous estimates for $\mathcal{F}(1)$ have been made in the framework of heavy-quark effective theory (HQET) and in lattice QCD. One recent compilation made for Particle Data Group [9] gives $\mathcal{F}(w) = 0.93 \pm 0.05$, while another compilation [10] gives $\mathcal{F}(1) = 0.913 \pm 0.042$.

TABLE 2. Results on $|V_{cb}|$ from experiments at the $\Upsilon(4S)$ and the Z. From Ref. [16]. A recent new preliminary CLEO measurement is discussed in the text.

Mode	$|V_{cb}|$	Comment
$\bar{B} \to D^{*+}\ell^-\bar{\nu}$ (from $Z \to b\bar{b}$)	$0.0367 \pm 0.0023 \pm 0.0018$	$\mathcal{F}(w)$ param. from dispersion relations
$\bar{B} \to D^{*+}\ell^-\bar{\nu}$ (from $\Upsilon(4S)$)	$0.0392 \pm 0.0030 \pm 0.0019$	Linear $\mathcal{F}(w)$ param. + correction
$\Gamma(b \to c\ell^-\bar{\nu})$ (Z)	$0.0408 \pm 0.0005 \pm 0.0025$	B^0, B^+, B_s, b-baryons
$\Gamma(B \to X_c\ell^-\bar{\nu})$ ($\Upsilon(4S)$)	$0.0400 \pm 0.0010 \pm 0.0024$	B^0, B^+ admixture

Experimentalists have attempted to measure the rate near $w = 1$, usually by performing a measurement over the full phase space and then performing an extrapolation to $w = 1$, guided by theoretical shapes. Figure 3 shows the w spectrum for $B \to D^*\ell^-\bar{\nu}$ decay measured by the OPAL experiment [14]. Table 2, taken from Ref. [16], compares the values of $|V_{cb}|$ obtained from several different methods. Heavy-quark expansion techniques have also been used by theorists to predict the inclusive semileptonic rate; these results are comparable in overall precision to those from $B \to D^*\ell^-\bar{\nu}$, although the experimental errors are somewhat smaller. An update [13] from CLEO for $B \to D^*\ell^-\bar{\nu}$ became available after this conference but is still based on only 3.3 M $B\bar{B}$ events. The preliminary results are $B(\bar{B}^0 \to D^{*+}\ell^-\bar{\nu}) = (5.66 \pm 0.29 \pm 0.33)\%$ and $|V_{cb}|\mathcal{F}(1) = 0.0424 \pm 0.0018 \pm 0.0019$.

The measurement of $|V_{ub}|$ is significantly more difficult. This quantity was first measured from the rate beyond the inclusive $B \to X_c\ell^-\bar{\nu}$ endpoint, but more recently it has been determined from measurements of $B \to \pi\ell^-\bar{\nu}$ and $B \to \rho\ell^-\bar{\nu}$ exclusive rates. Because the daughter quark in the $b \to u$ transition is light, the zero-recoil point of the daughter hadron does not provide a solid point for predicting the form factors. Recent results from CLEO [17] have been combined with an earlier study [18] and are summarized in Fig. 4, with the final value $|V_{ub}| = (3.25 \pm 0.14^{+0.21}_{-0.29} \pm 0.55) \times 10^{-3}$, where the errors are statistical, systematic, and theoretical. The $B \to \rho\ell^-\bar{\nu}$ signal yield is 216 ± 32 events. The measured branching fraction depends to some extent on which theoretical model is used to determine the experimental detection efficiencies, since the analysis uses cuts on the lepton energy and other quantities. An additional theoretical error on $|V_{ub}|$ arises from uncertainties in the theoretical predictions for the overall rate; this is now the dominant error for exclusive measurements. (It should be noted that the theoretical error quoted in the CLEO paper is not based on the spread among the $|V_{ub}|$ values, which is smaller.)

A third and very interesting approach to measuring $|V_{ub}|$ has been pursued by the LEP experiments, which can measure the $b \to u\ell^-\bar{\nu}$ rate inclusive rate over a reasonably broad region of phase space. Although these measurements have larger statistical errors, the theoretical uncertainties are reduced. The DELPHI collaboration [19] has used a technique in which the mass of the candidate hadronic system is required to be less than M_D; together with additional $b \to u\ell^-\bar{\nu}$ en-

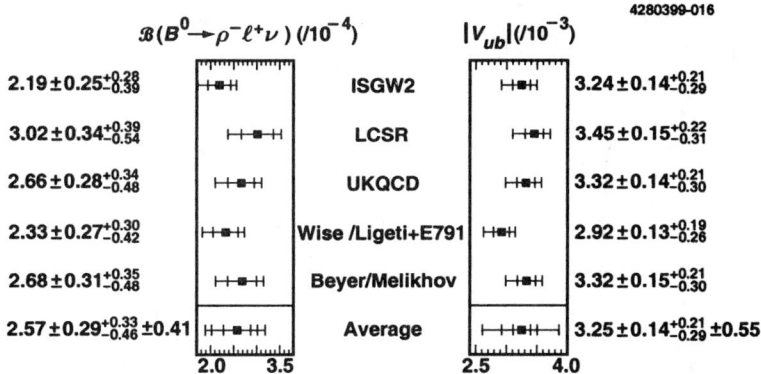

FIGURE 4. CLEO measurement of $B(B^0 \to \rho^-\ell^+\nu)$ and $|V_{ub}|$. The branching fraction depends on the shapes of kinematic distributions predicted by models through the efficiency of the selection cuts. There is an additional theoretical error on $|V_{ub}|$ due to uncertainty in the overall normalization of form factor predictions.

richment cuts (vertexing, π/K separation), this procedure yields a $b \to u\ell^-\bar{\nu}$ signal of 214 ± 56 events and $B(b \to u\ell^-\bar{\nu}) = (1.57 \pm 0.35 \pm 0.55) \times 10^{-3}$. The LEP Heavy Flavors Working Group has calculated [20] a LEP average value $|V_{ub}|/10^{-3} = 4.04^{+0.41}_{-0.46}(\text{expt})^{+0.43}_{-0.48}(b \to c)^{+0.24}_{-0.25}(b \to u) \pm 0.02(\tau_b) \pm 0.19(\text{HQS})$. Measurements of $|V_{ub}|$ will be an active area of research for years to come!

RARE B DECAYS

With of order 10 M $B\bar{B}$ events recorded, the B-factory experiments are sensitive to decay modes with branching fractions in the range 10^{-6} to 10^{-4}, depending on the number and types of particles produced in the final state and the branching fractions of the daughter particles. With this level of sensitivity, experiments are exploring new channels beyond the $b \to c$ transition, which is responsible for the dominant class of B decays. CLEO has been producing a steady stream of measurements of new rare decay modes (with BaBar and Belle soon to follow), and there is a large theoretical effort to understand and exploit this new information.

The current frontier of rare decays includes at least five different types of modes: (1) $b \to u$ transitions (e.g., $B \to \pi\pi$ and $B \to \rho\pi$), (2) strong (or gluonic) penguin decays (e.g., $B \to K\pi$), (3) electroweak penguin decays (e.g., $B \to K^*\gamma$, $b \to s\gamma$ inclusive, and $B \to K^{(*)}\ell^+\ell^-$), (4) leptonic decays ($B^+ \to \tau^+\nu_\tau$), and (5) $b \to c$ decays that are color suppressed due to so-called internal W emission. We will focus on the first three types of modes below; some useful reviews of penguins and other rare decays are given in the references [21–23].

Penguins were first observed in B decays by CLEO, which discovered [24] the radiative-penguin process $B \to K^*\gamma$ in 1993. This decay involves an internal loop associated with the $b \to tW^- \to s$ transition; the photon, which is distinctive due to its high energy, can be radiated from any of the charged particles. Since $B \to K\gamma$ is forbidden by angular momentum conservation, $B \to K^*\gamma$ is the simplest exclusive mode, and a handful of decays were originally seen in a sample of about 3M $B\bar{B}$ events. With a sample of 9.7M $B\bar{B}$ events, CLEO has updated and extended these measurements [25]

$$B(B^0 \to K^{*0}\gamma) = (4.55^{+0.72}_{-0.68} \pm 0.34) \times 10^{-5}$$
$$B(B^+ \to K^{*+}\gamma) = (3.76^{+0.89}_{-0.83} \pm 0.28) \times 10^{-5}$$
$$B(B \to K_2^*(1430)\gamma) = (1.66^{+0.59}_{-0.53} \pm 0.13) \times 10^{-5}.$$

The yields are $88.3^{+12.2}_{-11.5}$ events, $36.7^{+8.3}_{-7.6}$ events and $15.9^{+5.7}_{-5.1}$ events. The branching fractions are certainly within the range of theoretical expectations, but, due to the complications of hadronic effects, precise predictions are available only for the inclusive $b \to s\gamma$ (or $B \to X_s\gamma$) decay. After an enormous theoretical effort, the rate for $b \to s\gamma$ has been calculated [26–28] including all next-to-leading order terms, yielding $B(b \to s\gamma) = (3.28 \pm 0.33) \times 10^{-4}$. This value is consistent with measurements from CLEO [29] and ALEPH [30]:

$$B(b \to s\gamma) = (3.15 \pm 0.35 \pm 0.41) \times 10^{-4} \text{ (CLEO)}$$
$$B(b \to s\gamma) = (3.11 \pm 0.80 \pm 0.72) \times 10^{-4} \text{ (ALEPH)}.$$

Although the presence of a loop provides some sensitivity to higher mass particles such as SUSY partners, there is currently no evidence for an anomaly in the $b \to s\gamma$ rate.

The role of gluonic penguins in rare hadronic B decays has been a subject of intense interest for several years, and there is now strong evidence that such processes are important in a number of recently observed modes. The decay $B^- \to \bar{K}^0\pi^-$, for example, is naturally explained as a hadronic penguin: it cannot result from a $b \to u$ transition unless final-state interactions convert a $\bar{u}su\bar{u}$ system into a $\bar{u}sd\bar{d}$ system. This explanation is quite unlikely: the penguin amplitude is $\mathcal{O}(\lambda^2)$, while the tree diagram is $\mathcal{O}(\lambda^4)$, where $\lambda = \sin\theta_C$.

The news that gluonic-penguin contributions can be large is not entirely welcome in all cases. In particular, the decay $B^0 \to \pi^+\pi^-$ produces a CP eigenstate, and it appeared at first that this mode would be well suited to the determination of $\sin 2\alpha$ through a measurement of the time-dependent CP asymmetry. However, as Fig. 5 shows, both $b \to u$ tree and gluonic penguin diagrams can contribute to $B^0 \to \pi^+\pi^-$. This complicates the interpretation of the CP-asymmetry measurement, because there are two amplitudes in addition to the mixing amplitude. The quantitative impact of this complication is still not clear, because we do not yet know the relative size of the penguin and tree contributions.

We can, however, make some simple observations. Power counting in λ shows that the $b \to u$ tree process should contribute more to $\bar{B}^0 \to \pi^+\pi^-$ than to $\bar{B}^0 \to K^-\pi^+$, so that if this were the only amplitude present we would expect $B(\bar{B}^0 \to$

$\pi^+\pi^-) \gg B(\bar{B}^0 \to K^-\pi^+)$. (See the upper-left part of Fig. 5.) On the other hand, the penguin process (with intermediate t or c quarks) should contribute more to $\bar{B}^0 \to K^-\pi^+$ than to $\bar{B}^0 \to \pi^+\pi^-$, so $B(\bar{B}^0 \to K^-\pi^+) \gg B(\bar{B}^0 \to \pi^+\pi^-)$ would indicate a large gluonic-penguin contribution.

In a scenario where $B(\bar{B}^0 \to K^-\pi^+) \approx B(\bar{B}^0 \to \pi^+\pi^-)$, we could conclude that $\bar{B}^0 \to K^-\pi^+$ was predominantly penguin and that $\bar{B}^0 \to \pi^+\pi^-$ was predominantly a tree process. That is, the upper-left and lower-right decay diagrams in Fig. 5 must dominate. For, if the $\bar{B}^0 \to K^-\pi^+$ were mainly a tree process, then the $\bar{B}^0 \to \pi^+\pi^-$ tree contribution should be even larger, and we would observe $B(\bar{B}^0 \to \pi^+\pi^-) > B(\bar{B}^0 \to K^-\pi^+)$.

Table 3 lists just a few of the recent measurements [31–33] of hadronic rare modes by CLEO. These data indicate that $B(\bar{B}^0 \to K^-\pi^+)$ is significantly larger than $B(\bar{B}^- \to \pi^+\pi^-)$, so that the penguin contribution to $B^0 \to \pi^+\pi$ may not be negligible. In this case, it may be preferable to measure $\sin 2\alpha$ using the mode $B \to \rho\pi$, where a Dalitz-plot analysis can in principle be used to help understand the penguin contribution [34].

The interference between $b \to u$ tree and penguin diagrams is sensitive to the CKM angle γ, and direct CP violation is a possibility, although it is not possible to predict the values of the required strong phases. Searches for direct CP violation in B decays have so far yielded only limits [35]. There are numerous theoretical efforts [36–40] to better understand the dynamics of these rare decays. As we saw earlier, CP-violating phases can affect the partial decay widths. Using ratios of certain decay widths or the overall pattern of rare decay, constraints have been derived on γ with a varying degree of dependence on models. At present, our understanding of the strong interaction dynamics in these decays is not adequate to push such attempts very far, but this situation may well change over the next few years.

$D^0\bar{D}^0$ MIXING

The phenomenon of particle-antiparticle oscillations, or mixing, has been a powerful source of information on discrete symmetry violation and on particles at higher mass scales. Both $K^0\bar{K}^0$ and $B^0\bar{B}^0$ oscillations occur and have provided rough estimates of the charm and top quark masses, respectively. The Standard Model predictions for $D^0\bar{D}^0$ mixing are well below the current range of sensitivity. There are, however, many non-Standard Model predictions that give much higher mixing rates, and many of these have now been excluded by the new CLEO measurement [41] that I will discuss here.

Oscillations between a pseudoscalar meson P^0 and its antiparticle $\bar{P}^0$ can be described in terms of an effective hamiltonian matrix

$$\mathbf{H} = \begin{pmatrix} H_{11} & H_{12} \\ H_{21} & H_{22} \end{pmatrix} = \begin{pmatrix} M & M_{12} \\ M_{12}^* & M \end{pmatrix} - \frac{i}{2}\begin{pmatrix} \Gamma & \Gamma_{12} \\ \Gamma_{12}^* & \Gamma \end{pmatrix}.$$

TABLE 3. Selection of results [31–33] on hadronic rare B decays from CLEO. Limits are given at 90% C.L. The yields for modes with η or η' mesons are given separately for each final state reconstructed in Ref. [33].

Mode	$N_{\rm SIG}$	Signif. (σ)	$\mathcal{B}/10^{-6}$
$\pi^+\pi^-$	$20.0^{+7.7}_{-6.5}$	4.2	$4.3^{+1.6}_{-1.4}\pm 0.5$
$\pi^+\pi^0$	$21.3^{+9.7}_{-8.5}$	3.2	< 12.7
$K^\pm\pi^\mp$	$80.2^{+11.8}_{-11.0}$	11.7	$17.2^{+2.5}_{-2.4}\pm 1.2$
$K^\pm\pi^0$	$42.1^{+10.9}_{-9.9}$	6.1	$11.6^{+3.0+1.4}_{-2.7-1.3}$
$K^0\pi^\pm$	$25.2^{+6.4}_{-5.6}$	7.6	$18.2^{+4.6}_{-4.0}\pm 1.6$
$K^0\pi^0$	$16.1^{+5.9}_{-5.0}$	4.9	$14.6^{+5.9+2.4}_{-5.1-3.3}$
$B^- \to \pi^-\rho^0$	$29.8^{+9.3}_{-9.6}$	5.4	$10.4^{+3.3}_{-3.4}\pm 2.1$
$\bar{B}^0 \to \pi^\pm\rho^\mp$	$31.0^{+9.4}_{-8.3}$	5.6	$27.6^{+8.4}_{-7.4}\pm 4.2$
$B^- \to \pi^-\omega$	$28.5^{+8.2}_{-7.3}$	6.2	$11.3^{+3.3}_{-2.9}\pm 1.4$
$B^- \to K^-\rho^0$	$22.4^{+10.7}_{-9.1}$	3.7	< 17
$B^- \to \pi^-\bar{K}^{*0}$	$13.4^{+6.2}_{-5.2}$	3.6	< 16
$\bar{B}^0 \to K^\pm\rho^\mp$	$16.4^{+7.8}_{-6.6}$	3.5	< 32
$\bar{B}^0 \to \pi^0\bar{K}^{*0}$	$0.0^{+3.0}_{-0.0}$	0.0	< 3.6
$B^+ \to \eta'K^+$		16.8	$80^{+10}_{-9}\pm 7$
$B^0 \to \eta'K^0$		11.7	$89^{+18}_{-16}\pm 9$
$B^+ \to \eta K^{*+}$		4.8	$26.4^{+9.6}_{-8.2}\pm 3.3$
$B^0 \to \eta K^{*0}$		5.1	$13.8^{+5.5}_{-4.6}\pm 1.6$

By diagonalizing $\mathbf{H}$, we obtain the mass eigenstates as linear combinations of the two original basis states, P^0 and $\bar{P}^0$. The masses and lifetimes of these physical states are different; in the absence of CP violation these splittings are given by $\Delta M = \pm 2|M_{12}|$ and $\Delta\Gamma = \pm 2|\Gamma_{12}|$.

The elements of the mass matrix (M_{ij}) and the decay matrix (Γ_{ij}) describe distinctive classes of transitions between P^0 and $\bar{P}^0$: Γ_{12} is the amplitude for $\bar{P}^0 \to f \to P^0$ via on-shell intermediate states, while M_{12} describes transitions through off-shell, or virtual, intermediate states. The on-shell states are those that both P^0 and $\bar{P}^0$ can decay into, while off-shell states provide the sensitivity to physics at high mass scales. CP violation in mixing can arise from the inteference between on-shell and off-shell amplitudes, leading to $\Gamma(P^0 \to \bar{P}^0) \neq \Gamma(\bar{P}^0 \to P^0)$. This occurs when $M_{12} - \frac{i}{2}\Gamma_{12} \neq M_{12}^* - \frac{i}{2}\Gamma_{12}^*$. The dimensionless quantities $x = 2M_{12}/\Gamma$ and $y = \Gamma_{12}/\Gamma$ are commonly defined and we will use them below.

The process used to study mixing in CLEO is $e^+e^- \to c\bar{c}$, where the charm or anticharm quark fragments into a D^*: $c \to D^{*+} \to D^0\pi_s^+$. The charge of the "soft-pion" π_s^+ accompanying the D^0 is used to determine whether the intial state is D^0 or $\bar{D}^0$. In this analysis, neutral D mesons are reconstructed in $K^-\pi^+$ and $K^+\pi^-$ final states, so the normal Cabibbo-favored $D^0 \to K^-\pi^+$ decay produces an opposite-sign $\pi_s^+K^-$ charge correlation. If mixing occurs, $D^0 \to \bar{D}^0 \to K^+\pi^-$, it produces the "wrong-sign" charge correlation ($\pi_s^+K^+$). However, the direct doubly Cabibbo-suppressed decay (DCSD) $D^0 \to K^+\pi^-$ also gives this charge correlation

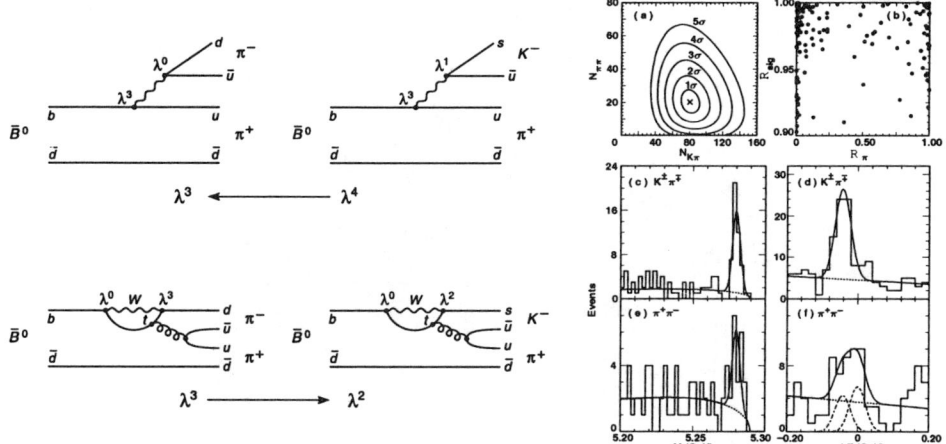

FIGURE 5. Left: Diagrams for $\bar{B}^0 \to \pi^+\pi^-$ and $\bar{B}^0 \to K^-\pi^+$. Both modes have contributions from tree $b \to u$ and penguin processes. The figure shows the dependence of each amplitude on $\lambda = \sin\theta_C$, as determined from the Wolfenstein parametrization of the CKM matrix. The figures at right show results from a recent analysis of $B \to \pi^+\pi^-$ and $B \to K^-\pi^+$. (a) shows the likelihood contours as a function of the number of $K\pi$ and $\pi\pi$ events, while (b) shows an event-by-event probability for an event to be signal (either $\pi\pi$ or $K\pi$ vs. the probability to be a $\pi\pi$ event. Signals are seen for these modes in the reconstructed B mass (c, e) and in the ΔE distributions (d, f), which compare the reconstructed energy to the expected energy (beam energy).

($\pi_s^+ K^+$), and it is distinguished from mixing using the decay time-dependence as described below. Semileptonic final states do not suffer from this ambiguity, but they have larger backgrounds and will be the subject of a future analysis.

Figure 6 shows the signal for the wrong-sign charge correlated events. Under the assumption of no mixing, this signal ($44.8^{+9.7}_{-8.7}$ events) leads to a wrong-sign fraction

$$R = \frac{\Gamma(D^0 \to K^+\pi^-)}{\Gamma(D^0 \to K^-\pi^+)} = (0.332^{+0.063}_{-0.065} \pm 0.040)\%, \qquad (10)$$

which is substantially more precise than the previous PDG value $R = (0.72 \pm 0.25)\%$.

To distinguish a potential mixing signal from DCSD decay, CLEO measures the wrong-sign rate as a function of the decay time, which can be determined from the distance between the D^0 production point and its decay vertex. In CLEO, this measurement is performed using a three-layer silicon vertex detector that surrounds the beam-pipe. The innermost layer of this device is only 2.35 cm from the beam axis, and the very small vertical size of the beam spot ($\sigma_y \approx 7~\mu$m) provides a good constraint on the D^0 production vertex. The wrong-sign rate is given by

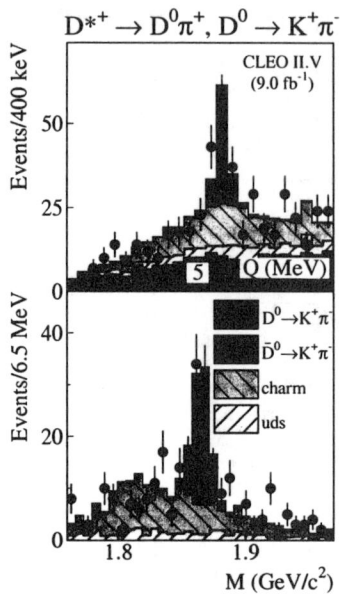 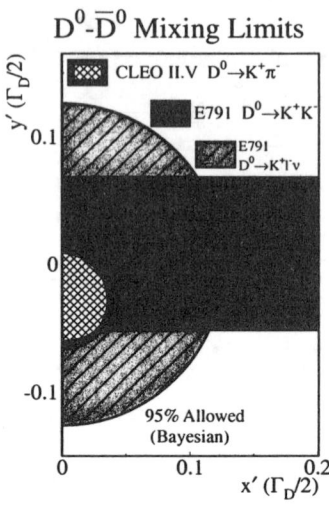

FIGURE 6. $D^0\bar{D}^0$ mixing results from CLEO. Left: data from the wrong-sign correlated events showing a signal corresponding to the D^{*+} decay (above) and the D^0 decay below. These are time-integrated signals; to distinguish between doubly Cabibbo suppressed decay and mixing CLEO analyzes the time dependence of the decay. Right: $D^0\bar{D}^0$ mixing limits from CLEO and Fermilab E791.

$$r_{ws}(t) = (R_D + \sqrt{R_D}y't + \frac{1}{4}(x'^2 + y'^2)t^2)e^{-t}, \tag{11}$$

where the decay time t is measurement expressed in units of the D^0 proper lifetime and R_D is the DCSD rate relative to the Cabibbo-favored rate. The quantities x' and y' in the equation allow for a possible strong phase δ between the Cabibbo-favored and the DCSD amplitudes: $x' = x\cos\delta + y\sin\delta$ and $y' = y\cos\delta - x\sin\delta$.

The fit to the proper lifetime distribution gives $x' = (0.0 \pm 1.5 \pm 0.2)\%$, or $|x'| < 2.9\%$ at 95% C.L. $((1/2)x'^2 < 0.041\%)$, and $y' = (-2.5^{+1.4}_{-1.6} \pm 0.3)\%$, or $-5.8\% < y' < 1.0\%$ at 95% C.L. Although there is no statistically significant indication for mixing, these limits are substantially more sensitive than previous results. Figure 6 shows the allowed region in the x'-y' plane. A extensive compilation of non-Standard Model $D^0\bar{D}^0$ mixing predictions shows [42] that many are excluded by these new limits. $D^0\bar{D}^0$ mixing studies are also now being performed by the FOCUS experiment [43].

TIME-DEPENDENT CP ASYMMETRIES FROM THE INTERFERENCE BETWEEN MIXING AND DECAY

The measurement of time-dependent CP asymmetries is the main goal of the new asymmetric-energy e^+e^- colliders PEP-II and KEKB, and their associated detectors, BaBar and Belle. The mechanism generating the interference was shown in Fig. 1: the processes $B^0 \to f$ and $B^0 \to \bar{B}^0 \to f$ can interfere, giving sensitivity to their relative phases. The final state f, which must be reachable from both B^0 and $\bar{B}^0$, can be an eigenstate of CP ($f = f_{CP}$), with eigenvalue η_{CP}. The measured asymmetry is defined by

$$\mathcal{A}_{CP} = \frac{\Gamma(B^0(t) \to f_{CP}) - \Gamma(\bar{B}^0(t) \to f_{CP})}{\Gamma(B^0(t) \to f_{CP}) + \Gamma(\bar{B}^0(t) \to f_{CP})}. \quad (12)$$

To measure the two rates as a function of time, one must be able to carry out the following procedure:

1. Reconstruct a sufficient number of B^0 or $\bar{B}^0$ decays in the final state f_{CP}. Examples of such final states are $J/\psi K_S^0$, $J/\psi K_L^0$, and $\pi^+\pi^-$.

2. Determine whether the meson was a B^0 or $\bar{B}^0$ at some well-defined reference time. Tagging the B flavor relies on measurements of particles produced in association with the B meson. At the $\Upsilon(4S)$, the initial system is $B^0\bar{B}^0$, and it evolves coherently until one of the B mesons decays. To a large extent, tagging relies on inclusive semileptonic decays ($b \to \ell^-$, $\bar{b} \to \ell^+$), in which the lepton charge is correlated with the B-meson flavor, and inclusive charge kaon production ($\bar{B}^0 \to D \to K^-$, $B^0 \to \bar{D} \to K^+$). For each tagging method, the mistag fraction must be carefully evaluated.

3. Measure the proper time interval between the decay time for $B^0(\bar{B}^0) \to f_{CP}$ and the reference time at which the flavor of the $B^0(\bar{B}^0)$ meson is tagged.

These three requirements place strong demands on both the accelerator and the detector. Figure 7 shows an example of the difference in decay rate vs. time for mesons that are initially B^0 and $\bar{B}^0$. The figure is limited to positive decay times, which means that the tag decay occurs prior to the decay to the CP eigenstate. If the time order of the two is reversed, it turns out that the asymmetry reverses sign as a consequence of Bose statistics and the p-wave nature of the $\Upsilon(4S)$ decay. If the decay time is not measured, the numerator in Eq. 12 integrates to zero, and there is no observable asymmetry! Measurement of the decay time is mandatory at the $\Upsilon(4S)$, while in other environments, e.g., hadron colliders, it is desirable but not strictly essential.

To determine the difference in decay time between the two neutral B mesons, one must measure the spatial separation between their decay vertices. The B mesons are produced nearly at rest in the $\Upsilon(4S)$ frame, so the $B\bar{B}$ system must be boosted relative to the detector frame. It is this requirement that has led to the enormous

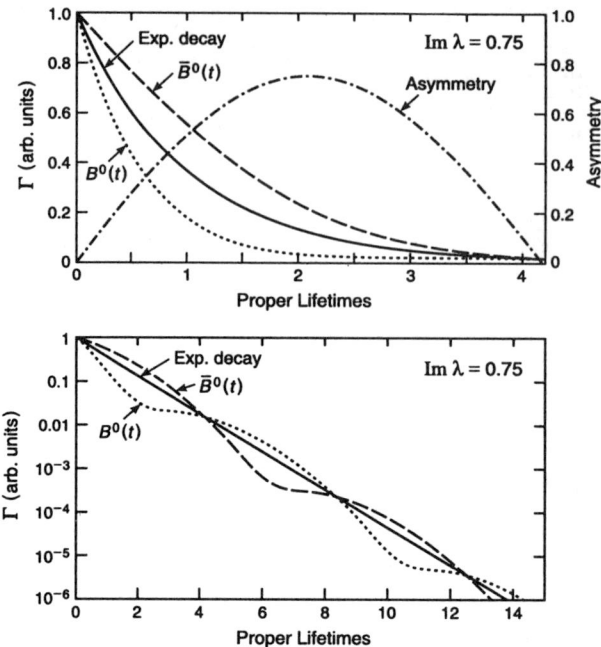

FIGURE 7. Example of a CP asymmetry resulting from the interference between B^0 mixing and decay as a function of proper lifetime. The size of the asymmetry is chosen arbitrarily but is consistent with some predictions for $\sin 2\beta$ based on the Standard Model and constraints from other measurements in B and K physics. Note that CP violation leads to a non-exponential decay.

effort of building asymmetric energy e^+e^- colliders. The unequal beam energies mean that the $\Upsilon(4S)$ has the boost $\beta\gamma = (E_{e^-} - E_{e^+})/2\sqrt{E_{e^-}E_{e^+}}$, so the mean vertex separation is $\Delta z = \beta\gamma c <\Delta t> \approx 260$ μm, since $<\Delta t> = \tau_B = 1.6$ ps. This is a much smaller distance scale than in kaon physics experiments, and it means that high-precision silicon strip vertex detectors play a large role in these measurements. The need for efficient tagging with a relatively low mistag rate implies that these experiments need good electron and muon identification and good π/K separation.

Figure 8 shows preliminary results obtained by Belle and BaBar that help to build an understanding of the detector systems needed to measure time-dependent CP asymmetries. The Belle data show their first measurement of the B^0 lifetime, while the BaBar data show a $B^0\bar{B}^0$ oscillation measurement using dilepton events. We can expect first results on CP asymmetries later this summer, but precision measurements will need to await more data.

With major e^+e^- and hadron-collider facilities coming on line, heavy-quark physics has a very bright future. At the next CIPANP meeting in three years, there should be enormous number of new results!

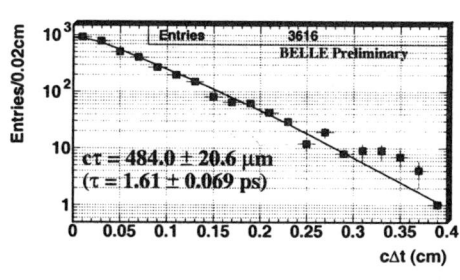 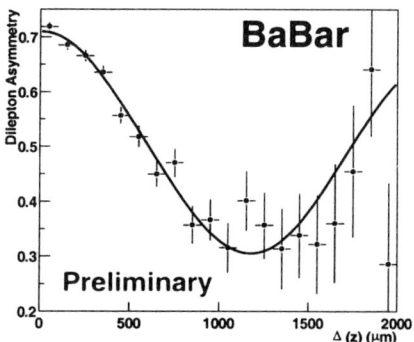

FIGURE 8. Progress from Belle and BaBar. Left: a preliminary measurement of the B lifetime from Belle; right: a preliminary of $B^0\bar{B}^0$ mixing from BaBar.

ACKNOWLEDGEMENTS

Many people provided valuable information that helped greatly in the preparation of this talk. It is a pleasure to thank Leo Piilonen from Belle and David Jaffe, Karl Ecklund, David Asner, Harry Nelson, and David Rubin from CLEO. I would also like to thank the conference organizers for a stimulating and collegial conference.

REFERENCES

1. B. Wicklund, Plenary Session talk at this conference.
2. R. Mohapatra, Plenary Session talk at this conference.
3. A. El-Khadra, Plenary Session talk at this conference.
4. J.H. Christenson, J.W. Cronin, V.L. Fitch, and R. Turlay, Phys. Rev. Lett. **13**, 138 (1964).
5. J.W. Cronin, Rev. Mod. Phys. **53**, 373 (1981).
6. Y. Nir, in *Proceedings of the Twentieth SLAC Summer Institute on Particle Physics: The Third Family and the Physics of Flavor*, Stanford, CA, July 13–24, 1992, SLAC-412.
7. A pedagogical discussion of CP violation and other aspects of B physics is contained in J.D. Richman, *Heavy Quark and CP Violation*, in *Probing the Standard Model of Particle Interactions*, Les Houches Session LXVIII, 28 July–5 September, 1997 (Elsevier, Amsterdam, 1999), ed. R.Gupta, A. Morel, E. De Rafael, and F. David, http://hep.ucsb.edu/driver/driver_houches12.ps (preprint).
8. M. Neubert, Phys. Rep. **245**, 259 (1994).
9. *Review of Particle Physics*, Eur. Phys. J. C **15**, 1 (2000).
10. BaBar Physics Book, P.F. Harrison and H.R. Quinn eds., SLAC-R-504 (1998).
11. J.E. Duboscq et al. (CLEO), Phys. Rev. Lett. **76**, 3898 (1996).

12. J. Bartelt *et al.* (CLEO), Phys. Rev. Lett. **82**, 3746 (1999).
13. J.P. Alexander *et al.*, CLEO CONF 00-03, ICHEP00 770, Submitted to XXXth International Conference on High Energy Physics, July 2000, Osaka, Japan.
14. G. Abbiendi *et al.* (OPAL), CERN-EP-2000-032, Phys. Lett. B**482**, 15 (2000).
15. J.D. Richman and P.R. Burchat, Rev. Mod. Phys. **67**, 893 (1995).
16. L. Gibbons and K. Honscheid, in *Review of Particle Physics*, Eur. Phys. J. C **15**, 581 (2000).
17. B.H. Behrens *et al.* (CLEO), Phys. Rev. D **61**, 052001 (2000).
18. J.P. Alexander, *et al.* (CLEO), Phys. Rev. Lett **77**, 5000 (1996).
19. P. Abreu *et al.* (DELPHI), CERN-EP-2000-030, submitted to Phys. Lett. B
20. Heavy Flavor Working Group, LEPHFS note 99-02, March 2000, and references therein.
21. A.J. Buras, Les Houches Lectures, in *Probing the Standard Model of Particle Interactions*, Les Houches Session LXVIII, 28 July–5 September, 1997 (Elsevier, Amsterdam, 1999), ed. R.Gupta, A. Morel, E. De Rafael, and F. David, hep/ph-9806471.
22. K. Lingel, T. Skwarnicki, J.G. Smith, Ann. Rev. Nucl. Part. Sci. **48**, 253 (1998).
23. A. Vainshtein, Int. J. Mod. Phys. A14, 4705 (1999).
24. R. Ammar *et al.* (CLEO), Phys. Rev. Lett. **71**, 674 (1993).
25. T.E. Coan *et al.* (CLEO), Phys. Rev. Lett. **84**, 5283 (2000).
26. K. Chetyrkin, M. Misiak, and M. Münz, Phys. Lett. B**400**, 206 (1997); Erratum– ibid., Phys. Lett. B**425**, 414 (1998).
27. A.J. Buras, A. Kwiatkowski and N. Pott, Phys. Lett. B**414**, 157 (1997); Erratum– ibid., Phys. Lett. B**434**, 459 (1998).
28. A.L. Kagan and Matthias Neubert, Eur. Phys. J. **C7**, 5 (1999).
29. S. Glenn *et al.* (CLEO), CLEO-CONF-98-17, submitted to the International Conference on High Energy Physics, Vancouver, Canada, 1998.
30. R. Barate *et al.* (ALEPH), Phys. Lett. B **429**, 169 (1998).
31. D. Cronin-Hennessy *et al.*, preprint CLNS-99-1650, hep-ex/0001010, submitted to Phys. Rev. Lett.
32. C.P. Jessop *et al.* (CLEO), preprint CLNS-99-1652, hep-ex/0006008, submitted to Phys. Rev. Lett.
33. S.J. Richichi *et al.* (CLEO), Phys. Rev. Lett. **85**, 520 (2000).
34. A.E. Snyder and H.R. Quinn, Phys. Rev. D **48**, 2139 (1993).
35. S. Chen *et al.* (CLEO), Phys. Rev. Lett. **85**, 525 (2000).
36. A. Ali, G. Kramer, C.D. Lu, Phys. Rev. D**58**, 094009 (1998).
37. A. Ali, G. Kramer, C.D. Lu, Phys. Rev. D**59**, 014005 (1999).
38. M. Beneke, G. Buchalla, M. Neubert, C.T. Sachrajda, hep-ph/0006124.
39. W.-S. Hou, J.G. Smith, F. Würthwein, hep-ex/9910014.
40. Y.-H. Chen, H.-Y. Cheng, B. Tseng, and K.-C. Yang, Phys. Rev. D **60**, 094014 (1999); hep-ph/9903453 v2.
41. R. Godang *et al.* (CLEO), Phys. Rev. Lett. **84**, 5038 (2000).
42. H.N. Nelson, submitted to the 1999 Lepton-Photon Symposium, hep-ex/9908021 (unpublished).
43. P. Sheldon, parallel session talk at this conference.

Structure Functions at High Q^2

Anthony T. Doyle*

*University of Glasgow[1]
Department of Physics and Astronomy, Glasgow, G12 8QQ*

Abstract. Progress in the study of structure functions in deep inelastic scattering is reviewed. A brief introduction to the formalism and the status of global analyses of parton distributions and their uncertainties is given. The review focuses on recent developments in the following areas: HERA results on F_2, F_L and charm; the resolution of the CCFR-NMC discrepancy; Tevatron jet cross-sections at different energies; and HERA high-Q^2 cross-sections and the measurement of xF_3.

Introduction

The neutral current (NC) cross-section $l(k)N(p) \to l(k')X(p')$ for a lepton (e, μ) with four-momentum k scattering off a nucleon with four-momentum p can be expressed as

$$\frac{d^2\sigma(l^\pm N)}{dx\, dQ^2} = \frac{2\pi\alpha^2}{xQ^4} \cdot [Y_+ F_2(x,Q^2) \mp Y_- xF_3(x,Q^2) - y^2 F_L(x,Q^2)]$$

where Q^2 is the four-momentum transfer squared, $x = Q^2/2p\cdot q$ is the Bjorken scaling variable, $y = p\cdot q/p\cdot k$ is the inelasticity variable and $Y_\pm = 1 \pm (1-y)^2$. The contribution from F_2 dominates the cross-section and is measured directly. To investigate sensitivity to F_L at large y or xF_3 at large Q^2, the reduced cross-section

$$\tilde{\sigma} \equiv \frac{xQ^4}{2\pi\alpha^2} \frac{1}{Y_+} \frac{d^2\sigma}{dxdQ^2} \equiv F_2 \mp \frac{Y_-}{Y_+} xF_3 - \frac{y^2}{Y_+} F_L$$

is adopted. The contribution from F_L is a QCD correction which is important at large y: the characteristic y dependence is utilised in the determination of F_L at low Q^2. The contribution from xF_3, due to γZ interference and Z exchange, violates parity conservation and enters for $Q^2 \simeq M_Z^2$: the change of sign for opposite charges of the incoming lepton beam enable xF_3 to be extracted. In the quark parton model

---

[1] supported by PPARC and the Royal Society.

(or in the DIS scheme of NLO QCD) F_2/x is the "charge-weighted" sum of the quark densities

$$F_2(x, Q^2) = x \sum_f A_f(Q^2) \cdot \Sigma(x, Q^2)$$

where $A_f(Q^2) = e_f^2 - 2v_l v_f e_f P_Z + (v_l^2 + a_l^2)(v_f^2 + a_f^2) P_Z^2$ contains vector and axial terms due to γ, γZ interference and Z exchange respectively, $P_Z = 1/(4 \sin^2 \theta_W \cos^2 \theta_W) \cdot Q^2/(Q^2 + M_Z^2)$ is the Z propagator term and $\Sigma(x, Q^2) = q_f(x, Q^2) + \bar{q}_f(x, Q^2)$ is the singlet summed quark and anti-quark distributions. Similarly,

$$xF_3(x, Q^2) = x \sum_i B_f(Q^2) \cdot q_{NS}(x, Q^2)$$

where $B_f(Q^2) = 2 a_l a_f e_i P_Z + 4 v_l a_l v_f a_f P_Z^2$ contains vector and axial terms due to $\gamma - Z$ and Z exchange respectively, and $q_{NS}(x, Q^2) = q_f(x, Q^2) - \bar{q}_f(x, Q^2)]$ is the non-singlet difference of the quark and anti-quark (valence quark) distributions.

The charged current (CC) cross-section $l^+(l^-)N \to \nu(\bar{\nu})X$ proceeds via W exchange and can be expressed as

$$\frac{d^2 \sigma^{CC}(l^\pm N)}{dx\, dQ^2} = \frac{G_F^2}{4\pi x} \left(\frac{M_W^2}{Q^2 + M_W^2} \right)^2$$

where G_F is the Fermi constant and a positive (negative) charged lepton beam picks out valence d (u) quarks.

Given a phenomenological input as a function of x, the parton distributions are evolved to different physical scales (Q^2) via the DGLAP evolution equations. The non-singlet contribution evolves as

$$\frac{\partial q_{NS}(x, Q^2)}{\partial t} = \frac{\alpha_s(Q^2)}{2\pi} P_{qq}^{NS} \otimes q_{NS}(x', Q^2)$$

where $t = \ln(Q^2/\Lambda^2)$ and the P_{ij}'s represent the NLO DGLAP splitting probabilities for radiating a parton with momentum fraction x from a parton with higher momentum x'. Quantities such as xF_3 provide a measure of $\alpha_s(Q^2)$ which is insensitive to the a priori unknown gluon distribution. Similarly, the singlet quark and gluon densities are coupled via the matrix equation

$$\frac{\partial}{\partial t} \begin{matrix} \Sigma(x, Q^2) \\ g(x, Q^2) \end{matrix} = \frac{\alpha_s(Q^2)}{2} \begin{matrix} P_{qq} & P_{qg} \\ P_{gq} & P_{gg} \end{matrix} \otimes \begin{matrix} \Sigma(x', Q^2) \\ g(x', Q^2) \end{matrix}$$

and quantities such as F_2 provide input for $\Sigma(x, Q^2)$ as well as coupled knowledge of $\alpha_s(Q^2)$ and the gluon, $g(x, Q^2)$. The gluon distribution, $g(x, Q^2)$, is determined through the scaling violations of F_2 at low x or constrained via jet or direct photon production at large x (where the gluon enters directly).

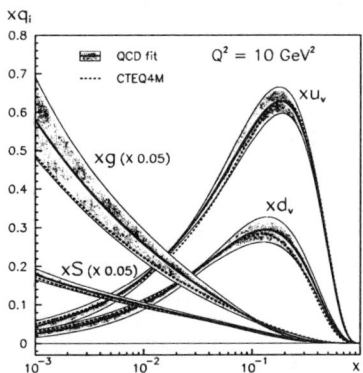

FIGURE 1. Parton momentum densities (full curves) at $Q^2 = 10$ GeV^2 including uncertainties (shaded bands) compared to the CTEQ4M set (dashed curves).

At the starting scale, Q_0^2, the light valence quarks (u_v and d_v) and the sea of quark and anti-quarks (S) as well as the gluon (g) are attributed a given functional form. The measured structure functions are then described by the convolutions of the parton densities with the appropriate NLO matrix elements. In Fig. 1 the latest analysis of the BCDMS, CCFR, E665, H1, NMC, SLAC and ZEUS DIS measurements by Botje [1] gives the parton densities and their associated uncertainties. The uncertainties are determined from a full treatment of the systematic uncertainties of the data, varying the strong coupling constant ($\Delta\alpha_S(M_Z) = \pm 0.005$), varying the renormalisation and factorisation scale uncertainty from $Q^2/2$ to $2Q^2$, and assessing effects due to uncertainties on the strange sea, nuclear effects and the charm threshold (treated here as a light quark above threshold). The analysis is noteworthy in that care has been taken to provide the correlation coefficients for each of the parton densities. The fit agress with the latest CTEQ and MRST fits, within the uncertainties of up to 10% on each parton density at $Q^2 = 10$ GeV^2.

HERA Structure Function Measurements

HERA $F_2(x, Q^2)$ measurements extend to: low y ($y_{\text{HERA}} \sim 0.005$) providing overlap with the fixed-target experiments; very low x ($x \lesssim 10^{-5}$) at low Q^2 exploring the transition region from soft to hard physics; high y ($y \to 0.8$) giving sensitivity to F_L; high $x \to 0.7$ probing sensitivity to electroweak effects in F_2 and xF_3 as well as constraining the valence quarks at large Q^2.

In Fig. 2 recent ZEUS F_2 data, as well as fixed target data from E665 and NMC, are plotted as function of Q^2 for fixed values of x. The observed scaling violations are intimately coupled with the rise of F_2 with decreasing x via the gluon density (in leading order $dF_2/d\ln Q^2 \sim xg(x)$ neglecting sea quark contributions). In order to study these scaling violations, a compact parameterisation $F_2(x) = A(x) + B(x)\log_{10}Q^2 + C(x)(\log_{10}Q^2)^2$ has been fitted, indicated by the full line. The F_2 points are observed to rise significantly at low x with a tendency to flatten at the lowest x values. Lines of fixed centre-of-mass energy $W^2 = Q^2(1/x - 1)$ are indicated by the dashed lines. The derivative $dF_2/d\log_{10} Q^2 = B(x) + C(x)\log_{10} Q^2$ is plotted in Fig. 3 for constant W as a function of Q^2 (upper) and x (lower plot), motivated by similar theoretical plots in an analysis of low-x data by Golec-Biernat and Wüsthoff [2], incorporating saturation effects. The turnover of the derivative occurs at higher Q^2 (or x) for increasing W, consistent with predictions from various models which go beyond the DGLAP formalism. The DGLAP formalism itself is however also sufficiently flexible to describe the data, provided that the gluon (proportional to the derivative) decreases significantly at low (x, Q^2). The behaviour of the HERA low (x, Q^2) data remains a challenge to our understanding of QCD: this presentation of the data represents one way in which to address the problem, although there are various physical interpretations of the observed behaviour.

The contribution of F_L enters as a QCD correction to the total DIS cross-section. As such it provides a method to calibrate the gluon at low x. H1 [3] have used two methods to extract F_L from the reduced cross-section $\tilde{\sigma} = F_2 - y^2/Y_+ \cdot F_L$ at high y. This is the region where the scattered electron energy is low: in the H1 analysis scattered positrons are measured down to $E_{e'} \simeq 5$ GeV and backgrounds reduced by requiring the associated track to have correct charge. F_L is determined as a function of $Q^2 \geq 4$ GeV^2 by measuring local derivatives of $\partial\tilde{\sigma}/\partial \log y$ and observing deviations from a straight line (or a QCD fit for $Q^2 > 10$ GeV^2) at high y. The extracted F_L data are given in Fig. 4, compared to the H1 NLO QCD fit to H1, NMC and BCDMS F_2 data as well as to F_L data from SLAC and NMC. The data are in agreement with the QCD expectations although the precision is limited by statistics and uncertainties due to the input F_2.

Another method to calibrate the gluon distribution is to extract the contribution due to charm. $D^* \to (K\pi)\pi_s$ measurements in DIS provide a significant test of the gluon density of the proton determined from the scaling violations of F_2. They also help to constrain theoretical uncertainties in the fits to F_2 where different prescriptions for charm production are adopted. The D^* data are extrapolated outside the measured range of η, p_T to obtain the charm cross-section and hence $F_2^{c\bar{c}}$. The ratio $F_2^{c\bar{c}}/F_2$ is plotted in Fig. 5 as a function of x for four Q^2 values, compared to the NLO QCD expectations, where the dashed curves correspond to the uncertainty due to the parton distributions from the NLO fit. There is an overall normalisation uncertainty of approximately 10% (not shown) due (mainly) to the charm hadronisation fraction to D^*. At high W (or low x), the charm contribution to F_2 is significant (up to 30%) and therefore a careful treatment of charm production is required in any analysis of low (x, Q^2) data. The threshold region of of $x \simeq 10^{-2}$, where the charm cross-section is changing rapidly, is also important in consideration of the CCFR-NMC discrepancy.

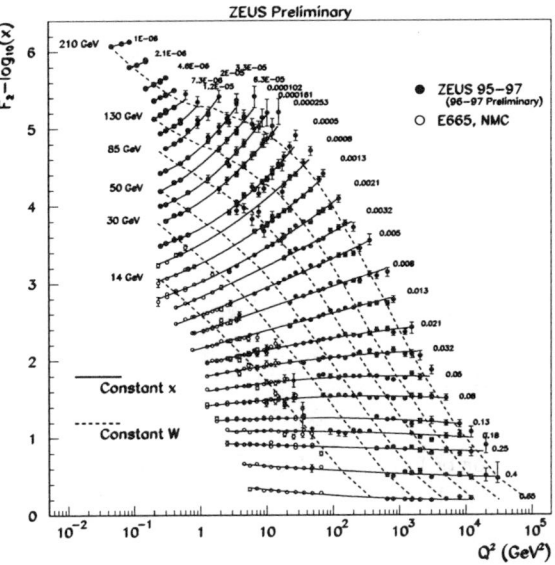

FIGURE 2. ZEUS, E665 and NMC $F_2 - \log_{10} x$ values versus Q^2. Fixed x values are indicated by the full lines and on the right hand side of the plot. Fixed W values are indicated by the dashed lines and on the left hand side of the plot.

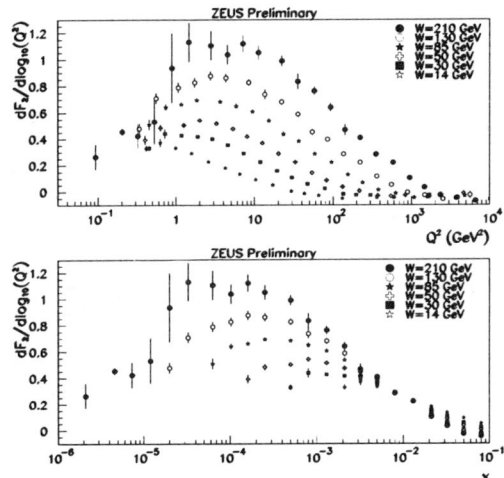

FIGURE 3. Derivatives of F_2 with respect to $\log_{10} Q^2$ for fixed W values indicated by the symbols. In the upper plot the data are plotted as a function of Q^2 and in the lower plot as a function of x.

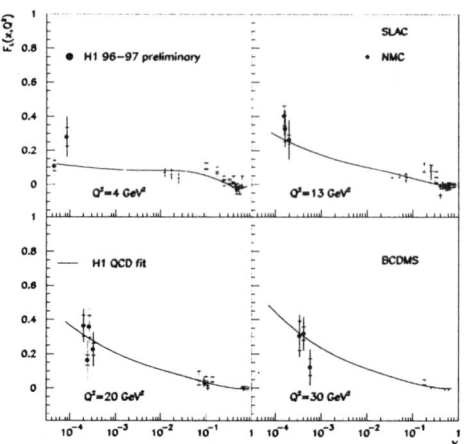

FIGURE 4. H1 determination of F_L versus x for four values of Q^2, compared to the NLO QCD fit expectation and data from BCDMS, NMC and SLAC.

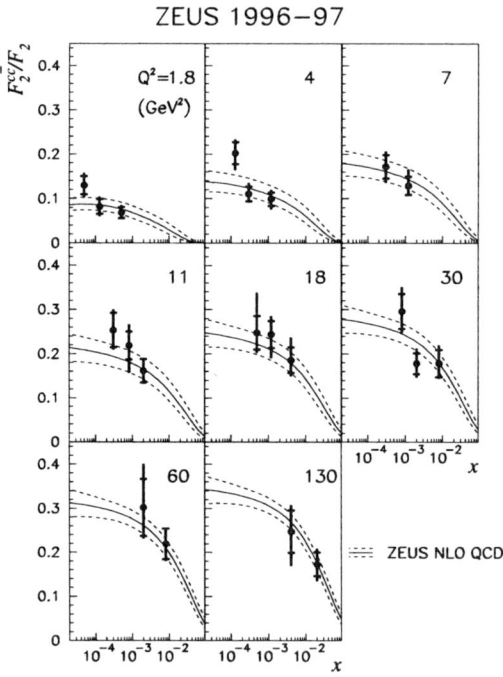

FIGURE 5. ZEUS ratio of $F_2^{c\bar{c}}/F_2$ as a function of x for four values of Q^2, compared to NLO QCD expectations.

Comparison of νN and lN Data

To leading order the muon (NMC) and neutrino (CCFR) structure functions are related by the "5/18ths rule" from quark counting. Here the strange sea enters as a correction, since $s \to c$ production is reduced in νN scattering due to the charm mass. The original CCFR analysis, where data were corrected using the dimuon result to constrain the strange sea, lay significantly ($\simeq 20\%$) above the NMC data for all values of Q^2 at $x < 0.1$. Advances in the H1 and ZEUS analyses of low y data, lead to a region of overlap where the HERA and fixed-target experiments can be compared with a precision of $\simeq 5\%$. In Fig. 6, a comparison of the CCFR (νN), H1, NMC and E665 (lN) data indicates that the discrepancy lies with the treatment of the CCFR data which is significantly above the lN data at low x.

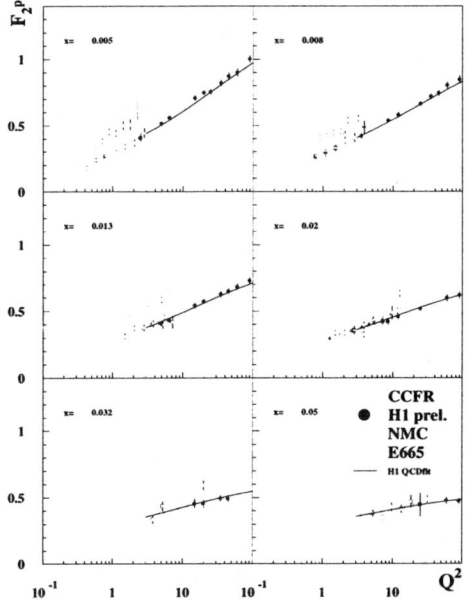

FIGURE 6. Original CCFR analysis compared to H1, E665 and NMC F_2 data. The F_2 data are plotted as a function of Q^2 for various x values. The full line is the H1 NLO QCD fit.

The HERA comparisons were made in parallel with a re-analysis of CCFR data by Yang and Bodek, reported at this conference [4]. The previous CCFR analysis used a slow rescaling variable $\xi = x(1 - m_c^2/Q^2)$, where m_c is the effective charm mass, in order to suppress charm production and correct F_2. In the latest CCFR analysis no explicit correction is made to the F_2 data for charm (or heavy target) effects. Various NLO treatments of charm are now available which allow data to be compared more precisely. The variable flavour scheme, adopted by CTEQ and MRST with different approaches to the threshold behaviour and fixed-order

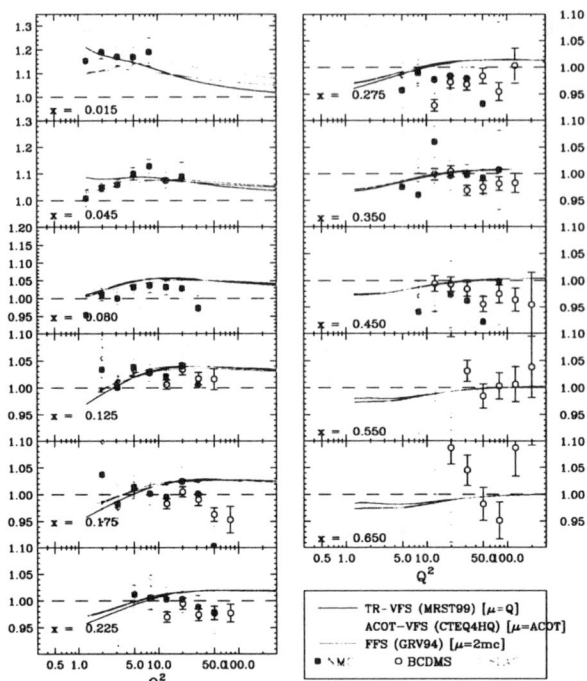

FIGURE 7. Model-independent CCFR analysis of the ratio of $5/18 F_2^{\nu N}/F_2^{lN}$ as a function of Q^2 in various x intervals, compared to NLO calculations incorporating various treatments of charm. The CCFR data is divided by NMC, BCDMS and SLAC data.

treatment, handle charm as another parton in the proton, a good approximation for $Q \gtrsim m_c$. The fixed flavour scheme (FFS), adopted by GRV, treats charm as generated dynamically in NLO QCD, a good approximation for $Q \simeq m_c$. The ratio of $5/18 F_2^{\nu N}/F_2^{lN}$ is plotted in Fig. 7. A clear departure from unity is observed at the lowest x values. The data are compared to NLO charm calculations from MRST, CTEQ and GRV which differ somewhat in shape, but provide a good description of the data ratios at all values of x. The difference in xF_3 from ν and $\bar{\nu}$ data ($\Delta x F_3^{\nu-\bar{\nu}}$, not shown) is also sensitive to charm production and provides an additional constraint on the calculations.

The significant discrepancy between $F_2^{\nu N}$ and F_2^{lN} data is thus resolved. The area is now open for more detailed comparisons of NLO charm methods, constraints on the strange sea from the ratio of CCFR to NMC and HERA data compared to CCFR dimuon data as well as an understanding nuclear shadowing corrections for νFe (CCFR) data. In addition new data from NuTeV tagging ν and $\bar{\nu}$ beams will provide more information on $\Delta x F_3^{\nu-\bar{\nu}}$, as well as improving the measurement of $\sin^2 \theta_W$.

Inclusive Jet Production at the Tevatron

The Tevatron collider provides a unique opportunity to study the properties of hard interactions in $p\bar{p}$ collisions at short distances. CDF and DØ have measured jet cross-sections at $\sqrt{s} = 1800$ GeV over ten orders of magnitude in $d^2\sigma/dE_T d\eta$ up to $E_T = 500$ GeV, half-way to the kinematic limit. These data indicate that the gluon should be enhanced at large x compared to direct photon measurements. One open area for further experimental and theoretical investigation is in this measurement at different energies. The ratio of the inclusive jet cross-section at $\sqrt{s} = 630$ GeV over $\sqrt{s} = 1800$ GeV is plotted in Fig. 8 as a function of scaled jet transverse energy $x_T = 2E_T/\sqrt{s}$. The ratio should lead to a partial cancellation of experimental and theoretical systematics [5]. However, it is clear that there is an experimental discrepancy at the lowest values of $x_T \lesssim 0.15$. In addition, the data lie below the NLO QCD predictions at all values of x_T. Two explanations have been proposed: first that different renormalisation scales could be used for the theoretical calculations at the two energies; second that a relatively small shift in jet energies by approximately 3 GeV (due to non-perturbative jet pedestal or underlying event effects) brings the theory in agreement. The forthcoming Run II will provide higher energy data in order to clarify these issues.

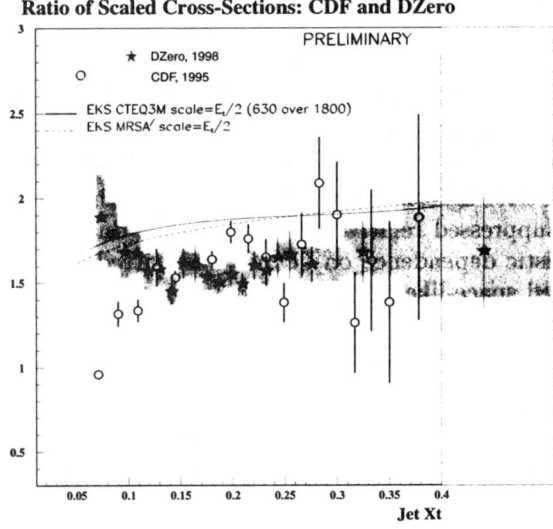

FIGURE 8. Ratio of CDF and DØ inclusive jet cross sections at $\sqrt{s} = 1800$ GeV to $\sqrt{s} = 630$ GeV, as a function of $x_T = 2E_T/\sqrt{s}$, compared to NLO QCD. The shaded band indicates the DØ energy scale uncertainty

High Q^2 Cross-Sections at HERA

The HERA collider provides a unique window to explore ep interactions at the highest energies, extending the range of momentum transfer Q^2 by about two orders of magnitude compared to fixed-target experiments. As the HERA luminosity increases we explore the region of $Q^2 \sim 10^4$ GeV^2, where electroweak effects play a rôle. It is in this unexplored kinematic region that we are sensitive to deviations from the standard model (SM). The theoretical uncertainties on the SM $e^\pm p$ NC and CC cross-sections were determined as discussed with respect to the parton densities of Fig. 1 and correspond to $\simeq$ 6-8% on the NC cross-sections and $\simeq$ 6-12% on the CC cross-sections at the highest accessible Q^2 values. The measured cross-sections generally agree with theory within these uncertainties and therefore represent a benchmark for the standard model. The example cross-sections, discussed below, are corrected to the electroweak Born level.

The e^+p charged-current cross-section [7,8] is sensitive to the valence d-quark distribution in the proton

$$\frac{d^2\sigma_{e^+p}}{dx\,dQ^2} \simeq \frac{G_F^2}{2\pi}\left(\frac{M_W^2}{Q^2+M_W^2}\right)^2 [\bar{u}+\bar{c}+(1-y)^2(d+s)].$$

In the upper plot of Fig. 9 the ZEUS cross-section is described by the SM, falling over four orders of magnitude. The ratio of the data to the SM, adopting the CTEQ4D PDF, is shown in the lower plot of Fig. 9 where good agreement is observed as a function of x for $Q^2 > 400$ GeV^2. Comparison of the data with the PDF uncertainties (shaded band) indicates that the data will help to constrain the d-quark densities at large x. The uncertainties on the data are large in this region, but an increase of the ratio of d/u quarks, such as that proposed by Yang and Bodek [6] results in better agreement with the data than the standard PDFs. The CC cross-section is suppressed relative to NC exchange due to the W propagator term: this characteristic dependence on Q^2 has been fitted to yield values for the mass of the exchanged space-like W-boson. The 'propagator-mass' fit value (with G_F fixed) of

$$M_W = 81.4^{+2.7}_{-2.6}(\text{stat.})\pm 2.0(\text{sys.})^{+3.3}_{-3.0}(\text{PDF})\ \text{GeV}$$

agrees with the PDG value of $M_W = 80.41 \pm 0.10$ GeV obtained from time-like production of W bosons at LEP and the Tevatron. The consistency of the data with the SM has been checked using the PDG values for G_F, α, M_Z and M_t, yielding a consistent value of

$$M_W = 80.50^{+0.24}_{-0.25}(\text{stat.})^{+0.13}_{-0.16}(\text{sys.})\pm 0.31(\text{PDF})^{+0.03}_{-0.06}(\Delta M_t, \Delta M_H, \Delta M_Z)\ \text{GeV}.$$

This result is in agreement with the value of $M_W = 80.35 \pm 0.21$ GeV obtained by NuTeV [9].

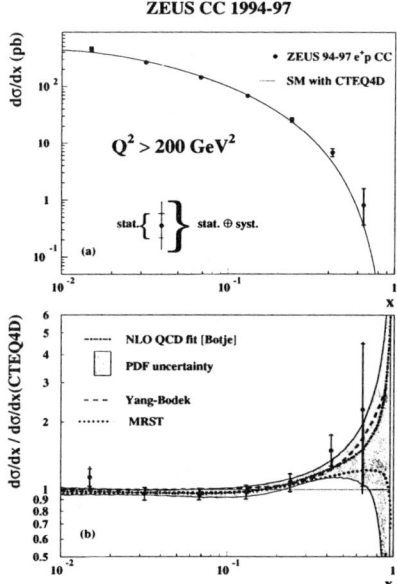

FIGURE 9. ZEUS charged current cross-section $d\sigma^{CC}(e^+p)/dx$ versus x for $Q^2 > 400$ GeV^2 and ratio with respect to the standard model prediction using the CTEQ4D PDF (lower plot). The curves and shaded band in the lower plot correspond to the NLO PDF fits described in the text.

The NC cross-sections are particularly sensitive to the valence u-quark distribution in the proton

$$\frac{d^2\sigma_{e^{\pm}p}}{dx\,dQ^2} \simeq \frac{2\pi\alpha^2}{xQ^4}[Y_+ F_2(x,Q^2) \mp Y_- xF_3(x,Q^2)].$$

Here, F_2 is the generalised structure function, incorporating γ and Z terms, which is sensitive to the singlet sum of the quark distributions $(xq + x\bar{q})$ and xF_3 is the parity-violating (Z-contribution) term which is sensitive to the non-singlet valence quark distributions $(xq - x\bar{q})$. The $e^{\pm}p$ data are now becoming sensitive to the difference in sign of the xF_3 contribution (due mainly to electroweak γZ interference effects) which suppresses (enhances) the e^+p (e^-p) NC cross-section by $\sim 30\%$ for $Q^2 > 10,000$ GeV^2. This is illustrated in Fig. 10 where the H1 reduced cross-section for e^+p (820 GeV) and e^-p (920 GeV) data is compared. A further step has been taken in the ZEUS analysis where the difference of the cross-sections (modified slightly due to the different beam energies) yields a first measurement of xF_3 [10]. In Fig. 11 the scaling of xF_3 is shown as a function of Q^2 compared to the CTEQ4D and MRST(99) PDFs: the statistical precision is currently limited, although the systematic precision on data and theory is good, boding well for a future determination of $\alpha_S(M_Z)$.

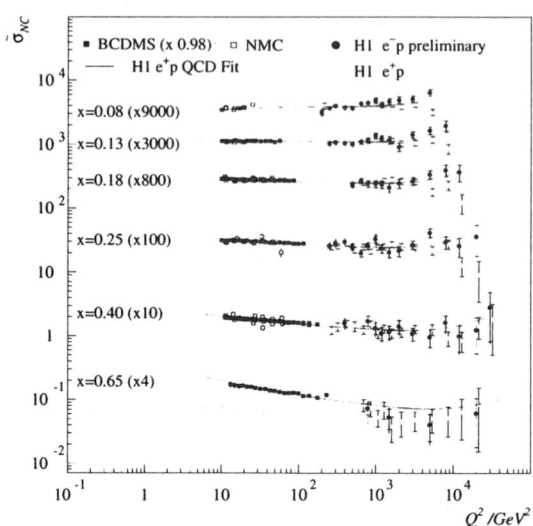

FIGURE 10. H1 neutral current reduced cross-section as a function of Q^2 for various values of x. The e^+p and e^-p data are compared to the NLO QCD fit expectations (full line) obtained from fitting the e^+p data.

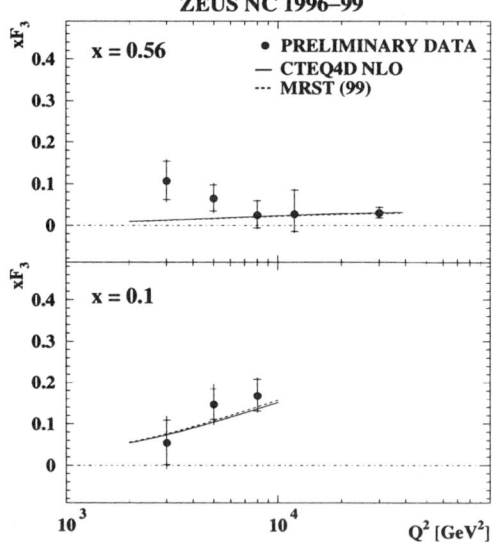

FIGURE 11. ZEUS xF_3 data extracted from the difference of the e^-p and e^+p neutral current cross-sections, compared to the CTEQ4D and MRST(99) PDF expectations.

Outlook: The first phase of HERA running from 1992-2000 is now complete. An upgrade of HERA during 2000-2001 will enable luminosities to be increased by a factor of 4-5. In the longer term, DESY has recognised a future programme for ep physics by ensuring that the proposed TESLA 500 GeV e linear collider is capable of combining with the HERA proton ring to provide the next generation of deep inelastic scattering measurements. The historical perspective of such an endeavour is illustrated in Fig. 12, where the resolved dimension (d [fm] $= 0.2/Q_{max}$ [GeV]) is plotted as a function of year. Progress on the path of DIScovery started with Rutherford, observed to be 30 years ahead of his time, and is planned to reach down to sub-attometre distances with the TERA collider.

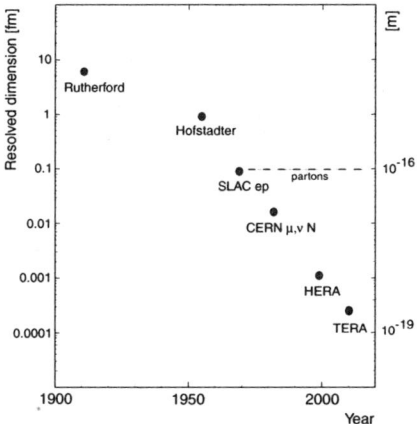

FIGURE 12. Resolved dimension as a function of year including the proposed TERA (=TESLA⊗HERA) collider.

Acknowledgements: It is a pleasure to thank the organisers for an excellent conference. Many thanks to Ian Bertram, Chris Cormack, Doris Eckstein, Brian Foster, Sergey Levonian, Jason Webb, Un-Ki Yang, Rik Yoshida and all the speakers in the structure functions parallel session for their help and advice.

REFERENCES

1. M. Botje, Eur. Phys. J. C14 (2000) 285.
2. K. Golec-Biernat and M. Wüsthoff, Phys. Rev. D59 (1999) 014017.
3. S. Levonian, CIPANP2000 proceedings.
4. U.K. Yang, CIPANP2000 proceedings.
5. R. Hirosky, CIPANP2000 proceedings.
6. U.K. Yang and A. Bodek, Phys. Rev. Lett. 82 (1999) 2467.
7. H1 Collab., C. Adloff et al., Eur. Phys. J. C13 (2000) 609.
8. ZEUS Collab., J. Breitweg et al., Eur. Phys. J. C12 (2000) 411.
9. CCFR/NuTeV Collab., K.S. McFarland et al., Eur. Phys. J. C1 (1998) 509.
10. C. Cormack, CIPANP2000 proceedings.

Polarized Structure Functions and the Spin Structure of the Nucleon

Manuella G. Vincter*

*Department of Physics, University of Alberta, Edmonton, AB, T6G 2J1, Canada
email:mvincter@phys.ualberta.ca*

Abstract. Recent polarized deep-inelastic scattering experiments have demonstrated that all of the components of the nucleon may perhaps contribute to its spin. This conclusion has been obtained through detailed measurements of polarized structure functions and direct measurements of polarized parton distributions. Both of these topics will be presented in this paper. However, there remains some open questions on nucleon spin structure such as the role of gluons and the third twist-2 structure function, the so-called transversity which is still an unknown quantity. The future of these topics in view of upcoming experiments will also be discussed.

INTRODUCTION

The nucleon possesses a rich substructure consisting of valence quarks, sea quarks, and gluons. Deep-inelastic scattering (DIS) has proven to be an ideal tool to probe this structure. It has provided precise information, via unpolarized structure functions, on the momentum distribution of quarks and gluons within the nucleon. However, the spin structure of the nucleon still is not completely understood. The first polarized deep-inelastic scattering experiments in the 1980's revealed that the quark spins contributed very little to the spin 1/2 of the nucleon; a result quite contrary to the expectations of the time. More recent measurements at SLAC, CERN, and DESY indicate that the quark spin contribution is approximately 30%. This number should be compared to the naive relativistic quark model which predicts a contribution of 60%. It is perhaps naive to expect that only quarks can contribute. In principle, all the partons within the nucleon may participate to the spin of the nucleon, S_z^N:

$$S_z^N = \frac{1}{2} = \frac{1}{2}(\Delta u_v + \Delta d_v + \Delta q_s) + \Delta G + L_q + L_g \qquad (1)$$

where $\Delta u_v, \Delta d_v$, and Δq_s are the spin contributions from valence and sea quark helicities, ΔG is the contribution from the gluon spin, and $L_q + L_g$ is the component from the orbital angular momentum of the partons in the nucleon. In this paper I shall review the status of the spin structure of the nucleon by discussing some recent measurements of polarized structure functions and quark and gluon distributions.

Ever since the early spin determinations in the 80's, several experiments have been built to better measure the spin structure of the nucleon. These include DIS experiments at CERN (EMC and SMC), at SLAC (E142, E143, E154, E155, E155x), at DESY (HERMES), and at JLAB (E91-023, E93-009, E94-010) as well as a real photon scattering experiment at MAMI (the GDH collaboration). I will present results from some of these experiments in this paper. Even though there have been numerous experiments so far, several key questions about nucleon spin structure still remain unanswered. These questions have motivated the next generation of spin experiments at CERN (COMPASS), at DESY (HERMES RUN2), at JLAB (E97-103, E97-110), and also the polarized proton-proton collider RHIC (PHENIX and STAR) at the Brookhaven National Lab. The sensitivity of these experiments to spin physics will also be discussed.

SPIN STRUCTURE TO LEADING TWIST

It is possible to interpret the spin structure of the nucleon by using the operator production expansion (OPE) in which the absorption of the virtual photon by the nucleon is described by a series of operators of increasing "twist" [1]. The leading order in $1/Q$, where Q is the four momentum-transfer of the process, is referred to as twist-2. All higher twists are suppressed by powers of $1/Q$. While leading twist describes the absorption of the virtual photon by an individual quark, all higher twists deal with correlations between the partons in the nucleon. After integrating over all the transverse momentum in the system, a quark inside the nucleon can be completely described by three functions in terms of the fraction of the nucleon's momentum x carried by the struck parton: $q(x)$, $\Delta q(x)$, $\delta q(x)$. The quark number density distribution $q(x)$ (or $\bar{q}(x)$ for anti-quarks) is a well known quantity related to the momentum distribution of the quark q in the nucleon. This distribution is the primary component of the unpolarized structure function $F_1(x) = 1/2 \sum_q e_q^2 [q(x) + \bar{q}(x)]$ where e_q is the fractional charge of the quark q in units of the elementary charge $|e|$. The second function is the helicity distribution, $\Delta q(x)$, which is the difference between the quark number density when the spin of the quark is aligned and anti-aligned with the spin of the host nucleon all of which aligned with the momentum of the virtual photon. $\Delta q(x)$ is related to the polarized structure function $g_1(x) = 1/2 \sum_q e_q^2 [\Delta q(x) + \Delta \bar{q}(x)]$. Finally, the third function, known as the quark transversity $\delta q(x)$, provides information on the quark's transverse polarization distribution and is similar to the quark helicity but in this case the quark and the nucleon's spin are transverse to the momentum of the photon. Again, this quark property is related to a third structure function $h_1(x) = 1/2 \sum_q e_q^2 [\delta q(x) + \delta \bar{q}(x)]$. Although $h_1(x)$ is on par in importance with $F_1(x)$ and $g_1(x)$ as all three functions are necessary to describe the nucleon structure to leading twist, due to the chiral odd nature of $h_1(x)$ it cannot be determined through inclusive deep-inelastic scattering and so as yet remains an unknown quantity. It is to be noted that due to relativistic and spin orbit effects, $\delta q(x) \neq \Delta q(x)$. In this paper, I will limit my discussion to two of these three quark distributions: $\Delta q(x)$ and $\delta q(x)$.

INCLUSIVE MEASUREMENTS

Polarized Structure Functions

The spin-dependent structure function $g_1(x, Q^2)$ is determined through inclusive polarized deep-inelastic scattering where only the scattered lepton is identified in the detector. It is extracted from an asymmetry constructed from cross sections when the beam and target's longitudinal polarization are parallel and anti-parallel to each other. World data from CERN, DESY, and SLAC on $g_1(x, Q^2)$ on the proton, deuteron, and the neutron (^{3}He) are shown in Fig. 1 [2]. One of the primary interests in the polarized structure function is its partonic interpretation in terms of parton distribution functions. In next-to-leading order (NLO) QCD, $g_1(x, Q^2)$ can be expressed in terms of three distributions $\Delta\Sigma$, Δq_{NS}, and ΔG [3]:

$$g_1(x, Q^2) = \frac{1}{2} \sum_q e_q^2 \left[\Delta\Sigma \otimes (1 + \frac{\alpha_s(Q^2)}{2\pi} \delta C_S) + \Delta q_{NS} \otimes (1 + \frac{\alpha_s(Q^2)}{2\pi} \delta C_{NS}) + \right.$$

$$\left. \Delta G \otimes (0 + \frac{\alpha_s(Q^2)}{2\pi} \delta C_g) \right] \quad (2)$$

where $\Delta\Sigma = \sum_q \Delta q$, $\Delta q_{NS} = \sum_q (e_q^2/\langle e^2 \rangle - 1)\Delta q$ are the singlet and non-singlet quark distributions and the δC's are the Wilson coefficients. In principle, $\Delta\Sigma$, Δq_{NS}, and $\Delta G(x, Q^2)$ can be uniquely determined in the DIS region due to their different Q^2 evolution. However, at NLO, the QCD splitting functions, coefficient functions, and parton distribution functions depend on normalization and factorization schemes. Two popular schemes are the $\overline{MS}$ [4] and AB [5] schemes, the main difference being whether or not the axial anomaly contributes to the first moments:

$$\Delta\Sigma_{AB} = \Delta\Sigma_{\overline{MS}}(Q^2) + N_f \frac{\alpha_s(Q^2)}{2\pi} \Delta G_{AB}(Q^2). \quad (3)$$

It is to be noted that some prefer the AB scheme as in this case $\Delta\Sigma_{AB} = (\Delta u + \Delta\bar{u}) + (\Delta d + \Delta\bar{d}) + (\Delta s + \Delta\bar{s})$ is independent of Q^2 and so can be more readily interpreted as the contribution of quark spins to the nucleon's. Several papers have been written on perturbative QCD (pQCD) fits to the Q^2 dependence of g_1. In particular, I discuss here the analysis from the SMC collaboration [6] shown in Fig. 2. Based on the Q^2 dependence of g_1 world data, they are able to determine well both the singlet ($\Delta\Sigma$) and non-singlet quark distributions (on the proton Δq_{NS}^p

and on the neutron Δq_{NS}^n). However, the gluonic contribution to the nucleon's spin (ΔG) is still not well established although it seems to be positive. At $Q^2 = 1$ GeV^2 in the AB scheme, they determine:

$$\Delta G_{(AB)} = 0.99^{+1.17}_{-0.31}{}^{+0.42}_{-0.22}{}^{+1.43}_{-0.45}$$
$$\Delta \Sigma_{(AB)} = 0.38^{+0.03}_{-0.03}{}^{+0.03}_{-0.02}{}^{+0.03}_{-0.05}.$$

It is to be noted that in the $\overline{MS}$ scheme, $\Delta\Sigma_{(\overline{MS})} = 0.19 \pm 0.05 \pm 0.04$. A comparison of this result in the two schemes seems to indicate that the axial anomaly cannot completely explain the low $\Delta\Sigma$ measured in the experiments and so something else other than quarks and gluons also contributes to the nucleon's spin. Since this pQCD fit was published, two new interesting measurements of g_1 at low x and low Q^2 were made by both the SMC [7] and HERMES [8] collaborations. Both results show no effects due to higher twist.

Several preliminary results obtained in the nucleon resonance region have emerged from Jefferson Lab Hall A (polarized electron ^{3}He scattering using the HRS spectrometer) and Hall B (polarized electron proton scattering using the CLAS spectrometer). These interesting determinations can provide insight on the quark structure of the resonances. Fig. 3 from Hall B [9] shows with good precision the Q^2 dependence of the double spin asymmetry $\vec{e}\vec{p} \to e\pi^+ n$ as a function of the invariant mass of the final hadronic system W. The Q^2 dependence of this asymmetry at a fixed W is an unexpected and as yet unexplained result.

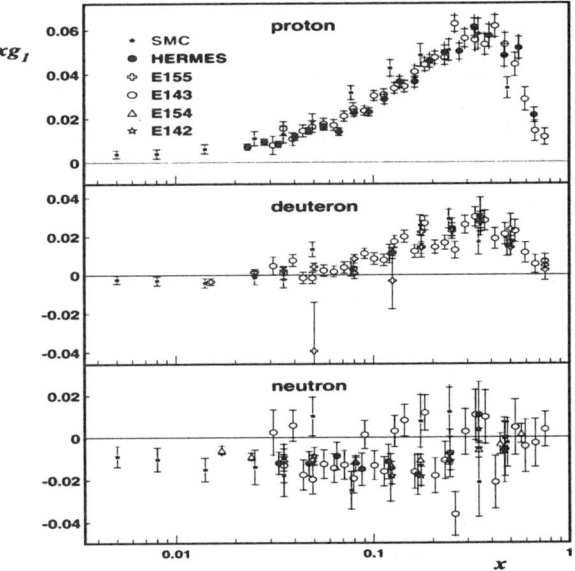

FIGURE 1. World data on x times the spin structure function $g_1(x)$ (CERN, DESY, and SLAC).

Sum Rules

The polarized cross section measurements and the structure functions derived from these quantities can be used to verify sum rules such as for example the Bjorken sum rule which relates the difference in the integral of g_1 on the proton and on the neutron to the ratio of the vector to axial vector couplings in neutron β decay. Another sum rule which has received considerable experimental attention recently

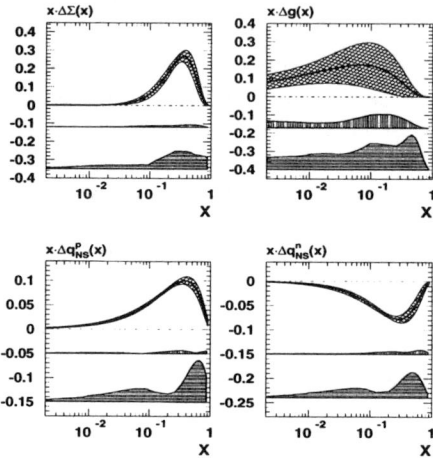

FIGURE 2. Results of an SMC pQCD fit to world data on polarized structure functions. Distributions are shown of the resulting quark singlet (top left), polarized gluon (top right), quark non-singlet on the proton (bottom left), and quark non-singlet on the neutron (bottom right). Cross-hatched band: statistical uncertainty, vertically-hatched band: experimental systematic uncertainty, and horizontally-hatched band: theoretical uncertainty.

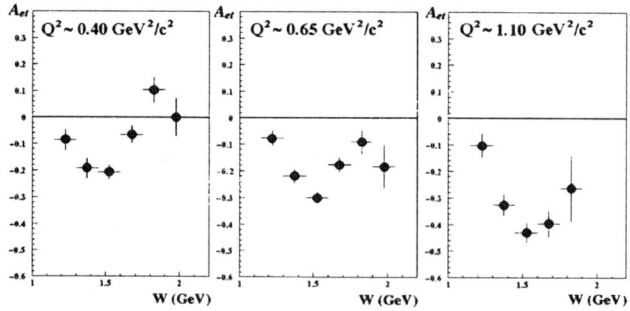

FIGURE 3. Q^2 dependence of the double spin asymmetry $\vec{e}\vec{p} \rightarrow e\pi^+ n$ as a function of the invariant mass of the final hadronic system W (from JLAB Hall B).

is the Gerasimov-Drell-Hearn (GDH) sum rule [10] which relates the anomalous contribution of the magnetic moment of the nucleon κ to the difference of the helicity dependent total photoabsorption cross section:

$$I = \int_{m_\pi}^{\infty} \frac{d\nu}{\nu} \left(\sigma_{3/2}(\nu) - \sigma_{1/2}(\nu) \right) = \frac{2\pi^2 \alpha}{M^2} \kappa^2 \quad (4)$$

where α is the fine structure constant, ν is the energy of the photon, and M is the mass of the nucleon. The predictions of this sum rule based on measurements of κ for the proton and the neutron are $I^{proton} = 204~\mu b$ and $I^{neutron} = 233~\mu b$. This sum rule has not yet been tested due to the inherent difficulties in measuring these cross sections with real photons. However, there has recently been a very interesting preliminary result from the GDH Collaboration at MAMI in Mainz. They use circularly polarized 200-800 MeV real photons on a polarized frozen butanol target to extract the cross section difference $\sigma_{3/2} - \sigma_{1/2}$ as a function of the photon energy (see Fig. 4 [11]). The contribution from the Δ resonance is quite prominent. The second bump includes contributions from single pion, double pion, and η production. The curve in this figure which attempts to take these contributions into account falls short of the experimental determination which is one of the reasons for the follow-up experiment at the Bonn accelerator ELSA which will probe the GDH sum in the region $E_\gamma = 550 - 3400$ MeV. The GDH integral over the measured region is $I^{proton} = 228 \pm 6~\mu b$(stat). An extrapolation to the unmeasured region is necessary before comparing this result to the GDH sum rule.

A renewed interest in the GDH sum also came about when it was realized that it could be generalized to the absorption of a virtual photon with energy ν and negative four momentum transfer squared Q^2

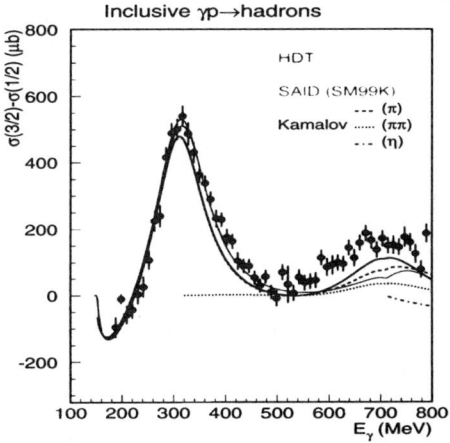

FIGURE 4. Helicity dependent cross section difference as measured with real photons at the GDH collaboration at MAMI. The curves are based on dispersion and multipole analyses.

$$I(Q^2) = \int_{m_\pi}^{\infty} \frac{d\nu}{\nu} \left(\sigma_{3/2}(\nu, Q^2) - \sigma_{1/2}(\nu, Q^2) \right). \tag{5}$$

The Q^2 dependence connects the static ground state properties of the nucleon with its helicity structure as measured in the resonance and DIS regions. HERMES has recently made a measurement of the generalized GDH sum for the first time in both the DIS region [12] and in the resonance region [13] and has made some interesting observations: 1) the contribution from the DIS region is at least as important as that from the resonance region even down to $Q^2 < 2$ GeV^2 and cannot be neglected in the GDH sum, 2) the resonance region contribution is strongly suppressed down to $Q^2 < 2$ GeV^2, and 3) in order to match with the GDH sum rule (from *real* photons), a sharp turn over in the GDH integral for the proton must occur somewhere between $Q^2 = 0 - 1$ GeV^2. It is to observe this sharp turn over that Jefferson Lab is measuring the GDH sum in the resonance region at low Q^2. More results are expected from both JLAB and HERMES within the next 1-2 years.

Higher Twist Effects in g_2

A second polarized structure function $g_2(x, Q^2)$ can be measured from a longitudinally polarized beam on a transversely polarized target: $\vec{e}\vec{N} \to eX$. There exists no simple quark parton model picture for g_2 but it is sensitive to higher twist effects. $g_2(x, Q^2)$ can be written as a purely twist-2 term (which is proportional to g_1) and the twist-3 term [14]

$$g_2(x, Q^2) = \underbrace{g_2^{WW}(x, Q^2)}_{\text{twist 2}} + \underbrace{\bar{g}_2(x, Q^2)}_{\text{mostly twist 3}}. \tag{6}$$

Higher twist effects can be seen as deviations in the measured $g_2(x, Q^2)$ from $g_2^{WW}(x, Q^2)$. There are some recent results from the E155x [15] collaboration at SLAC which has determined $g_2(x, Q^2)$ on both the proton and the deuteron. This collaboration has seen no evidence for higher twist effects nor any significant Q^2 dependence. JLAB will also soon make a precision determination of this quantity in the region of $x = 0.14 - 0.22$.

SEMI-INCLUSIVE MEASUREMENTS

Polarized Quark Distributions

The inclusive measurements presented above are becoming increasingly precise. However these are intrinsically limited in the information that they can provide as they are unable to give a full quark flavor decomposition of the spin. Further insight into the nucleon's structure may be obtained through semi-inclusive polarized deep-inelastic scattering where both the scattered lepton and hadrons produced from the struck quark are detected. This method of "flavor-tagging" by the detection of a leading hadron in a DIS event can be used to isolate individual components contributing to spin structure. Polarized quark distributions may be extracted from a semi-inclusive cross section asymmetry (longitudinally polarized leptons on a longitudinally polarized target). Results have been published by both SMC [16] and HERMES [17]. Fig. 5 shows the HERMES results based on data taken in 1995-7. Since these data were published, considerably more data has been taken by HERMES. The projection of the statistical precision of these new data is also shown in Fig. 5. A clean determination of the polarization of the quark sea should be possible including a first determination of the x dependence of the strange sea. The data to be taken during HERMES RUN2 should significantly add to this precision.

The spin program at the RHIC proton collider will also determine the polarized quark distributions to a high precision via the W boson production process. Fig. 6 shows the PHENIX single spin asymmetry sensitivity of W^+ and W^- production from which $\Delta q/q$ (where $q = u, d, \bar{u}, \bar{d}$) can be extracted [18]. STAR is expected to have a similar sensitivity. Combining both HERMES and RHIC results should clearly establish the quark helicity contribution to the spin of the nucleon and

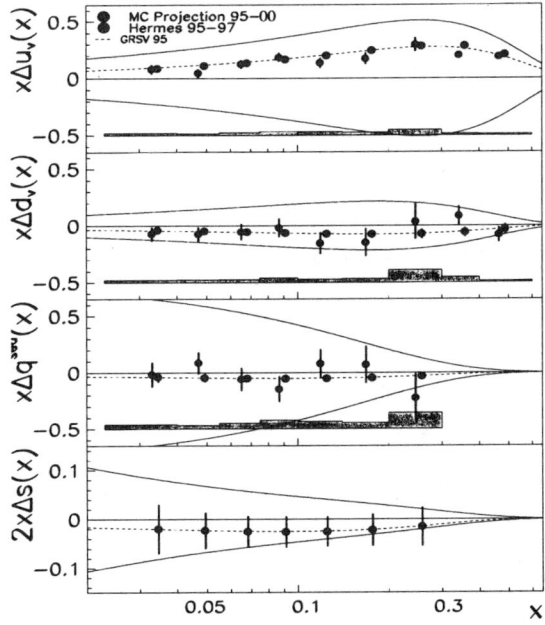

FIGURE 5. Polarized quark distributions at HERMES. Light colored points: published results based on 1995-7 data. Dark points: projected error bars at the end of 2000 data-taking period.

hopefully close the book on this topic.

Transversity

The third twist-2 quark distribution, δq, can only be determined through semi-inclusive processes and so is as yet an unknown quantity. This transversity distribution along with the transverse (Collins') fragmentation function $H_1^\perp$ can be accessed by looking at the spin dependence of the azimuthal angle ϕ of the leading pion (known as the Collins' effect [19]). Preliminary results have come from SMC [20] using unpolarized leptons on transversely polarized protons and deuterons. Transversity would manifest itself as a $\sin\phi$ moment in the asymmetry. Although their results are suggestive of such a moment, no clear determination has yet been established. A single spin asymmetry with a longitudinally polarized target is also sensitive to $\delta q \cdot H_1^\perp$. In this case, a $\sin\phi$ moment in the asymmetry has been observed by HERMES [21]. Fig. 7 [22] shows a clear sinusoidal behavior when a π^+ is observed in the semi-inclusive process; however, none is seen for the π^- asymmetry. This result was predicted by Collins and is due to u quark dominance. The $\sin\phi$ moment for π^+'s was also observed to increase with x indicating that sea quarks do not appear to play a significant role; a result also expected by Collins as transversity is expected have mostly valence properties. Much more precise results on transversity are expected to come in the near future from semi-inclusive DIS with transversely polarized targets from both HERMES and COMPASS and also from transversely polarized Drell-Yan scattering at RHIC.

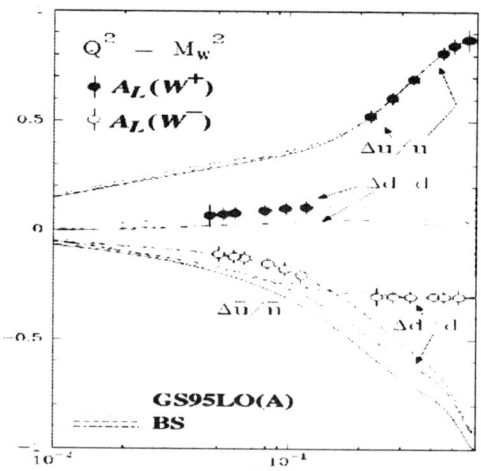

FIGURE 6. PHENIX single spin asymmetry sensitivity with W^+ (solid points) and W^- (open points) boson production compared to various predictions.

Polarized Gluons

One of the most significant remaining open question on the spin structure of the nucleon is the gluonic contribution. Results from inclusive physics as shown in Fig. 2 seem to indicate that the gluons are positively polarized. There are now some very good prospects for a more direct measurement of this quantity. RHIC's longitudinally polarized p-p collider is ideally suited for such a determination. One of the principle channels for a measurement of $\Delta G/G$ is the direct photon production where the away-side jet is also identified: $\vec{p}\vec{p} \to \gamma + jet + X$. If the quark-gluon compton process dominates then the spin asymmetry is directly related to $\Delta G/G$ and to the polarized quark asymmetry. Fig. 8 [23] shows the STAR predicted sensitivity to a determination of $x\Delta G$ for various parameterizations of the gluon polarization based on 20 weeks of data-taking. It is to be noted that data must be taken at two values of $\sqrt{s}$ (200 and 500 GeV) in order to cover the x range shown in the plot. An excellent measurement of ΔG should be possible although the high x region will not be completely constrained.

A determination of ΔG will also be made at COMPASS with the photon-gluon fusion mechanism using two channels. Open charm production is expected to give a sensitivity of $\delta(\Delta G/G) \approx 0.11$ in the accessible gluon momentum fraction region over 2.5 years of data-taking [24]. High-p_T hadron pair production's sensitivity is $\delta(\Delta G/G) \approx 0.05$ with one year of ^6LiD [25] targets. The high-p_T hadron pair method proposed in reference [25] has already been used at HERMES to determine $\Delta G/G$ [26]. Using PYTHIA as a model for processes occurring in the HERMES kinematical domain, three processes are seen to contribute significantly:

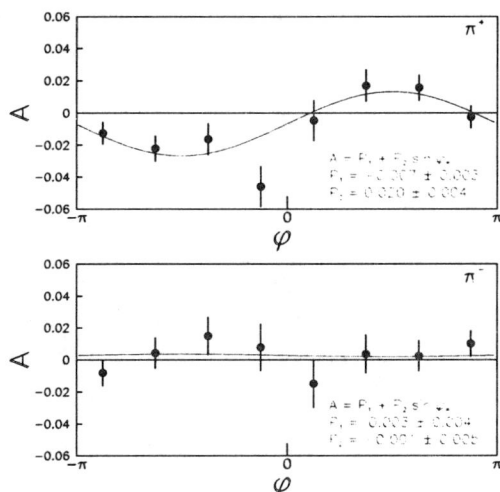

FIGURE 7. Single spin asymmetry with a longitudinally polarized target for the semi-inclusive production of π^+ (top) and π^- (bottom) mesons as a function of the Collins' angle ϕ.

QCD compton, vector meson dominance, and photon-gluon fusion (PGF) which is the process sensitive to ΔG. Monte Carlo studies have revealed that only PGF can produce the negative asymmetry observed in the data. From this, they have established a value of

$$\frac{\Delta G}{G} = 0.41 \pm 0.18(stat) \pm 0.03(sys) \qquad (7)$$

for h^+h^- pairs with the transverse momentum of one hadron $p_T^{h_1} > 1.5$ GeV/c and of the second hadron $p_T^{h_2} > 0.8$ GeV/c. The sign of this result is in agreement with the one obtained from pQCD fits of inclusive data.

CONCLUSION

Very precise measurements of polarized structure functions have emerged since the first experiments in the 1980's. These have enabled us to determine that more than just the quark spins contribute to the nucleon's spin structure. Efforts have now moved to semi-inclusive measurements for a full quark flavor decomposition and a determination of the gluon's role in nucleon spin. Some areas of nucleon

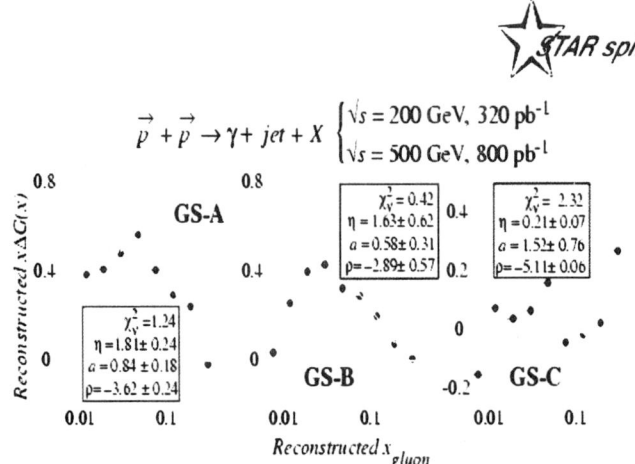

FIGURE 8. STAR sensitivity to a determination of $x\Delta G$ using various parameterizations of ΔG based on 20 weeks of data-taking.

structure, such as quark transversity, are still poorly understood. New results from spin experiments at BNL, CERN, DESY, and JLAB are expected within the next few years which will hopefully address all of these remaining unanswered questions.

ACKNOWLEDGMENTS

I would like to thank H.-J. Arends, L. Bland, V. Burkert, N. Saito, and B. Seitz for providing me with some of the figures in this paper.

REFERENCES

1. See for example Jaffe, R. and Ji, X. , *Phys. Rev.* **D43** 724 (1991).
2. HERMES compilation of world data from CERN, DESY, and SLAC.
3. Leader, E., hep-ph/0004106.
4. See for example Mertig, R. and van Neerven, W. L., *Z. Phys* **C70** 637 (1996); Vogelsang, W., *Phys. Rev.* **D54** 2023 (1996).
5. Ball, R. D. *et al., Phys. Lett.* **B378** 255 (1996).
6. The SMC Collaboration, *Phys. Rev.* **D58** 112002 (1998).
7. The SMC Collaboration, *Phys. Rev.* **D60** 072004 (1999).
8. The HERMES Collaboration, preliminary result. See R. Kaiser, these proceedings.
9. The E91-023 Collaboration, preliminary result. See A. Skabelin, these proceedings.
10. Gerasimov, S. B., *Sov. J. Nucl. Phys* **2** 430 (1966); Drell, S. D. and Hearn, A. C., *Phys. Lett.* **16** 908 (1966).
11. GDH Collaboration at MAMI, preliminary result. See W. Meyer, these proceedings.
12. The HERMES Collaboration, *Phys. Lett.* **B444** 531 (1998).
13. The HERMES Collaboration, preliminary result. See B. Seitz, these proceedings.
14. Cortes, J. L. *et al., Z. Phys.* **C55** 409 (1992); Wandura, S. and Wilczek, F., *Phys. Lett.* **B72** 195 (1977).
15. The E155x Collaboration, preliminary result. See D. McNulty, these proceedings.
16. The SMC Collaboration, *Phys. Lett.* **B420** 180 (1998).
17. The HERMES Collaboration, *Phys. Lett.* **B464** 123 (1999).
18. The PHENIX Collaboration. Curtesy of N. Saito.
19. Collins, J., *Nucl. Phys* **B396** 161 (1993).
20. The SMC Collaboration, preliminary result shown at DIS99.
21. The HERMES Collaboration, *Phys. Rev. Lett.* **84** 4047 (2000).
22. The HERMES Collaboration, final result (figure not published).
23. Bland, L. C., hep-ex/9907058.
24. Bradamante, F., *Proceedings of the 21st Course of the International School on Nuclear Physics*, Erice, September 1999.
25. Bravar, A., *et al., Phys. Lett.* **B421** 349 (1998).
26. The HERMES Collaboration, *Phys. Rev. Lett.* **84** 2584 (2000).

The Proton and Neutron Form Factors

Gerassimos G. Petratos

Kent State University, Kent, OH 44242, USA

Abstract.
A review of measurements of the nucleon elastic form factors is presented along with their interpretation and a brief theoretical background. All available data from Rosenbluth separations and experiments using polarized beams and polarized targets or recoil nucleon polarimeters are summarized. Future prospects for extending or improving the present measurements are outlined.

INTRODUCTION

Measurements of the proton and neutron elastic form factors are of fundamental importance in understanding their electromagnetic structure. The form factors are measured in elastic electron-nucleon scattering and contain all the information about the deviation of the proton and neutron charge and magnetization distributions from a pointlike structure. Large momentum transfer measurements can provide fundamental information on the structure of the nucleons in terms of their elementary constituents, the quarks, within the framework of quantum chromodynamics.

The electron-nucleon elastic scattering is described, to lowest order in the fine-coupling constant α, by the exchange of a single virtual photon. The differential cross section is given by the Rosenbluth [1] formula:

$$\frac{d\sigma}{d\Omega}(E,\theta) = \frac{\alpha^2 E'}{4E^3 \sin^4(\frac{\theta}{2})} \left[(F_1^2 + \kappa^2 \tau F_2^2) \cos^2(\frac{\theta}{2}) + 2\tau(F_1 + \kappa F_2)^2 \sin^2(\frac{\theta}{2}) \right], \quad (1)$$

where E is the incident electron energy, E' is the scattered electron energy, θ is the electron scattering angle, and κ is the anomalous magnetic moment of the nucleon. All the information about the nucleon structure is contained in the Dirac and Pauli form factors $F_1(Q^2)$ and $F_2(Q^2)$, respectively, with $Q^2 = 4EE' \sin^2(\theta/2)$ being the square of the four-momentum transferred to the nucleon. The kinematical factor τ is defined as $\tau = Q^2/4M^2$, where M is the nucleon mass. The Dirac form factor F_1 describes the distribution of charge and the normal part of the magnetic moment μ of the nucleon. The Pauli form factor describes the distribution of the anomalous part of the magnetic moment. The two form factors are normalized at $Q^2 = 0$ as follows: $F_{1p}(0) = 1$, $F_{2p}(0) = 1$ for the proton, and $F_{1n}(0) = 0$, $F_{2n}(0) = 1$ for the neutron.

The elastic cross section is given alternatively in terms of the Sachs [2] nucleon form factors $G_E(Q^2)$ and $G_M(Q^2)$:

$$\frac{d\sigma}{d\Omega}(E,\theta) = \frac{\alpha^2 E'}{4E^3 \sin^4(\frac{\theta}{2})} \left[\frac{G_E^2 + \tau G_M^2}{1+\tau} \cos^2(\frac{\theta}{2}) + 2\tau G_M^2 \sin^2(\frac{\theta}{2}) \right]. \quad (2)$$

The Sachs form factors $G_E(Q^2)$ and $G_M(Q^2)$ are referred to as the electric and magnetic form factors, respectively, because in the non-relativistic limit they are the Fourier transforms of the charge and magnetization distributions of the nucleons. They are normalized at $Q^2 = 0$ as follows: $G_{Ep}(0) = 1$, $G_{Mp}(0) = \mu_p = 2.79$ nm for the proton, and $G_{En}(0) = 0$, $G_{Mn}(0) = \mu_n = -1.91$ nm for the neutron.

The pioneering elastic electron-proton scattering experiments [3] at low Q^2 uncovered the empirical dipole law and form factor scaling: $G_{Mp}(Q^2)/\mu_p \simeq G_{Ep}(Q^2) \simeq (1 + Q^2/Q_o^2)^{-2} \equiv G_D(Q^2)$, where $Q_o^2 = 0.71$ (GeV/c)^2. The form factor scaling indicated that the proton charge and magnetization distributions have the same spatial dependence. The G_D dipole formula translated to an exponential spatial distribution with a rms radius of 0.8 fm. The low Q^2 measurements of the neutron magnetic form factor showed that $G_{Mn}(Q^2)/\mu_n \simeq G_D(Q^2)$, indicating that the magnetic moment distribution of the neutron is similar to that of the proton. The neutron electric form factor measurements showed that $G_{En}(Q^2) \simeq 0$, indicating a zero net neutron charge distribution, but also that the slope dG_{En}/dQ^2 is, for $Q^2 \simeq 0$, definitely > 0, indicating that the neutron charge distribution is not uniformly zero: the charge radius of the neutron is negative, i.e. there is a concentration of negative charge on the outside.

THEORETICAL OVERVIEW

The starting point in the theoretical description of the nucleon form factors is the Vector Meson Dominance (VMD) model. In this framework, the virtual photon couples to the nucleon through vector mesons (see Fig. 1), and the nucleon form factors are expressed in terms of photon-meson coupling strengths, $C_{\gamma V}$, and meson-nucleon vertex form factors, F_{VN}. VMD models [4,5] have given fair descriptions of the low Q^2 data but they do not incorporate a proper description of the nucleon form factors at large Q^2.

With the advent of the quark-parton model and Quantum Chromodynamics (QCD), it is believed that, at large Q^2, the nucleons must behave as systems of point-like quarks bound by gluon-exchanges and governed by the properties of the strong force. The first attempt at a quark-gluon description of the nucleon form factors was within the quark-dimensional scaling (QDS) framework [6,7]. QDS predicts that, at large Q^2, only the valence quark states are important in exclusive processes such as elastic electron-nucleon scattering. The underlying dynamical

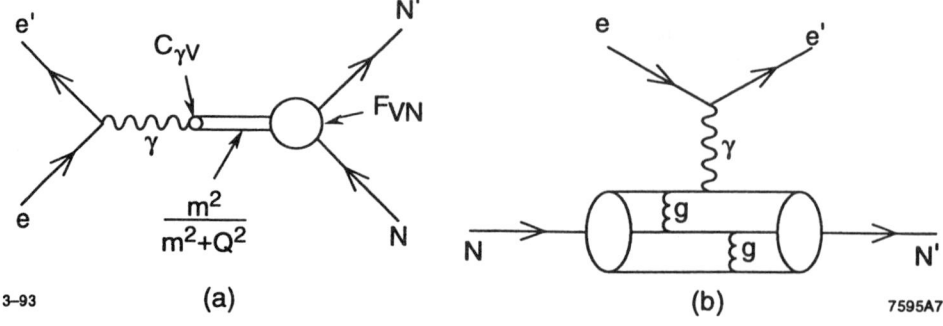

FIGURE 1. Elastic electron-nucleon scattering in a) the Vector Meson Dominance [4,5], and b) the hard-scattering pQCD [8] schemes.

mechanism is the hard rescattering of the quarks that constitute the nucleons, as shown in Fig. 1. In this case, a rough idea of the form factor Q^2 dependence can be gained by simply counting the number of gluon propagators. The magnetic form factor, for example, should scale asymptotically as $(Q^2)^{-(n-1)}$, where $n = 3$ is the number of the nucleon valence quarks.

The simple quark-counting rules were later substantiated within the framework of perturbative QCD (pQCD). Brodsky and Lepage demonstrated [8] that, at large Q^2, QCD effects produce only a logarithmic departure from the dimensional scaling power-laws, in agreement with the large Q^2 SLAC G_{Mp} data [9], as can be seen in Fig. 2. Within the hard-scattering scheme, the nucleon form factors are written as a convolution of distribution amplitudes, representing the scattering of the constituent quarks in a collinear approximation. Naive symmetric distribution amplitudes, where the three valence quarks share equally the nucleon's momentum, failed dramatically to account for the sign and normalization of the proton magnetic form factor. Agreement with the data can be achieved only with a distribution amplitude in which the momentum balance of the valence quarks in the proton is quite asymmetric [10].

The apparent success of pQCD for asymmetric distribution amplitudes is achieved at the expense of strong contributions from "soft" regions where one of the quark constituents carry a small fraction of the nucleon's momentum. This is, as has been pointed out by Isgur and Llewellyn-Smith [11], a very problematic situation for a calculation relying on perturbation theory. Although the hard scattering picture is likely to be the true asymptotic description of the nucleon form factors, it needs modifications at moderate momentum transfers.

An attempt to calculate the contribution to the nucleon form factors from soft non-perturbative processes has been provided by Nesterenko and Radyushkin [12]. They fixed the soft nucleon wave functions by employing QCD sum rules based on quark-hadron duality and decomposed the scattering process in a series of diagrams with gluon exchanges between the quarks. Their calculation showed that the form

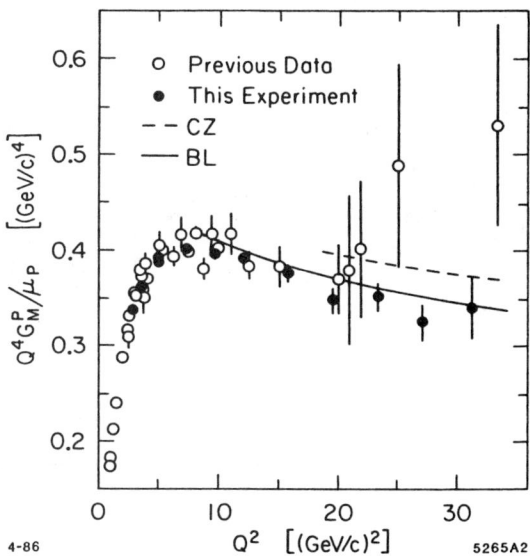

FIGURE 2. The proton magnetic form factor at large Q^2, extracted from SLAC cross section measurements [9] at forward scattering angles assuming form factor scaling. The theoretical curves are from Refs. [8] (BL) and [10] (CZ).

factors are dominated by "soft" processes over the Q^2 range of the existing data and they estimated that the scale of the transition to the hard two-gluon exchange is beyond the reach of present experiments.

In an effort to describe the moderate Q^2 region, Kroll and collaborators [13] proposed a generalization to the hard-scattering scheme by modeling nucleons as made up of quarks and diquarks. The diquarks are treated as quasi-elementary constituents which partly survive medium hard collisions. Their composite nature is taken into account by diquark form factors. The diquarks are viewed as an effective description of correlations in the nucleon wave function and constitute a model for non-perturbative effects. The model is designed such that, at large Q^2, when the diquarks dissolve into quarks, pQCD emerges.

Another approach [14,15] to describe the nucleon form factors at moderate momentum transfers is based on relativistic constituent-quark models. These calculations use a particular nucleon wave function model with two parameters: the effective quark mass m and a confinement scale a. For example, the model by Chung and Coester [14] assumes a simple exponential wave function of the form $\phi(M_o) = N \exp(-M_o^2/2a^2)$, where $M_o^2 = \sum_i \sqrt{m_i^2 + \vec{q}_i^2}$ with $\vec{q}_i$ being the quark relative momenta. The model of Schlumpf [15] assumes a symmetric also wave function of a particular form. The confinement scale a is of the order of 0.6 GeV and the quark mass that reproduces partially the available data is ~0.25 GeV, smaller than the conventional non-relativistic choice of 1/3 of the nucleon mass.

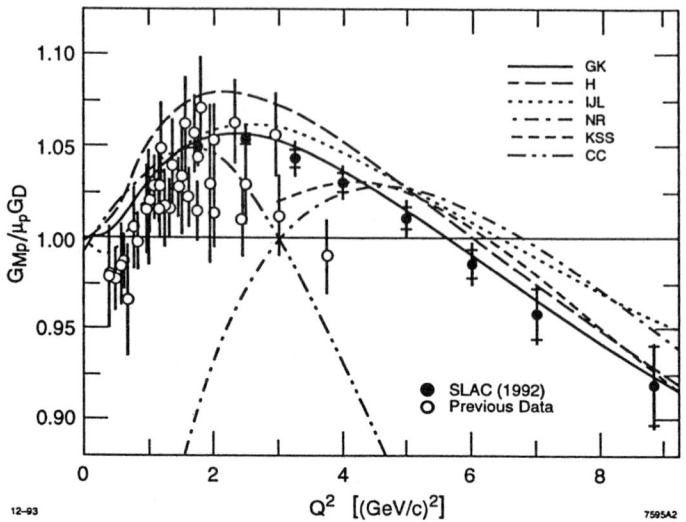

FIGURE 3. Proton magnetic form factor SLAC NE11 data, scaled by the dipole formula, along with previous data from Rosenbluth separations [17]. The GK, H, IJL, NR, KSS and CC curves are from Refs. [16], [5], [4], [12], [13] and [14], respectively.

Lastly, hybrid phenomenological approaches have attempted to describe the moderate Q^2 regime by synthesizing in a direct way the meson picture of VMD and the quark picture of pQCD using parametrizations of the form factors which properly combine these two pictures. A popular parametrization based on this approach is by Gary and Krümpelmann (GK) [16] that introduces a transition from meson dynamics to quark dynamics at ~ 5 $(\text{GeV}/c)^2$.

EXPERIMENTAL MEASUREMENTS

Rosenbluth separations

The majority of the existing data on the nucleon form factors have resulted from Rosenbluth separations in elastic electron-proton (e-p) and quasielastic electron-deuteron (e-d) scattering. The Rosenbluth separation method is based on cross section measurements at the same Q^2 at several different scattering angles. This technique allows the determination of both nucleon form factors as can be seen from Equations 1 and 2. A typical example based on this method is SLAC experiment NE11 that doubled the Q^2 range and improved the precision of previous measurements. The experiment [17,18] was performed at the SLAC End Station A and used beams from the Nuclear Physics Injector. Scattered electrons were detected in the SLAC 1.6 and 8 GeV/c magnetic spectrometers.

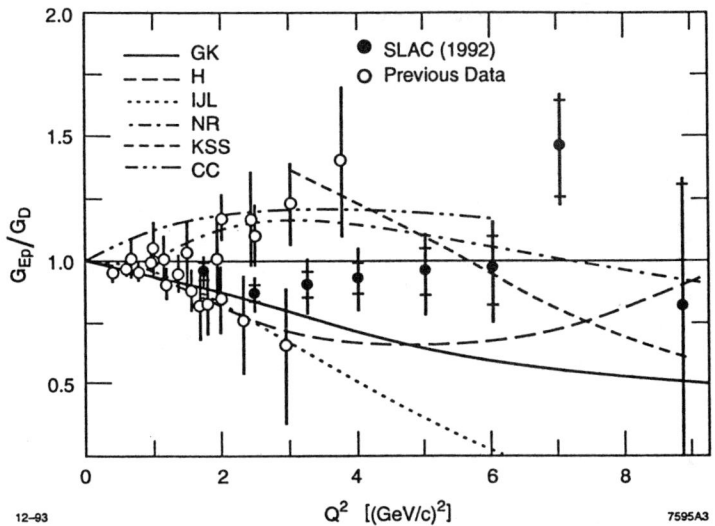

FIGURE 4. Proton electric form factor SLAC NE11 data, scaled by the dipole formula, along with previous data from Rosenbluth separations [17]. The GK, H, IJL, NR, KSS and CC curves are from Refs. [16], [5], [4], [12], [13] and [14], respectively.

The extracted proton form factors, scaled by the dipole formula $G_D(Q^2)$, are shown in Figs. 3 and 4 along with previous data from Rosenbluth separations [17]. The results at $Q^2 = 8.8$ $(GeV/c)^2$ were obtained by combining the NE11 backward angle data with previous forward angle cross section data [9] cross-normalized at the same $Q^2 = 5$ $(GeV/c)^2$ value. Also shown in Figs. 3 and 4 are VMD fits and theoretical calculations.

The neutron form factors were extracted by a Rosenbluth separation of the deuteron quasielastic response functions using the proton form factor measurements from the same experiment and a simplified form of the Durand-McGee non-relativistic Plane Wave Impulse Approximation quasielastic model [19]. Inelastic contributions were calculated using a Fermi-smearing model [20] to convolute measured proton resonance region data with the deuteron wave function. The extracted values for G_{Mn} from this experiment are shown in Fig. 5 along with previous data from Rosenbluth separations [18], and theoretical calculations and VMD fits.

The high luminosity and unique facilities of JLab can be used to push the extraction of G_{Mn} from inclusive e-d quasielastic scattering to its practical limit of about $Q^2 = 6.5$ $(GeV/c)^2$. To minimize the unknown contribution to the cross section from G_{En} and to eliminate the need for a Rosenbluth separation, it has been proposed and approved for the future to extract G_{Mn} by backward angle ($\sim 120°$) inclusive e-d quasielastic scattering [21]. This JLab measurement will be limited, as the SLAC NE11 one, by inherent theoretical uncertainties.

Precision G_{Mn} measurements

Precision G_{Mn} measurements can be obtained from quasielastic e-d scattering with detection of the recoil neutron in coincidence with the scattered electron [22,23]. A series of low Q^2 precision measurements has been performed during this decade using this method. The theoretical uncertainties in this method are further reduced by measuring in the same experiment quasielastic scattering with detection of the recoil proton in coincidence. The ratio of the neutron to proton quasielastic cross sections gives the G_{Mn}/G_{Mp} ratio, which is essentially free of theoretical uncertainties. This method is limited by the precision in the efficiency calibration of the detectors used to identify the recoil neutrons. The efficiency is usually determined using the $\gamma + p \rightarrow \pi + n$ reaction. Figure 6 shows all the G_{Mn} data to date from such precision measurements: MIT/Bates [24], NIKHEF [25], Bonn [26] and Mainz [27]. Although there are noticeable disagreements among the different data sets, G_{Mn} appears to be consistent with the dipole formula in the Q^2 range from 0 to 1 (GeV/c)^2. Measurements of G_{Mn} at larger momentum transfers, probably up to $Q^2 = 5$ (GeV/c)^2, are expected in the near future from a JLab experiment that will measure the neutron to proton quasielastic cross section ratio using the CLAS detector [28].

Experiments with polarized nucleon targets

Significant recent advances in high density polarized neutron targets allow for measurements of the neutron form factors by measuring the cross section asymmetry in quasielastic scattering of longitudinally polarized electrons from polarized neutrons [29]. For example, the asymmetry $A_\perp$, when the neutron spin is aligned perpendicular to the momentum transfer $\vec{q}$, is given by:

$$A_\perp = -P_e P_t \frac{2\sqrt{\tau(1+\tau)}\tan(\frac{\theta}{2}) \, G_{En}}{G_{En}^2 + \tau[1 + 2(1+\tau)\tan^2(\frac{\theta}{2})] \, G_{Mn}}, \quad (3)$$

where P_e and P_t are the electron and target nucleon polarizations. When the neutron spin is aligned parallel to $\vec{q}$, the asymmetry is:

$$A_\parallel = -P_e P_t \frac{2\tan(\frac{\theta}{2})\sqrt{1+\tau+(1+\tau)^2\tan^2(\frac{\theta}{2})}}{1 + 2(1+\tau)\tan^2(\frac{\theta}{2})}. \quad (4)$$

The latter equation is independent of the neutron form factors and serves as an excellent calibration check. Therefore, measurement of both asymmetries can provide the much sought G_{En}/G_{Mn} ratio as:

$$\frac{G_{En}}{G_{Mn}} = \frac{A_\perp}{A_\parallel}\sqrt{\tau + \tau(1+\tau)\tan^2(\frac{\theta}{2})}, \quad (5)$$

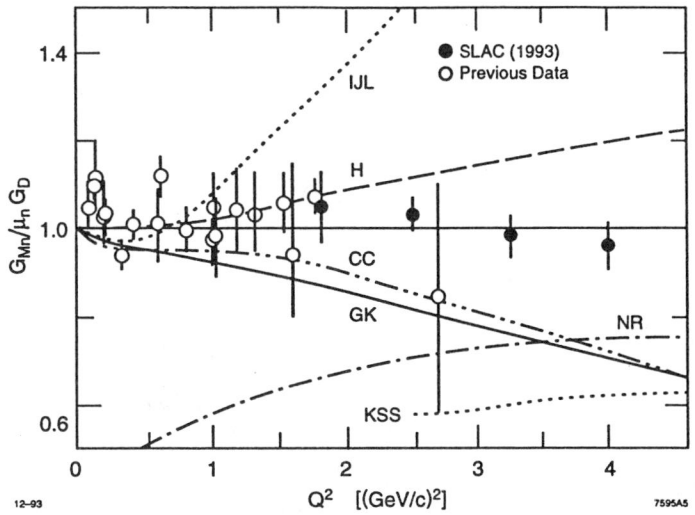

FIGURE 5. Neutron magnetic form factor SLAC NE11 data, scaled by the dipole formula, along with previous data from Rosenbluth separations [18]. The GK, H, IJL, NR, KSS and CC curves are from Refs. [16], [5], [4], [12], [13] and [14], respectively.

independently of the electron beam and target nucleon polarizations and their uncertainties.

The feasibility of this technique was demonstrated by two pilot MIT/Bates experiments using polarized ^{3}He targets (as effective polarized neutron targets) that measured G_{En} [30,31] and G_{Mn} [32]. The G_{Mn} result is shown in Fig. 6. Recent measurements at Mainz [33–35], using polarized ^{3}He targets, and at NIKHEF [36], using a vector polarized deuterium target, have provided precise G_{En} data showing a clear deviation from zero values (see Fig. 7). New data are becoming available or are expected, using polarized ^{3}He targets, from a JLab experiment [37] that has measured G_{Mn} in the range $0.1 \leq Q^2 \leq 0.6$ (GeV/c)^2, and from a NIKHEF measurement [38] of G_{En} at $Q^2 = 0.2$ (GeV/c)^2. Preliminary JLab G_{Mn} results are shown in Fig. 6. Measurements of G_{En} at larger momentum transfers, expected to reach $Q^2 = 1.5$ (GeV/c)^2, have started at JLab using polarized neutrons in a polarized deuterated ammonia (ND$_3$) target [39]. Preliminary JLab G_{En} results are shown in Fig. 7. The same technique using polarized protons in a polarized ammonia (NH$_3$) target can be used to provide precision G_{Ep} measurements in the future.

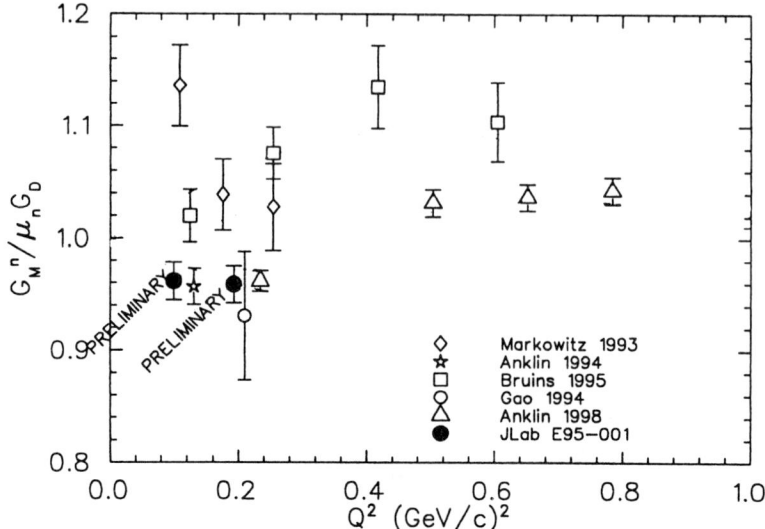

FIGURE 6. The neutron magnetic form factor from: a) electron-deuteron quasielastic scattering experiments with detection of recoil neutrons in coincidence: MIT/Bates (open diamonds) [24], NIKHEF (star) [25], Bonn (open squares) [26] and Mainz (open triangles) [27], and b) polarized electron-^{3}He scattering: MIT/Bates (open circle) [32] and preliminary JLab (solid circles) [37].

Experiments with a recoil nucleon polarimeter

Another way to measure the nucleon form factors using polarization observables is the recoil nucleon polarization technique [40,41]. This method requires a longitudinally polarized electron beam scattered off an unpolarized nucleon target, and a polarimeter to measure the polarization transferred to the recoil nucleon. The two components of the recoil nucleon polarization in the scattering plane, the transverse P_x and the longitudinal P_z, have been shown to be functions of the nucleon form factors:

$$P_x = -\frac{2}{I_o}\sqrt{\tau(1+\tau)}G_M G_E \tan\left(\frac{\theta}{2}\right), \quad P_z = \frac{E+E'}{MI_o}\sqrt{\tau(1+\tau)}G_M^2 \tan^2\left(\frac{\theta}{2}\right), \quad (6)$$

where $I_o = G_E^2 + \tau G_M^2/\epsilon$. The G_E/G_M ratio is then obtained from:

$$\frac{G_E}{G_M} = -\frac{P_x}{P_z}\frac{E+E'}{2M}\tan\left(\frac{\theta}{2}\right). \quad (7)$$

The P_x and P_z nucleon polarizations are measured from the azimuthal angular distribution $N(\Theta, \Phi)$ of the recoil nucleons scattered off the nucleon polarimeter. A Fourier analysis of $N(\Theta, \Phi)$ gives the two physical amplitudes $a(\Theta) = P_e A_c(\Theta) P_x$ and $b(\Theta) = P_e A_c(\Theta) P_z \sin\chi$ that determine the P_x/P_z ratio. Here, $A_c(\Theta)$ is the

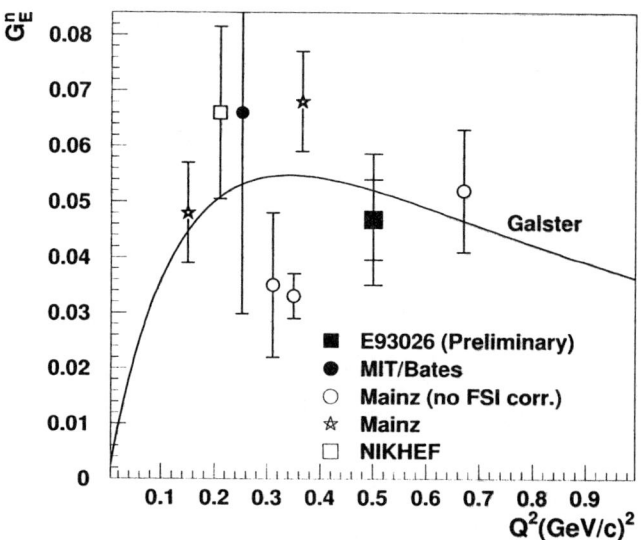

FIGURE 7. Precision measurements of the neutron electric form factor from MIT/Bates [42] (solid circle), Mainz (open circles/using polarized ^{3}He) [33–35], Mainz (stars/using a neutron polarimeter) [43,44], NIKHEF [36] (open square) and preliminary JLab [39] (solid square). The curve represents the Galster parametrization [46].

analyzing power of the polarimeter and χ is the spin precession angle of the nucleon in the magnetic spectrometer used for its detection.

The feasibility of this method was demonstrated by a pilot experiment also performed at MIT/Bates [42]. It measured the polarization of the recoil neutrons in quasielastic scattering of polarized electrons from deuterium and determined G_{En} at $Q^2 = 0.26$ (GeV/c)^2. A recent similar experiment at Mainz [43,44] has provided precision G_{En} data at $Q^2 = 0.15$ and 0.34 (GeV/c)^2. The data from the above experiments are shown in Fig. 7. It should be noted that all recent precise G_{En} results are within the limits determined from elastic electron-deuteron scattering measurements [46,47]. An extension of G_{En} measurements with a neutron recoil polarimeter, up to $Q^2 = 1.8$ (GeV/c)^2, is under way [45] at JLab. The latter experiment will also provide precision G_{Mn} measurements in the same Q^2 range.

The pilot MIT/Bates G_{Mn} measurement was followed by similar low Q^2 measurements of the G_{Ep}/G_{Mp} ratio using a recoil proton polarimeter [48]. The advent of JLab with its unique high luminosity and polarized electron beams allowed recently a precise measurement of the G_{Ep}/G_{Mp} ratio up to $Q^2 = 4$ (GeV/c)^2. The results from this experiment [49], showed for the first time a clear deviation from the previously assumed form factor scaling, as shown in Figure 8. The JLab G_{Ep}/G_{Mp} measurements are expected to be extended up to $Q^2 = 5.6$ (GeV/c)^2 in the near future [50] and be of crucial importance in testing pQCD predictions for this ratio.

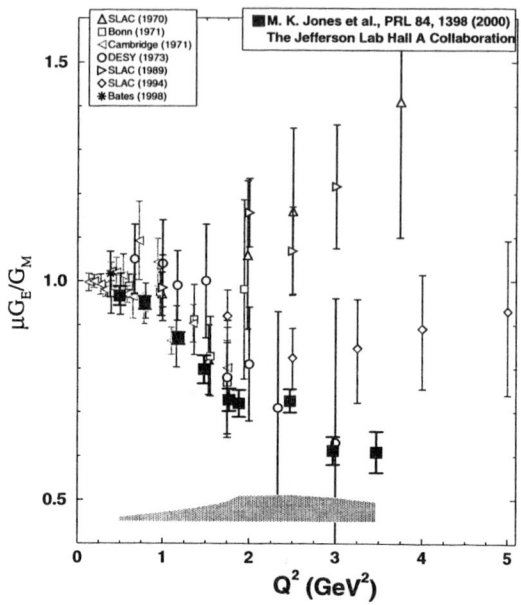

FIGURE 8. The ratio of the electric to magnetic proton form factors from JLab (solid squares) [49] and MIT/Bates (stars) [48] using the recoil proton polarization technique. Also shown are previous data from Rosenbluth separations (open symbols) [17].

SUMMARY

Rosenbluth separations in elastic electron-proton and quasielastic electron-deuteron scattering have provided a wealth of data on the electromagnetic form factors of the proton and neutron. These data have been unique in providing information on the nucleon charge and magnetization distributions, and in testing models of the nucleon structure. Recent advances in polarized nucleon targets and nucleon recoil polarimeters have made available, using polarized beams, precision data for G_{En} and G_{Mn}, and have improved the quality of the existing G_{Ep} data. Precision G_{Mn} data are becoming available from coincidence quasielastic electron-deuteron scattering experiments. Improved measurements of the nucleon form factors will be crucial in the quest for our understanding of the nucleon structure with a fundamental QCD prescription.

I would like to thank the CIPANP-2000 organizers for their kind invitation and to acknowledge the support of the U.S. National Science Foundation.

REFERENCES

1. M. N. Rosenbluth, *Phys. Rev.* **79**, 615 (1950).
2. F. J. Ernst, R. G. Sachs and K. C. Wali, *Phys. Rev.* **119**, 1105 (1960).

3. R. Hofstadter, *Ann. Rev. Nucl. Sci.* **7**, 231 (1957).
4. F. Iachello, A. Jackson and A. Lande, *Phys. Lett. B* **43**, 191 (1973).
5. G. Höhler et al., *Nucl. Phys. B* **114**, 505 (1976).
6. S. J. Brodsky and G. Farrar, *Phys. Rev. D* **11**, 1309 (1975).
7. V. Matveev, R. Muradyan and A. Tavkhelidze, *Lett. Nuovo Cimento* **7**, 719 (1973).
8. G. P. Lepage and S. J. Brodsky, *Phys. Rev. D* **22**, 2157 (1980).
9. A. F. Sill et al., *Phys. Rev. D* **48**, 29 (1993); and references therein.
10. V. L. Chernyak and I. R. Zhitnitsky, *Nucl. Phys. B* **246**, 52 (1984).
11. N. Isgur and C. Llewellyn-Smith, *Phys. Rev. Lett.* **52**, 1080 (1984); *Phys. Lett. B* **217**, 535 (1989).
12. V. A. Nesterenko and A. V. Radyushkin, *Phys. Lett. B* **128**, 439 (1983); *Sov. J. Nucl. Phys.* **39**, 811 (1984).
13. P. Kroll, M. Schürmann and W. Schweiger, *Z. Phys. A* **338**, 339 (1991).
14. P. L. Chung and F. Coester, *Phys. Rev. D* **44**, 229 (1991).
15. F. Schlumpf, *Mod. Phys. Lett. A* **8**, 2135 (1993); *J. Phys. G* **20**, 237 (1994).
16. M. Gari and W. Krümpelmann, *Z. Phys. A* **322**, 689 (1985); *Phys. Lett. B* **173**, 10 (1986).
17. L. Andivahis et al., *Phys. Rev. D* **50**, 5491 (1994); and references therein.
18. A. F. Lung et al., *Phys. Rev. Lett.* **70**, 718 (1993); and references therein.
19. L. Durand III, *Phys. Rev. Lett.* **6**, 631 (1961); I. McGee, *Phys. Rev.* **161**, 1640 (1967).
20. A. Bodek et al., *Phys Rev.* **20**, 1471 (1979); M. M. Sargsian, L. L. Frankfurt and M. I. Strikman, *Z. Phys. A* **335**, 431 (1990).
21. J. Gomez, G. Petratos et al., JLab Proposal E93-24.
22. P. Stein et al., *Phys. Rev. Lett.* **16**, 592 (1966).
23. W. Bartel et al., *Nucl. Phys. B* **58**, 429 (1973).
24. P. Markowitz et al., *Phys. Rev. C* **48**, R5 (1993).
25. H. Anklin et al., *Phys. Lett. B* **366**, 313 (1994).
26. E. Bruins et al., *Phys. Rev. Lett.* **75**, 21 (1995).
27. H. Anklin et al., *Phys. Lett. B* **428**, 248 (1998).
28. W. Brooks, M. Vineyard et al., JLab proposal E94-17.
29. T W. Donnelly and A. S. Raskin, *Ann. Phys.* **169**, 247 (1986).
30. C. E. Jones et al., *Phys. Rev. C* **44**, 571 (1991).
31. A. K. Thompson et al., *Phys. Rev. Lett.* **68**, 2901 (1992).
32. H. Gao et al., *Phys. Rev. C* **50**, R546 (1994).
33. M. Meyerhoff et al., *Phys. Lett. B* **327**, 201 (1994).
34. J. Becker et al., *Eur. Phys. J. A* **6**, 329 (1999),
35. D. Rohe et al., *Phys. Rev. Lett.* **83**, 4257 (1999).
36. I. Passchier et al., *Phys. Rev. Lett.* **82**, 4988 (1999).
37. D. Dutta et al., presented at this conference.
38. J. F. J. van den Brand et al., NIKHEF/AmPS Experiment E94-05.
39. M. Zeir et al., presented at this conference.
40. A. I. Akhiezer and M. P. Rekalo, *Sov. J. Part. Nucl.* **3**, 277 (1974).
41. R. G. Arnold, C. E. Carlson and F. Gross, *Phys. Rev. C* **21**, 1426 (1980).
42. T. Eden et al., *Phys. Rev. C* **50**, R1749 (1994).
43. M. Ostrick et al., *Phys. Rev. Lett.* **83**, 276 (1999).
44. C. Herberg et al., *Eur. Phys. J. A* **5**, 131 (1999).
45. S. Kowalski, R. Madey et al., JLab Proposal E93-38.
46. S. Galster et al., *Nucl. Phys. B* **32**, 221 (1971).
47. S. Platchkov et al., *Nucl. Phys. A* **510**, 740 (1990).
48. B. Milbrath et al. *Phys. Rev. Lett.* **80**, 452 (1998).
49. M. K. Jones et al., *Phys. Rev. Lett.* **84**, 1398 (2000).
50. E. Brash, M. Jones, C. Perdrisat, V. Punjabi et al., JLab Proposal E99-07.

Exotic Hadrons

Jim Napolitano

Department of Physics, Applied Physics, and Astronomy
Rensselaer Polytechnic Institute
Troy, NY 12180

This talk is dedicated to the memory of Nimai Chand Mukhopadhyay.

Abstract. This talk discusses ground and excited states of hadrons that are not explained by the quark model. This includes "extra" states, which are generally interpreted in terms of gluonic degrees of freedom or as multiquark configurations. It also includes "missing" states which are predicted by the quark model but are not yet observed. We concentrate on the more recent advances in this field, giving examples in both the meson and baryon sectors.

INTRODUCTION

We teach our students that by and large, the bound and excited states of hadrons are described by the quark model. Indeed, the model enjoyed early success by matching well with the observed spectrum of baryons [1] and mesons [2]. With the discovery of charm and beauty, the predictions were further solidified, especially with the realization that heavy quark systems were particularly well described with a quark potential well framework with additional spin degrees of freedom. The various disagreements with the light quark hadrons (i.e. bound states of u, d, and s quarks) were generally tossed off to a poor understanding of the long distance behavior of QCD, and more or less forgotten by the particle physics community.

As time went on, however, these disagreements in the light quark sector persisted even though our knowledge of strong QCD[1] grew. In fact, it became clear that these disagreements may in fact be *predicted* by picturing hadrons in terms of QCD at a fundamental level. That is, states should be observed (at some level) that manifest gluonic and multiquark degrees of freedom. It has taken some time to sort out these "legitimate" states from the flotsam and jetsam of decades of hadronic spectroscopy, but clear progress had been made in the past few years [3].

---

[1] By "strong QCD" we mean QCD in the long distance, strong coupling regime, as opposed to short distance "weak QCD" where perturbation theory applies.

This talk reviews the status of discrepancies between the quark model and the observed hadrons in the light quark sector. We include a discussion of "glueball" candidates, that is, states which are dominated by components having no quark degrees of freedom at all. We then describe recent results for mesons with manifestly "exotic" quantum numbers which cannot be constructed in any $q\bar{q}$ model. We use the term "oddball" to describe well-established states which have resisted description in terms of either the quark model or strong QCD, including the Roper Resonance in the baryon sector, the $a_0(980)$ and $f_0(980)$ mesons, and the old E/ι puzzle for mesons in the 1400 MeV region. We then discuss the $\xi(2230)$, whose status varies between glueball candidate and experimental artifact. We also briefly review states predicted by the quark model (and presumably by our present formulations of strong QCD) but not yet observed. We conclude with a discussion of the future of light hadron spectroscopy at Jefferson Laboratory and a few final remarks.

EXAMPLES OF EXOTIC STATES

In a broad sense, all of the discrepant states we discuss here are "exotic", and the term is frequently used in that sense. (Indeed, it is used in that way in the title of this talk.) In a more strict sense, it can mean states that are "manifestly exotic" by virtue of their quantum numbers. These examples are separated out in the discussion below.

The Scalar Glueball

Calculations of Strong QCD on the lattice now consistently obtain the lowest mass state of pure glue to be around 1700 MeV with $J^{PC} = 0^{++}$. For example, see [4,5]. The quark model predicts two lowest mass isoscalar scalar mesons, corresponding to $(L, S) = (1, 1)$ $q\bar{q}$ pairs, one being the isoscalar u, d combination (which we call $n\bar{n}$) and the other the $s\bar{s}$. These states would have masses not far from that of the glueball [3] and consequently all three states will likely mix.

Indeed, three states have been observed experimentally. These are the $f_0(1370)$ and $f_0(1500)$, studied thoroughly by the Crystal Barrel Collaboration [6] and others, and the $f_0(1710)$ most generally observed in J/ψ radiative decay. (There has been some controversy [3] as to the spin of the latter state, i.e. $J = 0$ or $J = 2$, and it has been traditionally been called the $f_J(1710)$. However, a detailed and unconstrained partial wave analysis [7] shows rather clearly that $J = 0$.) The $f_0(1370)$ is not only the lightest of the three, its decay modes indicate no significant coupling to $s\bar{s}$ and is likely predominantly $n\bar{n}$.

The decay patterns of the $f_0(1500)$ and $f_0(1710)$ are more complicated, both with observed decays to $\pi\pi$ and $K\bar{K}$. It is necessary to better understand the decay dynamics in order to determine the relative amounts of $s\bar{s}$ and glueball in each of these two states. For a recent discussion, see [8].

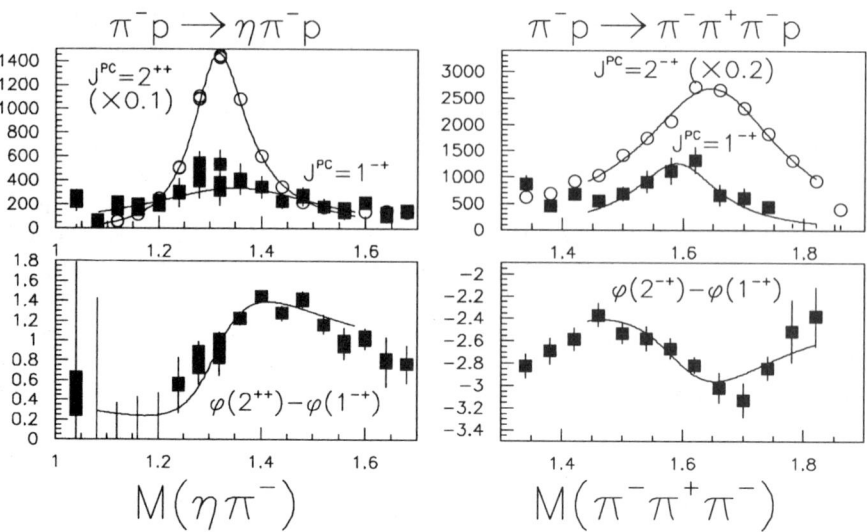

FIGURE 1. Evidence for mesons with exotic quantum numbers $J^{PC} = 1^{-+}$ from BNL experiment E852. The left side shows evidence for a state at 1400 MeV, decaying to $\eta\pi^-$. The right side, a state at 1600 MeV, decaying to $\pi^-\pi^+\pi^-$. In each case, the top plot compares the exotic wave with a more dominant conventional meson excitation, the $a_2(1320)$ on the left and the $\pi_2(1670)$ on the right. The bottom plot shows the phase motion between the two waves. The solid lines are simultaneous fits to Breit-Wigner resonance shapes for the intensity and the phase motion for the exotic and conventional state in each case.

Exotic Quantum Numbers

Mesons with manifestly exotic quantum numbers would provide unequivocal evidence of physics outside the quark model. Such states have been long sought, but until recently, evidence for them has been spotty at best. Now, however, two isovector states have been relatively well established with the exotic quantum numbers $J^{PC} = 1^{-+}$. These are the $\hat{\rho}(1400)$ [9–11], observed decaying to $\eta\pi$, and the $\hat{\rho}(1600)$ [12], observed decaying to $\rho\pi$. A summary of the data is shown in Fig 1. The isovector nature of these states implies that there is a $q\bar{q}$ component, but the quantum numbers insist that other degrees of freedom are present. Whether or not these other degrees of freedom are gluonic or not is still an open question. In fact, the resonant nature of the $\hat{\rho}(1400)$ has even been called into question [13], although it is not clear how the two different production mechanisms (i.e. peripheral production in E852 and $p\bar{p}$ annihilation in Crystal Barrel) can be reconciled otherwise.

There is recent evidence [14] for the $\hat{\rho}(1600)$ decaying to $\eta'\pi^-$, also from peripheral production in experiment E852. This is particularly noteworthy since unlike production in the $\eta\pi$ and 3π channels, where the exotic signal is a very small fraction of the total cross section, the exotic P-wave $\eta'\pi^-$ signal dominates in this case.

Oddballs

There are a number of well-established hadronic states that are very difficult to classify in terms of either the quark model, or in models which incorporate non-$q\bar{q}$ degrees of freedom. We review a few of these here.

The Roper Resonance

The lowest lying excitation[2] of the nucleon is the so-called Roper resonance. This is a $J^P = \frac{1}{2}^+$ state with mass 1440 MeV and a few hundred MeV wide. Its low mass is particularly mysterious, since in the quark model [1] the positive parity excitations of the nucleon are predominantly $2\hbar\omega$ excitations of the harmonic oscillator, and should lie above the negative parity single $\hbar\omega$ states. Indeed, the Roper is the only glaring exception to this rule. In fact, it was suggested quite some time ago [15] that it might be an example of a baryon with constituent gluons.

There has been recent work [16] that attempts to explain the Roper as a collective "breathing" mode of the nucleon. This is based on an extraction of the Roper signal using inelastic scattering of a 4.2 GeV α beam at small angles, $\theta_\alpha = 0.8°$. There is of course the problematic need to remove the substantial background from projectile excitation and other processes. Furthermore, at this beam energy there is a large variation in $-t$ (i.e. squared momentum transfer) as one crosses the wide Roper resonance. Nevertheless, the measurement and the interpretation are intriguing.

The $a_0(980)$ and $f_0(980)$

The $a_0(980)$ and $f_0(980)$ form an isovector/isoscalar pair of mesons that have been known for many years. Recently the mass and width of the $a_0(980)$ has been well measured in the $\eta\pi^\pm$ and $\eta\pi^0$ decay modes [17] and the $f_0(980)$ has been observed as a clear peak in peripheral $\pi^0\pi^0$ production at high $-t$ [18]. However, these states have always been a problem to fit neatly into the quark model. Their proximity to the $K\bar{K}$ threshold (at 990 MeV) and their propensity to decay to $K\bar{K}$, if kinematically allowed, has led several authors to suggest they are $K\bar{K}$ molecules. Other models are also proposed. See [3] for more details.

The decays $\phi \to \{a_0, f_0\}\gamma$ can be used to constrain the $s\bar{s}$ content of these states. Both the quark model wave function of the ϕ and the electromagnetic decay operator are relatively well understood, so the decay rates can lead to information on

---

[2] The well known $\Delta(1232)$ is best described as a three quark configuration with isospin $\frac{3}{2}$, in terms of the quark model. As such, it is a "ground state Δ" instead of an excited nucleon in our terminology.

the wave functions of the a_0 and f_0. In fact, recent measurements have been carried out at the CMD-2 [19] and SND [20,21] experiments using the VEPP-2M e^+e^- storage ring at Novosibirsk, and with the KLOE detector [22] and the DAΦNE ring at Frascati. These experiments certainly point to a significant $s\bar{s}$ content of the a_0 and f_0, and a new experiment [23] will shortly present results of comparable statistical uncertainty but with very different systematics. Unfortunately, however, the theoretical situation is not yet clear enough to distinguish molecular interpretations from other multiquark configurations.

Multiquarks, Molecules, and the E/ι Problem

Multiquark states, that is, states with $qq\bar{q}\bar{q}$ configurations, may very well be at the heart of understanding the a_0/f_0 situation and other problems. Whether or not these are clustered into mesonic molecules, such as $K\bar{K}$, is still a difficult question to answer. One reason is that other decay modes are possible which do not directly reflect the molecular nature of the state. For example, although both the $a_0(980)$ and $f_0(980)$ decay to $K\bar{K}$, we also have $a_0(980) \to \eta\pi$ and $f_0(980) \to \pi\pi$.

This situation is compounded in the case of $K^*\bar{K}$ (and hermitian conjugate) threshold at 1400 MeV. Since $K^* \to K\pi$, one expects to observe states from a $K\bar{K}\pi$ relative S-state in this region. One such combination arises if the $K\bar{K}$ manifests itself as an $a_0(980)$ which then forms an $L = 0$ molecule with the pion. This (isoscalar) molecule would have $J^{PC} = 0^{-+}$. On the other hand, an $L = 0$ $K^*\bar{K}$ molecular state would have $J^{PC} = 1^{++}$. Might one therefore expect *two* molecules in this system near 1400 MeV?

What makes all this very intriguing is that one also expects 0^{-+} and 1^{++} states in this region from the $q\bar{q}$ quark model itself! One of these is essentially the radially excited η' and the other is the mainly $s\bar{s}$ $(L, S) = (1, 1)$ axial vector meson, i.e. the f_1'. In other words, there might be four intermixed states in this region. In fact, there is evidence for two 0^{-+} and two 1^{++} states. See [3] and also [24]. The two 0^{-+} states appear in a number of different production mechanisms and are split by about 50 MeV, with the lower mass state decaying to $a_0\pi$ and the higher mass state to $K^*\bar{K}$.

On the other hand, the 1^{++} states each decay in multiple ways, but are very specific with respect to production reactions. The lower mass state, called $f_1(1420)$, is mostly prominent in peripheral reactions with pion beams and in (single tagged) $\gamma\gamma^*$ collisions. The $f_1(1510)$, however, appears clearly in peripheral reactions with kaon beams, in particular in the very clean results from the LASS experiment at SLAC [25]. No single reaction shows clear evidence for *both* 1^{++} states.

This is the essence of what was once known as the E/ι puzzle. Touted originally as evidence for a 0^{-+} glueball, it was seen in $K\bar{K}\pi$ final states resulting from radiative J/ψ decay, long assumed to be a "glueball factory". This is no longer a viable scenario, but the problem remains of disentangling the states in this region.

The $\xi(2230)$ and $\phi\phi$ Production

The $\xi(2230)$ (also known as the $f_J(2220)$) was originally observed by the SLAC Mark III collaboration in radiative J/ψ decay, and is a candidate for the lowest lying 2^{++} glueball. The BES collaboration continued these measurements and in fact reported multiple two body branching ratios for the state, but there is a difficulty. Their measurements, when combined with the non-observation of $\xi(2230)$ in $\bar{p}p$ annihilation in flight, suggest a very high lower bound on the radiative J/ψ branching ratio, and these results need to be confirmed. See [3] for more details.

However, an intriguing new result has been reported at this conference [26] from the JETSET collaboration. A partial wave analysis of the reaction $\bar{p}p \to \phi\phi$ leads to a narrow 2^{++} intensity, with phase motion consistent with a Breit-Wigner resonance. The mass and width are consistent with the $\xi(2230)$. (This is clearly *inconsistent* with old claims of glueballs in the $\phi\phi$ channel from peripheral pion production. See [3].) It remains to be seen how well this reconciles with existing measurements from BES, JETSET, and the Crystal Barrel.

Missing States

Meson and baryon spectroscopy is easily complicated by having to sift through a large number of overlapping states, so one does not expect to cleanly identify all the predicted excitations. However, some patterns emerge which suggest there might be cases where the quark model *overpredicts* the number of physical states. These "missing states" may be clues to the workings of strong QCD, or they may simply be artifacts of the experiments and the analyses done to date.

For example, a large number of the $L = 2$ meson excitations are missing [3]. A case in point is the missing $(L, S) = (2, 1)$ (i.e. 1^3D_2) isovector state ρ_2. Its mass is expected to be around 1700 MeV, close to the well known 1^3D_1 $\rho(1700)$ and 1^3D_3 $\rho_3(1690)$. The situation is similar for the 1^3D_2 isoscalars ω_2 and ϕ_2.

An older example is in the baryon sector. The problem here is that many of the positive parity $2\hbar\omega$ nucleon and Δ excitations are missing. This has been known for a long time, for example by Lichtenberg [27] who argued that the pattern of observed and missing states corresponded well with a di-quark model for baryons. Although there has been no evidence for di-quark clustering based on modern formulations of strong QCD, the problem and the patterns are very nearly unchanged over the past thirty years. It may simply be, however, that existing experiments have not measured reactions in which these states would have been found. This hypothesis is borne out by quark model calculations [28] and experiments are underway to search for these missing baryon states. See also [29,30].

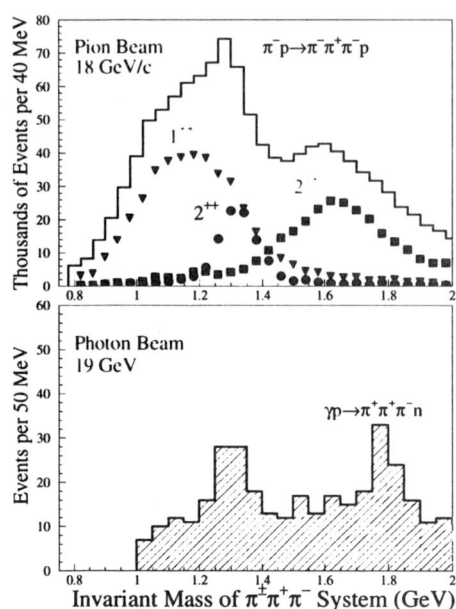

FIGURE 2. Production of the $\pi^{\pm}\pi^{+}\pi^{-}$ system with pion and photon beams. Pion production, upper panel, is dominated by three meson excitations, namely $a_1(1260)$, $a_2(1320)$, and $\pi_2(1670)$. It also shows a very large number of events, and this allows the quantum numbers of these excitations to be identified using partial wave analysis. Photoproduction, with either positive or negative charge exchange, are seen to be different from pion production, even with severely reduced statistics. Furthermore, they are different from each other. Both reactions seem to show production of $a_2(1320)$ with little indication of the other two states seen in pion production. The reaction $\gamma p \to \pi^{+}\pi^{+}\pi^{-}n$ suggests a new narrow state at ≈ 1775 MeV.

THE FUTURE: HALL D AT JEFFERSON LAB

Although hadron spectroscopy has been an active field for more than three decades, only in the past few years has clear evidence emerged for states beyond the quark model. This is largely due to technological advances that have allowed us to build sophisticated experiments, and to accumulate large data volumes and then incorporate computationally intensive analyses. Large data volumes are important if we are to carry out the sort of partial wave analyses that have proven so fruitful for E852 and the Crystal Barrel, among others.

Most of the spectroscopic data, particularly that which studies high mass states, has come from peripheral pion scattering, such as E852. In these reactions, states are produced via excitation of the (spin-0) pion. It is natural to expect a very different tower of excitations in peripheral reactions of a spin-1 hadronic beam. The photon provides such a beam, since its high energy peripheral hadronic reactions proceed through vector dominance. That is, peripheral photoproduction would look essentially at excitations built on (spin-1) ρ, ω, or ϕ ground states.

This simple idea is demonstrated in Fig. 2, which compares peripheral production of isovector (i.e. net nonzero charge) three pion systems. The upper plot shows results from a pion beam [3,12], while the lower is for a photon beam of about the same energy [31–33]. (The upper plot, in fact, is the same data set from E852 that yielded the exotic meson signal in the right panel of Fig. 1.)

An immediate conclusion is that the tower of excitations in peripheral photoproduction is indeed very different from that obtained with a pion beam. This is obvious simply from the shape of the invariant mass distributions. What appears to be the $a_2(1320)$ dominates the lower masses in photoproduction but is just one relatively small component in pion production. Furthermore, there is evidence for a relatively narrow high mass structure in photoproduction, but no such structure at all using the pion beam.

Of course, there is another major difference between the top and bottom plots in Fig. 2. There are more than three orders of magnitude more events for the experiment done with the pion beam. This results directly from the much lower intensity photon beams that have been available, as well as the inherently lower cross section for photoproduction. Furthermore, this large number of events is necessary for a thorough partial wave analysis, such as the one which is able to decompose the major contributing excited states to the upper plot in Fig. 2. Indeed, it is this superior statistical precision that allowed E852 to reliably extract the weak exotic signal shown on the right panel of Fig. 1.

A new experiment is being mounted at Jefferson Laboratory to study meson spectroscopy using peripheral photoproduction. This experiment, dubbed the "Hall D Project" [34–36], will use a linearly polarized photon beam with energy in the range 8 to 9 GeV, produced using coherent bremsstrahlung from the 12 GeV electron accelerator beam. A significant aspect of the project is the accelerator upgrade from its current beam energy maximum of 5.5 GeV. The photon beam intensity will exceed those previously available by more than four orders of magnitude, and the experimental facility is especially designed for large and continuous detector acceptance and efficiency, incorporating both charged and neutral particle detection.

CONCLUSIONS

To summarize, this talk puts forth four main points:

- **"Exotic hadrons"**, that is, excited states of strongly interacting particles that lie outside of the quark model, **exist.**
- **This is good.** Strong QCD predicts degrees of freedom that are not included in the quark model, and these should be observed.
- **Puzzles still remain.** The pattern of masses and decay properties of these exotic states is still emerging, and the comparison with strong QCD is yet to be satisfactorily made. Some states are still mysteries, and we do not yet have a consistent framework for dealing with multiquark systems.
- **New experiments are on the horizon.** Most notably, peripheral production with high energy, high intensity photon beams from Jefferson Lab, will yield new results in the next several years.

Acknowledgements. I would like to thank the conference organizers for their kind invitation, and the opportunity to present this physics to the larger community of high energy and nuclear physics. I am very grateful to my collaborators in E852 and on the Hall D project, as well as to my theoretical physics colleagues, all of whom have taught me so much over the years.

REFERENCES

1. Capstick, S., and Isgur, N., *Phys. Rev.* **D34**, 2809 (1986).
2. Godfrey, S., and Isgur, N., *Phys. Rev.* **D32**, 189 (1985).
3. Godfrey, S., and Napolitano, J., *Rev. Mod. Phys.* **71**, 1411 (1999).
4. Morningstar, C., and Peardon, M., *Phys. Rev.* **D60**, 134509 (1999).
5. Vaccarino, A., and Weingarten, D., *Phys. Rev.* **D60**, 114501 (1999).
6. Meyer, C., these proceedings.
7. Dunwoodie, W., SLAC-PUB-7163 (1997).
8. Close, F.E., and Kirk, A., *Phys. Lett.* **B483**, 345 (2000).
9. Thompson, D.R., et al., *Phys. Rev. Lett.* **79**, 1630 (1997).
10. Chung, S.U., et al., *Phys. Rev.* **D60**, 092001 (1999).
11. Abele, A., et al., *Phys. Lett.* **B423**, 175 (1998).
12. Adams, G.S., et al., *Phys. Rev. Lett.* **81**, 5760 (1998).
13. Donnachie, A. and Page, P.R., *Phys. Rev.* **D58**, 114012 (1998).
14. Cason, N., these proceedings.
15. Golowich, E., Haqq, E., Karl, G., *Phys. Rev.* **D28**, 160 (1983); Erratum *Phys. Rev.* **D33**, 859 (1986).
16. Morsch, H.P. and Zupranski, P., *Phys. Rev.* **C61**, 024002 (2000).
17. Teige, S., et al., *Phys. Rev.* **D59**, 012001 (1999).
18. Gunter, J., et al., arXiv:hep-ex/0001038 (2000).
19. Akhmetshin, R.R., et al., *Phys. Lett.* **B462**, 380 (1999).
20. Achasov, M.N., et al., *Phys. Lett.* **B485**, 349 (2000).
21. Achasov, M.N., et al., *Phys. Lett.* **B479**, 53 (2000).
22. Adinolfi, N., et al., arXiv:hep-ex/0006036 (2000).
23. Dzierba, A., *Nucl. Phys.* **A623**, 142c (1997).
24. Manak, J.J., et al., *Phys. Rev.* **D62**, 012003 (2000).
25. Aston, D., et al., *Phys. Lett.* **B201**, 573 (1988).
26. Jones, R.T., these proceedings.
27. Lichtenberg, D.B., *Phys. Rev.* **178**, 2197 (1969).
28. Capstick, S., and Roberts, W., *Phys. Rev.* **D49**, 4570 (1994).
29. Capstick, S., *Few Body Syst. Suppl.* **11**, 86 (1999).
30. Capstick, S., et al., *Phys. Rev.* **C59**, 3002 (1999).
31. Condo, G. et al., *Phys. Rev.* **D41**, 3317 (1990).
32. Condo, G. et al., *Phys. Rev.* **D48**, 3045 (1993).
33. Condo, G. et al., *Phys. Rev.* **D43**, 2787 (1991).
34. Dzierba, A., Meyer, C., and Swanson, E., *American Scientist* **88**, 406 (2000).
35. Dzierba, A., and Isgur, N., *CERN Courier* **September** (2000).
36. The Hall D website is http://dustbunny.physics.indiana.edu/HallD/.

The Phases of QCD in Heavy Ion Collisions and Compact Stars

Krishna Rajagopal[1]

Center for Theoretical Physics, Massachusetts Institute of Technology
Cambridge, MA 02139

Abstract. I review arguments for the existence of a critical point E in the QCD phase diagram as a function of temperature T and baryon chemical potential μ. I describe how heavy ion collision experiments at the SPS and RHIC can discover the tell-tale signatures of such a critical point, thus mapping this region of the QCD phase diagram. I then review the phenomena expected in cold dense quark matter: color superconductivity and color-flavor locking. I close with a snapshot of ongoing explorations of the implications of recent developments in our understanding of cold dense quark matter for the physics of compact stars.

The QCD vacuum in which we live, which has the familiar hadrons as its excitations, is but one phase of QCD, and far from the simplest one at that. One way to better understand this phase and the nonperturbative dynamics of QCD more generally is to study other phases and the transitions between phases. We are engaged in a voyage of exploration, mapping the QCD phase diagram as a function of temperature T and baryon number chemical potential μ. Because QCD is asymptotically free, its high temperature and high baryon density phases are more simply and more appropriately described in terms of quarks and gluons as degrees of freedom, rather than hadrons. The chiral symmetry breaking condensate which characterizes the vacuum phase melts away. At high temperatures, in the resulting quark-gluon plasma (QGP) phase all of the symmetries of the QCD Lagrangian are unbroken and the excitations have the quantum numbers of quarks and gluons. At high densities, on the other hand, quarks form Cooper pairs and

---

[1] Many thanks to the organizers for a conference which was stimulating precisely because it addressed so many facets of the intersection between nuclear and particle physics. I am grateful to M. Alford, B. Berdnikov, J. Berges, J. Bowers, E. Shuster, E. Shuryak, M. Stephanov and F. Wilczek for fruitful collaboration. I acknowledge helpful discussions with P. Bedaque, D. Blaschke, I. Bombaci, G. Carter, D. Chakrabarty, J. Madsen, C. Nayak, M. Prakash, D. Psaltis, S. Reddy, M. Ruderman, T. Schäfer, A. Sedrakian, D. Son, I. Wasserman and F. Weber. This work is supported in part by the U.S. Department of Energy (D.O.E.) under cooperative research agreement #DF-FC02-94ER40818 and by a DOE OJI Award and by the Alfred P. Sloan Foundation.

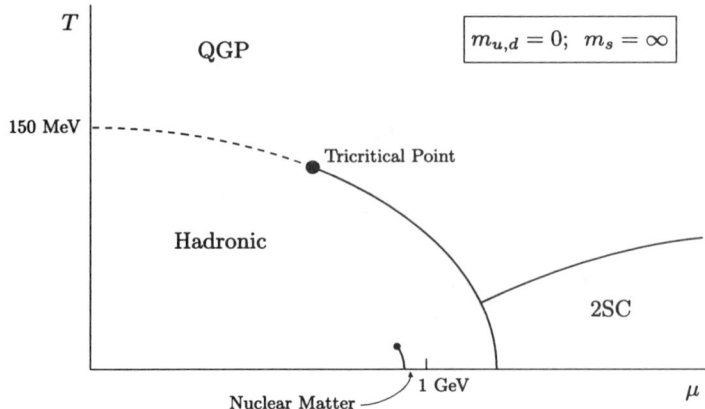

FIGURE 1. QCD Phase diagram for two massless quarks. Chiral symmetry is broken in the hadronic phase and is restored elsewhere in the diagram. The chiral phase transition changes from second to first order at a tricritical point. The phase at high density and low temperature is a color superconductor in which up and down quarks with two out of three colors pair and form a condensate. The transition between this 2SC phase and the QGP phase is likely first order. The transition on the horizontal axis between the hadronic and 2SC phases is first order. The transition between a nuclear matter "liquid" and a gas of individual nucleons is also marked; it ends at a critical point at a temperature of order 10 MeV, characteristic of the forces which bind nucleons into nuclei.

new condensates develop. The formation of such superconducting phases [1–5] requires only weak attractive interactions; these phases may nevertheless break chiral symmetry [5] and have excitations with the same quantum numbers as those in a confined phase [5–8]. These cold dense quark matter phases may arise in the core of neutron stars; mapping this region of the phase diagram requires an interplay between theory and neutron star phenomenology. We describe efforts in this direction in Section IV. A central goal of the experimental heavy ion physics program is to explore and map the higher temperature regions of the QCD phase diagram. Recent theoretical developments suggest that a key qualitative feature, namely a critical point which in a sense defines the landscape to be mapped, may be within reach of discovery and analysis as data is taken at several different energies [9,10]. The discovery of the critical point would transform this region of the map of the QCD phase diagram from one based only on reasonable inference from universality, lattice gauge theory and models into one with a solid experimental basis.

I THE CRITICAL POINT

We begin our walk through the phase diagram at zero baryon number density, with a brief review [11] of the phase changes which occur as a function of temperature. That is, we begin by restricting ourselves to the vertical axis in Figures 1

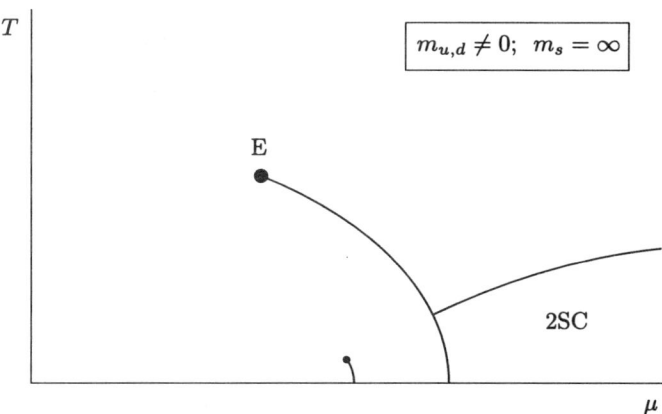

FIGURE 2. QCD phase diagram for two light quarks. Qualitatively as in Figure 1, except that the introduction of light quark masses turns the second order phase transition into a smooth crossover. The tricritical point becomes the critical endpoint E, which can be found in heavy ion collision experiments.

through 4. This slice of the phase diagram was explored by the early universe during the first tens of microseconds after the big bang and can be studied in lattice simulations. As heavy ion collisions are performed at higher and higher energies, they create plasmas with a lower and lower baryon number to entropy ratio and therefore explore regions of the phase diagram closer and closer to the vertical axis.

In QCD with two massless quarks ($m_{u,d} = 0$; $m_s = \infty$; Figure 1) the phase transition at which chiral symmetry is restored is likely second order and belongs to the universality class of $O(4)$ spin models in three dimensions [12]. Below T_c, chiral symmetry is broken and there are three massless pions. At $T = T_c$, there are four massless degrees of freedom: the pions and the sigma. Above $T = T_c$, the pion and sigma correlation lengths are degenerate and finite. In nature, the light quarks are not massless. Because of this explicit chiral symmetry breaking, the second order phase transition is replaced by an analytical crossover: physics changes dramatically but smoothly in the crossover region, and no correlation length diverges. Thus, in Figure 2, there is no sharp boundary on the vertical axis separating the low temperature hadronic world from the high temperature quark-gluon plasma. This picture is consistent with present lattice simulations [13,14], which suggest $T_c \sim 140 - 190$ MeV [15,14].

Arguments based on a variety of models [16,17,3,4,18,19] indicate that the chiral symmetry restoration transition is first order at large μ. (In Section III, we describe the color superconducting (2SC) phase of cold dense quark matter which occurs at values of μ above this first order transition; the fact that this is a transition in which two different condensates compete strengthens the argument that this transition is first order [18,20].) This suggests that the phase diagram features a critical point E at which the line of first order phase transitions present for $\mu > \mu_E$ ends, as

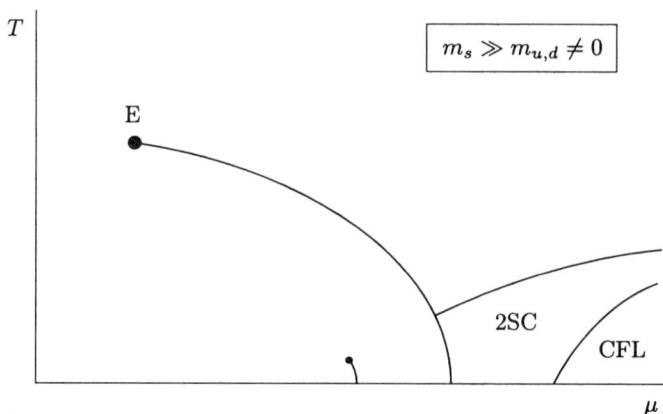

FIGURE 3. QCD phase diagram for two light quarks and a strange quark with a mass comparable to that in nature. The presence of the strange quark shifts E to the left, as can be seen by comparing with Figure 2. At sufficiently high density, cold quark matter is necessarily in the CFL phase in which quarks of all three colors and all three flavors form Cooper pairs. The diquark condensate in the CFL phase breaks chiral symmetry, and this phase has the same symmetries as baryonic matter which is dense enough that the nucleon and hyperon densities are comparable. The phase transition between the CFL and 2SC phases is first order.

shown in Figure 2.[2] At μ_E, the phase transition is second order and is in the Ising universality class [18,19]. Although the pions remain massive, the correlation length in the σ channel diverges due to universal long wavelength fluctuations of the order parameter. This results in characteristic signatures, analogues of critical opalescence in the sense that they are unique to collisions which freeze out near the critical point, which can be used to discover E [9,10].

Returning to the $\mu = 0$ axis, universal arguments [12], again backed by lattice simulation [13], tell us that if the strange quark were as light as the up and down quarks, the transition would be first order, rather than a smooth crossover. This means that if one could dial the strange quark mass m_s, one would find a critical m_s^c at which the transition as a function of temperature is second order [22,11]. Figures 2, 3 and 4 are drawn for a sequence of decreasing strange quark masses. Somewhere between Figures 3 and 4, m_s is decreased below m_s^c and the transition on the vertical axis becomes first order. The value of m_s^c is an open question, but lattice simulations suggest that it is about half the physical strange quark mass [23,24]. These results are not yet conclusive [25] but if they are correct then the phase diagram in nature is as shown in Figure 3, and the phase transition at low μ is a smooth crossover.

These observations fit together in a simple and elegant fashion. If we could vary m_s, we would find that as m_s is reduced from infinity to m_s^c, the critical point E

---

[2] If the up and down quarks were massless, E would be a tricritical point [21], at which the first order transition becomes second order. See Figure 1.

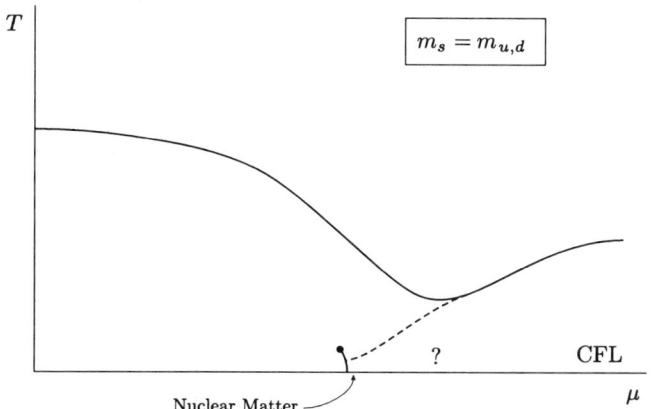

FIGURE 4. QCD phase diagram for three quarks which are degenerate in mass and which are either massless or light. The CFL phase and the baryonic phase have the same symmetries and may be continuously connected. The dashed line denotes the critical temperature at which baryon-baryon (or quark-quark) pairing vanishes; the region below the dashed line is superfluid. Chiral symmetry is broken everywhere below the solid line, which is a first order phase transition. The question mark serves to remind us that although no transition is required in this region, transition(s) may nevertheless arise as the magnitude of the gap increases qualitatively in going from the hypernuclear to the CFL phase. For quark masses as in nature, the high density region of the map may be as shown in Figure 3 or may be closer to that shown here, albeit with transition(s) in the vicinity of the question mark associated with the onset of nonzero hyperon density and the breaking of $U(1)_S$ [7].

in the (T, μ) plane moves toward the $\mu = 0$ axis [9]. This is shown in Figures 2-4. In nature, E is at some nonzero T_E and μ_E. When m_s is reduced to m_s^c, between Figure 3 and Figure 4, μ_E reaches zero. Of course, experimentalists cannot vary m_s. They can, however, vary μ. AGS collisions with center of mass energy $\sqrt{s} = 5$ AGeV create fireballs which freeze out near $\mu \sim 500 - 600$ MeV [26]. SPS collisions with $\sqrt{s} = 17$ AGeV create fireballs which freeze out near $\mu \sim 200 - 300$ MeV [26]. In time, we will also have data from SPS collisions with $\sqrt{s} = 9$ AGeV and from RHIC collisions with $\sqrt{s} = 56, 130$ and 200 AGeV and other energies.[3] By dialing $\sqrt{s}$ and thus μ, experimenters can find the critical point E.

II DISCOVERING THE CRITICAL POINT

We hope that the study of heavy ion collisions will, in the end, lead both to a quantitative study of the properties of the quark-gluon plasma phase at temperatures well above the transition and to a quantitative understanding of how to draw

---

[3] The first data from RHIC collisions at $\sqrt{s} = 56$ AGeV and $\sqrt{s} = 130$ AGeV have already appeared [27]. This bodes well for the analyses to come.

the phase transition region of the phase diagram. Probing the partonic matter created early in the collision relies on a suite of signatures including: the use of J/Ψ mesons, charmed mesons, and perhaps the Υ as probes; the energy loss of high momentum partons and consequent effects on the high-p_T hadron spectrum; and the detection of photons and dileptons over and above those emitted in the later hadronic stages of the collision. I will not review this program here. Instead, I focus on signatures of the critical point. The map of the QCD phase diagram which I have sketched so far is simple, coherent and consistent with all we know theoretically; the discovery of the critical point would provide an experimental foundation for the central qualitative feature of the landscape. This discovery would in addition confirm that in higher energy heavy ion collisions and in the big bang, the QCD phase transition is a smooth crossover. Furthermore, the discovery of collisions which create matter that freezes out near E would imply that conditions above the transition existed prior to freezeout, and would thus make it much easier to interpret the results of other experiments which study those observables which can probe the partonic matter created early in the collision.

We theorists must clearly do as much as we can to tell experimentalists *where* and *how* to find E. The "where" question, namely the question of predicting the value of μ_E and thus suggesting the $\sqrt{s}$ to use to find E, is much harder for us to answer. First, as we stress further in the next Section, *ab initio* analysis of QCD in its full glory — i.e. lattice calculations — are at present impossible at nonzero μ. We must therefore rely on models. Second, an intrinsic feature of the picture we have described is that μ_E is sensitive to the mass of the strange quark, and therefore particularly hard to predict. Crude models suggest that μ_E could be $\sim 600 - 800$ MeV in the absence of the strange quark [18,19]; this in turn suggests that in nature μ_E may have of order half this value, and may therefore be accessible at the SPS if the SPS runs with $\sqrt{s} < 17$ AGeV. However, at present theorists cannot predict the value of μ_E even to within a factor of two. The SPS can search a significant fraction of the parameter space; if it does not find E, it will then be up to the RHIC experiments to map the $\mu < 200$ MeV region.

Although we are trying to be helpful with the "where" question, we are not very good at answering it quantitatively. This question can only be answered convincingly by an experimental discovery. What we theorists *can* do reasonably well is to answer the "how" question, thus enabling experimenters to answer "where". This is the goal of a recent paper by Stephanov, myself and Shuryak [10]. The signatures we have proposed are based on the fact that E is a genuine thermodynamic singularity at which susceptibilities diverge and the order parameter fluctuates on long wavelengths. The resulting signatures are *nonmonotonic* as a function of $\sqrt{s}$: as this control parameter is varied, we should see the signatures strengthen and then weaken again as the critical point is approached and then passed.

The critical point E can also be sought by varying control parameters other than $\sqrt{s}$. Ion size, centrality selection and rapidity selection can all be varied. The advantage of using $\sqrt{s}$ is that we already know (by comparing results from the AGS and SPS) that dialing it changes the freeze out chemical potential μ, which is the goal in a search for E.

The simplest observables we analyze are the event-by-event fluctuations of the mean transverse momentum of the charged particles in an event, p_T, and of the total charged multiplicity in an event, N. We calculate the magnitude of the effects of critical fluctuations on these and other observables, making predictions which, we hope, will allow experiments to find E. As a necessary prelude, we analyze the contribution of noncritical thermodynamic fluctuations. We compare the noncritical fluctuations of an equilibriated resonance gas to the fluctuations measured by NA49 at $\sqrt{s} = 17$ AGeV [28]. The observed fluctuations are as perfect Gaussians as the data statistics allow, as expected for freeze-out from a system in thermal equilibrium. The width of the event-by-event distribution[4] of mean p_T is in good agreement with predictions based on noncritical thermodynamic fluctuations. That is, NA49 data are consistent with the hypothesis that almost all the observed event-by-event fluctuation in mean p_T, an intensive quantity, is thermodynamic in origin. This bodes well for the detectability of systematic changes in thermodynamic fluctuations near E.

One analysis described in detail in Ref. [10] is based on the ratio of the width of the true event-by-event distribution of the mean p_T to the width of the distribution in a sample of mixed events. This ratio was called $\sqrt{F}$. NA49 has measured $\sqrt{F} = 1.002 \pm 0.002$ [28,10], which is consistent with expectations for noncritical thermodynamic fluctuations.[5] Critical fluctuations of the σ field, i.e. the characteristic long wavelength fluctuations of the order parameter near E, influence pion momenta via the (large) $\sigma\pi\pi$ coupling and increase $\sqrt{F}$ [10]. The effect is proportional to $\xi_{\text{freezeout}}^2$, where $\xi_{\text{freezeout}}$ is the σ-field correlation length of the long-wavelength fluctuations at freezeout [10]. If $\xi_{\text{freezeout}} \sim 3$ fm (a reasonable estimate, as we describe below) the ratio $\sqrt{F}$ increases by $\sim 3-5\%$, ten to twenty times the statistical error in the present measurement [10]. This observable is valuable because data on it has been analyzed and presented by NA49, and it can therefore be used to learn that Pb+Pb collisions at 158 AGeV do *not* freeze out near E. The $3-5\%$ nonmonotonic variation in $\sqrt{F}$ as a function of $\sqrt{s}$ which we predict is easily detectable but is not so large as to make one confident of using this alone as a signature of E.

Once E is located, however, other observables which are more sensitive to critical effects will be more useful. For example, a $\sqrt{F_{\text{soft}}}$, defined using only the softest 10%

---

[4] This width can be measured even if one observes only two pions per event [29]; large acceptance data as from NA49 is required in order to learn that the distribution is Gaussian, that thermodynamic predictions may be valid, and that the width is therefore the only interesting quantity to measure.

[5] In an infinite system made of classical particles which is in thermal equilibrium, $\sqrt{F} = 1$. Bose effects increase $\sqrt{F}$ by $1-2\%$ [30,10]; an anticorrelation introduced by energy conservation in a finite system — when one mode fluctuates up it is more likely for other modes to fluctuate down — decreases $\sqrt{F}$ by $1-2\%$ [10]; two-track resolution also decreases $\sqrt{F}$ by $1-2\%$ [28]. The contributions due to correlations introduced by resonance decays and due to fluctuations in the flow velocity are each much smaller than 1% [10].

of the pions in each event, will be much more sensitive to the critical long wavelength fluctuations. The higher p_T pions are less affected by the σ fluctuations [10], and these relatively unaffected pions dominate the mean p_T of all the pions in the event. This is why the increase in $\sqrt{F}$ near the critical point will be much less than that of $\sqrt{F_{\text{soft}}}$. Depending on the details of the cuts used to define it, $\sqrt{F_{\text{soft}}}$ should be enhanced by many tens of percent in collisions passing near E. Ref. [10] suggests other such observables, and more can surely be found.

The multiplicity of soft pions is an example of an observable which may be used to detect the critical fluctuations without an event-by-event analysis. The post-freezeout decay of sigma mesons, which are copious and light at freezeout near E and which decay subsequently when their mass increases above twice the pion mass, should result in a population of pions with $p_T \sim m_\pi/2$ which appears only for freezeout near the critical point [10]. If $\xi_{\text{freezeout}} > 1/m_\pi$, this population of unusually low momentum pions will be comparable in number to that of the "direct" pions (i.e. those which were pions at freezeout) and will result in a large signature. This signature is therefore certainly large for $\xi_{\text{freezeout}} \sim 3$ fm and would not increase much further if $\xi_{\text{freezeout}}$ were larger still.

The variety of observables which should *all* vary nonmonotonically with $\sqrt{s}$ (and should all peak at the same $\sqrt{s}$) is sufficiently great that if it were to turn out that $\mu_E < 200$ MeV, making E inaccessible to the SPS, all four RHIC experiments could play a role in the study of the critical point.

In Ref. [31] we estimate how large $\xi_{\text{freezeout}}$ can become, thus making the predictions of Ref. [10] for the magnitude of various signatures more quantitative. The nonequilibrium dynamics we analyze is guaranteed to occur in a heavy ion collision which passes near E, even if local thermal equilibrium is achieved earlier at a higher temperature. If the system were to cool arbitrarily slowly, ξ would diverge at T_E. However, it would take an infinite time for ξ to grow infinitely large. Indeed, near a critical point, the longer the correlation length, the longer the equilibration time, and the slower the correlation length can grow. This critical slowing down means that the correlation length cannot grow sufficiently fast for the system to stay in equilibrium. We use the theory of dynamical critical phenomena to describe the effects of critical slowing down of the long wavelength dynamics near E on the time development of the correlation length. The correlation length does not have time to grow as large as it would in equilibrium: we find $\xi_{\text{freezeout}} \sim 2/T_E \sim 3$ fm for trajectories passing near E. Although critical slowing down hinders the growth of ξ, it also slows the decrease of ξ as the system continues to cool below the critical point. As a result, ξ does not decrease significantly between the phase transition and freezeout.

It is important to realize that one need not hit E precisely in order to find it. Our analysis [31] demonstrates that if one were to do a scan with collisions at many finely spaced values of the energy and thus μ, one would see signatures of E with approximately the same magnitude over a broad range of μ. The magnitude of the signatures will not be narrowly peaked as μ is varied. As long as one gets close enough to E that the equilibrium correlation length is $(2-3)/T_E$, the actual correlation length ξ will grow to $\sim 2/T_E$. There is no advantage to getting closer to E.

As described above, knowing that we are looking for $\xi_{\text{freezeout}} \sim 3$ fm allows us [31] to make quantitative estimates of the magnitude of the signatures of E described in detail in Ref. [10]. Together, the excess multiplicity at low momentum (due to post-freezeout sigma decays) and the excess event-by-event fluctuation of the momenta of the low momentum pions (due to their coupling to the order parameter which is fluctuating with correlation length $\xi_{\text{freezeout}}$) should allow a convincing detection of the critical point E. Both should behave nonmonotonically as the collision energy, and hence μ, are varied. Both should peak for those heavy ion collisions which freeze out near E, with $\xi_{\text{freezeout}} \sim 3$ fm.

We have learned much from the beautiful gaussian event-by-event fluctuations observed by NA49. The magnitude of these fluctuations are consistent with the hypothesis that the hadronic system at freezeout is in approximate thermal equilibrium. These and other data show none of the non-gaussian features that would signal that the system had been driven far from equilibrium either by a rapid traversal of the transition region or by the bubbling that would occur near a strong first order phase transition. There is also no sign of the enhanced, but still gaussian, fluctuations which would signal freezeout near the critical point E. Combining these observations with the observation of tantalizing indications that the matter created in SPS collisions is not well described at early times by hadronic models [32] suggests that collisions at the SPS may be exploring the crossover region to the left of the critical point E, in which the matter is not well-described as a hadron gas but is also not well-described as a quark-gluon plasma. This speculation could be confirmed in two ways. First, if the SPS is probing the crossover region then the coming experiments at RHIC may discover direct signatures of an early partonic phase, which are well-described by theoretical calculations beginning from an equilibrated quark-gluon plasma. Second, if $\sqrt{s} = 17$ AGeV collisions are probing the crossover region not far to the left of the critical point E, then SPS data taken at lower energies would result in the discovery of E. If, instead, RHIC were to discover E with $\mu_E < 200$ MeV, that would indicate that the SPS experiments have probed the weakly first order region just to the right of E. Regardless, discovering E would take all the speculation out of mapping this part of the QCD phase diagram.

III COLOR SUPERCONDUCTIVITY AND COLOR-FLAVOR LOCKING

I turn now to recent developments in our understanding of the low temperature, high density regions of the QCD phase diagram. First, a notational confession. It is conventional in the literature on cold dense quark matter to define μ as the *quark* number chemical potential, 1/3 the baryon number chemical potential used in Sections I and II. We make this change from here on. For example, neutron star cores likely have $\mu \sim 400 - 500$ MeV, corresponding to baryon number chemical potentials $\sim 1.2 - 1.5$ GeV in Figures 1-4.

The relevant degrees of freedom in cold dense quark matter are those which involve quarks with momenta near the Fermi surface. At high density, when the Fermi momentum is large, the QCD gauge coupling $g(\mu)$ is small. However, because of the infinite degeneracy among pairs of quarks with equal and opposite momenta at the Fermi surface, even an arbitrarily weak attraction between quarks renders the Fermi surface unstable to the formation of a condensate of quark Cooper pairs. Creating a pair costs no free energy at the Fermi surface and the attractive interaction results in a free energy benefit. Pairs of quarks cannot be color singlets, and in QCD with two flavors of massless quarks the Cooper pairs form in the (attractive) color $\bar{3}$ channel [1-4]. The resulting condensate creates a gap Δ at the Fermi surfaces of quarks with two out of the three colors and breaks $SU(3)_{color}$ to an $SU(2)_{color}$ subgroup, giving mass to five of the gluons by the Anderson-Higgs mechanism. In QCD with two flavors, the Cooper pairs are $ud - du$ flavor singlets and the global flavor symmetry $SU(2)_L \times SU(2)_R$ is intact. There is also an unbroken global symmetry which plays the role of $U(1)_B$. Thus, no global symmetries are broken in this 2SC phase. There must therefore be a phase transition between the 2SC and hadronic phases on the horizontal axis in Figure 1, at which chiral symmetry is restored. This phase transition is first order [3,18,33,20] since it involves a competition between chiral condensation and diquark condensation [18,20]. There need be no transition between the 2SC and quark-gluon plasma phases in Figure 1 because neither phase breaks any global symmetries. However, this transition, which is second order in mean field theory, is likely first order in QCD due to gauge field fluctuations [18], at least at high enough density [34].

In QCD with three flavors of massless quarks, the Cooper pairs *cannot* be flavor singlets, and both color and flavor symmetries are necessarily broken. The symmetries of the phase which results have been analyzed in [5,6]. The attractive channel favored by one-gluon exchange exhibits "color-flavor locking." A condensate of the form

$$\langle \psi_L^{\alpha a} \psi_L^{\beta b} \rangle \propto \Delta \epsilon^{\alpha\beta A} \epsilon^{abA} \qquad (1)$$

involving left-handed quarks alone, with α, β color indices and a, b flavor indices, locks $SU(3)_L$ flavor rotations to $SU(3)_{color}$: the condensate is not symmetric under either alone, but is symmetric under the simultaneous $SU(3)_{L+color}$ rotations.[6] A condensate involving right-handed quarks alone locks $SU(3)_R$ flavor rotations to $SU(3)_{color}$. Because color is vectorial, the combined effect of the LL and RR condensates is to lock $SU(3)_L$ to $SU(3)_R$, breaking chiral symmetry.[7] Thus, in quark

---

[6] It turns out [5] that condensation in the color $\bar{3}$ channel induces a condensate in the color 6 channel because this breaks no further symmetries [7]. The resulting condensates can be written in terms of κ_1 and κ_2 where $\langle \psi_L^{\alpha a} \psi_L^{\beta b} \rangle \sim \kappa_1 \delta^{\alpha a} \delta^{\beta b} + \kappa_2 \delta^{\alpha b} \delta^{\beta a}$. Here, the Kronecker δ's lock color and flavor rotations. The pure color $\bar{3}$ condensate (1) has $\kappa_2 = -\kappa_1$.

[7] Once chiral symmetry is broken by color-flavor locking, there is no symmetry argument precluding the existence of an ordinary chiral condensate. Indeed, instanton effects do induce a nonzero $\langle \bar{q}q \rangle$ [5], but this is a small effect [35].

matter with three massless quarks, the $SU(3)_{\text{color}} \times SU(3)_L \times SU(3)_R \times U(1)_B$ symmetry is broken down to the global diagonal $SU(3)_{\text{color}+L+R}$ group. All nine quarks have a gap. All eight gluons get a mass. There are nine massless Nambu-Goldstone bosons. There is an unbroken gauged $U(1)$ symmetry which plays the role of electromagnetism. Under this symmetry, all the quarks, all the massive vector bosons, and all the Nambu-Goldstone bosons have integer charges. The CFL phase therefore has the same symmetries as baryonic matter with a condensate of Cooper pairs of baryons [6]. Furthermore, many non-universal features of these two phases correspond [6]. This raises the possibility that quark matter and baryonic matter may be continuously connected [6], as shown in Figure 4.

The physics of the CFL phase has been the focus of much recent work [5–8,35–50]. Nature chooses two light quarks and one middle-weight strange quark, rather than three degenerate quarks as in Figure 4. A nonzero m_s weakens those condensates which involve pairing between light and strange quarks. The CFL phase requires nonzero $\langle us \rangle$ and $\langle ds \rangle$ condensates; because these condensates pair quarks with differing Fermi momenta they can only exist if they are larger than of order $m_s^2/2\mu$, the difference between the u and s Fermi momenta in the absence of pairing. If one imagines increasing m_s at fixed μ, one finds a first order unlocking transition [7,8]: for larger m_s only u and d quarks pair and the 2SC phase is obtained. Conversely, as m_s is reduced in going from Figure 2 to 3 to 4, the region occupied by the CFL phase expands to encompass regions with smaller and smaller μ [7,8]. For any $m_s \neq \infty$, the CFL phase is the ground state at arbitrarily high density [7]. For larger values of m_s, there is a 2SC interlude on the horizontal axis, in which chiral symmetry is restored, before the CFL phase breaks it again at high densities. For smaller values of m_s, the possibility of quark-hadron continuity [6] as shown in Figure 4 arises. It should be noted that when the strange and light quarks are not degenerate, the CFL phase may be continuous with a baryonic phase in which the densities of all the nucleons and hyperons are comparable; there are, however, phase transitions between this hypernuclear phase and ordinary nuclear matter [7].

The Nambu-Goldstone bosons in the CFL phase are Fermi surface excitations in which the orientation of the left-handed and right-handed diquark condensates oscillate out of phase in flavor space. The effective field theory describing these oscillations has been constructed [37,40,45]. Because the full theory is weakly coupled at asymptotically high densities, in this regime all coefficients in the effective theory describing the long wavelength meson physics are calculable from first principles. The decay constants $f_{\pi,K,\eta,\eta'}$ [40] and the meson masses $m_{\pi,K,\eta,\eta'}$ [40–44,46] are all now known. The meson masses depend on quark masses like $m^2 \sim m_q^2$ in the CFL phase (neglecting the small chiral condensate) [5], and their masses are inverted in the sense that the kaon is lighter than the pion [40]. The charged kaon mass $m_{K^\pm}^2 \sim m_d(m_u + m_s)\Delta/\mu$ is so light that it is likely less than the electron chemical potential, meaning that the CFL phase likely features a kaon condensate [49]. The dispersion relations describing the fermionic quasiparticle excitations in the CFL phase, which have the quantum numbers of an octet and a singlet of baryons, have also received attention [7,38]. So have the properties of the massive vector meson octet — the gluons which receive a mass via the Meissner-Anderson-Higgs mechanism [40,51,47]. We now have a description of the properties of the CFL phase and its excitations, in which much is known quantitatively if the value of the gap Δ is known.

It is interesting that both the 2SC and CFL phases satisfy anomaly matching constraints, even though it is not yet completely clear whether this must be the case when Lorentz invariance is broken by a nonzero density [52]. It is not yet clear how high density QCD with larger numbers of flavors [39] satisfies anomaly matching constraints. Also, anomaly matching in the 2SC phase requires that the up and down quarks of the third color remain ungapped; this requirement must, therefore, be modified once these quarks pair to form a $J = 1$ condensate, breaking rotational invariance [3].

Much effort has gone into estimating the magnitude of the gaps in the 2SC and CFL phases [2–5,7,8,18,20,35,53–68]. It would be ideal if this task were within the scope of lattice gauge theory as is, for example, the calculation of the critical temperature on the vertical axis of the phase diagram. Unfortunately, lattice methods relying on importance sampling have to this point been rendered exponentially impractical at nonzero baryon density by the complex action at nonzero μ. There are more sophisticated algorithms which have allowed theories which are simpler than QCD but which have as severe a fermion sign problem as that in QCD at nonzero chemical potential to be simulated [69]. This bodes well for the future. Given the present absence of suitable lattice methods,[8] the magnitude of the gaps in quark matter at large but accessible density has been estimated using two broad strategies. The first class of estimates are done within the context of models whose parameters are chosen to give reasonable vacuum physics [3,4,18,5,53,54,7,20,35,75,56]. These

---

[8] Note that quark pairing can be studied on the lattice in some models with four-fermion interactions and in two-color QCD [70]. The $N_c = 2$ case has also been studied analytically in Refs. [4,71]; pairing in this theory is simpler to analyze because quark Cooper pairs are color singlets. The $N_c \to \infty$ limit of QCD is often one in which hard problems become tractable. However, the ground state of $N_c = \infty$ QCD is a chiral density wave, not a color superconductor [72]. At asymptotically high densities color superconductivity persists up to N_c's of order thousands [73,74] before being supplanted by the phase described in Ref. [72]. At any finite N_c, color superconductivity occurs at arbitrarily weak coupling whereas the chiral density wave does not. For $N_c = 3$, color superconductivity is still favored over the chiral density wave (although not by much) even if the interaction is so strong that the color superconductivity gap is $\sim \mu/2$ [75]. The phase of $N_c = 3$ QCD with nonzero isospin density ($\mu_I \neq 0$) and zero baryon density ($\mu = 0$) *can* be simulated on the lattice [76]. Although not physically realizable, it is very interesting to consider because phenomena arise which are similar to those occurring at large μ and, in this context, these phenomena can be analyzed on the lattice. In this setting, therefore, lattice simulations can be used to test calculational methods which have also been applied at large μ, where lattice simulation is unavailable. Large μ_I physics features large Fermi surfaces for down quarks and anti-up quarks, Cooper pairing of down and anti-up quarks, and a gap whose g-dependence is as in (2), albeit with a different coefficient of $1/g$ in the exponent [76]. This condensate has the same quantum numbers as the pion condensate expected at much lower μ_I, which means that a hypothesis of continuity between hadronic — in this case pionic — and quark matter as a function of μ_I can be tested on the lattice [76].

methods yield results which are in qualitative agreement: the favored condensates are as described above; the gaps range between several tens of MeV up to as large as about 100 MeV; the associated critical temperatures (above which the diquark condensates vanish) can be as large as about $T_c \sim 50$ MeV.

The second strategy for estimating gaps and critical temperatures is to use $\mu = \infty$ physics as a guide. At asymptotically large μ, models with short-range interactions are bound to fail because the dominant interaction is due to the long-range magnetic interaction coming from single-gluon exchange [33,57]. The collinear infrared divergence in small angle scattering via one-gluon exchange (which is regulated by dynamical screening [57]) results in a gap which, at $\mu \to \infty$ where $g(\mu) \to 0$, takes the form [57]

$$\Delta \sim b\mu \, g(\mu)^{-5} \exp[-3\pi^2/\sqrt{2}g(\mu)] \, , \qquad (2)$$

whereas for a point-like interaction with four-fermion coupling g^2 the gap goes like $\exp(-1/g^2)$. Son's result (2) has now been confirmed using a variety of methods [61,58–60,62,63,66]. The $\mathcal{O}(g^0)$ contribution to the prefactor b in (2) is not yet fully understood [61,58,60,62–64,39,66–68]. Examination of the gauge-dependent (and g-dependent) contributions to b in calculations based on the one-loop Schwinger-Dyson equation reveals that they only begin to decrease for $g \lesssim 0.8$ [67]. This means that effects which have to date been neglected in all calculations (e.g. vertex corrections) are small corrections to b only for $\mu \gg 10^8$ MeV.

At large enough μ, the differences between u, d and s Fermi momenta decrease, while the result (2) demonstrates that the magnitude of the condensates *increases* slowly as $\mu \to \infty$. (As $\mu \to \infty$, the running coupling $g(\mu) \to 0$ logarithmically and the exponential factor in (2) goes to zero, but not sufficiently fast to overcome the growth of μ.) This means that the CFL phase is favored over the 2SC phase for $\mu \to \infty$ for any $m_s \neq \infty$ [7]. If we take the asymptotic estimates for the prefactor, quantitatively valid for $\mu \gg 10^8$ MeV [67], and apply them at accessible densities, say $\mu \sim 500$ MeV, it predicts gaps as large as about 100 MeV and critical temperatures as large as about 50 MeV [61]. Even though the asymptotic regime where Δ can be calculated from first principles with confidence is not accessed in nature, it is of great theoretical interest. The weak-coupling calculation of the gap in the CFL phase is the first step toward the weak-coupling calculation of other properties of this phase, in which chiral symmetry is broken and the spectrum of excitations is as in a confined phase. As we have described above, for example, the masses and decay constants of the pseudoscalar mesons can be calculated from first principles once Δ is known.

It is satisfying that two very different approaches, one using zero density phenomenology to normalize models, the other using weak-coupling methods valid at asymptotically high density, yield predictions for the gaps and critical temperatures at accessible densities which are in good agreement. Neither can be trusted quantitatively for quark number chemical potentials $\mu \sim 400 - 500$ MeV, as appropriate for the quark matter which may occur in compact stars. Still, both methods agree that the gaps at the Fermi surface are of order tens to 100 MeV, with critical temperatures about half as large.

$T_c \sim 50$ MeV is much larger relative to the Fermi momentum (say $\mu \sim 400 - 500$ MeV) than in low temperature superconductivity in metals. This reflects the fact that color superconductivity is induced by an attraction due to the primary, strong, interaction in the theory, rather than having to rely on much weaker secondary interactions, as in phonon mediated superconductivity in metals. Quark matter is a high-T_c superconductor by any reasonable definition. It is unfortunate that its T_c is nevertheless low enough that it is unlikely the phenomenon can be realized in heavy ion collisions.

IV COLOR SUPERCONDUCTIVITY IN COMPACT STARS

Our current understanding of the color superconducting state of quark matter leads us to believe that it may occur naturally in compact stars. The critical temperature T_c below which quark matter is a color superconductor is high enough that any quark matter which occurs within neutron stars that are more than a few seconds old is in a color superconducting state. In the absence of lattice simulations, present theoretical methods are not accurate enough to determine whether neutron star cores are made of hadronic matter or quark matter. They also cannot determine whether any quark matter which arises will be in the CFL or 2SC phase: the difference between the u, d and s Fermi momenta will be a few tens of MeV which is comparable to estimates of the gap Δ; the CFL phase occurs when Δ is large compared to all differences between Fermi momenta. Just as the higher temperature regions of the QCD phase diagram are being mapped out in heavy ion collisions, we need to learn how to use neutron star phenomena to determine whether they feature cores made of 2SC quark matter, CFL quark matter or hadronic matter, thus teaching us about the high density region of the QCD phase diagram. It is therefore important to look for astrophysical consequences of color superconductivity.

Equation of State: Much of the work on the consequences of quark matter within a compact star has focussed on the effects of quark matter on the equation of state, and hence on the radius of the star. As a Fermi surface phenomenon, color superconductivity has little effect on the equation of state: the pressure is an integral over the whole Fermi volume. Color superconductivity modifies the equation of state at the $\sim (\Delta/\mu)^2$ level, typically by a few percent [3]. Such small effects can be neglected in present calculations, and for this reason I will not attempt to survey the many ways in which observations of neutron stars are being used to constrain the equation of state [77].

Cooling by Neutrino Emission: We turn now to neutron star phenomena which *are* affected by Fermi surface physics. For the first 10^{5-6} years of its life, the cooling of a neutron star is governed by the balance between heat capacity and the loss of heat by neutrino emission. How are these quantities affected by the presence of a quark matter core? This has been addressed recently in Refs. [82,83], following earlier work in Ref. [84]. Both the specific heat C_V and the neutrino emission rate L_ν are dominated by physics within T of the Fermi surface. If, as in the CFL phase, all quarks have a gap $\Delta \gg T$ then the contribution of quark quasiparticles to C_V and L_ν is suppressed by $\sim \exp(-\Delta/T)$. There may be other contributions to L_ν [82], but these are also very small. The specific heat is dominated by that of the electrons, although it may also receive a small contribution from the CFL phase Goldstone bosons. Although further work is required, it is already clear that both C_V and L_ν are much smaller than in the nuclear matter outside the quark matter core. This means that the total heat capacity and the total neutrino emission rate (and hence the cooling rate) of a neutron star with a CFL core will be determined completely by the nuclear matter outside the core. The quark matter core is "inert": with its small heat capacity and emission rate it has little influence on the temperature of the star as a whole. As the rest of the star emits neutrinos and cools, the core cools by conduction, because the electrons keep it in good thermal contact with the rest of the star. These qualitative expectations are nicely borne out in the calculations presented by Page et al. [83].

The analysis of the cooling history of a neutron star with a quark matter core in the 2SC phase is more complicated. The red and green up and down quarks pair with a gap many orders of magnitude larger than the temperature, which is of order 10 keV, and are therefore inert as described above. Any strange quarks present will form a $\langle ss \rangle$ condensate with angular momentum $J = 1$ [85]. The resulting gap has been estimated to be of order hundreds of keV [85], although applying results of Ref. [86] suggests a somewhat smaller gap, around 10 keV. The blue up and down quarks can also pair, forming a $J = 1$ condensate which breaks rotational invariance [3]. The related gap was estimated to be a few keV [3], but this estimate was not robust and should be revisited in light of more recent developments given its importance in the following. The critical temperature T_c above which no condensate forms is of order the zero-temperature gap Δ. ($T_c = 0.57\Delta$ for $J = 0$ condensates [58].) Therefore, if there are quarks for which $\Delta \sim T$ or smaller, these quarks do not pair at temperature T. Such quark quasiparticles will radiate neutrinos rapidly (via direct URCA reactions like $d \to u+e+\bar{\nu}$, $u \to d+e^++\nu$, etc.) and the quark matter core will cool rapidly and determine the cooling history of the star as a whole [84,83]. The star will cool rapidly until its interior temperature is $T < T_c \sim \Delta$, at which time the quark matter core will become inert and the further cooling history will be dominated by neutrino emission from the nuclear matter fraction of the star. If future data were to show that neutron stars first cool rapidly (direct URCA) and then cool more slowly, such data would allow an estimate of the smallest quark matter gap. We are unlikely to be so lucky. The simple observation of rapid cooling would *not* be an unambiguous discovery of quark matter with small gaps; there are other circumstances in which the direct URCA processes occurs. However, if as data on neutron star temperatures improves in coming years the standard cooling scenario proves correct, indicating the absence of the direct URCA processes, this *would* rule out the presence of quark matter with gaps in the 10 keV range or smaller. The presence of a quark matter core in which all gaps are $\gg T$ can never be revealed by an analysis of the cooling history.

Supernova Neutrinos: We now turn from neutrino emission from a neutron star which is many years old to that from the protoneutron star during the first seconds of a supernova. Carter and Reddy [87] have pointed out that when this protoneutron star is heated up to its maximum temperature of order 30-50 MeV, it may feature a quark matter core which is too hot for color superconductivity. As the core of the protoneutron star cools over the coming seconds, if it contains quark matter this quark matter will cool through T_c, entering the color superconducting regime of the QCD phase diagram from above. For $T \sim T_c$, the specific heat rises and the cooling slows. Then, as T drops further and Δ increases to become greater than T, the specific heat drops rapidly. Furthermore, as the number density of quark quasiparticles becomes suppressed by $\exp(-\Delta/T)$, the neutrino transport mean free path rapidly becomes very long [87]. This means that all the neutrinos previously trapped in the now color superconducting core are able to escape in a sudden burst. If we are lucky enough that a terrestrial neutrino detector sees thousands of neutrinos from a future supernova, Carter and Reddy's results suggest that there may be a signature of the transition to color superconductivity present in the time distribution of these neutrinos. Neutrinos from the core of the protoneutron star will lose energy as they scatter on their way out, but because they will be the last to reach the surface of last scattering, they will be the final neutrinos received at the earth. If they are emitted from the quark matter core in a sudden burst, they may therefore result in a bump at late times in the temporal distribution of the detected neutrinos. More detailed study remains to be done in order to understand how Carter and Reddy's signature, dramatic when the neutrinos escape from the core, is processed as the neutrinos traverse the rest of the protoneutron star and reach their surface of last scattering.

R-mode Instabilities: Another arena in which color superconductivity comes into play is the physics of r-mode instabilities. A neutron star whose angular rotation frequency Ω is large enough is unstable to the growth of r-mode oscillations which radiate away angular momentum via gravitational waves, reducing Ω. What does "large enough" mean? The answer depends on the damping mechanisms which act to prevent the growth of the relevant modes. Both shear viscosity and bulk viscosity act to damp the r-modes, preventing them from going unstable. The bulk viscosity and the quark contribution to the shear viscosity both become exponentially small in quark matter with $\Delta > T$ and as a result, as Madsen [88] has shown, a compact star made *entirely* of quark matter with gaps $\Delta = 1$ MeV or greater is unstable if its spin frequency is greater than tens to 100 Hz. Many compact stars spin faster than this, and Madsen therefore argues that compact stars cannot be strange quark stars unless some quarks remain ungapped. Alas, this powerful argument becomes much less powerful in the context of a neutron star with a quark matter core. First, the r-mode oscillations have a wave form whose amplitude is greatest near the surface, not in the core. Second, in an ordinary neutron star there is a new source of damping: friction at the boundary between the crust and the neutron superfluid "mantle" keeps the r-modes stable regardless of the properties of a quark matter core [89,88].

Magnetic Field Evolution: Next, we turn to the physics of magnetic fields within color superconducting neutron star cores [90,91]. The interior of a conventional neutron star is a superfluid (because of neutron-neutron pairing) and is an electromagnetic superconductor (because of proton-proton pairing). Ordinary magnetic fields penetrate it only in the cores of magnetic flux tubes. A color superconductor behaves differently. At first glance, it seems that because a diquark Cooper pair has nonzero electric charge, a diquark condensate must exhibit the standard Meissner effect, expelling ordinary magnetic fields or restricting them to flux tubes within whose cores the condensate vanishes. This is not the case [91]. In both the 2SC and CFL phase a linear combination of the $U(1)$ gauge transformation of ordinary electromagnetism and one (the eighth) color gauge transformation remain unbroken even in the presence of the condensate. This means that the ordinary photon A_μ and the eighth gluon G^8_μ are replaced by new linear combinations

$$A^{\tilde{Q}}_\mu = \cos\alpha_0 \, A_\mu + \sin\alpha_0 \, G^8_\mu$$
$$A^X_\mu = -\sin\alpha_0 \, A_\mu + \cos\alpha_0 \, G^8_\mu \qquad (3)$$

where $A^{\tilde{Q}}_\mu$ is massless and A^X_μ is massive. This means that $B_{\tilde{Q}}$ satisfies the ordinary Maxwell equations while B_X experiences a Meissner effect. The mixing angle α_0 is the analogue of the Weinberg angle in electroweak theory, in which the presence of the Higgs condensate causes the A^Y_μ and the third $SU(2)_W$ gauge boson mix to form the photon, A_μ, and the massive Z boson. $\sin(\alpha_0)$ is proportional to e/g and turns out to be about $1/20$ in the 2SC phase and $1/40$ in the CFL phase [91]. This means that the $\tilde{Q}$-photon which propagates in color superconducting quark matter is mostly photon with only a small gluon admixture. If a color superconducting neutron star core is subjected to an ordinary magnetic field, it will either expel the X component of the flux or restrict it to flux tubes, but it can (and does [91]) admit the great majority of the flux in the form of a $B_{\tilde{Q}}$ magnetic field satisfying Maxwell's equations. The decay in time of this "free field" (i.e. not in flux tubes) is limited by the $\tilde{Q}$-conductivity of the quark matter. A color superconductor is not a $\tilde{Q}$-superconductor — that is the whole point — but it turns out to be a very good $\tilde{Q}$-conductor due to the presence of electrons: the $B_{\tilde{Q}}$ magnetic field decays only on a time scale which is much longer than the age of the universe [91]. This means that a quark matter core within a neutron star serves as an "anchor" for the magnetic field: whereas in ordinary nuclear matter the magnetic flux tubes can be dragged outward by the neutron superfluid vortices as the star spins down [92], the magnetic flux within the color superconducting core simply cannot decay. Even though this distinction is a qualitative one, it will be difficult to confront it with data since what is observed is the total dipole moment of the neutron star. A color superconducting core anchors those magnetic flux lines which pass through the core, while in a neutron star with no quark matter core the entire internal magnetic field can decay over time. In both cases, however, the total dipole moment can change since the magnetic flux lines which do not pass through the core can move.

Glitches in Quark Matter: The final consequence of color superconductivity we wish to discuss is the possibility that (some) glitches may originate within quark matter regions of a compact star [86]. In any context in which color superconductivity arises in nature, it is likely to involve pairing between species of quarks with differing chemical potentials. If the chemical potential difference is small enough, BCS pairing occurs as we have been discussing. If the Fermi surfaces are too far apart, no pairing between the species is possible. The transition between the BCS and unpaired states as the splitting between Fermi momenta increases has been studied in electron [93] and QCD [7,8,94] superconductors, assuming that no other state intervenes. However, there is good reason to think that another state can occur. This is the "LOFF" state, first explored by Larkin and Ovchinnikov [95] and Fulde and Ferrell [96] in the context of electron superconductivity in the presence of magnetic impurities. They found that near the unpairing transition, it is favorable to form a state in which the Cooper pairs have nonzero momentum. This is favored because it gives rise to a region of phase space where each of the two quarks in a pair can be close to its Fermi surface, and such pairs can be created at low cost in free energy. Condensates of this sort spontaneously break translational and rotational invariance, leading to gaps which vary periodically in a crystalline pattern. If in some shell within the quark matter core of a neutron star (or within a strange quark star) the quark number densities are such that crystalline color superconductivity arises, rotational vortices may be pinned in this shell, making it a locus for glitch phenomena.

We [86] have explored the range of parameters for which crystalline color superconductivity occurs in the QCD phase diagram, upon making various simplifying assumptions. For example, we focus primarily on a four-fermion interaction with the quantum numbers of single gluon exchange. Also, we only consider pairing between u and d quarks, with $\mu_d = \bar{\mu} + \delta\mu$ and $\mu_u = \bar{\mu} - \delta\mu$, whereas we expect a LOFF state when the difference between the Fermi momenta of any two quark flavors is near an unpairing transition. We find the LOFF state is favored for values of $\delta\mu$ which satisfy $\delta\mu_1 < \delta\mu < \delta\mu_2$ where $\delta\mu_1/\Delta_0 = 0.707$ and $\delta\mu_2/\Delta_0 = 0.754$ in the weak coupling limit in which $\Delta_0 \ll \mu$. (Here, Δ_0 is the 2SC gap that would arise if $\delta\mu$ were zero.) The LOFF gap parameter decreases from $0.23\Delta_0$ at $\delta\mu = \delta\mu_1$ (where there is a first order BCS-LOFF phase transition) to zero at $\delta\mu = \delta\mu_2$ (where there is a second order LOFF-normal transition). Except for very close to $\delta\mu_2$, the critical temperature above which the LOFF state melts will be much higher than typical neutron star temperatures. At stronger coupling the LOFF gap parameter decreases relative to Δ_0 and the window of $\delta\mu/\Delta_0$ within which the LOFF state is favored shrinks. The window grows if the interaction is changed to weight electric gluon exchange more heavily than magnetic gluon exchange.

The quark matter which may be present within a compact star will be in the crystalline color superconductor (LOFF) state if $\delta\mu/\Delta_0$ is in the requisite range. For a reasonable value of $\delta\mu$, say 25 MeV, this occurs if the gap Δ_0 which characterizes the uniform color superconductor present at smaller values of $\delta\mu$ is about 40 MeV. This is in the middle of the range of present estimates. Both $\delta\mu$ and Δ_0 vary as a function of density and hence as a function of radius in a compact star. Although it is too early to make quantitative predictions, the numbers are such that crystalline color superconducting quark matter may very well occur in a range of radii within a compact star. It is therefore worthwhile to consider the consequences.

Many pulsars have been observed to glitch. Glitches are sudden jumps in rotation frequency Ω which may be as large as $\Delta\Omega/\Omega \sim 10^{-6}$, but may also be several orders of magnitude smaller. The frequency of observed glitches is statistically consistent with the hypothesis that all radio pulsars experience glitches [97]. Glitches are thought to originate from interactions between the rigid crust, somewhat more than a kilometer thick in a typical neutron star, and rotational vortices in the neutron superfluid which are moving (or trying to move) outward as the star spins down. Although the models [98] differ in important respects, all agree that the fundamental requirements are the presence of rotational vortices in a superfluid and the presence of a rigid structure which impedes the motion of vortices and which encompasses enough of the volume of the pulsar to contribute significantly to the total moment of inertia.

Although it is premature to draw quantitative conclusions, it is interesting to speculate that some glitches may originate deep within a pulsar which features a quark matter core, in a region of that core in which the color superconducting quark matter is in a LOFF crystalline color superconductor phase. A three flavor analysis is required to determine whether the LOFF phase is a superfluid. If the only pairing is between u and d quarks, this 2SC phase is not a superfluid [3,7], whereas if all three quarks pair in some way, a superfluid *is* obtained [5,7]. Henceforth, we suppose that the LOFF phase is a superfluid, which means that if it occurs within a pulsar it will be threaded by an array of rotational vortices. It is reasonable to expect that these vortices will be pinned in a LOFF crystal, in which the diquark condensate varies periodically in space. Indeed, one of the suggestions for how to look for a LOFF phase in terrestrial electron superconductors relies on the fact that the pinning of magnetic flux tubes (which, like the rotational vortices of interest to us, have normal cores) is expected to be much stronger in a LOFF phase than in a uniform BCS superconductor [99].

A real calculation of the pinning force experienced by a vortex in a crystalline color superconductor must await the determination of the crystal structure of the LOFF phase. We can, however, attempt an order of magnitude estimate along the same lines as that done by Anderson and Itoh [100] for neutron vortices in the inner crust of a neutron star. In that context, this estimate has since been made quantitative [101,102,98]. For one specific choice of parameters [86], the LOFF phase is favored over the normal phase by a free energy $F_{\text{LOFF}} \sim 5 \times (10 \text{ MeV})^4$ and the spacing between nodes in the LOFF crystal is $b = \pi/(2|\mathbf{q}|) \sim 9$ fm. The thickness of a rotational vortex is given by the correlation length $\xi \sim 1/\Delta \sim 25$ fm. The pinning energy is the difference between the energy of a section of vortex of length b which is centered on a node of the LOFF crystal vs. one which is centered on a maximum of the LOFF crystal. It is of order $E_p \sim F_{\text{LOFF}} b^3 \sim 4$ MeV. The resulting pinning force per unit length of vortex is of order $f_p \sim E_p/b^2 \sim$ (4 MeV)/(80 fm^2). A complete calculation will be challenging because $b < \xi$, and is likely to yield an f_p which is somewhat less than that we have obtained by dimensional analysis. Note that our estimate of f_p is quite uncertain both because it is only based on dimensional analysis and because the values of Δ, b and F_{LOFF} are uncertain. (We hae a good understanding of all the ratios Δ/Δ_0, $\delta\mu/\Delta_0$, q/Δ_0 and consequently $b\Delta_0$ in the LOFF phase. It is of course the value of the BCS gap Δ_0 which is uncertain.) It is premature to compare our crude result to the results of serious calculations of the pinning of crustal neutron vortices as in Refs. [101,102,98]. It is nevertheless remarkable that they prove to be similar: the pinning energy of neutron vortices in the inner crust is $E_p \approx 1-3$ MeV and the pinning force per unit length is $f_p \approx (1-3 \text{ MeV})/(200-400 \text{ fm}^2)$. Perhaps, therefore, glitches occurring in a region of crystalline color superconducting quark matter may yield similar phenomenology to those occurring in the inner crust.

Perhaps the most interesting consequence of these speculations arises in the context of compact stars made entirely of strange quark matter. The work of Witten [103] and Farhi and Jaffe [104] raised the possibility that strange quark matter may be energetically stable relative to nuclear matter even at zero pressure. If this is the case it raises the question whether observed compact stars—pulsars, for example—are strange quark stars [105,106] rather than neutron stars. A conventional neutron star may feature a core made of strange quark matter, as we have been discussing above.[9] Strange quark stars, on the other hand, are made (almost) entirely of quark matter with either no hadronic matter content at all or with a thin crust, of order one hundred meters thick, which contains no neutron superfluid [106,107]. The nuclei in this thin crust are supported above the quark matter by electrostatic forces; these forces cannot support a neutron fluid. Because of the absence of superfluid neutrons, and because of the thinness of the crust, no successful models of glitches in the crust of a strange quark star have been proposed. Since pulsars are observed to glitch, the apparent lack of a glitch mechanism for strange quark stars has been the strongest argument that pulsars cannot be strange quark stars [108–110]. This conclusion must now be revisited.

Madsen's conclusion [88] that a strange quark star is prone to r-mode instability due to the absence of damping must also be revisited, since the relevant fluid oscillations may be damped within or at the boundary of a region of crystalline color superconductor.

The quark matter in a strange quark star, should one exist, would be a color superconductor. Depending on the mass of the star, the quark number densities increase by a factor of about two to ten in going from the surface to the center [106].

---

[9] Note that a convincing discovery of a quark matter core within an otherwise hadronic neutron star would demonstrate conclusively that strange quark matter is *not* stable at zero pressure, thus ruling out the existence of strange quark stars. It is not possible for neutron stars with quark matter cores and strange quark stars to both be stable.

This means that the chemical potential differences among the three quarks will vary also, and there could be a range of radii within which the quark matter is in a crystalline color superconductor phase. This raises the possibility of glitches in strange quark stars. Because the variation in density with radius is gradual, if a shell of LOFF quark matter exists it need not be particularly thin. And, we have seen, the pinning forces may be comparable in magnitude to those in the inner crust of a conventional neutron star. It has recently been suggested (for reasons unrelated to our considerations) that certain accreting compact stars may be strange quark stars [111], although the evidence is far from unambiguous [112]. In contrast, it has been thought that, because they glitch, conventional radio pulsars cannot be strange quark stars. Our work questions this assertion by raising the possibility that glitches may originate within a layer of quark matter which is in a crystalline color superconducting state.

Closing Remarks: The answer to the question of whether the QCD phase diagram does or does not feature a 2SC interlude on the horizontal axis, separating the CFL and baryonic phases in both of which chiral symmetry is broken, depends on whether the strange quark is effectively heavy or effectively light. This is the central outstanding qualitative question about the high density region of the QCD phase diagram. A central question at higher temperatures, namely where does nature locate the critical point E, also depends on the strange quark mass. Both questions are hard to answer theoretically with any confidence. The high temperature region is in better shape, however, because the program of experimentation described in Section II allows heavy ion collision experiments to search for the critical point E. Theorists have described how to use phenomena characteristic of freezeout in its vicinity to discover E; this gives experimentalists the ability to locate it convincingly. The discovery of E would allow us to draw the higher temperature regions of the map of the QCD phase diagram in ink. At high density, there has been much recent progress in our understanding of how the presence of color superconducting quark matter in a compact star would affect five different phenomena: cooling by neutrino emission, the temporal pattern of the neutrinos emitted by a supernova, the evolution of neutron star magnetic fields, r-mode instabilities, and glitches. Nevertheless, much theoretical work remains to be done before we can make sharp proposals for which astrophysical observations are most likely to help teach us how to ink in the boundaries of the 2SC and CFL regions in the QCD phase diagram. Best of all, though, and as in heavy ion physics, a wealth of new data is expected over the next few years.

REFERENCES

1. B. Barrois, Nucl. Phys. **B129**, 390 (1977); S. Frautschi, Proceedings of workshop on hadronic matter at extreme density, Erice 1978; B. Barrois, "Nonperturbative effects in dense quark matter", Cal Tech PhD thesis, UMI 79-04847-mc (1979).
2. D. Bailin and A. Love, Phys. Rept. **107**, 325 (1984), and references therein.

3. M. Alford, K. Rajagopal and F. Wilczek, Phys. Lett. **B422**, 247 (1998) [hep-ph/9711395].
4. R. Rapp, T. Schäfer, E. V. Shuryak and M. Velkovsky, Phys. Rev. Lett. **81**, 53 (1998) [hep-ph/9711396].
5. M. Alford, K. Rajagopal and F. Wilczek, Nucl. Phys. **B537**, 443 (1999) [hep-ph/9804403].
6. T. Schäfer and F. Wilczek, Phys. Rev. Lett. **82**, 3956 (1999) [hep-ph/9811473].
7. M. Alford, J. Berges and K. Rajagopal, Nucl. Phys. **B558**, 219 (1999) [hep-ph/9903502].
8. T. Schäfer and F. Wilczek, Phys. Rev. **D60**, 074014 (1999) [hep-ph/9903503].
9. M. Stephanov, K. Rajagopal and E. Shuryak, Phys. Rev. Lett. **81**, 4816 (1998) [hep-ph/9806219].
10. M. Stephanov, K. Rajagopal and E. Shuryak, Phys. Rev. **D60**, 114028 (1999) [hep-ph/9903292].
11. For a longer review, see K. Rajagopal, in Quark-Gluon Plasma 2, (World Scientific, 1995) 484, ed. R. Hwa [hep-ph/9504310].
12. R. Pisarski and F. Wilczek, Phys. Rev. **D29**, 338 (1984); F. Wilczek, Int. J. Mod. Phys. **A7**, 3911 (1992); K. Rajagopal and F. Wilczek, Nucl. Phys. **B399**, 395 (1993).
13. For reviews, see F. Karsch, hep-lat/9909006; E. Laermann Nucl. Phys. Proc. Suppl. **63**, 114 (1998); and A. Ukawa, Nucl. Phys. Proc. Suppl. **53**, 106 (1997).
14. A. Ali Khan et al, [CP-PACS Collaboration], hep-lat/0008011.
15. For example, S. Gottlieb et al., Phys. Rev. **D55**, 6852 (1997); F. Karsch, hep-lat/9909006.
16. A. Barducci, R. Casalbuoni, S. DeCurtis, R. Gatto, G. Pettini, Phys. Lett. **B231**, 463 (1989); S.P. Klevansky, Rev. Mod. Phys. **64**, 649 (1992); A. Barducci, R. Casalbuoni, G. Pettini and R. Gatto, Phys. Rev. **D49**, 426 (1994).
17. M. Stephanov, Phys. Rev. Lett. **76**, 4472 (1996); Nucl. Phys. Proc. Suppl. **53**, 469 (1997).
18. J. Berges and K. Rajagopal, Nucl. Phys. **B538**, 215 (1999) [hep-ph/9804233].
19. M. A. Halasz, A. D. Jackson, R. E. Shrock, M. A. Stephanov and J. J. Verbaarschot, Phys. Rev. **D58**, 096007 (1998) [hep-ph/9804290].
20. G. W. Carter and D. Diakonov, Phys. Rev. **D60**, 016004 (1999) [hep-ph/9812445].
21. For a review, see I. Lawrie and S. Sarbach in Phase Transitions and Critical Phenomena **9**, 1 (Academic Press, 1984), ed. C. Domb and J. Lebowitz.
22. F. Wilczek, Int. J. Mod. Phys. **A7**, 3911 (1992); K. Rajagopal and F. Wilczek, Nucl. Phys. **B399**, 395 (1993).
23. F. Brown et al, Phys. Rev. Lett. **65**, 2491 (1990).
24. JLQCD Collaboration, Nucl. Phys. Proc. Suppl. **73**, 459 (1999).
25. Y. Iwasaki et al, Phys. Rev. **D54**, 7010 (1996).
26. See, e.g., P. Braun-Munzinger, J. Stachel, J. P. Wessels and N. Xu, Phys. Lett. **B344**, 43 (1994); *ibid.* **B365**, 1 (1996); P. Braun-Munzinger and J. Stachel, Nucl.Phys. **A638**, 3 (1998).
27. B. B. Back et al, [PHOBOS Collaboration], to appear in Phys. Rev. Lett., hep-ex/0007036.

28. H. Appelshauser *et al.* [NA49 Collaboration], Phys. Lett. **B459**, 679 (1999).
29. A. Bialas and V. Koch, Phys. Lett. **B456**, 1 (1999).
30. St. Mrówczyński, Phys. Lett. **B430**, 9 (1998).
31. B. Berdnikov and K. Rajagopal, hep-ph/9912274.
32. Reviewed in U. Heinz and M. Jacob, nucl-th/0002042.
33. R. D. Pisarski and D. H. Rischke, Phys. Rev. Lett. **83**, 37 (1999) [nucl-th/9811104].
34. R. D. Pisarski, Phys. Rev. **C62**, 035202 (2000) [nucl-th/9912070].
35. R. Rapp, T. Schäfer, E. V. Shuryak and M. Velkovsky, Annals Phys. **280**, 35 (2000) [hep-ph/9904353].
36. D. K. Hong, M. Rho and I. Zahed, Phys. Lett. **B468**, 261 (1999) [hep-ph/9906551].
37. R. Casalbuoni and R. Gatto, Phys. Lett. **B464**, 111 (1999) [hep-ph/9908227].
38. M. Alford, J. Berges and K. Rajagopal, Phys. Rev. Lett. **84**, 598 (2000) [hep-ph/9908235].
39. T. Schäfer, Nucl. Phys. **B575**, 269 (2000) [hep-ph/9909574].
40. D. T. Son and M. A. Stephanov, Phys. Rev. **D61**, 074012 (2000) [hep-ph/9910491]; erratum, *ibid.* **D62**, 059902 (2000) [hep-ph/0004095].
41. M. Rho, A. Wirzba and I. Zahed, Phys. Lett. **B473**, 126 (2000) [hep-ph/9910550].
42. D. K. Hong, T. Lee and D. Min, Phys. Lett. **B477**, 137 (2000) [hep-ph/9912531].
43. C. Manuel and M. H. Tytgat, Phys. Lett. **B479**, 190 (2000) [hep-ph/0001095].
44. M. Rho, E. Shuryak, A. Wirzba and I. Zahed, Nucl. Phys. **A676**, 273 (2000) [hep-ph/0001104].
45. K. Zarembo, Phys. Rev. **D62**, 054003 (2000) [hep-ph/0002123].
46. S. R. Beane, P. F. Bedaque and M. J. Savage, Phys. Lett. **B483**, 131 (2000) [hep-ph/0002209].
47. D. H. Rischke, Phys. Rev. **D62**, 054017 (2000) [nucl-th/0003063].
48. D. K. Hong, hep-ph/0006105.
49. T. Schäfer, nucl-th/0007021.
50. M. A. Nowak, M. Rho, A. Wirzba and I. Zahed, hep-ph/0007034.
51. D. H. Rischke, Phys. Rev. **D62**, 034007 (2000) [nucl-th/0001040]; G. Carter and D. Diakonov, Nucl. Phys. **B582**, 571 (2000) [hep-ph/0001318].
52. F. Sannino, Phys. Lett. **B480**, 280 (2000) [hep-ph/0002277]; R. Casalbuoni, Z. Duan and F. Sannino, hep-ph/0004207; S. D. Hsu, F. Sannino and M. Schwetz, hep-ph/0006059.
53. N. Evans, S. D. H. Hsu and M. Schwetz, Nucl. Phys. **B551**, 275 (1999) [hep-ph/9808444]; Phys. Lett. **B449**, 281 (1999) [hep-ph/9810514].
54. T. Schäfer and F. Wilczek, Phys. Lett. **B450**, 325 (1999) [hep-ph/9810509].
55. N. O. Agasian, B. O. Kerbikov and V. I. Shevchenko, Phys. Rept. **320**, 131 (1999) [hep-ph/9902335].
56. B. Vanderheyden and A. D. Jackson, hep-ph/0003150.
57. D. T. Son, Phys. Rev. **D59**, 094019 (1999) [hep-ph/9812287].
58. R. D. Pisarski and D. H. Rischke, Phys. Rev. **D60**, 094013 (1999) [nucl-th/9903023]; Phys. Rev. **D61**, 051501 (2000) [nucl-th/9907041]; Phys. Rev. **D61**, 074017 (2000) [nucl-th/9910056];
59. D. K. Hong, Phys. Lett. **B473**, 118 (2000) [hep-ph/9812510]; Nucl. Phys. **B582**, 451 (2000) [hep-ph/9905523].

60. D. K. Hong, V. A. Miransky, I. A. Shovkovy and L. C. Wijewardhana, Phys. Rev. **D61**, 056001 (2000), erratum *ibid.* **D62**, 059903 (2000) [hep-ph/9906478].
61. T. Schäfer and F. Wilczek, Phys. Rev. **D60**, 114033 (1999) [hep-ph/9906512].
62. W. E. Brown, J. T. Liu and H. Ren, Phys. Rev. **D61**, 114012 (2000) [hep-ph/9908248]; Phys. Rev. **D62**, 054016 (2000) [hep-ph/9912409]; Phys. Rev. **D62**, 054013 (2000) [hep-ph/0003199]..
63. S. D. Hsu and M. Schwetz, Nucl. Phys. **B572**, 211 (2000) [hep-ph/9908310].
64. I. A. Shovkovy and L. C. Wijewardhana, Phys. Lett. **B470**, 189 (1999) [hep-ph/9910225].
65. N. Evans, J. Hormuzdiar, S. D. Hsu and M. Schwetz, Nucl. Phys. **B581**, 391 (2000) [hep-ph/9910313].
66. S. R. Beane, P. F. Bedaque and M. J. Savage, nucl-th/0004013.
67. K. Rajagopal and E. Shuster, hep-ph/0004074.
68. C. Manuel, hep-ph/0005040.
69. S. Chandrasekharan and U. Wiese, Phys. Rev. Lett. **83**, 3116 (1999) [cond-mat/9902128].
70. UKQCD Collaboration, Phys. Rev. **D59** (1999) 116002; S. Hands, J. B. Kogut, M. Lombardo and S. E. Morrison, Nucl. Phys. **B558**, 327 (1999) [hep-lat/9902034]; S. Hands, I. Montvay, S. Morrison, M. Oevers, L. Scorzato and J. Skullerud, hep-lat/0006018.
71. J. B. Kogut, M. A. Stephanov and D. Toublan, Phys. Lett. **B464**, 183 (1999) [hep-ph/9906346]; J. B. Kogut, M. A. Stephanov, D. Toublan, J. J. Verbaarschot and A. Zhitnitsky, Nucl. Phys. **B582**, 477 (2000) [hep-ph/0001171].
72. D. V. Deryagin, D. Yu. Grigoriev and V. A. Rubakov, Int. J. Mod. Phys. **A7**, 659 (1992).
73. E. Shuster and D. T. Son, Nucl. Phys. **B573**, 434 (2000) [hep-ph/9905448].
74. B. Park, M. Rho, A. Wirzba and I. Zahed, Phys. Rev. **D62**, 034015 (2000) [hep-ph/9910347].
75. R. Rapp, E. Shuryak and I. Zahed, hep-ph/0008207.
76. D. T. Son and M. A. Stephanov, hep-ph/0005225.
77. For a review, see H. Heiselberg and M. Hjorth-Jensen, Phys. Rept. **328**, 237 (2000) [nucl-th/9902033].
78. D. Blaschke, T. Klahn and D. N. Voskresensky, astro-ph/9908334.
79. D. Page, M. Prakash, J. M. Lattimer and A. Steiner, hep-ph/0005094.
80. C. Schaab *et al*, Astrophys. J. Lett **480** (1997) L111 and references therein.
81. T. Schäfer, hep-ph/0006034.
82. M. Alford, J. Bowers and K. Rajagopal, hep-ph/0008208.
83. G. W. Carter and S. Reddy, hep-ph/0005228.
84. J. Madsen, Phys. Rev. Lett. **81** (1998) 3311 and references therein.
85. L. Bildsten and G. Ushomirsky, astro-ph/9911155.
86. D. Blaschke, D. M. Sedrakian and K. M. Shahabasian, astro-ph/9904395.
87. M. Alford, J. Berges and K. Rajagopal, Nucl. Phys. **B571**, 269 (2000) [hep-ph/9910254].
88. For reviews, see J. Sauls, in Timing Neutron Stars, J. Ögleman and E. P. J. van den Heuvel, eds., (Kluwer, Dordrecht: 1989) 457; and D. Bhattacharya and G.

Srinivasan, in X-Ray Binaries, W. H. G. Lewin, J. van Paradijs, and E. P. J. van den Heuvel eds., (Cambridge University Press, 1995) 495.
89. A. M. Clogston, Phys. Rev. Lett. **9**, 266 (1962); B. S. Chandrasekhar, App. Phys. Lett. **1**, 7 (1962).
90. P. F. Bedaque, hep-ph/9910247.
91. A. I. Larkin and Yu. N. Ovchinnikov, Zh. Eksp. Teor. Fiz. **47**, 1136 (1964); translation: Sov. Phys. JETP **20**, 762 (1965).
92. P. Fulde and R. A. Ferrell, Phys. Rev. **135**, A550 (1964).
93. M. A. Alpar and C. Ho, Mon. Not. R. Astron. Soc. **204**, 655 (1983).
94. For reviews, see D. Pines and A. Alpar, Nature **316**, 27 (1985); D. Pines, in *Neutron Stars: Theory and Observation*, J. Ventura and D. Pines, eds., 57 (Kluwer, 1991); M. A. Alpar, in *The Lives of Neutron Stars*, M. A. Alpar et al., eds., 185 (Kluwer, 1995). For more recent developments and references to further work, see M. Ruderman, Astrophys. J. **382**, 587 (1991); R. I. Epstein and G. Baym, Astrophys. J. **387**, 276 (1992); M. A. Alpar, H. F. Chau, K. S. Cheng and D. Pines, Astrophys. J. **409**, 345 (1993); B. Link and R. I. Epstein, Astrophys. J. **457**, 844 (1996); A. Sedrakian and J. M. Cordes, Mon. Not. R. Astron. Soc. **307**, 365 (1999).
95. R. Modler *et al.*, Phys. Rev. Lett. **76**, 1292 (1996).
96. P. W. Anderson and N. Itoh, Nature **256**, 25 (1975).
97. M. A. Alpar, Astrophys. J. **213**, 527 (1977).
98. M. A. Alpar, P. W. Anderson, D. Pines and J. Shaham, Astrophys. J. **278**, 791 (1984).
99. E. Witten, Phys. Rev. **D30**, 272 (1984).
100. E. Farhi and R. L. Jaffe, Phys. Rev. **D30**, 2379 (1984).
101. P. Haensel, J. L. Zdunik and R. Schaeffer, Astron. Astrophys. **160**, 121 (1986).
102. C. Alcock, E. Farhi and A. Olinto, Phys. Rev. Lett. **57**, 2088 (1986); Astrophys. J. **310**, 261 (1986).
103. N. K. Glendenning and F. Weber, Astrophys. J. **400**, 647 (1992).
104. A. Alpar, Phys. Rev. Lett. **58**, 2152 (1987).
105. J. Madsen, Phys. Rev. Lett. **61**, 2909 (1988).
106. R. R. Caldwell and J. L. Friedman, Phys. Lett. **B264**, 143 (1991).
107. X.-D. Li, I. Bombaci, M. Dey, J. Dey, E. P. J. van den Heuvel, Phys. Rev. Lett. **83**, 3776 (1999); X.-D. Li, S. Ray, J. Dey, M. Dey, I. Bombaci, Astrophys. J. **527**, L51 (1999); B. Datta, A. V. Thampan, I. Bombaci, astro-ph/9912173; I. Bombaci, astro-ph/0002524.
108. D. Psaltis and D. Chakrabarty, Astrophys. J. **521**, 332 (1999); D. Chakrabarty, Phys. World **13**, No. 2, 26 (2000).

Effective Field Theory in Nuclear Physics[1]

Martin J. Savage

Department of Physics, University of Washington,
Seattle, WA 98915
and
Jefferson Lab., 12000 Jefferson Avenue, Newport News,
Virginia 23606.

Abstract. I review recent developments in the application of effective field theory to nuclear physics. Emphasis is placed on precision two-body calculations and efforts to formulate the nuclear shell model in terms of an effective field theory.

INTRODUCTION

A question that I have been asked many times is "*Why use Effective Field Theory in Nuclear Physics*"? The simple and somewhat glib answer to this question is that the only other option that one has to using an effective field theory (EFT) is to use the "*Theory of Everything*" (TOE), string theory or some derivative thereof. All other descriptions **must** be incomplete at some level and when precise predictions are compared with precise measurements, differences will become obvious. It is a daunting prospect for us (maybe only me) to use the TOE to compute low-energy hadronic processes, and in fact, it is quite silly to even consider such calculations. After all, we know that processes in QED can be computed to high precision which agree with experimental observations, without knowing anything about physics at the Planck scale, $M_{\rm pl}$.

The renormalizability of QED assures that ultra-violet divergences, arising from our lack of understanding of physics at short-distances, can be explicitly removed by a few constants (electric charge and fermion mass), allowing observables to be related to each other to arbitrary precision. In contrast, EFT's are non-renormalizable but are still predictive when a systematic power counting in small expansion parameters can be established. Relations between observables at a given

---

[1] NT@UW-00-018.

precision will involve a finite number of constants that are not dictated by the symmetries of the EFT alone. For instance, the standard model is a renormalizable field theory but processes at energies much less than the scale of electroweak symmetry breaking can be described by an non-renormalizable EFT of reduced symmetry, $SU(3)_c \otimes SU(2)_L \otimes U_Y(1) \rightarrow SU(3)_c \otimes U_{em}(1)$, where weak interactions are incorporated by higher-dimension four-fermi operators. Even with quantum effects, the theory provides a systematic expansion of observables in terms of q^2/M_W^2 and m_f^2/M_W^2, where m_f is a fermion mass, M_W is the mass of the weak gauge bosons, and q is the external momentum.

If one were interested in calculating the cross section for $np \rightarrow d\gamma$–radiative neutron capture by a proton to form a deuteron–directly from QCD, then a lattice calculation is the only technique available. The lattice calculation will provide an unambiguous cross section in terms of the quark masses and Λ_{QCD}, or equivalently in terms of other hadronic observables. A cartoon of a contribution to $np \rightarrow d\gamma$ in terms of perturbative quarks and gluons is shown in Figure 1. Sources of quarks and

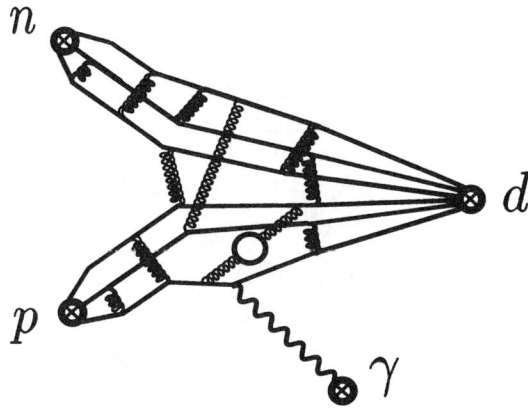

FIGURE 1. A "cartoon" of a contribution to $np \rightarrow d\gamma$ in QCD.

gluons that have non-zero overlap with the proton, neutron, deuteron and a source for the photon would used to generate the amplitude for $np \rightarrow d\gamma$. It is clear that a significant amount of work goes into forming the hadronic states themselves, let alone computing the interaction terms. In addition, as the deuteron has such a small binding energy and hence is quite extended compared to the nucleon, considerable effort will be required to generate the deuteron itself. Unfortunately, at this point in time the lattice community is not even close to being able to perform this multi-hadron calculation. Indeed, the deuteron itself remains to be generated in lattice calculations. [2] If all nuclear lengths scales were of order the chiral symmetry break-

---

[2] During the *Effective Field Theory* workshop to be held at the *Institute for Nuclear Physics* at the *University of Washington* during the summer of 2000, efforts will be made to estimate the computer resources necessary to determine the deuteron binding energy from lattice QCD [1].

ing scale Λ_χ then (very) naively lattice computations of nuclear observables would not be that much harder than computations in the single-nucleon sector. However, there are several low-energy length scales that play important roles in nuclear physics. Firstly Λ_χ, below which a hadronic description makes sense, and higher-dimension operators are induced that describe contributions from scales above Λ_χ. Secondly, the scale of the repulsive part of the nucleon-nucleon interaction, which is conventionally modeled by the exchange of vector mesons (far from mass-shell) and numerically is of order Λ_χ. Thirdly, the mass of the pion, which is much less than Λ_χ due to its special status as a pseudo-Goldstone boson. Finally, nuclear binding energies which are much smaller than one would naively guess.

If one is interested in this process at energies much less than Λ_χ, or the mass of the ρ-meson, but comparable to the mass of the pion, then it should be sufficient to use an EFT with only nucleons, pions and photons as dynamical degrees of freedom. All contributions from higher mass scales will be encapsulated in the infinite number of higher dimension operators that arise in the momentum and chiral expansions. Some diagrams that will contribute to $np \rightarrow d\gamma$ are shown in Figure 2. Unlike most EFT's that one encounters, higher dimension operators (dim-

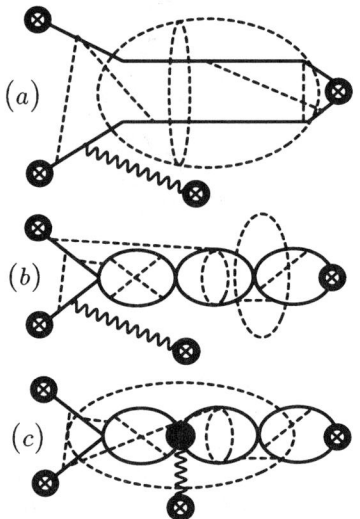

FIGURE 2. Contributions to $np \rightarrow d\gamma$ in a theory with nucleons, pions and photons. Diagram (a) shows a contribution from π-exchange alone, while diagram (b) shows a contribution from π-exchange and from short-distance interactions. Diagram (c) shows a contribution from a local, gauge invariant operator not constrained by nucleon-nucleon scattering data.

6) involving the nucleon field play a central role [2]-[6]. The fact that the deuteron is barely bound, with a binding energy of $B = 2.2$ MeV, requires a fine-tuning between pion exchange and short-distance physics. Naively, one would expect a binding energy set by f_π, the pion decay constant, much larger than one finds in

nature. Therefore, the class of diagrams shown in Figure 2(b) is not expected to be suppressed compared to those in Figure 2(a). In addition, contributions from diagrams shown in Figure 2(c) must be included. These arise from an insertion of operators that are gauge invariant by themselves, and are not related in any way to operators describing nucleon-nucleon scattering. They arise from short-distance physics and have a scale typically set by Λ_χ.

Continuing our descent in energy, if one is interested in this process at energies much less than the pion mass, m_π, then it should be sufficient to use an EFT with only nucleons and photons as dynamical degrees of freedom. All contributions from mass scales greater than m_π will be encapsulated in the infinite number of higher dimension operators that arise in the momentum expansion (chiral symmetry is explicitly broken, leaving isospin symmetry as the only relic of the flavor symmetries). Some diagrams that will contribute to $np \to d\gamma$ are shown in Figure 3. It is im-

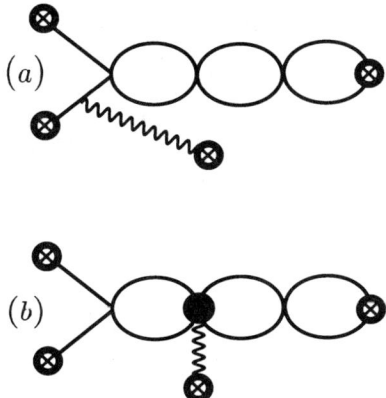

FIGURE 3. Diagrams that contribute to $np \to d\gamma$ in a theory with nucleons and photons. Diagram (a) shows a contribution from interactions between nucleons. Diagram (b) shows a contribution from a local, gauge invariant operator that does not contribute (at tree-level) to nucleon-nucleon scattering.

portant to realize that nucleon-nucleon scattering described by this EFT uniquely reproduces Effective Range Theory (ERT) [7]. However, for all other observables, such as those involving electroweak gauge fields, ERT (e.g. [7,8]) is seen to be an uncontrolled approximation to EFT (e.g. [9]).

In the following sections I attempt to indicate the status of EFT descriptions in the various energy regimes. Firstly, I will discuss low-energy $|\mathbf{p}| \ll m_\pi$ processes involving two and three nucleons, focusing on the recent high precision calculations that have been performed. Secondly, the issues, results and the present roadblocks to successfully describing the intermediate energy regime $|\mathbf{p}| \gtrsim m_\pi$ are presented. Finally, efforts to translate the understanding gained in EFT developments to many-nucleon systems are described. Such translation is necessary in order to achieve the ultimate goal of having a perturbative theory of nuclei that faithfully reproduces QCD.

$NP \to D\gamma$ AT LOW ENERGIES

During the past year there has been considerable focus placed on the radiative capture process $np \to d\gamma$. Firstly, it was pointed out [10] that the uncertainty in the cross section for $np \to d\gamma$ contributed significantly to the uncertainties in the predictions of Big-Bang-Nucleosynthesis (BBN) of light element abundances. This resulted from the lack of data in the energy region important for BBN from either $np \to d\gamma$ or $\gamma d \to np$. Further, available potential model calculations of these processes [11] are undocumented and error estimates are absent. Tools have recently been developed that allow for a $\sim 1\%$ calculation of the cross section with EFT [13,14]. This was facilitated in part by realizing that it is advantageous to get not only the location of the deuteron pole correct, but also the normalization of the deuteron s-state component. This had long been implemented in the methods of [15,16], and implicit in the construction of Weinberg [2,17], but was only recently implemented in the dimensionally regulated EFT [12]. Finally, there are experimental efforts to measure the small isoscalar $E2_S$ and $M1_S$ amplitudes contributing to $np \to d\gamma$ using polarized neutrons on polarized protons [18]. At the second workshop on Effective Field Theory in Nuclear Physics held at the *Institute for Nuclear Physics* at the *University of Washington* in 1999, Mannque Rho challenged the participants to compute the $E2_S$ and $M1_S$ amplitudes with the group whose predictions are verified experimentally winning a bottle of exceptional wine. With such a wonderful prize at stake, many workshop participants redirected their efforts to this project. This is now known as the *Rho-challenge*.

$np \to d\gamma$ for Big-Bang Nucleosynthesis [3]

As existing potential model calculations of $np \to d\gamma$ are undocumented and error estimates unavailable, a 5% uncertainty was assigned to the cross section as input into BBN codes [19]. It would be somewhat dismal if, after several decades of investigations into nuclear physics, the cross section for this process was uncertain at the 5% level. However, EFT calculations have demonstrated that the actual uncertainty is much less than 5%. EFT is a well defined method of calculation and estimates of the uncertainty in a given calculation can be made by considering the magnitude of higher order terms that have been omitted. The expression for the cross section for $np \to d\gamma$ valid at the $\sim 3\%$ level (with nonrelativistic kinematics) is

$$\sigma = \frac{4\pi\alpha \left(\gamma^2 + |\mathbf{P}|^2\right)^3}{\gamma^3 M_N^4 |\mathbf{P}|} \left[\, |\tilde{X}_{M1}|^2 + |\tilde{X}_{E1}|^2 \, \right] \quad . \tag{1}$$

---

[3] I thank Gautam Rupak for allowing me to present his results in this section

The isovector $M1_V$ and $E1_V$ amplitudes are

$$|\tilde{X}_{M1}|^2 = \frac{\kappa_1^2 \gamma^4 \left(\frac{1}{a_1} - \gamma\right)^2}{\left(\frac{1}{a_1^2} + |\mathbf{P}|^2\right)(\gamma^2 + |\mathbf{P}|^2)^2} \left[Z_d - r_0 \frac{\left(\frac{\gamma}{a_1} + |\mathbf{P}|^2\right)|\mathbf{P}|^2}{\left(\frac{1}{a_1^2} + |\mathbf{P}|^2\right)\left(\frac{1}{a_1} - \gamma\right)} - \frac{L_{np}}{\kappa_1} \frac{M_N}{2\pi} \frac{\gamma^2 + |\mathbf{P}|^2}{\frac{1}{a_1} - \gamma}\right]$$

$$|\tilde{X}_{E1}|^2 = \frac{|\mathbf{P}|^2 M_N^2 \gamma^4}{(\gamma^2 + |\mathbf{P}|^2)^4}\left[Z_d + \frac{M_N \gamma}{6\pi}\left(\frac{\gamma^2}{3} + |\mathbf{P}|^2\right)\overline{C}^{(P)}\right] \quad , \tag{2}$$

where κ_1 is the isovector nucleon magnetic moment, $a_1 = -23.714 \pm 0.013$ fm is the scattering length in the $^1S_0$ channel, $r_0 = 2.73 \pm 0.03$ fm is the effective range in the $^1S_0$ channel, $\gamma = \sqrt{M_N B}$ is the deuteron binding momentum, $\overline{C}^{(P)}$ is a number derivable from the nucleon-nucleon p-wave amplitudes, $Z_d = 1/(1 - \gamma \rho_d)$ is the residue of the $^3S_1$ nucleon-nucleon amplitude at the deuteron pole, and ρ_d is the effective range in the $^3S_1$ channel. There is also a contribution from an operator that is not related by gauge invariance to nucleon-nucleon scattering, L_{np}. This is the coefficient of a local operator involving four-nucleon fields and a magnetic photon. For incident neutrons with speed $|v| = 2200$ m/s the cross section for capture by protons at rest is measured to be $\sigma_{\text{cold}}^{\text{expt}} = 334.2 \pm 0.5$ mb [20]. The value of L_{np} is fixed by requiring that the expressions in eqs. (1) and (2) reproduce $\sigma_{\text{cold}}^{\text{expt}}$. The amplitudes in eq. (2) have been computed to one higher order by Rupak [14], including relativistic effects and the appearance of an E1 counterterm, providing a $\sim 1\%$ calculation of $np \to d\gamma$. Once L_{np} has been fixed, the photo-dissociation cross section $\gamma d \to np$ can be determined, and is shown in Figure 4. One finds that

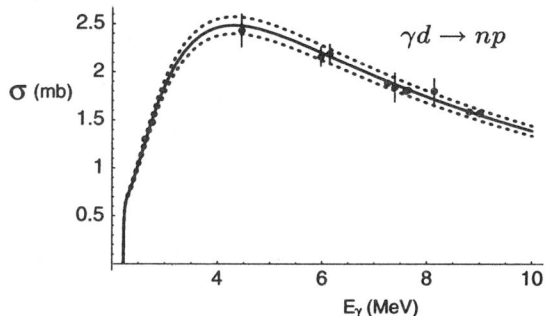

FIGURE 4. The photodissociation cross section for $\gamma d \to np$. The solid line results from eqs.(1) and (2) [13] with L_{np} determined by the cross section for cold $np \to d\gamma$. The dashed lines denote the theoretical uncertainty. Rupak has further reduced this uncertainty to below 1% [14].

this relatively simple analytic expression reproduces the data very well, once the counterterm L_{np} has been determined at a given energy. An idea of the convergence of the effective field theory calculation can be obtained from the numerical results of Rupak [14],

$$\sigma(2 \text{ MeV}) = 0.0218 \left(1 + 0.6389 + 0.0135 - 0.0053 - 0.0001 + ...\right) \text{ fm}^2$$

$$\sigma(20 \text{ keV}) = 0.1917 \, (1 + 0.1076 + 0.0001 + ...) \text{ fm}^2 \quad , \tag{3}$$

which are both seen to converge rapidly. The cross section at various energies

TABLE 1. $\sigma(np \to d\gamma)$ as a function of the nucleon center-of-mass energy, E. The asterisk denotes an input.

E (MeV)	total EFT σ (mb) [14]	ENDF σ (mb) [11]
1.264×10^{-8}	334.2 (*)	332.0
5.0×10^{-4}	1.668(0)	1.660
1.0×10^{-3}	1.172(0)	1.193
5.0×10^{-3}	0.4982(0)	0.496
1.0×10^{-2}	0.3324(0)	0.324
5.0×10^{-2}	0.1081(0)	0.108
0.100	0.06352(0)	0.0633
0.500	0.0341(1)	0.0345
1.00	0.0349(4)	0.0342

computed with EFT by Rupak [14], along with those from the on-line nuclear data center [11] are shown in Table 1. As expected the EFT calculation agrees with the numerical values from [11] at the $\sim 1\%$ level.

Isoscalar M1 and E2 Amplitudes in $np \to d\gamma$

As mentioned earlier, the *Rho Challenge* focused on the $M1_S$ and $E2_S$ isoscalar amplitudes that contribute to $np \to d\gamma$, so that predictions can be compared with the imminent measurement of polarization observables [18]. Two works were completed soon after the challenge was issued, one by Kubodera, Park, Min, and Rho [21], and one by Chen, Rupak and myself [22].

There are a couple of angular distributions that can be measured in $\vec{n} + \vec{p} \to d\gamma$, but if in addition, the polarization of the γ can be measured one finds that there is a different cross section for production of right-handed versus left-handed circularly polarized photons. Defining the asymmetry $A^\gamma(\theta)$ to be the ratio of the difference to the sum of these cross sections,

$$A^\gamma_{\eta_n}(\theta) = \eta_n \left[(P_\gamma(M1) + P_\gamma(E2)) \cos\theta + P_\gamma(E1) \sin^2\theta \right] \quad , \tag{4}$$

where the $P_\gamma(\Pi L)$ are combinations of the $M1_V$, $M1_S$, $E1_V$ and $E2_S$ amplitudes, and η_n is the neutron polarization vector.

The amplitudes and polarizations are computed in two very different ways. In [21] EFT wavefunctions are developed with a coordinate-space cut-off, which are then used to determine matrix elements of the various electric and magnetic multipole operators. Pions appear as dynamical degrees of freedom and determine the long-range part of the nucleon-nucleon interaction. In addition, counterterms are included via short-distance interactions (e.g. "delta-shell" and others) so that the magnetic and quadrupole moment of the deuteron are recovered. A very different construction is used in [22]. The EFT without pions is used and divergences are dimensionally regulated. As in [21], the four-nucleon-one photon counterterms are chosen to recover the deuteron magnetic and quadrupole moments, to give $P_\gamma(M1) = -7.1 \times 10^{-4}$, $P_\gamma(E2) = -3.5 \times 10^{-4}$ and a total of $P_\gamma = -1.06 \times 10^{-3}$ in the forward direction, approximately 2/3 of the experimentally determined value of [23] $P_\gamma^{\text{expt}} = -(1.5 \pm 0.3) \times 10^{-3}$. Given the large uncertainty in the calculation of the $M1_S$ amplitude, and the uncertainty of the measurement, the two are not inconsistent. $P_\gamma(M1) = -7.1 \times 10^{-4}$ agrees with the results of Burichenko and Kriplovich [24] of $P_\gamma(M1) = -7.0 \times 10^{-4}$ from a Reid soft-core calculation, but is somewhat less than their zero-range calculation of $P_\gamma(M1) = -9.2 \times 10^{-4}$. However, given the large uncertainty in the $M1_S$ amplitude of [22], both values are consistent. $P_\gamma(E2) = -3.5 \times 10^{-4}$ calculated in [22] agrees well with that computed in [21], and therefore these observables do not distinguish between the two EFT methods.

WEAK INTERACTIONS OF THE DEUTERON

Weak interaction processes involving the deuteron are central to current research efforts in nuclear physics. In addition to the accelerator based programs to elucidate the flavor structure of the nucleon, such as the SAMPLE experiments [25] at Bates, the interactions between neutrinos and the deuteron form the core of our efforts to learn about the neutrino and look beyond the standard model of electroweak interactions. Both in production, e.g. $pp \to de^!\nu_e$, and in detection at SNO (Sudbury Neutrino Observatory), e.g. $\nu_\mu d \to \nu_\mu np$, charged and neutral current weak interaction matrix elements between the deuteron and continuum states are required.

Much effort over the past few decades has been put into calculating the production mechanism $pp \to de^+\nu_e$, both from standard non-relativistic quantum mechanics [26], and from sophisticated potential model techniques [27]. Recently, EFT has been applied to this process by Kong and Ravndal [28] and by Park, Kubodera, Min and Rho [29] giving elegant expressions and numerical values of the weak capture cross section that are consistent with previous estimates[4]. The cross section depends somewhat on the value of a four-nucleon-one-weak-gauge-boson interaction, with coefficient $L_{1,A}$, as defined in [31].

The detection reactions, $\nu d \to np\nu$, $\nu_e d \to e^- pp$ and $\bar\nu_e d \to e^+ nn$ had been looked at by two groups, Ying, Haxton and Henley (YHH) [32] and Kubodera and Nozawa (KN) [33], using sophisticated potential models. The two sets of calculations differ at the 5% level, due to different treatments of meson-exchange currents (MEC). Recently, Butler and Chen [31] have determined the break-up cross sections with EFT. Only $L_{1,A}$ needs to be fixed in order to perform a $\sim 1\%$ calculation. The YYH and KN numerical results can be recovered with different

---

[4]) There has also been recent work in [30] that I have so far failed to comprehend.

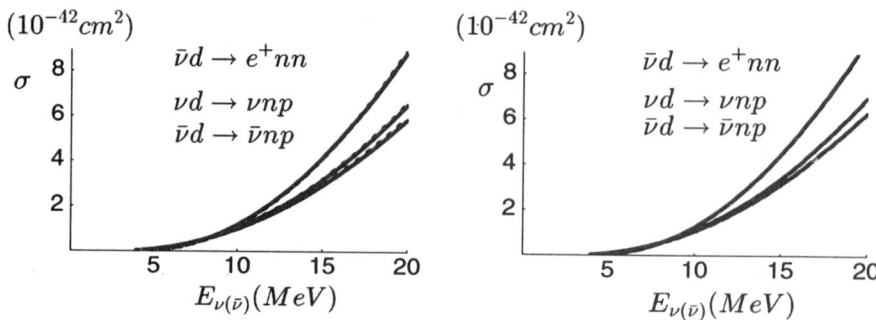

FIGURE 5. Inelastic $\nu(\bar{\nu})d$ cross-sections versus incident $\nu(\bar{\nu})$ energy. The solid curves in the left graph are KN results [33] while the dot-dashed curves, which lie on top of the solid curves, are NLO in EFT with $L_{1,A} = 6.3$ fm^3. The solid curves in the right graph are YHH results [32] while the dashed curves, which also lie on top of the solid curves, are NLO in EFT with $L_{1,A} = 1.0$ fm^3.

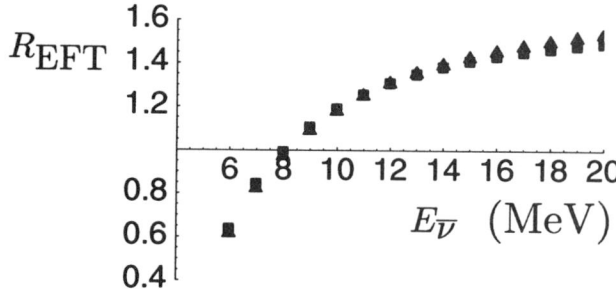

FIGURE 6. The ratio of charged current to neutral current cross sections in $\bar{\nu}d$ scattering versus incident $\bar{\nu}$ energy at NLO in EFT with $L_{1,A} = -20$ fm^3 (boxes) and 40 fm^3 (triangles).

choices of $L_{1,A}$, as shown in Figure 5[5], confirming that the difference between the two potential-model calculations is short-distance in origin. Therefore, to predict the break-up cross section with a precision of better than $\sim 5\%$, one has two options. Firstly, compute the β-decay of tritium, and use this to determine the counterterm $L_{1,A}$ in the EFT, or equivalently the MEC's in the potential models[6]. Secondly, one can perform an experiment to measure one of the break-up cross sections to high accuracy, and thereby extract $L_{1,A}$, or the MEC's. Such an experiment is currently under consideration [34]. An important input into determining if neutrinos are changing flavor as they move out of the sun and to the earth is the ratio of charged current to neutral current cross sections. Figure 6 shows that this ratio, unlike the individual cross sections, is relatively insensitive to the counterterm $L_{1,A}$.

---

[5] I thank Malcolm Butler and Jiunn-Wei Chen for allowing me to reproduce their figures
[6] This method of fixing the MEC's has already been implemented for $pp \to de^+\nu_e$ [27].

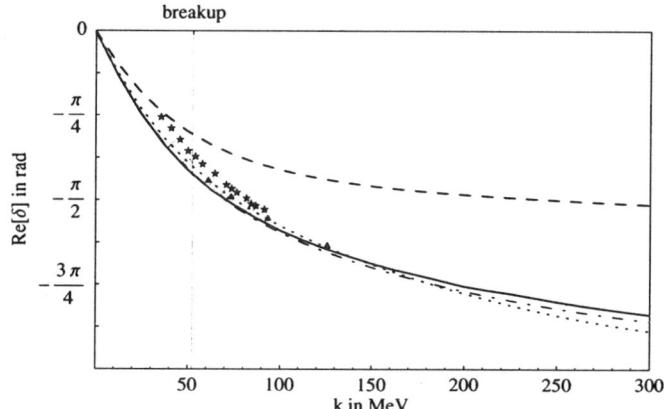

FIGURE 7. $Re(\delta)$ versus incident momentum for Nd scattering in the $J = \frac{3}{2}^+$ channel. The data are taken from the TUNL pd partial wave analysis [44]. The dashed, solid and dotted lines are the LO, NLO and NNLO EFT calculations.

LOW-ENERGY THREE-BODY PROCESSES

Significant progress has been made in understanding three-body systems with EFT [35,36]. A couple of years ago, Bedaque, Hammer and van Kolck[7] showed very clearly that low-energy Nd scattering in the $J = \frac{3}{2}^+$ channel could be described by EFT using contact interactions alone. One of the more impressive results was the calculation of $a_{\frac{3}{2}}$, the scattering length in the quartet s-wave channel. Bedaque and van Kolck calculated the first three terms to be $a_{\frac{3}{2}} = 5.01 + 1.0 + 0.32 + ...$ fm where the ellipses denote higher order contributions, that are estimated to be ± 0.1 fm. The calculated $a_{\frac{3}{2}} = 6.32 \pm 0.1$ fm agrees very well[8] with the experimental $a_{\frac{3}{2}}^{\text{expt}} = 6.35 \pm 0.02$ fm [39]. The first term had been computed in 1957 by Skornyakov and Ter Martirosian [40] while the second term was computed in 1991 by Efimov [41]. The third term had not been computed before and was determined unambiguously from the nucleon-nucleon scattering amplitude [35]. The results of extending this analysis to non-zero energy [35] can be seen in Figure 7 and are found to agree well with data. Scattering in higher partial waves has been examined by Bedaque, Gabbiani and Grießhammer [37] in the theory with only contact interactions between nucleons. A comparison between the EFT calculations[9], sophisticated potential model calculations and data for one partial wave is shown in Figure 8.

---

[7] I thank Paulo Bedaque, Hans-Werner Hammer and Bira van Kolck for allowing me to reproduce their figure.
[8] Subsequent "second generation" potential-model calculations agree with this result [38].
[9] I thank Paulo Bedaque, Fabrizio Gabbiani and Harald Grießhammer for allowing me to reproduce their figure.

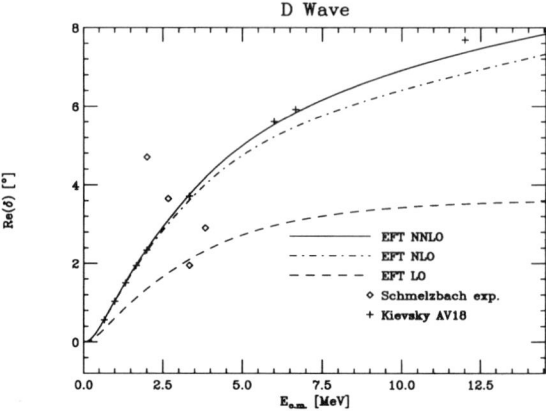

FIGURE 8. $Re(\delta)$ versus incident momentum for Nd scattering in the $L=2$ quartet channel. The dashed, dot-dashed and solid line is LO, NLO and NNLO. Calculations are shown as crosses [42], while the phase shift analysis is shown by open squares and diamonds [43].

Computations in the spin-doublet channel required much more development as local three-nucleon operators can contribute to the scattering amplitude [36]. The scale-dependence of the counterterm (more commonly known as the three-body force) is quite different from those that are familiar from perturbative field theory. Bedaque, Hammer and van Kolck showed that a single momentum independent counterterm could absorb all cut-off dependence from the leading operators resummed by the integral equation that describe three-body systems. The observed periodicity as a function of scale indicates that if a given calculation is performed with a given value of the cut-off, it is possible that the three-body "force" vanishes, while for a different value of the cut-off the three-body "force" may dominate. Further, the appearance of only one three-body counterterm required to render the scattering amplitude scale independent also naturally explains the *Phillips line* found in potential-models.

As a last comment on the three-body work, the techniques that have been developed in the nuclear physics setting are being applied to atomic systems, most notably scattering lengths and recombination rates in Bose condensates [45].

ISSUES AT HIGHER ENERGIES

The situation regarding the applicability of EFT's at higher momentum, near or above the pion mass, is much less clear. In Weinberg's scheme [2], pion exchange and local four-nucleon operators contribute to nucleon-nucleon scattering at the same order in the expansion parameter. A great calculational and conceptual simplification would arise if the exchange of pions between nucleons could be treated in perturbation theory in all partial waves, as suggested by Kaplan, Wise and myself (KSW) [6]. Unfortunately, one of the more disappointing results found during the past year is that the EFT with perturbative pions [6] appears not to be converging [46] (also, see earlier work by Cohen and Hansen [47]). Fleming, Mehen and Stewart [46] performed an analytic calculation of the NNLO amplitude for nucleon-nucleon scattering in the theory with pions and KSW power-counting [6]. They found large non-analytic contributions that appear to destroy the convergence of the series. In contrast, several computations were performed with Weinberg's power-counting [2] for the nucleon-nucleon potential which appears to give converging amplitudes. However, the formal inconsistency of Weinbergs power-counting remains. Amplitudes are not renormalization (cut-off) independent at any order in Weinberg's expansion, however, the cut-off dependence is found to be numerically small when renormalized at a typical strong interaction scale. Therefore, at this point in time there is no formally consistent, converging, perturbative EFT to describe nuclear interactions for momenta of order or higher than the mass of the pion.

To give you an idea of what has been attempted at these higher energies with both Weinberg and KSW power-counting, let me show you the results obtained for $\gamma d \to \gamma d$, deuteron Compton scattering. Three-orders in Weinberg's counting have been completed for $\gamma d \to \gamma d$ at higher energies [48]. The angular distribution of scattering photons at $E_\gamma = 69$ MeV is shown in Figure 9[10]. One can see from

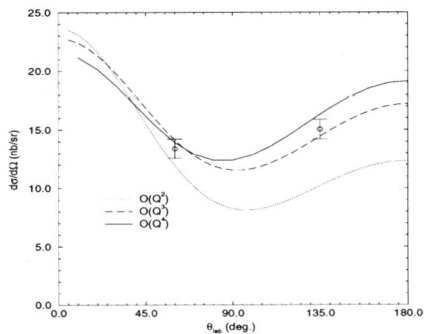

FIGURE 9. $\frac{d\sigma}{d\Omega}$ for $\gamma d \to \gamma d$ for incident photon energy $E_\gamma = 69$ MeV with Weinberg's power-counting [48]. The dotted, dashed and solid curves are LO, NLO and NNLO. Data is from [49].

Figure 9 that the expansion appears to be converging nicely to the experimental values. Similarly, $\gamma d \to \gamma d$ has been computed to two orders in KSW power-counting [50], the results of which can be seen in Figure 10. Very good agreement between data and the parameter free-prediction at NLO is found at $E_\gamma = 49$ MeV. The agreement is somewhat worse at $E_\gamma = 69$ MeV, and does not appear to be approaching the data in the same way that the calculation with Weinberg's counting

---

[10] I thank Daniel Phillips, Silas Beane and Bira van Kolck for allowing me to reproduce their figure.

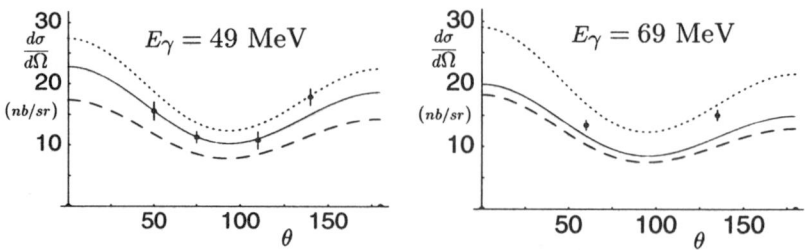

FIGURE 10. $\frac{d\sigma}{d\Omega}$ for γd Compton scattering at incident photon energies of $E_\gamma = 49$ MeV and 69 MeV determined in [50]. The dashed curves are LO. The solid (dotted) curves are NLO with (without) the graphs that contribute to the polarizability of the nucleon. Data is from [49].

appears to. It is clear that higher order calculations must be performed once the perturbative pions versus non-perturbative pions issue is understood, and further, it is clear that more precise data is required at low-energies. It is worth mentioning at this point that neither counting schemes, nor any theoretical calculation that exists at present comes close to reproducing the recent data at $E_\gamma = 95$ MeV [51].

There is much work still to be done in this area.

ON THE ROAD TO NUCLEI

In parallel to the efforts that I have described in the two- and three-body sectors, Haxton and collaborators [52] have been developing techniques to apply the ideas underpinning EFT to the nuclear shell model (efforts are ongoing by others [53] but I will not discuss their work in this talk).

Before I discuss this work I wish to show you the results of a relatively simple but demonstrative calculation by Phillips [54] (see also [55]). To show how the ideas of EFT can be translated into a potential-model mode of thinking, Phillips compared the deuteron quadrupole form factor, presented as T_{20}, computed with the Nijmegen93 potential [56] with that generated by an effective potential, $V_{\text{eff}}^L(r)$ (where L denotes the orbital angular momentum state) and local quadrupole moment counterterm. The effective potential consists of one-pion exchange at long distances and a square-well at short-distances, $V_{\text{eff}}^L(r) = V^{\text{OPE}}(r)$ for $r > R$, and $V_{\text{eff}}^L(r) = V_{0,L}$ for $r < R$. The values of the $V_{0,L}$ are chosen to reproduce the deuteron binding energy, and low-energy nucleon-nucleon scattering for each choice of R. Therefore, the long-distance behavior of the "true" potential and effective potential are identical. Further, the tail of the deuteron wave-function produced by the "true" potential and effective potential are identical. As one is brutalizing the nucleon-nucleon interaction at short-distances while preserving nucleon-nucleon scattering, it is expected that predictions for other observables, such as electromagnetic form factors, will deviate significantly from nature when probing distance scale comparable to or less than R, for reasonable values of R. In addition one expects

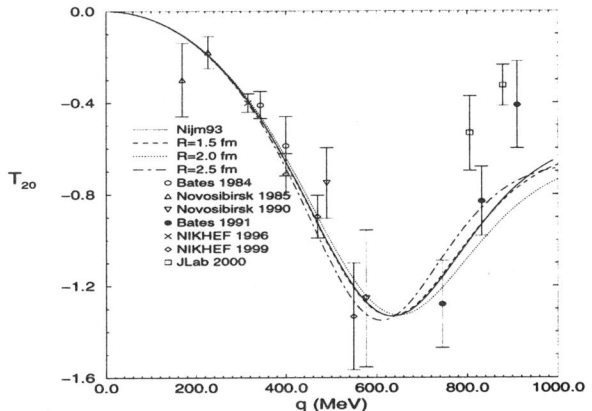

FIGURE 11. T_{20} computed with the Nijmegen93 potential and in the effective theory defined in the text for three values of the spatial cut-off R.

to find that the static moments differ somewhat from nature. It was shown in [9,57] that such short-distance modifications can be compensated by the inclusion of gauge invariant local operators. In the case of the deuteron quadrupole moment it is necessary to introduce a four-nucleon-one-quadrupole-photon operator, that is in no way related to the operators determining nucleon-nucleon scattering. These operators are induced at the chiral symmetry breaking scale, and must be included in any consistent calculation, and in fact, their omission is responsible for the discrepancy between all sophisticated potential model calculations of the deuteron quadrupole moment [58] and its experimental value. Phillips choose a value of this operator to reproduce the observed deuteron quadrupole moment for each value of R, and then predicted T_{20}^{eff} in the effective theory, the results of which are shown in Figure 11[11]. It is clear from Figure 11 that by fixing parameters in the effective theory to reproduce low-energy observables that, even with a nucleon-nucleon potential that has been brutalized at short-distances, one can essentially recover the "true" form factor over a quite impressive range of momentum transfers. This provides a very clear demonstration that T_{20} is determined largely by the tail of the deuteron wavefunction, the OPE tail of the nucleon-nucleon interaction and the deuteron quadrupole moment. For potential model calculations to accurately determine T_{20}, they must first recover the deuteron quadrupole moment, and to do so they must include a four-nucleon quadrupole operator [9,57].

A very similar exploration is ongoing by Haxton and collaborators [52]. They are attempting to construct a model-space-independent shell model, and are presently focusing on the deuteron to optimize their techniques. The underpinnings of EFT are basis independent, and as such should be able to be implemented in both a plane-wave basis (PWB) or a harmonic oscillator basis (HOB). For a continuum process, obviously the PWB is preferable, but for a bound state it seems reasonable

---

[11]) I thank Daniel Phillips for allowing me to reproduce his figure.

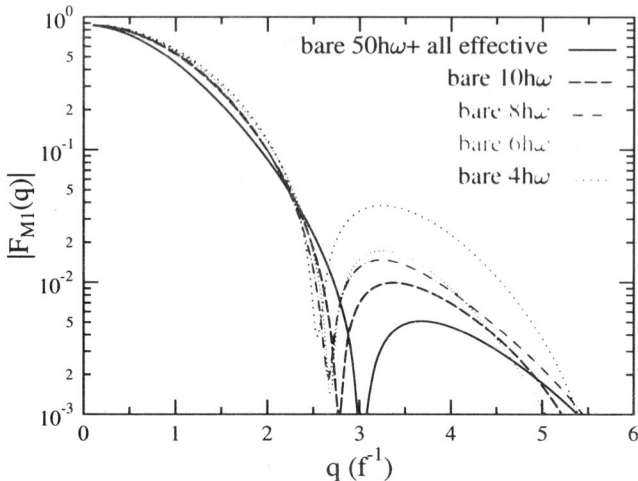

FIGURE 12. The deuteron $M1$ form factor.

that a bound state basis, such as the HOB, will be optimal. In an HOB basis it is natural to integrate out the levels of the HO step by step to allow for an easier calculation in a reduced model-space. As each HO level is removed, the hamiltonian and coefficients of gauge invariant operators are redefined to preserve observables. This is a discrete analog of the renormalization group (RG) implemented in the PWB. In Figure 12[12] the magnetic form factor of the deuteron $F_{M1}(q)$ is shown versus momentum transfer, where the $n = 50$ calculation provides the "true" calculation. The other curves correspond to the same calculation, but with the insertion of the bare $M1$ operator in each reduced model space. The solid line, however, is not just the $n = 50$ calculation but also the calculation from ALL reduced model-spaces when the renormalized $M1$ operator is inserted and NOT the bare operator. Clearly, a discrete RG can be implemented in an HOB.

One of the advantages of constructing a discrete RG for the nuclear shell model would be to greatly reduce the computer time required to compute matrix elements in a nucleus with $A \gg 2$. Presently, efforts are being made to reduce the shell-model space for the deuteron down from $n = 140$, which reproduces the deuteron binding energy perfectly (by construction), down to $n = 20$ or 30 and faithfully reproduce all deuteron observables. Part of the effort is to include the short-range part of the nucleon-nucleon interaction by local operators. If successful, this program will allow for high precision computations of nuclear properties, with greatly increased speed.

---

[12] I thank Wick Haxton for allowing me to reproduce his figure.

DISCUSSION

I have tried to give you an overview of a number of important developments of the last year or so. At low-energies, high precision calculations, $\sim 1\%$ have been performed in the two-body sector, and progress is being made toward calculations of similar precision in the three-body sector.

At somewhat higher energies, a formally consistent and converging EFT describing nucleon-nucleon interactions is yet to be uncovered. Weinberg's power-counting is formally ill-defined, yet gives numerical results that appear to be converging. In contrast, KSW power-counting is formally well-defined, yet appears not to be converging! I suspect, as do others, that some sort of union between the two power-countings may in fact be the correct one, but this is merely speculation.

Some of the more interesting developments this year were made in the implementation of EFT ideas in nuclear many-body calculations. There are indications that the EFT techniques may provide a means to compute properties of nuclei presently beyond reach. However, more work is required before any conclusions can be drawn.

REFERENCES

1. G. P. Lepage, Summary talk presented at the INT workshop on *Effective Field Theory in Nuclear Physics II*, edited by P.F. Bedaque, M.J. Savage, R. Seki and U. van Kolck, ISBN-981-02-4181-X.
2. S. Weinberg, *Phys. Lett.* **B251**, 288 (1990); *Nucl. Phys.* **B363**, 3 (1991); *Phys. Lett.* **B295**, 114 (1992).
3. C. Ordonez and U. van Kolck, *Phys. Lett.* **B291**, 459 (1992); C. Ordonez, L. Ray and U. van Kolck, *Phys. Rev. Lett.* **72**, 1982 (1994); *Phys. Rev.* **C53**, 2086 (1996); U. van Kolck, *Phys. Rev.* **C49**, 2932 (1994).
4. T.S. Park, D.P. Min and M. Rho, *Phys. Rev. Lett.* **74**, 4153 (1995); *Nucl. Phys.* **A596**, 515 (1996).
5. D.B. Kaplan, M.J. Savage and M.B. Wise, *Nucl. Phys.* **B478**, 629 (1996).
6. D.B. Kaplan, M.J. Savage and M.B. Wise, *Phys. Lett.* **B424**, 390 (1998); *Nucl. Phys.* **B534**, 329 (1998).
7. H. A. Bethe, *Phys. Rev.* **76**, 38 (1949); H. A. Bethe and C. Longmire, *Phys. Rev.* **77**, 647 (1950).
8. H. P. Noyes, *Nucl. Phys.* **74**, 508 (1965).
9. J.-W. Chen, G. Rupak and M. J. Savage, *Nucl. Phys.* **A653** 386 (1999).
10. S. Burles, K. M. Nollet, J. W. Truran and M. S. Turner, *Phys. Rev. Lett.* **82**, 4176 (1999).
11. ENDF online database at the NNDC Online Data Service, http://www.nndc.bnl.gov.
12. D. R. Phillips, G. Rupak, M. J. Savage *Phys. Lett.* **B473**, 209 (2000).
13. J.-W. Chen and M. J. Savage, *Phys. Rev.* **C60**, 065205 (1999).
14. G. Rupak, nucl-th/9911018.

15. T.-S. Park, D.-P. Min and M. Rho, *Phys. Rev. Lett.* **74**, 4153 (1995); *Nucl. Phys.* **A596** 515 (1996).
16. T.-S. Park, K. Kubodera, D.-P. Min, and M. Rho, *Astrophys. Jour.* **507**, 443 (1998); *Phys. Rev.* **C58**, R637 (1998).
17. C. Ordonez and U. van Kolck, *Phys. Lett.* **B291**, 459 (1992); C. Ordonez, L. Ray and U. van Kolck, *Phys. Rev. Lett.* **72**, 1982 (1994); *Phys. Rev.* **C53**, 2086 (1996); U. van Kolck, *Phys. Rev.* **C49**, 2932 (1994).
18. T. M. Muller, *Private Communication.*
19. M. S. Smith, L. H. Kawano and R. A. Malaney, *Astrophys. J. Suppl. Ser.* **85** 219 (1993).
20. A.E. Cox, S.A.R. Wynchank and C.H. Collie, *Nucl. Phys.* **74**, 497 (1965).
21. T.-S. Park, K. Kubodera, D.-P. Min, and M. Rho, *Phys. Lett.* **B472**, 232 (2000).
22. J.-W Chen, G. Rupak and M. J. Savage, *Phys. Lett.* **B464**, 1 (1999).
23. A. N. Bazhenov et al., *Phys. Lett.* **B289**, 17 (1992).
24. A. P. Burichenko and I. B. Khriplovich, *Nucl. Phys.* **A515**, 139 (1990).
25. D. T. Spayde et al. (SAMPLE Collaboration) *Phys. Rev. Lett.* **84**, 1106 (2000).
26. J. N. Bahcall and R. M. May, *Ap. J* **55**, 501 (1969).
27. R. Schiavilla et al., *Phys. Rev.* **C58** 1263 (1998).
28. X. Kong and F. Ravndal, nucl-th/0004038; *Nucl. Phys.* **A665** 137 (2000); *Nucl. Phys.* **A656** 421 (1999); *Phys. Lett.* **B470**, 1 (1999).
29. T.- S. Park, K. Kubodera, D.- P Min and M. Rho, *Nucl. Phys.* **A646**, 83 (1999).
30. A. N. Ivanov, H. Oberhummer, N. I. Troitskaya, and M. Faber nucl-th/9910021 and references therein.
31. M. N. Butler and J.-W Chen, nucl-th/9905059.
32. S. Ying, W. C. Haxton and E. M. Henley, *Phys. Rev.* **C45**, 1982 (1992); *Phys. Rev.* **D40**, 3211 (1989).
33. K. Kubodera and S. Nozawa, *Int. J. Mod. Phys.* **E3**, 101 (1994); Y. Kohyama and K. Kubodera, USC(NT)-report-92-1, unpublished.
34. F. Avignone, private communication.
35. P. F. Bedaque and H. W. Grießhammer, *Nucl. Phys.* **A671**, 357 (2000); P.F. Bedaque, H.W. Hammer and U. van Kolck, *Phys. Rev. Lett.* **82**, 463 (1999); *Nucl. Phys.* **A646**, 444 (1999); *Phys. Rev.* **C58**, R641 (1998). P.F. Bedaque and U. van Kolck, *Phys. Lett.* **B428**, 221 (1998).
36. P.F. Bedaque, H.W. Hammer and U. van Kolck, nucl-th/9906032.
37. F. Gabbiani, P. F. Bedaque and H. W. Grießhammer, nucl-th/9911034.
38. J. L. Friar, D. Huber, H. Witala and G. L. Payne, *Acta Phys. Polon.* **B31**, 749 (2000).
39. W. Dilg, L. Koester and W. Nistler, *Phys. Lett.* **B36**, 208 (1971).
40. G. V. Skornyakov and K. A. Ter-Martirosian, *Sov. Phys. JETP* **4**, 648 (1957).
41. V. Efimov, *Phys. Rev.* **C47**, 1876 (1993).
42. A. Kievsky, S. Rosati, W. Tornow and M. Viviani, *Nucl. Phys.* **A607**, 402 (1996).
43. E. Huttel, W. Arnold, H. Baumgart, H. Berg and G. Clausnitzer, *Nucl. Phys.* **A406**, 443 (1983); P. A. Schmelzbach, W. Grubler, R. E. White, V. Konig, R. Risler and P. Marmier, *Nucl. Phys.* **A197**, 273 (1972).
44. W. Tornow and H. Witala, "Proton-Deuteron Phase-Shift Analysis above the Deuteron Breakup Threshold", Technical Report TUNL XXXVI (1996-97).

45. P. F. Bedaque, E. Braaten and H. W. Hammer, cond-mat/0002365. P. F. Bedaque, H. W. Hammer, and U. van Kolck, *Nucl. Phys.* **A646**, 444 (1999).
46. S. Fleming, T. Mehen and I. W. Stewart, nucl-th/9911001; *Phys. Rev.* **C61**, 044005 (2000).
47. T. D. Cohen and J. M. Hansen, nucl-th/9908049; *Phys. Rev.* **C59**, 3047 (1999); *Phys. Rev.* **C59**, 13 (1999); T. D. Cohen, nucl-th/9904052
48. S. R. Beane, M. Malheiro, D. R. Phillips and U. van Kolck, *Nucl. Phys.* **A656**, 367 (1999); private communication.
49. M. A. Lucas, Ph. D. thesis, University of Illinois at Urbana-Champaign (1994)
50. J.-W. Chen, H. W. Grießhammer, M. J. Savage and R. P. Springer, *Nucl. Phys* **A644**, 245 (1998).
51. D. L. Hornidge *et al.*, *Phys. Rev. Lett.* **84**, 2334 (2000).
52. W. C. Haxton and C. L. Song, nucl-th/9907097; nucl-th/9906082.
53. E. Epelbaoum, W. Glockle and U.-G. Meissner, *Nucl. Phys.* **A671**, 295 (2000); *Few Body Syst. Suppl.* **10**, 479 (1999); *Phys. Lett.* **B439**, 1 (1998); E. Epelbaoum, W. Glockle, A. Kruger and U.-G. Meissner, *Nucl. Phys.* **A645**, 413 (1999); M. Lutz, 9906028; H. W. Hammer and R. J. Furnstahl, nucl-th/0004043.
54. D. R. Phillips, nucl-th/0004060.
55. D. R. Phillips and T. D. Cohen, *Nucl. Phys.* **A668**, 45 (2000).
56. V. G. J. Stoks, R. A. M. Klomp, C. P. F. Terheggen and J. J. de Swart, *Phys. Rev.* **C49**, 2950 (1994).
57. D. B. Kaplan, M. J. Savage and M. B. Wise, *Phys. Rev.* **C59**, 617 (1999).
58. R. Machleidt, nucl-th/0006014; R. B. Wiringa, V.G.J. Stoks, R. Schiavilla *Phys. Rev.* **C51**, 38 (1995).

Parity Violation and Nucleon Strangeness

Paul A. Souder[1]

Syracuse University, Syracuse, NY 13244

Abstract.
Parity violation in the scattering of polarized electrons is a unique tool for nuclear and particle physics that is presently being exploited at JLab, Mainz, MIT-Bates, and SLAC. One focus of these experiments is the measurement of the contribution of strange quarks to the elastic form factors of the nucleon. Recently, results on this topic from MIT-Bates and JLab have been reported. Future experiments, motivated by the search for strange form factors as well as testing the Standard Model and measuring the radius of the neutron distribution, are described.

INTRODUCTION

An open question in the structure of the nucleon is the detailed role of strangeness. It is known that about 3% of the momentum of the nucleon is carried by strange quarks [1], and analyses of data on spin structure functions [2–5] suggest that strange quarks also carry some of the spin. Based on early spin structure function data [6], Kaplan and Manohar [7] suggested that strange quarks may also contribute to the elastic nucleon form factors.

One way to observe the strange form factors is to measure the parity violating asymmetry in the scattering of polarized electrons from an unpolarized target [8–10]:

$$A^{PV} = \frac{\sigma_R - \sigma_L}{\sigma_R + \sigma_L}, \tag{1}$$

where $\sigma_{L(R)}$ is the cross section for the scattering of left(right) handed electrons. The cross section for unpolarized targets has contributions from three amplitudes, the electromagnetic amplitude f^γ, and the weak amplitudes f_V^Z, and f_A^Z shown in Fig. 1. The parity violating asymmetry isolates the weak amplitudes. In the Standard Model, the axial coupling of the electron to the Z is large, so f_V^Z is also large and provides a useful tool for measuring the weak vector current of the target.

---

[1] Work supported by the DOE under contract number DE-FG02-84ER40146

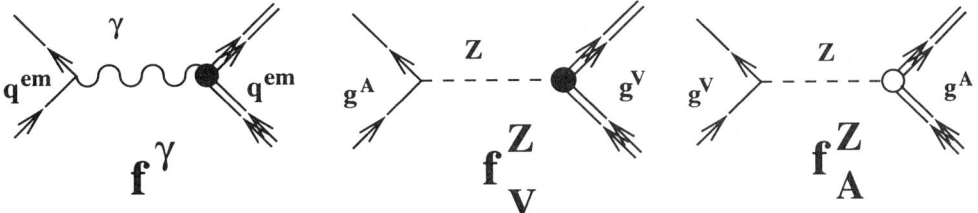

FIGURE 1. Feynman diagrams for electroweak scattering. Since $g^A \gg g^V$ in the Standard Model, f_V^Z is usually the largest parity-violating term.

On the other hand, g^V for the electron and thus f_A^Z is small, so experiments with kinematics chosen to emphasize this term have potential to measure electroweak radiative corrections or search for physics beyond the Standard Model.

A large number of experiments, listed in Table 1, have been completed [11–17] or are in progress [18–25] that exploit parity violation in polarized electron scattering. These experiments have a variety of physics goals as listed in the table. The most common goal recent work is the measurement of strange form factors.

THEORY OF STRANGE QUARKS IN ELASTIC SCATTERING

The spin-independent cross section for elastic scattering from the nucleon with four-momentum transfer Q^2 is

$$\frac{d\sigma}{d\Omega} = \left(\frac{d\sigma}{d\Omega}\right)_{Mott} \left[(F_1^{\gamma N})^2 + \tau(F_2^{\gamma N})^2 + 2\tau(F_1^{\gamma N} + F_2^{\gamma N})^2 \tan^2(\theta/2)\right], \quad (2)$$

where θ is the scattering angle, $\tau = Q^2/4M_N^2$, and M_N is the mass of the nucleon. The quantities $F_{1,2}^{\gamma N}$ are the Dirac elastic electromagnetic form factors, which are functions of Q^2. The form factors may be written in terms of contributions from individual quarks as follows:

$$F_{1,2}^{\gamma p} = \frac{2}{3} F_{1,2}^u - \frac{1}{3} F_{1,2}^d - \frac{1}{3} F_{1,2}^s \quad (3)$$

$$F_{1,2}^{\gamma n} = \frac{2}{3} F_{1,2}^d - \frac{1}{3} F_{1,2}^u - \frac{1}{3} F_{1,2}^s \quad (4)$$

where the second relation follows from charge symmetry. Weak scattering from the proton, in which the Z-boson is exchanged, is governed by the weak form factor

$$F_{1,2}^{Zp} = F_{1,2}^u - F_{1,2}^d - F_{1,2}^s - 4\sin^2\theta_W F_{1,2}^{\gamma p}. \quad (5)$$

Once the electromagnetic form factors for both the proton and neutron as well as the weak form factors F_i^{Zp} are measured, the strange form factors $F_{1,2}^s$ may be determined from the above equations.

It is more convenient to use the Sachs form factors: $G_E^i = F_1^i - \tau F_2^i$, and $G_M^i = F_1^i + F_2^i$. In terms of them, the parity-violating asymmetry for the proton is given by [26]:

$$A^{PV} = \left[\frac{-G_F M_p^2 \tau}{\pi \alpha \sqrt{2}}\right]\left\{(1 - 4\sin^2\theta_W) - \right. \tag{6}$$

$$\frac{[\varepsilon G_E^{p\gamma}(G_E^{n\gamma} + G_E^s) + \tau G_M^{p\gamma}(G_M^{n\gamma} + G_M^s)]}{\varepsilon(G_E^{p\gamma})^2 + \tau(G_M^{p\gamma})^2} -$$

$$\left. \frac{(1 - 4\sin^2\theta_W)\sqrt{\tau(1+\tau)}\sqrt{1-\varepsilon^2}G_M^{p\gamma}(-G_A^{(1)} + \frac{1}{2}F_A^s)}{\varepsilon(G_E^{p\gamma})^2 + \tau(G_M^{p\gamma})^2}\right\}$$

where $\varepsilon = [1 + 2(1+\tau)\tan^2(\theta/2)]^{-1}$ is the transverse polarization of the virtual photon exchanged, and F_A^s is the strange quark contribution to the axial form factor. Here the result is given explicitly in terms of the measured electromagnetic form factors and the unknown strange form factors.

TABLE 1. Survey of parity experiments using polarized electrons.

Experiment	Reaction	Physics Goals	A^{PV}
Completed Experiments			
SLAC E122 [11]	$\vec{e}D$ (DIS)	PV of Z	10^{-4}
Mainz [12]	$\vec{e}\,^9$Be QE	New Physics	10^{-5}
Bates [13]	$\vec{e}\,^{12}$C Elastic	New Physics	10^{-6}
SAMPLE-P [14]	$\vec{e}P$ Elastic	$G_M^s(0) = \mu_s$	10^{-5}
SAMPLE-D [15]	$\vec{e}D$ Elastic	G_A^e	10^{-5}
HAPPEX(JLab) [16,17]	$\vec{e}P$ Elastic	$G_M^s + 0.39 G_E^s$	10^{-5}
Approved Experiments			
Mainz [18]	$\vec{e}P$ Elastic	G_M^s, G_E^s	10^{-5}
G^0(JLab) [19]	$\vec{e}P$ Elastic	G_M^s, G_E^s	10^{-5}
^{4}He(JLab) [20]	$\vec{e}\,^4$He Elastic	G_E^s	10^{-5}
Moller(SLAC) [21]	$\vec{e}e$	New Physics	10^{-7}
HAPPEX II(JLab) [22]	$\vec{e}P$ Elastic	$G_M^s + 0.39 G_E^s$	10^{-6}
Pb(JLab) [23]	$\vec{e}$ Pb	Neutron Radius	10^{-6}
SAMPLE-D [24]	$\vec{e}D$ Elastic	G_A^e	10^{-5}
HAPPEX ^{4}He [25]	$\vec{e}\,^4$He Elastic	G_E^s	10^{-5}
Possible new experiments			
H(JLab)	$\vec{e}P$	New Physics	10^{-7}

The physics that may be learned from Eqn. 6 depends upon the kinematics. For small values of both τ and θ, the asymmetry is independent of form factors and is sensitive only to the fundamental weak interaction. A^{PV} is primarily sensitive to G_E^s and G_M^s for small θ but large τ, and primarily to G_M^s and G_A^e for large values of θ. Thus the experiments in Table 1 cover a wide range of kinematics.

The radiative corrections for the terms involving vector form factors are small and known [27]. On the other hand, the radiative correction for the axial term has substantial uncertainties [28]. The SAMPLE collaboration defines a quantity $G_A^e(T=0)$ which includes all radiative corrections and replaces $G_A^{(1)}$ in the above equation.

A number of papers have made predictions for the sizes of the form factors [29–42]. The predictions may be expressed in terms of the parameters ρ_s and μ_s which are the low Q^2 limits

$$G_E^s \to \tau \rho_s : \quad G_M^s \to \mu_s. \tag{7}$$

Here $G_E^s \to 0$ for $Q^2 \to 0$ because the net strangeness of the nucleon is zero. A summary of the predictions is given in Table 2. None of the predictions to date are particularly reliable, but they do give a reasonable flavor for the rough size that the strange form factors might possibly be. Reliable calculations are the goal of ongoing work. Lattice calculations are an especially promising approach.

EXPERIMENTAL METHODS

A number of experimental features are common to all polarized electron parity experiments. A typical apparatus is shown in Fig. 2. The heart of the experiments is the polarized electron source. Photoemission of circularly polarized laser light from a GaAs cathode produces the polarized electrons. The helicity of the electron beam is determined by the helicity of the light which is in turn determined by the high voltage on a Pockels cell. The polarized electrons are accelerated and then pass through a beam line. The beam line is highly instrumented to precisely measure any small correlations between beam parameters that influence the cross section, such as position and angle, with the helicity. Finally, the beam strikes the target and the scattered electrons are detected by a spectrometer. A wide variety of techniques are used for spectrometers to accommodate various kinematics for the different experiments. A computer monitors all the signals and records the data. In addition, the computer controls the Pockels cell to null any intensity asymmetry and controls the coils used to calibrate the sensitivity both of the monitors and of the spectrometers to beam parameters.

Since the measured asymmetries are small, between $10^{-4} - 10^{-7}$ depending on the experiment, it is essential to keep the effects of helicity correlations on the cross section due to any other beam parameter small. Innovative techniques have been developed [43] to maintain the helicity correlations at acceptable levels. The effect

TABLE 2. Some published theoretical estimates for ρ_s and μ_s.

Method	ρ_s	μ_s	Author
Pole fits	-2.1 ± 1.0	-0.31 ± 0.09	Jaffe [29]
	-2.9 ± 0.5	-0.24 ± 0.03	Hammer [30]
Kaon Loops	0.2	-0.03	Koepf [31]
	0.5 ± 0.1	-0.35 ± 0.05	Ramsey-Mulolf [32]
	0.3	-0.12	Ito [33]
Unquenched Quarks	0.6	0.04	Geiger [34]
Meson Exchange	0.03	0.002	Meissner [35]
Meson Cloud	–	0.-0.066	Ma [36]
NJL	3.0 ± 0.08	-0.15 ± 0.10	Weigel [37]
Skyrme	1.6	-0.13	Park [38]
	-0.7	-0.05	Park [39]
Chiral Bag Model	–	0.37	Hong [40]
Lattice QCD	1.7 ± 0.7	-0.36 ± 0.20	Dong [41]
Dispersion Relations	0.99	-0.42	Hammer [42]

of the beam parameters on the number of detected events may be calibrated by dithering steering coils in the beamline.

Recently, impressive progress has been made in the technology required for these challenging experiments. Polarized electron sources with intensities of $100\mu A$ and polarizations of $> 70\%$ have been achieved. Liquid hydrogen and deuterium targets have been developed [44] that can withstand the hundreds of watts that the high intensity beams deposit.

The helicity of the beam is reversed rapidly, typically between 15 and 300 times a second. For each window of beam with a given helicity, the flux in the detector D and the intensity I in the beam monitor are determined. For pulsed accelerators, including MIT-Bates and SLAC, the signals are integrated to accommodate the high instantaneous rates. For continuous wave accelerators including JLab and Mainz, either integrating or counting methods may be chosen. For each pair of pulses with opposite helicity, an asymmetry

$$A_{pair} = \frac{D^R/I^R - D^L/I^L}{D^R/I^R + D^L/I^L} \qquad (8)$$

is measured. For these experiments, A_{pair} can have ideal statistical behavior. One example, shown in Figure 3, shows that the distribution is Gaussian over many decades. The final raw experimental asymmetry A_{raw} is typically the average of $10^7 - 10^8$ values of A_{pair}.

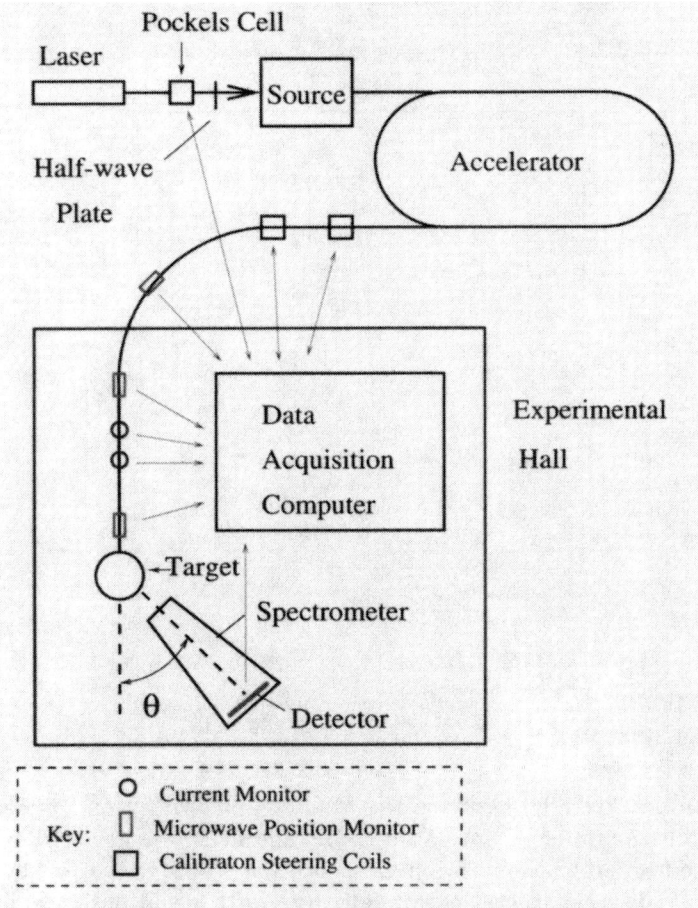

FIGURE 2. Generic Polarized Electron Parity Experiment

RECENT RESULTS

Results from the HAPPEX collaboration were presented at this conference by R. Holmes. The experiment took place at JLab using the high-resolution spectrometers in Hall A. The spectrometers provided excellent rejection of backgrounds. Electrons of energy 3.5 GeV scattered by 12.5° were detected. The average value of Q^2 was 0.477 $(GeV/c)^2$.

The HAPPEX experiment took place in two runs. The first was in 1998 [16] and the second in 1999 [17]. The 1998 run used a bulk GaAs crystal which provided a polarization was ∼40% and a current on target of ∼ 100μA. For the 1999 data, a strained GaAs crystal was used which produced a beam with ∼70% polarization and a current of typically 40μA.

For the 1998 run, helicity correlations in the beam parameters were still negli-

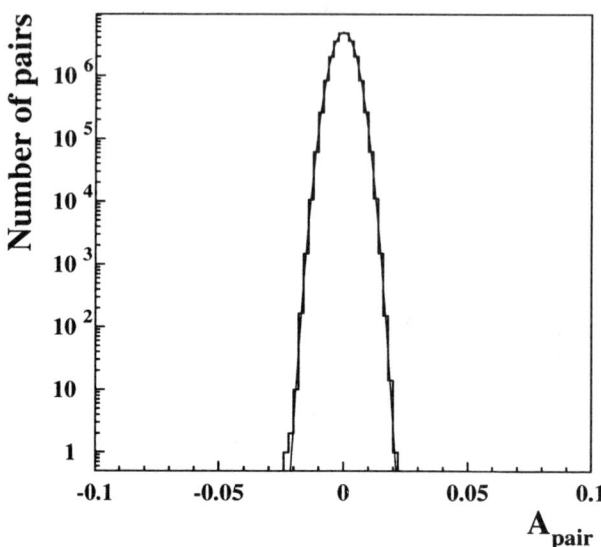

FIGURE 3. Distribution of asymmetries from individual pairs of pulses. Data are from the HAPPEX experiment.

gible. The intensity difference was less than 1 ppm. Position differences at the target were on the order of only a few nm. The helicity-correlated energy difference was also negligible; the position difference was <30 nm at a point on the beam line where the dispersion was ∼ 5m. The helicity correlations were larger in the 1999 run, mostly due to the large analyzing power of the strained GaAs crystal. However, the net effect of these correlations on the result was negligible.

The experimental asymmetry for the combined 1998 and 1999 was [17]

$$A^{PV} = -14.6 \pm 0.9 \pm 0.5 \text{ ppm} \tag{9}$$

where the first error is statistical and the second systematic.

Combining the asymmetry data with Eqn. 6 and with published data on electromagnetic form factors [45–52] gives the result

$$(G_E^s + 0.39 G_M^s)/(G_M^{p\gamma}/\mu_p) = 0.09 \pm 0.05 \pm 0.04 \tag{10}$$

where the first error is the experimental uncertainties added in quadrature and the second error arises from uncertainties in the electromagnetic form factor data. This number is consistent with negligible strange form factors. To study the data in the context of the models given above, we assume that the Q^2 dependence of the form factors is

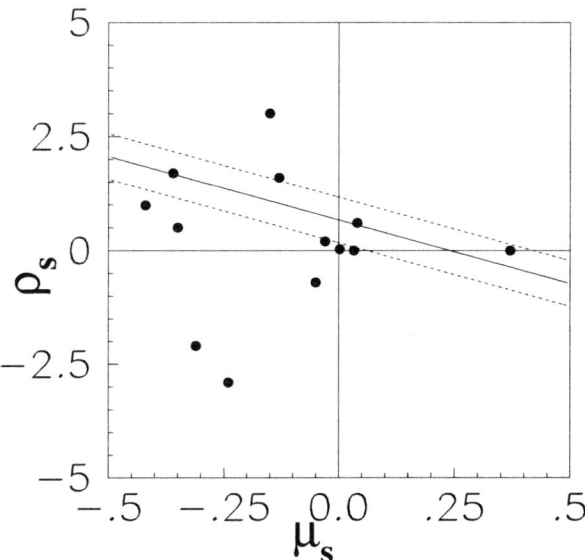

FIGURE 4: Band shows allowed region in the $\rho_s - \mu_s$ parameter space based on the HAPPEX data assuming that the approximation of Eqn. 7 is valid near $Q^2 = 0.48$ (GeV/c)^2.

$$G_E^s \sim \tau \rho_s G_M^{\gamma p}/\mu_p : \quad G_E^s \sim \mu_s G_M^{\gamma p}/\mu_p \qquad (11)$$

to obtain the band shown in Figure 4.

We note that a prerequisite for improving the limits on strange form factors, in addition to better parity data, is improved data on electromagnetic form factors. Several such experiments are in progress.

The SAMPLE collaboration at the MIT-Bates Laboratory is in the midst of a program of experiments measuring the asymmetry for electrons scattered from both hydrogen and deuterium by an angle of $\sim 130°$ and $Q^2 \sim 0.1$ (GeV/c)^2. A large solid angle of 1.5 sr is achieved with an array of ellipsoidal mirrors that focus the Čerenkov light produced by the scattered electrons in the air onto phototubes. Since the beam energy is only 200 MeV, inelastic events are below threshhold.

The result of the proton data is [14]

$$A^{PV} = -4.92 \pm 0.61 \pm 0.73 \text{ ppm} \qquad (12)$$

where the first error is statistical and the second is systematic. By using Eqn. 6, the results can be interpreted in terms of G_M^s and the radiatively corrected axial form factor G_A^e as shown in Figure 5. Also shown in the figure are the results from the recent deuterium run presented at this conference by T. M. Ito [15]. The slope

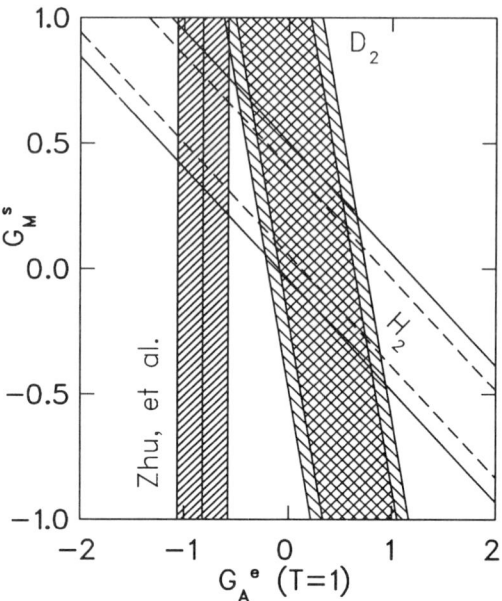

FIGURE 5. Results from the SAMPLE collaboration in terms of G_M^s and the radiatively corrected axial form factor $G_A^e(T=1)$. The bands labeled H_2 and D_2 are extracted from the data. The vertical band is a theoretical estimate for $G_A^e(T=1)$.

of the band from the deuterium data is steeper because of partial cancellation of G_M^s. The region where the bands intersect is consistent with $G_M^s = 0$ but slightly inconsistent with the theoretical prediction for G_A^e [28] shown as the vertical band in the figure. A new SAMPLE run [24] will provide additional deuterium data in the year 2001.

Future Experiments

A number of upcoming experiments will provide more data on strange form factors. An experiment at Mainz by the A4 collaboration [18] will measure elastic scattering from hydrogen at $\theta = 35°$ and $Q^2 = 0.23$ $(\text{GeV}/c)^2$. A unique feature of the experiment is the detector. It will count individual events with an array of 1022 tapered PbF_2 crystals. Identification of elastic events is achieved by the excellent resolution of the calorimeter. Initial data-taking is scheduled for the summer of 2000.

The G_0 experiment [19] is a major program at JLab to study strange form factors over a wide kinematic range. The unique feature is a large superconducting toroidal magnet that will serve as the spectrometer. In one configuration, the recoil protons from elastic scattering with $62° < \theta < 78°$ will be detected with plastic scintillators.

This kinematics, with a beam energy of 3 GeV, corresponds to electrons with $15° > \theta > 5°$ and $0.16 < Q^2 < 0.95$ (GeV/c)^2. In another configuration, electrons scattered with large angles will be detected. Runs with various beam energies between 0.3 and 0.9 GeV will cover the same Q^2 range as the forward data. The goal is to determine G_E^s and G_M^s separately over a large range of Q^2. Initial running is scheduled for the fall of 2001.

An extension of the HAPPEX experiment, called HAPPEX II [22], will measure elastic scattering from a hydrogen target at $Q^2 \sim 0.1 (\text{GeV}/c)^2$. The new run is motivated by the possibility that the strange form factors are proportional to Q^2 only at low values of Q^2, but fall off much faster at the HAPPEX kinematics. The small angle will be attained with septum magnets that allow the spectrometers in Hall A to reach angles as small as 6°.

Strange form factors may also be measured by elastic scattering from ^{4}He. Since the target is spinless, there is no axial hadronic contribution or uncertain radiative corrections. The asymmetry is given by [10,53]

$$A^{PV} = \frac{G_F Q^2}{\pi \alpha \sqrt{2}} \left[\sin^2 \theta_W + \frac{G_E^s}{2(G_E^p + G_E^n)} \right]. \tag{13}$$

It is sensitive only to G_E^s, whereas for hydrogen the asymmetry depends upon a combination of G_E^s and G_M^s. Two experiments have been approved at JLab, one at $Q^2 \sim 0.6$ (GeV/c)^2 and the other at $Q^2 \sim 0.1$ (GeV/c)^2.

Presently, experiments measuring parity violation with polarized electrons with goals other than measuring strange form factors are underway. One example is an experiment at JLab which will measure A^{PV} for elastic scattering from ^{208}Pb [23]. This will measure the radius of the neutron distribution in Pb with the same ease of interpretation as electromagnetic scattering data is for the charge radius [54,55]. The basic idea is that the Z couples mainly to neutrons in the same way that the photon couples to protons. The kinematics chosen is $\theta = 6°$ and $E=850$ MeV. The asymmetry will be measured to $\sim \pm 3\%$ in order to provide a 1% measure of the radius of the neutron distribution.

Another example is experiment E158 [21], which is presently being installed at SLAC. The goal of this project is to perform the most precise measurement of $\sin^2 \theta_W$ at low Q^2, and determine the running of $\sin^2 \theta_W$ [57]. The experiment is sensitive to physics beyond the Standard Model [56], such as a new force, such as a neutral current mediated by a new Z boson with a mass as high as 1 TeV. Another possibility is compositeness of the electron characterized by a new strong interaction with a scale $\Lambda < 15$ TeV.

The experiment, with a predicted asymmetry of only $A^{PV} = 0.32$ ppm, is particularly challenging. A 10 μA 48 GeV beam will scatter from a 1.5 m long liquid hydrogen target. A spectrometer based on quadrupoles will accept most of the possible solid angle for the desired kinematics. The plan is to obtain first data in the year 2001 and ultimately achieve a precision of 7% in the asymmetry.

CONCLUSIONS

Tremendous progress has been made recently in the field of the measurement of parity-violating asymmetries in the scattering of polarized electrons from unpolarized targets. A number of precise results have been published recently, and more precise data is expected soon. The experiments have already set significant limits on the size of strange elastic form factors of the nucleon. Future applications of parity violation also include tests of the Standard Model and a measurement of the radius of the neutron distribution in heavy nuclei.

REFERENCES

1. A. O. Bazarko et al., Z. Phys. C **65**, 189 (1995).
2. K. Abe et al., Phys. Lett. B **405**, 180 (1997).
3. B. Adeva et al., Phys. Rev. D **58**, 112002 (1998).
4. G. Alterelli et al., Acta Phys. Polon. B **29**, 1145 (1998).
5. E. Leader et al., Phys. Lett. B **462**, 189 (1999).
6. J. Ashman et al., Phys. Lett. B **206**, 364 (1988); Nucl. Phys. B **328**, 1 (1989).
7. D. B. Kaplan and A. Manohar, Nucl. Phys. B **310**, 527 (1988).
8. R. D. McKeown, Phys. Lett. B **219**, 140 (1989).
9. E. J. Beise and R. D. McKeown, Comments Nucl. Part. Phys. **20**, 105 (1991).
10. D. H. Beck, Phys. Rev. D **39**, 3248 (1989).
11. C. Y. Prescott et al., Phys. Lett. B **84**, 524 (1979).
12. W. Heil et al., Nucl. Phys. B **327**, 1 (1989).
13. P. A. Souder et al., Phys. Rev. Lett. **65**, 694 (1990).
14. D. T. Spayde et al., Phys. Rev. Lett. **84**, 1106 (2000).
15. T. M. Ito, these proceedings.
16. K. Aniol et al., Phys. Rev. Lett. **82**, 1096 (1999).
17. K. Aniol et al., nucl-ex/0006002.
18. Mainz proposal A4/1-93 (D. von Harrach, spokesperson).
19. JLab experiment 91-017 (D. Beck, spokesperson).
20. JLab experiment 91-004 (E. J. Beise, spokesperson).
21. SLAC experiment E158 (K. S. Kumar, spokesperson, E. W. Hughes and P. A. Souder, deputy spokespersons).
22. JLab experiment 99-115 (K. S. Kumar and D. Lhuillier, spokespersons).
23. JLab experiment 99-012 (R. Michaels and P. A. Souder, spokespersons).
24. MIT-Bates experiment 00-04, T. M. Ito, spokesperson.
25. JLab experiment 00-114 (D. S. Armstrong and R. Michaels, spokespersons).
26. M. J. Musolf et al., Phys. Rep. **239**, 1 (1994), and references therein.
27. Particle Data Group, C. Caso et al., Eur. Phys. J. C **3**, 1 (1998).
28. S. -L. Zhu, S. J. Puglia, B. R. Holstein, and M. J. Ramsey-Musolf, Phys. Rev. D **62**, 033008 (2000).
29. R. L. Jaffe, Phys. Lett. B **229**, 275 (1989).
30. H. -W. Hammer, Ulf-G. Meissner, and D. Drechsel, Phys. Lett. B **367**, 323 (1996).
31. W. Koepf, E. M. Henley, and J. S. Pollock, Phys. Lett. B **288**, 11 (1992).
32. M. J. Musolf and M. Burkhardt, Z. Phys. C **61**, 433 (1994).
33. H. Ito, Phys. Rev. C **52**, R1750 (1995).
34. P. Geiger and N. Isgur, Phys. Rev. D **55**, 299 (1997).

35. Ulf-G. Meissner, V. Mull, J. Speth, and J. W. Van Orden, *Phys. Lett. B* **408**, 381 (1997).
36. B.-Q Ma, *Phys. Lett. B* **408**, 387 (1997).
37. H. Weigel *et al.*, *Phys. Lett. B* **353**, 20 (1995).
38. N. W. Park, J. Schecter, and H. Weigel, *Phys. Rev. D* **43**, 869 (1991).
39. N. W. Park and H. Weigel, *Nucl. Phys. A* **541**, 453 (1992).
40. S-T. Hong, B-Y. Park, and D-P. Min, *Phys. Lett. B* **414**, 229 (1997).
41. S. J. Dong, K. F. Liu, and A. G. Williams, *Phys. Rev. D* **58**, 074504 (1998).
42. H.-W. Hammer and M. J. Ramsey-Musolf, *Phys. Rev. C* **60**, 045205 (1999).
43. T. Averett, C. E. Jones, R. D. McKeown, and M. Pitt, *Nucl. Instrum. Methods* **438**, 246 (1999)
44. E. J. Beise *et al.*, *Nucl. Instrum. Methods* **378**, 383 (1996).
45. M. K. Jones *et al.*, *Phys. Rev. Lett.* **84**, 1398 (2000).
46. R. C. Walker *et al.*, *Phys. Rev. D* **49**, 5671 (1994).
47. H. Anklin *et al.*, *Phys. Lett. B* **428**, 248 (1998).
48. E. E. W. Bruins *et al.*, *Phys. Rev. Lett.* **75**, 21 (1995).
49. C. Herberg *et al.*, *Eur. Phys. Jour. A* **5**, 131 (1999).
50. M. Ostrick *et al.*, *Phys. Rev. Lett.* **83**, 276 (1999).
51. I. Passchier *et al.*, *Phys. Rev. Lett.* **82**, 4988 (1999).
52. D. Rohe *et al.*, *Phys. Rev. Lett.* **83**, 4257 (1999).
53. M. J. Musolf, R. Schiavilla, and T. W. Donnelley *Phys. Rev. C* **50**, 2173 (1994).
54. C. J. Horowitz, *Phys. Rev. C* **57**, 3430 (1998).
55. C. J. Horowitz, S. J. Pollock, P. A. Souder, and R. Michaels, nucl-th/9912038.
56. K. S. Kumar, E. W. Hughes, R. Holmes, and P. A. Souder, Mod. Phys. Lett. **A10**, 2979 (1995).
57. A. Czarnecki and W. J. Marciano, *Phys. Rev. D* **53**, 1066 (1996).

Three Generations of Quarks and Leptons: Who Ordered that?

Douglas Bryman

Department of Physics and Astronomy,
University of British Columbia,
TRIUMF,
4004 Wesbrook Mall, Vancouver, B. C. Canada V6T 2A3
E-mail: doug@triumf.ca

Abstract. Ever more sensitive experiments dealing with flavor changing reactions in ultra-rare kaon and muon decays may provide new insight into the long-standing puzzle of the multiple flavors of quarks and leptons. The Standard Model prediction for $K^+ \to \pi^+ \nu \bar{\nu}$ is being challenged in an experiment at BNL. New experiments at BNL, KEK and FNAL are being developed to study the especially attractive CP-violating channel $K_L^0 \to \pi^0 \nu \bar{\nu}$. Very sensitive searches for lepton flavor violation in $\mu \to e\gamma$ and $\mu \to e$ conversion and in lepton number violation in neutrinoless double beta decay are also aiming for unprecedented levels of sensitivity. Experimental results and prospects for some important flavor-changing reactions are reviewed.

INTRODUCTION

The title recalls the famous conundrum expressed by I.I. Rabi upon hearing of the discovery of the muon, an apparently redundant heavy version of the electron. Now, nearly 70 years later, we live with precisely the same problem enhanced by three (and apparently only three) complete generations of both leptons and quarks and no additional insight into the origin of the replications of the basic constituents of matter. The quarks are related by the Standard Model (SM) Cabibbo-Kobayashi-Maskawa (CKM) mixing matrix where three generations are necessary to accommodate a complex CP-violating phase parameter. The lepton generation puzzle is manifested by three generations of "electrons" (e, μ, τ) and their associated neutrinos. Although the charged leptons have appeared to be isolated replications of each other except for mass, new evidence of oscillations in the neutrino sector is increasingly pointing to mixing phenomenology similar to that in the quark sector. Thus, the hypothesis of conservation of lepton flavor due to some presumed global symmetry does not seem to hold. The apparent absence of lepton flavor violation in reactions like $\mu \to e\gamma$ is therefore likely due to the smallness of neutrino mass. However, such reactions may still be generated at very high mass scales, as in many extensions of the SM producing measurable effects far greater than allowable just by tiny neutrino masses. Experiments searching for neutrinoless double beta decay may uncover evidence for the Majorana nature of the neutrino.

Flavor-changing reactions are being pursued vigorously in both the quark and lepton sectors (in addition to neutrino oscillations), in large part, because of the potential to provide new information on issues pertaining to the generation puzzle. In the quark sector experiments employing K and B mesons are focused on testing the validity of the CKM mixing matrix to provide a complete picture of quark mixing and, particularly, of CP violation. In many extensions of the SM there are additional phases which could lead to new sources of CP violation. Experiments involving kaon decays (which still provide the only observed evidence for CP violation) have recently reported interesting new findings, and further work extending the sensitivity boundaries by many orders of magnitude have been proposed. In particular, the presence of direct CP violation in $K \to \pi\pi$ decays seems to be confirmed by experiments measuring ϵ'/ϵ, consistent with SM predictions although leaving room for the possibility of new physics. Higher order decays $K^+ \to \pi^+ \nu \bar{\nu}$ and $K_L^0 \to \pi^0 \nu \bar{\nu}$ are potentially fertile testing grounds where such issues may be settled because hadronic effects are well known and the simple final states allow unusually precise calculations to be made. In addition, first results from the B factories are expected shortly to open up a rich new source of information on quark mixing and CP violation.

The existence of non-zero neutrino mass inferred from oscillations is a significant extension of the general problem of the wide spread of observed masses of the quarks and leptons ranging more than five orders of magnitude from the electron to the top quark. Now, we also have a huge variation of non-zero masses within each lepton generation. This deepens the unexplained mystery of the obvious symmetry between the quarks and leptons including the equality of the charge of the electron and proton and the matching pairs of doublets. In most extensions of the SM, lepton flavor violation (LFV) can occur naturally in decays of charged leptons at much higher levels then due to neutrino mass effects. Substantial progress on lepton flavor violating and lepton number violating searches continues to be made and there is no indication that the limits of achievable sensitivity have been exhausted.

In the following, I will discuss recent progress on measurements of some K and μ decays which bear directly on the questions of CP violation, quark mixing, and the search for new physics to resolve the generation puzzle.

DIRECT CP VIOLATION IN K DECAYS

In the SM, direct CP violation in $K \to 2\pi$ decays characterized by the parameter ϵ' is expected to be a small effect compared to the component due to mixing, i.e. $\epsilon' \ll \epsilon$. To obtain ϵ'/ϵ accurately, it is useful to measure the double ratio of decay rates

TABLE 1. Measurements of ϵ'/ϵ

		$\epsilon'/\epsilon(10^{-4})$
Previous Measurements c. 1993	FNAL E731 [4]	7.4 ± 5.9
	CERN NA31 [3]	23 ± 7
Latest/preliminary	FNAL KTEV [1]	28.0 ± 4.1
	CERN NA48 [2]	14.0 ± 4.3
"World Average"	(NA48) Unofficial!	19.3 ± 2.4

$$R = \frac{\frac{\Gamma(K_L \to \pi^0\pi^0)}{\Gamma(K_S \to \pi^0\pi^0)}}{\frac{\Gamma(K_L \to \pi^+\pi^-)}{\Gamma(K_S \to \pi^+\pi^-)}} \approx 1 - 6\frac{\epsilon'}{\epsilon}. \tag{1}$$

New values for ϵ'/ϵ based on measurements of R were reported recently by FNAL (KTEV-E832) [1] and CERN (NA48) [2] experiments. These results are shown in Table 1 along with the results of the previous round of experiments and the "unofficial" average of 19.3 ± 2.4 [2].

The value of ϵ' results from the difference in CP violation in the isospin $I = 0$ and $I = 2$ channels of $K \to 2\pi$ decays. The non-zero experimental value for ϵ' above therefore indicates the presence of direct CP violation due to one or both of the I= 0, 2 channels (ImA_0, ImA_2) and apparently rules out the pure superweak interpretation of CP violation in the K system. Typical calculations of ϵ'/ϵ (e.g. by Nir [5]) give

$$\epsilon'/\epsilon = 7.7^{+6.0}_{-3.5} \times 10^{-4}.$$

However, large hadronic matrix element uncertainties associated with strong interaction penguin diagrams, generally calculated from the lattice or with the $\frac{1}{N_c}$ expansion, indicate a possible range for ϵ'/ϵ from 0 to 25×10^{-4}. Thus, the large experimental value for ϵ'/ϵ measurements might be telling us something new but it is not presently possible to make that claim. If new physics scenarios such as SUSY enhanced Z_{ds} vertices [6] were contributing significantly to ϵ'/ϵ, such effects may show up unambiguously in measurements of $K^+ \to \pi^+\nu\bar{\nu}$ and $K_L^0 \to \pi^0\nu\bar{\nu}$ where precise SM predictions exist for comparison.

$K \to \pi\nu\bar{\nu}$ DECAYS

Theory

$K^+ \to \pi^+\nu\bar{\nu}$ and $K_L^0 \to \pi^0\nu\bar{\nu}$ present unique opportunities for evaluating the detailed predictions of higher order weak interactions in the SM and searching for new effects because the second order weak diagrams shown in Fig. 1 account for virtually the entire rates. $K^+ \to \pi^+\nu\bar{\nu}$ has comparable contributions in the SM due to t and c quark exchange but is free of long-distance (i.e. mesonic, photonic exchange) contributions down to the 10^{-13} level [7]. Thus, short distance

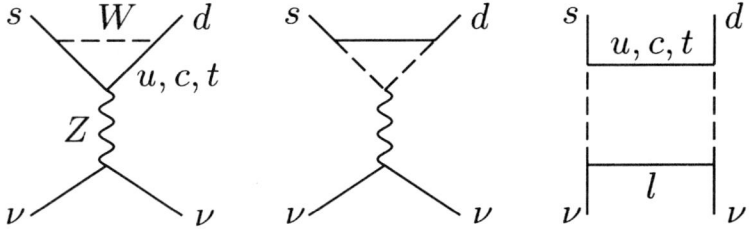

FIGURE 1. Second order weak Feynman diagrams leading to $K^+ \to \pi^+ \nu \bar{\nu}$ decay.

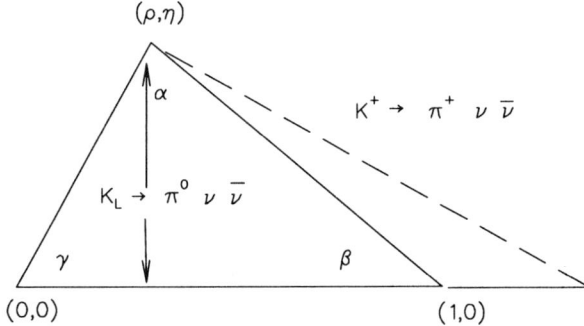

FIGURE 2. The CKM unitarity triangle. ρ and η are parameters in the Wolfenstein representation of the CKM matrix.

contributions (i.e. quark, lepton and gauge boson exchanges) dominate and SM or new physics effects, can be cleanly extracted.

$K^+ \to \pi^+ \nu \bar{\nu}$ is sensitive to important poorly known SM parameters, particularly the magnitude of V_{td}. Calculations by Buchalla, Buras and collaborators [8] beyond leading logarithms (using a two-loop renormalization group analysis for the c quark contribution and $O(\alpha_s)$ for t) indicate theoretical uncertainties no greater than 7% for typical SM parameters. The present range of branching ratios in the SM is $(0.6 - 1.5) \times 10^{-10}$.

$K_L \to \pi^0 \nu \bar{\nu}$ is also a flavor-changing neutral current process that is induced through the loop effects of Fig. 1. $K_L \to \pi^0 \nu \bar{\nu}$ is almost entirely due to direct CP violation where only the imaginary part of V_{td} survives in the amplitude. Using present estimates for SM parameters, the branching ratio for $K_L^0 \to \pi^0 \nu \bar{\nu}$ is expected [8] to lie in the range of $B(K_L \to \pi^0 \nu \bar{\nu}) = (3 \pm 2) \cdot 10^{-11}$.

Taken together $K^+ \to \pi^+ \nu \bar{\nu}$ and $K_L^0 \to \pi^0 \nu \bar{\nu}$ can be used to determine the unitarity triangle shown in Fig. 2. A clean measure of the height of the unitarity triangle is provided by the $K_L \to \pi^0 \nu \bar{\nu}$ branching ratio. $|V_{td}|$ (essentially the side from (ρ, η) to $(1,0)$) can be accurately determined by measuring $K^+ \to \pi^+ \nu \bar{\nu}$ after a small correction is applied for the c quark contribution as illustrated in Fig. 2 by the dashed line. A high statistics measurement of $K_L^0 \to \pi^0 \nu \bar{\nu}$ would be unsurpassed by techniques being pursued in the B meson system for examining CP violation in the SM. Precise determination of the CKM parameters in the K system would provide a basis for comparison with the comparable theoretically clean observables x_s/x_d, the ratio of B_s to B_d mixing, and $\sin(2\beta)$ from measurements in the B meson system. Since some non-SM approaches predict that different values of CKM parameters would be found in K and B experiments having the four measurements of quantities which are theoretically unambiguous in both systems is essential [9].

$K^+ \to \pi^+ \nu \bar{\nu}$ Experiment: E787 at BNL

The E787 group working at the 30 GeV Alternating Gradient Synchrotron (AGS) of Brookhaven National Laboratory (BNL) has reported evidence for the reaction $K^+ \to \pi^+ \nu \bar{\nu}$ based on the observation of a single clean event [10]. The experiment employs an advanced design low energy kaon beam produced by the intense AGS and a sophisticated detection apparatus located inside a 1 T, 3 m diameter solenoidal magnet. Definitive observation of $K^+ \to \pi^+ \nu \bar{\nu}$ requires suppression of all backgrounds to well below the sensitivity for the signal and reliable estimates of the residual background levels. Major background sources include the copious two-body decays $K^+ \to \mu^+ \nu_\mu$ ($K_{\mu 2}$) with a 64% branching ratio and $P = 236$ MeV/c and $K^+ \to \pi^+ \pi^0$ ($K_{\pi 2}$) with a 21% branching ratio and $P = 205$ MeV/c. The only other important background sources are scattering of pions in the beam and K^+ charge exchange (CEX) reactions resulting in decays $K_L^0 \to \pi^+ l^- \bar{\nu}$, where $l = e$ or μ. To suppress backgrounds, redundant kinematic and particle identification measurements were made and events with additional particles were eliminated.

The observation of the first $K^+ \to \pi^+ \nu \bar{\nu}$ candidate event was made in data taken in 1995. Recently, results from an additional data set of comparable sensitivity taken in 1996-97 was reported along with re-analysis of the 1995 data. Alas, no new events were found as indicated in Fig. 3 which shows the range in scintillator (R) vs. Energy (E) for the events surviving all other analysis cuts. Only events with measured momentum in the accepted region $211 \leq P \leq 230$ MeV/c are plotted. The rectangular box indicates the signal region specified as range $34 \leq R \leq 40$ cm of scintillator (corresponding to $214 \leq P_\pi \leq 231$ MeV/c) and energy $115 \leq E \leq 135$ MeV ($213 \leq P_\pi \leq 236$ MeV/c) which encloses the upper 16.2% of the $K^+ \to \pi^+ \nu \bar{\nu}$ phase space. The residual events below the signal region clustered at $E = 108$ MeV were due to $K_{\pi 2}$ decays where both photons had been missed. For the entire 1995-97 data set, the estimated background level was $b = 0.08 \pm 0.02$ events. If the observed event is due to $K^+ \to \pi^+ \nu \bar{\nu}$, the new branching ratio is $B(K^+ \to \pi^+ \nu \bar{\nu}) = 1.5^{+3.4}_{-1.2} \times 10^{-10}$ [11]. In addition, this data can be used to set a limit on the process $K^+ \to \pi^+ f$: $B(K^+ \to \pi^+ f) < 1.1 \times 10^{-10}$ for massless axions or familons ($m_f = 0$) [12].

The observation of an event with the signature of $K^+ \to \pi^+ \nu \bar{\nu}$ is consistent

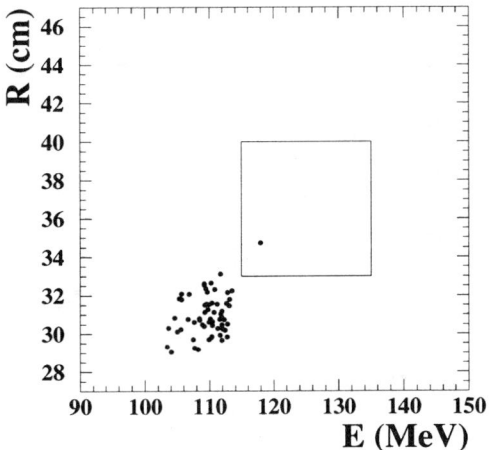

FIGURE 3. Range (R) vs. energy (E) distribution for the E787 1995-97 $K^+ \to \pi^+ \nu \bar\nu$ data set with the final cuts applied. The box enclosing the signal region contains a single candidate event.

with the expectations of the SM. Based on the result for $B(K^+ \to \pi^+ \nu \bar\nu)$, $|V_{td}|$ lies in the range $0.002 < |V_{td}| < 0.04$ (see ref. [10]). Additional data was acquired in 1998 from which E787 expects to obtain a single event sensitivity of $< 7 \times 10^{-11}$. During the next few years an extension of E787 (E949) will run at BNL aiming for a sensitivity that is an order of magnitude below the SM prediction. In the longer term, a proposed experiment at FNAL, "CKM", would attempt to collect 100 events of $K^+ \to \pi^+ \nu \bar\nu$ [36].

$K_L^0 \to \pi^0 \nu \bar\nu$ Experiments

Using the Dalitz decay $\pi^0 \to e^+ e^- \gamma$ mode for π^0 detection, the best present limit on $K_L^0 \to \pi^0 \nu \bar\nu$ was found by the KTEV group at FNAL to be $B(K_L^0 \to \pi^0 \nu \bar\nu) < 5.9 \times 10^{-7}$ [13]. While a significant improvement on previous results, there are still more than four orders of magnitude to go for an observation at the SM level. Further progress on the search for $K_L^0 \to \pi^0 \nu \bar\nu$ is anticipated at FNAL using the dominant $\pi^0 \to \gamma \gamma$ decay. Another $K_L^0 \to \pi^0 \nu \bar\nu$ experiment has been proposed at KEK [14] seeking a single event sensitivity of 2×10^{-11} employing a highly collimated "pencil" beam. The FNAL KTEV/KAMI group is also exploring the possibility of a new search for $K_L^0 \to \pi^0 \nu \bar\nu$ using a beam produced by the new main injector [15].

The KOPIO [16] group working at BNL has proposed a technique for measuring $K_L^0 \to \pi^0 \nu \bar\nu$ based on determining the K_L momentum by time-of-flight so that the detected π^0 can be reconstructed in the kaon center-of-mass system. Using the proposed detector sketched in Fig. 4 all possible kinematic measurements on the

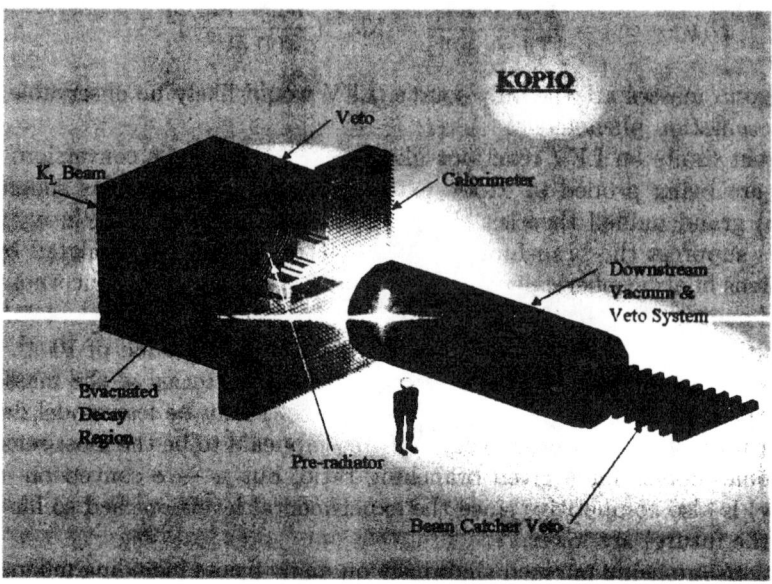

FIGURE 4. Proposed BNL KOPIO $K_L^0 \to \pi^0 \nu \bar{\nu}$ detector. The neutral beam of 1 GeV kaons enters from the left. In the forward detection region the photon detector system consists of a fine grained preradiator in which the photons are converted and the first e^+e^- pair is tracked, followed by an 18 radiation length (X_0) calorimeter.

photons from $\pi^0 \to \gamma\gamma$ decay are made. Energy, time, conversion point and angle of the individual photons are determined and a high efficiency veto system insures that no other particles are present in coincidence. Based on photon detection efficiency measurements made in E787 and using the extra information available from the kaon time-of-flight, it is expected that the inefficiency for detecting π^0s can be reduced to 10^{-8}, thereby suppressing the most dangerous background from $K \to \pi^0\pi^0$ decays to a level that is an order of magnitude below the expected signal. Using these and other special techniques, the goal of a $K_L^0 \to \pi^0 \nu \bar{\nu}$ signal in excess of 50 events may be reached in the presence of what otherwise may be insurmountable backgrounds.

Lepton Flavor Violation

Although LFV is absent in the SM with massless neutrinos, in a minimal extension of the theory incorporating small neutrino masses, reactions such as $\mu \to e\gamma$ may proceed although the branching ratios for this and other LFV reactions would be tiny [17]:

$$B(\mu \to e\gamma) = \frac{\Gamma(\mu \to e\gamma)}{\Gamma(\mu \to e\nu\bar{\nu})} = 10^{-39}(\frac{m_{\nu_1}^2 - m_{\nu_2}^2}{400 \text{ eV}^2})^2. \tag{2}$$

If small neutrino masses are the only source, LFV would likely be observable only in neutrino oscillation phenomena.

With present limits on LFV reactions like $\mu \to e\gamma$ and $\mu \to e$ conversion, high mass scales are being probed in exotic extensions of the SM such as supersymmetric SU(5) grand unified theories. However, progress is slow since heavy LFV force carriers suppress the branching ratios by factors of m_H^{-4}. Stimulated by recent suggestions of atmospheric neutrino oscillations [20], [21] models incorporating right-handed neutrinos at the GUT scale and $\Delta m^2 = m_{\nu_\mu}^2 - m_{\nu_\tau}^2 = 10^{-3} - 10^{-4}$ eV^2 suggest that the $\mu \to e\gamma$ branching ratio could have a *lower* bound of 10^{-14}. The uncertainties for predicting the rates of LFV processes in terms of the masses of supersymmetric partner particles like "selectrons" ($e_{\tilde{R}}$) may be less model dependent than for processes like proton decay. $\mu \to e\gamma$ appears to be the most sensitive process in some models for a given branching ratio, but $\mu \to e$ conversion (discussed below) is also competitive since the experimental levels reached so far (and perhaps, in the future) are lower.

LFV reactions are being pursued vigorously on many fronts including muon, tau, and kaon decays, and in ep collisions at HERA [18]. Table 2 gives some current results and goals for experiments in progress.[1]

TABLE 2. Limits on flavor violating and other exotic decays.

Reaction	Present limit	Reference	Proposed goal	Lab.
$\mu \to e\gamma$	1.2×10^{-11}	[22]	10^{-14}	PSI
$\mu \to 3e$	1×10^{-12}	[23]		
$\mu^- Ti \to e^- Ti$	6.1×10^{-13}	[24]	$< 10^{-16}$	BNL
$\mu^+ e^- \to \mu^- e^+$	8.2×10^{-11}	[25]		
$K_L \to \mu e$	5×10^{-12}	[28]		
$K^+ \to \pi^+ \mu^+ e^-$	2.8×10^{-11}	[30]		
$K_L \to \pi^0 \mu e$	3.2×10^{-9}	[29]	3×10^{-11}	FNAL
$K^+ \to \pi^+ X$	1.1×10^{-10}	[11]	10^{-11}	BNL
$\tau \to \mu\gamma$	3×10^{-6}	[31]		
$\tau \to e\gamma$	2.6×10^{-6}	[31]		
$\tau \to eee$	3×10^{-6}	[32]		

$\mu \to e\gamma$

$\mu \to e\gamma$ is the "Gold Standard" LFV process which is highly sensitive to new physics at heavy mass scales. The MEGA group working at LAMPF recently reported a final result for $\mu \to e\gamma$: $B(\mu \to e\gamma) < 1.2 \times 10^{-11}$ [22]. This is a factor 4 improvement in sensitivity over previous efforts at LAMPF. The big potential

---
[1] All limits presented in this paper will be at the 90% c.l. level.

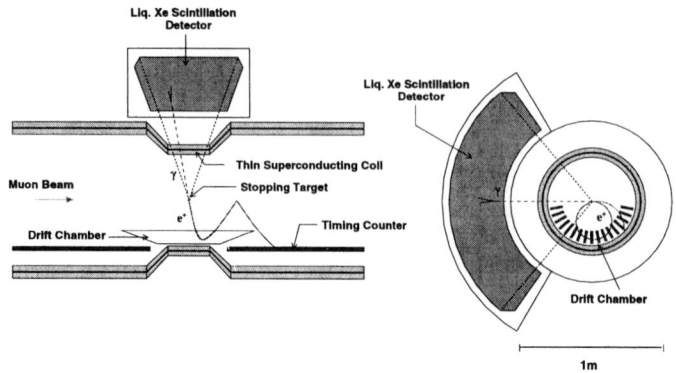

FIGURE 5. Proposed detector for studying $\mu \to e\gamma$ at PSI [33].

obstacle for improving on this hard-won result involves taming the accidental background (B(acc)) which arises from near simultaneous decays of two muons, one producing the positron and the other giving the photon from radiative decay. Since the accidental background goes approximately as B(acc) α δE_e δE_γ^2 δt $\delta \theta_{e\gamma}^2$ the gamma energy resolution and $e - \gamma$ angular resolution are key measurements.

A new $\mu \to e\gamma$ experiment has been proposed at PSI [33] aiming to reach a sensitivity of about 10^{-14}. The layout is shown in Fig. 5. The setup includes a "Super-K" geometry gamma calorimeter made of liquid Xe in which the gamma will be observed using scintillation light with $\delta E_\gamma = 1.4\%(FWHM)$ and $\delta\theta_{e\gamma}^2 = 17$ mrad. These proposed resolutions are each more than a factor of 3 better than obtained in the MEGA experiment and, if achieved, may enable the proposed sensitivity to be reached.

$\mu \to e$ Conversion

$\mu \to e$ conversion is another attractive process to pursue because the coherent capture is enhanced relative to ordinary nuclear muon capture and there are favorable experimental aspects which promote it above other LFV candidates. The signature of $\mu - e$ conversion is a monochromatic electron energy $E_{\max} = m_\mu - E_{\text{binding}} \approx 104$ MeV, which is far away from the end point of the free muon decay spectrum (53 MeV). However, the principal inherent background to $\mu - e$ conversion comes from muon decay in orbit around the nucleus where the energy can extend up to $E_{\max}$. Additional backgrounds come from cosmic rays and radiative capture of π^- contaminating the muon beam. These backgrounds may be reduced to a manageable level by good detector resolution and excellent suppression of π^-.

The SINDRUM2 group at PSI [24] reported an upper limit on the branching ratio for $\mu \to e$ conversion in Ti. No events were seen consistent with $\mu \to e$

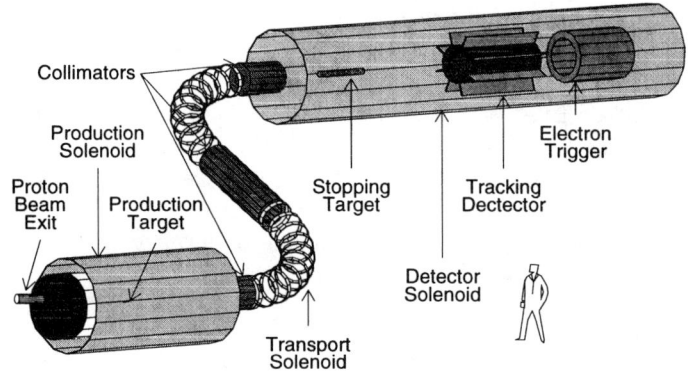

FIGURE 6. Proposed MECO detector for studying $\mu \to e$ conversion at BNL. [27]

conversion resulting in $B(\mu^-\text{Ti} \to e^-\text{Ti}) < 6.1 \times 10^{-13}$. This is about a factor of 7 improvement over the previous result from the TRIUMF TPC [26]. The PSI group, after experiencing some difficulties with pion contamination, is now concentrating on other targets particularly gold and expects to reach a level of 10^{-13} from running in 2000.

A new effort, MECO [27] at BNL, has been proposed to search for $\mu \to e$ conversion at the level of $< 10^{-16}$. This project developed from the original proposal of Lobashov et al. for the Moscow Meson Factory in which a high field gradient solenoid surrounds the pion production target and a muon beam produced by a pulsed proton beam is transported to a detector solenoid where the $\mu \to e$ conversion reaction is detected. Features of MECO shown in Fig. 6 include a bent transport solenoid which provides charge and momentum selection and a straw tube tracking detector.

LFV in Kaon Decays

$K_L^0 \to \mu e$ and $K \to \pi \mu e$ are LFV reactions which have been pursued in recent experiments at BNL and FNAL. $K_L^0 \to \mu e$ is potentially sensitive to axial-vector and pseudoscalar interactions and $K \to \pi \mu e$ can occur via vector or scalar currents. These reactions are especially attractive in non-SM approaches which have overall "generation" numbers conserved considering both the quark and lepton sectors. They are also sensitive to the presence of high mass scales in models with additional Higgs particles or leptoquarks.

Searches dealing with the processes $K_L^0 \to \pi^0 \mu^\pm e^\mp$ and $K^+ \to \pi^+ \mu^+ e^-$ are being made by the KTEV (E799) group at FNAL and the E865 group at BNL. The most recent result on $K_L^0 \to \pi^0 \mu^\pm e^\mp$ is $B(K_L^0 \to \pi^0 \mu^\pm e^\mp) < 3.2 \times 10^{-9}$ (see ref. [29]). The E865 group at BNL [30] has been studying $K^+ \to \pi^+ \mu^+ e^-$ using an upgraded beam and detector. They recently reported a new finding of B($K^+ \to \pi^+\mu^+e^-$) $< 3.9 \times 10^{-11}$ based on the observation of 3 events and an expected background from accidentals of 2.4 events. Combining this with their previous result gives an overall limit of B($K^+ \to \pi^+\mu^+e^-$) $< 2.8 \times 10^{-11}$.

The latest limit of B($K_L \to \mu e$) $< 5 \times 10^{-12}$ was reported [28] by the E871 group at BNL. This experiment used a magnetic spectrometer with a high rate straw chamber tracking apparatus and a beam plug inserted in the neutral beam upstream of the main detector to reduce accidentals. Although the experiment was not background limited, it was found that accidental backgrounds might limit future searches aiming for another order of magnitude sensitivity.

T-violation in $K_{\mu 3}$ Decay

CP violation outside the context of the SM may be required to explain the baryon asymmetry of the Universe. In some multi-Higgs models, exchange of a charged Higgs particles could interfere with SM processes in decays like $K \to \pi\mu\nu$ resulting in T-violating transverse muon polarization. Compared to expectations of zero in the SM, effects as large as $P_\mu^T \sim 10^{-3}$ are possible. The current limiting result is $P_\mu^T = 0.00185 \pm 0.0036$ (see ref. [34]). An experiment is in progress at KEK (E246) [35] aiming to substantially improve the sensitivity to P_μ^T using a technique involving stopped kaons (K^+), a toroidal spectrometer and an efficient muon polarimeter. The latest result is null: $P_\mu^T = (-0.0042 \pm 0.0049(stat) \pm 0.0009(sys))$ and the experiment is continuing.

$\beta\beta$ DECAY

Ordinary double beta decay ($\beta\beta\overline{\nu\nu}$) is a second order weak process involving the emission of two electrons and two neutrinos: $X_N^Z \to Y_N^{Z-2} e^- e^- \overline{\nu\nu}$. Following successful early measurements using geo-chemical techniques, $\beta\beta\overline{\nu\nu}$ was observed in direct counting experiments with several isotopes in a series of remarkable experiments over the last decade [37]. Recently, the Heidelberg-Moscow group reported $T_{1/2}^{76Ge}(\beta\beta\ \overline{\nu\nu}) = 1.77 \pm 0.01^{+0.13}_{-0.11} \times 10^{21}$ yr. based on an observation of 20,000 events [38].

However, the main quarry in such experiments is the hypothetical process of neutrinoless double beta decay $X_N^Z \to Y_N^{Z-2} e^- e^-$ ($\beta\beta$) which would represent a $\Delta L = 2$ change of lepton number. Such processes, completely outside the context of the SM, are potentially highly sensitive to new physics at mass scales ranging from sub-ev to TeV and beyond. In particular, $\beta\beta$ is potentially sensitive to the existence of Majorana neutrinos of effective mass $m_{eff} = \Sigma U_{ei}^2 m_i e^{i\phi_i}$ where U_{ei} is a mixing matrix element and ϕ_i is a phase. Based on the current limit of $T_{1/2}^{76Ge} > 1.6 \times 10^{25}$ yr., [39] it is concluded that $m_{eff} < 0.4 - 1$ eV where the range of the limit comes from the factor of three uncertainty generally assigned to the estimates of the nuclear matrix elements. Observation of $\beta\beta$ at greater sensitivity as planned for future experiments such as GENIUS [40] and NEMO [41] may provide a route to establishing absolute values for Majorana neutrino masses (whereas oscillation experiments are sensitive to mass differences). On the other hand, it would be difficult to distinguish this hypothesis for $\beta\beta$ from other $\Delta L = 2$ hypotheses due to exchange of high mass leptoquarks for which current experiments can set limits of $M_{LQ} > 200 - 300$ GeV.

CONCLUSION

Orders of magnitude gains are being made in searches and measurements for the most important rare kaon and muon decays. It is expected that a sensitivity below 10^{-10} will be attained shortly in the BNL E787 study of $K^+ \to \pi^+\nu\bar{\nu}$, representing nearly an order of magnitude larger data set than the one in which the first candidate event was found. The E949 extension of E787 which begins in late 2000 is aimed at a sensitivity of 10^{-11}. In addition, the "CKM" experiment has been proposed at FNAL with the goal of measuring 100 $K^+ \to \pi^+\nu\bar{\nu}$ events.

The KTEV group at FNAL has improved the sensitivity to $K_L^0 \to \pi^0\nu\bar{\nu}$ by two orders of magnitude in recent years. New measurements of $K_L^0 \to \pi^0\nu\bar{\nu}$ are being planned for BNL (KOPIO), KEK and FNAL (KAMI) aiming for a large sample of events at the 10^{-11} level. As a consequence of high theoretical precision, measurements of $K^+ \to \pi^+\nu\bar{\nu}$ and $K_L^0 \to \pi^0\nu\bar{\nu}$ with good experimental accuracy could test the SM origin of CP violation and may ultimately yield firm determinations of the CKM parameters ρ and η. Absence of $K^+ \to \pi^+\nu\bar{\nu}$ or $K_L^0 \to \pi^0\nu\bar{\nu}$ within the expected ranges or conflicts with other measures of quark mixing parameters would indicate new physics.

The search for lepton flavor violation and lepton number violation using decays of charged leptons and neutrinoless $\beta\beta$ decay continues unabated. New limits have been presented on $\mu \to e\gamma$ and $K^+ \to \pi^+\mu^+e^-$. Current and planned LFV experiments at BNL (MECO on $\mu \to e$ conversion) and ($\mu \to e\gamma$) at PSI aim for further large gains in sensitivity. Impressive limits have been placed on searches for double beta decay and new proposals also aim for improvements of orders of magnitude. Confirmed non-zero findings in these would be revolutionary.

REFERENCES

1. Alavi-Harati, A. et al., Phys. Rev. Lett. **83**, 22 (1999).
2. NA48 Collaboration, Barr, G.D., et al., CERN seminar Ceccucci, A., Feb. 29, 2000 (unpublished).
3. Gibbons, L.K., et al., Phys. Rev. **D55**, 6625 (1997).
4. Barr, G.D., et al., Phys. Lett. **B317**, 233 (1993)
5. Nir, Y., 27th SLAC Summer Inst. on Part. Phys., (1999), to be published; hep-ph/0011321.

6. Buras, A.J., *et al.*, hep-ph/9908371 (1999).
7. Rein, D. and Sehgal, L.M., *Phys. Rev.* **D39**, 3325, (1989); Hagelin, J.S. and Littenberg, L.S., *Prog. Part. Nucl. Phys.* **23**, 1 (1989); Lu, M. and Wise, M.B., *Phys. Lett.* **B324**, 461 (1994), and; Segre, G., *Phys. Rev.* **D61**, 077301-1 (2000).
8. Buchalla, G., Buras, A.J., *Phys. Rev.* **D54**, 6782(1996); Buchalla, G., Buras, A.J. and Lautenbacher, M.E., *Rev. Mod. Phys.* **68**, 1125 (1996); and, Buchalla, G., hep-ph/0002207 (2000).
9. Bergmann, S. and Perez, G., hep-ph/0007170 (2000).
10. Adler, S., *et al.*, *Phys. Rev. Lett.* **79**, 2204 (1997).
11. Adler, S., *et al.*, *Phys. Rev. Lett.* **84**, 3768 (2000).
12. Wilczek, F., *Phys. Rev. Lett.* **49**, 6782 (1982).
13. Alavi, A., *Phys. Rev.* **D61**, 072006 (2000).
14. see Inagaki, T., *Proc. Symp. on Flavor Changing Neutral Currents*, ed., Cline, D., *World Sci.* (1997).
15. Cheu, E., *et al.*, FNAL-PUB-97-321-E.
16. Chiang, I.H., *et al.*, BNL AGS proposal E926.
17. Gvozdev, A.A., *et al.*, *Phys. Lett.* **B345**, 490 (1995).
18. See Y. Sirois, Phys. Beyond the Std. Mod., Proc. Fifth International WEIN Symposium, eds. Herzeg, P., Hoffman, C. M. and Klapdor-Kleingrothaus, H.V., World Scientific, 382 (1999).
19. Barbieri, R. and Hall, L. *Phys. Lett.* **B338**, 212 (1994).
20. Kajita, T., Phys. Beyond the Std. Mod., Proc. Fifth International WEIN Symposium, *Ibid*, 70.
21. Hisano, J., Nomura, D., Yanigida, T., KEK-TH-548, hep-ph/9711348 (1997).
22. Brooks, M.L., *et al.*, *Phys.Rev.Lett.* **83**, 1521 (1999).
23. Belgardt, U., *et al.*, *Nucl. Phys.* **B299**, 1 (1988).
24. Kaulard, J., *et al.*, *Phys.Lett.B* **422**, 334 (1998).
25. Willmann, L., *et al.*, *Phys.Rev.Lett.* **82**, 49 (1999).
26. Ahmed, S., *et al.*, *Phys. Rev.* **D38**, 2102 (1988).
27. Molzon, W., Lepton and Baryon Number Violation, Proc. Lepton-Baryon98, ed. Klapdor-Kleingrothaus, H.V., 586 (1999); and Bachman, M. *et al.*, BNL proposal 940 (1997).
28. Ambrose, D., *et al.*, *Phys.Rev.Lett.* **81**, 5734 (1998).
29. Arisaka, K., *et al.*, *Phys. Lett.* **B432**, 230 (1998).
30. Appel, R., *et al.*, *Phys. Rev. Lett.*, to be published (2000).
31. Edwards, K., *et al.*, *Phys. Rev.* **D55**, 3919 (1997).
32. Bliss, D.W., *et al.*, *Phys. Rev* **D57**, 5903 (1998).
33. Barkow, L., *et al.*, PSI proposal R-99-05.1 (1999).
34. Blatt, S.R., *et al.*, *Phys. Rev.* **D27**, 1056 (1983).
35. Imazato, J., *et al.*, KEK E246. BNL E936 proposal (1997).
36. Coleman, R., *et al.*, FNAL proposal (1998).
37. See Moe, M. and Vogel, P., *Ann. Rev. Nucl. Part. Sci.* **44**, 247 (1994); Danevich, F.A., *et al.*, nucl-ex/0003001 (2000); Ejiri, H., *et al.*, *Nucl. Phys. A* **611**, 85 (1996); Alessandrello, A., *et al.*, *Phys. Lett.* **B433**, 156 (1998).
38. Heidelberg-Moscow collab., *Phys. Rev.* **D55**, 54 (1997); *Phys. Lett.* **B403**, 219

(1997).
39. Baudis, L., *et al.*, *Phys.Rev.Lett.* **83**, 41 (1999); see also Aalseth, C.E., *et al.*, *Phys.Rev.* **C59**, 2108 (1999).
40. See Klapdor-Kleingrothaus, H.-V., Phys. Beyond the Std. Mod., Proc. Fifth International WEIN Symposium, *Ibid*, 275.
41. Piquemal, F., NEMO III collaboration, *Nucl. Phys.* (Proc. Supl.) **B77**, 352 (1999).

Neutrino Oscillation Experiments

Kenzo Nakamura

High Energy Accelerator Research Organization (KEK)

Abstract. The present status of neutrino oscillation experiments and prospects of forthcoming experiments are reviewed. Particular emphasis is placed on the recent results from Super-Kamiokande atmospheric neutrino and solar neutrino observations.

INTRODUCTION

Neutrino oscillations provide the most sensitive means to explore the finite neutrino mass as well as the mixing in the neutrino sector. At present, there are three evidences of neutrino oscillations with different Δm^2: (i) $\Delta m^2 \sim 3 \times 10^{-3}$ eV^2 from atmospheric neutrino observations, (ii) $\Delta m^2 \sim 10^{-5}$ eV^2 or less from solar neutrino observations, and (iii) $\Delta m^2 \sim 1$ eV^2 from the LSND experiment. Recent neutrino oscillation experiments and planned forthcoming experiments mainly focus on further investigation of these evidences.

A simple observation is that these three different Δm^2 cannot be accommodated with only three neutrino species. Therefore, if all the three evidences are true, at least one sterile neutrino species, ν_s must exist because of the LEP constraint on the number of light, active neutrino species. Alternatively, one of the three evidences may not survive.

Among them, the evidence from atmospheric neutrino observations seems compelling. Most of the results were obtained by Super-Kamiokande (hereafter abbreviated as SuperK) [1–3], with supporting evidence from Kamiokande [4], IMB [5], MACRO [6,7], and Soudan 2 [8–10]. The observed mixing is nearly maximal. The SuperK data favors $\nu_\mu \to \nu_\tau$ oscillation and disfavors $\nu_\mu \to \nu_s$ oscillation as shown below [11].

Solar neutrinos have been observed by five experiments, Homestake chlorine experiment ($\nu_e\,^{37}\text{Cl} \to e^-\,^{37}\text{Ar}$) [12], SAGE [13] and GALLEX [14] gallium experiments ($\nu_e\,^{71}\text{Ga} \to e^-\,^{71}\text{Ge}$), and Kamiokande [15] and SuperK [16] νe scattering experiments. The solar neutrino fluxes measured by all these experiments are significantly lower than the standard solar model (SSM) predictions [17]. Astrophysics alone cannot explain these results. In terms of neutrino oscillations, a global analysis of the data obtained by all the experiments show various allowed regions in the two-flavor $\Delta m^2 - \sin^2 2\theta$ plane. Different regions are called the MSW Small Mixing

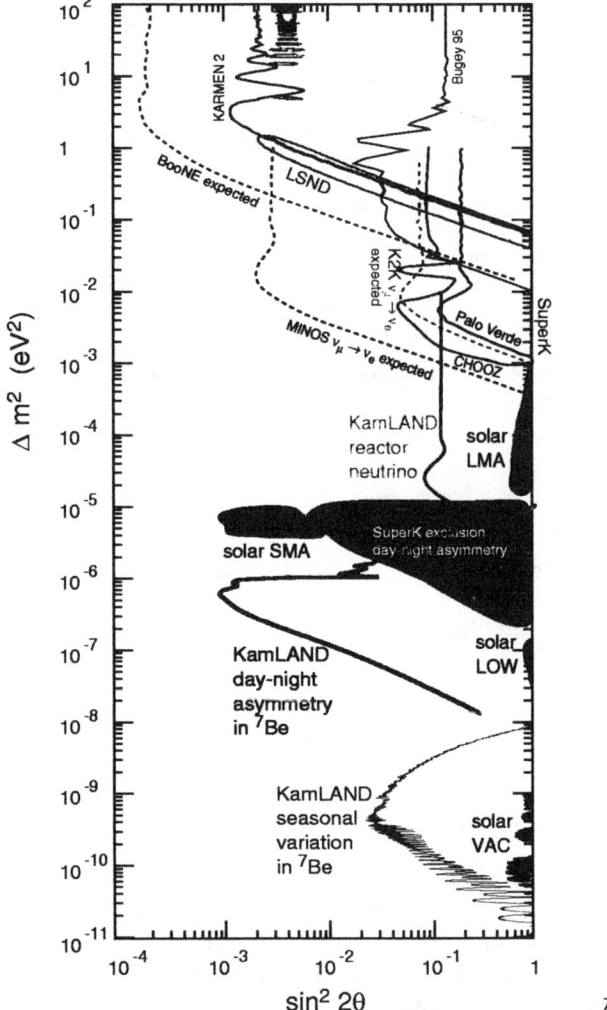

FIGURE 1. Two-flavor oscillation results from various experiments as well as the sensitivities of some of the forthcoming experiments, compiled by H. Murayama [21].

($\Delta m^2 \sim 5 \times 10^{-6}$ eV^2 and $\sin^2 2\theta \sim 3 \times 10^{-3}$), MSW Large Mixing($\Delta m^2 \sim 5 \times 10^{-5}$ eV^2 and $\sin^2 2\theta \sim 1$), MSW Low ($\Delta m^2 \sim 10^{-7}$ eV^2 and $\sin^2 2\theta \sim 1$), and Vacuum (or Just-So) ($\Delta m^2 \sim 10^{-10}$ eV^2 and $\sin^2 2\theta \sim 1$) solutions.

The LSND experiment at LAMPF reported evidence of neutrino oscillations for both $\bar{\nu}_\mu \to \bar{\nu}_e$ [18] and $\nu_\mu \to \nu_e$ [19]. The allowed (or "favored") (Δm^2, $\sin^2 2\theta$) regions are consistent for the two CP-conjugate oscillation channels. However, the Karmen 2 experiment [20] at ISIS, Rutherford Appleton Laboratory, found no evidence though it has similar, but slightly less, sensitivity to LSND.

Figure 1 neatly compiles the two-flavor oscillation results from important experiments as well as the sensitivities of some of the forthcoming experiments [21].

ATMOSPHERIC NEUTRINO RESULTS

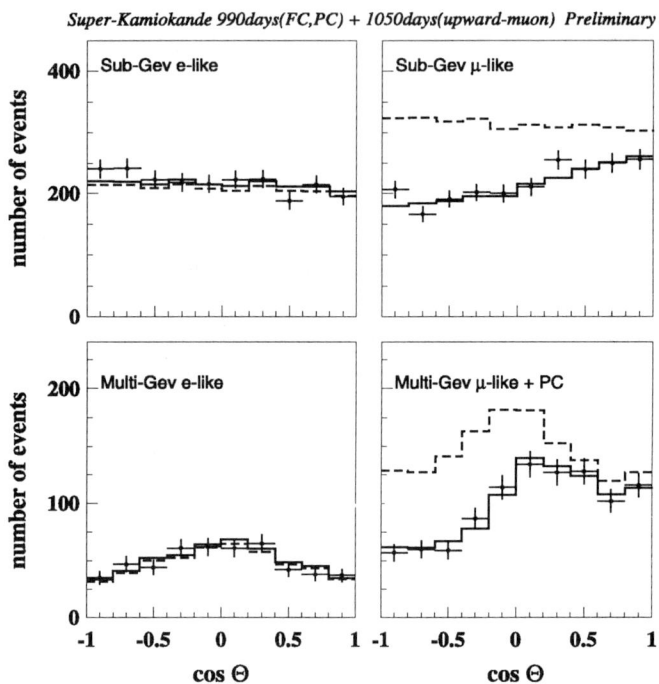

FIGURE 2. Zenith-angle distributions for sub-GeV e-like (upper left panel), sub-GeV μ-like (upper right panel), multi-GeV e-like (lower left panel), and multi-GeV (FC+PC) μ-like events (lower right panel) observed in SuperK. The horizontal axis shows cosine of the zenith angle; 1.0 corresponds to the downward direction and -1.0 the upward direction. The dashed histograms show the Monte Carlo prediction for the null oscillation hypothesis. The solid histograms show the Monte Carlo prediction for the hypothesis of $\nu_\mu \to \nu_\tau$ oscillations with $\sin^2 2\theta = 1$ and $\Delta m^2 = 2.8 \times 10^{-3}$ eV^2. For the oscillated histograms, the absolute normalization is adjusted to minimize χ^2.

The most recent results from SuperK atmospheric neutrino observations correspond to an exposure of 61 kton·year or 992 live days. Fully contained (FC) events are classified into sub-GeV events with $E_{\text{vis}} < 1.33$ GeV and multi-GeV events with $E_{\text{vis}} > 1.33$ GeV. Partially contained (PC) events are classified as multi-GeV events, though the minimum track length of about 2.5 m corresponds to muons with > 700 MeV/c momentum.

To study the atmospheric neutrino flavor ratio $(\nu_\mu + \bar\nu_\mu)/(\nu_e + \bar\nu_e)$, the double ratio $R = (\mu/e)_{\text{Data}}/(\mu/e)_{\text{MC}}$ is measured, where (μ/e) denotes the ratio of the numbers of μ-like to e-like neutrino interactions observed in the data or predicted by Monte Carlo (MC). The ratio of the data to MC is taken to cancel uncertainties in the neutrino flux and cross sections. The expected value for R is unity if there is agreement between the experiment and the theoretical prediction.

Figure 2 shows the zenith-angle dependence of sub-GeV e-like, sub-GeV μ-like, multi-GeV e-like, and multi-GeV μ-like (FC+PC) events. The dashed histograms show the Monte Carlo prediction for the hypothesis of null oscillation. These data indicate that the μ-like events show a strong deviation from expectation in the shape of the zenith-angle distribution. In particular, multi-GeV μ-like events show strong up/down asymmetry in contrast to the calculated up/down ratio of near unity. On the other hand, the zenith-angle distribution of e-like events is consistent with the prediction. These results update the published SuperK results [1].

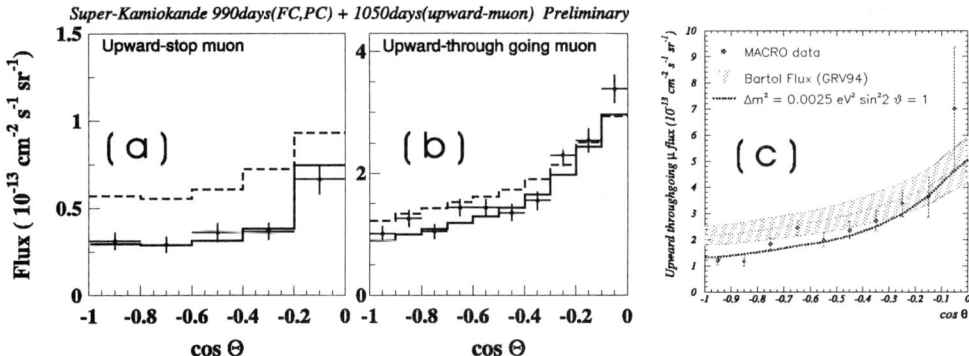

FIGURE 3. Zenith-angle distributions of upward-going muons. (a) SuperK results for upward stopping muons. (b) SuperK results for upward through-going muons. (c) MACRO results for upward through-going muons. In (a) and (b), the dashed histograms show the expected distribution for the hypothesis of null oscillation, while the solid histograms show the expectation for $\nu_\mu \to \nu_\tau$ with $\sin^2 2\theta = 1$ and $\Delta m^2 = 2.8 \times 10^{-3}$ eV^2. In (c), the shaded region shows the expectation and its uncertainties for the hypothesis of null oscillation, and the dotted line shows the expectation for $\nu_\mu \to \nu_\tau$ with $\sin^2 2\theta = 1$ and $\Delta m^2 = 2.5 \times 10^{-3}$ eV^2.

The fluxes of high-energy atmospheric ν_μs can be measured with underground detectors through the detection of muons produced by the charged-current interactions of ν_μs in the rock surrounding the detector. The SuperK Collaboration measured both the data of upward through-going muons and upward stopping muons that stop in the detector. Typical neutrino energies corresponding to upward through-going and upward stopping muon events in SuperK are 100 GeV and 10 GeV, respectively.

Figures 3 (a) and (b) show the SuperK results for the zenith-angle distributions of upward stopping and through-going muons. The observed zenith-angle distribution

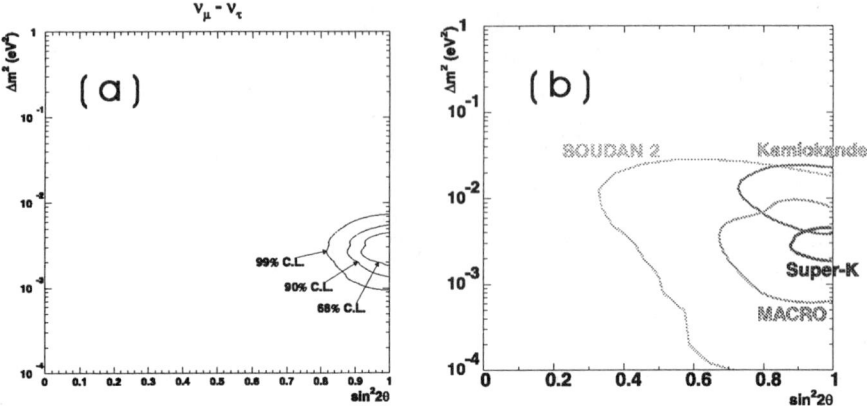

FIGURE 4. (a) The 68%, 90%, and 99% CL allowed regions for $\nu_\mu \to \nu_\tau$ are shown for the SuperK FC+PC atmospheric neutrino results. (b) The 90% CL allowed region for $\nu_\mu \to \nu_\tau$ obtained from a combined analysis of FC, PC, and upward through-going and stopping muon events measured by SuperK. The 90% CL allowed regions obtained from the Kamiokande, Soudan 2, and MACRO experiments are also shown.

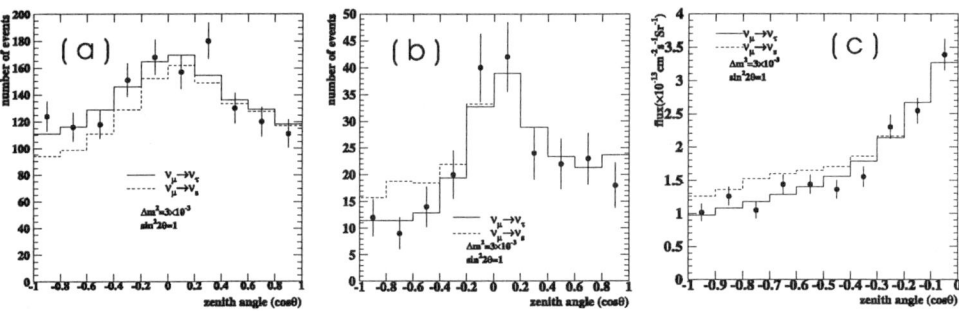

FIGURE 5. Zenith-angle distributions for (a) neutral-current enriched FC events, (b) high-energy PC events, and (c) upward through-going muons. The histograms show the Monte Carlo predictions for $\nu_\mu \to \nu_\tau$ and $\nu_\mu \to \nu_s$ hypotheses with $\Delta m^2 = 3 \times 10^{-3}$ eV^2 and $\sin^2 2\theta = 1$. The two predictions are normalized by a common factor so that the number of the observed events is the same as the number of the predicted events for $\nu_\mu \to \nu_\tau$.

for the upward through-going muons is steeper than the prediction with no neutrino oscillations. Also, the observed flux of the upward stopping muons is significantly smaller than the prediction with no neutrino oscillations. These results update the published SuperK results [2,3].

The upward-going muons were also measured by the MACRO experiment [6,7] at Gran Sasso. Figure 3(c) shows the zenith-angle distribution for the upward through-going muons observed by MACRO [7].

An oscillation analysis of the FC+PC atmospheric neutrino data from SuperK was made for the hypothesis of $\nu_\mu \to \nu_\tau$ using the flux model of Honda et al. [22]. The best fit was obtained at $\sin^2 2\theta = 1$ and $\Delta m^2 = 2.6 \times 10^{-3}$ eV^2 with χ^2_{min}/DOF (degree of freedom) = 45.4/67 by limiting the parameter space to the physical region ($\sin^2 2\theta \le 1$). On the other hand, the hypothesis of null oscillation gave χ^2_{min}/DOF = 191.0/69. The 68%, 90%, and 99% confidence level (CL) allowed regions in the ($\sin^2 2\theta, \Delta m^2$) plane are shown in Fig. 4(a).

A further oscillation analysis of the SuperK atmospheric neutrino data was made by including the upward through-going and stopping muons in addition to the FC and PC events. The obtained 90% CL allowed region for the hypothesis of $\nu_\mu \to \nu_\tau$ is shown in Fig. 4(b). In this figure, the 90% CL allowed regions reported by Kamiokande [23], Soudan 2 [10], and MACRO [7] are also shown. The best fit parameters from SuperK are $\sin^2 2\theta = 1$ and $\Delta m^2 = 2.8 \times 10^{-3}$ eV^2. The solid histograms in Figs. 2 and 3(a) and (b) are calculated with these parameter values. The 90% CL allowed intervals for the oscillation parameters obtained from the SuperK combined analysis are 2×10^{-3} eV$^2 < \Delta m^2 < 5 \times 10^{-3}$ eV^2 and $\sin^2 2\theta > 0.88$.

There is an alternative scenario that can explain the SuperK FC atmospheric neutrino events in terms of neutrino oscillations. As far as the FC events are concerned, the hypothesis of $\nu_\mu \to \nu_s$ gives an equally good fit to the zenith-angle distribution as the standard hypothesis of $\nu_\mu \to \nu_\tau$.

These two hypotheses may be discriminated by using the fact that for $\nu_\mu \to \nu_s$ (i) fewer neutral-current events should be observed than $\nu_\mu \to \nu_\tau$ because ν_s does not interact with matter, and (ii) matter effects suppress the oscillation probability [24] while there is no matter effects for $\nu_\mu \to \nu_\tau$. The difference of the oscillation probability is appreciable only for > 15 GeV neutrinos traveling through the Earth.

First, neutral-current enriched sample of events were selected from multi-ring FC events with $E_{vis} > 400$ MeV and the most energetic ring being e-like. A Monte Carlo study indicates that the fraction of neutral-current events in this sample is 29% for no oscillations. Figure 5(a) shows the zenith-angle distribution of these events with Monte Carlo predictions.

Next, matter effects were studied using PC events with $E_{vis} > 5$ GeV (for these events, the typical energy of the parent atmospheric neutrino is 20 GeV) and upward through-going muons (the typical energy of the parent neutrino is 100 GeV). Figures 5(b) and (c) show the zenith-angle distributions for the PC events and the upward through-going muons with Monte Carlo predictions.

A combined statistical analysis was performed using the SuperK neutral-current enriched FC events, high-energy PC events, and upward through-going muons, and the two hypotheses, $\nu_\mu \to \nu_\tau$ and $\nu_\mu \to \nu_s$ were tested. For neutral-current enriched FC events and high-energy PC events, an up/down ratio, up($-1 < \cos\theta < -0.4$)/down($0.4 < \cos\theta < 1$), is used as a discriminant to cancel some systematic errors. For upward through-going muons, a vertical/horizontal ratio, vertical($-1 < \cos\theta < -0.4$)/horizontal($-0.4 < \cos\theta < 0$), is used. Note that the matter effects

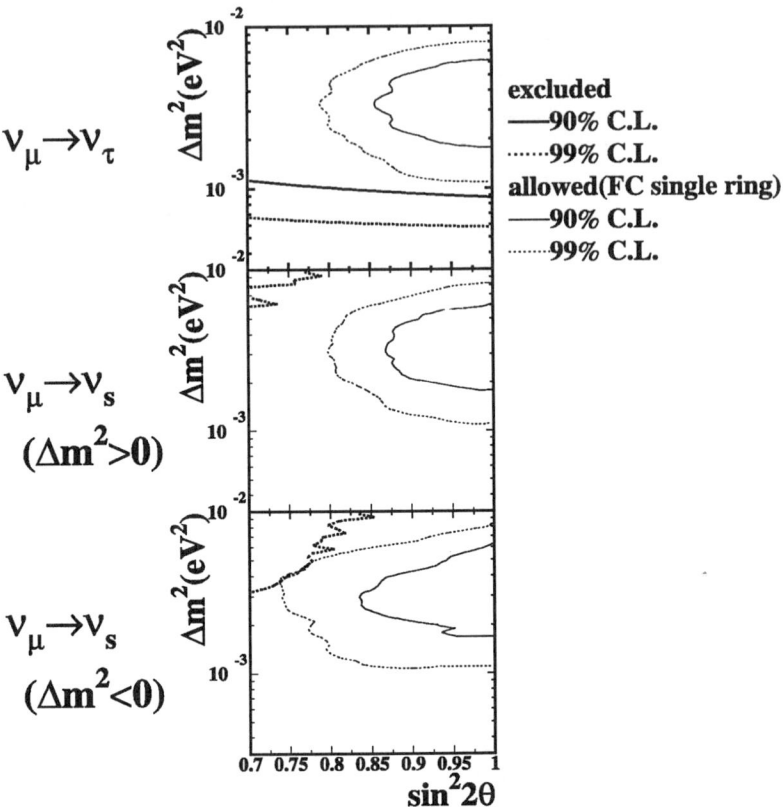

FIGURE 6. Excluded regions for three oscillation modes. (a) $\nu_\mu \to \nu_\tau$, (b) $\nu_\mu \to \nu_s$ with $\Delta m^2 > 0$, and (c) $\nu_\mu \to \nu_s$ with $\Delta m^2 < 0$. Also shown are the allowed regions from the analysis of the single-ring FC data.

depend on the sign of Δm^2, so that both positive and negative Δm^2 values were tested for $\nu_\mu \to \nu_s$. Figure 6 shows the excluded regions for each hypothesis. In this figure, the parameter regions allowed by the analysis of the SuperK FC data are also shown. The parameter regions allowed for the FC data are excluded at the 99% CL. Therefore, the SuperK results disfavor the hypothesis of $\nu_\mu \to \nu_s$ for the atmospheric muon neutrino oscillation.

SOLAR NEUTRINO RESULTS

The latest results of the solar neutrino flux measurements are listed and compared with the prediction of the Bahcall-Pinsonneault (BP98) SSM calculations [25] in Table 1. The SuperK result in Table 1 is obtained from 825 live days of solar neutrino observation, and updates the previous SuperK result from 300 live days of observation [16]. Compared to the SSM (standard solar model) prediction of BP98 [25], Data/SSM(BP98) = $0.475^{+0.008}_{-0.007} \pm 0.013$.

TABLE 1. Results of the solar-neutrino flux measurements compared with the BP98 [25] SSM calculation. The Homestake, GALLEX, and SAGE results and the corresponding calculations are given in terms of neutrino capture rates in units of SNU. The Kamiokande and SuperK results and the corresponding calculated results are given in terms of the ^{8}B solar-neutrino flux in units of $10^6 \text{cm}^{-2}\text{s}^{-1}$.

Experiment	Ref.	Result	BP98
Homestake	[12]	$2.56 \pm 0.16 \pm 0.16$	$7.7^{+1.2}_{-1.0}$
GALLEX	[14]	$77.5 \pm 6.2^{+4.3}_{-4.7}$	129^{+8}_{-6}
SAGE	[13]	$67.2^{+7.2+3.5}_{-7.0-3.0}$	129^{+8}_{-6}
Kamiokande	[15]	$2.80 \pm 0.19 \pm 0.33$	$5.15(1.00^{+0.19}_{-0.14})$
SuperK	[16]	$2.45 \pm 0.04 \pm 0.07$	$5.15(1.00^{+0.19}_{-0.14})$

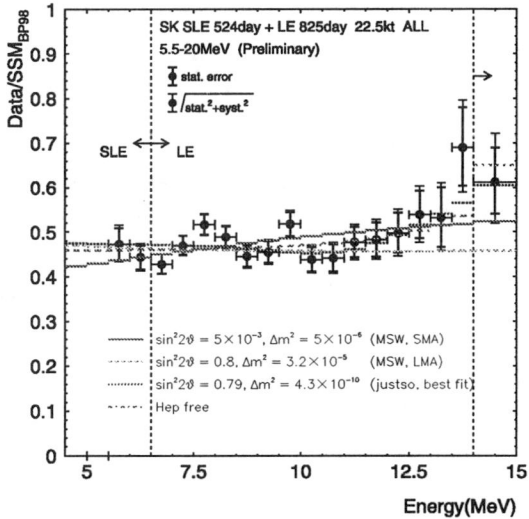

FIGURE 7. Recoil-electron energy spectrum relative to the SSM prediction, Data/SSM(BP98), observed by SuperK with 825 live days. The inner error bars are the statistical errors and the outer error bars are the quadratic sum of the statistical and systematic errors.

The SuperK solar neutrino observations produce not only the average flux but also other interesting data. Among them, the day-night flux difference and the recoil electron energy spectrum have important bearing on the discrimination of various possibilities to solve the solar neutrino problems.

The day-night flux difference, if any, would be caused by the regeneration of ν_es by the Earth. Should the MSW Large-Mixing solution be correct, an observable day-night flux difference would be expected. However, the SuperK results on the day and night fluxes are $(2.37\pm0.05^{+0.07}_{-0.06})\times 10^6$ cm^{-2} sec^{-1} and $(2.51^{+0.06}_{-0.05}\pm0.07)\times 10^6$ cm^{-2} sec^{-1}, respectively, and (Night-Day)/(Night+Day)/2 $= 0.065\pm0.031\pm0.013$. The significance of the observed day-night difference is about a 2σ level.

The SuperK recoil-electron energy spectrum normalized to the prediction of BP98 SSM calculation [25] is shown in Fig. 7 and compared with predictions for some ($\sin^2 2\theta$, Δm^2) values. Although errors are large, the data points at high-energy end are higher than the average. This might be explained by the larger contribution from "hep" neutrinos than the SSM prediction. However, to conclude anything, more statistics and better understanding of systematics are needed.

REACTOR EXPERIMENTS

Two reactor neutrino oscillation experiments, CHOOZ [26] and Palo Verde [27] investigated the Δm^2 range down to 10^{-3} eV^2 for $\bar{\nu}_e$ disappearance, $\bar{\nu}_e \to \bar{\nu}_x$, using liquid scintillator detectors. The CHOOZ detector is located in a 300-m water equivalent (mwe) underground laboratory at a distance of about 1 km from the Chooz power station in the Ardennes region of France. The Palo Verde detector is lacated at a distance of 750 - 890 m from three reactors at the Palo Verde Nuclear Generating Station near Phoenix, Arizona, but it has only 32 mwe overburden. Consequently, the Palo Verde experiment suffered from a large background rate, though the detector is subdivided.

The 90% CL excluded regions from these experiments are shown in Fig. 1. These results excludes the possibility that the atmospheric neutrino anomaly observed by Kamiokande is due to $\nu_\mu \to \nu_e$ oscillations, thus in concordance with the SuperK result.

In the context of three-flavor neutrino oscillations, the CHOOZ result gives a bound to the angle θ_{13}, $\sin^2\theta_{13} \leq 5\times 10^{-2}$.

ACCELERATOR EXPERIMENTS

LSND and Karmen 2

The LSND experiment searched for the appearance signal of the $\bar{\nu}_\mu \to \bar{\nu}_e$ oscillation. The 90% "favored" region from this experiment [18] is shown in Fig. 1. In addition, LSND found an evidence for $\nu_\mu \to \nu_e$ oscillations [19]. The oscillation probabilities are $(3.3\pm 0.9\pm 0.5)\times 10^{-3}$ for $\bar{\nu}_\mu \to \bar{\nu}_e$ and $(2.6\pm 1.0\pm 0.5)\times 10^{-3}$ for $\nu_\mu \to \nu_e$. The favored regions and oscillation probabilities for these CP-conjugate oscillation channels are consistent.

For the $\bar{\nu}_\mu \to \bar{\nu}_e$ oscillation, the Karmen 2 experiment has a sensitivity that nearly, but not entirely, covers the LSND favored region. However, Karmen 2 found no evidence, and the result of oscillation search [20] is shown in Fig. 1. At 90% CL, the region $0.2 < \Delta m^2 < 1$ eV^2 of the LSND favored region is not excluded yet.

CHORUS and NOMAD

CHORUS [28] and NOMAD [29] experiments at CERN searched for $\nu_\mu \to \nu_\tau$ oscillations by looking for the appearance of τ^-, in the context of massive neutrinos which might be a component of mixed dark matter. Their sensitivity goals are similar, $\sin^2 2\theta \sim 2 \times 10^{-4}$ for large Δm^2. At present, oscillations are excluded for $\sin^2 2\theta > 10^{-3}$ at 90% CL in the large Δm^2 limit. Also, $\Delta m^2 > 1$ eV^2 is excluded at $\sin^2 2\theta = 1$ (maximal mixing).

Status of K2K

The K2K (KEK-to-Kamioka) long baseline neutrino oscillation experiment aims at exploring neutrino oscillations $\nu_\mu \to \nu_x$ in the Δm^2 range of $10^{-2} \sim 10^{-3}$ eV^2, suggested by the atmospheric neutrino observations. ν_x may be ν_e, ν_τ, or ν_s, though the SuperK results strongly favor ν_τ. Muon neutrino beams are produced by the KEK 12-GeV Proton Synchrotron, and are shot to the SuperK located at a distance of 250 km from KEK. With a horn-focused neutrino beam with an average energy of 1.4 GeV, $\nu_\mu \to \nu_e$ is studied in the appearance mode, while $\nu_\mu \to \nu_\tau$ is studied in the disappearance mode. The goal of the experiment is to observe a few hundred charged-current events (in the case of no oscillations) in the 22.5 kton fiducial volume of SuperK with 10^{20} protons on the production target.

Initial K2K results are obtained from June and November runs in 1999. The total number of protons brought to the target amounted to 7.2×10^{18} for the runs used for the analysis. In SuperK, FC events which were correlated in time with 1.1 μsec neutrino beam pulses from KEK were searched for, and six events were found. Of these, three events had the vertex within the SuperK fiducial volume. Other 3 events had their vertex outside the fiducial volume. From the neutrino flux measured by the near detector and the near/far flux ratio deduced from the measurement of pion kinematics just after the magnetic horn, The expected number of SuperK FC events with vertex inside the 22.5 kton fiducial volume is estimated to be $12.3^{+1.7}_{-1.9}$ for no neutrino oscillation. For the SuperK FC events with vertex outside the fiducial volume, the expected number is $5.5^{+1.1}_{-1.2}$ for no neutrino oscillation. It looks that observed number of FC events with the vertex inside the SuperK fiducial volume is much smaller than the expectation for no oscillation. However, in order to conclude anything about neutrino oscillation, clearly more statistics are needed.

FORTHCOMING EXPERIMENTS

There are a number of forthcoming neutrino oscillation experiments which are at various stages; a new solar neutrino experiment SNO already started observation, some are under construction, and others are yet to be formally approved.

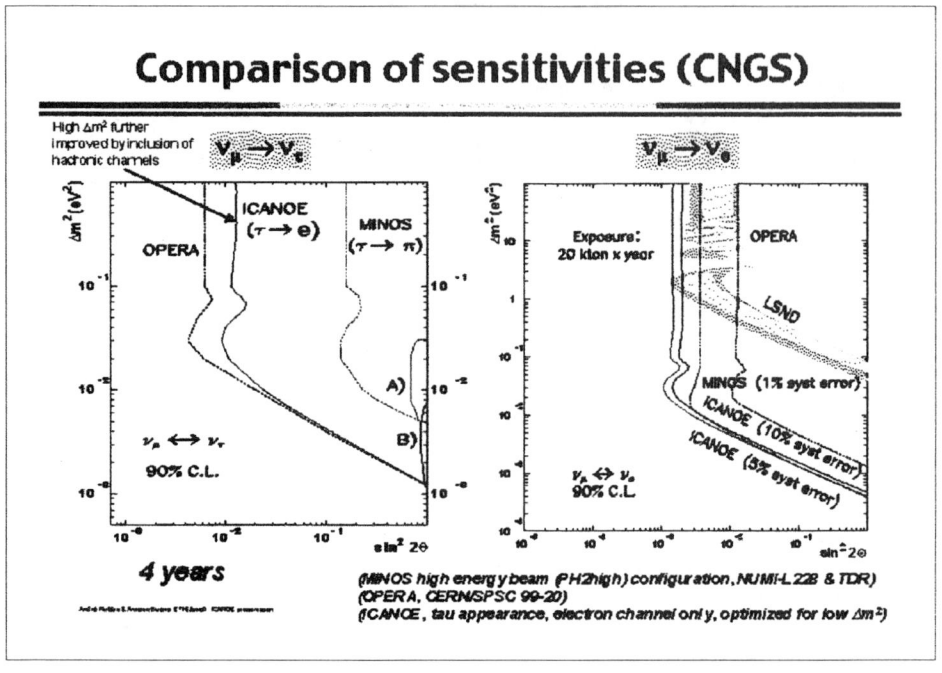

FIGURE 8. Comparison of ICANOE, OPERA, and MINOS sensitivities [30].

The main problems to be addressed by these forthcoming experiments are (i) to confirm the SuperK evidence for atmospheric ν_μ oscillations and to determine the oscillation mode by long baseline neutrino oscillation experiments, (ii) to solve the solar neutrino problem by identifying the unique solution, and (iii) to test the LSND evidence for $\bar{\nu}_\mu \to \bar{\nu}_e$ oscillations.

For the confirmation of the SuperK evidence for atmospheric ν_μ oscillations, K2K already started running. There are three other long baseline neutrino oscillation experiments with the baseline distance of 730 km; MINOS (Fermilab to the Soudan mine, under construction, expected to turn on in 2003), and OPERA and ICANOE (CERN to Gran Sasso, yet to be approved formally, expected to turn on in 2005). MINOS is a magnetized calorimeter experiment and OPERA is an emulsion experiment. The ICANOE detector is a combination of liquid argon TPC and a magnetized calorimeter. Though K2K is a $\nu_\mu \to \nu_x$ disappearance experiment, other three experiments have capabilities to measure the appearance of τ^-. Figure 8 compares the sensitivities of these experiments for both $\nu_\mu \to \nu_\tau$ and $\nu_\mu \to \nu_e$.

In the solar neutrino observations, the measurements of energy spectrum of the solar neutrinos and the day-night flux difference, and the measurement of solar-neutrino flux by utilizing neutral-current reactions are key issues. SuperK continues high-statistics measurements of the day-night flux difference and recoil electron spectrum. SNO, which uses 1,000 tons of heavy water D_2O, started data-taking at the end of 1999. SNO measures solar neutrinos through both inverse beta decay ($\nu_e d \to e^- pp$) and neutral-current interactions ($\nu_x d \to \nu_x pn$). In addition, νe scattering events will also be measured.

The Borexino experiment with 300 tons of ultra-pure liquid scintillator is under construction at Gran Sasso, and expected to turn on in 2001. The primary purpose of this experiment is the measurement of the ^{7}Be solar neutrino flux by lowering the detection threshold for the recoil electrons to 250 keV: if the vacuum oscillation is the solution, it causes seasonal variation of the ^{7}Be solar neutrino flux.

KamLAND, which is under construction at Kamioka and will be completed in 2001, is a multi-purpose neutrino experiment with 1,000 tons of ultra-pure liquid scintillator. One of the primary purposes of KamLAND is the search for neutrino oscillations by measuring neutrinos produced by power reactors. The sensitivity region of KamLAND (shown in Fig. 1) includes the MSW Large Mixing solution. It may be proved or disproved in 5 years of measurements. KamLAND can also observe the ^{7}Be solar neutrinos if the detection threshold can be lowered to a level similar to that of Borexino. The sensitivities to the solar-neutrino problem of the ^{7}Be solar neutrino observations are also shown in Fig. 1.

Finally, to test the LSND evidence for $\bar{\nu}_\mu \to \bar{\nu}_e$ and $\nu_\mu \to \nu_e$ oscillations, the MiniBooNE experiment at Fermilab is under construction: it is expected to turn on late in 2001. The MiniBooNE detector will be diluted liquid scintillator (total mass is 807 tons and fiducial mass is 445 tons) and will measure the electron appearance signature in the ν_μ beam from the Fermilab 8-GeV Booster. The expected sensitivity of MiniBooNE is shown as "BooNE expected" in Fig. 1.

CONCLUSIONS

The evidences for the neutrino oscillation from the SuperK atmospheric neutrino observations and from the solar neutrino experiments are becoming solid. To accommodate yet another evidence from LSND, a sterile neutrino species must be invoked in addition to the three known neutrino species. These evidences will be challenged by the forthcoming neutrino oscillation experiments. They, together with the ongoing experiments, eventually settle the current problems and may bring new surprizes. Once established, the neutrino mass will lead us not only to a new era of neutrino physics, but also to a wealth of new physics beyond the Standard Model.

REFERENCES

1. Fukuda Y. et al., *Phys. Rev. Lett.* **81**, 1562 (1998).
2. Fukuda Y. et al., *Phys. Rev. Lett.* **82**, 2644 (1999).
3. Fukuda Y. et al., *Phys. Lett. B* **467**, 185 (1999).
4. Fukuda Y. et al., *Phys. Lett. B* **335**, 237 (1994).
5. Casper D. et al., *Phys. Rev. Lett.* **66**, 2561 (1991); Becker-Szendy R. et al., *Phys. Rev. D*, **46**, 3720 (1992).
6. Ambrosio M. et al., *Phys. Lett. B* **434**, 451 (1998).
7. M. Spurio, *Nucl. Phys. B (Proc. Suppl.)* **85**, 37 (2000).
8. Allison W.W.M. et al., *Phys. Lett. B* **391**, 491 (1997).
9. Allison W.W.M. et al., *Phys. Lett. B* **449**, 137 (1999).
10. Mann W.A., hep-ex/9912007.
11. Fukuda S. et al., submitted to *Phys. Rev. Lett.*
12. Cleveland B.T. et al., *Astrophys. J.* **496**, 505 (1998).
13. Abdurashitov J.N. et al., *Phys. Rev. C* **60**, 0055801 (1999).
14. Hampel W. et al., *Phys. Lett. B* **447**, 127 (1999).
15. Fukuda Y. et al., *Phys. Rev. Lett.* **77**, 1683 (1996).
16. Fukuda Y. et al., *Phys. Rev. Lett.* **82**, 1810 (1999).
17. For a recent review of solar neutrinos, see Nakamura K., in *Reviews of Particle Physics (RPP 2000)*, *Eur. Phys. J. C* **15**, 366 (2000).
18. Athanassopoulos C. et al., *Phys. Rev. Lett.* **77**, 3082 (1996).
19. Athanassopoulos C. et al., *Phys. Rev. Lett.* **81**, 1774 (1998).
20. Jannakos T.E., *Nucl. Phys. B (Proc. Suppl.)* **85**, 84 (2000).
21. Murayama H., *Reviews of Particle Physics (RPP 2000)*, *Eur. Phys. J. C* **15**, 360 (2000).
22. Honda M. et al., *Phys. Rev. D* **52**, 4985 (1995).
23. Hatakeyama S. et al., *Phys. Rev. Lett.* **81**, 2016 (1998).
24. For example, see Lipari P., and Lusignoli M., *Phys. Rev. D* **58**, 073005 (1998).
25. Bahcall J.N., Basu S., and Pinsonneault M.H., *Phys. Lett. B* **433**, 1 (1998).
26. Apollonio M. et al., *Phys. Lett. B* **466**, 415 (1999).
27. Boehm F. et al., *Phys. Rev. Lett.* **84**, 3764 (2000).
28. Radicioni E., *Nucl. Phys. B (Proc. Suppl.)* **85**, 95 (2000).
29. Lupi A., *Nucl. Phys. B (Proc. Suppl.)* **85**, 91 (2000).
30. Taken from http://pcnometh4.cern.ch/LNGS_nov99.pdf.

Detection of Dark Matter

Maria C. P. Isaac

Center for Particle Astrophysics, 310 Le Conte Hall
University of California - Berkeley CA 94720

Abstract.
The identification of dark matter is still one of the major unsolved problems in physics. The core question of the problem concerns the nature of dark matter: is it baryonic, non-baryonic or if it is a mixture of normal matter and unknown particles, how much of which is in the universe. At last but not at least, a fundamental technological problem is how do we look for it. What are the techniques that will permit us to identify this major component of the universe? In this talk I will cover some well-known evidence for the existence of dark matter, discuss what would be its nature and give you a view of the current experimental panorama. Since time and space were limited, it was impossible to cover the whole subject in depth, and the view presented here of dark matter and its detection is somewhat incomplete.

EVIDENCE FOR DARK MATTER

Evidence for dark matter dates as far back as the 1930's, [1] [2]. The inconsistency between the visible amount of matter and its gravitational effects indicate that a large portion of the gravitationally bound matter is dark, does not absorb or emit light.

One of the strongest evidences for dark matter is in the rotation curves of galaxies. These curves clearly show the presence of matter at large distances from the galactic center, where there is no visible matter. Figure 1 (right) shows a fit of the experimental values of the rotational velocity as a function of the radius for NGC3198, with different mass contributions of the disk, the dark matter halo and HI regions [3]. The halo model assumed here is spherical. Clearly the mass of the halo dominates the gravitation potential at large radii.

Weak gravitational lensing uses background galaxies to map out the distribution of dark matter at large scales. Light from distant galaxies travel to us through many clumps of matter and dark matter that act as a lens scattering the light from the source. This phenomenon has been successful in revealing the existence of dark matter at very large scales. Figure 1 (left) is a well known image from the Hubble space telescope where the image of a distant blue galaxy is multiplied and deformed by the foreground cluster of galaxies. This technique is being used to map the distribution of dark matter, that is a function of its nature and global cosmological parameters [4].

Adding to the astrophysical challenges, a recent piece of the puzzle is actually changing the picture: The evidence for a cosmological constant or vacuum energy. Recent results from high redshift supernova type Ia searches [5] have indicated that we live in an ever-expanding universe with a finite cosmological constant. Other recent results from observations of fluctuations in the cosmic microwave background corroborate this view. Figure 3 shows the constraints in the Ω_Λ Ω_{matter} plane obtained by the MAXIMA collaboration [6] superimposed on the results from SN type Ia. These results indicate that around 60% of the energy density of the universe is completely unknown, while around 40% of it is in some form of matter, most of it of unknown nature[1]. The good news is that there is a wide variety of experimental efforts in many fronts that will lead us to an answer about the nature of this matter. I will try, in the remaining pages, give an overview of these efforts.

SEARCHES FOR BARYONIC DARK MATTER

One way to search for baryonic dark matter is to look for gravitational microlensing of compact objects when they pass in front of a light source. These compact objects would have masses ranging from $10^{-7} - 10^{-8}$ to 1 solar masses. Some of the current projects that search for MACHOS (Massive Compact Halo Objects) in the halo of our galaxy use stars in the Large Magellanic Cloud as light source. They are the MACHO Search and EROS. These experiments recently completed their life cycle and are actually building the case for non-baryonic dark matter, as we will see later.

We can describe the principles of gravitational microlensing as follows: the gravitational potential of an object of mass M will deflect light from a source if this object happens to be in the line of sight between a light source and the observer. This deflection will create multiple images, as in the case of distant galaxies and clusters of galaxies, but because of the distances involved, these multiple images will not be resolved (their separation is of the order of miliarcseconds) and the observer will effectively see an amplification of the light of the source. Moreover the duration of the event, how long the light from the source is magnified, is also a function of the mass of the object between observer and source. The resulting light curve from the source is achromatic and symmetric, defining a clear signal. The EROS and the MACHO collaboration have surveyed about 30 million stars combined. Details from the analysis of EROS can be found in [7] [8] and references therein. The MACHO collaboration results can be found in [9] [10]

The results from these two groups are very exciting. Due to the "meager crop" of events [8] we now have constraints in the fraction of the galactic halo made of compact objects. A halo made exclusively of MACHOs is ruled out and the favored

---

[1] The value of Ω represents the ratio between the present energy density in the universe and the critical density, $\rho_c = 3H_0/8\pi G$.

halo mass fraction is 20%. Besides the overall lack of events another interesting result is the lack of short events. Some candidates were observed with durations suggesting more massive objects, of the order of a solar mass, like white dwarfs. The likehood that white dwarfs are responsible for the dark matter in the galaxy is low, as pointed out in [11]. The observation of multi-TeV gamma rays by HEGRA puts strong constraints on the abundance of white dwarf progenitors, limiting the possibility of white dwarfs as baryonic dark matter. Figure 4 shows the 95% confidence level diagram on the halo mass fraction in the form of compact objects obtained from EROS and MACHO data. Details can be found in [8]. These results imply that another type of dark matter needs to contribute to the mass of the halo of the Milky Way.

Besides the observation that compact objects do not contribute significantly to the dark halo of the galaxy, there is new evidence from the primordial nucleosynthesis of light nuclei [12] that the total contributions from baryons to the energy density in the universe is $\Omega_{baryon} \sim 0.04$. These facts, coupled with growing experimental evidence, from a different number of experiments, that $\Omega_{matter} \sim 0.4$ indicate that the bulk of dark matter might then be "exotic", in a particle form currently unknown or unaccounted for.

SEARCHES FOR NON-BARYONIC DARK MATTER

In search for non-baryonic dark matter, a few candidates come to mind: neutrinos, axions and WIMPS (Weak Interacting Massive Particles). I will briefly cover neutrinos and axions and I will spend the remaining of the time discussing WIMPS and their detection.

Massive neutrinos would certainly contribute to the dark matter. In addition to contribute to the dark matter in the universe, non-zero neutrino masses would indicate physics beyond the Standard Model. In fact, since 1998 there is evidence for the existence of non-baryonic dark matter in the universe, in the form of massive neutrinos, from the results obtained by the Superkamiokande experiment on atmospheric neutrino oscillations [13].

To give a cosmologically interesting contribution to the energy density in the universe (Ω), a relatively narrow range of neutrino masses is allowed, namely $m_\nu \sim$ 1 − 30 eV, favoring $m_\nu \sim 10$ eV. The Superkamiokande evidence points towards a mass difference of the order of 10^{-3} eV^2. If this is the mass scale of neutrinos, then they are simply not heavy enough to contribute significantly to the dark matter in the universe. Another objection to massive neutrinos being the main component of dark matter is related to their fermionic nature. For example, dwarf galaxies are dominated by dark matter. Neutrinos alone could not be the dark matter in these galaxies, because such high density would violate the Pauli exclusion principle. Tremaine and Gunn [14] discuss the requirements for neutrino masses in dark matter-dominated galaxies. Another important objection to neutrinos as a major dark matter component come from arguments of structure formation. It is thus more than justifiable to search for other particles that would fulfill all the dark matter requirements.

Another candidate for dark matter is the axion. The axion was introduced to solve the strong CP problem of small electric dipole of the neutron. It was constrained by laboratory searches, stellar cooling and SN1987A to be very light, of the order of a few mili electronvolts. The current searches use the axion conversion into a photon in the presence of a magnetic field (Primakoff conversion) as means of detection. The frequency of the emitted photon is a function of the axion mass. The signal observed would be a small increase of power in the "axion mass" frequency. Two experiments are currently undertaking axion searches using this principle. These experiments are upgrading their set up and their sensitivities are beginning to probe axion models (see [15] and references therein for details and current limits). These experiments and their upgrades are quickly reaching the existing viable models for the axion. In the next years, we will be able to say more about axions and their contribution to dark matter.

WIMP Searches

WIMPS (Weak Interacting Massive Particles) are prime dark matter candidates. They would be massive, electrically neutral, weak interacting and long lived particles. If we assume that such particles have roughly the same gauge couplings as quarks and leptons, it is expected that large amounts of WIMPS were produced in the early universe, when thermal energies were high enough to allow for WIMP production in particle collisions. At these temperatures WIMPS are in thermal equilibrium with all other particles. As the universe expanded and cooled the WIMP production was "frozen". The abundance of a particle that falls out of thermal equilibrium after it has become non-relativistic is related to its annihilation rate at freeze-out. In the case of particles that are "cosmologically significant", this rate is similar to the one characteristic of the weak interaction. WIMPS are thus thought to have a cross section of the order of the weak interaction cross section. See [16] and [17] for a detailed explanation on the WIMP argument.

A promising candidate for WIMP is provided by the lightest supersymmetric particle, plausibly the neutralino χ. In most versions of the low-energy theory, there is a conserved multiplicative quantum number, R-parity: If R-parity is conserved, then the lightest supersymmetric particle is stable, since there is no kinematically allowed state with negative R-parity which it can decay to [2]. The lightest neutralino is a mixture of the supersymmetric partners of the photon, the Z and the two neutral CP-even Higgs bosons present in the minimal extension of the supersymmetric

---

[2] $R = (-1)^{3(B-L)+2S}$, where B is the baryon number, L, the lepton number and S the spin of the particle. $R = +1$ for ordinary particles and $R = -1$ for supersymmetric particles. Supersymmetric particles, thus can only be created or annihilated in pairs in reactions with ordinary particles. Neutralinos and anti-neutralinos produced in the early universe should have a relic abundance today.

standard model [18]. The neutralino is a very attractive dark matter candidate: it is electrically neutral, it is stable and it has gauge couplings and a mass that accommodates a large range of parameters in the supersymmetric sector yielding a relic density $\Omega_{matter} \sim 0.3$.

In this is experimental overview I will discuss WIMPs detection methods, mainly the direct detection methods but I will also touch on the indirect detection methods and their complementarity.

Direct detection methods are based on the elastic scattering of WIMPS and nuclei in the detector. An effective nucleon-WIMP elastic scattering cross section is obtained by summing the WIMP-quark scattering interaction over all quarks of the nucleon. In the low momentum transfer limit, the contributions of all nucleons are summed coherently over all nucleons, resulting in a WIMP-nucleus cross section. The estimation of the WIMP-nucleus cross section can be found in [16]. The treatment in the framework of the Minimal supersymmetric Standard Model can be found in [19].

The expected nuclear recoil spectrum and event rates depend on the cross section, the velocity distribution of WIMPS in the laboratory frame and the target nucleus. Table 1 [20] gives some values for different targets of the expected event rates and average energy deposition for different WIMP masses. Details on the calculation of rates and spectra can be found in [21].

Such low event rates and small energy depositions call for large target masses, very low background environments, low threshold and, if possible, some effective background rejection mechanism. Two of the possible background rejection mechanisms being used by direct search experiments are annual modulation and nuclear recoil discrimination.

"Conventional" WIMP detection methods rely on very large masses and low background environments. If background rejection mechanisms are available they are "statistical", not performed on an event-by-event basis. Conventional detection methods are represented by the use of standard germanium diodes (operating at 77K) and scintillating crystals. I will comment on some of these efforts.

The Heidelberg group is taking advantage of their years of experience in low radioactive background expertise and low background detector development to search for dark matter. This experience has been obtained with their experiments to search for neutrinoless double beta decay. Recently they have invested in an anticoincidence scheme for background event rejection based on having one Ge detector inside another. The prototype of the HDMS (Heidelberg Dark Matter Search) ran for about a year and achieved reasonable background rates [23]. Upper limits on WIMP-nucleon cross sections, obtained by the Heidelberg group, were until recently the most stringent ones for spin independent interaction. They were achieved after 0.53kgyr of measurement with one of the enriched ^{76}Ge detectors of the Heidelberg-Moscow experiment. A background level of 0.05events/kg/yr/keV in the energy region between 8.8 and 100 keV was reached [22]. The Heidelberg group has a very ambitious proposal of a 100 kg set up of bare Ge diodes in liquid nitrogen, that would serve the dual purpose of cooling down the detectors and providing an excellent shielding for external backgrounds. GENIUS, as it is called, is applying for funds in Europe.

The use of NaI detectors for WIMP searches relies on the large masses of detector that can be used. There are possible discrimination tools such as the use of pulse shape discrimination to differentiate between nuclear recoils and background, or annual modulation of the signal. The PSD technique takes advantage of the difference in the scintillation spectrum in nuclear recoil and electron recoil interactions. The UK Dark Matter collaboration, has been developing this technique for a number of years now. Their set up is located in a mine at Boulby in Yorkshire. The mine has large caverns at a depth of 1100 m, situated in salt rock. They have been plagued by an anomalous population of events that mimic the time constant distribution of nuclear recoils in NaI. The collaboration has developed a diagnostic array and has been searching for the origin of these events unsuccessfully, although recent results seem to indicate they are originated in the surface of the crystals, prompting the collaboration to start to use bare crystals in their experiment [24].

The DAMA (DArk MAtter) collaboration in Italy has recently published evidence for WIMP detection based on the annual modulation signal obtained after 4 years of data taking with 100 kg NaI array operating at the Gran Sasso Laboratory. The annual modulation signal discrimination is based on the fact that the Earth's velocity with respect to the dark matter halo of the galaxy is sinusoidally modulated by the Earth's orbit around the Sun. The laboratory frame velocity of the WIMPS changes during the year resulting on a variation on the recoil spectrum. The great advantages of this discrimination technique is the known phase of the modulated signal. The modulation amplitude, though, is of the order of a few percent of the total WIMP interaction rate, thus requiring an incredible stability of the experimental setup, since the signal is a small modulation on a large constant background. The DAMA collaborations uses a signal processing technique very similar to UK dark matter collaboration. This "pulse shape analysis" was used in the publication of their results in 1996. The recent result of 4 years of data, though, uses an "annual modulation only" analysis. Based on this observed annual modulation, the DAMA collaboration recently published an allowed region for WIMPS ([25] and references therein).

Other methods for dark matter detection rely heavily on discrimination techniques. Although a low background environment is still of great importance, being able to differentiate on an event by event basis between nuclear recoil and electron recoils can yield to background rejection levels >99%. Some types of cryogenic dark matter detectors operate on the principle that the ratio of ionization production to recoil energy in a semiconductor differs for nuclear and electron recoils. These detectors measure phonons and electron-hole pair production to determine the recoil energy and ionization yield for each event. Of course this technique can only be used at very low temperatures. Other detection techniques are used at low temperatures.

EDELWEISS, a cryogenic detector experiment, uses Ge crystals and records both phonon and ionization signals. The ionization readout is similar to conventional Ge detectors while the phonon readout is done with NTD thermistors. EDELWEISS is locate the Frejus Underground Laboratory, and during 1998-1999, the experiment was upgraded, and is currently pushing for lower background rates. Recent results can be found at [26].

The CRESST (Cryogenic Rare Event Search with Superconducting Thermometers) collaboration uses of a dielectric crystal with a small superconducting film, a transition edge thermometer, evaporated onto the surface. CRESST has its set up at the Gran Sasso. During phase I of the experiment they use tungsten films and sapphire absorbers, running near 15 mK. CRESST-I does not have background rejection capability but due to a very low threshold, is sensitive mainly to low mass WIMPS. For phase II of the experiment, scintillating crystals will be used as target materials. These absorbers permit background rejection via the simultaneous measurement of phonons (via the superconducting thermometers) and scintillation light. Nuclear recoils have a lower light yield compared to electron recoils (x-rays or electrons). A description of the CRESST experiment and future prospects can be found in [27]. The principles and description of the technique for background discrimination with scintillation can be found in [28].

CDMS (Cryogenic Dark Matter Search) also measures phonons and ionization on an event by event bases. The collaboration has initially used NTD thermistors as phonon sensors. More recently the CDMS collaboration has developed a solution to instrument large areas and be sensitive to athermal phonons. This is done coupling quasi particle traps with transition edge sensors [29].

Above 10 keV, CDMS detectors can discriminate between electron and nuclear recoil with an efficiency larger than 99%. Surface events are rejected with 95% efficiency. The CDMS I experiment is located at a depth of 16 mwe at the Stanford University campus. A cosmic ray veto (better than 99% efficient) rejects muon-coincident particles. A recent CDMS run with 96 live days with four 165g Ge detectors resulted the most sensitive dark matter limit to date [30]. In addition, the recent CDMS results signal the first significant stringent limit obtained by cryogenic detectors in dark matter searches. CDMS is building the second phase of the experiment in a deeper underground site in the Soudan mine. The new laboratory should become operational late in 2001 and CDMS will achieve a sensitivity 3 orders of magnitude lower than its current limit. Figure 5 shows the CDMS limit with the DAMA allowed region. Note that CDMS excludes at >84% C.L. the entire region allowed by DAMA.

Other WIMP detection methods are being explored. One interesting method is based on metastable gels. SDDs, Superheated Droplet Detectors, are not sensitive to minimum ionizing particles, can have high fluorine content (important for spin dependent WIMP interactions), and can be operated at moderate costs. An array of these detectors is being installed in an underground laboratory and will produce results very soon [31].

Liquid Xe is also a good dark matter detection medium. Being a high mass target, Xe is better suited for detection of high mass WIMPS. It gives larger recoil energy and it has potentially an excellent recoil discrimination via different modes of operation: liquid Xe is used with pulse shape discrimination; liquid/gas operation takes advantage of proportional scintillation in gas. A prototype chamber using the scintillation properties of gas is being assembled in the Boulby mine in the UK and should be operational by the end of 2000 [24]

The last detection technology I will mention is based on negative ion drift on a TCP. This technique can potentially discriminate between WIMPs and background based on directionality. The DRIFT (Directional Recoil Identification From Tracks) detector concept is being developed and a prototype is being installed in the Boulby mine in the UK [32]. The capability to obtain the direction of WIMPS is a powerful tool to discriminate nuclear recoils due to WIMPS and nuclear recoils due to a possible neutron background.

"Indirect" WIMP detection methods are based on the detection of WIMP-anti WIMP annihilation products, such as gamma rays and high energy neutrinos. Indirect WIMP detection methods are complementary to direct detection, in the sense that theory predicts high annihilation rates for low WIMP- nucleon interaction cross section. Obviously this is dependent on the nature of the WIMP and is valid for the prime WIMP candidate, the neutralino. Experiments such as MACRO already started to explore parameter space for supersymmetric models [33]. Other experiments such as GLAST and VERITAS can explore the indirect detection of WIMPS by gamma rays from the halo [34], while AMANDA, IceCube and other neutrino telescopes, would explore the indirect WIMP detection via a neutrino channel [35]. The advantage shared with gamma rays is that neutrinos keep their original direction. A high energy neutrino signal from the center of the Sun or Earth would be a way of discriminating between WIMP annihilation and a high energy neutrino background.

CONCLUSIONS

I hope I was able to give you a good picture of the current situation regarding the prospects of detection of dark matter. Non-baryonic dark matter is needed more than ever. The recent results in measurements of the primordial abundances of the light elements, combined with cosmic microwave background measurements seem to indicate that the major part of matter density in the Universe is in an unknown form. The experimental situation regarding axions and neutrinos is also becoming clearer, with positive results for neutrino oscillations from SuperK, new neutrino oscillation experiments coming online and axion experiments probing the most favored axion models.

Several directed detection experiments are starting to probe interesting regions of the parameter space of the minimal supersymmetric model. The incompatibility between the DAMA claim and the recent CDMS results do not alter the fact that experiments such as CDMS, GENIUS and DRIFT with its tracking capability, are potentially able to discover supersymmetry before the large colliders. Note that even though controversial, the annual modulation signal claimed by the DAMA experiment has been interpreted as being due to the neutralino [36].

The large neutrino telescopes and satellite experiments will contribute enormously, as they probe two important annihilation channels of supersymmetric WIMP particles. As a new generation of experiments and innovative techniques come online, we approach the solution of the 70-year old mystery of the nature of dark matter. Maybe it will be solved sooner than we thought!

(Thank you KG and SG!)

REFERENCES

1. J. H. Oort; Bull. Astr. Inst. Netherlands 6, 249, 1932
2. F. Zwicky; Helv. Phys. Acta 6, 110
3. J. Navarro; Astro-ph/9801073
4. D. M. Wittman et al.; Astro-ph/0003014
5. S. Perlmutter et al.; ApJ. 517, 565, 1999
6. A. Balbi et al.; Astro-ph/0005124 and ApJ Letters, in press.
7. E. Aubourg et. al; Nature 365, 623, 1993
8. T. Lasserre et al.; Astro-ph/0002253
9. C. Alcock et al.; Nature 365, 621, 1993
10. C. Alcock et al.; Astro-ph/0001272
11. K. Freese, B. Fields and D. Graff; Astro-ph/0002058
12. D. Tyler et al.; Astro-ph/0001318, Submitted to Physica Scripta
13. S. Hatakeyama et. al.; Phys. Rev. Lett., 81, 2016, 1998
14. S. Tremaine and J. E. Gunn; Phys. Rev. Lett. 42, 407, 1979
15. C.Hagmann et al.; Phys. Rev. Lett., 80, 2043, 1998
16. P. J. E. Peebles; *Principles of Physical Cosmology* Princeton University Press, 1993
17. E. W Kolb amd M. S. Turner; *The Early Universe*, Addison-Wesley, 1990
18. H.E. Haber and G.L. Kane, Phys. Rep. 117, 75, 1985
19. G. Jungman, M. Kamionkowski and K. Griest; Phys. Rep 267, 195, 1996
20. S. Golwala; Ph.D. Thesis, University of California, Berkeley, 2000.
21. J.D. Lewin nd P. F. Smith; Astropart. Phys., 6, 87, 1996.
22. L. Baudis et al.; Phys. Rep. 307, 291, 1998
23. B.Majorovits, for the HDMS collaboration, 4th International Symposium on Sources and Detection of Dark Matter/Energy in the Universe February 23-25, 2000, Marina del Rey, CA.
24. N. Spooner *ibid*.
25. R. Bernabei et al.; preprint ROM2/2000/01 AND INFN/AE-00/01.
26. A. Benoit et al.; Phys.Lett. B479, 8, 2000.
27. The CRESST Dark Matter Experiment: Status and Perspectives, M. Sisti et al., Proc. of the 8th Int. Workshop on Low Temperature Detectors LTD-8, Dalfsen, Netherlands, 15-20 Aug. 1999
28. CRESST-Collaboration; Astropart.Phys. 12, 107, 1999.
29. K. D. Irwin et al.; Review of Scientific Instruments 66, 5322, 1995
30. R. Abusaidi et al.; Phys.Rev.Lett. 84, 5699, 2000.

target	R_0 (kg^{-1}d^{-1}) M_δ (GeV c^{-2})			$E_0 r$ (keV) M_δ (GeV c^{-2})		
	10	100	1000	10	100	1000
H ($A = 1$)	3.9×10^{-6}	3.9×10^{-5}	3.9×10^{-7}	0.8	1.0	1.0
Si ($A = 28$)	7.8×10^{-2}	5.5×10^{-2}	8.1×10^{-3}	2.2	17.6	26.6
Ge ($A = 73$)	3.0×10^{-1}	5.4×10^{-1}	1.3×10^{-1}	1.2	25.9	64.1
I ($A = 127$)	5.8×10^{-1}	1.7×10^{0}	6.4×10^{-1}	0.8	26.7	101.7

TABLE 1. Rates (R_0) and average energy deposition ($E_0 r$) from different WIMP masses and different WIMP targets. R_0 and $E_0 r$ are indicative values of the rates and energy deposition expected.

31. J. I. Collar et al; Astro-ph/0005059
32. C. J. Martoff, 4th International Symposium on Sources and Detection of Dark Matter/Energy in the Universe February 23-25, 2000, Marina del Rey, CA.
33. The MACRO Collaboration; Nucl. Phys. B (Proc. Suppl.) 87, 108, 2000
34. L. Bergstrom and P. Ullio Nucl. Phys. B 504, 27, 1997.
35. F. Halzen; Comments Nucl. Part. Phys. 22, 155, 1997
36. R. Arnowitt and P. Nath; Phys. Rev. D60, 044002, 1999.

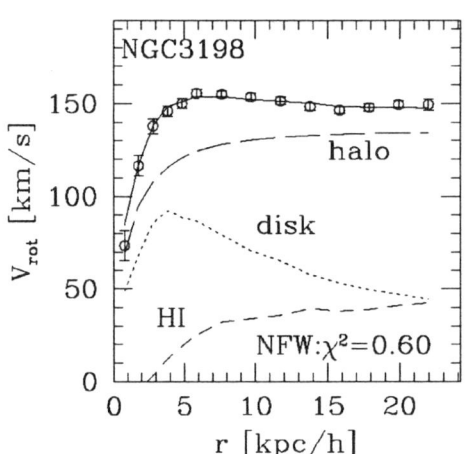 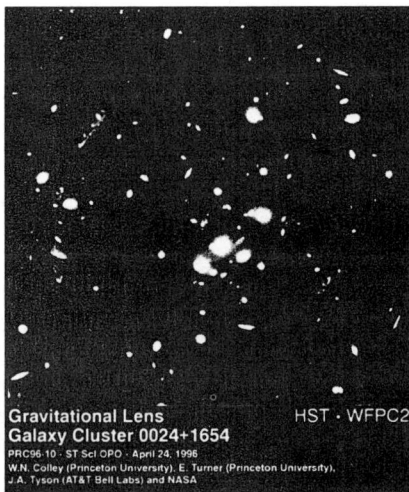

FIGURE 1. Right:Fit of the rotation curve of NGC3198 with the contributions from the gravitational potential of the disk, the halo and HI regions. Left:Evidence of dark matter at large scales: gravitational lensing.

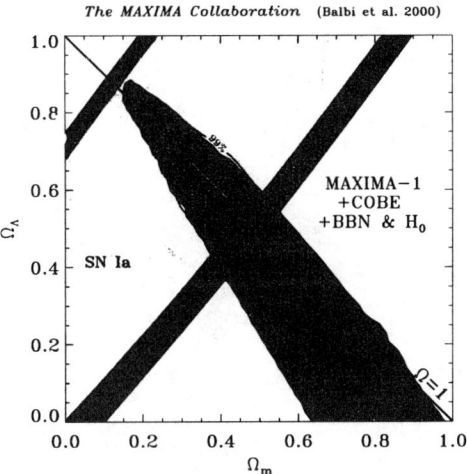

FIGURE 2. Constraints in the Ω_Λ Ω_{matter} plane by the MAXIMA collaboration. Closed contours are the confidence levels of the combined likehoods of CMB and SN Ia results. Here $\Omega_b h^2 = 0.0190 \pm 0.0024$ and $H_o = 65 \pm 7$ km s^{-1} Mpc^{-1} (from Ref. [6], with permission).

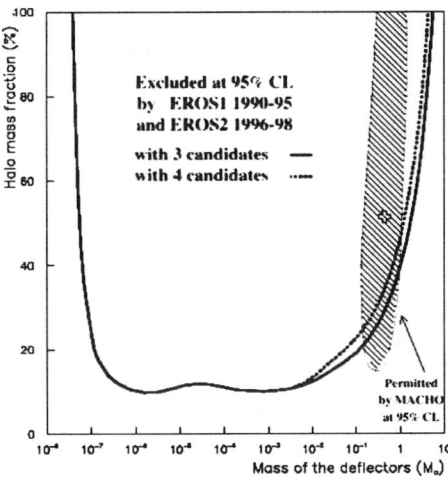

FIGURE 3. 95% confidence level diagram for a halo composed of compact objects.

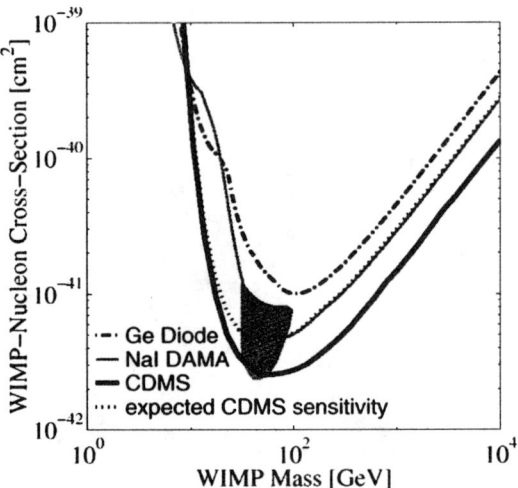

FIGURE 4. Spin independent WIMP-nucleon cross section as a function of WIMP mass. The regions above the curves are excluded at 90% CL. The solid dark curve shows the CDMS limit from ref. [30]. The shaded region is from ref. [25]

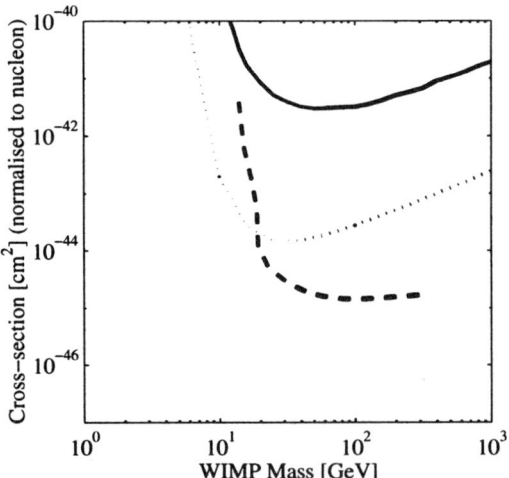

FIGURE 5. Projected exclusion limits for CDMS (dotted line) and GENIUS (dashed line), overlaid to different SUSY models (hatched regions. The current CDMS limit is the solid line on the plot. Regions above the curves are excluded.

New Particle Searches at LEP 200++

Robert A. McPherson

University of Victoria and Canadian Institute of Particle Physics
Dept. of Physics and Astronomy
University of Victoria
PO BOX 3055 MS7700
Victoria, BC, CANADA V8W 3P6

Abstract. The Standard Model of Particle Physics is a successful but experimentally incomplete theory, and the Large Electron Positron Collider running yearly at increasing energies and luminosities remains a unique facility for probing new physics possibilities. Searches for the Standard Model Higgs boson are presented and the signatures for the leading candidate of physics beyond the Standard Model, Supersymmetry, are explored. Constraints on models with large extra dimensions are also discussed.

INTRODUCTION

The Standard Model [1] of particle physics is an extremely successful theory, and has been tested to high precision at the Large Electron Positron Collider (LEP) and at other facilities [2]. Within the Standard Model (SM), gauge symmetry in the electro-weak interaction is spontaneously broken via the self-interactions of a scalar field doublet, the Higgs mechanism [3], and the model predicts the existence of a new particle, the Higgs boson. Fits to precision data prefer a Higgs boson with a mass of approximately 65 GeV, which is already excluded by direct searches as shown in Fig. 1. These indirect constraints predict that the Higgs mass is less than about 180 GeV at the 95% confidence level.

The Standard Model does, however, seem to have problems. The most severe may be the so called "fine tuning" of the Higgs mass due to the hierarchy of energy scales in the theory. The Higgs boson mass must receive radiative corrections due to its interactions with the Standard Model fermions and bosons, and the natural scale of those corrections is the scale of the highest energy to which the theory is valid. It the Standard Model is valid to the Planck mass, the natural scale for the correction to the higgs mass is $\mathcal{O}(10^{18})$ GeV, which is completely inconsistent with the model. Within the Standard Model, it is possible to fine tune the radiative corrections at about 1 part in 10^{20} to remove this problem.

There are several schemes available to "fix" the higgs mass without the severe fine tuning problem. The most popular by far is supersymmetry (SUSY) [4]. SUSY adds a new set of particles to the Standard Model, matching each of the Standard Model particles but with spins differing by a 1/2 unit. If SUSY were an exact symmetry, the new SUSY particles would have the same masses as their SM partners. Since this scenario is experimentally excluded, SUSY must be a broken symmetry. The general minimal supersymmetric extension to the Standard Models adds about 105 new parameters to the theory. This becomes an intractable problem, and instead it is typically assumed that SUSY is broken in some "hidden" sector of new particles, and is "communicated" (or mediated) to the "visible" sector of SM and SUSY particles by one of the known interactions. In this article, we will concentrate on scenarios inspired by "gravity-mediated" SUSY breaking, with a weakly interacting, stable lightest SUSY particle.

Another possibility for fixing the hierarchy problem is to cut off the theory at a relatively low energy scale of ~ 1 TeV. This can be done by assuming some composite substructure of the Standard Model particles. A rather new, and extremely popular, alternative is assuming that there are new, large extra dimensions at about the ~ 1 TeV scale [5–7], assumed to be the true "Planck Mass", M_D. The observed Planck scale, M_{Pl}, is large because gravity freely propagates in these extra dimensions, effectively diluting its influence on experimental observables. We explore several signatures from models with large extra dimensions at LEP.

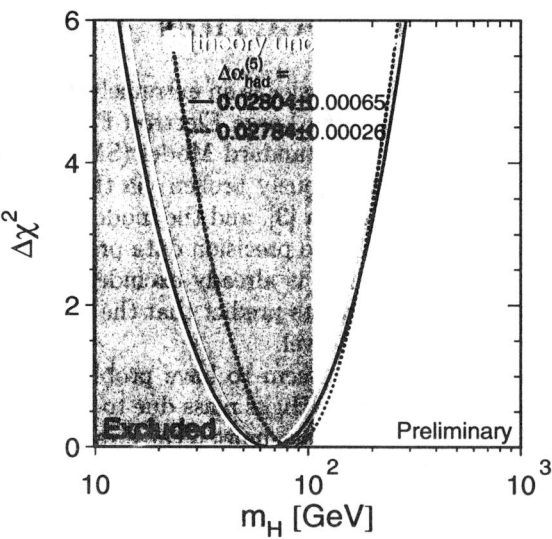

FIGURE 1. The χ^2 vs. the Higgs mass from precision electroweak data. The dark shaded region indicates the theoretical uncertainty. The light shaded region indicated the excluded region from direct searches, discussed in this article.

Large Electron Positron Collider

LEP is an e^+e^- collider at CERN, shown in Fig. 2. The results in this article concentrate on the high energy data taken over the past few years, summarized in Table 1. Results from the four general purpose detectors (ALEPH, DELPHI, L3 and OPAL) are presented in this article. The four experiments have similar sensitivities in the new particle searches considered here.

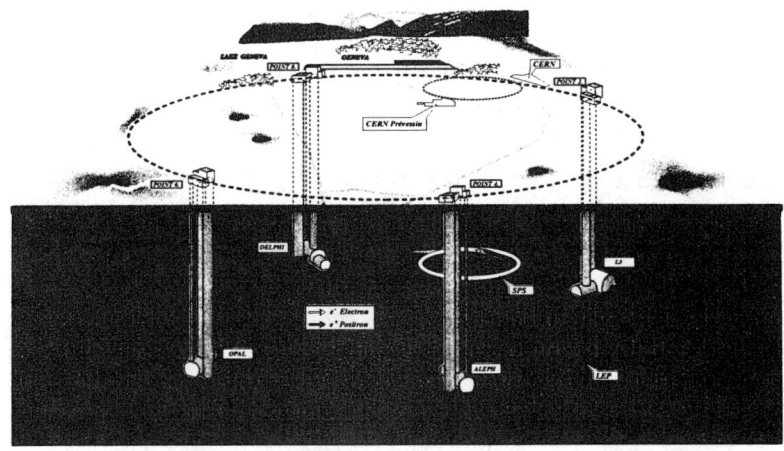

FIGURE 2. Schematic overview of the LEP e^+e^- Collider.

Stage	$\sqrt{s}$	Year	Luminosity
LEP 1	$\approx$ 91 GeV	1989-1995	175 pb^{-1}
LEP 1.5	130-140 GeV	1995	5 pb^{-1}
	161 GeV	1996	10 pb^{-1}
	172 GeV	1996	10 pb^{-1}
LEP 2	183 GeV	1997	55 pb^{-1}
	189 GeV	1998	180 pb^{-1}
	192–202 GeV	1999	230 pb^{-1}
	200–210! GeV	2000	???

TABLE 1. Summary of the energies and luminosities of the LEP Collider.

Standard Model Higgs

The Standard Model Higgs boson, H^0, would be predominantly produced at LEP via the process $e^+e^- \to Z^0H^0$, with a small contribution from the W^+W^- fusion process, both shown in Fig. 3. The Higgs predominantly decays to $b\bar{b}$;

i.e. two hadronic jets with b-flavoured hadrons. The different search channels are determined by the different decay modes of the Z^0 boson, shown in Fig. 4.

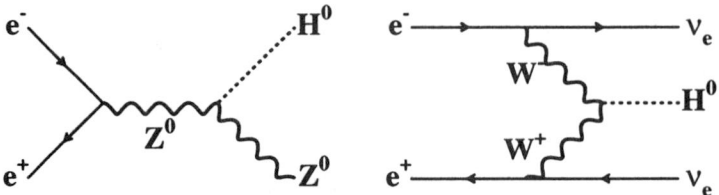

FIGURE 3. The dominant higgs production diagrams at LEP are "Higgs-strahlung" and "WW-fusion".

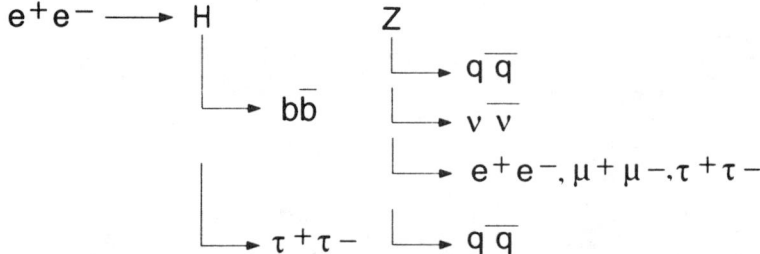

FIGURE 4. The principle Higgs decay modes. The channels considered are 4-jets, 2-jets with missing energy, and 2-jets with 2-leptons. In all cases, the jets are required to have identified b-hadrons.

The four LEP experiments have updated their Higgs searches using the data up the end of 1999 [8–11]. The results are summarized in Fig. 5, which show the reconstructed Higgs candidate masses of the selected events and background (the L3 analysis shown is optimized for a Higgs mass of 105 GeV). No significant excess is observed by any experiment. The experimental mass distributions are combined [12], and shown in Fig. 6. Again, the data is consistent with the background expectation.

Since no excess is observed, limits are derived on the Higgs mass. The confidence level for a Standard Model Higgs signal is plotted *vs.* the Higgs mass in Fig. 7. The limit is derived where the limited inferred from the data crosses the 5% confidence level curve. The four experiments combined exclude Higgs masses less than 107.9 GeV at the 95% confidence level, which is consistent with the median confidence level expected from background only of 109.1 GeV.

SUSY Searches

SUSY is not a well-constrained theory; adding the minimal number of new particles and fields to the Standard Model (the Minimal Supersymmetric Standard

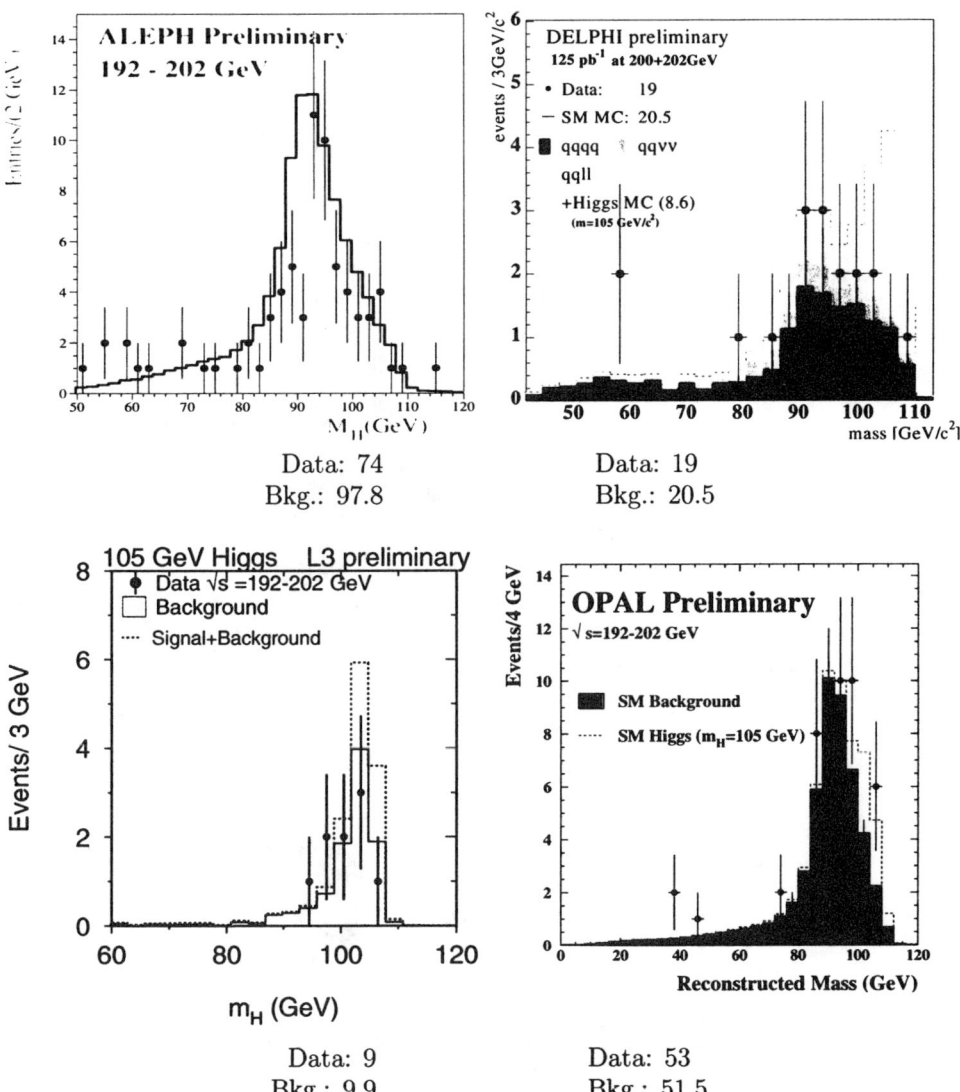

FIGURE 5. Selected higgs candidates from the data (black dots) and background expectation (histograms) from each of the four LEP experiments. The DELPHI, L3 and OPAL plots also show the expectation from a 105 GeV Higgs (dashed lines). There are fewer event in the L3 plot because they show the results from an analysis optimized for a 105 GeV Higgs.

Model, or MSSM) involves 105 additional parameters [13], and most of that parameter space is excluded by low energy experiments. Instead, restrictive models which are phenomenologically viable are constructed and used as benchmarks for

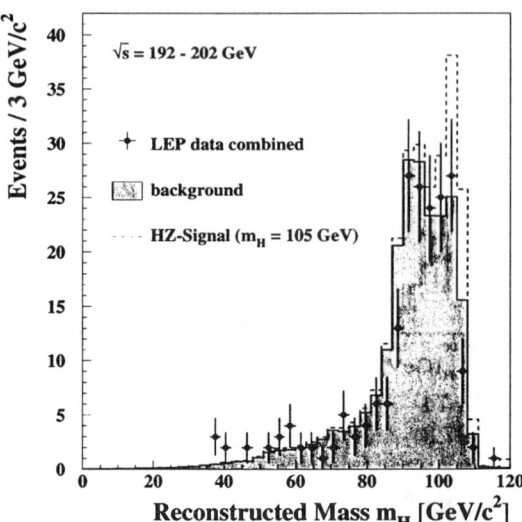

FIGURE 6. Selected higgs candidates from the four experiments combined. The points are the data, the solid line is the background expectation, and the dashed line is the additional contribution from a 105 GeV Higgs.

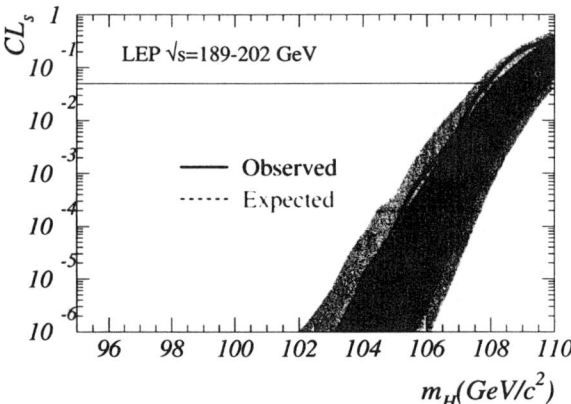

FIGURE 7. The confidence level for a Standard Model Higgs signal plotted *vs.* the Higgs mass. The limit inferred from the data is the solid line, which crosses the 5% confidence line at 107.9 GeV. The median expected limit from background only is shown by the dashed line, with a expected limit of 109.1 GeV. The shaded area is the 1-σ boundary on the expected limit.

searches. One such model is minimal supergravity, or mSUGRA, which contains only 4 parameters beyond the Standard Model. In fact, mSUGRA has restrictions which are not required by low energy phenomenology, such as requiring a common mass among the scalar fermions and higgs particles at the GUT scale. The LEP experiments use a somewhat less restrictive model, the "SUGRA-inspired" constrained MSSM, or CMSSM, using the parameters listed in Table 2, for interpreting their results. This review will only consider R-Parity conserving modes, implying that SUSY particles are produced in pairs and the lightest SUSY particle, the LSP, is neutral, weakly interacting, and stable. We will concentrate on models with a lightest neutralino LSP, the $\tilde{\chi}_1^0$.

Parameter	Description
M_2	EW-scale $SU(2)$ Gaugino Mass
	(GUT Unification gives M_1, M_3)
m_0	Common GUT-scale scalar mass
	(EW-scale masses from RGE's)
$\tan \beta$	v_2/v_1, Ratio of vev's of two higgs doublets
μ	Higgs mixing parameter
m_A	Pseudo scalar higgs mass
A_0	Common trilinear coupling

TABLE 2. Parameters of the constrained MSSM.

Even within the CMSSM, the experimental channel that will show the first signs of SUSY depends on what point in the parameter space nature has actually chosen. The "discovery channels" are listed for different extremes in the parameters in Table 3. The topologies all share the classic SUSY "missing energy" signature, since the observation is of the production of a particle which decays into the invisible $\tilde{\chi}_1^0$ which carries away energy and momentum. The analyses select events with significant missing energy, and then classify them according to the number of hadronic jets, identified leptons or photons, and event visible energy. If the mass difference between the produced SUSY particle and the $\tilde{\chi}_1^0$, ΔM, is large, the SUSY events will have large visible energy, but there are large backgrounds from the Standard Model processes $e^+e^- \rightarrow W^+W^-$ and $e^+e^- \rightarrow ZZ$. If ΔM is small, the dominant background is from "gamma-gamma" collision events, $e^+e^- \rightarrow e^+e^-\gamma^*\gamma^* \rightarrow e^+e^-f\bar{f}$, where f is a lepton or quark. In the intermediate ΔM region, backgrounds are lower and signal detection is more straightforward. The experiments optimize different selections in these different visible energy, or ΔM, regions.

Results from the four LEP experiments in the search for scalar lepton production are summarized in Table 4. A small excess is observed by each experiment in the search for scalar tau lepton production using the 1999 data at $\sqrt{s} = 192$–202 GeV. While the excess is consistent with a signature from new particle production, after detailed studies of the event kinematics, it is also consistent an upward fluctuation in Standard Model $e^+e^- \rightarrow W^+W^-$ production. Confirmation awaits more data.

Of particular interest in the CMSSM are the constraints on the mass of the LSP neutralino, $\tilde{\chi}_1^0$. Incorporating searches for scalar leptons, charginos, and neutralinos, the parameter space of the CMSSM is scanned. If every point in the parameter

Parameter Region		Discovery Channel	Signature
$\mu \ll M_2$ (Higgsino Region)	no $\tilde{\nu}$ coupling	$\tilde{\chi}_1^+ \tilde{\chi}_1^-, \tilde{\chi}_2^0 \tilde{\chi}_1$	Jets $+\not{E}_T$ Jets, Leptons $+\not{E}_T$
$\mu \gg M_2$ (Gaugino Region)	Large m_0	$\tilde{\chi}_1^+ \tilde{\chi}_1^-, \tilde{\chi}_2^0 \tilde{\chi}_1$	Leptons $+\not{E}_T$
	Small m_0	$\ell^+ \ell^-$ $\tilde{t}\tilde{t}, \tilde{b}\tilde{b}$	Leptons $+\not{E}_T$ Jets $+\not{E}_T$

TABLE 3. SUSY search channels

	Experiment	Data	SM Background	
$\tilde{e}$	ALEPH	42	48.1	-0.9σ
	DELPHI	68	59.3	$+0.4\sigma$
	L3	24	24.5	-0.1σ
	OPAL	–	–	–
$\tilde{\mu}$	ALEPH	39	43.4	-0.7σ
	DELPHI	13	20.4	-1.6σ
	L3	48	40.1	$+1.2\sigma$
	OPAL	–	–	–
$\tilde{\tau}$	ALEPH	46	34.2	$+2.0\sigma$
	DELPHI	25	19.2	$+1.3\sigma$
	L3	41	33.7	$+1.3\sigma$
	OPAL	71	56.8	$+2.0\sigma$
Sum	ALEPH	127	125.7	$+0.1\sigma$
	DELPHI	106	98.9	$+0.7\sigma$
	L3	113	98.3	$+1.5\sigma$
	OPAL	381	343.3	$+2.0\sigma$

TABLE 4. Numbers of events selected in the data and expected from Standard Model background processes from the four experiments in the search for scalar leptons, using the $\sqrt{s} = 192$–202 GeV data taken in 1999.

space giving a given $\tilde{\chi}_1^0$ is excluded at the 95% confidence level, than that neutralino mass is considered to be excluded. Example limits on the lightest neutralino mass vs. $\tan \beta$ are plotted in Fig. 8. The limits from the four LEP experiments are shown in Table 5.

Large Extra Dimensions

As discussed in the introduction, the possibility of large additional dimensions (beyond our known 3 spatial and 1 temporal ones) has been the subject of intense theoretical activity over the past 1–2 years. The models generally assume that our Standard Model "4-brane" lives in "bulk" of n additional dimensions, with a typical size R. If gravity is allowed to propagate in this bulk, the effective Planck scale, M_{Pl}, will be smaller than the true Planck scale, M_D. They are related by

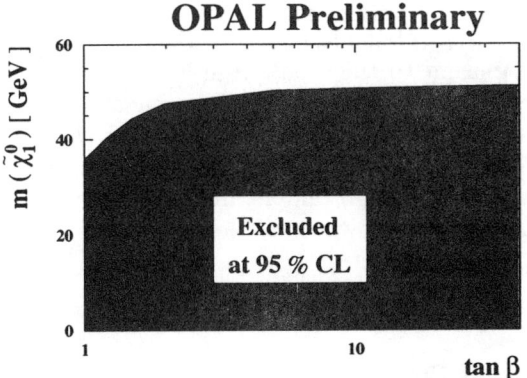

FIGURE 8. Limits on the lightest neutralino mass in the CMSSM plotted vs. $\tan\beta$, from the OPAL experiment. The light shaded region is excluded for all values of m_0, while the dark region is additionally excluded if $m_0 > 500$ GeV. For large values of m_0, the scalar fermions play no important role.

$\sqrt{s}=$	LSP $\tilde{\chi}_1^0$ limit (GeV)	
192-202 GeV	Large m_0	Any m_0
ALEPH	36.8	32.3
DELPHI	35.2	23.4
L3	–	37.5
OPAL	35.7	31.6

TABLE 5. 95% C.L. limits on the lightest neutralino mass in the CMSSM from the four LEP experiments. Limits are given for only large m_0, and also those valid for any value of m_0. For large values of m_0, the scalar fermions play no important role.

the equation:
$$M_{Pl}^2 = R^n \times M_D^{n+2}.$$

This principle benefit is that we can even lower M_D to the electroweak scale, and removed the hierarchy problem completely. Values of R below about 1 mm are not excluded by direct measurements. Assuming M_D is about 1 TeV, this implies that $n \geq 2$. If one assumes, motivated by string theory, that there are 11 additional dimensions, then $n \leq 7$ (for $n = 7$ and $M_D \sim 1$ TeV, R is about 1 fm, which is impossible to probe with current direct experiments).

In these models, Standard Model phenomenology is affected by the interaction of gravitons (G) with Standard Model processes. Some examples in e^+e^- collisions are shown in Figure 9. For example, the process $e^+e^- \to \gamma G$ gives an additional contribution to the signature of single photon production at LEP. The dominant Standard Model contribution to this final state is from $e^+e^- \to \gamma\nu\bar{\nu}$. The four LEP experiments have combined their single photon analyses, and the mass of the system recoiling off of the photon is plotted in Fig. 10 [14]. The data and background agree to within the quoted precision of the Monte Carlo generators used for the background estimates. Example limits on M_D for different values of n from the OPAL experiment are shown in Table 6 [15]. These results are based on the convention of Equation 2 in Ref. [7].

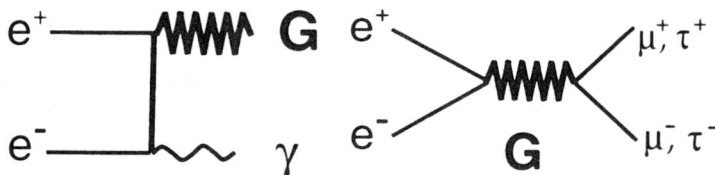

FIGURE 9. Example experimental signatures from gravity in extra dimensions.

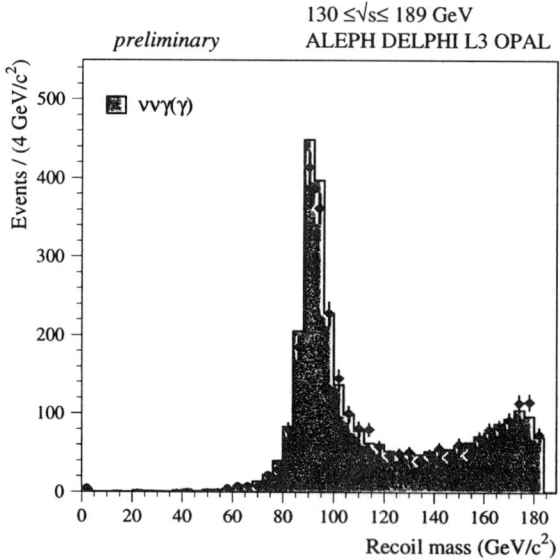

FIGURE 10. The mass of the invisible system recoil off of the observed photon in single photon events.

	OPAL 189 GeV
n	M_D Lower Limit (GeV) 95%C.L.
2	1086
3	862
4	710
5	605
6	528
7	470

TABLE 6. Limits on M_D from single photon events.

Conclusions

Many new particles searches have been performed at the LEP electron-positron collider. While no conclusive evidence for the Standard Model Higgs boson, or physics beyond the Standard Model, has been observed, there are features that require more data for investigation.

REFERENCES

1. S.L. Glashow, J. Iliopoulos and L. Maiani, Phys. Rev. **D2** (1970)1285;
   S. Weinberg, Phys. Rev. Lett. **19** (1967)1264;
   A. Salam, *Elementary Particle Theory*, ed. N. Svartholm (Almquist and Wiksells, Stockholm, 1968)367.
2. See, *eg.*, the article by H. Przysiezniak in these proceedings, and the references therein.
3. P.W. Higgs, Phys. Lett. **12** (1964) 321;
   F. Englert and R. Brout, Phys. Rev. Lett. **13** (1964) 321;
   G.S. Guralnik, C.R. Hagen and T.W.B. Kibble, Phys. Rev. Lett. **13** (1964) 585.
4. Y. Gol'fand and E. Likhtam, JETP Lett. **13** (1971) 323;
   D. Volkov and V. Akulov, Phys. Lett. **B46** (1973) 109;
   J. Wess and B. Zumino, Nucl. Phys. **B70** (1974) 39.
5. N. Arkani-Hamed, S. Dimopoulos, G. Dvali, Phys. Rev. **D59** (1999)86004.
6. J.L. Hewett, Phys. Rev. Lett. **82** (1999) 4765.
7. G.F. Giudice, R. Rattazzi, J.D. Wells, Nucl. Phys. **B544** (1999) 3.
8. ALEPH Collab., *Search for the neutral Higgs bosons of the Standard Model and the MSSM in e^+e^- collisions at centre-of-mass energies from 192 to 202 GeV*, ALEPH 2000-006 CONF 2000-003.
9. DELPHI Collab., *Searches for Neutral Higgs Bosons in e^+e^- collisions up to $\sqrt{s} = 202$ GeV*, DELPHI-2000-024 CONF 345.
10. L3 Collab., *Search for the SM Higgs boson in e^+e^- collisions at centre-of-mass energies up to 202 GeV*, L3 note 2511, March 6, 2000.

11. OPAL Collab., *Search for Neutral Higgs Bosons in e^+e^- Collisions at $\sqrt{s} \approx 192$-202 GeV*, OPAL Physics Note PN426, March 3, 2000.
12. The LEP Working Group for Higgs Boson Searches ALEPH, DELPHI, L3, and OPAL, CERN-EP-2000-055.
13. Nucl.Phys.Proc.Suppl. 62 (1998) 469-484.
14. LEP SUSY WG report LEPSUSYWG/99-05.1.
15. OPAL Collaboration, G. Abbiendi *et al.*, *Photonic Events with Missing Energy in e^+e^- Collisions at $\sqrt{s} = 189$ GeV*, CERN-EP-2000-050, Submitted to Eur.Phys.J.C.

What's the Matter? Physics Opportuntities and Future Facilities

Michael H. Shaevitz

Fermilab
Batavia, Illinois 60510
Columbia University
New York, New York 10027

Abstract. Particle and nuclear physics are coming into interesting times over the next decade. Many studies will probe the existence, interactions, and source of matter. This paper will cover investigations at exisiting and future facilities of neutrino mass, studies of QCD and the quark-gluon plasma, and the search for the Higgs boson along with other new types of matter.

I NEUTRINO MASS

Whether neutrinos have mass has been a key question in particle physics since the existence of neutrinos was first postulated. Theoretically, neutrinos are expected to be massive since they are the Standard Model partners of the massive charged leptons. Experimentally it has been shown that neutrino masses are very small. Some models, for example the See-Saw Model, have been developed that explain the smallness of neutrino masses through a mixing process with new isosinglet, heavy neutrinos usually called "sterile neutrinos". In this way, the measurement of a finite neutrino mass would give insights into new physics at the scale of these new heavy particle.

Neutrinos also have important cosmological consequences. There are of order 10^9 neutrinos per m^3 as relics of the big bang. Even a small mass (~1 eV) will effect models of cosmological evolution and relativistic neutrino matter seems needed to explain galaxy and cluster formation.

Direct neutrino mass measurements set limits many orders of magnitude smaller than the charged lepton masses but are now reaching their limits of sensitivity. To probe neutrino masses significantly below 1 eV requires indirect methods such as neutrino oscillation experiments. If the weak or flavor neutrino eigenstates are mixtures of several mass eigenstates with different masses, then one flavor of neutrino can oscillate to another as it propagates over a finite distance. With three generations of neutrinos, the flavor and mass eigenstates are related by a (3×3) unitary mixing matrix similar to the CKM mixing matrix of quarks.

$$\begin{pmatrix} \nu_1 \\ \nu_2 \\ \nu_3 \end{pmatrix} = \begin{pmatrix} c_{12}c_{13} & s_{12}c_{13} & s_{13}e^{-i\delta} \\ -s_{12}c_{23} - c_{12}s_{23}s_{13}e^{i\delta} & c_{12}c_{23} - s_{12}s_{23}s_{13}e^{i\delta} & s_{23}c_{13} \\ s_{12}s_{23} - c_{12}c_{23}s_{13}e^{i\delta} & -c_{12}s_{23} - s_{12}c_{23}s_{13}e^{i\delta} & c_{23}c_{13} \end{pmatrix} \begin{pmatrix} \nu_1 \\ \nu_2 \\ \nu_3 \end{pmatrix} \quad (1)$$

With three mass eigenstates, one can form three values for Δm^2, $\Delta m_{12}^2 = m_1^2 - m_2^2$, $\Delta m_{23}^2 = m_2^2 - m_3^2$, $\Delta m_{31}^2 = m_3^2 - m_1^2$, but only two are independent. At each Δm^2, there can be oscillations between all the neutrino flavors with different mixing angle combinations. (In Eq. 1, s_{ij} and cij refer to the sin and cos of the mixing angle θ_{ij}.) With active mixing between three or more flavors, the phase δ can lead to CP violation effects.

There are currently three experimental indications of neutrino oscillations, the solar neutrino deficit, the atmospheric muon neutrino deficit, and the LSND indication of $\bar{\nu}_\mu \to \bar{\nu}_e$ oscillations. These three results would correspond to oscillations at three different Δm^2 scales as shown in Figure 1. Over the next decade many new experiments will investigate these indications trying to answer the questions: 1) Are the indications really neutrino oscillations? 2) Which flavors are responsible for the oscillations? 3) What are the precise values for the oscillation parameters, Δm^2 and $\sin^2 2\theta$? Oscillations in the LSND region will be probed by the Mini-BooNE experiment at Fermilab and possibly the ORLaND experiment at the SNS facility in Oak Ridge. Accelerator studies of oscillations in the atmospheric region indicated by the Super-K and other will be made by the long-baseline, NuMI/Minos experiment at Fermilab and the CNGS experiments at CERN. Much more information on solar neutrino oscillations will come from the SNO experiment in Canada, the Borexino experiment the Gran Sasso Laboratory in Italy, and the Kamland reactor-based experiment in Japan.

II MUON STORAGE RING ν-FACTORY

The current list of neutrino oscillation experiments will mainly address the validity and parameters associated with the oscillations indications shown in Figure 1. A high intensity muon storage ring, on the other hand, could provide a unique, high-intensity neutrino beam facility for pursuing neutrino oscillation beyond these current measurements and for measuring the other parameters associated with the lepton mixing matrix given in Eq. 1.

The composition and spectra of the intense neutrino beam from a muon storage ring is selectable by the charge, momentum, and polarization of the stored muons, through the decays $\mu^- \to e^- \nu_\mu \bar{\nu}_e$ or $\mu^+ \to e^+ \nu_e \bar{\nu}_\mu$. There is no other comparable source of electron neutrinos and antineutrinos or neutrino beam with such well-understood fluxes. For muon energies above 20 GeV, the beam intensity from

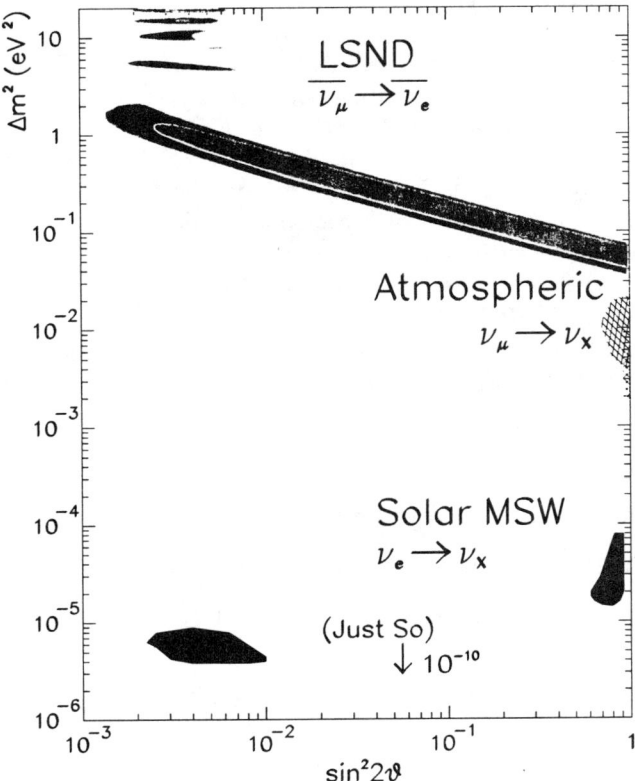

FIGURE 1. Three experimental indications of neutrino oscillations from solar, atmospheric and the LSND experiments.

a muon storage ring is much greater than conventional hadron focused neutrino beams. The combination of almost equal numbers of electron and muon neutrinos allows the study of all types of oscillation processes including $\nu_e \to \nu_{\mu \text{ or } \tau}$. Furthermore, the beam is highly collimated and very long baseline experiments from 2000 to 10,000 km are possible.

Two studies of a muon storage ring facility have recently been performed. The first was an accelerator study charged to develop a design concept for a muon storage ring and associated support facility that could support a compelling neutrino based research program. The goal of the study was to look at a design for a 50 GeV muon storage facility that would provide 2×10^{20} μ decays per year directed to a site 2900 km away such as Fermilab to SLAC or LBNL. The study came up with the conceptual design and identified the areas needing R&D. The conclusion of the study was: *"The result of this study clearly indicates that a neutrino source based on the concepts, which are presented here, is technically feasible."*

The second study looked at the physics motivation for a neutrino source based on a muon storage ring and how such a facility would allow the study of neutrino oscillations beyond the current set of experiments. The motivation for a ν-factory would be to extend the oscillation measurements to higher precision and to measure other parameters associated with the neutrino mixing matrix given in Eq. 1. These measurements will allow tests of the mixing and CP violation phenomenology predicted similar to studies of CKM quark mixing matrix. Specifically, a ν-factory could provide:

1. Unique access to ν_e oscillations to ν_μ or ν_τ

2. Measurements of Δm^2_{23}, θ_{23} to 1% and θ_{13} to the 10^{-4} level

3. A determination of the neutrino mass hierarchy by measuring the sign of Δm^2_{23}

4. Measurements of δ, the CP violation parameter, if solar or LSND parameters are favorable

The measurement of $\sin^2 2\theta_{13}$ can be made using the process $\nu_e \to \nu_\mu$ given the oscillation probability

$$P(\nu_e \to \nu_\mu) = \sin^2 \theta_{23} \sin^2 2\theta_{13} \sin^2 \left(1.27 \Delta m^2_{32} L/E_\nu\right)$$

By using the $\nu_e \to \nu_\mu$ signal, the measurement becomes a search for wrong-sign muons instead of a ν_e signal which has much higher background. For $\nu_e \to \nu_\mu$ oscillations, matter effects can change the oscillation probability as the neutrino propagates through the earth. This happens since the $\nu_e + e \to \nu_e + e$ process has both NC and CC contributions whereas the $\nu_\mu + e \to \nu_\mu + e$ has only a NC contribution. Through this mechanism, the oscillation probability becomes sensitive to the sign of $\Delta m^2_{32} = m^2_3 - m^2_2$ and therefore allows a measurement of the mass hierarchy, whether $m_3 > m_2$ or not. Furthermore, the difference between $\nu_e \to \nu_\mu$ and $\bar{\nu}_e \to \bar{\nu}_\mu$ oscillations is sensitive to CP violations (δ in Eq. 1). Figure 2 shows the

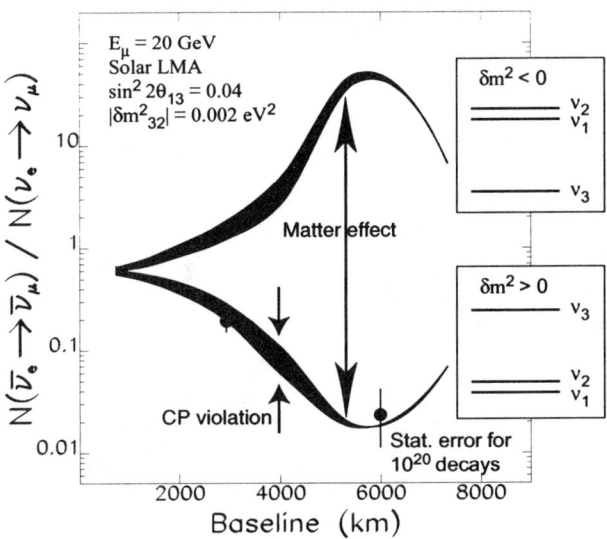

FIGURE 2. Predicted ratios of $\bar{\nu}_e \to \bar{\nu}_\mu$ to $\nu_e \to \nu_\mu$ rates at a 20 GeV neutrino factory. The upper(lower) band is for $\Delta m^2_{32} < 0$ ($\Delta m^2_{32} < 0$). The range of possible CP violation determines the widths of the bands. The statistical errors shown correspond to 10^{20} muon decays of each sign and a 50kt detector

ratio of number of $\bar{\nu}_e \to \bar{\nu}_\mu$ to $\nu_e \to \nu_\mu$ oscillations as function of baseline length for a 20 GeV muon storage ring with 10^{20} μ decays. There is a clear difference between the case with $\Delta m^2_{32} < 0$ (top) and $\Delta m^2_{32} > 0$ (bottom). The width of the curves represent the difference in rate for $\delta = 0$ and $\delta = \pi/2$. The black points with error bars represent measurements with the above statistics and clearly show that CP violation can be measured with a low-energy, modest-intensity ν-factory with the above parameters.

The physics study looked at the physics reach for a ν-factory as a function of muon energy and the number of muon decays per year. Figure 3 shows a summary of these results indicating that an entry level machine with modest energy and intensity could make probe the new oscillation modes and measure the mass hierarchy leading to a high intensity machine that could measure CP violation.

The ν-factory beams could also be used to improve many other types of neutrino measurements. A detector placed 50 to 100 meters from the storage ring would provide large event samples, i.e. over 20 million events in a 1 m liquid D_2 target, with well understood flux and a large ν_e component. In addition, these data can be used to improve electroweak measurements as well as search for exotic physics signatures from heavy neutral leptons, neutrino magnetic moment effects, anomalous

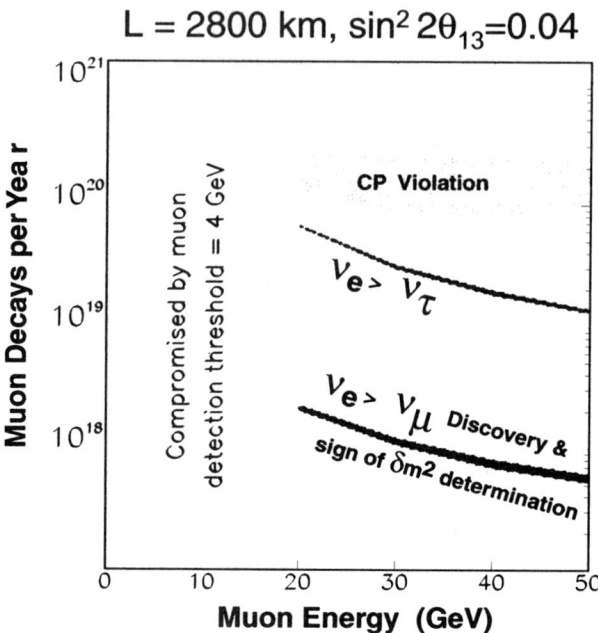

FIGURE 3. The required number of muon decays needed in a neutrino factory to observe $\nu_e \to \nu_\mu$ oscillations in a 50 kton detector and determine the sign of Δm^2 and the number of decays needed to observe $\nu_e \to \nu_\tau$ oscillations in a few kton detector, and ultimately put stringent limits on (or observe) CP violation in the lepton sector with a 50 kton detector.

τ production, and $D^0 - \overline{D}^0$ mixing studies.

III FUTURE QCD AND QGP MEASUREMENTS

Over the next decade, the luminosity upgrade for the HERA program, the energy upgrade for the TJ-Lab, and the ν-factory will provide high precision structure function and parton distribution measurements over a huge kinematic region covering $10^{-6} < x < 1$ and $0.1 < Q^2 < 10^5$ GeV^2 as shown in Figure 4. For neutrino QCD measurements, the ν-factory data sample will provide high precision determinations of parton distributions with enough statistics on light targets to yield A-dependence studies which rival and complement the charged lepton studies.

After the completion of the current HERA luminosity upgrade, ZEUS and H1 will be in a position to make a complete survey of the partons in the proton at low x. The heavy charm and bottom seas can be probed along with the strange sea from charged-current processes. Precision quark distribution measurements will allow the gluon density to be determined and $\gamma - Z$ interference effects can be used

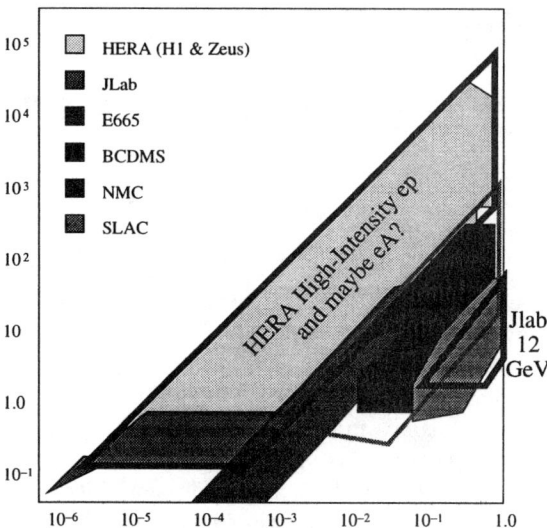

FIGURE 4. Kinematic range in x and Q^2 covered by various experimental programs.

to test the Standard Model. At TJ-Lab, studies can be done of the transition from the quasi-elastic to the deep-inelastic scattering (DIS) region. Nuclear effects for quarks and hadrons traversing the nuclear medium are accessible as are polarization effects associated with the nucleon spin components.

Important progress has been made recently on understanding the spin of the nucleon. The Bjorken sum rule has been firmly established but the Ellis-Jaffe sum rule seems to be violated. The decomposition of the nucleon spin seems to show that ~0.3 is in quarks with the strange quark giving approximately -0.1. The rest of the spin could therefore be associated with the gluons and antiquarks. The RHIC-Spin program will provide new opportunities to make spin studies with hadrons. Polarized Drell-Yan processes are sensitive to quark and antiquark spin components and gluon-fusion production measure the gluon spin contribution. Current DIS data suggest a gluon polarization at the level of $\Delta G = 1.8 \pm 0.6 \pm 1.3$.

The RHIC collider at Brookhaven with Pb + Pb collisions at 200 GeV/nucleon will bring QCD studies into a new regime, that of exploring the non-perturbative QCD vacuum and possibly quark-gluon plasma states. This will be followed by the future LHC collider at 5.5 TeV/ nucleon. These relativistic heavy ion collisions should reach densities and temperatures to produce a phase transition from normal hadron matter to a quark-gluon plasma. The phenomenology of the transition is difficult to model even qualitatively and, thus, many independent signatures will be needed to understand the transition. Signatures include deconfinement, chiral symmetry restoration, thermal radiation, jet quenching, strangeness/charm production. As in any exploration of a new physics regime, most likely there will be many surprises leading to possible new breakthroughs in our understanding of the application of QCD.

IV WHAT MAKES THE MATTER?

In the Standard Model (SM), the weak and electromagnetic interactions are unified and mediated by vector bosons associated with the SU(2)×U(1) gauge group. This unified interaction is broken by a process referred to as electroweak symmetry breaking (EWSB). In the SM, a single scalar Higgs boson (with mass < ~1 TeV) causes the EWSB and gives mass to the weak W and Z bosons as well all other particles. The comparison of the predictions of this theory to a wide range of electroweak measurements has been used to test the validity of the SM and to derive an upper bound on the Higgs mass of ~200 GeV. Theoretically, on the other hand, radiative corrections can easily make quantities such as the Higgs mass diverge unless one fine tunes the corrections very carefully. For this reason, various extensions to the SM have been proposed such as supersymmetry (SUSY) or technicolor which have extra Higgs bosons and/or other heavy particles that help contain the corrections.

For supersymmetry, every particle has a super-partner with opposite statistics that is the usual fermions have scalar partners and the gauge bosons have spin 1/2 (gaugino) partners. An unexpected consequence of including SUSY is that the weak, electromagnetic, and strong coupling constants seem to unify at a common scale of 10^{16} GeV. If SUSY is the source of the EWSB, then the lowest mass Higgs boson must have a mass below 180 GeV and the scale of the SUSY particles must be a few hundred GeV.

One of the main goals of particle physics over the next decade is to elucidate the source of the EWSB. The key experimental ingredients in this quest are to: 1) discover the Higgs boson(s), 2) verify the Higgs mechanism as the source of mass, and 3) measure the mass spectrum of new particles at the few GeV scale. Experimentally, hadron and electron colliders have and will provide the facilities for this quest as shown in Figure 5.

The Fermilab Tevatron $\bar{p}p$ collider at 2 TeV and integrated luminosities up to 30 fb^{-1} will explore the Higgs mass region up to about 180 GeV and search for SUSY particles with masses less than 400 GeV. In the later half of the decade, the Large Hadron Collider (LHC) will explore proton-proton collisions with a $E_{cm} = 14$ TeV and integrated luminosity samples up to 500 fb^{-1}. The LHC has great discovery potential and will be sensitive to new particles associated with supersymmetry and extra weak bosons associated with expanded theories. The LHC will probe SM Higgs boson masses up to 1 TeV and can cover almost the entire mass region associated with the Minimal Standard Supersymmetric Model. It is almost assured that some new physics signal will be discovered at the Tevatron and/or LHC.

After these discoveries, the next step will be to answer the questions associated with the mechanisms for EWSB. The hadron colliders will provide much informa-

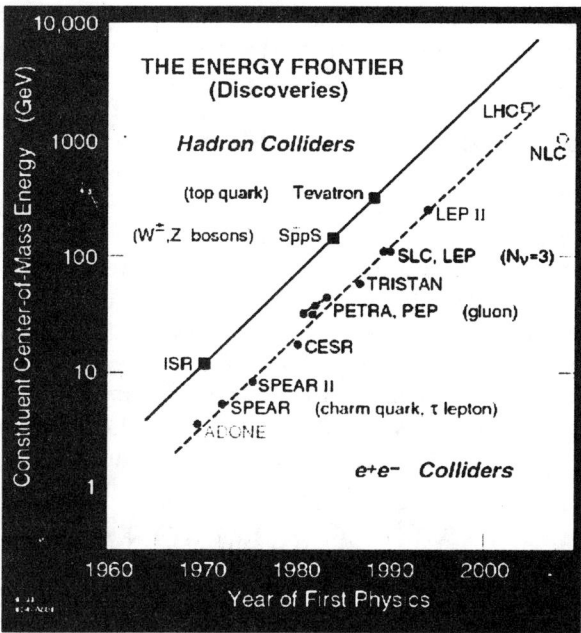

FIGURE 5. Timeline for hadron and electron colliders including major discoveries.

tion on these questions but e^+e^- linear colliders (LC) could offer a complementary probe with unique capabilities coming from the interaction of fundamental point particles. Several machines have been proposed: the Tesla collider is a 0.5-0.8 TeV superconducting machine about 20 km long, the Next-Linear-Collider has $E_{cm} = 0.5 - 1.5$ TeV from a 20km long warm RF acceleration structure, and CLIC is proposed to achieve ~3 TeV using a 40 km long two-beam acceleration scheme.

The LC machines can be used to accurately measure the mass, width, and spin of the Higgs boson. To test the Higgs mechanism, the LC would determine the couplings of the Higgs to various mass particles such as the bottom and charm quark, the τ-lepton, and the W/Z bosons. These couplings can also be used to discriminate a SM Higgs boson from one associated with the supersymmetric models. The Higgs potential, an important ingredient to EWSB in the SM, can also be determined from $e^+e^- \rightarrow ZHH$ and $\nu_e\nu_e HH$. In these ways, the LC will allow the precise investigations of the particles physics in the mass region up to a few TeV.

V FUTURE POSSIBILITIES AT THE ENERGY FRONTIER

To go beyond the few TeV mass region will take some breakthroughs in accelerator technology. One proposal is to push towards a hadron collider with a diameter of 200 km. Using low field magnets (~2 Tesla) would allow this machine to reach E_{cm} of 40 TeV which then could be upgraded with high field magnets (~12 Tesla) to 240 TeV. This machine, referred to as the Very Large Hadron Collider (VLHC) and would give discovery potential at the energy frontier after the LHC. If the Tevatron/LHC/LC sees the source of EWSB, then the VLHC can explore it in depth. On the other hand, if the new physics is beyond the LHC then the VLHC will be necessary to make progress.

Another option for a multi-TeV technology is the $\mu^+\mu^-$ collider. The muon collider uses beam particles are point constituents that do not have sizeable synchrotron radiation. Thus, relatively small circular rings can be used to reach energies of 3 to 5 TeV. In addition, the muon coupling to Higgs type particles is 40,000 times larger than electron machines. Much R&D is needed to develop systems to accelerate the muon quickly before it decays. This development has started in association with the muon storage ring ν factories discussed earlier and the technology of a muon collider may be an outgrowth of this program.

VI CONCLUSION

The future looks very interesting for nuclear and high-energy physics. With the many new facilities that are coming on-line over the next decade, it is almost a certainty that new discoveries will be uncovered. Our understanding of the source and interactions of matter may be fundamentally changed, giving new insights and theoretical understanding. This has been one of the prime goals of these fields and was put very eloquently by the late Director of Fermilab, Robert R. Wilson. "We have to understand in the simplest terms, what matter is, in order to understand who we are."

Intersections 2000: What's New in Hadron Physics [*]

James D. Bjorken

Stanford Linear Accelerator Center
Stanford University, Stanford, California 95409

Abstract. Hadron physics is that part of QCD dealing with hadron structure and vacuum structure, almost all of which is nonperturbative in nature. Some of the open problems in this field are outlined. We argue that hadron physics is a distinct subfield, no longer within particle physics, and not at all the same as classical nuclear physics. We believe that it needs to be better organized, and that a first step in doing so might be to establish hadron physics as a new division within the American Physical Society.

I. THE BIG PICTURE

The main portion of this talk deals with the subject of hadron physics: what it is, what some of its challenges are, and why I believe the hadron-physics community needs to identify itself more strongly and precisely in order to define and protect its long range experimental program. But before turning to that, it may be of use to put this subject in the context of the bigger picture of basic particle-physics goals. Another reason is that this conference has not been just about hadron physics. By my count about 60 percent of the parallel sessions dealt with hadron-physics issues, while only 45 percent of the plenary talks were on hadron physics, the remainder dealing with the bigger picture.

For most of the last twenty years, the Big Picture in particle physics has centered around the Big Three issues, namely Higgs, SUSY, and CP. I believe, as does most everyone, that the most important of these is the problem of mass and the nature of the Higgs sector, responsible for electroweak symmetry breaking. While this question seems timeless, having been around in an almost unchanged form for over two decades, our perspective of it has actually shifted somewhat. Thanks to the discovery at Fermilab of a very heavy top quark, and to the many beautiful precision electroweak measurements from CERN, SLAC, and elsewhere, the mass of the Higgs boson cannot be too large. This is encouragement that we will within

---

[*] Work supported by the Department of Energy under contract number DE–AC03–76SF00515.

a decade have a direct experimental handle on the question, from experiments at the Fermilab Tevatron and/or the CERN Large Hadron Collider.

How this turns out experimentally will have a profound influence on the future of the field. I like to contrast the options in terms of two extremes. One is the "desert" scenario, where the theory remains essentially what we now have all the way up to a very high mass scale, for example the Grand Unification scale of 10^{15} GeV or so. This can only occur if the Higgs boson, the only undiscovered particle that the desert scenario requires, has a mass of 160 ± 20 GeV. Other masses are ruled out by the requirements of vacuum stability and the absence of strong Higgs self-interactions (cf. Fig. 1). The other extreme is that of the supersymmetric extension of the standard model (SUSY, or MSSM), where each known particle has its superpartner, differing in spin by one-half unit, and with the superpartner masses typically less than a TeV. Also, the MSSM Higgs sector is larger, with at least one member expected to have mass less than about 130 GeV.

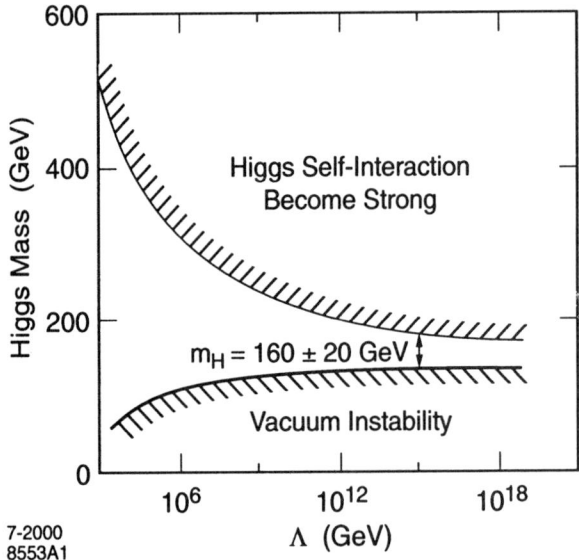

FIGURE 1. Values of m_{Higgs} and momentum scale for which the Standard Model exists, *i.e.* where electroweak perturbation theory converges. The upper region is forbidden because the self-interactions of the Higgs particle become strong. The lower region is forbidden because the vacuum itself becomes unstable.

If the SUSY scenario is correct, there will be full employment for experimentalists. They not only need to discover the superpartners, but also to determine the more than 100 extra fundamental parameters that characterize this extension of the standard model. On the other hand, if only the 160 GeV Higgs boson is seen, and nothing else new is found, this will be rather strong direct evidence for the desert scenario. In that extreme, there would be no reliable, landmark, higher-mass scales for new experimental facilities to aim for. It would become more difficult to justify multi-billion dollar future colliders were one to be unable to certify in advance new discoveries, in the way that has been done for previous facilities (W at the SPS, top at the TeVatron, Higgs at the SSC and LHC). For these reasons alone, I see the outcome of the Higgs search as a crucial turning point for the future of the field.

The desert scenario is rather unpopular because of the hierarchy problem, namely the problem of why the Higgs mass remains so low when the natural scale for it, via quadratically divergent radiative corrections, appears to be much larger. This has led to the demand that something be invented to cure the problem, the leading candidate being the MSSM. However, the cosmological constant suffers from a very similar situation, one for which a straightforward application of SUSY does not work. So it seems to me that a serious, viable approach to the Higgs hierarchy problem is to ignore it for the present, arguing that a much deeper source of the solution, at the level of what is required for the cosmological constant problem, is required. And once the hierarchy problem is ignored, the "desert" theory is really very consistent, with no trace of the quadratic divergence problem remaining, after renormalization, in the phenomenology.

While the Higgs situation has not changed all that much in the last two decades, this does not mean that the Standard Model has remained unchanged. Thanks to the strong evidence that neutrino oscillations really exist, we now have a New Standard Model to replace the venerable Old Standard Model. Instead of the twenty or so parameters characterizing the Old Standard Model, we now have thirty or so parameters for the New Standard Model, the exact number depending a bit on what one wants to include in the count. For sure there are three neutrino masses and four CKM-like mixing angles to determine. In addition perhaps the masses of the three heavy Majorana particles of the seesaw mechanism should be included as parameters, as well as a couple of phase factors seen at best only in double-beta-decay processes and the like.

There is, in addition to the new parameter count, a definite shift at the GUT level from $SU(5)$-like thinking to $SO(10)$-like thinking. All this to me represents a significant advance, despite the presence of the extra parameters that require explanation. As evidenced in this meeting, there clearly will be increasing emphasis on the neutrino sector in the future: it carries with it more than 30 percent of the parameters of the New Standard Model, and these parameters will be at least as difficult to determine as accurately as the CKM parameters, at present the focus of the B-physics program. But twenty years ago, the determination of CKM phases seemed to be a remote experimental possibility. Hopefully future progress on the neutrino front will parallel what is happening now in the realm of B physics.

II. HADRON PHYSICS: WHAT IS IT?

The main thrust of this talk has to do with hadron physics. I define it as the physics of hadron structure and of (strong-interaction) vacuum structure. This puts it as a subfield of quantum chromodynamics (QCD), just as chemistry, condensed-matter physics, and atomic physics are subfields of quantum electrodynamics (QED).

In their infancy those three subfields were part of elementary particle physics, but now are not. This was also the case for nuclear physics. And I think the same has already happened to hadron physics. Most elementary particle physicists, including those in positions of influence, do not pay much attention to the issues in this field [1]. And in fact most of the experimental research in this field is done within the nuclear physics community, even though it is a stretch to identify hadron physics with nuclear physics. Because experimental (and of course theoretical) hadron physics research spans all energy scales, this creates social and organizational problems, a subject I will return to later [2].

To deal with these social issues I think that it is fundamental and important to define in detail what hadron physics encompasses, what its long range goals are, and what experimental and theoretical programs are necessary to attain those goals. I cannot by myself articulate this here in full. Hadron physics is very a big subject and I am sure to make errors of omission, and to bias the subject matter toward my own particular interests. In fact a better method might be simply to peruse the contents of the proceedings of this conference. In the next section, I will simply catalog some open problems I find interesting, as examples of the huge challenges that are present in this subject, challenges which exist at quite fundamental levels. And because the subject matter of hadron physics rarely allows reliable perturbation-theory calculations, real progress requires a data-driven approach, characterized by close interaction between theory and experiment. In the final section I will return to the social issues confronting hadron physics.

III. SOME OPEN PROBLEMS IN HADRON PHYSICS

QCD, the basic theory of the strong interactions, is at short distances a perturbation theory of the pointlike quark and gluon constituents of hadrons. At very large distances QCD is a theory of pions and nucleons (and their strange counterparts), and is characterized by spontaneous symmetry breaking of the approximate chiral symmetry of QCD. At intermediate distance scales, there is the rich arena of *e.g.* hadron resonances, Regge trajectories, soft diffraction, and hadronization of the partons, just to name a few of many topics.

But the physics at all distance scales is linked, and it is hard to find a situation, even within the relatively clean, perturbative regime, where the nonperturbative effects do not enter. Our first example is chosen to illustrate this phenomenon.

1. Perturbing the Chiral Vacuum

A classic way of trying to understand the properties of a macroscopic system is to perturb it with a small, localized impurity and study its response. In the case of the chiral vacuum of long-distance QCD, a nice way of doing this is by putting a small color dipole, such as heavy onium, into the vacuum and examining its response. This implies creation at large distances of a very weak pion cloud around the onium. The importance of this cloud can be assessed by putting in another small dipole, and determining the long range force between them, due to essentially two-pion meson exchange. This has been done elegantly and cleanly by Fujii and Kharzeev [3], who find that this force dominates at separations greater than about 0.6 fermi. To be sure the potential energy associated with this effect is quite small, under 1 MeV.

Nevertheless, this effect can be amplified by putting the two dipoles into motion. At the qualitative level, one can see that the original clouds will be compressed into pancakes when the onia become extreme-relativistic. And the original rest energy of a pion cloud, however small, can be turned into an arbitrarily large amount of energy-momentum of a pionic pancake if an arbitrarily large boost is applied. Now boost the two onia in opposite directions and put them into collision, again at a large impact parameter. When the momentum density in each of the pancakes exceeds, say 1 GeV/fermi^2 in the overlap region, there will be ample amounts of cms energy available for particle production, and the two onia should act like light hadrons as far as their collision properties are concerned: in the jargon of the trade they will "exchange a soft Pomeron".

Now the study of onium-onium collisions at extremely high energies is a favorite playground of perturbative-QCD theorists. This is the so called BFKL regime, where much effort has gone into summing up Feynman diagrams to obtain a candidate phenomenology of "hard Pomeron exchange" [4]. Nevertheless, the above argument implies that, for a fixed size of the dipoles, however small, if one goes to high enough energy the perturbation theory approach is destined to fail, and the soft physics is destined to re-emerge. The smaller the dipoles are, the higher will be the energy scale at which this phenomenon occurs.

The pion-cloud argument is not inconsistent with BFKL ideology, which also anticipates a similar phenomenon occurring, due to "diffusion of gluon ladder transverse momenta into the infrared". However, I am not sure that the energy scale where the transition occurs is the same in the two approaches. But the bottom line remains the same: no matter how hard one works for cleanliness in short-distance QCD, the soft physics usually finds its way into the picture.

2. Foundations of Perturbative QCD

Perturbative QCD (pQCD) is a highly sophisticated and well developed subject. However, at a fundamental level there are, I believe, some real problems. The basic issue is that, despite the fact that it is commonly done, it is not legal in pQCD to put the quarks and gluons on mass shell, *i.e.* to treat them as asymptotic states that propagate to infinity. They clearly do not—there is no S-matrix for quarks and gluons. It is rather ironic that in the old days before QCD one had an S matrix formalism without a field theory, while now we need a field theory formalism without an S-matrix.

To see the problem more concretely, it is only necessary to look at the classic process of electron-positron annihilation into hadrons, in lowest order. The associated vacuum polarization amplitude is a quark loop (Fig. 2) which at large spacelike Q^2 is a safe short-distance calculation. Its absorptive part at timelike Q^2 is essentially the cross section of interest. It is not completely ultraviolet safe, requiring an energy average to make it safe. But even with that done, our problem emerges when one wants to look at the angular distribution of the quark-antiquark dijets which build the total cross section. To get that differential cross section, one typically calculates the absorptive part as if the quarks could be put on mass shell—which we must admit is illegal.

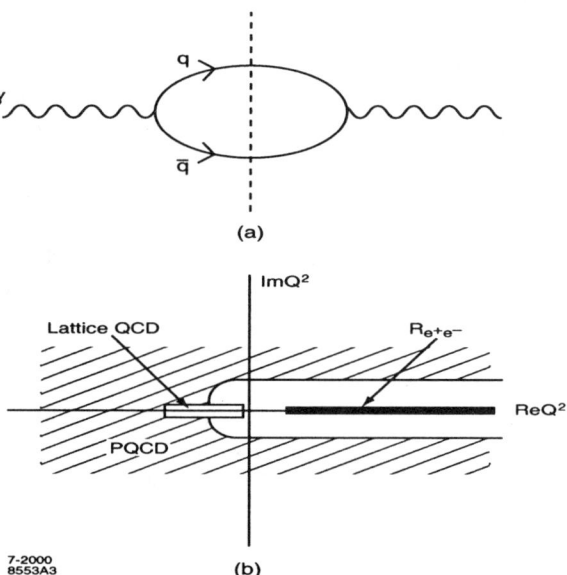

FIGURE 2. (a) The vacuum polarization amplitude whose absorptive part describes the quark-antiquark dijet final state in electron-positron annihilation. (b) The complex Q^2 plane appropriate to this process; only in the shaded region is perturbative QCD justifiable.

That this is not academic can be seen by imagining a QCD where only the bottom quark exists. With no light quarks available, the b and $\bar{b}$ quarks will (probably) be connected by an essentially unbreakable QCD string or flux tube. This leads (probably) to the conclusion that the final states will be a dense spectrum of excited onia, with no jets to be seen. The only way jets could occur is through glueball emission, and this appears to be a highly inefficient mechanism.

So the bottom line is that identification of final state jets with on-shell partons is a model assumption, which at present lacks a firm foundation. To be sure, it is an eminently reasonable assumption. But it would be better to have a sounder line of argument.

3. Parton Correlations and Multiplicities

How many quarks are there in a proton? "Three", says the spectroscopist. But the deep-inelastic community will (or should) answer "infinity". Both answers have their place, but connecting the two is still a problem. For example, there is a substantial sub-community of hadron physicists, in particular the practitioners of "exclusive QCD", who use a Fock-space description of the partons comprising a light-cone proton, with the "leading Fock-space component" having three and only three quarks in it. Now given that the average number of quarks is infinite (being essentially proportional to the integral over $\ln x$ of $F_2(x)$), this would mean that the multiplicity distribution of partons is quite peculiar, with the "Fock-space" piece of it of finite mean multiplicity (with how much total weight, please?), while the rest is of infinite multiplicity.

I am simply baffled that this inconsistency of approach seems not to be recognized at all as a problem. When I mention it to others, the response seems to be that I am the one with the problem. Maybe this is so. But maybe there is a clue in the example of pion clouds around onium in item 1. A partonic description of the collision process described there (especially if one of the onia is replaced by a spacelike photon, which also makes a splendid small color dipole) leads essentially to a parton distribution as shown in Fig. 3. The region to the right is essentially perturbative and probably amenable to the Fock-space, perturbative methodology. But present in addition are all the nonperturbative partons comprising the pion cloud. For phenomenology which concentrates on the large-x valence system, the cloud partons are presumably inconsequential. However, as one goes from the case of small color dipoles to realistic light hadrons, the role of the cloud partons becomes much more important, and in the light-quark limit one must work hard to justify their neglect.

Even leaving this issue aside, the parton multiplicity distribution itself is poorly understood, to say the least. Is it Poissonian, or KNO? How might one distinguish one from the other? And the correlations of the partons in the transverse plane are largely unknown. For example, are most of the infinite sea of wee partons inside the three constituent quarks, are they mostly outside, or are they mostly uncorrelated?

There are good reasons why such simple questions remain unanswered, and most of them have to do with the fact that it is very hard to get at them experimentally. Double parton—or multiple parton—collisions have the potential to provide information on the correlations. While some experimental work has been done already,

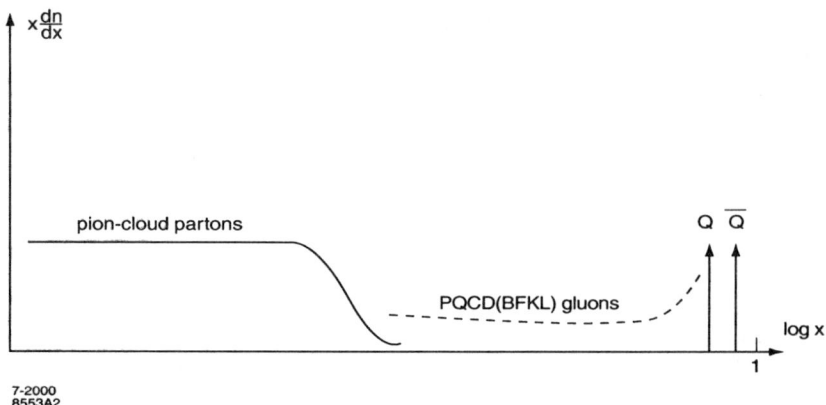

FIGURE 3. The parton distribution expected for a small, massive color dipole. The BFKL region where perturbative QCD may be applicable is of width $\sim \ln Q^2 d^2$, where d is the (color) dipole moment of the onium. A similar picture should apply for the structure function of a virtual photon with squared momentum $P^2 \sim d^{-2}$.

this subject will become increasingly practical at the LHC [5], provided that at least one of the several thousands of experimentalists working there will care enough to make the measurements and to do the analysis. Nevertheless, even in the absence of data, it still might be interesting to have the various theoretical options compete with each other at the Monte-Carlo simulation level, in order to search for sensitive experimental indicators.

4. Spectroscopy of Light Quarks

Thus far, our examples have swung from the very large distance regime to the very short distance regime, with minimal emphasis on the intermediate distance scale. That scale is the richest phenomenologically, and is certainly the crux region to understand, in order to obtain a command of what QCD is really about. And at the heart of the subject is the hadron spectrum, in particular the spectrum of hadrons built from light quarks. There is a long and distinguished history of hadron spectroscopy. It deserves at least half, and probably more than half, of the credit for the establishment of the standard-model quark picture of hadron structure. For me, a high point of hadron spectroscopy occurred in the mid 1970s with the very sophisticated measurements and analyses of baryon resonance spectra. An enormous body of work could be summarized by the SU(6) classification "**56**, L even; **70**, L odd", very consistent with a quark-diquark picture of baryon structure [6]. I am not sure how well this picture has survived the subsequent 25 years. But whatever the present situation is for baryons, there has never been such an easy summary of the situation regarding the meson multiplets. For a given choice of J^{PC}, the lowest lying multiplet may be in good shape. But as soon as one looks at higher excitations, there are missing states and there are extra states, as we heard at this meeting from Jim Napolitano [7]. Without question, there is a great need—and opportunity—for a new round of experiments, especially utilizing hadron beams.

At present multiparticle spectrometers such as MPS and Omega are being phased out, and the only replacement for the future is a new facility in a photon beam at Jlab. From the technical point of view, it seems to me that it should be possible with state-of-the-art technology (as used by *e.g.* ATLAS, CMS, CDF, D0, *etc.*) to create an "electronic bubble chamber" for hadron-induced processes with acceptance, resolution, and particle identification capability at least as good as the old bubble chambers, and with rate capability better by a millionfold. And the analysis power for partial wave analyses or amplitude analyses likewise should exceed what was done then by a millionfold. So a next generation attack across the board would seem to me to be natural and to be potentially extremely productive. It should be the case that a command of hadron resonance spectra up to and beyond a mass scale of 2 GeV should yield a great deal of understanding of the systematics of hadron structure.

5. Non-Singlet Regge Trajectories

One of the most profound and fertile topics in pre-QCD strong interactions was that of the Regge theory of complex angular momentum and of Regge poles. It is a subfield with a distinguished heritage, leading, via duality and the Veneziano amplitude, to the creation of string theory and the modern superstring industry.

In the context of hadron physics, Regge theory remains of great significance. Most importantly it works. The experimental evidence for the existence of Regge trajectories is in some cases extremely quantitative, in particular for the trajectories containing vector mesons. This is especially impressive for the very pure power-law energy dependences of K regeneration amplitudes, recently reviewed for engineering reasons by the KTeV collaboration [8].

Specific properties of the Regge trajectories are strong indicators of the underlying dynamics. In particular linearly rising trajectories suggest strongly a QCD string type of dynamics. This again places emphasis on spectroscopic measurements at high spin and at high mass scales. In any case the Regge picture systematizes in an important way the properties of the resonant states.

I think much more work could be done in this field. As best I can recall, it was only in the mid-seventies when the Regge behavior became well-established experimentally, more or less concurrent with the discovery of the ψ and the subsequent change in direction of the field as a whole away from that line of research. So a return visit and systematic study in hadron-hadron collisions, with modern detectors and analysis techniques, might be very productive.

One area of special importance is that of non-singlet Regge behavior in deep-inelastic processes. In some processes there is evidence for the expected Regge behavior in the scaling limit, *e.g.* the neutron-proton difference, in the F_3 for neutrino reactions, and in the integrand of the Adler neutrino sum rule. However, for the spin sum rule that bears my name, there seems to be evidence for slow convergence of the sum at small x, while Regge arguments would suggest that rapid convergence should be the case. Examination of the Regge limit of the GDH sum rule in photoproduction and its extension to low Q^2 electroproduction could be of considerable use. But aside from details, measurements and analyses at small x which incisively test for Regge asymptotics are rare in the contemporary deep inelastic scattering culture. Given the enormous amount of attention paid to deep inelastic scattering in general, I find this situation perplexing.

And just at the theoretical level, does QCD imply that Regge-pole contributions scale, or should they be of higher twist? Can the Regge residues be calculated from pQCD or something close to it? And while there is a great deal of attention (rightfully) paid by theorists to the vacuum Reggeon singularity (Pomeron), the structure of non-singlet Regge-pole trajectories should if anything be an easier problem. I believe they deserve a closer look by theorists.

6. Heavy Quark Spectroscopy

With the high statistics and superb quality of recent charm and bottom physics experiments, the opportunities for incisive spectroscopic studies have increased dramatically. Things have gotten to the point that the statistics of decays such as $D \to K\pi\pi$, or even $D \to 3\pi$ is so high that examination of the Dalitz plot yields useful information on spectroscopy of ordinary mesons made of light quarks [9].

But the heavy quark excitations are themselves very interesting. The onium systems are so clean that pQCD is the most appropriate starting point. And their final states are fertile territory for glueball searches. Especially interesting to me are the D and especially B mesons, where the machinery of heavy quark effective theory can be applied. What this boils down to is that everything having to do with the heavy quark is relatively trivial, computable within pQCD, leaving the nontrivial system something very close to a single constituent quark. In fact, a viable definition of a constituent quark is the B meson "without" the b quark. So the excitations of B's quite directly probe the properties of the single constituent quark: for example its couplings to pions and photons, its mass, its size, and (if the energy scale of the B^* excitations can be made large enough) any intrinsic excitations of its own.

While the electron-positron B factories are ill-suited for this kind of physics, the hadron-hadron colliders are very well-suited. And in addition to the intrinsic hadron-physics interest in the classification and study of B^* and other excitations of bottom hadrons, there is a good engineering reason to do so. B^*s can help distinguish a secondary B_d from a $B_{\bar{d}}$ in CP studies, freeing the experimentalist from having to find a second tagging B in the event.

7. Confinement, Instantons, and the Vacuum

From the point of view of theory, the belief that QCD is a viable theory at all distance scales, despite the intractability of perturbation theory in the infrared, rests on a hypothesis—that of confinement. There is pretty good evidence for this from lattice calculations, not to mention the experimental facts—which include the fact of our existence. Nevertheless, there is as yet no consensus amongst theorists as to the mechanism that is responsible for confinement. This is probably the leading outstanding problem in all of hadron physics.

The confinement problem is closely linked to the problem of vacuum structure. We have already alluded to the presence of chiral symmetry breaking, leading to a chiral condensate at large distance scales. In addition to that structure, there is the vacuum structure induced by the occurrence of gauge potentials with nontrivial topology, and the existence of instanton-induced transitions between vacua with differing gauge topologies. This creates both good news (existence of mass of the η' meson) and bad news (the possibility of CP violation in the strong interactions).

When the instantons were first discovered, their effects were in poor theoretical control. But at present the situation seems to be much better. The instanton size distribution seems to be sharply peaked about a value characterized by a momentum scale of 600 MeV. The density (in Euclidean space-time) is relatively low, so that the fraction of spacetime containing instanton fields is only a couple of percent or so. And there is a rather convincing line of argument that this instanton population distorts the Dirac sea of light quarks in just the right way to induce chiral symmetry breaking. It would be nice if the argument were to go further and account for confinement as well, but this seems to be much less likely.

It would also be nice to have a better handle experimentally on instanton-induced effects. Efforts have been made to search for signatures of instanton effects in multiparticle final states in collision processes, although this is very difficult and speculative territory [10]. I have my own favorite candidate for a "smoking-gun" instanton-induced effect, namely the leading decay modes of the 0^- η_c charmonium state. They are $\eta\pi\pi$, $\eta'\pi\pi$, and $\bar{K}K\pi$, each with about a 5% branching ratio, and each being a state naturally produced via the 't Hooft instanton-induced interaction:

$$\mathcal{L} \sim (\bar{c}c)(\bar{u}u)(\bar{d}d)(\bar{s}s) . \tag{1}$$

I think close theoretical and experimental attention to these modes might well be useful.

8. Equations of State and Quark-Gluon Plasma

Macroscopic properties of QCD are described by equations of state, which can be studied as a function of the parameters of the theory, including quark masses, number of colors, and of course temperature and chemical potentials. There is plenty of activity and theoretical progress, as described here by Krishna Rajagopal [11]. And of course the heavy ion program provides plenty of experimental impetus. The future looks bright indeed.

I believe that the heavy-ion programs, especially from RHIC and the LHC, will have important spinoffs into high energy physics in at least two respects. One is that if the goal of observation of quark-gluon plasma is achieved quantitatively, measurement of the critical temperature should provide a quite good, competitive value of Λ_{QCD}. More generally, the methods by which ion-ion collisions are studied, with the emphasis on space-time evolution and hydrodynamic flow, will be of great use in dealing with generic hadron-hadron collisions at Tevatron and especially LHC energies, where the number of parton-parton interactions per collision rivals the number of nucleon-nucleon interactions per Au-Au collision.

9. High Parton Densities at Very High Energies

The energetic proton carries with it a very large number of partons, in particular wee gluons, when it has momentum of a TeV and above. This was anticipated by pQCD theorists, and has been well established by the measurements at HERA. As we mentioned above, this has great implications for central, generic collisions at LHC energies, where the phenomenology of typical collisions is expected to differ sharply from that at low energies, essentially because opaque discs of dense gluons are coming into collision. Even at the partonic level there will be strong absorptive effects, copious minijet production, and possibly collective flow. There deserves to be at the LHC (as well as at the TeVatron) serious attention paid to the commonplace collisions as well as the high priority rare ones [12]. There is a frontier of new physics to be explored.

I cannot resist mentioning here a vaguely related, speculative application of high gluon-density physics for RHIC. Consider a RHIC Au-Au collision, not in their laboratory, cms frame, but in a frame where one of the Au nuclei is at rest. What that Au nucleus sees is a very energetic ion bearing down on it, carrying momentum of about 20 TeV per nucleon. There clearly is an enormous wee-gluon density that the rest-frame ion sees. With probability unity, each parton in each rest-frame nucleon gets hit by a gluon, and is "Compton-scattered" into a relativistic final state, with large longitudinal laboratory momentum.

What this essentially means is that everything that was in the rest-frame nuclear matter gets swept up by the projectile and carried away with it at the speed of light—the nuclear matter is stuck to the pancake. So in the volume originally occupied by the resting ion there is "nothing" left.

While this "nothing" is probably not vacuum, it cannot be highly excited either, at least that part of the "nothing" which will radiate secondaries more or less isotropically. This follows simply from conservation of $E - p_z$. In the initial state the value of $E - p_z$ is essentially A GeV, and this is spread over a large volume. So the density of $E - p_z$ in the final state must be low, no more than 140 MeV/fermi^3. This would seem to imply that at most only soft pions can be isotropically emitted from the "nothing". Perhaps these pions could be coherently emitted, a la Bose condensation or via a disoriented-chiral-condensate (DCC) mechanism. In the RHIC laboratory frame, this would imply a secondary "beam" of these pions emerging in the forward direction, with the same velocity as the incident ion beam. In other words, given 100 GeV/nucleon incident momentum, this cluster of pions would emerge with 14 GeV/pion, and with a transverse-momentum spread perhaps as low as 100 MeV. Such pions would be difficult to detect with the existing detectors, although it is not hard to envisage detectors which could do the job [13].

10. The Approach to Scaling

A very large amount of activity now exists in electron-nucleon and electron-ion scattering at intermediate energies, especially at Jlab, where a variety of very beautiful measurements are emerging. Here the basic challenge, as I see it, is to map out in detail the transition from manifestly long-distance descriptions (*e.g.* elastic e-p scattering at relatively small Q^2) to manifestly short distance descriptions, as used in the scaling region in deep-inelastic scattering. There are a variety of kinematic regions to experimentally explore, the two most important ones being high Q^2 at small inelasticity ("exclusive QCD") and low Q^2 at large inelasticity ("soft Pomeron physics"). Both are merged at moderate values of Q^2 and inelasticity with the physics of resonances and non-singlet Regge dynamics.

Among the theoretical challenges is the search for the best descriptive tools. Help comes from duality concepts, both of the Bloom-Gilman type as well as the Regge-resonance type, not to mention parton-hadron duality in its more general context. Sum rules are a powerful tool. While much more can be done with such tools, I think the time has come to search further for specific dynamical descriptions good enough to incorporate the sum rules (in particular a fully relativistic description) and other general features. An excellent example of what I have in mind is the chiral quark model of hadrons, as laid out by D. Diakonov, M. Polyakov, and their collaborators [14]. They start at the fundamental level of short distance QCD, integrate out instanton effects, and find a viable description of hadrons in the chiral sector which still is consistent with relativity and deep inelastic phenomenology. While the assumptions made are perhaps not under complete theoretical control, the credibility level is still very high, and capable in the future of going higher.

11. Diffraction

High energy hadrons, as extended objects, are nearly black discs. Their elastic scattering amplitudes provide clear evidence that this is the case. However, the shadows that hadrons cast in their high energy interactions are much more interesting and subtle than the shadow physics of elastic processes. In particular, inelastic diffraction is endemic, not only in hadron-hadron collisions, but also in electron-hadron collisions, where arguably 20 to 30 percent of all deep inelastic collisions lead to a diffractive final state (defined as a large final-state "rapidity gap", not exponentially suppressed, within which no secondary hadrons are to be found). In addition, diffractive final states are seen in a significant fraction of hard-collision events, namely events containing high-transverse-momentum jets, Ws, Zs, and/or leptons in the final state.

The favored descriptive tool for diffractive processes is that of the t-channel exchange of a Reggeon, the so-called Pomeron. Ingelman and Schlein, in a seminal work which created and thus far has defined the field of hard diffraction, suggested that this Pomeron should have a partonic description, like ordinary hadrons [15]. This concept has greatly helped to drive the field in a very productive manner, especially with regard to creation of a vital and exciting experimental program. Nevertheless, the foundations of the idea are speculative. And the field at present is confused. The data refuse to be easily integrated into the formalism. This was evidenced at this meeting in the excellent talk by Hatakeyama [16]. I personally think that it is time to retreat from the language of Pomeron structure functions, and to search, at the experimental level, for more general and reliable descriptive tools to organize and systematize the phenomenology.

In particular, it should not be taken as obvious that the t-channel exchange picture is the most appropriate language. It may be that it is better to emphasize more the s-channel, shadowy origins of the diffractive phenomena, perhaps in the style of Good and Walker's original description of diffraction dissociation [17], and/or of absorption models. And no matter what, one must acknowledge that the heart of the subject resides, from a diagrammatic point of view, in loop diagrams, not tree diagrams, and that quantum effects, as opposed to quasiclassical partonic visualizations, are essential no matter what descriptive viewpoint is adopted.

My favorite example for appreciating the subtlety of the phenomenon is that of high mass, soft inelastic diffraction. A typical final-state pseudorapidity distribution of secondary particles is illustrated in Fig. 4. First imagine that the reference frame is chosen such that zero rapidity is at position A. Then what one would see at early times is an ordinary multiparticle final state developing, with soft particles being produced at large angles at early times, and more energetic particles being produced at small angles at later times (the "inside-outside" cascade). But at some "macroscopic" late time (proportional to the energy of the fastest right-moving diffractively produced secondary), which can in practice be tens or hundreds of fermis, the leading right-moving emitter suddenly stops emitting: a quantum decision has been made that the right mover (neither a nucleon or "not a nucleon", but a quantum superposition of each possibility) projects to the right-moving nucleon state.

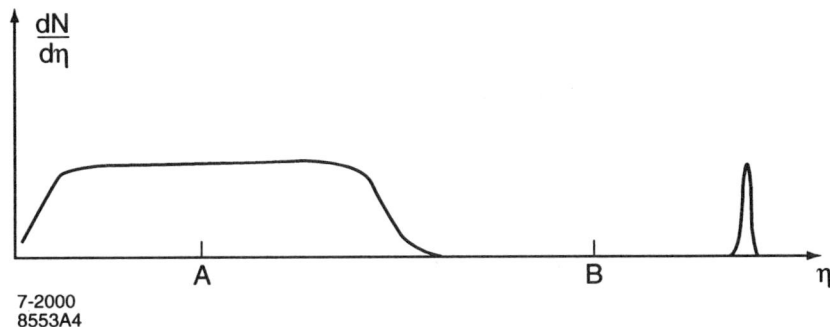

FIGURE 4. Typical pseudorapidity distribution of secondary hadrons in massive soft inelastic diffraction.

Now view the same process in a different reference frame, where zero pseudorapidity occurs at location B. In such a frame at early times there is *no* particle emission, as if the process were at most elastic scattering. But then, again at a "macroscopically" late time (proportional to the energy of the least energetic left-moving diffractive secondary—actually the same particle as before), the left mover (which has to be neither "not a nucleon" nor a nucleon, but a quantum superposition of each possibility) makes a quantum decision *not* to be the nucleon and to emit particles.

These two viewpoints represent the *same* physics, and a good picture of diffraction should be able to explain how and why they are the same. I find this a very interesting challenge, one which the present diagrammatic/partonic approach does not begin to touch. To do a really good job on diffractive physics may well be the last great frontier in hadron physics to be solved.

12. Hadronization Dynamics

A primary task of the hadron-physics subfield of multiparticle production is to understand and describe the transition from the multipartonic evolution at very short distance scales to the multihadronic final states observed experimentally. The prototypical reaction for doing this is electron-positron annihilation into hadrons. In that case, there is a rather satisfactory level of understanding, thanks both to the intrinsic cleanliness of the basic process, and to the large data set of complete events over a large energy scale.

Even so, it is interesting that two apparently competitive viewpoints, that of the nonperturbative QCD string, *a la* Lund, and that of the QCD partonic cascade, peacefully coexist. Quite sophisticated phenomena, such as the "string effect" in three-jet final states, can be described in either picture with comparable success.

Which is right? Conventional wisdom seems to be that most of the space-time evolution is in fact perturbative, with a rather quick transition to the final configuration of emergent hadrons. (This is basically the "preconfinement" picture). However, were the lightest quarks to have mass of a GeV or so, then QCD strings would not easily break, and the spacetime region of final-state evolution would be enlarged, with an interior boundary between perturbative evolution and stringy evolution (Fig. 5). In the heavy quark limit the future light cone gets filled with string. It might be of interest to study the hadronization phenomenology as function of quark masses in order to sort out perturbative from stringy effects.

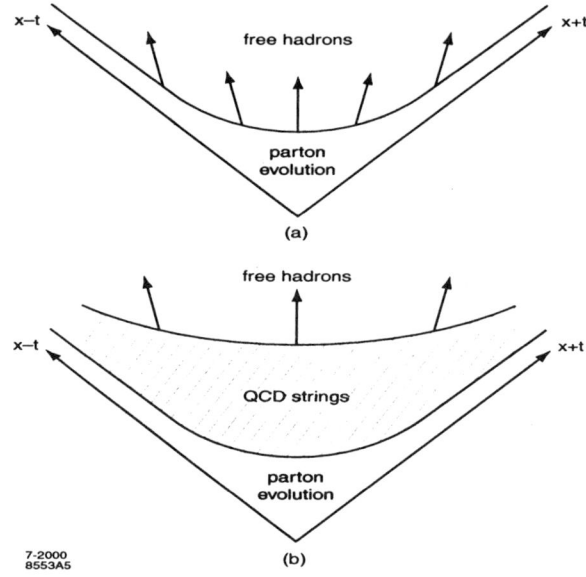

FIGURE 5. (a) Final-state evolution in dijet production in e^+e^- annihilation. (b) The same, in the case of the lightest quarks having masses ~ 1 GeV.

Hadronization phenomenology in hadron-hadron collisions is different and more difficult. The whole subject is in a much more primitive state, especially at collider energies. Not only is the theory much harder, but also there is a paucity of data. Much more experimental attention needs to be paid to the generic collision phenomenology. I think it is a necessary condition for significant progress to be made [18].

IV. THE FUTURE OF HADRON PHYSICS

In the introductory section, we argued that hadron physics has social problems that require it to be better defined and organized than at present. The main reason that drives this notion beyond the merely academic and that requires, in my opinion, some action is that this would facilitate a more rational pattern of funding and support, and that it would facilitate better access to the high-energy physics laboratories, built and managed for purposes other than exploring the details of hadron structure. In the previous cases of chemistry, atomic, and nuclear physics, those who chose to specialize in those fields instead of moving on to the higher energies and shorter distances of particle physics could rather easily do so. Experimental facilities fit, until quite recently, on university campuses. And the funding structures, including peer review systems, adiabatically evolved to adapt to changes of scientific emphases. But hadron physics now presents itself as a crossover field. It is beyond nuclear physics, although heavily populated by nuclear physicists. And much of it is within the energy scale of high energy physics, despite the fact that not many high energy physicists are practitioners.

Because of this, access of experimental hadron physicists to high energy laboratories is made especially difficult. If, in an austere fiscal situation, a hadron physics initiative of the highest quality is put in competition with, say, a quality next-generation neutrino experiment, there will likely be very little support within the high-energy community for the hadron-physics initiative. Indeed, were I myself wearing my high-energy physics hat, I would have a hard time too. Under these circumstances, it seems to me that the best way for hadron physics initiatives to be viable at high energy labs is that there be independent funding available, and that there be agreements with the high energy laboratories for a certain amount of access to collision regions, beam lines, luminosity, running time, infrastructure, *etc.* in return for appropriate contributions to the laboratory budgets. Hadron physics review structures at the program-committee level would be essentially independent of those of particle physics, although at higher policy levels there would necessarily be mixing of the communities.

Examples of this kind of setup exist. At SLAC the NPAS program allowed use of the linac for fixed target experiments at moderate energies of interest to the nuclear community. Despite its modest size, I have been told it was difficult to establish. The best example seems to me to be CERN, which for a long time has had nuclear physics as part of its program, a feature which is now expressed in the heavy ion initiatives at the SPS and at the LHC. The presence of the nuclear physics component at CERN has been important not only in providing a broader scientific base, with all the opportunities for cross-fertilization and diversification that that implies, but has also been useful in broadening its political base.

In the United States, Fermilab is a good example of a high energy laboratory where hadron-physics could be pursued much more aggressively than at present. One of the main-injector beam lines could well support that dream next-generation multiparticle spectrometer for spectroscopy, Regge dynamics, and correlation studies mentioned in item 4 of the previous section. Hyperon physics and charm physics are other possible options. Antiproton sources have been productive venues for charmonium and other studies, both at Fermilab and CERN, and there is more to be done. Indeed a workshop is scheduled for investigating such future options at Fermilab [19]. Finally, the C0 collision region of the TeVatron collider, the presumed home of the future BTeV B-physics initiative, is also a very attractive venue for studying hadron physics. The topics include charm physics, low p_T, diffraction, leading particle studies, and the study of collision dynamics, especially were full-acceptance detection of complete events available. And the future facilities under discussion, in particular muon colliders and/or neutrino sources based on muon-collider technology, are rich sources of hadron-physics spinoffs. While other spinoffs, such as K-decay physics, muon physics and deep inelastic neutrino reactions, have been discussed in this context, very little attention has been paid to the hadron-physics opportunities [20].

I think that a necessary condition for the situation to change is that the hadron physics community organizes itself better. It must not only identify itself and exhibit some political strength, but it must also define better what hadron physics comprises, what its fundamental scientific goals are, what the experimental programs are that deserve the greatest attention, and what the basic challenges to theory are. Since organizational changes within funding agencies and advisory structures are likely to be slow, it seems to me that the best opportunity for getting things going might be within the professional societies. In particular there perhaps should be a Division of Hadron Physics within the American Physical Society. It might provide the venue and organizational structures for achieving the above goals, and provide a basis for going further if, as I suspect is the case, it is deemed necessary to do so.

But most fundamental of all is that there exists a vital community of experimental and theoretical physicists just doing hadron physics, no matter what the obstacles. This meeting has been a splendid example that there are at present plenty of people doing just that. We should do everything we can to not only keep this field healthy, but to strengthen it. The scientific challenges will take quite some time to overcome, and in the meantime we must make every effort to acquire the means to overcome them.

V. ACKNOWLEDGMENTS

On behalf of the participants, it is a pleasure to thank the organizers, especially Stanley Kowalski, and Anne MacInnis, for all their hard work in making this such a splendid meeting.

REFERENCES

1. See for example, *Report of the Subpanel on Planning for the Future of U.S. High Energy Physics*, Gilman, F., chairman; Feb 1998-DOE/ER-0718,
   http://hepserve.fnal.gov:8080/doe-hep/hepap_reports.html, where there is no mention of hadron physics or even QCD in its recommendations.

2. An earlier discussion, by S. Heppelman and myself, can be found at http://sheppel.phys.psu.edu/FundQCD/.
3. Fujii, H. and Kharzeev, D., *Phys. Rev.* **D60**, 114039 (1999); hep-ph/9807383.
4. For a review, see Mueller, A., hep-ph/9911289.
5. See for example, Del Fabbro, A. and Treleani, D., *Phys. Rev.* **D61**, 077502 (2000); also hep-ph/0005025.
6. Hey, A., Litchfield, P., and Cashmore, R., *Nucl. Phys.* **B95**, 516 (1975).
7. Napolitano, J., these proceedings.
8. Briere, R., and Winstein, B., *Phys. Rev. Lett.* **75**, 402 (1995); erratum ibid **75**, 2070 (1995).
9. Malvezzi, S. and Napier, A., these proceedings.
10. Ringwald, A. and Schrempp, F., DESY-99-136, hep-ph/9909338.
11. Rajagopal, K., these proceedings.
12. "FELIX: A full Acceptance Detector at the LHC," Letter of Intent CERN/LHCC97-45 (1997); Eggert, K. and Taylor C., spokespersons (www.cern.ch/FELIX).
13. A discussion and sketch of detectors capable of doing this can be found in a note I have written for th eRHIC workshop, and in a study by Krasny, W.: http://quark.phy.bnl.gov/~raju/eRHIC.html.
14. Diakonov, D., hep-ph/9802298.
15. Ingelman, G. and Schlein, P., *Phys. Lett.* **B152**, 256 (1985).
16. Hatakeyama, K., these proceedings.
17. Good, M. and Walker, W., *Phys. Rev.* **120**, 1857 (1960).
18. More detailed discussion can be found in the FELIX Letter of Intent, Ref. [12].
19. See http://www.iit.edu/~bcps/hep/pbar2000.html for the particulars.
20. See the talk of Shaevitz, M., these proceedings. While one finds there some discussion of opportunities for parton-level deep-inelastic QCD, there is little if any discussion of opportunities at the hadron-physics level.

QCD Spectroscopy and Dynamics

A Study of the $\eta\eta\pi^-$ System Produced in the Reaction $\pi^- p \to p\pi^+\pi^-\pi^- 4\gamma$ at $18 GeV/c$

Paul Eugenio and Graham McNicoll
for the Brookhaven E852 Collaboration

Carnegie Mellon University
Pittsburgh, PA 15213, USA

Abstract. Preliminary results are reported on the partial wave analysis of the $\eta\eta\pi^-$ system in the reaction $\pi^- p \to p\pi^+\pi^-\pi^- 4\gamma$ at 18 GeV/c where both π^0 and η decay to $\gamma\gamma$. The data were obtained using the MultiParticle Spectrometer at Brookhaven National Laboratory. The $a_0(980)$ and $f_0(1500)$ are observed in the $\eta\pi^-$ and $\eta\eta$ mass spectra, respectively. The accepted $\eta\eta\pi$ mass spectrum exhibits a structure consistent with the $\pi(1800)$. The results of the PWA show the $\pi(1800)$ decaying to both $a_0(980)\eta$ and $f_0(1500)\pi$ decay modes.

INTRODUCTION

Experiment E852 at Brookhaven National Laboratory is an experiment in meson spectroscopy configured to detect both neutral and charged final states of 18 GeV/c $\pi^- p$ collisions in a search for meson states incompatible with the constituent quark model. The apparatus is located at the Multi–Particle Spectrometer facility (MPS) of Brookhaven's Alternating Gradient Synchrotron (AGS). A detailed description of the E852 apparatus is given in Reference [1].

An analysis effort of the Brookhaven E852 data is currently underway which focuses on the $\eta\eta\pi^-$ system produced in the reaction $\pi^- p \to p\pi^+\pi^-\pi^- 4\gamma$. The flux-tube model predicts characteristic decays of hybrid mesons to $L=0$, $L=1$ meson pairs [2,3]. Several states are predicted to decay into $\eta\eta\pi$. The $\pi(1800)$ recently reported by the VES Collaboration [4] to have a large decay to $\eta\eta\pi$ has been argued to be a hybrid meson based on its decay to flux-tube favored modes [5].

The data were collected during the 1995 Brookhaven E852 run where 265 million triggers of the type designed to enrich this exclusive final state sample were acquired. Out of these data, 45.6 thousand events were fully reconstructed and kinematically fitted to the hypothesis of $\pi^- p \to p\pi^+\pi^-\pi^-\pi^0\eta$. Events that also fitted a $\pi^- p \to p\pi^+\pi^-\pi^-\pi^0\pi^0$ hypothesis were eliminated.

FEATURES OF THE DATA

$\pi^+\pi^-\pi^-\pi^0\eta$ System

The accepted invariant mass distribution of the $\eta\pi^+\pi^-\pi^-\pi^0$ system is shown in Figure 1a. There are no obvious resonant-like structures in the distribution. The accepted invariant mass distribution of the $\pi^+\pi^-\pi^-\pi^0$ subsystem is shown in Figure 1b. This distribution exhibits a prominent enhancement at 1.3 GeV/c^2 with a width of 0.200 GeV/c^2. Assuming only isovector contributions, an overall charged 4π system must have an $I^G = 1^+$. The $b_1(1235)$ is the only known candidate for this structure.

Figure 1c displays the accepted $\pi^+\pi^-\pi^0$ invariant mass distribution. Two prominent narrow resonances are observed: at 0.550 GeV/c^2, an η signal; and at 0.785 GeV/c^2, an ω signal. A study of $\omega\eta\pi^-$ events shows that $b_1(1235)$ observed in the 4π spectrum shows up as an $\omega\pi^-$ intermediate state.

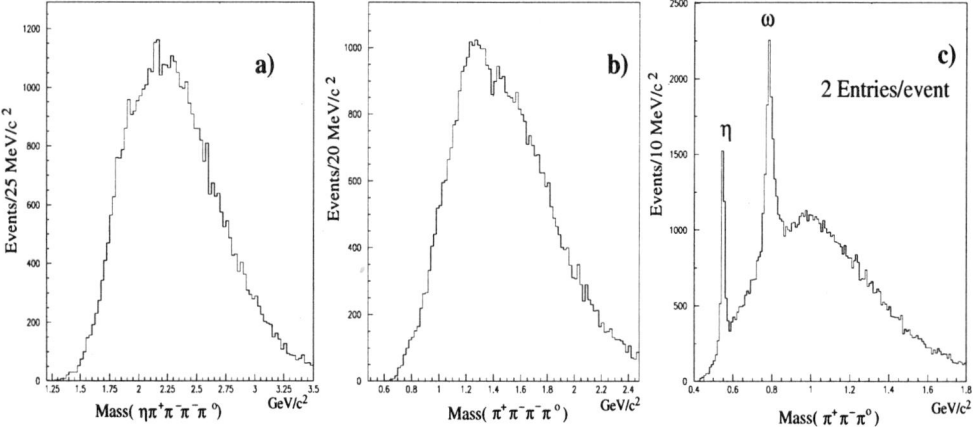

FIGURE 1. [a] The $\eta\pi^+\pi^-\pi^-\pi^0$ effective mass distribution, and [b] the $\pi^+\pi^-\pi^-\pi^0$ effective mass distribution from the reaction $\pi^-p \to p\pi^+\pi^-\pi^-\pi^0\eta$.

$\eta\eta\pi^-$ System

For an η selection of $0.500\,\text{GeV}/c^2 \leq \text{Mass}(\pi^+\pi^-\pi^0) \leq 0.580\,\text{GeV}/c^2$, about 5000 events were acquired for the study of the $\eta\eta\pi^-$ system. The non-$\eta \to 3\pi$ background has been estimated to be less than 10%. The accepted invariant mass distribution of the $\eta\pi$ subsystem is shown in Figure 2a. The main feature is a clear $a_0(980)$ signal. In addition, the distribution exhibits a shoulder-like structure in the region

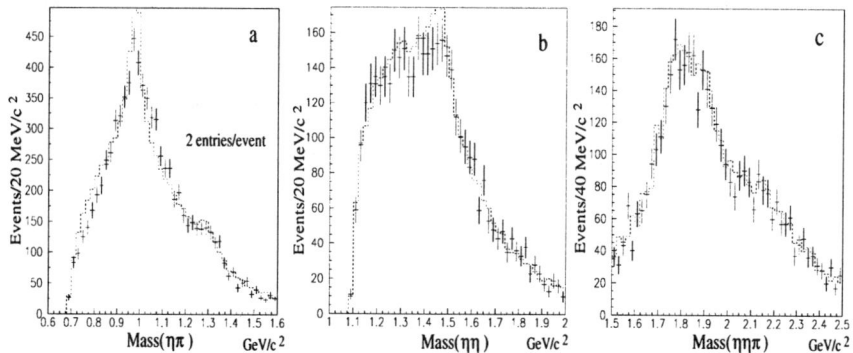

FIGURE 2. The accepted mass spectra for the $\eta\eta\pi^-$ analysis: a) Mass($\eta\pi^-$), Mass($\eta\eta$), and Mass($\eta\eta\pi^-$). The quality of the fit is shown by a comparison of the experimental data (with error bars) to Monte Carlo data weighted using the PWA fit results(dashed histogram).

of 1.3 GeV/c^2 suggesting the presence of some $a_2(1320)$ signal. The $\eta\eta$ subsystem can only couple to spin-even positive-parity isoscalar states(f_0, f_2, f_4, ...). Shown in Figure 2b is the accepted $\eta\eta$ invariant mass distribution. It exhibits a broad structure in the 1.4-1.5 GeV/c^2 region consistent with a $f_0(1500)$ signal. Figure 2c shows the accepted $\eta\eta\pi^-$ invariant mass distribution. The mass spectrum exhibits a resonant-like structure with a mass of 1.8 GeV/c^2 and a width of about 0.200 GeV/c^2.

PARTIAL WAVE ANALYSIS

A partial wave analysis of the data was performed in 50 MeV/c^2 wide mass bins from 1675 MeV/c^2 to 2175 MeV/c^2 and in the t region: $0 \leq |t| \leq 1.2(\text{GeV}/c^2)^2$ using the Brookhaven PWA program [6]. Partial waves for $L < 4$ and $|M| < 2$ were considered in the analysis. A good description of the data was achieved with the minimum set of 5 partial waves($J^{PC}M^\epsilon L(\text{isobar decay})$): $0^{-+}0^+S$ $a_0(980)\eta$, $0^{-+}0^+S$ $f_0(1500)\pi$, $2^{-+}0^+S$ $a_2(1320)\eta$, $2^{-+}0^+D$ $a_0(980)\eta$, and a flat non-interfering background. A comparison of the experimental data and the Monte Carlo data weighted using the fit results is shown in Figure 2. The $\pi(1800)$, $J^{PC} = 0^{-+}$, is observed decaying to $a_0(980)\eta$ and $f_0(1500)\pi$ (See the 0^{-+} intensities in Figures 3a,b). Both 0^{-+} intensities were fitted independently to a relativistic Breit-Wigner mass distribution resulting in: $Mass = 1884 \pm 19(stat)$ MeV/c^2, $\Gamma = 222 \pm 39(stat)$ MeV/c^2, $\chi^2/(9DoF) = 0.97$ for the $a_0(980)\eta$ decay mode, and $Mass = 1862 \pm 24(stat)$ MeV/c^2, $\Gamma = 166 \pm 46(stat)$ MeV/c^2, $\chi^2/(9DoF) = 0.76$ for the $f_0(1500)\pi$ decay mode[1]. The $2^{-+}0^+S$ $a_2(1320)\eta$ intensity shown in Figure

---

[1] The errors quoted do not take into account systematics.

3c exhibits a structure which peaks near 1900 MeV/c^2. This structure may be consistent with the $\pi_2(1670)$ being produced near the $a_2(1320)\eta$ mass threshold, but a 1900 MeV/c^2 resonance cannot yet be ruled out. The $2^{-+}0^+D$ $a_0(980)\eta$ wave is required in the fit and tends to contribute more at higher masses (See Figure 3d). Detailed studies of the relative phases, backgrounds, isobar parameterizations, and systematic errors are in progress.

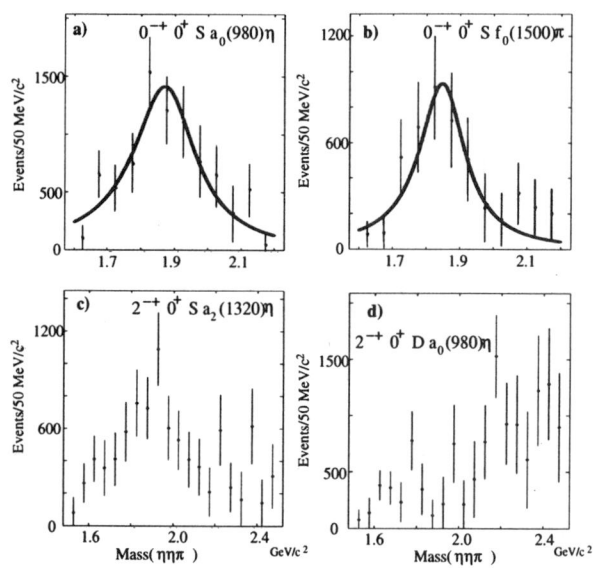

FIGURE 3. The partial wave intensities($J^{PC}M^\epsilon L(isobar\ decay)$): a) $0^{-+}0^+S\ a_0(980)\eta$, b) $0^{-+}0^+S\ f_0(1500)\pi$, c) $2^{-+}0^+S\ a_2(1320)\eta$, d) $2^{-+}0^+D\ a_0(980)\eta$ from the $\eta\eta\pi^-$ PWA.

REFERENCES

1. Brookhaven E852 Collaboration (S.U. Chung et al.), Phys. Rev., **D60**,92001, (1999);
2. N. Isgur and J. Paton, Phys. Rev., **D31**, 2910, (1985).
3. Close, F., and Page, P., Nucl. Phys. **B443** 33, (1995).
4. D. Amelin et al., Phys. Lett., **B356**, 595, (1995).
5. F. Close and P. Page , Phys. Rev., **D56**, 1584, (1997).
6. J.P. Cummings and D.P. Weygand , Brookhaven Report, **BNL-64637**, (1997).

Partial Wave Analysis Results from JETSET

Richard T. Jones

University of Connecticut, Storrs, CT 06269

Abstract. The Jetset experiment has conducted a partial-wave analysis of its data set for the reaction $p\bar{p} \to \phi\phi$. Of particular interest in this analysis is the decomposition of the cross section in the vicinity of the maximum, which coincides with the mass of the narrow $\xi(2230)$ whose existence and quantum numbers have been open to question for a long time. The PWA results are consistent with the existence of a 2^{++} resonance at 2.225 GeV/c^2 with a width of 30 MeV/c^2.

The JETSET experiment at CERN/LEAR has conducted a search for narrow resonances formed in proton-antiproton annihilation over the $\bar{p}$ momentum range 1.2 - 2.0 GeV/c. Of particular interest in this search is the $\xi(2230)$ state observed in radiative decays of the J/ψ decaying to K^+K^- and K_sK_s [1,2] and to $\pi^+\pi^-$ and $p\bar{p}$ [2]. An initial study indicated that this state is consistent with the quark model [3]; however a more complete calculation has recently challenged this conclusion [4]. Its appearance in final states containing strange mesons and also $p\bar{p}$ suggests that the ξ might be formed in the reaction $p\bar{p} \to \phi\phi$ which would normally be suppressed due to the OZI rule. The expected suppression of $\phi\phi$ was in fact not observed [5], can be seen from the mass plots shown in Fig. 1. An analysis of the total annihilation cross section to $4K^\pm$ revealed a dominance of $\phi\phi$ below 1.5 GeV/c, but only an upper limit could be placed on a possible enhancement in the $\xi(2230)$ region.

TABLE 1. Number of events per point that were included in the partial wave analysis. The b.g. count events is the estimate returned by the fit of the background under the $\phi\phi$ peak in the mass plot. The fit is done simultaneously in both angles and masses.

p(GeV)	events	b.g.	p(GeV)	events	b.g.	p(GeV)	events	b.g.
1.180	421	95	1.345	1594	589	1.465	2195	877
1.220	639	225	1.390	1847	585	1.505	1999	943
1.260	896	270	1.405	2668	886	1.700	2528	1592
1.300	1209	840	1.435	2243	868	1.900	2373	1666

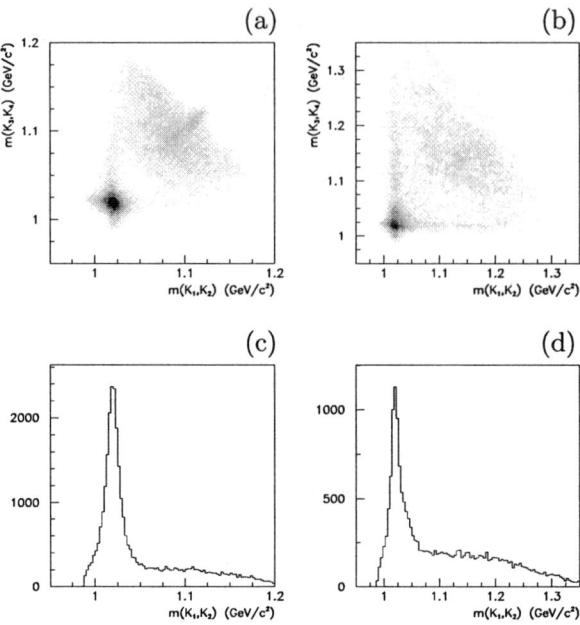

FIGURE 1. Invariant mass spectra for all events in the $4K^{\pm}$ sample (a) for $p < 1.5 \text{GeV}/c$ and (b) for $p > 1.5 \text{GeV}/c$, showing a clear enhancement at the mass of the $\phi(1020)$. Also shown is the mass of one kaon pair when that of the other pair is in the region of the ϕ (c) for $p < 1.5 \text{GeV}/c$ and (d) for $p > 1.5 \text{GeV}/c$.

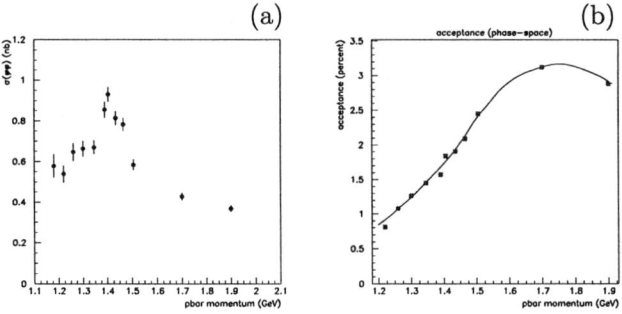

FIGURE 2. (a) Total cross section for $p\bar{p} \to \phi\phi \to 4K^{\pm}$ as a function of the antiproton beam momentum in the lab frame, which has been corrected for (b) the experimental acceptance determined from Monte Carlo under the naive assumption of uniform angular distributions for $\phi\phi$ production and decays.

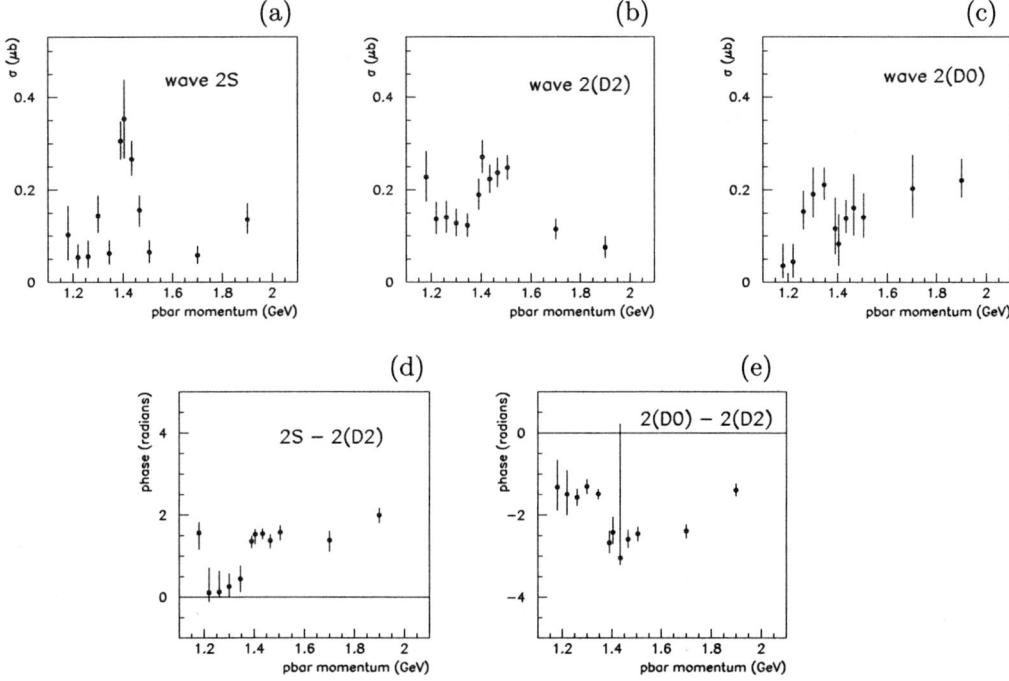

FIGURE 3. Partial-wave analysis results from the fit where only the three dominant waves were included: (a) the 2S wave, (b) the 2(D2) wave and (c) the 2(D0) wave, where 2S represents 2(S2) in the notation j(ls). The phases (d) and (e) are relative to wave 2(D2) which was chosen for this fit to be the reference wave.

A significant portion of the total data set was omitted from the analysis presented in Ref. [5] because of difficulties encountered in calculating the trigger acceptance below 1.4 GeV/c. Including all of the Jetset data in a single analysis leads to the total cross section shown in Fig. 2a, where the data have been rebinned to distribute about the same luminosity to each point. The acceptance plotted in Fig. 2b shows that the peak near 1.4 GeV/c is not an artifact of the acceptance correction. However the above-mentioned acceptance uncertainty makes a definite conclusion regarding the existence of a narrow peak near 1.4 GeV/c difficult to draw based upon the total $\phi\phi$ yield alone. For this reason these data have been submitted to a partial wave analysis to determine whether, independent of the shape of the total cross section, there is evidence in the angular distributions of resonance behavior near 1.4 GeV/c.

The PWA model used was a sum of a product of Breit-Wigners for the two ϕ's times the spin-averaged sum of angular distributions for all allowed $\phi\phi$ partial

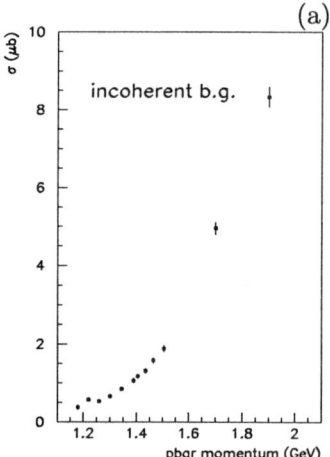 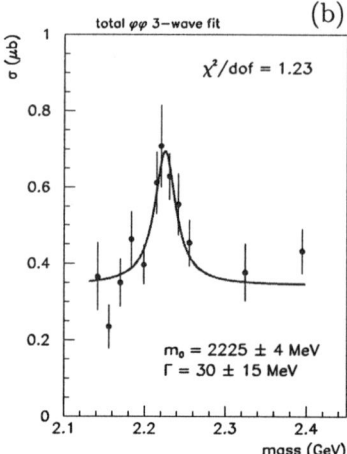

FIGURE 4. Cross section for (a) non-$\phi\phi$ background and (b) $\phi\phi$ production from the 3-wave fit. The background was normalized using phase-space acceptance, whereas the acceptance for $\phi\phi$ was calculated using the angular distributions returned by the fit. A fit is a simple Breit-Wigner plus a flat background.

waves, plus an isotropic $4K^\pm$ phase-space term for the background. In the simultaneous fit of mass and angular distributions, the discrimination between $\phi\phi$ and background comes from the mass distributions, while the separation of the intensity among the different $\phi\phi$ partial waves comes from the angular distributions. An independent fit was performed for each of the 12 points in Fig. 2. Table 1 shows the total number of events used for each fit, and the background estimate returned by the fit. A series of tests were done on the entire event sample, including all waves up to $J=4$ and $L=4$. It was found that across the entire momentum range the same three 2^{++} waves were dominant. Fits were then carried out with only these three $\phi\phi$ waves included, in addition to the incoherent $4K$ background. The results are shown in Fig. 3. The phases are taken relative to that of the 2(D2). The total cross sections for $\phi\phi$ and phase-space background are shown in Fig. 4.

REFERENCES

1. R.M. Baltrusaitis et al., Phys. Rev. Lett. **56** 107 (1986).
2. J.Z. Bai et al., Phys. Rev. Lett. **76** 3502 (1996).
3. S. Godfrey et al., Phys. Lett. **141B** 439 (1984).
4. H.G. Blundell et al., Phys. Rev. **D53** 3700 (1996).
5. C. Evangelista et al., Phys. Rev. **D57** 5370 (1998).

Light scalar mesons

Deirdre Black[1], Amir H. Fariborz and Joseph Schechter

Department of Physics, Syracuse University, Syracuse, NY 13244.

Abstract. In this talk we will present our work investigating the light scalar mesons. We begin with the lightest states (masses less than 1 GeV), which are needed within our framework to agree with experiment, and note that they have the quantum numbers of a nonet. Analysis of the isoscalar mixing pattern suggests that this nonet has a non-trivial $qq\bar{q}\bar{q}$ component. We extend our study to consider the conventional $q\bar{q}$ scalar candidates above 1 GeV and propose a simple mixing mechanism to explain some surprising features thereof.

INTRODUCTION

There has recently been much interest in the question of the light scalar meson spectrum, as is reflected in both the experimental and theoretical talks related to the subject at this meeting [1]. In the interests of time we will not present the details of the work of the Syracuse group on extracting evidence for scalar mesons from fits to $\pi\pi$ and πK scattering and $\eta' \to \eta\pi\pi$ decay, or our study of $a_0(980)$ and $a_0(1450)$ in $\pi\eta$ scattering and refer the reader to [2] for details of our $\frac{1}{N_c}$-inspired, chiral Lagrangian approach. Instead we focus on the possible family properties of such states and what can be inferred from this. We consider the scalar states [2]

$$\sigma(560), \quad \kappa(900), \quad a_0(980), \quad f_0(980), \quad K^*(1430) \quad a_0(1450). \qquad (1)$$

and do not discuss the heavier isoscalars in the 1-2 GeV mass range here. Now it can be seen that there are nine scalar states with masses less than 1 GeV which have the quantum numbers of an SU(3) flavor nonet, namely the isovector $a_0(980)$, the strange state $\kappa(900)$ and the isoscalars $\sigma(560)$ and $f_0(980)$. However if we compare them with a conventional $q\bar{q}$ nonet such as the vector mesons we see that the masses are not as we would expect - the $f_0(980)$ and $a_0(980)$ are degenerate (to be compared with the ϕ-ρ splitting) and also the σ is much lighter than all the other states (unlike its would-be analogue, the ω).

---

[1] Speaker.

LIGHT SCALAR NONET

Let us arrange the light scalar states in a nonet as follows:

$$N = \begin{bmatrix} N_1^1 & a_0^+ & \kappa^+ \\ a_0^- & N_2^2 & \kappa^0 \\ \kappa^- & \bar{\kappa}^0 & N_3^3 \end{bmatrix}, \tag{2}$$

where the isoscalar mixing convention is

$$\begin{pmatrix} \sigma \\ f_0 \end{pmatrix} \equiv \begin{pmatrix} \cos\theta_s & -\sin\theta_s \\ \sin\theta_s & \cos\theta_s \end{pmatrix} \begin{pmatrix} N_3^3 \\ \frac{N_1^1+N_2^2}{\sqrt{2}} \end{pmatrix}. \tag{3}$$

Now ideal mixing, following the original discussion of Okubo [4], may be neatly summarized using the following Lagrangian density:

$$\mathcal{L}_{mass} = -a\text{Tr}(NN) - b\text{Tr}(NN\mathcal{M}) \tag{4}$$

where $\mathcal{M}$ is the mass spurion which splits the strange quark mass from that of the light quarks. From this we can read off the sum rules

$$m^2(a_0) = m^2(\frac{N_1^1 + N_2^2}{\sqrt{(2)}}), \quad m^2(a_0) - m^2(\kappa) = m^2(\kappa) - m^2(N_3^3). \tag{5}$$

In the case of the vector mesons (so replacing $a_0 \to \rho$, $\kappa \to K^*$, $\sigma \to \omega$ and $f_0 \to \phi$) the solution of Eq. (5) corresponds to a mixing angle $\theta_s = \frac{\pi}{2}$ and natural $q\bar{q}$ assignment $\omega \sim \frac{u\bar{u}+d\bar{d}}{\sqrt{2}}$ and $\phi \sim s\bar{s}$.

We see that on the other hand the scalar masses seem to agree better with the solution $f_0 \sim \frac{N_1^1+N_2^2}{\sqrt{2}}$ and $\sigma \sim N_3^3$, which corresponds to $\theta_s = 0$. This assignment is more natural if we consider a nonet made from "dual" quarks and antiquarks, roughly $N_a^b \sim Q_a \bar{Q}^b$ where the dual quark Q_a is composed of two antiquarks $Q_a \sim \epsilon_{abc}\bar{q}^b\bar{q}^c$. Thus the alternative ideal mixing solution is $\sigma \sim u d \bar{u} \bar{d}$ and $f_0 \sim s\bar{s}\left(\frac{u\bar{u}+d\bar{d}}{\sqrt{2}}\right)$. The dynamical motivation for the existence of a light 4-quark nonet of scalar mesons comes from a calculation by Jaffe [5] in the MIT bag model where he showed that the binding energy of one such nonet would receive a large enhancement due to hyperfine interaction, approximating one-gluon exchange.

We add to Eq. (4) two terms $-c\text{Tr}(N)\text{Tr}(N)$ and $-d\text{Tr}(N)\text{Tr}(N\mathcal{M})$, which give deviations from ideal mixing, and solve for the constants a, b, c and d in terms of the four known masses of the scalar states. There are, as in the ideal mixing case, two solutions which turn out to be $\theta_s \approx -\frac{\pi}{2}$ like the conventional ideal mixing limit and $\theta_s \approx -20°$ which is closer to the "dual" ideal mixing scenario.

In order to distinguish between these two solutions we considered πK scattering and also the decay width of $f_0(980) \to \pi\pi$ again. The general trilinear scalar-pseudoscalar-pseudoscalar interaction, derived from a chiral invariant Lagrangian density, is

$$\mathcal{L}_{N\phi\phi} = A\epsilon^{abc}\epsilon_{def}N_a^d \partial_\mu \phi_b^e \partial_\mu \phi_c^f + B\text{Tr}(N)\text{Tr}(\partial_\mu\phi\partial_\mu\phi) + ... \qquad (6)$$
$$+ C\text{Tr}(N\partial_\mu\phi)\text{Tr}(\partial_\mu\phi) + D\text{Tr}(N)\text{Tr}(\partial_\mu\phi)\text{Tr}(\partial_\mu\phi),$$

which relates all of the trilinear couplings of the scalars in terms of the four constants A, B, C and D. In fact only A and B are needed for $\pi\pi$ and πK scattering, whilst C and D will also appear in processes involving η and η'.

It is important to note that since we are working with an effective Lagrangian description, using only the fact that the scalar mesons transform as an SU(3) nonet, we are making no a priori assumptions about the substructure of the nonet. When we fit the πK scattering data [6] again we obtained significantly better agreement with experiment for the case $\theta_s \approx -20°$. This means for example that σ is primarily N_3^3, which is suggestive of a significant $qq\bar{q}\bar{q}$ content.

NEXT-TO-LOWEST-LYING SCALAR MESON NONET

If the lightest scalar mesons are not conventional $q\bar{q}$ states, then it is natural to try and identify the members of the scalar $q\bar{q}$ multiplet predicted as p-wave states in the quark model. The 1998 PDG candidates for the I=1 and I=$\frac{1}{2}$ $q\bar{q}$ scalar states are $a_0(1450)$ & $K_0^*(1430)$. We notice that it is somewhat problematic to make the obvious $q\bar{q}$ assignments to these particles. First, presumably $a_0^+ \sim u\bar{d}$ and $K_0^* \sim u\bar{s}$ which based simply on counting the strange quarks would lead us to expect that $m_{a_0} < m_{K_0^*}$ which clearly is not the case. Moreover these states are heavy compared with other $q\bar{q}$ p-waves, although $\mathbf{L}\cdot\mathbf{S}$ forces should make the J=0 state lighter than the axial and tensor p-waves [for example compare with the $J^{PC} = 2^{++}$ analogues $a_2(1318)$ and $K_2^{*+}(1432)$]. Finally the SU(3) prediction using the interaction $\text{Tr}(N\partial_\mu\phi\partial^\mu\phi)$ is $\frac{\Gamma^{tot}(a_0(1450))}{\Gamma^{tot}(K_0^*(1430))} = 1.51$, whereas experimentally the ratio is 0.92 ± 0.12.

In order to explain these puzzles, we propose the mechanism which is illustrated in Fig. 1. We consider two nonets N and N' of $qq\bar{q}\bar{q}$ and $q\bar{q}$ states respectively which mix in order to produce the observed states. Recalling that the quark composition of the unmixed 4-quark isovectors and isospinors is, for example for the charged states $K_0^+ \sim u d\bar{d}\bar{s}$, $a_0^+ \sim u s\bar{d}\bar{s}$ (which counting strange quarks gives $m_{K_0} < m_{a_0}$), and that the bare $q\bar{q}$ scalars are p-wave states and so heavier than the s-wave nonet N, we see that the ordering of the prediagonalized states will be as in Fig. 1. This means that the mass difference between the isovectors is smaller than that between the strange states, and thus thinking from a perturbation theory point of view we expect that when mixed the $I=1$ states will be repelled more than the $K_0 - K_0'$ system, leading to the level crossing behaviour needed to yield the correct ordering of the observed spectrum in addition to pushing the heavy physical states up to higher masses than expected for pure $q\bar{q}$ scalar p-waves. Also, the mixing introduces different coupling patterns of $a_0(1450)$ and $K_0^*(1430)$ to their decay channels, which can explain their observed widths.

Quantitatively, using the simple mixing term

$$\mathcal{L}_{mixing} = \gamma \text{Tr}(NN') \qquad (7)$$

we input the known $a_0(980)$, $a_0(1450)$, $\kappa(900)$ and $K_0^*(1430)$ masses and find that there is indeed a range of values of γ for which the mechanism reproduces the expected ordering of the prediagonalized states. The value which gives a maximal splitting between the "bare" strange and non-strange $q\bar{q}$ states is $\gamma^2 = 0.33 \text{GeV}^4$, which corresponds to (in GeV) $m_{K_0} = 1.06$, $m_{a_0} = m_{a'_0} = 1.24$, $m_{K'_0} = 1.31$. This corresponds to maximal mixing of the isovectors and a mixing angle of about 30° for the isospinors.

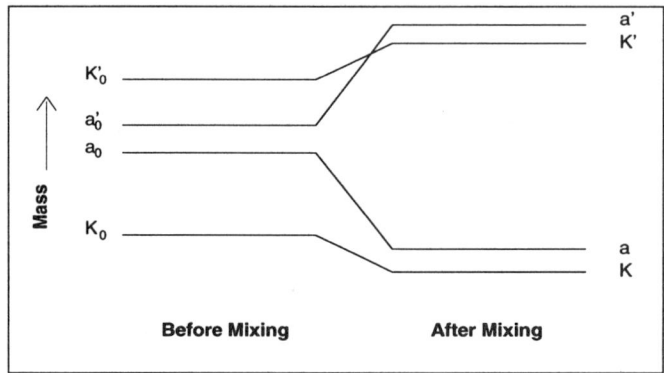

FIGURE 1. Mixing of scalar $qq\bar{q}\bar{q}$ states a_0 & K_0 with $q\bar{q}$ states a'_0 & K'_0 to produce observed isovector and isospinor states $a_0(980)$, $a_0(1450)$, $\kappa(900)$ & $K_0^*(1430)$ labelled a, a', K & K'.

REFERENCES

1. See plenary talk (this conference) on Exotic Hadrons by J. Napolitano, parallel session talks (QCD Spectroscopy and Dynamics) by K. Maltman, V. Yudichev, A. Sobol, C. Meyer, (Heavy Quark and Heavy Lepton Physics) by S. Malvezzi and A. Napier. See also references given in [2,3,7].
2. $\pi\pi$ scattering: M. Harada, F. Sannino, J. Schechter, Phys. Rev. **D54** 1991-2004 (1996), πK scattering: D. Black, A. H. Fariborz, F. Sannino., J. Schechter, Phys. Rev. **D58** 054019 (1998), $\eta' \to \eta\pi\pi$ decay: A. H. Fariborz, J. Schechter, Phys. Rev. **D60** 034002 (1999), $\pi\eta$ scattering: D. Black, A. H. Fariborz, J. Schechter, Phys. Rev. **D61** 074030 (2000).
3. D. Black, A. H. Fariborz, F. Sannino and J. Schechter, Phys. Rev. D 59, 074026 (1999).
4. S. Okubo, Phys. Lett. **5**, 165 (1963).
5. R.L. Jaffe, Phys. Rev. **D15**, 267 (1977).
6. D. Aston et al, Nucl. Phys. **B296**, 493 (1988).
7. D. Black, A. H. Fariborz and J. Schechter, Phys. Rev. D61 074001 (2000).

Scalar Meson Decay Constants and the Nature of the $a_0(980)$

K. Maltman*

Dept. Mathematics and Statistics, York Univ., 4700 Keele St., Toronto, ON Canada, and CSSM, Univ. of Adelaide, Adelaide, SA Australia[1]

Abstract. A form of QCD sum rules known to produce an accurate determination of the ρ decay constant is used to extract the $a_0(980)$, $a_0(1450)$ and $K_0^*(1430)$ decay constants. The resulting ratio of $a_0(980)$ to $K_0^*(1430)$ decay constants is shown to rule out the "loosely-bound-$K\bar{K}$-molecule", Gribov minion and "ordinary meson" scenarios for the $a_0(980)$. A recent conflicting sum-rule-analysis-based claim that the coupling of the $a_0(980)$ to the isovector scalar density is, in contrast, negligible, is shown to result from the overly-restrictive form employed for the spectral ansatz, and to be ruled out.

Scenarios proposed in the literature for the nature of the $f_0(980)$ and $a_0(980)$ (the loosely-bound $K\bar{K}$ molecule, crypto-exotic (four-quark), unitarized quark model (UQM), and Gribov minion pictures) differ significantly in their spatial extent. Processes proposed to distinguish between them (the $\gamma\gamma$ decay widths, and $\phi \to \gamma a_0, f_0$) suffer from difficult-to-quantify theoretical uncertainties associated with the necessity of modelling the non-trivial dynamics. In this paper we show how to determine the scalar meson decay constants and use this information to help distinguish between the different scenarios.

Because the various scenarios correspond to significantly different spatial extents, pointlike probes such as decay constants are ideal for distinguishing amongst them. Decay constant relations provide non-trivial information on $SU(3)_F$ classification and/or mixing in other contexts. For example, the scenario assigning the π and K to the same pseudo-Goldstone boson octet, in spite of the discrepancy in masses, $m_K/m_\pi \simeq 4.5$, requires $f_\pi \simeq f_K$, as observed experimentally. $f_{K^*} = 1.1 f_\rho \simeq f_\rho$ similarly confirms the assignment of the ρ and K^* to the same $SU(3)_F$ multiplet, and $f_\omega^{EM} \simeq f_\rho^{EM}/3$, (rather than $f_\omega^{EM} \simeq f_\rho^{EM}/\sqrt{3}$, as expected for a pure octet ω) near-ideal mixing in the vector meson sector. In this paper, we will use the $K_0^*(1430)$ as a "normal quark model" reference state and, using decay constants, in-

---

[1] Supported by the Natural Sciences and Research Engineering Council of Canada

vestigate which (if either) of the known a_0 resonances might belong to the same $SU(3)_F$ multiplet. The $K_0^*(1430)$ decay constant, $f_{K_0^*}$, is defined by $\langle 0|J_{us}|K^+\rangle = f_{K_0^*} m_{K_0^*}^2$, where $J_{us} = (m_s - m_u)\bar{s}u$, and can be read off from the $K_0^*(1430)$ peak value of the spectral function, ρ_{us}, of the correlator, $\Pi_{us}(q^2) \equiv i\int dx e^{iq\cdot x}\langle 0|T\left(J_{us}(x)J_{us}^\dagger(0)\right)|0\rangle$. Since $K\pi$ scattering is elastic up through the $K_0^*(1430)$ [1], $K\pi$ states saturate ρ_{us} below $s \sim 2$ GeV^2. Unitarity thus allows ρ_{us} to be expressed in terms of the timelike scalar $K\pi$ form factor, $f_{K\pi}(s)$. $f_{K\pi}$ follows from an Omnes representation using K_{e3} and $K\pi$ phase data as input [2,3]. The absence of a possible polynomial prefactor in the Omnes representation, and the high-s behavior of the phase must be assumed. Support for these assumptions is post facto: (1) finite energy sum rules (FESR) for Π_{us}, using ρ_{us} as generated above, produce very stable m_s values, and an extremely good hadronic/OPE match [4]; (2) the extracted m_s agrees with that obtained in a hadronic-τ-decay-based analysis [5]. One finds

$$f_{K_0^*} m_{K_0^*}^2 = 0.0842 \pm 0.0045 \; GeV^3 \; . \tag{1}$$

To determine the $a_0(980)$ and $a_0(1450)$ decay constants, we employ a form of FESR tested in the isovector vector channel and shown to produce a determination of f_ρ^{EM}, *using only the OPE, with the experimental α_s as dominant input*, accurate to within experimental errors [6,7]. The general FESR relation for a correlator Π is $\int_{s_{th}}^{s_0} ds\, w(s)\rho(s) = \frac{-1}{2\pi i}\oint_{|s|=s_0} ds\, w(s)\Pi(s)$, with $w(s)$ any analytic function, s_{th} the physical threshold, and $\rho(s)$ the corresponding spectral function. s_0 is to be chosen large enough that the OPE can be reliably employed on the RHS. Weight functions satisfying $w(s_0) = 0$, which cut out the region of the integral over the circle near the timelike real axis, produce sum rules very well satisfied in the isovector vector channel, even down to scales $s_0 \sim 2$ GeV^2 [6]. Since the $a_0(980)$ and $a_0(1450)$ are well separated, an incoherent 2-Breit-Wigner ansatz for the isovector scalar spectral function, using PDG values for the masses and widths, is reliable. The decay constants are fit in the FESR analysis of the correlator of the scalar density $(m_s-m_u)\bar{d}u$. One essentially uses analyticity, plus non-perturbative input (the known resonance positions and widths), to "measure" the decay constants in terms of α_s. The factor $m_s - m_u$ has been included so as to cancel in the ratio of a_0 to $K_0^*(1430)$ decay constants, leaving the ratio of the matrix elements of the $\bar{d}u$ and $\bar{s}u$ densities. Since these densities are members of an $SU(3)_F$ octet, the ratio will be 1 in the $SU(3)_F$ limit for an a_0 lying in the same multiplet as the $K_0^*(1430)$. On the OPE side of the FESR's, the dominant $D = 0$ part is known to 4-loops [2,8], and the small high D terms to $D = 6$ [2]. Instanton contributions are determined using the instanton liquid model [9]. A detailed description of the input and method of calculation can be found in Ref. [7].

Fitting the a_0 decay constants using the OPE as described above, one finds

$$f_{a_0} m_{a_0}^2 = 0.0447 \pm 0.0085 \; GeV^3 \; ,$$

$$f_{a_0'} m_{a_0'}^2 = 0.0647 \pm 0.0123 \; GeV^3 \;. \tag{2}$$

The errors are dominated by the estimate of the uncertainty associated with truncating the dominant $D = 0$ part of the OPE at 4-loop order. The quality of agreement between the OPE and the resulting hadronic side is shown in Figure 1 for $w(s) = (1 - s/s_0)(2 - s/s_0)$ (chosen to reduce the sensitivity to the less-well-known instanton contributions). The dotted line is the OPE side, the dashed-dotted line the hadronic side with Eqs. (2) as input. If the $a_0(980)$ is very diffuse compared to $K_0^*(1430)$ (the loosely-bound $K\bar{K}$ molecule scenario), one should find a much smaller decay constant; if very compact (the minion scenario), a much larger decay constant. The results of Eqs. (2) rule out both of these scenarios. The "normal meson" interpretation is also ruled out. Two additional possibilities need to be considered in more detail to make this conclusion definitive in the molecule case. The $a_0(980)$ spectral strength is proportional to the square of the decay constant. To see if this small-decay-constant scenario is plausible, we set the coefficient of the $a_0(980)$ Breit-Wigner to zero by hand and reoptimize the $a_0(1450)$ decay constant. The best fit obtained from this exercise is shown by the solid line in Figure 1; the match to the OPE side is clearly terrible. This failure cannot be cured by using a broad background, rather than narrow resonance contribution, in the region below the $a_0(1450)$. (The $\pi\eta$ matrix element of $\bar{d}u$ can be computed to leading order in the chiral expansion using Chiral Perturbation Theory (ChPT), and the corresponding background contribution to the spectral function obtained via unitarity. Setting the $a_0(980)$ resonance contribution to zero, and optimizing the $a_0(1450)$ decay constant in the presence of the resulting background, one obtains a "best" fit almost identical to that given by the solid line. Multiplying the ChPT-generated background contribution by a factor of 5 to allow (generously) for higher (chiral) order contributions, one obtains the "best" fit shown by the dashed line. Clearly no version of the loosely-bound molecule scenario corresponds to a good match to the OPE side, thus ruling out this scenario.) The relation between the $a_0(980)$ and $K_0^*(1430)$ decay constants given by the results above is, in contrast, exactly what one would expect in the UQM scenario if the $a_0(980)$ were a roughly equal admixture of a normal quark model meson core and a loosely bound two-meson component. A recent, conflicting claim, namely that the $a_0(980)$ coupling to $J'_{ud} = (m_u + m_d)\bar{d}u$ is small, and that the spectral distribution is dominated by a contribution with $m \simeq 1.5$ GeV [10], exists in the literature [10]. This claim is based on a Laplace sum rule analysis of the J'_{ud} correlator *assuming a spectral ansatz consisting of a single resonance plus an OPE-generated "continuum" beyond some "continuum threshold"*, s_0. The resonance mass and s_0 are fit in the sum rule analysis, whose validity, apart from the question of the suitability of the form of the spectral ansatz, relies only on analyticity and the applicability of the OPE, as in the sum rules above. *IF* these assumptions are valid, and *IF* the spectral function obtained in Ref. [10] is physical, then FESR's

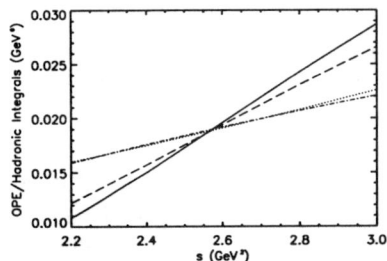
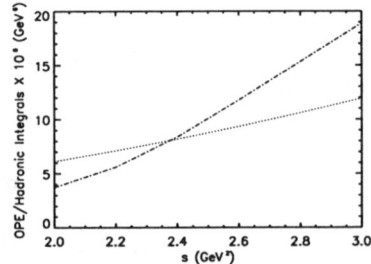

FIGURE 1. Left figure: OPE and hadronic sides for the $I = 1$ scalar sum rule for the various spectral ansatze. Right figure: Testing the "best fit" spectral solution of Ref. 10

analogous to those above must also be valid. Testing the spectral solution of Ref. [10] by means of the resulting FESR, one finds the results shown in Figure 2. The dotted line again represents the OPE side, the dashed-dotted line the hadronic side. One immediately sees that, although the solution of Ref. [10] may represent the "best" fit within the restricted form of the spectral ansatz employed, the quality of the OPE/hadronic match is, in fact, very poor, and, moreover, far inferior to that of the two-resonance form discussed above. The results of Ref. [10], and the apparent contradiction with the results obtained here are, therefore, an artifact of the overly-restrictive form of the spectral ansatz employed in Ref. [10].

REFERENCES

1. D. Aston et al., Nucl. Phys. B296 (1988) 493; W. Dunwoodie, private communication.
2. M. Jamin and M. Münz, Z. Phys. C66 (1995) 633.
3. P. Colangelo, F. De Fazio, G. Nardulli and N. Paver, Phys. Lett. B408 (1997) 340.
4. K. Maltman, Phys. Lett. B467 (1999) 195.
5. J. Kambor and K. Maltman, "The Strange Quark Mass From Flavor Breaking in Hadronic τ Decays", in preparation.
6. K. Maltman, Phys. Lett. B440 (1998) 367.
7. K. Maltman, Phys. Lett. B467 (1999) 14.
8. K.G. Chetyrkin, D. Pirjol and K. Schilcher, Phys. Lett. B404 (1997) 337.
9. E.V. Shuryak, Nucl. Phys. **B214** (1983) 237; A.E. Dorokhov, et al., J. Phys. **G23**.
10. F. Shi et al., hep-ph/9909475.

Pseudovector mesons, hybrids and glueballs

Leonid Burakovsky and Philip R. Page* [1]

Theoretical Division, MS-B283, Los Alamos National Laboratory, Los Alamos, NM 87545, USA

Abstract. We consider glueball– (hybrid) meson mixing for the low–lying four pseudovector states. The $h'_1(1380)$ decays dominantly to K^*K with some presence in $\rho\pi$ and $\omega\eta$. The newly observed $h_1(1600)$ has a D– to S–wave width ratio to $\omega\eta$ which does not enable differentiation between a conventional and hybrid meson interpretation. We predict the decay pattern of the isopartner conventional or hybrid meson $b_1(1650)$. A notably narrow $s\bar{s}$ partner $h'_1(1810)$ is predicted.

The pseudovector ($J^{PC} = 1^{+-}$) $s\bar{s}$ ground state has the interesting property that its OZI allowed decay to open strangeness, i.e. K^*K, which is *a priori* expected to be dominant, is severely suppressed by phase space. This not only makes the state anomalously narrow [1], but opens up the possibility that other decays could be significant. These can arise from $u\bar{u}$, $d\bar{d}$ components in the state, which can come from mixing with a glueball.

We solve Schwinger–type mass equations with linear masses, pioneered in refs. [2,3] and motivated in refs. [3–5]. In this approach the underlying nature of the meson, whether conventional or hybrid, is not specified. The primitive (bare) states are ideally mixed. Primitive isoscalar and isovector $u\bar{u}$, $d\bar{d}$ states are degenerate. In this work we further only allow $SU(3)$ symmetric glueball–meson coupling, with no meson–meson coupling. We restrict to ground state and first excited state mesons. It is known that such restriction is quite accurate if the glueball mass is far from those of the states [5], as is the case here.

The numerical input is as follows. The ratio between pseudovector and scalar glueball masses is evaluated in lattice QCD as 1.70 ± 0.05 [6] or 1.73 ± 0.09 [7]. Taking the world average scalar glueball mass as 1.6 GeV [4], this implies a (input) pseudovector glueball mass of 2.7 GeV. Our conclusions do not critically depend on this value. The primitive $u\bar{u} + d\bar{d}$ ground state is input as the b_1 mass [1]. The physical masses of $h_1(1170)$, $h'_1(1380)$ and the newly discovered $h_1(1600)$ at $1594 \pm 15^{+10}_{-60}$ MeV [8] are used as input. We further assume that the difference

---

[1] *E-mail:* burakov@lanl.gov, prp@lanl.gov

between the primitive $s\bar{s}$ and $u\bar{u} + d\bar{d}$ masses is the same for the ground states and excited states. Lastly, the primitive excited $u\bar{u} + d\bar{d}$ mass, the most uncertain input, is taken as 1650 ± 50 MeV. This is hence the assumed mass region for the yet undiscovered excited b_1 resonance.

The output of our analysis is as follows. The experimenally unobserved physical excited $s\bar{s}$ state ($|h_1'\rangle_2$) is predicted at 1810 ± 40 MeV. The difference between the primitive $s\bar{s}$ and $u\bar{u} + d\bar{d}$ masses, for both the ground and excited states, is 180 ± 10 MeV, yielding a primitive $s\bar{s}$ ground state ($|s\bar{s}\rangle_1$) at 1410 ± 10 MeV. This is consistent with 1445 ± 41 MeV derived from quark model relations[2]. The coupling, in the notation of refs. [2,3] is $g_1 = 0.19 \pm 0.01$ GeV for the ground states and $g_2 = 0.19 \pm 0.12$ GeV for the excited states. The accurate former value is larger than values found for scalar and tensor mesons [2,4].

The valence content of the physical mesons $|h_1\rangle_1$, $|h_1'\rangle_1$, $|h_1\rangle_2$ and $|h_1'\rangle_2$ is respectively given in terms of the primitive states by

$$\left(-0.22^{+0.02}_{-0.01}\right)|g\rangle + \left(0.06^{+0.03}_{-0.05}\right)|s\bar{s}\rangle_2 + \left(0.12^{+0.05}_{-0.09}\right)|u\bar{u}\rangle_2 + \left(0.17^{+0.01}_{-0.00}\right)|s\bar{s}\rangle_1 + \left(\mathbf{0.95^{+0.01}_{-0.01}}\right)|u\bar{u}\rangle_1,$$

$$\left(-0.13^{+0.02}_{-0.03}\right)|g\rangle + \left(0.06^{+0.03}_{-0.05}\right)|s\bar{s}\rangle_2 + \left(0.13^{+0.08}_{-0.11}\right)|u\bar{u}\rangle_2 + \left(\mathbf{0.96^{+0.02}_{-0.03}}\right)|s\bar{s}\rangle_1 + \left(-0.22^{+0.02}_{-0.03}\right)|u\bar{u}\rangle_1,$$

$$\left(-0.19^{+0.15}_{-0.08}\right)|g\rangle + \left(0.16^{+0.09}_{-0.16}\right)|s\bar{s}\rangle_2 + \left(\mathbf{0.94^{+0.07}_{-0.08}}\right)|u\bar{u}\rangle_2 + \left(-0.20^{+0.16}_{-0.11}\right)|s\bar{s}\rangle_1 + \left(-0.14^{+0.11}_{-0.06}\right)|u\bar{u}\rangle_1,$$

$$\left(-0.12^{+0.07}_{-0.01}\right)|g\rangle + \left(\mathbf{0.97^{+0.04}_{-0.03}}\right)|s\bar{s}\rangle_2 + \left(-0.21^{+0.22}_{-0.13}\right)|u\bar{u}\rangle_2 + \left(-0.06^{+0.03}_{-0.00}\right)|s\bar{s}\rangle_1 + \left(-0.05^{+0.03}_{-0.00}\right)|u\bar{u}\rangle_1.$$

The first three physical states are the experimental states $h_1(1170)$, $h_1'(1380)$ and $h_1(1600)$ [1].

Decays are now studied by using a finite width for the initial meson, and unless otherwise indicated, a narrow width approximation for the final mesons. Finite widths are implemented by smearing over relativistic Breit–Wigner shapes with Quigg – von Hippel energy dependent widths. Whenever a decay is OZI allowed from an ideally mixed initial state, we assume, for simplicity, that the initial state is 100% ideally mixed. OZI forbidden decays are implemented by using the (small) valence contents above to calculate connected decays [2].

The decays of conventional mesons are studied in the $^3P_0$ model using the methods, conventions and parameters of refs. [2,9]. Making the usual identification that the primitive ground state mesons are P–wave quark model states, we obtain the decay widths in Table 1. We note that although the experimentally observed K^*K mode is dominant, and similar to the total width of the state[3] [1], the $\rho\pi$

---

[2]) Combining $K(^1P_1) + K(^3P_1) = K(1270) + K(1400)$, $b_1 + (s\bar{s})_1 = 2K(^1P_1)$ and $K(^1P_1) - b_1 = K(^3P_1) - a_1$. Here all items are the corresponding masses. The $^1P_1$ and $^3P_1$ kaon masses before mixing are $K(^1P_1)$ and $K(^3P_1)$ respectively, and the primitive $s\bar{s}$ ground state mass is $(s\bar{s})_1$.

[3]) We find that mock meson phase space [13] gives a K^*K partial width of 191 ± 18 MeV, inconsistent with the experimental total width of the state [1]. For near threshold decays of this type mock meson phase space always gives a substantially larger width than relativistic phase space. Mock meson phase space results are hence not quoted for near threshold decays in the tables. We note that the K^*K partial width calculation in ref. [13] misses a flavour factor of two, and is hence unreliable.

TABLE 1. Partial decay widths of $h'_1(1380)$ in MeV in relativistic [9] and mock meson [13] phase space. The latter is in brackets. For conventional meson decays in relativistic phase space we allow the wave function parameter β, which is taken to be the same for the incoming and outgoing states in a decay, to vary between the reasonable values 0.35 and 0.45 GeV [9], giving rise to the error estimate. The dagger indicates that phase space is unreliable in the narrow resonance approximation for the final state, so that the width is calculated by smearing over a Breit–Wigner for all broad resonances involved, both in the initial and final states. Since the $|u\bar{u}\rangle_2$ component of the physical $h'_1(1380)$ has such a large uncertainty, we only employ the $|u\bar{u}\rangle_1$ component for OZI forbidden decays. However, omission of the $|u\bar{u}\rangle_2$ component could significantly affect widths and especially D/S-wave width ratios.

Mode	Wave	Width
K^*K †	S	137 ± 12
	D	1 ± 1
	D/S	$0.010^{+0.008}_{-0.004}$
$\rho\pi$	S	12 ± 3 (13)
	D	4 ± 3 (4)
	D/S	$0.4^{+0.4}_{-0.2}$ (0.4)
$\omega\eta$	S	2 ± 1 (2)
	D	0 (0)
	D/S	$0.01^{+0.01}_{-0.00}$ (0.01)
$b_1\pi$ †	P	0
Total		156

mode is detectable. It is not as large relative to K^*K as one might expect from the limited phase space of K^*K. Identification in $\rho\pi$ is complicated by the huge 360 ± 40 MeV width of the $h_1(1170)$ mainly in $\rho\pi$ [1]. This makes $\rho\pi$ an unattractive search channel for $h'_1(1380)$, since no viable production processes are known which strongly produce the dominant $s\bar{s}$ component in $h'_1(1380)$ as opposed to the dominant $u\bar{u} + d\bar{d}$ component in $h_1(1170)$. Although $\omega\eta$ is small, $h'_1(1380)$ has recently been observed in this mode [10]. Additional decay modes that have not been calculated but are expected to be small are decays to $h_1(\pi\pi)_S$ and direct three-body decays like $\pi^0\pi^+\pi^-$.

We proceed to analyse $h_1(1600)$. One has to allow for the possibility that the excited $u\bar{u}$, $d\bar{d}$ and $s\bar{s}$ states are hybrid mesons. The calculations for this possibility are performed in the Isgur–Paton flux-tube model with the standard parameters of ref. [11]. The results are displayed in Table 2. The $h_1(1600)$ is predicted to decay from most to least prevalent to $\rho\pi$ / $\rho(1450)\pi$, K^*K and $\omega\eta$ in all interpretations of the state. A minor feature that distinguishes interpretations is the relative size of the $\rho\pi$ and the $\rho(1450)\pi$ modes. The main distinguishing feature is the ratio of D-wave to S-wave widths, which is consistently larger for the meson than the hybrid interpretation. For the meson interpretation the S-wave width is suppressed due to a node in the amplitude, making it sensitive to the wave function parameter

TABLE 2. Partial decay widths of pure $u\bar{u} + d\bar{d}$ $h_1(1594)$ (with the experimental total Breit–Wigner width 384 MeV) in MeV for its interpretations as conventional and hybrid mesons. Hybrid meson decays are calculated in the IKP and PSS models [11]. Other conventions are as in Table 1. $h_1(1594) \to h_1(\pi\pi)_S$ is not estimated.

Mode	Wave	Meson	IKP Hybrid	PSS Hybrid
$\rho\pi$	S	14 ± 2 (13)	111 (96)	86 (74)
	D	126 ± 40 (97)	1 (1)	1 (1)
	D/S	9^{+1}_{-5} (7)	0.005 (0.004)	0.009 (0.008)
$\rho(1450)\pi$ †	S	31 ± 1	142	111
	D	6 ± 3	0	0
	D/S	0.2 ± 0.1	0.0002	0.0004
K^*K	S	15 ± 3 (17)	27 (31)	37 (42)
	D	17 ± 7 (17)	0 (0)	0 (0)
	D/S	$1.2^{+1.1}_{-0.6}$ (1.0)	0.0004 (0.0003)	0.0005 (0.0005)
$\omega\eta$	S	6 ± 2 (6)	19 (18)	24 (23)
	D	11 ± 5 (10)	0 (0)	0 (0)
	D/S	$1.8^{+1.8}_{-0.8}$ (1.6)	0.002 (0.001)	0.003 (0.002)
$b_1\pi$	P	0	136 (227)	0 (0)
Total		225	436	259

β employed. Table 2 shows that the D/S-wave width ratio in $\omega\eta$ is inconsistent with the experimental result $0.3^{+0.1}_{-0.[1-3]}$ [12] if $h_1(1600)$ is a conventional meson. In order to confirm this result, we perform three further checks. Firstly, we evaluate the ratio by taking the wave function parameter β to be different for different mesons participating in the decay. Varying β in the reasonable range $0.35 - 0.45$ GeV [9] confirms the result. Secondly, using the full valence content of $h_1(1600)$ above, and allowing decay via the ground state P-wave meson component, confirms the result. Thirdly, experimental data has few D-wave events above 1.8 GeV [8]. Restricting the Breit–Wigner smearing to invariant masses less than 1.8 GeV gives the nearest ratio to experiment in all these simulations, $0.9^{+1.2}_{-0.5}$. This ratio is still inconsistent with experiment. However, the small ratio predicted for a hybrid meson in Table 2 should be regarded as inconsistent with the observation of the D-wave in experiment. Thus current experiment does not enable differentiation between the conventional or hybrid meson interpretation of $h_1(1600)$, assuming that the state observed in experiment cannot be resolved into two separate states. Since the $^3P_0$ model has only been tested for a few D/S-wave width ratios [9], one needs further information. The total width $384 \pm 60^{+70}_{-100}$ MeV of $h_1(1600)$ is slightly more consistent with the hybrid interpretation. Future searches for $h_1(1600)$ should focus on obtaining the D/S-wave width ratio in the sizable $\rho\pi$ channel. The $b_1\pi$ mode distinguishes the two models of hybrid decay in Table 2.

We note that since the ρ Regge trajectory dominates the $\rho(1450)$ and b_1 trajectories, and $h_1(1600)$ has a healthy $\rho\pi$ coupling for all interpretations, one expects

TABLE 3. Partial decay widths of $b_1(1650)$ in MeV. Conventions are as in Table 2, including using the same total width for $b_1(1650)$ as for $h_1(1594)$. $b_1(1650) \to b_1(\pi\pi)_S$ is not estimated.

Mode	Wave	Meson	IKP Hybrid	PSS Hybrid
$\omega\pi$	S	4^{+2}_{-0} (4)	37 (30)	28 (22)
	D	48 ± 13 (35)	0 (0)	0 (0)
	D/S	11^{+0}_{-3} (9)	0.006 (0.005)	0.01 (0.01)
$\omega(1420)\pi$ †	S	11 ± 1	70	54
	D	7 ± 3	0	0
	D/S	$0.6^{+0.6}_{-0.2}$	0.0009	0.001
K^*K	S	13 ± 3 (14)	30 (32)	40 (42)
	D	23 ± 9 (22)	0 (0)	0 (0)
	D/S	$1.8^{+1.7}_{-0.9}$ (1.5)	0.0005 (0.0004)	0.0007 (0.0007)
$\rho\rho$ †	S	34 ± 6	0	0
	D	34 ± 15	0	0
	D/S	$1.0^{+0.8}_{-0.4}$		
$\rho\eta$	S	5 ± 1 (5)	20 (18)	25 (22)
	D	15 ± 6 (12)	0 (0)	0 (0)
	D/S	$3.1^{+2.6}_{-1.6}$ (2.6)	0.002 (0.002)	0.003 (0.003)
$a_0\pi$ †	P	8 ± 1	56	3
$a_1\pi$	P	11 ± 2 (16)	19 (30)	3 (5)
$a_2\pi$	P	82 ± 16 (132)	37 (60)	7 (12)
	F	3^{+4}_{-2} (4)	0 (0)	0 (0)
	F/P	$0.03^{+0.04}_{-0.01}$ (0.03)	0.005 (0.005)	0.0003 (0.0003)
$h_1\pi$	P	0	72 (108)	0 (0)
Total		296	341	160

the $h_1(1600)$ to be produced via natural parity exchange in the $\pi^- p$ collisions it has been observed in. This is confirmed in the experimental analysis [8], providing an independent check on our calculations. The non–observation of $h_1(1600)$ in unnatural parity exchange [8] may put bounds on its $b_1\pi$ coupling, discriminating between different hybrid decay models.

In Table 3 the widths for the isopartner state $b_1(1650)$ are calculated. The channels that distinguish between conventional and hybrid meson interpretations of the state, $\omega(1420)\pi$ and $\rho\rho$, are difficult to access experimentally. However, D/S–wave width ratios remain an excellent distinguishing feature. Possible search channels are $\omega\pi$ and $\rho\eta$.

The widths for the undiscovered excited $s\bar{s}$ state $h'_1(1810)$ are indicated in Table 4. It is interesting to note that the flux–tube model selection rule, which states that decays to $S + S$ states (K^*K, $\phi\eta$) are suppressed relative to $P + S$ states ($K_1(1270)K$) [11], is apparently violated. This is due to phase space. Whether the $h'_1(1810)$ is a conventional or hybrid meson, it is surprisingly narrow. Excellent

TABLE 4. Partial decay widths of pure $s\bar{s}$ $h_1'(1810)$ in MeV. Conventions are as in Table 2, except that h_1' has a total width of 100 MeV. $h_1'(1810) \to h_1'(1380)(\pi\pi)_S$ is not estimated.

Mode	Wave	Meson	IKP Hybrid	PSS Hybrid
K^*K	S	11 ± 9 (10)	47 (43)	47 (43)
	D	70 ± 30 (61)	0 (0)	0 (0)
	D/S	6^{+22}_{-4} (6)	0.004 (0.004)	0.009 (0.008)
$\phi\eta$	S	17 ± 5 (18)	22 (24)	56 (60)
	D	14 ± 7 (14)	0 (0)	0 (0)
	D/S	$0.8^{+1.3}_{-0.4}$ (0.8)	0.0004 (0.0004)	0.0006 (0.0006)
$K_1(1270)K$ †	P	1 ± 0	13	0
Total		113	82	103

search channels are K^*K and $\phi\eta$. The latter is especially interesting since it cannot come from a $u\bar{u} + d\bar{d}$ state via OZI allowed decay. Small OZI forbidden modes like $\rho\pi$ could also effect detection. A natural place to search for $s\bar{s}$ states is at Jefferson Lab, where the photon has a sizable coupling to $s\bar{s}$. Production is likely to be via diffractive exchange, as meson exchange involves OZI forbidden or evading processes.

We thank C. Amsler, P. Eugenio, E. Klempt and D. Weygand for useful discussions on their experimental data. This research is supported by the Department of Energy under contract W-7405-ENG-36.

REFERENCES

1. Particle Data Group (C. Caso et al.), *Eur. Phys. J.* **C3** (1998) 1.
2. L. Burakovsky, P.R. Page, *Eur. Phys. J.* **C12** (2000) 489.
3. L. Burakovsky, P.R. Page, *Phys. Rev.* **D62** (2000) 014011.
4. L. Burakovsky, P.R. Page, *Phys. Rev.* **D59** (1999) 014022, erratum – *ibid.*, 079902.
5. L. Burakovsky, P.R. Page, T. Goldman, *Phys. Lett.* **B467** (1999) 255.
6. C.J. Morningstar, M. Peardon, *Phys. Rev.* **D60** (1999) 034509.
7. M.J. Teper, hep-th/9812187.
8. P. Eugenio et al. (E852 Collab.), *in preparation*.
9. T. Barnes, F.E. Close, P.R. Page, E.S. Swanson, *Phys. Rev.* **D55** (1997) 4157.
10. C. Amsler, *Proc. of* 15th *Int. Conf. on Particles and Nuclei (PANIC '99)*, (June 1999, Uppsala, Sweden), eds. G. Fäldt et al. (North–Holland, Amsterdam, 2000), p. 92c.
11. P.R. Page, E.S. Swanson, A.P. Szczepaniak, *Phys. Rev.* **D59** (1999) 034016.
12. P. Eugenio, *private communication*.
13. R. Kokoski, N. Isgur, *Phys. Rev.* **D35** (1987) 907.

η Electroproduction in the $S_{11}(1535)$ Region with CLAS

James A. Mueller
For the CLAS Collaboration

Department of Physics and Astronomy, University of Pittsburgh, Pittsburgh, PA 15260, USA

Abstract. New cross sections measurements are presented for the reaction $ep \to ep\eta$ for total center-of-mass energy $1.49 < W < 1.95$ GeV, and four-momentum transfer $.5 < Q^2 < 1.5$ GeV^2. This data covers a much larger angular range than previous experiments allowing, for the first time, a full response function fit. This fit shows new indications of contributions beyond the $S_{11}(1535)$ baryon resonance. The photocoupling amplitude to the $S_{11}(1535)$ is also extracted from this data.

Photo and electroproduction experiments on the nucleon provide a clean probe of nucleon structure since Quantum Electrodynamics is well understood. The reaction $ep \to ep\eta$ is an especially clean reaction for studying excited states of the proton, since it selects out isospin $= 1/2$, N^* resonances. The cross section for this reaction appears to be dominated by the $S_{11}(1535)$ resonance, which has quantum numbers $IJ^P = \frac{1}{2}\frac{1}{2}^-$, and is reached from the nucleon through an electric dipole transition. This is the only baryon resonance known to have a branching ratio into ηp larger than a few percent. Past experiments [1] have established that the photocoupling amplitude ($A_{1/2}$) for the $S_{11}(1535)$ has a slower falloff with Q^2 indicating a more compact object than other N^* resonances. This feature is not well understood. In this paper, I describe new measurements of η electroproduction with nearly complete angular coverage. These data are used to look for additional contributions to this reactions beyond the $S_{11}(1535)$ and to extract new values for $A_{1/2}$ of the S_{11}.

This experiment was done using the CEBAF Large Acceptance Spectrometer (CLAS) [3] at the Thomas Jefferson National Accelerator Facility (TJNAF), nee CEBAF. The toroidal magnet coils separate CLAS into 6 largely identical sectors, each covering 54° in ϕ. The tracking chambers in CLAS measure angles and momenta of charged particles for lab polar angles in the range $8° < \theta < 142°$. Outside the DC scintillation counters (SC) provide time-of-flight measurements with which we can separate the charged hadrons into pions, kaons and protons. For lab angles $\theta < 48°$, threshold Cerenkov detectors (CC) and Electromagnetic Calorimeters

(EC) distinguish electrons from charged hadrons with high accuracy.

For this analysis, events were selected with an identified electron and proton. A fiducial cut on these particles was applied to avoid the complicated regions near the magnetic coils and the edges of the CC. Events were then binned according to Q^2, W, and the center-of-mass angles of the putative η ($\cos\theta_\eta^*$ and ϕ_η^*). The η yield was determined by fitting the distribution of missing mass recoiling against the e-p system. The fit is the sum of a peak at the η mass with a radiative tail plus a simple background function modified by the geometric acceptance for this reaction. The rms resolution for the missing mass peak is about 6 MeV. Acceptance for this reaction was calculated using a GEANT-based Monte Carlo simulation. The event generator included radiative effects using the peaking approximation, and the cross sections have been corrected for radiation. The acceptance varied from bin to bin with a high of 54%. Bins with acceptance less than 6% were not used in this analysis, leaving 87% of the bins for which cross sections were measured.

For each W and Q^2 bin, the differential cross sections were fit to a form that comes from an expansion of the response functions in terms of orthogonal polynomials.

$$\begin{aligned}
\frac{d^2\sigma}{d\Omega_\eta^*} &= \mathcal{PS}\left[R_T + \epsilon R_L + \sqrt{\frac{\epsilon(1+\epsilon)}{2}} R_{TL}\cos\phi_\eta^* + \epsilon R_{TT}\cos 2\phi_\eta^*\right] \\
&\approx \frac{|p_\eta^*|W}{KM_p}\Big[A + B\cdot\cos\theta_\eta^* + C\cdot P_2(\cos\theta_\eta^*) \\
&\quad + \left(D\cdot\sin\theta_\eta^* + E\cdot\sin\theta_\eta^*\cos\theta_\eta^*\right)\cos\phi_\eta^* + F\cdot\sin^2\theta_\eta^*\cos 2\phi_\eta^*\Big]
\end{aligned} \qquad (1)$$

Under the assumption that the cross section is dominated by the S_{11} partial wave, the A parameter represents the S-wave contribution, B and D come from S/P-wave interference, and C, D, and F are due to S/D-wave interference. Thus, nonzero values for B-F are evidence for nonresonant mechanisms or other N^* resonances. The results of these fits are shown in Figure 1. The terms representing R_{LT} and R_{TT} are small in agreement with theoretical expectations. A is the largest contribution and has, at low W, the expected Breit-Wigner shape due to an S-wave resonance close to threshold. C is slightly negative for low W, as is also seen in photoproduction, where it is due to the $D_{13}(1520)$. The most striking feature not seen in previous data is that the B parameter starts out negative at low W, but changes sign around 1.7 GeV. This could indicate variation of the relative phase between the S and P wave, or perhaps multiple canceling contributions whose relative magnitudes are changing. At these high W, however, the assumption of S-wave dominance is probably no longer valid. A more detailed partial wave analysis will be needed to disentangle these contributions.

By summing events over $\cos\theta_\eta^*$ and ϕ_η^*, adequate statistics are available to bin the events finer in both W and Q^2. These cross sections are presented in Figure 2. The prominent peak at $W \sim 1.5$ GeV is the $S_{11}(1535)$, as seen in the isotropic part of the angular distributions. Fits to a relativistic Breit-Wigner with an energy dependent width describe the low W region well, but there are deviations for $W > 1.65$ GeV.

It is interesting to note that this is the same energy region where the B parameter derived from the angular distributions shows a strong variation. Again, a full interpretation of this region is beyond the scope of this paper.

For each Q^2 bin, the peak cross section extracted from the fit can be used to extract $A_{1/2}$ with the assumption that $S_{1/2}$ is small. Consistent with PDG and Armstrong, et al., a value of the full width of 150 Mev and an $S_{11} \to \eta N$ branching fraction of 0.55 were used The results of this measurement of $A_{1/2}$ are shown in Figure 3 along with some previous results converted to be consistent with out choice of Γ and b_η. Four theoretical calculations [4] within the Constituent Quark Model are superimposed on the plot. Two of these are nonrelativistic, while the other two include features of relativity in the photon absorbtion and the 3-quark wave function.

The new η electroproduction measurement shown here is one of the first results of the CLAS at TJNAF. It has broader kinematic coverage than any previous experiment and covers the W region at and above where the $S_{11}(1535)$ is important.

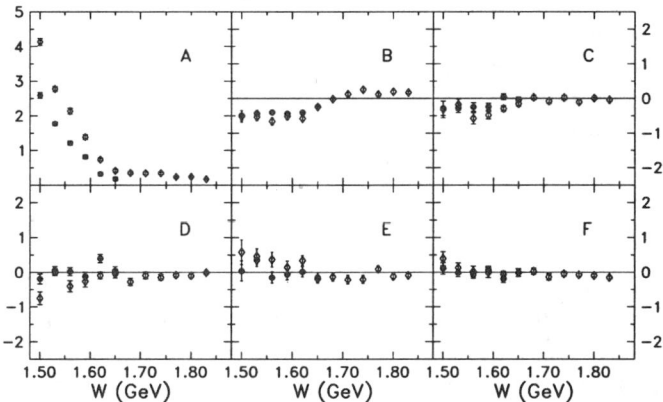

FIGURE 1. The results of the fit to the differential cross section using Equation 1. The open diamonds are for $Q^2 = 0.75$ GeV^2. The filled cirles are for $Q^2 = 1.25$ GeV^2.

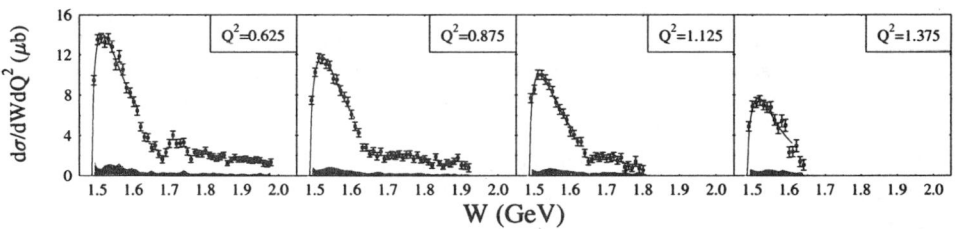

FIGURE 2. The integrated cross section measured for this experiment. The error bars on the points are statistical only. The size of the systematic uncertainty is indicated by the histogram at bottom of each plot.

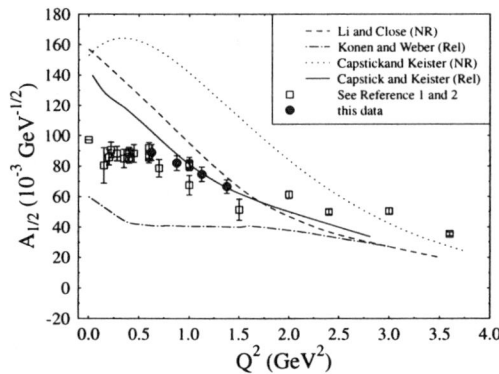

FIGURE 3. Values of the photon coupling amplitude, $A_{1/2}$ obtained from the integrated cross sections compared to previous experiments and a selected calculations. The previous reseults have been converted (see text).

The analysis presented here provides new evidence for the unusually slow falloff with Q^2 of the $S_{11}(1535)$ photocoupling amplitude. New Structure is seen above the $S_{11}(1535)$ both in the total cross section, and in the term in the partial wave analysis most naturally associated with interference between S and P-waves. There have been additional runs of this experiment since the data reported on here were collected. When analyzed, these data will improve the statistics for the current measurement by more than a factor of 5. There are now similarly large data sets for both lower and higher beam energy. With this new data we will be able to measure $A_{1/2}$ for $0.2 < Q^2 < 4.0$ GeV^2, covering the entire Q^2 range of fig(3). These data will also allow us higher statistics to further examine the features we have observed above the $S_{11}(1535)$. Polarized electron beams were also used during these runs, opening up the ability to measure additional response functions with different dependences on the underlining physics.

REFERENCES

1. F. Brasse et al., *Z. Phys. C* **22**, 33 (1984). H. Breuker et al., *Phys. Lett* **74B**, 409 (1978). F. Brasse et al., *Nucl. Phys.* **B139**, 37 (1978). P. Kummer et al., *Phys. Rev. Lett.* **30**, 873 (1973). U. Beck et al., *Phys. Lett* **51B**, 103 (1974). J. Alder et al., *Nucl. Phys.* **B91**, 386 (1975). C. S. Armstrong et al., *Phys. Rev.* **D60**, 052004 (1999).
2. B. Krusche et al., *Phys. Rev. Lett.* **74**, 3736 (1995).
3. W. Brooks, *Nucl. Phys.* **A663-664**, 1077 (2000).
4. Z. Li and F. Close, *Phys. Rev. D* **42**, 2207 (1990). W. Konen and H. J. Weber, *Phys. Rev. D* **41**, 2201 (1990). S .Capstick and B. D. Keister, *Phys. Rev. D* **51**, 3598 (1995).

CLAS Electro-Omega Production

V. Burkert[a], A. Coleman[b], H. Funsten[c], F. Klein[d], A. Larabee[e], B. Mecking[a], for the CLAS Collaboration

[a] JLAB, [b] Systems Planning and Analysis, Alexandria, Va, [c] WM College, [d] Miami International Universary, [e] Universary of Texas El Paso

Abstract. Electroproduction of $\omega(783)$ mesons from a proton target has been measured at CLAS in a search for so called "missing" baryon resonances. Scattered electrons were measured in coincidence with the recoiling proton and a π^+ from the ω decay. Missing mass techniques were applied to identify the outgoing ω and to reduce the contributions of $\rho(770)$ and 2π final states. The resulting ep missing mass distributions clearly show an ω peak superimposed on a predominantly 3-pion phase space. Preliminary analysis indicates that t distributions monotonically decrease for W > 2 GeV, as expected from pi-exchange and diffractive processes but for 1.8 GeV < W < 2 GeV the distributions are non monotonically decreasing.

CLAS Electroproduction of $\omega(783)$ mesons has been measured to search for "missing" N^*, $2\hbar\omega$ positive parity resonances predicted, [1–5], to lie in the mass region from 1.7 to ≈ 2.2 GeV. These have not been observed in nucleon resonance experiments, which predominately study πN elastic scattering; they may have weak πN coupling, $g_{\pi N}$, [1], ($\sigma_{\pi N\ elastic} \propto g_{\pi N}^4$).

These states may have significant coupling to γN and ωN states. The ω decay width is narrow, 8.4 MeV and it couples with a nucleon only to I=1/2 N^* resonances, eliminating Δ resonance contributions.

There are two predominant forward peaked Born electroproduction contributions, t - channel π exchange and vector - meson dominated diffractive scattering. A non monotonically decreasing t distribution could indicate possible non Born components.

There is evidence for non Born production in the ABBHHM collaboration photo-omega results [6]. A broad nonforward component occurs at W = 2 GeV but not at higher values of W.

CLAS DATA ANALYSIS

Events having at least $e'\ p\ \pi^+$ outgoing tracks were analyzed. For each event the squared missing masses $mm^2_{e'p}$ from outgoing $e'\ p$ tracks and $mm^2_{e'p\pi^+}$ from outgoing e' p π^+ tracks were obtained together with $W^2 = s$ from outgoing e' track and t from the outgoing p track. ΔW and Δt bins were set at 0.1 GeV and 0.35 GeV. Q^2 binning was not made.

$mm^2_{e'p\pi^+} > m^2_\pi$ event selection minimized two pion contributions from ρ mesons and two step sequential resonance pion decays. The resulting $mm_{e'p}(W,t)$ distribution displayed an ω peak superimposed on an underlying broad phase space distribution, $w(mm_{e'p}(W,t))$, characteristic of a 3π phase space. $mm_{e'p}(W,t)^{min} = 3m_\pi$. $mm_{e'p}(W,t)^{max}$ was given by the upper boundary in a $mm_{e'p} - t$ Chew-Low plot. $w(mm_{e'p}(W,t))$ was approximated [8]:

$$w(x) = x^2\sqrt{(1-x)} \quad x = \frac{mm_{e'p}(W,t)^{max} - 3M_\pi}{W - 3M_\pi - M_N} \quad (1)$$

$w(x)$ was multiplied by $(mm_{e'p})^K$ to account for possible distortion. It was convoluted over a W bin and the full Q^2 acceptance range ($\approx 1 \to 2\ Gev^2$). The result, together with an ω Gaussian, was fitted to the observed $w(mm_{e'p}(W,t))$ distributions. Three fit paramters, the Gaussian and background amplitudes and K, were obtained. (The Gaussian centroid was determined using all accepted events). CLAS aceptance was made using the GSIM-GEANT package and an ω event generator with uniform W,t distributions.

t distributions for the first (1988, 4 GeV, high torus field, 90 million eveents) CLAS run sequence generally displayed a Born-like monotonic fall off with t. However, two W bins, $W = 1.8 \to 1.9$ GeV and $W = 1.9 \to 2.0$ GeV displayed a significant rise in the distribution for $-1.9 \to t \to -2.3\ GeV^2$ corresponding to $\theta^* \approx 120°$. For both W bins this rise accounted for $\approx 30\%$ of the total ω yield.

Two additional 1998 runs were analyzed, (4 and 2.5 Gev, low torus field) produced t distributions having the same W,t behavior as the above.

1999 runs, having ≈ 200 million events, permitted halving the w bin size. Fits were made to $mm_{e'p}$ using a 2nd order polynomial in the ω peak region. t distributions in two W bins, $W_{av} = 1.825$ and $W_{av} = 1.875$ GeV displayed non monotonic t fall off, consistant with the 1988 data. t bumps appeared at $t \approx -1.9\ GeV^2$ for $W_{av} = 1.825$ and $t = -2.2\ GeV^2$ for $W_{av} = 1.875$, i.e., at $\theta^* \approx 125°$, approximately the same as for the 1988 run.

EXISTING N* RESONANCES

Three established ($\geq$ two star) N^* resonances exist in the $1.8 \to 2.0\ GeV$ region. Their masses and widths are inconsistant with the data.

Status	J^P	M (MeV)	Γ (MeV)	$\Gamma_{\pi N}/\Gamma_{Tot}$ %
**	3/2+	1900	500	26
**	7/2+	1990	200-500	6
**	5/2+	2000	100-500	8

CAPSTICK AND ROBERTS CALCULATIONS

An estimation the relative omega photoproduction rates for the seven "missing" N=2 positive parity states, using $A_{\gamma N}$ couplings of Capstick [4] and $A_{\omega N}$ of couplings of Capstick and Roberts [5] displays an intesting result. The 1910 MeV $\frac{3}{2}^+$ state carries almost the full electromagnetic strength of the group. Scaling to known $N^*(1520)$ couplings and $\gamma_V + p \to \pi + N^*(1520)$ cross section results in an ω yield from the $N^*(1910) \approx 2$ X a preliminary estimation of the observed ω yield.

DIFFRACTIVE COMPONENT

For W bins $W \geq 2.0$ GeV, t distributions exhibit a predominant diffractive, $(e^{b(\tau)t})$, t behavior (solid lines in Fig 1). Diffractive production hardens with $c\tau$, the fluctuation distance of a pointlike γ_V into hadron sized ω. From kinematics, averaged over acceptance bins
$c\tau^{average} \approx 0.5~fm$, a value lying in the transition region between a pure photon and a "saturated" vector meson. Diffraction should be "harder" than that for a saturated vector meson. The average measured slope parameter, $b(\tau^{average}) \approx 0.8~GeV^{-2}$, approximately equal to the $b(\tau)$ measured in vector meson electroproduction evaluated at $c\tau = 0.5~fm$ [7]. However within the τ region accessed in the experiment, $b(\tau)$ *increased* as τ *decreased*, i.e., $b(\tau)$ apparently softens as τ hardens. This could arise from a non negliable t channel π exchange contribution.

REFERENCES

1. Nathan Isgur *Proc. CEBAF/SURA 1984 Summer Workshop* (1984).
2. Nathan Isgur and Gabriel Karl, *Phys. Rev.* **D19**, 2563 (1978).
3. Roman Koniuk and Nathan Isgur , *Phys. Rev.* **D21**, 1868 (1980).
4. Simon Capstick, *Phys. Rev.* **D46**, 2864 (1992).
5. Simon Capstick and Winston Roberts, *Phys. Rev.* **D49**, 4570 (1993).
6. ABBHHM Collaboration, *Phys. Rev.* **175**, 1669 (1968).
7. D. G. Cassel et al, *Phys. Rev.* **D24**, 2787 (1982).
8. E. Byckling, K. Kajantie *Particle Kinematics* John Wiley & Sons (1973).

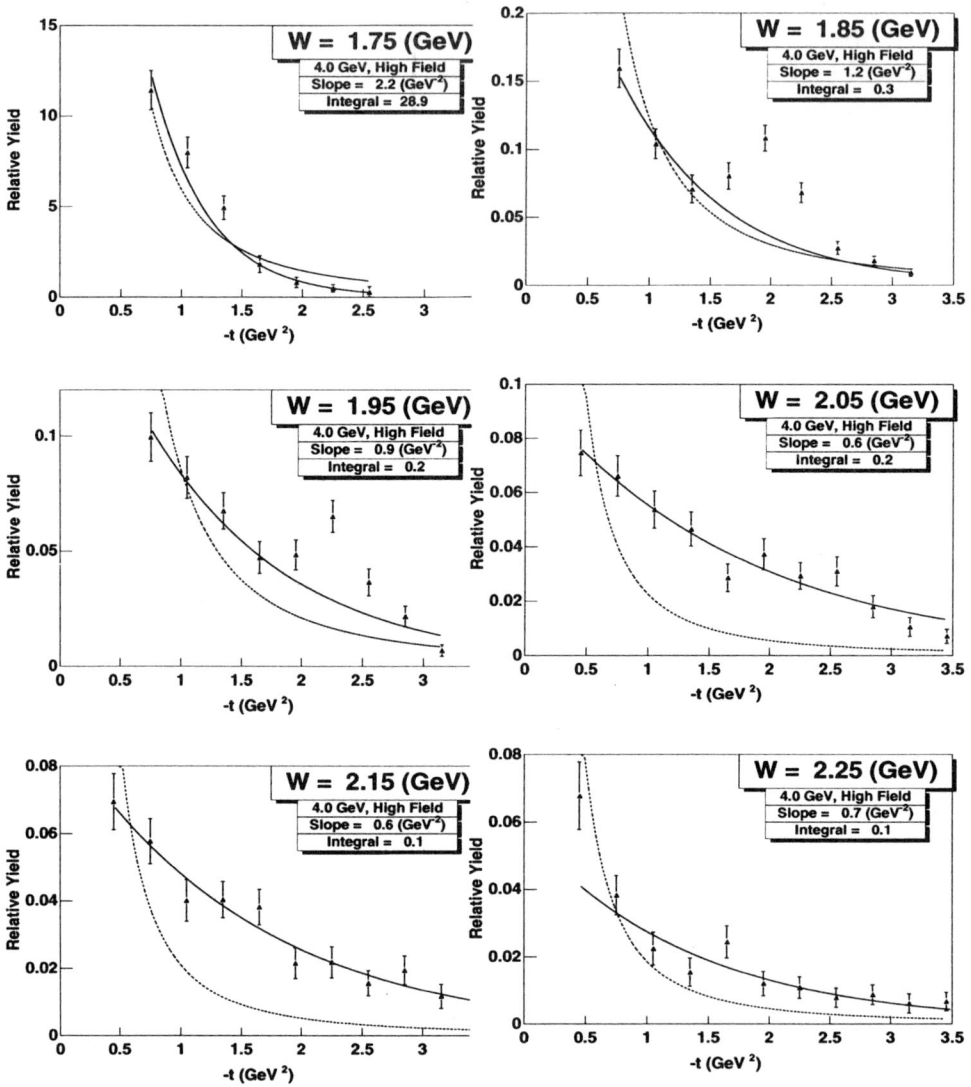

FIGURE 1. t distributions for the 4 GeV high field data. Points exhibiting a monotonic fall-off fitted by solid lines = exponential (diffractive), dashed lines = π t channel exchange.

Neutral Hyperon and H-Dibaryon Results from KTeV

Nickolas Solomey

Enrico Fermi Institute, The University of Chicago, Illinois 60637

Abstract. Results from the KTeV experiment at Fermilab using Ξ^0 hyperons are presented, especially the first form-factor (FF) measurement from its semi-leptonic decay. This decay, which is an important test of the Standard Model, was observed for the first time using the 1997 KTeV experiment. We also report on the analysis of a large samples of Ξ^0 weak radiative decays, and a search for the decay of a lightly bound H^0 dibaryon.

INTRODUCTION

The KTeV experiment is a neutral beam experiment produced by protons from the Tevatron accelerator at 800 GeV/c impinging on a target at a 4.8 mrad angle. The fiducial decay volume starts 90 m down stream of a target because of the space needed to collimate the neutral beam which is where most of the neutral particles decay and is also the location of the sweeping magnetics that eliminate the charged particles. The decay volume from 90 to 160 m from the target is an ultra high vacuum to reduce interactions and has scintillator ring counters to veto those events where the particles have left the fiducial volume. A spectrometer consisting of tracking chambers, an analysis magnet, electromagnetic calorimetry (CsI) [1], particle identification by transition radiation detectors (TRD) [2], and a muon counter system with 5 m of iron filter directly follows the decay volume.

Data was collected in 16 triggers for two different experimental configurations in 1997 and 1999. A rare kaon decay program E799 [3,4], and the search for direct CP violation E832 [5] are presented elsewhere in these proceedings. Presented here are part of the results from a neutral hyperon program that had three triggers in the E799 experiment configuration, and limited to the results from the 1997 run.

PHYSICS ANALYSIS

Physics results from the KTeV hyperon program are:

Semi-leptonic Decays: Neutral hyperon semi-leptonic decays accessible in KTeV are the beta-decay $\Xi^0 \to \Sigma^+ e^- \bar{\nu}_e$, and muon decay $\Xi^0 \to \Sigma^+ \mu^- \bar{\nu}_\mu$. They are important to study for their weak decay FF which give an understanding of their underlying structure. In the V-A formulation the transition amplitude for the beta decay is:

$$M = \frac{G}{\sqrt{2}} <\Sigma|J^\lambda|\Xi> \bar{u}_e \gamma_\lambda (1+\gamma_5) u_\nu \qquad (1)$$

The V-A hadronic current can be written as:

$$<\Sigma|J^\lambda|\Xi> = \mathcal{C} \, i \, \bar{u}(\Sigma) \, [\, f_1 \gamma^\lambda + f_2 \frac{\sigma^{\lambda\upsilon} \gamma_\upsilon}{M_\Xi} + f_3 q^\lambda \frac{M_e}{M_\Xi} +$$
$$[\, g_1 \gamma^\lambda + g_2 \frac{\sigma^{\lambda\upsilon} \gamma_\upsilon}{M_\Xi} + g_3 q^\lambda \frac{M_e}{M_\Xi}] \gamma_5 \,] u(\Xi) \qquad (2)$$

where $\mathcal{C}$ is the CKM matrix element, and q is the momentum transfer. There are 3 vector FF: f_1 (vector), f_2 (weak magnetism) and f_3 (an induced scaler); plus 3 axial-vector FF: g_1 (axial vector), g_2 (weak electricity) and g_3 (an induced pseudo-scaler). All six FF are real if time reversal invariance is valid. The quark model predicts a nonzero but small g_2 FF if SU(3) breaking is sizable, but the standard model assumes this FF is zero. The FF f_3 and g_3 for the beta decay, are expected to be large (i.e. $\frac{g_3}{g_1} \sim 8$), but it is multiplied by $\frac{M_e}{M_\Xi}$ making this term negligably small so as not to contribute any noticeable effect, but for the muon decay this may no longer be assumed. Furthermore, neither of these decays had previously been observed so measuring their branching ratio was also important as a test of the standard model, and in the case of the muon decay this could be the first place to look for a g_3 FF. The final results for the beta decay are a branching ratio of $(2.60 \pm 0.11 \pm 0.16) \times 10^{-4}$, based on 626 events where the first error is statistical and the second systematic, and the theoretical expected is 2.6×10^{-4}. For the muon decay its preliminary branching ratio is $(3.5 \, ^{+2.0}_{-1.0} \, ^{+0.5}_{-1.0}) \times 10^{-6}$ based on 5 events, again the first error is statistical and the second systematic, while the asymmetric error bars are from the small number of events and poisson statistics at the 68% C.I. [6]; theoretical expected is 2.6×10^{-6}.

A very clean sample of Ξ^0 beta decays, see figure 1 left, was obtained by using the TRD detector. The decays of the Σ^+ has a 98% analyzing power, and this fact makes its equivalent to a fully polarized beam. However, spin alignment magnetics gave the ability to control this and then test the technique on the much bigger normal mode decays: $\Xi^0 \to \Lambda^0 \pi^0$. By working in the Σ^+ reference frame all of the FF could be devised by measuring the angular distribution of the proton relative to the electron, neutrino (we typically use the reconstructed transverse neutrino direction) see figure 1 right, as well as test the technique by comparing the proton direction relative to the reconstructed Ξ^0. The final four FF are: $f_1 = 0.99 \pm 0.14$, $\frac{g_1}{f_1} = 1.24 \pm 0.27$, $\frac{f_2}{f_1} = 2.3 \pm 1.3$, and $\frac{g_2}{f_1} = -1.4 \pm 2.1$; here this analysis used the previously quoted branching ratio, and permitted the g_2 FF to float. The g_2 FF is

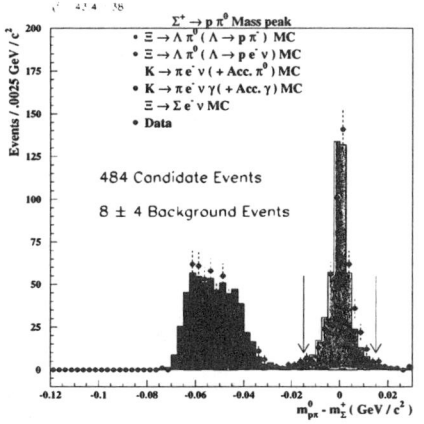
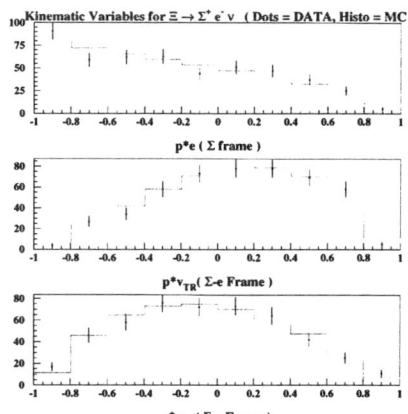

FIGURE 1. On the left is the Ξ^0 beta decay compared to its normalized MC and possible backgrounds, and on the right is the angular distribution in the Σ^+ reference frame of the angular asymmetry of the proton to electron, reconstructed neutrino and between the reconstructed neutrino and the electron.

consistent with zero and in another analysis it was constrained to be zero and the FF reanalyzed and they essentially remained unchanged.

Radiative Decays: Neutral hyperon radiative decays accessible in KTeV are: $\Xi^0 \to \Sigma^0 \gamma$ and $\Xi^0 \to \Lambda^0 \gamma$. Although these decays are not simple first-order processes but actually proceed threw complicated diagrams, they are important to study because they are difficult to calculate, but yet have high branching ratios. Furthermore, the angular emission of the high energy gamma relative to the hyperon polarization can be related to the underlying mechanism of its decay as a test of various theories. The decay $\Xi^0 \to \Sigma^+ \gamma$ in KTeV is identified when a second radiative decay of $\Sigma^0 \to \Lambda^0 \gamma$ and then for both radiative decays of the Ξ^0 through the charged particle decay of $\Lambda^0 \to p^+ \pi^-$. As with the semi-leptonic decays the self analyzing power with hyperon decays permits the polarization to be determined relative to the gamma angular distribution. The final results for the radiative decay $\Xi^0 \to \Sigma^0 \gamma$ is a branching ratio of $(3.34 \pm 0.05 \pm 0.11) \times 10^{-3}$, based on 4048 events, shown in figure 2 left, where the first error is statistical and the second systematic. The angular distribution of the gamma relative to the Ξ^0 polarization vector; is -0.65 ± 0.13, see figure 2 right. The radiative decay $\Xi^0 \to \Lambda^0 \gamma$ preliminary branching ratio is $(0.95 \pm 0.03 \pm 0.09) \times 10^{-3}$ based on 1105 events.

Dibaryon Search: The KTeV hyperon data was used to search for a six quark neutral hadron state which has two strange, two up and two down quarks that decays $H^0 \to \Lambda^0 p^+ \pi^-$. The data used for normalizing the decay was $\Xi^0 \to \Lambda^0 \pi^0$ with the decay of the $\Lambda^0 \to p^+ \pi^-$ and $\pi^0 \to e^+ e^- \gamma$. These two decays share

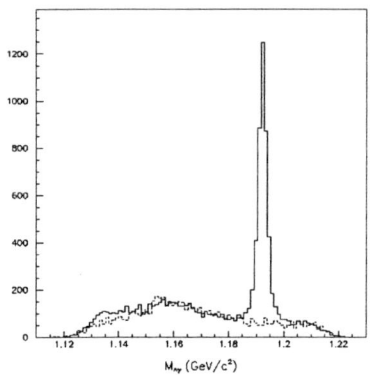 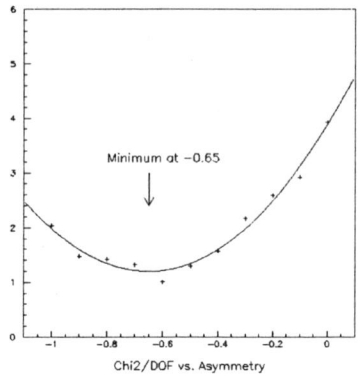

FIGURE 2. On the left is the $\Xi^0 \to \Sigma^0 \gamma$ decay signal compared to its normalized MC, and on the right the determination of the gamma angular asymmetry.

similar technical details for their analysis: two different charged particle vertices, a reconstructible Λ^0 mass peak and a stiff proton. Using the 1997 data from KTeV E799 run we have ruled out the existence of a H^0 dibaryon in the mass range from 2.194 to 2.231 GeV/c^2 for long lifetimes of 5×10^{-10} to 1×10^{-3} seconds at the 90% CI [7].

CONCLUSIONS

The results quoted here are the final measurements for these decays from the 1997 KTeV run. Three times more data exists from the 1999 run which is in the process of being analyzed. Future results are planned to make improvements in all decay modes, and some decays previously excluded by trigger requirements, such as: $\Xi^0 \to \Lambda^0 \pi^0$ where the decay $\pi^0 \to e^+ e^- \gamma$ will provide a sample of decays where the z vertex can be measured from the two electrons for use in a Ξ^0 mass and lifetime improved measurement, the decay $\Xi^0 \to \Sigma^0 \gamma$ with the special decay $\Sigma^0 \to \Lambda^0 e^+ e^-$, or the radiative decay $\Xi^0 \to \Lambda^0 \pi^0 \gamma$. All of these decays are especially well suited for the excellent abilities of π^0 and γ measurements with our high precession CsI electromagnetic calorimeter, and our unmatched $\pi^\pm/e^\pm$ rejection obtained with the TRD system.

REFERENCES

1. R.S. Kessler et al., Nucl. Instrum. Meth. A **368** (1996) 653.
2. N. Solomey, Nucl. Instrum. Meth. A **419** (1998) 637.
3. See A. Ledovskoy these proceedings.
4. See E. Halkiadakis these proceedings.
5. See A. Glazov these proceedings.
6. G.J. Feldman and R.D. Cousins, Phys. Rev. D **57** (1998) 3873.
7. A. Alavi-Harati et al., Phys. Rev. Lett. **84** (2000) 2593.

Fishing for Narrow Dibaryons in $pd \to pX$ Reaction

L.V. Fil'kov[a], V.L. Kashevarov[a], E.S. Konobeevski[b], M.V. Mordovskoy[b], S.I. Potashev[b], V.A. Simonov[b], V.M. Skorkin[b], and S.V. Zuev[b]

[a] *Lebedev Physical Institute, 117924 Moscow, Russia*
[b] *Institute for Nuclear Research, 117312 Moscow, Russia*

Abstract. An analysis of new experimental data, obtained at Linear Accelerator of INR, is carried out with the aim of searching for supernarrow dibaryons in the reactions $pd \to p + X$ and $pd \to p + pX_1$. Dibaryons with masses 1904±2, 1926±2, and 1942±2 MeV have been observed in missing mass M_X spectra. In missing mass M_{X_1} spectra, the resonancelike states $X_1 = \gamma + n$ at $M_{X_1} = 966 \pm 2$, 986±2, and 1003±2 MeV have been found. The analysis of the data obtained leads to the conclusion that the observed dibaryons are supernarrow dibaryons, the decay of which into two nucleons is forbidden by the Pauli exclusion principle.

In Ref. [1-3] the study of the reaction $pd \to pX$ was performed with the aim of searching for supernarrow dibaryons (SND), the decay of which into two nucleons is forbidden by the Pauli exclusion principle [4-6]. Such dibaryons with the mass $M < 2m_N + m_\pi$ can decay into two nucleons, mainly emitting a photon. The experiment was carried out at 305 MeV using the two-arm spectrometer TAMS. As was shown in Ref. [2,3], the nucleons and the deuteron from the decay of SND into γNN and γd have to be emitted in a narrow angle cone with respect to the direction of motion of the dibaryon. On the other hand, if a dibaryon decays mainly into two nucleons, then the expected angular cone of emitted nucleons must be more than 50°. Therefore, a detection of the scattered proton in coincidence with the proton (or the deuteron) from the decay of particle X at correlated angles allowed to suppress essentially the contribution of the background processes and to increase the relative contribution of a possible SND production. As a result, two narrow peaks in missing mass spectra have been observed at $M = 1905$ and 1924 MeV. The analysis of the angular distributions of the protons from the decay of particle X showed that the peak found at 1905 MeV most likely corresponds to a SND with isotopic spin equal to 1. In Ref. [3] arguments were presented for the resonance at $M = 1924$ MeV is a SND, too.

In the present paper we give the results of an analysis of new experimental data of $pd \to p + X$ and $pd \to p + pX_1$ reactions at 305 MeV. Experiment was performed using the spectrometer TAMS, the properties of which were described elsewhere [3]. CD_2 and C^{12} were used as targets. In this experiment, the scattered proton was detected in the left arm of the spectrometer TAMS at the angle $\theta_L = 70°$. The second charged particle (either p or d from the decay of X state) was detected in the right arm by three telescopes located at $\theta_R = 34°$, $36°$, and $38°$.

As follows from the present experiment, the main contribution into resonances observed here is given by processes where the second charged particle is a proton. The experimental missing mass M_X spectra obtained with the CD_2 target are shown in Figs. 1(a-c), where (a), (b), and (c) correspond to a detection of the proton from the decay of X states in the right arm detector at $\theta_R = 34°$, $36°$, and $38°$, respectively.

Three peaks at $M_X = 1904 \pm 2$, 1926 ± 2, and 1942 ± 2 MeV are observed in these spectra. The first two of them confirmed the values of the dibaryon mass obtained by us earlier [1–3] and the resonance at 1942 MeV is a new one. It is expected [3,6] that isoscalar SNDs contribute mainly into γd channel and isovector SNDs do into γNN one. As the main decay of the found dibaryons is observed into pX_1 channel, it is possible to assume that $X_1 = \gamma + n$ and all these states are isovector SNDs. The calculations for the SNDs $D(T = 1, J^P = 1^\pm)$ showed that the biggest contribution of such dibaryons must be at $\theta_R = 34°$ and $36°$. The contribution to spectrum at $38°$ is expected to be several times smaller. These predictions are in agreement with our experimental data. If the observed states are usual NN-

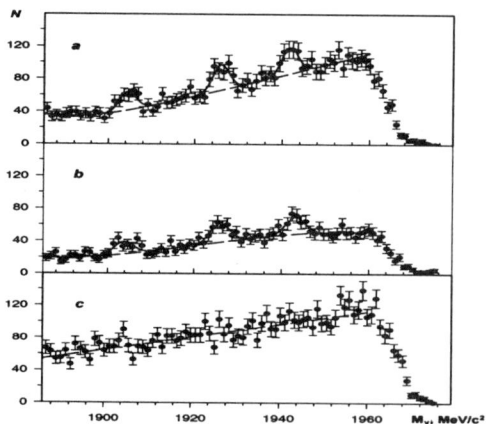

FIGURE 1. The missing mass M_X spectra for the reaction $pd \to p + X$; (a) – $\theta_R = 34°$, (b) – $\theta_R = 36°$, (c) – $\theta_R = 38°$,

coupled dibaryons decaying mainly into two nucleons then their contributions to the missing mass spectra in Fig.1(a), 1(b), and 1(c) would be nearly the same and would not exceed a few events. Hence, the peaks found most likely correspond to isovector SNDs.

The summary spectrum over angles $\theta_R = 34°$ and $36°$ is presented in Fig. 2a. This spectrum was interpolated by a second order polynomial (for the background) plus Gaussians (for the peaks). The numbers of standard deviations are 6.0, 7.0, and 6.3 SD for resonances at 1904, 1926, and 1942 MeV, respectively.

An additional information about the nature of the observed states is given by study of the missing mass M_{X_1} spectra of the reaction $pd \to p + pX_1$. If the state found is a dibaryon decaying mainly into two nucleons then X_1 is a neutron and

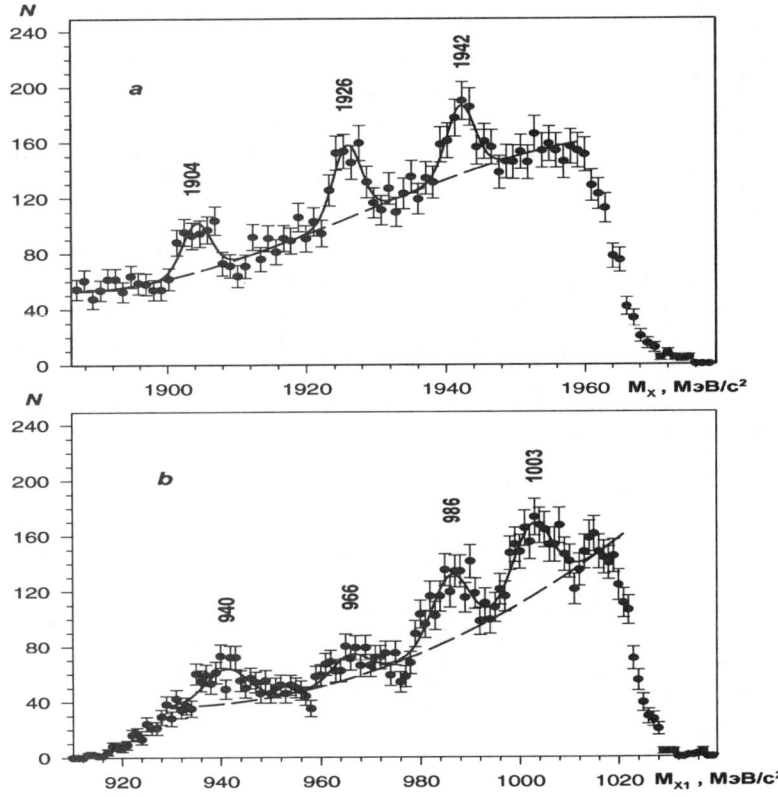

FIGURE 2. The missing mass M_X (a) and M_{X_1} (b) for the sum of angles of $\theta_R = 34°$ and $\theta_R = 36°$.

the mass M_{X_1} is equal to the neutron mass m_n. If the value of M_{X_1}, obtained from the experiment, differs essentially from m_n then $X_1 = \gamma + n$ and we have the additional indication that the observed dibaryon is SND.

The simulation of mass spectra for the reaction $pd \to p + pX_1$, where pX_1 are decay products of SNDs with masses 1904, 1926, and 1942 MeV, gave peaks at M_{X_1} =965, 987,and 1003 MeV, respectively. Fig. 2b demonstrates the missing mass M_{X_1} spectrum obtained from the experiment for the sum of the angles $\theta_R = 34°$ and 36°. As is seen from this figure, besides the peak at neutron mass, which caused by the process $pd \to p + pn$, resonance-like behavior of the spectrum is observed at 966 ± 2, 986 ± 2, and 1003 ± 2 MeV. These values of M_{X_1} coincide with the ones obtained from the simulation and differ essentially from the value of the neutron mass. Hence, for all states under study, $X_1 = \gamma + n$ and the dibaryons found are really SNDs.

It should be noted that a resonance-like behavior of X_1 at $M_{X_1} = 1003 \pm 2$ MeV corresponds to the resonance found in [7] and attributed to an excited nucleon state N^*. In this work, the authors brought out three such states with masses 1004, 1044, and 1094 MeV. Taking into account the found connection between the SNDs and the resonance-like states X_1, it is possible to assume that the peaks, observed in [7] are not the excited nucleons, but they are resonance-like states X_1 caused by possible existence and decay of SNDs with the masses 1942, 1982, and 2033 MeV, respectively.

The following conclusion can be made. As a result of the study of the reaction $pd \to pX$ and $pd \to p + pX_1$ three narrow peaks at 1904, 1926, and 1942 MeV have been observed in the missing mass M_X spectra. The analysis of the angular distributions of the protons from decay of X states showed that the peaks found can be explained as a manifestation of the SNDs, the decay of which into two nucleons is forbidden by the Pauli exclusion principle. The observation of the resonance-like structures in the missing mass M_{X_1} spectra at 966, 986, and 1003 MeV is an additional confirmation that the dibaryons found are really SNDs.

REFERENCES

1. Kashevarov V.L., Konobeevski E.S., Mordovskoy M.B., Potashev S.I., Skorkin V.M., and Fil'kov L.V., *Bulletin of Lebedev Phys. Inst.* No **11**, 36 (1998).
2. Fil'kov L.V., Kashevarov V.L., Konobeevski E.S., Mordovskoy M.V., Potashev S.I., and Skorkin V.M., *Phys. Atom. Nucl.* **62**, 2021, (1999).
3. Fil'kov L.V., Kashevarov V.L., Konobeevski E.S., Mordovskoy M.V., Potashev S.I., and Skorkin V.M., *Phys. Rev.* C **61**, 044004, (2000).
4. Mulders P.J., Aerts A.T., and de Swart J.J., *Phys. Rev.* D **21**, 2653, (1980).
5. Fil'kov L.V., *Sov. J. Nucl. Phys.* **47**, 437, (1988).
6. Akhmedov D.M. and Fil'kov L.V., *Nucl. Phys.* **A544**, 692, (1992).
7. Tatischeff B. et al. *Phys. Rev. Lett.* **79**, 601, (1997).

Radial Excitations of low–lying Baryons and the Structure of the Z^+ Penta–Quark

Herbert Weigel[1]

*Center for Theoretical Physics, Laboratory for Nuclear Science, and Department of Physics
Massachusetts Institute of Technology, Cambridge, Massachusetts 02139*

Abstract. Within the collective quantization scheme for chiral solitons we discuss states in higher dimensional representations of flavor $SU(3)$ and their relation to radially excited states in the octet. We also consider states which do not have counterparts of the same quantum numbers in the octet or decuplet and cannot be built from three quarks. We focus on the Z^+ penta–quark, presumably the lightest such state, by estimating its mass and decay width.

INTRODUCTION

In the collective quantization of chiral solitons higher dimensional representations of flavor $SU(3)$ play a major role[2]. Upon flavor symmetry breaking, members of these representations admix to the ordinary octet and decuplet states of the same quantum numbers to form the low–lying $\frac{1}{2}^+$ and $\frac{3}{2}^+$ baryons. Notably these higher dimensional representations also contain states with quantum numbers that are not found in the octet (decuplet) representation. In a quark model picture these baryons cannot be formed as simple three quark states, rather they consist of at least four quarks and an anti–quark. The $\overline{10}$ representation contains the presumably lightest such state, the Z^+, with the quantum numbers $Y = 2$, $I = 0$, $J = 1/2$. Details of the presented study are found in ref [2]. For earlier studies on the Z^+ see ref [3]. An experimental search for the Z^+ was proposed in ref [4].

COLLECTIVE QUANTIZATION

We consider a chiral Lagrangian in flavor $SU(3)$. The basic variable is the chiral field $U = \exp(i\lambda_a \phi^a/2)$ representing the pseudoscalar fields ϕ^a $(a = 0, \ldots, 8)$. Other fields may be included as well. For example, the specific model used later also contains a scalar meson. In general this chiral Lagrangian can be written as

---

[1] Heisenberg–Fellow
[2] See ref [1] for a review and references on chiral solitons in flavor $SU(3)$.

$$\mathcal{L} = \mathcal{L}_S + \mathcal{L}_{SB} \tag{1}$$

with flavor symmetric (S) and flavor symmetry breaking (SB) pieces. Denoting the (classical) soliton solution of (1) by $U_0(\vec{r})$ states with baryon quantum numbers are constructed by quantizing the flavor rotations

$$U(\vec{r}, t) = A(t) U_0(\vec{r}) A^\dagger(t), \qquad A(t) \in SU(3) \tag{2}$$

canonically. According to (1) the Hamiltonian for the collective coordinates $A(t)$ can be written as $H = H_S + H_{SB}$. For unit baryon number the eigenstates of H_S are the members of $SU(3)$ representations with the condition that the representation contains a state with identical spin and isospin quantum numbers (such as e.g. the nucleon or the Δ). The symmetry breaking piece, H_{SB} mixes states from different $SU(3)$ representations. For example, the state identified with the nucleon becomes

$$|N\rangle = |N, \mathbf{8}\rangle + c_{\overline{10}}^{(N)} |N, \overline{\mathbf{10}}\rangle + c_{27}^{(N)} |N, \mathbf{27}\rangle + \ldots , \tag{3}$$

where the coefficients are computed as matrix elements of H_{SB}. Another interesting question, of course, is what role do states play which diagonalize H and have dominant contributions from the higher dimensional representations such as

$$|N'\rangle = |N, \overline{\mathbf{10}}\rangle + c_8^{(N')} |N, \mathbf{8}\rangle + c_{27}^{(N')} |N, \mathbf{27}\rangle + \ldots , \tag{4}$$

$$|Z^+\rangle = |Z^+, \overline{\mathbf{10}}\rangle + c_{35}^{(Z^+)} |Z^+, \overline{\mathbf{35}}\rangle + \ldots ? \tag{5}$$

A naïve analysis predicts the N' to be 400 − 500MeV heavier than the nucleon which suggests to identify this state with the Roper (1440) resonance. However, such an interpretation contradicts the common folklore viewing the Roper as a radial excitation of the nucleon. In order to resolve this puzzle and also to provide a sensible interpretation to the state $|Z^+\rangle$, the radial degrees of freedom need to be quantized as well. For this end we introduce the corresponding collective coordinate $\xi(t)$ via [5,6]

$$U(\vec{r}, t) = A(t) U_0(\xi(t) \vec{r}) A^\dagger(t) . \tag{6}$$

Changing variables to $x(t) = [\xi(t)]^{-3/2}$ the flavor symmetric piece of the collective Hamiltonian can be diagonalized for a given $SU(3)$ representation of dimension μ

$$H_S = \frac{-1}{2\sqrt{m\alpha^3\beta^4}} \frac{\partial}{\partial x} \sqrt{\frac{\alpha^3 \beta^4}{m}} \frac{\partial}{\partial x} + V + \left(\frac{1}{2\alpha} - \frac{1}{2\beta}\right) J(J+1) + \frac{1}{2\beta} C_2(\mu) + s , \tag{7}$$

where J and $C_2(\mu)$ are the spin and (quadratic) Casimir eigenvalues associated with the representation μ. Note that $m = m(x), \alpha = \alpha(x), \ldots, s = s(x)$ are all functions of the scaling variable. We use (7) to build towers of radially excited states $|\mu, n_\mu\rangle$ with $n_\mu = 0, 1, \ldots$ denoting the number of nodes in the respective wave–functions. These states serve to diagonalize the Hamiltonian including symmetry breaking

$$H_{\mu,n_\mu;\mu',n'_{\mu'}} = \mathcal{E}_{\mu,n_\mu}\delta_{\mu,\mu'}\delta_{n_\mu,n'_{\mu'}} - \langle\mu,n_\mu|\tfrac{1}{2}\mathrm{tr}\left(\lambda_8 A\lambda_8 A^\dagger\right)s(x)|\mu',n'_{\mu'}\rangle , \qquad (8)$$

yielding the baryonic states $|B,m\rangle = \sum_{\mu,n_\mu} C^{(B,m)}_{\mu,n_\mu}|\mu,n_\mu\rangle$. The resulting spectrum [6] is depicted in figure 1 and compared to the known resonances [7]. We recognize a good agreement although the Roper is predicted about 90MeV too low. It also worth to note that this state turns out to dominantly be a radial excitation of the nucleon state in the octet representation, i.e. $|8,1\rangle$ [2].

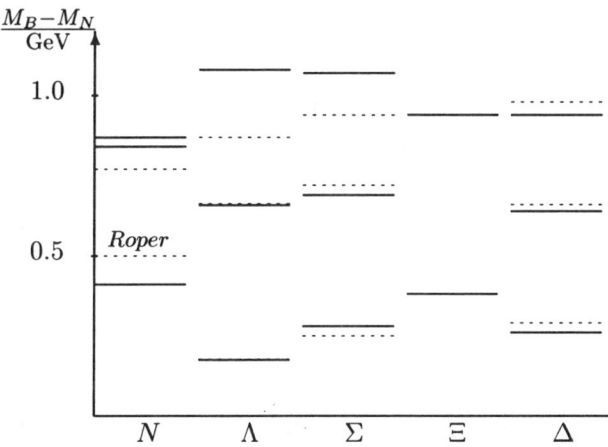

FIGURE 1 : The mass differences of the predicted baryons in the scaling mode treatment of the three flavor Skyrme model with a scalar field included (full lines) [6]; empirical data [7] are denoted by dotted lines. The model predictions for the ground states of the Λ and Ξ channels and the experimentally observed counterparts are (almost) indistinguishable.

For the baryon state we are interested in most, we find $M_{Z^+} - M_N \approx 630\mathrm{MeV}$. This state has a sizable admixture of the counterpart in the $\overline{35}$ representation.

ESTIMATE OF DECAY WIDTHS

In order to further specify the Z^+ we demand the width for the decay $Z^+ \to KN$. This is one of the strong decays of excited baryons which are described by

$$\Gamma_{B'\to B\phi} = \frac{3G^2_{B'\to B\phi}}{8\pi M_{B'}M_B}|\boldsymbol{p}_\phi|^3 . \qquad (9)$$

Here $\boldsymbol{p}_\phi$ is the momentum of the outgoing meson in the rest frame of B'. The coupling constants $G_{B'\to B\phi}$ are not easily accessible in chiral soliton models. Fortunately, it seems sufficient to estimate them from the appropriate flavor operator (For example, $a=3$ when B and B' have identical isospin projections, I_3.),

$$G_{B'\to B\phi} = C_\Delta\langle B'm'|x^{2n/3}\mathrm{tr}\left(\lambda_a A\lambda_3 A^\dagger\right)|Bm\rangle \qquad (10)$$

and allow for different forms for the scaling operators parameterized by the power n. The constant C_Δ is fixed such that the experimental width $\Gamma_{\Delta\to\pi N}=120\mathrm{MeV}$ is properly reproduced. From table 1 we observe that only the width of the decay $R\to N\pi$ has a significant dependence on the shape of the scaling function in (10).

TABLE 1. Decay widths (in MeV) and ratio of πN and $\pi \Delta$ coupling constants using the matrix elements (10). R denotes the Roper (1440) resonance. Experimental data are extracted from ref [7].

	e=5.0			e=5.5			Data
n	4	3	2	4	3	2	
$\Sigma^* \to \Sigma\pi$	1	1	1	2	2	2	4 ± 1
$\Sigma^* \to \Lambda\pi$	33	38	42	37	38	43	32 ± 4
$\Xi^* \to \Xi\pi$	5	7	10	7	9	11	10 ± 2
$R \to N\pi$	429	281	156	424	260	145	200 to 320
$R \to \Delta\pi$	4	2	2	9	6	3	50 to 80
$Z^+ \to NK$	118	121	124	130	124	126	?
$g_{\pi NN}/g_{\pi N\Delta}$	0.77	0.79	0.83	0.77	0.79	0.83	0.68

The width for $R \to \Delta\pi$ is underestimated because the respective phase space is small as a consequence of the too low prediction for the Roper mass. Otherwise the widths are reasonably well reproduced and we estimate $\Gamma_{Z^+ \to KN} \approx 120$MeV.

CONCLUSION

We have investigated the role of higher dimensional flavor $SU(3)$ representations for the description of excited baryon states. In particular the interplay between the states in these representations and the radial excitations of the octet (and decuplet) baryons has been discussed. A reasonable description of the spectrum and the decay widths has been achieved. The lightest of the exotic states in such a higher dimensional representation, the Z^+, was predicted to have a mass of about 1.57GeV and a width of about 120MeV for its decay into a nucleon and a kaon.

Acknowledgements

This work is supported in part by funds provided by the U.S. Department of Energy (D.O.E.) under cooperative research agreement #DF-FC02-94ER40818 and the Deutsche Forschungsgemeinschaft (DFG) under contract We 1254/3-1.

REFERENCES

1. H. Weigel, Int. J. Mod. Phys. **A11** (1996) 2419.
2. H. Weigel, Eur. Phys. J. **A2** (1998) 391.
3. H. Walliser, Nucl. Phys. **A548** (1992);
   D. Diakonov, V. Petrov and M. Polyakov, Z. Phys. **A359** (1997) 305.
4. M. Polyakov et al., nucl-th/9909048.
5. J. Schechter and H. Weigel, Phys. Rev. **D44** (1991) 2916.
6. J. Schechter and H. Weigel, Phys. Lett. **B261** (1991) 235.
7. C. Caso et al., (Particle Data Group), Eur. Phys. J. **C3** (1998) 1.

Mesons and Hybrids in a Relativistic Many Body Theory

Stephen R. Cotanch and Felipe J. Llanes-Estrada

Department of Physics, North Carolina State University, Raleigh NC 27695-8202

Abstract. Using a field theoretical, QCD inspired Hamiltonian formulated in the Coulomb gauge, relativistic many-body calculations are reported which reproduce the semiquantitative features of the observed meson and lattice glueball spectra. Dynamical chiral symmetry breaking is achieved utilizing a BCS vacuum ansatz, yielding gap equations and realistic quark (and gluon) constituent masses and condensates. The excited hadron states are obtained by diagonalizing the effective Hamiltonian in a truncated Fock space using both the TDA and RPA. The meson spectrum, for a variety of spin and parities, is reasonably well described. In general, the two approaches differ minimally, except for the π and η where only the RPA yields correct properties and a Goldstone boson in the chiral limit. Our unified model also describes hybrids and a three-body TDA calculation predicts exotic and non-exotic hybrid states all above 2 GeV. Our results are consistent with lattice and flux tube hybrid masses, suggesting that the recently observed 1^{-+} exotics below 2 GeV have an alternative, perhaps four quark, structure.

This work continues our comprehensive, many-body approach to hadron structure [1,2], and reports the first, quasiparticle three-body calculation for hybrid mesons. The formulation is based upon approximating the exact QCD Hamiltonian in the Coulomb gauge by an instantaneous kernel having linear confinement with slope, $\sigma = 0.18\ GeV^2$, specified by lattice. With the exception of the flavored current quark masses (here $m_u = m_d = 5\ MeV, m_s = 150\ MeV, m_c = 1.2\ GeV$), our theory entails only one predetermined parameter, yet is capable of providing a unified description of the quark, gluon and combined sectors for both ground (vacuum) and excited (hadron) states. The model also rigorously preserves chiral symmetry yet exhibits dynamical chiral symmetry breaking through constituent masses from the gap equation. In this paper we summarize our previous hadron results and detail a new hybrid meson application to confront interesting exotic $J^{PC} = 1^{-+}$ resonances recently observed at BNL [3].

We briefly describe our model (consult refs. [1,2] for additional information). Our QCD effective Hamiltonian is

$$H = \int d\mathbf{x} \Psi^\dagger(\mathbf{x})(-i\boldsymbol{\alpha}\cdot\nabla + \beta m)\Psi(\mathbf{x}) - \frac{1}{2}\int d\mathbf{x}d\mathbf{y}\rho^a(\mathbf{x})V_L(|\mathbf{x}-\mathbf{y}|)\rho^a(\mathbf{y})$$

$$-g_s \int d\mathbf{x} \Psi^\dagger(\mathbf{x})\vec{\alpha}\cdot\mathbf{A}(\mathbf{x})\Psi(\mathbf{x}) + Tr \int d\mathbf{x}(\mathbf{\Pi}\cdot\mathbf{\Pi} + \mathbf{B}\cdot\mathbf{B})$$

involving both quark, Ψ, and gluon, $\mathbf{A}$, fields with color density $\rho^a = \Psi^\dagger \frac{\lambda^a}{2}\Psi + f^{abc}\mathbf{A}^b \cdot \mathbf{\Pi}^c$. The linear interaction, $V_L = \sigma r$, is obtained from lattice studies and Regge phenomenology (for certain observables we supplement this with the canonical Coulomb potential $V_C = -\frac{\alpha_s}{r}$ with $\alpha_s = \frac{g_s^2}{4\pi}$).

We first address the ground state and set $g_s = 0$. The BCS ansatz produces uncoupled gap equations (Schwinger-Dyson equations) for both quark and gluon sectors. The generated dynamical mass is of order 100 MeV for light quarks and roughly 800 MeV for the gluons. These dressed quasiparticles, which in the vacuum form Cooper pairs ($^3P_0$ condensates), are the constituent degrees of freedom for the excited states, or hadron spectrum. Next, we separately apply the 1p-1h TDA truncation for the pure gluon and quark sectors. This generates glueballs [1] in broad agreement with lattice calculations and a meson spectrum [2] that adequately describes the known resonances, with a few exceptions, notably the pion. In particular, the TDA $\pi - \rho$ splitting is only ≈ 200 MeV which is insufficient, and motivated our improved RPA formulation. The RPA yields a better vacuum, $|RPA\rangle$, containing correlations beyond the BCS Cooper pairs (in particular, pion pairs), such that $Q(RPA)|RPA\rangle = 0$, where the RPA pion creator operator is $Q^\dagger = \sum_{ij}\left(X_{ij}q_i^\dagger \bar{q}_j^\dagger - Y_{ij}q_i\bar{q}_j\right)$. Here q_i, $\bar{q}_i$ are the rotated quasiparticle operators and X_{ij}, Y_{ij} are the RPA wavefunction components obtained from the coupled equations of motion. For the pion state, $|\pi\rangle$, the equations are obtained via $\langle\pi|[H,Q^\dagger]|RPA\rangle = M_\pi\langle\pi|Q^\dagger|RPA\rangle$. Significantly in the chiral limit, $m_q = 0$, and we compute $M_\pi = 0$, corresponding to a massless Goldstone boson. For a non zero quark mass, $m_q = 5$ MeV, the $\pi - \rho$ mass difference is now a more realistic 600 MeV indicating that chiral symmetry, in contrast to spin interactions, is the major source of mass splitting. Thus with only one parameter, σ, all the features of the meson spectrum, with the exception of the $\eta - \eta'$ channel which requires flavor mixing, are semi-quantitatively reproduced. Our results are displayed in figure 1.

Similarly for the glueball spectrum as detailed more completely in ref. [1]. Using both meson and glueball input, we now focus on the combined quark and glue sectors and formulate hybrid mesons as a three-body system ($q\bar{q}g$). We again use the BCS and TDA methods and diagonalize in an extended basis, $[[q \otimes \bar{q}]_8 \otimes g]_0|RPA\rangle$, now involving quark color octet states. The wave equation is studied variationally with alternative, carefully constructed, ansätze in different angular momentum channels incorporating the constraints from transversality in the Coulomb gauge, consistent with Yang's theorem. We consider only the lowest intermediate angular momenta, L_- ($q\bar{q}$ orbital), L_+ (($q\bar{q}$)g orbital) and S (total intrinsic quark spin) contributing to each J^{PC} channel. Our RPA calculation did not differ significantly, as the chiral color octet current is not conserved (in contrast to the color singlet charge) by our Hamiltonian (or by full QCD). In spite of the challenging difficulties attending a relativistic, non-local three body problem, the diagonalization is feasible on a supercomputer using a Monte Carlo multidimensional integration al-

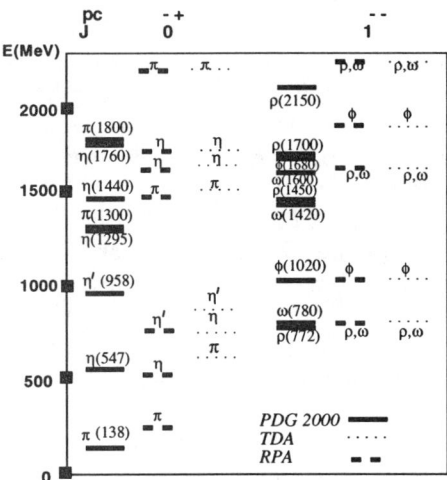

FIGURE 1. Calculations and data for the pseudoscalar and vector meson spectrum.

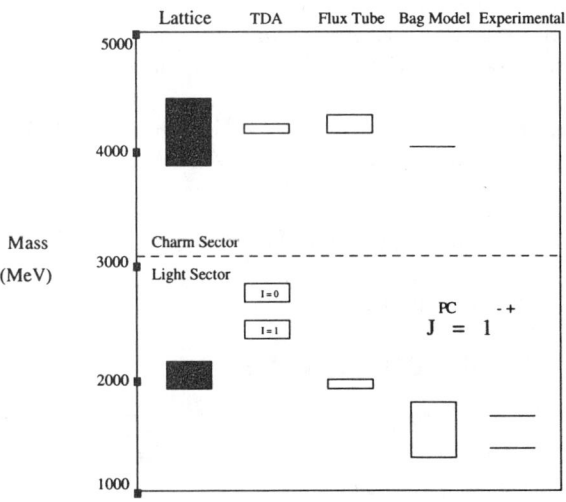

FIGURE 2. Comparison of different theoretical hybrid meson results in the exotic 1^{-+} sector.

gorithm and code Vegas. Further technical details will be reported in an upcoming extensive publication.

Our results are displayed in figure 2 and reveal a hybrid meson mass spectrum beginning in the low to mid 2 GeV range. The reasons why our hybrids are much heavier than their valence $q\bar{q}$ counterparts are the large dynamical gluon mass and the repulsive interaction between the quark pair in a color octet. Another noticeable feature of figure 2 is the $I = 1$ and 0 isospin splitting. This new effect, not present in color singlets, is due to quark and antiquark annihilation into a gluon for the spin 1, isospin 0 color octet state (analogous to annihilation in positronium). The process leads to a repulsive $q\bar{q}$ interaction, further raising the $I = 0$ states. In quark channels coupled to spin 0 we recover the isospin degenerate spectrum.

Note in particular the exotic states 1^{-+}, 0^{--}, 3^{-+} between 2 and 3 GeV. Each requires a p-wave, either L_+ or L_-. Hence, the lightest exotic appears around 2.4 GeV. This is somewhat higher, but still comparable to lattice. A comparison of the different theoretical models is also displayed in this figure along with the two controversial experimental states at 1.4 and 1.6 GeV [3]. Consensus is growing, reflected by this figure, that these resonances are not hybrids but rather four quark states (meson resonances or quark molecules) which would have a lighter mass spectrum.

We also display in figure 2 our charmed hybrid exotic prediction and compare to other models which give similar results. Clearly, any new experimental data would be very enlighting as an excellent opportunity appears at hand for documenting gluonic degrees of freedom.

In conclusion, we have described mesons, glueballs and hybrids reasonably well with a unified model containing only one predetermined parameter. Our hybrid results, especially for the interesting exotic states, generally agree with other theoretical approaches and suggests an alternative structure for the observed 1^{-+} states. Specifically, we submit these resonances may be four quark states. Our comprehensive approach readily extends to such systems and calculations are in currently in progress.

This work is supported by grants DOE DE-FG02-97ER41048 and NSF INT-9807009. Supercomputer time is provided by NERSC.

REFERENCES

1. A. P. Szczepaniak, E. S. Swanson, C.-R. Ji, and S. R. Cotanch, Phys. Rev. Lett. **76**, 2011 (1996).
2. F. J. Llanes-Estrada and S. R. Cotanch, Phys. Rev. Lett. **84**, 1102 (2000) and references therein.
3. G. S. Adams et al. (E852 Collaboration) Phys. Rev. Lett. **81**, 5760 (1998); D. R. Thompson et al. (E852 Collaboration) Phys. Rev. Lett. **79**, 1630 (1997).
4. C. McNeile, hep-lat/9904013 (1999).

Description of Glueballs in Effective Meson Lagrangians

M. K. Volkov[1] and V. L. Yudichev

Bogoliubov Laboratory of Theoretical Physics, Joint Institute for Nuclear Research, 141980 Dubna, Russia

Abstract. In the framework of a chiral quark model with dilaton, it is shown that $f_0(1500)$ is likely a scalar glueball state.

As it was shown in papers [1], in the framework of a nonlocal chiral quark model of Nambu–Jona-Lasinion (NJL) type, 19 scalar meson states with masses ranging from 0.4 GeV to 1.7 GeV can be interpreted as two scalar $q\bar{q}$ nonets: the ground and radially excited ones and a glueball. However, the glueball was not finally identified with one of the states $f_0(1500)$ or $f_0(1710)$, because only $q\bar{q}$ states were considered. To solve this problem, we suggest a way how to introduce the glueball into an effective meson Lagrangian, using the dilaton model (see, e.g., [2]). In our approach developed in [3], the principle of scale invariance of the effective meson Lagrangian is used in the chiral limit. The Lagrangian should also satisfy the Ward identity connected with the scale anomaly

$$\langle \partial_\mu S^\mu \rangle = \mathcal{C}_g - \sum_{q=u,d,s} m_q^0 \langle \bar{q}q \rangle, \qquad \mathcal{C}_g = \left(\tfrac{11}{24} N_c - \tfrac{1}{12} N_f\right) \left\langle \tfrac{\alpha}{\pi} G_{\mu\nu}^2 \right\rangle, \qquad (1)$$

where S^μ is the dilatation current, N_c is the number of colours; N_f, the number of flavors; $\langle \tfrac{\alpha}{\pi} G_{\mu\nu}^2 \rangle$ and $\langle \bar{q}q \rangle$ are the gluon and quark condensates; m_q^0, the current quark mass. To fulfill these requirements, one should start with an effective meson Lagrangian without glueball (a Lagrangian of that sort was considered by authors in [4]) and then introduce a dilaton field χ into each dimensional parameter: the four-quark interaction constant G, 't Hooft interaction constant K, the ultraviolet cutoff Λ, and constituent quark masses m, using the following rule:

$$G \to G\left(\chi_c/\chi\right)^2, \quad K \to K\left(\chi_c/\chi\right)^5, \quad \Lambda \to \Lambda\chi/\chi_c, \quad m \to m\chi/\chi_c, \qquad (2)$$

where χ_c is the vacuum expectation value (v.e.v.) of χ. *Current quark masses are not scaled, since the corresponding terms containing these masses explicitly violate scale invariance both in QCD and in effective meson models.* After this, the dilaton

---

[1] *Communicate to:* volkov@thsun1.jinr.ru

field appears in vertices of the effective meson Lagrangian, thereby allowing one to describe the mixing of the dilaton with quarkonia and the decays of the glueball into pseudoscalar mesons (π, K, η, and η').

We start with the chiral quark Lagrangian of NJL type with 't Hooft interaction describing the singlet-octet mixing of pseudoscalar and scalar mesons

$$L(\bar{q}, q) = \bar{q}(i\hat{\partial} - m^0)q + G/2 \sum_{a=1}^{9}[(\bar{q}\tau_a q)^2 + (\bar{q}i\gamma_5\tau_a q)^2] - \\ - K\{\det[\bar{q}(1+\gamma_5)q] + \det[\bar{q}(1-\gamma_5)q]\}, \quad (3)$$

where $\tau_a = \lambda_a$, $(a = 1, ..., 7)$ are the Gell-Mann matrices, $\tau_8 = (\sqrt{2}\lambda_0 + \lambda_8)/\sqrt{3}$, $\lambda_0 = \sqrt{2/3}\,\mathbf{1}$, $\tau_9 = (-\lambda_0 + \sqrt{2}\lambda_8)/\sqrt{3}$, with $\mathbf{1}$ being the unit matrix; m^0 is a current quark mass matrix with diagonal elements m_u^0, m_d^0, m_s^0 ($m_u^0 = m_d^0$).

Using the standard procedure [5], one can obtain an equivalent Lagrangian with four-quark vertices only

$$L = \bar{q}(i\hat{\partial} - \bar{m}^0)q + 1/2 \sum_{a,b=1}^{9}[G_{ab}^{(-)}(\bar{q}\tau_a q)(\bar{q}\tau_b q) + G_{ab}^{(+)}(\bar{q}i\gamma_5\tau_a q)(\bar{q}i\gamma_5\tau_b q)]. \quad (4)$$

Here $G_{ab}^{\pm}$ are new four-quark interaction constants, and $\bar{m}^0$ is a diagonal matrix composed of modified current quark masses. (For their definition see [3].)

After bosonization, the Lagrangian transforms to the form expressed in terms of already bosonic fields $\sigma = \sum_{a=1}^{9} \sigma_a \tau_a$, $\phi = \sum_{a=1}^{9} \phi_a \tau_a$

$$\mathcal{L}(\sigma, \phi) = \bar{L}_G(\sigma, \phi) + \mathrm{tr}\Big[1/4\,p^2(\sigma^2 + \phi^2) - 4mgI_1^\Lambda(m)\sigma + 2g^2 I_1^\Lambda(m)(\sigma^2 + Z\phi^2) \\ -m^2\sigma^2 + mg\sigma(\sigma^2 + Z\phi^2) - (g^2/4)((\sigma^2 + Z\phi^2)^2 - [\sigma - m, \phi]_-^2)\Big], \quad (5)$$

For $\bar{L}_G$, we have:

$$\bar{L}_G(\sigma, \phi) = -1/2 \sum_{a,b=1}^{9}(g_a\sigma_a - \mu_a + \bar{\mu}_a^0)\left(G^{(-)}\right)_{ab}^{-1}(g_b\sigma_b - \mu_b + \bar{\mu}_b^0) \\ - Z/2 \sum_{a,b=1}^{9} g_a\phi_a \left(G^{(+)}\right)_{ab}^{-1} g_b\phi_b, \quad (6)$$

where $\mu_a = 0$, $(a = 1, ..., 7)$, $\mu_8 = m_u$, $\mu_9 = -m_s/\sqrt{2}$ and $\bar{\mu}_a^0 = 0$, $(a = 1, ..., 7)$, $\bar{\mu}_8^0 = \bar{m}_u^0$, $\bar{\mu}_9^0 = -\bar{m}_s^0/\sqrt{2}$. Here m_u and m_s are constituent masses of u and d quarks to be introduced later. The constituent quark masses m_u and m_s appear as a result of spontaneous breaking of chiral symmetry. The integrals

$$I_n^\Lambda(m_a) = \frac{N_c}{(2\pi)^4}\int d_e^4 k \frac{\theta(\Lambda^2 - k^2)}{(k^2 + m_a^2)^n}, \quad (n = 1, 2;\ a = u, s), \quad (7)$$

are calculated in the Euclidean metric and regularized by a simple $O(4)$-symmetric ultra-violet cutoff Λ.

We also introduced Yukawa coupling constants g_a resulting from the renormalization of meson fields. They are expressed through logarithmically divergent integrals I_2 defined in [4,6]. The coefficient $Z \approx 1.44$ comes from π-A_1 transitions (A_1 is the axial-vector meson).

TABLE 1. The masses of physical scalar meson states σ_I, σ_{II}, σ_{III} and the values of the parameters χ_c, χ_0, bag constant B, and (bare) glueball mass m_g (in MeV) for two cases: 1) $M_{\sigma_{III}} = 1500$ MeV and 2) $M_{\sigma_{III}} = 1710$ MeV.

	σ_I	σ_{II}	σ_{III}	χ_c	χ_0	$B, [\text{GeV}^4]$	m_g
I	555	1075	1500	191	192	0.005	1485
II	555	1080	1710	167	168	0.005	1695

Let us introduce the dilaton field into Lagrangian (5) using rule (2)

$$\bar{\mathcal{L}}(\sigma,\phi,\chi) = \mathcal{L}(\chi) + L_{kin}(\sigma,\phi) + \bar{L}_G(\sigma,\phi,\chi) + \text{tr}\Big[-4mgI_1^\Lambda(m)\sigma\,(\chi/\chi_c)^3$$
$$+2g^2 I_1^\Lambda(m)(\sigma^2+\phi^2)(\chi/\chi_c)^2 - m^2 g^2 \sigma^2 (\chi/\chi_c)^2$$
$$+mg\chi/\chi_c\sigma(\sigma^2+\phi^2) - g^2/4((\sigma^2+\phi^2)^2 - [\sigma-m,\phi]_-^2)\Big], \quad (8)$$

where χ_c is v.e.v. of the dilaton field χ.

The pure dilaton Lagrangian $\mathcal{L}(\chi) = (\partial_\nu \chi)^2/2 - B(\chi/\chi_0)^4 \left[\ln(\chi/\chi_0)^4 - 1\right]$ contains a potential of a special form that has a minimum at $\chi = \chi_0$, and the parameter B represents the vacuum energy, when there are no quarks.

It is easy to check that Lagrangian (8) is scale-invariant in the chiral limit with exception of the dilaton potential and satisfies (1).

The gap equations are obtained from the requirement that first variations of the potential of effective Lagrangian are zero at the potential minimum. They look as follows

$$(m_u - \bar{m}_u^0)(G^{(-)})_{88}^{-1} - (m_s - \bar{m}_s^0)(G^{(-)})_{89}^{-1}/\sqrt{2} - 8m_u I_1^\Lambda(m_u) = 0, \quad (9)$$
$$(m_s - \bar{m}_s^0)(G^{(-)})_{99}^{-1} - \sqrt{2}(m_u - \bar{m}_u^0)(G^{(-)})_{98}^{-1} - 8m_s I_1^\Lambda(m_s) = 0, \quad (10)$$
$$4B(\chi_c^3/\chi_0^4)\ln(\chi_c/\chi_0)^4 + 2A/\chi_c = 0. \quad (11)$$

Here $A = \frac{1}{2}\sum_{a,b=8}^9 (\mu_a - \bar{\mu}_a^0)(G^{(-)})_{ab}^{-1}\mu_b^0$ is proportional to the current quark masses $\mu_b^0 \sim m_b^0$, and thereby, small.

Using (11), the Ward identity (1) written for the effective Lagrangian, and the "bare" mass of a glueball $m_g^2 = (4\mathcal{C}_g - A)/\chi_c^2$, we define parameters χ_c, χ_0, and B. The rest of parameters G, K, Λ, m_u, and m_s remain to be the same as for the model without dilaton and are defined in our earlier work [4]

$$m_u = 280 \text{ MeV}, \quad m_s = 420 \text{ MeV}, \quad \Lambda = 1.25 \text{ GeV},$$
$$G = 4.38 \text{ GeV}^{-2}, \quad K = 11.2 \text{ GeV}^{-5}. \quad (12)$$

With these values fixed and the value of the gluon condensate taken from [7], we obtain two sets of parameters χ_c, χ_0, and B for two candidates for the glueball.

The potential part of Lagrangian (8) which is quadratic in fields σ^r and χ' and which we denote as $L^{(2)}$ has the form

TABLE 2. The partial and total decay widths (in MeV) of the glueball and experimental values of total widths for two cases: $\sigma_{III} \equiv f_0(1500)$ and $\sigma_{III} \equiv f_0(1710)$.

	$\Gamma_{\pi\pi}$	$\Gamma_{K\bar{K}}$	$\Gamma_{\eta\eta}$	$\Gamma_{\eta\eta'}$	$\Gamma_{4\pi}$	Γ_{tot}	Γ_{tot}^{exp}
$f_0(1500)$	4	42	25	5	30	100	112
$f_0(1710)$	3	90	42	5	60	200	130

$$L^{(2)}(\sigma, \phi, \chi') = -1/2 g_8^2 \{[(G^{(-)})_{88}^{-1} - 8 I_1^\Lambda(m_u)] + 4 m_u^2\} \sigma_8^2$$
$$-1/2 g_9^2 \{[(G^{(-)})_{99}^{-1} - 8 I_1^\Lambda(m_s)] + 4 m_s^2\} \sigma_9^2$$
$$- g_8 g_9 (G^{(-)})_{89}^{-1} \sigma_8 \sigma_9 - 2 (C_g - A/4)(\chi'/\chi_c)^2 - \sum_{a,b=8,9} \mu_a^0 (G^{(-)})_{ab}^{-1} g_b \sigma_b \chi'/\chi_c. \quad (13)$$

From it one can extract the terms describing meson and glueball masses and the mixing term. After diagonalization that is performed numerically, one obtains the values of scalar isoscalar meson states that are shown in Table 1.

We also estimate the decays of the glueball into $\pi\pi$, $K\bar{K}$, $\eta\eta$, $\eta\eta'$, and 4π. The details are given in paper [3], here we present only the results collected in Table 2. In conclusion, we would like to say that, in our model, the mixing of the glueball with scalar isoscalar quarkonia and the amplitudes describing decays of a scalar meson into two pseudoscalar mesons are proportional to m_q^0 and vanish in the limit $m^0 \to 0$. Comparing our results with experimental data [8], one can see that the model decay widths of scalars better fit the experiment if one identifies the state $f_0(1500)$ as a glueball.

In a further work, we are going to consider also radially excited quarkonia together with the glueball.

REFERENCES

1. Volkov M. K. and Yudichev V. L., *Int. J. Mod. Phys. A* **14**, 4621–4640 (1999). Volkov M. K. and Yudichev V. L., *Fiz. Elem. Chast. At. Yadra.* **31**, 576–633 (2000); hep-ph/9906371.
2. Kusaka K., Volkov M. K., and Weise W., *Phys. Lett. B* **302**, 145–150 (1993); Ripka G. and Jaminon M., *Ann. Phys. N.Y.* **218**, 51–74 (1992); Cugnon J., Jaminon M., and Bosche B. Van den, *Nucl. Phys. A* **598**, 515–538 (1996); Jaminon M. and Bosche B. Van den, *Nucl. Phys. A* **619**, 285–294 (1997).
3. Ebert D., Nagy M., Volkov M. K., and Yudichev V. L., hep-ph/0007131;
4. Volkov M. K., Nagy M., and Yudichev V. L., *Nuovo Cim. A* **112**, 225–232 (1999).
5. Klevansky S. P., *Rev. Mod. Phys.*, **64**, 649–708 (1992).
6. Volkov M. K., *Sov. J. Part. and Nuclei* **17**, 186–203 (1986); Volkov M. K., *Phys. Part. Nucl.* **24**, 35–58 (1993).
7. Broadhurst D. J. et al., *Phys. Lett. B* **329**, 103–110 (1994); Geshkenbein B. V., *Phys. At. Nucl.* **58**, 1171 (1995); Narison S., *Phys. Lett. B* **387**, 162–172 (1996).
8. Caso C. et al., *Eur. Phys. J. C* **3**, 1–794 (1998).

Scalar and QCD String Confinement[1]

T.J. Allen*, M.G. Olsson†, and S. Veseli‡

*Physics Department, Hobart & William Smith Colleges, Geneva, NY 14456
†Physics Department, University of Wisconsin, Madison, WI 53706
‡Fermi National Accelerator Laboratory, P.O. Box 500, Batavia, IL 60510

Abstract. We compare scalar and string confinement mechanisms. Solutions for the massless quark case are found and discussed.

Although the concept of scalar confinement occupies a hallowed tradition in hadron physics, its origin from QCD has remained obscure. On the other hand, the QCD string provides a very plausible limit of QCD at large quark separation. In the first section we compare and contrast scalar and string confinement within a unified formalism [1]. In the first section we compare and contrast scalar and string confinement within a unified formalism. In the second part we examine the solutions for the ultra-relativistic limit of a massless quark confined by a scalar potential and by a QCD string [1].

FOUR-VECTOR INTERACTIONS: SCALAR AND QCD CONFINEMENT

QCD is intrinsically a vector-type theory and this is why it has been difficult to imagine scalar confinement in relation to QCD. Our first step will be to note that a four-vector interaction of a particular type is equivalent to a scalar potential. We begin with the invariant action for a mass m moving in a scalar potential and a four-vector potential,

$$S = -\int d\tau \left[m + \phi(x) - u^\mu A_\mu \right] , \qquad (1)$$

where u^μ is the four-velocity of m. Next consider a special four-vector potential

$$A^{\mu'}(x) = \left(\Phi(x), \mathbf{0} \right) . \qquad (2)$$

This four-vector is appropriate for an electric flux tube/string (in its rest frame). As pointed out by Buchmüller [2], this insures the Thomas-type spin-orbit interaction.

---

[1] Talk presented by M.G. Olsson

We then note that the invariant quantity $u^\mu A_\mu$ can be evaluated in the quark rest frame to give

$$-u^{\mu'} A_{\mu'} = -u^\mu A_\mu = \Phi(x). \tag{3}$$

Thus this particularly "string-like" four-potential is exactly equivalent to scalar confinement if we identify $\Phi(x)$ with $\phi(x)$. As discussed in more detail in Reference [1], this four-potential is not exactly string-like in that the energy is concentrated at the quark and that it should not depend on the radial velocity.

MASSLESS QUARK SPECTROSCOPY

For one fixed (heavy) and one massless quark, the squared Hamiltonian with scalar confinement is [1]

$$H^2_{\text{scalar}} = p^2 + (ar)^2. \tag{4}$$

This is a harmonic oscillator and hence the mass eigenstates are

$$M^2_{\text{scalar}} = 2a(J + 2n + 3/2), \tag{5}$$

where a is the "string tension," J and n are the angular and radial quantum numbers, respectively. An important aspect is the presence of "towers" of mass-degenerate states of the same parity. The spectroscopy of the scalar confinement result (5) is shown by the parallel lines in Fig. 1. The dots are exact numerical solutions [1,3] of the square root of Eq. (4).

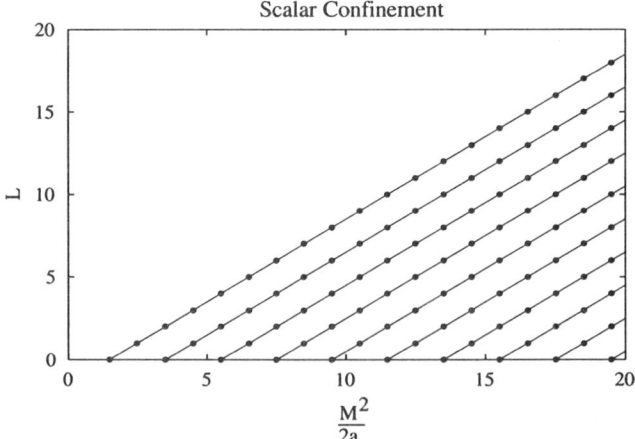

FIGURE 1. Regge structure and states in pure linear scalar confinement from numerical diagonization of the square root of H in (4). Solid lines are the harmonic oscillator result (5).

Next we examine the comparable solution for a QCD string [1,3]

$$H = W_r\gamma_\perp + ar\frac{\arcsin v_\perp}{v_\perp}, \tag{6}$$

$$\frac{J}{r} = W_r\gamma_\perp v_\perp + \frac{ar}{2v_\perp}\left(\frac{\arcsin v_\perp}{v_\perp} - \sqrt{1-v_\perp^2}\right), \tag{7}$$

where $v_\perp$ is the quark velocity perpendicular to the string and $W_r = \sqrt{p_r^2 + m^2}$. The massless quark limit is tricky since $W_r \to 0$ but $\gamma_\perp \to \infty$ for circular orbits. Considering the combination $H^2 - J^2/r^2$ and then setting $v_\perp \to 1$ in the interaction (since this is a smooth limit) we obtain

$$M^2 - \frac{a\pi J}{2} = p^2 + \left(\frac{a\pi r}{4}\right)^2, \tag{8}$$

where $p^2 = p_r^2 + J^2/r^2$. This would appear to be just a shifted harmonic oscillator but one must carefully examine the classical turning points. From (8) with $p_r = 0$ (and hence $v_\perp = 1$) we get

$$\frac{a\pi}{2}r_\pm = M \pm \sqrt{M^2 - a\pi J}. \tag{9}$$

These would seem to be the classical turning points. At these turning points [from (6)] we have (since $W_r\gamma_\perp \geq 0$)

$$M \geq \frac{a\pi r_\pm}{2} \tag{10}$$

or

$$M \geq M \pm \sqrt{M^2 - a\pi J}, \tag{11}$$

which is only satisfied for r_-. The larger turning point hence must be smaller than r_+. This larger turning point occurs at a finite value of p_r and hence is a "bounce." Its precise value depends on a more careful approximation, but if we take it as the circular orbit radius

$$\frac{a\pi}{2}r_0 = M, \tag{12}$$

we have a "half oscillator" which just doubles the radial excitation energies (by eliminating half the excitations). The half oscillator solution is obtained from (8) by replacing n by $2n + 1$, yielding

$$M^2 = a\pi\left(J + 2n + \frac{7}{4}\right). \tag{13}$$

This is very close to the scalar confinement result (5) although the physical process seems quite different. In Fig. 2, the solution $M^2 = a\pi(J + 2n + \frac{3}{2})$ is represented by parallel lines. The dots are the exact numerical solution [1,3] of the string equation (7).

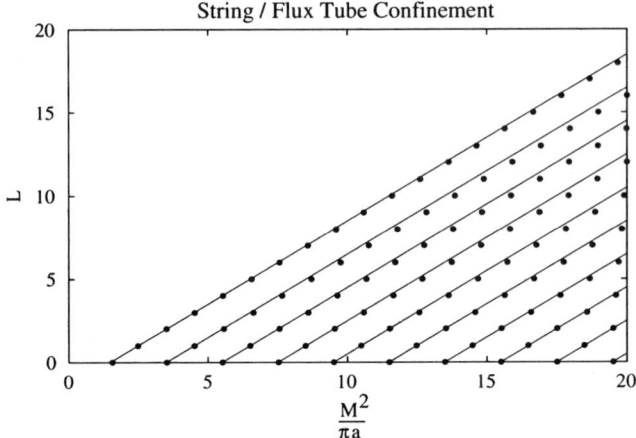

FIGURE 2. Regge structure of string confinement from numerical quantization of Eqs. (6) and (7). The lines are the solution (13) with intercept 3/2.

REFERENCES

1. Theodore J. Allen, M.G. Olsson, and Siniša Veseli, *From Scalar to String Confinement*, hep-ph/0001227, Phys. Rev. D (in press).
2. W. Buchmüller, Phys. Lett. **B112** 479 (1982).
3. M.G. Olsson and Siniša Veseli, Phys. Rev. **D51**, 3578 (1995).

Strong Isospin Breaking in CP-even and CP-odd $K \to \pi\pi$ Decays

C.E. Wolfe* and K. Maltman**

*Nuclear Theory Center, Indiana Univ., Bloomington, Indiana, USA[1]

**Dept. Mathematics and Statistics, York Univ., 4700 Keele St., Toronto, ON Canada, and CSSM, Univ. of Adelaide, Adelaide, SA Australia[2]

Abstract. Complete next-to-leading (chiral) order (NLO) expressions for the strong isospin-breaking (IB) contributions in $K \to \pi\pi$ are used to discuss (1) for CP-even, the impact on the magnitude of the $\Delta I = 1/2$ Rule, and (2) for CP-odd, the strong IB correction, Ω_{st}, for the gluonic penguin contribution to ϵ'/ϵ, with particular emphasis on the strong low-energy constant (LEC) and loop contributions, numerical values for which are model-independent at NLO.

In the presence of IB, the standard isospin decomposition of the $K^+ \to \pi^+\pi^0$, $K^0 \to \pi^+\pi^-, \pi^0\pi^0$ decay amplitudes, A_{+0}, A_{+-} and A_{00}, becomes [1]

$$A_{00} = \sqrt{1/3}A_0 e^{i\Phi_0} - [\sqrt{2/3}]A_2 e^{i\Phi_2},$$
$$A_{+-} = \sqrt{1/3}A_0 e^{i\Phi_0} + [1/\sqrt{6}]A_2 e^{i\Phi_2},$$
$$A_{+0} = [\sqrt{3}/2]A_2' e^{i\Phi_2'}. \tag{1}$$

Ignoring $\Delta I = 2$ electromagnetic (EM) channel coupling, Φ_I are the $\pi\pi$ phases. In general, EM- and strong-IB-induced $\Delta I = 5/2$ effects make $|A_2'| \neq |A_2|$. A_0, A_2 can be chosen real in the absence of CP violation.

Since $|A_0| \sim 22|A_2|$, IB "leakage" of the large octet ($I = 0$) amplitude into the $\Delta I = 3/2$ ($I = 2$) amplitude can be numerically significant. EM leakage has been computed to NLO in Ref. [1] (see also [2]); we compute the NLO strong leakage for both CP-even and gluonic-penguin-mediated CP-odd transitions. Strong cancellation between gluonic penguin (O_6) and electroweak penguin contributions make Standard Model predictions for ϵ'/ϵ sensitive to the latter [3].

At leading (chiral) order (LO), the ratio, $\delta A_2/A_0$, with δA_2 the strong octet leakage contribution to A_2, is unambiguous. The IC part of A_2, A_2^{IC}, is *decreased*

---

[1] Supported by DOE grant #DE-FG0287ER-40365
[2] Supported by the Natural Sciences and Engineering Research Council of Canada

TABLE 1. Strong octet and EM IB leakage contributions in units of 10^{-6} MeV. The isospin-conserving (IC) and LO IB fits yield $A_2 = A'_2 = -2.1 \times 10^{-5}$ MeV and -2.4×10^{-5} MeV, respectively.

Source	δA_2	$\delta A'_2$
(8)	$(-1.56 \pm 0.63) + (0.42 \pm 0.05)i$	$(-1.56 \pm 0.63) + (0.42 \pm 0.05)i$
(EM)	$(-1.27 \pm 0.40) - (1.28 \pm 0.02)i$	$(0.70 \pm 0.73) - (0.07 \pm 0.04)i$

by $[\Omega_{st}]_{LO} = 13\%$ once strong IB is included. The LO O_6 suppression in ϵ'/ϵ is $1 - \Omega_{st}$. Recent analyses of ϵ'/ϵ employ $\Omega_{st} = 0.25 \pm 0.08$ [3]. The difference reflects estimates of NLO effects mediated by an intermediate η' through the $K^0 \to \pi^0 \eta'$ transition [4]. This effect, in conventional ChPT (involving only the π, K and η as explicit degrees of freedom), represents one contribution to the CP-even (CP-odd) NLO weak LEC's, E_k^+ (E_k^-) [5], but does not exhaust such contributions.

Let us fix notation. At LO, the low-energy representations of the dominant CP-even $(+)$ octet operator and CP-odd $(-)$ gluonic penguin operator are [5]

$$c^{\pm} Tr \left[\lambda^{\pm} \partial_\mu U^\dagger \partial^\mu U \right], \qquad (2)$$

where $\lambda^+ = \lambda_6$, $\lambda^- = \lambda_7$, and $U = exp(i\lambda \cdot \pi)$, as usual. A complete set of NLO contributions is obtained by evaluating the sum of (i) graphs with a single LO weak vertex (proportional to $c^{\pm}$) and one of the external legs dressed by a single strong NLO vertex (proportional to one of the NLO strong LEC's, L_k) ("strong LEC contributions"); (ii) 1-loop graphs involving a single LO weak vertex and strong vertices only of LO ("loop contributions"); and (iii) tree graphs involving a single NLO weak vertex [5] ("NLO weak LEC contributions"). Only the sum of all three classes of contribution is renormalization-scale-independent and physically meaningful. From Eq. (2), it is immediately obvious that LO CP-even and CP-odd K^0 decay vertices are related through the substitution $c^+ \leftrightarrow ic^-$. The loop-plus-strong-LEC part of the NLO contribution to the *ratio* $\delta A_2/A_0$ is, therefore, the same for the CP-even and CP-odd cases. Because, moreover, the LO LEC's, $c^{\pm}$, cancel in the ratio, this contribution is, though scale-dependent, *model*-independent.

The strong LEC contributions to $\delta A_2/A_0$ have recently been discussed in Ref. [6]. They include an η'-mediated π^0-η mixing term proportional to L_7 which is, however, empirically, almost completely cancelled by an accompanying L_8 contribution [6]. The loop contributions have been evaluated in Ref. [7]. Since the full set of E_k^+ (E_k^-) cannot be determined empirically at present, one is forced to use model values. The models used are discussed below.

For the CP-even case, the leakage contributions, $\delta A_2 = \delta A'_2$, are given in Table 1 [7]. The corresponding EM contributions [1] are shown for comparison. The errors reflect uncertainties in the estimates of the NLO LEC's (see Ref. [7] for details, and references to the models employed for the E_k^+). Denoting the ratio of fitted LO 27-plet to octet weak LEC's obtained neglecting, or including, IB by r_{IC}, or r_{IB}, respectively, we find $R_{IB} \equiv r_{IB}/r_{IC} = 0.963 \pm 0.029 \pm 0.010 \pm 0.034$.

The errors correspond, respectively, to model dependence of the E_k^+, uncertainties in $B_0(m_d - m_u)$, and the EM uncertainties of Refs. [1]. The deviation from 1 is significantly smaller than at LO (where $R_{IB} = 0.870$). The true $\Delta I = 1/2$ rule enhancement, r_{IB}, can thus be approximated by r_{IC} to an accuracy of better than 10%. The $\Delta I = 5/2$ contribution (dominantly EM in character [7]), leads to $A_2/A_2' = 1.094 \pm 0.039 \neq 1$, and significantly exacerbates the phase discrepancy problem for the neutral K decays [7].

For the CP-odd case, $\Omega_{st} = \omega \operatorname{Im} \delta A_2 / \operatorname{Im} A_0$ ($\omega = \operatorname{Re} A_0 / \operatorname{Re} A_2 \simeq 22$). $\operatorname{Im} A_0$ is associated with O_6. At LO, $[\Omega_{st}]_{LO} = 0.13$. At NLO

$$\frac{\Omega_{st}}{[\Omega_{st}]_{LO}} = \left[1 + \frac{\operatorname{Im} \delta A_2^{(NLO;ND)}}{\operatorname{Im} \delta A_2^{(LO)}} - \frac{\operatorname{Im} A_0^{(NLO;ND)}}{\operatorname{Im} A_0^{(LO)}}\right] \equiv [1 + R_2 - R_0] \ . \tag{3}$$

The superscript $(NLO; ND)$ indicates the sum of non-dispersion NLO contributions (involving NLO weak and strong LEC's and the non-dispersive parts of loop graphs). Neither the NLO $I = 0$ isospin-conserving nor NLO $I = 2$ IB leakage E_k^- combinations are known. The NLO dispersive contributions create phases consistent with Watson's theorem. Although the positive $I = 0$ phases correspond to attractive FSI, the E_k^- terms may, nonetheless, make $\operatorname{Im} A_0$ smaller at NLO than the LO (see comments on Ref. [8] in Ref. [9] for a related discussion). If, however, NLO effects do enhance $Im A_0$ (decreasing the level of O_6-O_8 cancellation and increasing ϵ'/ϵ) Ω_{st} will be simultaneously suppressed, further amplifying this increase. The known NLO contributions (loops and strong LEC terms) give contributions $-0.24(-0.31)$ to R_2 and $-0.02(+0.42)$ to R_0, at scale $\mu = m_\eta(m_\rho)$ [10]. For the (model-dependent) E_k^- contributions, we considered the possibility of using 3 models: the weak deformation model (WDM) [11], the chiral quark model (ChQM) [12], and the scalar saturation model of Ref. [13] (SSM). Ref. [13] was the first to argue explicitly that NLO weak LEC contributions other than those mediated by the η' might be numerically important. A large negative contribution to $R_2 - R_0$ was obtained which, if correct, would significantly enhance the O_6 contribution to ϵ'/ϵ. In contrast, the two earlier models produce positive, or near-zero, E_k^- contributions to $R_2 - R_0$. Since there is a problem with the SSM estimate[3] we have used only the WDM and ChQM. Taking the WDM model E_k^- to correspond

---

[3] The model values, $\hat{E}_k^-$, of the E_k^- are obtained by (1) approximating the renormalized 4-quark operator O_6 as the product of renormalized scalar densities (the factorization approximation) and (2) dropping (divergent) seagull terms in the low-energy representation of this product. Two types of contribution to $\hat{E}_k^-$ result, $\left[\hat{E}_k^-\right]_1$ (proportional to the strong 6th order LEC's), and $\left[\hat{E}_k^-\right]_2$ (proportional to a product of 2 4th order strong LEC's). As an example, $\left[\hat{E}_1^-\right]_2 = 16\sqrt{2}G_F V_{us}^* V_{ud} \, Im \, (C_6) B_0^2 \, L_8^2$ [13]. In Ref. [13] it is assumed that the finite part of this expression is proportional to the square, $[L_8^r]^2$, of the finite part of L_8. This, however, is incorrect. In dimensional regularization, the Laurent expansion of L_8 begins at $O[1/(d-4)]$; the finite part of L_8^2 is thus actually $[L_8^r]^2 + 5L_8^{(-1)}/384\pi^2$, with $L_8^{(-1)}$ the coefficient of the $O[d-4]$ term in the L_8 Laurent expansion. $L_8^{(-1)}$ enters physical processes beginning at 6th order in the chiral expansion,

to a scale $\mu \sim m_\rho$, we find, combining with the strong LEC and loop contributions given above, $1 + R_2 - R_0 = 0.27$ in the WDM and 0.64 in the ChQM. Taking the more microscopic ChQM to provide a central value, and the difference between the WDM and ChQM to provide a *minimal* measure of model dependence in the theoretical result, we obtain, at NLO, $\Omega_{st} = 0.08 \pm 0.05$, significantly smaller than both the conventionally employed value 0.25 ± 0.08, and that, 0.16 ± 0.03, obtained in the estimate of Ref. [6], based on the strong LEC contributions only. The central value above, combined with conventional central values for the B-factors, leads to a $\sim 50\%$ increase in the predicted value for ϵ'/ϵ. To be conservative, we would propose using this lower central value with an even larger error estimate, in all future calculations of Standard Model values for ϵ'/ϵ.

REFERENCES

1. V. Cirigliano, J.F. Donoghue and E. Golowich, Phys. Lett. **B450**, 241 (1999); Phys. Rev. **D61** 093001 (2000); and Phys. Rev. **D61** 093002 (2000).
2. G. Ecker *et al.*, hep-ph/0006172.
3. See A. Buras, hep-ph/9908395 and references therein.
4. J.F. Donoghue, E. Golowich, B.R. Holstein and J. Trampetic, Phys. Lett. **B179**, 361 (1986), Erratum, *ibid.* **B188**, 511 (1987); A.J. Buras and J.M. Gerard, Phys. Lett. **B192**, 192 (1987).
5. J. Kambor, J. Missimer and D. Wyler, Nucl. Phys. **B346**, 17 (1990).
6. G. Ecker, G. Muller, H. Neufeld and A. Pich, Phys. Lett. **B477**,88 (2000).
7. C.E. Wolfe and K. Maltman, Phys. Lett. **B482**, 77 (2000).
8. E. Pallente and A. Pich, Phys. Rev. Lett. **84**, 2568 (2000).
9. A.J. Buras *et al.*, Phys. Lett. **B480**, 80 (2000).
10. C.E. Wolfe and K. Maltman, "Strong Isospin Breaking in ϵ'/ϵ at NLO in the Chiral Expansion", YU-PP-I/E-KM-7 and IU/NTC/00-04, July 2000.
11. G. Ecker, J. Kambor and D. Wyler, Nucl. Phys. **B394**, 101 (1993).
12. S. Bertolini, J. Eeg and M. Fabbrichesi, Nucl. Phys. **B449**, 197 (1995); S. Bertolini, J. Eeg, M. Fabbrichesi and E. Lashin, Nucl. Phys. **B514**, 63 (1998).
13. S. Gardner and G. Valencia, Phys. Lett. **B466**, 355 (1999).

---

and hence is on the same footing as the other 6th order strong LEC's appearing in $\left[\hat{E}_k^-\right]_1$ and retained in Ref. [13]. Since, it is inconsistent to drop the $L_k^{(-1)}$, and the SSM provides no means of estimating their values, the numerical estimates given in Ref. [13] are incomplete, and cannot be used.

Meson Spectra - Power Law Potential in the Dirac Equation

L. K. Sharma and J.O. Fiase

Department of Physics
University of Botswana
Private Bag 0022 ,Gaborone
Botswana
Tel:267 355 2141
e-mail: sharmalk@mopipi.ub.bw.

Abstract. A single mass spectra power law potential has been used in the relativistic Dirac equation to predict the spectra of both light and heavy mesons. Results for both light and heavy meson systems are in good agreement with the experimental data. Now that top quark has been dicovered and it's mass determined accurately, the same power law potential is also used to predict mass spectra of $t\bar{t}$(toponium?)systems. Since experimental data for the system is not yet available, it will be interesting to watch the experimental evolution of this system.

INTRODUCTION

A power law mass-dependent potential model is presented in the Dirac equation for calculating mass-spectra and decay width of both heavy and light mesons. As the mass of top quark has now been accurately measured [1], the mass-spectra of $t\bar{t}$(toponium?) bound states has also been predicted.

CALCULATIONS OF THE DIRAC BOUND - STATES

We choose a potential of the form:

$$V(r) = g_1 r^{m_0/2m_q} - V_0, \qquad (1)$$

where V(r) and r are measured in GeV and GeV^{-1} respectively.

Following Magyari[2] and with the choice of the vector fraction $g_v = \frac{1}{2}$, potential for the independent particle model of quarks is:

$$V'(r) = \frac{1}{2}V(r) = V_s(r) + V_v(r). \qquad (2)$$

Here each of the scalar and vector parts would equal $\frac{1}{4}V(r)$. Dirac equation can be written as ($\hbar = c = 1$):

$$(\vec{\alpha}.\vec{P} + m_q\beta)\psi(r) = [E' - V_v(r) - V_s(r)\beta]\psi(r), \qquad (3)$$

Equation (3) can be seperated into a system of the following coupled equations for the radial wavefunctions $\phi(r)$ and $\chi(r)$:

$$(E' - V_v(r) - V_s - m_q)\phi(r) + (\frac{k+1}{r} + \frac{d}{dr})\chi(r) = 0 \qquad (4)$$

$$(E' - V_v(r) + V_s + m_q)\chi(r) + (\frac{k-1}{r} - \frac{d}{dr})\phi(r) = 0$$

Setting $\phi(r) = \frac{U'(r)}{r}$ and $V_s(r) = V_v(r) = \frac{1}{4}\{g_1 r^{m_0/2m_q} - V_0\}$, the equation for large component $\phi(r)$ is:

$$\frac{d^2}{dr^2}U''(r) + [(E' + m_q)(E' - m_q - 2V_s(r)) - \frac{l(l+1)}{r^2}]U(r) = 0 \qquad (5)$$

Introducing a dimensionless variable $\rho = \frac{r}{r_0}$ and scale factor $r_0 = [\frac{(m_q+E')g_1}{2}]^{-1/(2+\frac{m_0}{m_q})}$,
equation (5) reduces to a Schroedinger like equation with a power law potential. The semi-classical solution of this equation gives following expression for the Dirac bound state masses:

$$(M_{nl})^{Dirac} = 2ax_{nl} + 2m_q - V_0, \qquad (6)$$

$a = (g_1)^{\frac{1}{(1+\frac{m_0}{m_q})}}$ and x_{nl} is the positive root of the following equation:

$$x_{nl}^{[1+\frac{4m_q}{m_0}]}(x_{nl} + b) = 2^{\frac{-4m_q}{m_0}}\epsilon_{nl}^{1+\frac{4m_q}{m_0}}, \qquad (7)$$

In equation (7)

$$\epsilon_{nl} = [(n+l/2-1/4)\frac{2\sqrt{\pi}\Gamma(\frac{3}{2}+\frac{2m_q}{m_0})}{\Gamma(1+\frac{2m_q}{m_0})}]^{\frac{2}{1+\frac{4m_q}{m_0}}} \qquad (8)$$

and

$$b = (2m_q - \frac{1}{2}V_0)/a \qquad (9)$$

RESULTS

In Table 1 energy levels of Ψ, Υ, Φ and ρ mesons have been computated and compared with the experimental data and with other potential models cited in the literature. The levels marked with asterisk denote input used to determine adjustable parameters. The results show considerable agreement with experimental data and that of other workers.

Table 1.

state	Data[3](MeV)	Calculated	Ref.[4]	Ref.[5]
		$c\bar{c}$ states		
1S	3096.88 ± 0.04	3097*	3114	3097
2S	3686.00 ± 0.09	3686.08	3690	3686
3S	4040 ± 10	4043.61	4040	4031
4S	4415 ± 6	4308.3	4320	4278
		$b\bar{b}$ states		
state	Data[3](MeV)	Calculated	Ref.[4]	Ref.[6]
1S	9460.37 ± 0.21	9460*	9460	9423
2S	10023.30 ± 0.31	10020.1	10000	10042
3S	10355.3 ± 0.5	10337.0	10320	10358
4S	10580 ± 3.5	10562.1	10570	10567
		$s\bar{s}$ states		
state	Data[3](MeV)	Calculated	Ref.[7]	Ref.[5]
1S	1019.413 ± 0.008	1020*	1098	1019.6
2S	1680 ± 20	1686.79	1616	1640.0
		$u\bar{u}$ states		
state	Data[3](MeV)	Calculated	Ref.[7]	Ref.[5]
1S	768.5 ± 0.6	767*	759	770
2S	1600	1600.61	1509	1402

Next calculations with the mass spectra for the charmed ($c\bar{q}$) and bottomed ($b\bar{q}$) mesons for the two lowest levels 1S and 2S have been made in Table 2.

Table 2.

meson	state	Data[3](MeV)	Calculated	Ref.[7]
$c\bar{u}$	$D*(1S)$	2010 ± 0.5	2010*	1990
	$D*(2S)$	—	2616.4	2575
$c\bar{s}$	$Ds*(1S)$	2112.4 ± 0.7	2110*	2102
	$Ds*(2S)$	–	2730.9	2685
$b\bar{u}$	$B*(1S)$	5324.8 ± 1.8	5325*	5311
	$B*(2S)$	—	59326	5930

Again reasonable agreement has been found with experimental data.

MESON SPECTRA OF $T\bar{T}$[TOPONIUM?] MESON

Motivated by the success of mass spectra of both light and heavy meson systems, we predict in Table 3. the energy spectra of $t\bar{t}$ system. Though the evidence of existence of this system has been recently confirmed [8] but there is no experimental data available as yet with which to compare our results. For the top mass we take the experimental value of $m_t = 176 GeV$ [1]. Note that V_0 is measured in GeV while g_1 is measured in $GeV^{(2m_t+1)/2m_t}$.

Table 3.

state	V_0, g_1	V_0, g_1	V_0, g_1	V_0, g_1	V_0, g_1
	5.0, 5.015	10.0, 10.060	20.0, 20.180	30.0, 30.323	40.0, 40.479
1S	352.00	352.000	352.00	352.00	352.00
2S	352.012	352.024	352.048	352.072	352.096
3S	352.018	352.037	352.074	352.110	352.147
4S	352.023	352.046	352.091	352.137	352.183
state	V_0, g_1	V_0, g_1	V_0, g_1	V_0, g_1	V_0, g_1
	60.0, 60.825	70.0, 71.009	80.0, 81.199	90.0, 91.395	100.0, 101.595
1S	352.00	352.000	352.00	352.00	352.00
2S	352.144	352.168	352.192	352.216	352.240
3S	352.221	352.258	352.295	352.332	352.368
4S	352.274	352.319	352.366	352.411	352.457

In Table 3. the calculations for different energy spectra with variation in the parameters V_0 and g_1 have been done for having an idea of the sensitivity of energy spacings for the $t\bar{t}$ bound - states with the given parameters. It is observed that the mass - spectra of the $t\bar{t}$ system is insensitive to variations in V_0 and g_1. Thus it can be conjectured from these modest analyses that the spectra of the $t\bar{t}$ system when eventually measured experimentally, will have a rather compressed energy spectra compared to that of the $c\bar{c}$ or $b\bar{b}$ systems. It will now be interesting to watch the experimental evaluation of $t\bar{t}$ as a test of our ideas on this meson structure.

REFERENCES

[1] Abe, F. et al.,Phys. Rev. Lett.**82**, 271(1999).
[2] Magyari, E. .Phys.Lett.**B95**, 295(1980).
[3] Caso, C. et al. Eur. Phy. J, **C3**,1(1998)
[4] Zhang, T. and Koniuk, R. ,Phys.Rev. **D43**,1688(1991).
[5] Jena, S.N., Phys.Rev. **D27**, 244(1983).
[6] Grant, A.K., etal, Phys.Rev. **D47**, 1981(1993).
[7] Crater, H. and Alstine, P.V., .Phys.Lett.**100B**, 166(1981).
[8] Abbott, B. et al.,Phys. Rev. Lett.**83**,1908(1999).

Measurement of the Pion Light-Cone Wave Function

Daniel Ashery [1]

School of Physics and Astronomy, Raymond and Beverly Sackler Faculty of Exact Sciences Tel Aviv University, Israel

Abstract. Diffractive dissociation of 500 GeV/c π^- mesons into di-jets from carbon and platinum targets is used to measure the momentum distribution of quarks in the pion and study color transparency effects. The measurements used data from Fermilab experiment E791. The results show that the $|q\bar{q}\rangle$ light-cone asymptotic wave function describes the data well for $Q^2 \sim 10$ (GeV/c)^2 or more. Evidence for color transparency comes from a measurement of the A-dependence of the yield of the diffractive di-jets.

INTRODUCTION

The Pion Momentum Wave Function

The internal momentum distribution of valence quarks in hadrons are fundamental to QCD. [1]. They are generated from the valence light-cone wave functions integrated over $k_t < Q^2$, where k_t is the intrinsic transverse momentum of the valence constituents and Q^2, the total momentum transfer squared. Even though these amplitudes were calculated about 20 years ago, there have been no direct measurements until those reported here.

The pion wave function can be expanded in terms of Fock states:

$$\Psi = a_1|q\bar{q}\rangle + a_2|q\bar{q}g\rangle + a_3|q\bar{q}gg\rangle + ... \quad (1)$$

Two functions have been proposed to describe the momentum distribution amplitude for the quark and antiquark in the $|q\bar{q}\rangle$ configuration. The asymptotic function was calculated using perturbative QCD (pQCD) methods [2–4], and is the solution to the pQCD evolution equation for very large Q^2 ($Q^2 \to \infty$):

$$\phi_{as}(x) = \sqrt{3}x(1-x). \quad (2)$$

---

[1] Representing the Fermilab E791 Collaboration. Supported in part by the Israel Science Foundation and the US-Israel Binational Science Foundation.

x is the fraction of the longitudinal momentum of the pion carried by the quark in the infinite momentum frame. The antiquark carries a fraction $(1-x)$. Using QCD sum rules, Chernyak and Zhitnitsky [5] proposed a function that is expected to be correct for low Q^2:

$$\phi_{cz}(x) = 5\sqrt{3}x(1-x)(1-2x)^2. \tag{3}$$

As can be seen from eqns. 2 and 3 and from Fig. 1, there is a large difference between the two functions. Measurements of form factors are insensitive to the wave function as these quantities are derived by integrating over the wave function and interpretation of the results is model dependent [6]. In this work we describe an experimental study that maps the momentum distribution of the valence $|q\bar{q}\rangle$ in the pion. This provides the first direct measurement of the pion light-cone wave function (squared). The concept of the measurement is the following: a high energy pion dissociates diffractively on a heavy nuclear target. This is a coherent process in which the quark and antiquark break apart and hadronize into two jets. If in this fragmentation process the quark momentum is transferred to the jet, measurement of the jet momentum gives the quark (and antiquark) momentum. Thus: $x_{measured} = \frac{p_{jet1}}{p_{jet1}+p_{jet2}}$. From simple kinematics and assuming that the masses of the jets are small compared with the mass of the di-jets, the virtuality and mass-squared of the di-jets are given by: $Q^2 \sim M_{DJ}^2 = \frac{k_t^2}{x(1-x)}$ where k_t is the transverse momentum of each jet. By studying the momentum distribution for various k_t bins, one can observe changes in the apparent fractions of asymptotic and Chernyak-Zhitnitsky (CZ) contributions to the pion wave function.

The basic assumption that the momentum carried by the dissociating $q\bar{q}$ is transferred to the di-jets was examined by Monte Carlo (MC) simulations of the asymptotic and (CZ) wave functions (squared). The MC samples were allowed to hadronize through the LUND PYTHIA-JETSET model [7] and then passed through simulation of the experimental apparatus (described in the next section) to simulate the effect of unmeasured neutrals and other experimental distortions. In Fig. 1 the initial distributions at the quark level are compared with the final distributions of the detected di-jets. As can be seen, the qualitative features of the distributions are retained. The results of this analysis come from comparing the observed x-distribution to a combination of the distributions shown, as examples, on the right of Figure 1.

The Color Transparency Effect

The Color-Transparency (CT) phenomenon is derived from the prediction that color fields cancel for physically small color singlet systems of quarks and gluons [8]. This effect of color neutrality (or color screening) is expected to lead to the suppression of initial and final state interactions for a small sized system or point-like configuration (PLC) formed in a large angle hard process [9]. Observation

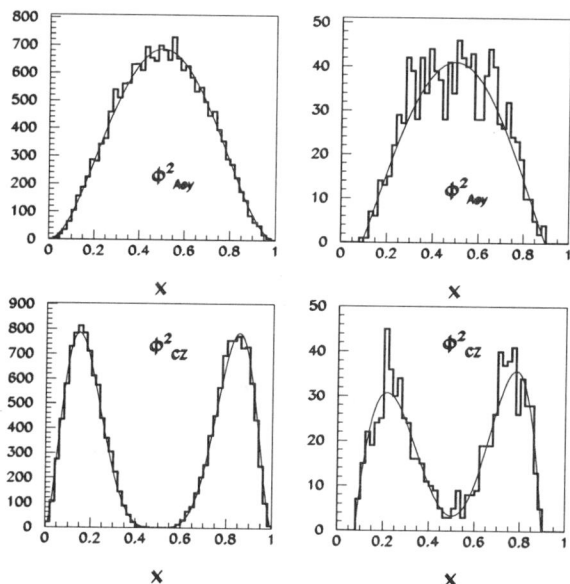

FIGURE 1. Monte Carlo simulations of squares of the two wave functions at the quark level (left) and of the reconstructed distributions of di-jets as detected (right). ϕ^2_{Asy} is the asymptotic function (squared) and ϕ^2_{CZ} is the Chernyak-Zhitnitsky function (squared). The di-jet mass used in the simulation is 6 GeV/c^2 and the plots are for 1.5 GeV/c $\leq k_t \leq$ 2.5 GeV/c.

of CT requires that a PLC is formed and that the energies are high enough so that expansion of the PLC does not occur [10] (the frozen approximation). Under conditions of $k_t > 1.5$ GeV/c, which translates to $Q^2 \sim 10$ (GeV/c)^2 and $\langle r \rangle \sim 0.1 fm$, observation of these effects can be expected. Bertsch et al. [4] proposed that the small $|q\bar{q}\rangle$ component will be filtered by the nucleus. They predicted an $A^{1/3}$ dependence for the *integrated* cross section. Frankfurt et al. [11] predicted an A^2 dependence for the *differential* cross section at $t = 0$ and $k_t > 1.5$ GeV/c. This prediction is studied in the present work.

EXPERIMENTAL RESULTS

Fermilab experiment E791 recorded 2×10^{10} events from interactions of a 500 GeV/c π^- beam with carbon and platinum targets. Details of the experiment are given in [12]. Only about 10% of the E791 data was used for the analysis presented here. The data were analysed by selecting events in which 90% of the beam momentum was carried by charged particles. Jets were identifies using the JADE jet-finding algorithm [13]. To insure clean selection of two-jet events, a minimum k_t of 1.25 GeV/c was required and their relative azimuthal angle, which for pure di-jets should be 180° was required to be within 20° of this value.

Diffractive di-jets were identified through the $e^{-bq_t^2}$ dependence of their yield (q_t^2 is the square of the transverse momentum transferred to the nucleus and $b = \frac{<R^2>}{3}$ where R is the nuclear radius). Figure 2 shows the q_t^2 distributions of di-jet events from platinum and carbon. The different slopes in the low q_t^2 coherent region reflect the different nuclear radii. Events in this region come from diffractive dissociation of the pion.

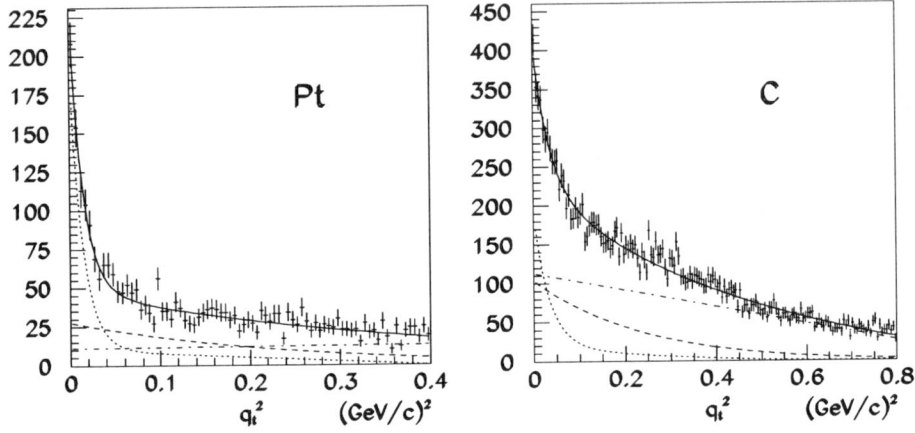

FIGURE 2. q_t^2 distributions of di-jets with $1.5 \leq k_t \leq 2.0$ GeV/c for the platinum and carbon targets. The lines are fits of the MC simulations to the data: coherent dissociation (dotted line), incoherent dissociation (dashed line), background (dashed-dotted line), and total fit (solid line).

The Pion Wave Function

For measurement of the wave function we used data from the platinum target as it has a sharp diffractive distribution and low background. We used events with $q_t^2 < 0.015$ GeV/c^2. For these events, the value of x was computed from the measured longitudinal momentum of each jet. A background, estimated from the x distribution for events with larger q_t^2 was subtracted. The analysis was carried out in two windows of k_t: 1.25 GeV/c $\leq k_t \leq$ 1.5 GeV/c and 1.5 GeV/c $\leq k_t \leq$ 2.5 GeV/c. The resulting x distributions are shown in Fig. 3. In order to get a measure of the correspondence between the experimental results and the calculated light-cone wave functions, we fit the results with a linear combination of squares of the two wave functions (right side of Fig. 1). This assumes an incoherent combination of the two wave functions and that the evolution of the CZ function is slow (as stated in [5]).

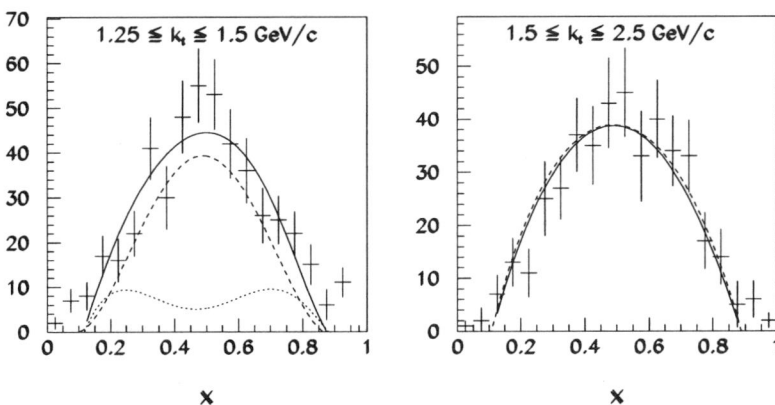

FIGURE 3. The x distribution of diffractive di-jets from the platinum target for $1.25 \leq k_t \leq 1.5$ GeV/c (left) and for $1.5 \leq k_t \leq 2.5$ GeV/c (right). The solid line is a fit to a combination of the asymptotic and CZ wave functions. The dashed line shows the contribution from the asymptotic function and the dotted line that of the CZ function.

k_t (GeV/c)	a_{as}	$\Delta_{a_{as}}^{stat}$	$\Delta_{a_{as}}^{sys}$	$\Delta_{a_{as}}$	a_{cz}	$\Delta_{a_{cz}}^{stat}$	$\Delta_{a_{cz}}^{sys}$	$\Delta_{a_{cz}}$
1.25 - 1.5	0.64	±0.12	+0.07 -0.01	+0.14 -0.12	0.36	∓0.12	-0.07 +0.01	-0.14 +0.12
1.5 - 2.5	1.00	±0.10	+0.00 -0.10	+0.10 -0.14	0.00	∓0.10	-0.00 +0.10	-0.10 +0.14

TABLE 1. Asymptotic (a_{as}) and CZ (a_{cz}) wave functions contributions to a fit of the data.

The results of the fits are given in Fig. 3 and in Table 1 in terms of the coefficients a_{as} and a_{cz} representing the contributions of the asymptotic and CZ functions, respectively. The results for the higher k_t window show clearly that the asymptotic wave function describes the data very well. This shows that for $k_t > 1.5$ GeV/c, which translates to $Q^2 \sim 10$ (GeV/c)^2, the pQCD approach that led to construction of the asymptotic wave function is reasonable. The distribution in the lower window is consistent with a significant contribution from the CZ wave function or may indicate contributions due to other non-perturbative effects.

The k_t dependence of diffractive di-jets is another observable that can show how well the perturbative calculations describe the data. As shown in [11] assuming interaction via two gluon exchange and ϕ_{as} would lead to $\frac{d\sigma}{dk_t} \sim k_t^{-6}$. The results, corrected for experimental acceptance, are shown in figure 4(a) fitted by k_t^n for $k_t > 1.25$ GeV with n = $-9.2 \pm 0.4(stat) \pm 0.3(sys)$ and $\chi^2/dof = 1.0$. This slope is significantly larger than expected. However, the region above $k_t \sim 1.8$ GeV/c can be fitted (Fig. 4(b)) with n = -6.5 ± 2.0 with $\chi^2/dof = 0.8$, consistent with the predictions. This would support the evaluation of the light-cone wave function at large k_t as due to one gluon exchange. The lower k_t region can be fitted with the non-perturbative Gaussian function: $\psi \sim e^{-\beta k_t^2}$ [14], with

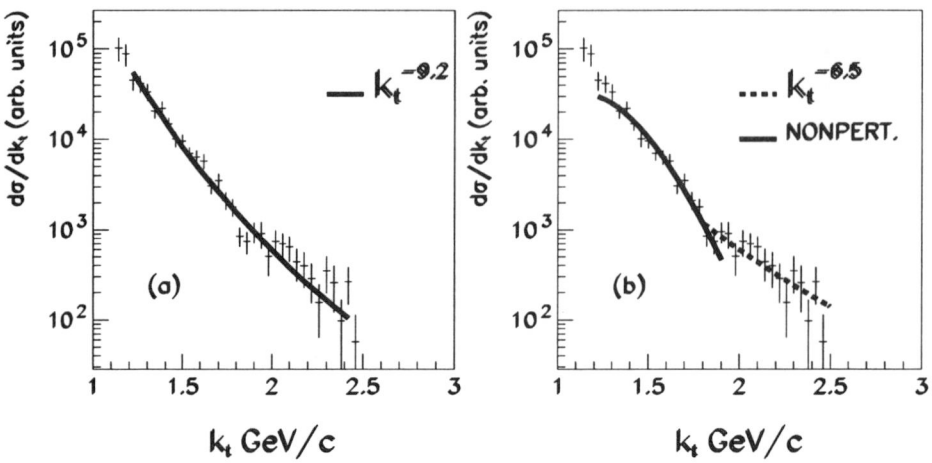

FIGURE 4. Comparison of the k_t distribution of acceptance-corrected data with fits to cross section dependence (a) according to a power law, (b) based on a nonperturbative Gaussian wave function for low k_t and a power law, as expected from perturbative calculations, for high k_t.

$\beta = 1.78 \pm 0.05(stat) \pm 0.1(sys)$ and $\chi^2/dof = 1.1$. Model-dependent values in the range of 0.9 - 4.0 were used [14]. These results are consistent with the measurements of the wave function that indicated noticeable non-perturbative effects up to $k_t \sim 1.5$ GeV/c.

The Color Transparency Effect

To study the CT effect we measure the A-dependence of the diffractive di-jet yield. The coherence length is estimated using $2p_{lab} = 1000$ GeV/c and $M_J \sim 5$ GeV/c^2. The result is $l_c \sim 10$ fm, larger than the platinum nuclear radius. In order to correct for experimental acceptance, we generate MC simulations of diffractive di-jets using the asymptotic wave function and di-jet masses of 4,5, and 6 GeV/c. The simulated coherent q_t^2 distributions of the di-jets represent the nuclear form factors of carbon (R=2.44 fm) and platinum (R=5.27 fm) [15] and the incoherent dissociation is simulated according to the nucleon radius (R=0.8 fm) [15] truncated at $q_t^2 < 0.015$. A combination of these simulations was used to fit the data (Fig. 2). We derive the numbers of produced di-jet events in the data for each target in three k_t bins by integrating over the diffractive terms in the fits. Using the resulting yields and the known target thicknesses, we determine the ratio of cross

k_t bin GeV/c	α	$\Delta\alpha_{stat}$	$\Delta\alpha_{sys}$	$\Delta\alpha$	α (CT)
1.25 − 1.5	1.64	±0.05	+0.04 −0.11	+0.06 −0.12	1.25
1.5 − 2.0	1.52	±0.09	±0.08	±0.12	1.45
2.0 − 2.5	1.55	±0.11	±0.12	±0.16	1.60

TABLE 2. The exponent in $\sigma \propto A^\alpha$, experimental results for coherent dissociation and the Color-Transparency (CT) predictions.

sections for diffractive dissociation on platinum and carbon (the two targets were subjected to essentially the same beam flux). The exponents α are then calculated using the cross section dependence $\sigma \propto A^\alpha$. The results are listed in Table 2 and compared with CT theoretical predictions [11]. The results are consistent with those expected from color-transparency calculations and clearly inconsistent with α values for incoherent scattering observed in other hadronic interactions.

REFERENCES

1. S.J. Brodsky, hep-ph/9908456.
2. S.J. Brodsky and G.P. Lepage, Phys. Rev. **D22**, 2157 (1980); S.J. Brodsky, G.P. Lepage, Phys. Scripta **23**, 945 (1981); S.J. Brodsky, Springer Tracts in Modern Physics **100**, 81 (1982).
3. A.V. Efremov and A.V. Radyushkin, Theor. Math. Phys. **42**, 97 (1980).
4. G. Bertsch, S.J. Brodsky, A.S. Goldhaber, and J. Gunion, Phys. Rev. Lett. **47**, 297 (1981).
5. V.L. Chernyak and A.R. Zhitnitsky, Phys. Rep. **112**, 173 (1984).
6. G. Sterman and P. Stoler, Ann. Rev. Nuc. Part. Sci. **43**, 193 (1997).
7. H.-U. Bengtsson and T. Sjöstrand, Comp. Phys. Comm. **82**, 74 (1994); T. Sjöstrand, PYTHIA 5.7 and JETSET 7.4 Physics and Manual, CERN-TH.7112/93, (1995).
8. F. E. Low, Phys. Rev. **D12**, 163 (1975); S. Nussinov, Phys. Rev. Lett **34**, 1286 (1975).
9. A.H. Mueller in Proceedings of the Seventeenth Rencontre de Moriond, Les Arcs, France (1982) ed. J. Tran Thanh Van (Editions Frontieres, Gif-sur-Yvette, France, 1982) Vol. I, p13; S.J. Brodsky in Proceedings of the Thirteenth Int'l Symposium on Multiparticle Dynamics, ed. W. Kittel, W. Metzger, and A. Stergiou (World Scientific, Singapore, 1982) p963.
10. S.J. Brodsky and A.H. Mueller, Phys. Lett. **B206**, 685 (1998).
11. L.L. Frankfurt, G.A. Miller, and M. Strikman, Phys. Lett. **B304**, 1 (1993).
12. E791 Collaboration, E.M. Aitala *et al.*, EPJdirect **C4**, 1 (1999).
13. JADE collaboration, W. Bartel *et al.*, Z. Phys. **C33**, 23 (1986).
14. R. Jakob and P. Kroll, Phys. Lett. **B315**, 463 (1993).
15. Atom. data Nucl. Data tabl. **14**, 479 (1974).

The Transverse Quark Distribution and Proton Electromagnetic Form Factors in Skew Distribution Formalism

John P. Ralston[a], Pankaj Jain[b] and Roman V. Buniy[a]

[a] *Department of Physics and Astronomy, University of Kansas, Lawerence, KS 66045, USA*
[b] *Physics Department, I.I.T. Kanpur, India 208016*

Abstract. Skew density matrices can be diagonalized to yield probability interpretation. The power-counting prediction of perturbative QCD is found consistent with recent CEBAF data on $F_2(Q^2)/F_1(Q^2)$.

Perturbative QCD can be applied to hard exclusive processes in many ways. Use of "skew" or "off-diagonal" parton distributions [1] has generated attention. Skew distributions are matrix elements sharing features with density matrices. The diagonal elements of a positive density matrix are interpreted as probabilities. Density matrices depend on the basis: probability in one frame will appear to be an interference of amplitudes in another frame. This occurs with quark spin amplitudes [2], where transverse polarization ("transversity") distributions appear as interference in the helicity basis, and helicity distributions appear as interference in the transverse polarization basis.

Here we find a probabilistic interpretation for skew distributions involved in the proton electromagnetic form factors. Then a simple, kinematic angular momentum argument resolve a recent puzzle from data for the electromagnetic form factors $F_1(Q^2), F_2(Q^2)$. For about 30 years it was thought that $F_2(Q^2)/F_1(Q^2) \sim 1/Q^2$ was a prediction of hard scattering, and in particular, of $pQCD$. We made the same prediction on the basis of skew distributions in 1993 [3]. Recent $CEBAF$ data [4] shows that $F_2/F_1 \sim 1/\sqrt{|Q^2|}$. Our prediction missed some instructive points, and the $CEBAF$ result is expected, if proper kinematics is simply taken into account.

Diagonalization: Consider $2 \to 2$ quark-proton scattering with momentum transfer $\Delta^\mu, \Delta^2 < 0$. The proton is described by states $|p - \Delta/2, s>$ coming "in" and $|p + \Delta/2, s'>$ going "out". The scattering matrix element is

$$\Phi_{\alpha,\beta}^{s,s'}(p;k,k',\Delta)\delta(k+p-k'-p-\Delta)$$
$$= \int d^4z d^4z' \exp(-ikz + ik'z') <p - \Delta/2, s|\psi_\beta(z)\bar{\psi}_\alpha(z')|p + \Delta/2, s'> .$$

Here α, β are Dirac indices of the quark fields ψ, and the in- and out-quarks have momenta k, k' with $k = x(p - \Delta/2) + \vec{k}_T$, $k' = x'(p + \Delta/2) + \vec{k}'_T$. A time-ordering symbol is dropped, as only one time ordering contributes: see Diehl and Gousset [1] Now $\Phi(\Delta)$ can be decomposed into terms symmetric and antisymmetric in Δ; the symmetric part is Hermitian. This leads to a density matrix we can diagonalize.

To diagonalize we make a series of coordinate transformations. Make a Lorentz transformation to the frame: $p = (p^+, 0, m^2/2p^+)$; $\Delta = (0, \Delta_T, -t/2p^+)$; $\Delta_T^2 = -t$. Now Δ^μ becomes entirely transverse as $p^+ \to \infty$. The partons have 4-vectors $k^\mu = xp^\mu + k_T^\mu, x(p + \Delta)^\nu + k_T^{\prime\nu}$, yielding

$$x = x'; \quad k'_T = k_T + \Delta_T(1 - x).$$

The matrix element is now diagonal in x. This might seem impossible, because the x, x' dependence of skew distributions is thought to be invariant: but x is not Lorentz-invariant sideways. The convolution in $\vec{k}_T$ is diagonalized by conjugate transverse spatial coordinates $\vec{b}$. Then $\int d^2b \, \Phi(s, s'; \vec{b}/(1-x), x) e^{i\vec{b}\cdot\vec{\Delta}}$ can be inverted by Fourier transform to find the integrand, diagonal in everything but spin, which also can be made diagonal by familiar helicity or transverse bases.

The choice of frames and diagonalization was used in an independent early introduction of off-diagonal distributions [3]. Due to a kinematics goof we missed the factor $1/(1 - x)$. A review by Brodsky and Lepage [5] was useful.

Interpretation: The electromagnetic form factors (with $-Q^2 = \Delta_T^2$) are found by following the Feynman rules, which includes multiplication by the quark charge e_q, tracing Φ with γ^μ, and doing the integral $\int dx \, d^2b$. At large $-Q^2$ this is dominated by $b^2 \sim 1/|Q^2|$ by Fourier analysis. The "short distance" implied by large Q is far more general than the more problematic "quark-counting" argument. We simply have *one* quark located by the hard momentum Q, while the asymptotic short distance theory [5] assumes that *all possible Fock states* are separated by asymptotically short distance. We remain within the framework of *pQCD*, of course, while choosing a more general factorization method.

Since the form factor $F_1(Q^2)$ is known from data, we invert the Fourier transform to solve for a positive-definite diagonal element of the density matrix, namely a probability:

$$P(\lambda, \lambda; \vec{b}) = \frac{1}{e_q^2} \int d^2Q_T \, e^{-i\vec{b}\cdot\vec{Q}_T} F_1(Q^2). \tag{1}$$

Positivity is assured because $F_1(Q^2)$ is monotonically decreasing. The quantity $P(\lambda, \lambda; \vec{b})$ is on a similar footing to the usual parton distributions. Indeed, the usual parton distributions are functions of x integrated over k_T, while $P(b)$ depends on $b = |\vec{b}|$ and has been integrated over x.

A fit to the Fourier transform of $F_1(Q^2)$ has been performed. A profile of the transverse ($1/(1-x)$-weighted) probability to find a quark (mostly of the up-type) was shown at the meeting. We expect that this invariant quantity will be useful in many studies involving the transverse coordinate.

Orbital Angular Momentum: Quark orbital angular momentum is a fascinating subject of great interest in the proton spin puzzle. In the case at hand, we are lucky to have two quarks evaluated at the same $x, \vec{b}$ points, so that difficulties of gauge invariance are minimal. Indeed, $P(b)$ is gauge invariant by definition in terms of observable quantities.

We expand the operators in terms of solutions to $\nabla^2 \psi = 0$. By usual methods [2] the operators are evaluated inside the proton state, letting the correlation Φ be expressed by c-number "wave functions". Partons are below threshold in a form factor, so we need an expansion for orbital angular momentum for spacelike k:

$$\psi(z^-, \vec{b}, z^+ = 0) = \sum_n \int dx \, \psi_n(x) I_n(b) e^{-in\phi} e^{ixp^+ z^-}. \tag{2}$$

Here $e^{-in\phi}$ are light-cone $SO(2)$ orbital angular momentum basis functions; $I_n(b)$ are modified Bessel functions more usually seen as $J_n(b)$ for a timelike basis.[1] We are interested in the F_2 form factor, associated with $i\sigma_{\mu\nu} Q^\nu / 2m$, which represents proton helicity-flip at large Q^2. Since we cannot change the helicity of a quark with a hard scattering (the near-perfect chiral symmetry of $pQCD$), the proton can only flip its spin to make F_2 by transferring a unit of *orbital* angular momentum in the quark [3]. (Hoodbhoy and Ji [6] subsequently verified the same result.)

Note we are not attempting here to derive the functional dependence of $P(\lambda, \lambda; \vec{b})$: in our approach this comes from data. Our approach is to relate each power of b or angular momentum to further suppression by powers of $1/Q$. The argument for short distance is kinematically compelling here (if controversial in the quark-counting method), so for large Q the Fourier transform is dominated by $I_0(b) \sim b^0$ if this channel is allowed. But $I_0(b)$ represents the s-wave component, with zero angular momentum, and so this channel is open only to the helicity non-flip, namely

$$F_1 \sim \delta_{\lambda\lambda'} \int dx \, d^2b \, \bar{\psi}_0(x) \Gamma \psi_0(x) \left[I_0 \left(b/(1-x) \right) \right]^2 \exp(i\vec{Q}_T \cdot \vec{b}). \tag{3}$$

Here Γ represents the necessary Dirac matrices.

Conversely, the only possibility for the helicity-flip F_2 is to use powers of b to conserve the angular momentum, and suffer the corresponding power-suppression in Q. On the basis of this power counting, it was reasoned [3] that by angular momentum selection rules, the integrals over b would vanish unless two representations of the same angular momentum matched up, giving the previous prediction $F_2(Q^2)/F_1(Q^2) \sim 1/Q^2$ for $Q^2 > GeV^2$. Now consulting the formulas this is simply not true. There is a factor of $e^{i\vec{Q}_T \cdot \vec{b}_T}$ carrying angular momentum:

$$F_2 \sim \delta_{\lambda, -\lambda'} \int dx \, d^2b \, \bar{\psi}_0(x) \Gamma \psi_1(x) I_0 \left(b/(1-x) \right) I_1 \left(b/(1-x) \right) e^{i|Q_T|b\cos(\phi - \phi_Q)} e^{i\phi} + cc. \tag{4}$$

---

[1] If one is concerned about $b \to \infty$, then Green functions can be expanded in series of $I_n(b_<) K_n(b_>)$ where $b_<(b_>)$ is the smaller (larger) of two b arguments. We only need short distance.

Physically, the probe $\vec{Q}$ breaks the rotational symmetry of the problem.

We reiterate that this analysis is entirely within the context of pQCD. In pQCD one takes some matrix elements from the data, and makes predictions for others. What can we predict here? From the power-counting cited, we have

$$\frac{F_2(Q^2)}{F_1(Q^2)} = \frac{<\bar{\psi}_1(x)\psi_0(x)>}{\sqrt{|Q^2|}<b\bar{\psi}_0(x)\psi_0(x)>}, \tag{5}$$

where the braces represent the integrals. The fact that $(Q/GeV)F_2(Q^2)/F_1(Q^2)$ is not far from unity in the CEBAF data indicates that the proton wave functions for quark angular momenta 1 and 0 are not too different in magnitude. Constituent quark, or non-relativistic models are ruled out, but those models were never capable of capturing the Fock space description of the skew distribution. It will be interesting to continue these studies in the context of the larger proton spin puzzle.

Acknowledgments: Work supported by DOE grant number DE-FG02-98ER41079.

REFERENCES

1. J. Bartles, Zeit. Phys. C **12** (1982) 263; B. Geyer et al, Zeit. Phys. C **26** (1985) 521; T. Braunschweig et al, Zeit. Phys. C **33** (1987) 521; F. Dittes et al, Phys. Lett. B **209** (1988) 325; I. Balitsky and V. Braun, Nucl. Phys. B **311** (1989) 1541; P. Jain, J. P. Ralston and B. Pire, Proceedings of the DPF92 meeting, Fermilab, November 10-14 (1992), hep-ph/9212243; X. Ji, Phys. Rev. D **55** (1997) 7114; A. Radyushkin, Phys. Lett. B **380** (1996) 417; Phys. Rev. D 56 (1997) 5524; M. Diehl and T. Gousset, Phys. Lett. B **428**, 359 (1998).
2. J. P. Ralston and D. E. Soper, Nucl. Phys. B **152**, 109 (1979).
3. P. Jain and J. P. Ralston, *Future Directions in Particle and Nuclear Physics at Multi-GeV Hadron Beam Facilities* Proceedings of the Workshop held at BNL, 4-6 March, 1993, hep-ph/9305250.
4. M. K. Jones et al, Phys. Rev. Lett. **84** 1398 (2000).
5. S. J. Brodsky and G. P. Lepage, *Perturbative Quantum Chromodynamics*, Ed. by A.H. Mueller, World Scientific, 1989.
6. P. Hoodbhoy and X. Ji, Phys. Rev. Lett. **76**, 740 (1996).

A New Measurement of the Energy Dependence of Nuclear Transparency for Large Momentum Transfer $^{12}C(p,2p)$ Scattering

A. Leksanov[a], J. Alster[b], G. Asryan[c], Y. Averichev[h],
D. Barton[d], V. Baturin[a,e], N. Bukhtojarova[d,e], A. Carroll[c],
A. Schetkovsky[a,e], S. Heppelmann[a], T. Kawabata[f], A. Malki[b],
Y. Makdisi[c], E. Minina[a], I. Navon[b], H. Nicholson[g],
A. Ogawa[a], Y. Panebratsev[h], E. Piasetzky[b], S. Shimanskiy[h],
A. Tang[i], J.W. Watson[i], H. Yoshida[f], D. Zhalov[a]

[a] *Physics Department, Pennsylvania State University, University Park, PA 16801, USA*
[b] *School of Physics and Astronomy, Sackler Faculty of Exact Sciences, Tel Aviv University, Ramat Aviv 69978, Israel*
[c] *Yerevan Physics Institute, Yerevan 375036, Armenia*
[d] *Collider-Accelerator Department, Brookhaven National Laboratory, Upton, NY 11973, USA*
[e] *Petersburg Nuclear Physics Institute, Gatchina, St. Petersburg 188350, Russia*
[f] *Department of Physics, Kyoto University, Sakyoku, Kyoto, 606-8502, Japan*
[g] *Department of Physics, Mount Holyoke College, South Hadley, MA 01075, USA*
[h] *J.I.N.R., Dubna, 141980, Russia*
[i] *Department of Physics, Kent State University, Kent, OH 44242, USA*

Abstract. We present a new measurement of the energy dependence of nuclear transparency from AGS experiment E850, performed using the EVA solenoidal spectrometer, upgraded since 1995. Using a secondary beam from the AGS accelerator, we simultaneously measured pp elastic scattering from hydrogen and $(p,2p)$ quasi-elastic scattering in carbon at incoming momenta of 5.9, 8.0, 9.0, 11.7 and 14.4 GeV/c. This incident momentum range corresponds to a Q^2 region between 4.8 and 12.7 (GeV/c)^2. The detector allowed us to do a complete kinematic analysis for the center-of-mass polar angles in the range $85° - 90°$. We report on the measured variation of the nuclear transparency with energy and compare the new results with previous measurements.

Color Transparency(CT) is the predicted reduction in the initial and final state interactions, which may take place when a large transverse momentum(p_t), quasi-exclusive scattering, involving hadrons, occurs in nuclear matter. We study nuclear transparency for $(p,2p)$ quasi-elastic scattering. It can be informally defined as

$$T_r = \frac{\frac{d\sigma}{dt}(pp \text{ quasi-elastic in nucleus})}{Z\frac{d\sigma}{dt}(pp \text{ elastic in hydrogen})} \quad . \tag{1}$$

Theoretical predictions for the variations of this quantity with energy are very model dependent. For example, Glauber-based calculations show no energy dependence above 5 GeV with a value around thirty percent, whereas QCD based models predict an increase of T_r with energy, reaching eventually an asymptotic value of unity [1]. Interpretations of fluctuating energy dependence of the transparency [2] at the values of Q^2 above 10 (Gev/c)^2 have been presented in the models described in Refs. [3] and [4].

Many widely accepted pictures of CT dynamics are based on the assumption that selection of small-size Fock state configurations occurs in certain hard scattering processes and that these small configurations have a reduced cross-section for reinteraction with the surrounding nucleus. The protons remain in these configurations during a time period which is different for different energies and large on the scale of nuclear processes [5].

We performed our measurement using the EVA apparatus located in the C1 line of the AGS at BNL. Two differential Cerenkov counters were used for the incident particle identification. The beam flux measurement was performed using a scintillating-fiber beam hodoscope located in the upstream part of the apparatus. We used solid carbon and polyethylene targets, whose dimensions were 5.1 cm × 5.1 cm × 6.6 cm. The AGS provided a secondary beam with approximately 1-2×10^7 particles per 2 second spill and about 20 spills per minute. EVA is a magnetic spectrometer with tracking capabilities. It was built around a solenoidal superconducting magnet which produced nearly uniform field in the beam direction allowing for charged particle transverse momentum(p_t) measurements. It further allowed us to reject low p_t tracks at the trigger level.

The tracking with this apparatus used an array of four cylindrical straw tube drift chambers with charge division, providing us with 3D position information. The chambers were used both for triggering and for final track reconstruction. The signals from two fan shaped hodoscopes of scintillating counters served as input to the first level trigger, which started the digitization of the event [6], [7], [8], [9], [10].

The variable we present in this paper is the *transparency ratio* T_{CH}, which we define to be the ratio of $C(p,2p)$ quasi-elastic(QE) to pp elastic cross-section divided by the Z of the nucleus(6 in our case) for a restricted region of kinematics near that of pp elastic scattering. Within the impulse approximation this quantity is related to nuclear transparency via

$$T_{CH} = \frac{C}{ZH} = T_r \int_{\alpha_1}^{\alpha_2} \frac{d\alpha}{\alpha} \int\int d\vec{p}_{Ft} n(\alpha, \vec{p}_{Ft}) \frac{\frac{d\sigma}{dt}(s)}{\frac{d\sigma}{dt}(s_0)} \quad . \tag{2}$$

Here $n(\alpha, \vec{p}_{Ft})$ is the momentum distribution of the target proton in the nucleus, $\alpha = \frac{E_F - p_{Fz}}{m} \simeq \frac{s}{s_0}$ near $\alpha = 1$, s and s_0 are the Mandelstam variables for C and H events correspondingly, $\vec{p}_{Ft} = (p_{Fx}, p_{Fy})$ is the transverse component of the target nucleon momentum, $\frac{d\sigma}{dt}(s)$ is the differential pp cross-section for the process inside the nucleus, $\frac{d\sigma}{dt}(s_0)$ - for the hydrogen.

In the presentation of these measurements, we attempt to limit the influence of incomplete knowledge of the spectral function. This was accomplished by applying an unrestrictive cut on transverse Fermi momentum but a tight cut on longitudinal nuclear motion so that $\frac{d\sigma}{dt}(s) \simeq \frac{d\sigma}{dt}(s_0)$, neglecting small variations in cross-section.

Our apparatus was capable of measuring all three components of momenta of both outgoing particles. Therefore the kinematic variables of interest were calculated from the data. The following cuts ensured that the conditions were satisfied: $|p_{Fy}| < 0.3$ GeV/c, $|p_{Fx}| < 0.5$ GeV/c and $0.95 < \alpha_0 < 1.05$. The variable α_0, defined as

$$\alpha_0 = 1 - \frac{2p\cos\frac{\theta_1-\theta_2}{2}\cos\frac{\theta_1+\theta_2}{2} - p_{inc\,z}}{m} \qquad (3)$$

with $p = \sqrt{(\frac{E_{beam}+m}{2})^2 - m^2}$, is an analog of α, which was defined to be independent of momentum variables and varies only with the opening angle in the Lab frame. The value of $|\alpha - \alpha_0|$ in the region of interest is well below one percent [7], [8].

To extract the number of events from C and CH_2 targets, we used the missing mass variable defined as:

$$M^2_{miss} = E^2_{miss} - \vec{p}^2_F \,. \qquad (4)$$

As it follows from equation (4), M^2_{miss} distribution for the reconstructed QE events should display a peak around zero. From the events that satisfy the cuts we select candidates for QE by demanding exactly two charged tracks in the final state. We also select events with more then two tracks to model the behavior of the background. The signal is then extracted from these distributions after background subtraction.

The hydrogen signal was measured as the normalized difference between the yields from the CH_2 and C targets. Then the transparency ratio can be defined as

$$T_{CH} = \frac{\beta\gamma C}{3(CH_2 - \beta\gamma C)} \,. \qquad (5)$$

Here C and CH_2 are the numbers of extracted events. $\beta = 0.4535$ is the ratio of number of CH_2 molecules to the number of C atoms in the corresponding targets, γ is the beam normalization factor $\gamma = \frac{beam_C}{beam_{CH_2}}$. The beam normalization factor is determined from the flux measurements using scaler readings and comparing yields for the same target positions and different target configurations.

We performed our measurements at incoming momenta of 5.9 and 7.5 GeV/c in 1994 [11] and 5.9, 8.0, 9.0, 11.7 and 14.4 GeV/c in 1998. These corresponded to the following values of momentum transfer Q^2: 4.8 and 6.2 (GeV/c)^2 in 1994 and 4.8, 6.6, 7.6, 10.1 and 12.7 (GeV/c)^2 in 1998. The preliminary result of our experiment, the energy dependence of the transparency ratio, integrated over the

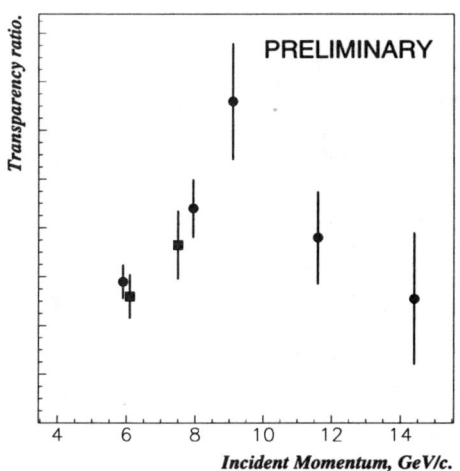

FIGURE 1. The measured dependence of the transparency ratio, T_{CH}, on the incident momentum, P_{inc}. Boxes represent 1994 results and circles - 1998. No systematic errors are included.

center-of-mass scattering angle in the region $85° - 90°$, is presented in the figure 1. We can draw the following conclusions from it. First, the ratio definitely varies with energy with strong deviations from Glauber calculations. Second, we clearly see the rise of the ratio for the incident momenta below 10 GeV/c. Third, we also observe the fall of the transparency above 10 GeV/c. Finally, our results seem to be in a good agreement with the energy dependence reported in 1988 by E834 [2]

REFERENCES

1. G.R. Farrar et al., *Phys. Rev. Lett.* **61**, 686 (1988).
2. A.S. Carroll et al., *Phys. Rev. Lett.* **61**, 1698 (1988).
3. B. Pire and J. Ralston, *Phys. Rev. Lett.* **61**, 1823 (1988).
4. C.F. de Teramond and S.J. Brodsky, *Phys. Rev. Lett.* **60**, 1924 (1988).
5. M. Strikman, L. Frankfurt and G.A. Miller, *Comm. Nucl. Part. Phys.* **21**, 1 (1992).
6. J. Wu et al., *Nucl. Instr. and Meth.* **349**, 183 (1994).
7. I. Mardor, *PhD Thesis*, unpublished, Tel-Aviv University, 1997.
8. Y. Mardor, *PhD Thesis*, unpublished, Tel-Aviv University, 1997.
9. S. Durrant, *PhD Thesis*, unpublished, The Pennsylvania State University, 1994.
10. A. Leksanov, *PhD Thesis*, in preparation, The Pennsylvania State University, 2000.
11. I. Mardor et al., *Phys. Rev. Lett.* **81**, 5085 (1998).

Longitudinal Momentum Fraction X_L for Two High P_t Protons in pp→ppX Reaction

D. Zhalov[e], S. Heppelmann[e], J. Alster[a], G. Asryan[c,b],
Y. Averichev[h], D. Barton[c], V. Baturin[e,d], N. Bukhtoyarova[c,d],
A. Carroll[c], T. Kawabata[f], A. Leksanov[e], Y. Makdisi[c], A. Malki[a],
E. Minina[e], I. Navon[a], H. Nicholson[g], A. Ogawa[e], Yu. Panebratsev[h],
E. Piasetzky[a], A. Schetkovsky[e,d], S. Shimanskiy[h], A. Tang[i],
J.W. Watson[i], H. Yoshida[f]

[a] *School of Physics and Astronomy, Sackler Faculty of Exact Sciences, Tel Aviv University, Ramat Aviv 69978, Israel*
[b] *Yerevan Physics Institute, Yerevan 375036, Armenia*
[c] *Collider-Accelerator Department, Brookhaven National Laboratory, Upton, NY 11973, USA*
[d] *Petersburg Nuclear Physics Institute, Gatchina, St. Petersburg 188350, Russia*
[e] *Physics Department, Pennsylvania State University, University Park, PA 16801, USA*
[f] *Department of Physics, Kyoto University, Sakyoku, Kyoto, 606-8502, Japan*
[g] *Department of Physics, Mount Holyoke College, South Hadley, MA 01075, USA*
[h] *J.I.N.R., Dubna, 141980, Russia*
[i] *Department of Physics, Kent State University, Kent, OH 44242, USA*

Abstract. We present an analysis of new data from Experiment E850 at BNL. We have characterized the inclusive cross section near the endpoint for pp exclusive scattering in Hydrogen and in Carbon with incident beam energy of 6 GeV. We select events with a pair of back-to-back hadrons at large transverse momentum. These cross sections are parameterized with a form $\frac{d\sigma}{dX_L} \sim (1 - X_L)^p$, where X_L is the ratio of the longitudinal momentum of the observed pair to the total incident beam momentum. Small value of p may suggest that the number of partons participating in the reaction is large and reaction has a strong dependence on the center-of-mass energy. We also discuss nuclear effects observed in our kinematic region.

There are two main questions which can be addressed in the study of the Longitudinal Momentum Fraction carried by a pair of large P_t final state particles in the pp→ppX reactions. We learn first something about the number of partons involved in the interaction. Concluding, as we will, that the number of partons involved is large, we discuss a mechanism for picking up a momentum boost from the spectator nucleus in events where the observed momentum fraction is equal and larger than the incoming momentum.

We defined the Longitudinal Momentum Fraction variable X_L as:

$$X_L = \frac{P_{L_1} + P_{L_2}}{P_{L_{inc}}}. \tag{1}$$

Here $P_{L_{inc}}$ is the momentum of the beam coming along the Z-axis, P_{L_1} and P_{L_2} are the longitudinal components of the two detected out-going particles having large transverse momenta. We have argued that the two detected out-going particles in the pp→ppX reaction are protons [1]. We then expect the shape of the cross section, as X_L approaches unity, to reflect the number of partons participating in the collision. If the cross section is parameterized as

$$\frac{d\sigma}{dX_L} \sim (1 - X_L)^p \tag{2}$$

then a large p is expected for single parton collisions, reflecting the endpoint behavior of structure functions and fragmentation functions [2]. As the number of participating partons increases, we expect the power p to be reduced.

In the experiment discussed here, we measure p for selected classes of two track events. Our result, that p is in the range of 1-2, indicates that many partons are participating in the hard process. It can be expected that the cross section may be falling very rapidly with center-of-mass (cm) energy for these events, perhaps similar to the pp elastic energy dependence.

The cm energy dependence of the underlying pp→ppX cross section is affected by the number of partons participating in the interaction. The underlying cross section falls with energy as a power of energy, and that power increases as the number of partons increases. It is known that for exclusive scattering, the cross section has a dependence on the cm energy given by (s is a Mandelstam variable) $\sigma \sim s^{-m}, m = (n_1 + n_2 + n_3 + n_4 - 2)$ for a process with $n_1 + n_2$ partons $\rightarrow n_3 + n_4$ partons [3].

With this observation about the nature of the selected pp→ppX events, we can further conclude that interesting nuclear effects would also be expected. The nucleus can change the cm energy of the incident proton and the nuclear proton by virtue of nuclear fermi momentum. If the nucleus contributes a target proton which is moving in the direction of the beam particle, the cm energy is reduced. Because the increase in cross section associated with a reduction in cm energy is very large, we can expect a strong bias for the nucleus to contribute momentum to the scattering process in the forward direction. The typical magnitude of this nuclear contribution to the final state momentum may be much larger than typical fermi momentum seen in other processes.

Since the nucleus tends to add positive momentum to the two final state particles, in some sense, the stopping power for these events is apparently negative. When one compares the same pp→ppX processes measured in nuclear matter with that measured in Hydrogen, the effect of the nucleus is to move strength in the X_L

distribution to larger X_L. If a ratio, like a transparency ratio of the cross section in nuclear matter to that measured in Hydrogen is calculated, the ratio will be artificially enhanced.

The data used for this analysis was collected on two targets ^{12}C and CH_2 using the EVA Spectrometer (E850 Exp., BNL, AGS). The CH_2 target enabled us to extract cross sections for free proton (Hydrogen) scattering. The acceptance of the detector for two high P_t tracks in the r-z plane was centered around the opening angle for elastic pp→pp scattering at 90° in the cm frame [4]. Charged particle acceptance in the azimuthal angle was nearly 2π.

In Fig. 1, we present the X_L distribution for Carbon and Hydrogen. In one case we applied a 'semi-elastic' cut (two and only two charged tracks in our charged track acceptance) which allowed us to look at the X_L distribution for pp→pp events with a background of pp→pp+X^0 and pp→pp+$X^{charged}$. The quasi-elastic signals have been extensively analyzed elsewhere [4,5].

Secondly, we select a more inclusive data set made up of events with two high P_t tracks but any number of additional lower momentum charged tracks in the pp→ppX process.

We have fitted X_L distribution with a form (here A and p are the fit constants):

$$\frac{d\sigma}{dX_L} = A * (1 - X_L)^p + Gaussian \tag{3}$$

The polynomial part of the fit was restricted for X_L <1 while the Gaussian is included to account for the quasi-elastic contributions near $X_L \simeq 1$.

The peak corresponding to the Gaussian part of the fit is actually shifted for Carbon events from $X_L = 1$ towards larger X_L because of the s^{-10} dependence of the pp→pp reaction's cross section.

We have determined the power of the polynomial in the fit to be in the range 1.1-1.8 depending on the tightness of the cuts around the region of the elastic kinematics. According to our interpretation, we conclude that the number of quarks participating in the process is large. Consequently, the power m of s^{-m} dependence of the cross section would be expected to be large as well.

In Fig. 1, the effective ratio for scattering in Carbon vs Hydrogen is shown. While the ratio is well in excess of Glauber levels, our interpretation is that the Carbon distribution is enhanced due to the strong energy dependence of the cross section.

A recent study of neutron correlations from this experiment (E850 collaboration at BNL) has shown an anomalously large yield of neutrons going into the backward hemisphere in pp→ppX reactions compared to the yields reported by other experiments [1,6]. This result was observed with a data sample similar to the one described above. A possible interpretation of this phenomenon is that strong s-dependence of the cross-section for this class of events may select events with large positive fermi momentum. The nucleus must balance this momentum by some kind of recoil in the backward direction. The idea that most of the recoil momentum is absorbed by a single recoiling neutron leads to the prediction of enhanced backward neutron production [7]. Since a large yield of backward neutrons has been observed in the pp→ppX data set described here, the picture seems to have merit.

Keeping this in mind, we looked at the X_L distribution for Carbon and Hydrogen (Fig. 1). In the case of exclusive events with two high P_t particles, we can clearly see the shift in the position of the Gaussian peak in the Carbon sample. This effect is known to be caused by the s^{-10} dependence of the cross section. We have compared the more inclusive data set distributions for Carbon and Hydrogen for pp→ppX and interpreted the enhanced ratio as evidence for a similar shift in the Carbon distribution with respect to the Hydrogen for this more inclusive sample as well.

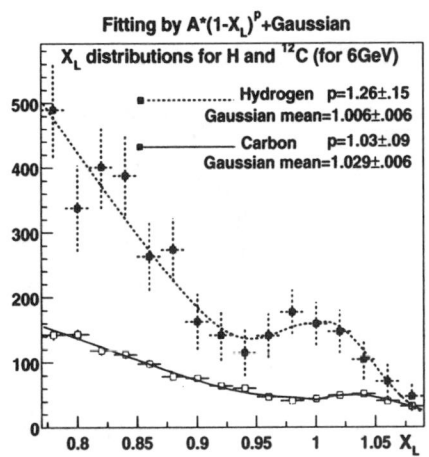
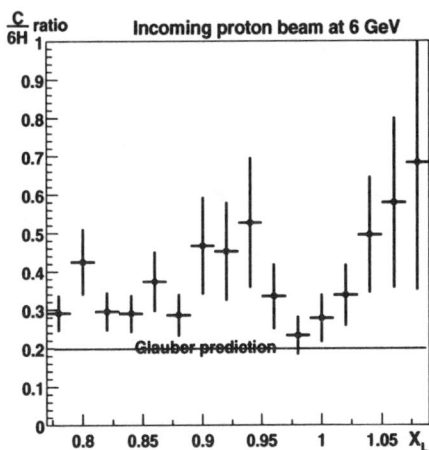

FIGURE 1. LEFT: X_L distributions for Carbon nucleus and 6*(Hydrogen) with two and only two tracks cut. RIGHT: Ratio of the Carbon signal to the 6*(Hydrogen) signal with two and only two tracks cut.

REFERENCES

1. Malki A. et al, *to be published*.
2. Brodsky S. et al, *Nucl. Phys.* **B441**, 197 (1995).
3. Brodsky S., Farrar G., *Phys. Rev. Lett.* **31**, 1153 (1973).
4. Leksanov A. et al, in *Proceedings of CIPANP2000, Quebec, May 22-28, 2000*.
5. Carroll A. et al, *Phys. Rev. Lett.* **61**, 1698 (1988).
6. Malki A. et al, in *Proceedings of CIPANP2000, Quebec, May 22-28, 2000*.
7. Frankfurt L.L., Strikman M.I., *Phys. Lett.* **B69**, 87 (1977).

Dynamics of glueball and $q\bar{q}$ production in the central region of pp collisions

Andrei Sobol[a,b] representing the WA102 collaboration[1]

[a] *Laboratoire de la Physique des Particules, Annecy-le-Vieux, France*
[b] *Institute for High Energy Physics, Protvino, Russia*
E-mail: sobol@mx.ihep.su

Abstract. A strong dependence of meson production with different J^{PC} on the angle between the transverse momentum vectors of the outgoing protons is observed. The ϕ and t dependences of several resonances with $J^{PC} = 0^{\pm+}, 1^{++}, 2^{\pm+}$ can be described by a model of double Pomeron exchange for the soft Pomerons acting as non-conserved vector currents. The 0^{++} and 2^{++} sector reveals a systematic behaviour in the data that appears to distinguish between $q\bar{q}$ and non-$q\bar{q}$ or glueball candidates.

INTRODUCTION

The WA102 fixed target experiment [1] at the CERN Omega Spectrometer studies centrally produced exclusive final states formed in the reaction

$$pp \to p_f X p_s$$

at 450 GeV/c, where subscripts f and s refer to the fastest and slowest particles in the laboratory frame respectively and X represents the central system. Such reactions are predicted to be a source of gluonic final states via double Pomeron exchange (DPE) [2]. There has been an intensive experimental programme in the last two years by the WA102 collaboration which has produced a large and detailed data set in meson spectroscopy [3]. A strong dependence on the angle between the transverse momentum vectors of the outgoing protons was observed in meson production with different J^{PC}.

---

[1] D.Barberis, F.G.Binon, F.E.Close, K.M.Danielsen, S.V.Donskov, B.C.Earl, D.Evans, B.R.French, T.Hino, S.Inaba, A.Jacholkowski, T.Jacobsen, G.V.Khaustov, J.B.Kinson, A.Kirk, A.A.Kondashov, A.A.Lednev, V.Lenti, I.Minashvili, J.P.Peigneux, V.Romanovsky, N.Russakovich, A.Semenov, P.M.Shagin, H.Shimizu, A.V.Singovsky, A.Sobol, M.Stassinaki, J.P.Stroot, K.Takamatsu, T.Tsuru, O.Villalobos Baillie, M.F.Votruba, Y.Yasu.

THE EFFECT OF NON FLAT AZIMUTHAL ANGLE

The azimuthal angle ϕ is defined as the angle between the p_T vectors of the p_f and p_s in the pp centre of mass. Naively it may be expected that this angle would be flat irrespective of the resonances produced. The experimentally observed ϕ dependences are clearly non flat [3] and considerable variations are found between resonances with different J^{PC}s. Figures 1(a), 2(a), 3(a), 4(a-d) show the ϕ dependences for resonances with $J^{PC} = 0^{-+}$ (the η'), with $J^{PC} = 2^{-+}$ (the $\eta_2(1645)$), with $J^{PC} = 1^{++}$ (the $f_1(1285)$), two with $J^{PC} = 0^{++}$ (the $f_0(1370)$ and $f_0(1500)$) and two with $J^{PC} = 2^{++}$ (the $f_2(1270)$ and $f_2(1950)$) respectively.

THEORETICAL STUDY ON ϕ DEPENDENCE

Several theoretical papers have been published on ϕ dependence [4,5]. All agree that the exchanged particle (it can be Pomeron) must have $J > 0$ and that $J = 1$ is the simplest explanation. Using $\gamma^*\gamma^*$ collisions as an analogy Close and Schuler have calculated the ϕ dependencies for the production of resonances with different J^{PC}s [5]. In their model for DPE the Pomeron acts as non-conserved vector current and the cross section of central production is expressed as follows:

$$\frac{d\sigma}{dt_1 dt_2 d\phi'} \sim G_E^{p\,2}(t_1) G_E^{p\,2}(t_2) F^2(t_1, t_2) A(t_1, t_2, \phi'),$$

where ϕ' is the angle between two pp scattering planes in the Pomeron-Pomeron centre of mass[2]; $G_E^p(t)$ is the proton-Pomeron form factor which is described using the Donnachie Landshoff formalism [6]; $A(t_1, t_2, \phi')$ is the prediction for the interaction of two Pomerons acting like non-conserved vector currents; $F^2(t_1, t_2)$ is the Pomeron-Pomeron-meson form factor which for collisions of two Pomerons with a given polarisation (transverse T or longitudinal L) is parametrized as: $e^{-(b_T t_i + b_L t_j)}$, $i, j = 1, 2$.

0^{-+} – the η'. According to Close-Schuler model

$$A(t_1, t_2, \phi') = t_1 t_2 \sin^2(\phi'), \quad F^2(t_1, t_2) = e^{-b_T(t_1+t_2)}.$$

Fig.1 (a) shows a good agreement of model prediction with experimental ϕ dependence. In order to describe the t dependence, fig.1 (b), one need $b_T = 0.5\ GeV^{-2}$.

1^{++} – the $f_1(1285)$. For 1^{++} states the model predicts that $J_z = \pm 1$ should dominate, which has been found to be correct [3], and then

$$A(t_1, t_2, \phi') = (\sqrt{t_1} - \sqrt{t_2})^2 + 4\sqrt{t_1 t_2} \sin^2(\phi'/2), \quad F^2(t_1, t_2) = e^{(-b_T t_i + b_L t_j)}.$$

---

[2] The WA102 collaboration measures the azimuthal angle ϕ in the pp c.m. frame. To compare the experimental result with model predictions the angle ϕ' is transformed to ϕ.

As can be seen from fig.3 (a) the ϕ distribution is in good agreement with prediction. Fixing b_T from the previous case we can determine b_L from the t distribution, fig.3 (b), to be $b_L = 3\ GeV^{-2}$. In addition we have a parameter free prediction of the variation of the ϕ distribution as a function of $|t_1 - t_2|$. Fig. 3 (c) and (d) show the output of the Monte Carlo superimposed on the ϕ for the $f_1(1285)$ for $|t_1 - t_2| \leq 0.2\ GeV^{-2}$ and $|t_1 - t_2| \geq 0.4\ GeV^{-2}$ respectively. The agreement between the data and the prediction is very good.

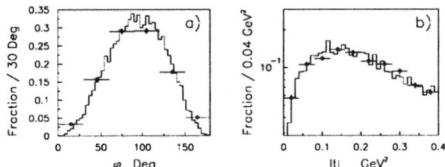

FIGURE 1. (a) The ϕ and (b) t distributions for the η' for the data (dots) and the model predictions from Monte-Carlo (histogram).

FIGURE 2. (a) The ϕ and (b) t distributions for the $\eta_2(1645)$ for the data (dots) and the Monte-Carlo (histogram).

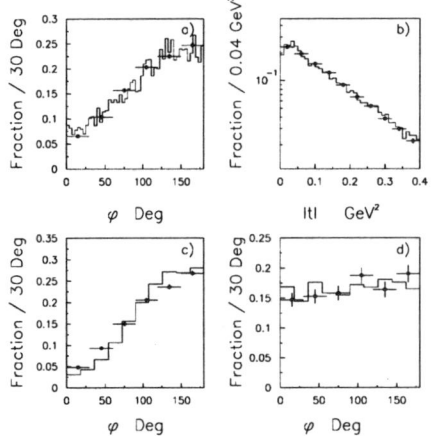

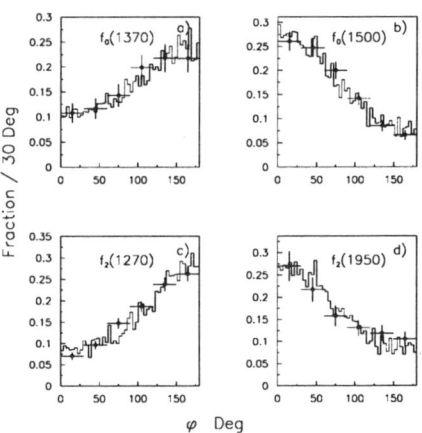

FIGURE 3. (a) The ϕ and (b) t distributions for the $f_1(1285)$ for the data (dots) and the Monte-Carlo (histogram). (c) and (d) the ϕ distributions for $|t_1 - t_2| \leq 0.2$ and $|t_1 - t_2| \geq 0.4$ GeV^2 respectively.

FIGURE 4. The ϕ distributions for the (a) $f_0(1370)$, (b) $f_0(1500)$, (c) $f_2(1270)$ and (d) $f_2(1950)$ for the data (dots) and the Monte-Carlo (histogram).

<u>2^{-+} – the $\eta_2(1645)$.</u> The 2^{-+} states are predicted to be produced polarized and helicity 2 suppressed. This has been found experimentally to be true [3]. In this case

$$A(t_1, t_2, \phi') = (\sqrt{t_1} - \sqrt{t_2})^2 + 4\sqrt{t_1 t_2}\cos^2(\phi'/2), \quad F^2(t_1, t_2) = e^{(-b_T t_i + b_L t_j)}.$$

Parameters b_T and b_L are fixed now. Fig.2 shows the remarkable agreement between the data and parameter free model predictions.

$J^{PC} = 0^{++}$ and 2^{++}. For 0^{++} and helicity zero amplitude of 2^{++} (which experimentally is found to dominate [3]) the model prediction is

$$F^2(t_1, t_2) A(t_1, t_2, \phi') = t_1 t_2 [e^{(-\frac{b_L(t_1+t_2)}{2})} + \frac{\sqrt{t_1 t_2}}{\mu^2} e^{(-\frac{b_T(t_1+t_2)}{2})} \cos(\phi')]^2.$$

The overall ϕ dependences for the $f_0(1370)$, $f_0(1500)$, $f_2(1270)$ and $f_2(1950)$ can be described by varying the quantity μ^2. Results are shown in fig.4. It is clear that these ϕ dependences discriminate two classes of resonances in the 0^{++} sector and also in the 2^{++}. The $f_0(1370)$ and $f_2(1270)$ can be described using $\mu^2 \approx -0.5$ GeV^2, for the $f_0(1500)$ and $f_2(1950)$ $\mu^2 = +0.7$ GeV^2. If this is the long sought discriminator between $q\bar{q}$ and non-$q\bar{q}$ states or glueballs is for future.

REFERENCES

1. T.A.Armstrong et al., *Nucl.Instrum.Methods* **A274**, 165 (1989); F.Antinori et al., *Nuovo Cimento* **A107**, 1857 (1994).
2. D.Robson, *Nucl.Phys.* **B130**, 328 (1977); F.E.Close, *Rep.Prog.Phys.* **51**, 833 (1988).
3. D.Barberis et al., *Phys.Lett.* **B413**, 217-224 (1997); *Phys.Lett.* **B413**, 225-231 (1997); *Phys.Lett.* **B422**, 399-404 (1998); *Phys.Lett.* **B427**, 398-402 (1998); *Phys.Lett.* **B432**, 436-442 (1998); *Phys.Lett.* **B436**, 204-210 (1998); *Phys.Lett.* **B440**, 225-232 (1998); *Phys.Lett.* **B446**, 342-348 (1999); *Phys.Lett.* **B453**, 305-315 (1999); *Phys.Lett.* **B453**, 316-324 (1999); *Phys.Lett.* **B453**, 325-332 (1999); *Phys.Lett.* **B462**, 462-470 (1999); *Phys.Lett.* **B467**, 165-170 (1999); *Phys.Lett.* **B471**, 429-434 (2000); *Phys.Lett.* **B471**, 435-439 (2000); *Phys.Lett.* **B471**, 440-448 (2000); *Phys.Lett.* **B474**, 423-426 (2000); *Phys.Lett.* **B479**, 59-66 (2000).
4. P.Castoldi et al., *Phys.Lett.* **B425**, 359 (1998); J.Ellis and D.Kharzeev, hep-ph/9811222; N.I.Kochelev, hep-ph/9902203; N.I.Kochelev, T.Morii and A.V.Vinnikov, hep-ph/9903279.
5. F.E.Close and G.Schuler, *Phys.Lett.* **B464**, 279 (1999); F.E.Close, A.Kirk and G.Schuler, *Phys.Lett.* **B477**, 13 (2000).
6. A.Donnachie and P.V.Landshoff, *Phys.Lett.* **B231**, 189 (1983).

Exotic meson production in the reaction $\pi^- p \to \eta' \pi^- p$ at 18 GeV/c

Neal M. Cason for the E852 Collaboration[1]

Department of Physics, University of Notre Dame, Notre Dame, IN 46556

Abstract.
The $\eta'\pi^-$ system has been studied in the reaction $\pi^-p \to \eta'\pi^-p$ at 18 GeV/c. A partial wave analysis of 6040 kinematically-identified events shows that the reaction is dominated by natural parity exchange. Production of an exotic isovector state $\pi_1(1600)$ is observed in the $I^G(J^{PC}) = 1^-(1^{-+})$ wave. The mass and width of the state are estimated via simultaneous mass-dependent fits of the $1^-(1^{-+})$ and $1^-(2^{++})$ waves. The $a_2^-(1320)$ and a wide structure at 1.8 GeV/c^2 are observed in the $1^-(2^{++})$ wave. An amplitude analysis in the high-mass region (above 1.8 GeV/c^2) which includes the $1^-(4^{++})$ wave is also presented.

Brookhaven E852 was designed to search for exotic mesons using the MPS detector with upgrades [1–5] for both triggering and for the detection of charged particles and photons. We have previously reported on the observation of $I^G(J^{PC}) = 1^-(1^{-+})$ exotic states in the $\eta\pi^-$ system at 1.4 GeV/c^2 (the $\pi_1(1400)$) [6,7] and in the $\rho^0\pi^-$ system (the $\pi_1(1600)$) [8] at 1.6 GeV/c^2. In this paper we report on the search for similar states in the $\eta'\pi^-$ system using data from the reaction $\pi^-p \to \eta'\pi^-p$ at 18 GeV/c. (Note that a P-wave resonance in the $\eta'\pi^-$ system has $I^G(J^{PC}) = 1^-(1^{-+})$ and would be manifestly exotic.)

Preliminary E852 results on the $\eta'\pi^-$ system have been published previously [9–12]. Here we report on data taken in 1995 using a trigger requiring a recoil proton, three forward charged tracks, and energy deposit in the lead glass detector. Of the 165 million triggers, a final data sample of 6,040 events satisfied all quality cuts and kinematic requirements for this analysis (see ref. [12]).

Shown in Fig. 1 are the 2γ, $\eta\pi^+\pi^-$ and $\eta'\pi^-$ effective mass distributions for the data. Clear evidence for the η is seen in the 2γ mass distribution. These photon pairs were kinematically constrained to be consistent with an η. The resulting $\eta\pi^+\pi^-$ mass distribution shows a clear η' signal. After constraining the $\eta\pi^+\pi^-$ system to be consistent with the η', the resulting $\eta'\pi^-$ mass distribution shown in the figure contains structure in the 1.6 GeV/c^2 region, the nature of which will

---

[1] The E852 collaborating institutions are: Brookhaven, Indiana, IHEP Protvino, Massachusetts Dartmouth, Moscow State, Northwestern, Notre Dame, and Rensselaer Polytechnic Institute.

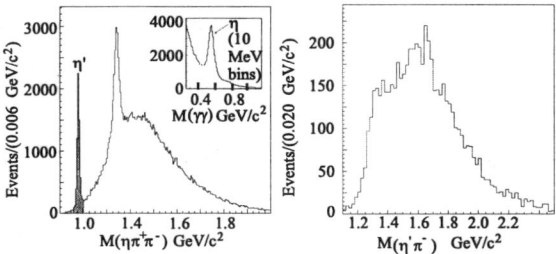

FIGURE 1. Distributions of the 2γ effective mass (inset), the $\eta\pi^+\pi^-$ effective mass (left) and the $\eta'\pi^-$ effective mass (right).

become clear from the following partial wave amplitude (PWA) analysis.

Amplitude analyses were carried out using the standard E852 PWA package (see ref. [7] for a complete description).[2] The waves used in the fit included, for natural-parity exchange ($\epsilon = +1$), P, D, and (in some cases) G waves; and for unnatural-parity exchange, ($\epsilon = -1$) S, P, and D waves. We also assume $|M| \leq 1$. (Note that for $\epsilon = +1$, $M = 0$ is forbidden.) In this paper, we denote the natural-parity waves as P_+, D_+, and G_+; and the unnatural-parity waves as S_0, P_0, and D_0 for the $M = 0$ waves and P_- and D_- for the $|M| = 1$ waves.

Shown in Fig. 2 are results of amplitude analyses of the data between 1.0 GeV/c^2 and 2.5 GeV/c^2. We have carried out analyses in 50 and 100 MeV/c^2 bins, as well as analyses with and without the G_+ amplitude. The G_+ amplitude is not needed below an $\eta'\pi^-$ mass of about 1.7 GeV/c^2, and is not shown below 1.7 GeV/c^2. Furthermore, the G-wave intensity is small so that we show its intensity and phase-difference plots in 100-MeV/c^2 bins as opposed to the binning of 50-MeV/c^2 shown for the other waves. The unnatural-parity exchange waves (not shown) are all quite small although they are required to obtain good fits to the data.

The main features of the amplitudes are: the strong production of the P_+ wave in Fig. 2a; structure in Fig. 2b in the $a_2(1320)$ region and at higher mass; and significant structure in the phase difference plots of Figs. 2d, e, and f.

Several preliminary mass-dependent analyses, using Breit-Wigner parameterizations of the amplitudes, were carried out in order to see if the amplitudes were consistent with resonance production. We have carried out fits in the low-mass region (below 1.8 GeV/c^2) fitting the P_+ and D_+ waves (intensities and phase differences) as well as fits extending above 2.0 GeV/c^2 including the G_+ wave and its phase difference with both the P_+ and D_+ waves. We find a satisfactory description of the data if we assume the presence of a single exotic P-wave resonance, a pair

---

[2] The PWA formalism defines the amplitudes in the reflectivity basis, so they are characterized by the naturality (ϵ) of the exchanged particle, the orbital angular momentum of the $\eta'\pi^-$ system (L), and the projection of the angular momentum L along the beam direction (M).

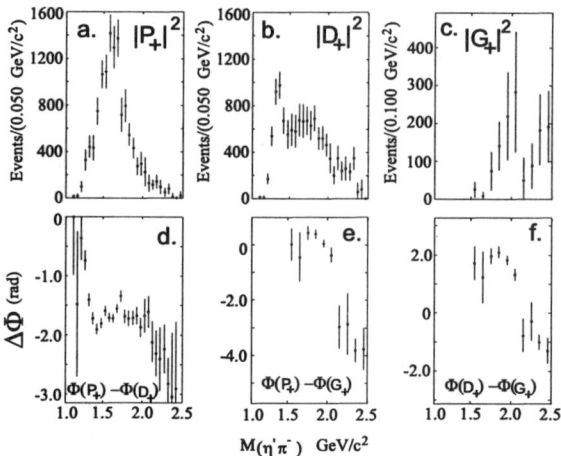

FIGURE 2. Natural-parity exchange waves from the mass-independent amplitude analysis.

of D-wave resonances, and a G-wave resonance. A preliminary fit leads to values for the mass and width of the 1^{-+} exotic state of 1.598 ± 0.009 GeV/c^2 and 0.380 ± 0.022 GeV/c^2 respectively. This is most likely the same state (the $\pi_1(1600)$) observed previously [8] decaying to $\rho\pi$. The fitted parameters of the lower-mass D-wave resonance are consistent with those of the $a_2(1320)$; and the parameters of the G-wave state (M = 2.03 ± 0.04 GeV/c^2 and $\Gamma = 0.32 \pm 0.14$ GeV/c^2) are consistent with production of the $a_4(2040)$.

Fits were carried out with two P-wave resonances. The fits did improve somewhat when the $\pi_1(1400)$ was included but since a satisfactory fit was attained without this second state, we can only say that our data is consistent with its production but the fit does not require it. The second D-wave resonance in the fits is very broad ($\sim$500-600 MeV/c^2 wide) and peaks around 1.8 GeV/c^2. Although this may be a resonant state, the very broad width suggests that it represents the production of a slowly varying non-resonant D-wave system.

Shown in Fig. 3 are the results of amplitude analyses obtained when the data is divided into a low-momentum-transfer sample and a high-momentum-transfer sample. It is interesting that the t-dependence for production of the $\pi_1(1600)$ state is different from that for $a_2(1320)$ production. This is clear because there is a significant decrease in the high-t $a_2(1320)$ production (compare the $a_2(1320)$ signals in Fig. 3c and d) whereas production of the $\pi_1(1600)$ is approximately the same in the two samples (compare the $\pi_1(1600)$ signals in Fig. 3a and b). Since both the $a_2(1320)$ and the $\pi_1(1600)$ are produced by natural-parity exchange, (presumably ρ- and Pomeron-exchange), one might speculate that the different valence structure

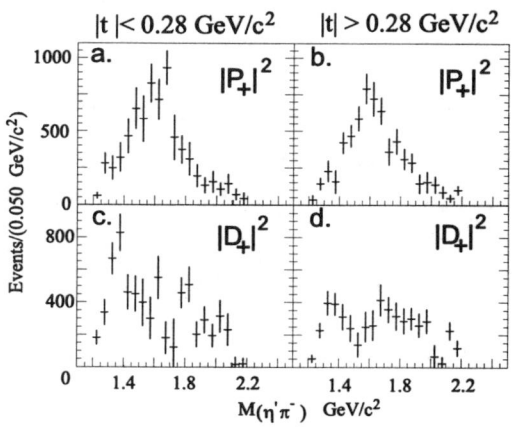

FIGURE 3. P_+ and D_+ intensity distributions for low- and high-$|t|$ production.

for the two states leads to significantly different couplings to the ρ and Pomeron.

In summary, we have observed an $I^G(J^{PC}) = 1^-(1^{-+})$ exotic meson, the $\pi_1(1600)$ decaying to $\eta'\pi^-$. Preliminary values of the mass and width are 1.598 ± 0.009 GeV/c^2 and 0.380 ± 0.022 GeV/c^2 respectively. The state is produced with a broader momentum-transfer distribution than that for the $a_2(1320)$.

REFERENCES

1. B. Brabson et al., Nucl. Instr. & Meth A **332**, 419 (1993).
2. R. R. Crittenden et al., Nucl. Instr. & Meth A, **387**, 377 (1997).
3. T. Adams et al., Nucl. Instr. & Meth A **386**, 617 (1996).
4. Z. Bar-Yam et al., Nucl. Instr. & Meth A **342**, 398 (1994).
5. S. Teige et al., Phys. Rev. **D59**, 012001 (1999).
6. D.R. Thompson et al., Phys. Rev. Lett. **79**, 1630 (1997).
7. S.U. Chung et al., Phys. Rev. D **60**, 092001 (1999).
8. G.S. Adams et al., Phys. Rev. Lett. **81**, 5760 (1998).
9. N.M. Cason et al., Proceedings of the VIth International Conference on Hadron Spectroscopy, Manchester, England, eds. M.C. Birse, G.D. Lafferty, and J.A. McGovern, World Scientific, Singapore, 55 (1996).
10. D. Stienike, "An analysis of the $\pi^-\eta'$ system produced in the reaction $\pi^-p \to \eta'\pi^-p$ at 18 GeV/c", Ph.D thesis, University of Notre Dame, Notre Dame, IN (1998).
11. D. Ryabchikov et al., Proceedings of the VIIth International Conference on Hadron Spectroscopy, Upton, N.Y., eds. S.U. Chung, H.J. Willutzki, AIP Conference Proceedings 432, 527 (1997).
12. E.R. Tatar, "A study of the $\eta'\pi^-$ system produced in the reaction $\pi^-p \to \eta'\pi^-p$ at 18 GeV/c", Ph.D thesis, University of Notre Dame, Notre Dame, IN (2000).

Search for the Tensor Glueball Candidate $\xi(2230)$ in an Antiproton-Proton Formation Experiment

Wilhelm Roethel, for the CB Collaboration

Department of Physics, Northwestern University, Evanston, IL 60208

Abstract. Results of a search for the tensor (2^{++}) glueball candidate $\xi(2230)$ are presented. A scan of the formation of $\xi(2230)$ in the mass region $\sqrt{s} = 2222.7 - 2239.7$ MeV/c^2 was made using the Crystal Barrel Detector at LEAR (CERN). $\pi^0\pi^0$ and $\eta\eta$ final states were investigated and no indication for the formation of ξ was found. Breit-Wigner fits, assuming a resonance width of 10-20 MeV/c^2 yield the 95% confidence level upper limits $B(\bar{p}p \to \xi) B(\xi \to \pi^0\pi^0) < 6 \cdot 10^{-5}$ and $B(\bar{p}p \to \xi) B(\xi \to \eta\eta) < 4 \cdot 10^{-5}$.

INTRODUCTION

In 1986 the Mark III collaboration reported evidence for a narrow resonance, named $\xi(2230)$, in the invariant mass spectrum of K^+K^- and $K_S K_S$ in the reactions $J/\Psi \to \gamma K^+K^-$ and $J/\Psi \to K_S K_S$ [1]. The resonance was found to have a mass around 2230 MeV/c^2 and a width of ≈ 20 MeV/c^2. This unusually narrow width and $J^{PC} = (\text{even})^{++}$ led to various interpretations of the $\xi(2230)$ as a state beyond the simple $q\bar{q}$ model. In particular, the glueball hypothesis is of interest since various theoretical models predict a mass of the 2^{++} glueball around 2200-2400 MeV/c^2 [2]. The DM2 Collaboration, however, repeated the measurement of Mark III and found no indication for a narrow $\xi(2230)$ [3]. Instead they quoted upper limits for the product branching ratios of $B(J/\Psi \to \gamma\xi) B(\xi \to K^+K^-)$ and $B(J/\Psi \to \gamma\xi) B(\xi \to K_S K_S)$ at 90% CL of 1/2 the reported values of Mark III. Other searches for the $\xi(2230)$ included peripheral production and $\bar{p}p$ annihilation. No indication for a narrow resonance was found in these experiments. It is worth noting that evidence for a somewhat broader resonance was reported in the reactions $\pi^- p \to n X$ (with $X = \eta\eta'$, $K_S K_S$) [4] and $K^- p \to \Lambda X$ (with $X = K^+K^-$, $K_S K_S$) [5]. In summary the nature of the $\xi(2230)$ remained in question.

In 1996 the BES Collaboration reported new evidence for the $\xi(2230)$ in their analysis of radiative J/Ψ decays [6]. Not only did they find a narrow signal in the K^+K^- and $K_S K_S$ spectra in agreement with the measurement of Mark III, but also claimed additional decay modes of $\xi \to \pi^+\pi^-$, $\pi^0\pi^0$ and $p\bar{p}$, all with product branching ratios in the range of $1.5 \cdot 10^{-5}$ - $5.6 \cdot 10^{-5}$. The nearly flavor blind decay of the $\xi(2230)$, a signature of a glueball, and the apparent coupling of ξ to $p\bar{p}$, again stimulated a search for the resonance in antiproton-proton annihilation.

EXPERIMENTAL SETUP AND DATA SELECTION

Using the Crystal Barrel Detector at LEAR (CERN), the region around the $\xi(2230)$ was scanned in 9 steps from 2222 MeV/c^2 to 2240 MeV/c^2. The Crystal Barrel Detector is a modern type detector allowing the full reconstruction of neutral and charged particles over a solid angle of almost 4π [7]. At each scan point an average of $6.5 \cdot 10^5$ events were taken with a trigger requiring no charged tracks in the final state and the energy sum of the 1380 CsI crystals in the calorimeter in an appropriate window.

$\pi^0\pi^0$, $\eta\eta$ and $\eta\pi^0$ events were selected using the following cuts: (a) no charged tracks in the Drift Chamber (b) $E_{tot} > 1.27 \cdot P_{tot} + m_p/2$, (c) exactly 4$\gamma$s and (d) the invariant mass of the two 2γ pairs within the π^0 or η region. Optionally, a kinematic fit on the selected final state was made. The selection provides a data sample of about 11,000 $\pi^0\pi^0$, 400 $\eta\eta$ and 3,500 $\eta\pi^0$ events at each energy point. The background was estimated by Monte Carlo simulation and was found to contribute less than 0.1% for $\pi^0\pi^0$ and $\eta\pi^0$, and $\approx$ 3% for $\eta\eta$.

The data were normalized directly to the number of antiprotons by monitoring the incoming $\bar{p}$ with beam counters, and recording the DAQ live-time. The differential cross section $d\sigma_X/d\Omega$ for a final state X, was then calculated from the number of selected events in the solid angle $d\Omega$, N_X, with

$$\frac{d\sigma_X}{d\Omega} = \frac{N_X \epsilon_X}{N_{\bar{p}}} \cdot N_{target},$$

where ϵ_X is the overall efficiency in the solid angle $d\Omega$, and $N_{target} = 5.33 \cdot 10^6$ μb, as obtained from the target geometry and the density of LH$_2$.

The integrated cross sections over the range of $\cos\theta = 0$ to 0.85 are shown in Fig. 1. The distributions are consistent with straight lines as the fit results in Fig. 1 demonstrate. No indication of an enhancement, indicative of a narrow resonance, can be found. Possible systematical effects of the normalization can be eliminated by using the ratios $\sigma(\pi^0\pi^0)/\sigma(\pi^0\eta)$ and $\sigma(\eta\eta)/\sigma(\pi^0\eta)$. Since $\eta\pi^0$ has isospin 1, a glueball (I=0) can not decay into this final state. The results are shown in the last two panels of Fig. 1. Again no signal of a narrow resonance can be found.

To obtain upper limits for the product branching ratios $B(\bar{p}p \to \xi)B(\xi \to \pi^0\pi^0)$ and $B(\bar{p}p \to \xi)B(\xi \to \eta\eta)$ the excitation function was fitted to a linear background + a Breit-Wigner

$$\sigma_X = (A + \sqrt{B}) \cdot \frac{5}{4} \left(\frac{4\pi(\hbar c)^2}{s - 4m_p^2}\right) \frac{4 B_{in} B_{out}}{1 + [2[\sqrt{s} - M_R]/\Gamma]^2},$$

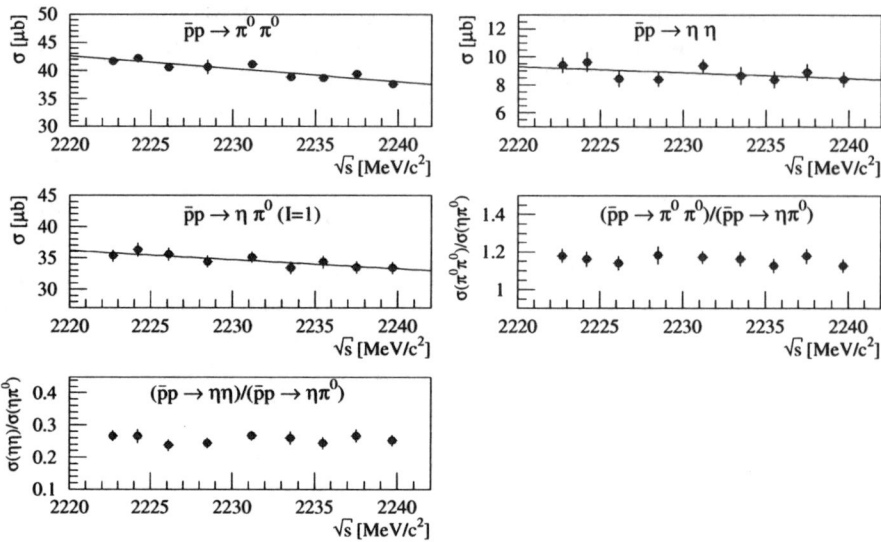

FIGURE 1. Integrated cross sections over the range $\cos\theta \leq 0.85$ for $\bar{p}p \to \pi^0\pi^0$, $\eta\eta$ and $\eta\pi^0$. The solid lines show the results of the straight line fits giving χ^2 values of 1.38, 0.69 and 0.34, respectively. The last two panels show the the ratios of $\sigma(\pi^0\pi^0)/\sigma(\pi^0\eta)$ and $\sigma(\eta\eta)/\sigma(\pi^0\eta)$. Again the distribution is consistent with a straight line.

keeping M_R and Γ fixed and using the results of the straight line fit above for the background parameterization. Figure 2 shows results for the upper limits for $B_{in} B_{out}$ at 95% confidence level, for assumed widths of 10 and 20 MeV/c^2 as a function of the resonance mass M_R. Values for the upper limits of

$$B(\bar{p}p \to \xi)B(\xi \to \pi^0\pi^0) < 6 \cdot 10^{-5}, \quad 95\% \text{ CL} \quad \text{and}$$
$$B(\bar{p}p \to \xi)B(\xi \to \eta\eta) < 4 \cdot 10^{-5}, \quad 95\% \text{ CL}.$$

are found. Combining these results with the product branching ratios reported by BES for $B(J/\Psi \to \gamma\xi) B(\xi \to p\bar{p})$ and $B(J/\Psi \to \gamma\xi) B(\xi \to \pi^0\pi^0)$[1] leads to a lower limit for the branching ratio of $B(J/\Psi \to \gamma\xi) > 0.29\%$, a comparatively large value, and upper limits for the decay of $\xi(2230)$ into $p\bar{p}$ and $\pi^0\pi^0$ of 0.51% and 1.2%. Using the other product branching ratios published by BES leads to upper limits of $\xi \to \pi^+\pi^-$, K^+K^- and $K_S K_S$ of 2.3%, 1.1% and 0.9%, respectively.

---

[1] For the BES branching ratio for $\pi^0\pi^0$ the weighted average of the product branching ratios $B(\pi^+\pi^-)$ and $B(\pi^0\pi^0)$ was used.

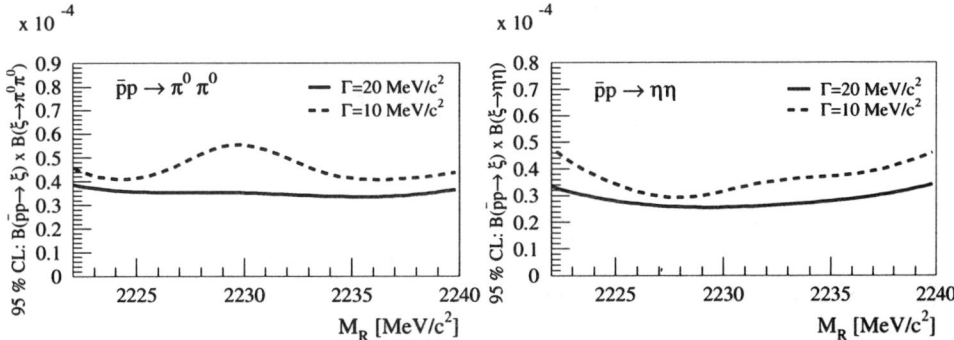

FIGURE 2. 95% confidence level upper limits for the product branching ratios $B(\bar{p}p \to \xi) B(\xi \to \pi^0\pi^0)$ and $B(\bar{p}p \to \xi) B(\xi \to \eta\eta)$ as a function of the resonance mass M_R with assumed widths of 10 and 20 MeV/c².

CONCLUSION

The result of the measurements $\bar{p}p \to \pi^0\pi^0$ and $\bar{p}p \to \eta\eta$ show no evidence for the existence of a narrow resonance in the mass region 2222.7 to 2239.7 MeV/c². 95% confidence levels of $6\cdot 10^{-5}$ and $4\cdot 10^{-5}$ for the formation of ξ in $\bar{p}p$ annihilation and its decay into $\pi^0\pi^0$ and $\eta\eta$, respectively, have been established. This leads to the conclusion that either the $\xi(2230)$ does not exist in the mass range covered by this measurement, or the coupling of ξ to $\bar{p}p$ is much weaker than the BES measurements seem to indicate.

REFERENCES

1. R. M. Baltrusaitis et al., Mark III Collaboration, Phys. Rev. Lett. **56** (1986) 107.
2. e.g. Flux Tube Model: N. Isgur and J. Paton, Phys. Rev. **D31** (1985) 2910.
   Bag Model: R. Jaffe and K. Johnson, Phys. Lett. **60B** (1976) 201.
   QCD Sum Rules: M. Shifman et al., Nucl. Phys. **B147** (1979) 385, 448.
   Recent predictions from lattice gauge calculations assign a mass of $\approx 2400 \pm 25 \pm 120$ MeV/c² for the 2^{++} glueball (C. Morningstar and M. Peardon, Phys. Rev. **D60** (1999) 034509).
3. J. E. Augustin et al., DM2 Collaboration, Phys. Rev. Lett. **60** (1988) 2238.
4. D. Alde et al., IHEP-IISN-LANL-LAPP Collaboration, Phys. Lett **177B** (1986) 120; B. Bolonkin et al., Nucl. Phys. **B309** (1998) 426.
5. D. Aston et al., Phys. Lett. **215** (1988) 199; Nucl. Phys. **301** (1988) 525;
6. J. Z. Bai et al., BES Collaboration, Phys. Rev. Lett. **76** (1996) 3502; Phys. Rev. Lett. **81** (1998) 1179.
7. E. Aker et al., Crystal Barrel Collaboration, Nucl. Instr. Meth. **A321** (1992) 69.

4π Decays of Scalar Mesons

Curtis A. Meyer*
for the Crystal Barrel Collaboration

U.C. Berkeley, Bochum, Bonn, Budapest, R.A.L., CERN, Hamburg, Karlsruhe, Queen Mary, U.C.L.A., München, LPNHE Paris VI,VII, Carnegie Mellon*, Strassburgh, Zúrich

Abstract. The Crystal Barrel Experiment has studied $\bar{p}N \to 5\pi$ reactions at rest to look for 4π decays of scalar mesons. Both the $f_0(1370)$ and the $f_0(1500)$ are seen to decay dominately to 4π final states.

The Crystal Barrel Experiment is a nearly 4π detector for both charged particles and photons [1]. It ran at the CERN Low Energy Antiproton Ring, (LEAR) from 1989 to 1996 studying antiproton annihilations on hydrogen and deuterium both at rest and in flight. The data reported here come from both $\bar{p}p$ and $\bar{p}d$ annihilations at rest on liquid targets. The analysis looked for events with 5πs in the final state with the specific goal of studying the 4π decays of scalar mesons via reaction 1.

$$\bar{p}N \to X\pi \to (\pi\pi\pi\pi)\pi \tag{1}$$

In this paper, we report on results from three different 5π final states, as follows, as well as the annihilation rates into these final states.

$$\bar{p}n \to \pi^- 4\pi^0 \quad \text{BR} = (0.67 \pm 0.10) \times 10^{-2} \tag{2}$$
$$\bar{p}n \to \pi^+ 2\pi^- 2\pi^0 \quad \text{BR} = (3.15 \pm 0.47) \times 10^{-2} \tag{3}$$
$$\bar{p}p \to 5\pi^0 \quad \text{BR} = (0.71 \pm 0.11) \times 10^{-2} \ [2]. \tag{4}$$

A partial wave analysis, (PWA), of these final states has been performed, and a set of solutions which are internally consistent between the different data sets have been found. The PWA performed assuming that the reactions can be described in terms of the isobar model where the transition to the final 5π state is described in terms of a series of two-body processes. The analysis allows for three different general decay chains:

$$\bar{p}N \to A_1\pi \to [A_2 A_3]\pi \to [(\pi\pi)(\pi\pi)]\pi$$
$$\bar{p}N \to B_1\pi \to [B_2\pi]\pi \to [(B_3\pi\pi)]\pi \to [(\pi\pi)\pi\pi]\pi$$
$$\bar{p}N \to C_1 C_2 \to [(C_3\pi)](\pi\pi) \to [(\pi\pi)\pi](\pi\pi)$$

There are three features which can be seen in the invariant mass plots shown in Figure 1. The first is a clear peak above phase space in the $4\pi^\circ$ invariant mass. This is found to be a combination of both the $f_0(1370)$ and the $f_0(1500)$ decaying into $4\pi^\circ$. The second is a small shoulder in the $\pi^-2\pi^\circ$ plot which is found to be caused by the $\pi(1300) \to \rho\pi$. The last is the $\rho^-(770)$ seen in the $\pi^-\pi^\circ$ invariant mass.

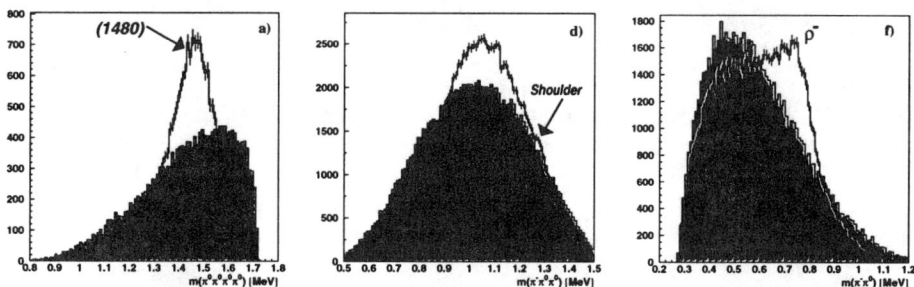

FIGURE 1. A comparison between the data (line) and phase space (solid) distributions for (right) $4\pi^\circ$, (middle) $\pi^-2\pi^\circ$ and (left) $\pi^-\pi^\circ$ invariant masses from the $\pi^-4\pi^\circ$ data set.

We start by fitting the $\pi^-4\pi^\circ$ data, which is simplest in terms of the 4π combinations. We then use the results from the fit to fit both the $5\pi^\circ$ and the $\pi^+2\pi^-2\pi^\circ$ data. We have found a set of results with are consistent with all three final states. A key element of the fits is the need for two f_0 states to explain the peak in the $4\pi^\circ$ mass spectrum. A single resonance fit is significantly worse than the two-resonance solution. In the 2 and 3, both states are seen decaying into two pairs of S-wave dipions, $(\pi\pi)_s(\pi\pi)_s$, which we will write as $\sigma\sigma$. These σ are not resonant states, but rather describe the S-wave $\pi\pi$ scattering amplitude and phase from threshold up to about $1\,\text{GeV}/c^2$ [4]. They are extremely important in describing these data, and both the $f_0(1370)$ and the $f_0(1500)$ are seen to decay into $\sigma\sigma$. In Figure 2 are shown the mass projections for the fits to reaction 2.

The projections do an excellent job with all projections except the $\pi^-\pi^\circ$, where there appears to be insufficient ρ^-. This can be fixed by incoherently protonium P-wave annihilation to any final state involving a ρ^- at about the 7% level. However, there is no sensitivity to the either the protonium initial state, or the final state — it just needs to have ρ^- in it. The components needed in the fit to $\pi^-4\pi^\circ$ are given in table 1. Using these results, it is possible to fit reaction 3 using the same parameters. The result is an excellent description as shown in Figure 3. Not only are the masses and widths of the f_0 states the same, but the relative production and decays are also the same.

With the addition of data from reaction 4, it is possible to also include $\rho\rho$, $a_1(1260)\pi$ and $\pi(1300)\pi$ final states of both the $f_0(1370)$ and the $f_0(1500)$. Note that while we checked for $\sigma\pi$ decays of both the $a_1(1260)$ and the $\pi(1300)$, the fit did not need any of these, the $\rho\pi$ decays were however quite significant. The results

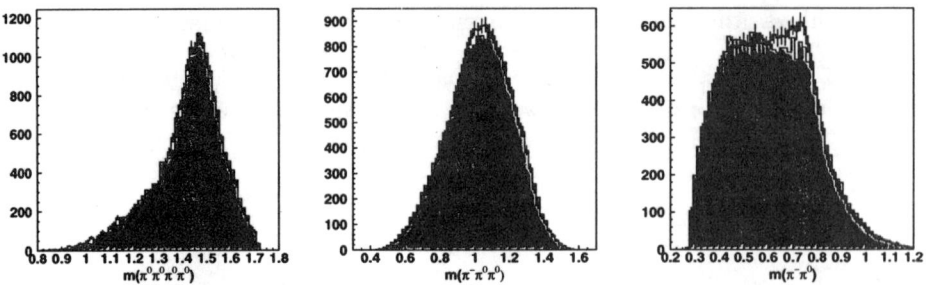

FIGURE 2. A comparison between the data (points) and the fit (solid) for several mass distributions. (right) $4\pi^\circ$, (middle) $\pi^-2\pi^\circ$ and (left) $\pi^-\pi^\circ$ invariant masses from the $\pi^-4\pi^\circ$ data set.

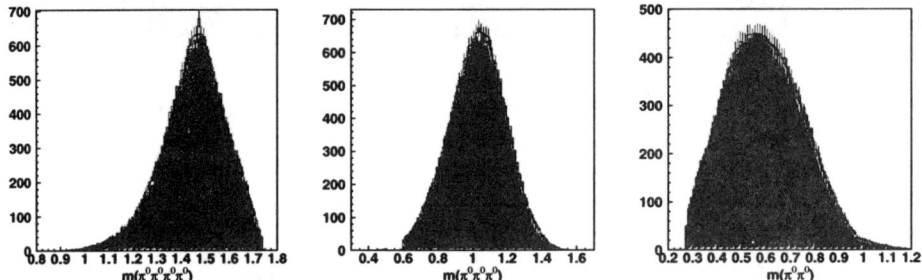

FIGURE 3. A comparison between the data (points) and the fit (solid) for several mass distributions. (right) $4\pi^\circ$, (middle) $3\pi^\circ$ and (left) $2\pi^\circ$ invariant masses from the $5\pi^\circ$ data set.

TABLE 1. The fit results for the $\pi^-4\pi^\circ$ data.

Amplitude	Parameters	Fraction
$f_0^{(1)}\pi, f_0 \to (\sigma\sigma)$	$m = 1395 \pm 40\,\text{MeV}/c^2$ $\Gamma = 275 \pm 55\,\text{MeV}/c^2$	$69.9 \pm 4\%$
$f_0^{(2)}\pi, f_0 \to (\sigma\sigma)$	$m = 1490 \pm 30\,\text{MeV}/c^2$ $\Gamma = 140 \pm 40\,\text{MeV}/c^2$	$9.7 \pm 0.7\%$
$\pi(1300)\sigma, \pi(1300) \to \rho\pi$	$m = 1375 \pm 40\,\text{MeV}/c^2$ $\Gamma = 268 \pm 50\,\text{MeV}/c^2$	$5.4 \pm 0.4\%$
$\rho(1450)\pi, \rho(1700)\pi$	$m = 1435\,\text{MeV}/c^{2\dagger}, \Gamma = 325\,\text{MeV}/c^{2\dagger}$	$3.6 \pm 0.5\%$
	$m = 1700\,\text{MeV}/c^{2\dagger}, \Gamma = 235\,\text{MeV}/c^{2\dagger}$	$9.7 \pm 1.0\%$

for the 4π decays of both the $f_0(1370)$ and the $f_0(1500)$ are given in table 2. It should be pointed out that these rates assume simple isospin scaling to account for the unmeasured final states. Still, it is clear that the 4π decay of the $f_0(1370)$ is the dominant decay of this particle, as seen in an earlier analyses [5]. In addition, the total of all 4π decays accounts for about $\frac{1}{2}$ of all $f_0(1500)$ decays. Similar results have been reported recently by WA102 [9], however they see no $\sigma\sigma$ decay for the

$f_0(1370)$ whereas the *sigmaσ* is critical in the description of these data.

TABLE 2. A summary of all measured decay rates of the $f_0(1370)$ and the $f_0(1500)$. What are reported are the product branching ratio for $\bar{p}p \to f_0\pi \to$ final state in units of 10^{-4}. The two-pseudoscalar rates are taken from [6], [7], [8].

	$\sigma\sigma$	$\rho\rho$	$\pi(1300)\pi$	$a_1(1260)\pi$	4π
$f_0(1370)$	93.2 ± 27.1	88.3 ± 33.8	8.0 ± 3.6	44.7 ± 17.7	234.3
$f_0(1500)$	12.2 ± 7.2	6.8 ± 5.1	9.5 ± 7.1	11.3 ± 7.5	39.7

	$\pi\pi$	$\eta\eta$	$\eta\eta'$	$K\bar{K}$	Σ
$f_0(1370)$	19.2 ± 7.2	0.4 ± 0.2		$7.0 \pm 1.6\text{--}18.8 \pm 4.0$	32.5
$f_0(1500)$	24.6	1.91 ± 0.24	1.61 ± 0.06	4.52 ± 0.36	32.6

In summary, we have looked at the 4π decays of the $f_0(1370)$ and the $f_0(1500)$. We find that for the $f_0(1370)$, these decays account for most of the decay width, while for the $f_0(1500)$, they represent about $\frac{1}{2}$ of the total decay widths.

REFERENCES

1. E. Aker, *et al.*, (The Crystal Barrel Collaboration), Nucl. Instrum. Meth. **A321**, 69, (1992).
2. A. Abele, *et al.*, (The Crystal Barrel Collaboration), Phys. Lett. **B380**, 453, (1996).
3. A. Abele, *et al.*, (The Crystal Barrel Collaboration), Submitted to Phys. Lett. B, (2000).
4. K. L. Au, D. Mornington and M. R. Pennington, Phys. Rev. **D35**, 1633, (1987).
5. C. Amsler, *et al.*, (The Crystal Barrel Collaboration), Phys Lett. **B322**, 431, (1994).
6. A. Abele, *et al.*, (The Crystal Barrel Collaboration), Phys. Lett. **B385**, 425, (1996).
7. A. Abele, *et al.*, (The Crystal Barrel Collaboration), Nucl. Phys. **A609**, 562, (1996).
8. A. Abele, *et al.*, (The Crystal Barrel Collaboration), Phys. Rev. **D57**, 3860, (1998).
9. D. Barberis, *et al.* (The WA102 Collbaortaion), Phys. Lett. **B471**, 440, (2000).

Studies of Coulomb Gauge QCD

Adam P. Szczepaniak* and Eric S. Swanson[†]

*Physics Department and Nuclear Theory Center
Indiana University, Bloomington, Indiana 47405-4202*
[†] *Department of Physics and Astronomy, University of Pittsburgh, Pittsburgh PA 15260 and Jefferson Lab, 12000 Jefferson Ave, Newport News, VA 23606.*

Due to freedom in choosing a gauge there exist many ways of defining the QCD Hamiltonian. The Coulomb gauge is particularly suitable if one wishes to have a probabilistic interpretation in terms of a Fock space decomposition of hadronic wave functions and to find a connection between QCD and quark model degrees of freedom [1]. In the Coulomb gauge all degrees of freedom, including the two component gluon field, are physical since the Coulomb law constraint is build into the Hamiltonian. This also leads to a nonabelian Coulomb potential, which is a direct interaction term between colour charge densities and, in the heavy quark limit, leads to a confining force [2]. Here we will discuss how the nonabelian Coulomb kernel exhibits confinement already at the mean field level. In the heavy quark limit residual interactions between heavy quarks and transverse gluons are spin dependent *i.e.* relativistic and can be calculated using the Foldy-Wouthuysen transformation. This makes the Coulomb gauge suitable for studying the nonrelativistic limit [3]. Finally it is possible to use standard mean field techniques to define quasiparticle excitations, which, as we discuss below, have similar properties to what is usually assumed about constituent quarks in the light quark sector.

There are many different confinement scenarios. For example it has often been assumed that the gluon propagator in a covariant, gauge should be more singular that $1/q^2$ for small momenta, $q^2 \to 0$. The $1/q^4$ behaviour in particular would lead to linear confinement at large distance between static colour sources. If on the other hand one thinks of confinement as suppressing propagation of colour nonsinglet objects then the gluon propagator ought to vanish as $q^2 \to 0$. This apparent conflict can be resolved if by appropriate gauge choice, *e.g.* the Coulomb gauge, the gluon field is split into physical components which should not be allowed to propagate and residual components which provide confinement and can be enhanced at low momentum. This is the scenario which we will explore here.

The pure gauge QCD Hamiltonian in Coulomb gauge may be written as $H = H_g + H_C$, where the terms are given by $H_g = \text{Tr}\int d^3x \left[\mathcal{J}^{-1}\mathbf{\Pi}\cdot\mathcal{J}\mathbf{\Pi} + \mathbf{B}\cdot\mathbf{B}\right]$ and $H_C = \frac{1}{2}g^2 \int d^3x d^3y \, \mathcal{J}^{-1}\rho^a(\mathbf{x})K^{ab}(\mathbf{x},\mathbf{y})\mathcal{J}\rho^b(\mathbf{y})$ The second term is the instantaneous nonabelian Coulomb interaction for QCD with the kernel K given by $K^{ab}(\mathbf{x},\mathbf{y}) = \langle \mathbf{x}, a|(\nabla\cdot\mathbf{D})^{-1}(-\nabla^2)(\nabla\cdot\mathbf{D})^{-1}|\mathbf{y}, b\rangle$. Our goal is to find a suitable basis (or Fock space) in which the gluon (and later quark) fields are expanded and in which the Schrodinger equation, $H\Psi = E\Psi$ for colour singlet hadrons can be reliably solved in terms of a few Fock space components. In order to proceed, the Hamiltonian has to be regularized to remove UV divergences. In the absence of bare quark masses the UV cutoff Λ is the only mass parameter in the theory and therefore should not be removed. Independence of physical observables on the cutoff parameter is achieved through dimensional transmutation by making the coupling constant and coefficient multiplying various counterterms Λ-dependent. To assure that the cutoff is local, *i.e.* modifies the Hamiltonian only at relative separations $O(1/\Lambda)$, the field products should be point-split. As long as observables to be studied with H involve momenta smaller then the cutoff, one can keep only a finite number of counterterms which are ordered by inverse powers of Λ. To $O(1/\Lambda^2)$ these are $c_{-2}(\Lambda)\Lambda^2 \int d^3x A^2/2$, $c_0(\Lambda) \int d^3x A\nabla^2 A/2$, $(Z^{-1}(\Lambda) - 1) \int d^3x \Pi^2/2$, and $(Z(\Lambda) - 1) \int d^3x \mathbf{B}^2/2$, where c_{-2}, c_0 and Z are dimensionless coefficients which should be determined by fitting to physical observables [4]. We have employed the quasiparticle basis to represent single gluon states built on top of the BCS vacuum which contains two-body correlations as viewed in the perturbative basis (defined as eigenstates of $H_0 = H(\alpha_s = 0)$). The strength of these two body correlations is determined by minimizing the Hamiltonian density in the BCS vacuum. This results in a set of coupled self-consistent equations [4]. One of them, the gap equation, determines the overlap between quasiparticle and bare fields and is given in terms of a single function of gluon momentum $\omega(q)$. The other equation gives the expectation value of the Coulomb kernel $\tilde{V}_c$ which to a good approximation can is given by an algebraic equation,

$$\tilde{V}_c(q) = \frac{4\pi\tilde{\alpha}_s(\Lambda)}{q^2\left(1 - \frac{\tilde{\alpha}_s(\Lambda)\beta_c}{4\pi}\log\left[\frac{\Lambda^2}{\omega^2(q)}\right]\right)} \qquad (1)$$

Two of the four parameters c_{-2}, c_0, Z and α_s are fixed by $\tilde{V}_c$. First, requiring that $\tilde{V}_c$ behaves as $(3/2)b/q^4$ *i.e.* confines as $q \to 0$ with a strength determined by the slope of the linear potential $b \sim 0.18$ GeV^2 second requiring that the potential is renormalization group invariant, *i.e.* $d\tilde{V}_c/d\Lambda = 0$. The first condition implies that $m^2(0) \equiv (\omega^2(q) - q^2)_{q=0} = 3\beta_c b/(8\pi) = 0.257$ GeV^2 and the second fixes α, $m^2(0) = \Lambda^2 \exp(-4\pi/\alpha(\Lambda)\beta_c)$, *i.e* it demands that $m(0)$ plays the role of Λ_{QCD}. The third renormalization condition is obtained by fixing $\omega(q)$ at some finite value of q. The wave function renormalization, Z enters the single gluon dispersion relation, $E(q)$ which determines the strength of the one-body quasiparticle operator. Due to confinement, however, the single gluon energy is IR singular and does not have a physical meaning. Alternatively one could use the ground state glueball mass to fix Z. Our renormalization conditions for the other three constants could also be replaced by conditions on truly physical observables (glueball masses) rather than on $\omega(q)$ which does not have a physical meaning. The condition for the potential is however physical since in mean-field approximation the same potential represents the energy of the static quark-antiquark source which is finite and can, for example be measured on the lattice. These lattice simulations [5] do lead to the

$-\alpha/r + br$ form for the potential, consistent with our $\tilde{V}$. This is to be contrasted with lattice computations of the Landau-gauge gluon propagator [6] which saturates at low momentum, $\sim 1/(q^\alpha + (M_g)^\alpha)$, $\alpha \sim 2$. The gluon mass is determined to be $M_g \sim 800$ MeV. This again illustrates the inconsistency of a confinement scenario based on the behaviour of the covariant gluon propagator with a scenario based on the Wilson loop or noncovariant separation of gluon degrees of freedom.

Even though transverse gluons (with momenta below the cutoff) do not propagate in our scenario, they can effect the spectrum through mixing between colour singlet $Q\bar{Q}$ bound states and colour singlet *hybrid*, $Q\bar{Q}g$ states induced, for example by the magnetic interaction $\psi^\dagger \vec{\alpha} \cdot \vec{A}\psi$. Furthermore the effects of transverse gluons with momenta larger then the cutoff are implicitly accounted for by the local counterterms. The short distance contribution from transverse gluons is for example responsible for the majority of the hyperfine interactions. In the quark model it is this term which is responsible for the π-ρ mass splitting [7]. In QCD however, one should be able to derive pion properties based on the (approximate) chiral symmetry of the underlying Hamiltonian. The main question is whether long distance interactions alone (through spontaneous chiral symmetry breaking) are responsible for the $\pi - \rho$ mass splitting. If this were the case the short distance nature of the spin-spin interaction and its pQCD origin would be invalid for light quarks.

To address this issue we go back to the full Hamiltonian (including the quark sector) and approximate the Coulomb kernel by a local binding interaction, $c_c(\Lambda)\psi^\dagger\psi\psi^\dagger\psi/\Lambda^2$, and add a $O(1/\Lambda^2)$ spin-dependent counterterm, $\propto \psi^\dagger\vec{\alpha}\psi \cdot \psi^\dagger\vec{\alpha}\psi$ [8]. In order for the central potential to reproduce the features of $\tilde{V}_c$ it is necessary to establish the relation between the bare quark mass, $m_0(\Lambda)$ and Λ. For a heavy $Q\bar{Q}$ system the Coulomb potential dominates and the average momenta of the quarks in the bound state are $O(\alpha(\Lambda)m_0(\Lambda))$. Since in the local approximation $\langle p \rangle \sim \Lambda$ in this case one should set $\Lambda \sim c_c(\Lambda)m_0(\Lambda)$. Similarly to reproduce the proper momentum distribution arising from a purely linear potential one should set $\Lambda \sim (bm_0(\Lambda))^{\frac{1}{3}} + \sqrt{b} \sim (\Lambda_{QCD}^2 m_0(\Lambda))^{1/3} + O(\Lambda_{QCD})$. In the quark model the hyperfine splitting is determined by $\alpha/m^2 \langle p^3 \rangle$ where α/m^2 comes from the one-gluon-exchange potential and $\langle p^3 \rangle$ is from the wave function. This leads to a $\alpha^4 m$ behaviour of the vector-pseudoscalar mass splitting for Coulombic bound states and b/m for linearly confined systems. In our approach the constituent mass m is determined from the gap equation, however, for heavy quarks the difference between m and m_0 is negligible. In Fig. 1 we show the vector-pseudoscalar mass splitting as a function of m_0. The strengths of $c_c(\Lambda)$ and $c_h(\Lambda)$ are fixed to give $m_0(\Lambda = \sqrt{b}) \sim 5$ MeV and to reproduce the π and ρ meson masses. As $m_0(\Lambda)$ increases c_c and c_h are held fixed and Λ increases as $\Lambda \sim c_c(\Lambda)m_0(\Lambda) + (\Lambda_{QCD}^2 m_0(\Lambda))^{1/3} + O(\Lambda_{QCD})$ for the Coulomb + linear case and as $\Lambda \sim (\Lambda_{QCD}^2 m_0(\Lambda))^{1/3} + O(\Lambda_{QCD})$ for the confined case. The two short-dashed lines show the evolution of the pseudoscalar and vector masses. The solid line is the splitting for the Coulomb + linear case and the long-dashed line is the splitting for the linear case. In the former the linear growth with m_0 is recovered; in the

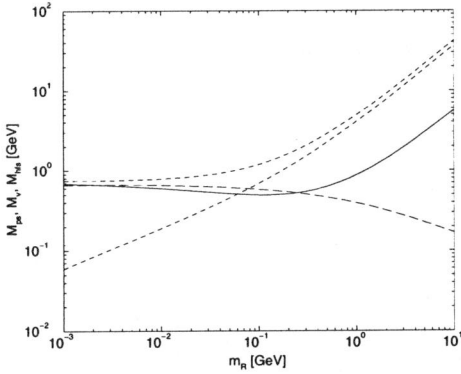

FIGURE 1. Unflavoured pseudoscalar and vector meson masses and hyperfine splittings as a function of the renormalized quark mass.

confined case the splitting turns out to be proportional to $\Lambda^2/m \sim m_0^{1/3}$. This is to be contrasted with the $\Lambda^3/m_0^2 \sim 1/m_0$ behaviour of the quark model. The difference is due to additional spin-dependent forces (coming from mixing with negative energy states) which are the ones responsible for the chirally symmetric behaviour of the pseudoscalar mass, $m_\pi^2 \sim m_0$ for small quark masses. In the light quark limit the $\rho - \pi$ mass difference turns out be predominantly driven by the hyperfine interaction ($\sim 80\%$), and as a function of the constituent mass (which in the light sector saturates at ~ 380 MeV does indeed follow the $1/m$ behaviour of the quark model hyperfine splitting. As a function of the bare mass however, the dependence changes for m_0 below ~ 500 MeV as mixing with negative energy states increases due to chiral symmetry. These findings confirm that effective constituent quarks with residual hyperfine interactions inversely proportional to the constituent quark mass are a good approximation to QCD dynamics in the light quark sector.

The authors would like to thank Nathan Isgur and Anthony Thomas for discussions. This work was supported in part by US Department of Energy grants under contracts DE-FG02-96ER40944, DE-AC05-84ER40150 (ES), and DE-FG02-87ER40365 (AS).

REFERENCES

1. Szczepaniak A.P., and Swanson E.S., *Phys. Rev.* **D55**, 1578 (1997).
2. Szczepaniak A.P., and Swanson E.S., *Phys. Rev.* **D55**, 3987 (1997).
3. Feinberg F.L., *Phys. Rev.* **D17**, 2659 (1978).
4. Szczepaniak A.P, and Swanson E.S., hep-ph/0005083.
5. Juge K.J., Kuti J., and Morningstar C.J., *Nucl. Phys. Proc. Suppl.* **63**, 326 (1998).
6. Leinweber D.B. *et al.*, *Phys. Rev.* **D58**, 031501 (1998).
7. Godfrey S., and Isgur N., *Phys. Rev.* **D32**, 189 (1985).
8. Szczepaniak A.P., Swanson E.S, hep-ph/0006306.

Relativistic Heavy Ions

Pb-Pb at 158GeV Results from a Theorist's View

Sangyong Jeon

Nuclear Science Division
Lawrence Berkeley National Laboratory

Abstract. The objective of this talk is to give a short critical review of Pb-Pb at 158 GeV results obtained by CERN SPS. The emphasis are given to the strangeness enhancement and the anomalous J/ψ suppression. I argue that the results obtained so far does not support a firm conclusion.

INTRODUCTION

As the CERN heavy ion program is closing down and the RHIC is starting up, there is no doubt that assessing the results of the CERN program is of great importance. The big question is whether or not we witnessed the creation of the Quark-Gluon Plasma (QGP) at CERN. The answer given by the CERN can be broadly categorized in two ways.

1. The energy density of about $3\,\text{GeV}/\text{fm}^3$ and the temperature higher than about 170 MeV are achieved. Therefore there must have been a phase transition.

2. There are features of the data which cannot be explained by the conventional models. Therefore, something new must have been created.

Both of these evidences are indirect and subject to large uncertainties. The estimates of energy density and the temperature depend very much on the assumptions such as the validity of the Bjorken model and the validity of the ideal hydrodynamics. It would be nice if one can make a measurement free of model assumptions. However, this amounts to solving the full QCD for two colliding relativistic heavy ions. Unfortunately, current status of lattice gauge theory doesn't allow us to do that as yet. Hence, the estimates of energy density and the temperature has to be taken with caution.

As for features that cannot be explained by conventional models, the difficulty is again the similar: Unless we solve the full QCD, we have to rule out *all* possible hadronic scenarios. In other words, the theoretical uncertainties are still large.

Nevertheless, one shouldn't be too pessimistic. There can be consequences of QGP that are dramatically different from any conceivable hadronic scenario. The CERN announcement particularly emphasized the direct photon measurement from WA98, the dilepton measurement from CERES, the strangeness enhancement from WA97 and the J/ψ suppression from NA50. In this talk, I'll mainly discuss the latter two.

STRANGENESS ENHANCEMENT

The strangeness enhancement as a signal of QGP was suggested some time ago [2] on the basis of the thermodynamics of free quarks and gluons. Today, we also have lattice calculations indicating that at temperature above 170 MeV, the entropy due to the strange quarks is roughly the same as the entropy due to a light quark species [3]. If such strangeness abundance is reflected in the final particle spectrum, this counts as a factor of about 4 enhancement over pp or e^+e^- results. A smaller factor of 2 enhancement in overall strangeness production is indeed observed by various CERN experiments such as NA49 and WA97.

The fact that there is an enhancement alone does not constitute as a proof of QGP. In many purely hadronic models such as RQMD [4] rescattering among secondaries can reproduce the overall strangeness enhancement factor of 2. What is hard for all these models is to reproduce the Ξ and Ω enhancement seen by WA97 [5]. The Ω particles actually show a factor of 10 enhancement over the extrapolated pA result.

However, there are problems. The experimental acceptance is limited to high $p_T (> 700$ MeV) and mid-rapidity. Hence, strictly speaking, the dramatic enhancement depend critically on how one extrapolates to lower p_T and all rapidity. In view of this fact, it is not inconceivable that purely hadronic processes such as the Baryon Junction [6] or a non-equilibrium production mechanism [7] are behind this enhancement. Basically, what one needs are a modest overall enhancement due to secondary scatterings and a modest enhancement in baryon stopping.

J/ψ SUPPRESSION

The original argument of the suppressed J/ψ production in the presence of QGP goes back to Satz and Matusi [8]. They argued that that any suppression of J/ψ in nuclear collisions is the signal of QGP formation. However, the observation of suppression in pA collisions made the modification of this original idea unavoidable. The new ideas are most clearly stated in Ref. [9]. The authors found that to fit the pA data the nuclear absorption cross-section should be twice the value expected from the photo-production of J/ψ. The larger cross-section can be explained if one assumes that the pre-resonance $c\bar{c}$ state dominates at the initial stage. The claim here is then that Glauber model calculation of J/ψ suppression with 7 mb nuclear absorption cross-section is not enough to explain the suppression observed by NA50.

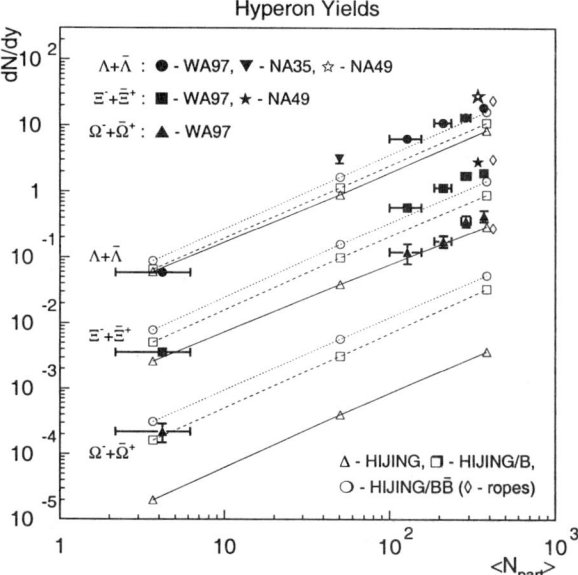

FIGURE 1. Hyperon yield calculation from Ref. [6] shown here with data from the NA35, the NA49 and the WA97 collaborations.

The observation of J/ψ suppression by NA50 has generated a lot of activities and many alternative scenarios have been suggested. In the analysis presented in Ref. [9], the authors ignored two major effects: Energy loss of the projectiles and the nuclear absorption by secondary particles (comovers). Most of the alternative explanations one way or another exploit these two effects. There are too many papers to mention all in this limited space. Let me just mention the works of two groups. The energy loss effect of the projectile was comprehensively studied in Ref. [10] using LEXUS and without comovers. There we found that the J/ψ yields can be explained up to S-U collisions but we cannot explain not Pb-Pb results without comovers. Perhaps the most comprehensive cascade calculation including the effect of comovers has been carried out in Ref. [11] using UrQMD. The conclusion of that paper is that hadronic scenario is certainly not ruled out.

CONCLUSION

It is of course impossible to summarize decades of theoretical and experimental work given the limitation of a short talk. The basic message is this: Before claiming the discovery of a 'new physics', the 'old physics' has to be thoroughly investigated and found wanting. Unfortunately, we are not yet at that stage with the two most prominent signals of QGP – the strangeness enhancement and J/ψ suppression. Certainly, the formation of QGP at CERN SPS is not ruled out, but firm supporting evidences are still lacking.

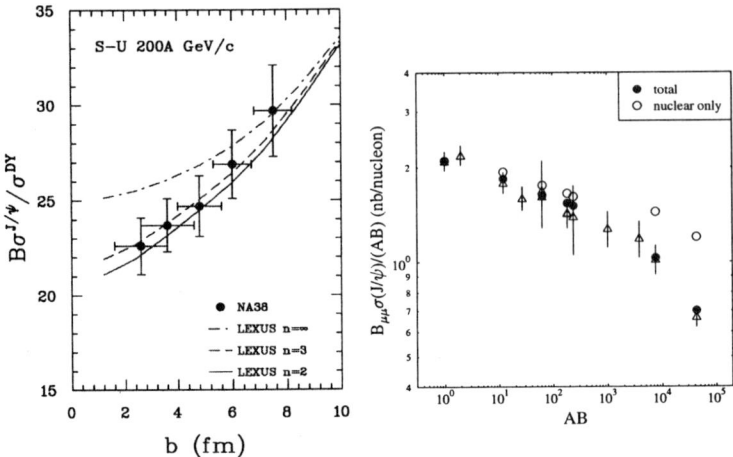

FIGURE 2. The left figure taken from Ref. [10] shows the effect of energy loss and coherence effect. The nuclear absorption cross-section is set to 3.6mb. The right figure taken from [11] is UrQMD calculation of the J/ψ cross-section in heavy ion collisions together with data.

REFERENCES

1. D. Bucher [WA98 Collaboration], Nucl. Phys. **A661**, 510 (1999). Proceedings, QM99.
2. J. Rafelski and B. Muller, Phys. Rev. Lett. **48**, 1066 (1982).
3. S. Gottlieb et al., Phys. Rev. **D55**, 6852 (1997) [hep-lat/9612020].
4. F. Wang, H. Liu, H. Sorge, N. Xu and J. Yang, heavy-ion collisions," nucl-th/9909001.
5. F. Antinori et al. [WA97 Collaboration], Eur. Phys. J. **C11**, 79 (1999).
6. S. E. Vance and M. Gyulassy, Phys. Rev. Lett. **83**, 1735 (1999) [nucl-th/9901009]. See also the contribution in this proceedings.
7. A. Capella and C. A. Salgado, model," Phys. Rev. **C60**, 054906 (1999) [hep-ph/9903414].
8. T. Matsui and H. Satz, Phys. Lett. **B178**, 416 (1986).
9. D. Kharzeev, C. Lourenco, M. Nardi and H. Satz, collisions," Z. Phys. **C74**, 307 (1997) [hep-ph/9612217].
10. C. Gale, S. Jeon and J. Kapusta, relativistic heavy ion collisions," hep-ph/9912213.
11. C. Spieles, R. Vogt, L. Gerland, S. A. Bass, M. Bleicher, H. Stocker and W. Greiner, Phys. Rev. **C60**, 054901 (1999) [hep-ph/9902337].

Searching for QGP: the J/ψ probe in the NA50/CERN experiment

Presented by Sérgio Ramos[6,b]

NA50 Collaboration

M.C. Abreu[6,a], B. Alessandro[10], C. Alexa[3], R. Arnaldi[10], M. Atayan[12], C. Baglin[1], A. Baldit[2], M. Bedjidian[11], S. Beolè[10], V. Boldea[3], P. Bordalo[6,b], A. Bussière[1], L. Capelli[11], L. Casagrande[6,c], J. Castor[2], T. Chambon[2], B. Chaurand[9], I. Chevrot[2], B. Cheynis[11], E. Chiavassa[10], C. Cicalò[4], T. Claudino[6], M.P. Comets[8], N. Constans[9], S. Constantinescu[3], N. De Marco[10], A. De Falco[4], G. Dellacasa[10,d], A. Devaux[2], S. Dita[3], O. Drapier[11], L. Ducroux[11], B. Espagnon[2], J. Fargeix[2], P. Force[2], M. Gallio[10], Y.K. Gavrilov[7], C. Gerschel[8], P. Giubellino[10], M.B. Golubeva[7], M. Gonin[9], A.A. Grigorian[12], J.Y. Grossiord[11], F.F. Guber[7], A. Guichard[11], H. Gulkanyan[12], R. Hakobyan[12], R. Haroutunian[11], M. Idzik[10,e], D. Jouan[8], T.L. Karavitcheva[7], L. Kluberg[9], A.B. Kurepin[7], Y. Le Bornec[8], C. Lourenço[5], P. Macciotta[4], M. Mac Cormick[8], A. Marzari-Chiesa[10], M. Masera[10], A. Masoni[4], M. Mehrabyan[12], M. Monteno[10], A. Musso[10], P. Petiau[9], A. Piccotti[10], J.R. Pizzi[11], F. Prino[10], G. Puddu[4], C. Quintans[6], S. Ramos[6,b], L. Ramello[10,d], P. Rato Mendes[6], L. Riccati[10], A. Romana[9], I. Ropotar[5], P. Saturnini[2], E. Scomparin[10] S. Serci[4], R. Shahoyan[6,f], S. Silva[6], M. Sitta[10,d], C. Soave[10], P. Sonderegger[5,b], X. Tarrago[8], N.S. Topilskaya[7], G.L. Usai[4], E. Vercellin[10], L. Villatte[8], N. Willis[8].

[1] LAPP, CNRS-IN2P3, Annecy-le-Vieux, France. [2] LPC, Univ. Blaise Pascal and CNRS-IN2P3, Aubière, France. [3] IFA, Bucharest, Romania. [4] Università di Cagliari/INFN, Cagliari, Italy. [5] CERN, Geneva, Switzerland. [6] LIP, Lisbon, Portugal. [7] INR, Moscow, Russia. [8] IPN, Univ. de Paris-Sud and CNRS-IN2P3, Orsay, France. [9] LPNHE, Ecole Polytechnique and CNRS-IN2P3, Palaiseau, France. [10] Università di Torino/INFN, Torino, Italy. [11] IPN, Univ. Claude Bernard Lyon-I and CNRS-IN2P3, Villeurbanne, France. [12] YerPhI, Yerevan, Armenia.

a) also at UCEH, Universidade de Algarve, Faro, Portugal; b) also at IST, Universidade Técnica de Lisboa, Lisbon, Portugal; c) now at CERN; d) Università del Piemonte Orientale, Alessandria and INFN-Torino, Italy; e) now at Fac. Physics and Nuclear Techniques, Univ. Mining and Metallurgy, Cracow, Poland; f) on leave of absence of YerPhI, Yerevan, Armenia.

Abstract. The J/ψ production in 158 A GeV Pb-Pb interactions is studied as a function of centrality. Its pattern, after a first sharp variation around E_T= 40 GeV, continues to fall down and exhibits a curvature change around E_T= 90 GeV. This trend excludes any conventional hadronic model and is in agreement with a deconfined quark-gluon phase scenario.

CP549, *Intersections of Particle and Nuclear Physics: 7th Conference*, edited by Z. Parsa and W. J. Marciano
© 2000 American Institute of Physics 1-56396-978-5/00/$17.00

INTRODUCTION

A phase transition from ordinary hadronic matter to a new state of deconfined quarks and gluons is predicted for temperatures above $T_c \sim 150$ MeV by non-perturbative Quantum ChromoDynamics [1] and is thus expected to be achieved in heavy ion collisions at the CERN-SPS. In this context, experiment NA50 searches for a specific signal for deconfinement, the predicted suppression of charmonia production, the $c\bar{c}$ bound states being prevented to be formed by the colour screening potential in the very dense medium [2].

I EXPERIMENTAL SETUP AND DATA SELECTION

The NA50 apparatus mainly consists of a muon spectrometer, a segmented active target and three independent centrality detectors: an electromagnetic calorimeter which measures the neutral transverse energy (E_T) produced in the interaction, a zero degree calorimeter measuring the very forward hadronic energy (E_{ZDC}) of the spectator nucleons and a silicon strip multiplicity detector [3].

Data taking conditions in all periods were the same, apart from the total thickness of the target and the number of its sub-targets. The segmented lead target was made, in 1996, of 7 sub-targets with a total thickness of 12 mm corresponding to 30 % interaction length; due to its low efficiency to fight re-interactions in Pb-Pb central collisions, its total thickness was reduced to 3 mm (7 % λ_I) in 1998, with only one sub-target (thus improving detectors' resolution).

The active target algorithm being inefficient for peripheral collisions, a contour-cut selection (correlation $E_T - E_{ZDC}$) [4] was applied to the huge 1996 statistics in order to recuperate low E_T events. But, because of the re-interactions problem at high E_T in 1996, the 1998 run was devoted to the detailed study of the suppression pattern in central collisions, and as the in-target over off-target events ratio was much less favorable, no attempt was made to recover the lost low E_T events (thus preventing any contamination of the final sample by Pb-air events).

The kinematical domain used, $2.92 \leq y_{lab} \leq 3.92$ and $|\cos \theta_{CS}| < 0.5$ leads, in the mass region of interest, to acceptances of the order of 15%. The muon pairs selected for the analysis satisfy the standard NA50 criteria [3].

II ANALYSIS AND RESULTS

The physical ingredients taken into account, besides J/ψ and ψ' resonances and the Drell-Yan mechanism, are the background and the open charm ($D\bar{D}$) contributions. The background is mainly due to uncorrelated π and K decays into muons. It is computed from the like-sign mass distributions using $N_{BG} = 2\sqrt{N^{++}N^{--}}$.

The opposite sign muon pair invariant mass distribution (Fig. 1) is fitted according to the following procedure in order to determine the amounts of its different components. The shapes of the muon pairs originating from J/ψ and ψ' decays and from the Drell-Yan process are obtained from a simulation of the NA50 detector using the same reconstruction and selection criteria as used for the real data. Drell-Yan contribution is calculated at the leading order and uses the MRSA structure functions, which take into account the $\bar{u}/\bar{d}$ asymmetry as measured by NA51 experiment [5]. The $D\bar{D}$ shape is taken from Pythia and its normalisation is obtained by fitting the data in the range $2.2 < M_{\mu\mu} < 2.9$ GeV/c^2. Finally, the amplitudes of J/ψ, ψ' and Drell-Yan contributions are obtained from a fit to the experimental data above $M_{\mu\mu} > 2.9$ GeV/c^2.

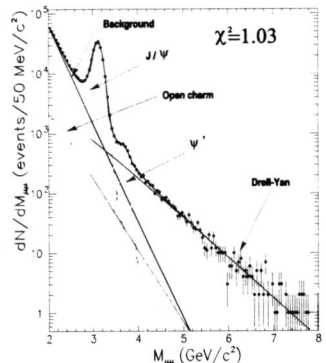

FIGURE 1. Fit to the mass spectrum

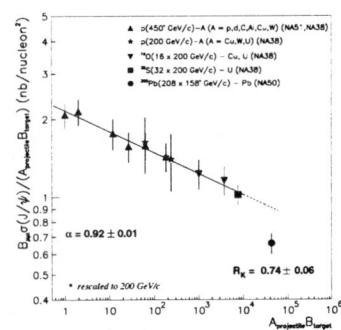

FIGURE 2. ψ/AB as a function of $A_{\text{proj}} \cdot B_{\text{tgt}}$

A systematic study of Drell-Yan behaviour, ranging from p-p and some p-A systems to S-U and Pb-Pb, proves that its cross-section behaves normally and is proportional to the number of elementary nucleon-nucleon collisions (i.e., the product A·B) [6]. Thus, it can be used as a reference in the study of J/ψ production.

Fig. 2 shows the J/ψ cross-section per nucleon-nucleon collision as a function of $A_{\text{proj}} \cdot B_{\text{tgt}}$: p-A collisions at 450 GeV/c (rescaled to 200 GeV/c), and proton, oxygen and sulphur induced reactions at 200 GeV/c. The common trend observed from p-p to S-U allows a single parametrization; using a power law, one obtains:

$$\sigma^\psi \propto (A\,B)^\alpha \quad \Rightarrow \quad \alpha = 0.92 \pm 0.01 \quad .$$

The Pb-Pb point (which error bar includes a 7 % systematic error contribution) is several standard deviations under the expected value. One concludes that J/ψ is anomalously suppressed in lead-lead, as compared to other reactions.

The J/ψ cross-section per nucleon-nucleon collision can also be evaluated from the ratio of the J/ψ to Drell-Yan cross-sections. This ratio is almost free of systematic errors, which are common to both samples (only 1.5% left). Fig. 3 shows it, for a large range of systems (from p-p and several p-A to S-U and Pb-Pb), as a function L, the path length of the pre-resonant $c\bar{c}g$ state in nuclear matter (which allows a common description of the very different interacting systems) [4].

A simple exponential parametrization applied to the lighter systems (previous NA38 [7–9] and NA51 [5] data ranging from p-p to S-U): $B_{\mu\mu}\sigma^\psi/\sigma^{DY} \propto e^{-\rho L \sigma_{abs}}$

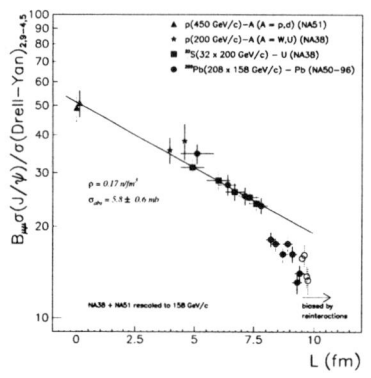
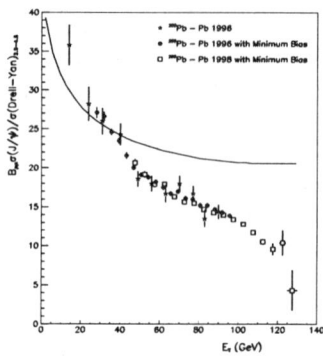

FIGURE 3. ψ/DY as a function of L

FIGURE 4. ψ/DY as a function of E_T

gives an absorption value of the $c\bar{c}g$ state in nuclear matter of $\sigma_{abs} = 5.8 \pm 0.6$ mb (full calculation gives 6.4 ± 0.5 mb). Whereas the more peripheral Pb-Pb points lie on the absorption curve, the more central ones show a sudden 20% drop at L = 8 fm, suggesting the onset of another J/ψ suppression mechanism.

The ratio of J/ψ to Drell-Yan cross-sections has also been studied as a function of E_T, for the different data taking periods, in Pb-Pb collisions. Two methods are used. Drell-Yan (and J/ψ) is extracted from the standard fit procedure described above, or it may be evaluated from the very large minimum bias sample (thus reducing significantly statistical errors) obtained with the independent ZDC calorimeter trigger, by a procedure described in [10].

The 1996 minimum bias analysis confirms the results of the standard analysis. They are shown in Fig. 4 in the domain where re-interactions are negligible. The 1998 data, showed here through the minimum bias method, are in agreement with 1996 analyses and reaches higher E_T values. The solid line presented in the figure corresponds to the ordinary nuclear absorption fit to lighter systems data (from p-p to S-U) using $\sigma_{abs} = 6.4$ mb. The overall displayed distribution exhibits, besides a 20% drop at 40 GeV (the one showed at L = 8 fm), an inflexion point at 90 GeV followed by a steady steep decrease.

III DISCUSSION AND CONCLUSION

The J/ψ pattern as a function of centrality is showed again in Fig. 5 together with conventional hadronic models (see [11] and Refs. therein) which assume charmonia absorption by comoving hadrons. It clearly rules out any hadronic model, as all them show a smooth decrease, with shapes similar to the absorption curve fitted to our data, without any drops or inflexion points. On the other hand, our data agree with the trend expected in the framework of quark-gluon plasma formation.

In order to further describe the J/ψ suppression pattern and in view of putting together all available systems, from p-p to Pb-Pb, we studied it as a function of the energy density ϵ, estimated by the Bjorken's model (for p-A collisions the RQMD cascade model [12] calculation was used), corrected for the rapidity interval of our E_T measurement ($1.1 < y < 2.3$). The result is shown in Fig. 6 (see also [11]).

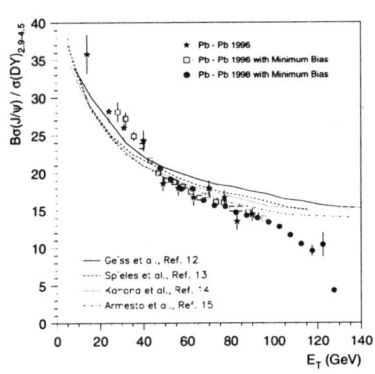

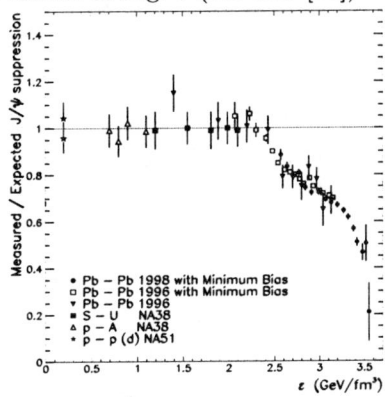

FIGURE 5. ψ/DY as a function of E_T: Comparision with hadronic models

FIGURE 6. ψ suppression as a function of ϵ

The data represent the measured J/ψ to Drell-Yan cross-section ratios (for p-A the direct J/ψ cross-sections per nucleon-nucleon collision were used) normalized to our fitted absorption curve to lighter systems [7–9].

The stepwise suppression pattern shows a sudden drop at 2.3 GeV/fm^3 and an inflexion point at 3.1 GeV/fm^3 followed by a steady steepy decrease. It rules out any conventional hadronic model and can be explained in the framework of quark-gluon plasma formation as successive melting of χ, then ψ, charmonium states.

REFERENCES

1. C. Bernard et al., *Phys. Rev. Lett.* **D 54** 4586 (1996).
2. T. Matsui and H. Satz, *Phys. Lett.* **B 178** 416 (1986).
3. M.C. Abreu et al. (NA50 Coll.), *Phys. Lett.* **B 410** 327 (1997).
4. M.C. Abreu et al. (NA50 Coll.), *Phys. Lett.* **B 410** 337 (1997).
5. M.C. Abreu et al. (NA51 Coll.), *Phys. Lett.* **B 438** 35 (1998).
6. S. Ramos et al. (NA50 Coll.), *Proc. of ICHEP98*, Vancouver, Canada (1998).
7. M.C. Abreu et al. (NA38 Coll.), *Phys. Lett.* **B 444** 516 (1998).
8. M.C. Abreu et al. (NA38 Coll.), *Phys. Lett.* **B 449** 128 (1999).
9. M.C. Abreu et al. (NA38 Coll.), *Phys. Lett.* **B 466** 408 (1999).
10. M.C. Abreu et al. (NA50 Coll.), *Phys. Lett.* **B 450** 456 (1999).
11. M.C. Abreu et al. (NA50 Coll.), *Phys. Lett.* **B 477** 28 (2000).
12. H. Sorge et al., *Phys. Lett.*, **B 243** 7 (1990); *Phys. Lett.* **B 271** 37 (1991).

J/Ψ, Ψ' and Drell-Yan nuclear dependence in 800 GeV/c p - A collisions.

M.A. Vasiliev[i,1], M.E. Beddo[g], C.N. Brown[c], T.A. Carey[f],
T.H. Chang[g,2], W.E. Cooper[c], C.A. Gagliardi[i], G.T. Garvey[f],
D.F. Geesaman[b], E.A. Hawker[i,f], X.C. He[d], L.D. Isenhower[a],
D.M. Kaplan[e], S.B. Kaufman[b], D.D. Koetke[j], W.M. Lee[d],
M.J. Leitch[f], P.L. McGaughey[f], J.M. Moss[f], B.A. Mueller[b],
V. Papavassiliou[g], J.C. Peng[f], G. Petitt[d], P.E. Reimer[f,b],
M.E. Sadler[a], W.E. Sondheim[f], P.W. Stankus[h], R.S. Towell[a,f],
R.E. Tribble[i], J.C. Webb[g], J.L. Willis[a], G.R. Young[h]
(FNAL E866/NuSea Collaboration[3])

Abilene Christian[a], ANL[b], FNAL[c], Georgia State[d], Illinois Institute of Tech.[e], LANL[f],
New Mexico State[g], ORNL[h], Texas A & M[i], Valparaiso[j]

Abstract. Precise measurements of J/Ψ, Ψ' and Drell-Yan production in 800 GeV/c p-A collisions are reported. The observed J/Ψ and Ψ' suppression is smallest and nearly constant for $-0.1 < x_F < 0.25$ and increases at larger values of x_F. Suppression is also strongest at small p_T. A substantial difference in J/Ψ and Ψ' suppression at low x_F is observed for the first time in p-A collisions. The behavior of the Drell-Yan ratios is well described at small x_2 by an existing fit to the shadowing observed in deep-inelastic scattering. The cross-section ratios as a function of x_1 set tight limits on the energy loss of quarks passing through a cold nucleus.

Strong suppression of the yields for J/ψ's produced in heavy nuclei relative to light nuclei has been observed in proton (E772) [1], pion (NA3) [2] and heavy-ion (NA50) induced collisions [3]. It is essential to understand these processes before J/Ψ suppresion can be used as an unambiguous signature for production of a quark-gluon plasma. The kinematic dependencies of this suppression are strong, especially with Feynman-x (x_F) and transverse momentum (p_T). Broad coverage

---

[1] On leave from Kurchatov Institute, Moscow 123182, Russia.
[2] Present address: University of Illinois, Urbana, IL 61801
[3] This work supported in part by the U.S. Department of Energy

in these kinematic variables is required to be able to unravel the sources of the suppression. It is also important in understanding these mechanisms to contrast suppression of vector meson production with similar studies of the Drell-Yan process and open charm production. In Drell-Yan scattering only initial-state interactions are important since the dimuon in the final state does not interact strongly with the medium. This makes the Drell-Yan process an ideal tool to study energy loss of fast quarks in nuclear matter – a subject of considerable theoretical interest [4–7] with significant implications for the physics of relativistic heavy-ion collisions. Here we report measurements made during Fermilab E866 of J/ψ, Ψ' and Drell-Yan production for proton-nucleus collisions on Be, Fe, and W targets.

The experiment used the same 3-dipole magnetic spectrometer that was described in [8]. An 800 GeV/c extracted proton beam averaging 3×10^{11} protons per 20 s spill bombarded one of three solid targets or an empty target frame. The Be, Fe and W targets were 3% – 19% of an interaction length thick.

Over 3×10^6 J/Ψ and 10^5 Ψ' with $-0.10 < x_F < 0.93$ and $p_T \leq 4$ GeV/c and 1.3×10^5 Drell-Yan pairs with $4.0 < M < 8.4$ GeV/c^2, $0.01 < x_2 < 0.12$, $0.21 < x_1 < 0.95$, $0.13 < x_F < 0.93$ and $p_T \leq 4$ GeV/c survived analysis cuts. Random pairs were subtracted from the data using simulated opposite-charge random dimuons constructed by mixing single-muon tracks that were obtained simultaneously from prescaled single-muon triggers.

We present our results for J/Ψ and Ψ' suppression [9] in terms of α, where α is obtained by assuming the cross section dependence on nuclear mass, A, to be of the form $\sigma_A = \sigma_N \times A^\alpha$, where σ_N is the cross section on a nucleon. Previous experiments such as E772 have had a limited acceptance in p_T which varied with x_F. Since the value of α depends strongly on p_T, this can cause a distortion of the apparent shape of α versus x_F. The improvements in the E866/NuSea trigger allowed a much broader p_T acceptance than in these earlier measurements. However, for the lowest values of x_F at each spectrometer setting our p_T acceptance still becomes somewhat restricted. For the results presented here, we have corrected the values of $\alpha(x_F)$ using a detailed simulation of our acceptance and a differential cross section shape versus p_T derived from our data. The resulting dependence of α on x_F is shown in Fig. 1. The systematic uncertainty of 1% in the corrected α is dominated by the p_T acceptance correction. α for the J/ψ is largest at values of x_F of 0.25 and below, but strongly decreases at larger values of x_F. For the ψ', α is smaller than for the J/ψ for $x_F < 0.2$, remains relatively constant up to x_F of 0.5 (becoming slightly larger than for the J/ψ), and then falls to values consistent with those for the J/ψ for $x_F > 0.6$. The significance of the overall J/ψ, ψ' difference for $x_F < 0.2$ is about 4 sigma with respect to the statistical and relative systematic uncertainties. This difference is consistent with less accurate results obtained by NA38 for p-A at 450 GeV/c [10], but is inconsistent with the quoted NA38 result that also included the p-p and p-d data from NA51. Although slightly larger α values for the ψ' than for the J/ψ can be seen near $x_F = 0.55$, we should point out that if instead we emphasize the velocity of the $c\bar{c}$ and plot α versus rapidity, then the agreement is quite good in this region.

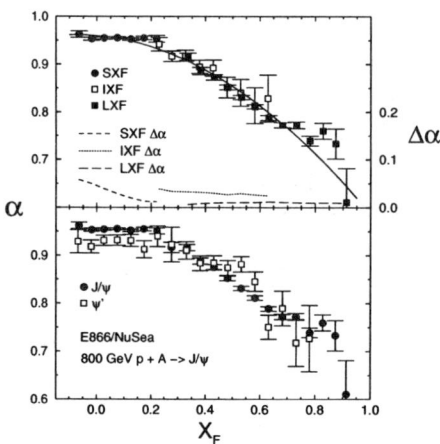

FIGURE 1. α for the J/ψ versus x_F for the three different data sets (top) and for the J/ψ and ψ' after the data sets are combined (bottom). Values are corrected for the p_T acceptance, as discussed in the text.

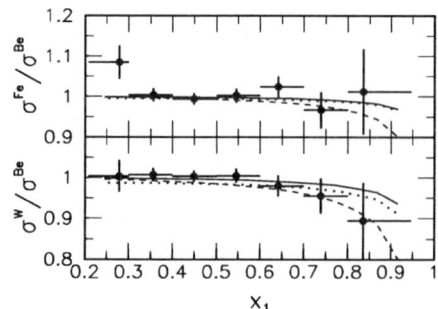

FIGURE 2. Drell-Yan cross section ratios per nucleon corrected for shadowing. The solid curves are the best fit using energy loss model [4], and the dashed curves show the upper limits. The dotted curves show the upper limits using the energy loss models [5–7]

The x_1 dependence of the Drell-Yan cross-section ratios provides the best direct measure of the energy loss of the incident quarks in the nuclear medium. However, as shown in [11], shadowing at small x_2 explains a substantial fraction of the apparent variation in the cross-section ratios versus x_1. This must be removed before one can isolate a nuclear dependence due to energy loss. Figure 2 show the cross-section ratios vs. x_1, corrected for shadowing by weighting each event with the calculated ratio of the Drell-Yan cross sections per nucleon for deuterium and nucleus A at the same (x_1, x_2, Q^2), using the EKS98 [12] fit to shadowing in deep-inelastic scattering and the MRST [13] parton distribution functions.

Several groups have studied energy loss of partons in nuclei. Their results can be expressed in terms of the average change in the incident-parton momentum fraction, Δx_1, as a function of the target nuclear mass A. Gavin and Milana [4] analyzed the E772 Drell-Yan data for energy loss based on the parameterization $\Delta x_1 = -\kappa_1 x_1 A^{1/3}$. Brodsky and Hoyer [5] argued that the energy loss found by Gavin and Milana was too large since the time scale for QCD bremsstrahlung was too short to allow for multiple contributions to the energy loss. Brodsky and Hoyer assumed initial-parton energy loss is described by $\Delta x_1 \approx -\frac{\kappa_2}{s} A^{1/3}$, where s is the square of the nucleon-nucleon center-of-mass energy. They found an upper limit for the gluon radiation from the uncertainty relation. They also noted that elastic scattering should make a similar contribution to the energy loss. Overall, they concluded that energetic partons should lose $\lesssim 0.5$ GeV/fm in nuclei. Baier et al. [6,7] found that the energy loss of energetic partons depends on a characteristic length and the broadening of the squared transverse momentum of the parton. For finite nuclei, both factors vary as $A^{1/3}$, so Baier et al. predict $\Delta x_1 \approx -\frac{\kappa_3}{s} A^{2/3}$. Baier et al. also predict that energy loss may be different in hot and cold nuclear matter. Given these energy loss expressions, it is possible to obtain empirical values for the κ's by performing simultaneous fits to the Fe/Be and W/Be Drell-Yan cross-section ratios versus x_1. Curves corresponding to the fits are included in Fig. 2. When assuming the form [4], we find $\kappa_1 = 0.0004 \pm 0.0009$. This implies that the observed fractional energy loss of the incident quarks is $< 0.14\%$/fm at 1σ. For the energy loss forms [5] and [6,7] the best fits imply essentially zero energy loss. We find the 1σ upper limits to be $\kappa_2 < 0.75$ GeV^2 and $\kappa_3 < 0.10$ GeV^2. The κ_2 limit indicates that the incident quarks lose energy at a constant rate of < 0.44 GeV/fm. The κ_3 limit implies that the observed energy loss of the incident quarks within the model of Baier et al. is $\Delta E < 0.046$ GeV/fm$^2 \times L^2$, where L is the quark propagation length through the nucleus. This is very close to the lower value given by Baier et al. for cold nuclear matter [14]. In all three cases, the quoted errors include both statistics and the overall normalization uncertainty, with the latter dominating.

Baier et al., also predict a quantitative connection between energy loss and p_t-broadening, $-dE/dz = \frac{\alpha_s N_c}{4} \times \Delta < p_t^2 >$. Using this we can extract energy loss indirectly from the Drell-Yan dimuon p_t spectra. In order to determine absolute values of $< p_t^2 >$, a high statistics Monte-Carlo simulation of the p_t acceptance of the E866 spectrometer was done. Our preliminary result for energy loss from p_t broadening is $\Delta E = \left(0.008 \pm 0.002 \frac{GeV}{fm^2}\right) \times L^2$, which is consistent with the limit found for this model above.

REFERENCES

1. D.M. Alde et al., Phys. Rev. Lett **66**, 133 (1991).
2. M.J. Corden et al., Phys. Lett. **B110**, 415 (1982).
3. M.C. Abreu et al., Phys. Lett.**B450**, 456 (1999).
4. S. Gavin and J. Milana, Phys. Rev. Lett. **68**, 1834 (1992).
5. S.J. Brodsky and P. Hoyer, Phys. Lett. **B298**, 165 (1993).
6. R. Baier et al., Nucl. Phys. **B484**, 265 (1997).
7. R. Baier et al., Nucl. Phys. **B531**, 403(1998).
8. E. Hawker et al., Phys. Rev. Lett. **80**, 3715 (1998).
9. M.J. Leitch et al., Phys. Rev. Lett. **84**, 3256 (2000).
10. M.C. Abreu et al., Phys. Lett. **B444**, 516 (1998).
11. M.A. Vasiliev et al., Phys. Rev. Lett. **83**, 2304 (1999).
12. K.J. Eskola, V.J. Kolhinen and P.V. Ruuskanen, Nucl. Phys. **B535**, 351 (1998); K.J. Eskola, V.J. Kolhinen and C.A. Salgado, Eur. Phys. J. **A3**, 351 (1998).
13. A.D. Martin et al., Eur. Phys. J. **C4**, 463 (1998).
14. Note that [7] indicates that the estimate 0.02 GeV/fm^2 given in [6] should be increased by a factor of 2.

An Overview of AGS Physics Results

John G. Lajoie

Iowa State University, Department of Physics and Astronomy, Ames, IA 50011

Abstract. As the study of relativistic heavy-ion collisions moves forward into the RHIC era, it is instructive to reflect on the past decade of the AGS heavy-ion program. I will present selected physics results from the AGS program and emphasize the role that they have played in our understanding of hot hadronic matter at high net baryon density. This regime is separate and unique from that to be studied at RHIC, and recent studies indicate that it may occupy a region in the QCD phase diagram worthy of study in its own right [1]. Finally, I will emphasize some of the outstanding features of the AGS data that may defy simple explanation as the product of hot, hadronic matter.

INTRODUCTION

Relativistic heavy-ion collisions at the AGS have provided the nuclear physics community with an excellent opportunity to study hadronic matter at nonzero temperature and high baryon density. This corresponds to a region of the nuclear matter phase diagram that is less well understood theoretically than the regime of high temperature and low net baryon density that will be studied at RHIC. In this talk I will very briefly review a few selected aspects of the experimental program at the AGS in an attempt to address the following questions: Does the interacting matter achieve equilibrium? How much stopping is there at the AGS? Can we build a qualitative (if not quantitative) understanding of the fireball created in heavy-ion collisions at the AGS? Do we understand strangeness enhancement and the production of strange antibaryons at the AGS?

THERMAL MODELS AND COLLECTIVE EXPANSION

The canonical comparison made to address the question of thermal equilibration in heavy-ion collision is typically to a thermal model calculation. Despite the simplicity of the underlying model, thermal calculations do a remarkably good job of predicting particle ratios measured at the AGS over many orders of magnitude. Of course, a comparison of particle ratios integrated over all phase space is independent of source dynamics such as flow, etc. These models indicate that the nuclear matter at the AGS is thermally equilibrated with a temperature between 120-140 MeV and a baryon chemical potential between 500-600 MeV [2]. In contrast, the same thermal models indicate that at the SPS equilibrium is achieved with T≈170 MeV and a baryon chemical potential of between 200-300 MeV.

However, it has been noted both at the AGS and SPS that the transverse momentum spectra of particles are ordered my mass such that the momentum spectra of heavier particles are significantly more "stiff" than those for lighter species [3]. This ordering arises naturally when random thermal motion is superimposed on a collective expansion of the system - for example, the nuclear fireball is "exploding" outward. If this is the case, then while the thermal models described above can do a fair job of predicting particle ratios, we cannot expect the measured transverse momentum spectra of particles to directly represent the equilibrium temperature. It has been shown that by choosing a particular equation of state and initial condition a unique temperature and velocity profile at thermal freezeout can be determined [4]. Under these assumptions we find that the gross features of measured particle spectra at the AGS are consistent with emission from a thermal fireball with a mean transvers expansion velocity of .3-.4 times the speed of light. It is interesting to note that under this analysis collisions at both the SPS and the AGS seem to undergo thermal freezeout at a constant energy per hadron of about 1 GeV.

STOPPING AT THE AGS

Given the the gross features of particle emission at the AGS appear to be consistent with emission from a thermal source, we can ask how much of a memory of the inital state there is in the final system. For example, how much of the longitudinal energy of the nucleons in the initial state nuclei is converted to thermal energy in the final state. One way to answer this question is to examine the distribution of protons in the final state as a function of rapidity. This has been done by the E917 collaboration for a variety of collisions energies and centralities [5], demonstrating that the most central collisions show a high degree of stopping as evidenced by a central plateau in the proton dN/dy spectra. However, the stopping is not complete, and thermal models require a mean longitudinal expansion velocity of the emitting source of about 0.5 times the speed of light in order to describe the rapidity distribution of particle emission [6]. Using a Bjorken picture, which may be justified by the rapidity plateau observed in the proton dN/dy, one can estimate the nucleon and energy densities at 1 fm/c to be 1.1 $/fm^3$ and 1.4 GeV/fm^3 respectively [2]. These values are approximately one order of magnitude higher than thermal models and HBT source size estimates [7] would indicate at freezeout.

STRANGE PARTICLE PRODUCTION

Recent measurements of the excitation function for kaon production at the AGS by the the E866/E917 collaborations may indicate that strangeness enhancement at the AGS may be purely an effect of hadronic rescattering [8]. Measurements indicate a smooth evolution of the K^+/π^+ ratio above threshold, and no sudden jump in this ratio as might be expected if there were a transition to a deconfined plasma state. In addition, the double ratio $(K^+/\pi^+)_{AuAu}/(K^+/\pi^+)_{pp}$ indicates that strangeness enhancement is strongest near threshold, where hadronic rescattering should play the largest role in enhancing strange particle production.

I STRANGE ANTIBARYONS

The study of antimatter produced in heavy-ion collisions is interesting becaise it involves the interplay between mechanisms of enhanced production [11–13] and the large antinucleon annihilation cross section [11,14] which should deplete the antiprotons observed in the final state. Strange antibaryons play a unique role by combining two features that are predicted to be enhanced if the collision forms a deconfined plasma state.

Despite the fact that it appears that there may be no surprises in strange particle production at the AGS, the production of strange antibaryons continues to present

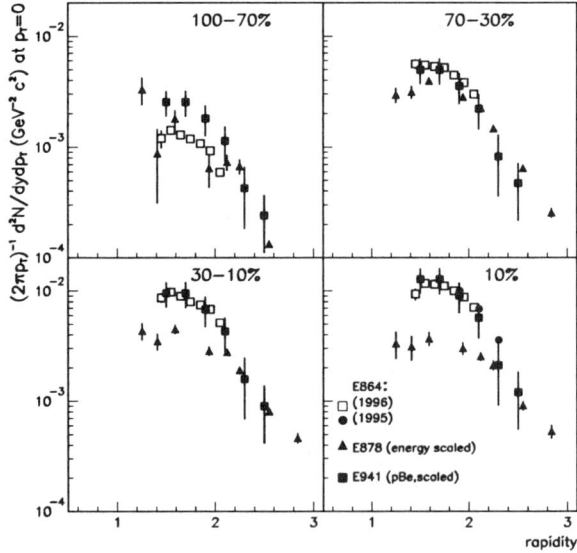

FIGURE 1. A comparison of measured antiproton production in E878, E864 (Au+Au) and E941 (p+Be, scaled by first collisions) at 12 GeV/c.

somewhat of a mystery. Early indications from the E859 collabortion of direct measurements of antilambda production demonstrated a very high production of antilambdas relative to antiprotons in Si+Au collisions [9,10]. The combination of subsequent measurements from E864 and E878 demonstrated a similarly high ratio in Au+Au collisions, and that this ratio grows as a function of collision centraility, from $\overline{\Lambda}/\overline{p} > 0.02$ for peripheral collsions to $\overline{\Lambda}/\overline{p} > 2.3$ for 10% central collsions [15,16]. This has been further confirmed by direct measurements of antilamba production in Au+Au collisions by the E866/E917 collaboration [17].

A very high $\overline{\Lambda}/\overline{p}$ ratio is difficult to understand withing the context of a thermal model, especially within the constraints of the measured K/π ratios. In addition, simple cascade models which include reduced annihilation cross sections for the $\overline{\Lambda}$ and conversion processes such as $\overline{p} + K^+ \to \pi^+ + \overline{\Lambda}$ have been unable to produce the centrality dependence of the data [18].

The situation is further confused when recent data from the E941 collaboration for antiproton production in p+Be collisons at 12 GeV/c are compared with the E864 and E878 data, as shown in Figure 1. The difference between the E864 and E878 data shown on the plot is due to the different experimental acceptance for antiprotons from antihyperon feeddown, while the E941 p+Be data is scaled by the number of first collsions estimated in each Au+Au centraility bin using a simple Glauber model. While the E941 scaled p+Be data is somewhat higher than the Au+Au data for peripheral collsions, it grows to match the E864 data for higher centrality bin. This leads to the somewhat surprising conclusion that the combination of annihilation and enhanced production of antiprotons and antihyperons (as measured in the E864 data) scales with the number of first collsions. The importance of annihilation has been underscored by recent measurements of antiproton antiflow by the E877 collaboration [19].

A complete understanding of the relative production of antiprotons and antihyperons is one of the remaining open questions of the AGS experimental program.

II CONCLUSIONS

The regime of nuclear matter studied at the AGS is unique in that it represents the highest baryon density collisions available to the experimenal heavy-ion community. While gross features of these collsions appear consistent with a simple model of a thermal fireball combined with longitudinal (incomplete stopping) and transverse expansion, there are features that defy explanation within such a simple picture. One of these features discussed here, strange antibaryon production, is particularly interesting in light of the combination of Au+Au and p+A measurements from a variety of AGS experiments.

REFERENCES

1. Stephanov M., Rajagopol K., Shuryak E.*Phys. Rev. Lett.* **81**, 22 (1998).

2. Stachel J., *Nucl. Phys.* **A610**, 509c (1996).
3. E864 Collaboration, *Phys. Rev.* **C61**, 6 (1999).
4. Braun-Munzinger P., Stachel J., *Nucl. Phys.* **A638**, 3c (1998).
5. E917 Collaboration, nucl-ex/0003007v2.
6. Braun-Munzinger P., Stachel J., N. Xu, *Phys. Lett.* **B344**, 43 (1994).
7. E877 Collaboration, *Phys. Rev. Lett.* **78**, 2916 (1997).
8. E917 Collaboration, *Phys. Lett.* **B476**, 8 (2000).
9. Cole B., *Nucl. Phys.* **A590**, 509c (1995).
10. Yeudong Wu, proceedings of HIPAGS '96
11. Koch P., Dover C., *Phys. Rev.* **C40**, 145 (1989).
12. Koch P. et. al., *Phys. Lett.* **A8**, 737 (1988).
13. Koch V. et. al., *Phys. Lett.* **B265**, 29 (1991).
14. Kahana S., Dover C., *Phys. Rev.* **C47**, 1356 (1993).
15. E864 Collaboration, *Phys. Rev.* **C59**, 2699 (1999).
16. E878 Collaboration, *Phys. Rev.* **C56**, 1521 (1997).
17. E917 Collaboration, nucl-ex/9904010.
18. Wang, Welke G., Bellweid R., Pruneau C., nucl-ex/9807036.
19. E877 Collaboration, nucl-ex/0004002.

Flow and HBT Measurements in the E895 Experiment

M.A. Lisa*, E895 Collaboration[†]

The Ohio State University, Columbus, Ohio 43210
[†]*Stony Brook, U.C. Davis, LBNL, Columbia, BNL, Purdue, Harbin Institute, Kent State, Carnegie Mellon, St. Mary's College, Univ. Auckland, Ohio State*

Abstract. There has been considerable interest in the energy dependence of sideward flow in non-central heavy ion collisions in the transition region between Bevalac/SIS (~1 AGeV) and top AGS (~10 AGeV) energies, as a suggested "dip" in flow may signal the onset of Quark Gluon Plasma formation at these energies. The E895 Collaboration has measured the excitation function of sideward flow. While the data are smooth as a function of energy, displaying no dramatic dips, transport models are severely challenged as they try to reproduce the details of the flow. At the same time, E895 has mapped out the azimuthal dependence of two-pion correlations, and observes a new effect that provides unprecedented information on the *spatial* orientation of the hot matter created in these collisions. This new information should provide further insight into the nature of directed flow at these energies.

Heavy ion collisions at non-zero impact parameter generate a highly excited and compressed system with a built-in azimuthal anisotropy. This anisotropy provides an important handle through which to study in detail the response of the system to the pressure; this response, in turn, depends heavily on the underlying dynamics. The 1st-order azimuthal anisotropy in momentum space, directed flow, has been studied extensively over a wide range of collision energies [1]. Directed flow at AGS energies has attracted considerable attention, due to speculation based on one-fluid hydrodynamic models [2] that a pronounced minimum in the excitation function of directed flow might signal the onset of QGP formation at these energies.

The E895 Collaboration at the AGS has used a large acceptance Time Projection Chamber to measure a variety of physical observables in Au+Au collisions at 2, 4, 6, and 8 AGeV, including sideward and elliptic flow of nucleons [3,4] and two-pion correlations (HBT) [5,6,18]. Here, we review the E895 proton sideward flow results, and make a new connection between these momentum-space anisotropies and the *coordinate*-space anisotropies which may be measured via HBT.

Following the standard technique, the estimated reaction plane azimuth for an event is the orientation of $\mathbf{Q} = \sum_j w \, \mathbf{p}_j^\perp / p_j^\perp$, where j runs over all baryonic fragments in the event, $\mathbf{p}^\perp$ is momentum in the plane perpendicular to the projectile direction, and we use the weighting factor $w = y_j' / \max(|y_j'|, 0.8)$, where $y_j' = y_j^{lab}/y^{mid} - 1$ [3,7]. The centrality of a collision is estimated from the charged particle multiplicity; the flow analysis presented here is for collisions estimated to correspond to impact parameters between 5 and 7 fm.

The mean proton transverse momentum projected onto the reaction plane generally increases as a function of y. Shapes of $\langle p_x(y) \rangle$ are normally close to linear over an interval centered on midrapidity, and a function $Fy' + Cy'^3$ typically fits the $\langle p_x(y') \rangle$ distribution over the y' region dominated by participant fragments. We use the fitted linear coefficient F (or $F_y = F/y^{\text{mid}}$, the corresponding slope for unnormalized rapidity) to characterize the overall strength of the sideward flow. We average the fitted coefficient F with and without imposing $C = 0$, and the difference generally dominates the systematic uncertainty in the slope (which is large compared with the statistical error). Figure 1 presents both F and F_y as functions of beam energy, along with the same quantities for comparable centrality, as measured in the same detector at lower energy [8], and in E877 [9] at maximum AGS energy. All experimental flow signals are corrected for finite resolution in determining the reaction plane using the prescription described in Ref. [10].

One gains a somewhat different view by examining the so-called number flow, in which one does not weight by the particle's p_T, but instead examines the anisotropic distribution of track azimuths ϕ relative to the reaction plane. These anisotropies generally can be well described by the truncated Fourier expansion $dN/d\phi \approx v_0 \left[1 + 2v_1 \cos \phi + 2v_2 \cos 2\phi\right]$. The coefficient v_2 represents elliptic flow [1], already reported for the E895 data [4]. Figure 2 presents measured v_1 coefficients for protons as a function of rapidity.

The data exhibit no pronounced dip or non-smooth behavior as a function of energy that might indicate a sudden change in the underlying dynamics. However, flow at these energies remains quite sensitive to the physics, as indicated by the wide variation in flow predictions of popular models (RQMD [11], ART [12], uRQMD [13], BEM [14]) shown in Figures 1 and 2, as the physical assumptions in the model are varied. None of the models is close to simultaneously reproducing E895 sideward [3] and elliptic [4] flow observations.

Clearly, fresh information on sideward flow is needed. Indeed, it has recently been shown [15,16] that HBT analyses, correlated with the event-wise reaction plane, can easily distinguish *coordinate-space* anisotropies of the emitting source. In the Bertsch-Pratt ("out-side-long") decomposition of the relative momentum q between the pair of pions, the correlation function for a given emission angle ϕ with respect to the reaction plane, is fit with with the standard Gaussian parameterization [17] $C(\mathbf{q}, \phi) = 1 + \lambda(\phi) \exp\left[-\sum q_i q_j R_{ij}^2(\phi)\right]$ where $i, j = o, s, l$. In contrast to previous analyses, *all six* radius parameters are relevant here [16,15,17].

E895 pion correlation systematics as a function of beam energy, m_T, centrality, and rapidity, have been reported [5,6]. Full details of our recent ϕ-sensitive analysis for non-central collisions may be found in Ref. [18].

In these non-central collisions, the initial transverse geometry of the participant zone reflects the overlap of the two spherical nuclei, with a larger extent perpendicular to than parallel the impact parameter vector. A freeze-out configuration that retains this shape leads to equal-amplitude 2nd order oscillations in the squared radii R_o^2, R_s^2, R_{os}^2; this is seen in the experimental data of Figure 4, which corre-

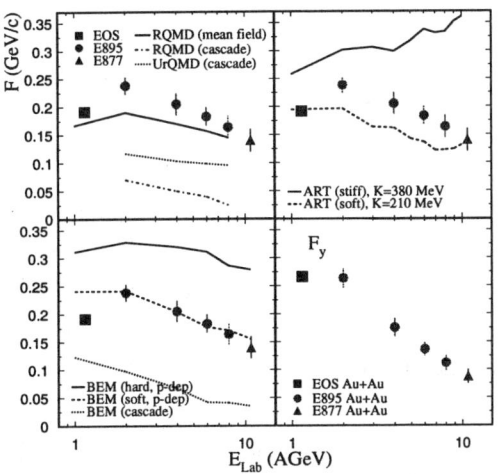
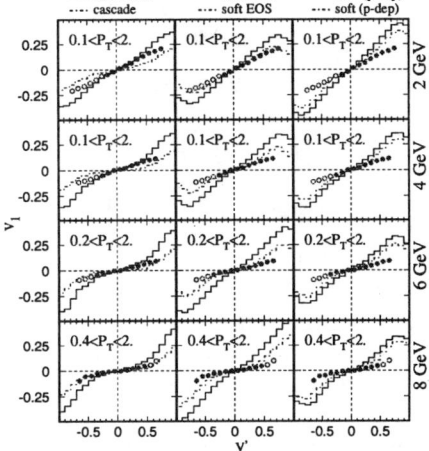

FIGURE 1. E895 sideward flow results compared to model predictions.

FIGURE 2. E895 and model v_1 coefficients as a function of y.

sponds to low p_T (< 350 MeV/c) π^- at midrapidity from 4 AGeV Au+Au collisions at impact parameter $b \approx 4-8$ fm.

More striking are the *first*-order oscillations of the terms R_{ol}^2 and R_{sl}^2, which may be directly related [15] to the spatial "tilt" of the pion-emitting source away from the beam axis. All energies which we have analyzed display oscillations similar to those in Figure 4. The first-order oscillations there indicate a tilt angle which is positive (i.e. in the same direction as proton sideward flow) and large– about 40°.

RQMD calculations agree qualitatively with the data [18]. The pion freezeout configuration predicted by the model is shown in Figure 3. The coordinate-space positive tilt is clear in the contours. The momentum-space *negative* tilt is indicated with arrows (note that the bulk of of the pions come from around $z = 0$); this is the pion "anti-flow" which is also observed in the data [19]. This picture makes it clear that tilt angles are *not* the same as *momentum*-space "flow angles" [1], and that the ϕ-sensitive HBT analysis accesses new information– the size, shape, and orientation of the pion-emitting source. For example, the observation in the data that the flow and tilt angles point on opposite sides of the beam axis confirms conclusively that pion flow is not hydrodynamic in nature; model-dependent information regarding pion rescattering and absorption may also be gained from the tilt angles [15].

According to RQMD, nearly all of these low-p_T pions arise from Δ decay. Thus, ϕ-sensitive HBT images the flowing baryonic matter itself. This information should prove especially useful in model comparisons, as the coordinate-space configuration of the flowing matter is sensitive to interesting underlying physics [18,20].

In summary, E895 has measured detailed excitation functions of sideward flow and pion HBT at the AGS. The data show no obvious anomalous structure, but transport calculations are unable to consistently reproduce the details of sideward

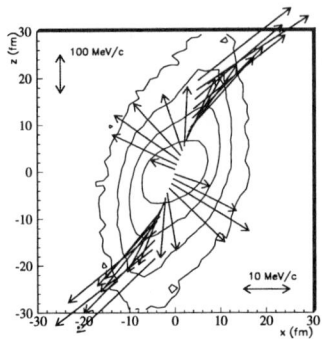

FIGURE 3. RQMD pion freeze-out in coordinate (contours) and momentum (arrows) space.

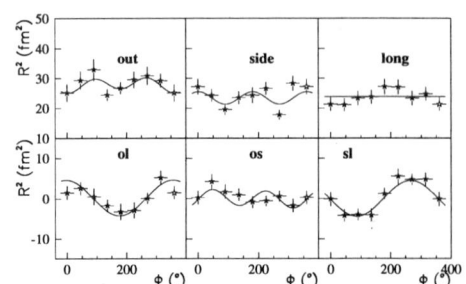

FIGURE 4. HBT radii versus ϕ relative to the reaction plane for semicentral Au+Au collisions at 4 AGeV.

flow. New information on the coordinate-space aspects comes from azimuthally sensitive HBT, which provides the shape and orientation of the emitting source. It is hoped that the 6-dimensional picture obtained by combining flow and HBT information provides further insight into the dynamics of non-central collisions.

REFERENCES

1. W. Reisdorf and H.G. Ritter, Annu. Rev. Nucl. Part. Sci. **47**, 663 (1997).
2. D.H. Rischke, S. Bernard, and J.A. Maruhn, Nucl. Phys. **A595**, 346 (1995); D.H. Rischke and M. Gyulassy, *ibid* **A597**, 701 (1996).
3. E895 Collaboration, H. Liu, et al., Phys. Rev. Lett. **84**, 5488 (2000).
4. E895 Collaboration, C. Pinkenburg, et al., Phys. Rev. Lett. **83**, 1295 (1999).
5. E895 Collaboration, M.A. Lisa, et al., Phys. Rev. Lett. 84, 2798 (2000).
6. E895 Collaboration, M.A. Lisa, et al., Nucl. Phys. **A661**, 444c (1999) (QM99).
7. P. Danielewicz and G. Odyniec, Phys. Lett. **157B**, 146 (1985).
8. EOS Collaboration, M.D. Partlan et al., Phys. Rev. Lett. **75**, 2100 (1995).
9. E877 Collaboration, J. Barrette et al., Phys. Rev. **C56**, 3254 (1997).
10. A. Poskanzer and S. Voloshin, Phys. Rev. **C58**, 1671 (1998).
11. H. Sorge, Phys. Rev. **C52** 3291 (1995).
12. B.-A. Li and C.M. Ko, Phys. Rev. C **58**, R1382 (1998).
13. S.A. Bass et al., Prog. Part. Nucl. Phys. **41**,225 (1998).
14. P. Danielewicz et al., Phys. Rev. Lett. **81**, 2438 (1998).
15. M.A. Lisa, U. Heinz, and U.A. Wiedemann, in press, Phys. Lett. **B**; nucl-th/0003022.
16. U.A. Wiedemann, Phys. Rev. C **57**, 266 (1998).
17. U.A. Wiedemann and U. Heinz, Phys. Rep. **319**, 145 (1999).
18. E895 Collaboration, M.A. Lisa, et al., nucl-ex/0007022 (2000).
19. EOS Collaboration, J. Kintner et al., Phys. Rev. Lett. **78**, 4165 (1997); W. Caskey, Ph.D. thesis, University of California at Davis (1999)
20. J. Brachmann, et al., Phys.Rev. C61 (2000) 024909

Collective Evolution of Hot QCD Matter from the QGP to Freeze-Out

Adrian Dumitru* and Steffen A. Bass[†]

*Physics Department, Columbia University, 538W 120th Street, New York, NY 10027
[†]Nat. Supercond. Cycl. Lab., Michigan State University, East Lansing, MI 48824-1321

Abstract. We present results on the evolution of $\langle p_t \rangle$ with energy, for hadrons emerging from hadronization of a quark-gluon plasma possibly produced in high-energy heavy-ion collisions. We find that BNL-RHIC energy corresponds to the previously predicted plateau of $\langle p_t \rangle$, reminiscent of the assumed first-order QCD phase transition. Heavy hadrons are the best messengers of the transverse flow stall.

Theoretical arguments [1] and lattice QCD studies [2] suggest that at high temperatures, roughly $T \sim 200$ MeV, the thermodynamically stable state of QCD is different from that at $T = 0$, i.e. the theory of strong interactions exhibits a phase transition to a new state called "Quark Gluon Plasma" (QGP). It is the only phase transition of a fundamental theory accessible to experiments under defined laboratory conditions. In the QGP the effective number of relativistic degrees of freedom is expected to be substantially larger than below the phase transition temperature T_c. The search for signatures of that phase transition is one of the primary goals of high-energy heavy-ion collision experiments.

In particular, if the transition is first order with non-vanishing latent heat, it is supposed to leave fingerprints on the hydrodynamical expansion pattern [3,4]. Isentropic expansion would proceed through phase coexistence, where the pressure p is independent of the energy density ϵ, such that the isentropic speed of sound $c_s^2 = dp/d\epsilon = 0$. Thus, during that phase coexistence stage energy density gradients in the system do not reflect in pressure gradients, and the average transverse flow velocity does not increase substantially. As the p_t spectra of the emitted hadrons are determined by the transverse boost velocity and the temperature at freeze-out, one expects a plateau of $\langle p_t \rangle$ of charged hadrons for some range of beam energies where the central region "sweeps" through the mixed phase [3].

However, the early studies of that effect suffered from some simplifications which render comparisons to existing experimental data, and extrapolation to higher energies, unrealistic. First, the latent heat of the transition (and therefore the space-time volume of the mixed phase) was overestimated because only pions were assumed to contribute to the entropy of the hadronic phase. Second, final-state decays of hadronic resonances (like $\rho \to \pi\pi$ or $\Delta \to \pi N$) which contribute more than 50% to the measured pion multiplicity were not taken into account. Finally, the most serious problem arises from the arbitrariness of the decoupling hypersurface, which can not be determined within hydrodynamics but affects the results considerably.

To reduce the uncertainties and the number of free parameters one therefore has to improve on the treatment of the post-hadronization stage by considering microscopic transport [5,6], e.g. within the relativistic Boltzmann equation with collision kernel as determined by the known hadronic cross sections. The expansion and subsequent hadronization of the high-temperature QGP state can be modelled within relativistic hydrodynamics, assuming that the system evolves locally through a phase coexistence. Thus, in the space-time region inbetween the initial-time (purely space-like) hypersurface and the hadronization hypersurface (the boundary between mixed phase and hadronic phase, which has both space-like and time-like parts) we solve the continuity equation for the energy-momentum tensor,

$$\partial_\mu T^{\mu\nu} = 0 \;, \text{ with } T^{\mu\nu} = (\epsilon + p)u^\mu u^\nu - pg^{\mu\nu} \;. \tag{1}$$

In the QGP we assume a gas of quasi-free u, d, s quarks and gluons; for simplicity, the only contribution from interactions to the QGP pressure is due to the temperature independent vacuum pressure (bag pressure), see e.g. [1].

On the hadronization hypersurface, we switch to the Boltzmann equation,

$$p \cdot \partial f_i(x^\mu, p^\nu) = \mathcal{C}_i \;, \tag{2}$$

where $f_i(x^\mu, p^\nu)$ denotes the phase-space distribution function of species i. The binary collision approximation is used to construct the collision kernel $\mathcal{C}_i$ [7], which treats almost all hadrons from [8] explicitly. All those hadrons are also taken into account in the equation of state of the mixed phase entering eq. (1). The switch from hydrodynamics to microscopic transport is performed by matching the energy momentum tensors and conserved currents on the hadronization hypersurface. The evolution towards freeze-out is thus determined by the relevant hadronic cross-sections and decay rates, competing with the local expansion rate $\partial_\mu u^\mu$. The initial condition for central collisions of Pb/Au nuclei and CERN-SPS energy is that hydrodynamic flow sets in on the $\tau_i = 1$ fm proper time hypersurface, and the entropy per net baryon is $s/n_B = 45$, which reproduces a variety of measured final-state hadron multiplicities as well as the p_t spectra of π, K, p, Λ, Ξ, and Ω, see [5,6]. For the higher BNL-RHIC energy, $\sqrt{s} = 200A$ GeV, the parameters were assumed to be $\tau_i = 0.6$ fm and $s/n_B = 205$, which yields $dN_{ch}/dy \approx 800$ at $y = 0$. Fig. 1 depicts $\langle p_t \rangle$ for a variety of hadrons, for CERN-SPS, BNL-RHIC, and CERN-LHC energies. One observes that at each energy $\langle p_t \rangle$ increases with m. Reminiscent of the collective expansion before decoupling, all hadrons flow approximately with the same transverse velocity, and therefore their momentum essentially increases in proportion to their mass. This effect is seen in data obtained for Pb+Pb collisions at the SPS [9]. The spectra of heavy hadrons are much less affected by the random thermal motion at decoupling, $v_{th} \propto \sqrt{T/m}$, and are therefore better suited to measure the transverse flow build up during the evolution. One also observes that despite the five times higher entropy per baryon assumed to be achieved at RHIC energy, the average transverse momenta are predicted by this model to be similar to the lower SPS energy. This is mainly due to the first-order phase transition featuring a region of energy densities where c_s^2 is small (it is not exactly zero in this calculation due to the finite conserved baryon charge); and to some extent also due to earlier decoupling of the hadrons emerging from the hadronization of the plasma at RHIC energy [5].

If the initial energy density in the central region continues to grow towards higher energy as predicted by models assuming saturation of the transverse density of gluons [10] (see however [11]), the hydrodynamical expansion will lead to very large $\langle p_t \rangle$ at LHC energy, see Fig. 1. Thus, RHIC-energy is possibly right in the plateau of $\langle p_t \rangle$, if the concept of a phase mixture is applicable to high-energy collisions.

Fig. 2 shows the freeze-out volume of the pions at central rapidity, calculated as the average transverse area $\pi \langle r_t \rangle^2$ times the average length of the central rapidity slice, which is equal to proper time τ. The space-time volume is essentially increasing linearly with the multiplicity. Thus, the pion density at freeze-out is not affected much by the presence of the phase transition, unlike the average transverse momentum. When extrapolating each line to $dN_\pi/dy = 0$ we obtain a "hollow" volume V_0, which increases with p_t. This is caused by the radial flow: high p_t pions can not be emitted from the center, where the collective velocity field vanishes for symmetry reasons, and where the pions have only random thermal momenta.

Acknowledgements: A.D. thanks the organizers for the invitation to CIPANP2000 and acknowledges support from a DOE Research Grant, Contract

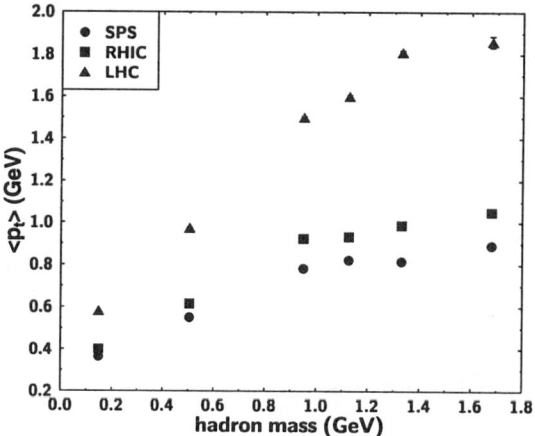

FIGURE 1. $\langle p_t \rangle$ versus hadron mass m for π, K, p, Λ, Ξ, and Ω. Dots for $\sqrt{s} = 18A$ GeV, squares for $\sqrt{s} = 200A$ GeV, and triangles for $\sqrt{s} = 5500A$ GeV.

No. De-FG-02-93ER-40764. S.A.B. is supported by NSF grant PHY-00-70818.

REFERENCES

1. e.g., Shuryak E., *Phys. Rept.* **61**, 71 (1980); McLerran L., *Rev. Mod. Phys.* **58**, 1021 (1986).
2. e.g., Oevers M., Karsch F., Laermann E., and Schmidt P., *Nucl. Phys. Proc. Suppl.* **73**, 465 (1999).
3. Kataja M., Ruuskanen P., McLerran L., von Gersdorff H., *Phys. Rev. D* **34**, 794 and 2755 (1986).
4. Blaizot J., and Ollitrault J., *Phys. Rev. D* **36**, 916 (1987); Hung C., and Shuryak E., *Phys. Rev. Lett.* **75**, 4003 (1995); Rischke D., and Gyulassy M., *Nucl. Phys.* **A608**, 479 (1996); Brachmann J., et al., nucl-th/9908010; nucl-th/9912014.
5. Dumitru A., Bass S., Bleicher M., Stöcker H., and Greiner W., *Phys. Lett.* **B460**, 411 (1999).
6. Bass S., and Dumitru A., *Phys. Rev. C* **61**, 064909 (2000).
7. Bass S., et al., *Prog. Part. Nucl. Phys.* **41**, 225 (1998).
8. Caso C., et al., *Eur. Phys. J.* **C3**, 1 (1998).
9. Bearden I., et al., [NA44 Collaboration], *Phys. Rev. Lett.* **78**, 2080 (1997).
10. Krasnitz A., and Venugopalan R., *Phys. Rev. Lett.* **84**, 4309 (2000); Eskola K., Kajantie K., Ruuskanen P., and Tuominen K., *Nucl. Phys.* **B570**, 379 (2000).
11. Dumitru A., and Gyulassy M., hep-ph/0006257.

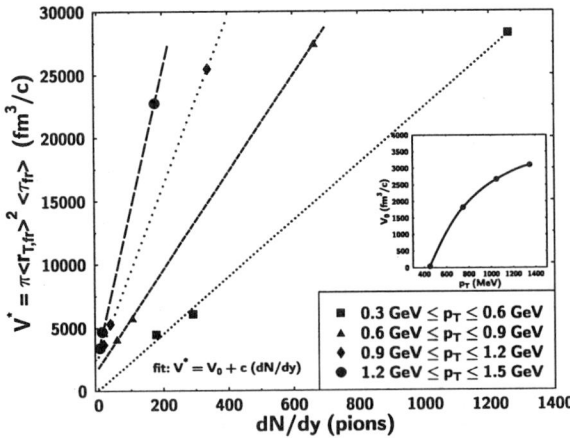

FIGURE 2. Freeze-out volume of the pions as a function of the pion rapidity density at central rapidity, for various p_t cuts. Increasingly high p_t pions are only emitted from a "shell", the radius of the hollow core increasing with p_t, as shown in the inset.

Strange and multi-strange baryon production at SPS as a probe of QGP formation

Presented by R. CALIANDRO for the WA97 Collaboration:

F. Antinori[e,i], H. Bakke[b], W. Beusch[e], I.J. Bloodworth[d], R.Caliandro[a], N. Carrer[i],
D. Di Bari[a], S. Di Liberto[k], D. Elia[a], D. Evans[d], K. Fanebust[b], R.A. Fini[a], J. Ftáčnik[f],
B. Ghidini[a], G. Grella[l], H. Helstrup[c], A.K. Holme[h], D. Huss[g], A. Jacholkowski[a],
G.T. Jones[d], J.B. Kinson[d], K. Knudson[e], I. Králik[f], V. Lenti[a], R. Lietava[e],
R.A. Loconsole[a], G. Løvhøiden[h], V. Manzari[a], M.A. Mazzoni[k], F. Meddi[k],
A. Michalon[m], M.E. Michalon-Mentzer[m], M. Morando[i], P.I. Norman[d], B. Pastirčák[f],
E. Quercigh[e], G. Romano[l], K. Šafařík[e], L. Šándor[e,f], G. Segato[i], P. Staroba[j],
M. Thompson[d], T.F. Thorsteinsen[b,†], G.D. Torrieri[d], T.S. Tveter[h], J. Urbán[f],
O. Villalobos Baillie[d], T. Virgili[l], M.F. Votruba[d] and P. Závada[j].

[a] Dipartimento I.A. di Fisica dell'Università e del Politecnico di Bari and Sezione INFN, Bari, Italy
[b] Fysisk institutt, Universitetet i Bergen, Bergen, Norway
[c] Høgskolen i Bergen, Bergen, Norvay
[d] School of Physics and Astronomy, University of Birmingham, Birmingham, UK
[e] CERN, European Laboratory for Particle Physics, Geneva, Switzerland
[f] Institute of Experimental Physics, Slovak Academy of Sciences, Košice, Slovakia
[g] GRPHE, Université de Haute Alsace, Mulhouse, France
[h] Fysisk institutt, Universitetet i Oslo, Oslo, Norway
[i] Dipartimento di Fisica dell'Università and Sezione INFN, Padua, Italy
[j] Institute of Physics, Academy of Sciences of the Czech Republic, Prague, Czech Republic
[k] Dipartimento di Fisica dell'Università "La Sapienza" and Sezione INFN, Rome, Italy
[l] Dipartimento di Scienze Fisiche "E.R. Caianiello" dell'Università and Sezione INFN, Salerno, Italy
[m] Institut de Recherches Subatomiques, IN2P3/ULP, Strasbourg, France
† Deceased

Abstract.
The WA97 experiment studies strangeness production at central rapidity in Pb-Pb, p-Pb and p-Be collisions at CERN SPS. A comprehensive study, including the most recent results on yields and transverse momentum spectra of strange and multi-strange particles, is presented.

I INTRODUCTION

The study of strange particle production in heavy ion collisions at high energy has the ultimate goal of finding evidence for a new state of matter: the Quark Gluon Plasma (QGP). If a QGP state is formed, the strangeness production is

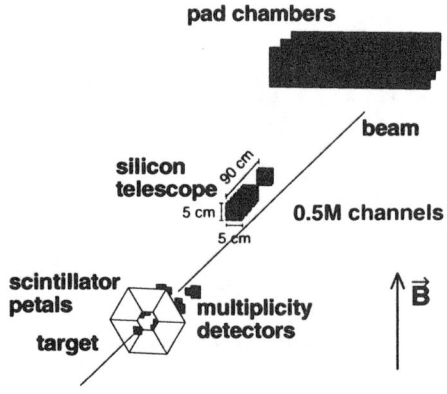

FIGURE 1. The WA97 set-up.

expected to equilibrate in a few fm/c [1], thanks to the partial restoration of the chiral symmetry and, at SPS energies, to Pauli blocking. If, instead, no QGP state is formed, the production of strange and expecially multi-strange baryon in hadronic rescattering is much more difficult and chemical equilibration cannot be reached within the lifetime of the interacting system. Therefore, the strangeness signature of QGP is an enhancement of strange particle production in nucleus-nucleus collisions versus nucleon-nucleon or nucleon-nucleus collisions, where only hadronic processes occur.

The main features of the strangeness signal can be summarized in the following points:

- the enhancement is expected to increase with the strangeness content, according to statistical hadronization [2];

- the strange baryons and antibaryon abundances are expected to be close to hadronic thermal and chemical equilibrium;

- expansion flow effects should be present in the hyperon transverse momentum spectra.

II THE WA97 EXPERIMENT

The WA97 experiment is designed to study strange particle production at central rapidity in Pb-Pb interactions and, for comparison, in p-Pb and p-Be collisions at the same beam momentum.

The WA97 set-up, sketched in fig. II, is described in [3], [4]. The 158 A GeV/c lead beam from the CERN SPS was incident on a 1% interaction length lead target. An 8% interaction length target was used for the proton induced runs. A multiplicity detector, consisting of silicon strips, samples the charged particle multiplicity for the off-line centrality analysis. A silicon telescope, made of pixel and microstrip planes, allows track reconstruction with high efficiency and resolution up to the most central Pb-Pb events. It was placed slightly above the beam line and inclined (pointing to the target), in order to accept particles at central rapidity and transverse momentum p_T greater than a few hundreds MeV/c. The apparatus was placed inside the 1.8 T magnetic field of the CERN Omega magnet.

The Λ, Ξ^- and Ω^- byperons and their antiparticles were identified by reconstructing their decays into charged particles. We also studied K_s^0 and negatives (h^-). The details of the analysis, i.e. the extraction of the signals and the weighting procedure are discussed in [4], [5].

III RESULTS

A Transverse mass distributions

The momentum spectra of identified hyperons has been parametrized as

$$\frac{d^2N}{dm_T dy} = f(y)\, m_T \, \exp\left(-\frac{m_T}{T}\right) \quad (1)$$

where $m_T = \sqrt{p_T^2 + m^2}$ is the transverse mass, the rapidity function $f(y)$ is assumed to be constant in the limited acceptance region of the experiment and the inverse slope parameter T has been extracted by a maximum likelihood fit.

In fig. 2 the inverse slope parameters for identified hadrons in the Pb-Pb system are shown as a function of their rest mass, along with results from other CERN SPS heavy-ion experiments. The figure shows a linear increase of T with the particle mass, which can be interpreted as a result of the global expansion flow. However Ωs, and possibly also Ξs, deviate from the general trend, indicated by the straight line.

This behaviour has been reproduced by a RQMD simulation and understood as an early decoupling of Ωs and possibly Ξs [6]. This means that the multi-strange production is very little affected by the later stages of the reaction.

B Particle yields

Particle yields were extrapolated to the common region covering full p_T and one unit of rapidity centered at midrapidity (y_{cm}) by integrating eq. (1):

$$Y = \int_m^\infty dm_T \int_{y_{cm}-0.5}^{y_{cm}+0.5} \frac{d^2N}{dm_T dy} dy$$

The yields have been studied as a function of the centrality of the collision, measured by the number of wounded nucleons, i.e. the nucleons participating to the collision (N_{part}). In Pb-Pb collisions, N_{part} has been estimated from the measured

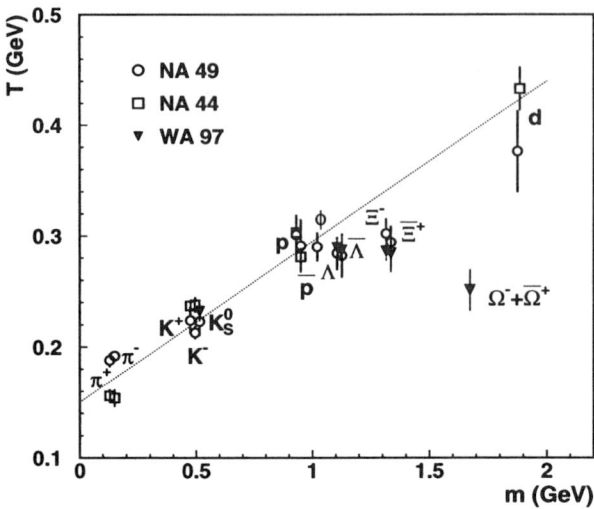

FIGURE 2. Dependence of inverse slope parameter T versus the particle mass.

TABLE 1. Values of the average number of participants.

	p-Be	p-Pb	Pb-Pb (4 classes)
$\langle N_{part} \rangle$	2.5	4.75	120 205 289 351

charged particle multiplicity, using the Wounded Nucleon Model [7], whereas in proton induced interactions it corresponds to minimum bias collisions. The resulting values of the average number of participants are reported in table 1, where the Pb-Pb centrality range has been divided into four classes.

The particle yields as a function of $\langle N_{part} \rangle$ are plotted in fig. 3. The particles are divided in two groups: those with at least one valence quark in common with the nucleon (plotted on the left) and those with no valence quark in common with the nucleon (plotted on the right). We have kept the two groups separate since they are empirically know to exhibit different production features. From the figure, it can be seen that all hyperon yields show a steady increase with centrality, from p-Be to p-Pb up to very central Pb-Pb collisions.

In fig. 4 the yields per participant relative to the p-Be ones are shown. They are compared to a yield curve, drawn in full line, proportional to the number of participants and intersecting the p-Be common point. It can be noted that all the yields increase with centrality from p-Be to p-Pb scaling with the number of participants, while from p-nucleus to Pb-Pb the increase is much faster than what expected from scaling with the number of participants. Moreover, it can be seen that, in both the considered groups, the enhancement increases with the strangeness content of the particle, up to a factor about 15 for Ωs.

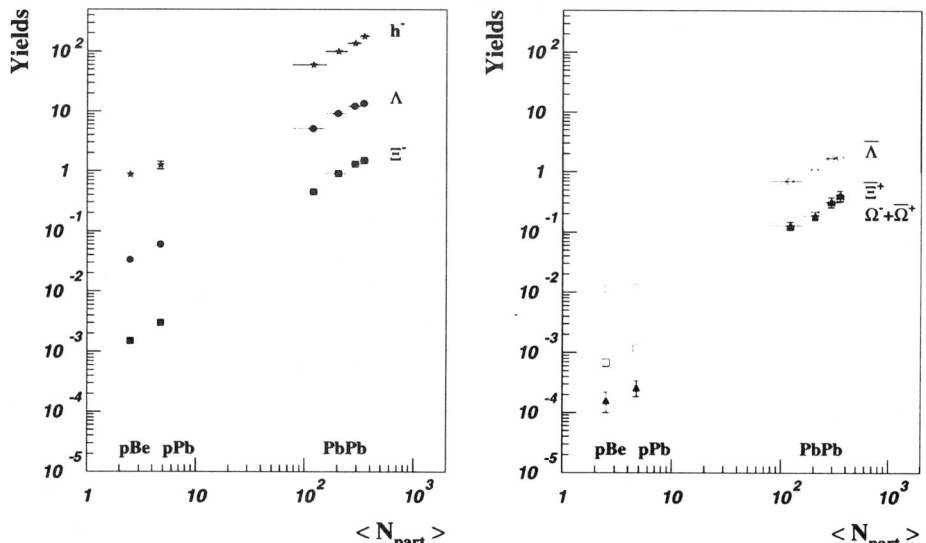

FIGURE 3. Particle yields as a function of the number of participants.

Fig. 4 also shows that the Pb-Pb yields increase proportionally to N_{part} in the covered centrality range, corresponding to $\langle N_{part} \rangle > 100$. This suggests a saturation of the enhancements, namely it seems that all the reported enhancements reach their maximum values for $\langle N_{part} \rangle$ below 100 and then remain constant.

Recently, thermal fits to particle ratios using our data have been performed [8]: they show that hyperon yields in Pb-Pb are consistent with thermal and chemical equilibrium.

Moreover, we showed in a recent analysis [9] that two of the most used hadronic models, VENUS 4.12 and RQMD 2.3, underestimate the measured multi-strange enhancements.

IV CONCLUSIONS AND OUTLOOK

The WA97 experiment has produced a comprehensive set of data on the production of strange and multi-strange hadrons at SPS. The emerging picture can be summarized as follows:

- the particle yields per participant at central rapidity are found to be enhanced when going from p-A to Pb-Pb collisions. The enhancement increases with the strangeness content of the particle up to a factor 15 for $\Omega^- + \overline{\Omega}^+$;

- the mechanism responsible for the enhancement is already saturated at $\langle N_{part} \rangle \sim 100$;

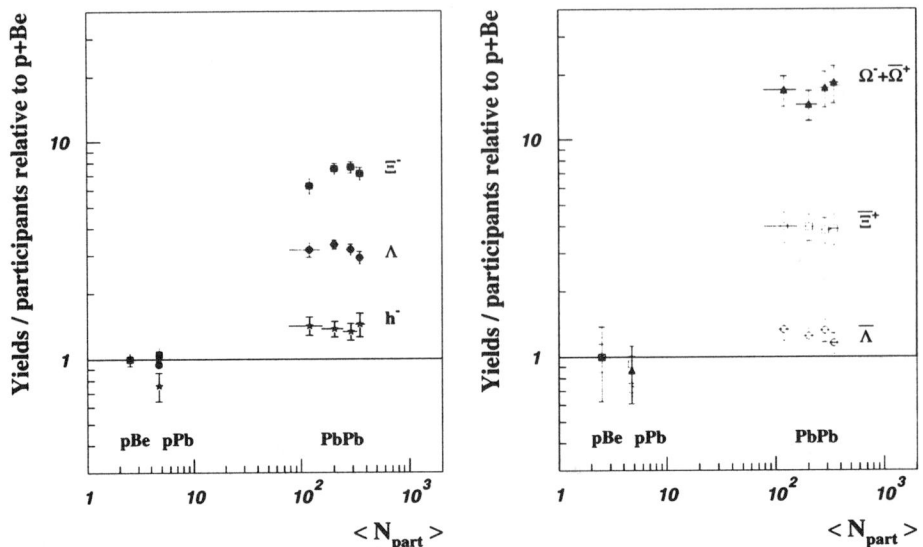

FIGURE 4. Particle yields per participant relative to the p-Be yields, as a function of the number of participants. The line refers to a yield proportional to the number of participants crossing the p-Be common point.

- hadronic models fail to simultaneously reproduce multi-strange enhancements in Pb-Pb versus p-Be and p-Pb data.

This picture provides strong evidence for production of deconfinded matter in central Pb-Pb collisions at SPS.

The question now is: Where deconfinement sets in? This questions will be addressed by the forthcoming NA57 experiment [10], which will take data at extendend centrality and at lower beam energy.

REFERENCES

1. J. Rafelski and B. Müller, Phys. Rev. Lett. **48** (1982) 1066.
2. A. Bialas, Phys. Lett. **B262** (1998) 449.
3. F. Antinori et al., Nucl. Phys. **A590** (1995) 139c.
4. E. Andersen et al., Phys. Lett. **B433** (1998) 209.
5. E. Andersen et al., Phys. Lett. **B449** (1999) 401.
6. H. van Hecke, H. Sorge, N. Xu, Phys. Rev. Lett. **81** (1998) 5764.
7. F. Antinori et al., CERN-EP-2000-002, submitted to Eur. Phys. J. C.
8. P. Braun-Munzinger, I. Heppe, J. Stachel, Phys. Lett. **B465** (1999) 15.
9. F. Antinori et al., Eur. Phys. J. **C 11** (1999) 79.
10. R. Caliandro et al., CERN/SPSLC/96-40 SPSLC/P300 20 August 1996.

Meson Mixing and Dilepton Production in Heavy Ion Collisions

A.K. Dutt-Mazumder, C. Gale[1], and O. Teodorescu

Physics Department, McGill University
3600 University St., Montreal, Quebec H3A 2T8, Canada

Abstract. We study the possibility of $\rho - a_0$ mixing via N-N excitations in dense nuclear matter. This mixing is found to induce a peak in the dilepton spectra at an invariant mass equal to that of the a_0. We calculate the cross section for dilepton production through mixing and we compare its size with that of $\pi - \pi$ annihilation. In-medium masses and mixing angles are also calculated. Some preliminary results of the mixing effect on the dilepton production rates at finite temperature are also presented.

Electromagnetic radiation, especially lepton pairs, constitutes a class of valuable probes in the context of heavy ion collisions. This owes to the fact that the leptons couple to hadrons via vector mesons and therefore hadronic processes involving $\ell^+\ell^-$ in the final channel are expected to reveal their properties in the dilepton spectra. Furthermore, the $\ell^+\ell^-$ pairs suffer minimum final state interactions and are thus likely to bring information to the detectors essentially unscathed. We will only consider dielectrons in this work.

Several experiments have measured, or are planning to measure, the lepton pairs produced in nucleus-nucleus collisions. They have been carried out by the DLS at LBL [1], and by HELIOS [2] and CERES [3] at CERN. Two new initiatives that will focus on electromagnetic probes will be PHENIX at RHIC [4] and HADES at GSI [5]. The density-dependent characteristics of vector mesons can also be highlighted through experiments performed at TJNAF [6].

We explore here the possibility of ρ-a_0 mixing via n-n excitations in nuclear matter. This is a pure density-dependent effect and is forbidden in free space on account of Lorentz symmetry. We show that such a mixing opens up a new channel in dilepton production and will modify the spectrum in the ϕ mass region. The details of the calculation will only be sketched here. The interested reader is invited to consult [7].

---

[1] Speaker.

The interaction Lagrangian we will use can written as

$$\mathcal{L}_{int} = g_\sigma \bar\psi \phi_\sigma \psi + g_{a_0} \bar\psi \phi_{a_0,a} \tau^a \psi + g_{\omega NN} \bar\psi \gamma_\mu \psi \omega^\mu$$
$$+ g_\rho [\bar\psi \gamma_\mu \tau^a \psi + \frac{\kappa_\rho}{2m_n} \bar\psi \sigma_{\mu\nu} \tau^a \partial^\nu] \rho_\alpha^\mu , \qquad (1)$$

where ψ, ϕ_σ, ϕ_{a_0}, ρ and ω correspond to nucleon, σ, a_0, ρ and ω fields, and τ_a is a Pauli matrix. The values used for the coupling parameters are obtained from Ref. [8].

The polarization vector through which the a_0 couples to ρ via the n-n loop is given by

$$\Pi_\mu(q_0, |\vec q|) = 2i g_{a_0} g_\rho \int \frac{d^4k}{(2\pi)^4} \mathrm{Tr}[G(k) \Gamma_\mu G(k+q)]. , \qquad (2)$$

where 2 is an isospin factor and the vertex for ρ-nn coupling is

$$\Gamma_\mu = \gamma_\mu - \frac{\kappa_\rho}{2m_n} \sigma_{\mu\nu} q^\nu . \qquad (3)$$

$G(k)$ is the in-medium nucleon propagator [9].

With the evaluation of the trace and after a little algebra Eq. (2) could be cast into a suggestive form:

$$\Pi_\mu(q_0, |q|) = \frac{g_\rho g_{a_0}}{\pi^3} 2q^2 (2m_n^* - \frac{\kappa q^2}{2m_n}) \int_0^{k_F} \frac{d^3k}{E^*(k)} \frac{k_\mu - \frac{q_\mu}{q^2}(k \cdot q)}{q^4 - 4(k \cdot q)^2} . \qquad (4)$$

It easily can be seen that it obeys current conservation: $q^\mu \Pi_\mu = 0 = \Pi_\nu q^\nu$. This implies that only one component of Π_μ is independent. In fact, only the longitudinal component of the ρ meson couples to the scalar meson while the transverse mode remains unaltered.

In presence of mixing the combined meson propagator might be written in a matrix form where the dressed propagator would no longer be diagonal:

$$\mathcal{D} = \mathcal{D}^0 + \mathcal{D}^0 \Pi \mathcal{D} \qquad \mathcal{D}^0 = \begin{pmatrix} D^0_{\mu\nu} & 0 \\ 0 & \Delta_0 \end{pmatrix} . \qquad (5)$$

The noninteracting propagator for the a_0 and ρ are given respectively by

$$\Delta_0(q) = \frac{1}{q^2 - m_{a_0}^2 + i\epsilon} , \qquad D^0_{\mu\nu}(q) = \frac{-g_{\mu\nu} + \frac{q_\mu q_\nu}{q^2}}{q^2 - m_\rho^2 + i\epsilon} . \qquad (6)$$

The mixing is characterised by the polarization matrix which contains non-diagonal elements

$$\Pi = \begin{pmatrix} \Pi^\rho_{\mu\nu}(q) & \Pi_\nu(q) \\ \Pi_\mu(q) & \Pi^{a_0}(q) \end{pmatrix} . \qquad (7)$$

Π^{a_0} and $\Pi^{\rho}_{\mu\nu}$ refer to the diagonal self-energies of the a_0 and ρ meson induced by the n-n polarization.

The ρ meson being a vector, one can write the longitudinal and transverse polarization as $\Pi_L = -\Pi_{00} + \Pi_{33}$ and $\Pi_T = \Pi_{11} = \Pi_{22}$. To determine the collective modes and their dispersion relation, one can define the dielectric function and look for its zeroes [10]. The left panel of Fig. 1 shows the relevant dispersion curves with

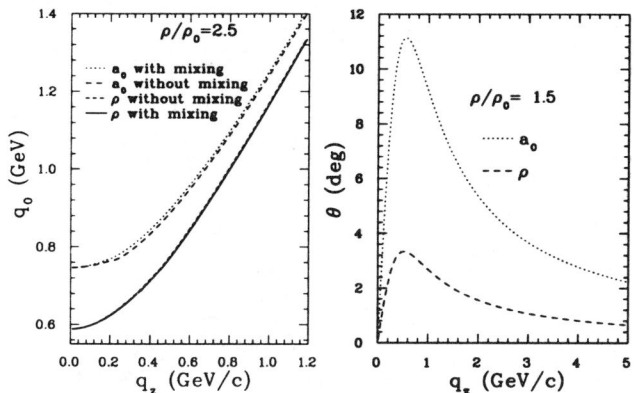

FIGURE 1. The dispersion curve and mixing angle at $\rho=2.5\rho_0$.

and without mixing at density $\rho=2.5\rho_0$. As only the L mode mixes with the scalar mode, we do not consider the T mode. The later in fact is the same as presented in Ref. [11] for the ρ meson, and in Ref. [12] for $\sigma - \omega$ mixing. The effect of mixing on the pole masses, as evident from Fig. 1, are found to be small. However, the mixing could be large when the mesons involved go off-shell.

To calculate the mixing angle, one diagonalises the mass matrix [11] with the mixing and obtains

$$\theta_{mix} = \frac{1}{2} \arctan\left(\frac{2\Pi^{\rho a_0}_{mix}}{m_{a_0}^2 - m_\rho^2 - \Pi^\rho_L + \Pi^{a_0}}\right) \quad (8)$$

In Eq. (8) $\Pi^{\rho a_0}_{mix} = M_i/|\vec{q}|\Pi_0$ which increases with density. Π_0 is the zero'th component of Eq. (4). The momentum dependence for a density of 1.5 times higher than the normal nuclear matter density is shown in the right panel of Fig. 1. This shows that for momenta beyond $|\vec{q}| \approx 0.2$ GeV/c the mixing is quite appreciable. It should also be noted that the mixing angle vanishes at $|\vec{q}| = 0$ as it should.

The ρ-a_0 mixing opens a new channel, $\pi + \eta \to e^+ + e^-$, in dense nuclear matter through n-n excitations. The cross-section for this process is expressed in terms of the mixing amplitude (Π_0) [7] and is plotted in Fig. 2.

One can notice in Fig. 2 that the process, $\pi + \eta \to e^+ + e^-$, at densities higher than ρ_0, not only enhances the overall production of lepton pairs but also induces an additional peak near the ϕ mass region. The contribution at the a_0 mass is

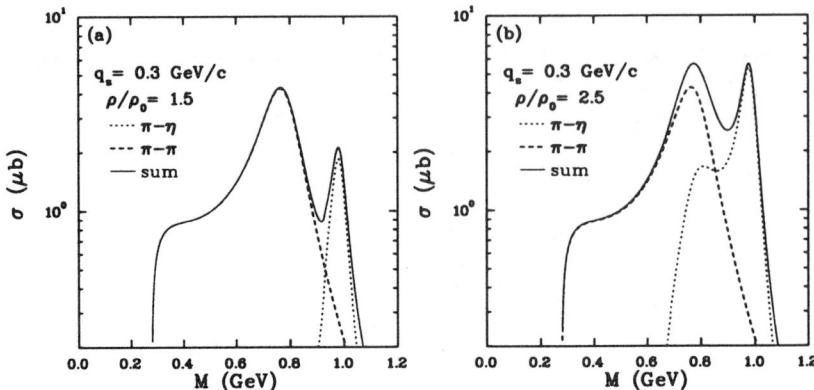

FIGURE 2. Dilepton production cross section for $\pi + \eta \to e^+ + e^-$ through matter-induced $\rho - a_0$ mixing, and for $\pi + \pi \to e^+ + e^-$.

comparable to that of $\pi + \pi \to e^+ + e^-$ near the ρ peak, for densities higher than ρ_0. Fig. 2 also shows that as the density goes even higher the dilepton yield arising out of the mixing also increases further.

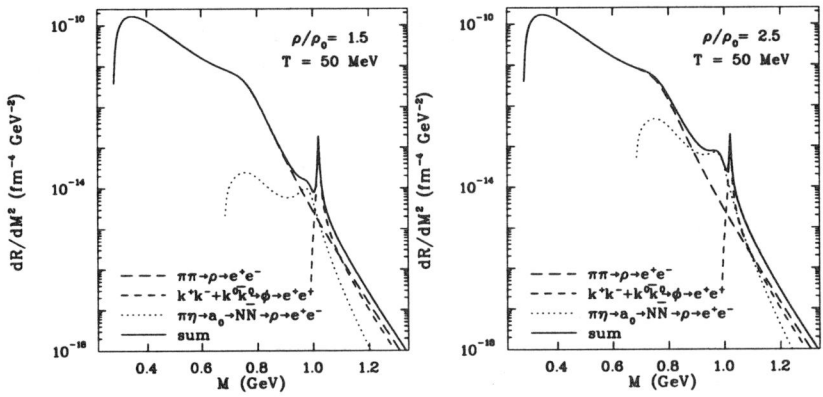

FIGURE 3. Dilepton production rates for T=50 MeV.

Further studies are in progress to assess finite temperature effects and to incorporate the necessary many-body machinery. We will show here only some preliminary results for the dilepton production rate including the effect of $\rho - a_0$ mixing. We work the independent particle approximation of kinetic theory. We assume a thermal gas of mesons at $T = 50 MeV$. Fig. 3 shows the dilepton production rate as compared with the standard $\pi\pi$ and $K\bar{K}$ annihilation. One can observe from Fig 3 that even at density $\rho/\rho_0 = 1.5$ the contribution of the mixing wins over the π-π

annihilation rate in the vicinity of M = 1.0 GeV. Naturally at higher density this goes up as evident from the right panel of Fig. 3.

We have highlighted the possibility of ρ-a_0 mixing in dense nuclear matter. We observe the appearance of an additional peak at a dilepton invariant mass that corresponds to that of the a_0. With sufficient experimental resolution, this effect could be observed. Maybe not as an individual peak, but probably more realistically as a shoulder in the ϕ spectrum. This feature would then be exclusively temperature and density-dependent and would thus be the reflection of a genuine in-medium effect.

This work was supported in part by the Natural Sciences and Engineering Research Council of Canada and in part by the Fonds FCAR of the Québec Government.

REFERENCES

1. See, for example, R. J. Porter et al., Nucl. Phys. **A638**, 499 (1998), and references therein.
2. M. Masera for the HELIOS collaboration, Nucl. Phys. **A590**, 93c (1995).
3. P. Wurm for the CERES collaboration, Nucl. Phys. **A590**, 103c (1995).
4. D. P. Morrison et al., Nucl. Phys. **A638**, 560 (1998).
5. J. Friese et al., Prog. Part. Nucl. Phys. **42**, 235 (1999).
6. M. Kossov et al., TJNAF proposal No. PR-94-002 (1994).
7. O. Teodorescu, A. K. Dutt-Mazumder, and C. Gale, Phys. Rev. C **61**, 051901 (2000).
8. R. Machleidt, Adv. Nucl. Phys. **19**, 198 (1989).
9. See, for example, Brian D. Serot and John D. Walecka, Adv. Nucl. Phys. **16**, 1 (1986).
10. S. A. Chin, Ann. Phys. 108, 301 (1977)
11. A. K. Dutt-Mazumder, B. Dutta-Roy, and A. Kundu, Phys. Lett. **399B**, 196 (1997).
12. K. Saito, K. Tsushima, A. W. Thomas, and A. G. Williams, Phys. Lett. **433B**, 243 (1998).

Measurements of Hadronic Observables with the Solenoidal Tracker at RHIC

Rene Bellwied for the STAR Collaboration

*Physics Department, Wayne State University,
Detroit, MI 48201, USA*

Abstract.
The capabilities of the STAR detector at RHIC regarding the measurement of hadronic observables will be described. Special emphasis will be given to the determination of event-by-event observables. Many of the hadronic particle measurements described in this short document may shed light on the occurance of a phase transition to a plasma phase in ultrarelativistic heavy ion collisions.

I INTRODUCTION

The recent announcement of indirect evidence for the formation of a new state of matter in several CERN heavy-ion experiments has led to renewed interest towards the field of relativistic heavy-ion collisions. With the advent of first collisions at the RHIC collider at BNL this paper is meant as a compilation of the hadronic physics observables measurable with STAR in the near future. STAR took first events at 30+30 A GeV/c by mid-June 2000, and first events at 65+65 A GeV/c by the end of June 2000. Fig. 1 shows a typical event taken about two days before the writing of this manuscript (first week of July).

As expected, the very first measurements reported from RHIC will be based on hadronic observables. Hadronic observables in relativistic heavy ion collisions have been used widely to determine the collision centrality (charged particle multiplicity) and to characterize the freeze-out properties of the final state (slope parameter, particle abundances etc.). Thermal model descriptions of hadronic measurements at CERN, in particular the strange particle abundances and ratios, have concluded that only the inclusion of a phase transition into any model calculation will allow a proper description of the measured results [1], [2].

STAR is perfectly equipped to measure hadronic particle spectra, abundances, and ratios at energies that are up to ten times larger than the CERN-SPS energies. In the following I will first detail the STAR detector setup, then show the STAR measurement capabilities, and end with a short description of the STAR hadronic physics program.

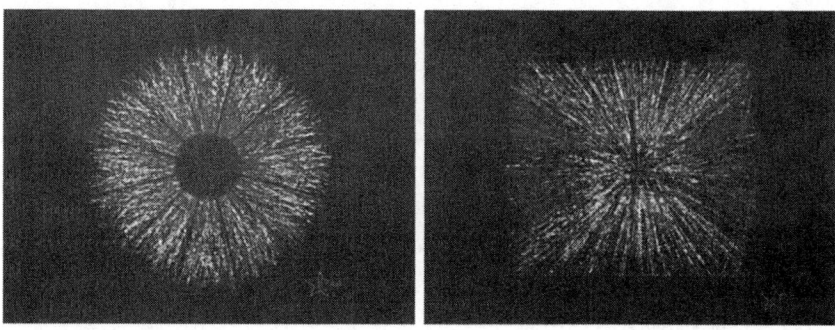

FIGURE 1. Front and Side View of the track distribution in the STAR-TPC based on a central Au-Au Collisions at RHIC at 65+65 A GeV/c.

II STAR DETECTOR SETUP AND PERFORMANCE CAPABILITIES

The main detector component of STAR is a large 2π Time Projection Chamber (TPC), which covers a pseudo-rapidity range from $\eta = -1.5$ to 1.5. The TPC is surrounded by a large conventional solenoidal magnet that generates a field up to 0.5 T. The TPC allows the precise measurement of the momenta of charged particles and it has limited particle identification capabilities through the measurement of energy loss in the TPC gas. Additional detector components have been added or will be added to the TPC to improve its performance in some specific areas. At smaller radius a Silicon Vertex Tracker (SVT) will be added to allow the reconstruction of secondary vertices, in particular to measure short lived strange and multi-strange baryons (Λ, Ξ^-, Ω^-, and their anti-particles). This device will also extend the low transverse momentum coverage of the TPC for all charged particles. In forward direction (η=2-4) a Forward Time Projection Chamber (FTPC) will be added to extend the phase space coverage for charged particle reconstruction. At mid-rapidity small detector segments of a Ring Imaging Cherenkov Counter (RICH) and a Time Of Flight patch (TOFp) will be added to extend the momentum range for identified charged particles beyond momenta up to which the energy-loss method can be applied. Finally, the TPC will be surrounded in 2π by an Electro-Magnetic Calorimeter (EMC) to enable the measurement of the kinematics of neutral hadrons and photons through conversion into lepton pairs and the subsequent measurement of the pair energy.

With the combined tracking system of SVT and TPC reconstruction efficiencies of charged particles range above 90% from a transverse momentum of 100 MeV/c up to the highest measurable momenta (above 50 GeV/c). The achievable momentum resolution is about 2% for momenta below 5 GeV/c and degrades only slowly to higher momenta. At 50 GeV/c the resolution is still below 10%. The calorimeter resolution is measured to be 18% $/\sqrt{E}$, and particle identification capabilities are summarized in the following table.

TABLE 1. Momentum range (in Gev/c) of particle identified charged hadrons depending on the detector component used in the analysis

Detector	pion range	kaon range	proton range
SVT/TPC	0.07-0.6	0.07-0.6	0.7-1.0
TOFp	0.2-1.6	0.6-1.8	0.6-3.0
RICH	0.6-3.0	1.1-3.0	1.5-5.0

III STAR PHYSICS PROGRAM

The first set of measurements to arise from STAR after RHIC turn-on will be based on global observables, in particular the charged particle multiplicity as a function of centrality. STAR has two independent centrality detectors, namely a Central Trigger Barrel (CTB), which surrounds the TPC and which measures charged particle hit density, and a Zero Degree Calorimeter (ZDC) which measures the energy deposition in very forward direction and which is common to all RHIC experiments.

The next measurements in STAR will be the transverse momentum and rapidity distributions for all primary charged particles (π, k, p, d, and their anti-particles). In addition to providing a basic measurement of the kinematics of the emitted particles, which is a crucial part of every theory prediction, these kinematic spectra also contain information of the energy loss of partons in the nuclear medium [3].

In subsequent measurements STAR will add the capability to reconstruct short-lived hadrons via measuring their decay particles ($K_s^0, \Lambda, \Xi^-, \Omega^-$).

The next set of measurements will include the more sophisticated analyses, e.g. two particle interferometry based on the method suggested by Hanbury-Brown and Twiss (HBT) to measure the spatial extension of a particle emitting source [4].

An even more ambitious program which requires calorimetry and good particle reconstruction in the TPC is the measurement of the leptonic decay of mesons and vector mesons (π^0, Φ, J/ψ, ψ', and the Y). Although not the primary goal of STAR, these measurements will have some complementarity and some overlap with the similar program measured by the PHENIX detector at RHIC [5].

All of the analyses described above were based on event samples that range in statistics from around 100,000 events up to several million events required to obtain the necessary statistics for a significant measurement. Some observables, in particular the ones based on pions, might be measurable on an event-by-event basis. We expect around 1500 charged pions per event within the coverage of the TPC. This sample is statistically significant and the potential event-by-event observables will be non-statistical charged hadron fluctuations (topologies) in momentum and/or rapidity space, which are postulated through a variety of QGP induced mechanisms, e.g. chiral symmetry restoration [6]. Recently it was also postulated that charge fluctuations in momentum space could potentially be interpreted as a sign of parity violation induced by QGP formation [7]. In addition we will attempt to measure charged pion ratios, hadron/pion ratios and two pion interferometry on an event-by-event basis. The hope is that only a subset of events actually exhibits the phase transition characteristics and thus event-by-event analyses should be more sensitive to the QGP signals than inclusive measurements. Furthermore by correlating signals on an event-by-event basis the separation of QGP events from non-QGP events should be event stronger [8].

IV CONCLUSIONS

The latest major achievement of relativistic heavy ion physics was the recent announcement of the first successful collisions of Gold beams at the Relativistic Heavy Ion Collider (RHIC) at Brookhaven National Laboratory (BNL). This announcement coincided with the announcement of indirect evidence for a Quark Gluon Plasma formation in fixed target relativistic heavy-ion collisions at the CERN accelerator in Geneva.

STAR will be the primary device at RHIC to measure hadronic observables (ratios, slopes, $<p_T>$, spectral shapes, rapidity distributions, non-statistical phase space fluctuations). Its large solid angle coverage will allow high precision measurements event-by-event as well as on an inclusive basis. Although it is generally agreed that no single hadronic parameter will serve as a 'smoking gun' for the QGP, the wealth of data deduced from STAR will provide the basis for a many-parameter, correlated approach to the determination of the equation of state. Hadronic measurements lead to complete information of the three dimensional thermodynamic phase space of the state of nuclear matter (as a function of the temperature T and the chemical potentials μ_B and μ_S). Correlations between hadronic observables might yield conclusive evidence for a phase transition to a quark gluon plasma.

Acknowledgements: I would like to thank the following group of STAR physicists and theorists for providing their simulations to me: Sean Gavin, David Hardtke, Gerd Kunde, Qun Li, Curtis Lansdell, Tom LeCompte, Mike Lisa, Bill Llope, Claude Pruneau, Vladimir Rykov, Thomas Ullrich, Xin-Nian Wang, Nu Xu.

REFERENCES

1. J. Letessier and J. Rafelski, Phys.Rev. C59 (1999) 947
2. U. Heinz, Nucl.Phys. A661 (1999) 140c
3. X.N. Wang, Phys.Rev. C61 (2000) 064910
4. R. Hanbury-Brown and R.Q. Twiss, Phil.Mag.Nature 178 (1956) 1046
5. K. Shigaki for the PHENIX Collaboration, contribution to these proceedings
6. J.D. Bjorken, Acta Phys.Polon. B28 (1997) 2773
7. D. Kharzeev and R. Pisarski, Phys.Rev. D61 (2000) 111901
8. S. Gavin and C. Pruneau, Phys.Rev. C61 (2000) 004901

Physics via Lepton Channels at RHIC/PHENIX

Kenta Shigaki* for the PHENIX Collaboration

High Energy Accelerator Research Organization (KEK), 1-1 Oho, Tsukuba, Ibaraki 305-0801, Japan

Abstract. The first physics data taking has started at Relativistic Heavy Ion Collider (RHIC) at Brookhaven National Laboratory. The physics goal of RHIC is to investigate QCD in extreme conditions and scales, and ultimately to create, to observe, and to understand the nature of a deconfined quark-gluon plasma phase. Experimental approaches and physics prospects via lepton channels at RHIC are discussed along with strategies and capabilities of the PHENIX experiment.

INTRODUCTION

Heavy ion programs at CERN SPS suggested possible onsets of the QCD phase transition [1]. It has been generally agreed that various signatures have to be combined to understand the whole picture of relativistic heavy ion reactions: hadrons to probe the boundary conditions of collision dynamics, photons to trace the evolution of the system, and leptons to probe the early stage of the collisions. Among these the particular importance of lepton channels was endorsed.

Relativistic Heavy Ion Collider (RHIC) at Brookhaven National Laboratory, which started its first physics operation in 2000, will create a much higher energy density and clearer environment due to the center-of-mass collision energy 12 times higher than at SPS. It has comprehensive and complementary programs to achieve conclusive experimental results on the physics goals to investigate QCD in extreme conditions and scales, and ultimately to create, to observe, and to understand the nature of a new phase of matter predicted by QCD, *i.e.* a deconfined quark-gluon plasma phase. In particular the PHENIX experiment [2,3] uniquely features systematic studies of penetrating probes via single- and di-lepton channels in conjunction with photon and hadron measurement in the single experiment. We will discuss experimental approaches and physics prospects via lepton channels at RHIC along with strategies and capabilities of PHENIX.

PHENIX EXPERIMENT AT RHIC

The basic strategy of PHENIX is to utilize a wide variety of probes with the single detector and trigger system. It is designed for sensitivity to as many signatures of the QCD phase transition as possible and to essentially all the time scales of relativistic heavy ion reactions. The signatures in the scope of PHENIX include strangeness and heavy flavor production, jet quenching, color Debye screening, chiral symmetry restoration via vector meson properties and perhaps disoriented chiral condensation, and thermal radiation from hot hadron gas. With the emphasis on penetrating probes, *i.e.* real and virtual photons and single- and di-leptons, PHENIX is characterized with the high particle identification capability, high momentum and mass resolution, and broad kinematic coverage. The high rate capability with selective multi-level triggering plays another vital role to access the rare physics probes.

As shown in Figure 1, PHENIX consists of four arms of magnetic spectrometers along with a set of global detectors for event characterization: two central arms to measure photons, electrons and hadrons, and two forward arms for muon measurement.

PHENIX measures electrons in the two central arms covering a pseudo-rapidity range $-0.35 < \eta < 0.35$ and azimuthally $d\phi = \pi/2$ each. Charged particles are

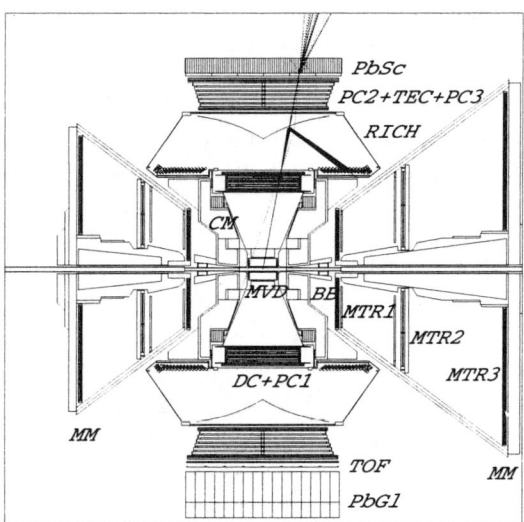

FIGURE 1. A plan view of the experimental apparatus of PHENIX and a simulated single-electron event. BB: beam/beam counter. MVD: silicon multiplicity and vertex detector. DC: drift chamber. PC1–3: pad chambers. TEC: time expansion chamber. RICH: ring imaging Čerenkov counter. TOF: time-of-flight wall. PbSc: lead-scintillator calorimeter. PbGl: lead-glass calorimeter. MTR1–3: muon trackers. CM: central magnet. MM: muon magnets.

tracked with drift chambers, pad chambers and time expansion chambers with a help from ring imaging Čerenkov counters, a time-of-flight wall and electromagnetic calorimeters. The ring imaging Čerenkov counters serve as the primary electron identification device providing hadron rejection at 10^4 level for single tracks and at 10^3 level in central Au+Au environment. The calorimeters provide additional hadron rejection via measured energy/momentum ratio especially at high momentum, and the time expansion chambers via dE/dx measurement at relatively low momentum. The combined electron identification capability stays above 10^4 up to the threshold momentum of pions in the Čerenkov counter, which is 3.5 GeV/c with CO_2 radiator and 4.7 GeV/c with C_2H_6. In the momentum range the momentum resolution dp/p is better than 2% in r.m.s..

Muons are measured in the two forward arms with full azimuthal coverage. Their pseudo-rapidity coverage is $1.2 < \eta < 2.4$ for the north arm and $1.2 < \eta < 2.2$ for the south. They are equipped with tracking chambers in the magnetic field followed by muon identifiers with the energy threshold at 2 GeV. The left panel in Figure 2 shows a simulated di-muon mass spectrum demonstrating the high mass resolution to separate not only J/ψ and $\psi(2S)$ but even $\Upsilon(1S)$ and $\Upsilon(2S+3S)$.

PHYSICS VIA LEPTON CHANNELS

Quarkonium has been proved at SPS as a promising probe of the QCD phase transition. NA50 observed $\psi(2S)$ strongly suppressed [4] and more interestingly a multi-step-like behavior of J/ψ possibly due to dissolution of χ_c and J/ψ [5]. If the quarkonium suppression is scaled to the energy density in the reaction system, J/ψ is expected to dissolve from semi-peripheral collisions in Au+Au reaction at the RHIC energy. On the other hand there exist many uncertainty factors to be cleared before understanding the so-called "anomalous" suppression, e.g. elemen-

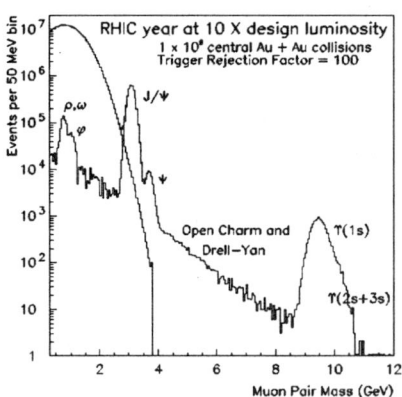 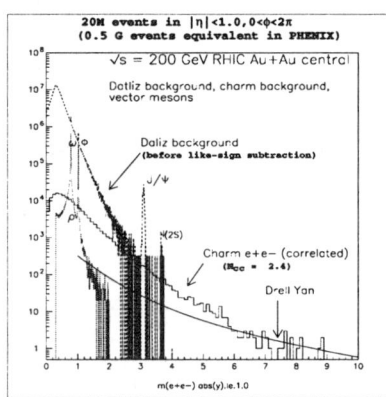

FIGURE 2. Simulated di-muon (left) and di-electron (right) spectra in PHENIX forward and central arms, respectively.

TABLE 1. PHENIX acceptance and resolution for quarkonium measurement.

	central arms	forward arms
pseudo-rapidity coverage	$-0.35 < \eta < 0.35$	$1.2 < \eta < 2.4$ (north)
		$-2.2 < y < -1.2$ (south)
J/ψ acceptance	0.8% of $B_{ee}\sigma$	4.3% of $B_{\mu\mu}\sigma$ (per arm)
	(4% of $B_{ee}\sigma$ in $|y|<0.5$)	
Υ acceptance	1.7% of $B_{ee}\sigma$	3.0% of $B_{\mu\mu}\sigma$ (per arm)
	(5% of $B_{ee}\sigma$ in $|y|<0.5$)	
J/ψ mass resolution	20 MeV	105 MeV
Υ mass resolution	160 MeV	180 MeV

tary J/ψ production cross section, initial state suppression or gluon shadowing, and nuclear absorption. PHENIX attacks these problems to understand the whole picture of quarkonium production/suppression in relativistic heavy ion reactions through systematic measurement of quarkonia.

On the accelerator side, RHIC is capable of colliding any combination of ions from proton to ^{197}Au in the independent two rings, and at any energy from the injection energy to 500 GeV per nucleon per rigidity (200 A GeV for heavy ions). It allows us p+p and p+A measurement as a function of $\sqrt{s}$ as the baseline. Note that the elementary cross section of J/ψ production at the RHIC energy presently has uncertainty of a factor of ~ 2.

The basics of PHENIX capability in quarkonium measurement is summarized in Table 1. The central and forward arms look into kinematic regions with substantially different local energy densities, as demonstrated in Figure 3 using the rapidity density of produced particles as a measure. Simultaneous access to the two regions provides a good test whether the quarkonium suppression is a function of local energy density. The transverse momentum coverage for di-electron also spans widely from 0 to above 5 GeV/c. As another key systematics, PHENIX can identify several members of J/ψ and Υ families, namely J/ψ, $\psi(2S)$, $\Upsilon(1S)$ and $\Upsilon(2S+3S)$ separately (see Figure 2). It is also capable to measure many reference channels such as di-lepton continuum from charm and Drell Yan, single leptons and single photons.

The expected statistics for one year of RHIC Au+Au running are shown in Table 2 with following assumptions: (1) 1.6 nb^{-1} of integrated luminosity, corresponding to 27 weeks of running with a 50% experimental duty factor at the RHIC nominal Au+Au luminosity of 2×10^{26} cm^{-2} s^{-1}, and (2) 400 nb of $B_{ll}\sigma_{NN}(J/\psi)$ at $\sqrt{s} = 200$ GeV. Nuclear absorption or any further suppression is not included in the calculation. NA50 estimates the factor in Pb+Pb reactions at $\sqrt{s} = 17$ GeV to be ~ 0.31 for J/ψ in minimum-bias events, ~ 0.27 for J/ψ in central events, and ~ 0.06 for $\psi(2S)$ [4,5]. High statistics data from the high rate capability with selective triggering open a possibility of further systematics, e.g. quantitative evaluation of χ_c contribution to J/ψ and detailed study of centrality dependence of quarkonium behavior.

TABLE 2. Expected quarkonium statistics in PHENIX for one year of RHIC running. See text for details of assumptions.

	central arms		forward arm (per arm)	
	10 % central	minimum-bias	10 % central	minimum-bias
J/ψ †	~ 70,000	~ 200,000	~ 350,000	~ 1,000,000
ψ (2S) †	~ 1,400	~ 4,000	~ 7,000	~ 20,000
Υ †	~ 80	~ 250	~ 160	~ 500
Drell Yan (> 4 GeV)	~ 250	~ 700	~ 1,500	~ 4,000

† Nuclear absorption and "anomalous" suppression factor are not included.

Open heavy flavor production is a probe of the initial state and also serves as a good reference to quarkonium production. PHENIX addresses it via single leptons at high momentum, high-mass di-leptons, and more sophisticatedly electron-muon coincidence. As an example the right panel in Figure 2 shows a simulated di-electron spectra at $\sqrt{s} = 200$ GeV. Contribution of open charm will dominate single electrons above 2 GeV/c and di-electrons above the J/ψ mass.

A change of properties of light vector mesons, such as mass, decay width and/or branching ratio, is proposed as a probe of chiral symmetry restoration. NA45 reported enhancement of low-mass di-electrons which could be attributed to enhancement and melting of ρ [6]. PHENIX addresses it with the high electron identification capability and resolution. With the mass resolution better than 5 MeV in r.m.s., ϕ and ω will be clearly separated as shown in Figure 2. Without like-sign subtraction signal-to-noise ratios of ~ 1/10 and ~ 1/15 will be achieved for ϕ and ω, respectively.

Thermal di-lepton is another probe of interest to access radiation from the hot gas state as a complement to direct photon measurement. At RHIC there is a possible invariant mass window to look into thermal di-electrons from 1 to 2 GeV.

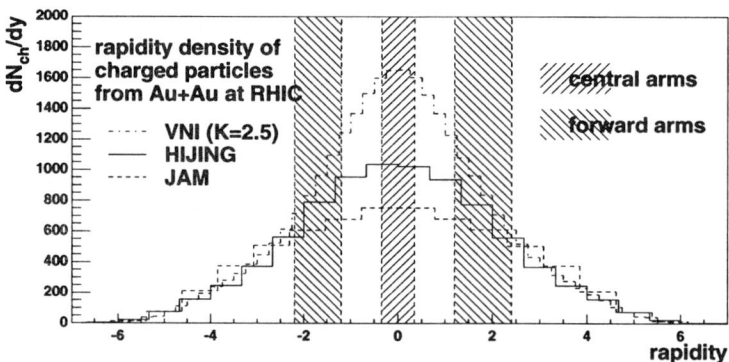

FIGURE 3. Expected produced particle density, as a measure of local energy density, in kinematic coverage of PHENIX central and forward arms.

UPCOMING PHYSICS

In the programs at RHIC in 2000, the first priority is placed to record at least 1 μb^{-1} of Au+Au reactions at $\sqrt{s} = 130$ GeV. First physics results via photon, electron and hadron channels will come out of PHENIX. It should be stressed that PHENIX is capable of accessing most of the physics probes and essentially all the time scales of the reaction already in the first year: initial hard processes via jet, hard photon and hadrons at high transverse momentum; deconfinement via heavy quarkonium; chiral restoration via light vector mesons; thermalization via soft photons, non-resonant di-electrons, single electrons from open charm and photons at high transverse momentum; hadronization via hadron spectra, strangeness and interferometry; and final-state hydro-dynamics via transverse energy distribution and produced particle density.

In the following year, Au+Au running at the full energy of $\sqrt{s} = 200$ GeV is planned, along with p+p running to characterize the baseline and for spin physics programs. PHENIX will also add the muon measurement capability. A greater sensitivity to rare probes from relativistic heavy ion reactions, as well as first results on spin physics, is expected in 2001.

SUMMARY AND CONCLUDING REMARKS

RHIC, a unique and versatile facility to investigate QCD in extreme conditions and scales, has started its operation. The PHENIX experiment at RHIC is suited for systematic studies of relativistic heavy ion reactions, notably via lepton channels with the high particle identification capability, high resolution, broad kinematic coverage and high rate/statistics capability. Utilizing a wide variety of physics probes, it covers many signatures of the QCD phase transition and accesses essentially all the time scales of relativistic heavy ion reactions. The longstanding search for the QCD phase transition and a deconfined quark-gluon plasma phase is expected to reach conclusive experimental results. Systematic study of quarkonium production/suppression is one of the promising clues. First physics results via photon, electron and hadron channels will come out of PHENIX in 2000, followed by more systematic data including muon channels in 2001.

REFERENCES

1. *http://press.web.cern.ch/Press/Releases00/PR01.00EQuarkGluonMatter.html.*
2. *http://www.phenix.bnl.gov/.*
3. PHENIX Collaboration, D. P. Morrison *et al.*, *Nucl. Phys. A* **638** (1998) 565c.
4. NA50 Collaboration, M. C. Abreu *et al.*, *Nucl. Phys. A* **638** (1998) 261c.
5. NA50 Collaboration, M. C. Abreu *et al.*, *Phys. Lett. B* **477** (2000) 28.
6. NA45 Collaboration, G. Agakichiev *et al.*, *Phys. Lett. B* **422** (1998) 405.

Results from pQCD for A+A collisions at RHIC & LHC energies

K. Tuominen[1]

*Department of Physics, University of Jyväskylä,
P.O. Box 35, FIN-40351, Jyväskylä, Finland*

Abstract. This talk will discuss how to compute initial quantites in heavy ion collisions at RHIC (200 AGeV) and at LHC (5500 AGeV) using perturbative QCD (pQCD) by including the next-to-leading order (NLO) corrections and a dynamical determination of the dominant physical scale. The initial numbers are converted into final ones by assuming kinetic thermalization and adiabatic expansion.

The whole heavy ion physics community has entered into exciting and intense period as the first results from RHIC are expected to appear, and the era of LHC seems no more very distant. However, despite the solid status of QCD as the theory of strongly interacting matter, many uncertainties remain in the predictions for the outcome of the current experiments and various methods have been applied. Further experimental data will (hopefully) single out the best candidates for the correct approach.

As the collision energy is increased from that of the SPS, larger intrinsic scales are generated and the applicability of perturbative QCD (pQCD) becomes possible. The initial particle production is expected to be dominated by minijets, i.e. partons with $p_T \sim 1\ldots 2\text{GeV} \gg \Lambda_{QCD}$ [1]. By assuming independent multiple semi-hard parton-parton collisions the average energy carried by minijets with $p_T \geq p_0$ at the rapidity interval ΔY in a central $\mathbf{b} = 0$ AA collision is given in leading-order (LO) by [2]

$$\overline{E}_{AA}^T = T_{AA}(\mathbf{0})\sigma\langle E_T\rangle$$
$$= T_{AA}(\mathbf{0}) \sum_{q,\bar{q},g} \int_{p_0,\Delta Y} dp_t dy_1 dy_2 x_1 f_{i/p}(x_1, Q^2) x_2 f_{j/p}(x_2, Q^2) \frac{d\hat{\sigma}^{ij \to jk}}{d\hat{t}} p_T, \quad (1)$$

where $T_{AA}(\mathbf{b})$ is the standard nuclear overlap function and p_0 is the smallest transverse momentum scale to be considered. Collinear factorization is assumed to

---

[1] kimmo.tuominen@phys.jyu.fi

hold and any effects beyond it are neglected. This perturbative minijet approach suffers from two major sources of uncertainties:

(i) The next-to-leading order (NLO) corrections, which have not been known prior to [3,4], but rather have been simulated by *ad hoc* K-factors.

(ii) The determination of which value of p_0 to use at, say, RHIC or LHC energies. Should one stick to some constant universal value of p_0 or will this parameter possess some nontrivial $\sqrt{s}$- and A-dependence?

In the following we will provide answers to both of these questions and combine them to obtain numerical estimates of average transverse energies and charged particle multiplicities at RHIC and LHC energies.

As is evident from formula (1), to deal with the first uncertainty, the relevant quantity to compute in NLO pQCD is the $\sigma\langle E_T \rangle$, the first moment of the perturbative E_T distribution in a pp-collision. The infrared safe NLO computation of this quantity has been presented in [4], where the computation was formulated via the subtraction algorithm of S. Ellis, Z. Kunszt and D. Soper [5].

The E_T in central rapidity region is defined to be the total p_T entering this region and originating from hard subprocessess in which at least an amount of $2p_0$ of transverse momentum is released. The numerical results are shown in fig. 1(a).

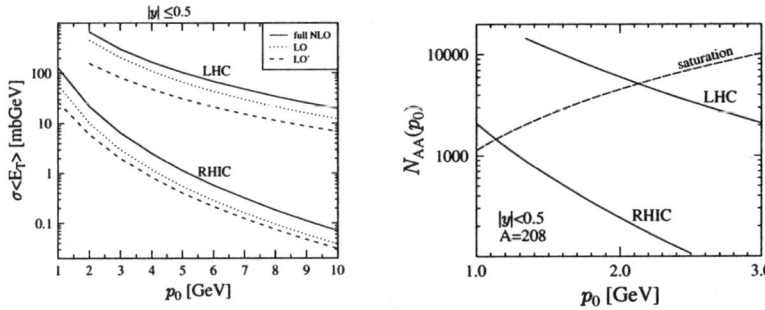

FIGURE 1. (a) The NLO $\sigma\langle E_T \rangle$. The LO' stands for the leading order result evaluated with 2-loop α_s and NLO parton distributions, whereas the LO stands for the leading order result evaluated with 1-loop α_s and LO parton distributions. The rapidity interval is chosen to be the central unit, and the parton distributions used are those of GRV-94 -set [6]. (b) The average number of QCD-quanta produced with $p_T \geq p_0$ and $|y| \leq 0.5$ as a function of p_0. The saturation scale p_{sat} is determined by the points of intersection with the dashed curve $(p_0^2 R_A^2)$ labelled "saturation".

To analyze the implications of these numbers, let us define two K-factors: $K =$ (full NLO)/LO and $K' =$ (full NLO)/LO', the first of these measuring the deviation of the NLO results from the consistent LO calculation and the latter measuring rather the relative difference between two subsequent terms of the perturbation series. The magnitude of these K-factors is due to a new kinematical region which manifests itself only at NLO. If one rejects this new domain, the resulting K-factors would be a factor of two smaller than those obtained from fig. 1(a), but then a significant amount of perturbatively calculable E_T would be neglected.

For a detailed description of the calculation and issuses such as the scale choices, see ref. [4] and references therein. For the purposes to be considered here, the sufficient observation is that we now have control over the magnitude of NLO corrections which are stable relative to LO results even at few GeV scales, thus signalling the applicability of pQCD in this domain.

Turning to the uncertainty (ii), then, it is clear that as p_0 is decreased the cross section, as well as the uncertainty, grows. At certain value of saturation, $p_0 = p_{\text{sat}}$ the system becomes very dense and new physics enters [7]. For large nuclei and large collision energies this may happen already in the perturbative domain, which we concluded on the basis of the NLO analysis to include also the few GeV region. Then the corresponding values for the number of particles as well as for the amount of E_T are easily produced via the perturbative computation at $p_0 = p_{\text{sat}}$, which effectively accounts for the contributions of all scales, since the partons with $p_T \gg p_{\text{sat}}$ are rare and those with $p_T \ll p_{\text{sat}}$, altough numerous, contribute negligibly to total E_T.

Various ways to determine the actual magnitude of p_{sat} can be conjectured. A simple geometric criterion has been presented in [8]. This is based on the idea that if one assigns an effective area π/p_0^2 to each gluon produced, then at certain value of p_0 the total area of $N_{AA}(p_0, \sqrt{s}\Delta Y)$ gluons produced will exceed the effective transverse area πR_A^2 of the nucleus. Therefore one can iterate the equation $N_{AA} = p_{\text{sat}}^2 R_A^2$ to determine p_{sat} for given A and $\sqrt{s}$, see fig.1(b). On the basis of the NLO analysis, we take here $K = 2$ to account for the NLO corrections and also implement nuclear shadowing via the EKS98 parametrization [9].

All the initial quantities then computed can well be fitted by a scaling law of a type $CA^b(\sqrt{s})^b$. In particular one finds that

$$
\begin{aligned}
p_{\text{sat}}/\text{GeV} &= 0.208 A^{0.128}(\sqrt{s})^{0.191} \\
\epsilon_i/(\text{GeV}/\text{fm}^3) &= 0.103 A^{0.504}(\sqrt{s})^{0.786} \\
n_i \cdot \text{fm}^3 &= 0.370 A^{0.383}(\sqrt{s})^{0.574}
\end{aligned}
\quad (2)
$$

where the particle and energy densities are evaluated at p_{sat}, and the whole production process is then considered to take place at $\tau_0 = 1/p_{\text{sat}}$.

These, however, are just the initial numbers at 0.2 (0.1) fm/c at RHIC (LHC), and the major problem is how to get from these to the ones at later instants and to finally arrive at experimentally visible quantities. Let us therefore assume, not completely without reason (see [8]), that the system is initially thermalized in the sense that it possesses a correct ratio of energy per particle as far as the dominant gluonic particle content is considered, and expands conserving the total entropy $S \approx 3.6 N_i$. As the final particles consist dominantly of pions, for which $S \approx 4 N_f$, we find that $N_f = 0.9 N_i$. As the initial volume is $V_i = \pi R_A^2 \Delta Y/p_{\text{sat}}$, formulae (2) give

$$N_f = 1.245 A^{0.92}(\sqrt{s})^{0.383}. \quad (3)$$

After the conference the very first measurements by PHOBOS collaboration have been announced [10]. According to them, the charged particle multiplicity at midrapidity is $dN/d\eta = 408 \pm 12 \pm 30$ at 56 AGeV and $555 \pm 12 \pm 35$ at 130 AGeV. The framework described here gives $N_{ch} = 2/3 N_f = 370$ and 530 per unit η respectively, when taking into account that number of participants was reported to be 330 for $\sqrt{s} = 56$ AGeV and 343 for $\sqrt{s} = 130$ AGeV, and that $dN/dy = 1.15 dN/d\eta$.

The final E_T in this scenario is obtained by means of hydrodynamics [8,11] as

$$E_T = N_f \times [0.39 + 0.061 \ln(N_f/A)]. \quad (4)$$

Numerical values then per unit η are $E_T = 260$ GeV for $\sqrt{s} = 56$ AGeV and $E_T = 390$ GeV for $\sqrt{s} = 130$ AGeV using the multiplicites computed above and the quoted participant numbers. Measurements of these final transverse energies are awaited to appear soon. These measurements will then allow us to draw conclusions on the issues such as the true degree of thermalization in the system.

Acknowledgements: I wish to thank K.J. Eskola, K. Kajantie and P.V. Ruuskanen for collaboration. The financial support from the Academy of Finland as well as from the organizers of the CIPANP2000 conference is gratefully acknowledged.

REFERENCES

1. Blaizot J.P., and Mueller A.H., Nucl. Phys. **B289** 847 (1987);
   Kajantie K., Landshoff P.V. and Lindfors J., Phys. Rev. Lett. **59** 2527 (1987).
2. Eskola K.J., Kajantie K. and Lindfors J., Nucl. Phys. **B323** 37 (1989).
3. Leonidov A. and Ostrovsky D., Eur. Phys. J. **C11** 495 (1999).
4. Eskola K.J. and Tuominen K., hep-ph/0002008.
5. Ellis S.D., Kunszt Z. and Soper D.E., Phys. Rev. **D40** 2188 (1989);
   Ellis S.D., Kunszt Z. and Soper D.E., Phys. Rev. Lett. **64** 2121 (1990);
   Kunszt Z. and Soper D.E., Phys. Rev **D46** 192 (1992).
6. M. Glück, E. Reya and A. Vogt, Z.Phys. **C67** 433 (1995).
7. Gribov L.V., Levin E.M. and Ryskin M.G., Phys. Rept. **100** 1 (1983);
   Mueller A.H. and Qiu J., Nucl. Phys. **B268** 427 (1986).
8. Eskola K.J., Kajantie K., Ruuskanen P.V. and Tuominen K., Nucl. Phys. **B570** 379 (2000). hep-ph/9909456.
9. Eskola K.J., Kolhinen V.J. and Ruuskanen P.V., Nucl. Phys. **B535** 351 (1998). hep-ph/9802350;
   Eskola K.J., Kolhinen V.J. and Salgado C.A., Eur. Phys. J. **C9** 61 (1999). hep-ph/9807297.
10. PHOBOS collaboration, Back B.B. et al., hep-ex/0007036.
11. Kataja M., Ruuskanen P.V., McLerran L. and von Gersdorff H., Phys. Rev. **D34** 2755 (1986).

From Crêpes to Pancakes in the MV Model

Gregory Mahlon

Department of Physics, McGill University
3600 University Street, Montréal, Québec H3A 2T8, Canada

Abstract. The McLerran-Venugopalan model provides a framework which allows one to compute the gluon distribution function of a very large nucleus from the equations of QCD, provided that the longitudinal momentum fraction, x_F, is sufficiently small. The source of color charge for this computation may be thought of as a crêpe moving along the z axis at the speed of light. We refine the MV model by allowing for the presence of non-trivial longitudinal correlations between the color charges that comprise the nucleons. We find that a consistent treatment forces us to consider a pancake-like source which moves at slightly less than the speed of light. Our calculation allows us to consider larger values of x_F than were allowed in the original MV model.

Several years ago, McLerran and Venugopalan realized that for large enough nuclei at small enough values of the longitudinal momentum fraction x_F, it ought to be possible to compute the gluon distribution function using QCD [1]. Based on this observation, the framework known as the McLerran-Venugopalan model (MV model) was subsequently developed [2–5]. Recently, it was shown [6] that the infrared divergences present in the MV model may be cured by capturing the physics of confinement via a color-neutrality condition to be imposed on the charge density correlation function used as input to the MV model. In this talk, I will describe work [7] on extending the MV model to larger values of x_F.

The MV model as originally formulated in Refs. [1–5], is restricted to very small longitudinal momentum fractions $x_F \lesssim A^{-1/3}/(ma)$, where A is the number of nucleons, m is the nucleon mass, and $a \sim \Lambda_{\text{QCD}}^{-1}$ is the nucleon radius. In this regime, the longitudinal resolution of the gluons is so poor that they probe distances which are much longer than the (Lorentz-contracted) thickness of the nucleus. Thus, all of the quarks inside the nucleus effectively have the "same" value of the longitudinal coördinate x^-. At each value of the transverse coördinate x describing this crêpe-shaped (*i.e.* very thin) nucleus, the color charges from a large number of valence quarks must be summed, resulting in a color charge density which is in a high-dimensional representation of the gauge group. This is a necessary condition for a classical treatment to be valid. In going to larger values of x_F, we find that the longitudinal resolution of the gluons improves, and we begin to see the longitudinal structure of the nucleus. In order to include this longitudinal structure, we are naturally led to a pancake-shaped geometry (*i.e.* one with a finite non-zero thickness). The details of this fully 3-dimensional calculation are contained in Ref. [7].

For a classical treatment to be valid, not only must the color charge be in a large representation of the gauge group, but the gauge coupling α_s should also be weak. McLerran and Venugopalan [1] argue that the running coupling ought to be evaluated at the scale μ^2, which is set by the charge-squared per unit transverse area. For large enough nuclei, $\mu^2 \gg \Lambda_{QCD}^2$, implying that $\alpha_s(\mu^2) \ll 1$.

Thus, we begin our computation of the gluon distribution function for a large nucleus by solving the *classical* Yang-Mills equations describing a pancake-shaped distribution of color charge moving along the z axis at nearly the speed of light. The result is a non-linear expression for the vector potential $A(x^-, \boldsymbol{x})$ in terms of the charge density $\rho(x^-, \boldsymbol{x})$. In the spirit of the Weizsäcker-Williams approximation [8], we extract the gluon number density from the two-point correlation function, $\langle A(x^-, \boldsymbol{x}) A(x'^-, \boldsymbol{x}') \rangle$. We replace the quantum mechanical average implied by the angled brackets with a classical average over an ensemble of nuclei. This ensemble is specified by inputting the two-point charge-density correlator $\langle \rho \rho \rangle \sim \mathcal{D}(x^-, \boldsymbol{x})$. Furthermore, we assume that the correlations are Gaussian. Confinement is incorporated into the calculation at this stage in the form of a color neutrality condition on $\mathcal{D}$ [6]. When $\mathcal{D}$ satisfies the color neutrality condition, the two-point correlation function is infrared finite, and may be Fourier-transformed to momentum space, producing a gluon number density $dN/dx_F d^2\boldsymbol{q}$ which is differential not only in x_F, but in the transverse momentum $\boldsymbol{q}$ as well.

In the limit $A^{-1/3} \ll 1$, the result of this rather lengthy calculation reads[1]

$$\frac{dN}{dx_F d^2\boldsymbol{q}} = 3AC_F \frac{2\alpha_s}{\pi^2} \frac{1}{x_F} \int d^2\boldsymbol{\Delta}\, e^{i\boldsymbol{q}\cdot\boldsymbol{\Delta}} \mathcal{L}(x_F, \boldsymbol{\Delta}) \frac{\exp[N_c \mathcal{X}_\infty L(\boldsymbol{\Delta})] - 1}{N_c \mathcal{X}_\infty L(\boldsymbol{\Delta})}. \qquad (1)$$

Although complicated in appearance, Eq. (1) is made up of several easily-understood parts. The prefactor shows that at lowest order the number of gluons is simply proportional to the number of quarks. The lowest order result is governed by the function

$$\mathcal{L}(x_F, \boldsymbol{\Delta}) \equiv \frac{1}{2} \int \frac{d^2\boldsymbol{q}}{4\pi^2} e^{-i\boldsymbol{q}\cdot\boldsymbol{\Delta}} \frac{q^2 \tilde{\mathcal{D}}(x_F, \boldsymbol{q})}{[q^2 + (x_F m)^2]^2}, \qquad (2)$$

and is what would be obtained by considering an Abelian theory. The non-Abelian corrections to this result are contained in the exponential factor, which depends on two quantities. First, the the spatial dependence is determined by

$$L(\boldsymbol{\Delta}) \equiv \int \frac{d^2\boldsymbol{q}}{4\pi^2} \frac{\tilde{\mathcal{D}}(0, \boldsymbol{q})}{q^4} \left[e^{-i\boldsymbol{q}\cdot\boldsymbol{\Delta}} - 1 \right]. \qquad (3)$$

---

[1] Eq. (1) has been written assuming cylindrical geometry for the nucleus. For a full discussion of the (rather weak) geometric dependence of the result, see Ref. [7].

The strength of the non-Abelian corrections is set by the factor

$$\mathcal{X}_\infty = \frac{1}{\pi R^2} \frac{3Ag^4 C_F}{N_c^2 - 1} \sim 8\pi \alpha_s^2 A^{1/3} \Lambda_{\text{QCD}}^2. \tag{4}$$

These effects are most prominent in very large nuclei. For uranium we have $\mathcal{X}_\infty \sim 5$ or $6\Lambda_{\text{QCD}}^2$. To obtain $\mathcal{X}_\infty = 20\Lambda_{\text{QCD}}^2$ (as is employed in the plots below) requires of order 10^4 nucleons. Finally, we note that if we set $x_F = 0$ in Eqs. (1)–(3), we explicitly reproduce the original MV result [5].

The functions appearing in Eqs. (2) and (3) are not finite at $q = 0$ unless the charge density correlator $\mathcal{D}$ satisfies the requirement of color neutrality. As explained in Ref. [6], a key consequence of confinement is the appearance of color neutral nucleons. Mathematically, this consequence may be implemented as a constraint on $\mathcal{D}$:

$$\int d\Delta^- d^2\Delta \; \mathcal{D}(\Delta^-, \Delta) = 0, \quad \text{or} \quad \tilde{\mathcal{D}}(0,0) = 0. \tag{5}$$

Eq. (5) ensures that widely-separated nucleons are essentially uncorrelated. In order for (5) to be true, $\mathcal{D}$ must contain a length scale. Not surprisingly, this scale turns out to be the nucleon radius $a \sim \Lambda_{\text{QCD}}^{-1}$. When combined with the assumption of rotational symmetry in the transverse plane, Eq. (5) is sufficient to render the functions contributing to the gluon number density completely infrared finite: the integrals get cut off at $q \sim a^{-1} \sim \Lambda_{\text{QCD}}$.

Aside from the neutrality condition (5), the correlation function $\mathcal{D}$ is unspecified. In order to illustrate the general features of the gluon number density (1), it is convenient to choose the following form for $\tilde{\mathcal{D}}$:

$$\tilde{\mathcal{D}}(x_F, q) = 1 - \frac{1}{1 + (aq)^2 + (x_F m a)^2}. \tag{6}$$

In the context of Kovchegov's nuclear model [9], the correlation function given in Eq. (6) corresponds to quarks which are distributed within the nucleons according to the weight

$$|\psi(\vec{r})|^2 = \frac{1}{2\pi^2 a^3} \frac{a}{r} K_1\left(\frac{r}{a}\right). \tag{7}$$

At large distances, the modified Bessel function produces a Yukawa-like behavior in this probability distribution.

Fig. 1 exhibits the behavior of the fully differential gluon number density as a function of the transverse momentum q^2. We see that there is saturation: the number of gluons increases as $q^2 \to 0$ to some maximum value and then stops growing. Qualitatively, these distributions are very much like those obtained by Muller from the point of view of onium scattering [10,11].

The gluon distribution function resolved at the scale Q^2 is related to the fully differential gluon number density by

$$g_A(x_F, Q^2) \equiv \int_{|q| \leq Q} d^2q \, \frac{dN}{dx_F d^2q}. \tag{8}$$

In Fig. 2 we multiply the fully differential distribution by q^2: on the semi-log scale used this produces a "true" representation of where the important contributions to Eq. (8) are. We see that the very low q^2 region does not play a significant role provided Q^2 is not too small. From the plot we see that for $Q^2 \to \infty$, the gluon distribution function becomes insensitive to the presence or absence of the non-Abelian corrections. In fact, we can prove under rather general circumstances that

$$\int d^2q \left\{ \left. \frac{dN}{dx_F d^2q} \right|_{\text{all orders}} - \left. \frac{dN}{dx_F d^2q} \right|_{\text{lowest order}} \right\} = 0, \tag{9}$$

independent of $\mathcal{D}$ [6].

In Fig. 3, we exhibit $x_F g_A(x_F, Q^2)$ as a function of x_F for several different values of Q^2. At low values of Q^2, the addition of the non-Abelian corrections reduces the number of gluons from the Abelian result, while by the time $Q^2 = 2500 \Lambda^2_{\text{QCD}}$ is reached, the effect of the non-Abelian terms is negligible. Although we have plotted out to $x_F = 1$, the calculation is not reliable beyond $x_F \sim 0.25$ [7]. For small x_F the gluon distribution function exhibits the pure $1/x_F$ dependence characteristic of the original MV model.

Finally, we present Fig. 4, which illustrates the deviation of the gluon distribution function from the naïve expectation that for A nucleons we should obtain A times

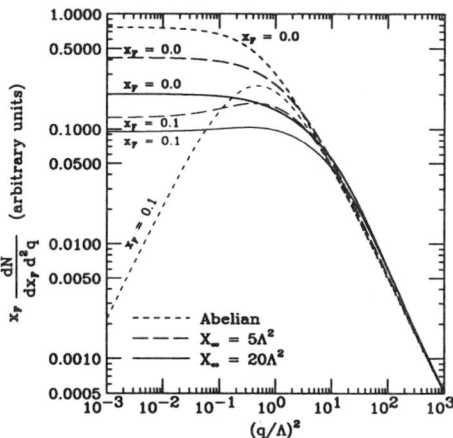

FIGURE 1. Fully differential gluon number density plotted versus q^2 at $x_F = 0.0$ and 0.1. The three curves are for $\mathcal{X}_\infty = 0$ (Abelian), $5\Lambda^2_{\text{QCD}}$, and $20\Lambda^2_{\text{QCD}}$. The gluon density saturates at small q^2.

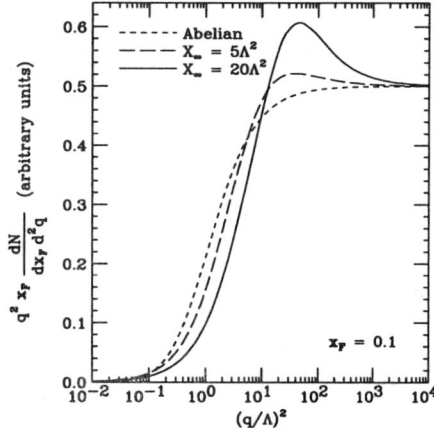

FIGURE 2. Fully differential gluon number density at $x_F = 0.1$ multiplied by q^2 versus q^2 for $\mathcal{X}_\infty = 0$ (Abelian), $5\Lambda^2_{\text{QCD}}$, and $20\Lambda^2_{\text{QCD}}$. The area under each curve accurately reflects the contributions to $g_A(x_F, Q^2)$.

the result for a single nucleon. We see that at low values of Q^2 the distribution function grows less rapidly than the number of nucleons as A is increased, whereas at large Q^2, the simple scaling expectation holds.

REFERENCES

1. L. McLerran and R. Venugopalan, *Phys. Rev.* **D49**, 2233 (1994).
2. L. McLerran and R. Venugopalan, *Phys. Rev.* **D49**, 3352 (1994).
3. L. McLerran and R. Venugopalan, *Phys. Rev.* **D50**, 2225 (1994).
4. A. Ayala, J. Jalilian-Marian, L. McLerran, and R. Venugopalan, *Phys. Rev.* **D52**, 2935 (1995).
5. J. Jalilian-Marian, A. Kovner, L. McLerran, and H. Weigert, *Phys. Rev.* **D55**, 5414 (1997).
6. C.S. Lam and G. Mahlon, *Phys. Rev.* **D61**, 014005 (2000).
7. C.S. Lam and G. Mahlon, "Longitudinal Resolution in a Large Relativistic Nucleus: Adding a Dimension to the McLerran-Venugopalan Model," hep-ph/0007133.
8. C. Weizsäcker and E. Williams, *Z. Phys.* **88**, 244 (1934).
9. Yu. Kovchegov, *Phys. Rev.* **D54**, 5463 (1996).
10. A. Mueller, *Nucl. Phys.* **B335**, 115 (1990).
11. A. Mueller, *Nucl. Phys.* **B558**, 285 (1999).

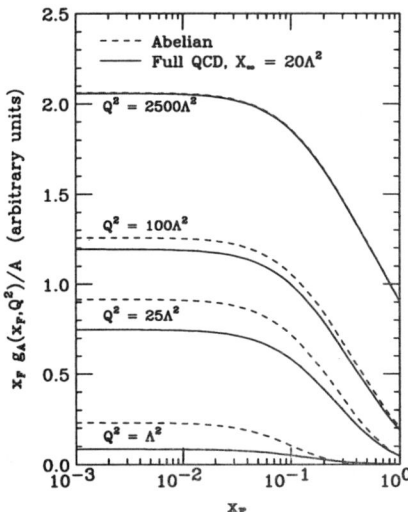

FIGURE 3. The gluon structure function $x_F g_A(x_F, Q^2)$ plotted versus x_F for several values of Q^2. The dashed curves illustrate the effect of ignoring the non-Abelian corrections, which are included in the solid curves.

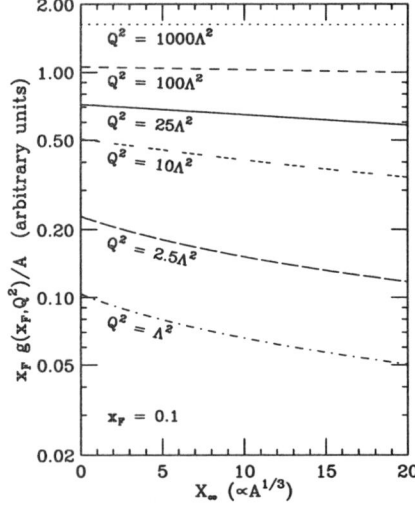

FIGURE 4. Nuclear dependence of the gluon structure function $x_F g_A(x_F, Q^2)$. For sufficiently small Q^2 the number of gluons grows more slowly than the number of nucleons as A is increased.

QCD and Nuclear Structure

Measurement of the Neutron Electric Form Factor G_E^n in $\vec{D}(\vec{e},e'n)p$ Quasi-Elastic Scattering at $Q^2 = 0.5 (GeV/c)^2$

M. Zeier[a], for the E93-026 Jlab collaboration

[a] *Department of Physics, University of Virginia, Charlottesville, VA, USA*

Abstract. We have determined the electric form factor of the neutron G_E^n from the reaction $\vec{D}(\vec{e},e'n)p$ using a longitudinally polarized electron beam and a polarized deuterium target at Jefferson Lab's Hall C. The knocked out neutron was detected in coincidence with the electron in a shielded neutron detector. The beam-target asymmetry of quasi-elastically scattered electrons was measured for opposite orientations of the beam helicity which allowed the extraction of G_E^n. This method is insensitive to the deuteron structure and avoids longitudinal/transverse Rosenbluth separation, both potential sources of large systematic errors. We present the results of a preliminary analysis for G_E^n at $Q^2 = 0.5 (GeV/c)^2$.

INTRODUCTION

G_E^n is related to the charge distribution of the neutron and is thus of fundamental importance for our understanding of the structure of nucleons. It provides a unique test for QCD and meson theory.

Recently it has become technically possible to perform double polarization experiments which measure asymmetries and detect the neutron in coincidence with the electron. This type of experiment avoids Rosenbluth separation and subtraction of the proton contribution and is less sensitive to the structure of the deuteron. Thus it avoids the main systematic difficulties of both elastic and inelastic inclusive cross section measurements, which were designed in the past to determine G_E^n.

The experiment described here used a longitudinally polarized electron beam in combination with a polarized deuterium target. To minimize the influence of the magnetic form factor G_M^n and to be most sensitive to G_E^n the direction of the target polarization was chosen to be perpendicular to the three momentum transfer $\vec{q}$ and to lie in the scattering plane. For a free neutron then the electron-neutron asymmetry A_n is given as

$$A_n = \frac{-2\sqrt{\tau(1+\tau)}\tan(\Theta_e/2)\, G_E^n G_M^n}{(G_E^n)^2 + \tau(1+2(1+\tau)\tan^2(\Theta_e/2))(G_M^n)^2}; \qquad \tau = \frac{Q^2}{4M^2}$$

with electron scattering angle Θ_e and neutron mass M.

The electron–deuteron asymmetry A_{ed}^V has been calculated by Arenhövel [1] for electrodisintegration of the deuteron for different kinematic regions. The quasielastic region with a neutron emission in the direction of $\vec{q}$ shows negligible sensitivity to potential models, final state interactions and subnuclear degrees of freedom (meson exchange currents, isobar contributions).

EXPERIMENT E93-026

The experiment was performed in Hall C at Thomas Jefferson National Accelerator Facility (Jlab) in Newport News, Virginia, USA. The measurements took place from August to October 1998 and will be continued in 2001. The kinematics measured were $Q^2 = 0.5 (GeV/c)^2$ and $Q^2 = 1 (GeV/c)^2$. Results are presented here only for $Q^2 = 0.5 (GeV/c)^2$.

The longitudinally polarized electron beam had an energy of 2.723 GeV. The beam helicity was reversed every second. The average beam polarization was 76% and was determined with an accuracy of 1% using a high precision moller polarimeter [2].

The polarized target [3] consisted of a superconducting magnet operating at 5 Tesla and a ^4He evaporation refrigerator operating at 1 K. Deuterized ammonia ($^{15}ND_3$), doped by irradiation was used as target material. The polarization was achieved dynamically by means of micro waves. A NMR circuit was used to determine the magnitude of polarization which averaged to 21% over the whole experiment.

A raster system was used to distribute the beam over the entire target surface and chicane magnets were installed before and after the target to compensate for the influence of the target magnet on the electron beam. Scattered electrons were detected in quasielastic kinematics in Hall C's standard high momentum spectrometer (HMS) while hadrons were detected in coincidence in a neutron detector enclosed in a concrete hut.

The neutron detector consisted of plastic scintillator detectors equipped with photo-multiplier tubes on each end. 79 Bars with a cross section of 10×10 cm^2 were arranged in 5 planes covering an area of 160×160 cm^2 at a distance of 400 cm from the target. 39 paddles with a thickness of 1 cm were arranged in two planes in front of the bars to veto the dominant proton contribution and to allow for a clean identification of the neutrons.

ANALYSIS AND RESULTS

Events were reconstructed using the standard HMS reconstruction code extended for the effects of beam raster and target magnetic field. On the neutron detector side a tracking algorithm was developed for proper particle identification. After event reconstruction and event selection charge and dead-time normalized yields were produced for each beam helicity state: $N^{\uparrow\downarrow}$. From these yields asymmetries ϵ were calculated and with the knowledge of target (P_T) and beam (P_B) polarization and the dilution factor (f) it was possible to extract experimental deuteron asymmetries A_{ed}^V,

$$\varepsilon = \frac{N^{\uparrow} - N^{\downarrow}}{N^{\uparrow} + N^{\downarrow}} = P_B \cdot P_T \cdot f \cdot A_{ed}^V.$$

The dilution factor f accounts for scattering from nuclei other than deuterium and was determined from Monte Carlo simulations which were calibrated on data taken with a ^{12}C target. Additionally the experimental asymmetries were corrected for radiative effects and accidental background. Arenhövel provided theoretical calculations of A_{ed}^V on a kinematical grid and for different values of G_E^n. Using Monte Carlo technique we averaged the theoretical calculations over our experimental acceptance and compared it with the experimental A_{ed}^V. Figure 1 presents the experimental A_{ed}^V along with the theoretical A_{ed}^V averaged over our acceptance for three values of G_E^n in units of the Galster parametrization [4]. The asymmetries are plotted vs. θ_{nq} which is the lab angle between the $\vec{q}$ vector and the neutron track.

Our experimental value for G_E^n was then determined by fitting a linear interpolation of the theoretical curves to the data with the result $G_E^n = 0.90 \times (G_E^n)_{Galster}$ This procedure was repeated for other kinematic variables than θ_{nq} and produced consistent results.

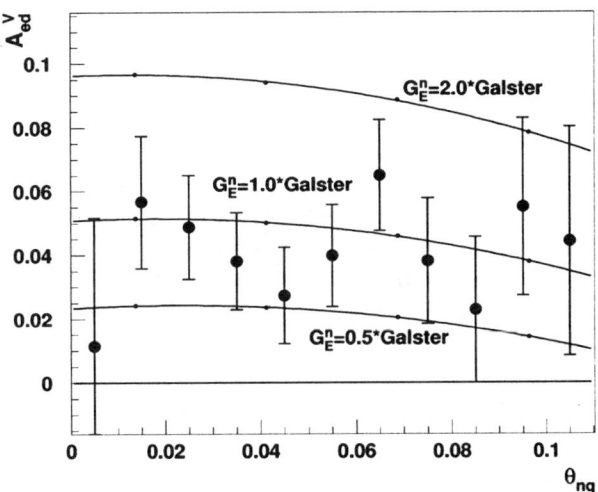

FIGURE 1. Deuteron asymmetries A_{ed}^V are plotted vs. θ_{nq}. The curves represent theoretical calculations averaged over our kinematical acceptance for different values of G_E^n. The points are measured A_{ed}^V for single bins of θ_{nq}.

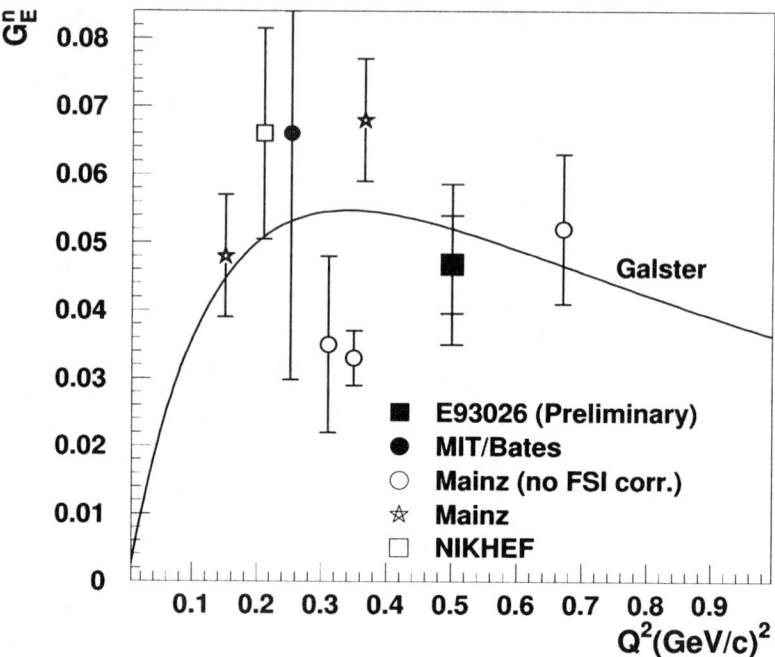

FIGURE 2. G_E^n World data from double polarization experiments. The solid line represents the Galster parameterization. The error bar of the E93026 point includes a 20% systematic error for now. The open circles are results from Mainz with polarized ^3He target and do not include corrections for final state interactions.

Our preliminary result for G_E^n at $Q^2 = 0.5 (GeV/c)^2$ is:

$$G_E^n = 0.0468 \pm 0.0072_{stat}$$

Systematic studies are still in progress. The systematic error will be dominated by the target polarization (5%) and dilution factor (>4%) and is expected to be smaller than 10%. For now we quote a 20% systematic error. Figure 2 shows our experimental result together with other double polarization results from Mainz, NIKHEF and MIT/Bates.

REFERENCES

1. Arenhövel, H. et al, *Z. Phys.* **A331**, 123 (1988).
2. Hauger, M. et al., *Nucl. Instr. Meth.* **A** in print.
3. Crabb, D. et al., *Nucl. Instr. Meth.* **A356**, 9 (1995).
4. Galster, S. et al., *Nucl. Phys.* **B32**, 221 (1971).

Methods for the Nonperturbative Approximation of Form Factors and Scattering Amplitudes

John R. Hiller

Department of Physics
University of Minnesota Duluth
Duluth Minnesota 55812

Abstract. Methods are described for the nonperturbative calculation of wave functions and scattering amplitudes in light-cone quantization. Form factors are computed from the boost-invariant wave functions, which appear as coefficients in a Fock-state expansion of the field-theoretic eigenstate. A technique is proposed for calculating scattering amplitudes from matrix elements of a T operator between such composite-particle eigenstates.

INTRODUCTION

To benefit from the recent progress on the calculation of field-theoretic bound states in light-cone quantization [1,2], we explore methods by which form factors and scattering amplitudes can be extracted nonperturbatively. In the case of form factors, this is relatively straightforward; well-known formulas [3] yield the form factors as overlap integrals of Fock-state wave functions. For scattering amplitudes, the way is less certain. One possible method [4] is discussed briefly here. Others have been considered by Kröger [5], Ji and Surya [6], and Fuda [7].

The formulations given are in terms of light-cone coordinates [8,1], where $x^+ \equiv t+z$ plays the role of time and the conjugate variable $p^- \equiv E - p_z$ is the light-cone energy. The light-cone three-momentum is $\underline{p} = (p^+ \equiv E + p_z, \mathbf{p}_\perp)$. An eigenstate $|P,\sigma\rangle$ of the light-cone Hamiltonian operators $\mathcal{P}^\pm$, $\mathcal{P}_\perp$ and helicity σ is written as a Fock-state expansion

$$|P,\sigma\rangle = \sum_n \int [dx][d^2k_\perp] \psi_{P,\sigma}^{(n)}(x,\mathbf{k}_\perp)|n : \underline{p}_i\rangle, \tag{1}$$

with

$$\int [dx][d^2k_\perp] = \int \delta(1 - \sum_i x_i) \prod_i \frac{dx_i}{\sqrt{x_i}} 16\pi^3 \delta(\sum_i \mathbf{k}_{\perp i}) \prod_i \frac{d^2k_{\perp i}}{16\pi^3} \tag{2}$$

and where the $\psi^{(n)}$ are wave functions for n particles, $x_i \equiv p_i^+/P^+$ are longitudinal momentum fractions, and $\mathbf{k}_{\perp i} = \mathbf{p}_{\perp i} - x_i \mathbf{P}_\perp$ are relative transverse momenta. Use of light-cone coordinates brings several advantages, including boost invariance of the wave functions.

The eigenvalue problem $\mathcal{P}|P,\sigma\rangle = P|P,\sigma\rangle$ for fixed σ determines the wave functions as solutions of a coupled set of integral equations. A method frequently applied to these equations is discrete light-cone quantization (DLCQ) [9,1], which approximates the integrals by the trapezoidal rule and computes the wave functions on an equally spaced momentum grid. Any bound-state property can then, in principle, be calculated from these wave functions. The grid is parameterized by a longitudinal resolution K and transverse resolution $N_\perp$, such that longitudinal momentum fractions are multiples of $1/K$ and transverse momenta have as many as $2N_\perp + 1$ values in each direction. The value of $N_\perp$ is associated with a cutoff Λ^2 on the invariant mass of each constituent and with the choice of transverse momentum scale $\pi/L_\perp$.

FORM FACTORS

For a spin-1/2 fermion, the two form factors can be obtained from matrix elements of the plus component of the electromagnetic current J

$$F_1(Q^2) = \frac{1}{2}\langle P+Q,\sigma|J^+(0)/P^+|P,\sigma\rangle, \qquad (3)$$

$$-\left(\frac{Q_x - iQ_y}{2M}\right)F_2(Q^2) = \frac{1}{4\sigma}\langle P+Q,\sigma|J^+(0)/P^+|P,-\sigma\rangle. \qquad (4)$$

These can be reduced to overlap integrals [3]

$$F_1(Q^2) = \sum_n \sum_j e_j \int [dx][d^2k_\perp]\psi^{(n)*}_{P+Q,1/2}(x,\mathbf{k}'_\perp)\psi^{(n)}_{P,1/2}(x,\mathbf{k}_\perp), \qquad (5)$$

$$-\left(\frac{Q_x - iQ_y}{2M}\right)F_2(Q^2) = \sum_n \sum_j e_j \int [dx][d^2k_\perp]\psi^{(n)*}_{P+Q,1/2}(x,\mathbf{k}'_\perp)\psi^{(n)}_{P,-1/2}(x,\mathbf{k}_\perp), \qquad (6)$$

in the frame where the photon momentum Q is written $(0, 2Q\cdot P/P^+, \mathbf{Q}_\perp)$ and

$$\mathbf{k}'_{\perp i} = \begin{cases} \mathbf{k}_{\perp i} - x_i \mathbf{Q}_\perp, & i \neq j \\ \mathbf{k}_{\perp j} + (1-x_j)\mathbf{Q}_\perp, & i = j. \end{cases} \qquad (7)$$

For the model studied by Brodsky, Hiller, and McCartor [2], an explicit calculation of F_1 has been done [10]. In this model, a bare fermion acts as a source and sink for bosons of mass μ. The lowest massive eigenstate is a fermion dressed by a boson cloud. The theory is regulated by a Pauli–Villars boson [11] with an imaginary coupling, and renormalized by fits of physical quantities to "data." Because no spin-flip interactions are included, F_2 is zero. Results for F_1 are shown in Fig. 1. The large-momentum-transfer value of F_1 is the bare fermion probability and therefore is not zero.

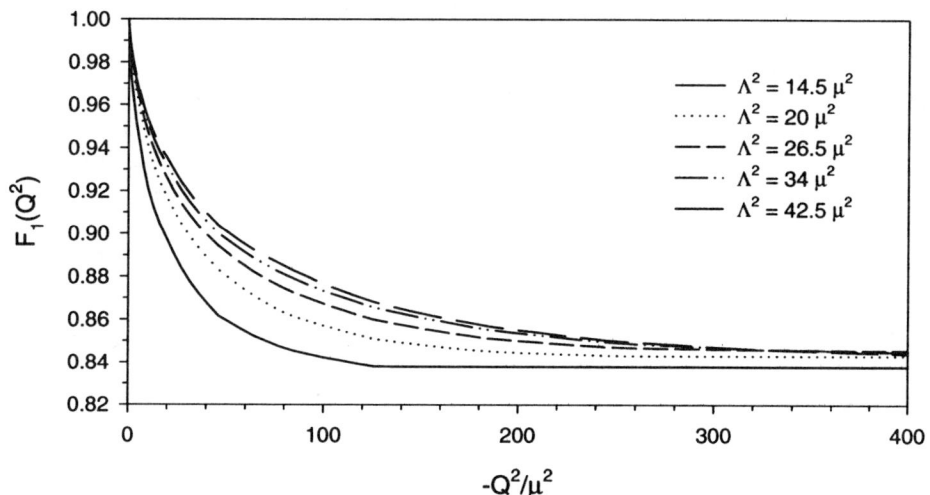

FIGURE 1. The form factor F_1 for fixed longitudinal resolution $K = 9$ and transverse scale $L_\perp = 2\pi/\mu$, and for a particular set of model parameters. Various cutoffs Λ^2 are considered, with the transverse resolution $N_\perp$ ranging from 5 to 9.

SCATTERING AMPLITUDES

The center-of-mass cross section for two-body scattering $(A + B \to C + D)$ is [12]

$$\frac{d\sigma}{d\Omega_{\text{cm}}} = \frac{1}{2E_A 2E_B v_{\text{rel}}} \frac{|\vec{p}_C||\mathcal{M}_{fi}|^2}{16\pi^2 E_{\text{cm}}}, \tag{8}$$

where $\mathcal{M}_{fi}$ is the invariant amplitude obtained from the S matrix

$$S_{fi} = \langle f|i\rangle + (2\pi)^4 \delta^{(4)}(p_f - p_i) i \mathcal{M}_{fi} = \delta_{CD,AB} - 2\pi i \delta(s_{AB} - s_{CD}) T_{\text{LC}fi}, \tag{9}$$

with $s_{AB} = \frac{m_A^2 + p_{A\perp}^2}{p_A^+/P^+} + \frac{m_B^2 + p_{B\perp}^2}{p_B^+/P^+}$. The T matrix for scattering of composites is given by [4,13]

$$T_{\text{LC}fi} = P^+ T_{fi} = \langle C|V_D^\dagger \frac{1}{s_{AB} + i\epsilon - H_{\text{LC}}} V_B|A\rangle + \langle C|DV_B|A\rangle. \tag{10}$$

Here $|A\rangle$ and $|C\rangle$ are composite-particle eigenstates of the light-cone Hamiltonian H_{LC}, and the operator V_B is defined by

$$V_B = [H_{\text{LC}}, B^\dagger] - \frac{m_B^2 + p_{B\perp}^2}{p_B^+/P^+} B^\dagger, \tag{11}$$

with $B^\dagger$ the creation operator for the B particle, i.e. $|B\rangle = B^\dagger|0\rangle$. This construction generalizes one presented some time ago by Wick [13]. Details can be found in Ref. [4]. Given numerical solutions for the composite-particle eigenstates, obtained with DLCQ, the most difficult remaining task is the estimation of the matrix element of $(s + i\epsilon - \bar{H}_{\mathrm{LC}})^{-1}$. For this type of matrix element, the recursion method of Haydock [14] has worked well. A nonrelativistic application is described in [4]; an application to the field-theoretic model studied in [2] is in progress.

ACKNOWLEDGMENTS

This work was supported in part by the Minnesota Supercomputing Institute through grants of computing time and by the Department of Energy, contract DE-FG02-98ER41087.

REFERENCES

1. For a review, see Brodsky, S.J., Pauli, H.-C., and Pinsky, S.S., *Phys. Rep.* **301**, 299-486 (1997).
2. Brodsky, S.J., Hiller, J.R., and McCartor, G., *Phys. Rev.* D **58**, 025005 (1998); **60**, 054506 (1999).
3. Drell, S.D., and Yan, T.-M., *Phys. Rev. Lett.* **24**, 181-185 (1970); Brodsky, S.J., and Drell, S.D., *Phys. Rev.* D **22**, 2236-2243 (1980).
4. Hiller, J.R., "Nonperturbative Calculation of Scattering Amplitudes," to appear in the proceedings of the fourth workshop on Continuous Advances in QCD, Minneapolis, Minnesota, May 12-14, 2000, hep-ph/0007231.
5. Kröger, H., *Phys. Rep.* **210**, 45-109 (1992).
6. Ji, C.-R., and Surya, Y., *Phys. Rev.* D **46**, 3565-3575 (1992).
7. Fuda, M.G., and Zhang, Y., *Phys. Rev.* C **51**, 23-37 (1995) and references given therein.
8. Dirac, P.A.M., *Rev. Mod. Phys.* **21**, 392-399 (1949).
9. Pauli, H.-C., and Brodsky, S.J., *Phys. Rev.* D **32**, 1993-2000 (1985); **32**, 2001-2013 (1985).
10. Hiller, J.R., "On the Use of Discrete Light-Cone Quantization to Compute Form Factors," in *The Transition from Low to High Q Form Factors*, edited by G. Strobel and D. Mack, TJNAF Workshop Proceedings, 1999, pp. 193-199, hep-ph/9909471.
11. Pauli, W., and Villars, F., *Rev. Mod. Phys.* **21**, 434-444 (1949).
12. Peskin, M.E., and Schroeder, D.V., *An Introduction to Quantum Field Theory*, Addison-Wesley, Reading, MA, 1995, pp. 99-108.
13. Wick, G.C., *Rev. Mod. Phys.* **27**, 339-362 (1955).
14. Haydock, R., in *Solid State Physics*, edited by H. Ehrenreich, F. Seitz, and D. Turnbull, Vol. 35, Academic Press, New York, 1980, pp. 215-294.

Rare Pion Double Radiative Capture Reactions on Hydrogen and Deuterium

M.D. Hasinoff[1], D.S. Armstrong[2], J. Clark[2], T.P. Gorringe[3],
M. Kovash[3], S. Tripathi[3], D.H. Wright[4] and P.A. Żołnierczuk[3]

[1] *Dept of Physics & Astronomy, Univ. of British Columbia, Vancouver, B.C., V6T 1Z1*
[2] *College of William and Mary, Williamsburg, Virginia, 23187*
[3] *Dept. of Physics & Astronomy, University of Kentucky, Lexington, Kentucky, 40506*
[4] *TRIUMF, Vancouver, B.C., Canada, V6T 2A3*

Abstract. The rare 2-photon radiative capture reaction has been observed for the first time on hydrogen and deuterium using the RMC high acceptance cylindrical pair spectrometer at TRIUMF. Our preliminary branching ratios are 3.8×10^{-5} for hydrogen and 1.6×10^{-5} for deuterium. Our π^-p data confirms the predicted dominance of the $\pi\pi \to \gamma\gamma$ annihilation mechanism. Moreover, since crossing symmetry relates $\pi\pi \to \gamma\gamma$ to $\gamma\pi \to \gamma\pi$ this threshold $(\pi, 2\gamma)$ reaction might also provide new information on the electric polarizability of the pion. Our π^-d data shows no evidence for the predicted $d_1^*(1920)$ dibaryon.

INTRODUCTION

The $(\pi^-, 2\gamma)$ absorption mode of pionic atoms was first proposed by Ericson and Wilkin [1] as a means of studying the distribution of the virtual π^+ cloud inside the nucleus, in analogy with positron annihilation in a solid. The dominant reaction mechanism was predicted to be the annihilation of the real, stopped π^- on a soft, virtual π^+ with an estimated branching ratio of $\sim 5 \times 10^{-6}$. The first $(\pi, 2\gamma)$ measurements [2,3] were performed on Be and C targets about 20 years ago at CERN and TRIUMF with the measured branching ratios being about 3× larger than these first estimates. More detailed calculations by Christillin and Ericson [4] and by Gil and Oset [5] disagree about the importance of the relative contributions of the bremsstrahlung and annihilation terms. The former agrees with the data in the forward angle region but does not reproduce the peak at backward angles, while the latter reproduces the backward angle peak but is 4× too large at small $\gamma\gamma$ angles.

An understanding of the elementary process $\pi^-p \to \gamma\gamma n$ will be a very valuable benchmark for the nuclear $(\pi, 2\gamma)$ process and any possible renormalization of the virtual pion field inside the nuclear medium. Several authors have studied the $\gamma\gamma n$ capture mode of pionic hydrogen [6]. The most recent calculation using a pseudoscalar-coupled π-N theory predicts a branching ratio of 5.1×10^{-5} with the $\pi\pi$ annihilation diagram becoming dominant at small photon opening angles and contributing about 2/3 of the overall total rate. This is nearly $10\times$ smaller than the experimental upper limit reported [7] from JINR, Dubna based on 2 ± 10 events above a random background.

Since the $\pi^-p \to \gamma\gamma n$ reaction is predicted to be dominated by the annihilation diagram $\pi\pi \to \gamma\gamma$, by invoking crossing symmetry, this reaction can be considered as the transition of a real π to a virtual π via Compton scattering, ie., $\gamma\pi \to \gamma\pi$. Hence this elementary $\pi p \to \gamma\gamma n$ reaction can be interpreted as a probe of pion Compton scattering which is expected to be quite sensitive to both QCD loop corrections and the electromagnetic polarizability, $\alpha_E^{\pi^\pm}$ of the pion. The current experimental situation for $\alpha_E^{\pi^\pm}$ is rather confusing with large uncertainties and discrepancies in the data(see Table 1). No measurement is yet accurate enough to test the χPT prediction $(2.7\pm 0.4)\times 10^{-4} fm^3$ which is based solely on the values of the pion decay constant, f_π and the pion charge radius, r_π.

TABLE 1. Pion electric polarizability, $\alpha_E^{\pi^\pm}$.

Reaction	$\alpha_E^{\pi^\pm}$ ($\times 10^{-4} fm^3$)	Experiment	Reference
$\pi A \to \gamma\pi A$	$6.8 \pm 1.4 \pm 1.2$	Serpukhov	[8]
$\gamma p \to \gamma\pi n$	20 ± 12	Lebedev	[9]
$\gamma\gamma \to \pi\pi$	$19.1 \pm 4.9 \pm 5.6$	PLUTO	[10]
$\gamma\gamma \to \pi\pi$	2.2 ± 1.6	MARK II	[10]

A comparison of the hydrogen and deuterium $(\pi, 2\gamma)$ capture rates might indicate the onset of the effects of nuclear binding on the virtual pion cloud surrounding the bound nucleon. In addition there has been a speculation by Gerasimov [11] that the $d_1^*(1920)$ dibaryon reported by the DIB-2gamma collaboration [12] might be observable through the reactions

$$\pi^-d \to d_1^*(1920)\gamma_1, \quad d_1^*(1920) \to \gamma_2 nn$$

with $E_{\gamma_1} \sim 90$ MeV and $E_{\gamma_2} \sim 30$ MeV. Treating the d_1^*-dibaryon as an N-Δ bound state ($J^\pi=1^+$,T=1) Gerasimov estimated the branching ratio due to d_1^* production to be $\sim 0.2\%$, or roughly $100\times$ more probable than the non-resonant $\pi p \to \gamma\gamma n$ reaction.

EXPERIMENTAL DETAILS

The $(\pi, 2\gamma)$ experiments reported here were performed using the large acceptance RMC pair spectrometer on the M9A beamline at TRIUMF over the past 15 months. An 82 MeV/c pion beam ($\Delta p/p=10\%$) was collimated and stopped in a $LH_2(LD_2)$ target (16 cmϕ $\times$ 15 cm length) placed at the centre of a large solenoidal magnet. The outgoing photons were converted in a 1mm thick, 35 cm long cylindrical Pb converter located at a radius of 12.7 cm. The e^+e^- pair was then bent by a uniform magnetic field of 1.2 kG and tracked by a 1 cm dual spiral cathode proportional chamber and a large drift chamber [13] with 3 axial plus 1 stereo layers. A segmented array of inner and outer scintillators was used to provide the fast 1st level trigger. The very copious back-to-back photons from stopped π^0 production (on hydrogen) were partially rejected in hardware using both scintillator and drift chamber hit pattern information.

RESULTS

A typical $\gamma\gamma$ event is shown in Fig.1. The points in layer 3 do not necessarily lie on the circular track since this layer is a stereo layer at 7deg. A cut on the beam counter ADC value was used to reject the large number of random 2 photon events produced when 2 pions are captured in the same beam burst. Pions in adjacent bursts were rejected by the inner C-counter timing.

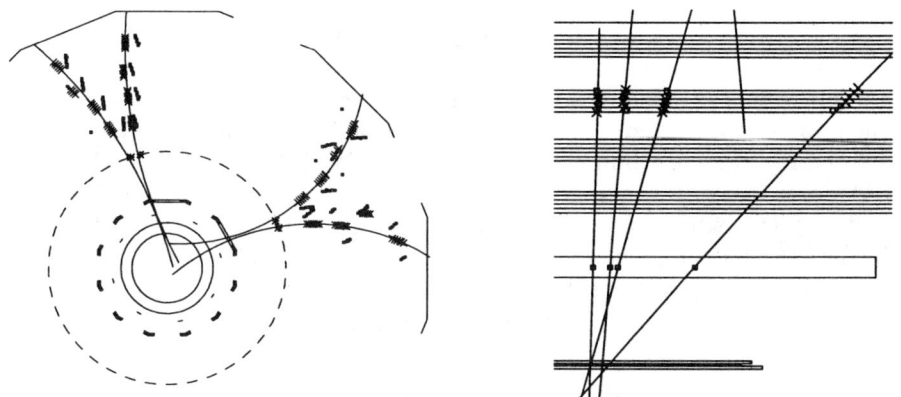

FIGURE 1. End and side views of a typical 2 photon event.

The events with an opening angle $\cos\theta \leq -0.2$ in Fig. 2 have been removed due to the large number of π^0 events still remaining in the spectrum. The drop in the spectrum at small forward angles is created by an acceptance cutoff when both γ's pass through the same inner scintillator. In total we have observed ~1000 good $(\pi, 2\gamma)$ events for hydrogen and ~500 good events for deuterium. Our preliminary branching ratios are 3.8×10^{-5} for hydrogen and 1.6×10^{-5} for deuterium. The hydrogen $(\pi, 2\gamma)$ value is in approximate agreement with Beder's calculation [6]. The relative $2\gamma/1\gamma$ branching ratios are 1.0×10^{-4} and 0.6×10^{-4}, respectively; these are about 10× smaller than in Be or C. Our current overall uncertainty is ~20%; we expect our final uncertainty to be <10%.

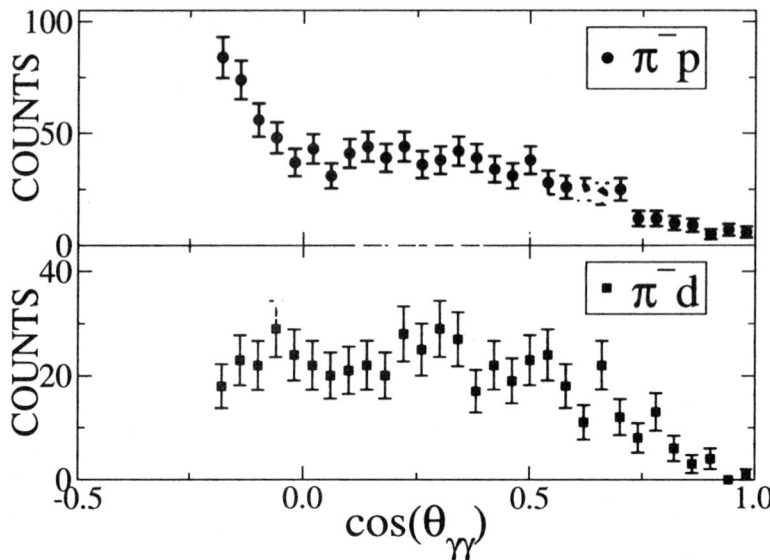

FIGURE 2. Opening angle distributions for pionic hydrogen (circles) and deuterium (squares).

CONCLUSIONS

Our preliminary results confirm that the $\pi\pi \to \gamma\gamma$ annihilation graph is dominant in the $\pi^- p \to \gamma\gamma n$ reaction. Hence this double radiative capture mode on hydrogen is a potential probe of pion Compton scattering. We have found no evidence for the production of a $d_1^*(1920)$-dibaryon in the $\pi^- d \to \gamma\gamma nn$ reaction. We are currently considering the extension of these $(\pi, 2\gamma)$ measurements to other nuclear targets. A higher statistics hydrogen experiment awaits an evaluation of the sensitivity to the pion's electric polarizability.

REFERENCES

1. Ericson, T.E.O., and Wilkin, C., *Phys. Lett.* **57B**, 345 (1975).
2. Deutsch, J., et al., *Phys. Lett.* **80B**, 347 (1979).
3. Mazzucato, E., et al., *Phys. Lett.* **96B**, 43 (1980).
4. Christillin, P., and Ericson, T.E.O., *Phys. Lett.* **87B**, 163 (9179).
5. Gil, A., and Oset, E., *Phys. Lett.* **B346**, 1 (1995). **15**, 558 (1972).
6. Beder, D., *Nucl. Phys.* **B156**, 482 (1979) and references therein.
7. Vasilebsky, I., et al., *Nucl. Phys.* **B9** 673 (1969).
8. Antipov, A., et al., *Z.Phys.* **C24**, 39 (1984).
9. Aibergenov, T., et al., *Czech. J. Phys.* **B36**, 948 (1986).
10. Babusci, D. et al., *Phys. Lett.* **B277**, 138 (1992).
11. Gerasimov, S., nucl-th/9812077 and nucl-th/9808070.
12. Khrykin, A., et al., πN *Newslett.* **13**, 250 (1997).
13. Wright, D.H., et al., *Nucl. Inst. & Meth.*, **A320**, 249 (1992).

Ortho-Para Transition in Muonic Molecular Hydrogen[1]

Jessica H.D. Clark*, D.S. Armstrong*, T.P. Gorringe[†],
M.D. Hasinoff[‡], P.M. King*, T.J. Stocki[‡], S. Tripathi[†],
D.H. Wright[§], P.A. Żołnierczuk [†]

College of William and Mary, Williamsburg, VA, USA, 23185
[†] *University of Kentucky, Lexington, KY, USA, 40506*
[‡] *University of British Columbia, Vancouver, BC, Canada V6T 1Z1*
[§] *TRIUMF, Vancouver, BC, Canada V6T 2A3*

Abstract. Knowledge of the transition rate between the ortho- and para-molecular states of muonic molecular hydrogen is critical for the extraction of the induced pseudoscalar coupling of the proton, g_p, from muon capture experiments in hydrogen. A measurement of this rate has recently been completed at TRIUMF, and a status report on the experiment is presented here.

INTRODUCTION

Negative muon capture ($\mu^- + p \to \nu_\mu + n$) can be used to probe nucleon structure. The strong interaction present in this semi-leptonic process induces extra structure within the vector and axial weak currents. Of these semi-leptonic weak coupling constants, the induced pseudoscalar coupling, g_p, is the least well-known, theoretically and experimentally. Using heavy baryon chiral perturbation theory and the chiral Ward identities of QCD, g_p has been calculated to be 8.44 ± 0.23 [1]. More traditional calculations using PCAC yield

$$g_p(q^2 = -0.88 m_\mu^2) = \frac{2 M m_\mu g_a}{q^2 + m_\pi^2} = 8.56 \quad (1)$$

where M is the mass of the proton, m_μ is the mass of the muon, g_a is the axial coupling constant, and m_π is the mass of the pion. Experiment, however, has not yet reached this level of precision.

Since the muon capture branching ratio is $\sim 10^{-3}$, experiments commonly utilize liquid hydrogen targets [2]. However, with the higher proton density comes an

---

[1] Research supported by NSF (U. Kentucky), NSERC (Canadian collaborators), Jeffress Memorial Trust, Clare Boothe Luce graduate fellowship, William and Mary Endowment Fund.

increase in the probability of the formation of muonic molecules. Due to the spin-dependence of the weak interaction, interpretation of muon capture data requires knowledge of the relative hyperfine states, singlet and triplet, of the μp system. The higher energy orthomolecular state is comprised of ~75% singlet and ~25% triplet atomic states, whereas the paramolecular state reverses this relative population of states. The capture rate from a singlet state is ~60 times faster than the capture from a triplet state [2]. Therefore, transitions between the ortho and para states will change the fraction of muons that capture in the system. Muon capture experiments in liquid hydrogen require precise knowledge of the transition rate between the ortho and para states, λ_{op}.

There has only been one measurement of λ_{op}, which yielded the value $\lambda_{op} = (4.1 \pm 1.4) \times 10^4$ s^{-1} [3]. This is significantly different from the theoretical value [4], $\lambda_{op} = (7.1 \pm 1.2) \times 10^4$ s^{-1}.

This uncertainty in the value of λ_{op} has affected the interpretation of the two most precise measurements of g_p, as shown in Fig. 1. The radiative muon capture (RMC)

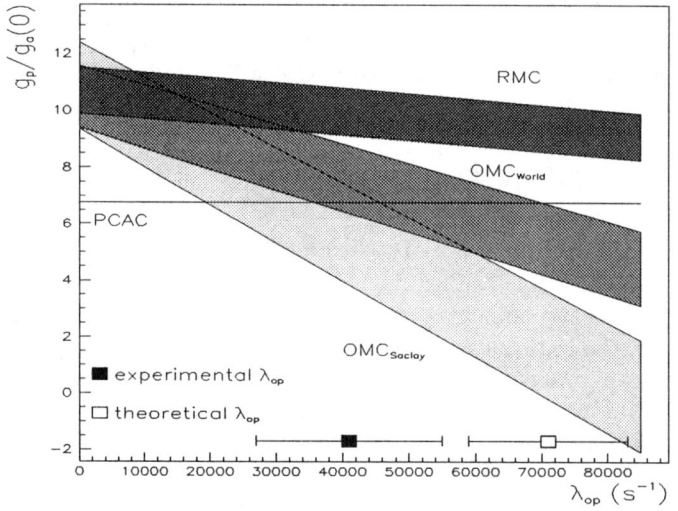

FIGURE 1. The pseudoscalar coupling g_p from radiative (RMC) and ordinary (OMC) muon capture experiments vs. λ_{op}. Also included is the average of all OMC measurements, OMC$_{World}$. The PCAC prediction for $g_p/g_a(0)$ is indicated.

experiment [5] is less sensitive to λ_{op} than the ordinary muon capture (OMC) experiment [3]. Also indicated is the world average of all OMC measurements in both gas and liquid targets. The value of λ_{op} chosen clearly affects each experiment's agreement with predicted values for g_p and with each other.

EXPERIMENTAL TECHNIQUE

The present experiment measures λ_{op} by obtaining the time difference between a muon arrival in the target and the detection of the monoenergetic, 5.2 MeV neutron produced by muon capture on a proton. Due to the presence of λ_{op}, the time distribution is modified from a simple exponential and is approximated by

$$\frac{dn}{dt} = Ae^{-t/\tau_{\mu^-}} \left[e^{-\lambda_{p\mu p} t}(\lambda_s - \lambda_{om}) + e^{-\lambda_{op} t}(\lambda_{om} - \lambda_{pm}) + e^{-\lambda_{pm} t}\lambda_{pm} \right] \quad (2)$$

where λ_s, λ_{om}, λ_{pm} are the muon capture rates in the singlet μp, the ortho $p\mu p$, and the para $p\mu p$ states respectively, $\lambda_{p\mu p}$ is the $p\mu p$ formation rate, and τ_{μ^-} is the muon lifetime. This relation assumes negligible contamination from deuterium or higher-Z species, ignores the fleeting existence of isolated μp atoms in the atomic triplet state, and assumes that the $p\mu p$ is formed only in the ortho state [6].

Five BC501A equivalent liquid scintillator neutron detectors surround a 2.7 liter ultra-pure, isotopically enriched liquid hydrogen target. Two plastic scintillators serve as the beam telescope, and there are five scintillators to detect charged particles that enter the neutron detectors. In order to differentiate between gammas and neutrons, the signals from the detectors are fed into pulse shape discrimination (PSD) modules. Since this experiment measures a time spectrum and not the absolute neutron yield, the difficult task of determining detector efficiencies is not needed.

This experiment differs from the previous λ_{op} measurement [3] by utilizing multihit timing electronics and working at a facility with a CW, rather than a pulsed, beam. However, a major similarity is the presence of many backgrounds. The following list enumerates the backgrounds and their predicted effects on the measurement:

1. Cyclotron and cosmic ray neutrons: large, but time-independent– shows up as a large flat background to the time spectrum but does not distort the shape.

2. Muon capture on target impurities (deuterium and high-Z elements): negligible, < 1 ppm deuterium, < 1 ppb higher-Z impurities.

3. Muon capture on target walls (Au, Ag): negligible, with 0-500 ns timing cut since Au and Ag captures cause a rapid muon disappearance rate [2].

4. Muon capture in beam and veto scintillators: $\sim 10\%$ of signal, measured during empty target runs.

5. "Fake" neutrons– photo-nuclear events and two-photon pile-up events: $\sim 10\%$ of signal, measured with a μ^+ beam, in which the muon capture process is "turned off" but the dominant muon decay bremsstrahlung photons remain.

CONCLUSIONS

A total of 2.0×10^{10} muon stops in the target were recorded during the run. The 5.2 MeV neutron signal from a typical liquid scintillator is shown in Fig. 2. Electron-rejection and PSD cuts have been applied, and neutron events occuring within 0.5 and 2.0 microseconds of a muon stop in the target have been subtracted. Clearly seen is the characteristic "box" extending downwards from ~5 MeV to the low energy cut-off at ~3.5 MeV, indicating the presence of muon capture neutrons.

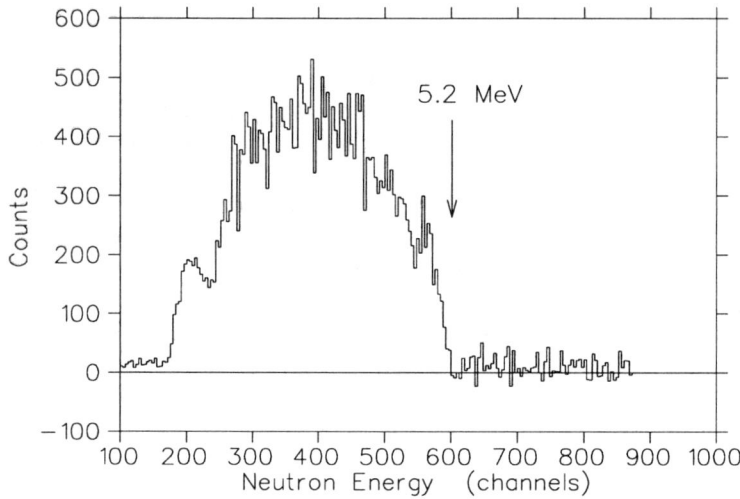

FIGURE 2. Typical neutron energy spectrum (backgrounds subtracted).

The cuts used to select the capture neutrons for the neutron time distribution are currently being optimized. The error on this measurement of λ_{op} is expected to be largely statistics dominated. Final results are expected in Fall 2000 and can be expected to make a major impact on our knowledge of the proton's induced pseudoscalar coupling.

REFERENCES

1. V. Bernard, N. Kaiser, Ulf-G. Meissner, *Phys. Rev. D* **50**, 6899 (1994).
2. N. Mukhopadhyay, *Phys. Rep.* **30C**, 1 (1977).
3. G. Bardin et al., *Phys. Lett.* **104B**, 320 (1981).
4. D.D. Bakalov et al., *Nucl. Phys.* **A384**, 302 (1982).
5. G. Jonkmans et al., *Phys. Rev. Lett.* **77**, 4512 (1996); D.H. Wright, et al., *Phys. Rev. C* **57**, 373 (1998).
6. L.I. Ponomarev and M.P. Faifman, *Sov. Phys. JETP* **44**, 886 (1976).

Creation and Decay of η-Mesic Nuclei

G.A. Sokol[1], T.A. Aibergenov, A.V. Koltsov, A.V. Kravtsov,
Yu.I. Krutov, A.I. L'vov, L.N. Pavlyuchenko,
V.P. Pavlyuchenko, S.S. Sidorin

P.N. Lebedev Physical Institute, Leninsky Prospect 53, Moscow 117924, Russia

Abstract.
First experimental results on photoproduction of η-mesic nuclei are analyzed. In an experiment performed at the 1 GeV electron synchrotron of the Lebedev Physical Institute, correlated $\pi^+ n$ pairs arising from the reaction

$$\gamma + {}^{12}C \to N + {}_\eta(A-1) \to N + \pi^+ + n + (A-2)$$

and flying transversely to the photon beam have been observed. When the photon energy exceeds the η-meson production threshold, a distribution of the $\pi^+ n$ pairs over their total energy is found to have a peak in the subthreshold region of the internal-conversion process $\eta p \to \pi^+ n$ which signals about formation of η-mesic nuclei.

The idea that a bound state of the η-meson and a nucleus (the so-called η-mesic nucleus) can exist in Nature was put forward long ago by Peng [1] who relied on the first estimates of the ηN scattering length $a_{\eta N}$ obtained by Bhalerao and Liu [2]. Owing to Re $a_{\eta N} > 0$, an average attractive potential exists between slow η and nucleons. This can result in binding ηA systems, provided the life time of η in nuclei is long enough [3]. Modern calculations [4,5] predict a rather strong ηN attraction which is sufficient for binding η in all nuclei with $A \geq 4$.

The very first experiments on searching for the η-mesic nuclei performed at BNL [6] and LAMPF [7] gave negative results. Meantime, studies of the reactions $p(d,{}^3He)\eta$ [8,9], ${}^{18}O(\pi^+,\pi^-){}^{18}Ne$ [10], and $\vec{d}(d,{}^4He)\eta$ [11] suggest that a quasi-bound ηA state is formed in these reactions [12,13].

In the present work we report on first results concerning formation of η-mesic nuclei in photoreactions. A very efficient trigger for searching for η-mesic nuclei [15] consists in detecting decay products of the η-mesic nucleus, viz. a πN pair produced in the reaction $\eta N \to \pi N$ inside the nucleus. Here η itself is produced at an earlier stage, in the reaction $\gamma N \to \eta N$ in our case. Both these reactions are mediated by the $S_{11}(1535)$ resonance which affects also a propagation of the

---

[1] E-mail: gsokol@x4u.lebedev.ru

intermediate η in the medium (via multiple ηN rescattering) and leads to capturing slow η into a bound state (Fig. 1). Formation of the bound state of the η and the nucleus becomes possible when the momentum of the produced η is small (typically less than 150 MeV/c). This requirement suggests photon energies $E_\gamma = 650$–850 MeV as most suitable for creating η-nuclei.

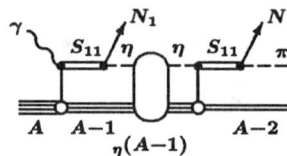

FIGURE 1. Mechanism of creation and decay of η-mesic nuclei.

πN pairs emerging from η-mesic nucleus decays have an opening angle $\langle \theta_{\pi N} \rangle = 180°$ and specific kinetic energies of their components (though smeared by the Fermi motion), $\langle E_\pi \rangle \simeq 300$ MeV, $\langle E_n \rangle \simeq 100$ MeV. Among four possible isotopic combinations $\pi^+ n$, $\pi^- p$, $\pi^0 n$, $\pi^0 p$ the first one is quite suitable for measuring energies of the particles.

Accordingly, in an experiment performed at the 1 GeV electron synchrotron of the Lebedev Physical Institute, correlated $\pi^+ n$ pairs arising from the reaction

$$\gamma + {}^{12}C \to N + {}_\eta(A-1) \to N + \pi^+ + n + (A-2) \tag{1}$$

have been searched for. An experimental setup (Fig. 2) consisted of a carbon target $\varnothing 4$ cm × 4 cm and two time-of-flight scintillator spectrometers having a time resolution of $\delta\tau \simeq 0.1$ ns. A plastic anticounter A of charged particles (of the 90% efficiency), placed in front of the neutron detectors, and dE/dx layers, placed between start and stop detectors in the pion spectrometer, were used for a better identification of particles.

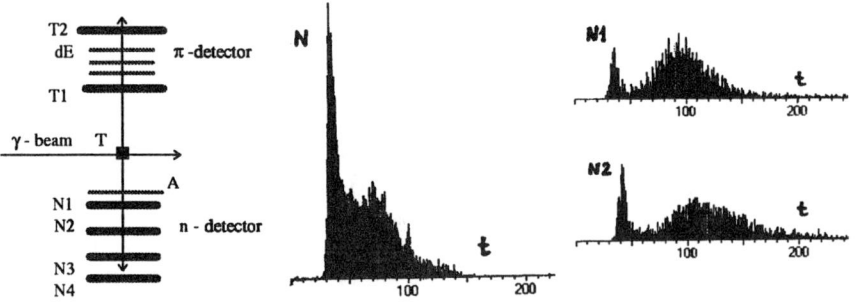

FIGURE 2. Layout of the experimental setup. Shown also time-of-flight spectra in the π (left) and n (right) spectrometers.

Strategy of measurements was as follows. Two bremsstrahlung-beam energies were used, $E_{\gamma \max} = 650$ MeV and 850 MeV, i.e., well below and well above η production threshold on the free nucleon which is 707 MeV. The first, "calibration" run was performed at 650 MeV with the spectrometers positioned at angles $\theta_n = \theta_\pi = 50°$ around the beam. In that run, this was a quasi-free photoproduction $\gamma p \to \pi^+ n$ which dominated the observed yield of the $\pi^+ n$ pairs. Then, at the same "low" energy 650 MeV, the spectrometers were positioned at $\theta_n = \theta_\pi = 90°$ (the "background" run). In such a kinematics, the quasi-free production did not contribute and the observed count was presumably dominated by double-pion production. At last, the third run (the "effect+background" run) was performed at the same 90°/90° position, however with the higher photon beam energy of 850 MeV, at which η mesons are produced too.

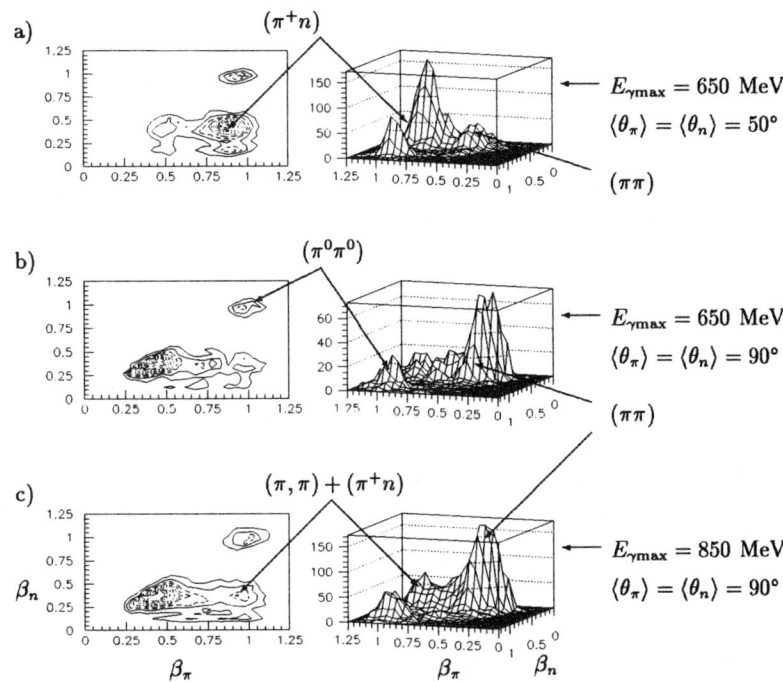

FIGURE 3. Raw $\pi^+ n$ event distributions over the particle velocities β_n and β_π for the "calibration" run (a), "background" run (b), and "effect+background" run (c).

In accordance with measured velocities of particles detected by the spectrometers, all candidates to the $\pi^+ n$ events were separated into three classes: fast-fast (FF), fast-slow (FS), and slow-slow (SS) events. The FF events mostly correspond to $\pi^0 \pi^0$ production which results in hitting detectors by photons or e^+/e^-. The FS events mostly emerge from the $\pi^+ n$ pairs. Comparing yields and time spectra in these runs (and, in particular, using the SS events for extrapolating and subtracting a background), we have found a clear excess of the FS events which appeared when the photon energy exceeded η production threshold. The total cross section of photoproduction of such excess pairs, averaged over the photon energy range 650–850 MeV, was found to be about $\sigma_{tot}(\pi^+ n) \simeq 10$ μb. See Ref. [14] for more details. In the present work a further analysis of the excess FS events is done and their energy characteristics are determined.

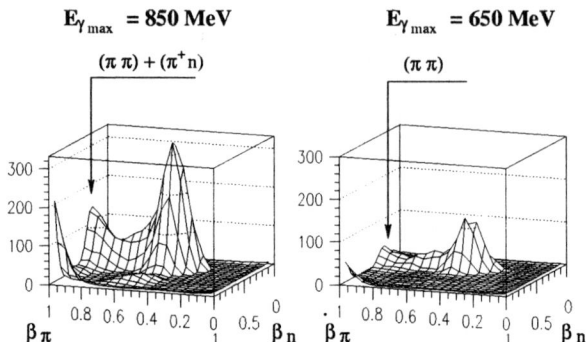

FIGURE 4. Corrected two-dimensional distributions over the velocities β of the $\pi^+ n$ events with the end-point energy of the bremsstrahlung spectrum $E_{\gamma \max} = 850$ and 650 MeV.

In order to find kinetic energies of the neutron and pion, the velocities $\beta_i = L_i/ct_i$ of both the particles have to be determined. They are subject to fluctuations stemming from errors δt_i and δL_i in the time-of-flight t_i and the flight base L_i. Such fluctuations are clearly seen in the case of the ultra-relativistic FF events which have experimentally observed velocities close but not equal to 1 (see Fig. 3). Therefore, an experimental β-resolution of the setup can be directly inferred from the FF events. Then, using this information and applying an inverse-problem statistical method described in Ref. [16], one can unfold the experimental spectrum, obtain a smooth velocity distribution in the physical region $\beta_i < 1$ (Fig. 4), and eventually find a distribution of the particle's kinetic energies $E_i = M_i[(1 - \beta_i^2)^{-1/2} - 1]$. Finding E_i, we introduced corrections related with average energy losses of particles in absorbers and in the detector matter. It is worth to say that the number of the $\pi^+ n$ FS events visibly increases when the photon beam energy becomes sufficient for producing η mesons.

Of the most interest is the distribution of the $\pi^+ n$ events over their total energy $E_{tot} = E_n + E_\pi$, because creation and decay of η-mesic nuclei is expected to produce a relatively narrow peak in E_{tot} of the width $\sim$ 50–70 MeV (see, e.g., [14,17]). Such a peak was indeed observed: see Fig. 5, in which an excess of the FS events appears when the photon energy exceeds the η-production threshold. Subtracting a smooth background, we have found a 1-dimensional energy distribution of the $\pi^+ n$ events presumably coming from (bound) η decaying in the nucleus, see Fig. 6. The experimental width of this distribution is about 100 MeV, including the apparatus resolution. Its center lies by $\Delta E = 40$ MeV below the energy excess $m_\eta - m_\pi = 408$ MeV in the reaction $\eta N \to \pi N$, and it is well below the position of the $S_{11}(1535)$ resonance too. Up to effects of binding of protons annihilated in

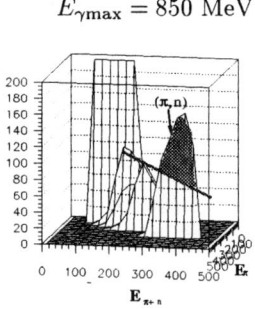
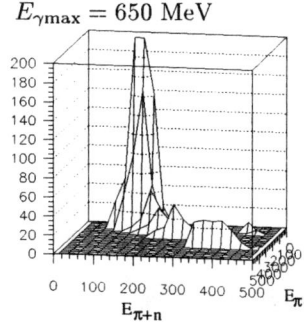

FIGURE 5. Distribution over the total kinetic energy of the $\pi^+ n$ pairs for the "effect+background" run (the left panel) and for the "background" run (the right panel) obtained after unfolding the raw spectra.

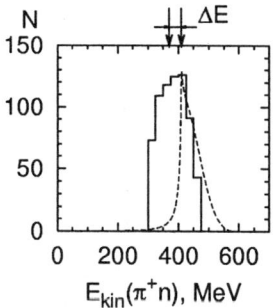

FIGURE 6. Distribution over the total kinetic energy of the $\pi^+ n$ pairs after subtraction of the background. Arrows indicate threshold in the reaction $\eta N \to \pi N$, i.e. 408 MeV, and the weighted center of the histogram. For a comparison, a product of free-particle cross sections of $\gamma N \to \eta N$ and $\eta N \to \pi N$ [5] is shown with the dashed line (in arbitrary units).

the decay subprocess $\eta p \to \pi^+ n$, the value ΔE characterizes the binding energy of η in the nucleus. The width of that peak is determined both by the width of the η-bound state and by the Fermi motion.

Whereas the fixed opening angle $\theta_{\pi n} = 180°$ chosen in the kinematics with $\theta_n = \theta_\pi = 90°$ selects $\pi^+ n$ pairs carrying a low total momentum in the direction of the photon beam, an independent check of the transverse momentum $p_\perp = p_\pi - p_n$ is meaningful. The corresponding distribution is shown in Fig. 7. On the top of a background, there is a narrower peak in $p_\perp$ having a width compatible with the Fermi momentum of nucleons in the nucleus.

In conclusion, an excess of correlated $\pi^+ n$ pairs with the opening angle $\langle \theta_{\pi N} \rangle = 180°$ has been experimentally observed when the energy of photons exceeded the η-production threshold. A distribution of the pairs over their total kinetic energy was found to have a peak lying below threshold of the elementary process $\pi N \to \eta N$.

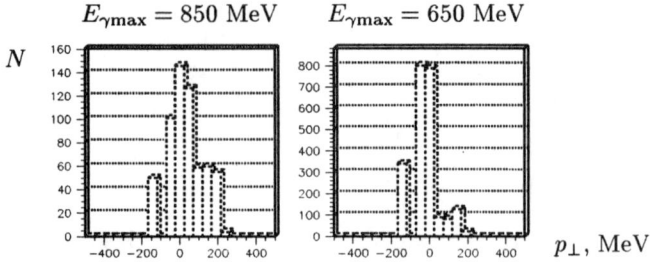

FIGURE 7. Distribution over the total transverse momentum $p_\perp$ of $\pi^+ n$ pairs for the "effect+background" run (the left panel) and for the "calibration" run (the right panel).

A narrow peak is also found in the pair's distribution over their total transverse momentum. All that suggests that these $\pi^+ n$ pairs arise from creation and decay of captured bound η in the nucleus, i.e., they arise through the stage of formation of an η-mesic nucleus.

The work was supported by RFBR grant 99-02-18224.

REFERENCES

1. J.C. Peng, AIP Conference Proceedings **133**, 255 (1985).
2. R.S. Bhalerao and L.C. Liu, *Phys.Rev.Lett.* **54**, 865 (1985).
3. L.C. Liu and Q. Haider, *Phys. Rev. C* **34** (1986) 1845.
4. M. Batinić et al., *Phys. Rev. C* **57** (1998) 1004.
5. A.M. Green and S. Wycech, *Phys. Rev. C* **55** (1997) R2167.
6. R.E. Chrien et al., *Phys. Rev. Lett.* **60** (1988) 2595.
7. B.J. Lieb and L.C. Liu, LAMPF Progress Report, LA-11670-PR (1988).
8. J. Berger et al., *Phys. Rev. Lett.* **61** (1988) 919.
9. B. Mayer et al., *Phys. Rev. C* **53** (1996) 2068.
10. J.D. Johnson et al., *Phys. Rev. C* **47** (1993) 2571.
11. N. Willis et al., *Phys. Lett.* **B 406** (1997) 14.
12. C. Wilkin, *Phys. Rev. C* **47** (1993) R938.
13. L. Kondratyuk et al., Proc. Int. Conf. "Mesons and Nuclei at Intermediate Energies", Dubna 1994 (World Scientific, Singapore, Eds. M.Kh. Khanhasayev and Zh.B. Kurmanov), p. 714.
14. G.A. Sokol et al., *Fizika B* **8** (1998) 81.
15. G.A. Sokol and V.A. Tryasuchev, *Kratk. Soobsh. Fiz.* [Sov. Phys. – Lebedev Institute Reports] 4 (1991) 23.
16. V.P. Pavlyuchenko, *Vopr. Atom. Nauki i Tekhn, ser. Tekh. Fiz. Eksp.* 1/13 (1983) 39.
17. A.I. L'vov, nucl-th/9810054 and Proc. 7th Int. Conf. "Mesons and Light Nuclei", Pruhonice, Prague, Czech Republic, 1998 (World Scientific, eds. J. Adam et al.), p. 469.

η and η' Mesons in the Improved Ladder Bethe-Salpeter Approach

Makoto Takizawa*, Kenichi Naito**, Yukio Nemoto[†]
and Makoto Oka[††]

*Showa Pharmaceutical University, Machida, Tokyo 194-8543, Japan
**Radiation Laboratory, the Institute of Physical and Chemical Research,
Wako, Saitama 351-0198, Japan
[†]Yukawa Institute of Theoretical Physics, Kyoto University, Kyoto 606-8502, Japan
[††]Department of Physics, Tokyo Institute of Technology, Meguro, Tokyo 152-8551, Japan

Abstract. We study the properties of the η and η' mesons using the improved ladder Schwinger-Dyson and Bethe-Salpeter equations in which the one gluon exchange kernel with the one-loop running coupling constant is used. The effect of the $U_A(1)$ anomaly is introduced by the $U_A(1)$ breaking 6-quark flavor-mixing interaction. The masses of η and η' mesons are reproduced by a relatively weak flavor-mixing interaction, for which the chiral symmetry breaking is dominantly induced by the soft-gluon exchange interaction. The decay constants are calculated and the anomalous PCAC relation is numerically checked.

INTRODUCTION

It is known that the classical QCD Lagrangian is invariant under the $U_L(3) \times U_R(3)$ symmetry except for the quark mass term and this symmetry is broken down to the $U_V(3)$ spontaneously by the strong interaction between quarks in the low-energy QCD. It is also known that the $U_A(1)$ symmetry is broken by the quantum anomaly. Therefore the number of the Nambu-Goldstone bosons associated with the spontaneous breaking of the chiral symmetry: $SU_L(3) \times SU_R(3) \times U_V(1) \to U_V(3)$ is eight, namely, π^+, π^-, π^0, K^+, K^-, K^0, $\bar{K}^0$ and η. The masses of these Nambu-Goldstone bosons are considered to be due to the quark masses. The heavy mass of η' is due to not only quark masses but also the $U_A(1)$ anomaly. Since the energy scales of the spontaneous breaking of chiral symmetry, strange quark mass and $U_A(1)$ anomaly are similar, one should treat these effects on an equal footing. In order to understand the dynamics of the low-energy QCD, it is important to study the properties of the η and η' mesons.

As for the η meson, its mass, decay constant and radiative decays are extensively studied [1] in the framework of the three-flavor Nambu-Jona-Lasinio (NJL) model

FIGURE 1. The diagram for the SD equation.

with the $U_A(1)$ breaking 6-quark flavor-mixing interaction proposed by Kobayashi and Maskawa [2] which is similar to 't Hooft instanton induced interaction [3]. It is shown that η-meson mass, $\eta \to \gamma\gamma$, $\eta \to \gamma\mu^-\mu^+$ and $\eta \to \pi^0\gamma\gamma$ decay amplitudes are in good agreement with the experimental values when the $U_A(1)$ breaking is rather strong. A shortcoming of this approach is that the η' mass has unphysical imaginary part associated with the unphysical decay channel $\eta' \to \bar{q}q$.

We study [4] the properties of the η and η' mesons using the improved ladder Schwinger-Dyson (SD) and Bethe-Salpeter (BS) equations in which the one gluon exchange kernel with the one-loop running coupling constant is used. In this approach Aoki et al. [5,6] have shown that the chiral symmetry is broken dynamically, and that the pion decay constant f_π, the quark condensate $\langle \bar{q}q \rangle$ and the masses of the lowest lying vector and axial-vector mesons are reproduced rather well in the chiral limit. It has been also shown that the asymptotic behavior of the QCD is reproduced in this approach. Recently, this approach has been extended to the finite current quark mass case [7] and reasonable values of the pion mass, pion decay constant and the quark condensate are obtained with a rather large Λ_{QCD}. The effect of the $U_A(1)$ anomaly is introduced by the $U_A(1)$ breaking 6-quark flavor-mixing interaction proposed by Kobayashi and Maskawa [2]. The instanton size effects are taken into account by the form factor of the interaction vertices, which guarantees the right asymptotic behavior of the solution of the SD and BS equations. In this approach, dynamical quark mass depends on its momentum and increases in the low-q^2 region. The dynamical quark mass is large enough to prevent the η' from decaying into $\bar{q}q$ pair unphysically.

IMPROVED LADDER SD AND BS APPROACH

Using the Cornwall-Jackiw-Tomboulis (CJT) effective action formulation [8] we derive the rainbow like SD equation for the quark propagator on the nonperturbative vacuum and the ladder like BS equation for the η, η' mesons. The diagrams for these equations are shown in Fig. 1 and 2. FM means the $U_A(1)$ breaking 6-quark flavor-mixing interaction. We employ the Landau gauge gluon propagator with the Higashijima-Miransky type running coupling constant [9]. For the η and η' mesons, one has to solve eight coupled channel BS equation, two in flavor space and four in Dirac space.

Using the same model parameters of the running coupling constant used in Ref. [7], we have solved the SD and BS equations numerically. The strength of FM interaction is determined so as to reproduce the η and η' meson masses. Our

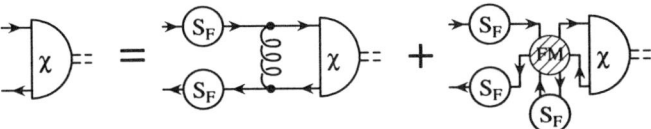

FIGURE 2. The diagram for the BS equation.

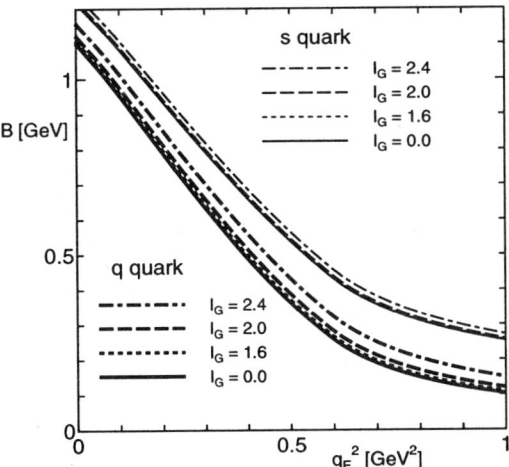

FIGURE 3. q_E^2 dependences of the solutions of the SD equation in the $SU(2)$ chiral limit with $I_G = 0.0, 1.6, 2.0, 2.4$ [GeV^{-1}].

numerical results are as follows. The dynamical quark masses at $q^2 = 0$ are $B_{u,d}(q^2 = 0) = 1.17$ [GeV] and $B_s(q^2 = 0) = 1.24$ [GeV] and the q_E^2 depandence is shown in Fig. 3. I_G is the strength parameter of the FM interaction. The quark condensates are $\langle \bar{u}u \rangle_R = \langle \bar{d}d \rangle_R = -(0.276)^3$ [GeV^3] and $\langle \bar{s}s \rangle_R = -(0.236)^3$ [GeV^3], the pion mass is $M_\pi = 152$ [MeV] and the pion decay constant is $f_\pi = 103$ [MeV]. Here we have not performed the precise parameter fittings so as to reproduce the observed pion mass and decay constant since solving the BS equation of the non-local interaction requires the rather large computer resources.

As for the η and η' mesons, we obtain $M_\eta = 511$ [MeV] and $M_{\eta'} = 1060$ [MeV]. Since the flavor structure of the BS amplitude in the η, η' channel depends on the relative and total momenta, one can only define the decay constants associated with the octet and singlet axial-vector current for the η or η' mesons, i.e., f_8^η, f_0^η, $f_8^{\eta'}$ and $f_0^{\eta'}$. Our results are $f_8^\eta = 113$ [MeV], $f_0^\eta = 19$ [MeV], $f_8^{\eta'} = -36$ [MeV] and $f_0^{\eta'} = 122$ [MeV]. In order to see effects of the flavor mixing, we introduce the flavor singlet-octet mixing angles for the η and η' mesons, $-f_0^\eta/f_8^\eta = \tan\theta_\eta$, $f_8^{\eta'}/f_0^{\eta'} = \tan\theta_{\eta'}$. These quantities may indicate the averaged flavor contents of the η and η' mesons. Our numerical results are $\theta_\eta = -9.5$ [deg] and $\theta_{\eta'} = -16.4$

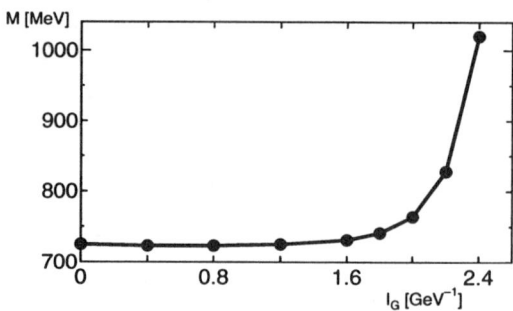

FIGURE 4. I_G dependence of the mass of the η' meson with $m_{qR} = 0$ and $m_{sR} = 100$ [MeV].

[deg]. It shows that the momentum dependences of the flavor structures of the η and η' mesons are not so small and the momentum independent treatment of the $\eta - \eta'$ mixing angle is rather questionable. If one switch off the FMI, the dynamical quark mass becomes $B_{u,d}(q^2 = 0) = 1.11$ [GeV], so the contribution of FMI to the dynamical chiral symmetry breaking is about 5%. We plot the η' meson mass as a function of the $U_A(1)$ breaking parameter I_G in Fig. 4. The effect of the mixing of the up and down quark component seems to be negligible and the η' mass grows rapidly from $I_G \sim 2.0$ [GeV^{-1}].

In summary, we have solved the improved ladder SD and BS equations which include the $U_A(1)$ breaking flavor mixing interaction. Using the same model parameters of the running coupling constant used in Ref. [7], we have obtained reasonable values of M_π, M_η, $M_{\eta'}$, f_π and $\langle \bar{q}q \rangle$ with a relatively weak flavor-mixing interaction (FMI), for which the chiral symmetry breaking is dominantly induced by the soft-gluon exchange interaction. It is in contrast with the NJL model results, where about 1/3 of the dynamical quark mass is due to FMI.

REFERENCES

1. Takizawa M., Nemoto Y., and Oka M., *Phys. Rev. D* **54**, 4083 (1997).
2. Kobayashi M., and Maskawa T., *Prog. Theor. Phys.* **44**, 1422 (1970).
3. 't Hooft G., *Phys. Rev. D* **14**, 3432 (1976).
4. Naito K.. Nemoto Y., Takizawa M., Yoshida K., and Oka M., *Phys. Rev. C* **61**, 065201 (2000).
5. Aoki K.-I., Bando M., Kugo T., Mitchard M.G., and Nakatani H., *Prog. Theor. Phys.* **84**, 683 (1990).
6. Aoki K.-I., Kugo T., and Mitchard M.G., *Phys. Lett. B* **206**, 467 (1991).
7. Naito K., Yoshida K., Nemoto Y., Oka M., and Takizawa M., *Phys. Rev. C* **59**. 1722 (1999).
8. Cornwall J.M., Jackiw R., and Tomboulis E., *Phys. Rev. D* **10**, 2428 (1974).
9. Hgashijima K., *Phys. Rev. D* **29**, 1228 (1984);
   Miransky V.A., *Sov. J. Nucl. Phys.* **38**, 280 (1984).

Subthreshold K^+-Production Studies with ANKE at COSY-Jülich*

S. Barsov[a], U. Bechstedt[b], G. Borchert[b], W. Borgs[b], M. Büscher[b],
M. Debowski[c], W. Erven[b], R. Eßer[d,1], P. Fedorets[e], D. Gotta[b],
M. Hartmann[b], H. Junghans[b], A. Kacharava[f,i], B. Kamys[g],
F. Klehr[b], H.R. Koch[b], V.I. Komarov[f], V.P. Koptev[a], P. Kulessa[b],
A. Kulikov[f], V. Kurbatov[f], G. Macharashvili[f,i], R. Maier[b],
S. Merzliakov[f], S. Mikirtychiants[a], H. Müller[c], A. Mussgiller[h],
M. Nioradze[i], H. Ohm[b], A. Petrus[f], D. Prasuhn[b], K. Pysz[j],
F. Rathmann[k], B. Rimarzig[c], Z. Rudy[g], R. Schleichert[b],
Chr. Schneider[c], H. Schneider[b], O.W.B. Schult[b], H. Seyfarth[b],
K. Sistemich[b], H.J. Stein[b], H. Ströher[b], I. Zychor[l]
for the ANKE collaboration[2]

[a] *Petersburg Nuclear Physics Institute, 188350 Gatchina, Russia*
[b] *Forschungszentrum Jülich, 52425 Jülich, Germany*
[c] *Forschungszentrum Rossendorf, 01314 Dresden, Germany*
[d] *Universität zu Köln, 50923 Köln, Germany*
[e] *Institute of Theoretical and Experimental Physics, 117259 Moscow, Russia*
[f] *Joint Institute for Nuclear Research, 141980 Dubna, Russia*
[g] *Jagellonian University, 30059 Cracow, Poland*
[h] *Fachhochschule München, 80335 München, Germany*
[i] *HEP Institute of Tbilisi State University, 380086 Tbilisi, Georgia*
[j] *Institute of Nuclear Physics, 31342 Cracow, Poland*
[k] *Universität Erlangen-Nürnberg, 91058 Erlangen, Germany*
[l] *The Andrzej Soltan Institute for Nuclear Studies, 05400 Swierk, Poland*

Abstract. The spectrometer ANKE at the synchrotron COSY-Jülich allows to momentum analyze ejectiles emitted from an internal target at forward angles $\varphi \simeq 0°$ with large angular acceptance ($\Delta\Omega \simeq 50$ msr). In first measurements the K^+-production in pA collisions at beam energies from 1.0 GeV, far below the nucleon-nucleon threshold of 1.58 GeV, up to 2.3 GeV was investigated. The experimental challenge at 1.0 GeV is to measure the K^+-production with cross sections of about 10^{-34} cm^2 at a background cross sections level of 10^{-27} cm^2. Data on the target-mass dependence of the differential cross section for kaon momenta $p_{K^+} = 150...510$ MeV/c and three different targets, C, Cu and Au, are presented and compared with theoretical predictions.

[1)] Now working at: BICRON Vertriebs GmbH, 42929 Wermelskirchen, Germany.
[2)] For a complete collaboration list see [1].

I THE ANKE SPECTROMETER

The spectrometer ANKE [2–4] is located in one of the two straight sections of the race-track shaped accelerator ring COSY [5]. Its three dipole magnets guide the circulating beam out of its nominal path towards the target and back into the standard orbit. The spectrometer dipole D2 deflects forward-emitted ejectiles from proton-induced reactions on strip targets of C, Cu and Au from the circulating beam and allows the determination of their emission angles and momenta with the K^+-detection system, see Fig. 1 a). The 20 cm gap of the dipole D2 provides a vertical acceptance in the range of $3.5° < |\vartheta| < 5.5°$; the horizontal acceptance amounts to $\varphi \sim -12° \ldots 12°$.

The K^+-detection system consists of: *i)* 23 Start counters close to the 500 μm thick Al vacuum window at the side of D2. *ii)* 15 telescopes along the focal surface of D2 to detect positively charged particles in the momentum range from 110 to 525 MeV/c at $B_{D2} = 1.3$ T. The telescopes, see Fig. 1 b), comprise an active part (Stop-, ΔE-, Veto scintillation- and lucite Cherenkov counters (only telescopes 9–15)) and two passive Cu degraders. *iii)* 2 multi-wire proportional chambers MWPC1 and MWPC2 which are used to reconstruct ejectile trajectories. They have each three sensitive anode planes with vertically and $\pm 30°$ inclined tungsten wires.

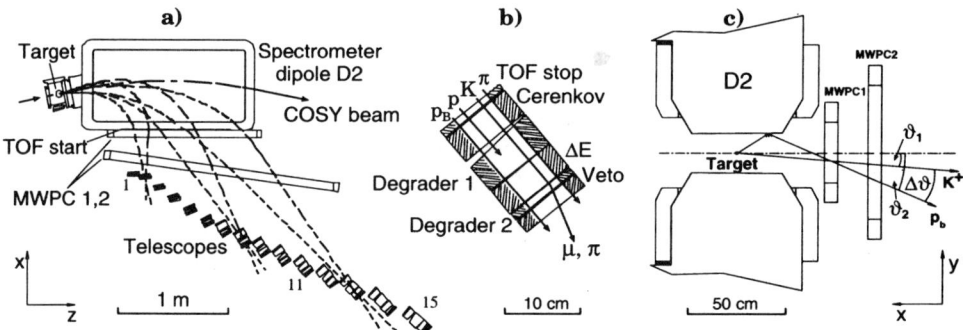

FIGURE 1. a) Top view of the ANKE spectrometer. b) Setup of telescope #14. c) Side view of ANKE; rescattered particles p_b have angles (ϑ_2) different from target particles (ϑ_1) and may cross MWPC1 at the same point.

The K^+-identification procedure is briefly described in the following — see [2,3] for more details: Pions and protons from the target are strongly suppressed by appropriate time of flight (TOF) criteria between the Start- and Stop counter and the energy-loss cuts in these counters. The first degrader optimizes the energy-loss ratio of kaons to pions in the ΔE counter and stops protons originating from the target. The second degrader prevents kaons from reaching the Veto counter. This offers the possibility to use the K^+-decay as a criterion for the discrimination against background: the K^+-decay products, mainly monoenergetic pions or muons, can be detected with a characteristic delay with respect to the signal in the Stop counter of the same telescope. Exploiting the information from the scintillation counters one can suppress nearly all the background from the target. The remaining background, mostly rescattered protons p_b (Fig. 1 c)), can effectively be discriminated with wire chambers since particles from the target at a vertical angle ϑ are grouped in a narrow peak in the $\Delta\vartheta$ distribution, mainly defined by multiple Coulomb scattering. Similar distributions for rescattered particles turn out to be flat.

II A-DEPENDENCE OF THE K^+ CROSS SECTION

The measurements were performed at proton beam energies of $T = 1.0$, 1.2, 1.5, 1.8, 2.0 and 2.3 GeV with carbon targets. At $T = 1.0$ and 2.3 GeV the kaon production was also studied with gold and copper targets. In Fig. 2 the number of identified kaons in the different telescopes from the three targets are presented. Absolute values of the cross sections will be obtained by normalization of N_{K^+} to the known pion production cross section [6–8].

In addition, it is also necessary to correct the spectra for the angular-momentum acceptances and the detection efficiencies of the telescopes and wire chambers. The corresponding measurements were carried out and are currently evaluated. These calibration constants are equal for all measurements and do not influence the cross-section ratios for kaon production from different targets. The corresponding ratios for pion production are known with an accuracy of 3–10% [6–8].

To improve the statistical accuracy of the cross-section ratios, the corresponding spectra of the different telescopes have been summed up. Fig. 3 shows the distribution for $\Delta\vartheta$ for the C target at $T = 2.3$ GeV and for C, Cu and Au at 1.0 GeV. At 2.3 GeV there is almost no background left after the scintillation-counter information is applied. At 1.0 GeV the kaon-to-background ratio decreases with increasing target mass number. Table 1 compares the momentum-integrated ratios of cross sections with theoretical expectations. At 2.3 GeV theoretical models [9–12] predict that K^+-production takes place in a single proton-nucleon collision. In this

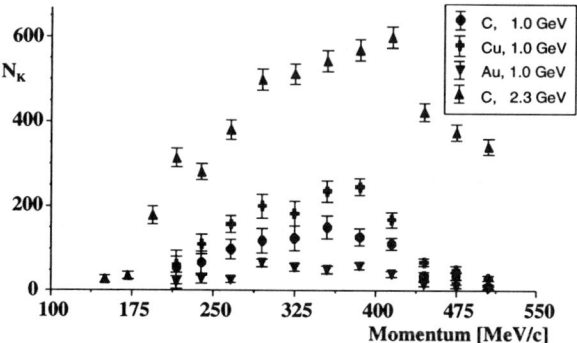

FIGURE 2. Number of identified kaons for Au, Cu and C targets at 1.0 GeV and for C at 2.3 GeV. Statistical errors are indicated. Each data point corresponds to the number of kaons in one telescope, the first telescope has been omitted because of a too high background level.

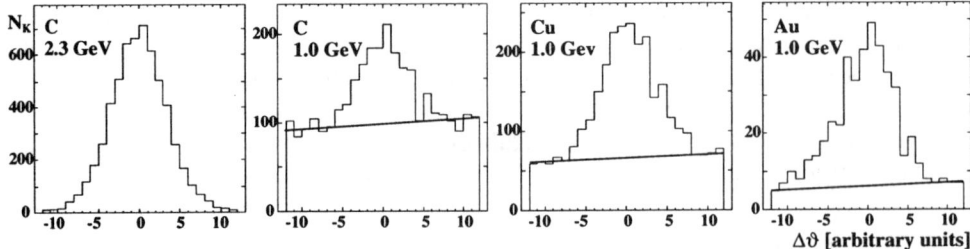

FIGURE 3. Distribution of $\Delta\vartheta$ for the C target at 2.3 GeV and for the targets C, Cu and Au at 1.0 GeV beam energy. The distribution of background can be described with straight lines. Its slopes have been determined in a detailed analysis.

case a scaling of the total K^+-production cross section with $A^{2/3}$ is expected from a simple geometrical model. The experimental results are in good agreement with that model. At 1.0 GeV the K^+-production should occur mainly via two-step mechanisms where in a first proton-nucleon collision an intermediate pion is produced. Such mechanisms imply a scaling of the K^+ cross section with A. The ratio Cu/C is in accordance with this assumption while the ratio Au/C does not seem to fit the model. Currently detailed calculations are in progress to find out whether the rescattering of the kaons in the heavy gold target leads to a reduced K^+ yield.

TABLE 1. Comparison of measured cross-section ratios with theoretical predictions for different targets and different energies.

Energy [GeV]	$(Cu/C)_{exp}$	$(Cu/C)_{theory}$	$(Au/C)_{exp}$	$(Au/C)_{theory}$
1.0	5.6±0.5	5.3	9.2±1.2	16.3
2.3	3.6±0.4	3.0	6.6±0.6	6.5

REFERENCES

1. See http://ikpd15.ikp.kfa-juelich.de:8085/doc/Anke.html
2. S. Barsov et al., *Nucl. Phys.*, **A663–664**, 1107 (2000)
3. S. Barsov et al., *Nucl. Phys.*, **A675**, 230 (2000)
4. S. Barsov et al., *submitted to NIM A*, (May 2000)
5. R.A. Maier, *Nucl. Phys. News* **7** No. 4, (1997)
6. D.R.F. Cochran et al., *Phys. Rev.* **D6**, 3085 (1972)
7. J. Papp et all., *Phys. Rev. Lett.* **34**, 601 (1975)
8. V.V. Abaev et al., *J. Phys.* **G14**, 903 (1988)
9. V.P. Koptev et al., *JETP* **67**, 2177 (1988)
10. W. Cassing et al., *Phys. Lett.* **B238**, 25 (1990)
11. A. Sibirtsev et al., *Z. Phys.* **A347**, 191 (1994)
12. E.Ya. Paryev, *Eur. Phys. J.* **A5**, 307 (1999)

*Supported by: BMBF, DFG, State of North Rhine Westfalia; Russian Ministry of Science, RFFI (99-02-04034, 99-02-18179a); Polish Committee for Scientific Research (2 P03B 101 19); INTAS.

ANTIKAON PRODUCTION AND MEDIUM EFFECTS IN PROTON–NUCLEUS REACTIONS AT SUBTHRESHOLD BEAM ENERGIES

E.Ya. Paryev

Institute for Nuclear Research, Russian Academy of Sciences, Moscow 117312, Russia

Abstract. The inclusive K^- meson production in proton–nucleus collisions in the subthreshold energy regime is analyzed in the framework of an appropriate folding model for incoherent primary proton–nucleon production process, which takes properly into account the struck target nucleon momentum and removal energy distribution (nucleon spectral function) as well as nuclear mean–field potential effects on the one-step antikaon creation process. A detailed comparison of the model calculations of the K^- differential cross sections for the reaction $p + Be^9$ at subthreshold energies with the first experimental data obtained at the ITEP proton synchrotron is given.

Kaon and antikaon properties in dense matter are a subject of considerable current interest in the nuclear physics commynity [1]. The knowledge of these properties is important for understanding both chiral symmetry restoration in dense nuclear medium and neutron star properties. Subthreshold kaon and antikaon production in heavy–ion collisions in which the high densities are accessible is apparently best suited for the studying of their properties in dense matter. The dropping K^- mass scenario will lead to a substantial enhancement of the K^- yield in heavy–ion collisions at subthreshold incident energies due to in–medium shifts of the elementary production thresholds to lower energies. Antikaon enhancement in nucleus–nucleus interactions has been recently observed by the KaoS Collaboration at SIS/GSI [2, 3]. This phenomenon has been attributed to the in–medium K^- mass reduction [4]. Of special question is the validity of extrapolation of extracted in [4] an "empirical" kaon and antikaon dispersion relations from densities of $(2-3)\rho_0$ to the density of ordinary nuclei. This can be clarified from the study of subthreshold K^+ and K^- production in proton–induced reactions. The advantage of these reactions is that a possible kaon and antikaon mass changes (up to 5% and 20% for K^+ and K^-, respectively), although smaller than those in heavy–ion collisions, can be better controlled due to their simpler dynamics compared to the case of nucleus–nucleus interactions. Therefore, the information obtained from the proton–induced reactions will supplement that deduced from heavy–ion collision studies.

Another a very important information that can be extracted from the study of K^+ and K^- meson production in pA–collisions at subthreshold incident energies concerns such intrinsic properties of target nuclei as Fermi motion, high momentum components of the nuclear wave function.

The inclusive K^+ production in proton–nucleus reactions at bombarding energies less than threshold energies in a collision of free nucleons has been extensively studied both experimentally and theoretically in recent years. Up to now, there have been, however, no data on subthreshold K^- production in proton–nucleus collisions. Recently, such experimental data have been obtained at the ITEP proton synchrotron [5]. The main goal of the present work is to analyze these data within the spectral function approach.

Apart from participation in the elastic scattering an incident proton can produce a K^- directly in the first inelastic pN–collision due to nucleon Fermi motion. Since we are interested in a few GeV region (up to 3 GeV), we have taken into account the following elementary process which has the lowest free production threshold (2.99 GeV for kinematical conditions of the experiment [5] in which the rather "hard" K^- mesons with momentum of 1.28 GeV/c at the laboratory angle of 10.5^0 have been detected):

$$p + N \to N + N + K + K^-, \tag{1}$$

where K stands for K^+ or K^0 for the specific isospin channel. In the following calculations, we will include the medium modification of the final hadrons (nucleons, kaon and antikaon) participating in the production process (1) by using their in–medium masses m_h^* determined below. The kaon and antikaon masses in the medium $m_{K^\pm}^*$ can be obtained from the mean–field approximation to the effective chiral Lagrangian [4, 6], i.e.:

$$m_{K^\pm}^*(\rho_N) = m_K + U_{K^\pm}(\rho_N), \tag{2}$$

where m_K is the rest mass of a kaon in free space, the $K^\pm$ optical potentials $U_{K^\pm}(\rho_N)$ are proportional to the nuclear density ρ_N

$$U_{K^\pm}(\rho_N) = U_{K^\pm}^0 \frac{\rho_N}{\rho_0} \tag{3}$$

and

$$U_{K^+}^0 = 22 \ MeV, \quad U_{K^-}^0 = -126 \ MeV. \tag{4}$$

The effective mass m_N^* of secondary nucleons produced in the reaction (1) can be expressed via the scalar mean–field potential $U_N(\rho_N)$ as follows [7]:

$$m_N^*(\rho_N) = m_N + U_N(\rho_N), \tag{5}$$

where m_N is the bare nucleon mass. The potential $U_N(\rho_N)$ was assumed to be proportional nuclear density

$$U_N(\rho_N) = U_N^0 \frac{\rho_N}{\rho_0} \qquad (6)$$

with the depth at nuclear saturation density ρ_0 relevant [7] for the momentum range of outgoing nucleons for the most part of kinematical conditions of the experiment [5] on subthreshold antikaon production

$$U_N^0 = -34 \ MeV. \qquad (7)$$

We have taken into account in the calculation of the K^- production cross section from the one–step process (1) also the influence of the nuclear optical potential $V_0 \approx 40 \ MeV$ on the incoming proton [7]. The total energy of the struck target nucleon N just before the collision (1) can be easily expressed [7] through the respective recoil and excitation energies of the residual $(A-1)$ system. In our method, the K^- production cross section for pA reactions from primary reaction channel (1) can be expressed [7] as the respective integral of the in–medium inclusive elementary K^- production cross section and nucleon spectral function $P(\mathbf{p}_t, E)$ over the struck target nucleon momentum $\mathbf{p}_t$ and removal energy E. The nucleon spectral function $P(\mathbf{p}_t, E)$ represents the probability to find in the nucleus a nucleon with momentum $\mathbf{p}_t$ and removal (binding) energy E. The expressions for the high momentum–energy part (correlated part) of the nucleon spectral function as well as for the single particle (uncorrelated) part of the one adopted in [7] were used in our calculations of K^- production in pBe collisions.

Figure 1 shows a comparison of the calculated invariant cross section for the production of K^- mesons with momentum of $1.28 \ GeV/c$ at the laboratory angle of 10.5^0 from primary $pN \rightarrow NNKK^-$ channel with the data from the experiment [5] for $p + Be^9 \rightarrow K^- + X$ reaction at the various bombarding energies. One can see that:

1) our model for primary antikaon production process, based on nucleon spectral function, fails completely (especially at "low" beam energies, dash–dotted line) to reproduce the experimental data at subthreshold beam energies without allowance for the influence of the corresponding nuclear mean–field potentials on the one–step production process (1);

2) a simultaneous inclusion of potentials for final nucleons, kaon and antikaon (dashed line with two dots) leads to an enhancement of the K^- yield by about a factors of 1.6 and 3, respectively, at "high" and "low" incident energies as well as to a reasonable well description of the experimental data except for the four lowest data points;

3) the previous scenario is hardly distinguishable from the one with employing only the attractive outgoing nucleon effective potential (dashed line with three

dots), what indicates that the simultaneous application of kaon and antikaon potentials unaffects the K^- yield and it is mainly governed by the nucleon mean–field potential;

4) our calculations with including simultaneously both attractive antikaon (3), (4) and nucleon (6), (7) effective potentials (solid line in Figure 1) reproduce quite well the experimental data in the energy region $\epsilon_0 \geq 2.4\,GeV$, but nevertheless underestimate the data at lower bombarding energies as in the considered above cases with the different scenarios for the in–medium masses of hadrons produced in the primary production process (1);

5) the antikaon yield from the one–step K^- production mechanism is entirely governed by the correlated part of the nucleon spectral function only in the far subthreshold energy region (at bombarding energies of $\epsilon_0 \leq 2.4\,GeV$), what intimates that the internal nucleon momenta greater than the Fermi momentum are needed for K^- production in direct process (1) at given kinematics and these beam energies.

The results presented in Figure 1 indicate, as was also noted above, that the one–step production process (1) misses the experimental data in the energy region far below the free threshold (at beam energies $\epsilon_0 \leq 2.4\,GeV$) even when the influence of the nuclear density–dependent mean–field potentials (3), (4), (6), (7) has been included. But the K^- creation due to first chance pN–collisions (1) in this energy region occurs, as is evident from the foregoing, when the incident protons collide with the short–range two–nucleon (or multinucleon) correlations inside the target nucleus, what means that the local baryon density around the spatial creation points of hadrons in these collisions can be high. Therefore, the antikaon production in the far subthreshold energy region should be evaluated more likely for the density–independent potentials with depths (4) and (7) taken at normal nuclear density ρ_0 than for the density–dependent fields (3) and (6) where the local average nuclear density is involved. The results of such calculations obtained for the one–step reaction channel (1), as was shown in [8], reproduce quite well the data in this energy region. Finally, it should be noted that our calculations of the antikaon yield from the secondary pion–nucleon production processes with including the influence of the different in–medium scenarios on it underestimate essentially the data and calculated cross sections from primary process (1) [8], what implies the dominance of the one–step K^- production mechanism for the considered antikaon production at all beam energies of interest.

Taking into account the considered above, one may conclude that the determination of the K^- potential in nuclear matter from the measurements of the primary–proton energy dependence of the double differential cross sections for production of "hard" antikaons on light target nuclei in the subthreshold energy regime appears to be difficult. It is apparent that the K^- potential has a strong effect on the K^- yields at low antikaon momenta (see, also, [8–10]). Therefore, the measurements of the differential cross sections for subthreshold "soft" (low momentum) K^- production on different target nuclei will allow as one may hope to shed light on the antikaon potential in nuclear medium. Such measurements might be conducted at, for example, the accelerator COSY using proton beam in the COSY-ANKE detector system.

REFERENCES

1. Cassing, W., and Bratkovskaya, E.L., *Phys. Rep.* **308**, 65 (1999).
2. Barth, R., et al., *Phys. Rev. Lett.* **78**, 4007 (1997).
3. Laue, F., et al., *Phys. Rev. Lett.* **82**, 1640 (1999).
4. Li, G.Q., et al., *Nucl. Phys.* **A625**, 372 (1997).
5. Akindinov, A.V., et al., *Preprint ITEP* **41-99**, Moscow (1999).
6. Li, G.Q., and Ko, C.M., *Phys. Rev.* **C54**, 1897 (1996).
7. Paryev, E.Ya., *Eur. Phys. J.* **A5**, 307 (1999).
8. Paryev, E.Ya., *Eur. Phys. J.A*, in the press.
9. Sibirtsev, A., and Cassing, W., *Nucl.Phys.* **A641**, 476 (1998).
10. Kiselev, Yu.T., for the FHS Collaboration: *J. Phys. G.* **25**, 381 (1999).

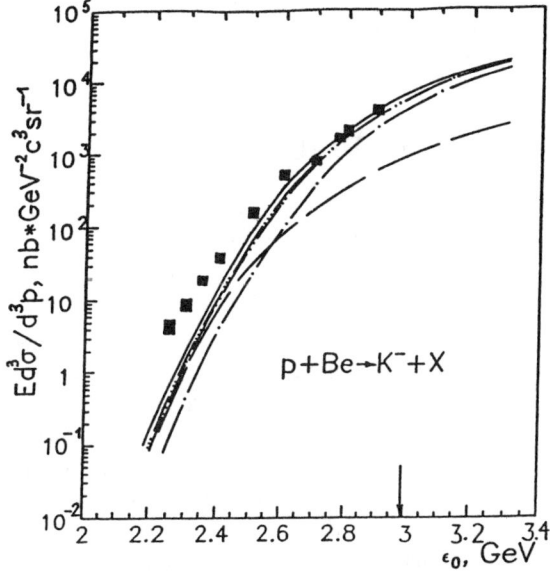

Figure 1. Lorentz invariant cross sections for the production of K^- mesons with momentum of 1.28 GeV/c at the lab angle of 10.5^0 in $p + Be^9$ reactions as functions of the laboratory kinetic energy ϵ_0 of the proton. The experimental data (full squares) are from the experiment [5]. The curves are our calculation with the density-dependent potentials. The dashed lines with one, two, three dots and the solid line are calculations for primary production process (1) with the total nucleon spectral function at $U_N(\rho_N) = 0$, $U_{K^+}(\rho_N) = 0$, $U_{K^-}(\rho_N) = 0$; $U_N(\rho_N) = -34(\rho_N/\rho_0)MeV$, $U_{K^+}(\rho_N) = 22(\rho_N/\rho_0)MeV$, $U_{K^-}(\rho_N) = -126(\rho_N/\rho_0)MeV$; $U_N(\rho_N) = -34(\rho_N/\rho_0)MeV$, $U_{K^+}(\rho_N) = 0$, $U_{K^-}(\rho_N) = 0$ and $U_N(\rho_N) = -34(\rho_N/\rho_0)MeV$, $U_{K^+}(\rho_N) = 0$, $U_{K^-}(\rho_N) = -126(\rho_N/\rho_0)MeV$, respectively. The long-dashed line denotes the same as the dashed line with two dots, but it is supposed in addition that the total nucleon spectral function is replaced by its correlated part. The arrow indicates the threshold for the reaction $pN \rightarrow NNKK^-$ occuring on a free nucleon at the kinematics under consideration.

Measurement of Proton Polarization in the $d(\vec{\gamma}, \vec{p})n$ Reaction

Steffen Strauch for the JLab Hall-A Collaboration

Rutgers, The State University of New Jersey, Piscataway, New Jersey, 08854, USA

Abstract. Recoil proton polarization was measured in deuteron photodisintegration at $90°_{cm}$ for photon energies up to 2.5 GeV. The induced polarization p_y is consistent with zero above 1 GeV, consistent with expectations from quark models and in disagreement with meson-baryon calculations. Polarization transfer observables are non-zero, indicating that hadron helicity is not conserved.

INTRODUCTION

Whether a meson-baryon or a quark-gluon picture is more appropriate to describe high momentum transfer exclusive reactions is a fundamental issue in nuclear physics. Deuteron photodisintegration in the energy range between a few hundred MeV and a few GeV provides the possibility of investigating this issue by searching for two signatures of QCD effects. The first signature is the scaling of the reaction cross section. Constituent counting rules as derived from perturbative QCD (pQCD) predict that the differential cross section scales with s, the center of mass energy squared, as $d\sigma/dt \propto s^{2-n}$ [1], where n is the total number of pointlike constituents involved in the initial and final state. For deuteron photodisintegration $n = 13$. The second signature is the observation of hadron helicity conservation (HHC) [2]. Its assumption leads to distinct predictions for spin observables. The main focus at higher energies has been on cross section measurements, most recently at SLAC and Thomas Jefferson National Accelerator Facility Hall C [3]. There is no unambiguous clear interpretation of the data, as illustrated in Fig. 1. At 90° and for energies up to 1.6 GeV the meson exchange model (MEC) of the Bonn group [4] (not shown) is in very good agreement with the data. Effective QCD models like the quark gluon string (QGS) model [5] and the QCD rescattering model [6] give a reasonably good description at higher energies in their respective ranges of validity; see Ref. [7] for a discussion of further models. At 90° and 69° the cross section seems to follow the scaling law, even though pQCD is not expected to work at these low energies.

Polarization observables provide further constraints on the models. There are 12 independent amplitudes. The induced polarization p_y and the polarization transfer

C_x are proportional to the imaginary and real parts of the same combination of amplitudes, which are products of helicity conserving and helicity non-conserving amplitudes [8], so both polarization observables vanish in case of HHC. Under additional assumptions on amplitudes at $\theta_{cm} = 90°$ the polarization transfer C_z also vanishes [9]. Predictions for polarizations are available for the meson exchange model [4] and are underway for the QGS and QCD rescattering model.

EXPERIMENT

Experiment E89-019 was done at Jefferson Lab Hall A [7]. Longitudinally polarized electrons were incident on a copper radiator, producing a beam of circularly polarized photons. The photons and electrons then impinged on a liquid deuterium target. Beam energies were chosen at eight settings between 0.5 and 2.5 GeV. Also ep elastic scattering data were taken for calibrations. Protons from the target were detected at a center of mass angle of 90° in one of the Hall A high resolution spectrometers, equipped with a focal plane polarimeter (FPP). The photon energy was reconstructed from the measured proton momentum and scattering angle. Background contributions from the target windows and from electrodisintegration were subtracted. Events with pion production were excluded by accepting only protons with the highest momenta. Polarized protons lead to azimuthal asymmetries after scattering in the carbon analyzer of the FPP. These distributions were analyzed by means of a maximum likelihood method to obtain the induced and transferred polarization components. Spin precession in the spectrometer magnetic fields and the helicity of the electron beam were taken into account. The carbon analyzing power used in the analysis was taken from ep calibration data at every setting and cross-checked with previous work [10]. These calibration data also provide a measure of the small instrumental asymmetries.

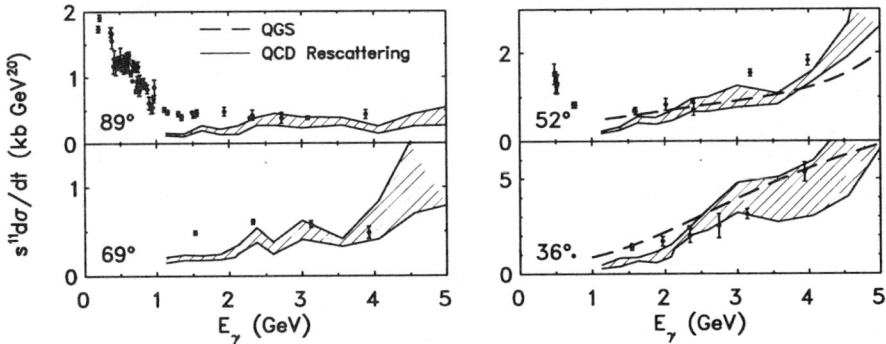

FIGURE 1. $s^{11} d\sigma/dt$ vs E_γ for different center of mass angles. Data shown are taken from Ref. [3]. Calculations shown are the QGS model [5] (dashed), and the QCD rescattering calculations [6] (hatched).

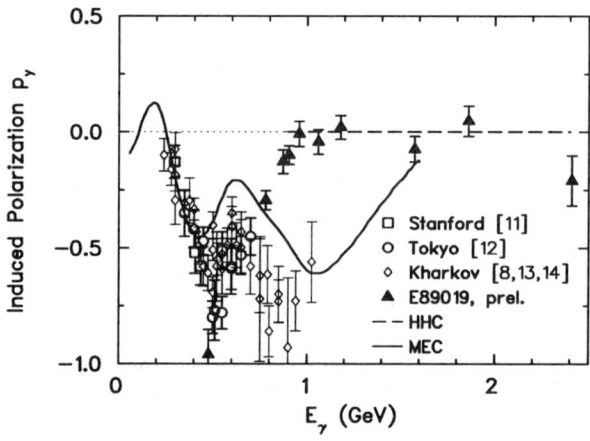

FIGURE 2. Induced polarization p_y in the $\vec{\gamma}d \to \vec{p}n$ reaction at $\theta_{cm} = 90°$.

RESULTS

The preliminary result for the normal component p_y of the induced polarization is shown as triangles as a function of the photon energy in Fig. 2. Our data agree with earlier low energy measurements [8,11–13] but disagree with the higher energy Kharkov results [14]. Our data are also compared to available theoretical predictions. The solid curve shows the result of the meson-baryon calculation of the Bonn group [4], which is in fair agreement with the data in the delta region but there is no evidence for the predicted structure related to higher mass resonances. This is an unexpected result if the meson-baryon picture is applicable. For $E_\gamma \geq 1$ GeV, the cross sections scale and p_y is consistent with HHC; all amplitudes having the same phase could also make p_y vanish. The HHC prediction can further be tested with the polarization transfer observables C_x and C_z. Our preliminary results are shown in Fig. 3; neither other data nor calculation exist. The data are non-zero and indicate that hadron helicity is not conserved in the investigated energy range. It appears likely that a non-perturbative quark model is necessary to describe the data. The knowledge of deuteron photodisintegration will be deepened in future experiments [15], which will measure an angular distribution of polarization observables at high photon energies, and will extend the $\theta_{cm} = 90°$ data to above 3 GeV photon energy.

ACKNOWLEDGMENTS

I acknowledge many fruitful discussions with, and the help of, my collaborators [7], especially R. Gilman. This work was supported in part by the United States National Science Foundatation and DOE contract DE-AC05-84ER40150 under which the Southeastern Universities Research Association (SURA) operates the Thomas Jefferson National Accelerator Facility.

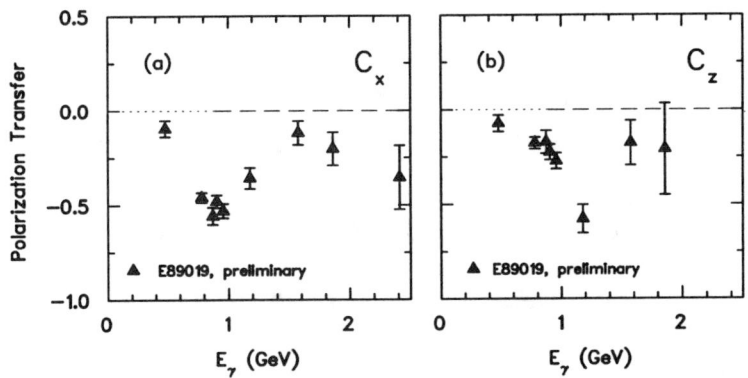

FIGURE 3. Polarization transfer in the $\vec{\gamma}d \to \vec{p}n$ reaction at $\theta_{cm} = 90°$, C_x (a) and C_z (b).

REFERENCES

1. S.J. Brodsky and G.R. Farrar, *Phys. Rev. Lett.* **31**, 1153 (1973).
2. G.P. Lepage and S.J. Brodsky, *Phys. Rev. D* **22**, 2157 (1980).
3. C. Bochna et al., *Phys. Rev. Lett.* **81**, 4576 (1998) and references therein.
4. Y. Kang et al., *Abstracts of the Particle and Nuclear Intersections Conference*, (MIT, Cambridge, MA 1990); Y. Kang, Ph.D. dissertation, Bonn (1993).
5. L.A. Kondratyuk et al., *Phys. Rev. C* **48**, 2491 (1993).
6. L.L. Frankfurt et al., *Phys. Rev. Lett.* **84**, 3045 (2000) and private communication.
7. TJNAF Experiment E89-019, R. Gilman, R.J. Holt, and Z.-E. Meziani, spokespeople; K. Wijesooriya et al., to be published.
8. V.P. Barranik et al., *Nucl. Phys. A* **451**, 751 (1986).
9. S.I. Nagornyĭ, Yu.A. Kasatkin, and I.K. Kirichenko, *Yad. Fiz.* **55**, 345 (1992), [*Sov. J. Nucl. Phys.* **55**, 189 (1992)].
10. B. Bonin et al., *Nucl. Inst. Meth. A* **288**, 379 (1990); M.W. McNaughton et al., *Nucl. Inst. Meth. A* **241**, 435 (1985);
11. F.F. Liu et al., *Phys. Rev.* **165**, 1478 (1968).
12. T. Kamae et al., *Phys. Rev. Lett.* **38**, 468 (1977); T. Kamae et al., *Nucl. Phys. B* **139**, 394 (1978); H. Ikeda et al., *Phys. Rev. Lett.* **42**, 1321 (1979); H. Ikeda et al., *Nucl. Phys. B* **172**, 509 (1980).
13. A.S. Bratashevskiĭ et al., *Nucl. Phys. B* **166**, 525 (1980); A.S. Bratashevskiĭ et al., *Yad. Fiz.* **31**, 860 (1980), [*Sov. J. Nucl. Phys.* **31**, 444 (1980)]; A.S. Bratashevskiĭ et al., *PZETF* **35**, 489 (1982), [*JETP Lett.* **35**, 605 (1982)]; A.S. Bratashevskiĭ et al., *Yad. Fiz.* **43**, 785 (1986), [*Sov. J. Nucl. Phys.* **43**, 499 (1986)]. A.A. Zybalov et al., *Nucl. Phys. A* **533**, 642 (1991). V.B. Ganenko et al., *Z. Phys. A* **341**, 205 (1992).
14. A.S. Bratashevskiĭ et al., *PZETF* **34**, 410, (1981); A.S. Bratashevskiĭ et al., *PZETF* **36**, 174 (1982), [*JETP Lett.* **36**, 216 (1982)]; A.S. Bratashevskiĭ et al., *Yad. Fiz.* **44**, 960 (1986), [*Sov. J. Nucl. Phys.* **44**, 619 (1986)].
15. TJNAF Experiment E00-007 and E00-107, R. Gilman, R.J. Holt, and Z.-E. Meziani, spokespeople.

Single π^0 Electroproduction from CLAS Data at Jefferson Lab

K. Joo* for the CLAS collaboration

University of Virginia, Charlottesville, Virginia 22903, U.S.A.

Abstract. New measurements of the electroproduction of the $\Delta(1232)$ resonance through the $p(e,e'p)\pi^o$ reaction have been performed. The data were taken with the CEBAF Large Acceptance Spectrometer (CLAS) at Jefferson Lab using incident electron energies of 1.6, and 2.4 GeV. Cross sections were measured simultaneously with continuous coverage over a large range of four-momentum transfer $Q^2 = (0.3\text{-}1.2 \text{ GeV}^2)$. Decay angular distributions in the $p\pi^o$ center-of-mass were obtained over the full range of $\cos\theta_{c.m.}$ and $\phi_{c.m.}$. The high statistical accuracy of this data set is expected to provide strong constraints on dynamic models of the $N \to \Delta$ transition form factors.

INTRODUCTION

Single π^0 electroproduction in the Δ resonance provides information that is sensitive to models describing the internal motion of baryon constituents. One of the crucial tests of our understanding of the $\Delta(1232)$ resonance is to determine the electric and scalar quadrupole transition multipoles E_{1+} and S_{1+} and the magnetic dipole M_{1+}. In $SU(6)$ symmetric quark models, the $\gamma N\Delta$ transition is mediated by a single quark spin flip in the nucleon ground state leading to M_{1+} dominance and $E_{1+} = S_{1+} \equiv 0$, while helicity conservation in pQCD requires $E_{1+} = M_{1+}$ as $Q^2 \to \infty$. In various constituent quark models, a tensor force from the interquark hyperfine interaction, leads to a d-state admixture in the baryon ground state wave function. As a result small but finite values for the ratios, $R_{EM} = E_{1+}/M_{1+}$ and $R_{SM} = S_{1+}/M_{1+}$, are predicted.

The determination of R_{EM} and R_{SM} in the region of the $\Delta(1232)$ resonance has been the aim of a considerable number of experiments and theoretical activities in the past. Even though theoretical models have become more refined, most previous measurements have large systematic, statistical errors and significant kinematic limitations. A new program using CLAS at Jefferson Lab/Hall B has been inaugurated to vastly improve the systematic and statistical precision and cover a wide kinematic range in four momentum transfer Q^2 and invariant mass W, as well as the full angular range of the resonance decay into the $p\pi^o$ final state. In this report, some of the first preliminary results using CLAS will be presented.

EXPERIMENTAL SETUP AND DATA ANALYSIS

The data were taken using an electron beam at 100 % duty factor with energies of 1.645 and 2.445 GeV and an average current of 2.5 nA incident on a 4 cm hydrogen target. Scattered electrons and protons were detected by CLAS. A schematic view of CLAS is shown in Figure 1. CLAS is a toroidal magnetic spectrometer divided into six identical sectors. Each sector has three drift chambers to determine the momentum and charge of the charged particles, 48 time-of-flight scintillator paddles to determine the masses of the hadrons, a threshold Čerenkov detector to distinguish between electrons and negatively charged pions, and a calorimeter to detect electromagnetic showers from electrons and photons.

To obtain differential cross sections, CLAS geometrical acceptance, tracking efficiency and resolution were simulated using a GEANT model of the detector geometry which incorporated the magnetic field map and the surveyed positions of detector elements (including target position relative to coils). Software fiducial cuts were used to define the solid angle for electrons and hadrons, excluding regions of low Čerenkov efficiency or multiple scattering from magnetic coils. Radiative corrections were also calculated using a Monte-Carlo method based on the Mo and Tsai formula [1], without using the peaking approximation.

Once differential cross sections were obtained, R_{EM} and R_{SM} were extracted by an energy independent multipole analysis. In the region of the $\Delta(1232)$ resonance, one may expect that only $s-$ and $p-$ waves with $J \leq 3/2$ contribute. Therefore, the partial wave expansion in the differential cross section contains seven unknown multipoles, E_{0+}, M_{1-}, E_{1+}, M_{1+}, S_{0+}, S_{1+}, and S_{1-}. Furthermore, one can retain only terms containing M_{1+} assuming that the $\gamma N\Delta$ transition is dominated by a M_{1+} magnetic dipole transition. By utilizing these two approximations, R_{EM} and R_{SM} can be determined unambiguously by mapping the angular dependence of the differential cross sections.

PRELIMINARY RESULTS AND OUTLOOK

The data presented here were measured in Spring 1998 and represent roughly 5-10% of the total events collected so far. Preliminary results for R_{EM} and R_{SM} are shown in Figure 2 for Q^2=0.40, 0.52, 0.65, 0.75 and 0.91 GeV^2 compared to recent model calculations [2–4] and previous measurements [5–7]. The errors shown throughout are statistical only. R_{EM} is small and negative with a weak Q^2 dependence, while R_{SM} exhibits a strong Q^2 dependence with a trend towards increasingly negative values. The trend of the data is qualitatively described by quark models that include pion degrees of freedom. These results are in contrast to previous data which give ambiguous results for the sign of R_{EM} and show no clear Q^2 dependence of R_{SM}.

Analysis of the full data set using both polarization observables and cross sections is underway. It is expected to give results as little model-dependent as possible.

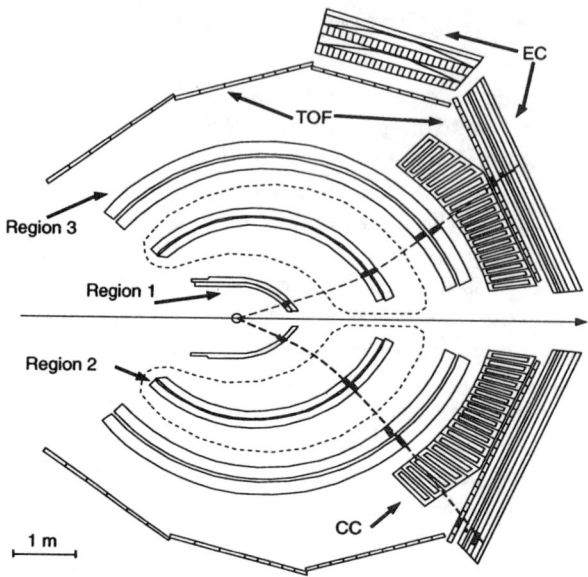

FIGURE 1. Horizontal midplane cut through the CLAS detector at beam line elevation showing two charged particles traversing the drift chambers (Region 1,2,3) in opposite sectors. Outside of Region 3 are time-of-flight (TOF) counters, calorimeters (EC), and Čerenkov counters (CC). The 2 T toroidal magnetic field is contained within the boundary surrounding Region 2.

ACKNOWLEDGMENTS

This work was supported in part by DOE contract DEFG02-97ER41018. Thanks go to Lee Cole Smith and Ralph Minehart of the University of Virginia for their help in the data analysis, and to the staffs at Jefferson Lab/Hall B and the CLAS collaboration for their hard work in supporting the experiment.

REFERENCES

1. L. W. Mo and Y. Tsai, *Rev. Mod. Phys.* **41**, 205 (1969)
2. A. Silva, D. Urbano, T. Watabe, M. Fiolhais and K. Goeke, hep-ph/9904326
3. S.S. Kamalov and Shin Nan Yang, *Phys. Rev. Lett.* **83**, 4494 (1999)
4. T. Sato and T.-S.H. Lee, *Phys. Rev.* C **54**, 2660 (1996) and T.-S.H. Lee, private communication, 2000.
5. R. Beck *et al*, *Phys. Rev. Lett.* **78**, 606 (1997)
6. G. Blanpied *et al*, *Phys. Rev. Lett.* **79**, 4337 (1997)
7. C. Mertz *et al*, nucl-ex/9902012, H. Schmieden, nucl-ex/9909006

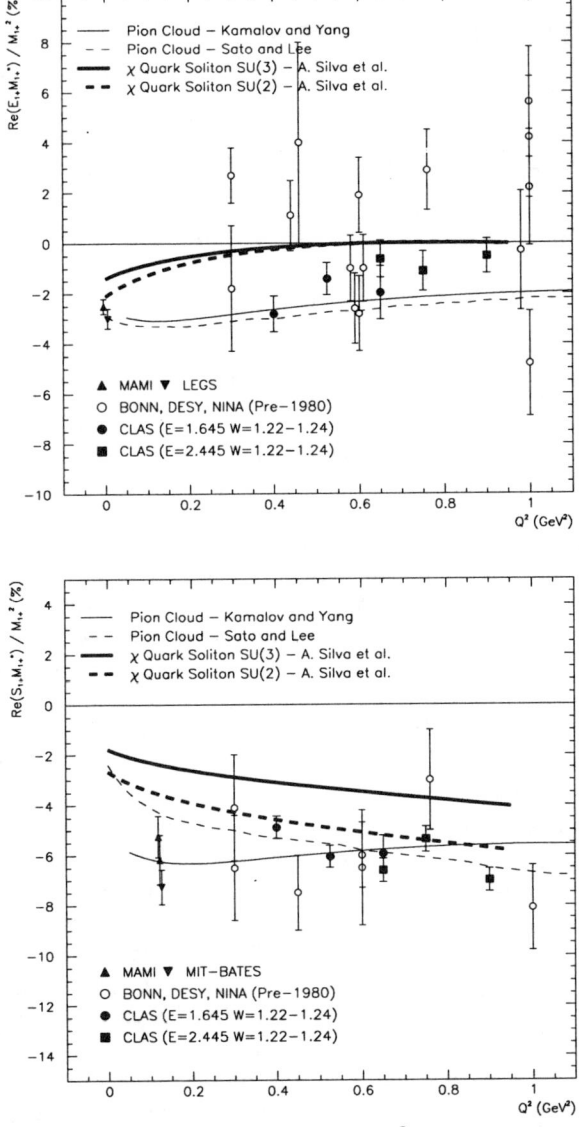

FIGURE 2. Preliminary multipole ratios vs. Q^2 at electron beam energies E=1.645, 2.445 GeV. Curves are from recent model calculations[2,3,4]. Top: Electric quadrupole $\mathrm{Re}(E^*_{1+}M_{1+})/|M_{1+}|^2$ compared to previous measurements[5,6]. Bottom: Coulomb quadrupole $\mathrm{Re}(S^*_{1+}M_{1+})/|M_{1+}|^2$ compared to previous measurements[7].

JLab Measurements of the Deuteron Electric and Magnetic Form Factors

Gerassimos G. Petratos [1]

Kent State University, Kent, OH 44242, USA

Abstract.
Large-momentum transfer JLab measurements of the deuteron electric and magnetic form factors are reported. The data are compared to theoretical models based on the relativistic impulse approximation with the inclusion of meson-exchange currents, and to predictions of quark-dimensional scaling and perturbative QCD.

Measurements of the deuteron elastic form factors are crucial in understanding the electromagnetic structure of the nuclear two-body system. In particular, they offer unique opportunities to test models of the short-range nucleon-nucleon interaction, meson-exchange currents (MEC), as well as the possible influence of explicit quark degrees of freedom [1-2]. The existing data [3-7] are described fairly well by calculations [8-10] based on the impulse approximation with the inclusion of MEC. At large momentum transfers, the elastic form factors are expected to be calculable with only quarks and gluons. Quark-dimensional scaling (QDS) [11] and perturbative QCD (pQCD) [12] predict, that the "electric" form factor, $A(Q^2)$, should fall with increasing four-momentum transfer as $(Q^2)^{-10}$. Previous SLAC data [4] exhibit, for $Q^2 > 2$ (GeV/c)^2, some evidence for such scaling behaviour.

Using the Continuous Electron Beam Accelerator and Hall A Facilities of Thomas Jefferson National Accelerator Facility (JLab) we have a) extended the kinematical range of $A(Q^2)$ by measuring [13] the elastic electron-deuteron cross section for $0.7 \leq Q^2 \leq 6.0$ (GeV/c)^2, and b) performed precision measurements, by means of Rosenbluth separation, of the "magnetic" form factor, $B(Q^2)$, in the range $0.7 \leq Q^2 \leq 1.4$ (GeV/c)^2. Electron beams of 100% duty factor, 5-120 μA intensity and 0.5-4.4 GeV energy, were scattered off a 15 cm long liquid deuterium target that provided a record high luminosity. Scattered electrons were detected in the electron High Resolution Spectrometer (HRS), which was equipped with a set of

---

[1] *Representing the Jefferson Lab Hall A Collaboration*: Blaise Pascal, California State LA, Duke, Florida International, Florida State, Gent, Georgia, Grenoble, Hampton, Harvard, INFN, JLab, Kent State, Kentucky, Kharkov, Maryland, MIT, New Hampshire, Norfolk State, North Carolina Central, Old Dominion, Orsay, Princeton, Regina, Rutgers, Saclay, Stony Brook, Syracuse, Temple, Tohoku, Virginia, William and Mary, Yamagata and Yerevan.

drift chambers, two scintillation hodoscopes, a Čerenkov counter and a lead-glass calorimeter. To suppress backgrounds and separate elastic from inelastic processes, recoil deuterons were detected in coincidence with the scattered electrons in the hadron HRS, using a set of drift chambers and a pair of scintillation hodoscopes.

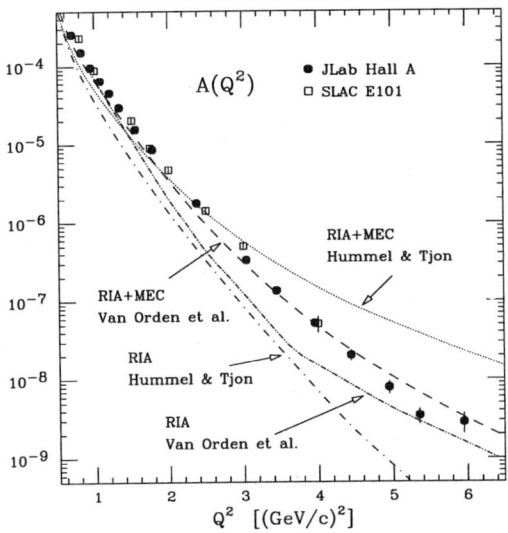

FIGURE 1. The deuteron electric form factor $A(Q^2)$ from this experiment compared to relativistic impulse approximation calculations [8-9]. Also shown are previous SLAC data [4].

Figure 1 shows our $A(Q^2)$ data, extracted from forward-angle cross section measurements, together with previous SLAC data [4] and theoretical calculations. The error bars represent statistical and systematic uncertainties added in quadrature. Our data agree very well with the SLAC data in the range of overlap and exhibit a smooth fall-off with Q^2 with no apparent diffractive structure. The double dot-dashed and dot-dashed curves in Fig. 1 represent the relativistic impulse approximation (RIA) calculations of Van Orden, Devine and Gross (VDG) [8] and Hummel and Tjon (HT) [9], respectively. The VDG curve is based on a manifestly covariant calculation that uses the Gross equation and assumes that the virtual photon is absorbed by an off-mass-shell nucleon or a nucleon that is on-mass-shell right before or after the interaction. The HT curve is based on a one-boson-exchange quasi-potential approximation of the Bethe-Salpeter equation where the two nucleons are treated symmetrically by putting them equally off their mass shell with zero relative energy. In both cases the RIA appears to be lower than the data. Both groups have augmented their models by including the $\rho\pi\gamma$ MEC contribution. The difference in the two models is indicative of the size of theoretical uncertainties. Our $A(Q^2)$ data are described best by the VDG model which includes the $\rho\pi\gamma$ meson-exchange current correction.

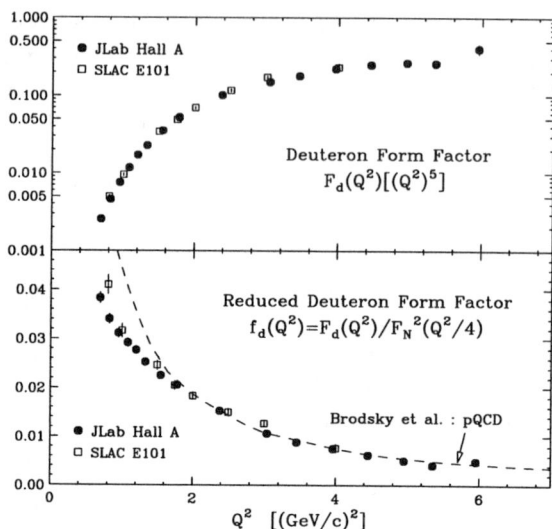

FIGURE 2. JLab and SLAC data [4] on the deuteron form factor $F_d(Q^2)$ times $(Q^2)^5$ (top) and the reduced deuteron form factor $f_d(Q^2)$ (bottom). The curve is the asymptotic pQCD prediction of Ref. [12] for $\Lambda=100$ MeV, arbitrarily normalized to the data at $Q^2 = 4$ (GeV/c)^2.

Figure 2 (top) shows values for the "deuteron form factor" $F_d(Q^2) \equiv \sqrt{A(Q^2)}$ multiplied by $(Q^2)^5$: our data exhibit a behavior consistent with the power law of QDS and of pQCD. Figure 2 (bottom) shows values for the "reduced" deuteron form factor [14] $f_d(Q^2) \equiv F_d(Q^2)/F_N^2(Q^2/4)$ where the two powers of the nucleon form factor $F_N(Q^2) = (1+Q^2/0.71)^{-2}$ remove in a minimal and approximate way the effects of nucleon compositeness [14]. Although several authors have questioned the validity of QDS and pQCD at the momentum transfers of this experiment [15], the data appear to follow, for $Q^2 > 2$ (GeV/c)^2, the asymptotic pQCD prediction of Brodsky et al. [12].

Preliminary data for $B(Q^2)$ are shown in Figure 3. They have been extracted from a Rosenbluth separation of forward-angle and backward-angle (144.5°) cross section measurements. The error bars represent statistical and systematic uncertainties added in quadrature. Also shown are the VDG and HT theoretical calculations. Our data are in agreement with previous measurements [7] and they are described best by the HT RIA model which includes the $\rho\pi\gamma$ MEC correction.

In conclusion, we have measured the deuteron electric and magnetic form factors up to large momentum transfers. The results have significantly improved the precision of the existing data in the range of overlap. RIA+MEC calculations are consistent with our data but fail to provide simultaneously a quantitative description of both form factors. The $A(Q^2)$ results are also indicative of a scaling behaviour consistent with predictions of quark-dimensional scaling and perturbative QCD.

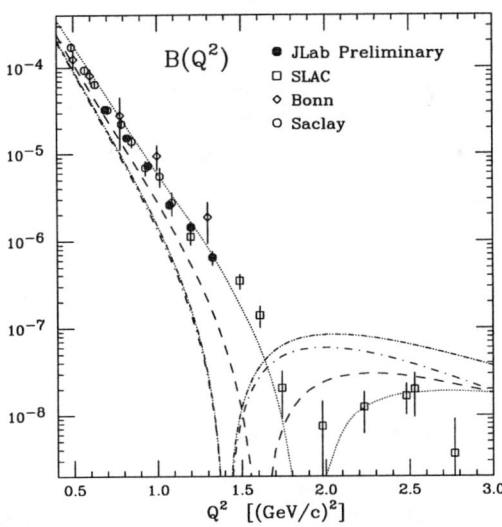

FIGURE 3. The deuteron magnetic form factor $B(Q^2)$ from this experiment compared to RIA calculations [8-9] (for meaning of curves see Fig. 1). Also shown are previous data [7].

We acknowledge the outstanding support of the staff of the Accelerator and Physics Divisions of JLab that made this experiment possible. This work was supported in part by the U.S. Department of Energy and National Science Foundation.

REFERENCES

1. J. Carlson and R. Schiavilla, *Rev. Mod. Phys.* **70**, 743 (1998).
2. C.E. Carlson, J.R. Hiller, R.J. Holt, *Annu. Rev. Nucl. Part. Sci.* **47**, 395 (1997).
3. J.E. Elias et al., *Phys. Rev.* **177**, 2075 (1969).
4. R.G. Arnold et al., *Phys. Rev. Lett.* **35**, 776 (1975).
5. R. Cramer et al., *Z. Phys. C* **29**, 513 (1985).
6. S. Platchkov et al., *Nucl. Phys. A* **510**, 740 (1990); and references therein.
7. P.E. Bosted et al., *Phys. Rev. C* **42**, 38 (1990); and references therein.
8. J.W. Van Orden, N. Devine and F. Gross, *Phys. Rev. Lett.* **75**, 4369 (1995).
9. E. Hummel and J.A. Tjon, *Phys. Rev. Lett.* **63**, 1788 (1989).
10. R.B. Wiringa, V.G.J. Stoks and R. Schiavilla, *Phys. Rev. C* **51**, 38 (1995); R. Schiavilla and D.O. Riska, *Phys. Rev. C* **43**, 437 (1991).
11. S.J. Brodsky and G.R. Farrar, *Phys. Rev. Lett.* **31**, 1153 (1973); V.A. Matveev, R.M. Muradyan and A.N. Tavkhelidze, *Lett. Nuovo Cimento* **7**, 719 (1973).
12. S.J. Brodsky, C-R. Ji and G.P. Lepage, *Phys. Rev. Lett.* **51**, 83 (1983).
13. L. C. Alexa et al., *Phys. Rev. Lett.* **82**, 1374 (1999).
14. S.J. Brodsky and B.T. Chertok, *Phys. Rev. D* **14**, 3003 (1976).
15. N. Isgur and C.H. Llewellyn Smith, *Phys. Rev. Lett.* **52**, 1080 (1984); *Phys. Lett. B* **217**, 535 (1989); G.R. Farrar et al., *Phys. Rev. Lett.* **74**, 650 (1995).

Precise Electro-Pion Production Experiments in the Delta Region

A.M.Bernstein

M.I.T

Cambridge, Mass., U.S.A.

Abstract. High precision, exclusive measurements of the $\gamma^* N \to \Delta$ transition have been pursued at the MIT/Bates laboratory during in the last several years. These involve polarized beams, out-of-plane detection and focal plane polarimetry in the $H(\vec{e}, e'p)\pi^0$ and $H(\vec{e}, e'\pi^+)n$ channels. The goal is a precise determination of the quadrupole amplitudes in the $\gamma^* N \to \Delta$ transition due to non-spherical interquark interactions and the isolation of competing background mechanisms such as Born terms and tails of higher resonances. Similar studies are being pursed intensively at many laboratories. The recent data from Bates are compared to theoretical calculations.

INTRODUCTION

The quadrupole amplitudes in the $\gamma^* N \to \Delta$ transition is a fundamental issue in hadronic physics. In a constituent-quark picture it is due d-state admixture in the quark wave function of the nucleon which is a consequence of a spin-spin tensor color-hyperfine interaction [1,2]. In pion cloud [3] and dynamical models of the πN system, the effect of the pionic cloud will also cause the appearance of quadrupole amplitudes [4,5].

Three multipoles (M1, E2 and C2) can contribute to the $\gamma^* N \to \Delta$ electro-excitation. M1 is the dominant magnetic-dipole (quark spin-flip) amplitude. The quadrupole amplitudes ,E2 and C2, usually quantified in terms of the Coulomb-to-Magnetic (CMR = C2/M1) and Electric-to-Magnetic (EMR = E2/M1) ratios. Various models of the nucleon predict few % values for these ratios at low momentum transfers. However since most models do not calculate exclusive reaction cross sections their comparison to experimental data is qualitative only.

The measurement of the quadrupole amplitude is difficult to accomplish since it is much smaller than the leading term, and must be isolated from the non-resonant, coherent "background" processes such as the Born terms and tails of higher resonances [6]. Finally, the isospin decomposition of resonant and non-resonant pieces is also required.

Recently, photoproduction experiments with polarized photon beams yielded high precision results which allowed for a multipole analysis and constrained the transverse electric amplitude [7,8]. To this date, relatively little information is available from electroproduction experiments to permit a clear identification of the quadrupole scalar amplitude and of the Q^2-evolution of both the EMR and the CMR ratios. However, many electro-production experiments are planned or underway in practically every electron facility and a new generation of data will soon emerge in this sector as well.

THE OOPS $\gamma^* N \to \Delta$ PROGRAM

The Out-Of-Plane Spectrometer ("OOPS") collaboration[1] is presently pursuing an extensive program to measure the quadrupole resonant terms in the $\gamma^* N \to \Delta$ transition at a Q^2 values of 0.127 and 0.071 $(\text{GeV}/c)^2$ [11–13] through the isolation of the interference nuclear response functions contributing to the cross section [9]. Measurements are done over a wide range of hadronic invariant masses (W) and center-of-mass angles (θ_{pq}).

The unique features of the OOPS set-up include an extended out-of-plane detection capability and the possibility to perform simultaneous measurements with virtually identical spectrometers at optimally chosen positions, thus significantly minimizing systematic errors [10].

All of the necessary equipment for simultaneous measurements with four out-of-plane spectrometer modules and an electron arm is now in place.

In a typical OOPS configuration each of the spectrometers is precisely aligned to given azimuthal angles ϕ_{pq} on the proton emission cone for constant θ_{pq} centered along the direction of the momentum transfer $\vec{q}$. Each OOPS module [14] consists of a light-weight (16 ton) dipole and quadrupole combination with a 12° vertical bending angle, offering good angular (1.3 mrad) and energy ($\Delta E/E = 10^{-2}$) resolutions up to a maximum momentum of 850 MeV/c.

Results and discussion

In our first pilot run the in-plane cross section was measured with an unpolarized electron beam. In a subsequent measurement using a focal plane polarimeter and the recoil polarization P_n was also measured [16]

These data were followed by measurements with a 850 MeV, 35 % polarized beam. The recoil protons were detected in two OOPS modules positioned at $\phi = \pi/4$ and $\phi = 3\pi/4$. Measurements of the $H(\vec{e}, e'p)\pi^0$ channel were performed at $W = 1172$ MeV for $\theta_{pq} = 63°$ and $W = 1232$ MeV ($\theta_{pq} = 51°$), at $Q^2 = 0.127$ GeV^2 [17]. In the same set-up, the $H(\vec{e}, e'p)\pi^+$ cross section and helicity asymmetries were

---

[1] Arizona State University; Bates Linear Accelerator Center; California State University at Los Angeles; Florida State University; National and Capodistrian University of Athens; Massachusetts Institute of Technology; Old Dominion University; Tohoku University; University of Illinois at Urbana-Champaign; University of Mainz; University of Massachusetts at Amherst.

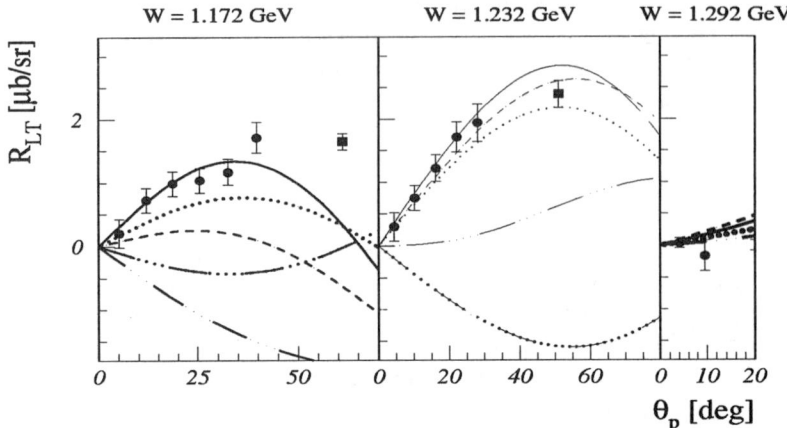

FIGURE 1. The R_{LT} structure function for the $H(e,e'p)\pi^0$ reaction at $Q^2 = 0.127$ (GeV/c)^2 and different W. Data from Refs. [15] (circles) and [17] (squares) respectively. The solid, short-dash and dotted curves are from [18], [19] and [4], respectively. The long-dash and dot-dash curves are the results from [18] and [4], respectively, for a CMR=0.

measured after the insertion of suitably designed proton absorbers in OOPS. In Figs. 1 and 2 we show our data on the total cross section and the interference structure function R_{LT}.

The parallel kinematics cross section $\sigma_{\parallel}$ is dominated by R_T, which contains $|M_{1+}|^2$. The response functions which are most sensitive to the S_{1+} multipole are R_L and R_{LT}; the latter can be conveniently extracted from the left/right (relative to $\vec{q}$) space asymmetry. Finally, R'_{LT} and the recoil induced polarization P_n are sensitive to the background contributions, since without such contributions S_{1+} and M_{1+} would have the same phase and thus response function would vanish.

From our results it is clear that no model completely agrees with the data. The Sato-Lee dynamical model

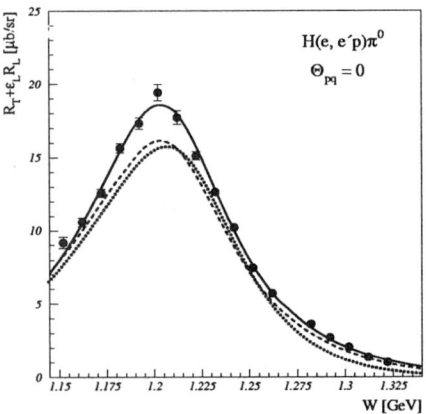

FIGURE 2. The $H(e,e'p)\pi^0$ cross section in parallel kinematics. Data from [15], curve labels as in Fig. 1

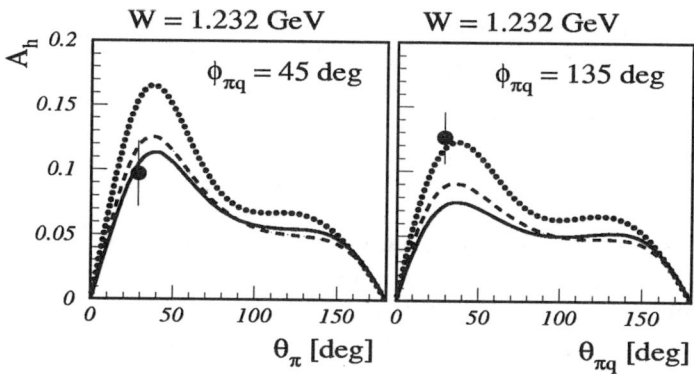

FIGURE 3. The preliminary result for the beam-helicity asymmetry in the $H(e,e'n)\pi^+$ at $Q^2 = 0.127$ (GeV/c)^2. Curves as in Fig 1.

[4], which couples a bare $\gamma^* N \to \Delta$ amplitude consistent with the CQM to pion and nucleon degrees of freedom in a gauge-invariant fashion, under-predicts the cross section. The original version of the Mainz unitary model MAID [19](with parameters fixed before our data) does not agree with our experiment. However one virtue of this model is that its parameters can be readily adjusted to fit the data. This has been done and the new results [18] with C2/M1 = (-6.5 ± 0.2) % do fairly well on resonance and have improved considerably the description of R_{LT} below resonance but still miss the behavior of the new data [17] at larger angles and of the old data [15] above resonance. Their value for the recoil polarization at $\theta_{pq} = 0$ and on resonance is -0.52 whereas the measured value is $P_n = (-0.397 \pm 0.055 \pm 0.009)$ [16]. The data presented here have sufficient sensitivity to significantly constrain the longitudinal couplings.

The sizeable values of the $R_{LT'}$ and P_n observables show clearly that background and non-resonant contributions play an important role. In addition, their values imply as well that an analysis that would retain only the small terms that interfere linearly with M_{1+} is not sufficiently accurate since the neglected strength is at least as large as the quadrupole signal. For this reason it is clearly very important to obtain a precise theoretical description of R_{LT} below and above the peak of the resonance.

CONCLUSIONS AND FUTURE PROSPECTS

High precision electroproduction data from Bates on the $\gamma^* N \to \Delta^+(1232)$ transition typical of the new generation of data now emerging show sensitivity to quadruple amplitudes and unprecedented discriminative power among nucleon models. The agreement between theory and experiment has not yet reached the desired level: it is qualitatively good, indicating the substantial progress that has been achieved in recent years. A clear preference for a negative quadrupole transition amplitude resulting in a CMR of about -6.5 % has been obtained. Precise statements on the underlying physics of interest will need to await further development of theory and additional data, spanning wider angular and invariant mass ranges.

In the upcoming months a new series of measurements will be undertaken. These will be the first measurements with the fully developed(four module) OOPS system and a high duty-factor extracted beam from the Bates South Hall Ring. We plan to pursue measurements of R_{LT} up to large angles at $W = 1172, 1232$ and 1292 MeV. Simultaneously, we will extract the EMR at $Q^2 > 0$ by measuring both the $(R_T + \epsilon_L R_L)$ and R_{TT} response function. We also plan to extend our coverage of the π^+ channel [20].

REFERENCES

1. N. Isgur, G. Karl and R. Koniuk Phys. Rev. **D85**, 239 (1982).
2. S. Capstick and G. Karl Phys. Rev. **D41**, 2767 (1990).
3. M. Fiolhais et al., Phys. Lett. **B373**, 229 (1996). D. H. Lu et al., Phys. Lett. **C55**, 3108 (1997).
4. T. Sato and T.-S. H. Lee, Phys. Rev. **C54**, 2660 (1996). T.-S. H.Lee, private communication.
5. S.N.Yang, J. Phys. G11, L205(1985). S.S. Kamalov and S.N. Yang, Phys. Rev. Lett. **83**, 4494 (1999).
6. A.M.Bernstein, S. Nozawa, and M.A.Moinester, Phys. Rev. C47,1274(1993).
7. R. Beck et al., Phys. Rev. Lett. **78**, 606 (1997).
8. G. Blanpied et al., Phys. Rev. Lett. **79**, 4337 (1997).
9. A.S. Raskin and T.W. Donnelly, Annals of Physics **191**, 78 (1989). D. Drechsel and L. Tiator, J. Phys. G: Nucl. Part. Phys. **18**, 449 (1992).
10. C.N. Papanicolas et al Nucl. Phys **A497**, 509c (1989).
11. Bates proposal 87-09 (1987), C.N. Papanicolas, spokesman.
12. Bates proposal 97-04 (1997), A. M. Bernstein and M. O. Distler, spokesmen.
13. Bates proposal 97-05 (1997), N. I. Kaloskamis and C.N. Papanicolas, spokesmen.
14. J. B. Mandeville et al Nucl. Instr. Meth. **A344**, 583 (1994); S. M. Dolfini et al Nucl. Instr. Meth. **A344**, 571(1994)
15. C. Mertz et al., submitted to Phys. Rev. Lett.; nucl-ex/9902012.
16. G. A. Warren et al., Phys. Rev. **C58**, 3722 (1998).
17. C. Kunz, Massachusetts Institute of Technology, Ph.D. Thesis, unpublished (2000).
18. http://www.kph.uni-mainz.de/MAID/maid2000/maid2000.html; S.S. Kamalov, private communication.
19. D. Drechsel et al., *Nucl. Phys.* **A645**, 145 (1999).
20. A similar, recent summary of the $\gamma^* N \to \Delta$ program of the OOPS group has been presented by T. Botto and C.N. Papanicolas at the XVIth Few Body Conference, Taipei, 2000.

Backward emitted high-energy neutrons in hard reactions of p and π^+ on carbon

A. Malki[a], E. Piasetzky[a], J. Alster[a], G. Asryan[c,b], Y. Averichev[h],
D. Barton[c], V. Baturin[e,d], N. Bukhtoyarova[c,d], A. Carroll[c],
S. Heppelmann[e], T. Kawabata[f], A. Leksanov[e], Y. Makdisi[c],
E. Minina[e], I. Navon[a], H. Nicholson[g], A. Ogawa[e],
Yu. Panebratsev[h], A. Schetkovsky[e,d], S. Shimanskiy[h], A. Tang[i],
J.W. Watson[i], H. Yoshida[f], D. Zhalov[e]

[a] School of Physics and Astronomy, Sackler Faculty of Exact Sciences, Tel Aviv University, Ramat Aviv 69978, Israel
[b] Yerevan Physics Institute, Yerevan 375036, Armenia
[c] Collider-Accelerator Department, Brookhaven National Laboratory, Upton, NY 11973, USA
[d] Petersburg Nuclear Physics Institute, Gatchina, St. Petersburg 188350, Russia
[e] Physics Department, Pennsylvania State University, University Park, PA 16801, USA
[f] Department of Physics, Kyoto University, Sakyoku, Kyoto, 606-8502, Japan
[g] Department of Physics, Mount Holyoke College, South Hadley, MA 01075, USA
[h] J.I.N.R., Dubna, 141980, Russia
[i] Department of Physics, Kent State University, Kent, OH 44242, USA

Abstract. Beams of protons and pions of 5.9 GeV/c were incident on a C target. Neutrons emitted into the back hemisphere, in the laboratory system, were detected in (triple) coincidence with two emerging p_t >0.6 GeV/c particles. We present the momentum spectra of the backward going neutrons. We also integrated the spectra and determined the fraction of the hard scattering events which are in coincidence with at least one neutron emitted into the back hemisphere, with momenta above 0.32 GeV/c. Contrary to the earlier measurements which found that only a small fraction (of the order of 10%) of the total inelastic cross section for light nuclei was associated with backward going nucleons, we find that about half of the events are of this nature. We speculate that the reason for the large difference is due to the strong dependence of the hard-scattering reaction upon the total center of mass energy (s) and short range nucleon correlations in nuclei.

We present results from a measurement which was performed during 1998 with the EVA spectrometer at the AGS accelerator of Brookhaven National Laboratory. The spectrometer consisted of a super-conducting solenoidal magnet. The scattered particles were tracked by 4 sets of 4-layer straw tube cylindrical drift chambers which surrounded the beam axis cylindrically. We triggered the spectrometer on two positively charged particles which emerged at polar angles which corresponds to about $90°$ scattering in the pp center of mass. Three scintillator arrays measured the direction and energy of neutrons, in coincidence with these two particles.

In Fig 1. we present the measured invariant momentum spectra $(E_n/p_n) \times \frac{dY_3}{Y_2 d(p_n^2)}$ in arbitrary units for pion and proton incident beams, where E_n and p_n are the energy and momentum of the neutron detected in the backward hemisphere ($90° < \theta_n < 130°$).

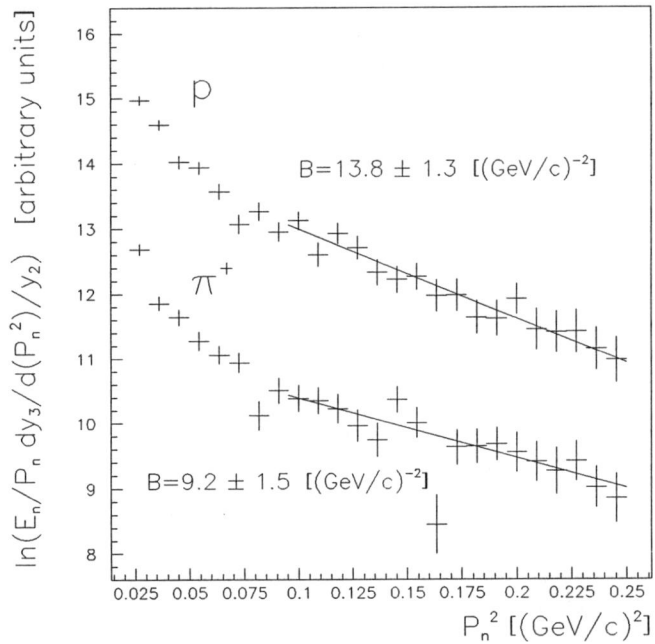

FIGURE 1. Proton and pion induced neutron invariant momentum spectra.

We call Y_2 the yield of events with two high transverse momentum charged particles with $p_t > 0.6$ GeV/c, each and no other charged particles seen in the detector. We applied software cuts to allow better determination of the target position from the track reconstruction and to obtain a better separation between incident protons and pions. The effective yield of triple coincidence events which fulfills all the conditions of the Y_2 events and, in addition, have a single neutron in the scintillator bars is indicated by Y_3. The effective yield Y_3 includes corrections to efficiency and attenuation which depend on the neutron momentum. Above $p_n^2 > 0.1$ (GeV/c)^2 the points are fitted to a straight line. The slopes (Bs) we measured in this experiment agree, within the measured uncertainties, with the slopes obtained for the emission of backward going nucleons from nuclei in collisions with high energy (approximately > 1 GeV) hadrons, real and virtual photons, muons, neutrinos and antineutrinos [1].

We also integrated the spectra and determined the fraction of the hard scattering events which are in coincidence with at least one neutron emitted into the back hemisphere, with momenta above 0.32 GeV/c.

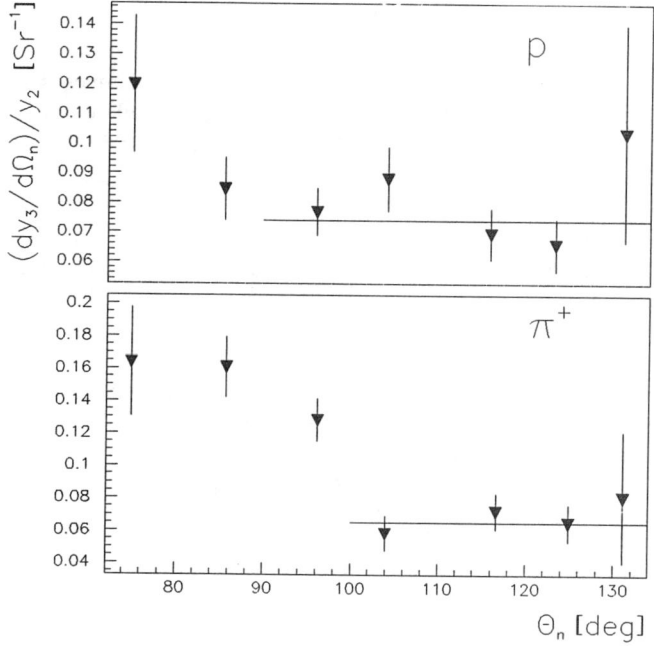

FIGURE 2. The relative yield per solid angle $\frac{dY_3}{Y_2 d\Omega_n}$ of backward going neutrons above 0.32 GeV/c as a function of the neutron angle. The lines represent fits to a constant which is used to estimate the total backward emission yield.

Contrary to the earlier measurements which found that only a small fraction (of the order of 10%) of the total inelastic cross section for light nuclei was associated with backward going nucleons, we found that about half of the events are of this nature.

We speculate that the reason for the difference is the strong dependence of the hard-scattering reaction upon the total center of mass energy (s) and its sensitivity to the short range nucleon correlations in nuclei. The reaction favors nucleons in the nuclei with a large momentum in the direction of the incident particle (small s) [2]. The high momentum protons most likely have a correlated partner at short range which are the neutrons that dominate the backward going yield [3]. This interpretation is consistent with the conclusion we offered for the triple coincidence measurement of the reaction ^{12}C(p,2p+n) [4] and with forthcoming papers [5]. The speculations need to be checked with detailed calculations.

This research was supported by the U.S. - Israel Binational Science foundation, the Israel Science Foundation founded by the Israel Academy of Sciences and Humanities, the NSF grants PHY-9501114, PHY-9722519 and the U.S. Department of Energy grant DEFG0290ER40553.

REFERENCES

1. G.A. Leksin in Proceeding of the XVIII International conference on High Energy Physics, Tbilisi, 1976, ed. N. N. Bogolubov et al. ; Yu D. Bayukov *et al.*, Sov. J. Nucl. Phys. 18 (1974) 639; Sov. J. Nucl. Phys. 42 (1985) 116, and 238; Sov. J. Nucl. Phys. 34 (1981) 437; Yu D. Bayukov *et al.*, Sov. J. Nucl. Phys. 41 (1985) 101; ITEP-5-1985 (1985) 67. ; V.I. Komarov *et al.*, Nucl. Phys. A326(1979)297; S. Frankel *et al.*, Phys. Rev. Lett. 36 (1976) 642; K.V. Alanakyan *et al.*, Sov. J. Nucl. Phys. 25 (1977) 292; M. R. Adams *et al.*, Phys. Rev. Lett. 74 (1995) 5198; J. P. Berge *et al.*, Phys. Rev. D18 (1978) 1367; V.I. Efremenko *et al.*, Phys. Rev. D22 (1980) 2581; E. Matsinos *et al.*, Z. Phys. C44 (1989) 79.
2. G. R. Farrar *et al.*, Phys. Rev. Lett. 62 (1989) 1095.
3. L.L. Frankfurt and M.I. Strikman, Phys. Rep. 76 (1981) 214; *ibid*, 160 (1988) 235; L.L. Frankfurt and M.I. Strikman, Phys. Lett. B69 (1977) 87.
4. J. Aclander *et al.*, Phys. Lett. B453 (1999) 211.
5. A.Malki *et al.*, to be published; A.Tang *et al.*, to be published.

n-p Short-Range Correlations from (p,2p + n) Measurements

A. Tang[a], J. W. Watson[a], J. Alster[b], G. Arsyan[d,c], Y. Averichev[i],
D. Barton[d], V. Baturin[f,e], N. Bukhtoyarova[d,e], A. Carroll[d],
S. Heppelmann[f], T. Kawabata[g], A. Leksanov[f], Y. Makdisi[d],
A. Malki[b], E. Minina[f], I. Navon[b], H. Nicholson[h], A. Ogawa[f],
Yu. Panebratsev[i], E. Piasetzky[b], A. Schetkovsky[f,e],
S. Shimanskiy[i], H. Yoshida[g], D. Zhalov[f]

[a] Dept. of Physics, Kent State Univ., Kent, OH 44242, U.S.A.
[b] School of Physics and Astronomy, Sackler Faculty of Exact Sciences, Tel Aviv University, Ramat Aviv 69978, Israel
[c] Yerevan Physics Institute, Yerevan 375036, Armenia
[d] Collider-Accelerator Department, Brookhaven National Laboratory, Upton, NY 11973, USA
[e] Petersburg Nuclear Physics Institute, Gatchina, St. Petersburg 188350, Russia
[f] Physics Department, Pennsylvania State University, University Park, PA 16801, U.S.A.
[g] Dept. of Physics, Kyoto Univ., Sakyoku, Kyoto, 606-8502, Japan
[h] Dept. of Physics, Mount Holyoke College, South Hadley, MA 01075, U.S.A.
[i] J.I.N.R., Dubna, Moscow 141980, Russia

Abstract. Recently, a new technique for measuring short-range NN correlations in nuclei (NN SRCs) was reported by the E850 collaboration, using data from the EVA spectrometer at the AGS at Brookhaven Nat. Lab. In this talk, we will report on a larger set of data from new measurement by the collaboration, utilizing the same technique. This technique is based on a very simple kinematic approach. For quasi-elastic knockout of protons from a nucleus (^{12}C(p,2p) was used for the current work), we can reconstruct the momentum $\mathbf{p}_f$ of the struck proton in the nucleus before the reaction, from the three momenta of the two detected protons, $\mathbf{p}_1$ and $\mathbf{p}_2$ and the three momentum of the incident proton, $\mathbf{p}_0$:

$$\mathbf{p}_f = \mathbf{p}_1 + \mathbf{p}_2 - \mathbf{p}_0$$

If there are significant n-p SRCs, then we would expect to find a neutron with momentum $-\mathbf{p}_f$ in coincidence with the two protons, provided $\mathbf{p}_f$ is larger than the Fermi momentum k_F for the nucleus (~220 McV/c for ^{12}C). Our results reported here confirm the earlier results from the E850 collaboration.

For the past half century the dominant model for the structure of nuclei, especially light nuclei, has been the nuclear shell model. In the shell model, the long-range ($\sim$ 2 fm) part of the N-N force, in combination with the Pauli principle, produces an average potential in which the nucleons undergo nearly independent motion, and the residual interactions can be treated by perturbation theory. However the N-N interaction is also highly repulsive at short-range ($\sim$ 0.4 fm) and it has long been a goal of nuclear physics to observe the effects of this short-range repulsion. These effects are most easily pictured in terms of momentum correlations rather than in terms of spatial correlations. When two nucleons in a nucleus interact at short range, they must have large relative momenta (because of their strong repulsion at short range). Following such a collision they will have equal and opposite momenta in their two-body c.m. frame. Typically, to obtain high enough relative momenta to probe the N-N repulsive core, one would expect the two-body c.m. frame to coincide roughly with the c.m. frame for the nucleus as a whole.

Recently, Aclander et al.[1] described new technique for observing such short-range correlations using data taken with the EVA spectrometer[2,3] at the AGS. This technique is based on a very simple kinematic approach. For the quasi-elastic knockout of protons from nuclei, e.g. ^{12}C(p,2p) in [1] and for this work, we can reconstruct (event by event) the three momentum $\mathbf{p}_f$ that each struck proton had before the reaction:

$$\mathbf{p}_f = \mathbf{p}_1 + \mathbf{p}_2 - \mathbf{p}_0 \tag{1}$$

where $\mathbf{p}_0$ is the momentum of the incident proton and $\mathbf{p}_1$ and $\mathbf{p}_2$ are the momenta of the two detected protons. The question we then ask is whether or not there is a coincident neutron with $\mathbf{p}_n \approx -\mathbf{p}_f$. To answer this question, we deployed 36 neutron detectors to look for triple coincidences of the kind ^{12}C(p,2p+n).

The EVA spectrometer is designed to detect proton pairs from quasielastic collisions with $\theta_{cm} \approx 90°$. Because the cross section for this geometry fall steeply with the Mandelstam variable s, the (p,2p) reaction preferentially occurs for nuclear protons with $\mathbf{p}_f$ in the forward going lab direction. Therefore most of our 36 neutron detectors were placed in the backward laboratory hemisphere. Figure 1 shows the layout of the 36 neutron detectors relative to EVA. The detectors in arrays 1 and 2 had dimensions 10 cm x 12.5 cm x 1m. The detectors in array 3 had dimensions 25 cm x 10 cm x 1 m.

Our initial objective with the new, more extensive, triple coincidence measurement with EVA in 1998 was to confirm the results from the 1994 data reported in [1]. To this end we applied the following cuts:

(1). There should be two (and only two) high $\mathbf{p}_t$ positive tracks - kinematics dictates that these are both protons.

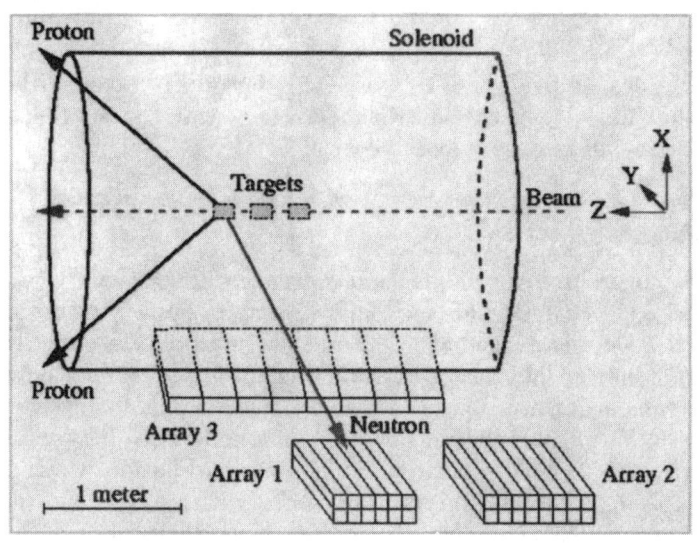

FIGURE 1. Layout of the experiment

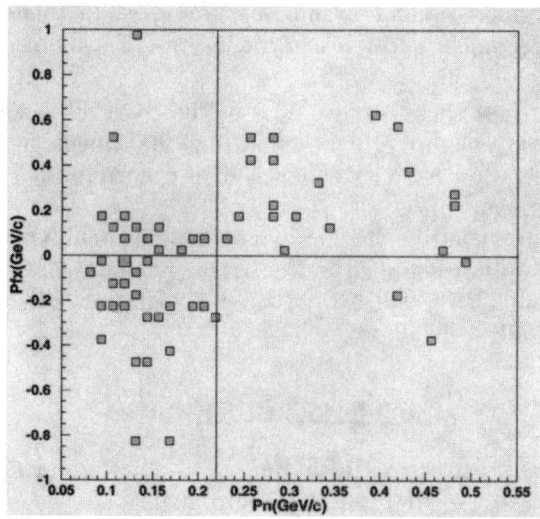

FIGURE 2. Pfx vs. Pn with cuts 1,2,3 and 4 for ^{12}C(p,2p+n) at 5.9 GeV/c

(2). The missing energy, E_{miss} should be appropriate for quasi-elastic scattering (within our resolution): $-0.2 < E_{miss} < 0.8$ GeV.

(3). $0.05 < p_n < 0.55$ GeV/c.

There is another cut we can apply to more fully reproduce the conditions of the 1994 run. For that data, only the straw-tube sectors near the midplane of EVA were working. So we can impose a fourth cut:

(4). The two detected protons were limited to a plane parallel to the neutron detectors within $\pm 25°$.

Figure 2 shows our preliminary momentum correlation results for $^{12}C(p,2p+n)$ at 5.9 GeV/c with cuts (1), (2), (3) and (4). Figure 2 is a plot of $\mathbf{p}_{fx}$, the reconstructed x component of $\mathbf{p}_f$, vs. the measured $\mathbf{p}_n$. Since the neutrons are detected largely going "downward" in the laboratory we would expect $\mathbf{p}_{fx}$ to be "upward" for correlated high-momentum n-p pairs. We see in Fig. 2 that for events with $\mathbf{p}_n > 0.22$ GeV/c (the Fermi momentum for ^{12}C) that indeed $\mathbf{p}_{fx}$ is predominantly "upward". For $\mathbf{p}_n < 0.22$ GeV/c there is no evident correlation, which is also as expected. For $\mathbf{p}_n \geq 0.22$ GeV/c, the ratio of events with $\mathbf{p}_{fx} \geq 0$ and $\mathbf{p}_{fx} < 0$ in Fig. 2 is 18/3. This result is completely consistent with the ratio of 17/1 reported in [1] and provides strong confirmation of that result.

With cut (4) we are selectively rejecting events with an approximately "coplanar" geometry where all four transverse momenta $\mathbf{p}_{t1}$, $\mathbf{p}_{t2}$, $\mathbf{p}_{tf}$ and $\mathbf{p}_{tn}$ all lie roughly in the same plane with $\Delta\phi = |\phi_2 - \phi_1| \approx 180°$. The type of events selected by (4) are then of the "non co-planar" type where $\Delta\phi$ differs significantly from 180°. This preferential selection of non co-planar events has an unintended benefit, in that our reconstruction of $\mathbf{p}_f$ has much better resolution for non co-planar events than for co-planar events.

The data in Figs. 2 represents 10% to 20% of the total data recorded in 1998. As we continue the analysis, we will be exploring the kinematic and geometric constraints on where we find strong evidence of n-p correlations - such as were reported in [1] and are seen in Fig. 2.

This research was supported by the U.S. - Israel Binational Science foundation, the Israel Science Foundation founded by the Israel Academy of Sciences and Humanities, the NSF grants PHY-9501114, PHY-9722519 and U.S. Department of Energy grant DEFG0290ER40553.

REFERENCES

1. J. Aclander et al., Phyl. Lett. B453 (1999) 211.
2. I. Mardor, Ph.D. Dissertation, Tel Aviv University, 1997.
3. Y. Mardor, Ph.D. Dissertation, Tel Aviv University, 1998.

PROGRESS IN PERTURBATIVE COLOR TRANSPARENCY

John P. Ralston[a], Pankaj Jain[b], Bijoy Kundu[c], Jim Samuelsson[d]

[a] Department of Physics and Astronomy, University of Kansas, Lawrence, KS 66045, USA
Email: ralston@ukans.edu
[b] Department of Physics, IIT Kanpur, Kanpur-208 016, India
E-mail: pkjain@iitk.ac.in
[c] Department of Physics, University of Virginia, Charlottesville VA, USA
E-mail: bkundu@jlab.org
[d] Department of Theoretical Physics, University of Lund, Lund S-22362, Sweden
E-mail: jim@thep.lu.se

Abstract. A brief overview of the status of color transparency experiments is presented. We report on the first complete calculations of color transparency within a perturbative QCD framework. We also comment on the underlying factorization method and assumptions. Detailed calculations show that the slope of the transparency ratio with Q^2, and the effective attenuation cross sections extracted from color transparency experiments depend on the x distribution of wave functions.

I OVERVIEW

Several experiments indicate that color transparency [1] and nuclear filtering [2–4] have been observed at large nuclear number A. The first color transparency experiment of Carroll et al [5] convincingly showed that interference effects in proton-proton scattering were filtered away in nuclear targets. Fits to the attenuation cross section in nuclear targets show values significantly below the Glauber theory values [6]. The FNAL E-665 experiment [7] also proved consistent with filtering effects [8]. The observation of increasing longitudinal final state polarization in $\gamma^* A \to \rho A$ as a function of Q^2 is noteworthy. We still await confirmation of predicted longitudinal polarization increasing as a function of A [4].

Electron beam experiments remain controversial, with few signals of interesting Q^2 dependence [9]. A basic feature of γ^*-initiated reactions is that most events are knocked out from the back side of the nucleus. This minimizes the resolving power of such experiments to measure the size of propagating states. The A dependence is a particularly useful tool [6] to measure effective attenuation cross sections. O'Neill et al [10] showed that effective attenuation cross sections extracted from $A(e, e'p)$

SLAC data were smaller than Glauber theory calculations by a statistically significant amount. However, the precision of the data [9] was insufficient to establish a large effect, and model dependence in the choice of the normalization of hard scattering is another complication. Reports on new $(e, e'p)$ beam experiments from CEBAF are expected shortly.

Progress on the theory front has come from looking deeper at the basic factorization methods [3], and doing the work of labor-intensive calculations [11]. The asymptotic factorization scheme of Lepage and Brodsky [12] is inadequate. An integration over the transverse separation of quarks is needed in the description. We call this "impact parameter factorization", which is needed to describe the interactions with the nuclear target, which otherwise vanish prematurely in the pure short-distance scheme. The impact parameter method was originally found necessary to regulate Landshoff and Sudakov effects in pp scattering [13]. We adapted it to describe color transparency and nuclear filtering [3,11]. Impact-parameter factorization has subsequently become very popular for the description of free-space form factors [14,15], which remain controversial [16,17] at laboratory Q^2 values.

II THE CALCULATIONS

Elsewhere [11] we report details of calculations of hard-pion knockout, $\gamma^* A \to \pi A'$, and hard nucleon knockout, $\gamma^* A \to p A'$. These are exploratory concept studies, designed to see how $pQCD$ predicts color transparency and filtering with few parameters. Since all the details except for experimental acceptances have been incorporated, the calculations are also fully quantitative predictions of the type needed to compare to experiments.

The case of pionic transparency deserves special mention. First, the pion is the cleanest theoretical laboratory one would desire. A short-distance wave function is known from experiments on pion decay, without relying on the sometimes circular logic of schemes such as QCD sum rules. Second, a pion is ultra-relativistic at energies as low as a few GeV. This helps strengthen the approximations made in $pQCD$. Finally, the pion has only two quarks in its valence state, and one transverse separation b, reducing the complexity of the calculations.

Working in configuration (impact-parameter b) space the expression for a $\gamma^* -$ meson form-factor becomes:

$$F_\pi(Q^2) = \int dx_1 dx_2 \frac{d^2\vec{b}}{(2\pi)^2} \mathcal{P}(x_2, b, P_2, \mu) \tilde{H}(x_1, x_2, Q^2, \vec{b}, \mu) \mathcal{P}(x_1, b, P_1, \mu). \quad (1)$$

Here $\mathcal{P}(x, b, P, \mu)$ represent the Fourier transforms of the wave functions, including Sudakov factors; $\tilde{H}(x_1, x_2, Q^2, \vec{b}, \mu)$ represents the hard scattering kernel from perturbation theory. The impact parameter $\vec{b}$ is conjugate to $\vec{k}_{T1} - \vec{k}_{T2}$, μ is the renormalization scale, and P_1, P_2 are the initial and final momenta of the meson.

The nuclear medium modifies the quark wave function by an interaction kernel f_A, which is called the nuclear filtering amplitude. An eikonal form [3] appropriate for f_A is: $f_A(b; B) = exp(-\int_z^\infty dz' \sigma(b) \rho(B, z')/2)$. Here $\rho(B, z')$ is the nuclear number density at longitudinal distance z' and impact parameter B relative to the nuclear center. We parametrize $\sigma(b)$ as kb^2 for our calculations. Finally, we must include the probability to find a target at position B, z inside the nucleus. Then the wave functions $\mathcal{P}_A$ appropriate for the nuclear target are [3]

$$\mathcal{P}_A(x, b, P, \mu) = f_A(b; B) \mathcal{P}(x, b, P, \mu).$$

Putting together the factors, the process of knocking out a hadron from inside a nuclear target has an amplitude M given by

$$M = \int_0^\infty d^2B (\Pi dx_i d^2 b_i) \int_{-\infty}^{+\infty} dz \rho(B, z) \times F_\pi(x_1, x_2, b, Q^2) \times f_A(b, B) \quad (2)$$

The analysis for the proton is similar but vastly more complicated. A 9 dimensional integration over the various x_i, b_i coordinates is performed by Monte Carlo. Sudakov effects are set to depend on the maximum of the three quark separation distances, $b_{max} = max(b_1, b_2, b_3)$.

We find that the physics is not described by a free-space hard scattering, followed by some model of propagation with or without "expansion", which is the ansatz of most competing groups. The integrations over the transverse quark variable extend over the whole volume of the nucleus. There is no easy decoupling into a simple product of "hard" times "nuclear" effects. Color transparency truly probes the internal structure of hadrons.

We found uncertainties in the nuclear correlations at the 10% level to be a major concern, in some cases exceeding the theoretical uncertainties from the rest of the calculation [11]. Primary new results include a discovery that the *slope* of the transparency ratio with Q^2 depends strongly on the $x-$ dependence of wave functions (distribution amplitudes). This mysterious effect was traced to the fact that central wave functions are more effective in maintaining short distance. Endpoint dominated distributions tend to exacerbate long-distance effects, which are found not to produce transparency. Nuclear filtering was observed to depend on the choice of wave functions as consistent. Thus both the slope of transparency ratios, and the magnitude of effective attenuation cross sections extracted from data, are probes of $x-$ dependence. Extensive details and nearly a dozen plots are given in Ref. [11].

Acknowledgments: This work was supported by BRNS grant No. DAE/PHY/96152, the Crafoord Foundation, the Helge Ax:son Johnson Foundation, DOE Grant number DE-FGO2-98ER41079, the KU General Research Fund and NSF-K*STAR Program under the Kansas Institute for Theoretical and Computational Science.

REFERENCES

1. S. J. Brodsky and A. H. Mueller, *Phys. Lett.* B **206**, 685 (1988).

2. J. P. Ralston and B. Pire, *Phys. Rev. Lett.* **61**, 1823 (1988).
3. J. P. Ralston and B. Pire, *Phys. Rev. Lett.* **65**, 2343 (1990).
4. P. Jain, B. Pire and J. P. Ralston, *Phys. Rep.* **271**, 67 (1996).
5. A. S. Carroll *et. al.*, *Phys. Rev. Lett.* **61**, 1698 (1988).
6. P. Jain and J. P. Ralston, *Phys. Rev.* D **48**, 1104 (1993).
7. M. R. Adams *et al Phys. Rev. Lett.* **74**, 1525 (1995).
8. B. Z. Kopeliovich, J. Nemchick, N. N. Nikolaev, and B. G. Zakharov, *Phys. Lett.* B **309**, 179 (1993); *Phys. Lett.* B **324**, 469 (1994).
9. N. Makins *et. al.*, *Phys. Rev.* D **12**, 163 (1994).
10. T. G. O' Neill *et. al*, *Phys. Lett.* B **351**, 87 (1995).
11. B. Kundu, J. Samuelsson, P. Jain and J. P. Ralston, hep-ph/9812506.
12. S. J. Brodsky and G. P. Lepage, *Phys. Rev.* D **24**, 2848 (1981).
13. J. Botts and G. Sterman, *Nucl. Phys.* **B325**, 62 (1989).
14. H-n. Li and G. Sterman, *Nucl. Phys.* **B381**,129 (1992).
15. H-n. Li, *Phys. Rev.* D **48**, 4243 (1993).
16. B. Kundu, H. N. Li, J. Samuelsson and P. Jain, *Eur. Phys. J.* **C8**, 637 (1999), hep-ph/9806419.
17. J. Bolz, R. Jakob, P. Kroll, M. Bergmann, and N.G. Stefanis, *Z. Phys.* C **66**, 267 (1995).

Testing the Spin-Dependence of the In-Medium Nucleon-Nucleon Interaction Using Polarization Observables in $(\vec{p}, \vec{p}')$ Scattering at Intermediate Energies

Francesca Sammarruca* and Edward J. Stephenson[†]

University of Idaho, Moscow, Idaho 83844
[†]*IUCF, Bloomington, Indiana 47408*

Abstract. We examine medium modifications of the nucleon-nucleon interaction through proton-nucleus inelastic scattering at intermediate energy. Predictions from conventional calculations reveal problems with the spin-dependence of the effective interaction. Attempts to incorporate modifications of meson spectral functions are presented and discussed.

INTRODUCTION

The main motivation for this work is to examine the behavior of hadrons in nuclear matter. This topic has been receiving considerable attention, particularly in the context of dilepton production in relativistic heavy-ion collisions [1].

Modifications of hadron spectral properties are typically discussed within the framework of hot nuclear matter near the chiral phase transition. However, QCD-based theories predict that a precursor of this transition could already be observable in nuclear systems at normal nuclear densities [2]. Furthermore, recent experimental determinations of the ρ-meson mass in light nuclei [3] provide additional motivation for looking into the question of whether changes induced in the QCD vacuum by increasing density/temperature have relevance for the dynamics of nuclei under ordinary conditions.

Motivated in part by these considerations, we have done a systematic investigation of proton-induced reactions on nuclei, in particular inelastic proton-nucleus scattering to unnatural-parity states. Polarization observables for these states are sensitive to the spin dependent terms in the effective interaction that have been associated with possible non-conventional mechanisms that would alter the tensor force in the medium [4].

RESULTS OF CONVENTIONAL CALCULATIONS

Proton-nucleus scattering can be used to test our understanding of the nucleon-nucleon (NN) force in the nuclear medium. High precision polarization data are now available for a number of discrete nuclear states around 200 MeV.

Our model for the effective interaction is based on a one-boson-exchange (OBE) potential that describes free-space two-body data very well, an important prerequisite for reliable (p,p') predictions. Medium effects are calculated within the Dirac-Brueckner approach and are consistent with nuclear matter saturation properties. Our model is also satisfactory for calculations of elastic scattering and natural-parity inelastic transitions [5], which are very sensitive to the established many-body effects, such as those included in conventional Brueckner (BHF) or Dirac-Brueckner (DBHF) theory.

To test specifically the spin dependence of the effective interaction, the high-spin stretched transitions are particularly appropriate due to their unnatural parity, or "spin-flip," character. Because of their high spin, there is usually only one particle-hole transition that can contribute, which simplifies the structure. This choice is important because the lack of sensitivity of stretched states to conventional medium effects allows us to maximize the prospect of seeing new physics beyond the conventional.

With our baseline model described in Ref. [5], we have performed a DWBA calculation of the 6^- states in ^{28}Si and the 4^- states in ^{16}O [6]. For the isovector case, the cross section is well constrained by electron scattering formfactor measurements.

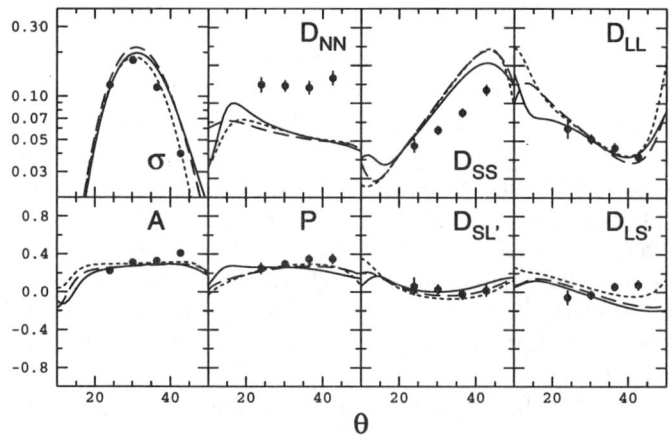

FIGURE 1. Measurements and predictions for cross section and spin observables for the 6^-, T=1 state at 14.35 MeV in ^{28}Si. The curves are based on the free (short dash), BHF (long dash), and DBHF (solid) interactions.

Calculations also reproduce reasonably well the angular distributions of A, P, $D_{SL'}$, and $D_{LS'}$. But substantial disagreement is seen for some of the diagonal D_{ii}, in particular D_{NN} which is too negative (see Fig. 1). It can be easily shown that bringing these values closer to zero requires a reduction of the spin-longitudinal contribution to the effective interaction [6].

Similar features are observed for the 4^-, T=1 transition in ^{16}O, thus the discrepancies are systematic. Problems also exist in the isoscalar channel (not shown here) where the data call for a reduction of the spin-orbit component relative to the tensor [6].

BEYOND CONVENTIONAL EFFECTS

Because the established medium effects contained in our model do not provide an adequate description of these states, we have been exploring the possibility of less conventional mechanisms. One way to reduce the tensor attraction in the OBE potential, as required by the present data in the isovector channel, is to increase the repulsive influence of the ρ-meson by reducing its mass with increasing density (Brown-Rho scaling [7]). However, calculations done within the meson-mass rescaling model of Ref. [8], which incorporates meson mass scaling in a DBHF framework, did not result in a significant improvement of the predictions [6].

Of course, a mass-rescaling scenario is useful only if the hadron spectral function in the medium exhibits a well-defined quasiparticle peak. In such a case, it would make sense to define a density-dependent, in-medium effective mass. A novel approach has recently been proposed by Friman et al. to explore systematically meson properties in nuclear matter [9]. By fitting meson-nucleon scattering and meson production data, they extracted empirically the meson-nucleon scattering amplitude, which can be related to the meson spectral function through a low-energy theorem, to leading order in density. The resulting spectral functions for ρ and ω are shown [9] to exhibit a two-peak structure, with the low-lying mode carrying about 20% of the strength at nuclear matter density. This suggests a simple scenario where the vector meson spectral functions are parametrized in terms of two quasiparticles, with density-dependent masses and strengths [10]. Preliminary results are shown in Fig. 2, and indicate that a two-state model for vector mesons in the medium may be capable of addressing some of the problems with the spin observables mentioned above. This model will have to be confronted with a larger body of data and the saturation properties of nuclear matter.

CONCLUSIONS

We have presented predictions for proton-nucleus scattering to unnatural-parity states at intermediate energy. The results from conventional calculations of medium effects indicate that the spin-dependence of the in-medium effective interaction is not yet modeled in a satisfactory manner. Models of meson-mass rescaling were

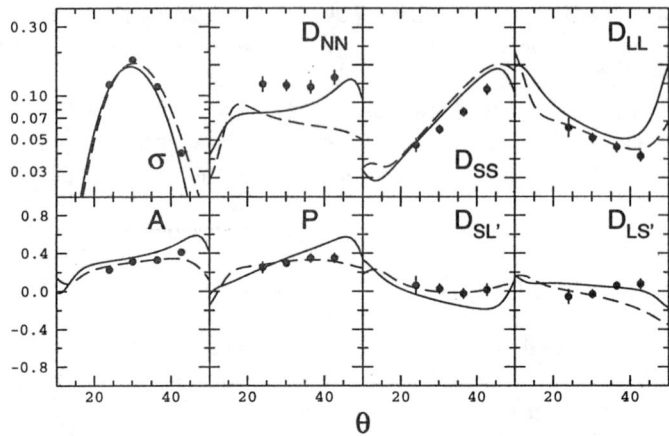

FIGURE 2. Same observables as in Fig. 1. The solid curves are the predictions of the two-state model for vector mesons [7-8] and are compared with our standard DBHF calculation (long dash).

not able to successfully address the problems. Alternative ways to model the vector meson spectral function in nuclear matter which incorporate fragmentation of the meson strength over a broader mass range shows a greater potential to explain the polarization data.

REFERENCES

1. G.Q. Li, C.M. Ko, G.E. Brown, and H. Sorge, *Nucl. Phys.* **A611**, 539 (1996).
2. T. Hatsuda and T. Kunihiro, *Phys. Rep.* **247**, 221 (1994).
3. G.J. Lolos et al., *Phys. Rev. Lett.* **80**, 241 (1998).
4. E.J. Stephenson et al., *Phys. Rev. Lett.* **78**, 1636 (1997).
5. F. Sammarruca, E.J. Stephenson, and K. Jiang, *Phys. Rev.* C **60**, 064610 (1999).
6. F. Sammarruca et al., *Phys. Rev.* C **61**, 014309 (2000).
7. G.E. Brown and M. Rho, *Phys. Lett.* **66**, 2720 (1990).
8. R. Rapp, R. Machleidt, J.W. Durso, and G.E. Brown, *Phys. Rev. Lett.* **82**, 1827 (1999).
9. B. Friman, M. Lutz, and G. Wolf, "From Meson-Nucleon Scattering to Vector Mesons in Nuclear Matter", in *Hadrons in Dense Matter*, edited by M. Buballa, W. Nörenberg, B.-J. Schaefer, and J. Wambach, Gesellschaft für Schwerionenforschung (GSI), Darmstadt, 2000, pp.161-169.
10. Y. Kim, R. Rapp, G.E. Brown, and M. Rho, nucl-th/9902009.

Flavor Asymmetry of the Nucleon Sea: Unambiguous Role of Goldstone Bosons

Anthony W. Thomas[a], W. Melnitchouk[a,b] and Fernando M. Steffens[c]

[a] *Special Research Centre for the Subatomic Structure of Matter,*[1]
University of Adelaide, SA 5005 Australia
[b] *Jefferson Lab,*[2] *12000 Jefferson Avenue, Newport News, VA 23606 USA*
[c] *Instituto de Fisica – USP,*[3] *C.P. 66 318, 05315-970, Sao Paulo, Brazil*

Abstract. We consider the non-analytic behaviour of $\bar{d}(x) - \bar{u}(x)$ and $s(x) - \bar{s}(x)$ as a function of the quark mass. Using the fact that such non-analytic behaviour is a unique characteristic of Goldstone boson loops, we establish two features of the nucleon sea in a model independent fashion. First we establish the unambiguous role of the pion cloud of the nucleon in the observed excess of $\bar{d}$ over $\bar{u}$ quarks in the proton. Second, we establish that $s(x)$ cannot in general be equal to $\bar{s}(x)$.

$SU(2)$ FLAVOR SYMMETRY VIOLATION

The role of the pion cloud of the nucleon (required by the spontaneous breaking of chiral symmetry in QCD) in generating an excess of $\bar{d}$ over $\bar{u}$ sea quarks in the proton was predicted in 1983 [1]. Since the experimental discovery that indeed $\bar{d} > \bar{u}$ [2], many authors have presented analyses of the data in terms of the pion cloud [3]. On the other hand, there has been a contrary view that one should not mention "long range" concepts such as the pion when dealing with parton distributions – instead one should only discuss the problem in terms of quarks and gluons. As a result of this controversy, the fundamental importance of the pion cloud in connection with flavor symmetry violation is not universally appreciated.

Here we use the unique feature of Goldstone boson loops, namely that they give rise to a dependence on the quark mass which is not an analytic function [4], to establish the unambiguous role of pion loops in the measured difference $\bar{d} - \bar{u}$.

The leading twist contribution to the n-th moment of the $\bar{d}(x) - \bar{u}(x)$ difference is given by [1,5]:

$$\left(\overline{D} - \overline{U}\right)^{(n)} = \int_0^1 dx\, x^n \left(\bar{d}(x) - \bar{u}(x)\right) = \frac{2}{3} V_\pi^{(n)} \cdot f_{\pi N}^{(n)}. \tag{1}$$

---

[1] Supported by the Australian Research Council and the University of Adelaide.
[2] Supported by DOE contract DE-AC05-84ER40150.
[3] Supported by FAPESP (96/7756-6, 98/2249-4).

Here $V_\pi^{(n)}$ is the n-th moment of the valence pion structure function and $f_{\pi N}^{(n)}$ is the n-th moment of the pion distribution function in the nucleon (or the $N \to \pi N$ splitting function):

$$f_{\pi N}^{(n)} = \int_0^1 dy \, y^n \, f_{\pi N}(y) \,. \tag{2}$$

The momentum dependence of the pion distribution function is given by [5,6]:

$$f_{\pi N}(y) = \left(\frac{3g_{\pi NN}^2}{16\pi^2}\right) y \int_{t_{min}}^{\mu^2} dt \, \frac{t}{(t+m_\pi^2)^2} \,, \tag{3}$$

where $t = -k_\mu k^\mu$ (k_μ is the four-momentum of the pion), with a minimum value $t_{min} = M^2 y^2/(1-y)$, and $g_{\pi NN}$ is the πNN coupling constant. (Since the non-analytic structure of pion loops does not depend on the short-distance behavior, we have for simplicity introduced an ultra-violet cut-off, μ, to regulate the integral in Eq. (3). One could have equally well used a form factor for the πNN vertex [7], or a more elaborate regularization procedure.)

We stress that this contribution to $\bar{d} - \bar{u}$ is a leading twist contribution to the structure function of the nucleon. The hard scattering involves the constituents of the pion itself, while the momentum of the pion is typical of those met in chiral models of nucleon structure, namely a few hundred MeV/c. The fact that the momentum associated with the pion is low is the reason one can discuss the LNA structure of $\bar{d} - \bar{u}$. There may, of course, be other terms which contribute to the physical difference between $\bar{d}$ and $\bar{u}$, which cannot be expressed in the factorized form of Eq. (1), such as interactions of the spectator quark in the pion with the recoil nucleon. However, the LNA behavior of $\bar{d} - \bar{u}$ is entirely determined by the one pion loop. It cannot be altered by such contributions and is strictly model independent. Evaluating the n-th moment of the distribution in Eq. (3), the LNA chiral log contribution from a pion loop is [8]:

$$f_{\pi N}^{(n)}\Big|_{LNA} = \left(3M^2 g_A^2/(4\pi f_\pi)^2\right) \times \begin{cases} (-1)^{n/2}((n+4)/(2n+4)) \, (m_\pi/M)^{n+2} \log(m_\pi^2/\mu^2) \\ \quad (n \text{ even}), \\ (-1)^{(n+1)/2}((n+5)/2) \, (m_\pi/M)^{n+3} \log(m_\pi^2/\mu^2) \\ \quad (n \text{ odd}). \end{cases} \tag{4}$$

We note that the PCAC relation has been used to express the πNN coupling constant in terms of the axial charge g_A (both g_A and the nucleon mass, M, are taken in the chiral SU(2) limit). For the $n = 0$ moment, conservation of baryon number requires that $V_\pi^{(0)} = 1$, which leads directly to our result for the 1st moment:

$$\left(\bar{D} - \bar{U}\right)_{LNA}^{(0)} \equiv \int_0^1 dx \left(\bar{d}(x) - \bar{u}(x)\right)_{LNA} = \frac{2g_A^2}{(4\pi f_\pi)^2} m_\pi^2 \log(m_\pi^2/\mu^2). \tag{5}$$

The LNA contributions to the $n > 0$ moments are suppressed in the chiral limit by additional powers of m_π^n. The scale dependence of $V_\pi^{(n)}$ for $n > 0$ introduces a Q^2 dependence into the higher moments of $\bar{d} - \bar{u}$.

While the current analysis aims only at establishing the model-independent, chiral behavior of flavor asymmetries, without necessarily trying to explain the asymmetries quantitatively, we observe that with a mass scale $\mu \sim 4\pi f_\pi \sim 1$ GeV, the magnitude of the LNA contribution (at the physical pion mass) to the $n = 0$ moment of $\bar{d} - \bar{u}$ is quite large – of order 0.2. This is the same order of magnitude as the latest experimental value for the asymmetry, $(\overline{D} - \overline{U})^{(0)} \approx 0.1 - 0.15$ [2].

SU(3) CASE

Following the original suggestion of Signal and Thomas [9] that virtual kaon loops are a source of non-perturbative strangeness in the nucleon [10], we turn to the chiral behaviour of the strange component of the sea of the nucleon. In this case both the kaon and the associated hyperon carry non-zero strangeness. The different momentum distributions of $\bar{s}$ quarks in the kaon and s quarks in the hyperon then lead to different s and $\bar{s}$ distributions as a function of x (as well as to non-zero values for strange electromagnetic form factors [10]).

The n-th moment of the $s - \bar{s}$ difference arising from a one-kaon loop can be written:

$$\left(s - \bar{s}\right)^{(n)} = \int_0^1 dx\, x^n \left(s(x) - \bar{s}(x)\right) = V_\Lambda^{(n)} \cdot f_{\Lambda K}^{(n)} - V_K^{(n)} \cdot f_{K\Lambda}^{(n)}, \qquad (6)$$

where $f_{K\Lambda}^{(n)}$ is the n-th moment of the $N \to K\Lambda$ splitting function [9]. The corresponding moment of the Λ distribution, $f_{\Lambda K}^{(n)}$, can be evaluated from $f_{K\Lambda}^{(n)}$ through the symmetry relation between the splitting functions:

$$f_{\Lambda K}(y) = f_{K\Lambda}(1 - y) . \qquad (7)$$

The requirement of zero net strangeness in the nucleon implies the vanishing of the $n = 0$ moment, $(S - \bar{S})^{(0)} = 0$, which follows from Eq. (7) and strangeness number conservation, $V_\Lambda^{(0)} = V_K^{(0)} = 1$. However, for higher moments this is no longer the case, so that in general $(S - \bar{S})^{(n)}$ will be non-zero for $n > 0$. In particular, the LNA components of the strange distributions will be given by:

$$f_{K\Lambda}^{(n)}\Big|_{LNA} = \frac{27}{25} \frac{M^2 g_A^2}{(4\pi f_\pi)^2} (M_\Lambda - M)^2 (-1)^n \frac{m_K^{2n+2}}{\Delta M^{2n+4}} \log(m_K^2/\mu^2), \qquad (8)$$

where we have used SU(6) symmetry to relate $g_{KN\Lambda}$ to g_A/f_π. Furthermore, while the LNA part of the n-th moment of $\bar{s}$ is of order $m_K^{2n+2} \log m_K^2$, from Eq.(7) the LNA contribution to the n-th moment of s will be of order $m_K^2 \log m_K^2$. Since the LNA terms in the chiral expansion are model-independent, and in general not cancelled by other contributions, this result establishes the fact that the process of dynamical symmetry breaking in QCD implies that the s and $\bar{s}$ distributions must have a different dependence on Bjorken x.

CONCLUSION

We have derived the leading non-analytic chiral behavior of the flavor asymmetries in the sea of the proton which are associated with Goldstone boson loops. In particular, we have established unambiguously that the study of flavor asymmetries in the nucleon sea should yield direct information on dynamical chiral symmetry breaking in QCD.

REFERENCES

1. A.W. Thomas, *Phys. Lett.* **B 126**, 97 (1983).
2. P. Amaudraz et al., *Phys. Rev. Lett.* **66**, 2712 (1991); A. Baldit et al., *Phys. Lett.* **B 332**, 244 (1994); E.A. Hawker et al., *Phys. Rev. Lett.* **80**, 3715 (1998).
3. E. M. Henley and G. A. Miller, *Phys. Lett.* **B251**, 453 (1990); A. Signal, A. W. Schreiber and A. W. Thomas, *Mod. Phys. Lett.* **A6**, 271 (1991); S. Kumano, *Phys. Rev.* **D43**, 3067 (1991); S. Kumano and J. T. Londergan, *Phys. Rev.* **D44**, 717 (1991); W.-Y. P. Hwang, J. Speth and G. E. Brown, *Z. Phys.* **A339**, 383 (1991).
4. S. Weinberg, *Physica (Amsterdam)* **96 A**, 327 (1979); J. Gasser and H. Leutwyler, *Ann. Phys.* **158**, 142 (1984); E. Jenkins, M. Luke, A.V. Manohar and M.J. Savage, *Phys. Lett.* **B 302**, 482 (1993); V. Bernard, N. Kaiser and U.-G. Meißner, *Int. J. Mod. Phys.* **E 4**, 193 (1995).
5. For reviews see: J. Speth and A.W. Thomas, *Adv. Nucl. Phys.* **24**, 83 (1998); S. Kumano, *Phys. Rep.* **303**, 183 (1998).
6. J.D. Sullivan, *Phys. Rev.* **D 5**, 1732 (1972).
7. S. Théberge, G.A. Miller and A.W. Thomas, *Phys. Rev.* **D 22**, 2838 (1980); A.W. Thomas, *Adv. Nucl. Phys.* **13**, 1 (1984).
8. A. W. Thomas, W. Melnitchouk and F. M. Steffens, hep-ph/0005043.
9. A.I. Signal and A.W. Thomas, *Phys. Lett.* **B 191**, 206 (1987).
10. X. Ji and J. Tang, *Phys. Lett.* **B 362**, 182 (1995); S.J. Brodsky and B.Q. Ma, *Phys. Lett.* **B 381**, 317 (1996); P. Geiger and N. Isgur, *Phys. Rev.* **D 55**, 299 (1997); W. Melnitchouk and M. Malheiro, *Phys. Rev.* **C 55**, 431 (1997).

Hadron Structure Functions in the Bosonized Nambu–Jona–Lasinio Model [1]

Herbert Weigel[2,a)] and Leonard Gamberg [b)]

[a)] *Center for Theoretical Physics, Laboratory for Nuclear Science, and Department of Physics Massachusetts Institute of Technology, Cambridge, Massachusetts 02139*
[b)] *Department of Physics and Astronomy University of Oklahoma, 440 West Brooks Norman, Ok 73019*

Abstract. We outline a consistent regularization procedure to compute hadron structure functions within bosonized chiral quark models. We impose the Pauli–Villars scheme, which reproduces the chiral anomaly, to regularize the bosonized action. Applying the Bjorken limit to the Compton amplitude to extract structure functions that are consistent with the scaling laws and sum rules of deep inelastic scattering.

INTRODUCTION

The bosonized action of chiral quark models can be cast in the form

$$\mathcal{A}[S,P] = -iN_C \mathrm{Tr}_\Lambda \log\left[i\slashed{\partial} - (S+i\gamma_5 P)\right] - \frac{1}{4G}\int d^4x\, \mathrm{tr}\,\mathcal{V}(S,P)\,. \tag{1}$$

Here $\mathcal{V}$ is a local potential respectively for scalar and pseudoscalar fields S and P. In the Nambu–Jona–Lasinio (NJL) model [1] one has $\mathcal{V} = S^2 + P^2 + 2\hat{m}_0(S+iP)$. From the gap–equation we obtain the VEV, $\langle S \rangle = m$ which is identified as the constituent quark mass and parameterizes the dynamical chiral symmetry breaking. The cut–off Λ indicates the regularization of the quadratically divergent quark loop.

DIS off hadrons is parameterized by the hadronic tensor $W^{\mu\nu}(q)$ with q being the momentum transmitted to the hadron by the photon. $W^{\mu\nu}(q)$ is obtained from the hadron matrix element of the commutator $[J^\mu(\xi), J^\nu(0)]$. The major concern in regularizing the functional (1) is to maintain the chiral anomaly. We achieve this by splitting this functional into γ_5–even and odd pieces and only regularize the former as the γ_5–odd part is conditionally finite. Details of this presentation are published in [2]. For related work see refs [3–5].

[1)] This work is supported in parts by funds provided by the U.S. Department of Energy (D.O.E.) under cooperative research agreements #DF–FC02–94ER40818 and #DE–FG0398ER41066 and by the Deutsche Forschungsgemeinschaft (DFG) under contract We 1254/3-1.
[2)] Heisenberg–Fellow.

REGULARIZATION OF THE COMPTON TENSOR

For the evaluation of the hadronic tensor we start from the absorptive part of the forward virtual Compton amplitude, $T^{\mu\nu}$

$$W^{\mu\nu}(q) = \frac{1}{2\pi}\Im\left(T^{\mu\nu}(q)\right) \quad \text{with} \quad T^{\mu\nu}(q) = \int d^4\xi\, e^{iq\cdot\xi} \langle p,s|T\left(J^\mu(\xi)J^\nu(0)\right)|p,s\rangle. \quad (2)$$

(We denote the momentum of the hadron by p and eventually its spin by s.) The time–ordered product is unambiguously obtained from the regularized action

$$T\left(J^\mu(\xi)J^\nu(0)\right) = \frac{\delta^2}{\delta v_\mu(\xi)\,\delta v_\nu(0)} \operatorname{Tr}_\Lambda \log\left[i\partial\!\!\!/ - (S+i\gamma_5 P) + \mathcal{Q}\,v\!\!\!/\right]\Big|_{v_\mu=0}, \quad (3)$$

with $\mathcal{Q}$ being a flavor matrix. In order to extract the leading twist pieces of the structure functions, we consider $W^{\mu\nu}(q)$ in the Bjorken limit: $q^2 \to -\infty$ with $x = -q^2/p\cdot q$ fixed.

To specify the regularization of the functional trace in (3) we define

$$i\mathbf{D} = i\partial\!\!\!/ - (S+i\gamma_5 P) + v\!\!\!/\mathcal{Q} \quad \text{and} \quad i\mathbf{D}_5 = -i\partial\!\!\!/ - (S-i\gamma_5 P) - v\!\!\!/\mathcal{Q}$$

and separate the functional trace into (un–)regularized γ_5–even (odd) pieces,

$$\operatorname{Tr}_\Lambda \log\left[i\partial\!\!\!/ - (S+i\gamma_5 P) + \mathcal{Q}\,v\!\!\!/\right] = -i\frac{N_C}{2}\sum_{i=0}^{2} c_i \operatorname{Tr}\log\left[-\mathbf{D}\mathbf{D}_5 + \Lambda_i^2 - i\epsilon\right]$$

$$-i\frac{N_C}{2}\operatorname{Tr}\log\left[-\mathbf{D}\left(\mathbf{D}_5\right)^{-1} - i\epsilon\right]. \quad (4)$$

With the conditions $c_0 = 1$, $\Lambda_0 = 0$, $\sum_{i=0}^{2} c_i = 0$ and $\sum_{i=0}^{2} c_i\Lambda_i^2 = 0$ the double Pauli–Villars regularization renders the functional in (3) finite.

PION STRUCTURE FUNCTION

DIS off pions is characterized by a single structure function, $F(x)$. For its computation we have to specify the pion matrix element in the Compton amplitude (2), whence we introduce the pion field $\vec{\pi}$ via[3]

$$S + iP\gamma_5 = m\left(U\right)^{\gamma_5} = m\exp\left(i\frac{g}{m}\gamma_5\vec{\pi}\cdot\vec{\tau}\right). \quad (5)$$

Due to the separation into $\mathbf{D}$ and $\mathbf{D}_5$ the computation of the Compton amplitude for virtual pion–photon scattering differs from the evaluation of the 'handbag' diagram in the Bjorken limit: isospin violating dimension–five operators emerge. Fortunately all isospin violating pieces cancel yielding

---

[3] The coupling g and the constituent quark mass m are related by the pion decay constant. In the chiral limit the relation is linear $m = gf_\pi$.

$$F(x) = \frac{5}{9}(4N_C g^2)\frac{d}{dm_\pi^2}\left\{m_\pi^2 \sum_{i=0}^{2} c_i \frac{d^4k}{(2\pi)^4 i}\left[-k^2 - x(1-x)m_\pi^2 + m^2 + \Lambda_i^2 - i\epsilon\right]^{-2}\right\}.$$

The cancellation of the isospin violating pieces is a feature of the Bjorken limit: insertions of the pion field on the propagator carrying the infinitely large photon momentum can be ignored. Furthermore this propagator can be taken to be the one for non–interacting massless fermions. Thus the Pauli–Villars cut–offs can be omitted for this propagator which leads to the scaling behavior of the structure function.

NUCLEON STRUCTURE FUNCTIONS

Baryons emerge as solitons of the meson fields [6]. We specify the soliton by

$$U(\vec{x},t) = A(t)\exp\left(i\vec{\tau}\cdot\hat{r}\Theta(r)\right)A^\dagger(t). \tag{6}$$

The chiral angle $\Theta(r)$ is determined from the stationary condition for constant A. Subsequently we quantize the collective coordinates A to generate nucleon states.

As we take the quark propagator with the infinite photon momentum to be free and massless it is sufficient to differentiate

$$\frac{N_C}{4i}\sum_{i=0}^{2} c_i \mathrm{Tr}\left\{\left(-\mathbf{D}^{(\pi)}\mathbf{D}_5^{(\pi)} + \Lambda_i^2\right)^{-1}\left[\mathcal{Q}^2 \slashed{v}\left(\slashed{\partial}\right)^{-1}\slashed{v}\mathbf{D}_5^{(\pi)} - \mathbf{D}^{(\pi)}(\slashed{v}\left(\slashed{\partial}\right)^{-1}\slashed{v})_5 \mathcal{Q}^2\right]\right\}$$
$$+\frac{N_C}{4i}\mathrm{Tr}\left\{\left(-\mathbf{D}^{(\pi)}\mathbf{D}_5^{(\pi)}\right)^{-1}\left[\mathcal{Q}^2 \slashed{v}\left(\slashed{\partial}\right)^{-1}\slashed{v}\mathbf{D}_5^{(\pi)} + \mathbf{D}^{(\pi)}(\slashed{v}\left(\slashed{\partial}\right)^{-1}\slashed{v})_5 \mathcal{Q}^2\right]\right\}, \tag{7}$$

with respect to the photon field v_μ. We have introduced the $(\ldots)_5$ description

$$\gamma_\mu\gamma_\rho\gamma_\nu = S_{\mu\rho\nu\sigma}\gamma^\sigma - i\epsilon_{\mu\rho\nu\sigma}\gamma^\sigma\gamma^5 \quad \text{and} \quad (\gamma_\mu\gamma_\rho\gamma_\nu)_5 = S_{\mu\rho\nu\sigma}\gamma^\sigma + i\epsilon_{\mu\rho\nu\sigma}\gamma^\sigma\gamma^5$$

to account for the unconventional appearance of axial sources in $\mathbf{D}_5$ [2]. Substituting (6) into (7) and computing the functional trace using basis states obtained from the Dirac Hamiltonian in the background of $U(\vec{x},t)$, we find analytical results for the structure functions. For details and the verification of sum rules we refer to [2]. Here we solely report that the structure function entering the Gottfried sum rule does not undergo regularization. This is surprising since in the parton model this structure function differs from the one entering the Adler sum rule only by the sign of the anti–quark distribution. The latter structure function, however, gets regularized, in agreement with the quantization rules for the collective coordinates.

Unfortunately numerical results for the full structure functions, *i.e.* including the properly regularized vacuum piece, are not yet available. However, we have verified that in the Pauli–Villars regularization scheme the axial charges are saturated to 95% or more by their valence quark contributions once the self–consistent soliton is substituted. This provides sufficient justification to adopt the valence quark contribution to the polarized structure functions as a reliable approximation [3]. In Fig. 1 we compare the model predictions for the linearly independent polarized structure functions to experimental data [9].

Fig. 1: Model predictions for the polarized proton structure functions xg_1 and xg_2. The curves labeled 'RF' denote the results as obtained from the valence quark contribution to (7). These undergo a projection to the infinite momentum frame 'IMF' [7] and a leading order 'LO' DGLAP evolution [8]. Data are from SLAC–E143 [9].

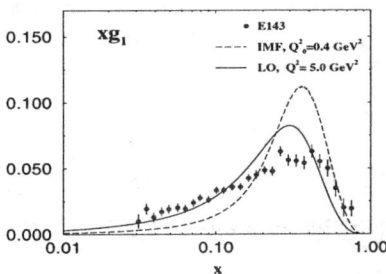
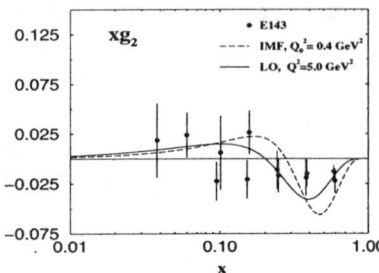

The evolution of the structure function g_2 to the momentum scale of the experiments requires the separation into twist–2 and twist–3 components [8]. We observe that the model results for the polarized structure functions, which we argued to have reliably approximated, agree reasonably well with the experimental data.

ACKNOWLEDGMENTS

We thank our collaborators E. Ruiz Arriola, O. Schröder and H. Reinhardt for valuable contributions.

REFERENCES

1. Y. Nambu and G. Jona–Lasinio, Phys. Rev. **122** (1961) 345; **124** (1961) 246.
2. H. Weigel, E. Ruiz Arriola and L. Gamberg, Nucl. Phys. **B560** (1999) 383.
3. H. Weigel, L. Gamberg and H. Reinhardt, Mod. Phys. Lett. **A11** (1996) 3021, Phys. Lett. **B399** (1997) 287, Phys. Rev. **D55** (1997) 6910; L. Gamberg, H. Reinhardt and H. Weigel, Phys. Rev. **D58** (1998) 054014; O. Schröder, H. Reinhardt and H. Weigel, Phys. Lett. **B439** (1998) 398; H. Weigel, Nucl. Phys. **A670** (2000) 92.
4. D. I. Diakonov et al., Nucl. Phys. **B480** (1996) 341, Phys. Rev. **D56** (1997) 4069.
5. M. Wakamatsu and T. Kubota, Phys. Rev. **D57** (1998) 5755, Phys. Rev. **D60** (1999) 034020.
6. R. Alkofer, H. Reinhardt and H. Weigel, Phys. Rep. **265** (1996) 139; C. V. Christov et al., Prog. Part. Nucl. Phys. **37** (1996) 91.
7. L. Gamberg, H. Reinhardt and H. Weigel, Int. J. Mod. Phys. **A13** (1998) 5519.
8. G. Altarelli, P. Nason and G. Parisi, Phys. Lett. **B320** (1994) 152, **B325** (1994) 538 (E); A. Ali, V.M. Braun and G. Hiller, Phys. Lett. **B226** (1991) 117.
9. K. Abe et al., Phys. Rev. **D58** (1998) 112003.

Pseudospin Symmetry in Nuclei, Spin Symmetry in Hadrons

P.R. Page, T. Goldman, J.N. Ginocchio* [1]

Theoretical Division, MS-B283, Los Alamos National Laboratory, Los Alamos, NM 87545, USA

Abstract. Ginocchio argued that chiral symmetry breaking in QCD is responsible for the relativistic pseudospin symmetry in the Dirac equation, explaining the observed approximate pseudospin symmetry in sizable nuclei. On a much smaller scale, it is known that spin-orbit splittings in hadrons are small. Specifically, new experimental data from CLEO indicate small splittings in D-mesons. For heavy-light mesons we identify a cousin of pseudospin symmetry that suppresses these splittings in the Dirac equation, known as spin symmetry. We suggest an experimental test of the implications of spin symmetry for wave functions in electron-positron annihilation. We investigate how QCD can give rise to two different dynamical symmetries on nuclear and hadronic scales.

Recently, Isgur [1] has re-emphasized the experimental fact that spin-orbit splittings in meson and baryon systems, which might be expected to originate from one-gluon-exchange (OGE) effects between quarks, are absent from the observed spectrum. He conjectures that this is due to a fairly precise, but accidental, cancellation between OGE and Thomas precession effects, each of which has "splittings of hundreds of MeV" [1]. Taking the point of view that precise cancellations reflect symmetries rather than accidents, we have examined what dynamical requirements would lead to such a result.

Below, we first elucidate the experimental evidence for small spin-orbit splittings. Then we identify the symmetry that suppresses these splittings. We show how this symmetry leads to a proposed experimental test in electron-positron annihilation. The existence of pseudospin symmetry in nuclei is also indicated. Finally, we investigate how QCD can give rise to two different dynamical symmetries on nuclear and hadronic scales.

We study systems made from one light quark and one heavy antiquark. Because the light quark is expected to move relativistically, and the heavy quark can be treated as stationary, we examine the relativistic Dirac Equation. The equation includes the possibility of the OGE and Thomas precession effects mentioned before.

---

[1] *E-mail:* prp@lanl.gov, tgoldman@lanl.gov, gino@t5.lanl.gov

The dynamics are approximated as a light quark moving in a relativistic vector and scalar potential of a heavy antiquark.

In the limit where the antiquark is infinitely heavy, the angular momentum of the light quark, j, is separately conserved. The states can be labelled by l_j, where l is the orbital angular momentum of the light quark. In non–relativistic models of mesons the splitting between $l_{l+\frac{1}{2}}$ and $l_{l-\frac{1}{2}}$ levels, e.g. the $p_{\frac{3}{2}}$ and $p_{\frac{1}{2}}$ or $d_{\frac{5}{2}}$ and $d_{\frac{3}{2}}$ levels, can *only* arise from spin-orbit interactions [1]. In Fig. 1 it is seen that the $p_{\frac{1}{2}}$ level corresponds to two degenerate broad states with different total angular momenta $J = j \pm \frac{1}{2}$, where $\frac{1}{2}$ is the spin of the heavy antiquark. For example in the case of D-mesons the two states are called D_0^* and D_1'. There are also two degenerate narrow $p_{\frac{3}{2}}$ states D_1 and D_2^*. The degenerate states separate as one moves slightly away from the heavy antiquark limit.

For the D–mesons, the CLEO collaboration claims a broad $J^P = 1^+$ state at $2461^{+41}_{-34} \pm 10 \pm 32$ MeV [2], belonging to the $p_{\frac{1}{2}}$ level, in close vicinity to the D_2^* at 2459 ± 2 MeV, belonging to the $p_{\frac{3}{2}}$ level, indicating a remarkable $p_{\frac{3}{2}}$-$p_{\frac{1}{2}}$ spin-orbit degeneracy of -2 ± 50 MeV.

For the K-mesons, the $p_{\frac{1}{2}}$ level is at 1409 ± 5 MeV, with $p_{\frac{3}{2}}$ nearby at 1371 ± 3 MeV, corresponding to a $p_{\frac{3}{2}}$-$p_{\frac{1}{2}}$ splitting of -38 ± 6 MeV. The splitting between the higher-lying $d_{\frac{5}{2}}$ and $d_{\frac{3}{2}}$ levels is -4 ± 14 MeV or 41 ± 13 MeV, depending on how the states are paired into doublets. These results indicate a near spin-orbit degeneracy if the strange quark can be treated as heavy, although it has certainly not been established that such a treatment is valid.

For B-mesons, L3 has performed an analysis, using input from theoretical models and heavy quark effective theory, to determine that the $p_{\frac{3}{2}}$-$p_{\frac{1}{2}}$ splitting is 97 ± 11 MeV [3]. Note that this is *not* a model-independent experiment.

There is currently no agreement within either lattice QCD, or models, on the magnitude, and even sign, of the $p_{\frac{3}{2}}$-$p_{\frac{1}{2}}$ splitting in D, D_s, B and B_s-mesons [4].

In order to more quantitatively measure the spin-orbit splitting, define $r = (p_{\frac{3}{2}} - p_{\frac{1}{2}})/((4p_{\frac{3}{2}}+2p_{\frac{1}{2}})/6-s_{\frac{1}{2}})$, where all entries refer to masses. The experimental data on D, K and B mesons give respectively $r = 0.00\pm0.10$, -0.06 ± 0.00 and 0.23 ± 0.04. For the Dirac equation with arbitrary vector and scalar Coulomb potentials, the only potential for which the relevant analytic solutions are known, $-0.7 \lesssim r \lesssim 0.6$. It is hence evident that the spin-orbit splittings extracted from experimental results are indeed small.

The Dirac hamiltonian describing the motion of the light quark is

$$H = \vec{\alpha} \cdot \vec{p} + \beta(m + V_S) + V_V + M, \tag{1}$$

where we have set $\hbar = c = 1$, $\vec{\alpha}$, β are the usual Dirac matrices, $\vec{p}$ is the momentum, and m and M are the masses of the light and heavy (anti)quarks respectively.

If the vector potential $V_V(r)$ is equal to the scalar potential plus a constant potential U, which is independent of the spatial location $r = |\vec{r}|$ of the light quark

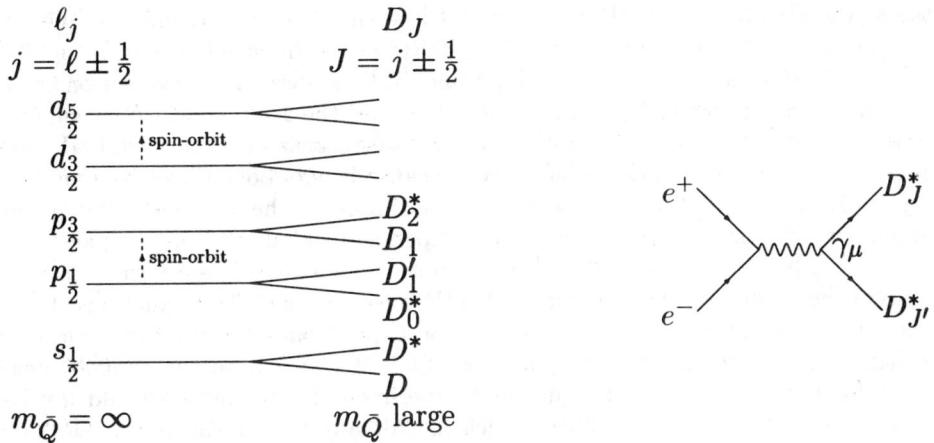

FIGURE 1. Notation (left) and electron-positron annihilation (right).

relative to the heavy one, $V_V(r) = V_S(r) + U$, then the hamiltonian is invariant under a spin symmetry [5], $[H, \hat{S}_i] = 0$. The generators of that symmetry are,

$$\hat{S}_i = \begin{pmatrix} \hat{s}_i & 0 \\ 0 & \tilde{\hat{s}}_i \end{pmatrix} \qquad (2)$$

where $\hat{s}_i = \sigma_i/2$ are the usual spin generators, σ_i the Pauli matrices, and $\tilde{\hat{s}}_i = U_p \hat{s}_i U_p$ with $U_p = \frac{\vec{\sigma}\cdot\vec{p}}{p}$. Thus Dirac eigenstates can be labeled by the orientation of the spin, even though the system may be highly relativistic, and the eigenstates with different spin orientation will be degenerate. This means the states with the same orbital angular momentum will be degenerate. Thus, for example, the $p_{1/2}$ and $p_{3/2}$ states will be degenerate, yielding exact spin-orbit degeneracy.

Thus, we have identified a symmetry in the heavy-light quark system which produces spin-orbit degeneracies independent of the details of the potential. If this potential is strong, the heavy-light quark system will be very relativistic; that is, the lower component for the light quark will be comparable in magnitude to the upper component of the light quark. In non-relativistic models spin-orbit degeneracy naturally arises in the absence of spin-dependent forces. It is remarkable that non-relativistic behaviour of energy levels can arise for such fully relativistic systems.

If future experiments determine that spin-orbit splittings are small not only for the lowest excited states in mesons but are small throughout the meson spectrum, this experimental fact dictates that the effective QCD vector and scalar potentials between a quark and antiquark are approximately equal up to a constant, which would be a significant observation about the nature of non-perturbative QCD.

It is possible to show that the radial wave functions in momentum space of the upper and lower components of the Dirac wave function are separately equal for both eigenstates with the same orbital angular momentum, e.g. the $p_{1/2}$ and $p_{3/2}$

states [4]. This has explicitly been verified in a fit to the spectrum of K-mesons [4]. The prediction can be tested in the following experiment, depicted in Fig. 1.

The annihilation $e^+e^- \to D_0^* D_0^*$, $D_0^* D_2^*$ and $D_2^* D_2^*$ allows for the extraction of the D_0^* and D_2^* electromagnetic static form factors and the D_0^* to D_2^* electromagnetic transition form factor. The photon interaction γ_μ ensures that all radial wave functions of the light quark are accessed, because it acts both diagonally and off-diagonally, and because it does *not* act on the (infinitely) heavy quark. When spin symmetry is realised, there are only two independent radial momentum space wave functions, which should enable the prediction of one of the three form factors in terms of the other two. This should enable the verification of the predictions of spin symmetry. The proposed experiment can be carried out at the Beijing Electron Positron Collider at an energy of approximately 1 GeV above the $\psi(4040)$ peak in the final state $DD\pi\pi$. An equivalent experiment for K-mesons would involve detection of the $KK\pi\pi$ final state, which has already been measured at DM2. If B-mesons also exhibit spin symmetry, one can do equivalent experiments around 1 GeV above the $\Upsilon(4S)$ peak at the SLAC, KEK or CESR B-factories.

If $\hat{s}_i$ and $\hat{\tilde{s}}_i$ are interchanged in Eq. 2 one finds a symmetry of the hamiltonian in Eq. 1, $[H, \hat{S}_i] = 0$, when $V_V(\vec{r}) = -V_S(\vec{r}) + U$. This symmetry, called pseudospin symmetry, has been identified as being responsible for the pseudospin degeneracies observed in nuclei, for example the particle or hole states in double magic ^{208}Pb [6]. These are nuclei obtained from ^{208}Pb by subtracting or adding either a neutron or a proton.

What may the QCD origins of pseudospin and spin symmetry on different hadronic scales be? It has been argued by considering QCD sum rules in an infinite nuclear medium that the ratio of the scalar and vector self-energies of the nucleon is $-\sigma_N/(8m) \approx -1$, where σ_N is the nucleon σ-term related to chiral symmetry breaking, and m is the light quark current mass [7]. These scalar and vector self-energies in an infinite nuclear medium inform about the scalar and vector potentials of the nucleon, with exactly the relation needed for the validity of pseudospin symmetry. The case for how spin symmetry can arise from QCD has been made in ref. [4].

We are supported by the Department of Energy under contract W-7405-ENG-36.

REFERENCES

1. N. Isgur, nucl-th/9908028.
2. S. Anderson *et al.* (CLEO Collab.), *Nucl. Phys.* **A663-664** (2000) 647.
3. M. Acciari *et al.* (L3 Collab.), *Phys. Lett.* **B465** (1999) 323-334.
4. P.R. Page, T. Goldman, J.N. Ginocchio, hep-ph/0002094.
5. J.S. Bell and H. Ruegg, *Nucl. Phys.* **B98** (1975) 151.
6. J.N. Ginocchio, *Phys. Rep.* **315** (1999) 231.
7. T.D. Cohen *et al.*, *Prog. Part. Nucl. Phys.* **35** (1995) 221.

Extension of the Delta-hole Approach to Higher Baryon-hole Excitations in Nuclei

V.V.Balashov, A.V.Bibikov, V.K.Dolinov, M.M.Kaskulov

Institute of Nuclear Physics, Moscow State University,
Moscow 119899, Russia

Abstract.
Exchange interaction between baryon-hole excitations of different sorts can play an important role in nuclear reactions in the baryon resonances region. Calculations are presented for the coherent η-meson photoproduction from ^{12}C in the region of the $D_{13}(1520)$ and $S_{11}(1535)$ resonances.

Extending the Delta-hole model to higher baryon-hole resonances we start from its basic theoretical concepts [1-3]. On the other hand, taking into account the growing role of exclusive experiments in physics of baryon excitation in nuclei [4-9], we perform parallel investigations on the Δ-h model itself to adapt it (and, in perspective, its extended versions) to these new trends.

In [10,11] we introduced a concept of the density matrix of the isobar produced in nuclei and applied it to investigate polarization characteristics of the Δ-isobar in nuclei under pion and photon absorption. Our latest calculations relate to the production and decay of the Delta-isobar in the charge exchange reaction $(^3He, t\pi)$. They give an important information on the distribution of the $(^3He, t\pi)$ events over the wide scale of the excitation energy of the target nucleus in the final state.

To study higher baryon resonances we begin with the Roper resonance $P_{11}(1440)$. Taking into account the strong selectivity of the (α, α') reaction relative to the Roper resonance excitation we investigate this reaction on helium target. The idea is to detect both the scattered and the recoil α-particles in coincidence: the first one - to measure the transferred energy distribution and the other - to exclude the acompanying process of coherent pion production via the virtual excitation of the Δ-isobar in the projectile which breaks up the recoil α-particle (in the (α, α') reaction on hydrogen [12,13] this mechanism is dominating among those masking the Roper resonance excitation). Preliminary calculations in our $(\alpha, 2\alpha)$ case show a resonable scale of the expected cross section and provide with enough information on energy and angular distributions of both α-particles. In progress are detailed investigations of other aspects of this problem.

In [14] we have suggested a multiconfiguration baryon-hole model where the baryon-hole excitations of different sorts are considered as an unified basis and are coupled together via pion, ρ-meson and other meson exchange and the short-range correlations of the Landau-Migdal type. Exploring this approach, we present here our calculations for the process of coherent η-meson photoproduction from nuclei in the $D_{13}(1520)$ and $S_{11}(1535)$ resonance region intensively investigated in many theoretical groups [15-18].

The effective Lagrangians of corresponding elementary processes are used in their nonrelativistic form:

$$L_{\eta N D_{13}} = \frac{g_{\eta N D_{13}}}{m_\eta} \bar{\Psi}^\mu_{D_{13}} \gamma_5 \Psi_N \partial_\mu \varphi_\eta \to \delta H_{\eta N D_{13}} = i\frac{g_{\eta N D_{13}}}{2m_\eta M}(\vec{S}^\dagger \cdot \vec{k})(\vec{\sigma} \cdot \vec{k}),$$

$$L_{\eta N S_{11}} = -ig_{\eta N S_{11}} \bar{\Psi}_N \Psi_{S_{11}} \varphi_\eta \to \delta H_{\eta N S_{11}} = ig_{\eta N S_{11}},$$

$$L^{(1)}_{\gamma N D_{13}} = i\frac{eg^{(1)}_{\gamma N D_{13}}}{2M} \bar{\Psi}^\mu_{D_{13}} \gamma^\nu \Psi_N F_{\mu\nu} \to$$

$$\to \delta H^{(1)}_{\gamma N D_{13}} = -\frac{eg^{(1)}_{\gamma N D_{13}}}{2M} E_\gamma (\vec{S}^\dagger \cdot \vec{\epsilon}_{\lambda\vec{k}}) - i\frac{eg^{(1)}_{\gamma N D_{13}}}{4M^2}(\vec{S}^\dagger \cdot \vec{k})(\vec{\sigma} \times \vec{k})\vec{\epsilon}_{\lambda\vec{k}},$$

$$L^{(2)}_{\gamma N D_{13}} = \frac{eg^{(2)}_{\gamma N D_{13}}}{4M^2} \bar{\Psi}^\mu_{D_{13}} \partial^\nu \Psi_N F_{\mu\nu} \to \delta H^{(2)}_{\gamma N D_{13}} = \frac{eg^{(2)}_{\gamma N D_{13}}}{4M} E_\gamma (\vec{S}^\dagger \cdot \vec{\epsilon}_{\lambda\vec{k}}),$$

$$L_{\gamma N S_{11}} = \frac{eg_{\gamma N S_{11}}}{4M} \bar{\Psi}_{S_{11}} \gamma_5 \sigma_{\mu\nu} \Psi_N F^{\mu\nu} \to \delta H_{\gamma N S_{11}} = \frac{eg_{\gamma N S_{11}}}{2M} E_\gamma (\vec{\sigma} \cdot \vec{\epsilon}_{\lambda\vec{k}}).$$

with vertex constants of strong and electromagnetic interactions taken as in [18]:

$$g_{\gamma p S_{11}} = 0.73 \qquad g^{(1)}_{\gamma p D_{13}} = 5.46 \qquad g^{(2)}_{\gamma p D_{13}} = 5.76$$
$$g_{\gamma n S_{11}} = -0.62 \qquad g^{(1)}_{\gamma n D_{13}} = -0.97 \qquad g^{(2)}_{\gamma n D_{13}} = 0.66$$
$$g_{\eta N S_{11}} = 2.1 \qquad g_{\eta N D_{13}} = 6.76$$

One can see here a very strong suppression of the isoscalar photoexcitation amplitude $\gamma + N \to N(1535)S_{11}$. So, inspite of the dominant coupling of the $N(1535)S_{11}$ resonance with the $N + \eta$ channel and the very strong evidence of this resonance in the η-photoproduction cross section on hydrogen, the role of this resonance in coherent η-meson photoproduction from zero-isospin nuclei seems to be small. Interaction between the $D_{13}(1520)$-hole and $S_{11}(1535)$-hole excitations changes the situation radically. Schematically it looks as if the photon absorption takes place in the $D_{13}(1520)$-hole channel, then this excitation is transferred into the $S_{11}(1535)$-hole states, from which the excited nucleus decays to the coherent η-production channel. Being absolutely excluded in the elementary photoproduction process on an isolated nucleon, this, in a sense, two-step mechanism turns out to work actively in nuclei.

In the calculations presented here we introduce the $D_{13}(1520)$-hole and $S_{11}(1535)$-hole configuration mixing via the virtual η-exchange:

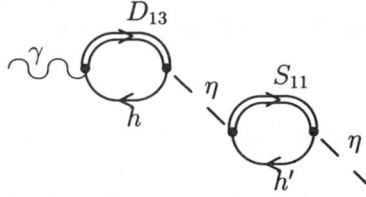

Fig.1 illustrates its effect schematically. It relates to a pure resonant approach where the coherent η-meson photoproduction from carbon nucleus is considered in the very near-threshold region as if it takes place exclusively via excitation and decay of baryon-hole excitations. The cross section turns out to be strongly enhanced due to the $D_{13}(1520)$-hole and $S_{11}(1535)$-hole interaction.

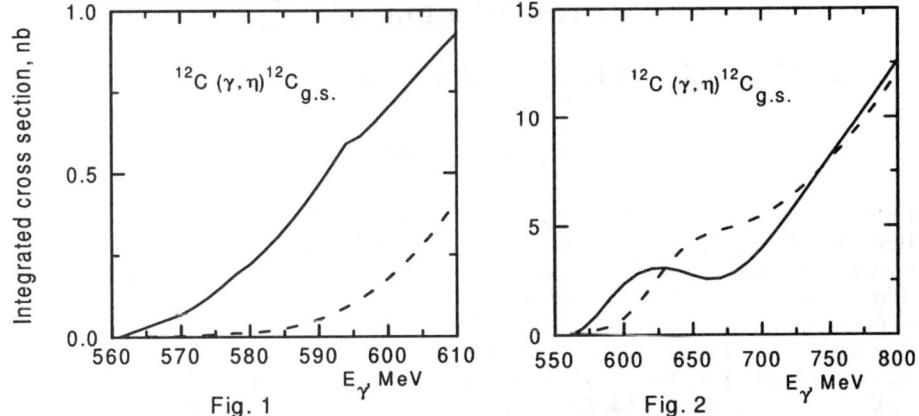

FIGURE 1. Integrated cross section for the reaction $^{12}C(\gamma,\eta)^{12}C_{g.s.}$ calculated with (solid line) and without (dashed line) the $D_{13}(1520)$-hole and $S_{11}(1535)$-hole interaction taken into account.

FIGURE 2. The same as in Fig.1 but calculated with taking into account both the resonant and direct photoproduction processes.

Realistic calculations performed with taking into account also the direct ω-exchange process

confirms predictions of the pure resonant approach (Fig.2). Looking forward to possible experimental investigations of the process under consideration, note a nontrivial, nonmonotonic energy dependence of the integrated cross section in the

$D_{13}(1520)$ and $S_{11}(1535)$ resonance region. More calculations, including those for other targets, are in progress and will be presented elsewhere.

Conclusion

Exchange interaction between baryon-hole states of different sorts can play an important role in a wide class of nuclear reactions with excitation of baryon resonances in nuclei. Its mechanism must be investigated in much more detail and, to our opinion, should be taken into consideration in any version of theoretical description of nuclear reactions in the higher baryon resonances region.

Acknowledgments

The work was supported by the INTAS-RFBR grant N 1345.

References

1. Kisslinger L.S., Wang W.L., *Ann.Phys.* (N.Y.) **66**, 374 (1976).
2. Hirata M., Koch J.H., Lenz F., Moniz E.J., *Ann.Phys.* **120**, 205 (1979).
3. Oset E. and Weise W., *Nucl.Phys.* **A368**, 375 (1981).
4. Glavanokov I.V. et al., *Yad.Phys.* **29**, 1455 (1979).
5. Chiba J. et al., *Phys.Rev.Lett* **67**, 1982 (1991).
6. Hennino T. et al., *Nucl.Phys.* **A527**, 399 (1992).
7. Pham L.D. et al., *Phys.Rev.* **C46**, 621 (1992).
8. Hicks K. et al., *Phys.Rev.* **C55**, R12 (1997).
9. van Uden M.A. et al., *Phys.Rev.* **C58**, 3462 (1998).
10. Balashov V.V., Bibikov A.V., Vostroknutova O.N., *Phys. of Atom. Nuclei* **58**, 1150 (1995); *Yad. Fiz. (Russian Nucl.Phys.)* **58**, 1229 (1995).
11. Balashov V.V., Bibikov A.V., Vostroknutova O.N., *Int Conf. "Mesons and Nuclei at Intermediate Energies"*, Dubna, JINR, 1994.
12. Morsch H.P. et al., *Phys.Rev.Lett.* **69**, 1336 (1992).
13. Hirenzaki S., Fernandez de Cordoba P., Oset E., *Phys.Rev.* **C53**, 277 (1996).
14. Balashov V.V., Bibikov A.V., Dolinov V.K., *16 Euro. Conf. on FEW BODY PROBLEMS IN PHYSICS, Autrans, France, 1-6 June 1998, Abstracts Booklet, Grenoble 1998, p.34*.
15. Bennhold C. and Tanabe A., *Nucl.Phys.* **A530**, 625 (1991).
16. Benmerrouche M., Mukhopadhyay N.C. and Zhang J.F., *Phys.Rev.* **D51**, 3237 (1995).
17. Fix A. and Arenhovel H., *Nucl.Phys.* **A620**, 457 (1997).
18. Peters W., Lenske H., Mosel U., *Nucl.Phys.* **A642**, 506 (1998).

Structure and Production of Lambda Baryons

J.T. Londergan[1], C. Boros[2] and A.W. Thomas[2]

[1] *Dept. of Physics and Nuclear Theory Center, Indiana University Bloomington, IN 47405, USA;* [2] *CSSM, University of Adelaide Adelaide 5005, Australia.*

Recently we have obtained experimental evidence for Λ production in semi-inclusive electron-induced reactions. As emphasized by Burkardt and Jaffe [1,2], the naive quark structure of the Λ is that of a s quark which couples to a $u - d$ pair with $S = T = 0$. If we accept arguments based on $SU(3)$ or $SU(6)$ quark symmetry, then we should be able to relate spin-dependent amplitudes for Λ production to known nucleon structure quantities. These arguments were used by de Florian et al. [7,8] to predict the Λ polarization measured in these experiments. In this contribution, we describe a simple model which predicts significant violation of $SU(3)$ quark symmetry in hyperon production. This model is used to analyze Λ production experiments.

Semi-inclusive hyperon production cross sections are proportional to the fragmentation functions for quarks into a hyperon. The Adelaide group [3] have produced a method for extracting leading twist quark distributions from quark model wavefunctions. This method leads to the relation

$$q^{\uparrow\downarrow}(x) = 2P^+ \sum_n |\langle n; p_n|\psi_+^{\uparrow\downarrow}(0)|B; PS\rangle|^2 \delta(P^+(1-x) - p_n^+) \quad (1)$$

Eq. 1 gives the quark distributions inside a baryon B, for quarks whose helicity is parallel ($q^{\uparrow}(x)$) or antiparallel ($q^{\downarrow}(x)$) to the baryon helicity. This equation includes a sum over a complete set of states n, with momentum $p_n^+ = \sqrt{M_n^2 + p_n^2} + p_{nz}$. These are the states which remain following a hard collision which removes a single quark from (or adds an antiquark to) a baryon. The contribution to a quark distribution $q(x)$ from an intermediate state with mass M_n peaks at

$$x_{max} \approx 1 - M_n/M \ . \quad (2)$$

The effective mass M_n in Eq. 1 depends on the helicity and flavor of the struck quark. This in turn produces a spin-flavor dependence in the resulting parton distributions.

Such spin-flavor effects can be very significant in the case of hyperon parton distributions, as was pointed out by Alberg et al. [4]. For large x the dominant contributions come from two-quark intermediate states. The diquark effective masses for non-strange quarks can be estimated from fitting $N - \Delta$ splitting;

one obtains a scalar effective mass $M_2(S=0) \sim 600$ MeV, and a vector effective mass $M_2(S=1) \sim 800$ MeV. The corresponding effective masses for two-quark states containing one s quark can be obtained from $\Lambda - \Sigma$ splitting, giving $M_2'(S=0) \sim 890$ MeV and $M_2'(S=1) \sim 1010$ MeV. The simplest quark model picture of the Λ couples an s quark to a ud pair with $S=0$. Arguments based on $SU(3)$ symmetry generally assume equal distributions for all three quarks in the Λ. Inserting the relevant states into Eq. 2, we see that removing a strange quark from the Λ should produce a peak in the s distribution at $x_{max} \approx 0.48$. If a u quark in the Λ is struck, this leaves behind a sd pair which is most probably in an $S=1$ state. The resulting vector diquark produces a peak in the u quark distribution at $x_{max} \approx 0.12$. From this argument, one expects the s quark distribution in the Λ hyperon to be *much harder* than the corresponding u or d distributions. Applying these arguments to spin distributions, one expects the Δu quark distribution in the Λ to be positive at large x, even if the average of this distribution over all x is zero [5].

Boros et al. [5] have extended this method to calculate fragmentation functions of quarks into hyperons. Applying the same arguments which were used to determine parton distributions, one can generate fragmentation functions $D(z)$ from quark model wavefunctions through the relation

$$\frac{1}{z} D_{q\Lambda}^{\uparrow\downarrow}(z) = P^+ \sum_n |\langle 0|\psi_+^{\uparrow\downarrow}(0)|\Lambda(PS); np_n\rangle|^2 \delta(P^+(\frac{1}{z}-1) - p_n^+) \qquad (3)$$

Eq. 3 describes the process by which a quark fragments into a Λ hyperon plus an n-quark final state. The contribution to the fragmentation from a state with mass M_n will produce a peak at a value $z_{max} \approx 1/(1 + M_n/M_\Lambda)$. Therefore, the fragmentation function at large z will be dominated by contributions from quark fragmentation leading to a Λ plus a pair of antiquarks ($n=2$). Because of quark mass differences, and spin-flavor effects in quark interactions, one expects that at large z the fragmentation of s quarks into Λ hyperons will be much larger than that for u or d quarks. This differs significantly from models based on $SU(3)$ quark symmetry, which generally assume that u, d, and s quarks have identical fragmentation into Λ's.

The results of this flavor dependence are shown in Fig. 1, which shows Λ production arising from e^+e^- annihilation at the Z resonance. Fig. 1a shows the spin-averaged production cross sections, which are proportional to the fragmentation functions of quarks into a Λ particle. The dot-dashed curve is the contribution of the s quark fragmentation producing two-quark final states; this contribution, termed "valence" fragmentation, is calculated in the MIT bag model by Boros et al. [5] at the bag scale, and evolved to M_Z^2. The dotted curve is the corresponding contribution from u and d valence fragmentation. The dashed curve is the contribution from quark fragmentation leading to $n=4$ and higher states. This "sea" contribution is fit to the data, which has been taken at both LEP and SLD facilities [6]. This simple model gives quite a good fit to the experiment; at large z note that the cross sections are dominated by s quark fragmentation.

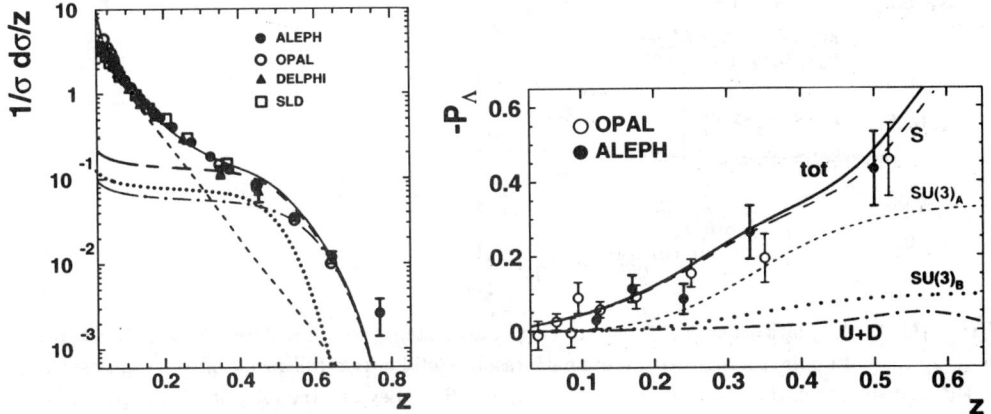

FIGURE 1. e^+e^- annihilation at the Z resonance. (a) Inclusive Λ production cross section. Dash-dot (dotted) curves: s quark (u,d quark) "valence" fragmentation into Λ. Dashed line: "sea" fragmentation contribution. (b) Λ polarization. Solid curve: full calculation; dashed curve: s quark contribution. Dashed, dotted curves: results of assuming $SU(3)$ symmetry for unpolarized quark fragmentation.

The resulting longitudinal Λ polarization is shown in Fig. 1b. At large z the polarization is large and negative; it is predicted that $P_\Lambda \rightarrow -0.94$ in the limit $z \rightarrow 1$. The solid curve is the calculation of Boros et al., while the dotted and dashed curves are the results assuming $SU(3)$ symmetry for the spin-averaged fragmentation. Curve $SU(3)_A$ assumes the naive quark model result, $\Delta D_{s\Lambda} = D_{s\Lambda}; \Delta D_{u\Lambda} = \Delta D_{d\Lambda} = 0$. The curve $SU(3)_B$ results from using the parton spin distributions measured for nucleons, and converts these into the corresponding Λ parton distributions using $SU(3)$. This produces hyperon polarizations which differ radically from the data. De Florian et al. [7] obtain closer agreement with the polarized data using $SU(3)$-based arguments, however their spin-averaged fragmentation functions are freely varied to reproduce the hyperon production cross sections. Λ polarization can also be measured in semi-inclusive DIS, where one detects charged particles such as mesons, resulting from polarized $\vec{e}$ scattering from an unpolarized nucleon. The Λ polarization is given by

$$\vec{P}_\Lambda = \hat{e}_3 P_e \frac{y(2-y)}{1+(1-y)^2} \frac{\sum_q e_q^2 q_N(x,Q^2) \Delta D_{q\Lambda}(z,Q^2)}{\sum_q e_q^2 q_N(x,Q^2) D_{q\Lambda}(z,Q^2)} \qquad (4)$$

At large z, Λ production is dominated by u quark fragmentation, since u quark distributions are dominant in the proton valence region. The u quark contribution is further enhanced by the coupling of the virtual photon, which is proportional to the squared charge of the quark. We predict positive values of P_Λ in semi-inclusive DIS

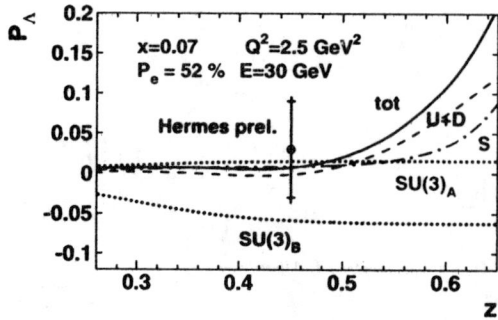

FIGURE 2. Λ polarization in $\vec{e} - p$ scattering. Parameters listed are those of HERMES expt, Ref. [9]; the data point shows their preliminary result. Solid curve: full result of Ref. [5]; dashed: $u + d$ contrib; dot-dashed: s contrib. Dotted curves: $SU(3)$ symmetry assumed for unpolarized quark fragmentation.

at large z, in contradiction to SU(3)-based models which predict zero or negative P_Λ [8], as shown in Fig. 2. The HERMES collaboration [9] has preliminary results for P_Λ at a single z value. This result is slightly positive, but with significant error bars. At present, the experiment disagrees with the $SU(3)B$ prediction (the notation is that of Fig. 1), but does not rule out any model. However, our model predicts that P_Λ will increase very rapidly with increasing z, contrary to the $SU(3)$-based models. Thus reasonably precise measurements at substantially larger z could distinguish between these models of semi-inclusive DIS.

The author would also like to thank C. Benesh, J-C Peng, and F. Steffens for useful discussions. This research was supported in part by the NSF under research contract PHY97-22706, and by the Australian Research Council.

REFERENCES

1. Burkardt, M. and Jaffe, R.L., *Phys. Rev. Lett.* **70**, 2537-2540 (1993).
2. Jaffe, R.L., *Phys. Rev.* **D54**, 6581-6585 (1996).
3. Signal, A.I. and Thomas, A.W., *Phys. Rev.* **D40**, 2832-2843 (1989); Schreiber, A.W. et al., *Phys. Rev.* **D42**, 2226-2236 (1990); *Phys. Rev.* **D44**, 2653-2662 (1991).
4. Alberg, M. et al., *Phys. Lett.* **B389**, 367-373 (1996); *Nucl. Phys.* **A644**, 93-104 (1998).
5. Boros, C., Londergan, J.T. and Thomas, A.W., *Phys. Rev.* **D61**, 014007:1-11 (2000).
6. Ackerstaff, K. et al. (OPAL Collaboration), *Eur. Phys. J.* **C2**, 49 (1998); **C8**, 241 (1999); Abe, K. et al. (SLD Collaboration) *Phys. Rev.* **D59**, 052001:1-33 (1999); Buskulic, D. et al. (ALEPH Collaboration) *Z. Phys.* **C64** 361 (1994); *Phys. Lett.* **B374**, 319-330 (1996).
7. De Florian, D., Stratmann, M. and Vogelsang, W., *Phys. Rev.* **D57**, 5811-5824 (1998).
8. De Florian, D., Stratmann, M. and Vogelsang, W., *Phys. Rev. Lett.* **81**, 530 (1998).
9. Airepetian, A. et al. (HERMES Collaboration), hep-ex/9911017 (1999).

Relativistic Calculations for Incoherent η-Photoproduction

I. R. Blokland and H. S. Sherif

Department of Physics, University of Alberta
Edmonton, Alberta, Canada T6G 2J1

Abstract. We present and discuss the results of our calculations using a relativistic model for incoherent η-photoproduction on nuclei. We find that the incoherent cross sections are dominated by contributions from the $S_{11}(1535)$ and $D_{13}(1520)$ resonances. We find that isovector transitions are favored over isoscalar transitions. This can lead to incoherent η-photoproduction cross sections that are often larger than those of the coherent process.

INTRODUCTION

The study of η meson photoproduction reactions is primarily motivated by the promise of learning more about the isospin-$\frac{1}{2}$ nucleon resonances to which the η meson couples so selectively. Using nuclear targets, η-photoproduction can be used to investigate the modification of hadron properties in the nuclear medium. Furthermore, the study of the final-state interactions of η mesons with nuclei is closely connected with very recent experimental evidence of the existence of η-mesic nuclei [1].

Although inclusive [2,3] and coherent [4–8] η-photoproductions reactions have been well studied in recent years, the incoherent reaction has been largely ignored [4], partly because of an aversion to nuclear structure complications, and partly due to an expectation of low cross sections. Nevertheless, in order to understand the underlying mechanisms of the η-photoproduction process better, the recent experimental results [9–13] must eventually be complemented by measurements of exclusive quasifree, coherent, and incoherent processes.

In the following sections, we present a brief overview of our study of incoherent η-photoproduction reactions. Further details can be obtained from a forthcoming publication [14] and from the references provided at the end of this contribution.

THEORY

Our model for incoherent η-photoproduction is built upon an effective Lagrangian description of the elementary reaction [15], wherein the phenomenological parameters are constrained by the recent experiments studying this reaction. We use a relativistic mean field approach [16,17] and the nuclear shell model to describe the dynamics of the nucleons within nuclear matter. Following the procedure in [18], we have derived expressions for the reaction amplitude and observables for an incoherent η-photoproduction reaction.

We would now like to show how we decompose the reaction amplitude into isoscalar and isovector components. The total reaction amplitude arises as a sum over the individual amplitudes of the protons and neutrons

$$\begin{aligned} S &= \sum_p S_p + \sum_n S_n \\ &= \sum_N \left(\frac{1}{2} + \tau_z\right) S_p + \sum_N \left(\frac{1}{2} - \tau_z\right) S_n \\ &= \sum_N \left\{\left[\frac{1}{2}(S_p + S_n)\right] + [S_p - S_n]\tau_z\right\} \\ &= \sum_N \{S_0 + S_1 \tau_z\} \end{aligned} \quad (1)$$

With the assumption that the reaction proceeds primarily through the formation of an S_{11} resonance, the Mainz data for η-photoproduction on the deuteron [10] indicates that $S_n = (-0.80)S_p$. In the context of the equation above, this implies that the amplitude for an isovector transition, S_1, is significantly larger than the amplitude for an isoscalar transition, S_0. This is one of the major reasons why our calculations suggest that incoherent η-photoproduction cross sections are not inherently smaller than those of the coherent process on $N = Z$ target nuclei.

RESULTS

We have performed calculations for incoherent η-photoproduction cross sections on ^{12}C, ^{16}O, and ^{40}Ca. Qualitatively, the results are quite similar for the three nuclei. Some of the ^{12}C calculations will be presented in our forthcoming publication [14]. In this work, we will show calculations for the ^{40}Ca$(\gamma, \eta)^{40}$Ca$^*(5^-1; 8.55)$ reaction, under the assumption that the nuclear excited state is well described by a $1d_{3/2}^{-1} - 1f_{7/2}$ configuration.

Figure 1 shows the differential cross sections for this reaction when $E_\gamma = 750$ MeV. The various dotted lines indicate the contributions from the individual reaction channels. Note that the η mesons are predominantly produced in the forward direction. The $S_{11}(1535)$ and $D_{13}(1520)$ resonances provide the most significant contributions and the relative phases of these contributions are such that the total differential cross section is enhanced.

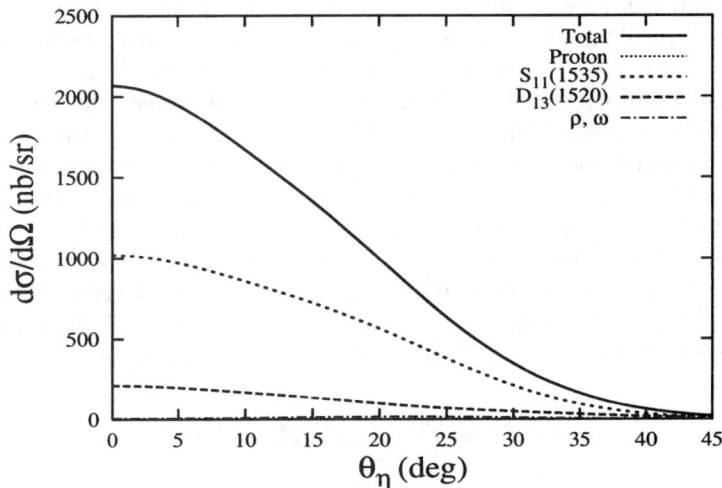

FIGURE 1. Contributions of the different reaction channels to the differential cross section for the $^{40}\text{Ca}(\gamma,\eta)^{40}\text{Ca}^*(5^-1;8.55)$ reaction at $E_\gamma = 750\,\text{MeV}$.

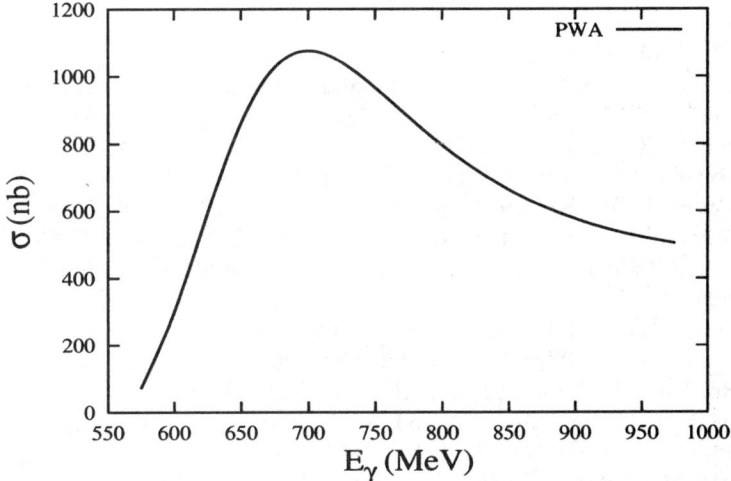

FIGURE 2. Total cross section, as a function of the incident photon energy E_γ, for the $^{40}\text{Ca}(\gamma,\eta)^{40}\text{Ca}^*(5^-1;8.55)$ reaction in the Plane Wave Approximation.

In Figure 2, we show the total cross section for this reaction as a function of the energy of the incident photon. Unlike the corresponding graphs for the coherent and quasifree processes, which would increase monotonically from threshold, this curve rises sharply to a maximum and then decreases. As a result, searches for incoherent processes should stay within 100 MeV or so of threshold. Both figures show calculations carried out in the Plane Wave Approximation. Distortion effects, arising from final-state interactions of the η meson with the recoiling nucleus typically reduce the cross sections by about a factor of two [14].

Based on Figure 2, for $E_\gamma = 700$ MeV, we would expect the cross section for this particular incoherent reaction to be a few hundred nb, after accounting for final-state interactions and realistic nuclear spectroscopy. By way of comparison, the coherent cross section on ^{40}Ca at this energy has been calculated variously as 10 nb [6] and 50 nb [8]; the measured inclusive η-photoproduction cross section is 50 μb [11]. Considering that there are several nuclear excited states that can be expected to contribute to an overall incoherent η-photoproduction cross section, we suggest that under certain circumstances, the incoherent η-photoproduction cross sections are significantly larger than those of the coherent process.

ACKNOWLEDGEMENTS

This work was supported in part by the Natural Sciences and Engineering Research Council of Canada.

REFERENCES

1. See G. Sokol, these proceedings.
2. F. X. Lee, L. E. Wright, C. Bennhold, and L. Tiator, Nucl. Phys. A **603**, 345 (1996).
3. M. Hedayati-Poor and H. S. Sherif, Phys. Rev. C **58**, 326 (1998).
4. C. Bennhold and H. Tanabe, Nucl. Phys. A **530**, 625 (1991).
5. A. Fix and H. Arenhövel, Nucl. Phys. A **620**, 457 (1997).
6. W. Peters, H. Lenske, and U. Mosel, Nucl. Phys. A **642**, 506 (1998).
7. J. Piekarewicz, A. J. Sarty, and M. Benmerrouche, Phys. Rev. C **55**, 2571 (1997).
8. L. J. Abu-Raddad, J. Piekarewicz, A. J. Sarty, and M. Benmerrouche, Phys. Rev. C **57**, 2053 (1998).
9. B. Krusche et al., Phys. Rev. Lett. **74**, 3736 (1995).
10. B. Krusche et al., Phys. Lett. B **358**, 40 (1995).
11. M. Röbig-Landau et al., Phys. Lett. B **373**, 45 (1996).
12. J. Ajaka et al., Phys. Rev. Lett. **81**, 1797 (1998).
13. A. Bock et al., Phys. Rev. Lett. **81**, 534 (1998).
14. I. R. Blokland and H. S. Sherif, *submitted for publication*.
15. M. Benmerrouche, N. C. Mukhopadhyay, and J. F. Zhang, Phys. Rev. D **51**, 3237 (1995).
16. J. D. Walecka, Ann. Phys. (N.Y.) **83**, 491 (1974).
17. B. D. Serot and J. D. Walecka, in *Advances in Nuclear Physics*, edited by J. W. Negele and E. Vogt (Plenum, New York, 1986), Vol. 16.
18. M. Hedayati-Poor and H. S. Sherif, Phys. Rev. C **56**, 1557 (1997).

Phase Transitions in Finite Density Baryonic Matter

Oliver Schwindt and Niels R. Walet

Dept. of Physics, UMIST, P.O. Box 88, Manchester M60 1QD, UK

Abstract.
We discuss the calculation of the phase diagram for the 2D Skyrme model, as a first step towards that of the 3D model. This should give us access to the behaviour of baryonic matter at finite densities and temperatures.

Understanding the phase structure of baryonic matter at finite temperatures and densities is an important and difficult problem. Unfortunately, at present only limited information can be extracted directly from QCD, and one has to resort to effective field theories. One such theory is provided by the Skyrme model [1] which was first used in the sixties by Skyrme, and was later justified by 't Hooft [2] and Witten [3] who analysed the large number of colours limit of QCD. This analysis implied that baryons would emerge as solitons in the large N_c limit.

The Skyrme Lagrangian is

$$\mathcal{L} = \frac{1}{2}(\partial_\mu \phi_k)^2 - \left[\frac{1}{4}(\partial_\mu \phi_k)^2 (\partial_\nu \phi_l)^2 - \frac{1}{4}(\partial_\mu \phi_k \partial_\mu \phi_l)^2\right] - m_\pi^2(1 - \phi_1), \qquad (1)$$

where roman indices run from 1 to 4 and the greek indices from 0 to 3. The fields obey the unitarity constraint $\sum_k \phi_k^2 = 1$. The field ϕ_1 is proportional to the sigma field, and (ϕ_2, ϕ_3, ϕ_4) is proportional to the pion field. The first term in the Lagrangian is the non-linear sigma term, the second term is the quartic Skyrme term. The balance between these two terms gives rise to stable soliton solutions, in this case called Skyrmions. The final term is the pion mass term, which allows for a more realistic pion, and breaks chiral symmetry. The mathematical reason for the emergence of soliton solutions is the interpretation of the field ϕ as a mapping from $\mathcal{S}^3$ to $\mathcal{S}^3$, which has a non-trivial winding number. The topological solitons are then interpreted as baryons.

The Skyrme model has been used to describe finite density zero temperature baryonic matter. Crystalline matter is observed for all densities, with a phase transition between a phase with broken chiral symmetry at low densities $\langle \phi_k \rangle \propto \delta_{k1}$ and a chirally symmetric phase at high densities, $\langle \phi_k \rangle = 0$.

The zero temperature limit of such models could well be misleading, and it is an interesting question as to what structure the model has at a finite temperature. Before solving this problem, we first look at the 2D analogue, for which the lagrangian is identical to equation (1), except that now the roman indices run from 1 to 3 and the greek indices from 0 to 2. Solutions are now stabilized by the balance between the non-linear sigma term and the mass term, and this might also make a difference in the properties of the model.

In order to study the thermodynamics we discretize the model on a spatial grid, and evaluate the partition function

$$\mathcal{Z} = \int \prod_p d^3\phi_p d^3\dot{\phi}_p \delta(\vec{\phi}_p \cdot \vec{\dot{\phi}}_p) \delta(\vec{\phi}_p^2 - 1) \exp(-\beta \mathcal{E}_p), \qquad (2)$$

where the labels p run over all the lattice points. Here $\mathcal{E}_p$ is the energy density and the two δ functions, $\delta(\vec{\phi}_p^2 - 1)$ and $\delta(\vec{\phi}_p \cdot \vec{\dot{\phi}}_p)$, impose the constraints arising from unitarity. The density of the system is determined by the initial topological charge in the system. (We use a box with periodic boundary conditions, so topological charge cannot leak out of the system). The function $\mathcal{Z}$ cannot be evaluated analytically, and we use a standard metropolis algorithm to evaluate it numerically. The relevant techniques are similar to those discussed in Ref. [4], where simulated annealing was used to find zero temperature solutions.

For low temperatures and low densities small numbers of Skyrmions bind together. For high densities and low temperatures, the Skyrmions merge into a lattice where it is not possible to identify individual Skyrmions, see Figure 1. The lattice is determined most easily from the structure of the "holes" with zero baryon density. There are two such minima present per baryon. In the low density phase, the average field $\langle \phi_k \rangle$ is polarized in the direction of the sigma field, but for high densities, the average field is zero. Therefore, a mean-field chiral symmetry exists in

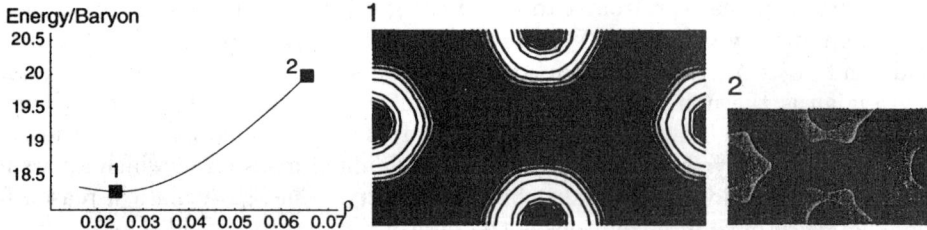

FIGURE 1. High density phase at zero temperature. *Left:* Energy per baryon against density. *Middle:* A baryon density plot of the lowest energy per baryon number attainable, i.e. the "natural" density. Lighter shades represent higher baryon density. *Right:* A baryon density plot at high density. The baryon number is more evenly distributed throughout the structure. The middle and right plot each show a unit cell, in correct proportion to each other.

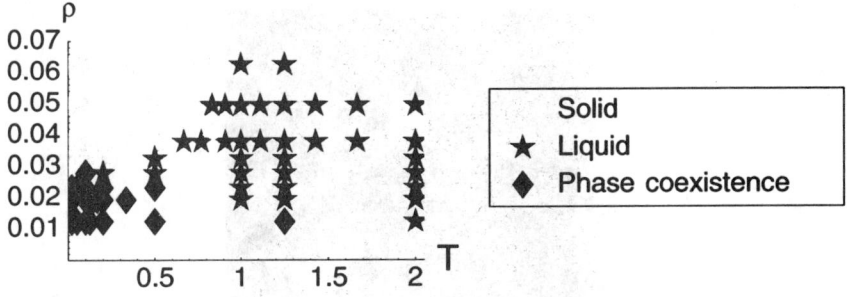

FIGURE 2. The Phase Diagram.

the high density phase and not in the low density phase. Thermodynamics systems are in equilibrium when maximizing entropy and minimizing energy. The lowest energy per baryon that can be achieved occurs in the high density phase (at zero temperature) and this is therefore the "natural" density of the system.

At finite temperatures, a solid phase, a liquid phase, and a phase coexistence region are observed (see Fig. 2). Along the lines first proposed by Klebanov [5], we analyse the grid-averaged fields $\langle \phi_k \rangle$. In the solid and fluid phases we seem to find $\langle \phi_k \rangle = 0$, which would imply that both are chirally symmetric. The phase coexistence region is similar to a crystal of nuclei. The liquid and solid phases were distinguished by their pseudotime-averaged baryon densities: periodic for a solid and zero for a fluid.

An example of a solid is given in figure 3. On the left we show a pseudotime-averaged baryon density plot and on the right a snapshot where thermal fluctua-

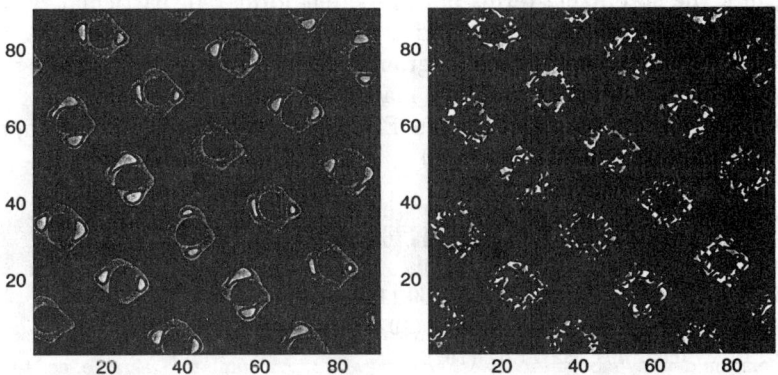

FIGURE 3. The solid region. Lighter shades represent higher baryon density. *Left:* Pseudotime averaged baryon density plot. *Right:* Snapshot of the baryon density.

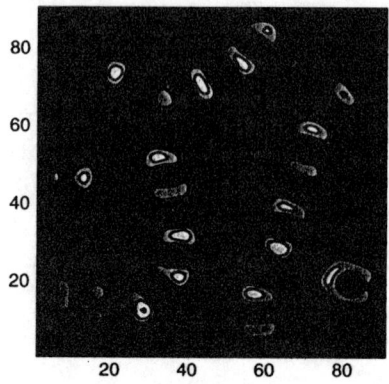

FIGURE 4. Pseudotime averaged baryon density plot in the phase coexistence region. Lighter shades represent high baryon density.

tions are evident. The approximate triangular crystal structure can be observed although defects are present, which are probably created by the periodical boundary conditions.

A pseudotime-average of a liquid shows an almost constant baryon density- because structures move. In a snapshot, the individual baryons are visible.

In the phase coexistence region, one observes series of multi-solitons that don't change with time, i.e. a solid structure with large regions of vacuum. These can either be interpreted as a crystal of nuclei, or as a percollating network. An example is shown in figure 4.

The 2D baby Skyrme model shows a rich phase diagram. A similar diagram will be created for the 3D Skyrme model, and the phase diagrams are expected to differ because the stabilizing terms have different forms. In particular, we would like to know at which temperatures and densities the phase transitions occur and compare these with the QCD phase diagram in Ref [6]. We would also like to know the natural crystal structure of Skyrme matter.

OS acknowledges support through an EPSRC studentship. NRW is supported through an EPSRC grant (GR/L22331).

REFERENCES

1. Skyrme, T.H.R., *Nucl. Phys.* **31** 556-569 (1962).
2. 't Hooft, G., *Nucl. Phys.* **B72** 461-473 (1974).
3. Witten, E., *Nucl. Phys.* **B160** 57 (1979).
4. Hale, M., Schwindt, O., and Weidig,T., hep-th/0002058 (2000), *Phys. Rev.* **E** in press.
5. Klebanov, I., *Nucl. Phys.* **B262** 133-143 (1985).
6. Rajagopal, K., *Nucl. Phys.* **A661** 150-161 (1999); these proceedings.

The Density Dependence of Charge Symmetry Breaking

C. J. Horowitz

Nuclear Theory Center, 2401 Milo B. Sampson Lane, Bloomington, Indiana 47405

J. Piekarewicz

Department of Physics, Florida State University, Tallahassee, FL 32306

Abstract.
We study the density dependence of charge symmetry breaking by calculating the energy difference as a function of density between adding one neutron or one proton to uniform nuclear matter. Isospin violation from electromagnetic interactions are expected to increase with density as the average distance between quarks decreases. In contrast, isospin violation from the up-down quark mass difference should decrease with density as the contributions from the small quark masses are suppressed by the increasing Fermi momentum. The Okamoto-Nolen-Schiffer (ONS) anomaly in the binding energy of mirror nuclei was studied at high density by adding a single neutron or a proton to a quark-gluon plasma. In the limit of very high density we find an anomaly equal to two-thirds of the Coulomb exchange energy of a proton. This effect is dominated by quark electromagnetic interactions—rather than by the up-down quark mass difference and has the same sign as that observed in normal nuclei.

There are two—apparently independent—sources of isospin violation in the Standard Model: the electromagnetic interaction between quarks and the "down-up" quark-mass difference Δm. Often these two sources make comparable contributions. This complicates the calculation of many isospin-violating or charge-symmetry-breaking (CSB) observables.

There are a number of measured isospin-breaking observables, such as the difference in the analyzing powers between the proton and the neutron in np elastic scattering [1]. There have been many calculations of these observables that include a large number of possible contributions (see for example [2]). It would be useful to have an organizational scheme that, at least in principle, would allow one to separate electromagnetic and quark mass contributions. This could make CSB a more powerful tool for elucidating the structure of strongly interacting systems.

In this paper we focus on the energy difference between a system with either one extra proton or one extra neutron added to uniform nuclear matter. This is closely related to the Okamoto-Nolen-Schiffer (ONS) anomaly—which is the long-standing discrepancy between the calculated and the measured binding-energy differences of mirror nuclei [2–4]. By studying the *density dependence* of this energy difference we are able to identify a high-density limit where CSB is dominated by electromagnetic effects rather than by the quark-mass difference. In this limit the calculation of CSB observables may simplify.

The ONS anomaly can be calculated or described on several levels. Perhaps the simplest is the observation by B. A. Brown [5] that the magnitude of the anomaly is approximately equal to the Coulomb exchange energy. If one adds an extra proton to a nucleus in a simple Hartree-Fock picture, there will be both a direct (Hartree) and an exchange (Fock) Coulomb interaction with the other protons. If one—arbitrarily—neglects the Fock term, then one obtains a better agreement with experiment. Although this observation by Brown is not a dynamical explanation, it is an interesting characterization of the size of the anomaly. Could there be something "wrong" with the exchange term? In this paper we study EM effects involving the Coulomb exchange interactions of composite nucleons.

To clarify the importance of EM and Δm terms we consider a high-density limit of the ONS anomaly. Consider uniform nuclear matter at very high density. We assume that an electron gas makes the system electrically neutral. Thus the direct Coulomb interaction vanishes. Yet Coulomb exchange effects are still present.

Now we add to this system either a single neutron or a single proton and compute the change in energy $\Delta E \equiv E(Z, N+1, \rho) - E(Z+1, N, \rho)$. Here $E(Z, N, \rho)$ denotes the energy of the system with Z protons and N neutrons at a baryon density ρ. At zero density the energy difference reduces to the mass difference ΔM between the neutron M_n and the proton M_p. In the absence of isospin breaking ($\alpha = \Delta m \equiv 0$) we expect $\Delta E(\rho)$ to vanish at all densities. Thus, if one expands $\Delta E(\rho)$ in a Taylor series about this isospin-symmetric point, one obtains—to lowest order in α and Δm—the following expression:

$$\Delta E(\rho) \approx a(\rho)\alpha + b(\rho)\Delta m . \tag{1}$$

This equation defines the two coefficients in the Taylor expansion through $a(\rho) = \partial \Delta E/\partial \alpha|_{\alpha=\Delta m=0}$, $b(\rho) = \partial \Delta E/\partial \Delta m|_{\alpha=\Delta m=0}$. Note that we assume that the average light quark mass $\overline{m} \equiv (m_u + m_d)/2 \neq 0$ is kept fixed while Δm is varied. Here m_u, m_d are *current* quark masses with $\Delta m = m_d - m_u$. The ONS anomaly is closely related to the *density dependence* of $a(\rho)$ and $b(\rho)$.

The electromagnetic coefficient $a(\rho)$ is sensitive to the Coulomb interactions among quarks. These are expected to increase as the average distance between quarks decreases. Thus $a(\rho)$ should grow with density. Of course, $a(\rho)$ is also sensitive to other electromagnetic effects. However, in the very high density limit (see below) the only momentum scale is set by the quark Fermi momentum k_F and the only length scale is $1/k_F$. Thus, in this limit $a(\rho)$ is expected to be proportional to k_F on purely dimensional grounds.

In contrast, the quark-mass term $b(\rho)$ is expected to decrease with density on very general kinematic grounds—at least in the limit of very high density. Any contribution of a small quark mass $m \ll k_F$ to the energy is expected to be of order m^2/k_F, because $(k_f^2 + m^2)^{1/2} \approx k_f + m^2/k_F$, and this is suppressed at large momentum scales k_F. Therefore $b(\rho)$ should be suppressed by $1/k_F$ at high density. One of the main results of this paper is the expectation that

$$a(\rho) \propto k_F , \quad b(\rho) \propto \overline{m}/k_F , \tag{2}$$

in the limit of very high density.

The general nature of the arguments of the last two paragraphs is important. Here we illustrate it with a simple model where the system is either a free Fermi gas of nucleons or a (nearly) free Fermi gas of quarks. We start with the former. An added proton to a free Fermi gas of elementary nucleons will have a Coulomb exchange energy equal to $V_p = -e^2 \frac{k_F}{\pi}$. Here k_F is the Fermi momentum of the system and e is its electric charge. In contrast, in this model an added neutron has zero Coulomb exchange energy: $V_n = 0$.

Next we consider a quark-gluon plasma. We assume—because of asymptotic freedom—that at very high density the system is nearly a free Fermi gas of quarks. This is because the strong coupling $\alpha_S(k_F^2)$ becomes small at the large momentum scale characterized by k_F. When a proton is added to such a system we assume that it will dissociate into two up and one down quark. Therefore, the Coulomb exchange energy of these three quarks is

$$V_p^{(q)} = -\left(\sum_{i=1}^{3} e_i^2\right) \frac{k_F}{\pi} , \tag{3}$$

where e_i denotes the quark electric charge and k_F is the quark Fermi momentum. The sum of the squares of the valence charges in a proton is $(4/9 + 4/9 + 1/9)e^2 = e^2$. Because of this "numerical accident" the quark Coulomb exchange energy is equal to the Coulomb exchange energy of an elementary proton.

An interesting difference arises when we add a neutron. In a quark-gluon plasma the Coulomb exchange energy is no longer zero because a neutron is made up of charged constituents. The sum of the squares of the valence quark charges in a neutron is $(4/9 + 1/9 + 1/9)e^2 = 2e^2/3$. Thus, the neutron Coulomb energy is fully two thirds of that of a proton: $V_n^{(q)} = -\frac{2}{3}e^2 k_F/\pi$. This "unexpected" neutron exchange energy produces an ONS anomaly.

We should also consider the contribution from the up-down quark mass difference. However, in the high-density limit, all contributions from Δm are suppressed by the large Fermi momentum. For example, the difference in the Fermi energy of free down and up quarks is: $\sqrt{k_F^2 + m_d^2} - \sqrt{k_F^2 + m_u^2} \approx (m_d^2 - m_u^2)/2k_F$.

Our results may only be valid in the limit of very high density. Unfortunately, model calculations are needed to determine quantitatively how high a density is required. There have been a variety of model-calculations of the ONS anomaly [6–8] which find significant contributions at normal nuclear densities. We too have performed a model-calculation of the Coulomb energy as a function of density for pure neutron matter in a string-flip quark model [9]. In this model the wave function is explicitly antisymmetric even for the exchange of quarks from different nucleons. The Coulomb energy of neutron matter arises in the model because of the electric charge of the neutron constituents. We find a significant increase in this energy at a density of 0.08 neutrons per fm^{-3} [10].

In this paper we suggest that isospin-violating contributions to a variety of processes may be classified, at least in principle, according to their density dependence. In the high-density limit the contribution from the quark-mass difference Δm should be suppressed relative to electromagnetic effects. In future work one should study the Coulomb energy of neutron matter, characterize the density dependence of $a(\rho)$ and $b(\rho)$ at low density, and study the approach to the high-density limit.

A high-density limit of the Okamoto-Nolen-Schiffer anomaly was studied by adding a single neutron or a proton to a quark-gluon plasma. In this limit we find: (1) that there is an ONS anomaly and (2) that it is dominated by EM effects between quarks with Δm being unimportant. Further, (3) the magnitude of the anomaly is simply related to the proton Coulomb exchange energy. The anomaly arises in part from an attractive Coulomb exchange energy for an added neutron because of the electric charge of its quark constituents.

In conclusion, we have studied the density dependence of charge symmetry breaking by computing the difference in energy between adding a single neutron or a single proton to uniform nuclear matter. On very general grounds we expect that isospin violation from electromagnetic interactions should increase with density as the average distance between quarks decreases. In contrast, the contribution to isospin violation from the quark-mass difference should decrease with density because of simple kinematical considerations.

We thank Susan Gardner for useful conversations. This work was supported in part by DOE grants DE-FG02-87ER40365 and DE-FG05-92ER40750.

REFERENCES

1. L.D. Knutson et al., Nucl. Phys. **A508**, 185c (1990); Phys. Rev. Lett. **66**, 1410 (1991); S.E. Vigdor et al., Phys. Rev. C **46**, 410 (1992).
2. G. A. Miller, B. M. K. Nefkens and I. Slaus, Phys. Rep. **194** (1990) 1.
3. K. Okamoto, Phys. Lett. **11** (1964) 150.
4. J. A. Nolen and J. P. Schiffer, Ann. Rev. Nucl. Sci. **19** (1969) 471.
5. B. Alex Brown, Phys. Rev. **C58** (1998) 220.
6. Thomas Schafer, Volker Koch and Gerald E. Brown, Nuc. Phys. **A562** (1993) 644.
7. S. Nakamura, et al., Phys. Rev. Lett. **76** (1996) 881.
8. K. Tsushima, K. Saito and A. W. Thomas, Phys. Lett. **B465** (1999) 36.
9. F. Lenz et al., Ann. of Phys. (N.Y.) **170** (1986) 65; C. J. Horowitz and J. Piekarewicz, Nuc. Phys. **A536** (1992) 669.
10. C. J. Horowitz and J. Piekarewicz, to be published.

Elastic Form Factors and Charge Densities of Medium and Heavy Nuclei and the Proton Occupancies of Shell States

Ilia S. Gulkarov* and Bishan P. Nigam†

Mathematics/Computer Science Department, Paradise Valley Community College, Phoenix, AZ 85032, USA

†*Department of Physics and Astronomy, Arizona State University, Tempe, AZ 85287 USA*

Abstract. The charge density distributions (CDD) and form factors of the medium and heavy nuclei were calculated using the radial wave functions of the harmonic oscillator assuming that the occupation numbers of shell states different from 0 or 1. The occupation numbers of shell states were found from experimental data on electron scattering. The modified shell model gives a good description of the model independent CDD of nuclei.

The aim of this investigation is the systematic study of the proton occupancies of the shell states of nuclei. Information from the charge densities of nuclei may be translated into information about single-particle occupancies. Such analysis allows us to arrive at a more definite conclusion regarding the occupation numbers and the shell structure of the atomic nuclei.

The charge densities of nuclei from ^{12}C to ^{208}Pb [1-4] are calculated on the basis of a modified shell model with fractional numbers of the shell states. The proton occupancies of these nuclei were determined by comparison with the experimental charge densities and were found to be different from 0 or 1.

The radial distribution of the protons in the nucleus my be obtained from

$$\rho(r) = \sum_{n\ell j} (2j+1) \, R_{n\ell j}(r) / 4\pi,$$

where $R_{n\ell j}(r)$ is the radial single-particle oscillator wave functions. The shell of all nuclei we divided into three groups: *core, eroding and gaining shells*. Some of the protons redistribute from eroding shells into gaining shells. At the same time the number of protons in the core shells are constant. The result of calculation of charge density of ^{150}Nd is shown in figure 1.

The charge densities were calculated with n_{2s} is constant and various values of 2p (fig. a) and 2p is constant and various values of 2s (fig.b). It can be seen that the CDD changes considerably upon introducing fractional occupation numbers for the 2s and 2p shells and good description of ρ(r) in the range 0< r <4 fm can be obtained.

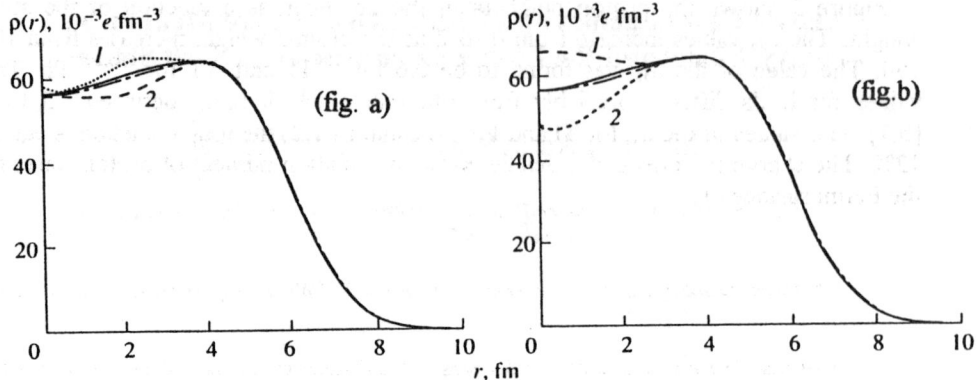

Figure 1. Charge densities in ^{150}Nd.

We also calculated the differences of the CDD for the ^{16,18}O, ^{32,34}S, ^{40,48}Ca, ^{142,146,150}Nd, and ^{206,208}Pb isotopes. Addition of neutrous to the nucleus changes the distribution of protons in the shell: the charge is distributed from the interior of the nucleus to its surface. In all probability, this is because a part of charge is transferred to the added neutrons, so that the neutron CDD increases in nuclear matter [5].

The difference in charge densities of the ^{206,208}Pb are shown in figure 2.

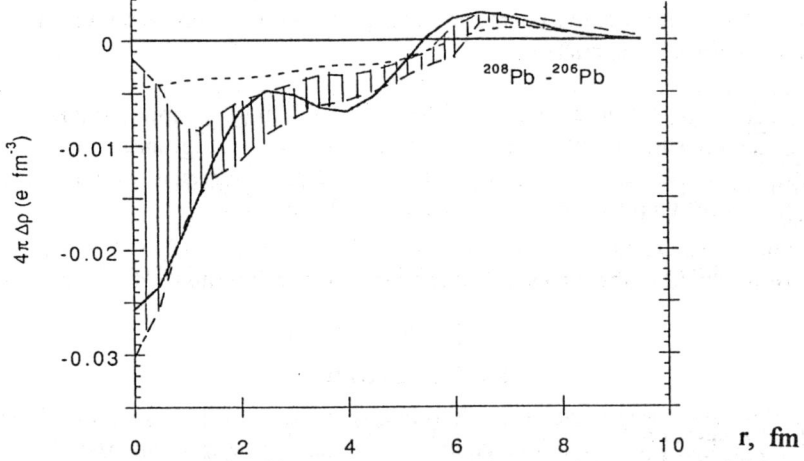

Figure 2. The difference of the charge densities of nuclei ^{208}Pb – ^{206}Pb.

The addition of two neutrons to the ^{206}Pb has essentially changed the distribution of protons in shells: the charge distributes from the central part of the nucleus to its surface. The dashed curve was calculated in the simple shell model, the solid curve in the modified shell model and the hatched curves are the experimental data. The charge which is moved from the inner region to the outer region of the ^{208}Pb by addition of the two neutrons to ^{206}Pb correspond to 0.24 e of one proton charge.

Figure 3 shows the proton numbers in the 2s shells as a function of the atomic weight. The n_{2s} values increase from 0 to 2 as the atomic weight increases from 10 to 150. The value of the n_{3s} was found to be 0.6 for ^{206}Tl, and ≈ 1 for ^{206,208}Pb. These values for leads differ appreciably from other data which range between 1.3 to 1.6 [6.7]. The values of the n_{2s} for Tl and Pb are equal to 1.2, the total depletion is equal to 12%. The charge radii depend crucially on the occupation number of proton states near the Fermi surface [8].

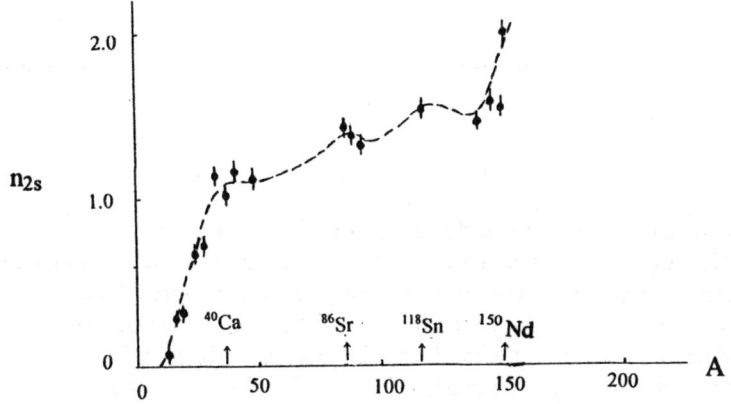

Figure 3. The quantities n2s versus atomic weight.

The present work provides a clear argument that nucleon-nucleon correlations are important for the correct description of the charge densities and form factors of nuclei. The modified shell model allows us to find the occupation numbers of shell states from experimental data on electron scattering. A systematic investigation of the occupation numbers of the entire region of nuclei allow us to arrive at a more definite conclusion regarding these numbers. The shell model is still interesting and the analysis on the basis of modified model gives new information regarding the shell structure of nuclei.

REFERENCES

1. Gulkarov, I.S, *Sov. J. Nucl. Phys.* **51**, 97-102 (1990)
2. Gulkarov, I.S. and Nigam, B.P., *Phys. Review* **C52**, 663-668 (1996)
3. Gulkarov, I.S. and Nigam, B.P., *Physics of Atomic Nuclei* **60**, 890-896 (1997)
4. Gulkarov, I.S. and Nigam, B.P., *Intern. J. of Modern Phys.* **E7**, 367-377 (1998)
5. Bunatyan, G.G., *Sov. J. Nucl. Phys.* **48**, 820-326 (1988)
6. Hasse, R.W., Friman B.L, Berdichevsky D., *Phys. Letters* **B181**, 5-12 (1986)
7. Quint E.N.M. et al, *Phys. Rev. Lett.* **57**, 186-189 (1986); **58**, 1088-109 (1987)
8. Bhattacharya, R., *Zeit. Phys.* **A351**, 137-144 (1995)

Lepton-Hadron and Hadron-Hadron Scattering

Structure Functions at Very High Q^2 From HERA

Christopher M. Cormack*

For the H1 and ZEUS Collaborations

Rutherford Appleton Laboratory, Chilton, Didcot, Oxford, OX11 0QX, United Kingdom

Abstract. Measurements of Deep-Inelastic Neutral and Charged current interactions are presented in lepton proton scattering at HERA. The measurements are obtained from taken during 1996 to 1999 and consists of 30 pb^{-1} of e^+p and 16 pb^{-1} of e^-p data. The addition of the new high statistics electron data with the positron data allows the first extraction of the parity violating structure function xF_3 and tests of high-Q^2 electroweak effects of the heavy bosons Z^0 and W are observed and found to be consistent with the Standard Model expectation.

INTRODUCTION

Deep–Inelastic scattering (DIS) of leptons off nucleons has played a fundamental role in understanding the structure of matter. HERA the first electron-proton (ep) collider ever built is capable of investigating the two main contributions to DIS, Neutral Current (NC) interactions, $ep \to eX$ and Charged Current (CC) interactions, $ep \to \nu X$. In the Standard Model a photon (γ) or a Z^0 boson is exchanged in a NC interaction, and a $W^\pm$ boson is exchanged in a CC interaction. DIS can be described by the four-momentum transfer squared Q^2, Bjorken-x and inelasticity y defined as:

$$Q^2 = -q^2 \equiv (k - k')^2 \tag{1}$$

$$x = \frac{Q^2}{2p \cdot q} \tag{2}$$

$$y = \frac{p \cdot q}{p \cdot k} \tag{3}$$

with $k(k')$ and p being the four-momentum of the incident (scattered) lepton and proton respectively. The centre-of-mass energy $\sqrt{s}$ of the ep interaction is given by $s \equiv (p + k)^2 = Q^2/xy$.

In this paper measurements are presented of the NC and CC cross-sections at high $Q^2 > 200\,\text{GeV}^2$ for both electron and positron scattering from both the H1 and ZEUS experiments. The integrated luminosity in the e^+p data sample is $35.6\,\text{pb}^{-1}$ ($30\,\text{pb}^{-1}$) and in the e^-p data sample is $16\,\text{pb}^{-1}$ ($16\,\text{pb}^{-1}$) for the H1 (ZEUS) collaborations.

I EXPERIMENTAL SETUP

The H1 and ZEUS detectors are described elsewhere [1], [2]. Both detectors are nearly hermetic multi-purpose apparatus built to investigate ep interactions. The measurements in both experiments rely primarily on the calorimetry, tracking and the luminosity detectors.

In both experiments the Charged Current event selection is based on the observation of large P_T^{miss}, which is assumed to be the transverse momentum carried by the outgoing neutrino. For NC events it is based on the identification of a scattered electron (positron), further fiducial (NC) and kinematic cuts (CC and NC) are then applied. Details of the selection procedure can be found in [2,3].

II CROSS SECTIONS

The electroweak Born-level NC DIS cross section $d^2\sigma^{NC}/dxdQ^2$ for the reaction $e^\pm p \to e^\pm X$ can be written as:

$$\frac{d^2\sigma^{e^\pm p}_{Born}}{dxdQ^2} = \frac{2\pi\alpha^2}{xQ^4}\left[Y_+ F_2(x,Q^2) \mp Y_- xF_3(x,Q^2) - y^2 F_L(x,Q^2)\right] \quad (4)$$

For unpolarised beams, the structure functions F_2 and xF_3 can be decomposed taking into account Z^0 exchange as:

$$F_2 \equiv F_2^{em} - v\frac{\kappa_w Q^2}{(Q^2+M_Z^2)}F_2^{\gamma Z} + (v^2+a^2)\left(\frac{\kappa_w Q^2}{Q^2+M_Z^2}\right)F_2^Z \quad (5)$$

$$xF_3 \equiv -a\frac{\kappa_w Q^2}{(Q^2+M_Z^2)}xF_3^{\gamma Z} + (2va)\left(\frac{\kappa_w Q^2}{Q^2+M_Z^2}\right)^2 xF_3^Z \quad (6)$$

where M_Z is the mass of the Z^0, $\kappa_w = 1/(4\sin^2\theta_w \cos^2\theta_w)$ is a function of the Weinberg angle (θ_w) and v and a are the vector and axial couplings of the electron to the Z^0. For unpolarised beams, F_2 is the same for electron and positron scattering, while xF_3 changes sign.

The NC "reduced cross-section" $\tilde\sigma$ is defined from the measured as:

$$\tilde\sigma_{NC}(x,Q^2) \equiv \frac{1}{Y_+}\frac{xQ^4}{2\pi\alpha^2}\frac{d^2\sigma^{NC}}{dxdQ^2} \quad (7)$$

The Born double differential CC cross-section for $e^+p \to \bar\nu X$ can be written in leading order QCD as:

$$\left(\frac{d^2\sigma_{CC}^{e^+p}}{dxdQ^2}\right)_{Born} = \frac{G_F^2}{2\pi}\left(\frac{M_W^2}{M_W^2 + Q^2}\right)^2 [(\bar{u}+\bar{c}) + (1-y)^2(d+s)] \qquad (8)$$

for e^-p scattering the coupling is predominantly to the u type quarks for e^+p scattering the electroweak coupling is predominantly to the d type quarks.

$$\left(\frac{d^2\sigma_{CC}^{e^-p}}{dxdQ^2}\right)_{Born} = \frac{G_F^2}{2\pi}\left(\frac{M_W^2}{M_W^2 + Q^2}\right)^2 [u+c + (1-y)^2(\bar{d}+\bar{s})] \qquad (9)$$

The double differential CC "reduced" cross-section $\tilde{\sigma}_{CC}$ is defined as:

$$\tilde{\sigma}_{CC} = \frac{2\pi x}{G_F^2}\left(\frac{M_W^2 + Q^2}{M_W^2}\right)^2 \frac{d^2\sigma_{CC}}{dxdQ^2} \qquad (10)$$

III RESULTS AND INTERPRETATION

The NC and CC single differential cross-sections and $d\sigma_{NC}/dQ^2$, $d\sigma_{CC}/dQ^2$ are shown in fig 1, also shown are the Standard Model expectations given from NLO QCD fits to the data by both the CTEQ [5] and the H1 [7] collaborations. The Standard Model is seen to give a good description of the data.

The measurement of he NC cross-section spans more than two orders of magnitude in Q^2 and the cross-section falls with Q^2 by about 6 orders of magnitude for both the e^+p and the e^-p cross section. For Q^2 values above $1000\,\text{GeV}^2$ the e^-p cross section can be seen to be significantly above the e^+p cross section. This is consistent with the Standard Model picture in which one expects constructive interference of the photon and Z^0 for e^-p interactions and destructive interference for e^+p interactions.

Due to the propagator mass term and the different coupling the CC cross-section is smaller than the NC cross-section and it falls less steeply, by about 3 orders of magnitude, between $Q^2 = 300$ and $15\,000\,\text{GeV}^2$ for both the e^-p and e^+p cross section. The shape and magnitude of the NC and CC cross-sections are well described by the Standard Model expectation. The e^-p cross section is at relatively low Q^2 to be approximately twice as big as the e^+p cross section, which is consistent with the Standard Model picture in which the u quark density probed is expected to be approximately twice the d quark density.

Cross Sections Shown in fig. 1 is the single differential CC cross section $d\sigma_{CC}/dx$ for $Q^2 > 200\,\text{GeV}^2$ for e^+p scattering from the ZEUS collaboration [10]. The ratio of the measured cross section $d\sigma_{CC}/dx$. At high x the e^+p CC cross-section depends mainly on the d quark density which is less constrained than the u quark density. All data agree well with the Standard Model expectation using PDFs extracted from CTEQ4D and those from the ZEUS NLO fit [11]. The possibility of a larger d/u ratio than previously assumed has been of interest in recent years, for example

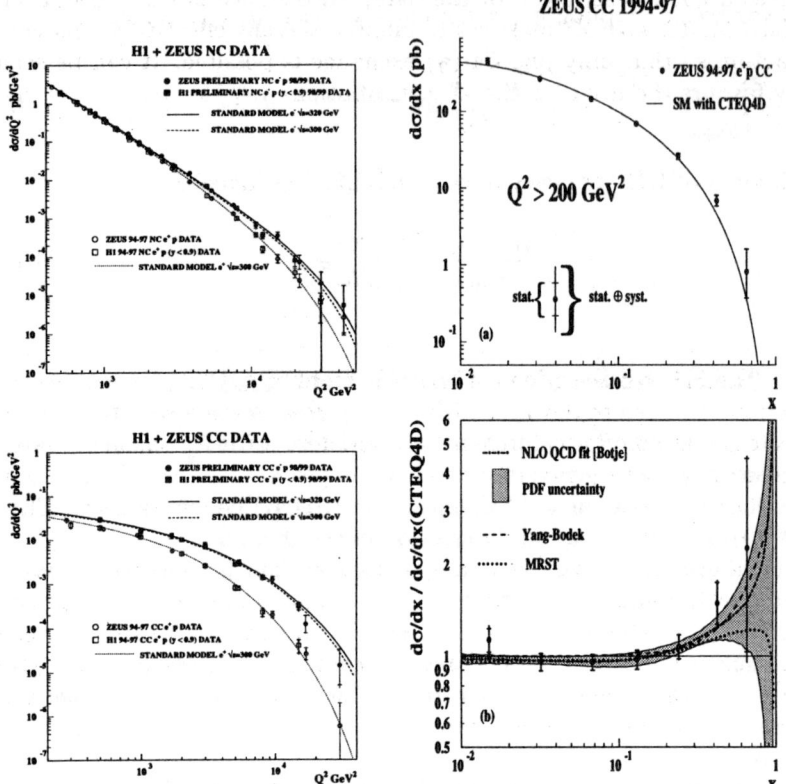

FIGURE 1. a) The high-Q^2 $e^\pm$ NC cross section $d\sigma_{e^\pm p}/dQ^2$ from the ZEUS and H1 collaborations along with the Standard Model predictions using the CTEQ4D parton momentum distributions. b) The high-Q^2 CC cross section $d\sigma_{e^\pm p}/dQ^2$ from the ZEUS and H1 collaborations along with the Standard Model predictions using the CTEQ4D parton momentum distributions. c) The ZEUS CC high-Q^2 structure function $d\sigma/dx$ with the Standard Model prediction using CTEQ4D. d) The ratio of the measured ZEUS CC high-Q^2 structure function $d\sigma/dx$ with the Standard Model prediction using CTEQ4D. Also shown are the predictions from Yang-Bodek and the ZEUS NLO QCD fit.

see [12,13]. Modification [12] of PDFs with an additional term $\delta(d/u)$ yields $d\sigma/dx$ close to the NLO QCD fit.

The NC cross-section $d\sigma_{NC}/dx$ for $Q^2 > 10\,000\,\text{GeV}^2$ is shown in Fig. 2 along with the expectations from the Standard Model with a Z^0 mass of set to the PDG value, which gives a good description of the data. Also shown is the expectation from the Standard Model with Z^0 mass set to infinity, thereby effectively removing the weak interaction so that only photon (γ) exchange is possible. It can be seen the data clearly favours the need for the Z^0 contribution.

IV DOUBLE DIFFERENTIAL NC AND CC CROSS-SECTIONS

The double differential reduced NC cross-section is shown in Fig. 2 for fixed x as a function of Q^2 for both e^+ and e^- data. Also shown are the expectations of the Standard Model which give a good description of the data. For Q^2 values below $1\,000\,\text{GeV}^2$ the e^+ and e^- cross-sections are within errors equal, for higher values of Q^2 the cross sections are seen to deviate, with the e^-p cross-section seen to increase above the e^+p cross section. This difference is consistent with the change in sign of the parity violating structure function xF_3.

The ZEUS collaboration go a step further and combine the e^+ and e^- cross sections, where $30\,\text{pb}^{-1}$ of e^+p data collected at a centre of mass energy of $\sqrt{s} = 300\,\text{GeV}$ were combined with $16\,\text{pb}^{-1}$ of e^+p data collected at a centre of mass energy of $\sqrt{s} = 300\,\text{GeV}$. To reduce statistical fluctuations several bins from the double differential binning [9]. Figure 2 shows xF_3 at fixed values of Q^2 as a function of x. The Standard Model expectation evaluated with either the CTEQ4D or MRST(99) parton density functions are seen to give a good description of the data.

The reduced cross sections as functions of x and Q^2 are displayed in Figs. 2 along with the prediction of the H1 NLO QCD fit [7]. The relative increase in the e^-p cross section over the e^+p cross section is consistent with a larger u quark density relative to that of the d quark.

The absolute magnitude of the CC cross section, described by equation 8 is determined by the Fermi constant G_F and the PDFs, while the Q^2 dependance of the CC cross-section includes the propagator term $[M_W^2/(M_W^2 + Q^2)]$ which produces substantial damping of the cross section. The ZEUS collabiration have fitted the measured differential cross section, $d\sigma/dQ^2$, treating G_F and M_W as free parameters, yields

$$G_F = (1.171 \pm 0.034((\text{stat.})^{+0.026}_{-0.015}(\text{syst.})^{+0.016}_{-0.015}(\text{PDF})) \times 10^{-5}\,\text{GeV}^{-2} \quad (11)$$

and

$$M_W = 80.8^{+4.9}_{-4.5}(stat.)^{+5.0}_{-4.3}(syst.)^{+1.4}_{-1.3}(PDF)\,\text{GeV} \quad (12)$$

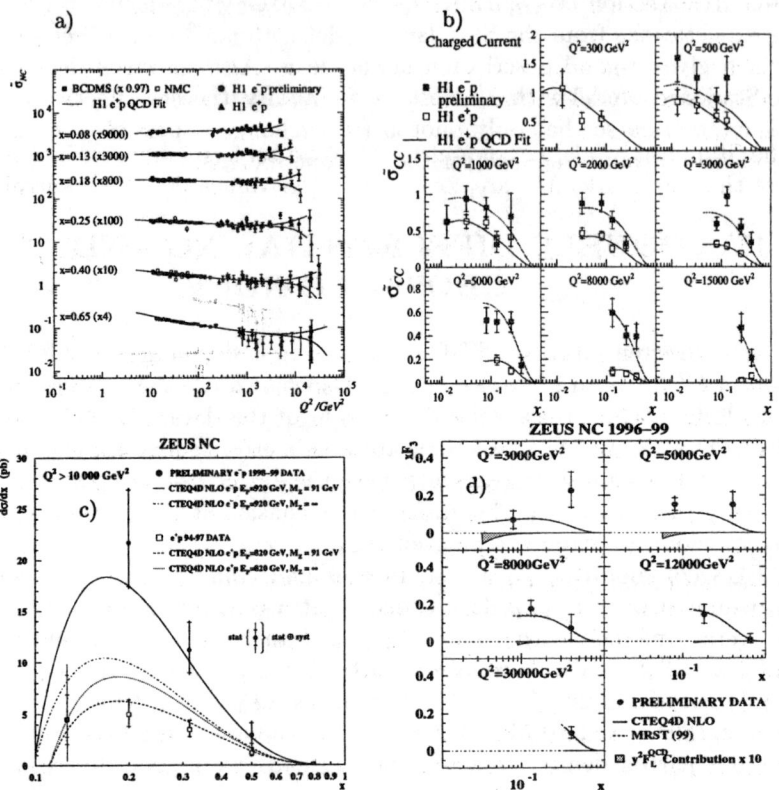

FIGURE 2. a) The high-Q^2 NC e^-p (circles) e^+p (triangles) reduced cross sections $d^2\sigma_{e^{\pm}p}/dxdQ^2$ from the H1 collaboration along with the Standard Model predictions from the H1 NLO QCD fit. b) The high-Q^2 CC reduced cross section $\tilde{\sigma}_{CC}$ from the ZEUS and H1 collaborations along with the Standard Model predictions using the CTEQ4D parton momentum distributions. c) The high-Q^2 $e^{\pm}p$ cross section $d\sigma_{e^{\pm}p}/dx$ and the Standard Model predictions using the CTEQ4D PDFs for $Q^2 > 10\,000\,\text{GeV}^2$. Also shown are the theoretical predictions without the weak interaction were absent ($M_Z = \inf$). d) The Structure function xF_3, extracted by the ZEUS collaboration. the data were obtained from the 1998/99 e^-p data set recorded at $\sqrt{s} = 318\,\text{GeV}$ and the e^+p data set taken $\sqrt{s} = 300\,\text{GeV}$, plotted at fixed Q^2 between $3\,000\,\text{GeV}^2$ and $30\,000\,\text{GeV}^2$ as a function of x together with the Standard Model predictions using the CTEQ4D and MRST(99) PDFs [6].

The central values are obtained using the CTEQ4D PDFs. Further fits to the data were also performed [10], the 'propagator-mass' fit to the measured differential cross-section, $d\sigma/dQ^2$, with G_F fixed to the value $G_F = 1.16639 \times 10^{-5}$ GeV^{-2} yields the result

$$M_W = 81.4^{+2.7}_{-2.6}(stat.) \pm 2.0(syst.)^{+3.3}_{-3.0}(PDF) \text{ GeV} \qquad (13)$$

Using the Standard Model constraint α, M_Z, and all fermion masses, other than the mass of the top quark, M_t, are set to the PDG values [4]. The central result of the fit was obtained with $M_t = 175$ GeV and the mass of the Higgs boson $M_H = 100$ GeV. The χ^2 function is evaluated along the line given by the SM constraint, gives the following result.

$$M_W = 80.50^{+0.24}_{-0.25}(stat.)^{+0.13}_{-0.16}(syst.) \pm 0.31(PDF)^{+0.03}_{-0.06}(\Delta M_t, \Delta M_H, \Delta M_Z) \text{ GeV} \qquad (14)$$

V HERA UPGRADE

From September 2000, the luminosity of the HERA collider will be increased by a factor of five. At the same time longitudinal lepton beam polarisation will be provided for the collider experiments H1 and ZEUS. Over a six year running period it is anticipated that the total luminosity of 1 000 pb^{-1} will be delivered. This large data volume will allow F_2^{NC} to be extracted with an accuracy of 3% over the kinematic range $2 \times 10^{-5} < x < 0.7$ and $2 \times 10^{-5} < Q^2 < 5 \times 10^4$ GeV^2. With this accuracy it is anticipated that it will be possible to determine α_S from the scaling violations of F_2^{NC} with a precision of < 0.003. The gluon distribution will also be determined from such a fit with a precision of 3% for $x = 10^{-4}$ and $Q^2 = 20$ GeV^2.

There will also be a significant benefit to the CC cross-section measurement, with the potential of a clean determination of the u and d quark densities.

With the introduction of polarised beams it will be possible to obtain a measurement of the vector and axial-vector couplings of the u-quark, ν_u and a_u respectively, obtained in a fit in which ν_u and a_u are allowed to vary. With a luminosity of 250 pb^{-1} per charge, polarisation combination and taking the vector and axial-vector couplings of the u and d-quarks as free parameters gives a precision of 13 %, 6 %, 17 % for ν_u, a_u, ν_d and a_d respectively. By comparing the NC couplings of the c and b-quarks obtained at LEP, it will be possible to test the universality of the NC couplings.

VI SUMMARY AND OUTLOOK

The latest Deep-Inelastic Neutral and Charge current cross-sections have been presented from the H1 and ZEUS experiments. The Standard Model is seen to give a good description of the data in all cases. The high luminosity data at high Q^2 from both experiments has enabled the first tests of electroweak effects in both Neutral and Charged current interactions. In NC interactions the data are seen to be consistent with the effect of γZ° interference and has allowed the first extraction of the parity violating structure function xF_3 at high Q^2.

With the factor of five increase in yearly luminosity expected from the HERA upgrade further high precision tests of the Strong and Weak Interaction will be made. The future CC measurements will allow precision determinations of the u and d quark densities and a determination of the vector and axial-vector couplings of the quarks from the NC interaction, providing important information for future high energy hadron colliders such as the LHC.

REFERENCES

1. The H1 Detector H1 Collab., I. Abt et al. Nucl. Instr. Meth.
2. The ZEUS Detector, Status Report 1993, DESY (1993). **A336** (1993) 460.
3. H1 Collab., C. Adloff et al., Eur Phys J **C 13** (2000) 609-639
4. Particle Data Group, C. Caso et al., Eur. Phys. J. **C3** (1998) 1.
5. H.L. Lai et al., Phys. Rev. **D55** (1997) 1280.
6. A.D. Martin, R.G. Roberts, W.J. Stirling, and R.S. Thorne, Eur. Phys. J. **C4** (1998) 463.
7. H1 Collab., S. Aid et al., Nucl. Phys. **B470** (1996) 3.
8. ZEUS Collab., J. Breitweg et al. Euro. Phys **C 11** (1999) 3, 427-445
9. ZEUS Collab., Abstract 1049 *XXX International Conference on High-Energy Physics, Osaka,* July 2000
10. ZEUS Collab., J. Breitweg et al. Euro. Phys **C 12** (2000) 1, 53-68
11. M. Botje, *A QCD analysis of HERA and fixed target structure function data,* DESY 99-038, NIKHEF-99-011 (in preparation).
12. U.K. Yang and A. Bodek, *Phys. Rev. Lett.* **82** (1999) 2467.
13. W. Melnitchouk and A.W. Thomas, *Phys. Lett.* **B377** (1996) 11; W. Melnitchouk and J.C. Peng, *Phys. Lett.* **B400** (1997) 220.

New Measurements of Nucleon Structure Functions from the CCFR/NuTeV Collaboration

A. Bodek,[7] U. K. Yang,[7] T. Adams, [4] A. Alton,[4] C. G. Arroyo,[2]
S. Avvakumov,[7] L. de Barbaro,[5] P. de Barbaro, [7] A. O. Bazarko,[2]
R. H. Bernstein,[3] T. Bolton, [4] J. Brau,[6] D. Buchholz,[5] H. Budd,[7]
L. Bugel,[3] J. Conrad,[2] R. B. Drucker,[6] B. T. Fleming,[2]
J. A. Formaggio,[2] R. Frey,[6] J. Goldman,[4] M. Goncharov,[4]
D. A. Harris,[7] R. A. Johnson,[1] J. H. Kim,[2] B. J. King,[2] T. Kinnel,[8]
S. Koutsoliotas,[2] M. J. Lamm,[3] W. Marsh,[3] D. Mason,[6]
K. S. McFarland, [7] C. McNulty,[2] S. R. Mishra,[2] D. Naples,[4]
P. Nienaber,[3] A. Romosan,[2] W. K. Sakumoto,[7] H. Schellman,[5]
F. J. Sciulli,[2] W. G. Seligman,[2] M. H. Shaevitz,[2] W. H. Smith,[8]
P. Spentzouris,[2] E. G. Stern,[2] M. Vakili,[1] A. Vaitaitis,[2] V. Wu,[1]
J. Yu,[3] G. P. Zeller,[5] and E. D. Zimmerman[2]

(The CCFR/NuTeV Collaboration)

[1] *Univ. of Cincinnati, Cincinnati, OH 45221;* [2] *Columbia University, New York, NY 10027*
[3] *Fermilab, Batavia, IL 60510* [4] *Kansas State University, Manhattan, KS 66506*
[5] *Northwestern University, Evanston, IL 60208;* [6] *Univ. of Oregon, Eugene, OR 97403*
[7] *Univ. of Rochester, Rochester, NY 14627;* [8] *Univ. of Wisconsin, Madison, WI 53706*
Presented by Arie Bodek

Abstract. We report on the extraction of the structure functions F_2 and $\Delta x F_3 = x F_3^{\nu} - x F_3^{\bar{\nu}}$ from CCFR ν_μ-Fe and $\bar{\nu}_\mu$-Fe differential cross sections. The extraction is performed in a physics model independent (PMI) way. This first measurement for $\Delta x F_3$, which is useful in testing models of heavy charm production, is higher than current theoretical predictions. The F_2 (PMI) values measured in ν_μ and μ scattering are in good agreement with the predictions of Next to Leading Order PDFs (using massive charm production schemes), thus resolving the long-standing discrepancy between the two sets of data.

Deep inelastic lepton-nucleon scattering experiments have been used to determine the quark distributions in the nucleon. However, the quark distributions determined from muon and neutrino experiments were found to be different at small values of x, because of a disagreement in the extracted structure functions. Here, we report on a measurement of differential cross sections and structure functions from CCFR ν_μ-Fe and $\bar{\nu}_\mu$-Fe data. We find that the neutrino-muon difference is resolved by extracting the ν_μ structure functions in a physics model independent way.

The sum of ν_μ and $\bar{\nu}_\mu$ differential cross sections for charged current interactions on an isoscalar target is related to the structure functions as follows:

$$F(\epsilon) \equiv \left[\frac{d^2\sigma^\nu}{dxdy} + \frac{d^2\sigma^{\bar{\nu}}}{dxdy}\right] \frac{(1-\epsilon)\pi}{y^2 G_F^2 M E_\nu} = 2xF_1[1+\epsilon R] + \frac{y(1-y/2)}{1+(1-y)^2}\Delta x F_3.$$

Here G_F is the Fermi weak coupling constant, M is the nucleon mass, E_ν is the incident energy, the scaling variable $y = E_h/E_\nu$ is the fractional energy transferred to the hadronic vertex, E_h is the final state hadronic energy, and $\epsilon \simeq 2(1-y)/(1+(1-y)^2)$ is the polarization of the virtual W boson. The structure function $2xF_1$ is expressed in terms of F_2 by $2xF_1(x,Q^2) = F_2(x,Q^2) \times \frac{1+4M^2x^2/Q^2}{1+R(x,Q^2)}$, where Q^2 is the square of the four-momentum transfer to the nucleon, $x = Q^2/2ME_h$ (the Bjorken scaling variable) is the fractional momentum carried by the struck quark, and $R = \frac{\sigma_L}{\sigma_T}$ is the ratio of the cross-sections of longitudinally- to transversely-polarized W-bosons. The $\Delta x F_3$ term, which in leading order $\simeq 4x(s-c)$, is not present in the μ-scattering case. In addition, in a ν_μ charged current interaction with s (or $\bar{c}$) quarks, there is a threshold suppression originating from the production of heavy c quarks in the final state. For μ-scattering, there is no suppression for scattering from s quarks, but more suppression when scattering from c quarks since there are two heavy quarks (c and $\bar{c}$) in the final state.

In previous analyses of ν_μ data, light-flavor universal physics model dependent (PMD) structure functions were extracted by applying a slow rescaling correction to correct for the charm mass suppression in the final state. In addition, the $\Delta x F_3$ term (used as input in the extraction) was calculated from a leading order charm production model. These resulted in a physics model dependent (PMD) structure functions. In the new analysis reported here, slow rescaling corrections are not applied, and $\Delta x F_3$ and F_2 are extracted from two parameter fits to the data. We compare the values of $\Delta x F_3$ to various charm production models. The extracted physics model independent (PMI) values for F_2^ν are then compared with F_2^μ within the framework of NLO models for massive charm production.

The CCFR experiment collected data using the Fermilab Tevatron Quad-Triplet wide-band ν_μ and $\bar{\nu}_\mu$ beam. The raw differential cross sections per nucleon on iron are determined in bins of x, y, and E_ν ($0.01 < x < 0.65$, $0.05 < y < 0.95$, and $30 < E_\nu < 360$. GeV). Figure 1 (a) shows typical differential cross sections at $E_\nu = 150$ GeV. Next, the raw cross sections are corrected for electroweak radiative effects, the W boson propagator, and for the 5.67% non-isoscalar excess of neutrons over protons in iron (only important at high x). Values of $\Delta x F_3$ and F_2 are extracted from the sums of the corrected ν_μ-Fe and $\bar{\nu}_\mu$-Fe differential cross sections at different energy bins according to Eq. (1). It is challenging to fit $\Delta x F_3$, R, and $2xF_1$ using the y distribution at a given x and Q^2 because of the strong correlation between the $\Delta x F_3$ and R terms, unless the full range of y is covered by the data. Covering this range (especially the high y region) is hard because of the low acceptance. Therefore, we restrict the analysis to two parameter fits. Our strategy is to fit $\Delta x F_3$ and $2xF_1$ (or equivalently F_2) for $x < 0.1$ where the $\Delta x F_3$ contribution is relatively large, while constraining R using the $R_{world}^{\mu/e}$ QCD inspired empirical fit to all available R from electron- and μ-scattering data. The $R_{world}^{\mu/e}$ fit is also in good agreement with NMC R^μ data at low x, and with the most recent NNLO QCD calculations (including target mass effects) of R by Bodek and Yang

For $x < 0.1$, R in neutrino scattering is expected to be somewhat larger than R for muon scattering because of the production of massive charm quarks in the final state. A correction for this difference is applied to $R_{world}^{\mu/e}$ using a leading order slow rescaling model to obtain an effective R for neutrino scattering, R_{eff}^ν. The difference between $R_{world}^{\mu/e}$ and R_{eff}^ν is used as a systematic error. Because of the positive correlation between R and $\Delta x F_3$, the extracted values of F_2 are rather insensitive to the input R. If a large input R is used, a larger value of xF_3 is extracted from the y distribution, thus yielding the same value of F_2. In contrast, the extracted values of $\Delta x F_3$ are sensitive to the assumed value of R, which is reflected in a larger systematic error. The values of $\Delta x F_3$ are sensitive to the energy dependence of the neutrino flux ($\sim y$ dependence), but are insensitive to the absolute normalization. The uncertainty on the flux shape is estimated by using the constraint that F_2 and xF_3 should be flat over y (or E_ν) for each x and Q^2 bin.

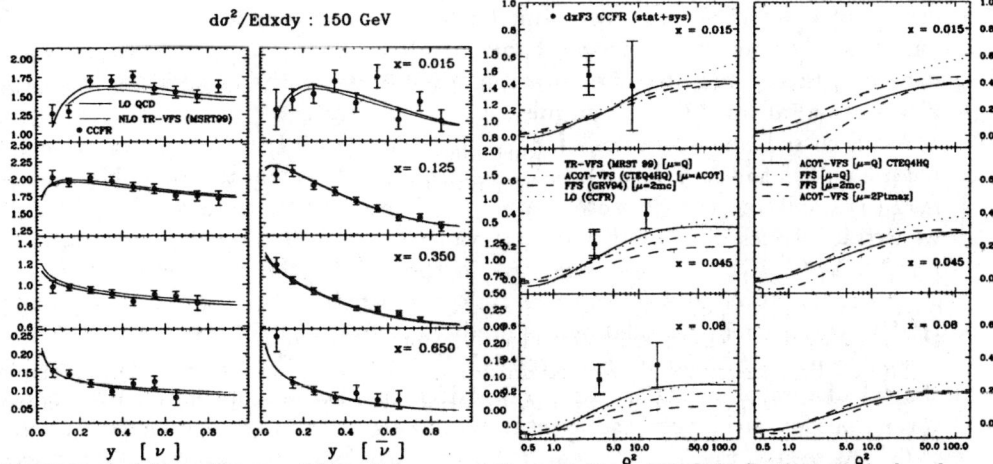

FIGURE 1. (a) Typical raw differential cross sections at $E_\nu = 150$ GeV (both statistical and systematic errors are included). (b) $\Delta x F_3$ data as a function of x compared with various schemes for massive charm production: RT-VFS(MRST), ACOT-VFS(CTEQ4HQ), FFS(GRV94), and LO(CCFR), a leading order model with a slow rescaling correction (left); Also shown is the sensitivity of the theoretical calculations to the choice of scale (right).

Because of the limited statistics, we use large bins in Q^2 in the extraction of

$\Delta x F_3$ with bin centering corrections from the NLO Thorne & Roberts Variable Flavor Scheme (TR-VFS) calculation with the MRST PDFs. Figure 1 (b) shows the extracted values of $\Delta x F_3$ as a function of x, including both statistical and systematic errors, compared to various theoretical methods for modeling heavy charm productions within a QCD framework. The three-flavor Fixed Flavor Scheme (FFS) assumes that there is no intrinsic charm in the nucleon, and all scattering from c quarks occurs via the gluon-fusion diagram. The concept behind the Variable Flavor Scheme (VFS) proposed by ACOT is that at low scale, μ, one uses the three-flavor FFS scheme, and above some scale, one changes to a four-flavor calculation and an intrinsic charm sea (which is evolved from zero) is introduced. The concept in the RT-VFS scheme is that it starts with the three-flavor FFS scheme at a low scale, becomes the four-flavor VFS scheme at high scale, and interpolates smoothly between the two regions. Shown are the predictions from the TR-VFS scheme (as corrected after DIS-2000 and implemented with MRST PDFs), with their suggested scale $\mu = Q$, and the predictions of the other two NLO calculations, ACOT-VFS (implemented with CTEQ4HQ and the recent ACOT suggested scale $\mu = m_c$ for $Q < m_c$, and $\mu^2 = m_c^2 + cQ^2(1 - m_c^2/Q^2)^n$ for $Q < m_c$ with $c = 0.5$ and $n = 2$), and the FFS (implemented with the GRV94 PDFs and GRV94 recommended scale $\mu = 2m_c$). Also shown are the predictions from $\Delta x F_3 \simeq 4Ks(x, Q^2)$ from a leading order model (LO(CCFR)) Buras-Gaemers type fit to the CCFR dimuon data (here K is a slow rescaling correction). Figure 1 (b) (right) also shows the sensitivity to the choice of scale. The data do not favor the ACOT-VFS(CTEQ4HQ) predictions if implemented with an earlier suggested scale of $\mu = 2Pt_{max}$. With reasonable choices of scale, all the theoretical models yield similar results. However, at low Q^2 our $\Delta x F_3$ data are higher than all the theortical models. The difference between data and theory may be due to an underestimate of the strange sea (or gluon distribution) at low Q^2, or from missing NNLO terms.

As discussed above, values of F_2 (PMI) for $x < 0.1$ are extracted from two parameter fits to the y distributions. In the $x > 0.1$ region, the contribution from $\Delta x F_3$ is small and the extracted values of F_2 are insensitive to $\Delta x F_3$. Therefore, we extract values of F_2 with an input value of R and with $\Delta x F_3$ constrained to the TR-VFS(MRST) predictions. As in the case of the two parameter fits for $x < 0.1$, no corrections for slow rescaling are applied. Fig. 2 (a) shows our F_2 (PMI) measurements divided by the predictions from the TR-VFS(MRST) theory. Also shown are F_2^μ and F_2^e from the NMC divided by the theory predictions. In the calculation of the QCD TR-VFS(MRST) predictions, we have also included corrections for nuclear effects, target mass and higher twist corrections at low values of Q^2. As seen in Fig. 2, both the CCFR and NMC structure functions are in good agreement with the TR-VFS(MRST) predictions, and therefore in good agreement with each other. A comparison using the ACOT-VFS(CTEQ4HQ) predictions yields similar results.

In the previous analysis of the CCFR data, the extracted values of F_2 (PMD) at the lowest $x = 0.015$ and Q^2 bin were up to 20% higher than both the NMC data and the predictions of the light-flavor MRSR2 PDFs. (see figure 2 (b)).

About half of the difference originates from having used a leading order model for $\Delta x F_3$ versus using our new measurement. The other half originates from having used the leading order slow rescaling corrections, instead of using a NLO massive charm production model, and from improved modeling of the low Q^2 PDFs (which changes the radiative corrections and the overall absolute normalization to the total neutrino cross sections).

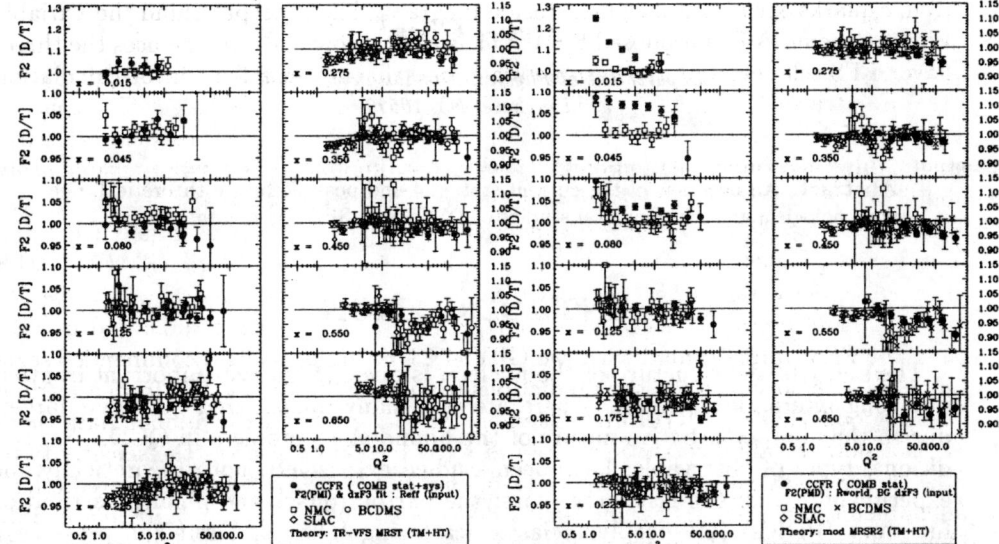

FIGURE 2. (a) Left side: The ratio (data/theory) of the F_2^ν (PMI) data divided by the predictions of TR-VFS(MRST) (with target mass and higher twist corrections). Both statistical and systematic errors are included. Also shown are the ratios of the F_2^μ (NMC) and F_2^e (SLAC) to the TR-VFS(MRST) predictions. (b) Right side: The ratio (data/theory) of the previous F_2^ν (PMD) data (and also F_2^μ (NMC) and F_2^e (SLAC)) divided by the predictions of the MRSR2 light-flavor PDFs (with target mass and higher twist corrections).

In conclusion, the F_2 (PMI) values measured in neutrino-iron and muon-deuterium scattering show good agreement with with the predictions of Next to Leading Order PDFs (using massive charm production schemes), thus resolving the long-standing discrepancy between the two sets of data. The first measurements of $\Delta x F_3$ are higher than current theoretical predictions.

Deeply Virtual Compton Scattering and Skewed Parton Distribution

Zhang Chen[1]

Department of Physics, Manhattanville College
Purchase, NY 10577

Abstract. An overview of the current status of and possible future theoretical, phenomenological and experimental studies of DVCS and SPD's is presented.

INTRODUCTION

The study of the structure of the nucleon is one of the most important frontiers in strong interaction physics. There are still many unanswered questions largely due to the non-perturbative nature of the bound state problem in QCD. Two traditional types of observables have been studied extensively, both theoretically and experimentally, for the last forty years: the parton (quark and gluon) distribution functions (PDF's) (via deeply inelastic scattering (DIS) or Drell-Yan processes), and elastic form factors of the nucleon. In the past few years, studies of a new type of nucleon observable, the skewed parton distributions (SPD's), have flourished(eg, [1]). The SPD's generalize and interpolate between the ordinary PDF's and elastic form factors and therefore contain rich structural information. They can be measured in (exclusive) diffractive processes in which the nucleon recoils elastically after receiving a non-zero momentum transfer in the so-called deeply virtual limit[2].

THEORY: FROM DIS & PDF'S TO DVCS & SPD'S

Ordinary PDF's in DIS are accessed through an optical theorem that relates the imaginary part of a forward (Compton) scattering amplitude to the cross section. Through operator product expansion (OPE) one factorizes the hard scattering from the soft physics. The hard scattering can be calculated order by order in perturbation theory while the soft part is parametrized as PDF's. They are essentially matrix elements of light-cone bilocal (quark and gluon) operators between equal momentum (symmetric) states. The factorization scale dependence of the matrix

---

[1] *email:* chenz@phys.mville.edu
[2] $Q^2 \to \infty$ while keeping Bjorken variable $x = x_B = \frac{Q^2}{2p \cdot q}$ fixed, and $Q^2 \gg t = -(p'-p)^2$ where p (p') is the initial(recoil) nucleon momentum.

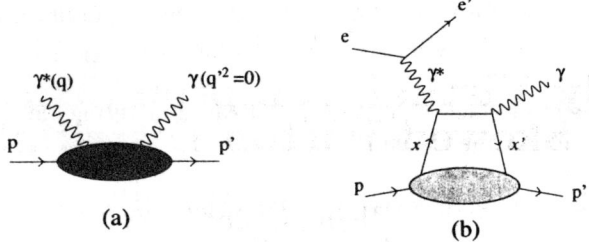

FIGURE 1. The DVCS process

elements/PDF's is governed by a renormalization group equation. The cross section is related to (usually a linear combination of) PDF's.

The SPD's are, on the other hand, essentially matrix elements of the same light-cone bilocal operators between *different* momentum (asymmetric) states. They can be accessed thus via deeply virtual processes like diffractive vector meson production ($ep \to ep' VM$) [2–5] and deeply virtual Compton scattering (DVCS) [6–8]. DVCS(See fig.1. (b) is the lowest order "handbag" diagram in an e-p collision, where x and x' denote the longitudinal momentum fractions of the interacting parton,) is a non-forward process signified by non-zero t with the longitudinal momentum (fraction) transfer $x-x' \equiv \zeta = x_B$ and also in general a non-zero transverse momentum transfer.

One still has valid factorization theorem [4,9] and OPE [6,10] (fig.2) in the non-forward case. That is, the DVCS amplitude can be factorized into a hard scattering part calculable perturbatively (the upper part of the OPE diagram and also a crossed term) and a soft part (lower part of the diagram) parametrized as SPD's.

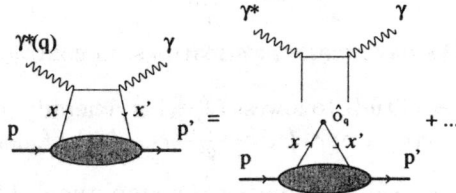

FIGURE 2. Non-forward OPE (only the lowest $q - q$ diagram is shown)

SPD's, like usual PDF's, depend on factorization scale. Its QCD evolution is governed again by a renormalization group equation through a kernel that has been worked out to next-to-leading-order (NLO) [11].[3] The QCD Q^2 evolution of the SPD's has been studied extensively and has been shown to exhibit characteristics of both the DGLAP evolution of usual PDF's and the ERBL evolution of meson distribution amplitudes, depending on different kinematic regions(eg, [1,10]). In particular, in the small x_B, ie, small $\xi = \frac{x_B}{2(1-x_B/2)}$ region SPD's are closely connected with usual PDF's [4]. For $x \gg \xi$, SPD's $\approx$ PDF's, while for $x \sim \xi$, in leading

---

[3] Also worked out is the NLO correction to hard scattering in DVCS.

$\log \frac{1}{x} \sim \log \frac{1}{\xi}$, SPD's $\to$ forward PDF's [2] and at large factorization scale the Q^2 evolution tends to wash out effects of the asymmetry ξ (eg, [12]). The region $x < \xi$ is least well-known.

SPD's are indeed form factors of the non-forward scattering amplitude $\gamma^* p \to \gamma p'$, eg, take quarks (similar for gluons)

$$FT \; \langle p', s'|\overline{\psi}(0)\gamma^\mu \psi(z)|p, s\rangle = H(x,\xi,t) \; \bar{u}(p',s')\gamma^+ u(p,s)$$
$$+ E(x,\xi,t) \; \bar{u}(p',s') \frac{i\sigma^{+\nu}(p'-p)_\nu}{2M} u(p,s) + \cdots , \quad \text{(Quark Spin Sum)}$$
$$FT \; \langle p', s'|\overline{\psi}(0)\gamma^+\gamma_5 \psi(z)|p, s\rangle = \widetilde{H}(x,\xi,t) \; \bar{u}(p',s')\gamma^+\gamma_5 u(p,s)$$
$$+ \widetilde{E}(x,\xi,t) \; \bar{u}(p',s') \frac{(p'-p)^+ \gamma_5}{2M} u(p,s) + \cdots , \quad \text{(Difference)}$$

where $H(x,\xi,t)$, $E(x,\xi,t)$, $\widetilde{H}(x,\xi,t)$, $\widetilde{E}(x,\xi,t)$ are skewed quark distributions with $2\xi = \frac{x_B}{1-x_B/2}$ [4] and FT denotes Fourier Transform.

While in general two SPDs correspond to one usual PDF, in the forward limit of $p = p'$ ($\xi \to 0$ and $t \to 0$), SPD's are reduced to the normal distributions: $H(x,0,0) = q(x)$, $\widetilde{H}(x,0,0) = \Delta q(x)$, where $q(x)$ and $\Delta q(x)$ are the conventional forward quark and quark helicity distributions (Similar equations hold for gluons). At the same time, the first moments of these SPD's are related to nucleon form factors of corresponding EM or Axial currents by the sum rules:

$$\int_{-1}^{1} dx H(x,\xi,t) = F_1(t) , \quad \int_{-1}^{1} dx E(x,\xi,t) = F_2(t) ,$$
$$\int_{-1}^{1} dx \widetilde{H}(x,\xi,t) = G_A(t) , \quad \int_{-1}^{1} dx \widetilde{E}(x,\xi,t) = G_P(t) .$$

At the same time, SPD's have many new features, in contrast to the usual PDF's.

- The non-forward amplitude to lowest $O(\alpha_s)$ is generally related to a integration over the SPD's of type $Amp \sim \int dx \frac{1}{x-\zeta+i\epsilon} f(x,\zeta,t)$ (f denotes SPD's).

- Cross section is obtained by squaring the amplitude. There is not an optical theorem, nor a simple probability interpretation. SPD's can be viewed as overlap of wavefunctions between different parton numbers (Fock states) [13]

- SPD's are interference/correlation functions of different wave functions/ probability amplitudes. The extreme case with $x > 0$ and $x' < 0$ can be understood by re-interpret x' line as antiparton with momentum fraction $-x'$ resulting in "extracting $q\bar{q}$-pair" from the nucleon.

SPD(via DVCS) is the only place so far probing the orbital angular momentum of partons [7]. Moments of the SPD's provide information on quark and gluon

---

[4] Notations in the literature differ. That of Ji's is used and mom. fractions refer to $\frac{1}{2}(p+p')$.

contributions to nucleon's spin because they are closely related to the form factors of the (QCD) energy-momentum tensor. For example, one can measure the skewed quark distribution in spin-averaged experiments and extract form factors of the tensor. Extrapolating to $t = 0$ one can obtain the total (spin+orbital) quark contribution to the nucleon spin $J_q(0)$ by taking $t = 0$ in Ji's sum rule:

$$\frac{1}{2}\int dx\, x\, (H_q(x,\xi,t) + E_q(x,\xi,t)) = J_q(t).$$

There is a similar sum rule for gluon and one finds $J_q(0) + J_g(0) = \frac{1}{2}$.

PHENOMENOLOGY AND EXPERIMENTS

Using known usual PDF's as input, models of SPD's have been studies by making ansatz that fulfills the (some of which very non-trivial) general properties of SPD's(eg, [1,14]). There are also studies on contributions from meson exchange in the ERBL region ($x > 0, x' < 0$, on SPD's as overlap of proton wavefunctions for $x > \xi$ [15] and on different dynamical models of SPD's, including bag model, instanton vacuum and quark soliton model (eg, [16]).

Experiments usually access integrals of SPD's with parton momenta that are multi-variable and can not be directly obtained from cross sections. Thus two main issues remain, one being experimental difficulty, the other extracting SPD's from data.

The key background to DVCS is the QED Bethe-Heitler(BH) process(Compton scattering), which depend on EM form factor, and its interference with DVCS. At higher Q^2 the background is smaller, but so is the signal. This makes measuring DVCS cross section very difficult experimentally. The first experimental measurement come from HERA [17], where the preliminary on 96/97 data has shown evidence of the DVCS signal. However, the interference also makes possible exploring DVCS at the amplitude level, measuring its imaginary and real part independently.[5] There has been a lot of data on vector meson production (eg, [19,20]), but the extraction of SPD's is still a pending task.[6] Table 1 is a very brief overview of the current experimental status.

Independent of the actual form of SPD's, factorization predicts that in the scaling limit of $Q^2 \to \infty; x_B, t$ fixed, DVCS $\sim \frac{1}{Q^0}$ and VM production $\sim \frac{1}{Q}$. There are also

---

[5] A proposal for JLAB Hall A [18] proposed measuring cross section difference for leptons of opposite helicities, which is proportional to the interference of the imaginary part of the DVCS amplitude with a known BH weight. This quantity turns out to be a *linear* combination of SPD's, thus if scaling is reached at the kinematic region with 6Gev beam, it would be a measurement of SPDs' contribution. Also since the higher twist effects is only down $\frac{1}{Q}$ and has a different angular distribution (thus not masked by leading-twist) it can give an estimation of these effects.

[6] Another reason of the interest in exclusive diffractive VM at small-x is that the cross section is predicted to be proportional to the *square* of gluon density [2]. This could be a direct measurement of the glue in the proton, rather than getting it from evolution as in DIS.

TABLE 1. Overview of Status of SPD-related Exps, DVCS and VM Production

Process(es)	DATA	Proposed
DVCS	HERA [17] ($e^+p \to e^+\gamma p'$)	COMPASS, HERMES, JLAB [18]
Meson Prod.	HERMES [19], HERA [20]	JLAB, COMPASS

helicity selection rules stating that for DVCS only $\gamma^*(T) \to \gamma(T) \sim \frac{1}{Q^0}$ and $\gamma^*(L) \to \gamma(T)$ is power suppressed while for VM only $\gamma^*(L) \to VM(L) \sim \frac{1}{Q}$ and all else are power suppressed. This is due to helicity conservation and introduces helicity parton distributions (eg, helicity flip gluon distribution that has no counterpart in DIS) [21], while being higher-twist gives the suppression. Therefore by looking at the angular distributions one have a window on higher-twist effects [22] since it is not masked by leading twist because of different helicity.

Other SPD-related experimental processes been proposed include diffractive di-jets (photo-production) and crossed DVCS (eg, $\gamma^*\gamma \to \pi^+\pi_-$) [23], etc.

ACKNOWLEDGEMENTS

I thank Xiangdong Ji and Macus Diehl for helpful discussions. This research is partly supported by a grant from Manhattanville College.

REFERENCES

1. Ji, X., *et al*, *Phys. Rev.* **D56**, 5511(1997); Ji, X., *J. Phys.* **G24**, 1181(1998); Radyushkin, A., *Phys. Rev.* **D56**, 5524(1997); Musatov I. & Radyushkin, A., *Phys. Rev.* **D61** 074027(2000).
2. Brodsky, S., *et al*, *Phys. Rev.* **D50**, 3134(1994).
3. Radyushkin, A., *Phys. Lett.* **B385**, 333(1996).
4. Collins, J., *et al*, *Phys. Rev.* **D56**, 2982(1997).
5. Frankfurt, L., *et al*, *Phys. Lett.* **B418**, 345(1998).
6. Mueller, D., *et al*, *Fortschr. Phys.* **42** 101 (1994).
7. Ji, X., *Phys. Rev. Lett.* **78**, 610(1997); *Phys. Rev.* **D55**, 7114(1997).
8. Radyushkin, A., *Phys. Lett.* **B380**, 417(1996).
9. Collins, J. & Freund, A., *Phys. Rev.* **D59**, 074009(1999).
10. For example, Chen, Z., *Nucl. Phys.* **B525**, 369(1998).
11. Belitsky, A., *et al*, *Phys. Lett.* **B474** 163(2000).
12. Shuraev, A., *et al*, *Phys. Rev.* **D60**, 014015(1999).
13. Brodsky, S. & Hwang, D., *Nucl. Phys.* **B543**, 239(1999).
14. Mankiewicz, L., *et al*, *Eur. Phys. J.* **C10** 307(1999); Guichon, P., *et al*, *Phys. Rev.* **D60**, 094017(1999).
15. Diehl, M., *et al*, *Eur. Phys. J.* **C8** 409(1999).
16. Petrov, V., *et al*, hep-ph/9712203, *Phys. Rev.* **D57**, 4325(1998); Penttinen, M., *et al*, *Phys. Rev.* **D62** 014024(2000).

17. Saul, P., *et al*, for Zeus Collaboration, hep-ex/0003030; similar results from H1 collaboration also presented at CIPANP2000.
18. Sabatie, F., *et al*, Proposal to Jefferson Lab PAC 18 (Hall A).
19. For example, Airapetian, A., *et al*, for HERMES collaboration, hep-ex/0004023.
20. For example, Crittenden, J., hep-ex/9910068.
21. Diehl, M., *Eur. Phys. J.* **C4** 497(1998); Hoodbhoy, P. & Ji, X., *Phys. Rev.* **D58**, 054006(1998).
22. Anikin, I., *et al*, hep-ph/0003203; Penttinen, M., *et al*, hep-ph/0006321.
23. Golec-Biernat, K., *et al*, *Phys. Rev.* **D58**, 094001(1998); Diehl, M., *et al*, *Phys. Rev. Lett.* **81**, 1782(1998).

Study of Neutron Spin Structure Functions at Low Q^2 with Polarized ^{3}He

Seonho Choi[1]

Temple University, Philadelphia, PA 19122, USA

Abstract. The recently completed experiment E94-010 at Jefferson Lab studies the neutron spin structure functions at low momentum transfer (Q^2) values. Using a polarized ^{3}He target and polarized electron beam, we have measured the asymmetries and cross sections for $^3\vec{\mathrm{He}}(e,e')$ from the elastic to the deep inelastic region. The covered Q^2 ranges from 0.03 to 1.1 GeV^2. From the data, the Q^2 evolution of the spin structure functions for ^{3}He and neutron, and of the Gerasimov-Drell-Hearn (GDH) sum rule has been studied, and the preliminary results are presented.

INTRODUCTION

The study of the spin structure of the nucleon has been of central interest in the last decade. Since the nucleon shows different features depending on the length scale at which it is observed, it is crucial to explore the spin structure over a wide range of the length scale. At a very small length scale, or high Q^2, the Bjorken sum rule holds [1]. On the other hand, at $Q^2 = 0$, the GDH sum rule [2,3] is valid which relates the spin dependent total cross section for hadron photoproduction to the anomalous magnetic moment of the nucleon. In the intermediate Q^2 region, it is possible to generalize the GDH sum rule for the absorption of the virtual photons, giving a Q^2 dependence. Recently, it has been shown by Ji and Osborne that this generalized GDH sum rule is closely connected to the Bjorken sum rule at high Q^2 limit [4]. Thus, the GDH sum rule is a crucial test of our understanding of the spin structure of the nucleon at all the values of Q^2.

GENERALIZED GDH SUM RULE

Originally, the GDH sum rule was derived for real photons:

$$\int_{\mathrm{thr}}^{\infty} \frac{\sigma_{1/2} - \sigma_{3/2}}{\nu} d\nu = -\frac{2\pi^2 \alpha}{M^2} \kappa^2, \tag{1}$$

---

[1] *For the Jefferson Lab E94-010 Collaboration*

where $\sigma_{1/2}$ and $\sigma_{3/2}$ are the spin dependent cross sections for the absorption of real photons, and κ is the anomalous magnetic moment of the nucleon.

The integral on the left side of Eq. 1 can be easily generalized for the virtual photon case[2]:

$$I_{GDH} \equiv 8\pi^2 \alpha \int G_1(\nu, Q^2) \frac{d\nu}{\nu}, \qquad (2)$$

where α is the electromagnetic coupling constant and $G_1(\nu, Q^2)[= g_1(x, Q^2)/M\nu]$ is the spin structure function. Then, this integral can be related to the forward virtual Compton scattering amplitude, $S_1(Q^2)$, through

$$I_{GDH} = 2\pi^2 \alpha S_1(Q^2), \qquad (3)$$

yielding a generalized GDH sum rule valid for any values of Q^2.

At small Q^2, $S_1(Q^2)$ can be calculated using Chiral Perturbation Theory and the results of this experiment will be a valuable test of that theory. At large Q^2, it can also be calculated with a higher order QCD expansion (twist expansion). For the intermediate Q^2 regions, more theoretical efforts (such as lattice QCD calculation) are necessary.

EXPERIMENT

We have measured the inclusive cross section for the scattering of polarized electrons from a polarized ^{3}He target in Hall A of the Jefferson Lab. An electron beam with average polarization of 70% and current up to 15μA was scattered off a high density (10 atm in a 40 cm long glass cell) polarized ^{3}He target with 30 to 40% polarization. The scattered electrons were detected by the two essentially identical spectrometers sitting at 15.5° on both sides of the beam.

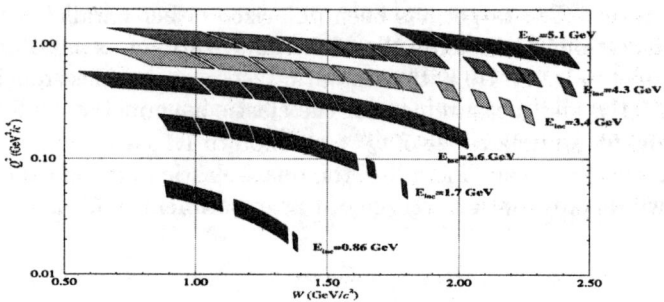

FIGURE 1. Kinematic coverage of JLab E94-010 experiment.

---

[2] Actually, there are several different generalizations, all of which agrees for the real photon case at $Q^2 = 0$. We followed Ji and Osborne's definition.

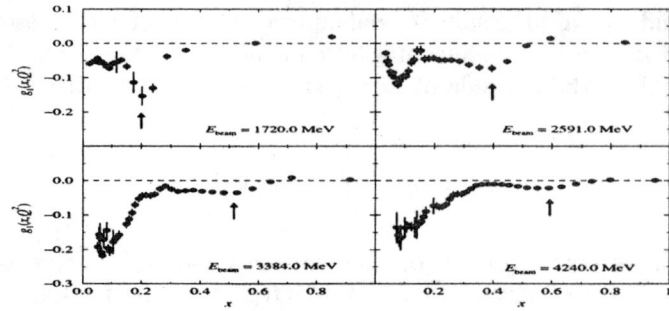

FIGURE 2. Preliminary results on the spin dependent structure function $g_1^{^3He}(x,Q^2)$.

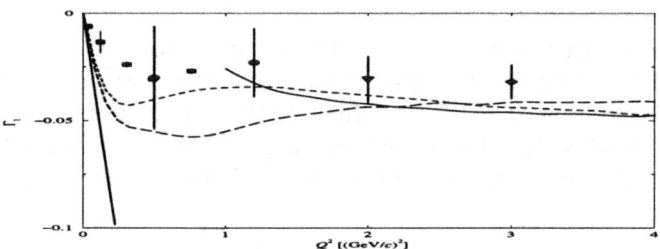

FIGURE 3. Integral of $g_1(x,Q^2)$ for the neutron. The present data (squares) are plotted together with data from SLAC E142 (inverted triangle) and E143(circle). The curves correspond to the evolution[7] of the deep-inelastic results due to changing α_s (solid), the predictions of Burkert and Ioffe[8] (dotted), the model of Soffer[9] (long dash), and the GDH approach to $Q^2 = 0$ (solid).

The ^{3}He target was polarized using spin exchange principle. Rb atoms were polarized by optical pumping and the polarization was transferred to ^{3}He via spin exchange collisions. The target has been polarized either parallel or perpendicular to the beam direction. The magnitude of the polarization was monitored with three independent methods: NMR with adiabatic fast passage, Electron Paramagnetic Resonance (EPR), and measurement of the elastic asymmetry.

The experiment covers a range of Q^2 from 0.03 to 1.1 GeV^2, and from the elastic to the deep inelastic regime, including the quasi-elastic and the resonance regions. Figure 1 shows the kinematic coverage of this experiment in Q^2 and invariant mass W.

PRELIMINARY RESULTS

The differential cross sections for each combination of electron and ^{3}He spin directions have been deduced from the experimental data. Overall consistency has been checked comparing ^{3}He elastic cross sections and asymmetries with the world data.

From the polarized beam asymmetries with the target polarization either parallel or perpendicular to the beam direction, the structure functions $g_1^{^3He}(x,Q^2)$ and $g_2^{^3He}(x,Q^2)$ have been deduced. Figure 2 shows the structure function $g_1^{^3He}(x,Q^2)$ for four beam energies. We can observe that $g_1^{^3He}(x,Q^2)$ for Δ resonance (arrow mark) is negative and decreases in magnitude as the beam energy (or Q^2) increases.

In Fig. 3, the integral of $g_1(x,Q^2)$, $\Gamma_1 = \int_0^1 g_1(x,Q^2)dx$ has been compared with the existing data from SLAC E142 [5] and E143 [6], and a few calculations. The preliminary results are consistent with the existing data but with much more improved statistical precision[3].

SUMMARY AND FUTURE

We have measured spin dependent cross sections and asymmetries for $^3\vec{\text{He}}(e,e')$. The preliminary results are compatible with the existing data and have significantly improved statistical precision. Detailed analysis of the systematic errors is in progress, and final results will be available shortly. An extension to even smaller Q^2 ranges and higher energies is planned in the future at Jefferson Lab [11]. A more detailed study of the GDH sum rule and the spin structure of the nucleon will enrich our understanding of the internal structure of the nucleon.

REFERENCES

1. Bjorken, J. D., *Phys. Rev.* **148**, 1467 (1966).
2. Gerasimov, S. B., *Sov. J. Nucl. Phys.* **2**, 430 (1966).
3. Drell, S. D., and Hearn, A. C., *Phys. Rev. Lett.* **16**, 908 (1966).
4. Ji, X., and Osborne, J., hep-ph/9905010 (1999).
5. E142 Collaboration, Anthony, P. L. et al., *Phys. Rev. Lett.* **71**, 959(1993); *Phys. Rev.* **D 54**, 6620 (1996)
6. E143 Collaboration, Abe, K. et al., *Phys. Rev.* **D 58**, 112003 (1998).
7. Larin, S. A., and Vermaseren, J. A. M., *Phys. Lett.* **B259**, 345 (1991), and references therein; Larin, S. A., *ibid.* **334**, 192 (1994).
8. Burkert, V. D., and Ioffe, B. L., *Phys. Lett.* **B296**, 223 (1992); *JETP* **78**, 619 (1994).
9. Soffer, J., and Teryaev, O. V., *Phys. Rev.* **D 51**, 25 (1995); **56**, 7458 (1997).
10. degli Atti, C., and Scopetta, S., *Phys. Lett.* **B44**, 223 (1997).
11. Chen, J. P., Cates, G., and Garibaldi, F., *JLab Proposal E97-110*.

---

[3] Due to the small components of S' and D waves in the ground state of the ^{3}He, there are corrections to be made to get $g_1^{neutron}$ from $g_1^{^3He}$ [10]. These corrections have been ignored in this comparison. For our results, the integral has been performed with $g_1^{^3He}(x,Q^2)$ at constant beam energy within the measured region of x, and the results are plotted versus average Q^2 value.

Parton Distributions from 1+1 QCD

Varghese John, Govind S. Krishnaswami [1] and S. G. Rajeev

Department of Physics and Astronomy,
University of Rochester, Rochester, New York 14627

Abstract. We study the large N_c limit of a previously introduced reformulation of 2d QCD, HadronDynamics. This model is used for an effective description of the baryon in Deep Inelastic Scattering, when transverse momenta of partons are ignored. This allows us to determine the non-perturbative initial condition for Q^2 evolution equations: x_B dependence of structure functions at an initial Q_0^2. After including effects of transverse momenta via the DGLAP evolution, we compare our prediction with experimental measurements of xF_3 and find good agreement. We have only two parameters: the initial scale Q_0 and the fraction of baryon momentum carried by valence quarks.

The Q^2 evolution of the hadronic structure functions of Deep Inelastic Scattering (DIS) can be determined within the framework of perturbative QCD [1,4]. However, the initial condition for the DGLAP equations, the dependence on x_B at an initial Q_0^2 is non-perturbative and we do not yet have a reliable way of calculating it from QCD. Therefore, we have had to rely on parametrizations of experimental data for these initial conditions [4]. Here we study Rajeev's bi-local reformulation of 2 dimensional QCD, HadronDynamics [2]. This model is proposed as an effective description of the baryon at a low value of Q^2, where the transverse momenta of partons are ignored. The solutions of this model are used as initial conditions for DGLAP evolution (which can be thought of as including corrections due to transverse momenta [1]). The iso-spin averaged valence quark distribution determined this way compares well with experimental measurements of the structure function $xF_3(x_B, Q^2)$. Our only parameters are Q_0^2 and the fraction of baryon momentum carried by valence quarks, f.

We take as our effective hamiltonian for the leading 'relevant' interactions of the quarks, the hamiltonian of 2d Hadron Dynamics [2]. This is the same as the hamiltonian of 2d QCD in the null gauge $A_- = 0$, expressed in terms of the gauge invariant meson field $\hat{M}_b^a(x,y) = \frac{2}{N_c} : \chi_{ba}(x) \chi^{\dagger a a}(y) :$, where x and y lie along a null line. The longitudinal gluons induce a linear potential between quarks. In the large N_c limit, the ground state of the baryon (Baryon Number = $-\frac{1}{2} tr M = 1$) is

---

[1] speaker, govind@pas.rochester.edu. Supported by DOE grant No. DE–FG02-91ER40685

determined by minimizing the energy

$$\frac{E[M]}{N_c} = -\frac{1}{4}\int [p + \frac{\mu_a^2}{p}]\tilde{M}_a^a(p,p)\frac{dp}{2\pi} + \frac{\tilde{g}^2}{8}\int M_b^a(x,y)M_a^b(y,x)|x-y|dxdy. \quad (1)$$

μ_a^2 is related to the current quark masses m_a by a finite renormalization: $\mu_a^2 = m_a^2 - \frac{\tilde{g}^2}{\pi}$ [5] and $\tilde{g}^2 = g^2 N_c$. Though the hamiltonian is quadratic, this is an interacting theory since the phase space is a curved manifold due to the quadratic constraint arising from the Pauli principle for quarks, $(\epsilon + M)^2 = 1$ (ϵ is the hilbert transform).

The diagonal entries of the (normal ordered) density matrix, $-\frac{1}{2}\tilde{M}_a^a(p,p)$ and $-\frac{1}{2}\tilde{M}_a^a(-p,-p), 0 \leq p \leq P$ are the quark and anti-quark probability densities. Experimentally, it is inferred that gluons carry about half the baryon momentum at low Q_0^2 [4]. Since we have ignored the transversely polarized gluons, we require that the valence quarks $-\frac{1}{2}\sum_{a=u,d}(M_a^a(p,p) - M_{\bar{a}}^{\bar{a}}(-p,-p)), p > 0$ carry only a fraction f of the total baryon momentum P.

We have found the *exact* minimum of this variational principle in the chiral and large N_c limit [3]. The minimum occurs for a configuration consisting of valence quarks alone: $\tilde{M}_1(p,q) = -2\tilde{\psi}(p)\tilde{\psi}^*(q)$ for $p,q \geq 0$ and zero otherwise. Indeed, $\tilde{\psi}(p)$ can be regarded as the valence quark wave function, and in the ground state $\tilde{\psi}(p) = \sqrt{\frac{2\pi}{fP}}\exp(\frac{-p}{2fP})$, where $\bar{P}$ is the mean baryon momentum per color. For finite N_c, we find that the functional form $\tilde{\psi}(p) = Cp^a(1-\frac{p}{P})^b, 0 \leq p \leq P$ provides a good approximation to the ground state with $b = \frac{N_c}{2f} - 1 + a(\frac{N_c}{f} - 1)$ and $a \to \sqrt{\frac{3}{\pi}\frac{m}{\tilde{g}}}$ for small current quark masses.

The above purely valence quark configuration is the exact minimum in the chiral and large N_c limits. To estimate the anti-quark and sea quark content of the baryon away from chiral symmetry, we must allow for more general configurations. The simplest baryon number one configuration that departs from the valence quark approximation and satisfies the quadratic constraints can be expressed as:

$$M_3 = -2\psi \otimes \psi^\dagger + 2\zeta_-^2[\psi_- \otimes \psi_-^\dagger - \psi_+ \otimes \psi_+^\dagger] + 2\zeta_-\zeta_+[\psi_- \otimes \psi_+^\dagger + \psi_+ \otimes \psi_-^\dagger]. \quad (2)$$

Here ψ, ψ_+ are orthonormal and non-vanishing only for positive momenta; they are the valence and sea quark wave functions. ψ_- vanishes for positive momenta and describes anti-quarks. $0 \leq \zeta_- \leq 1$ measures the deviation from the purely valence quark configuration. $\zeta_+ = \sqrt{1-\zeta_-^2}$. Baryon number is given by $B = \int_0^\infty \{|\tilde{\psi}(p)|^2 + \zeta_-^2[|\tilde{\psi}_+(p)|^2 - |\tilde{\psi}_-(-p)|^2]\}\frac{dp}{2\pi}$. From our previous result we expect ζ_- to vanish as $\frac{m^2}{\tilde{g}^2} \to 0$. For physically relevant current quark masses ($\frac{m^2}{\tilde{g}^2} \sim 10^{-3}$), our estimates [3] show that the corrections due to anti and sea-quarks is negligible; for instance they carry less than a percent of baryon momentum.

Thus, in the chiral limit, our approximation for the valence quark probability distribution is $V(x_B, Q_0^2) = C(1-x_B)^{\frac{N_c}{f}-2}$ (For $f = 0$ this agrees with numerical computations of Hornbostel et. al. [6]). We can compare this prediction

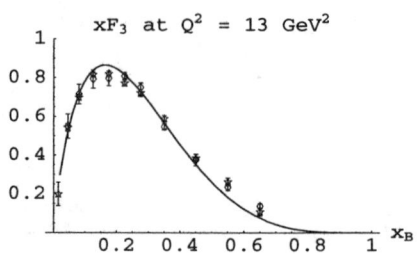

FIGURE 1. Predicted xF_3 compared with measurements by CDHS($\Diamond$) and CCFR($\star$).

with experimental data after including corrections due to transverse momenta by evolving the distribution to higher Q^2 via the DGLAP equation. The valence quark parton distribution function is given by $\phi^V(x_B, Q_0^2) = \nu(Q_0^2) V(x_B, Q_0^2)$, where $\nu(Q_0^2) = \int_0^1 dx_B \phi^V(x_B, Q_0^2)$ is determined by solving the integrated DGLAP equation with initial condition $\nu(\infty) = N_c = 3$ from $Q^2 = \infty$ to Q_0^2. Within the leading logarithmic approximation, $\phi^V(x_B, Q^2)$ is the same as $F_3(x_B, Q^2)$ averaged over neutrino and anti-neutrino scattering on an isoscalar target. Then $\nu(Q_0^2) = \int_0^1 dx_B F_3(x_B, Q_0^2)$, is given by the GLS sum rule [7]. We compare our prediction for xF_3 with measurements by the CDHS and CCFR collaborations in Fig. 1 and find good agreement for a choice of parameters $Q_0^2 \sim 0.4 GeV^2$ and $f \sim \frac{1}{2}$. Thus HadronDynamics is a successful way of deriving structure functions from collinear QCD.

REFERENCES

1. G. Altarelli and G. Parisi, Nucl. Phys. B126, 298 (1977);
2. S. G. Rajeev, in *1991 Summer School in High Energy Physics and Cosmology* Vol.II, World Scientific, Singapore (1992); S. G. Rajeev, Int. J. Mod. Phys. A31(1994) 5583; S.G. Rajeev, hep-th/9905072 (1999). S.G. Rajeev, Nucl. Phys. B (Proc. Suppl.) 86 (2000) 86.
3. G. S. Krishnaswami and S. G. Rajeev, Phys. Lett. B441 (1998) 449; G. S. Krishnaswami, Undergraduate Thesis, University of Rochester hep-ph/9911538 (1999); V. John, G. S. Krishnaswami and S. G. Rajeev, 'An Interacting Parton Model for Quark and Anti-quark distributions in the Baryon', Phys. Lett. B, to be published.
4. CTEQ Collab., G. Sterman et al., Rev. Mod. Phys. **67**, (1995) 157.
5. G. 't Hooft, Nucl. Phys. B75 (1974) 461.
6. K. Hornbostel, S.J. Brodsky and H.C. Pauli, Phys. Rev. **D41** (1990) 3814.
7. D. J. Gross and C. H. Lewellyn-Smith, Nucl. Phys. B14, 337 (1969)

Photon and Pion Production at High p_T

Marek Zieliński

University of Rochester, Rochester, New York 14627

Abstract. We present a study of high-p_T photon and pion production in hadronic interactions, focusing on a comparison of the yields with expectations from next-to-leading order perturbative QCD (NLO pQCD). We examine the impact of phenomenological models of k_T smearing (which approximate effects of additional soft-gluon emission) on absolute predictions for photon and pion production and their ratio.

Single and double direct-photon production in hadronic collisions at high transverse momenta (p_T) have long been viewed as an ideal testing ground for the formalism of pQCD. A reliable theoretical description of the direct-photon process is of special importance because of its sensitivity to the gluon distribution in a proton through the quark–gluon scattering subprocess ($gq \to \gamma q$). The gluon distribution, $G(x)$, is relatively well constrained for $x < 0.1$, but much less so at larger x [1]. In principle, fixed-target direct-photon production can constrain $G(x)$ at large x, and such data have therefore been incorporated in several modern global parton distribution function (PDF) analyses [2–4].

However, both the completeness of the NLO description of the direct-photon process, as well as the consistency of results from different experiments, have been questioned [4–11]. The inclusive production of hadrons provides a further means of testing the predictions of the NLO pQCD formalism. Deviations have been observed between measured inclusive direct-photon and pion cross sections and NLO pQCD calculations. Examples of such discrepancies are shown in Fig. 1 where ratios of data to theory are displayed as a function of $x_T = 2p_T/\sqrt{s}$ for photon and pion data. (Unless otherwise indicated, all NLO calculations [12–16] in this paper use a single scale of $\mu = p_T/2$, CTEQ4M PDFs [2], and BKK fragmentation functions for pions [17].) It has been suggested that part of the deviations from theory for both photons and pions can be ascribed to higher-order effects of initial-state soft-gluon radiation [6–8].

Given the scatter of the data shown in Figs. 1, it may be instructive to consider measurements of the γ/π^0 ratio over a wide range of $\sqrt{s}$ [18]. Both experimental and theoretical uncertainties tend to cancel in such a ratio, and the ratio should also be less sensitive to incomplete treatment of gluon radiation. A compilation of comparisons between data and theory, shown for simplicity without their uncer-

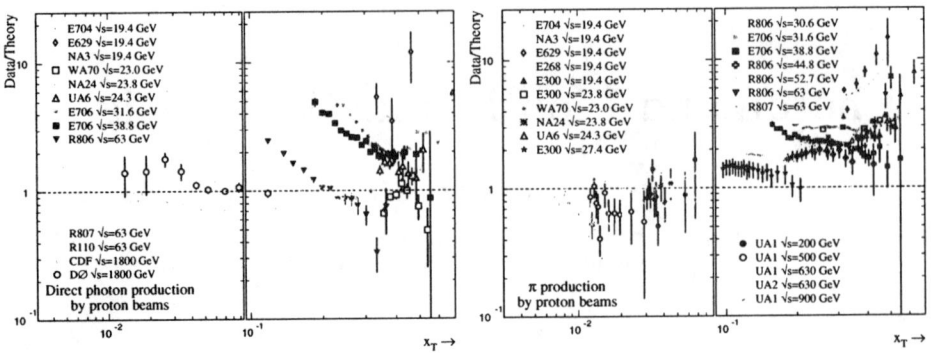

FIGURE 1. Comparison of direct-photon (left) and pion (right) data to NLO pQCD vs x_T.

tainties, is presented in Fig. 2. In this figure, the ratios of data to theory for the γ to π^0 measurements have been approximated as a constant value at high-p_T, and the results plotted as a function of $\sqrt{s}$ (see [18] for details). The figure suggests an energy dependence in the ratio of data to theory for γ/π^0 production. There is, however, an indication of substantial differences between the experiments at low $\sqrt{s}$ (where the observed γ/π^0 is smallest), which makes it difficult to quantify this trend. Recognizing the presence of these differences is especially important because thus far only the low energy photon experiments have been used in PDF fits to extract the gluon distribution.

The differences between many of the data sets and pQCD, seen in Fig. 1, may be due to the impact of the effective parton transverse momentum, k_T. In hadronic hard-scattering processes, there is generally a substantial amount of k_T in the initial state resulting from gluon emission [8]. The presence of k_T impacts the final state and has been observed in measurements of Drell-Yan, diphoton, and heavy quark production; the amount of k_T expected from NLO calculations is not sufficient to describe the data. The effective values of $\langle k_T \rangle$/parton for these processes vary from ≈ 1 GeV/c at fixed target energies, as illustrated in Fig. 2 for diphoton distributions from E706 [19], to 3–4 GeV/c at the Tevatron Collider — the growth is approximately logarithmic with center-of-mass energy [8]. The size of the $\langle k_T \rangle$ values, and their dependence on energy, argue against a purely "intrinsic" non-perturbative origin. Rather, the major part of this effect is generally attributed to soft-gluon emission. While the importance of including gluon emission through the resummation formalism has long been recognized and calculations have been available for some time for Drell-Yan [20], diphoton [21,22], and W/Z production [21], they have only recently been developed for inclusive direct-photon production [23–28].

In the absence of a rigorous theoretical treatment of the impact of gluon emission on high-p_T inclusive production, a more intuitive phenomenological approach has proved successful [8]. The soft gluon radiation was parametrized in terms of an effective $\langle k_T \rangle$ that provided an additional transverse impulse to the outgoing

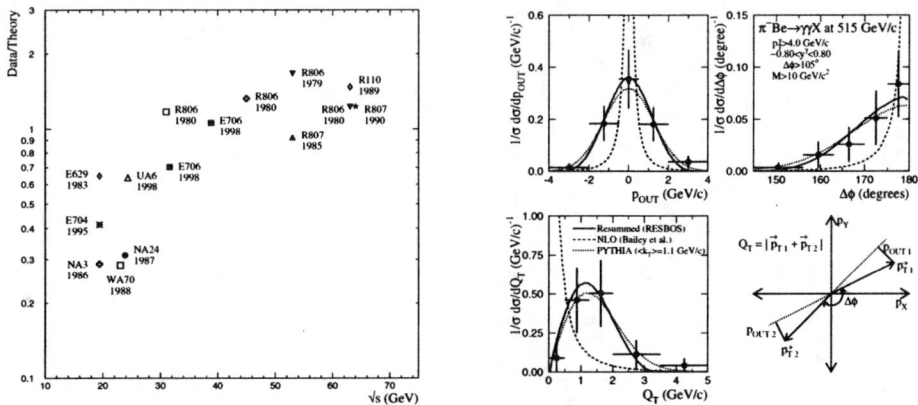

FIGURE 2. Left: The ratio of data for the γ/π^0 measurements to NLO theory, as a function of $\sqrt{s}$. Right: Diphoton data from E706 [19] compared to NLO [16] (dashed), resummed [21] (solid), and Pythia [30] (dotted) calculations, using GRV PDF [3].

partons. Because of the steeply falling cross section in p_T, such a $\langle k_T \rangle$ can shift the production of final-state particles from lower to higher values of p_T, effectively enhancing the cross section.

As described in [8], a leading-order (LO) pQCD calculation [29] has been used to generate K-factors (ratios of calculations for any given $\langle k_T \rangle$ to the result for $\langle k_T \rangle = 0$) for inclusive cross sections. These p_T-dependent factors have been then applied to the NLO pQCD calculations. The enhancements that would be expected for direct-photon production from parton-showering models [30,31] have also been investigated [18]. These programs do not provide sufficient smearing at fixed-target energies because shower development is constrained by cut-off parameters that ensure the perturbative nature of the process. Consequently, these calculations allow additional input k_T for Gaussian smearing, and are often used that way in comparisons to data. The respective corrections have been obtained using default settings for other program parameters and an input $\langle k_T \rangle$ of 1.2 GeV/c for the smearing, relative to these same settings with $\langle k_T \rangle = 0$, and then applied to NLO pQCD calculations. The resulting comparisons to data from E706 [7] are displayed in Fig. 3 (similar results hold for pion production, not shown). The observed differences should be kept in mind when comparing these models to data for k_T-sensitive quantities.

Recently, there has been significant progress in more rigorous resummed pQCD calculations for single direct-photon production [23–28]. Substantial corrections to fixed-order QCD calculations are expected from soft-gluon emission, especially in regions of phase space where gluon emission is restricted kinematically. At large x, there is a suppression of gluon radiation due to the rapidly falling parton distributions and a complete description of the cross section in this region requires the resummation of "threshold" terms. Two recent threshold-resummed pQCD

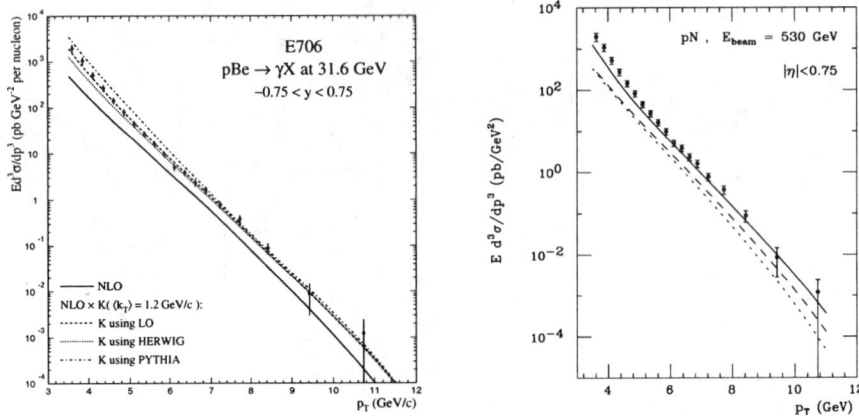

FIGURE 3. Left: Comparison between the E706 direct-photon data at $\sqrt{s} = 31.6$ GeV [7] and the NLO pQCD calculation (solid), and the NLO theory enhanced by K-factors obtained using the LO calculation [29] (dashed), HERWIG [31] (dotted), and PYTHIA [30] (dash-dotted). Right: Same data compared to recent QCD calculations. The dotted line represents the full NLO calculation [14], while the dashed and solid lines, respectively, incorporate purely threshold resummation [23] and joint threshold and recoil resummation [28].

calculations for direct photons [23,24] exhibit far less dependence on QCD scales than found in NLO theory. These calculations agree with the NLO prediction for the scale $\mu \approx p_T/2$ at low p_T (without inclusion of explicit k_T or recoil effects), and show an enhancement in cross section at high p_T.

A method for simultaneous treatment of recoil and threshold corrections in inclusive single-photon cross sections is being developed [28] within the formalism of collinear factorization. This approach accounts explicitly for the recoil from soft radiation in the hard-scattering subprocess, and conserves both energy and transverse momentum for the resummed radiation. The possibility of substantial enhancements from higher-order perturbative and power-law nonperturbative corrections relative to NLO are indicated at both moderate and high p_T for fixed-target energies, similar to the enhancements obtained with the simple k_T-smearing model discussed above. Figure 3 (right) displays the results of an example calculation [28] based on this approach compared with direct-photon measurements from E706.

While there is still no resummation calculation for inclusive pion production, the trend of recent developments in direct-photon processes has led to an increased appreciation of the importance of the effects of multiple gluon emission, and to the emergence of tools for incorporating these effects. These latest theoretical developments encourage optimism that the long-standing difficulties in developing an adequate description of these processes can eventually be resolved, making possible a global re-examination of parton distributions with an emphasis on the determination of the gluon distribution from the direct-photon data [32].

ACKNOWLEDGMENTS

This work has been done in collaboration with L. Apanasevich, M. Begel, C. Bromberg, T. Ferbel, G. Ginther, J. Huston, S. Kuhlmann, P. Slattery, and V. Zutshi. We also wish to thank P. Aurenche, C. Balázs, S. Catani, M. Fontannaz, N. Kidonakis, J. F. Owens, E. Pilon, G. Sterman, W. Vogelsang and J. Womersley for helpful discussions.

REFERENCES

1. Huston, J., et al., *Phys. Rev.* **D58**, 114034 (1998).
2. Lai, H. L., et al., *Phys. Rev.* **D55**, 1280 (1997).
3. Gluck, M., Reya, E., and Vogt, A., *Z. Phys.* **C53**, 127 (1992).
4. Martin, A. D., et al., *Eur. Phys. J.* **C4**, 463 (1998).
5. Baer, H., and Reno, M., *Phys. Rev.* **D54**, 2017 (1996).
6. Huston, J., et al., *Phys. Rev.* **D51**, 6139 (1995).
7. Apanasevich, L., et al., *Phys. Rev. Lett.* **81**, 2642 (1998).
8. Apanasevich, L., et al., *Phys. Rev.* **D59**, 074007 (1999).
9. Aurenche, P., et al., *Eur. Phys. J.* **C9**, 107 (1999).
10. Aurenche, P., et al., *Eur. Phys. J.* **C13**, 347 (2000).
11. Kimber, M. A., Martin, A. D., and Ryskin, M. G., *Eur. Phys. J.* **C12**, 655 (2000).
12. Aurenche, P., et al., *Phys. Lett.* **140B**, 87 (1984).
13. Berger, E. L., and Qiu, J.-W., *Phys. Rev.* **D44**, 2002 (1991).
14. Gordon, L. E., and Vogelsang, W., *Phys. Rev.* **D48**, 3136 (1993); **D50**, 1901 (1994).
15. Aversa, F., et al., *Phys. Lett.* **B210**, 225 (1988); **B211**, 465 (1988).
16. Bailey, B., Owens, J. F., and Ohnemus, J., *Phys. Rev.* **D46**, 2018 (1992).
17. Binnewies, J., Kniehl, B. A., and Kramer, G., *Phys. Rev.* **D52**, 4947 (1995).
18. Apanasevich, L., et al., hep-ph/0007191.
19. Begel, M., *Nucl. Phys.* **B79**, 244 (1999); Ph.D. thesis, Univ. of Rochester, 1999.
20. Altarelli, G., et al., *Nucl. Phys.* **B246**, 12 (1984).
21. Balázs, C., et al., *Phys. Rev.* **D57**, 6934 (1998).
22. Chiappetta, P., Fergani, R., and Guillet, J., *Phys. Lett.* **B348**, 646 (1995).
23. Catani, S., et al., *JHEP* **03**, 025 (1999).
24. Kidonakis, N., and Owens, J. F., *Phys. Rev.* **D61**, 094004 (2000).
25. Laenen, E., Oderda, G., and Sterman, G., *Phys. Lett.* **B438**, 173 (1998).
26. Lai, H.-L., and Li, H.-N., *Phys. Rev.* **D58**, 114020 (1998).
27. Li, H.-N., *Phys. Lett.* **B454**, 328 (1999).
28. Laenen, E., Sterman, G., and Vogelsang, W., *Phys. Rev. Lett.* **84**, 4296 (2000).
29. Owens, J. F., *Rev. Mod. Phys.* **59**, 465 (1987).
30. Sjöstrand, T., *Comput. Phys. Commun.* **82**, 74 (1994).
31. Marchesini, G., et al., *Comput. Phys. Commun.* **67**, 465 (1992).
32. Sterman, G., and Vogelsang, W., hep-ph/0002132.

Measurement of Drell-Yan Cross Sections and Flavor Asymmetry in the Nucleon Sea

J.C. Webb[h], T.C. Awes[i], M.E. Beddo[h], M.L. Brooks[f], C.N. Brown[c],
J.D. Bush[a], T.A. Carey[f], T.H. Chang[h,1], W.E. Cooper[c],
C.A. Gagliardi[j], G.T. Garvey[f], D.F. Geesaman[b], E.A. Hawker[j,f],
X.C. He[d], L.D. Isenhower[a], D.M. Kaplan[e], S.B. Kaufman[b],
P.N. Kirk[g], D.D. Koetke[k], G. Kyle[h], D.M. Lee[f], W.M. Lee[d],
M.J. Leitch[f], N. Makins[b,1], P.L. McGaughey[f], J.M. Moss[f],
B.A. Mueller[b], P.M. Nord[k], V. Papavassiliou[h], B.K. Park[f],
J.C. Peng[f], G. Petitt[d], P.E. Reimer[f,b], M.E. Sadler[a], J. Selden[h],
W.E. Sondheim[f], P.W. Stankus[i], T.N. Thompson[f], R.S. Towell[a,f],
R.E. Tribble[j], M.A. Vasiliev[j,2], Y.C. Wang[g], Z.F. Wang[g],
J.L. Willis[a], D.K. Wise[a], G.R. Young[i]

(FNAL E866/NuSea Collaboration)

Abilene Christian University[a], Argonne National Laboratory[b], Fermi National Accelerator Laboratory[c], Georgia State University[d], Illinois Institute of Technology[e], Los Alamos National Laboratory[f], Louisiana State University[g], New Mexico State University[h], Oak Ridge National Laboratory[i], Texas A&M University[j], Valparaiso University[k]

Abstract.
Measurements of Drell-Yan cross sections from the bombardment of liquid deuterium and hydrogen with an 800 GeV proton beam are reported. The relative cross section ($\sigma_{pd}/2\sigma_{pp}$) directly probes the ratio of anti-down ($\bar{d}$) to anti-up ($\bar{u}$) quarks in the nucleon sea, while absolute measurement of the deuterium cross section constrain $\bar{d}+\bar{u}$. Results are compared with previous experiments and next-to-leading order calculations based on current sets of parton distribution functions.

Recent experiments [1–5] have demonstrated a large asymmetry between the up and down anti-quark distributions in the nucleon sea. Although no known

---

[1] Present address: University of Illinois, Urbana, IL 61801
[2] On leave from Kurchatov Institute, Moscow 123182, Russia.

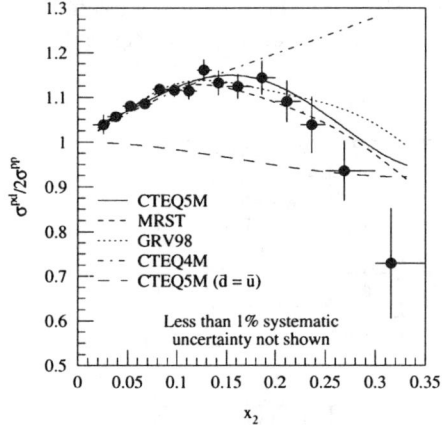

FIGURE 1. The ratio of deuterium to hydrogen cross sections vs. x. The curves are the ratios of next-to-leading order cross sections based on MRST, CTEQ5 and GRV98 parton distributions.

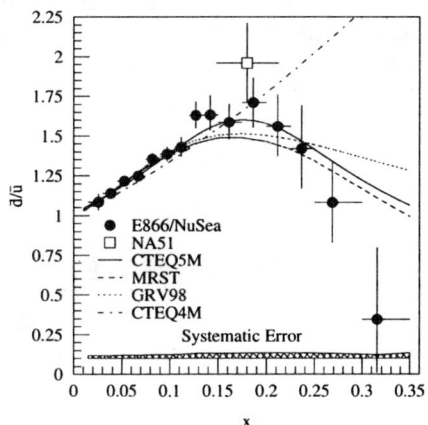

FIGURE 2. $\bar{d}/\bar{u}(x)$ vs x, extracted from the ratio of Drell-Yan cross sections. Statistical and systematic errors are shown. The NA51 data point is also shown.

symmetry demands the equality of $\bar{d}$ and $\bar{u}$, a large asymmetry was not expected from a perturbatively generated sea.

Fermilab E866 [4] was the first experiment to measure the Bjorken-x dependence of $\bar{d}/\bar{u}$. That previous work was based on a subset of the E866 data which had relatively large masses of the muon pair. A complete analysis [6] of the cross section ratio $\sigma_{pd}/2\sigma_{pp}$ is presented here. Progress has also been made in absolute measurements of the deuterium and hydrogen cross sections. We present here preliminary results for deuterium, which help justify one of the assumptions made in the extraction of $\bar{d}/\bar{u}$.

The apparatus used in E866 was the E605 spectrometer [7]. It was modified from its previous configuration by the replacement of the station 1 detectors with new hodoscope and drift chambers, reconfiguration of the hadronic absorbing wall, and an upgrade to the trigger system [8].

Experiment 866 measured the dimuon yields from an 800 GeV proton beam incident on liquid hydrogen and deuterium targets. Figure 1 shows the ratio of deuterium to hydrogen cross sections, versus the momentum fraction of the antiquark. Also shown are next-to-leading order (NLO) calculations of the ratio using different parton distributions. Parameterizations which predate E866 do not show the observed drop off above $x \approx 0.2$.

An iterative process was used to extract $\bar{d}/\bar{u}$ from the cross section ratio. This process calculated $\sigma_{pd}/2\sigma_{pp}$ at NLO based on the MRST [9] partons[3], and com-

---

[3] Results using CTEQ5 [10] and GRV98 [11] parton distributions were consistent within the stated systematic error.

pared it with the measured ratio. We then adjusted $\bar{d}/\bar{u}$, holding $\bar{d}+\bar{u}$ fixed at the MRST values until the calculated and measured cross section ratios agreed. The result is shown in figure 2.

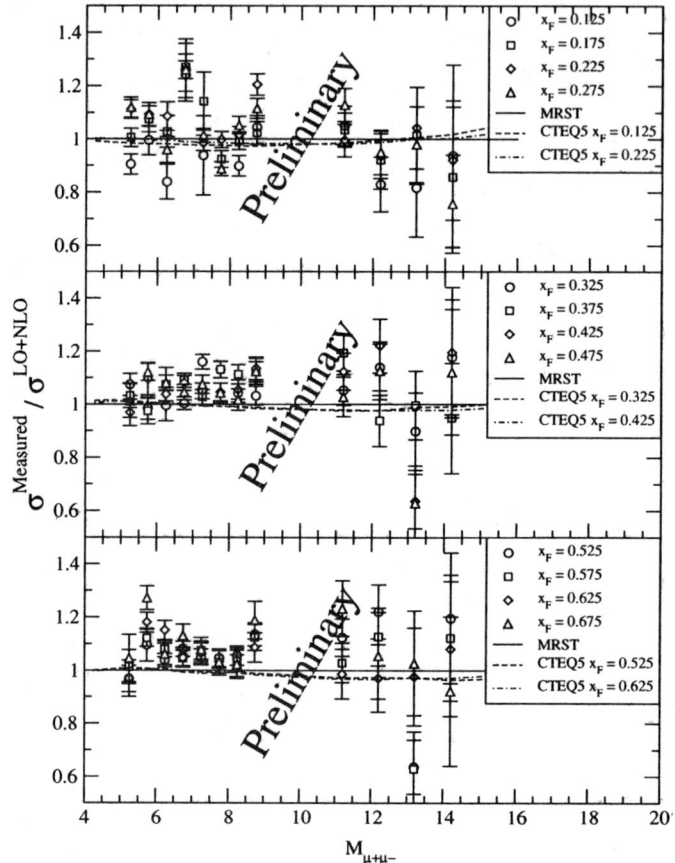

FIGURE 3. Preliminary measurement of $d^2\sigma/dMdx_F$ for $pd \to \mu^+\mu^- X$. The cross sections are divided by next-to-leading order calculations based on MRST parton distributions. Error bars represent statistical uncertainties only: there is an additional 10% uncertainty in the normalization and another 10% uncertainty in the acceptance.

The Drell-Yan cross section for $pd \to \mu^+\mu^- X$ is given by

$$M^2 \frac{d^2\sigma}{dM^2 dx_F} = \frac{4\pi\alpha^2}{9M^2} \frac{x_1 x_2}{x_1+x_2} \frac{4u(x_1)+d(x_1)}{9}[\bar{d}(x_2)+\bar{u}(x_2)] \qquad (1)$$

where $x_{1(2)}$ is the momentum fraction carried by the beam(target) quark(antiquark), M is the mass of the muon pair, and Feynman-x is $x_F = x_1 - x_2$. We have also taken the limit of large x_F, enabling us to identify the quark with the beam and the anti-quark with the target. Since the valence distributions are, in principle, well constrained by deep inelastic scattering measurements, equation 1 is a probe of $\bar{d}+\bar{u}$.

Figure 3 shows the measured cross section for deuterium divided by NLO calculations using MRST parton distributions. The data and predictions agree well within the stated systematic errors. Thus, the parameterizations of $\bar{d}+\bar{u}$ in the global fits are consistent with our data. A previous measurement of the cross section [12] also agrees with our measurement[4].

Although previous experiments demonstrated that $\bar{d} > \bar{u}$, E866 was the first experiment to measure the x dependence of $\bar{d}/\bar{u}$. This measurement has had a major impact on the global fits to the parton distributions, and has provided tests of non-perturbative models of the origin of the nucleon sea [13]. From the standpoint of the absolute cross sections, the parton distributions appear to be on a firm footing. Deviations from the NLO predictions are less than 5%, which is within our systematic error.

REFERENCES

1. P. Amaudruz et al., Phys. Rev. Lett. **66**, 2712 (1991); M. Arneodo et al., Phys. Rev. D **50**, R1 (1994).
2. M.R. Adams et al., Phys. Rev. Lett. **75**, 1466 (1995).
3. A. Baldit et al., Phys. Lett. B **332**, 244 (1994).
4. E.A. Hawker et al., Phys. Rev. Lett. **80**, 3715 (1998)
5. HERMES Collaboration, K. Ackerstaff et al., Phys. Rev. Lett. **81**, 5519 (1998).
6. R.S. Towell, P.L. McGaughy, et al., in preparation.
7. The E605 Collaboration, G. Moreno et al., Phys. Rev. D **47**, 2815 (1991).
8. C.A. Gagliardi, E.A. Hawker, R.E. Tribble, D.D. Koetke, P.M. Nord, et al.Nucl. Inst. Methods A **418**, 322 (1998).
9. A.D. Martin, R.G. Roberts, W.J. Stirling and R.S. Thorne, Eur. Phys. J. C **4**, 463 (1998).
10. H.L. Lai et al., Eur. Phys. J. C **12**, 375 (2000).
11. M. Glück, E. Reya, and A. Vogt, Eur. Phys. J. C **5**, 461 (1998).
12. P.L. McGaughey, et al., Phys. Rev. D **50**, 3038 (1994.); Erratum, Phys. Rev. D **60**, 119903 (1999).
13. FNAL E866/NuSea Collaboration, J.C. Peng, et al., Phys. Rev. D **58**, 092004 (1998).

---

[4]) Except in the region of large x_F and small mass, where E772 lies about 10-20% above the E866 measurement, though this is within the quoted systematic errors.

Challenges in the Global QCD Analysis of Parton Structure of Nucleons

Wu-Ki Tung

Michigan State University and Fermi National Laboratory [1]

Abstract. We briefly summarize the current status of global QCD analysis of the parton structure of the nucleon and then highlight the open questions and challenges which confront this endeavor on which much of the phenomenology of the Standard Model and the search of New Physics depend.

INTRODUCTION

All calculations of high energy processes with initial hadrons, both for precision study of the Standard Model (SM) or for exploring New Physics, require parton distribution functions (PDFs) as an essential input. The reliability of these calculations, which underpins future theoretical and experimental progress, depends substantially on having the best estimate of PDFs, and on understanding the uncertainties of the PDFs.

Global QCD analysis makes use of a large number of experimental inputs from many different high energy physical processes, treated in a uniform theoretical framework, typically next-to-leading order (NLO) QCD, in order to extract the non-perturbative parton distribution functions (PDFs) at some low momentum scale (Q_0). These unknown distributions are usually given by some suitably chosen parametrization form; and the parameters are determined phenomenologically in the global analysis. The most comprehensive efforts are represented by the widely used MRST and CTEQ distributions. The number of experimental data points are of the order of 1300, from about 15 different experiments; the number of theoretical parameters are typically around $15 \sim 25$.

This talk highlights the open questions and challenges which confront global QCD analysis of parton distributions due to the many sources of experimental and theoretical uncertainties, as well as possible choices of methodology, all which complicate this important task. Unfortunately, space limitation for this written version restricts most topics to a simple list, allowing neither detailed discussions

---

[1] On Sabbatical Leave

(including most of the figures shown during the talk) nor citations of most of the relevant papers on the topics touched.

OPEN ISSUES AND CHALLENGES

We begin with general remarks on challenges to experiment, theory and phenomenology; then turn to specific open physics issues in current global analysis.

General Challenges

Experimental input to global QCD analysis comes from the following physical processes: deep inelastic scattering (DIS), Drell-Yan pair production and asymmetry measurements, W/Z-, direct photon-, and Jet- production in hadron collisions. The accuracy of the data sets and the degree of details available on systematic errors vary very widely among the experiments. This makes the relative weighing of the data sets in the global analysis and the assessment of uncertainties a very complex and difficult task.

An important challenge to experimentalists is to establish common practices of reporting detailed errors, even when they are non-gaussianly distributed, and to present this information in convenient formats for phenomenological analysis as well as for future reference. Cf. "manifesto" in [1]. Furthermore, it is important that experimental measurements should be analyzed in a way as independent of theoretical models as possible, so as to avoid creating (unsuspected) artificial effects. A good example is that the long-standing "disagreement" between the neutral-current and (model-dependent) charge-current DIS F_2 determination has been cleared up by a recent model-independent re-analysis of the CCFR data. [2]

On the theory side, NLO QCD calculation of hard cross-sections form the foundation of current global analysis, but they are not adequate for treating all measured hard processes. Improved PQCD theory is needed for processes where NLO results show excessive renormalization and factorization scale dependence. Examples are direct photon and heavy quark production. Power law (higher-twist) corrections need to be brought under control. Both of these areas are related to the active research efforts devoted to the *resummation* of logarithmic effects of various kinds (small x-, p_T-, threshold-, ... etc.). Finally, for experiments performed on nuclear targets – deuteron (in lieu of the neutron), as well as heavy nuclei (mainly for neutrino DIS) – nuclear corrections need to be understood in order to obtain reliable information on the parton structure of the nucleon.

To obtain dependable parton distributions and estimates of their uncertainties, the global analysis efforts must, beyond using the most up-to-date theory and experiment, include systematic exploration of the dependence on the parametrization of the non-perturbative parton distribution functions, incorporate reliable methods of incorporating systematic errors, and provide practical ways to deliver the best

PDFs and ways to propagate their uncertainties to physical variables to the users throughout the theory and experimental communities.

We now turn to the open issues in current PDF analysis.

Breaking of SU(2) flavor symmetry

For studying the breaking of SU(2) flavor symmetry – the u vs. d and $\bar{u}$ vs. $\bar{d}$ differences – the relevant experiments are: DIS: $F_2^n - F_2^p$, F_2^n/F_2^p, F_2^{nc}/F_2^{cc}; Drell-Yan charge asymmetry: $\sigma^{pd}/2\sigma^{pp}$; and W-production: lepton y-asymmetry. The unresolved issues are: (i) "Correct" deuteron correction (relevant for $F^d \to F^n$ conversion); Heavy nuclear target corrections (relevant for extracting $F_{2,3}^{\nu N}$ from $F_{2,3}^{\nu A}$); other power-law (e.g. target mass) corrections; Small x: What is the origin of SU(2) flavor symmetry breaking of the sea?; Large x: Should the ratio $d(x)/u(x) \to 0$ as $x \to 1$?; and Charge symmetry violation: Is there any evidence for $u_p(x) \neq d_n(x)$? There are extensive discussions of these issues in the literature. There is no clear consensus on most of them.

The challenge, to both particle and nuclear physicists, is therefore to clean these problems up as far as possible – and model-independently to the extent possible. One way is to use nucleon targets as much as possible, e.g. use neutral and charged current events at HERA, using both electron and positron beams. With progress along this line, we can learn nuclear physics from these measurement, rather then invoking nuclear model calculations to extract PDFs.

Strange Sea Size, Shape and Charge Symmetry

Two combinations of total inclusive DIS structure function measurements are sensitive to the strange quark distribution. At the naive parton level:

$$\Delta F_3^{CC} = F_3^{\nu N} - F_3^{\bar{\nu} N} \approx 4(s-c) \quad \text{and} \quad \frac{5}{6}F_2^{CC} - 3F_2^{NC} \approx (s-c)$$

and, in addition, a *direct* measurement on inclusive structure function with tagged charm yields naively $F_2^{\text{charm}} \approx s$. These relations must be supplemented by contributions involving gluon initial state to be well-defined in the QCD framework.

Most current global analyses use as input the relation, $s(x) = 2\kappa/(\bar{u}(x) + \bar{d}(x))$ with $\kappa \simeq 0.5$, inferred from analysis of the charm production cross-section, primarily from the CCFR experiment. There is no clear evidence that this result is consistent with the two combinations of total inclusive structure functions given above. In fact, there are recent re-analyses of the earlier CDHS and CHARM data which reached different conclusions. The challenge here is to present the neutrino charm production data in the model-independent form of F_2^{charm}, so that they can be incorporated into the general global analysis. Then one can conclusively test the consistency of all available data, and determine the strange quark distribution independent of assumptions. Another open question is whether the commonly accepted assumption $s(x) = \bar{s}(x)$ is valid or not. This too can be tested experimentally if all available data are incorporated in a comprehensive analysis.

The Gluon Distribution

The gluon distribution contributes to all high energy processes. It is important for all SM, SUSY, and other New Physics calculations. Yet, it is among the least well determined of the parton distributions. Most relevant measurements for its determination are: $dF_2(x,Q)/d\ln Q$ in DIS; Inclusive and differential jet production cross-section; Direct Photon Production; and Heavy Quark Production.

Comprehensive analysis of the Q^2 dependence of the DIS structure functions, with the aim of determining $G(x,Q)$, has been carried out by the experimental groups, particularly BCDMS, NMC, H1, and ZEUS. Bands of uncertainties on $G(x,Q)$ have been presented. Inclusive jet production cross-section [3] which is directly sensitive to $G(x,Q)$ have been used by the CTEQ collaboration in their global analysis of PDFs; MRST plans to do the same in the future.

In contrast, the "classical" process for gluon determination – direct photon production – has been beset with uncontrollable theoretical uncertainties so far, in spite of recent experimental advances [4]. The current CTEQ analysis does not use these data; the MRST group also plans to drop this process from their input because the results on $G(x,Q)$ are totally dictated by the theoretical assumptions made in the treatment of direct photon data. In principle, heavy quark production cross-sections are also very sensitive to $G(x,Q)$. Unfortunately, the theoretical uncertainty in NLO QCD for this process is also not very reliable.

The primary challenge here is clearly to the theorists: we need to gain control over the theoretical accuracy of direct photon and heavy quark production calculations. Much research has recently been directed toward solving this problem by incorporating the resummation of relevant logarithms which arise in these multi-scale problems in QCD.

Heavy Quarks

An independent question concerns the heavy quark distributions $c(x,Q)$ and $b(x,Q)$. Are they purely "radiatively generated" by gluon splitting, or are there "intrinsic" charm and bottom partons inside the nucleon? In order to ask this question in a meaningful way, one needs a theoretical framework in which the relevant concepts are well-defined. This is not the case, for instance, in most papers on "NLO calculation" of heavy quark production cross-sections using the "fixed-flavor number scheme" in which there is no charm/bottom parton at all, by definition! General QCD formalisms for including heavy quark partons have been developed in the past 15 years. The various (theoretically equivalent but practically limited) sub-schemes have proven to be relatively robust in comparison with DIS data, but problems persist with hadron collider data. Comprehensive phenomenology has yet to be carried out using the general formalism.

UNCERTAINTIES OF PARTON DISTRIBUTION FUNCTIONS

The reliability of all calculations of high energy processes, both within and beyond the Standard Model, which underpins both future theoretical and experimental progress, depends substantially on understanding the uncertainties of the PDFs. This highlights one of the most important challenges of global QCD analysis.

It might appear, since some χ^2 minimization procedure is used to perform the global analysis, one could simply assess the uncertainties of the PDFs, and propagate the experimental errors to the predicted physical quantities, according to standard textbook methods. However, in practice, because a large body of data from many diverse experiments is included, using a theoretical model with many parameters and with many sources of uncertainty of its own, the standard error matrix method and the general tools for evaluating the error become inadequate.

Several approaches to this problem have recently been proposed, with rather different emphases on the rigor of the statistical method, scope of experimental input, and attention to the various practical complications. [1] Our group at Michigan State has initiated one of these efforts, with the emphasis on developing practical methods which enable the utilization of the full constraints of the global data – in contrast to others which often turn out to be applicable only to specific sets of data, usually DIS structure functions. This work is enabled by distinctive improvements and extensions of standard general purpose methods and tools [5]. We give here a glimpse of this new development.

The uncertainties of the PDFs are embodied in those of the phenomenological parameters $\{a_i, i = 1, ..., n\}$ which are used to specify the non-perturbative PDFs at a low (confinement) scale, say Q_0. PDFs and their uncertainties at higher scales are obtained by QCD evolution. The Error Matrix of the a_i parameters is the inverse of the Hessian, the matrix comprising the second derivative of the χ^2 function with respect to $\{a_i\}$. For many reasons, the complex system of global QCD analysis results in very disparate rate of change of χ^2 along different directions in the parameter space, resulting in five orders of magnitude difference in the values of the eigenvalues of the Hessian matrix. This renders the standard methods for calculating the Hessian, such as the widely used program MINUIT, numerically unstable, hence the results unusable. This fact motivated us to devise an iterative procedure to calculate the Hessian, which results in uniform accuracy in all directions of the parameter space, and hence stable and reliable results all around. Comparative tests show that this method renders the traditional error matrix formalism applicable where it fails badly in the standard way of calculation. Once a reliable Hessian matrix is available, along with a set of "best-fit" PDFs, one can calculate both the best estimate (prediction) of any physical quantity of interest and the uncertainty of that quantity associated with PDFs – the latter by folding the error matrix into the gradient vector of the relevant physical quantity with respect to the parameters $\{a_i\}$, assuming the linear approximation works.

The above "standard method" relies on the validity of linear approximations of all relevant quantities with respect to $\{a_i\}$. However, in the final analysis, it is the uncertainties of physical quantities, say $\{X\}$, over the entire parameter space that really count – it is desirable to directly quantify the uncertainties of physical variables without the intermediate steps involving $\{a_i\}$ and the associated approximations. We found that this challenge can be met by using the well-known mathematical method of the Lagrange undetermined multiplier. The basic idea is to map out the allowed range of variation of, say a given X, with respect to the overall χ^2 value of the global fit by performing many unconstrained fits to the global data sets using the new minimization function $\chi^2(a_i) + \lambda X(a_i)$ with a series of values of λ, the Lagrange undetermined multiplier variable. This procedure can be generalized to several physical variables, using one Lagrange multiplier for each.

The Lagrange multiplier method yields the most robust estimate of the uncertainty on X; and it can be used to test the efficacy of the Hessian method. We have performed such studies [5], and found the two methods to be quite consistent with each other, thus lending credibility to both. The development of these methods marks the beginning of a systematic study of uncertainties associated with PDFs.

CONCLUSION

We have given an overview of the current status of, and the challenges facing, the Global QCD Analysis of Parton Structure of Nucleons. Unfortunately, due to space limitation, this brief narrative cannot be supported by the figures shown during the talk, and by key references to the topics covered. Nonetheless, we hope the reader can get a sense of the importance of the endeavor, of where things stand now, and where the field is going.

Acknowledgement: I thank my CTEQ and MSU colleagues, particularly J. Pumplin and D. Stump for stimulating discussions and collaboration.

REFERENCES

1. Contribution of the *Parton Distributions Working Group* to the Proceedings of Fermilab RunII Workshop on *QCD and Gauge Boson Physics*, hep-ph/0006300.
2. See A. Bodek, contribution to these proceedings.
3. See M. Zielinski, contribution to these proceedings.
4. See R. Hirosky, contribution to these proceedings.
5. J. Pumplin, D. Stump, and W.K. Tung, *Multivariate Fitting and the Error Matrix in Global Analysis of Data*, hep-ph/0008191.

Diffractive Physics at the Tevatron

K. Hatakeyama
For the CDF and DØ collaborations

*The Rockefeller University, 1230 York Avenue,
New York, NY, 10021, U.S.A.*

Abstract. Experimental results of hard single diffraction and double pomeron exchange studies at the Fermilab Tevatron $p\bar{p}$ collider are presented. Single diffraction results are compared with predictions from phenomenological models and expectations from results obtained in diffractive deep inelastic scattering experiments at the DESY ep collider HERA. Double pomeron exchange results are compared with corresponding single diffraction results to test factorization.

INTRODUCTION

Recent results from studies of hard single diffraction (SD) [1,2] and double pomeron exchange (DPE) [3] events produced in $p\bar{p}$ collisions at the Fermilab Tevatron collider by the CDF and DØ collaborations are presented. Hard SD events contain a hard scattering and a leading (anti)proton associated with a rapidity gap, which is defined as a pseudorapidity [4] region devoid of particles. In hard DPE events, both incoming proton and antiproton are scattered quasielastically and are separated by rapidity gaps from a hard scattering occurring in the central region. The rapidity gap is presumed to be due to the exchange of a Pomeron ($I\!P$), which is a color-singlet state with vacuum quantum numbers. In this framework, a $p\bar{p}$ hard SD event can be expressed as $\bar{p}+p \rightarrow [\bar{p}'+I\!P]+p \rightarrow \bar{p}'+(jj, W, b\bar{b}, ...)+X$, and similarly, a $p\bar{p}$ hard DPE event as $\bar{p}+p \rightarrow [\bar{p}'+I\!P]+[p'+I\!P] \rightarrow \bar{p}'+p'+(jj, W, b\bar{b}, ...)+X$. These processes are illustrated in figure 1.

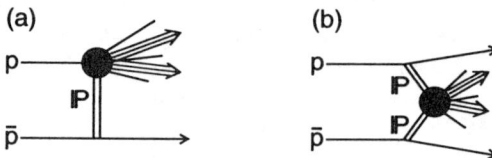

FIGURE 1. Diagrams of (a) single diffractive dijet production and (b) dijet production in double pomeron exchange in a $p\bar{p}$ interaction.

The question which we seek to answer is whether hard diffraction processes obey QCD factorization, i.e. can be expressed in terms of the parton-parton cross section convoluted with a universal diffractive (anti)proton structure function. The diffractive structure function depends not only on Q^2 and the Bjorken scaling variable, x, but also on the fractional momentum loss of the (anti)proton, ξ, and the four momentum transfer squared at the p-$I\!P$ ($\bar{p}$-$I\!P$) vertex, t. In this talk, factorization for hard diffraction processes is tested by comparing $p\bar{p}$ hard SD results with predictions from phenomenological models, expectations from results obtained in diffractive deep inelastic scattering (D-DIS) experiments, and results obtained in $p\bar{p}$ hard DPE studies.

HARD DIFFRACTION WITH A RAPIDITY GAP

The CDF collaboration has measured the fraction of events that contain forward rapidity gaps in W, dijet and b-quark production at $\sqrt{s} = 1800$ GeV [5]. Recently, the DØ collaboration has also reported a measurement of the gap fraction in forward and central dijet events at $\sqrt{s} = 630$ and 1800 GeV [1]. In the DØ analysis, rapidity gaps are identified by using the LØ forward scintillator arrays ($2.3 < |\eta| < 4.3$), and a portion of the forward calorimeters ($3.0 < |\eta| < 5.2$). The measured gap fractions are shown in Table 1. The gap fractions predicted with four different pomeron structure functions, (1) "hard gluon", $s(\beta) \propto \beta(1-\beta)$; (2) "flat gluon", $s(\beta) \propto$ constant; (3) "soft gluon", $s(\beta) \propto (1-\beta)^5$; and (4) "quark", the quark analog of (1), are also shown in Table 1, where $\beta(=x/\xi)$ is the momentum fraction of the Pomeron carried by the interacting parton. Although the quark structure is in general agreement with the data, it has previously been shown to yield a higher rate than that measured for diffractive W production [5]. The lower half of Table 1 provides valuable information, since the Monte Carlo normalization cancels in the ratio of gap fractions for the same $\sqrt{s}$. A gluonic Pomeron containing significant soft and hard components, combined with a reduced (renormalized [6]) pomeron flux factor, could reasonably describe all the data samples.

TABLE 1. The measured and predicted gap fractions and their ratios.

Sample	Data	Hard Gluon	Flat Gluon	Soft Gluon	Quark
		Gap Fraction			
1800 GeV $|\eta| > 1.6$	$(0.65 \pm 0.04)\%$	$(2.2 \pm 0.3)\%$	$(2.2 \pm 0.3)\%$	$(1.4 \pm 0.2)\%$	$(0.79 \pm 0.12)\%$
1800 GeV $|\eta| < 1.0$	$(0.22 \pm 0.05)\%$	$(2.5 \pm 0.4)\%$	$(3.5 \pm 0.5)\%$	$(0.05 \pm 0.01)\%$	$(0.49 \pm 0.06)\%$
630 GeV $|\eta| > 1.6$	$(1.19 \pm 0.08)\%$	$(3.9 \pm 0.9)\%$	$(3.1 \pm 0.8)\%$	$(1.9 \pm 0.4)\%$	$(2.2 \pm 0.5)\%$
630 GeV $|\eta| < 1.0$	$(0.90 \pm 0.06)\%$	$(5.2 \pm 0.7)\%$	$(6.3 \pm 0.9)\%$	$(0.14 \pm 0.04)\%$	$(1.6 \pm 0.2)\%$
		Ratio of Gap Fraction			
630/1800 $|\eta| > 1.6$	1.8 ± 0.2	1.7 ± 0.4	1.4 ± 0.3	1.4 ± 0.3	2.7 ± 0.6
630/1800 $|\eta| < 1.0$	4.1 ± 0.9	2.1 ± 0.4	1.8 ± 0.3	3.1 ± 1.1	3.2 ± 0.5
1800 $|\eta| > 1.6/|\eta| < 1.0$	3.0 ± 0.7	0.88 ± 0.18	0.64 ± 0.12	$30. \pm 8.$	1.6 ± 0.3
630 $|\eta| > 1.6/|\eta| < 1.0$	1.3 ± 0.1	0.75 ± 0.16	0.48 ± 0.12	$13. \pm 4.$	1.4 ± 0.3

DIFFRACTIVE DIJETS WITH A LEADING ANTIPROTON

The CDF collaboration has studied SD dijet production at $\sqrt{s} = 1800$ GeV [2] by using events triggered on a leading antiproton detected in a Roman Pot spectrometer (RPS). In this study, the diffractive structure function of the antiproton is measured and compared with expectations from results obtained in diffractive DIS experiments at HERA to test factorization. In leading order QCD, the ratio $R_{ND}^{SD}(x, Q^2, \xi)$ of diffractive to non-diffractive (ND) dijet rates is equal to the ratio of the diffractive to ND structure functions. The relevant diffractive structure functions integrated over t can be written as $F_{jj}^D(x, Q^2, \xi) = x[g^D(x, Q^2, \xi) + \frac{4}{9} q^D(x, Q^2, \xi)]$ where $g^D(x, Q^2, \xi)$ and $q^D(x, Q^2, \xi)$ are respectively the diffractive gluon and quark parton densities, and $\frac{4}{9}$ is a color factor. Therefore, the diffractive structure function can be obtained by multiplying the ratio $R_{ND}^{SD}(x, Q^2, \xi)$ by the known non-diffractive structure function $F_{jj}(x, Q^2)$. The value of x is evaluated from the detected jets as $x = \sum_{i=1}^{2(3)} E_T^i e^{-\eta^i}/\sqrt{s}$, where the sum is carried out over the two leading (highest transverse energy) jets plus a third jet if $E_T^{jet3} > 5$ GeV. By changing variables, $x \to \beta\xi$, the obtained diffractive structure function $F_{jj}^D(x, Q^2, \xi)$ can be transformed to $F_{jj}^D(\beta, Q^2, \xi)$.

Figure 2 shows the measured $F_{jj}^D(\beta)$ for $0.035 < \xi < 0.095$, $|t| < 1$ GeV^2 and $E_T^{jet1,2} > 7$ GeV. The dashed (dotted) curve is the expectation for $F_{jj}^D(\beta)$ from fit 2 (fit 3) of the H1 diffractive parton densities [7] at $Q^2 = 75$ GeV^2, which approximately corresponds to the $\langle E_T^{jet} \rangle^2$ of the CDF data. The measured $F_{jj}^D(\beta)$ distribution and expectations from the H1 analysis of diffractive DIS events disagree both in normalization and shape, which indicates a breakdown of factorization. The discrepancy in normalization, defined as the ratio of the integral over β of data to expectation, is $D = 0.06 \pm 0.02$ ($D = 0.05 \pm 0.02$) for fit 2 (fit 3).

DIJET PRODUCTION IN DOUBLE POMERON EXCHANGE

Dijet production in DPE has been studied by the CDF collaboration at $\sqrt{s} = 1800$ GeV [3] using events triggered on a leading antiproton and requiring a forward rapidity gap on the leading proton side.

As mentioned earlier, in leading order QCD, the ratio $R_{ND}^{SD}(x_{\bar{p}})$ of the SD to ND dijet events as a function of $x_{\bar{p}}$ is equal to the ratio of the diffractive to ND structure functions of the antiproton. Similarly, the ratio $R_{SD}^{DPE}(x_p)$ of the DPE to SD dijet rates as a function of x_p is equal to the ratio of the diffractive to ND structure functions of the proton. The variables x_p and $x_{\bar{p}}$ are the Bjorken scaling variables for the proton and antiproton, respectively. Therefore, factorization for hard diffraction processes can be tested by comparing $R_{ND}^{SD}(x_{\bar{p}})$ with $R_{SD}^{DPE}(x_p)$. A

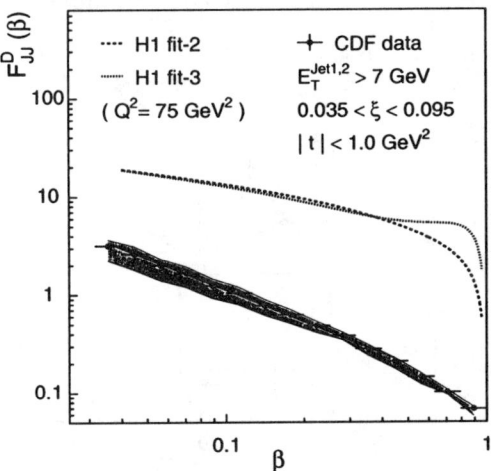

FIGURE 2. Data β distribution (points) compared with expectations from the parton densities of the proton extracted from diffractive deep inelastic scattering by the H1 collaboration. The straight line is a fit to the data of the form β^{-n} in the region $\beta < 0.5$. The lower (upper) boundary of the filled band represents the data distribution obtained by using only the two leading jets (up to four jets of $E_T > 5$ GeV) in evaluating β. The dashed (dotted) lines are expectations from the H1 fit 2 (fit 3). The systematic uncertainty in the data normalization is $\pm 25\%$.

deviation of the double ratio $D = R_{ND}^{SD}(x_{\bar{p}})/R_{SD}^{DPE}(x_p)$ from unity would indicate a breakdown of factorization.

In figure 3, the ratio $R_{SD}^{DPE}(x_p)$ is compared with $R_{ND}^{SD}(x_{\bar{p}})$ as a function of $x (\equiv x_p = x_{\bar{p}})$. The data are restricted to the regions $7 < E_T^{jet1,2} < 10$ GeV, $|t_{\bar{p}}| < 1$ GeV^2, $0.035 < \xi_{\bar{p}} < 0.095$, and for DPE $0.01 < \xi_p < 0.03$, where $\xi_{\bar{p}}$ (ξ_p) is the fractional momentum loss of the antiproton (proton), and $t_{\bar{p}}$ is the four momentum transfer squared at the $\bar{p}$-$\mathbb{P}$ vertex. Both ratios are normalized per unit ξ. Although the factorization test has to be performed at the same x and ξ values, the $\xi_{\bar{p}}$ and ξ_p regions in the data do not overlap. The $\xi_{\bar{p}}$ dependence of the ratio $\tilde{R}_{ND}^{SD}$ is examined in the insert of figure 3, where the tilde over the R indicates the weighted average in the region of x within the vertical dashed lines in the main figure. The ratio $\tilde{R}_{ND}^{SD}$ is found to be flat in the region $0.035 < \xi_{\bar{p}} < 0.095$. The ratio of $\tilde{R}_{ND}^{SD}$ extrapolated down to $\xi_{\bar{p}} = 0.02$ to be compared with $\tilde{R}_{SD}^{DPE}$ in the region $0.01 < \xi_p < 0.03$ is $D \equiv \tilde{R}_{ND}^{SD}/\tilde{R}_{SD}^{DPE} = 0.19 \pm 0.07$, which indicates a breakdown of factorization.

CONCLUSION

The CDF and DØ collaborations have studied hard single diffraction and double pomeron exchange events produced in $p\bar{p}$ collisions at the Fermilab Tevatron

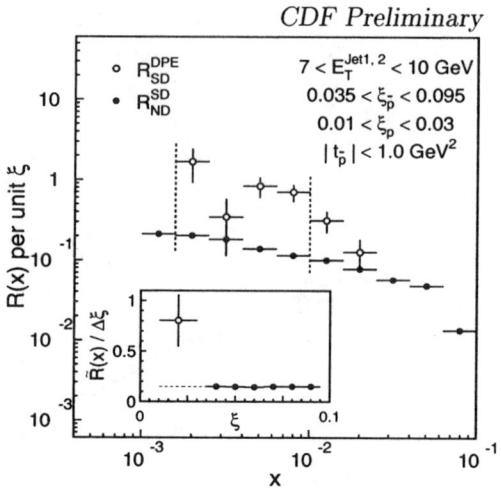

FIGURE 3. Ratios of DPE to SD (SD to ND) dijet event rates per unit ξ, shown as open (filled) circles, as a function x-Bjorken of partons in the p ($\bar{p}$). The errors are statistical only. The SD/ND ratio has a normalization uncertainty of $\pm 20\%$. The insert shows $\tilde{R}(x)$ per unit ξ versus ξ, where the tilde over the R indicates the weighted average of the $R(x)$ points in the region of x within the vertical dashed lines, which mark the DPE kinematic boundary (left) and the value of $x = \xi_p^{min}$ (right).

collider. A breakdown of factorization for hard diffraction processes is observed in comparing $p\bar{p}$ hard single diffraction results with expectations from results obtained in diffractive deep inelastic scattering experiments, or with results from a study of $p\bar{p}$ hard double pomeron exchange events. These studies provide further insight into the mechanism of hard diffraction, which will undoubtedly help in establishing the phenomenology of hadronic diffraction processes.

REFERENCES

1. B. Abbott et al., (DØ Collaboration), FERMILAB-PUB-99/375-E.
2. T. Affolder et al., (CDF Collaboration), Phys. Rev. Lett. **84**, 5043 (2000).
3. T. Affolder et al., (CDF Collaboration), FERMILAB-PUB-00/098-E.
4. The pseudorapidity, η, is defined as $\eta = -\ln(\tan\frac{\theta}{2})$, where θ is the polar angle between a particle and the proton beam direction.
5. F. Abe et al., (CDF Collaboration), Phys. Rev. Lett. **78**, 2698 (1997); **79**, 2636 (1997); T. Affolder et al., Phys. Rev. Lett. **84**, 232 (2000).
6. K. Goulianos, Phys. Lett. B **358**, 379 (1995); B **363**, 268 (1995).
7. C. Adloff et al., (H1 Collaboration), Z. Phys. C **76**, 613 (1997)

A Ring Imaging Čerenkov for HERMES

Dirk De Schepper [1], ANL

In 1998 a dual radiator ring imaging Čerenkov (RICH) was installed at HERMES. This detector will significantly enhance the hadron identification capability of the spectrometer. A concept drawing of one of the two RICH halves is shown below. The 5.5 cm thick aerogel tile wall is located directly behind the entrance window. The rest of the volume is filled with fluorobutane (C_4F_{10}) gas at room temperature and pressure. A large segmented mirror at the rear of the detector images the Čerenkov light onto the photon detector. The box is fitted with gas connections and pressure regulators which provide a continuous controlled flow of recirculating gas, regulated at a small overpressure with respect to atmosphere. The aerogel is incased in a gas-tight container, through which dry nitrogen is flowing.

The photon detector consists of a hexagonal array of 1934 PMTs wrapped in 100 μm of μ-metal and held in a soft steel matrix to shield them from the residual field of the spectrometer magnet. The gas seal is provided by a quartz window, which was glued to a soft steel insert, which itself was glued into the matrix. A mylar spacer is inserted between the photocathode and the window. To decrease the dead space between the PMT's, an aluminized foil cone is inserted in front of each PMT. This foil sticks out above the soft steel insert, effectively minimizing the dead space between the PMTs. The readout electronics are slightly modified PCOS4 modules, which digitize the information from each PMT, registring only when a PMT response exceed the threshold. This reduces the data transfer to a manageable size.

---

[1] for the HERMES collaboration

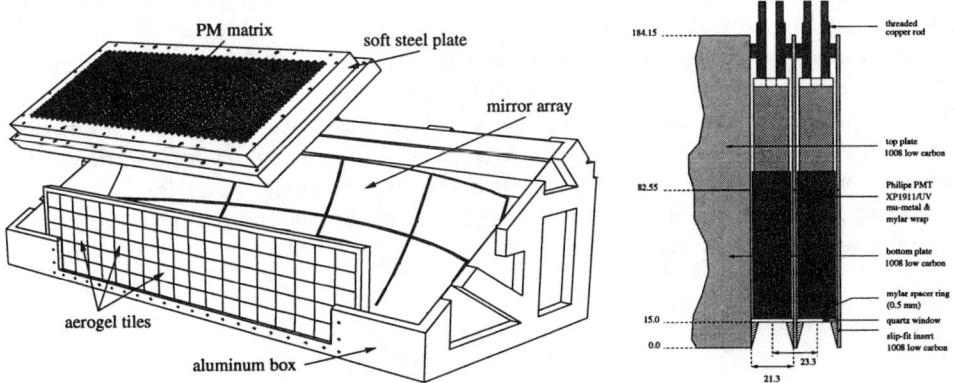

FIGURE 1. Schematic view of the RICH counter and the photon detector design.

Table 1 lists estimates for the essential contributions to the single photon resolution. $\Delta\theta_{em}$ results from the uncertainty in the emission vertex along the track in the radiator. The detector granularity determines $\Delta\theta_{pix}$, which follows from the size of the pixels. These two contributions affect both the aerogel and the gas angles. The chromatic aberration $\Delta\theta_{chr}$ derives from the variation of the index of refraction with respect to the wavelength and is only important for the resolution in the aerogel. These quantities do not add up quadratically and are only listed separately to indicate their relative importance. Their combined effect is calculated in the MC and is listed as $\Delta\theta_{MC}$.

	aerogel	C_4F_{10}
$\Delta\theta_{em}$	1.8 mrad	2.2 mrad
$\Delta\theta_{pix}$	5.6 mrad	5.2 mrad
$\Delta\theta_{chr}$	2.5 mrad	-
$\Delta\theta_{MC}$	7.1 mrad	7.2 mrad

Table 1 Fundamental contributions to the single photon resolutions

	aerogel	C_4F_{10}
$\Delta\theta_{tile}$	~3.0 mrad	-
$\Delta\theta_{nvar}$	~1.1 mrad	-
$\Delta\theta_{press}$	-	~1.0 mrad
$\Delta\theta_{exp}$	7.6 mrad	7.5 mrad
$\Delta\theta_{mirr}$	~2.4 mrad	~2.2 mrad

Table 2 Additional contributions to the single photon resolutions

Additional contributions to the single photon resolution are listed in table 2. A significant effect is the irregular shape of the aerogel tile surface near the edge. This gives rise to the $\Delta\theta_{tile}$ contribution. $\Delta\theta_{nvar}$ accounts for variations in the index of refraction between the aerogel tiles. Its value is determined from the measured indexes of refraction. Pressure and temperature fluctuations change the density of the gas, and therefore its index of refraction. This contribution, $\Delta\theta_{press}$, was determined from the observed maximum pressure fluctuations. The experimental value $\Delta\theta_{exp}$ was determined for high energy single lepton tracks that are not affected by any tile edge or mirror acceptance effects. Events with a large background were excluded. The reconstructed angle spectra were fitted with a Gaussian plus a linear background. This selection of the data means that $\Delta\theta_{tile}$ does not contribute, and the effect of the background is minimized. Since the data were taken in a period of stable atmospheric pressure, $\Delta\theta_{press}$ is negligible too. $\Delta\theta_{nvar}$ will still contribute to the aerogel resolution.

There remains the contribution of the mirror imperfections, in particular the non-sphericity of the mirror array and the diffuse reflection component of its surface. This is combined in the $\Delta\theta_{mirr}$ contribution, which is calculated from the difference of the experimental resolution and the quadratic sum of the relevant contributions to the resolution. This value agrees with the expectations. The fact that it is somewhat larger for the aerogel is in qualitative agreement with the fact that the aerogel rings are larger and therefore more sensitive to the outlying parts of the mirror.

Overview of Exotic Strange Quark Matter Search Experiments

James L. Nagle*

Columbia University, Department of Physics, New York, NY 10027

Abstract. A brief overview is given of experimental searches for strange quark matter. Recent searches for a bound H-dibaryon state and larger strangelets are summarized, and a projection on the future of such searches is presented.

INTRODUCTION

QCD allows for bound states of quarks in color-singlet configurations (hadrons such as baryons and mesons). However, nuclei are not single hadrons, but instead bound states of individual nucleons where the nucleons represent the relevant degrees of freedom (not quarks and gluons). Quark matter, where the quarks are the relevant degrees of freedom, composed of only up and down quarks is known to be unstable (will decay strongly) since normal nuclei do not spontaneously transition into two flavor quark matter.

Chin and Kerman [1] predicted in 1979 that strange quark matter (SQM) having a roughly equal number of up, down and strange quarks with $A < 10$ might be metastable ($\tau < 10^{-4}$ seconds). These SQM states have a reduced Fermi energy due to the additional quark flavor and a reduced Coulomb repulsion since an equal u,d,s population yields an electrically neutral state. Thus SQM states could range in size from $A=2$ (referred to as the H-dibaryon) to $A=3-200$ (referred to as strangelets) to A greater than 10^6 (bulk size). Witten proposed in 1984 that SQM in bulk could even be the ground state of nuclear matter and could exist as remnants of the Big Bang [2].

Since the normal world around us is not dominantly composed of SQM, there has been much speculation under what conditions in nature or in the laboratory SQM would exist and be detectable.

Early Universe and Strange Stars

As Witten proposed, shortly after the Big Bang, nuggets of SQM could have formed and if perfectly stable would still exist today in the universe. These SQM nuggets were proposed as a possible source of dark matter. There have been numerous experiments searching for such states that have employed a variety of techniques (including Rutherford type scattering and electrostatic charged-particle separation). All searches have yielded null results with limits significantly below predicted yields in the case of stable SQM. A recent overview is given in [3].

There is also the possibility that the core of dense stars is composed of SQM. SQM would alter the equation of state of the star's core and thus might yield measurable effects in the star's spin down. It has recently been postulated that the millisecond pulsar SAX J1808.4-3658 may be a strange star [4]. Searches for more candidates and other more striking signatures are ongoing.

Coalescence of Strange Baryons

In an effort to produce SQM states in the laboratory, the great challenge is to bring together enough strange quarks into a small volume where they might fuse to create SQM. Since even in the highest energy reactions and/or the largest size heavy ion reactions the total strangeness produced is relatively small, we are limited to searches for the H-dibaryon and strangelets ($A < 100$).

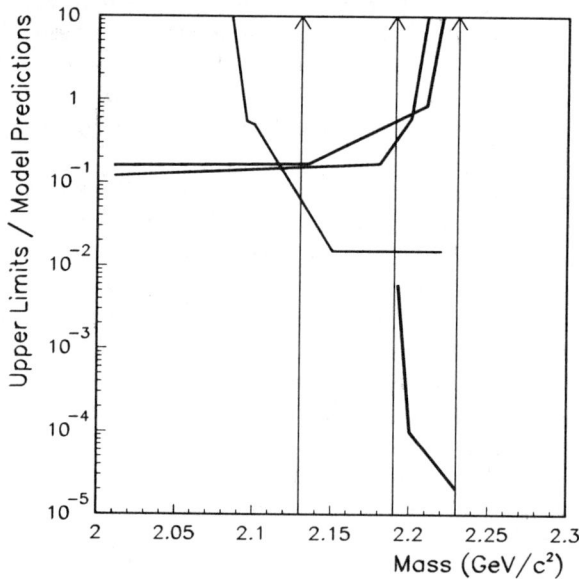

FIGURE 1. Experimental H-dibaryon upper limits divided by theoretical estimates of the production level, plotted as a function of H mass. The vertical lines indicate the different decay mode thresholds discussed in the text. The curves are from experiments E224, E836, E888, KTeV (seen from upper left to lower right).

There have been many searches for the H-dibaryon state, in part because the required two units of strangeness can be produced in close proximity in double strangeness exchange reactions (with kaon beams), in elementary particle collisions, in proton-nucleus reactions and in heavy ion reactions. Donoghue et al. [5] have estimated the possible lifetimes and decay modes for the H. If the H has a mass greater than that of two Λs, then the H is unbound and may only be a resonance similar to the d^* in proton-proton interactions. If the H has a mass greater than $\Lambda + p + \pi$, then it will undergo a weak decay into these three particles with a short lifetime of order $\tau_H = \tau_\Lambda/2$. This particular state would be difficult to distinguish from a two Λ bound state. If the H is even more bound then it might decay into $\Sigma^- + p$, $\Sigma^0 + n$, or $\Lambda + n$ with an expected lifetime of order 10^{-8} seconds. Below this binding, it would only be able to decay via a $\Delta S = 2$ reaction which would result in a very long lifetime of 10^5 seconds.

Many experiments have searched for the H in various decay modes and lifetime windows. Four of the most recent and sensitive searches find null results and are summarized in Figure 1. The two upper curves (green and black) are the results for double strangeness exchange reaction searches from experiment E224 at KEK and E836 at Brookhaven National Lab. Their current limits are approximately an order of magnitude below theoretical estimates for the H production in these reactions. Experiment E888 at Brookhaven searched for the H via its $\Lambda + n$ decay channel in proton-nucleus collisions. They originally published two candidates [6], but have subsequently shown them to be consistent with background [7]. Their null observation is shown as the blue curve with limits approximately two orders of magnitude below predictions.

In this case it is particularly interesting to investigate the theoretical prediction used in calculating this curve. In a calculation by Cousins [8], they predict an H cross section for these proton-nucleus reactions of 1.0 μb, as compared to the experimental upper limit of $\sigma_H < 60 nb$. However, experiment E888 is only sensitive to H production forward of mid-rapidity ($x > 0$). Shown in Figure 2 is a calculation we have done for p+p and p+Pt reactions using the transport model RQMD with a coalescence after-burner [9]. One can see that the Λ production is shifted substantially to backward rapidities in p+Pt, and this is reflected even more strongly in a backward shift of the H production. Thus, since E888 measures only forward of $y = 1.9$, we believe they have overestimated their sensitivity (as suggested by Cole [10]). We calculate the E888 limit as only $\sigma_H < 1.2\mu b$. However, in our model, if we calculate the expected yield with only the first N-N reaction contributing we get a predicted H production of $2\mu b$ (in rough agreement with Cousins), but for a real p-Pt reaction we see a substantial enhancement in the predicted H production ($\sigma_H = 40\mu b$). This new comparison certainly needs further study.

The most exciting new result is a limit from the KTeV collaboration [11] in the $\Lambda + p + \pi^-$ channel. Their extremely low limit is shown in Figure 1. We have re-checked the production model they use from Rotondo [12], and find good agreement using RQMD+Coalescence within a factor of two. It is notable that the KTeV limits are sensitive for lifetimes $> \tau_\Lambda/2$. It is possible they could then miss a very weakly bound H. Experiment E896 at Brookhaven covers this lifetime range, and so those results will be quite useful.

It has been proposed that if one can observe double-Λ hypernuclei that decay by sequential $\Delta S = 1$ reactions, then the H-dibaryon cannot be deeply bound. There are three candidates for such states from three different experiments, but no consensus about their correct identification. Recently experiment E906 at Brookhaven has reported a possible sample of tens of $^{4}_{\Lambda\Lambda}H$ states [13]. If these are confirmed, then anything but the most weakly bound H are ruled out.

Strangelet Searches in Heavy Ion Reactions

SQM with $|S| > 2$ can only be produced in the laboratory in relativistic heavy ion reactions. It is well known that strangeness production is enhanced in these reactions relative to p-p, and thus as the system cools off it is possible that multiple strange and non-strange baryons can coalescence into either multi-strange hypernuclei or strangelets (depending on their relative stability). It has also been hypothesized that if a deconfined plasma of quarks and gluons is formed that it could cool off into a strangelet. This latter mechanism is extremely speculative and does not provide a clear quantitative dynamical picture. Strangelets are searched for by looking for anomalously low Z/A ratios, outside the range possible for nuclear isotopes.

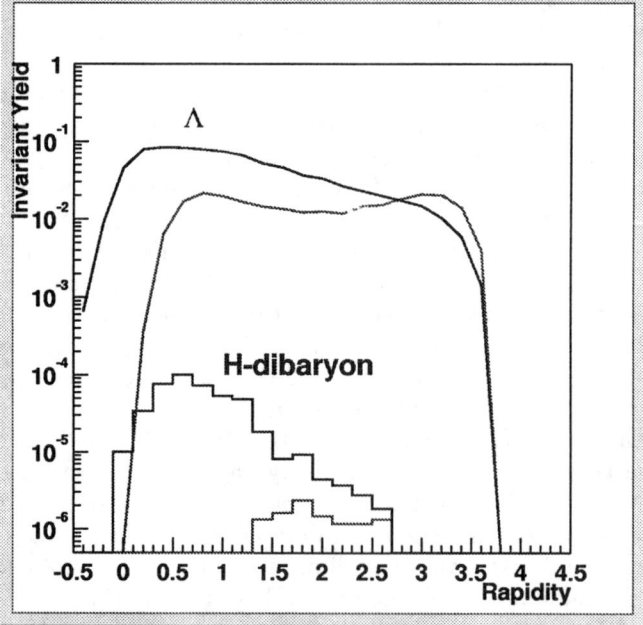

FIGURE 2. RQMD+Coalescence model predictions for Λ and H-dibaryons in p+p and p+Pt collisions. The relative normalization between the Λ and H is arbitrary.

To date, the most sensitive experiment at Brookhaven E864 [14] and the most sensitive at the CERN-SPS NA52 [15] have analyzed most of their existing data samples and have no candidates for SQM. The upper limits are of order 10^{-8} to 10^{-9} per central Au+Au collision, and final limits should be published soon. It is important to note that these limits are only for SQM with lifetimes $> 50ns$, and if the SQM have shorter lifetimes the experiments have no sensitivity. These limits rule out strangelet production via coalescence only for $A < 5-6$, and it is difficult to make a quantitative statement about the plasma production scenario (though many published predictions are ruled out).

CONCLUSIONS

There have been a wide array of experiments searching for different states of strange quark matter. To date, searches for the H-dibaryon, strangelets and SQM in bulk have yielded null results. In particular, the null results for the H-dibaryon may be close to ruling out any deeply bound state, but we must await the final results from a few crucial experiments.

ACKNOWLEDGMENTS

I would like to thank Frank Rotondo, Ram Ben-David, Adam Rusek, Bob Cousins, Josh Klein, Sebastian White, Bill Zajc and many others for useful discussions and insights.

REFERENCES

1. S.Chin and A.Kerman, Phys. Rev. Lett. 43, 1292 (1979).
2. E.Witten, Phys. Rev. D 30, 272 (1984).
3. R.Klingenberg et al., J. Phys. G: Nucl. Part. Phys. 25, R273 (1999).
4. X.-D.Li et al., Phys. Rev. Lett. 83, 3776 (1999).
5. J.F.Donoghue et al., Phys. Rev. D 34, 3434 (1986).
6. J.Belz et al., Phys. Rev. Lett. 76, 3277 (1996).
7. J.Belz et al., Phys. Rev. C 56, 1164 (1997).
8. R.Cousins et al., Phys. Rev. D 56, 1673 (1997).
9. J.L.Nagle et al., Phys. Rev. C 53, 367 (1996).
10. B.Cole et al., Phys. Lett. B 350, 147 (1995).
11. A. Alavi-Harati et al., Phys. Rev. Lett. 84, 2593 (2000).
12. F.Rotondo et al., Phys. Rev. D 47, 3871 (1993).
13. A. Rusek et al. Presentation at AGS User's Meeting (2000).
14. T.A.Armstrong et al., Phys. Rev. Lett. 79, 3612 (1997).
15. R. Klingenberg et al., Nucl. Phys. A, 306c (1996).

Calculations of the Quark Masses and Smallest Eigenvalues of Wilson-Dirac Operator in Domain-Wall Formalism

Valeriya Gadiyak, Xiangdong Ji, Chulwoo Jung

Department of Physics
University of Maryland
College Park, Maryland 20742

Abstract. In the domain-wall formulation of chiral fermion, the finite separation between domain-walls (L_s) induces an effective quark mass (m_{eff}) which complicates the chiral limit. In this work, we study the size of the effective mass as the function of L_s and the domain-wall height m_0 by calculating the smallest eigenvalue of the hermitian domain-wall Dirac operator in the topologically-nontrivial background fields. Our numerical result is consistent with a previous study of the effective mass from the GMOR relation.

Chiral symmetry and its explicit and/or spontaneous breaking are important aspects of strong interaction phenomenology. Chiral dynamics dominates the low-energy hadron structure and interactions. The chiral phase transition at finite temperature has been sought after experimentally for a long time. On the theoretical frontier massless fermions defy the naive nonperturbative treatments. In the last few years, Kaplan and Shamir's domain-wall construction [1,2] and Narayanan and Neuberger's overlap fermion formalism [3] have emerged as promising approaches to simulating massless quarks. In this paper, we aim to study the effectiveness of the domain-wall approach [4]-[10].

Following previous studies, we adopt Shamir's version of the domain wall fermion formulation [2], in which the 5-dimensional Wilson fermion is first introduced. The finite fifth dimension with L_s lattice sites extends from $s = 0$ to $s = L_s - 1$. Dirichlet boundary condition on the quark fields is applied to the four-dimensional slices at $s = -1$ and $s = L_s$. The fifth component of the gauge potential is identically zero and the other four components are the same at every s slice. These chiral fermions appear as the surface modes at $s = 0$ and $s = L_s - 1$ with opposite chirality. For the finite wall separation ($L_s \neq \infty$), the chiral mode on the $s = 0$ wall couples with the one with the opposite chirality on the $s = L_s - 1$ wall in an exponentially small way. Because of this coupling, a finite residual fermion mass is produced. The goal of this paper is to understand the size and dependence of this induced fermion mass on m_0 (the wall height) and L_s when realistic background gauge fields are introduced.

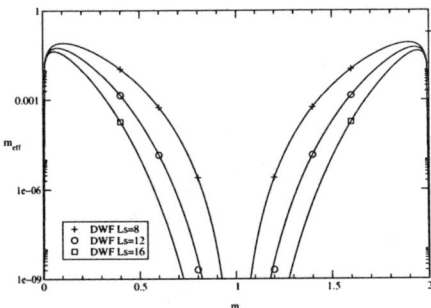 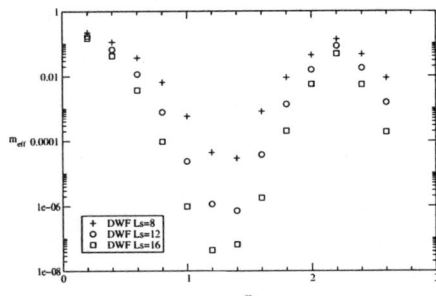

FIGURE 1. Effective quark mass induced by domain-walls for the free field configuration. L_s is the number of lattice sites in the fifth direction.

FIGURE 2. Effective quark mass induced by domain walls in the smooth instanton field. L_s is the number of lattice sites in the fifth direction.

In the absence of the gauge potentials, the effective mass can be defined in terms of the pole of the free Green's function. For a large L_s, it has a simple analytical form [2],

$$m_{\text{eff}} = m_0(2-m_0)(1-m_0)^{L_s} . \qquad (1)$$

One can also obtain m_{eff} by either diagonalizing the hermitian domain-wall Dirac operator $H_{DW} = \gamma_5 P_s D$ or $D^\dagger D = H_{DW}^2$ [2,4], where D is the domain-wall Dirac operator and P_s is the reflection along the fifth dimension. On a lattice with the periodic boundary condition, the lowest four-momentum of a fermion is zero, and the lowest eigenvalue of $\gamma_5 P_s D$ is just m_{eff}. We have computed m_{eff} on an 8^4 lattice with $L_s = 8, 12$, and 16, and the result is shown in Fig. 1 together with the exact answer from Eq. (1).

In the presence of a realistic gauge potential, the effective quark mass result from the finite wall separation may depend on how it is defined. Here we explore the effective mass in the following way.

The domain-wall formulation can be regarded as an approximation to the overlap formalism [3]. Edwards, Heller, and Narayanan have done extensive studies of the topological properties of lattice gauge configurations [11,12]. In a topologically-nontrivial background thus defined, the domain-wall Dirac operator with a finite L_s has small eigenvalues, nonvanishing only because of the finite wall separation. We *define* the smallest eigenvalues of the hermitian domain-wall operator H_{DW} as the wall-induced effective fermion mass.

As a first nontrivial example, we performed calculations in a smooth instanton field configuration on an 8^4 lattice. The lowest eigenvalue of the hermitian domain-wall Dirac operator H_{DW} is shown in logarithmic scale in Fig. 2. The overall profile of the eigenvalue as a function of m_0 is similar to the free case in Fig. 1. For a fixed m_0, the effective fermion mass decreases exponentially as L_s increases from 8 to 12 and 12 to 16.

A smooth instanton on the lattice is far from a typical equilibrium gauge configuration entering in the Feynman path integral. A more realistic study of the induced quark masses requires gauge configurations with quantum fluctuations fully included.

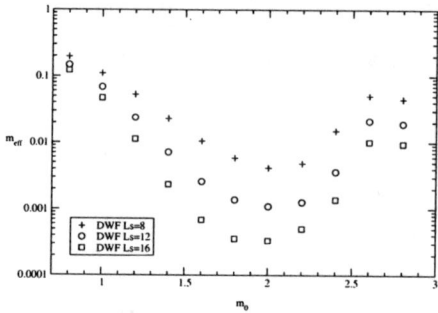

FIGURE 3. Effective quark mass induced by the domain walls in the topological configuration.

In the following, we work on a set of Monte Carlo configurations generated on a four-dimensional 8^4 lattice and with $\beta = 6$.

In Fig.3, we have shown the smallest eigenvalue of the hermitian domain-wall Dirac operator H_{DW}. Over a large region of m_0, the effective quark mass decreases exponentially in L_s, as is clear from the approximate equal spacings between pluses, circles, and squares.

The important point about Fig. 3 is that the magnitude of the effective mass is much enhanced relative to the case of the smooth instanton configuration.

Since the effective quark mass depends on particular gauge fields, it is useful to get an average over an ensemble of gauge configurations. For this purpose, we have generated a set of 150 configurations on a 8^4 lattice at $\beta = 6.0$. By studying the spectral flow of the hermitian Wilson-Dirac operators, we found 12 topologically-nontrivial configurations. We find the average effective quark mass over the gauge ensemble is $8(4) \times 10^{-4}$ at $L_s = 16$.

Our result can also be compared with a previous study of the wall-induced quark mass using the Gell-Mann-Oakes-Renner (GMOR) relation [13] on a $16^3 \times 32$ lattice at the same value of β.

We finally return to the central question of the domain-wall fermion formalism: How large an L_s is needed for a practical simulation? The answer depends on the size of quantum fluctuations. For a large value of β such as the case we have presented, the configurations are relatively smooth; one can work with $L_s = 12$ or 16 and keep the induced quark mass under control.

On the other hand, at $\beta = 5.7$, one needs to have very large L_s (30 to 40) to keep m_{eff} small.

To summarize, we have studied the induced quark mass resulted from the finite domain wall separation by diagonalizing the hermitian domain-wall Dirac operator in topologically nontrivial configurations. We find the quantum fluctuation strongly enhances the domain-wall effects. However, the effective mass does show an exponential decay as a function of L_s. Our result on an 8^4 lattice with $\beta = 6$ is consistent with the effective fermion masses from the GMOR relation, although a detailed analysis shows that the two definitions of the effective mass are not the same. Finally, we comment on the size of L_s needed in a practical Monte Carlo simulation.

We thank N. Christ, R. Edwards and J. Negele for useful discussions related to the subject of this paper. The numerical calculations reported here were performed on the Calico Alpha Linux Cluster at the Jefferson Laboratory, Virginia. This work is supported in part by funds provided by the U.S. Department of Energy (D.O.E.) under cooperative agreement DOE-FG02-93ER-40762.

REFERENCES

1. D. B. Kaplan, Phys. Lett. B **288**, 342 (1992).
2. Shamir, Y., Nucl. Phys. **B406**, 90 (1993);
   Furman, V., and Shamir, Y., Nucl. Phys. **B439**, 54 (1995).
3. Narayanan, R., and Neuberger, H., Phys. Lett. B **302**, 62 (1993);
4. Vranas, P., Nucl. Phys. B (Proc. Supp.) **63**, 605 (1998);
   Vranas, P., et al., hep-lat/9911002.
5. Blum, T., and Soni, A., Phys Rev. D **56**, 174 (1997);
6. Blum, T., Soni, A., and Wingate, M., Phys. Rev. D **60**, 114507 (1999).
7. Chen, P., et al., Phys. Rev. D **59**, 054508 (1999).
8. Chen, P., et al., Nucl. Phys. B (Proc. Suppl.) **73**, 204, 207, 405 (1999).
9. Ali Khan, A., et al., Nucl. Phys. B (Proc. Suppl.) **83-84**, 591 (2000).
10. Aoki, S., Izubuchi, T., Kuramashi, Y., and Taniguchi, Y., Nucl. Phys. B (Proc. Suppl.) **83-84**, 624 (2000).
11. Edwards, R. G., Heller, U. M., and Narayanan, R., Nucl. Phys. **B522**, 285 (1998).
12. Edwards, R. G., Heller, U. M., and Narayanan, R., Nucl. Phys. **B535**, 403 (1998);
    Edwards, R. G., Heller, U. M., and Narayanan, R., Phys. Rev. D **60**, 077502 (1999).
13. Fleming, G. R., Nucl. Phys. B (Proc. Suppl.) **83-84**, 363 (2000);
    Christ, N. H., private communications.

Hyperon Electroproduction and Strangeness Physics

Stephan L. Mintz

Physics Department
Florida International University
Miami, Florida 33199

Abstract. We calculate the differential cross section for the weak production of Λ and Σ^0 hyperons via electron scattering from protons for a range of incident electron energies from 0.5 to 6.0 GeV. The calculations presented are phenomenologically based and make use of $SU(3)$ and $SU(2)$ relations. We obtain contributions from individual form factors to the differential cross sections and calculate the expected event rates for these reactions. We show that Λ production should be observable and that Σ^0 production, though smaller might be observable. Finally we discuss what might be learned about the structure of the weak, strangeness changing current from these two processes as well as how these processes might serve as tests of $SU(3)$ relations.

INTRODUCTION

The advent of medium energy electron accelerators such as the continuous electron beam accelerator at the Thomas Jefferson National Accelerator Facility have made the study of weak strangeness changing electron induced processes such as $e^- + p \longrightarrow \nu_e + \Lambda$ and $e^- + p \longrightarrow \nu_e + \Sigma^0$ possible. There has been a continuing interest in observing these processes[1,2] and so in this paper we present results for the differential cross sections for both of these reactions.

The reason for the interest in these reactions is several fold. First the weak strangeness changing neutral current has been studied at low and intermediate energies in only a limited number of cases such as the decay reactions $\Lambda \longrightarrow p + e^- + \bar{\nu}_e$ and $\Sigma^- \longrightarrow p + e^- + \bar{\nu}_e$ as well as in K decays. Questions still remain concerning its isospin structure, and how well its matrix elements obey $SU(3)$ relations. At facilities such as TJNAF, electron energies from 0.5 GeV to 6.0 GeV are available thus making systematic analysis possible.

In this paper we attempt as phenomenological a calculation as possible. We make no use of internal models for the proton or hyperons. Instead we make use of measured form factors along with $SU(3)$ and $SU(2)$ relations to obtain the differential cross sections. These work well for the decays mentioned above but have not been tested at higher energies.

MATRIX ELEMENTS

The first order transition matrix element for the production of hyperons from the scattering of electrons from protons may be written as:

$$<\nu_e\Lambda|H_w|e^-p> = \frac{G}{\sqrt{2}}\sin\theta_C \bar{u}_\nu\gamma^\lambda(1-\gamma_5)u_e <\Lambda|J_\lambda^\dagger(0)|p>. \quad (1)$$

We have written this equation for the Λ hyperon but it holds equally well for the Σ^0 hyperon. We note here that the hadronic current may be written as $J_\mu(0) = V_\mu(0) - A_\mu(0)$ where $V_\mu(0)$ is the vector current and $A_\mu(0)$ is the axial current. The matrix elements of the vector and axial vector currents may be written as:

$$<\Lambda|V_\mu^\dagger(0)|p> = \bar{u}_f[\gamma_\mu F_V(q^2) + \frac{iF_M(q^2)\sigma_{\mu\nu}q^\nu}{2m_p} - \frac{F_S(q^2)q_\mu}{2m_p}]u_i \quad (2)$$

$$<\Lambda|A_\mu^\dagger(0)|p> = \bar{u}_f[\gamma_\mu\gamma_5 F_A(q^2) + \frac{q_\mu\gamma_5 F_P(q^2)}{m_\pi} + \frac{iF_E(q^2)\sigma_{\mu\nu}q^\nu\gamma_5}{2m_p}]u_i \quad (3)$$

where we have written the matrix elements for the Λ case but which hold equally well for the Σ^0 case. The structure of the particles is contained in the six form factors in the two above equations. However because we are interested in electron induced processes we will not be able to observe form factors F_P and F_S. Thus our matrix elements will depend only upon $F_V, F_M, F_A,$ and F_E.

We may make use of well known $SU(3)$ relations to obtain:

$$F_r = \frac{-1}{\sqrt{6}}(3\tilde{F}_r + \tilde{D}_r) \quad (4)$$

for the Λ case and

$$F_r = \frac{1}{\sqrt{2}}(\tilde{F}_r - \tilde{D}_r) \quad (5)$$

for the Σ^0 case and where r stands for V, M, A or E. The quantities $\tilde{D}_V, \tilde{D}_M, \tilde{F}_V$ and $\tilde{F}_M$ can be obtained from proton and neutron data[3]. From this data we obtain the form factors F_V and F_M. Because all form factors will turn out to be consistent with a dipole fit, we write the general form only once as:

$$F_r(q^2) = \frac{F_r(0)}{(1 - q^2/M_r^2)^2} \tag{6}$$

where $r = V, M, A$, or E. For the form factors F_V and F_M we find $F_V(0) = 1.2247$ and $M_V = .98\ GeV/c^2$ and $F_M(0) = 1.793/2m_p$ and $M_M = .71\ GeV/c^2$ for the Λ case and $F_V(0) = -.707$ and $M_V = .98\ GeV/c^2$ and $F_M(0) = -1.437/2m_p$ and $M_M = .71\ GeV/c^2$ for the Σ^0 case.

The axial current form factors have been determined in beta decay measurements. For the Λ case[4] it has been found to be $F_A(0)/F_V(0) = 0.718 \pm 0.015$ and is consistent with a dipole fit given by Eq.(6) with $F_A(0) = .8793$ and $M_A = 1.25\ GeV/c^2$. The Σ^0 case is more difficult because there is no data for the weak decay of Σ^0. From the isospin relation $[I^+, J_\lambda] = 0$ we obtain $<p|[I^+, J_\lambda]|\Sigma^0> = 0$ which leads to the relation:

$$\frac{1}{\sqrt{2}} <n|J_\lambda|\Sigma^-> = <p|J_\lambda|\Sigma^0> \tag{7}$$

and from the measured value[4] for $F_A^{\Sigma^-}/F_V^{\Sigma^-}$ we find $F_A^{\Sigma^0}/F_V^{\Sigma^0} = F_A^{\Sigma^-}/F_V^{\Sigma^-} = -.34 \pm .017$ and obtain for the Σ^0 case $F_A(0) = .24$ and $M_A = 1.25\ GeV/c^2$.

Now only F_E remains. We make use of a theoretical estimate[5] to obtain this quantity. It is again given by Eq.(6) with $F_E(0) = .705/2m_p$ and $M_M = M_E$. These values are for the Λ case but we also use them as an upper limit for the Σ^0 case since the form factors here are generally smaller. It will turn out however that the F_E form factor makes too small a contribution to the cross section to be observed.

RESULTS AND DISCUSSION

We are now able to calculate the differential cross sections for the reactions $e^- + p \longrightarrow \Lambda + \nu_e$ and $e^- + p \longrightarrow \Sigma^0 + \nu_e$. We have done so for electron energies from 0.5 GeV to 6.0 GeV. In figure 1 we show representative differential cross sections for both processes for incident electrons with a laboratory energy of 4.0 GeV.

We have also calculated the contributions from the form factors to the differential cross sections. For the Λ case, in the observable (low q^2) region around the maximal Λ outgoing laboratory angle, the contributions from F_A and F_V become almost identical. All of the other form factors in this region are too small to observe.

The case is very different for the Σ^0 processes. Here in the region of the maximal angle only F_V makes a substantial contribution. The axial vector form factor is substantially supressed. Thus a measurement of the differential cross section for this process is essentially a measurement of F_V.

Finally we consider the possibility of observing these reactions. For the Λ case, for parameters appropriate to TJNAF,i.e. a 15 cm liquid hydrogen target and a current of 200 μA one obtains approximately 300 events per hour assuming observation of the differential cross section near its maximal value. For the Σ^0 case the same parameters yield 50 to 60 events per hour due to the smaller differential cross section at maximum. Experimentalists at TJNAF believe that the Λ case would be observable but the observability of the Σ^0 case is less clear. The backgrounds though difficult are manageable for both processes. The observation of either process would enable a test of the efficacy of $SU(3)$ relations in predicting the strangeness changing weak current form factors. However the Σ^0 process would also enable a search for $I = 3/2$ contributions to the weak current. Because Λ is an $I = 0$ particle, an $I = 3/2$ current cannot connect it with a proton. However Σ^0 is an $I = 1$ particle which can be connected to a proton by an $I = 3/2$ current.

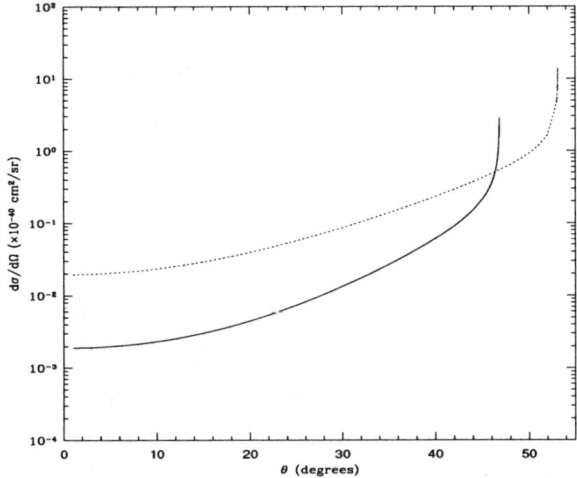

FIGURE 1. Plot showing a comparison of differential cross sections ($\times 10^{-40}$) versus hyperon angle for the weak production of hyperons by electron scattering from protons. The dashed line is for the Λ case and the solid line is for the Σ^0 case. The results are for 4.0 GeV electrons and all results are in the laboratory frame.

REFERENCES

1. Private communication, J.M. Finn.
2. Private communication, K. Baker.
3. Mintz S.L., *Nucl. Phys.* **A 657**,303(1999).
4. Particle Data Group,*Eur. Phys. J.* **C 3**,1(1998).
5. Hwang W-Y.P.and Henley E.M.,*Phys. Rev.* **D 38**, 798(1988).

Precision Measurement of the Neutron Magnetic Form Factor from $^3\vec{He}(\vec{e},e')$

D. Dutta[1]

Laboratory for Nuclear Science and Department of Physics,
Massachusetts Institute of Technology,
77 Massachusetts Ave., Cambridge, MA 02139

Abstract. A precision measurement of the inclusive quasielastic transverse asymmetry $A_{T'}$ from $^3\vec{He}(\vec{e},e')$ was completed recently in Hall A at Jefferson Lab (E95-001). The preliminary results on the neutron magnetic form factor at low Q^2 are presented here.

INTRODUCTION

Electromagnetic form factors are of fundamental importance for an understanding of the underlying structure of nucleons. While the proton form factors are known with good precision over a large range of four-momentum transfer squared, Q^2, the corresponding data for the neutron are of inferior quality due to the lack of free neutron targets. Until recently, most data on G_M^n had been deduced from elastic and quasielastic electron-deuteron scattering experiments. For inclusive measurements, this procedure requires the subtraction of a large proton contribution and suffers from large theoretical uncertainties due to the deuteron model employed and corrections for final-state interactions (FSI) and meson-exchange currents (MEC). The proton subtraction is avoided in coincidence $d(e,e'n)$ experiments [1], and the sensitivity to nuclear structure can be greatly reduced by measuring the cross section ratio $d(e,e'n)/d(e,e'p)$ at quasielastic kinematics. Several recent experiments [2-4] have employed the latter technique to extract G_M^n with uncertainties of <2% in the Q^2 range of 0.1 to 0.8 (GeV/c)^2. While this precision is excellent, the results of these experiments [1-4] are not fully consistent.

An alternative approach to a precision measurement of G_M^n is through the inclusive quasi-elastic reaction $^3\vec{He}(\vec{e},e')$. In comparison to deuterium experiments, this technique employs a different target and relies on polarization degrees of freedom. It is thus subject to completely different systematics.

---

[1] For the Jefferson Lab Hall A, E95-001 collaboration.

INCLUSIVE QUASIELASTIC SCATTERING OF POLARIZED ELECTRONS FROM POLARIZED ^{3}HE TARGETS

The polarized ^{3}He nucleus is a good candidate for an effective neutron target because its ground-state wave function is dominated by the S-state in which the proton spins cancel and the nuclear spin is entirely due to the neutron.

The spin-dependent contribution to the $^3\vec{\mathrm{He}}(\vec{e},e')$ cross section is completely contained in two nuclear response functions, a transverse response $R_{T'}$ and a longitudinal-transverse response $R_{TL'}$. These appear in addition to the spin-independent longitudinal and transverse responses R_L and R_T. $R_{T'}$ and $R_{TL'}$ can be isolated experimentally by forming the spin-dependent asymmetry A defined as $A = \frac{\sigma^{h+} - \sigma^{h-}}{\sigma^{h+} + \sigma^{h-}}$, where $\sigma^{h\pm}$ denotes the cross section for the two different helicities of the polarized electrons. In terms of the nuclear response functions, A can be written [5] as,

$$A = \frac{-(\cos\theta^* \nu_{T'} R_{T'} + 2\sin\theta^* \cos\phi^* \nu_{TL'} R_{TL'})}{\nu_L R_L + \nu_T R_T} \qquad (1)$$

where the ν_k are kinematic factors and θ^* and ϕ^* are the polar and azimuthal angles of target spin with respect to the 3-momentum transfer vector $\vec{q}$. The response functions R_k depend on Q^2 and the electron energy transfer ω. By orienting the target spin at $\theta^* = 0°$ corresponding to the spin direction along $\vec{q}$, one can select the transverse asymmetry $A_{T'}$ (proportional to $R_{T'}$). For the quasielastic $^3\vec{\mathrm{He}}(\vec{e},e')$ process, $A_{T'}$ is most sensitive to $(G_M^n)^2$ [6-9].

JLAB EXPERIMENT E95-001

The experiment was carried out in Hall A at JLab in early 1999 using a longitudinally polarized continuous-wave electron beam of 10 μA current. A high pressure polarized ^{3}He gas target utilizing spin-exchange optical pumping of rubidium at a density of 2.5×10^{20} nuclei/cm^3 was employed. The average beam (target) polarization for the experiment was approximately 70% and 30%, respectively.

Six kinematic points were measured corresponding to $Q^2 = 0.1\text{-}0.6$ (GeV/c)^2 (0.1). An incident electron beam energy of 0.778 GeV was employed for the two lowest Q^2 values of the experiment and the remaining points were completed at an incident beam energy of 1.727 GeV. The scattered electrons were detected in the two Hall A High Resolution Spectrometers, HRSe and HRSh. Both spectrometers were configured to detect electrons in single-arm mode. The HRSe was set for quasielastic kinematics while the HRSh detected elastically scattered electrons. Since the elastic asymmetry can be calculated with high precision at low Q^2 using the well-known elastic form factors of ^{3}He [10], the elastic measurement allows accurate monitoring of the product of the beam and target polarizations, $P_t P_b$.

To extract G_M^n for the two lowest Q^2 kinematics, the transverse asymmetry data were averaged over a 30 MeV bin around the quasi-elastic peak. A full Faddeev calculation including MEC [15] was employed to generate $A_{T'}$ as a function of G_M^n in the same ω region. The G_M^n at $Q^2 = 0.1$ and 0.2 $(\text{GeV}/c)^2$ were extracted by comparing the measured asymmetries with the predictions of the full calculation. The extracted preliminary values of G_M^n are shown in Fig. 1 as solid circles. The uncertainties shown are the quadrature sum of the statistical and experimental systematic uncertainties.

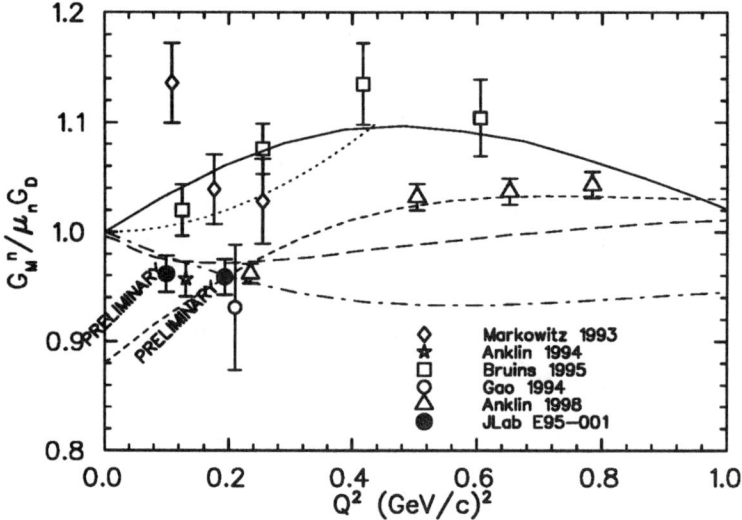

FIGURE 1. The neutron magnetic form factor G_M^n in units of the standard dipole parameterization, $\mu_n G_D$, in the low Q^2 region, as determined in several recent measurements: Markowitz et al. [1] (diamonds) using $d(e,e'n)$; Anklin et al. [2] (star), Bruins et al. [3] (squares), and Anklin et al. [4] (triangles) using the ratio $d(e,e'n)/d(e,e'p)$; and Gao et al. [16] (circle) using $^3\vec{\text{He}}(\vec{e},e')$. The preliminary results from JLab experiment E95-001 are shown as solid circles with total experimental uncertainties. The solid curve is a cloudy bag model calculation [17] and the dotted curve is the minimal vector dominance model calculation [18]. The short and long dashed curves are the non-relativistic and relativistic quark model calculations [19,20], respectively. The dash-dotted curve is a calculation [21] based on a fit of the proton data using dispersion theoretical arguments.

SUMMARY

The inclusive transverse asymmetry $A_{T'}$ from the quasi-elastic $^3\vec{\text{He}}(\vec{e},e')$ process has been measured with high precision at Q^2-values of 0.1 to 0.6 $(\text{GeV}/c)^2$ from JLab experiment E95-001. Using a full Faddeev calculation which includes FSI and MEC the neutron magnetic form factor G^n_M was extracted at Q^2 of 0.1 and 0.2 $(\text{GeV}/c)^2$. Full calculations are at present not available for $Q^2 \geq 0.3$ $(\text{GeV}/c)^2$ to allow the extraction of G^n_M with high precision at higher Q^2. Theoretical efforts are currently underway to extend the full calculation to higher Q^2 [22].

ACKNOWLEDGMENTS

This work is supported in part by the U.S. Department of Energy under contract numbers DE-FC02-94ER40818 and DE-AC05-84ER40150.

REFERENCES

1. P. Markowitz et al., Phys. Rev. C **48** (1993) R5.
2. H. Anklin et al., Phys. Lett. **B336** (1994) 313.
3. E.E.W. Bruins et al., Phys. Rev. Lett. **75** (1995) 21.
4. H. Anklin et al., Phys. Lett. **B428** (1998) 248.
5. T.W. Donnelly and A.S. Raskin, Ann. Phys. 169, 247 (1986).
6. R.-W. Schulze and P.U. Sauer, Phys. Rev. C **48** (1993) 38.
7. C. Ciofi degli Atti, E. Pace and G. Salmè, Phys. Rev. C 51 (1995) 1108; C. Ciofi degli Atti, E. Pace and G. Salmè, in *Proceedings of the 6th Workshop on Perspectives in Nuclear Physics at Intermediate Energies*, ICTP, Trieste May 1993, (World Scientific); C. Ciofi degli Atti, E. Pace and G. Salmè, Phys. Rev. C **51**, 1108 (1995).
8. S. Ishikawa et al., Phys. Rev. C **57** (1998) 39.
9. J. Golak, private communication.
10. A. Amroun et al., Nucl. Phys. **A579** (1994) 596.
11. A. Kievsky, E. Pace, G. Salmè and M. Viviani, Phys. Rev. C **56**, 64 (1997).
12. G. Höhler et al., Nucl. Phys. **B114**, 505 (1976).
13. J. Golak et al., Phys. Rev. C **51**, 1638 (1995).
14. D.O. Riska, Phys. Scr. **31**, 471 (1985).
15. V.V. Kotlyer, H. Kamada, W. Glöckle and J. Golak, Few-Body Syst. **28**, 35 (2000).
16. H. Gao et al., Phys. Rev. C **50** (1994) R546; H. Gao, Nucl. Phys. **A631**, 170c (1998).
17. D.H. Lu, A.W. Thomas and A.G. Williams, Phys. Rev. C **57**, 2628 (1998).
18. U.-G. Meißner, Phys. Rep. **161**, 213 (1988).
19. E. Eich, Z. Phys. C **45**, 627 (1988).
20. F. Schlumpf, J. Phys. G **20** 237, 1994.
21. P. Mergell, U.-G. Meißner and D. Drechsel, Nucl. Phys. **A596**, 367 (1996).
22. W. Glöckle, private communication.

Heavy Quark and Heavy Lepton Physics

D-Meson Dalitz Fit from Focus

Sandra Malvezzi [1]

Istituto Nazionale Fisica Nucleare
Via Celoria 16 – 20133 Milano – Italy

Abstract. Preliminary results of Dalitz analyses of D-meson to three-pseudoscalar decays by Focus are presented. The role of FSI, which create phase shifts between interfering resonant channels, can be studied in different decay modes and the annihilation contribution measured in the charm sector via the $D_s \to 3\pi$ decay. Particular attention was devoted to the parametrization of the $f_0(980)$ coupled-channel Breit-Wigner: the mass and width of this state have been extracted from our $D_s \to 3\pi$ signal via the K-matrix formalism and then used to fit the $KK\pi$ final state.

INTRODUCTION

The analysis of substructures should be considered in the context of probing hadronic-decay dynamics. A complete lifetime-hierarchy is already experimentally established [2,3], posing severe constraints on theoretical models and indicating hadronic-sector corrections. The role of final-state interactions (FSI) can be gauged from isospin-amplitude interference; phase shifts between different isospin amplitudes are often near 90° [4], indicating the general importance of FSI. Amplitude analysis in the non-leptonic sector emerges as a natural extension to the three-body decays, where the phase-shifts between interfering amplitudes for various resonant channels can be measured. Moreover, a still unsolved problem in charm decay is the reliable estimate of non-spectator contributions. The best candidate to occur via annihilation is $D_s \to 3\pi$; nonetheless, the presence of resonant channels with an $s\bar{s}$ quark content suggests spectator processes rather than W annihilation, requiring a Dalitz-plot analysis to disentangle direct three-body decays from resonances.

I THE AMPLITUDE FORMALISM: THE $KK\pi$ STATE

The $D_s, D^+ \to KK\pi$ Dalitz plots in $m^2_{KK}, m^2_{K\pi}$, obtained by Focus and shown in Fig. 1 are particularly instructive. The D_s Dalitz is highly dominated by the $\phi\pi$ and $\bar{K}^{*0}K$ channels while the D^+ has a significant additional contribution appearing as a uniform event intensity. Depopulation of the central region of the ϕ and $\bar{K}^{*0}$

---

[1] for a complete list of the Focus Collaboration see the paper by E. Vaandering, these proceedings

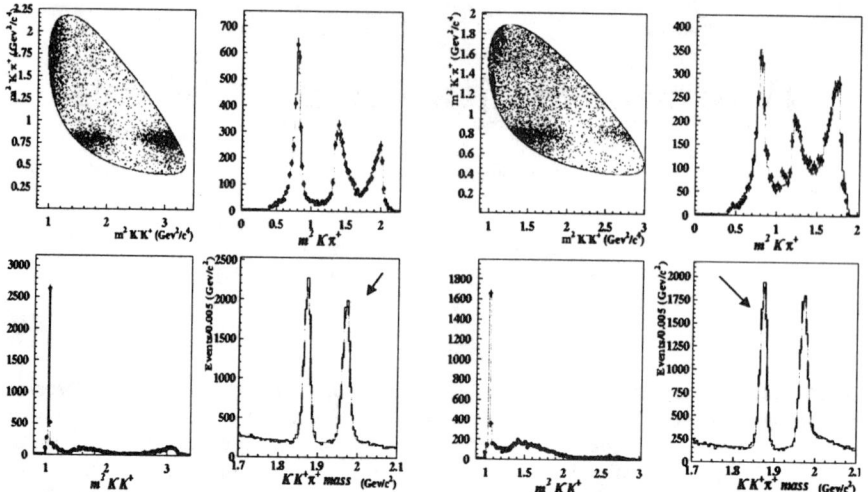

FIGURE 1. $KK\pi$ Dalitz plot, projection fit and mass distribution for D_s and D^+

is due to a node in the function, reflecting angular-momentum conservation; the $\cos\theta_{\text{helicity}}$ (Fig. 2) behaviour arises from the vector nature of the mesons (ϕ, $\overline{K}^{*0}$) decaying into two pseudoscalars. The Dalitz plots are fit to a coherent sum of resonances, quasi-two-body channel, of the type

$$D \to r + c$$
$$\hookrightarrow a + b,$$

described by a decay function (see also Fig. 3),

$$\mathcal{M} = F_D\, F_r \times |\bar{c}|^J |\bar{a}|^J\, P_j(\cos\theta^r_{ac}) \times BW(m_{ab}), \tag{1}$$

the product of two vertex form factors (Blatt-Weisskopf momentum-dependent factors), a Legendre polynomial of order J, representing the angular decay wave function, and a relativistic Breit-Wigner (BW).

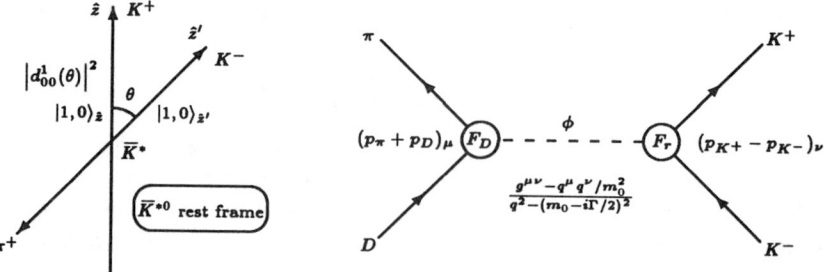

FIGURE 2. Helicity angle definition.

FIGURE 3. $D \to KK\pi$ decay diagram.

The total decay amplitude is a sum of functions of the form (1), each multiplied by a factor $a_r e^{i\delta_r}$, where the modulus a gives the strength of the channel and δ is the phase shift (for a more complete description see ref. [1]). Under the factorization assumption, the bare amplitudes are real. Phases are induced via FSI by virtue of Watson's theorem, which states that the physical amplitudes are related to the bare amplitudes by the square root of a strong-interaction S-matrix, describing hadron-hadron scattering. The asymmetry, not present in the D_s case, may be interpreted as an interference effect (Fig. 4) with a nearly constant amplitude (a broad scalar), which may be written as $\cos\delta + i\sin\delta$. The interference of these two amplitudes contributes to the intensity in the form

$$2\text{Re}\left\{(\cos\delta + i\sin\delta)^* \frac{\cos\theta_{KK}}{m_r^2 - m_{K\pi}^2 - i\Gamma m_r}\right\}$$

$$= \frac{2(m_r^2 - m_{K\pi}^2)\cos\theta_{KK}\cos\delta}{(m_r^2 - m_{K\pi}^2)^2 + \Gamma^2 m_r^2} + \frac{2\Gamma M_r \cos\theta_{KK}\sin\delta}{(m_r^2 - m_{K\pi}^2)^2 + \Gamma^2 m_r^2}. \qquad (2)$$

Along the $\bar{K}^{*0}$ band in the direction of increasing m_{KK}^2, $\cos\theta$ changes from $+1$ to -1 and both terms in eq. 2 switch sign. Along a line of constant m_{KK}^2 the first term, which dominates for a real amplitude ($\delta \sim 0°$), switches sign on passing through the resonance pole (thus cancelling the interference), while the second interference term, which dominates for a relatively imaginary amplitude does not change sign under such motion. Thus, the patterns for real and imaginary phases are different. The interference seen in the D^+ best resembles the pattern for a relatively imaginary amplitude $\delta \sim 90°$. The fully coherent Dalitz analysis of $D^+ \to KK\pi$ confirms this, giving a phase shift between $K^*(1430)$ and $\bar{K}^{*0}(892)$ of $\sim 70°$ (for a complete discussion of the fit results see Table 2 and the next section).

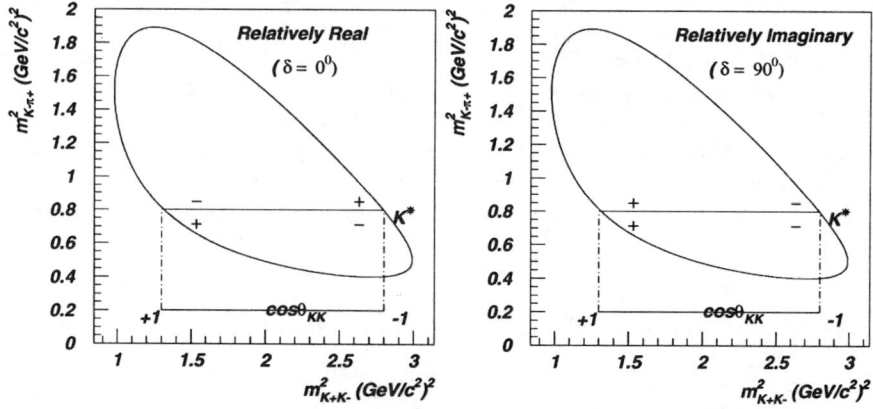

FIGURE 4. The interference pattern of $D^+ \to KK\pi$ in the $\bar{K}^{*0}$ band.

II THE $D_S, D^+ \to 3\pi$ DECAY AND THE $f_0(980)$

To gauge the rate of non-spectator diagrams in the charm sector, the $D_s \to 3\pi$ final state can be analyzed. Inspection of the Dalitz plot (see Fig. 5) reveals the decay to be dominated by the scalar $f_0(980)$ and by unusual resonances ($f_2(1270)$ and $f_0(1500)$) coupling not only to $s\bar{s}$, but also to $u\bar{u}$ and $d\bar{d}$, which appear to balance the strangeness by absorbing the initial strange quark and emitting a dipion.

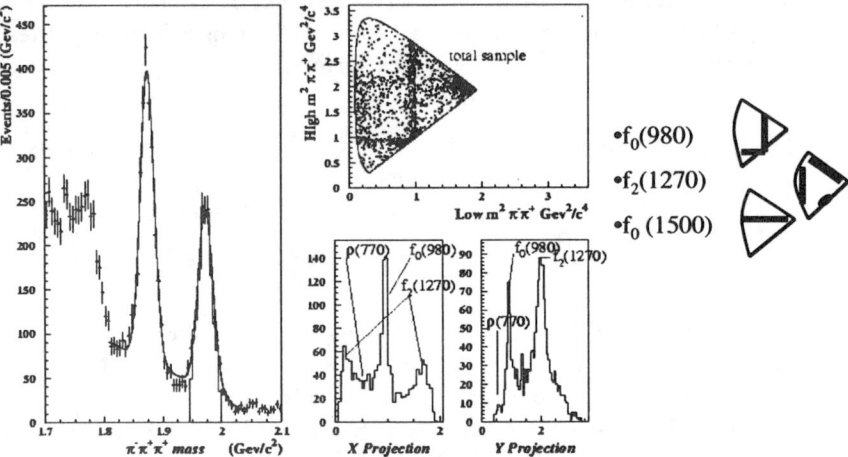

FIGURE 5. $D_s \to 3\pi$ mass distribution and Dalitz Plots, the resonance features are highlighted.

The quality and the statistics of the Focus data (the $D_s \to 3\pi$ yield is about 13 times that of E687 with the same signal-to-noise ratio), allowed us to study the scalar f_0. The nature of this state has been a puzzle for many years: it not being clear whether it is a KK molecule, an amalgam of resonances, a glue-ball or a normal BW-like structure. Despite considerable experimental effort, its parameters are not yet precisely established. The Focus data reveal it to be a narrow state, rendering inadequate the old WA76 parametrization used in E687. Systematic precision in the resonance definition led us to examine the general basis of scattering theory; the K-matrix formalism [5] revealed itself an elegant and convenient representation of the S-(T-) scattering operator, expressing unitarity and analyticity of the amplitudes. The $f_0(980)$ appears as a "regular" resonance in the $\pi\pi$ system, the relevant Breit-Wigner denominator, for m close to m_r, being

$$m_r^2 - m^2 - im_r\Gamma_r. \qquad (3)$$

In the resonance approximation the K-matrix mass m_0 and width Γ_0 are

$$\begin{aligned} m_0^2 &= m_r^2 + \left(\frac{\gamma_{KK}}{\gamma_{\pi\pi}}\right)^2 \left[\frac{|\rho_{KK}(m_r)|}{\rho_{\pi\pi}(m_r)}\right] m_r\Gamma_r \\ \Gamma_0 &= \frac{m_r\Gamma_r}{m_0\rho_{\pi\pi}(m_r)\gamma_{\pi\pi}^2}, \end{aligned} \qquad (4)$$

where $\rho_{\pi\pi}$ and ρ_{KK} are the phase-space terms, $\gamma_{\pi\pi}$ and γ_{KK} are coupling constants normalized to $\gamma_{\pi\pi}^2 + \gamma_{KK}^2 = 1$. We fitted our $D_s \to 3\pi$ data to find the best K-matrix mass and width to then employ in the $KK\pi$ channel; we found $m_0 = 982\,\mathrm{MeV}$ and $\Gamma_0 = 142\,\mathrm{MeV}$, corresponding to values for the K-matrix partial widths of $\Gamma_{\pi\pi} = 54\,\mathrm{MeV}$ and $\Gamma_{KK} = 9\,\mathrm{MeV}$. With the K-matrix formalism a valuable improvement in our likelihood function and in the fit quality was obtained. The Dalitz plot fit and results are shown in Fig. 6 and Table 1.

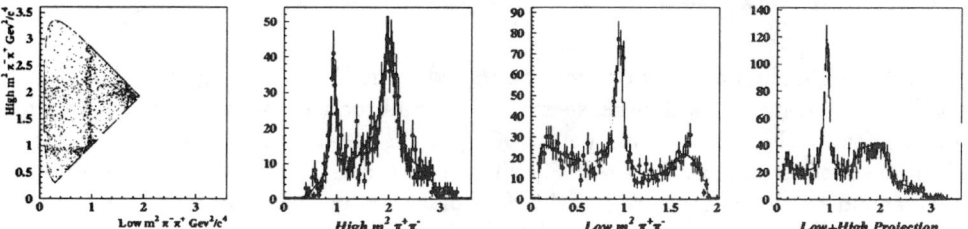

FIGURE 6. $D_s \to 3\pi$ mass distribution and Dalitz Plot.

TABLE 1. D_s and $D^+ \to 3\pi$ fit fractions and phases.

	D_s		D^+	
	fit fraction	phases	fit fraction	phases
NR	0.19 ± 0.04	$233 \pm 7°$	0.10 ± 0.08	$0°$ (fixed)
$\rho(770)$	0.02 ± 0.01	$39 \pm 40°$	0.34 ± 0.04	$84 \pm 14°$
$f_2(1270)$	0.09 ± 0.01	$152 \pm 8°$	0.11 ± 0.02	$171 \pm 15°$
$f_0(980)$	0.82 ± 0.04	$0°$ (fixed)	0.06 ± 0.02	$238 \pm 21°$
$S_0(1475)$	0.24 ± 0.04	$254 \pm 6°$	0.01 ± 0.01	$273 \pm 34°$
$\rho(1450)$	0.04 ± 0.01	$210 \pm 2°$	0.04 ± 0.01	$0 \pm 30°$
$f_0(400)$	–	–	0.26 ± 0.01	$-64 \pm 16°$
$f_0(1300)$	–	–	0.01 ± 0.01	$-19 \pm 80°$

In Table 1 the results of the $D^+ \to 3\pi$ are also reported and the fit is shown in Fig. 7. Our preliminary results seem to indicate the presence of a wide scalar low-mass state, here called $f_0(400)$, otherwise known as σ.

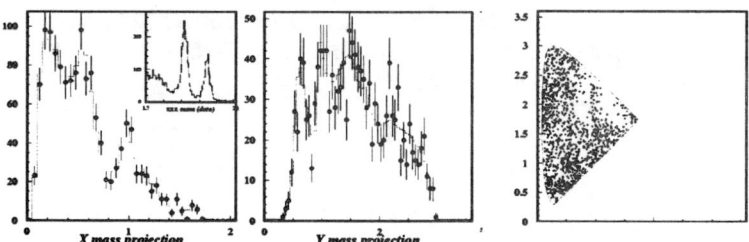

FIGURE 7. $D^+ \to 3\pi$ mass distribution and Dalitz Plot.

TABLE 2. D_s and $D^+ \to KK\pi$ fit fractions and phases.

	D_s		D^+	
	fit fraction	phases	fit fraction	phases
$K^*(892)$	0.445 ± 0.009	$0°$ (fixed)	0.285 ± 0.009	$0°$ (fixed)
$\Phi(1020)$	0.450 ± 0.009	$-220 \pm 4°$	0.285 ± 0.009	$-154 \pm 5°$
$K^*(1430)$	0.086 ± 0.008	$118 \pm 8°$	0.537 ± 0.025	$68 \pm 3°$
$f_j(1710)$	0.023 ± 0.004	$107 \pm 6°$	–	–
$a_0(980)$	0.106 ± 0.008	$128 \pm 5°$	0.030 ± 0.009	$123 \pm 6°$

Once the f_0 mass and width are determined from $D_s \to 3\pi$, they may be used to fit the $KK\pi$ state. This channel seems to prefer a broader resonance, which we tried to identify as the coupled-channel $a_0(980)$, again parametrized via the K-matrix. A good set of resonances for the $KK\pi$ fit is reported in Table 2; the results in Fig. 1 correspond to this choice. When the f_0 replaces a_0 in the amplitude function the resulting fit is slightly worse. Fits with the f_0 and a_0 considered simultaneously are under study to understand the true pattern of interference between the two almost completely overlapping resonances and the sensitivity to their discrimination.

In conclusion, the amplitude analysis of charm decays offers an opportunity to probe hadronic dynamics and the role of non-spectator diagrams. Moreover, the high statistics and quality of the Focus data make Dalitz plots a good laboratory to characterize not yet well established resonances, such as the $f_0(980)$ (maybe the $\sigma(400)$). The K-matrix provides a robust formalism to deal with coupled-channel Breit-Wigner and reproduces the phase-shift movement at the opening of inelastic thresholds and the pole position in the Riemann sheets. Finally we believe that the $D^+ \to KK\pi$ Dalitz studies may provide new handles to measure CP violation in charm hadronic decays via phase-shift differences in the charge split samples (D^+ and D^-), posing it as an original, independent approach to the problem.

REFERENCES

1. P.L. Frabetti et al., *Phys. Lett.* **B351**, 591 (1995).
2. P.L. Frabetti et al., *Phys. Lett.* **B357**, 678 (1995).
3. S. Malvezzi, "Charm lifetime" in *Sixth Int. Symp. on Heavy Flavours Physics* (Pisa, 1995), eds. F. Costantini and M. Giorgi, *Nuovo Cim.* **109A**, 727 (1996).
4. T. Browder, K. Honsheid and D. Pedrini, *Ann. Rev. Nucl. Part. Sci.* **46**, 395 (1996).
5. S.U. Chung et al., *Ann. Physik.* **4**, 404 (1995).

Extracting m_s From Flavor Breaking in Hadronic τ Decays

K. Maltman[*] and J. Kambor[**]

[*]*Dept. Mathematics and Statistics, York Univ., 4700 Keele St., Toronto, ON Canada, and CSSM, Univ. of Adelaide, Adelaide, SA Australia*[1]

[**]*Institut für Theor. Physik, Univ. Zürich, CH-8057 Zürich, Switzerland*[2]

Abstract. New finite energy sum rules (FESR's) for extracting m_s from hadronic τ decay data are constructed which (1) significantly reduce potential theoretical uncertainties present in existing sum rule analyses and (2) remove problems associated with both the poor convergence of the OPE representation of the longitudinal part of the us vector and axial vector correlators and the large statistical errors in the us spectral data above the K^* region.

The ratio of the hadronic τ decay rate through the $f=ij=ud, us$ vector (V) or axial vector (A) current to the corresponding electronic decay rate, $R_\tau^{V/A;ij} = \Gamma[\tau^- \to \nu_\tau \text{hadrons}_{V/A;ij}(\gamma)]/\Gamma[\tau^- \to \nu_\tau e^- \bar{\nu}_e(\gamma)]$, can be written [1]:

$$\frac{R_\tau^{V/A;ij}}{[|V_{ij}|^2 S_{EW}]} = 12\pi^2 \int_0^1 dy\, (1-y)^2 \left[(1+2y)\, \rho_{V/A;ij}^{(0+1)}(s) - 2y\, \rho_{V/A;ij}^{(0)}(s)\right]$$

$$= 6\pi i \oint_{|y|=1} dy\, (1-y)^2 \left[(1+2y)\, \Pi_{V/A;ij}^{(0+1)}(s) - 2y\, \Pi_{V/A;ij}^{(0)}(s)\right], \quad (1)$$

with $y = s/m_\tau^2$, V_{ij} the $f=ij$ CKM matrix element, S_{EW} an electroweak correction, and $\rho_{V/A;ij}^{(J)}(s)$ the spectral function of $\Pi_{V/A;ij}^{(J)}(s)$, $\Pi_{V/A;ij}^{(J)}(s)$ ($J=0,1$) being the spin J part of the $f=ij$ vector (V) or axial vector (A) correlator. The second line follows from the first as a consequence of the general FESR relation, valid for any Π without kinematic singularities, and any $w(s)$ analytic in the region of the contour, $\int_{s_{th}}^{s_0} ds\, w(s)\rho(s) = \frac{-1}{2\pi i}\oint_{|s|=s_0} ds\, w(s)\Pi(s)$. Experimental data thus allows access to the ud-us spectral difference, and hence to integrals of the corresponding correlator difference, which, for large

---
[1]) Supported by the Natural Sciences and Research Engineering Council of Canada
[2]) Supported by the Schweizerischer Nationalfonds and the EEC-TMR program under Contract No. CT 98-0169

enough s_0, are dominated by the $D=2$ term of the OPE representation, proportional to m_s^2. For such s_0 one may, therefore, hope to extract m_s in terms of appropriately weighted integrals of the experimental spectral data.

A complicating factor in any attempt to determine m_s based on this observation is the non-convergence of the OPE representation of the longitudinal $((J)=(0))$ integral at scales $\leq m_\tau^2$ [2]. The current inability to make an experimental longitudinal/transverse separation above $s \sim 1$ GeV^2 thus, at present, precludes a reliable analysis using sum rules with significant longitudinal contributions. Recent analyses [3,4], which either work with the spectral data without making a longitudinal subtraction, or attempt to place loose experimental bounds on the longitudinal contribution, employ "spectral weights" (for the transverse $((0+1))$ case, defining $y = s/s_0$, these are $w_\tau^k(y) \equiv [(1-y)^2(1+2y)](1-y)^k)$. In this paper we work with the "transverse" $(V+A)_{ud}^{(0+1)} - (V+A)_{us}^{(0+1)}$ difference, and try to find alternate weight choices which improve the reliability of the analysis. The $D=2$ term in the corresponding OPE is

$$\left[\Pi_{V+A;ud}^{(0+1)} - \Pi_{V+A;us}^{(0+1)}\right]_{D=2} = \frac{3}{2\pi^2}\frac{m_s(Q^2)}{Q^2}\left[1 + \frac{7}{3}a + 19.9332a^2 + \cdots\right] \quad (2)$$

where $a = \alpha_s(Q^2)/\pi$. Details of the treatment of $D=4,6,8$ contributions may be found in Ref. [5].

On the spectral side we employ the ALEPH ud and us distributions [3,6]. The longitudinal subtraction is performed using sum rule methods [5]; for the weights employed in our analysis the integrated longitudinal subtraction represents $< 1\%$ of the integrated $D=2$ OPE ud-us difference, making the impact of any uncertainties associated with this procedure negligible.

The weights employed in the ud-us FESR's of our analysis have been chosen so as to reduce both theoretical and experimental difficulties. On the experimental side, we seek to (1) de-emphasize us spectral contributions from the region above the K^*, since the ALEPH determination of $\rho_{V+A;us}$ has $\sim 20-30\%$ statistical errors in this region [3] and (2) reduce the strong cancellation in the ud-us difference, which otherwise greatly magnifies the impact of experimental uncertainties. (See Ref. [5] for a detailed discussion of the second point.) Weights which fall more strongly with s above $s \sim 1$ GeV^2 decrease the level of ud-us cancellation and simultaneously suppress high-s contributions, decreasing the impact of both the longitudinal subtraction and experimental errors on $\rho_{V+A;us}$.

On the theoretical side, the goal is to control an important potential theoretical systematic uncertainty. For the weights w_τ^k, the contour improvement prescription [7] is known to produce a significant improvement in the convergence of the known terms of the integrated $D=2$ series. The smallness of the last $(O(\alpha_s^2))$ known term, however, turns out to result from strong cancellations between contributions from different regions of the circular part of

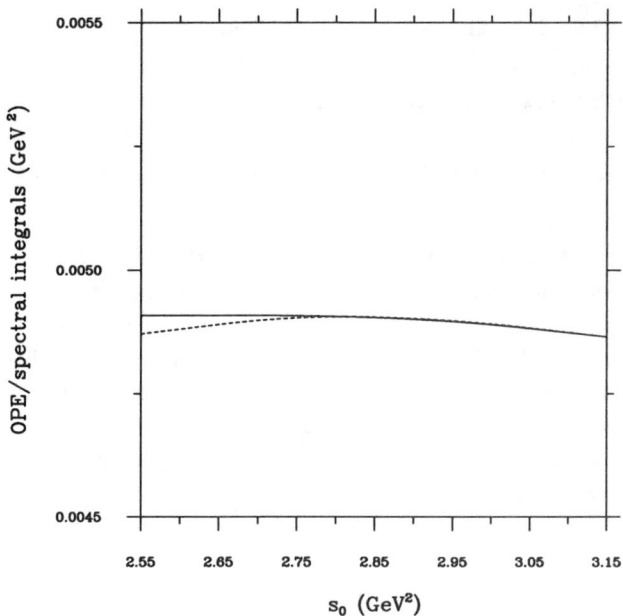

FIGURE 1. The agreement between the OPE and hadronic sides of the FESR corresponding to the weight, $w_{20}(y)$ for $2.55 \text{ GeV}^2 \leq s_0 \leq m_\tau^2$. The solid line is the OPE side, using the values of m_s and the $D=8$ contribution obtained in the fitting procedure described in the text and, in more detail, in Ref. [5]. The dashed line is the hadronic side, obtained using the ALEPH spectral data from which the longitudinal component has been subtracted as described in Ref. [5].

the contour, $|s| = s_0$ [5]. Since, assuming continued geometric growth of the coefficients, similar cancellations do not persist to higher orders [5], an estimate of the truncation error based on the size of the $O(\alpha_s^2)$ term is unreliable. We have constructed 3 alternate polynomial weights which both avoid such "accidental" cancellations and emphasize regions of the plane for which the convergence of the $D=2$ series is optimal. The result is a very strong suppression of possible higher order $D=2$ contributions [5]. The explicit forms of the weights, as well as details of this improvement, are given in Ref. [5].

The results of our analysis are as follows. First, all 3 new weights yield consistent, and stable, values of m_s in the window $2.55 \text{ GeV}^2 < s_0 < m_\tau^2$. An illustration of this fact is given, in Table 1, for the weight $w_{10}(y) = [1-y]^4[1+y]^2[1+y^2][1+y+y^2] = 1-y-y^2+2y^5-y^8-y^9+y^{10}$, where $y = s/s_0$, which is favorable from a theoretical point of view because the absence of y^3, y^4 terms removes $D=8, 10$ contributions to the integrated OPE. Second, for our two

TABLE 1. The extracted value of $m_s(1\text{ GeV}^2)$ in MeV as a function of s_0 for the weight w_{10} having no $D = 8, 10$ contributions.

s_0 (GeV^2):	2.35	2.55	2.75	2.95	3.15
$m_s(1\text{ GeV}^2)$ (MeV):	153.2	159.0	162.2	163.4	163.2

other weights, which do not share this property, the $D = 8$ contributions, which are determined self-consistently, are also stable in this window. These conditions are not satisfied for the FESR's based on the spectral weights, w_τ^k. We choose, for our final analysis, that weight among the three constructed above (called w_{20} in Ref. [5]) which leads to the smallest fractional statistical error. The match between the OPE and hadronic sides of the corresponding FESR which results once m_s and the $D = 8$ contribution have been optimized is shown in Figure 1, and is clearly excellent. Our final numerical result for m_s, in the $\overline{MS}$ scheme, is

$$m_s(1\text{ GeV}^2) = 158.6 \pm 18.7 \pm 16.3 \pm 13.3 \text{ MeV} , \quad (3)$$

which, using four-loop running, corresponds to

$$m_s(4\text{ GeV}^2) = 115.1 \pm 13.6 \pm 11.8 \pm 9.7 \text{ MeV} . \quad (4)$$

The first error is statistical, the second due to the uncertainty in $|V_{us}|$, and the third theoretical, with the latter dominated by our estimate of the error associated with truncating the $D = 2$ series at $O(\alpha_s^2)$. Improvements in the accuracy of the us spectral data, such as will be possible at BaBar, will serve to significantly reduce the first error and, simultaneously, allow use of weights other than w_{20} which produce reduced theoretical truncation errors.

REFERENCES

1. See, e.g., A. Pich, Nucl. Phys. Proc. Suppl. 39BC (1995) 326.
2. K. Maltman, Phys. Rev. D58 (1998) 093015.
3. R. Barate et al. (The ALEPH Collaboration), Eur. Phys. J. C11 (1999) 599.
4. A. Pich and J. Prades, Nucl. Phys. Proc. Suppl. 74 (1999) 309; JHEP 9910 (1999) 004.
5. J. Kambor and K. Maltman, hep-ph/0005156.
6. R. Barate et al. (The ALEPH Collaboration), Eur. Phys. J. C4 (1998) 409.
7. F. Le Diberder and A. Pich, Phys. Lett. B286 (1992) 147, B289 (1992) 165.

Electroweak Couplings of the τ from LEP

T. Paul

Northeastern University
Boston MA, 02115, USA

Abstract. This note reviews LEP measurements of the couplings of the τ lepton to the electroweak gauge bosons.

Introduction

τ lepton production and decay has been studied at LEP for nearly ten years, and among the profusion of results are fairly detailed analyses of the couplings of the τ to the electroweak gauge bosons. This note briefly discusses a few of these measurements. The majority of the results discussed are based on the LEPI samples of $e^+e^- \to \tau^+\tau^-$ events.

Weak neutral current couplings of the τ

Studies of $\tau-Z$ couplings have been largely based on detailed spin analyses of the τ. Unlike the other leptons, τ's decay inside the LEP detectors and measurements of the energy and angular distributions of the decay products can be used to infer the τ polarization state. The longitudinal component of the polarization is sensitive to the vector and axial-vector couplings of both the τ and the electron to the Z, and thus contributes to the precision electroweak measurements discussed elsewhere [1]. In addition to this, correlations among transverse and normal spin components have been extracted from distributions of the aplanarity of τ decay products [2]. Though these measurements are not competitive with longitudinal polarization in terms of sensitivity to vector and axial vector couplings, transverse-normal spin correlations are potentially sensitive to CP violating effects.

Transverse and normal spin components have also been employed to search for anomalous $Z - \tau - \tau$ couplings, which may be introduced to the theory through an effective Lagrangian:

$$\mathcal{L}^{eff}_{int} = \frac{1}{2}\frac{eF_2^w}{2m_\tau}\bar{\psi}\sigma^{\mu\nu}\psi Z_{\mu\nu} - \frac{i}{2}\frac{eF_3^w}{2m_\tau}\bar{\psi}\sigma^{\mu\nu}\gamma_5\psi Z_{\mu\nu} \qquad (1)$$

where the so-called weak electric and magnetic dipole moments are given in terms of the form factors, the electric charge, and τ mass as $d_\tau^w = eF_3^w/2m_\tau$ and $a_\tau^w = F_2^w$ respectively. In the standard model, loops endow d_τ^w and a_τ^w with values that are non-zero, but too tiny to measure at LEP [4]; conversely, unexpectedly large values would signal something new. The LEP experiments have determined [3] $Re(d_\tau^w) = (-0.34 \pm 1.50) \times 10^{-18}$ e·cm, $Im(d_\tau^w) = (0.07 \pm 0.46) \times 10^{-17}$ e·cm, $Re(a_\tau^w) = (0.0 \pm 2.8) \times 10^{-3}$, and $Im(a_\tau^w) = (-1.0 \pm 5.9) \times 10^{-3}$.

Charged current couplings of the τ

The first question one may wish to pose concerning the $\tau - W$ coupling is whether its strength is the same as the weak coupling enjoyed by the other leptons. The rate for τ decays is given by,

$$\Gamma(\tau \to \ell \nu_\tau \bar{\nu}_\ell) = \frac{\mathcal{B}_\ell}{\tau_\tau} = \frac{G_F^2 m_\tau^2}{192\pi^3} f(x_\ell^2) r_{RC} \qquad (2)$$

where ℓ is an electron or muon, $\mathcal{B}_\ell$ indicates the branching fraction, $f(x_\ell^2)$ is a phase space factor and r_{RC} contains effects of radiative corrections. The Fermi coupling, G_F, is proportional to the product of the coupling constants, g_ℓ and g_τ; under lepton universality these are assumed to be identical. From equation 2 and the corresponding expression for muon decay, one can write the ratios g_τ/g_μ and g_τ/g_e in terms of the leptonic branching fractions and the lifetimes and masses of the τ and muon. LEP is well suited to the τ lifetime measurement because of the large boost of the τ's. The figure at the right shows an ALEPH measurement of the distribution of the distance from the e^+e^- interaction point to the τ decay vertex for the 3 pion mode; the lifetime is clearly visible. All four LEP experiments have performed measurements of the τ lifetime and leptonic branching fractions [5], and combining them leads to $g_\tau/g_\mu = 0.9996 \pm 0.0025$, $g_\tau/g_e = 0.9989 \pm 0.0024$, which are evidently consistent with the universality hypothesis. In addition, the rates for on-shell W decays to the three different leptons have been measured at LEP2 [1], again supporting universality, though with less precision than the aforementioned measurements.

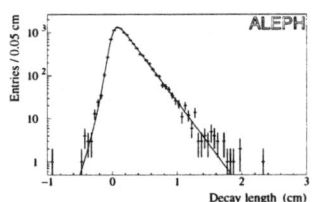

The Lorentz structure of the charged current is also accessible at LEP. The usual approach has been to propose a general derivative-free four-fermion contact interaction:

$$\mathcal{M} = \frac{4G_F}{\sqrt{2}} \sum_{\substack{\gamma=S,V,T \\ \lambda,\iota=R,L}} g^\gamma_{\lambda\iota} <\bar{\ell}_\lambda |\Gamma^\gamma|(\nu_\ell)_n><(\bar{\nu}_\tau)_m|\Gamma_\gamma|\tau_\iota> . \qquad (3)$$

where γ labels the interaction type as vector, scalar or tensor, and λ and ι indicate whether the incoming and outgoing fermions are right or left handed. In the standard model, the charged current includes only vector interactions between left-handed leptons, that is $g_{LL}^V = 1$ and all other constants are zero. The well-known Michel parameters are bilinear combinations of these constants. The shapes of the energy distributions of the final state leptons are sensitive to four of the Michel parameters [6] and the τ polarization, while semileptonic decay spectra can be described in terms of the τ polarization together with a single parameter, ξ_{had}, which turns out to be the average helicity of the ν_τ.

The LEP experiments have performed global fits to the joint and single-sided leptonic and semileptonic decay distributions [1] to extract the Michel parameters [7]. The combined results are given in table 1.

TABLE 1. Michel parameter measurements from LEP shown separately for electronic and muonic decays as well as for the universality assumption. The electronic mode has no sensitivity to η.

Parameter	e	μ	assuming universality	SM expectation
ρ	0.751 ± 0.020	0.796 ± 0.032	0.747 ± 0.008	0.75
η			0.030 ± 0.029	0
ξ	1.015 ± 0.079	1.050 ± 0.110	0.969 ± 0.028	1
$\xi\delta$	0.800 ± 0.060	0.791 ± 0.060	0.744 ± 0.021	0.75
ξ_{had}			-0.995 ± 0.006	-1

Anomalous electromagnetic moments of the τ

Reminiscent of the anomalous weak dipole moments discussed previously, electromagnetic form factors can be introduced at the $\tau - \tau - \gamma$ vertex by replacing the γ^μ in the current by a more general Lorentz invariant form,

$$F_1(q^2)\gamma^\mu + (iF_2(q^2)/2m_\tau)\sigma^{\mu\nu}q_\nu - F_3(q^2)\sigma^{\mu\nu}\gamma^5 q_\nu \qquad (4)$$

If $q^2 = 0$ and if the τ is on-shell at both sides of the $\tau - \tau - \gamma$ vertex, then two of these form factors correspond to the electric and magnetic dipole moment anomalies for the τ: $F_3(0) = d_\tau/Q_\tau$ and $F_2(0) = a_\tau = (g_\tau - 2)/2$ where Q_τ is the τ charge. A non-zero value of d_τ is forbidden by T invariance, while the standard model predicts $a_\tau = 0.0011773(3)$ [8]. Though this expectation for a_τ is beyond current experimental reach, new phenomena could increase its value [9].

Large values of a_τ or d_τ would alter the energy and angular distributions of the final state particles in $e^+e^- \to \tau^+\tau^-\gamma$ events, generally enhancing production of high energy, isolated photons [10]. Such distributions have been measured [11], and the figure shows an L3 measurement of the distributions of the photon energy, the angle between the photon and the nearest τ decay product, and an example of what the distribution would look like for a large a_τ. There is

---

[1] Fitting joint distributions allows disentanglement of the effects of τ polarization from those of the ξ and $\xi\delta$ parameters.

no evidence in the data for anything unexpected, and the following limits have been placed: $-0.052 < a_\tau < 0.058$ and $-3.7 \times 10^{-16} \text{e} \cdot \text{cm} < d_\tau < 3.7 \times 10^{-16} \text{e} \cdot \text{cm}$.

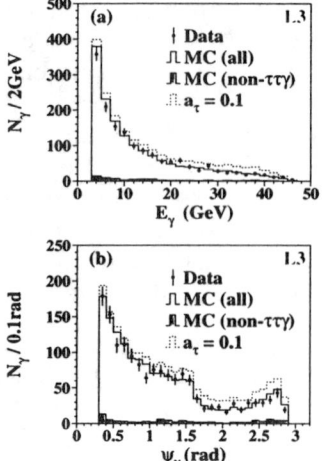

Conclusions

Thus far the τ appears identical to the other leptons in all respects except its mass and apparently distinct (and conserved) lepton number. This note has described some of the LEP contributions to measurements of τ electroweak couplings; these support the universality hypothesis at the per mil level in some cases. Though the LEP era is now drawing to a close, we can, in the near term, look to the b factories to continue the enterprise.

REFERENCES

1. H. Przysiezniak, these proceedings; LEP Electroweak Working Group, CERN-EP-2000-016.
2. ALEPH Collab.,*Phys.Lett.*B405, 191 (1997); DELPHI Collab., *Phys.Lett.*B404, 194 (1997); R. Volkert, DESY-ZEUTHEN-97-04 (1997).
3. A. Zalite in proceedings of Tau '98 *Nucl. Phys. Proc. Suppl.*76, 237 (1999), and references therein.
4. E.R. Cohen and B.N. Taylor, *Rev. Mod. Phys*59, 1121 (1987); V.W. Hughes and T. Kinoshita, *Comm. Nucl. and Part. Phys.* 14, 341 (1985).
5. ALEPH Collab., ALEPH99-014; DELPHI Collab., *Eur. Phys. J.*C10, 201 (1999); OPAL Collab., *Phys.Lett.*B447, 134 (1999); B. Stugu, in proceedings of Tau '98 *Nucl.Phys.Proc.Suppl.*76, 101 (1999); ALEPH Collab., *Phys. Lett.*B414, 362 (1997); DELPHI Collab., DELPHI99-133; L3 Collab., *Phys. Lett.*B479, 67 (2000); OPAL Collab., *Phys. Lett.*B374, 341 (1996).
6. K. Mursula and F. Scheck, *Nucl. Phys.*B253, 189 (1985); C. Nelson, *Phys. Rev.*D40, 123 (1989), erratum *Phys. Rev.*D41, 2327 (1990); W. Fetscher, *Phys. Rev.*D42, 1544 (1990); R. Alemany et al., *Nucl. Phys.*B379, 3 (1992).
7. ALEPH Collab., ALEPH98-073; L3 Collab., *Phys. Lett.*B438,405 (1998); OPAL Collab., *Eur. Phys. J.*C8, 3 (1999); DELPHI Collab., CERN-EP-2000-027;
8. M.A. Samuel, G. Li and R. Mendel, *Phys. Rev. Lett.*67, 688 (1991), erratum *ibid* 69, 995 (1992); F. Hamzeh and N.F. Nasrallah, *Phys. Lett.*B373, 221 (1996).
9. D.J. Silverman, G.L. Shaw, *Phys. Rev.*D27, 1196 (1983); J.L. Hewett, T.G. Rizzo, *Phys. Rev.*D56, 5709 (1997).
10. S.S. Gau, T. Paul, J. Swain and L. Taylor, *Nucl. Phys.*B523, 439 (1998); T. Paul, J. Swain and Z. Wąs, *Comput. Phys. Commun.*124, 243 (2000);
11. L3 Collab., *Phys.Lett.*B434, 169 (1998); OPAL Collab., *Phys.Lett.*B431, 188 (1998).

Recent Charmed Baryon Results from FOCUS

Eric W. Vaandering[1]
for the FOCUS Collaboration

University of Colorado
Department of Physics
Boulder Colorado 80309-0390

Abstract. Using data from the FOCUS experiment at Fermilab, we present measurements of the Σ_c mass difference measurements $M(\Sigma_c^0 - \Lambda_c^+)$ and $M(\Sigma_c^{++} - \Lambda_c^+)$. We find $M(\Sigma_c^0 - \Lambda_c^+) = 167.38 \pm 0.21 \pm 0.13$ MeV/c^2 and $M(\Sigma_c^{++} - \Lambda_c^+) = 167.35 \pm 0.19 \pm 0.12$ MeV/c^2. We measure the isospin mass splitting $M(\Sigma_c^{++} - \Sigma_c^0)$ to be $-0.03 \pm 0.28 \pm 0.11$ MeV/c^2. We also present a preliminary measurement of the relative branching ratio $\Xi_c^+ \to pK^-\pi^+ / \Xi_c^+ \to \Xi^-\pi^+\pi^+$ which we find to be $0.19 \pm 0.04 \pm 0.03$.

INTRODUCTION

In this paper we present two recent charmed baryon results from FOCUS. We present the most accurate, to date, measurements of the Σ_c^{++} and Σ_c^0 masses[2] and a measurement of the relative branching ratio for the Cabibbo suppressed decay $\Xi_c^+ \to pK^-\pi^+$.

FOCUS collected data using the Wideband photon beamline during the 1996–1997 Fermilab fixed-target run and is an upgraded version of FNAL-E687 [1]. The FOCUS experiment utilizes a forward multiparticle spectrometer to study charmed particles produced by the interactions of high energy photons ($\langle E \rangle \approx 180$ GeV) with a segmented BeO target [2].

MEASUREMENT OF Σ_c MASS DIFFERENCES

Many experiments [3–10] have measured the mass differences of the Σ_c^0 and Σ_c^{++} baryons with respect to the Λ_c^+. Only FNAL-E791 [3] and CLEO II [4] have measured the mass differences with respect to the Λ_c^+ to a total (statistical and systematic) precision of less than 0.5 MeV/c^2. Some of these previous measurements

---

[1] Present address: Vanderbilt University, Nashville, TN 37235
[2] Throughout this paper, charge conjugate states are implicitly included unless stated otherwise.

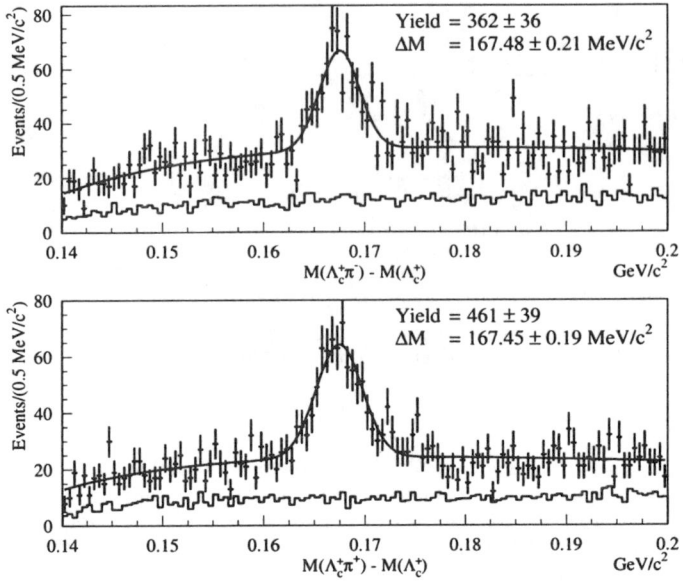

FIGURE 1. Mass difference distributions for $M(\Sigma_c^0 - \Lambda_c^+)$ and $M(\Sigma_c^{++} - \Lambda_c^+)$. The dotted histograms are obtained by combining pions with events from Λ_c^+ sidebands. These mass difference values have not been corrected with the mass calibration adjustment described in the text.

have suggested that the Σ_c multiplet is unique in that the masses of the isospin states *increase* with the quark substitution $d \to u$.

To reconstruct the decays $\Sigma_c \to \Lambda_c^+ \pi^\pm$, we first obtain a sample of Λ_c^+ baryons using the decay mode $\Lambda_c^+ \to pK^-\pi^+$. Λ_c^+ candidates are distinguished from background hadronic interactions primarily by requiring that the production and decay vertices are distinct using a candidate driven vertexing algorithm [1]. We apply a minimum detachment requirement which requires that the measured separation of the two vertices divided by the error on that measurement is greater than 6. Additional kinematic and Čerenkov requirements are placed on the $\Lambda_c^+ \to pK^-\pi^+$ candidate events. We find $12\,410 \pm 180$ $\Lambda_c^+ \to pK^-\pi^+$ candidates.

Σ_c candidates are reconstructed by combining the Λ_c^+ candidates in the peak region with a charged pion. To remove any systematic effects due to the reconstruction of the Λ_c^+, we compute and plot the invariant mass difference (ΔM).

The resulting distributions are then fit with the background function $N(1 + \alpha(\Delta M - m_\pi)\Delta M^\beta)$ where N, α, and β are allowed to vary. A Gaussian fitting function is used to represent the Σ_c signals. A small component attributed to $\Lambda_{c1}^{*+}(2625) \to \Lambda_c^+ \pi^+ \pi^-$ decays is also included. The invariant mass distributions, fits, and fit values are shown in Figure 1. We obtain samples of 362 ± 36 $\Sigma_c^0 \to \Lambda_c^+ \pi^-$ and 461 ± 39 $\Sigma_c^{++} \to \Lambda_c^+ \pi^+$ decays.

From other studies we estimate that the measured $\Sigma_c - \Lambda_c^+$ mass differences are 0.10 ± 0.05 MeV/c^2 above the true values. The final values quoted are adjusted for this shift and a systematic uncertainty is incorporated into the systematic error. Our measurement of the $\Sigma_c^{++} - \Sigma_c^0$ mass difference is unchanged by this shift.

We consider other sources of systematic errors such as effects introduced by our analysis cuts, fitting methods, and reconstruction algorithms. Considering both the statistical and systematic errors and applying the shift due to the momentum miscalibration, we find final values of $M(\Sigma_c^0 - \Lambda_c^+) = 167.38 \pm 0.21 \pm 0.13$ MeV/c^2, $M(\Sigma_c^{++} - \Lambda_c^+) = 167.35 \pm 0.19 \pm 0.12$ MeV/c^2, and $M(\Sigma_c^{++} - \Sigma_c^0) = -0.03 \pm 0.28 \pm 0.11$ MeV/c^2, where the first errors are statistical and the second are systematic. We find no evidence for a mass difference between the two Σ_c states. These results are in excellent agreement with preliminary results from CLEO II.V [11].

MEASUREMENT OF BR($\Xi_c^+ \to pK^-\pi^+ / \Xi_c^+ \to \Xi^-\pi^+\pi^+$)

From a theoretical standpoint the three body decays of charmed baryons are difficult to study due to several factors, especially the presence of strong final state interactions. Here we report additional evidence of the only observed Cabibbo suppressed decay of the Ξ_c^+ baryon: $\Xi_c^+ \to pK^-\pi^+$. This observation is a confirmation of the findings [12] of the SELEX experiment.

Tight Čerenkov cuts were placed on all the decay daughters, with the most restrictive cuts being placed on the proton and the loosest on the pion. A tight detachment cut required that the measured separation of the two vertices divided by the error on that measurement is greater than 8.[3]

A significant fraction of the background under $\Xi_c^+ \to pK^-\pi^+$ is from the charmed meson reflections $D^+/D_s^+ \to K^+K^-\pi^+$ and $D^+ \to K^-\pi^+\pi^+$. An additional background arises from $\phi(1020) \to K^+K^-$ where the K^+ is misidentified as a proton. To reduce these backgrounds, tighter Čerenkov cuts are applied to Ξ_c^+ candidates kinematically consistent with these other decays. An illustration of the relative strengths of these backgrounds *before* the application of tighter Čerenkov cuts is shown in Figure 2.

To measure the relative branching ratio we select the normalization mode $\Xi_c^+ \to \Xi^-\pi^+\pi^+$ with similar analysis cuts. The signals for both decay modes of the Ξ_c^+ are shown in Figure 3. Both distributions are fit with a polynomial background function and an unbounded Gaussian distribution for the signal. We obtain samples of 173 ± 32 $\Xi_c^+ \to pK^-\pi^+$ decays and 134 ± 14 $\Xi_c^+ \to \Xi^-\pi^+\pi^+$ decays. Our Monte Carlo efficiencies for the two decay modes are found to be $1.351 \pm 0.015\%$ and $0.201 \pm 0.005\%$ respectively (statistical errors only).

Extensive systematic studies and consistency checks were performed on the branching ratio measurement. These studies included separation into two independent run periods, a study of particles and antiparticles, and studies of different

---

[3] Due to the longer lifetime of the Ξ_c^+, such a cut is less destructive than for the $\Lambda_c^+ \to pK^-\pi^+$ decay channel.

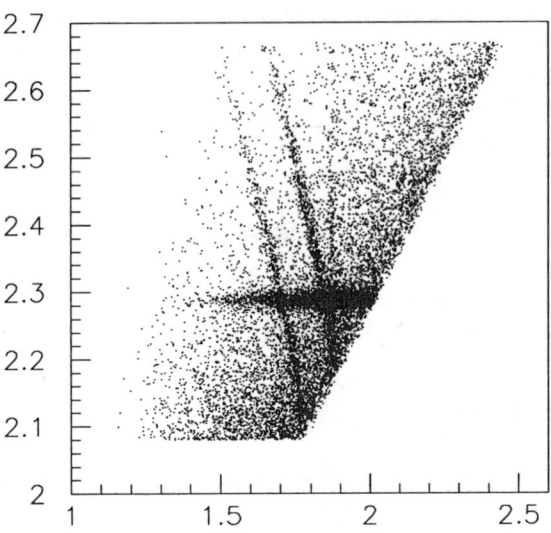

FIGURE 2. Charmed meson reflections in $\Xi_c^+ \to pK^-\pi^+$. The vertical axis shows the invariant mass of $\Xi_c^+ \to pK^-\pi^+$ candidates while the horizontal axis shows the same candidates reconstructed as the final state $\pi^+K^-\pi^+$. The dark diagonal bands are reflections from the misidentified decays $D^+/D_s^+ \to K^+K^-\pi^+$. The light vertical band is from $D^+ \to K^-\pi^+\pi^+$. The dark horizontal band is $\Lambda_c^+ \to pK^-\pi^+$ and the decay $\Xi_c^+ \to pK^-\pi^+$ is barely visible.

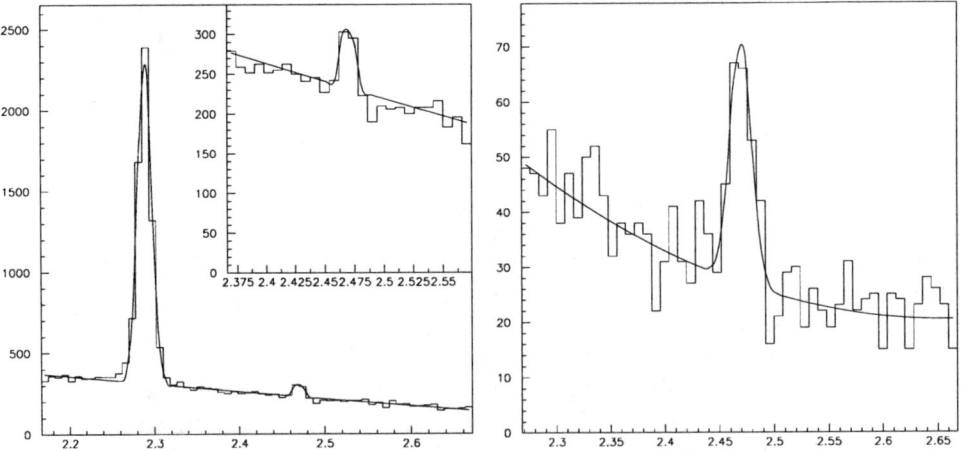

FIGURE 3. The plot on the left shows the $pK^-\pi^+$ invariant mass distribution. The decay $\Lambda_c^+ \to pK^-\pi^+$ is clearly visible and in the inset the decay $\Xi_c^+ \to pK^-\pi^+$ is also visible. The plot on the right shows the normalizing mode $\Xi_c^+ \to \Xi^-\pi^+\pi^+$ with similar analysis cuts.

fitting and binning methods. We also study the stability of the result as a function of the analysis cuts. From these studies we estimate an upper limit of the systematic error to be 3%.

In conclusion we preliminarily determine the relative branching ratio of the Cabibbo suppressed decay channel $\Xi_c^+ \to pK^-\pi^+$ to be

$$\frac{\text{BR}(\Xi_c^+ \to pK^-\pi^+)}{\text{BR}(\Xi_c^+ \to \Xi^-\pi^+\pi^+)} = 0.19 \pm 0.04 \pm 0.03$$

where the first error is statistical and the second is systematic.

ACKNOWLEDGEMENTS

We wish to acknowledge the assistance of the staffs of Fermi National Accelerator Laboratory, the INFN of Italy, and the physics departments of the collaborating institutions. This research was supported in part by the U. S. National Science Foundation, the U. S. Department of Energy, the Italian Istituto Nazionale di Fisica Nucleare and Ministero dell'Università e della Ricerca Scientifica e Tecnologica, the Brazilian Conselho Nacional de Desenvolvimento Científico e Tecnológico, CONACyT-México, the Korean Ministry of Education, and the Korean Science and Engineering Foundation.

REFERENCES

1. P. L. Frabetti et al., Nucl. Instrum. Meth. **A320**, 519 (1992).
2. J. M. Link, The FOCUS spectrometer and hadronic decays in FOCUS, in *Heavy Quarks at Fixed Target*, edited by H. W. K. Cheung and J. N. Butler, pages 261–269, American Institute of Physics, 1998.
3. E. M. Aitala et al., Phys. Lett. **379B**, 292 (1996).
4. G. Crawford et al., Phys. Rev. Lett. **71**, 3259 (1993).
5. P. L. Frabetti et al., Phys. Lett. **365B**, 461 (1995).
6. A. N. Aleev et al., Dubna JINR - 3(77)-96 (96/05,rec.Nov.) 31-46.
7. H. Albrecht et al., Phys. Lett. **211B**, 489 (1988).
8. J. C. Anjos et al., Phys. Rev. Lett. **62**, 1721 (1989).
9. M. Diesburg et al., Phys. Rev. Lett. **59**, 2711 (1987).
10. T. Bowcock et al., Phys. Rev. Lett. **62**, 1240 (1989).
11. J. Yelton, These proceedings.
12. S. Y. Jun et al., Phys. Rev. Lett. **84**, 1857 (2000).

Recent CLEO Results on Charmed Baryon Spectroscopy

John Yelton

University of Florida
yelton@phys.ufl.edu
Speaking on behalf of the
CLEO Collaboration

Abstract. The CLEO detector has collected more than 14 fb^{-1} of data running at and near the $\Upsilon(4s)$ resonance. Here we present new results on the measurement of the masses and widths of the Σ_c^{++} and Σ_C^0 baryons, a new measurement of the mass of the Σ_c^+, the discovery of a new state, the Σ_c^{*+}, and the first CLEO observation of the charmed, doubly-strange baryon, the Ω_c^0. All results are preliminary.

INTRODUCTION

A The Detector

The CLEO II and CLEO II.V detector configurations have recorded around $15 fb^{-1}$ of e^+e^- annihilation data in the last decade. Although CLEO and CESR were designed primarily with a view to doing B physics, many of our publications are on the subject of charm physics. In particular, CLEO has reported the first observation of around 12 different charmed baryon state. In this talk I will present some of our latest results on charmed baryons.

The CLEO detector is well known [1,2]. The main difference between the two configurations is that CLEO II.V includes a silicon vertex detector. This is particularly instrumental in measuring the angle of soft pions, and this makes a notable improvement in mass difference measurements.

We consider charmed baryons as consisting of a heavy c quark and a light diquark. The three quarks combine into a specific state of J^P. The light quark degrees of freedom also have a specific J^P which we refer to as J^P_{LIGHT}. Transitions between states must obey the quantum mechanical decay rules for transitions between specific J^P states, both for the total J^P and also J^P_{LIGHT}.

Compared with mesons, there are more states. More importantly there are more *narrow* states, as the excitation energy associated with one unit of orbital angular momentum is reduced as the moment of inertia of a state rises.

CP549, *Intersections of Particle and Nuclear Physics: 7th Conference,* edited by Z. Parsa and W. J. Marciano
© 2000 American Institute of Physics 1-56396-978-5/00/$17.00

To find charmed baryons at CLEO we generally look at high momenta where backgrounds are lower. This restricts us to continuum charm events rather than $B \to c$. We build the baryons out of protons, kaons and pions identified by time of flight and dE/dx measurements, and hyperons identified by their detached vertices.

I Σ_C^{++} AND Σ_C^0 WIDTHS

The Σ_c^{++} and Σ_c^0 baryons have been well established for several years. It has always been expected, by analogy with the non-charmed Σ baryons, that they should have natural widths of the order of 2 MeV. Furthermore, their natural widths can be calculated by scaling from the Σ_c^* widths that have been measured [3], albeit roughly, by CLEO. We now have high statistics and good enough resolution to see widths of this order. First we reconstruct a large sample of Λ_c^+ decays (if we look at $x_p > 0.5$ we have around 58,000 with a signal:noise of 1:1.2). We add a pion and take the mass difference. We take care to use a fitting technique that optimizes the resolution (the pion is constrained to come from the event vertex). Figure 1 shows the resultant mass difference plot. The solid fit is a p-wave Breit-Wigner convoluted with a double-Gaussian resolution function. The dashed line is what we would expect with no natural width - clearly inconsistent with the data. We can extract masses and widths of the Σ_c^0 and Σ_c^{++}. The widths (see the table) are in agreement with theoretical predictions [4]. The masses are in good agreement with the particle data group values and the FOCUS results shown earlier in this conference.

II Σ_C^+ AND Σ_C^{*+} RECONSTRUCTION

Theoretically these particles are really identical to their isospin partners. However experimentally they are more challenging as they decay via a *neutral* transition pion. Using the same Λ_c^+ sample as before, and π^0's with momentum greater than 150 MeV (optimized using Monte Carlo), we make the $M(\Lambda_c^+\pi^0) - M(\Lambda_c^+)$ mass difference plot (Figure 2). It can either be fit with two Gaussian signals to get yields of the Σ_c^+ of 661^{+63}_{-60} events and Σ_c^{*+} of 327^{+78}_{-73} events, or with Breit-Wigners smeared with Gaussian resolution functions (shown in Figure 2) which is the fit we use for the final mass measurement. This observation constitutes the discovery of the Σ_c^{*+}, the last of the seven $L = 0$ $\Sigma_c - \Lambda_c$ baryons to be found; as such it fills one more missing piece in the jigsaw. The masses and widths are tabulated in the table. Note that our new number for the Σ_c^+ mass is less than the previously published CLEO number [5]. This is in agreement with models that predict that the singly charged state is slightly lower mass than its isospin partners. Now the Σ_c^{*+} has been found, we can ask if this is the end of spectroscopy in the $\Sigma_c - \Lambda_c^+$ sector. The question is not whether there are more states made - there clearly are - the question is how many of them are narrow and well-defined enough to be detected. There are a whole series of $L = 1$ particles to be discovered. Some of them may be narrow. Most of them will decay eventually to the Λ_c^+'s [1]. It seems that a very large fraction of Λ_c^+'s found in e^+e^- annihilation are actually daughters of excited states.

III Ω_C OBSERVATION

The Ω_c, which consists of css quarks, has proved the most difficult of the ground state singly-charmed baryons to detect. Early results from CERN consisted of only a few events [6]. ARGUS then found a small signal [7] in $\Xi^- K^- \pi^+ \pi^+$ which was directly contradicted by CLEO limits [8], unfortunately unpublished. E-687 found two signals, one rather weak [9], but then a very good looking one [10] in $\Sigma_c^+ K^- K^- \pi^+$ that remains the best signal to date. WA-89 found peaks but never published a mass value.

Why is it so difficult to find the Ω_c? Of course its large mass makes it difficult to be produced. In e^+e^- annihilation, you need to consider also the mass of the particles that are produced along with it, that need to conserve quantum numbers (e.g. it must often be produced in conjunction with a Ξ^-). Unlike the Λ_c^+ case, there are not a huge number of excited states funneling into the Ω_c^0; some of its excited states will probably decay into $\Xi_c K$ combinations. Lastly, in fixed target machines, its short lifetime makes it difficult to detect the displaced vertices.

It is always possible to find a peak if you look in enough decay modes. We take the philosophy that we choose the five most likely modes based upon a) guessing what decay modes seem likely based upon analogy with other charmed baryons, b) choosing modes where we have good efficiency, and c) choosing modes with reasonably small backgrounds. The modes we choose are shown in Figure 3. There are x_p cuts placed on the data ($x_p > 0.5$ for an $x_p > 0.6$ for the others as their backgrounds are higher.) We also look in $\Sigma^+ K^- K^- \pi^+$ as that is the mode that has been seen by E-687, but consider that plot on a different basis to the others as we would not *a priori* consider as a favoured decay mode. The best of the hints of signals we can see is in $\Omega^- \pi^+$. If we add our five decay modes together (Figure 4), then a signal really looks good, and the yield in $\approx 40 \pm 9$ events. We do not fit this plot because the 5 decay modes have differing detector resolutions. Instead we do a simultaneous maximum likelihood fit to the five modes to extract our best mass value of $M = 2694.7 \pm 1.6 \pm 2.0$ MeV, and a yield of 40.4 ± 9.0 events. This mass is a little lower than the E-687 value, but consistent with it. We do not find a statistically significant signal in $\Sigma^+ K^- K^- \pi^+$ but find only a loose limit of 4.8 at the 90% confidence level, for the ratio of this decay to that of $\Omega^- \pi^+$. It is very exciting after so many years of looking to be able to find the Ω_c^0, however the data is still not overwhelming.

---

[1] Baryon-D meson decays will be competitive for some of the high mass states

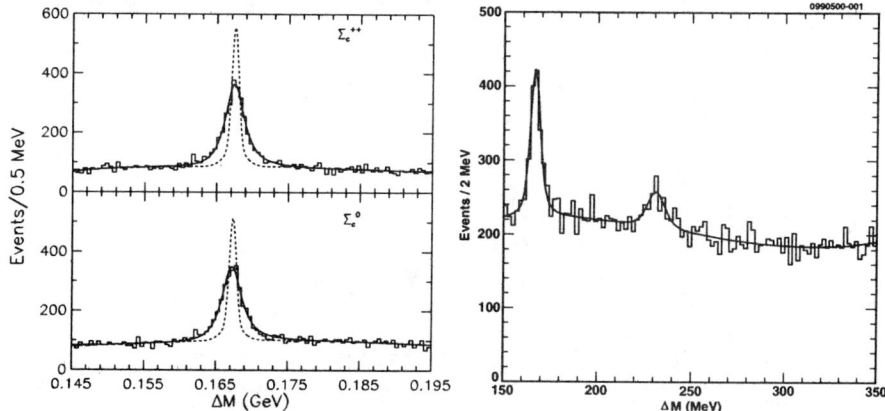

FIGURE 1. $M(\Lambda_c^+\pi^+) - M(\Lambda_c^+)$ and $M(\Lambda_c^+\pi^-) - M(\Lambda_c^+)$. The solid line is a fit to the data. The dashed line assumes no intrinsic width.

FIGURE 2. $M(\Lambda_c^+\pi^0) - M(\Lambda_c^+)$. There are clear peaks for the Σ_c^+ and Σ_c^{*+}. The second peak has not previously been reported.

Particle	$M - M(\Lambda_c^+)$ (MeV)	Γ (MeV)
(Σ_c^0)	$167.2 \pm 0.1 \pm 0.2$	$2.4 \pm 0.2 \pm 0.4$
(Σ_c^+)	$166.4 \pm 0.3 \pm 0.3$	< 4.6
(Σ_c^{++})	$167.4 \pm 0.1 \pm 0.2$	$2.5 \pm 0.2 \pm 0.4$
(Σ_c^{*+})	$231.0 \pm 1.1 \pm 2.0$	< 17

IV CONCLUSIONS

Using the large CLEO II and CLEO II.V data samples we have shown new results on the Σ_c^0 and Σ_c^{++} masses and widths, and the Σ_c^+ mass. We have shown the evidence for the discovery of the Σ_c^{*+} (this conference is the first time this has been shown). We have shown the first CLEO evidence for the Ω_c^0, which is a valuable addition to the published literature. I stress that all the results shown are preliminary. CLEO is not finished! There will be plenty of new results from CLEO II and CLEO II.V data and now CLEO III is taking data. The spectroscopy of charmed baryons is exceptionally rich and we can expect many results from all the B factories in particular in the next 10 years.

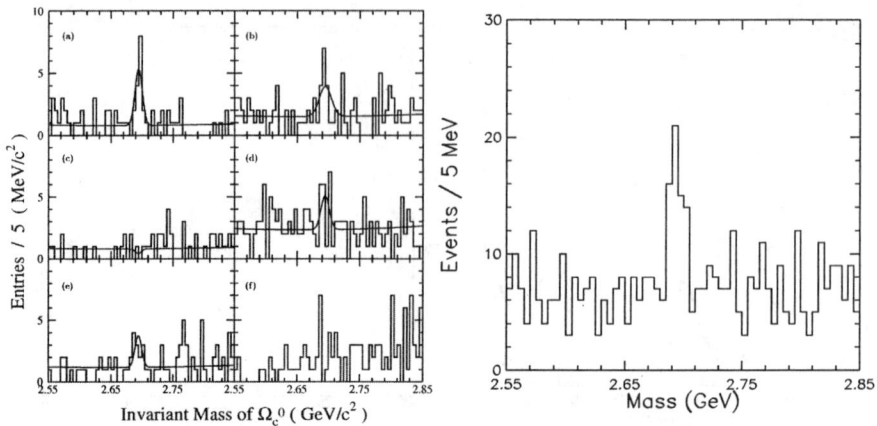

FIGURE 3. The above plot shows simultaneous fits to the five Ω_c^0 modes: (a) $\Omega^-\pi^+$, (b) $\Omega^-\pi^+\pi^0$, (c) $\Omega^-\pi^+\pi^-\pi^+$, (d) $\Xi^0 K^-\pi^+$, (e) $\Xi^- K^-\pi^+\pi^+$. The mode (f) $\Sigma^+ K^- K^-\pi^+$ has not been included in the fit.

FIGURE 4. Mass distributions for the five selected modes of the Ω_c^0 added together. The peak is about 40 ± 9 events.

REFERENCES

1. Y. Kubota et al., Nucl. Instr. and Meth. A **320** 66 (1992).
2. T. Hill et al., Nucl. Instr. and Meth. A **418** 32 (1998).
3. G. Brandenberg et al., Phys. Rev. Lett. **78** 2304 (1997).
4. J. Rosner et al., Phys. Rev. D **52**, 6461 (1995), D. Pirjol and T. Yan, Phys. Rev. D **56**, 5483 (1997), S. Tawfiq et al.,, Phys. Rev. D **58**, 5483 (1997).
5. G. Crawford et al., Phys. Rev. Lett. **71**, 3259 (1993).
6. P. Biagi et al., Z. Phys. C **28**, 175 (1985).
7. H. Albrecht et al., Phys. Lett. B **288** 367 (1992).
8. M. Battle et al., CLEO CONF 93-9.
9. P. Frabetti et al., Phys. Lett. B **300** 190 (1993).
10. P. Frabetti et al., Phys. Lett. B **338** 106 (1994).

Review of Top Quark Physics

Ulrich Heintz

Physics Department, Boston University, Boston, MA 02215

Abstract. The status of measurements of the properties of the top quark, including the $t\bar{t}$ production cross section, the top quark mass, and other properties are summarized for data from Run 1 (1992-1996) of the Fermilab Tevatron.

INTRODUCTION

According to the standard model, the building blocks of matter are the three generations of fermions. The first experimental evidence for the third generation dates from 1975, when the τ-lepton was discovered. The first experimental evidence for the third quark generation came in 1977 with the discovery of the b quark, which was measured to behave like a member of a doublet with weak isospin $-1/2$. The absence of flavor changing neutral currents required the existence of its partner with weak isospin $= +1/2$ — the top quark.

The search for the top quark lasted for two decades, until the CDF and DØ experiments at the Fermilab Tevatron finally observed it in 1995 [1]. It turned out to be much more massive than expected, massive enough to decay to a W-boson and a b-quark. The standard model predicts that the width of the top quark is 1.55 GeV, assuming it always decays to Wb. That means its lifetime is about 4×10^{-25} s, shorter than the time scale of the strong interactions of about 10^{-24} s. Thus, the top quark decays before it hadronizes, which makes its phenomenology fundamentally different from any other quark. Since the top quark does not form hadrons, we can study many of its properties directly and precisely.

From 1992 to 1996 (Run 1), the Fermilab Tevatron collided protons and antiprotons at a center of mass energy of 1.8 TeV and luminosities up to 2×10^{31} cm^{-2}s^{-1}. Two detectors (DØ [2] and CDF [3]) each accumulated about 125 pb^{-1} of data.

TOP QUARK PRODUCTION AND DECAY

Most frequently, top quarks are produced in $t\bar{t}$ pairs via the strong interaction. Single top quarks are also expected to be produced via the weak interaction.

In the standard model, the top quark decays almost exclusively to a W-boson and a b-quark if $m_t > M_W + m_b$. Its decay channels are defined by the decay modes of the W boson. The dilepton channels (both Ws decay to $e\nu$ or $\mu\nu$) make up about 5% of all $t\bar{t}$ decays. The lepton+jets channels (one W decays to $e\nu$ or $\mu\nu$, the other to jets) make up 30% of all $t\bar{t}$ decays. In 44% of the decays both Ws decay to jets. In the remaining 21% of decays one or both Ws decay to $\tau\nu$. The all-hadronic decays and decays involving τ-leptons are more difficult to isolate from background than decays involving e or μ.

Table 1 lists the number of events found in the DØ and CDF data and the expected backgrounds, broken down in the various decay channels. Different analyses were used in the lepton+jets channel. Top quark production can be tagged by the characteristics of the topology of events containing such massive particles (used by DØ) or by the presence of b-quarks among the decay products. Both experiments tag b-quarks by their decay muons. CDF also tags b-quarks by the presence of a displaced secondary vertex from the decay of a long lived particle. All channels combined, DØ measures a cross section of 5.9 ± 1.7 pb [4] and CDF $6.5\pm^{1.7}_{1.4}$ pb [5]. Predictions based on QCD calculations range between 4.7 and 6.2 pb [6].

TABLE 1. Number of $t\bar{t}$ candidate events observed and background events expected in the absence of $t\bar{t}$ production.

experiment	DØ		CDF	
dilepton channel	$ee, e\mu, \mu\mu$		$ee, e\mu, \mu\mu$	$e\tau, \mu\tau$
number of events	9		9	4
background	2.6±0.6		2.4±0.5	2.0±0.4
lepton+jets channel	topological	lepton tag	vertex tag	lepton tag
number of events	19	11	29	25[a]
background	8.7±1.7	2.4±0.5	8.0±1.0	13.2±1.2
all jets channel				
number of events	41		187	
background	24±2.4		151±10	

[a] 11 events are in common with the vertex tagged sample

TOP QUARK MASS

The top-quark mass affects predictions of the standard model via radiative corrections. For example, the dominant radiative corrections to the W boson mass involve top-quark and Higgs-boson loops. Thus, the W mass can be expressed as the tree-level value plus a function of the top-quark and Higgs-boson masses. Precise measurements of these masses can be used to constrain the mass of the Higgs boson within the framework of the standard model.

In Run I, the most precise mass measurement came from the lepton+jets channel. A total of 18 quantities define the six-particle final state of $t\bar{t}$ decay. In the lepton+jets channel, 17 can be measured in the event and only the momentum of the neutrino along the beam axis is unknown. There are three constraints: The invariant masses of the charged lepton and neutrino pair, and the two jets from the

TABLE 2. Summary of mass measurements. The two uncertainties are statistical and systematic.

experiment	DØ	CDF	combined
lepton+jets	173.3±5.6±5.5 GeV [8]	176.1±5.1±5.3 GeV [9]	
dilepton	168.4±12.3±3.6 GeV [7]	167.4±10.3±4.8 GeV [10]	
all-hadronic		186.0±10.0±5.7 GeV [11]	
combined	172.1±7.1 GeV	176.1±6.6 GeV	174.3±5.1 GeV [12]

other W-boson decay must both equal the W boson mass. The invariant mass of the decay products of the top and the antitop quarks must be equal. Thus we can perform a 2-C kinematic fit to determine the mass of the top quark. There is an up to 24-fold combinatoric ambiguity in the assignment of the four leading jets to the four quarks in the final state. The kinematic fit is performed for all permutations, and the best permutations are used to specify a fitted top-quark mass. A maximum likelihood fit to the spectrum of the fitted mass values of background and signal predictions for a range of hypothesized top-quark masses yields a measurement of the top quark mass. The largest systematic uncertainties are due to the jet energy scale calibration, and the Monte Carlo event generator.

In the dilepton channel, only the sum of the transverse momenta of the two neutrinos can be inferred from the measurements. Thus, of the six quantities needed to define the neutrino momenta, only two are measured and four are unknown. The same three constraints as for the lepton+jets channel apply. Therefore, the kinematics are underconstrained and a kinematic fit is not possible. In one analysis, a dynamical likelihood method is employed [7]. A weight function that characterizes how likely the event occurs for a range of top-quark masses is assigned to each event. These weight functions are then compared to MC predictions to determine the most likely top-quark mass. With the present small sample, the statistical uncertainty in the result is large. However, this channel has somewhat smaller systematic uncertainties than the lepton+jets channel. It may therefore become competitive with the lepton+jets channel when larger data samples are available.

In the all-hadronic channel the final state is measured completely, and a 3-C kinematic fit can be performed. The disadvantage of this channel is the large background from multijet production.

Table 2 summarizes all the measured values. The large value of the mass of the top quark means that its Yukawa coupling is near one. This has led to suggestions that the top quark may have a special connection to electroweak symmetry breaking. A comparison of predictions from the standard model with measurements indicates that the range of mas values is now restricted to M_H <170 at 95% confidence level [13].

OTHER RESULTS FROM RUN I

Both experiments are searching for production of single top quarks via the electroweak processes $q\bar{q} \to t\bar{b}$ and $qg \to qt\bar{b}$. CDF has set an upper limit of 13.5 pb at 95% confidence level on their combined cross section. DØ has set 95% confidence level upper limits of 39 pb on the first process and 58 pb on the second [14]. The standard model predicts 2.4 pb combined [15] [16]. These processes involve the Wtb vertex, and their cross section therefore provides a measurement of $|V_{tb}|$ and the width of the top quark.

The p_T spectrum of top quarks produced in $p\bar{p}$ collisions is a sensitive probe of new physics, e.g., new strong interactions [17]. DØ sees good agreement with the standard model [8]. The high momentum tail of the spectrum is most sensitive to new physics, and CDF has set a model independent upper limit on the fraction of top quarks produced with $p_T > 225$ GeV of 0.114 at 95% confidence level.

Models of electroweak symmetry breaking through strong dynamics predict the existence of resonances that decay to $t\bar{t}$ with masses of several hundred GeV. They would manifest themselves as peaks in the $t\bar{t}$ invariant mass spectrum. Both DØ [8] and CDF have searched for such signals but have not seen any evidence for such resonances.

The standard model predicts that the orientations of the spins of the top and antitop quarks are correlated. The degree of correlation is characterized by a correlation coefficient κ. DØ has analyzed the sample of dilepton events and finds $\kappa > -0.25$ at 68% confidence level, in agreement with the standard model value of 0.88 [18].

If the top quark has spin $1/2$ and pure $V-A$ coupling to the W boson, it decays to W bosons with helicity 0 or -1 in the ratio $F_0/F_{-1} = m_t^2/2m_W^2$. Thus, about 75% of the W bosons should have helicity 0. This fraction can be measured using the p_T distribution of leptons from $t\bar{t}$ events. CDF determines $F_0=0.91\pm0.39$. Fixing F_0 to the standard model value CDF fits a $V+A$ component that would result in W bosons with helicity $+1$. CDF finds $F_{+1}=0.11\pm0.16$ [19], consistent with the standard model.

FUTURE PROSPECTS

Run 2 of the Tevatron is scheduled to begin in March 2001. The accelerator is expected to deliver $p\bar{p}$ collisions at a center-of-mass energy of 2 TeV and a luminosity of up to 2×10^{32}cm^{-2}s^{-1}. Both detectors have been upgraded and expect to accumulate 2 fb^{-1} in the first 2 years, and as much as 15 fb^{-1} by 2006. The expected $t\bar{t}$ event yields per experiment for 2 fb^{-1} are over 200 dilepton events, 1800 lepton+jet events with 4 jets, 1400 with 3 jets and one b-tag, and 450 with 4 jets and 2 b-tags.

With these data samples, each experiment will measure the top quark mass to about 2 – 3 GeV in the lepton+jets channel. Together with the expected measurement of the W boson mass to about 40 MeV, this will constrain the Higgs boson mass to 80% of its value. The error will be reduced because of the larger data samples and the additional capabilities of the detectors. For example, the calibration of the jet p_T scale can be improved by using $Z \to b\bar{b}$ decays and hadronic W decays from $t\bar{t}$ events. Requiring two tagged b-jets in the events will greatly reduce the combinatoric uncertainties. The dilepton channel will also contribute significantly due to its smaller systematic uncertainties.

Using the Run 2 data sample, the observation of single top-quark production will be possible and the measurements of many properties of the top quark, that are now limited in precision by the small event sample from Run 1, will be much improved, to enable a meaningful comparison with standard model predictions.

CONCLUSION

DØ and CDF have studied the top quark using the data collected during Run 1 of the Tevatron. The $t\bar{t}$ pair production cross section agrees with expectations, and the top quark mass is 174.3±5.1 GeV. Run 2 provides the unique opportunity to study the top quark with greater precision. It will provide large data samples to address the question whether its properties agree with the standard model.

REFERENCES

1. CDF, Phys. Rev. Lett. 74, 2626 (1995); DØ, Phys. Rev. Lett. 74, 2632 (1995).
2. DØ, Nucl. Instr. Meth. A338, 185 (1994).
3. CDF, Nucl. Instr. Meth. A271, 387 (1988); Phys. Rev. D50, 2966 (1994).
4. DØ, Phys. Rev. Lett. 79 1203 (1997).
5. CDF, Phys. Rev. Lett. 80 2773 (1998) and recent updates.
6. Berger, Contopanagos, Phys. Lett. B361, 115 (1995); Phys. Rev D54, 3085 (1996); Bonciani et al., Nucl. Phys. B529, 424 (1998); Laenen et al., Phys. Lett. B321, 254 (1994); Catani et al., Phys. Lett. B378, 329 (1996).
7. DØ, Phys. Rev. Lett. 80, 2063 (1998); Phys. Rev D60, 052001 (1999).
8. DØ, Phys. Rev. Lett. 79 1197 (1997); Phys. Rev. D58, 52001 (1998).
9. CDF, Phys. Rev. Lett. 80 2767 (1998).
10. CDF, Phys. Rev. Lett. 82, 271 (1999).
11. CDF, Phys. Rev. Lett. 79, 1992 (1997).
12. DØ and CDF, FERMILAB-TM-2084 (1999).
13. Pepe-Altarelli, IVth Rencontres du Vietnam, Hanoi, July 2000.
14. DØ, submitted to Phys. Rev. Lett., Fermilab-Pub-00/188-E.
15. Stelzer et al., Phys. Rev. D56, 5919 (1997).
16. Smith, Willenbrock, Phys. Rev. D54, 6696 (1996).
17. Hill and Parke, Phys. Rev. D49, 4454 (1994).
18. DØ, Phys. Rev. Lett. 85, 256 (2000).
19. CDF, Phys. Rev. Lett. 84, 216 (2000).

Bound States in NRQCD and NRQED and the Renormalization Group[1]

Iain W. Stewart

Department of Physics, University of California at San Diego,
9500 Gilman Drive, La Jolla, CA 92093

Abstract.
The application of renormalization group techniques to bound states in non-relativistic QED and QCD is discussed. For QED bound states like Hydrogen and positronium, the renormalization group allows large logarithms of the velocity, $\ln v$ (or equivalently $\ln \alpha$'s), to be predicted in a universal and simple way. The series of $(\alpha \ln \alpha)$'s are shown to terminate after a few terms. For QCD one can systematically sum infinite series of the form $[\alpha_s \ln \alpha_s]^k$, and answer definitively the question "α_s at what scale?".

For Coulombic bound states of two fermions, the relevant scales include the mass of the fermions, m, their momentum, $p \sim mv$, and their energy, $E \sim mv^2$. Since $v \ll 1$ it is useful to calculate properties of these bound states in a double expansion in v and α, with terms $\alpha/v \sim 1$ kept to all orders. However, the expansion may still involve large logarithms of v or α, which appear through factors of $\ln(p/m)$, $\ln(E/p)$, and $\ln(E/m)$. In this talk I discuss how the renormalization group can be used to systematically predict and sum powers of $\alpha \ln v$ [1–6].

At a given order in QED, terms involving $\ln \alpha$ typically give the largest contributions, and the precision of experiments make their prediction quite important. In Ref. [5] it was shown for the first time that $\ln \alpha$'s in the Lamb shift, hyperfine splittings, and annihilation decay widths can be predicted with the renormalization group. This is in contrast to the usual method of computing these logarithms by evaluating matrix elements at the scale m. The calculations are simple enough that we can simultaneously treat Hydrogen, muonium ($\mu^+ e^-$), and positronium ($e^+ e^-$).

For the $t\bar{t}$ system near threshold, the relevant scales are $m_t \sim 175\,\mathrm{GeV}$, $m_t v \sim 30\,\mathrm{GeV}$, and $m_t v^2 \sim 5\,\mathrm{GeV}$, which are all $\gg \Lambda_{\mathrm{QCD}}$ and can be treated perturbatively. In QCD there is a strong dependence of the coupling on the scale; $\alpha_s(m_t)$ is much different from $\alpha_s(m_t v^2)$. The renormalization group allows us to handle this complication, or equivalently it allows us to systematically sum terms

---

[1] UCSD/PTH 00-19

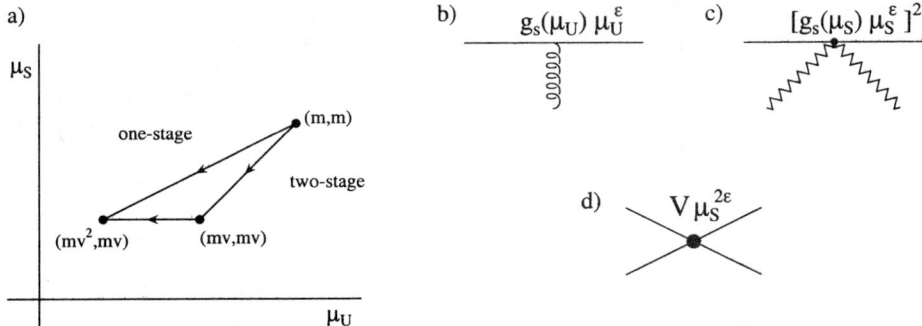

FIGURE 1. a) Paths in the (μ_U, μ_S) plane for one-stage and two-stage running. b),c),d) Examples of the μ_U and μ_S dependence of the Feynman rules.

$(\alpha_s \ln v)^k$. Predictions for the running coefficients of the $t\bar{t}$ potentials and the $t\bar{t}$ production current will be discussed below [2,4].

Effective theories for non-relativistic QED and QCD (NRQED and NRQCD) allow the double expansion in v and α to be performed in a simple way [7]. However, application of the renormalization group in these theories is complicated by the presence of two low energy scales, p and E, which are coupled by the equations of motion, $E = p^2/(2m)$. If one attempts to lower the cutoff on the energy to $E \lesssim \Lambda$, then one can still excite larger momenta $p \lesssim \sqrt{m\Lambda}$. Using dimensional regularization one can deal with this coupling of scales by using a velocity renormalization group [1], which has a subtraction velocity ν rather than the usual subtraction momentum μ. Running in one-stage from $\nu = 1$ to $\nu = v$ simultaneously lowers the subtraction point for momenta, $\mu_S \equiv m\nu$, to the scale mv, and the subtraction point for energy, $\mu_U \equiv m\nu^2$, to mv^2. In Fig. 1a) this one-stage approach is contrasted with the alternative two-stage approach where one first runs from $\mu = m$ to mv and then runs from $\mu = mv$ to mv^2. In Ref. [6] it was shown that for QED bound states, the most obvious method of two-stage running fails to reproduce terms involving $(\ln \alpha)^k$ with $k \geq 2$. This occurs because in the two-stage method the coupling between the energy and momentum is ignored.

To calculate observables at the hard scale m, the way the effective theory is formulated does not matter too much as long as it has a consistent power counting in v. Below m, only on-shell degrees of freedom are kept in the effective theory, ie. degrees of freedom which fluctuate near their mass shell. For example, a potential gluon exchanged between two quarks has energy $\sim mv^2$ but momentum $\sim mv$, and is therefore far offshell. Instead of a potential gluon the effective theory has a four quark operator

$$\mathcal{L}_p = - \sum_{\mathbf{p},\mathbf{p}'} V(\mathbf{p}, \mathbf{p}') \, \mu_S^{2\epsilon} \, \psi_{\mathbf{p}'}^\dagger \psi_{\mathbf{p}} \, \chi_{-\mathbf{p}'}^\dagger \chi_{-\mathbf{p}}, \tag{1}$$

where $\psi_{\mathbf{p}}$ ($\chi_{\mathbf{p}}$) destroys a quark (antiquark) with momentum $\mathbf{p}$ and spin and color

TABLE 1. QED $\ln \alpha$'s which follow from the leading order (LO) and next-to-leading order (NLO) anomalous dimensions in Ref. [5]. $\mu^+ e^-$ predictions include $1/m_\mu$ dependence. (h.f.s is hyperfine splitting and $\Delta\Gamma/\Gamma$ is the $e^+ e^-$ decay width correction.)

$\alpha^8 \ln^3 \alpha$	Lamb shift	H	agrees with [8,9], disagrees with [10,11][a]
	(no h.f.s., no $\Delta\Gamma/\Gamma$)	$\mu^+ e^-$, $e^+ e^-$	new predictions
$\alpha^7 \ln^2 \alpha$	h.f.s.	H, $\mu^+ e^-$, $e^+ e^-$	agree with [12,8,13]
	Lamb shift	H, $\mu^+ e^-$, $e^+ e^-$	needs running of $V^{(1)}$
$\alpha^3 \ln^2 \alpha$	$\Delta\Gamma/\Gamma$	$e^+ e^-$ ortho and para	agree with [8]
$\alpha^6 \ln \alpha$	Lamb shift, h.f.s.	H, $\mu^+ e^-$, $e^+ e^-$	needs ρ_s, $V^{(1)}(1)$
$\alpha^2 \ln \alpha$	$\Delta\Gamma/\Gamma$	$e^+ e^-$ ortho and para	agree with [14]
$\alpha^5 \ln \alpha$	Lamb shift	H, $\mu^+ e^-$, $e^+ e^-$	agree

[a] There is a growing consensus that the value in Ref. [8] is correct [15].

labels are suppressed. $V(\mathbf{p}, \mathbf{p}')$ is the potential

$$V(\mathbf{p}, \mathbf{p}') = \frac{U_c}{\mathbf{k}^2} + \frac{U_k}{|\mathbf{k}|} + U_2 + U_s \mathbf{S}^2 + \frac{U_r (\mathbf{p}^2 + \mathbf{p}'^2)}{2\mathbf{k}^2} + U_t \left(\sigma_1 \cdot \sigma_2 - \frac{3 \mathbf{k} \cdot \sigma_1 \mathbf{k} \cdot \sigma_2}{\mathbf{k}^2} \right)$$
$$- \frac{i \mathbf{U}_\Lambda \cdot (\mathbf{p}' \times \mathbf{p})}{\mathbf{k}^2} + U_3 |\mathbf{k}| + U_{3s} \mathbf{S}^2 |\mathbf{k}| + \frac{U_{rk} (\mathbf{p}^2 + \mathbf{p}'^2)}{2|\mathbf{k}|} + \ldots, \quad (2)$$

where the $U_i(\nu)$'s are running coefficients. In QCD the color singlet and octet coefficients have different values. We have absorbed in the U_i's the dependence on the fermion masses; only the momentum dependence is important for the power counting. For the QQ and $Q\bar{Q}$ potentials in an arbitrary Lie gauge group the matching coefficients at one-loop to order v^2 can be found in Ref. [3].

The effective Lagrangian has quarks with $(E, p) \sim (mv^2, mv)$ interacting with soft gluons which have $(E, p) \sim (mv, mv)$, and ultrasoft gluons which have $(E, p) \sim (mv^2, mv^2)$ (see Refs. [1–4]). To implement the velocity renormalization group we renormalize the Lagrangian and compute anomalous dimensions. This procedure is fairly simple since it can be done treating the Coulomb potential perturbatively. In Figs. 1b),c),d) the μ's which appear in some typical interactions are shown. Fig. 1b) shows a quark interacting with a single ultrasoft gluon. Physically, the fact that the ultrasoft mode involves the coupling $g(\mu_U)$ makes sense; due to the multipole expansion the scale mv^2 is the only scale it sees. Fig. 1c) shows a soft gluon scattering off a quark, and Fig. 1d) shows an insertion of the potential. For these interactions the parameter $\mu_S \sim mv$ appears.

Below the electron mass the electromagnetic coupling in NRQED does not run, but coefficients in the potential do. For fermions with mass and charge $(m_1, -e)$ and (m_2, Ze) we find the anomalous dimensions:

$$\nu \frac{dU_2}{d\nu}\bigg|_{LO} = \frac{2\alpha}{3\pi} \left(\frac{1}{m_1} + \frac{Z}{m_2} \right)^2 U_c + \frac{14 Z^2 \alpha^2}{3 m_1 m_2},$$

TABLE 2. LO and LL values of coefficients of the $t\bar{t}$ potential in QCD. Here $U_\Lambda = SU_\Lambda$ and the values are in units of the top mass m_t.

	$U_k^{(s)} m_t$	$U_r^{(s)} m_t^2$	$U_2^{(s)} m_t^2$	$U_s^{(s)} m_t^2$	$U_\Lambda^{(s)} m_t^2$	$U_t^{(s)} m_t^2$
$\nu = 1$	-0.36	-1.81	0	0.60	0.15	2.71
$\nu = v$	-0.03	-1.49	0.63	0.53	0.16	3.11

$$\nu \frac{dU_{2+s}}{d\nu}\bigg|_{\rm NLO} = \rho_{c22}\, U_c \left(U_{2+s}^2 + 2 U_{2+s} U_r + \frac{3}{4} U_r^2 - 9 U_t^2 {\bf S}^2 \right) + \rho_{ccc}\, U_c^3$$
$$+ \rho_{cc2}\, U_c^2 (U_{2+s} + U_r) + \rho_{ck}\, U_c U_k + \rho_{k2}\, U_k (U_{2+s} + U_r/2)$$
$$+ \rho_{c3}\, U_c \left(U_3 + U_{3s} {\bf S}^2 + \frac{U_{rk}}{2} \right) + \rho_s \frac{Z^3 \alpha^3}{m_1 m_2}, \qquad (3)$$

where $U_{2+s} = U_2 + U_s {\bf S}^2$ and the ρ_i's are mass dependent numbers [5]. Solving these equations gives the results summarized in Table 1.[2] Taking the matrix element of the leading log (LL) value of $U_2(\nu)$ gives the $\alpha^5 \ln\alpha$ Lamb shifts for Hydrogen, muonium, and positronium. Furthermore, the LO anomalous dimension is independent of ν so there are no higher terms, $\alpha^{k+4} \ln^k \alpha$ for $k \geq 2$. At next-to-leading log (NLL) order the most logarithms are generated by the $\rho_{c22} U_c U_2(\nu)^2$ term which gives the $\alpha^8 \ln^3 \alpha$ Lamb shifts. Thus, there are no terms $\alpha^{k+5} \ln^k \alpha$ for $k \geq 4$. Hyperfine splittings are generated by the ${\bf S}^2$ terms in Eq. (3), and the ortho and para-positronium widths are generated by imaginary terms which enter through the matching condition for $U_{2+s}(1)$.

NRQCD is better at generating logarithms than NRQED since the running of α_s causes all potential coefficients to run. For $t\bar{t}$ the change in the color singlet couplings from $\nu = 1$ to $\nu = v = 0.15$ are shown in Table 2. The largest changes occur in the spin independent couplings $U_k^{(s)}(\nu)$, $U_r^{(s)}(\nu)$, and $U_2^{(s)}(\nu)$. It is these couplings which depend on $\alpha_s(m_t \nu^2)$ since their anomalous dimensions have contributions from ultrasoft diagrams. In Fig. 2a) we plot the two-loop running of $U_k^{(s)}(\nu)$ whose value changes by an order of magnitude between $\nu = 1$ and $\nu = v$. The full theory $t\bar{t}$ production current gets matched onto a current in the effective theory $\bar{t} \gamma^i t = c_1 \psi_{\bf p}^\dagger \sigma^i \chi_{-\bf p}^* + \ldots$. At NLL order the running potentials mix into the running of the production current coefficient $c_1(\nu)$ [1]. Fig. 2b) plots the running of $c_1(\nu)$ at NLL from Ref. [4]. Summing the logarithms improves the convergence by reducing the size of the NLO matching coefficient by a factor of 2. It would be interesting to see if a similar improvement in the convergence takes place for the rather large NNLO matching correction found in Ref. [16].

This work was supported in part by the Department of Energy under grant DOE-FG03-97ER40546, by the National Science Foundation under award NYI PHY-9457911, and by NSERC of Canada.

---

[2] Note that the $\alpha^7 \ln^2 \alpha$ Lamb shift requires the LL running of $V^{(1)}$ since this potential mixes into U_2, and the $\alpha^6 \ln\alpha$ Lamb shift requires the NLL running of U_k. These will be discussed in a future publication.

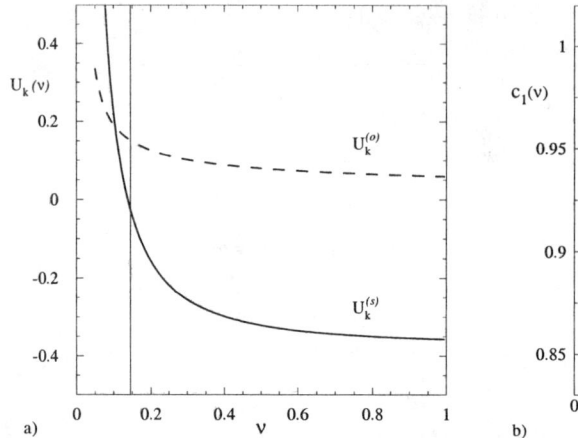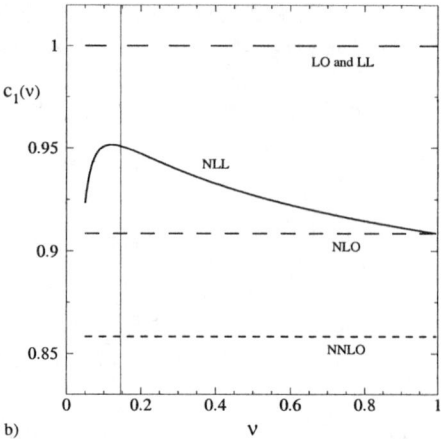

FIGURE 2. For $t\bar{t}$ production near threshold the running of the color singlet and octet $1/|\mathbf{k}|$ potentials are shown in a), and the NLL value of the production current is given in b) (solid line) [4]. In b) the large, medium, and small dashes are the LO, NLO, and NNLO [16] matching results. In a) and b), $\nu = 1$ is the scale $\mu = m_t$ and the solid vertical line is the Coulombic velocity.

REFERENCES

1. M.E. Luke, A.V. Manohar, and I.Z. Rothstein, Phys. Rev. **D61**, 074025 (2000).
2. A.V. Manohar and I.W. Stewart, Phys. Rev. **D62**, 014033 (2000).
3. A.V. Manohar and I.W. Stewart, hep-ph/0003032, to appear in PRD.
4. A.V. Manohar and I.W. Stewart, hep-ph/0003107.
5. A.V. Manohar and I.W. Stewart, hep-ph/0004018, to appear in PRL.
6. A.V. Manohar, J.Soto, and I.W. Stewart, Phys. Lett. **B486**, 400 (2000).
7. W.E. Caswell and G.P. Lepage, Phys. Lett. **167B**, 437 (1986); G.T. Bodwin, E. Braaten and G.P. Lepage, Phys. Rev. **D51**, 1125 (1995), Erratum ibid. **D55**, 5853 (1997); P. Labelle, Phys. Rev. **D58**, 093013 (1998); M. Luke and A.V. Manohar, Phys. Rev. **D55**, 4129 (1997); B. Grinstein and I.Z. Rothstein, Phys. Rev. **D57**, 78 (1998); M. Luke and M.J. Savage, Phys. Rev. **D57**, 413 (1998); A. Pineda and J. Soto, Nucl. Phys. Proc. Suppl. **64**, 428 (1998); M. Beneke and V.A. Smirnov, Nucl. Phys. **B522**, 321 (1998).
8. S.G. Karshenboim, Sov. Phys. JETP **76**, 541 (1993).
9. I. Goidenko et al., Phys. Rev. Lett. **83**, 2312 (1999).
10. S. Mallampalli and J. Sapirstein, Phys. Rev. Lett. **80**, 5297 (1998).
11. V.A. Yerokhin, hep-ph/0001327.
12. P. Labelle, Ph. D. thesis, Cornell University, 1994 (unpublished).
13. K. Melnikov and A.S. Yelkhovsky, Phys. Lett. **B458**, 143 (1999).
14. W.E. Caswell and G.P. Lepage, Phys. Rev. **A20**, 36 (1979); I.B. Khriplovich and A.S. Yelkhovsky, Phys. Lett. **B246**, 520 (1990).
15. J. Sapirstein, private communication.

16. For QED see: A.H. Hoang, Phys. Rev. **D56**, 5851 (1997); Phys. Rev. **D56**, 7276 (1997). For QCD see: A. Czarnecki and K. Melnikov, Phys. Rev. Lett. **80**, 2531 (1998); M. Beneke, A. Signer and V.A. Smirnov, Phys. Rev. Lett. **80**, 2535 (1998).
17. M. Beneke, A. Signer and V. A. Smirnov, Phys. Lett. **B454**, 137 (1999).

Extracting $|V_{ub}|$ from Cut Spectra

Ira Z. Rothstein

Department of Physics
Carnegie Mellon University
Pittsburgh Pa 15213

Abstract. In this talk I review recent progress made in extracting $|V_{ub}|$ from the cut electron energy and hadronic mass spectra of inclusive B meson decays.

INTRODUCTION

It is an unfortunate fact that experimental cuts can take a nice clean theoretical prediction and turn it into a troublesome mess. A perfect example of this scenario arises in the extraction of V_{ub} from inclusive B decays. In principle this extraction should be straightforward. One measures the inclusive rate for semi-leptonic decays into non-charmed states and compares the result to the theoretical prediction for the total rate, which is under good theoretical control [1]. Of course, the snag is that there is no simple way, at least at this time, to measure the total inclusive rate to charmless states. Thus, some cut must be applied to reject the charmed final states. Perhaps the simplest choice, from an experimentalist viewpoint, is to cut on the electron energy, rejecting all events with $E_l < (m_B^2 - m_D^2)/(2m_B)$. This is the oldest method for extracting V_{ub} [2]. Unfortunately, the theoretical prediction for the integrated cut spectrum is rather complicated.

The problem arises from the fact that the cut introduces a new scale into the problem. Without the cut there are only two scales of relevant physics, namely, the quark mass m_b (suitably defined) and the QCD scale Λ_{QCD}. Suppose we cut on a scaled kinematic variable, such as the electron energy $x = 2E_l/m_b$. If we cut near the endpoint, $x \simeq 1$, then the scale $(1-x)m_b$ can introduce a large (small) dimensionless ratio into the calculation $1/(1-x)$. This can lead to power law, as well as logarithmic, amplifications of what normally would be small effects.

The calculation of the total inclusive rate can be derived from first principles [1] within a systematic expansion in $\alpha(m_b)$ and Λ/m_b [1]. The aforementioned amplifications in the cut rate arise as corrections of the form $\Lambda/(m_b(1-x))$ and

---

[1] Even in the total rate for semi-leptonic decays one still has "mild" local-duality assumptions arising from the fact that the contour approaches the real axis at a point.

$\alpha_s Log^2(1-x)$. Physically, the reasons for these enhancements are clear. The non-perturbative corrections arise from the fact that near the endpoint the spectrum becomes very sensitive to the fermi motion of the heavy quark. The logarithmic corrections are due to exclusivity of the cut rate. In the limit where x approaches one there is no room for gluon radiation, thus leading to the usual infra-red divergences of the Bloch-Nordstrom type.

Of these two types of enhancements it is really the power corrections that make the calculation more difficult. The reason for this is that the effects of Fermi motion are incalculable. In the past the Fermi motion was modeled, leading to extractions of V_{ub} for which it was not really possible to make meaningful theoretical error estimates. This is not to say that those extractions will not lead to a number which in the end could turn out to be "correct". Or that the old error band is not reasonable. However, given that we are now entering age of precision CKM measurements the theoretical has more of an obligation to make more systematic estimates of their errors.

Fortunately, there is a way around having to model the Fermi motion. The relevant point is that the effect of the Fermi motion on the decay spectra are universal. Thus, if we can extract the necessary information about the end point spectrum from one decay we can use it to make a prediction for another decay spectrum. On top of the Fermi motion issue, the end point spectrum is also plagued by the above mentioned large logs. These logarithms can jeopardize the convergence of the perturbative expansion in α_s. However, these large corrections can be resummed using modified renormalization group arguments, thus leading to a controlled approximation.

In this talk I will review progress made in making predictions for the integrated cut spectra, within a systematic expansion in α_s and Λ/m_b. In particular, I will talk about predictions for the endpoint electron energy as well as the end point of the hadronic invariant mass spectrum.

I THE CUT ELECTRON SPECTRUM

The electron energy endpoint spectrum can be written as

$$\frac{d\Gamma}{dE} = \int_{2E-m_b}^{\bar{\Lambda}} dk_+ f(k_+) \frac{d\Gamma_p}{dE}(m_b^*), \tag{1}$$

where E is the charged lepton energy in semi-leptonic B decay. m_b^* is the effective mass which accounts for the residual momentum k_+, such that $m_B^* = m_b + k_+$ and the structure function, which accounts for the Fermi motion, is given by

$$f(k_+) = \langle B(v) \mid \bar{b}_v (iD_+)^n b_v \mid B(v) \rangle. \tag{2}$$

A similar expression can be given for the radiative decay spectrum. Thus, in principle, one could measure one end point spectrum, deconvolve (1), and then extract $f(k_+)$ to make a prediction for the shape of another spectrum. However, this would be a rather Herculean task which I would wish on neither friend nor foe. Fortunately, this analysis can be avoided by first taking the Mellin transform of the spectrum. However, before discussing how to do this, I will first discuss the breakdown of the perturbative expansion near the end point.

As I mentioned above, near the end point large infrared logs arise which jeopardize the convergence of the the perturbative expansion. The pattern of these logs is extensively discussed in [5,6]. The upshot is that if we take the quantity $L \equiv \alpha_s Log(1-x)$ to be of order one, then the Mellin transform of the perturbative expansion can be reorganized into the form

$$M[\frac{d\Gamma}{dx}; N] = H(\alpha_s) exp\left(Log(N)g_1(\chi) + g_2(\chi) + \alpha_s g_3(\chi) + \ldots\right) \quad (3)$$

The function H is independent of the large logs and χ is given by $\alpha_s \beta_0 Log(N)$ [2]. The functions g_1 and g_2 were calculated in [4,5] and are related to similar functions used in deep-inelastic scattering [10]. Now the point is that when the structure function is included in the rate the Mellin transform turns the convolution in (1) into a product. Thus we may remove all dependence on the structure function by first taking the ratio of the moments of two spectra and then taking the inverse Mellin transform. This is exactly what was done for the electron energy spectrum in ref. [7]. In this reference it was shown that following this procedure $|V_{ub}|^2/|V_{ts}^* V_{tb}|^2$ may be extracted from the relation

$$\frac{|V_{ub}|^2}{|V_{ts}^* V_{tb}|^2} = \frac{3\alpha C_7(m_b)^2}{\pi(1 + 3\bar{\Lambda}/M_B)} \int_{x_B^c}^1 dx_B \frac{d\Gamma}{dx_B}$$
$$\times \left\{ \int_{x_B^c}^1 dx_B \int_{x_B}^1 du_B \, u_B^2 \frac{d\Gamma^\gamma}{du_B} K\left[x_B; \frac{4}{3\pi\beta_0} \log(1 - \alpha_s \beta_0 l_{x_B/u_B})\right] \right\}^{-1}, \quad (4)$$

where we have rewritten our result in terms of the hadronically scaled variable $x_B = 2E/M_B$, $\bar{\Lambda} = M_B - m_b$ and the experimental cut x_B^c. The function $K(x;y)$ in the convolution is the integral over the neutrino energy x_ν, taking into account the x dependence up to $\mathcal{O}((1-x)^2)$, and is given by

$$K(x;y) = 6\left\{\left[1 + \frac{4\alpha_s}{3\pi}\left(1 - \psi^{(1)}(4+y)\right)\right]\frac{1}{(y+2)(y+3)}\right.$$
$$\left. - \frac{\alpha_s}{3\pi}\left[\frac{1}{(y+2)^2} - \frac{7}{(y+3)^2}\right] - \frac{4\alpha_s}{3\pi}\left[\frac{1}{(y+2)^3} - \frac{1}{(y+3)^3}\right]\right\}$$
$$- 3(1-x)^2. \quad (5)$$

C_7 is the Wilson coefficient of the O_7 operator which dominates the radiative decay and $l_{x/u} = -\log[-\log(x/u)]$.

---

[2] One should consider $\alpha_s Log(1-x)$ as parametrically scaling like χ.

Note that the experimental cut on the radiative decay is actually lower (around 2.1 GeV at present) then the cut on the electron energy spectrum. Thus the integral over the radiative spectrum is known. This result is complete up to next to leading order in the infrared logs (i.e. includes g_1 and g_2 in (3)). The leading errors are of order α^2, $\alpha_s(1-x)$ and $(\Lambda^2/((1-x)m_b^2)$ [3], which lead to errors of order %10 in extractions of V_{ub}. It should always be remembered that when it comes to inclusive decays there are always questions regarding the errors due to duality errors which are unquantifiable within QCD. This is especially true in cut spectra since we are looking in a limited region of phase space where we are including only a limited number of hadrons in the final state. For this reason it is imperative that we see convergence among as many extractions as possible.

II THE CUT HADRONIC MASS SPECTRUM

It is also possible to remove the background from charmed transitions by cutting on the hadronic invariant mass [8]. While this choice presents a greater experimental challenge, it benefits from the fact that, unlike the electron spectrum, most of the $B \to X_u e\nu$ decays are expected to lie within the region $s_H < M_D^2$. Furthermore, it is believed that even though both the invariant mass region $s_H < M_D^2$ and electron energy regions $M_B/2 > E_e > (M_B^2 - M_D^2)/(2M_B)$ receive contributions from hadronic final states with invariant mass up to M_D, the cut mass spectrum will be less sensitive to local duality violations. This belief rests on the fact that the contribution of large mass states is kinematically suppressed for the electron energy spectrum in the region of interest.

We may write down a closed form expression for V_{ub} using the same strategy discussed above for the electron energy spectrum. First, the rate is factorized in terms of hadronic variables [9], then, after taking the Mellin transform, a ratio is taken with the Mellin transformed radiative spectrum, thus eliminating our ignorance of the Fermi motion. The final expression for V_{ub} in terms of the cut hadronic rate and the cut radiative rate is rather lengthy and can be found in ref. ([11]).

The quantifiable errors in this extraction are of the same order as those discussed above for the electron spectrum. Finally, I should mention that once the neutrino reconstruction can be performed with sufficient accuracy, it is also possible to make a cut on the leptonic invariant mass spectrum [12,13]. This method has the great advantage that it is less sensitive to the structure function.

III CONCLUSIONS

We have successfully circumvented the need to use quark models to extract the CKM matrix CKM. This is a first, crucial, step towards being able to make "precision" extractions of V_{ub} using inclusive decays. But we should not fool ourselves

---

[3] This last error was not made explicit in (7)

into thinking that our footing is rock solid. I don't believe that the term "precision" as understood in a "LEP" sense will ever really apply in the CKM sector, simply due to the fact that hadronic physics is just too hard. It might very well be that one or more of the extractions deviates from the others. Which is to say our experimental colleagues should appreciate the fact that local duality lurks in the background and is not well understood. But we are certainly doing a lot better then we were before when we relied on quark models.

REFERENCES

1. J. Chay, H. Georgi, and B. Grinstein, Phys. Lett. **B247** (1990) 399; M. Voloshin and M. Shifman, Sov. J. Nucl. Phys. **41** (1985) 120.
2. Review of Particle Physics. Eur. Phys. J. **C15** (2000).
3. M. Neubert, Phys. Rev. **D49** (1994) 3392; T. Mannel and M. Neubert, Phys. Rev. **D50** (1994) 2037; I.I. Bigi, M.A. Shifman, N.G. Uraltsev, and A.I. Vainshtein, Int. J. Mod. Phys. **A9** (1994) 2467.
4. G.P. Korchemsky and G. Sterman, Phys. Lett. **B340** (1994) 96.
5. R. Akhoury and I.Z. Rothstein, Phys. Rev. **D54** (1996) 2349.
6. A.K. Leibovich and I.Z. Rothstein, Phys. Rev. **D61** (2000) 074006.
7. A.K. Leibovich, I. Low, and I.Z. Rothstein, Phys. Rev. **D61** (2000) 053006.
8. J. Dai, Phys. Lett. **B333** (1994) 212. V. Barger, C.S. Kim, and R.J.N. Phillips, Phys. Lett. **B251** (1990) 629; J. Dai, Phys. Lett. **B333** (1994) 212; C. Greub and S.J. Rey, Phys. Rev. **D56** (1997) 4250; A. Falk, Z. Ligeti and M.B. Wise, Phys. Lett. **B406** (1997) 225.
9. A.K. Leibovich, I. Low, and I.Z. Rothstein, Phys. Rev. **D62** (2000) 014010.
10. For a review of resummation techniques and a collection of references see, G. Sterman, in *QCD and Beyond, Proceedings of the Theoretical Advanced Study Institute in Elementary Particle Physics (TASI 95)*, ed. D.E. Soper (World Scientific), 1996, hep-ph/9606312.
11. A.K. Leibovich, I. Low, and I.Z. Rothstein, Phys. Lett. **B486** (2000) 86.
12. C. Bauer, Z. Ligeti and M. Luke, Phys. Lett. **B479** (2000) 401.
13. C. Bauer these proceedings.

Search for New Physics in Rare B Decays

H. Landsman* and Y. Rozen*

*Physics Department, Technion - Israel Institute of Technology,[1]
32000 Haifa, Israel

Abstract. A search for the decay $B^{\pm} \rightarrow K^{\pm}K^{\pm}\pi^{\mp}$ was performed using data collected by the OPAL detector at LEP. These decays are strongly suppressed in the Standard Model but could occur with a higher branching ratio in supersymmetric models, especially in those with R-parity violating couplings. No evidence for a signal was observed and a 90% confidence level upper limit of 1.29×10^{-4} was set for the branching ratio. This upper limit was used to set a limit on the R-parity violating couplings.

INTRODUCTION

Rare b decays offer an opportunity to discover new physics beyond the Standard Model (SM). The process $b \rightarrow ss\bar{d}$, induced by a box diagram, is predicted to be exceedingly small in the SM, of the order of 10^{-11} [1]. However, in the minimal supersymmetric standard model (MSSM) [2], this transition can be induced by the squark-gaugino (or higgsino) box diagrams at a level of $10^{-7} - 10^{-8}$. An alternative mechanism for this channel in supersymmetric models is through R-parity violating couplings [3]. These two possibilities appear to be the only ones that will produce significant enhancement of this decay within supersymmetric models [1]. he decay $B^- \rightarrow K^-K^-\pi^+$ (charge conjugation is assumed throughout this paper), either as a direct three-body decay or through a K*-like resonance, is a clear signature of this process. This document describes the first search for the decay $B^- \rightarrow K^-K^-\pi^+$ carried by the OPAL collaboration [4].

EVENT SELECTION

We used data collected at LEP by the OPAL detector [5] between 1990 and 1995 running at center-of-mass energies in the vicinity of the Z^0 peak. Hadronic Z^0

---

[1] This research is supported by the Israeli Science Foundataion and by the Fund for the promotion of research at the Technion.

decays were selected using the number of charged tracks and the visible energy in each event as in Reference [6]. This selection yielded 4.41 million hadronic events.

We searched the hadronic event sample for the decay $B^-\to K^-K^-\pi^+$ by combining three charged tracks to form a B meson candidate. All three track combinations were considered. All tracks were required to have a momentum of at least 2 GeV/c and to be in the same jet. Two of the tracks were required to have the same charge and were assigned the mass of a kaon. A third track, with an opposite charge, was assumed to be the pion. Tracks were required to satisfy selection criteria based on the measured rate of energy loss due to ionisation (dE/dx). The dE/dx selection criteria are 44% efficient while rejecting 98.5% of the background.

The three tracks were fitted to a common vertex and the decay length, the distance from the e^+e^- interaction point to the reconstructed secondary vertex, was calculated. Candidates where the secondary vertex is in the hemisphere opposite to the candidate's jet were rejected. This criterion left 55% of the remaining background events, but kept 96% of the signal events.

Since the hadronic data sample consisted mostly of non-$b\bar{b}$ events, we suppressed these events by means of a b-tagging algorithm [7]. Events were accepted if any of the jets were tagged. The b-tagging selection was found to be 79% efficient, while rejecting 80% of the remaining background.

The final selection was based on an artificial neural network (ANN) designed to select $B^-\to K^-K^-\pi^+$ events while rejecting background events. The neural network used seven input parameters: the momenta of the three tracks; the B candidate momentum; the ratio of B candidate energy to the jet energy; the decay length; and the vertex probability, the probability of the three tracks to originate from a common vertex which is calculated using the track parameters. The neural network retains 74% of the signal events and rejects 97% of background events when selecting candidates with an ANN output above 0.9.

Candidates were accepted if their invariant mass was in the region: 5.10 GeV/c^2 < $M_{KK\pi}$ < 5.46 GeV/c^2, which corresponds to twice the mass resolution around the nominal B^- mass. Only one candidate per jet was accepted, based on the largest neural network output for candidates in a jet. Figure 1(a) shows the invariant mass distribution of the $K^-K^-\pi^+$ candidates. No enhancement is seen in the signal region, where 17 events were observed. Monte Carlo studies indicated that 88% of the background at this stage consists of $b\bar{b}$ events.

If the decay chain $B^-\to K^-K^-\pi^+$ is assumed to be direct (i.e., without an intermediate K^* resonance), then the mass of the $K^-\pi^+$ system can be exploited to further reduce background where the pion and one of the kaons are from the decay of a K^* resonance. The mass of the $K^-\pi^+$ system was added as an input to the neural network, and the training procedure of the ANN was repeated. Figure 1(b) shows the invariant mass distribution of the $K^-K^-\pi^+$ candidates passing the selection. Here too, no enhancement is seen in the signal region and the observed 14 events are used to determine an upper limit on the branching ratio.

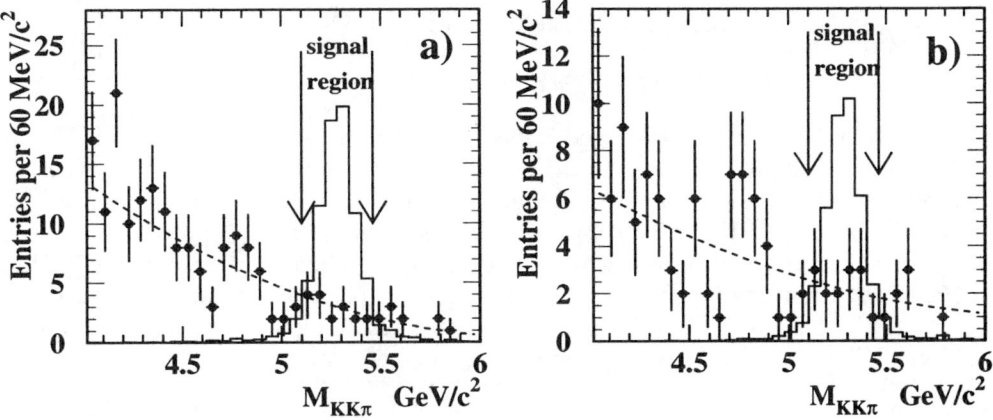

FIGURE 1. Invariant mass distribution of the $K^-K^-\pi^+$ candidates after all selection criteria were applied a) via intermediate K^* resonance and b) with direct production. The dots represent the data, the solid line shows the expected signal shape from Monte Carlo events after all the selection criteria were applied with arbitrary normalisation, and the dashed line is the expected background.

Background Estimation

The background was estimated by fitting a second-order polynomial to the invariant mass of a combinatorial background, obtained by releasing the ANN cut, and then normalising the shape to the mass side-bands of Figure 1. Monte Carlo studies indicated that the background shape is not altered by this procedure. Alternatively, we repeated this procedure by releasing each of the selection criteria separately and by obtaining the shape from Monte Carlo. All the alternative fits gave a consistent result. We also took the number of events within the signal region in each of the above cases and scaled it to the appropriate sample size. Here too, all estimates were consistent.

As we are setting upper limits, the conservative approach is to estimate the number of signal events using the lowest background estimate. The lowest estimate were 17.5 and 14.1 events for the two cases.

Limit Determination

The above numbers were used to determine N^{90}, the 90% C.L. upper limit on the number of signal events. We obtained $N^{90} = 7.8$ and $N^{90} = 7.4$ events for the resonant/direct decay, respectively.

To calculate an upper limit on the branching ratio we used:

$$\mathrm{Br}(B^- \to K^-K^-\pi^+) \leq \frac{N^{90}}{\epsilon\, N_B},$$

where N_B is the number of charged B mesons in the sample and ϵ is the efficiency for Monte Carlo simulated events of the process $B^- \to K^-K^-\pi^+$ to survive the selection procedure. With 4.41 million hadronic Z^0 decays, using $\mathrm{Br}(b \to B^\pm) = 0.397^{+0.018}_{-0.022}$ and $\Gamma_{b\bar{b}}/\Gamma_{\mathrm{had}} = 0.2170 \pm 0.0009$ [8], we obtained $N_B = 759\,800^{+34\,600}_{-42\,200}$.

The conservative approach when setting upper limits is to use the model giving the lowest efficiency for the signal. If one assumes resonance production, then the lowest efficiency, 8.11±0.19%, is obtained when assuming the signal decay channel is via $K^{*0}(892)$. The lowest efficiency for non-resonant decay, was found to be 11.3±0.2%.

RESULTS

When incorporating systematic uncertainty of 8.5 %, according to the method outlined in reference [9], we obtained with $N^{90}_{\mathrm{res.}} = 7.8$ events and $N^{90}_{\mathrm{no\ res.}} = 7.4$:

$$\mathrm{Br}(B^- \to K^-K^-\pi^+) \leq 1.29 \times 10^{-4}\ @\ 90\%\ \mathrm{C.L.}$$

$$\mathrm{Br}(B^- \to K^-K^-\pi^+)\ \mathrm{non-resonance} \leq 8.79 \times 10^{-5}\ @\ 90\%\ \mathrm{C.L.}$$

SUMMARY

A search for the decay of charged B mesons to $K^-K^-\pi^+$ was conducted by the OPAL collaboration. This decay channel is strongly suppressed in the Standard Model, but may be large in R-parity violating models. Hence, this decay mode may serve as a probe for new physics beyond the Standard Model. No evidence has been observed for such a decay. Upper limits on the branching ratio have been set of 1.29×10^{-4}, or of 8.79×10^{-5} if one assumes that the decay is not via a K^* resonance, both at 90% confidence level.

Using these limits, and the estimate that $\frac{B^- \to K^-K^-\pi^+}{b \to s s \bar{d}} \approx \frac{1}{4}$ [1], we can put new limits on the contribution of R-parity violating couplings in this process. Starting from Equation 9 of reference [1],

$$\Gamma = \frac{m_b^5 f_{\mathrm{QCD}}^2}{512(2\pi)^3} \times \left(\left| \Sigma_{n=1}^3 \frac{1}{m_{\tilde{\nu}_n}^2} \lambda'_{n32}\lambda'^*_{n21} \right|^2 + \left| \Sigma_{n=1}^3 \frac{1}{m_{\tilde{\nu}_n}^2} \lambda'_{n12}\lambda'^*_{n23} \right|^2 \right) \quad (1)$$

where m_b is the mass of the b quark, $f_{\mathrm{QCD}} = (\alpha_s(m_b)/\alpha_s(m_{\tilde{\nu}_n}))^{24/23}$, $m_{\tilde{\nu}_n}$ is the mass of the sneutrino involved and λ' is a dimensionless coupling. As an example, with $m_b = 4.5$ GeV/c^2, $f_{\mathrm{QCD}} \simeq 2$, $m_{\tilde{\nu}_n} = 100$ GeV/c^2 as in [1] and $\tau_{B^-} = 1.65$ ps we obtain:

$$\sqrt{|\Sigma_{n=1}^3 \lambda'_{n32}\lambda'^*_{n21}|^2 + |\Sigma_{n=1}^3 \lambda'_{n12}\lambda'^*_{n23}|^2} < 5.9 \times 10^{-4},$$

which can be compared to the previous limit of 0.1 obtained from K and B mixing.

REFERENCES

1. K. Huitu, D.-X. Zhang, C.-D. Lü and P. Singer, Phys. Rev. Lett. **81** (1998) 4313.
2. H.P. Nilles, Phys. Rep. **110** (1984) 1;
   H.E. Haber and G.L. Kane, Phys. Rep. **117** (1985) 75.
3. P. Fayet, in *"Unification of the Fundamental Particle Interactions"*, eds. S. Ferrara, J. Ellis and P. Van Nieuewenhuizen, Plenum Press (1980) 727.
4. OPAL Collab., G. Abbiendi *et al.*, Phys. Lett. **B 476** (2000) 233.
5. OPAL Collab., K. Ahmet *et al.*, Nucl. Instr. Meth. **A 305** (1991) 275; OPAL Collab., P.P. Allport *et al.*, Nucl. Instr. Meth. **A 324** (1993) 34;
6. OPAL Collab., G. Alexander *et al.*, Z. Phys. **C 52** (1991) 175.
7. OPAL Collab., R. Akers *et al.*, Z. Phys. **C 66** (1995) 19.
8. *A Combination of Preliminary Electroweak Measurements and Constraints on the Standard Model,* ALEPH, DELPHI, L3 and OPAL collaborations, the LEP Electroweak Working Group and the SLD Heavy Flavour and Electroweak Groups, CERN-EP/99-015.
9. R.D. Cousins and V.L. Highland, Nucl. Instr. Meth. **A 320** (1992) 331.

$b \to s\ell^+\ell^-$ decays in and beyond the Standard Model [1]

Gudrun Hiller*

Stanford Linear Accelerator Center, Stanford University, Stanford, CA 94309, USA

Abstract. We briefly review the status of rare radiative and semileptonic $b \to s(\gamma, \ell^+\ell^-)$, $(\ell = e, \mu)$ decays. We discuss possible signatures of new physics in these modes and emphasize the role of the exclusive channels. In particular, measurements of the Forward-Backward asymmetry in $B \to K^*\ell^+\ell^-$ decays and its zero provide a clean test of the Standard Model, complementary to studies in $b \to s\gamma$ decays. Further, the Forward-Backward CP asymmetry in $B \to K^*\ell^+\ell^-$ decays is sensitive to possible non-standard sources of CP violation mediated by flavor changing neutral current Z-penguins.

INTRODUCTION

Flavor changing neutral current (FCNC) b decays do not occur at tree level in the Standard Model (SM). Being loop induced, they feel scales of order $\mathcal{O}(m_W, m_t)$ and in principle much higher ones, making them important probes of the flavor sector of the SM and beyond.

Rare radiative $b \to s\gamma$ decays proceed via so-called electromagnetic penguins. They have been measured in exclusive $B \to K^*\gamma$ [1] and inclusive $B \to X_s\gamma$ [2,3] decays. In dilepton channels $b \to s\ell^+\ell^-$ $(\ell = e, \mu)$, we identify two additional structures in the Feynman diagrams: boxes and Z-penguins. None of the dilepton modes has been detected to date, but we expect large data samples from operating B-factories (CLEO,BaBar,Belle), dedicated B-physics programmes at colliders (Tevatron Run II,Hera-B) and LHC-B in the long term. Corresponding $b \to d$ transition amplitudes are CKM suppressed $V_{td}^*/V_{ts}^* \propto \lambda \sim 0.22$.

The existing best bound in the dimuon channels for inclusive decays is $\mathcal{B}(B \to X_s\mu^+\mu^-) < 5.8 \cdot 10^{-5}$ at 90% C.L. [4], which is one order of magnitude above the NLO SM expectation $\mathcal{B}(B \to X_s\mu^+\mu^-)_{SM} = 5.7 \pm 1.1 \cdot 10^{-6}$ [5]. Note that the NNLO calculation in $b \to s\ell^+\ell^-$ is only partially available [6]. Corresponding bounds for the exclusive channels are $\mathcal{B}(B^+ \to K^+\mu^+\mu^-) < 5.2 \cdot 10^{-6}$, $\mathcal{B}(B^0 \to K^{*0}\mu^+\mu^-) < 4.0 \cdot 10^{-6}$ at 90% C.L. [7] and their respective SM predictions are $\mathcal{B}(B \to K\mu^+\mu^-)_{SM} = 5.9 \pm 2.1 \cdot 10^{-7}$, $\mathcal{B}(B \to K^*\mu^+\mu^-)_{SM} = 2.0 \pm 0.7 \cdot 10^{-6}$

---

[1] Work supported by the Department of Energy, Contract DE-AC03-76SF00515

with the dominant theoretical uncertainty resulting from hadronic matrix elements, which are estimated here using Light cone sum rules [8]. Currently, the exclusive $B \to K^*\mu^+\mu^-$ decay has the most interesting bound, which is only a factor of 2 away from the SM prediction. Despite larger theoretical uncertainty than in the inclusive cases, rare exclusive decays are more accessible experimentally in the near future and have observables (e.g. existence and position of the zero of the Forward-Backward asymmetry discussed below), which are as clean as the respective ones in the inclusive modes.

MODEL INDEPENDENT ANALYSIS
PHOTON AND Z PENGUINS

The calculational tool for the description of $b \to s(\gamma, \ell^+\ell^-)$ decays is the low energy effective Hamiltonian $\mathcal{H}_{eff} \sim G_F V_{ts}^* V_{tb} \sum_{i=1}^{10} C_i(\mu) O_i(\mu)$ [9]. This enables the analysis of relevant observables in a model independent way, with the goal being to extract the Wilson coefficients C_i from data [10]. The major player is the $bs\gamma$ vertex $O_7 \sim m_b \bar{s}_L \sigma_{\mu\nu} b_R F^{\mu\nu}$. Its effective coupling strength C_7^{eff} is related to the branching ratio $\mathcal{B}(B \to X_s \gamma) \sim |C_7^{eff}|^2$ thus $0.25 \le |C_7^{eff}| \le 0.37$ [2,8] which is in good agreement with the SM value $C_7^{eff}|_{SM} = -0.31$ at $\mu = m_b$ at leading log. We see that the $b \to s\gamma$ data fix the modulus of C_7^{eff}, but not its sign (phase).

In $b \to s\ell^+\ell^-$ decays, in addition to O_7, also 4-Fermi operators involving dileptons contribute, given by $O_9 \sim \bar{s}_L \gamma^\mu b_L \bar{\ell}\gamma_\mu \ell$, $O_{10} \sim \bar{s}_L \gamma^\mu b_L \bar{\ell}\gamma_\mu \gamma_5 \ell$. Due to the charge assignments of lepton-Z couplings the Z-penguin contribution to C_9 is suppressed with respect to C_{10} by $(\bar{\ell}\ell Z|_V)/(\bar{\ell}\ell Z|_A) = -1 + 4\sin^2\theta_W \sim -0.08$. We thus identify C_{10} as a measure of the sZb coupling modulo the box contribution [11].

Decomposition of the $B \to K^*\mu^+\mu^-$ branching ratio yields $\mathcal{B} = a|C_7^{eff}|^2 + b|C_9|^2 + c|C_{10}|^2 + dC_7^{eff}C_9 + eC_7^{eff} + fC_9 + g$ [8]. Using the CDF bound [7] on this mode and allowing C_7^{eff} to have both SM-like and SM-opposite signs gives the present best bound on the strength of generic FCNC Z-penguins of $|C_{10}| \le 10$, which is a factor of 2-3 larger than the SM value $C_{10}|_{SM} = -4.7$ [11]. Scenarios with non-standard Z-penguins arise in many extensions of the SM like such as supersymmetry, 4th generation and Z' [11]. Another interesting possibility to test the sZb vertex arises in $b \to s\nu\bar{\nu}$ decays, since here no photon penguins contribute.

EXCLUSIVE $B \to K, K^*\ell^+\ell^-$ DECAYS

Supersymmetric effects in inclusive $b \to s\ell^+\ell^-$ decays have been studied in [12,13]. The reach of a new physics search in the dilepton invariant mass distribution in $B \to K^*\mu^+\mu^-$ decays is exemplified in Fig. 1 [8]. Supergravity (SUGRA) (dotted) and a supersymmetric scenario with non-minimal sources of flavor violation in the mass insertion approximation (dashed), can be well discriminated from the SM (solid) and its hadronic uncertainties (shaded area); the upper curves contain resonant $c\bar{c}$ background via $B \to K^*\Psi^i \to K^*\mu^+\mu^-$, lower ones are pure short-distance contributions [14]. Note that the dashed curve saturates the experimental bound in this channel. Similar findings are valid for $B \to K\mu^+\mu^-$ decays, which however show less sensitivity to C_7^{eff} as the photon pole at $s = 0$ is absent. In either case $C_7^{eff} > 0$ (opposite-to-SM sign) enhances the rates through constructive interference of C_7^{eff} with C_9.

Forward-Backward Asymmetry

The Forward-Backward asymmetry in $B \to K^*\ell^+\ell^-$ decays results from V/A interference in the lepton pair $A(s) \sim C_{10}(C_7^{eff} + \beta(s)Re(C_9^{eff}))$, shown in Fig. 2 [8]; see [11] for a discussion of the sign of A, which is opposite to the Forward-Backward asymmetry $\bar{A}$ of the CP conjugate channel in the CP conserving limit. In the SM ($C_7^{eff} < 0$, solid curve), A has a zero around $s_0 \sim 3\text{GeV}^2$, which would disappear if C_7^{eff} would have the opposite sign (long-short-dashed curve). The existence of a zero in the Forward-Backward asymmetry in $B \to K^*\ell^+\ell^-$ decays below the J/Ψ resonance is an important test of the SM and is independent of hadronic matrix elements [8]. Positive C_7^{eff} occurs generically in supersymmetric theories [12], but only for large $\tan\beta$ in relaxed and/or minimal SUGRA [13].

Further, the position of the zero s_0 has very small hadronic uncertainties [15]. In the limit where the final hadron has large energy, i.e., small dilepton mass, the

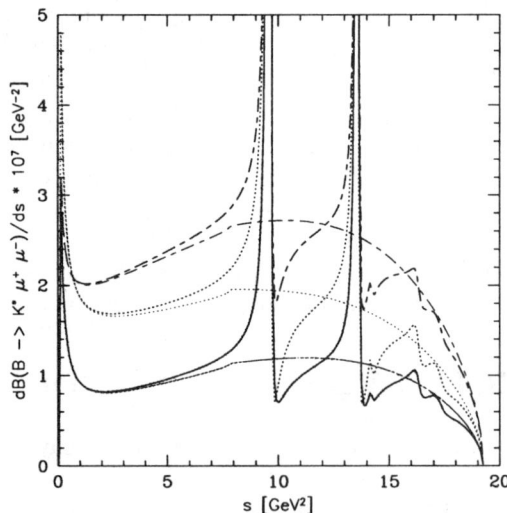

FIGURE 1. Dilepton invariant mass spectrum in $B \to K^*\mu^+\mu^-$ decays. Figure taken from [8].

Large Energy Effective Theory [16] is applicable and here all form factors cancel out in the ratios which determines s_0 [8].

A recently proposed observable, the Forward-Backward CP asymmetry FB_{CP} in $B \to K^* \ell^+ \ell^-$ decays probes the phase of C_{10} or of the sZb vertex [11], respectively. Defined as $FB_{CP} \equiv (A + \bar{A})/(A - \bar{A}) = ImC_{10}/ReC_{10} ImC_9^{eff}/ReC_9^{eff}(1 + \ldots)$, its magnitude scales with $ImC_9^{eff} = ImY$, where $Y(s)$ contains contributions from resonant and non-resonant $c\bar{c}$ states [14]; $ImY = 0$ below threshold, so it is sizeable only in the high dilepton mass region. Integration over $m_{\psi'}^2 < s \leq (m_B - m_{K^*})^2$ yields $\Delta FB_{CP} = (3 \pm 1)\% ImC_{10}/ReC_{10}$, which can be large in the case of $\mathcal{O}(1)$ phases. Hadronic uncertainties are not small, but the SM background is below 10^{-3} and any effect above this would be due to a non-SM source of CP violation [11].

RADIATIVE RARE B-DECAYS

The SM branching ratio in $B \to X_s \gamma$ decays is known at NLO with 10 % accuracy $\mathcal{B}(B \to X_s \gamma) = (3.32 \pm_{0.11}^{0.00} \pm_{0.08}^{0.00} \pm_{0.25}^{0.26}) \cdot 10^{-4}$ [17]. This is in good agreement with data from LEP $\mathcal{B}(B \to X_s \gamma) = (3.11 \pm 0.80 \pm 0.72) \cdot 10^{-4}$ [3], and CLEO, $\mathcal{B}(B \to X_s \gamma) = (3.15 \pm 0.35 \pm 0.32 \pm 0.26) \cdot 10^{-4}$ [2].

A promising observable in $b \to s\gamma$ decays where dramatic signals of possible physics beyond the SM could show up is the CP asymmetry in the rate $a_{CP} \equiv (\Gamma(\bar{B} \to X_s \gamma) - \Gamma(B \to X_{\bar{s}} \gamma))/(\Gamma(\bar{B} \to X_s \gamma) + \Gamma(B \to X_{\bar{s}} \gamma))$ [18]. It is very tiny

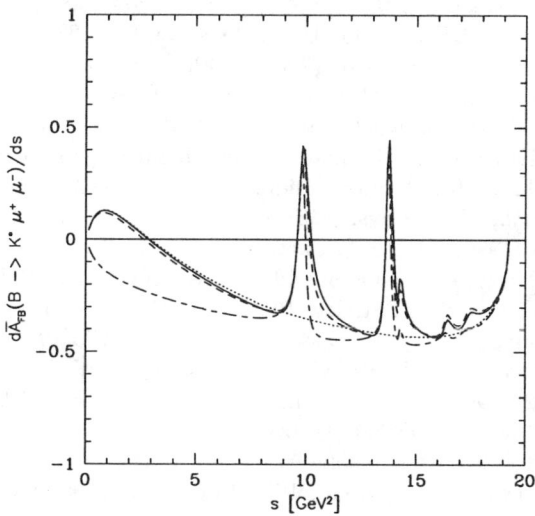

FIGURE 2. Forward-backward asymmetry for $B \to K^* \mu^+ \mu^-$. Figure taken from [8].

in the SM since $a_{CP} \propto \alpha_s(m_b)\eta\lambda^2$, η, λ are Wolfenstein parameters, thus $a_{CP} \leq 1\%$ [18]. However, large effects of (10-50) % are possible in scenarios with an enhanced chromo-magnetic dipole operator C_8 in the bsg vertex. CP asymmetry in exclusive $B \to K^*\gamma$ decays has a less clean prediction due to strong phases, however, in the SM, $a_{CP}^{B \to K^*\gamma} \leq \mathcal{O}(1\%)$ holds [19]. Any significant deviation from this would signal new physics. In both inclusive $-0.09 < a_{CP}^{B \to X_s\gamma} < 0.42$ [2] and exclusive cases $a_{CP}^{B \to K^*\gamma} = (8 \pm 13 \pm 3)\%$ [1], the measurements are not conclusive yet.

SUMMARY

Theoretically clear signatures of possible new physics can be experimentally isolated in $b \to s\gamma$ and $b \to s\ell^+\ell^-$ decays in the near future via the observables $(A(s_0), FB_{CP}, a_{CP})$. It is exciting to see whether the SM passes this next round of FCNC tests.

REFERENCES

1. T.E.Coan et al., (CLEO Collaboration), Phys.Rev.Lett. 84 (2000) 5283.
2. M.S. Alam et al., (CLEO Collaboration), Phys. Rev. Lett.74(1995)2885; S. Ahmed et.al. (CLEO Collaboration), CLEO CONF 99–10 (hep–ex/9908022).
3. R. Barate et al., (ALEPH Collaboration) Phys. Lett. B429 (1998) 169.
4. S. Glenn et al., (CLEO Collaboration), Phys. Rev. Lett. 80 (1998) 2289.
5. A. Ali, L.T. Handoko, G. Hiller and T. Morozumi, Phys. Rev. D55 (1997) 4105.
6. C.Bobeth, M. Misiak and J. Urban, Nucl. Phys. B574 (2000) 291.
7. T. Affolder et al., (CDF Collaboration), Phys.Rev.Lett. 83 (1999) 3378.
8. See, A. Ali et al., Phys. Rev. D61 (2000) 074024, and references therein.
9. A.J. Buras et al., Nucl. Phys. B424(1994)374; A.J. Buras and M. Münz, Phys. Rev. D52(1995)186; M. Misiak, Nucl. Phys. B393(1993)23; Err. ibid. B439 (1995) 461.
10. A. Ali, G.F. Giudice and T. Mannel, Zeitschrift für Physik C67(1995)417.
11. See, G. Buchalla, G. Hiller and G. Isidori, hep-ph/0006136, and references therein.
12. S. Bertolini et al., Nucl. Phys. B353(1991)591;P. Cho, M. Misiak and D. Wyler, Phys. Rev. D54(1996)3329;J.L. Hewett and J.D. Wells, Phys. Rev.D55(1997)5549; E. Lunghi, A. Masiero, I. Scimemi and L. Silvestrini, Nucl. Phys. B568 (2000) 120.
13. T. Goto, Y. Okada and Y. Shimizu, Phys. Rev. D58(1998)094006.
14. A. Ali, T. Mannel and T. Morozumi, Phys. Lett. B273(1991)505; F. Krüger and L.M. Sehgal, Phys. Lett. B380(1996)199; Z. Ligeti, I.W. Stewart and M.B. Wise, Phys. Lett. B420(1998)359; A. Ali and G. Hiller, Phys. Rev. D60 (1999) 034017.
15. G. Burdman, Phys. Rev. D57(1998)4254.
16. J. Charles et al., Phys. Rev. D60(1999)014001.
17. See, F. Borzumati and C. Greub, Phys.Rev. D59 (1999) 057501, and references therein.
18. A.L.Kagan and M.Neubert, Phys.Rev. D58 (1998) 094012.
19. C. Greub, H. Simma and D. Wyler,Nucl. Phys. B434 (1995) 39.

Radiative B decays and B → τν at CLEO

Hubert Schwarthoff

The Ohio State University
174 W 18th Avenue
Columbus, OH 43210

Abstract.
We have investigated exclusive, radiative B meson decays in 9.7 million $B\bar{B}$ decays accumulated with the CLEO detector. The B → K*(892) γ branching fractions are determined to be BF(B^0 → K^{*0}(892),γ) = $(4.55^{+0.72}_{-0.68} \pm 0.34) \times 10^{-5}$ and BF(B^+ → K^{*+}(892) γ) = $(3.76^{+0.89}_{-0.83} \pm 0.28) \times 10^{-5}$. We have searched for CP asymmetry in B → K*(892) γ decays and measure $A_{CP} = +0.08 \pm 0.13 \pm 0.03$. We also report the first observation of the decay B → K^*_2(1430) γ with a branching fraction of $(1.66^{+0.59}_{-0.53} \pm 0.13) \times 10^{-5}$. No significant evidence for the decays B→ $\rho, \omega \gamma$ is found, and we limit BF(B→ $\rho/\omega \gamma$)/BF(B→ K*(892) γ) < 0.32 at 90% c.l. We also find no evidence for the non-penguin decays $B_0 \to \phi \gamma$ and $B_0 \to D^{*0} \gamma$. Non-observation of the latter decay provides evidence that weak radiative B decays are dominated by the b → sγ short-distance mechanism of the Standard Model.

In the same dataset we have investigated the decay B → τν. No significant evidence is found, and we limit BF(B → τν) < 8.4×10^{-4} at 90% c.l.

INTRODUCTION

The radiative decay of the b quark into an s quark and a single photon, b → sγ, was first observed by CLEO in an exclusive measurement in 1993 [1], and an inclusive measurement has been made in 1995 [2]. The graph for this decay involves a loop ("penguin") diagram. In the Standard Model this loop is dominated by a t quark (see Fig. 1), giving access to the CKM matrix elements V_{ts} and V_{tb}. It constitutes a simple flavor changing neutral current, which is not directly allowed by the Standard Model. The decay B → K* γ is assumed to be dominated by the b → sγ transition. In case there are processes involved that are not described by the Standard Model, the loop can contain other particles, such as a supersymmetric charged Higgs boson.

Another radiative quark decay can be described with the W exchange diagram, also shown in Fig. 1. It is the process involved in the decay B → $D^{*0} \gamma$ and could provide a significant background for the inclusive measurement of b → sγ. In the description of the Standard Model it is strongly suppressed.

The data presented here have been recorded with the CLEO detector at the symmetric 10.58 GeV e^+e^- collider CESR at Cornell University, NY, USA. The center of mass energy is chosen so that a large number of all collisions create an $\Upsilon(4s)$ meson in resonance, which decays mostly into a pair of B mesons. The rest of the collisions produce non-resonant $q\bar{q}$ pairs. One third of the data were recorded at a center of mass energy of approximately 60 MeV below the resonance (*off-resonance*) and used for background studies.

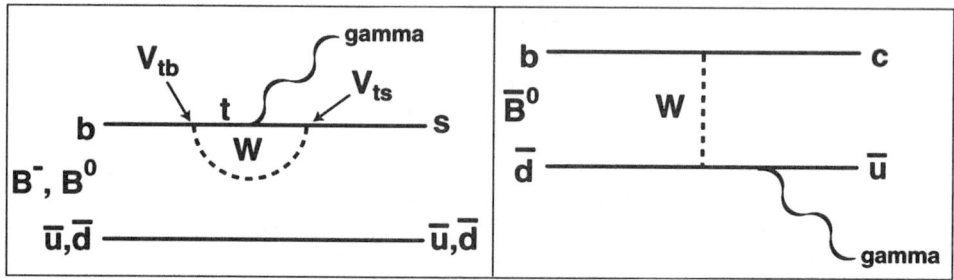

FIGURE 1. Feynman graphs for radiative B meson decays. Left: $b \to s\gamma$ ("penguin"), Right: W exchange diagram.

EXCLUSIVE RADIATIVE B DECAY MEASUREMENTS AND SEARCHES

The general signature of a radiative B decay in the CLEO detector is a high energy photon, typically between 2.1 GeV and 2.7 GeV, accompanied by the daughter particle of the radiative decay and the decay products of the other B meson from the $\Upsilon(4s)$.

The background to the radiative B decays is dominated by photons from initial state radiation in $e^+e^- \to q\bar{q}$ events. Other sources are photons coming from π^0 and η decays. Generally the background can be suppressed by studying the angle between the reconstructed B meson thrust axis and the rest of the event, Θ_T. Monte Carlo studies show that the cosine of this angle is flatly distributed for signal events, while the off-resonance background tends to peak at small values of Θ_T.

Only the barrel part of the crystal calorimeter is used for detection of the signature high energy photon: $|\cos\Theta_\gamma| < 0.71$, where Θ_γ is the angle between the beam axis and the candidate photon. This facilitates additional background suppression and allows more precise simulation studies, since the crystal calorimeter modeling is optimal in this region. A veto on photons that can be combined with another photon to have the π^0 or η mass further reduces background.

Reconstruction channels

The following decay channels are used in the analysis. Reference to the charge conjugate states is implicit unless stated otherwise.

- B → K*(892) γ and B → K*$_2$(1430) γ: K*+ → K+π⁰, K$_s^0$π+ and K*⁰ → K+π−, K$_s^0$π⁰
- B → ρ, ω γ: ρ⁰ → π+π− and ρ+ → π+π⁰ and ω → π+π−π⁰
- B → φ γ: φ → K+K−
- B → D*⁰ γ: D*⁰ → D⁰ π⁰, D⁰ γ and D⁰ → K+ π−, K− π+ π⁰, K− π+ π− π+

The background suppression can be enhanced for all these channels using event shape cuts as well as constraints on the variables ΔE and M_B.

$$\Delta E = |E(R) + E_\gamma - E_{beam}|$$

E(R) is the energy of the reconstructed signature hadron K*, ρ, ω, or φ, and E_{beam} is the beam energy, and

$$M_B = \sqrt{E_{beam}^2 - p_B^2}$$

M_B is the *Beam-constrained B mass*, where p_B is the momentum of the reconstructed B meson candidate. Typical cut values are $|\Delta E| < 300$ MeV (for K*) and $5.2\,\text{GeV} < M_B < 5.3\,\text{GeV}$.

Analysis results on the exclusive decays

Following the above mentioned background suppression procedure the results for $B^0 \to K^* \gamma$ are quoted in Table 1, and for the remaining exclusive channels in Table 2. They have been published in [3,4]. Quoting these numbers we assume that equal numbers of B^0 and B^+ mesons are produced at CLEO, and that the decay rates of $B^+ \to K_2^{*+}(1430)$ and $B^0 \to K_2^{*0}(1430)$ are equal.

TABLE 1. Branching fraction measurements from the CLEO analysis of 9.7 Million exclusive radiative B meson decays into K*. The K*$_2$ result is an average over neutral and charged modes.

Decay channel	Signal yield (number of events)	Branching fraction in 10^{-5} (90% c.l.)
$B^0 \to K^{*0}(892)\gamma$	$88.3^{+12.2}_{-11.5}$	$4.55^{+0.72}_{-0.68} \pm 0.34$
$B^+ \to K^{*+}(892)\gamma$	$36.7^{+8.3}_{-7.6}$	$3.76^{+0.89}_{-0.83} \pm 0.28$
$B \to K^*_2(1430)\gamma$	$15.8^{+5.7}_{-5.1}$	$1.66^{+0.59}_{-0.53} \pm 0.13$

The measurement of the branching fraction for $B \to K^*_2(1430)\gamma$ is the first observation of this decay. The signal has a significance of 3.3 σ, including systematic errors.

TABLE 2. Results from the CLEO analysis of 9.7 Million exclusive radiative B meson decays for the studied channels. All branching fraction numbers are upper limits.

Decay channel	Yield (events)	Estimated background events	Branching fraction (90% c.l.)
$B \to \rho^0 \gamma$	24	$9.3^{+0.6}_{-0.5}$ (continuum) 5.4 ± 0.8 (K*)	$< 1.7 \times 10^{-5}$
$B \to \rho^+ \gamma$	10	5.2 ± 0.4 (continuum) 2.6 ± 0.6 (K*)	$< 1.3 \times 10^{-5}$
$B \to \omega \gamma$	5	2.7 ± 0.1	$< 0.92 \times 10^{-5}$
$B \to \phi \gamma$	1	1.2 ± 0.1	$< 0.33 \times 10^{-5}$
$B \to D^{*0} \gamma$	0		$< 5.0 \times 10^{-5}$

For all other inclusive searches no significant signal was observed. While the analysis of all inclusive channels suffers from background from $e^+e^- \to q\bar{q}$ (continuum), it is possible to misidentify a charged K as a π in the decay $B \to \rho\gamma$. This introduces additional background from the decay $B \to K^*\gamma$. Table 2 contains the numbers of such background events, along with the numbers for continuum background.

The analysis of $B \to D^{*0}\gamma$ is similar to the procedure for $B \to K^*_2(1430)\gamma$. Apart from the continuum background, the main background comes from $B^+ \to D^{*0}\rho^+$.

Conclusions from the results for radiative B decays

The ratio of the two branching fractions BF($B \to K^*_2(1430)\gamma$) and BF($B \to K^*(892)\gamma$) has been predicted by theory to lie between 3.0 and 4.9 [6], and, in a more recent paper [5], to be 0.37 ± 0.1. The ratio from the measurements presented in this paper, $0.39^{+0.15}_{-0.13}$, agrees well with [5].

Furthermore, theory predicts the ratio of the two branching fractions

$$\frac{\text{BF}(B \to \rho\gamma)}{\text{BF}(B \to K^*\gamma)} = \xi \left|\frac{V_{td}}{V_{ts}}\right|^2$$

where ξ is the ratio of the form factors for $B \to \rho\gamma$ and $B \to K^*\gamma$: $0.58 < \xi < 0.91$ from model calculations ([7–9]). Assuming BF($B \to \rho\gamma$) = BF($B^+ \to \rho^+\gamma$) = 2 BF($B^0 \to \rho^0\gamma$) = 2 BF($B^0 \to \omega\gamma$), we can deduct a limit on the CKM matrix elements: $|\frac{V_{td}}{V_{ts}}| < 0.79$ (95% c.l.). At the same time the best limit on this quantity has been presented from a combination of data from the LEP experiments, SLD, and CDF: $|\frac{V_{td}}{V_{ts}}| < 0.24$ (95% c.l.). We expect a considerable improvement from the new data taken with the CLEO III detector.

The low upper limit on the branching fraction for the decay $B \to D^{*0}\gamma$, along with a low limit on $B \to \phi\gamma$, leads us to the interpretation that radiative B decays are in fact dominated by the radiative penguin diagram $b \to s\gamma$. This conclusion supports the assumption that the W exchange is indeed strongly suppressed, as described in the Standard Model.

THE SEARCH FOR B → τν

The second topic of this presentation is the analysis of the process $B^+ \to \tau^+ \nu_\tau$. The branching fraction for this B decay can be expressed in a simple way:

$$\text{BF}(B^+ \to \ell^+ \nu) = \frac{G_F^2 m_B m_\ell^2}{8\pi}(1 - \frac{m_\ell^2}{m_B^2}) f_B^2 |V_{ub}|^2 \tau_B,$$

where ℓ is a charged lepton, G_F the Fermi coupling constant, m_B the B meson mass, m_ℓ the lepton mass, and V_{ub} the CKM matrix element for b to u quark transitions. It is largest for B → τν and, if measured, allows the determination of $f_B |V_{ub}|$.

To obtain a high signal yield a large number of decay modes for the second B meson in the $\Upsilon(4s)$ event is used: B → D^0, $D^{*0}(n\pi)^+$, n = 1-5, where $D^{*0} \to D^0 \pi^0$, $D^0 \gamma$. 46% of the τ decays are reconstructed through the channels τ → $e \nu_e \nu_\tau$, $\mu \nu_\mu \nu_\tau$, $\pi \nu_\tau$, requiring that only a single charged track be present after full reconstruction of the other B meson.

The analysis finds 6 events in the on-resonance data. An extended likelihood fit of the Δ E distribution, after selecting events with a beam constrained B mass within 2.5 standard deviations of the true B meson mass, yields 0.96 signal events, which leads to an upper limit on the branching fraction:

$$\text{BF}(B \to \tau^+ \nu_\tau) < 8.4 \times 10^{-4} \; (90\% \text{c.l.})$$

This upper limit is still above the Standard Model prediction ($2-10 \times 10^{-5}$, [10] [11]).

REFERENCES

1. CLEO Collaboration, R. Ammar et al., *Phys. Rev. Lett.* **71**, 676 (1993).
2. CLEO Collaboration, M.S. Alam et al., *Phys. Rev. Lett.* **74**, 2885 (1995).
3. CLEO Collaboration, T.E. Coan et al., *Phys. Rev. Lett.* **84**, 5283 (2000).
4. CLEO Collaboration, M. Artuso et al., *Phys. Rev. Lett.* **84**, 4292 (2000).
5. S. Veseli, and M.G. Olsson, *Phys. Lett.* **B 367**, 309 (1996).
6. A. Ali, T. Ohl, and T. Mannel, *Phys. Lett.* **B 298**, 195 (1993).
7. A. Ali, V.M. Braun, and H. Simma, *Z. Phys.* **C 63**, 437 (1994).
8. S. Narison, *Phys. Lett.* **B 327**, 354 (1994).
9. J.M. Soares, *Phys. Rev.* **D 49**, 283 (1994).
10. Using $F_B = 185^{+35}_{-25}$ MeV from Lattice QCD calculations, T. Draper, *Nucl. Phys. Proc. Suppl.* **73**, 43 (1999).
11. Using $V_{ub} = 3.25^{+0.61}_{-0.64} \times 10^{-3}$ from B. H. Behrens, et al., *Phys. Rev.* **D61**, 43 (1999).

Neutral Meson Decays to Four Leptons in KTeV

Eva Halkiadakis

Rutgers University Physics Department
Piscataway, New Jersey 08854

Abstract. The Fermilab Experiment 799-II searches for a variety of rare neutral K_L decays, with particular emphasis of those relevant to CP violation. We report preliminary and published findings from a number of rare decay analyses. In each case, the sensitivity of E799-II is at least an order of magnitude greater than that of any previously published result.

INTRODUCTION

We present preliminary branching ratio measurements of the rare decays $\pi^0 \to e^+e^-e^+e^-$ and $K_L \to e^+e^-e^+e^-$ from data taken during the 1997 run of KTeV/E799-II. We also present the first unambiguous evidence for the decay $K_L \to e^+e^-\mu^+\mu^-$. The invariant mass distribution of the lepton pairs probes the *meson* $\to \gamma^*\gamma^*$ form factors. The measurement of the $K_L^0 \to \gamma^*\gamma^*$ form factor is needed to better understand the long-distance contribution to $K_L \to \mu^+\mu^-$ [1]. In addition, we present analyses of the angular distributions of the leptons which are sensitive to P and CP violating components in the matrix elements.

THE KTEV DETECTOR

E799-II and E832 (KTeV), neutral kaon experiments, took data in the 1996-97 Tevatron fixed target run. The primary goal of E832 was to measure $\text{Re}(\frac{\epsilon'}{\epsilon})$ to a precision of $\sim 10^{-4}$. E799-II was a high sensitivity experiment designed to search for rare K_L decays.

A detailed description of the detector can be found in [2].

THE DECAY $\pi^0 \to e^+e^-e^+e^-$

Fully reconstructed $K_L \to \pi^0\pi^0\pi^0$ decays were used to study π^0 decays, such as $\pi^0 \to e^+e^-e^+e^-$. Figure 1 (left) shows the invariant mass of $e^+e^-e^+e^-$ coming from $K_L \to \pi^0\pi^0\pi^0$ decays. The sharp peak shows $\pi^0 \to e^+e^-e^+e^-$ events (or double Dalitz events), where the other two π^0's decayed to two photons. The broad distribution under the peak comes from $K_L \to \pi^0\pi^0\pi^0$ decays where two of the π^0's Dalitz decay ($e^+e^-\gamma$ in the final state) and the third π^0 decays to two photons. These two-single-Dalitz events (2D) are used for normalization to the double-Dalitz events (DD). In order to distinguish between these two decays we can pair the photons and electrons for each decay by creating a χ^2 based on how the π^0's decay. The pair with the lowest χ^2 for each decay, $K_L \to \pi^0\pi^0_D\pi^0_D$ and $K_L \to \pi^0\pi^0\pi^0_{DD}$, is the one that is chosen. The smaller of the two determines if an event is a $K_L \to \pi^0\pi^0\pi^0_{DD}$ or if it is a $K_L \to \pi^0\pi^0_D\pi^0_D$. This technique yields about 12000 $\pi^0 \to e^+e^-e^+e^-$ events with very low background. The KTeV preliminary result for the branching ratio is $\frac{\Gamma(\pi^0 \to e^+e^-e^+e^-)}{\Gamma(\pi^0 \to \gamma\gamma)} = (3.31 \pm 0.04(stat) \pm 0.22(sys)) \times 10^{-5}$. The systematic error is dominated by the external error due to the measurement of the $BR(K_L \to \pi^0\pi^0_D\pi^0_D)$ in the PDG.

The theoretical expectation is 3.42×10^{-5} [3]. The previous measurement, a bubble chamber experiment, yielded 146 events [4].

We have tested parity, P, in the decay $\pi^0 \to e^+e^-e^+e^-$ by studying ϕ, the angle between the planes of the two e^+e^- pairs. This distribution for our $\pi^0 \to e^+e^-e^+e^-$ events can be seen in figure 1 (left). Due to the ambiguity in the pairing of the e^+e^-, we have chosen the pair that minimizes the product the invariant masses of the e^+e^- pairs. Also, in order to have two well defined planes we have placed a cut on the invariant mass of the two e^+e^- pairs, M_{ee1} and M_{ee2}, requiring that M_{ee1} and M_{ee2} both be greater than $10 MeV$. We fit the Kroll-Wada formula [5]

$$\frac{d\Gamma(\pi^0 \to e^+e^-e^+e^-)}{d\phi} \propto (1 + \alpha\cos(2\phi)), \qquad (1)$$

where α is a constant whose expected magnitude is calculated to be 0.250 ± 0.003, with the $M_{e^+e^-}$ cut mentioned above. For an odd(even) P eigenstate α is negative(positive). Fitting to the data gives the preliminary result $\alpha = -0.239 \pm 0.030$. The previous measurement [4], with 112 events and both opening angles $> 3°$, yielded $\alpha = -0.12 \pm 0.15$. The KTeV measurement is more sensitive to α by a factor of 10.

THE DECAY $K_L \to e^+e^-e^+e^-$

We have detected 430 $K_L \to e^+e^-e^+e^-$ events, as can be seen in figure 2 (left). These events contain very little background, on the order of a few events, which comes primarily from photons converting in material in the detector. The KTeV preliminary result for the branching ratio is $BR(K_L \to e^+e^-e^+e^-) = (4.14 \pm 0.27 \text{ (stat)} \pm 0.31 \text{ (sys)}) \times 10^{-8}$. The decay $K_L \to \pi^0\pi^0_D\pi^0_D$ was used as a normalization.

The theoretical expectation for the $K_L \to e^+e^-e^+e^-$ branching ratio is 3.68×10^{-8} [3]. Other theoretical models predict values of the same order [6], [7]. The previous measurement yielded 27 events [8].

We have tested charge-parity, CP, in the decay $K_L \to e^+e^-e^+e^-$ again by studying ϕ, the angle between the planes of the two e^+e^- pairs. This distribution for our $K_L \to e^+e^-e^+e^-$ events can be seen in figure 2 (middle). The pairing of the e^+e^- is done as described in the previous section and we also require that M_{ee1} and M_{ee2} both be greater than $10 MeV$. Again we fit the ϕ distribution to the formula in equation 1. This time the expected magnitude of α is 0.247 ± 0.005. For an odd(even) CP eigenstate α is negative(positive). Fitting to the data gives the preliminary result $\alpha = -0.220 \pm 0.110$. The previous measurement with 27 events found $\alpha = -0.21 \pm 0.29$ [8]. The KTeV measurement is more sensitive to α by a factor of 3.

A visual inspection of the invariant mass of the e^+e^- pairs indicates evidence of a non-flat form factor in $K_L \to e^+e^-e^+e^-$, see figure 2 (right). We can fit for the form factor parameter α_{K^*}, the Bergström, Massó, and Singer (BMS) parameterization of the $K_L \to \gamma^*\gamma$ form factor using an extended Vector Meson Dominance model [9]. The BMS model predicts $|\alpha_{K^*}| = 0.2 - 0.3$. The PDG world average is $\alpha_{K^*} = -0.28 \pm 0.08$. This will be the first measurement of $K_L^0 \to \gamma^*\gamma^*$ from $K_L \to e^+e^-e^+e^-$.

THE DECAY $K_L \to e^+e^-\mu^+\mu^-$

We have detected 38 $K_L \to e^+e^-\mu^+\mu^-$ events, as can be seen in figure 3 (left). These events contain very little background, on the order of less than an event, which comes primarily from photons converting in material in the detector. This is the first unambiguous observation of this decay. The previous experiment observed one event [10]. $K_L \to e^+e^-\mu^+\mu^-$ is ideal for studying the angular distribution between the lepton pairs (CP test) and probing the $K\gamma^*\gamma^*$ vertex. The invariant mass distribution of e^+e^- and $\mu^+\mu^-$ pairs can be seen in figure 3 (right). With these 38 events, there is no compelling disagreement between data and Monte Carlo [11]. In addition, the branching ratio may be able to say something about CP violation in this decay [6], [7], [11].

REFERENCES

1. G. Bélanger and C. Q. Geng, Phys. Rev. D **43**, 140 (1991).
2. J. Adams, et al., Phys. Rev. Lett. **80**, 4123 (1998).
3. T. Miyazaki and E. Takasugi, Phys. Rev. D **8**, 2051 (1973).
4. N. P. Samios, et al., Phys. Rev. **126**, 1844 (1962).
5. N. M. Kroll and W. Wada, Phys. Rev. **98**, 1355 (1955).
6. C. Quigg and J. D. Jackson, UCRL-18487 (1968).
7. L. Zhang and L. J. Goity, Phys. Rev. D **57**, 7031 (1998).

8. P. Gu, et al., Phys. Rev. Let. **72**, 3000 (1994).
9. L. Bergström et al., Phys. Rev. Let. **131**, 229(1983).
10. P. Gu, et al., Phys. Rev. Let. **76**, 4312 (1996).
11. Z. Uy, Phys. Rev. D **43**, 802 (1991).

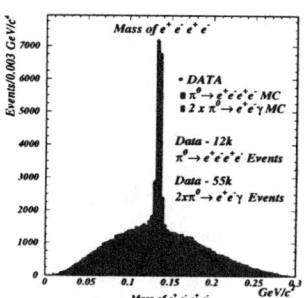
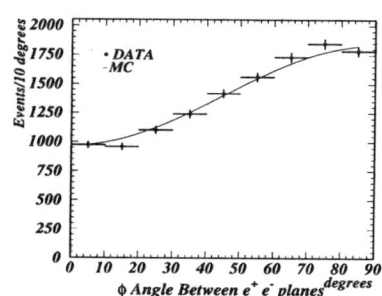

FIGURE 1. Left: Invariant mass of $e^+e^-e^+e^-$. The sharp peak shows $\pi^0 \to e^+e^-e^+e^-$ events and the broad distribution under the peak comes from $K_L \to \pi^0 \pi_D^0 \pi_D^0$. The dots are the data and the filled histograms are the Monte Carlo. Right: ϕ distribution for $\pi^0 \to e^+e^-e^+e^-$ events. The dots are the data and the line is the Monte Carlo.

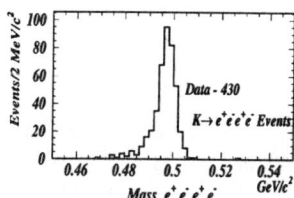
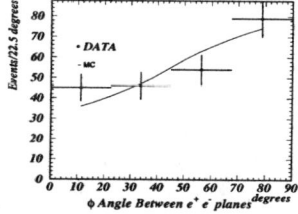
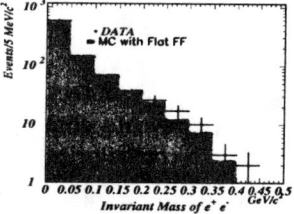

FIGURE 2. Left: Invariant mass of $e^+e^-e^+e^-$ for $K_L \to e^+e^-e^+e^-$ events. Middle: ϕ distribution for $K_L \to e^+e^-e^+e^-$ events. The dots are the data and the line is the Monte Carlo. Right: Invariant mass of e^+e^- for $K_L \to e^+e^-e^+e^-$ events.

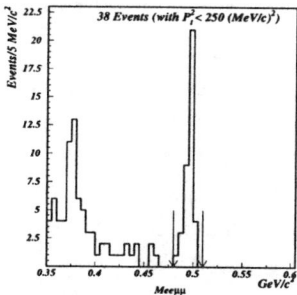
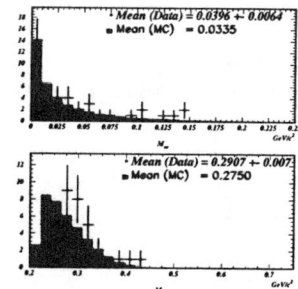

FIGURE 3. Left: Invariant mass of $e^+e^-\mu^+\mu^-$ for $K_L \to e^+e^-e^+e^-$ events. Right: Invariant masses of e^+e^- and $\mu^+\mu^-$ for $K_L \to e^+e^-\mu^+\mu^-$ events.

B Meson Decays to Charmonia at CLEO

Alexey Ershov
for the CLEO Collaboration

Harvard University,
42 Oxford St.,
Cambridge, MA 02138
E-mail: ershov@huhepl.harvard.edu

Abstract. We present the recent CLEO results on B meson decays to charmonia. New measurements of the B^0 decay modes useful for $\sin 2\beta$ measurement are presented. We describe a search for direct CP violation in $B^\pm \to J/\psi K^\pm$ and $B^\pm \to \psi(2S) K^\pm$ decays. First observations of the $B \to J/\psi \phi K$ and $B \to \eta_c K$ decays are reported. We present a new measurement of the ratio of charged to neutral B meson production at the $\Upsilon(4S)$. Finally, we describe a precise measurement of the B^0 and B^+ meson masses from $B \to \psi^{(\prime)} K$ decays.

I INTRODUCTION

The data were collected at the Cornell Electron Storage Ring (CESR) with two configurations of the CLEO detector called CLEO II [1] and CLEO II.V [2]. We use 9.2 fb^{-1} of e^+e^- data taken at the $\Upsilon(4S)$ resonance and 4.6 fb^{-1} taken 60 MeV below the $\Upsilon(4S)$ resonance. The data sample corresponds to 9.7×10^6 $B\bar{B}$ meson pairs. Two thirds of the data were collected with the CLEO II.V detector.

II B^0 DECAYS USEFUL FOR $\sin 2\beta$ MEASUREMENTS

This analysis is described in detail in Ref. [3]. We study the decays $B^0 \to J/\psi K_S^0$, $B^0 \to \chi_{c1} K_S^0$, and $B^0 \to J/\psi \pi^0$. The latter two modes are observed for the first time. We have developed a $K_S^0 \to \pi^0 \pi^0$ detection technique and applied it to the reconstruction of the decay $B^0 \to J/\psi K_S^0$. In the standard model, the measurement of the CP asymmetry in $B^0(\overline{B}^0) \to J/\psi K_S^0$ decays determines $\sin 2\beta$, where $\beta \equiv \mathrm{Arg}\left(-V_{cd}V_{cb}^*/V_{td}V_{tb}^*\right)$ [4]. A measurement of $\sin 2\beta$ with $B^0(\overline{B}^0) \to \chi_{c1} K_S^0$ decays is as theoretically clean as one with $B^0(\overline{B}^0) \to J/\psi K_S^0$. If the penguin ($b \to d c \bar{c}$) amplitude is negligible compared to the tree ($b \to c \bar{c} d$) amplitude, then the measurement of the CP asymmetry in $B^0(\overline{B}^0) \to J/\psi \pi^0$ decays allows a theoretically clean extraction of $\sin 2\beta$. The asymmetries measured with $J/\psi K_S^0$ and

$J/\psi \pi^0$ final states should have exactly the same absolute values but opposite signs, thus providing a useful check for charge-correlated systematic bias in B-flavor tagging. If the ratio of penguin to tree amplitudes is not too small, then comparison of the measured asymmetries in $J/\psi K_S^0$ and $J/\psi \pi^0$ modes may allow a resolution of one of the two discrete ambiguities ($\beta \to \beta + \pi$) remaining after a $\sin 2\beta$ measurement [5]. The results are listed in Table 1.

TABLE 1. Number of signal candidates, estimated background, product of secondary branching fractions ($\mathcal{B}_s$), detection efficiency, and measured branching fraction.

Decay mode	Signal candidates	Total background	$\mathcal{B}_s$ (%)	Efficiency (%)	Branching fraction ($\times 10^{-4}$)
$B^0 \to J/\psi K^0$					$9.5 \pm 0.8 \pm 0.6$
$K_S^0 \to \pi^+\pi^-$	142	0.3 ± 0.2	4.04 ± 0.06	37.0 ± 2.3	$9.8 \pm 0.8 \pm 0.7$
$K_S^0 \to \pi^0\pi^0$	22	1.1 ± 0.3	1.85 ± 0.03	13.9 ± 1.1	$8.4^{+2.1}_{-1.9} \pm 0.7$
$B^0 \to \chi_{c1} K^0$	9	0.9 ± 0.3	1.10 ± 0.07	19.2 ± 1.3	$3.9^{+1.9}_{-1.3} \pm 0.4$
$B^0 \to J/\psi \pi^0$	10	1.0 ± 0.5	11.8 ± 0.2	31.4 ± 2.2	$0.25^{+0.11}_{-0.09} \pm 0.02$

III SEARCH FOR DIRECT CP VIOLATION IN $B^\pm \to J/\psi K^\pm$ AND $B^\pm \to \psi(2S) K^\pm$ DECAYS

This analysis is described in detail in Ref. [6]. The CP-violating charge asymmetry in $B^\pm \to \psi^{(\prime)} K^\pm$ decays is defined as a branching fraction asymmetry

$$\mathcal{A}_{CP} \equiv \frac{\mathcal{B}(B^- \to \psi^{(\prime)} K^-) - \mathcal{B}(B^+ \to \psi^{(\prime)} K^+)}{\mathcal{B}(B^- \to \psi^{(\prime)} K^-) + \mathcal{B}(B^+ \to \psi^{(\prime)} K^+)}.$$

The relative weak phase between the tree and penguin $b \to c\bar{c}s$ quark transition amplitudes is very small in the standard model [4]. Therefore, the CP asymmetry in $B^\pm \to J/\psi K^\pm$ decays is firmly predicted in the standard model to be much smaller than the 4% precision of our measurement. A CP asymmetry of $\mathcal{O}(10\%)$ in $B^\pm \to J/\psi K^\pm$ decays is possible in a specific two-Higgs doublet model described in Ref. [7]; such a large asymmetry could be measured with our current data. We have fully reconstructed 534 $B^\pm \to J/\psi K^\pm$ and 120 $B^\pm \to \psi(2S) K^\pm$ decays with very low background. We measure the CP-violating charge asymmetry to be $(+1.8 \pm 4.3[\text{stat}] \pm 0.4[\text{syst}])\%$ for $B^\pm \to J/\psi K^\pm$ and $(+2.0 \pm 9.1[\text{stat}] \pm 1.0[\text{syst}])\%$ for $B^\pm \to \psi(2S) K^\pm$. Our results are consistent with the standard model expectations and provide the first experimental test of the assumption that direct CP violation is negligible in $B \to \psi^{(\prime)} K$ decays.

IV OBSERVATION OF THE DECAY $B \to J/\psi \phi K$

This analysis is described in detail in Ref. [8]. The decay $B \to J/\psi \phi K$ can occur only if an additional $s\bar{s}$ quark pair is created in the decay chain besides the quarks produced in the weak $b \to c\bar{c}s$ transition. The $B \to J/\psi \phi K$ transition most likely proceeds as a three-body decay. Another possibility is that the $B \to J/\psi \phi K$ decay proceeds as a quasi-two-body decay in which the J/ψ and ϕ mesons are daughters of a hybrid charmonium state [9]. We have fully reconstructed 8 $B^+ \to J/\psi \phi K^+$ candidates and 2 $B^0 \to J/\psi \phi K_S^0$ candidates. The total background is estimated to be 0.5 ± 0.2 events. We measure the branching fraction $\mathcal{B}(B \to J/\psi \phi K) = (8.8^{+3.5}_{-3.0}[\text{stat}] \pm 1.3[\text{syst}]) \times 10^{-5}$, which is approximately an order of magnitude smaller than $\mathcal{B}(B \to J/\psi K)$. The $J/\psi \phi$ invariant mass is above the DD^{**} production threshold for all 10 $B \to J/\psi \phi K$ candidates thus disfavoring the hybrid charmonium dominance scenario [9].

V OBSERVATION OF THE DECAY $B \to \eta_c K$

This analysis is described in detail in Ref. [10]. We have searched for the exclusive B decays to η_c meson, the spin-0 charmonium ground state. We reconstruct the η_c in the decay modes $\eta_c \to \phi\phi \to K^+K^-K^+K^-$ and $\eta_c \to K_S^0 K^\pm \pi^\mp$. The signal yield is extracted by a maximum likelihood fit in several discriminating variables. The statistical significance of the signal is 5.2 standard deviations for the $B^+ \to \eta_c K^+$ decay and 4.8 for $B^0 \to \eta_c K_S^0$. We measure branching fractions $\mathcal{B}(B^+ \to \eta_c K^+) = (0.69^{+0.26}_{-0.21} \pm 0.08 \pm 0.20) \times 10^{-3}$ and $\mathcal{B}(B \to \eta_c K^0) = (1.09^{+0.55}_{-0.42} \pm 0.12 \pm 0.31) \times 10^{-3}$, where the first error is statistical, the second is systematic, and the third is from the η_c branching fraction uncertainty. Within the uncertainties, the decay $B \to \eta_c K$ occurs at the same rate as the $B \to J/\psi K$ decay.

VI CHARGED AND NEUTRAL B MESON PRODUCTION AT THE $\Upsilon(4S)$

This analysis is described in detail in Ref. [11]. Measurements of exclusive B-decay branching fractions from e^+e^- colliders operating at the $\Upsilon(4S)$ resonance assume equal production rates of charged and neutral B-meson pairs [12]. Any physics based upon comparisons of absolute decay rates of charged and neutral B mesons will profit from a more precise knowledge of the B-production ratio $f_{+-}/f_{00} \equiv \Gamma(\Upsilon(4S) \to B^+B^-)/\Gamma(\Upsilon(4S) \to B^0\bar{B}^0)$. We study the decays $B \to J/\psi K^{(*)}$, which are isospin conserving transitions. The decays $B^+ \to J/\psi K^{(*)+}$ and $B^0 \to J/\psi K^{(*)0}$ must therefore have equal partial widths and we can extract $\frac{f_{+-}}{f_{00}} \times \frac{\tau_{B^+}}{\tau_{B^0}} = \frac{\mathcal{N}(B^+ \to J/\psi K^{(*)+})}{\mathcal{N}(B^0 \to J/\psi K^{(*)0})}$, where $\mathcal{N}$ is the efficiency-corrected signal yield. We measure the rates for $B^0 \to J/\psi K^{(*)0}$ and $B^+ \to J/\psi K^{(*)+}$ decays and use the world-average B-meson lifetime ratio to extract the relative widths $f_{+-}/f_{00} = $

$1.04 \pm 0.07[\text{stat}] \pm 0.04[\text{syst}]$. This is the most precise measurement of f_{+-}/f_{00} to date.

VII MEASUREMENT OF THE B^0 AND B^+ MESON MASSES FROM $B \to \psi^{(\prime)} K$ DECAYS

This analysis is described in detail in Ref. [13]. The previous measurements of the B meson masses at e^+e^- colliders operating at $\Upsilon(4S)$ energy were obtained from the fits to the distributions of the beam-constrained B mass, defined as $M_{\text{bc}} \equiv \sqrt{E_{\text{beam}}^2 - p^2(B)}$, where $p(B)$ is the B candidate momentum. Substitution of the beam energy for the measured energy of the B meson candidate results in a significant improvement of the mass resolution. The precision of the measurement of the absolute B^0 and B^+ masses, however, is limited by the systematic uncertainties in the absolute beam energy scale and in the correction for initial state radiation. In this measurement, we have fully reconstructed 135 $B^0 \to \psi^{(\prime)} K_S^0$ and 526 $B^+ \to \psi^{(\prime)} K^+$ candidates with very low background. We fit the $\psi^{(\prime)} K$ invariant mass distributions of these B meson candidates and measure the masses of the neutral and charged B mesons to be $M(B^0) = 5279.1 \pm 0.7[\text{stat}] \pm 0.3[\text{syst}]$ MeV/c^2 and $M(B^+) = 5279.1 \pm 0.4[\text{stat}] \pm 0.4[\text{syst}]$ MeV/c^2. The precision is a significant improvement over previous measurements.

REFERENCES

1. Y. Kubota et al., *Nucl. Instrum. Methods* A **320**, 66 (1992).
2. T.S. Hill, *Nucl. Instrum. Methods* A **418**, 32 (1998).
3. P. Avery et al. (CLEO Collaboration), *Phys. Rev.* D **62**, 051101 (2000).
4. For a recent review see Y. Nir, lectures given at 27th SLAC Summer Institute on Particle Physics, (SSI 99), Stanford, California, Report IASSNS-HEP-99-96, hep-ph/9911321.
5. Y. Grossman and H. R. Quinn, *Phys. Rev.* D **56**, 7259 (1997).
6. G. Bonvicini et al. (CLEO Collaboration), *Phys. Rev. Lett.* **84**, 5940 (2000).
7. G. Wu and A. Soni, Report BNL-HET-99/40, hep-ph/9911419; K. Kiers, A. Soni, and G. Wu, *Phys. Rev.* D **59**, 96001 (1999).
8. C.P. Jessop et al. (CLEO Collaboration), *Phys. Rev. Lett.* **84**, 1393 (2000).
9. F.E. Close et al., *Phys. Rev.* D **57**, 5653 (1998).
10. K.W. Edwards et al. (CLEO Collaboration), Report No. CLNS 00/1680, hep-ex/0007012 (submitted to *Phys. Rev. Lett.*).
11. J.P. Alexander et al. (CLEO Collaboration), Report No. CLNS 00/1670, hep-ex/0006002 (submitted to *Phys. Rev. Lett.*).
12. C. Caso et al. (Particle Data Group), *Eur. Phys. J.* C **3** 1 (1998).
13. S.E. Csorna et al. (CLEO Collaboration), *Phys. Rev.* D **61**, 111101 (2000).

Lifetimes and Inclusive Decay Rates of c and b Hadrons

M.B. Voloshin

Theoretical Physics Institute, University of Minnesota, Minneapolis, MN 55455
and
Institute of Theoretical and Experimental Physics, Moscow, 117259

Abstract. I give a brief report on the status of understanding the differences of inclusive decay rates among the charmed and beauty hadrons. Also is pointed out the relation between the differences of the decay rates for heavy baryons and the decays $\Xi_Q \to \Lambda_Q \pi$.

The differences in inclusive weak decay rates between hadrons containing the same heavy quark Q (c or b) are very prominent for the charmed hadrons, and are observable, although small, for the b hadrons [1]. The decrease of the relative magnitude of these differences from charm to beauty is in full agreement with the idea that they are associated with nonperturbative effects suppressed by inverse powers of the heavy quark mass. The systematic description [2–4] of these effects is provided by an application of the operator product expansion (OPE) to the effective operator, describing the total decay rates,

$$L_{eff} = 2\,\text{Im}\left[i\int d^4x\, e^{iqx}\, T\{L_W(x), L_W(0)\}\right],\tag{1}$$

where L_W is the appropriate part of the weak interaction Lagrangian responsible for the particular type of decay of the heavy quark. The inclusive decay rate of a hadron H_Q, containing the heavy quark Q is then given by [1]

$$\Gamma_H = \langle H_Q | L_{eff} | H_Q \rangle.\tag{2}$$

At large mass m_Q of the heavy quark one can expand the operator L_{eff} at $q^2 = m_Q^2$ in powers of m_Q^{-1} in terms of local operators. The general expression for first three terms in this expansion reads as

$$L_{eff} = L_{eff}^{(0)} + L_{eff}^{(2)} + L_{eff}^{(3)} = \tag{3}$$

$$c^{(0)}\frac{G_F^2\, m_Q^5}{64\,\pi^3}\left(\overline{Q}Q\right) + c^{(2)}\frac{G_F^2\, m_Q^3}{64\,\pi^3}\left(\overline{Q}\sigma^{\mu\nu}G_{\mu\nu}Q\right) + \frac{G_F^2\, m_Q^2}{4\,\pi}\sum_i c_i^{(3)}\,(\overline{q}_i\Gamma_i q_i)(\overline{Q}\Gamma_i' Q),$$

---

[1] The non-relativistic normalization is used here for the *heavy* quark states: $\langle Q|Q^\dagger Q|Q\rangle = 1$.

where the superscripts denote the power of m_Q^{-1} in the relative suppression of the corresponding term in the expansion with respect to the leading one, $G_{\mu\nu}$ is the gluon field tensor, q_i stand for light quarks, u, d, s, and, finally, Γ_i, Γ'_i denote spin and color structures of the four-quark operators. The coefficients $c^{(a)}$ depend on the specific part of the weak interaction Lagrangian L_W, describing the relevant underlying quark process.

The leading first term in the expansion (3) describes the perturbative 'parton' decay rate, and does not depend on the flavor of light quarks, accompanying the heavy quark in the hadron. The same is true for the second, chromomagnetic term [5], which depends only on the correlation of the heavy quark spin with the rest of the constituents. The dependence on the flavor of the light quarks arises only in the third term, containing four-quark operators [2–4,6]. It is believed that the contribution of the third term dominates the observed differences of the inclusive decay rates among the charmed hadrons and should also be dominant in similar differences for the b hadrons. The main problem in describing quantitatively the effects of this term on the decay rates is a poor knowledge of the matrix elements of the relevant four-quark operators over hadrons. For mesons one can use, as a reference, the estimate of these matrix elements in the limit of factorization [3,4,7], thus expressing the matrix elements in terms of the pseudoscalar mesons annihilation constants, f_D, f_B. In the case of heavy baryons there is no similar 'guideline', and one has to rely on the existing experimental data on e.g. lifetimes of the charmed baryons, in order to extract the matrix elements, and then use them in predicting differences in other inclusive decay rates of charmed and b baryons.

The analysis of the spectator effects in decays of baryons is simplified for the flavor SU(3) (anti)triplets of the baryons (Λ_Q, $\Xi_Q^{(u)}$, and $\Xi_Q^{(d)}$), where the spin of the hyperon is carried by the spin of the heavy quark, which is not correlated with spin degrees of freedom of the light components. Using this observation and the SU(3) symmetry, one can express the effects of the four-quark operators in the decay rates of the baryons in the triplet through four independent matrix elements. Two SU(3) nonsinglet matrix elements x and y are defined as

$$x = \left\langle \frac{1}{2} (\overline{Q}\gamma_\mu Q)\left[(\overline{u}\gamma_\mu u) - (\overline{s}\gamma_\mu s)\right] \right\rangle_{\Xi_Q^{(d)} - \Lambda_Q}, \tag{4}$$

$$y = \left\langle \frac{1}{2} (\overline{Q}_i \gamma_\mu Q_k)\left[(\overline{u}_k \gamma_\mu u_i) - (\overline{s}_k \gamma_\mu s_i)\right] \right\rangle_{\Xi_Q^{(d)} - \Lambda_Q},$$

with the notation for the differences of the matrix elements: $\langle \mathcal{O} \rangle_{A-B} = \langle A|\mathcal{O}|A\rangle - \langle B|\mathcal{O}|B\rangle$, and two flavor singlet matrix elements are

$$x_s = \frac{1}{3} \langle H_Q|(\overline{Q}\gamma_\mu Q)\left((\overline{u}\gamma_\mu u) + (\overline{d}\gamma_\mu d) + (\overline{s}\gamma_\mu s)\right)|H_Q\rangle$$

$$y_s = \frac{1}{3} \langle H_Q|(\overline{Q}_i \gamma_\mu Q_k)\left((\overline{u}_k \gamma_\mu u_i) + (\overline{d}_k \gamma_\mu d_i) + (\overline{s}_k \gamma_\mu s_i)\right)|H_Q\rangle, \tag{5}$$

where H_Q stands for any heavy hyperon in the (anti)triplet.

Using explicit expressions [4,7–9] for the relevant parts of the third term in the effective Lagrangian (3) one can find in terms of only x and y the differences among the charmed baryons of the inclusive rates for the following decays: the dominant CKM unsuppressed nonleptonic decays ($\Delta S = \Delta C$), the once CKM suppressed ones ($\Delta S = 0$), and the semileptonic decays [8]. Combining these one can find the differences of the total decay rates within the triplet of the charmed hyperons and thus extract the values of x and y from the data on the lifetimes. The relations for the total decay widths in terms of the matrix elements x and y normalized at $\mu = m_c$ read as

$$\Gamma(\Xi_c^0) - \Gamma(\Lambda_c) =$$
$$\frac{G_F^2 m_c^2}{4\pi} \cos^2\theta_c \left\{ -x \left[\cos^2\theta_c\, C_+ C_- + \tfrac{\sin^2\theta_c}{4}(5C_+^2 + 5C_-^2 + 6C_+ C_-) \right] + 3y \left[\cos^2\theta_c\, C_+ C_- + \tfrac{\sin^2\theta_c}{4}(6C_+ C_- - 3C_+^2 + C_-^2) + \tfrac{2}{3} \right] \right\}, \quad (6)$$

and

$$\Gamma(\Xi_c^+) - \Gamma(\Lambda_c) = \frac{G_F^2 m_c^2}{4\pi} \left\{ -x\,\tfrac{\cos^4\theta_c}{4}(5C_+^2 + 5C_-^2 - 2C_+ C_-) - 3y\left[\tfrac{\cos^4\theta_c}{4}(3C_+^2 - C_-^2 + 2C_+ C_-) + \tfrac{2}{3}(\cos^2\theta - \sin^2\theta)\right] \right\}, \quad (7)$$

where θ_c is the Cabibbo angle, and C_+ and C_- are the QCD renormalization coefficients for nonleptonic weak Lagrangian at $\mu = m_c$: $C_-(m_c) = C_+^{-2}(m_c) = [\alpha_s(m_c)/\alpha_s(m_W)]^{12/25}$.

Comparing the formulas (6) and (7) with the data on the lifetimes, one finds [9] $x = -(0.04 \pm 0.01)\,GeV^3\,(1.4\,GeV/m_c)^2$ and $y(m_c) = (0.02 \pm 0.01)\,GeV^3\,(1.4\,GeV/m_c)^2$ where, as explicitly indicated, the matrix element x does not depend on the normalization point, while the result for y is given for $\mu = m_c$. The extracted values significantly differ from the naive constituent quark model estimate $y = -x \approx f_D^2\, M_D/12 \approx 0.006\,GeV^3$ used in earlier analyses.

The extracted values of x and y can be used for predicting differences within the heavy hyperon triplet of other inclusive decay rates (e.g. semileptonic and the Cabibbo suppressed nonleptonic), as well as for predicting the similar differences among the b baryons. Given the correlation of errors in x and y it is better to express the predictions directly in terms of the total decay rates of the charmed baryons. In this way one finds [10,11] a large expected difference of the semileptonic decay rates between either of the Ξ_c baryons and Λ_c:

$$\Gamma_{sl}(\Xi_c) - \Gamma_{sl}(\Lambda_c) \approx 0.13\,\Gamma(\Xi_c^0) + 0.065\,\Gamma(\Xi_c^+) - 0.195\,\Gamma(\Lambda_c) \approx 0.59 \pm 0.32\,ps^{-1}, \quad (8)$$

which in fact implies that in absolute terms the semileptonic decay rate for the charmed cascades is by a factor of 2 to 3 larger than that of Λ_c: $\Gamma_{sl}(\Xi_c) = (2 - 3)\,\Gamma_{sl}(\Lambda_c)$. For the total decay rates of the b baryons one finds the prediction that $\Gamma(\Lambda_b) \approx \Gamma(\Xi_b^0)$, and

$$\Gamma(\Lambda_b) - \Gamma(\Xi_b^-) \approx 0.015\,\Gamma(\Xi_c^0) - 0.016\,\Gamma(\Xi_c^+) + 0.001\,\Gamma(\Lambda_c) \approx 0.11 \pm 0.03\,ps^{-1}\ . \quad (9)$$

The central value here corresponds to about 14% of the total decay rate of Λ_b.

The singlet matrix elements x_s and y_s (cf. eq.(5)) are related to the shift of the average decay rate of the baryons in the triplet: $\overline{\Gamma}_Q = [\Gamma(\Lambda_Q) + \Gamma(\Xi_Q^{(u)}) + \Gamma(\Xi_Q^{(d)})]/3$. It can be found that the average shift due to the four-quark operators in the expansion (3) is proportional to the combination $x_s - 3\,y_s$ for both the charmed and the b baryons. Thus these unknown matrix elements cancel in the ratio, and one finds

$$\delta^{(3)}\overline{\Gamma}_b = \frac{|V_{bc}|^2}{\cos^4\theta_c}\frac{m_b^2}{m_c^2}\frac{(\tilde{C}_+ - \tilde{C}_-)^2}{C_+^2 + C_-^2}\left[\frac{\alpha_s(m_c)}{\alpha_s(m_b)}\right]^{5/18}\delta_{nl}^{(3)}\overline{\Gamma}_c \approx 0.0025\,\delta_{nl}^{(3)}\overline{\Gamma}_c\ , \quad (10)$$

where $\tilde{C}_\pm$ are the same renormalization coefficients as $C_\pm$, however normalized at $\mu = m_b$. This equation shows that relatively to the charmed baryons the overall shift of the decay rates in the b baryon triplet is greatly suppressed by the ratio $(\tilde{C}_+ - \tilde{C}_-)^2/(C_+^2 + C_-^2)$, which parametrically is of the second order in α_s, and numerically is only about 0.12. A conservative estimate of the shift of the average nonleptonic decay rate for the charmed baryons due to the four quark operators is $\delta_{nl}^{(3)}\overline{\Gamma}_c < 3\,ps^{-1}$, for which Eq.(10) predicts $\delta^{(3)}\overline{\Gamma}_b < 0.008 \pm 0.001\,ps^{-1}$, which is only about 1% of the total decay rate of Λ_c. The shift of the Λ_b decay rate with respect to the average width $\overline{\Gamma}_b$ due to the non-singlet operators is one third of the splitting (9), i.e. about 5%. Adding to this the 1% shift of the average width and another 1% difference from the meson decays due to the suppression of the latter by the m_b^{-2} chromomagnetic effects, one concludes that at the present level of theoretical understanding it looks impossible to explain a more than 10% enhancement of the total decay rate of Λ_b relative to B_d, where an ample 3% margin is added for the uncertainties of higher order terms in OPE as well as for higher order QCD radiative effects in the discussed corrections. In other words, the expected pattern of the lifetimes of the b hyperons in the triplet, relative to B_d, is $\tau(\Xi_b^0) \approx \tau(\Lambda_b) < \tau(B_d) < \tau(\Xi_b^-)$, with the "best" theoretical estimate of the differences to be about 7% for each step of the inequality.

An additional test of the four-quark matrix elements over heavy baryons is possible in the weak decays $\Xi_Q \to \Lambda_Q \pi$ [12], where the heavy quark is not destroyed, but rather it is the strange quark that decays. Due to absence of spin correlations between the heavy quark and the light quarks, and due to that the light system (diquark) in the baryons has quantum numbers $J^P = 0^+$, the pion transitions are purely S wave (in the heavy quark limit). The S wave amplitudes can be well approximated by their chiral limit, i.e. at zero four-momentum of the pion, where they are described by the PCAC reduction formula: $\langle \Lambda_Q \pi_i(p=0)|H_W|\Xi_Q\rangle = \sqrt{2}\,\langle \Lambda_Q|[Q_i^5, H_W]|\Xi_Q\rangle/f_\pi$, where π_i is the pion triplet in the Cartesian notation, Q_i^5 is the corresponding isotopic triplet of axial charges, and $f_\pi \approx 130\,MeV$ is the pion decay constant. The Hamiltonian H_W is the non-leptonic strangeness-changing weak interaction Hamiltonian:

$$H_W = \sqrt{2} G_F \cos\theta_c \sin\theta_c \left\{ (C_+ + C_-) \left[(\bar{u}_L \gamma_\mu s_L)(\bar{d}_L \gamma_\mu u_L) - (\bar{c}_L \gamma_\mu s_L)(\bar{d}_L \gamma_\mu c_L) \right] + \right.$$
$$\left. (C_+ - C_-) \left[(\bar{d}_L \gamma_\mu s_L)(\bar{u}_L \gamma_\mu u_L) - (\bar{d}_L \gamma_\mu s_L)(\bar{c}_L \gamma_\mu c_L) \right] \right\} \ . \tag{11}$$

The applicability of PCAC ensures that the considered decays obey the $\Delta I = 1/2$ rule, so that $\Gamma(\Xi_Q^{(d)} \to \Lambda_Q \pi^-) = 2\Gamma(\Xi_Q^{(u)} \to \Lambda_Q \pi^0)$. The terms in the Hamiltonian (11) without the charmed quark give equal contribution to the transitions between the c baryons and the b baryons, while the terms with the c quark, corresponding to the weak scattering $s\,c \to d\,c$ contribute only to the transitions between the charmed baryons. Using the PCAC relation and a flavor SU(3) rotation, the contribution of the latter terms to the transition amplitude can be expressed through the matrix elements x and y (cf. Eq.(4)) and thus through the differences of the total widths of the charmed baryon. The resulting relation [12] reads as

$$\langle \Lambda_c \pi^-(p=0) | H_W | \Xi_c^0 \rangle - \langle \Lambda_b \pi^-(p=0) | H_W | \Xi_b^- \rangle \approx \tag{12}$$
$$-\frac{\sqrt{2}\pi \cos\theta_c \sin\theta_c}{G_F m_c^2 f_\pi} \left[0.45\,\Gamma(\Xi_c^0) - 0.04\,\Gamma(\Xi_c^+) - 0.41\,\Gamma(\Lambda_c) \right] \approx -(5.4 \pm 2) \times 10^{-7} \ ,$$

which can be also written as a triangle inequality for the rates:

$$\sqrt{\Gamma(\Xi_b^- \to \Lambda_b \pi^-)} + \sqrt{\Gamma(\Xi_c^0 \to \Lambda_c \pi^-)} \geq \sqrt{0.9 \times 10^{10}\,s^{-1}} \ . \tag{13}$$

This inequality shows that the rate of at least one of these decays should be at the level of $0.01\,ps^{-1}$, which is a considerable rate for an exclusive decay channel of either charmed or b hardons.

This work and participation in the conference is supported in part by DOE under the grant number DE-FG02-94ER40823.

REFERENCES

1. Particle Data Group, *Eur. Phys. J* **C3**, 1 (1998).
2. M.A. Shifman and M.B. Voloshin, (1981) unpublished, presented in V.A. Khoze and M.A. Shifman, *Sov. Phys. Usp.* **26**, 387 (1983).
3. M.A. Shifman and M.B. Voloshin, *Sov. J. Nucl. Phys.* **41**, 120 (1985).
4. M.A. Shifman and M.B. Voloshin, *Sov. Phys. JETP* **64**, 698 (1986).
5. I.I. Bigi, N.G. Uraltsev, and A.I. Vainshtein, *Phys. Lett.* **B293**, 430 (1992); [E: **297**, 477 (1993)].
6. N. Bilić, B. Guberina, and J. Trampetić, *Nucl. Phys.* **B248**, 261 (1984).
7. M. Neubert and C.T. Sachrajda, *Nucl. Phys.* **B483**, 339 (1997).
8. M.B. Voloshin, *Phys. Rep.* **320**, 275 (1999).
9. M.B. Voloshin, *Phys. Rev.* **D61**, 074026 (2000).
10. M.B. Voloshin, *Phys. Lett.* **B385**, 369 (1996).
11. B. Guberina and B. Melić, *Eur.Phys.J.* **C2**, 697 (1998).
12. M.B. Voloshin, *Phys. Lett.* **B476**, 297 (2000).

Spin Physics

Measurement of Polarized Quark Distributions of the Nucleon at HERMES

Marc Beckmann[a] on behalf of the HERMES Collaboration

[a] Fakultät für Physik, Albert–Ludwigs–Universität Freiburg, D–79104 Freiburg i. Brsg., Germany

Abstract. Double spin asymmetries of the semi–inclusive cross sections for the production of charged hadrons have been measured by the HERMES experiment in deep–inelastic scattering of polarized positrons on polarized hydrogen and ^{3}He targets, in the kinematic range $0.023 < x < 0.6$, and $1\ \text{GeV}^2 < Q^2 < 10\ \text{GeV}^2$. Polarized quark distributions are extracted as a function of x for up $(u+\bar{u})$ and down $(d+\bar{d})$ flavors, as well as for valence and sea quarks. In the measured range, the up quark polarization is positive and the polarization of the down quarks is negative. The polarization of the sea is compatible with zero. The moments of the polarized quark distributions are compared to predictions based on $SU(3)_f$ symmetry and to a prediction from lattice QCD. New data taken on a polarized deuterium target together with the identification of charged kaons by a RICH detector will eventually enable the extraction of the strange quark polarization in the sea.

Semi–inclusive polarized deep–inelastic scattering (DIS) provides a useful tool to measure the polarized quark distributions $\Delta q(x, Q^2) \equiv q^{\uparrow}(x, Q^2) - q^{\downarrow}(x, Q^2)$ for the different quark types $q = u, \bar{u}, d, \bar{d}, s, \bar{s}$. Here, $q^{\uparrow(\downarrow)}(x, Q^2)$ denotes the number density of quarks with spin aligned parallel (anti–parallel) to the nucleon's spin.

Assuming factorization, the DIS cross section asymmetry A_1^h can be written in LO QCD as

$$A_1^h(x, Q^2) \stackrel{g_2=0}{=} \frac{\sum_q e_q^2\, \Delta q(x, Q^2) \int_{0.2}^1 D_q^h(Q^2, z)\, dz}{\sum_q e_q^2\, q(x, Q^2) \int_{0.2}^1 D_q^h(Q^2, z)\, dz} \times \underbrace{\frac{1 + R(x, Q^2)}{1 + \gamma^2}}_{\equiv\, \mathcal{C}_R}. \qquad (1)$$

In this expression, the $q(x, Q^2)$ denote the unpolarized quark distribution functions, $\gamma^2 = Q^2/\nu^2$, and $R(x, Q^2) = \sigma_L/\sigma_T$ is the photo absorption cross section ratio for longitudinal and transverse virtual photons. The fragmentation functions $D_q^h(Q^2, z)$ are integrated over the range $0.2 - 1$ in z. To account for the longitudinal component of the photo absorption cross section, which is included in the experimentally determined parameterizations of $q(x, Q^2)$, but not in $\Delta q(x, Q^2)$, the correction factor $\mathcal{C}_R$ is introduced. In Eq. (1) contributions from the spin structure function g_2

are omitted, which was found to be very small in measurements at SLAC [1,2]. Provided the fragmentations functions are known, Eq. (1) can be used to extract the quark polarizations $\Delta q(x)/q(x)$ from a set of measured asymmetries.

The HERMES experiment [3] utilizes the longitudinally polarized 27.5 GeV positron beam of the HERA storage ring at DESY, incident on polarized pure atomic gas targets, which are internal to the HERA beam line. In 1995 an optically pumped target of polarized ^{3}He atoms was used, while in 1996/97 data was taken on a target of polarized hydrogen atoms from an atomic beam source. The HERMES detector is an open geometry forward spectrometer with good particle identification accomplished by the combined information from a preshower counter, a transition radiation detector, a lead glass calorimeter and a threshold Čerenkov detector. The measured inclusive and semi–inclusive asymmetries on polarized ^{3}He and proton targets can be found in Ref. [5]. In total, 2.2 (2.3) million DIS events taken on a ^{3}He (proton) target were included in this analysis.

For a set of measured inclusive and semi–inclusive asymmetries on different targets, Eq. (1) can be rewritten in matrix form

$$\vec{A}(x) = \mathcal{P}(x)\,\vec{Q}(x), \qquad (2)$$

where the elements of the vector $\vec{A}(x)$ contain the measured asymmetries, and the vector $\vec{Q}(x)$ holds the polarizations of the different quark flavors to be extracted. The elements of the matrix $\mathcal{P}(x)$ depend on the fragmentation functions, the unpolarized parton densities, and the cross section ratio $R(x, Q^2)$. The fragmentation functions were determined from the LUND string fragmentation model implemented in the JETSET 7.4 [6] Monte Carlo package. The LUND string breaking parameters have been tuned to fit the hadron multiplicities measured at HERMES. For the unpolarized parton distributions the parameterizations of Ref. [7] were used.

Due to the low weight of the individual sea quark polarizations in Eq. (1), they were combined, using the assumption that the polarization of all sea quarks is flavor symmetric:

$$\frac{\Delta q_s}{q_s} \equiv \frac{\Delta u_s}{u_s} = \frac{\Delta \bar{u}}{\bar{u}} = \frac{\Delta d_s}{d_s} = \frac{\Delta \bar{d}}{\bar{d}} = \frac{\Delta s}{s} = \frac{\Delta \bar{s}}{\bar{s}}. \qquad (3)$$

With this assumption the flavor decomposition of the quark polarization was obtained by solving Eq. (2) for the vector $\vec{Q}(x) = \left(\frac{\Delta u+\Delta\bar{u}}{u+\bar{u}}, \frac{\Delta d+\Delta\bar{d}}{d+\bar{d}}, \frac{\Delta s+\Delta\bar{s}}{s+\bar{s}}\right)$, where the polarization of the strange quarks $(\Delta s + \Delta \bar{s})/(s+\bar{s}) = \Delta q_s/q_s$ equals the polarization of all sea quarks due to Eq. (3). For values of $x > 0.3$ the sea polarization was fixed to zero and the resulting influence on the up and down quark polarizations was included in their systematic uncertainties. The extracted polarizations of the up, down and sea quarks are shown in Fig. 1 as a function of x. The polarization of the up quarks is positive everywhere in the measured range and the down quarks are polarized in the opposite direction, while the extracted polarization of the sea quarks is zero within the combined uncertainties.

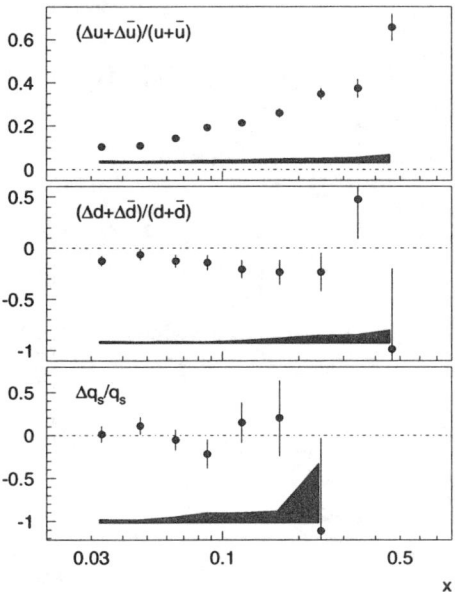

FIGURE 1. The polarization of the up, down, and sea quarks as a function of x. The error bars represent the statistical uncertainties while the shaded bands indicate the systematic uncertainties.

The polarized quark distributions $\Delta q(x)$ are obtained by multiplying the extracted quark polarizations with the unpolarized quark distributions $q(x)$ from Ref. [7]. Fig. 2 shows the polarized quark distributions separately for valence and sea quark flavors at a scale of 2.5 GeV^2. The Δu_v distribution is positive over the measured x-range and the Δd_v distribution is mainly negative. For comparison, earlier results by the SMC collaboration [4] and a parameterization [8] of the polarized quark distributions are also shown in Fig. 2. The agreement with the SMC results is good and both data sets are consistent with the parameterization.

First and second moments of the polarized quark distributions were extracted. The results from the SMC and the HERMES experiments are compatible in the common measured range $0.023 < x < 0.6$. Using extrapolations in the unmeasured regions, the HERMES results are furthermore in agreement with analyses assuming SU(3)$_f$ flavor symmetry [5]. Significant deviations are found for the first and second moments of the Δu_v distribution from predictions by quenched lattice QCD calculations at similar values of Q^2, while the corresponding results for the moments of the Δd_v distribution are in agreement. From the HERMES data, the spin carried by the different quark flavors is found to be $\Delta u + \Delta \bar{u} = 0.57 \pm 0.04$, $\Delta d + \Delta \bar{d} = -0.25 \pm 0.08$, and $\Delta s + \Delta \bar{s} = -0.01 \pm 0.05$ at a scale of $Q^2 = 2.5$ GeV^2.

The HERMES results presented here were obtained from the data taken in the years 1995 to 1997 and have been published in Ref. [5]. Since 1998 HERMES is taking data on a polarized deuterium target with a currently accumulated statistics

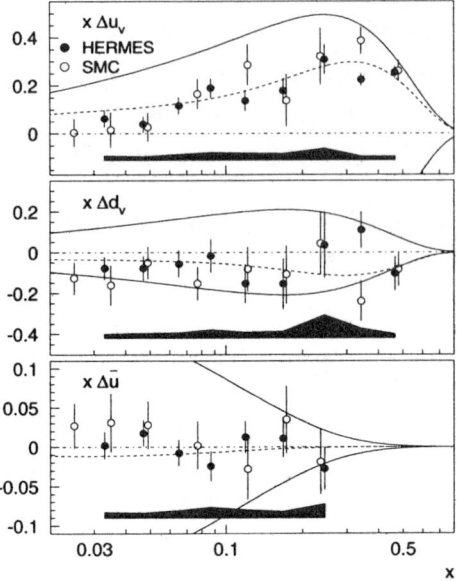

FIGURE 2. The polarized valence and sea quark distributions $x\,\Delta u_v(x)$, $x\,\Delta d_v(x)$, and $x\,\Delta\bar{u}(x)$ at $Q^2 = 2.5$ GeV^2. The HERMES data points (filled circles) are compared to SMC data [4] (open points), which have been evolved to the same scale, and to a parameterization [8] given by the dashed lines. The solid lines indicate the positivity limits on the polarized quark distributions.

of about 5.7 million DIS events, which will substantially reduce the statistical uncertainties on the polarizations of the up and down quarks. Furthermore, the replacement of the threshold Čerenkov counter by a RICH [9] counter in 1998 allows the analysis of semi–inclusive kaon asymmetries. This will substantially improve the sensitivity on the sea quark polarization and eventually enable an extraction of the thusfar unmeasured polarization of the strange quarks.

REFERENCES

1. Abe, K. *et al.*, *Phys. Rev.* **D 58**, 112003 (1998).
2. Anthony, P.L. *et al.*, *Phys. Lett.* **B 458**, 529–535 (1999).
3. Ackerstaff, K. *et al.*, *Nucl. Instr. Meth.* **A 417**, 230–265 (1998).
4. Adeva, B. *et al.*, *Phys. Lett.* **B 420**, 180–190 (1998).
5. Ackerstaff, K. *et al.*, *Phys. Lett.* **B 464**, 123–134 (1999).
6. Sjöstrand, T., *Comp. Phys. Comm.* **82**, 74–89 (1994).
7. Lai, H.L. *et al.*, *Phys. Rev.* **D 55**, 1280–1296 (1997).
8. Glück, M. *et al.*, *Phys. Rev.* **D 53**, 4775–4786 (1996).
9. De Schepper, D., these proceedings.

Large Flavor Asymmetry of Polarized Antiquarks: Prediction & Possible Tests in semi-inclusive DIS and Polarized DY

P. Schweitzer*, B. Dressler*, K. Goeke*, M.V. Polyakov*[†], C. Weiss*

Institut für Theoretische Physik II, Ruhr-Universität Bochum, D-44780 Bochum, Germany
[†]*Petersburg Nuclear Physics Institute, Gatchina, St. Petersburg 188350, Russia*

Abstract. Calculations of polarized parton distributions in the large N_c-limit predict a large antiquark flavor asymmetry $\Delta\bar{u}(x)-\Delta\bar{d}(x)$ in the nucleon. The framework for the calculations is the chiral quark–soliton model. The predicted large flavor asymmetry in the polarized quark sea does not contradict the recent semi-inclusive DIS data from HERMES. However a $\Delta\bar{u}(x)-\Delta\bar{d}(x)$ as predicted by the model would have a large effect on the longitudinal double spin asymmetry measurable in the process of Drell–Yan lepton pair production in polarized nucleon–nucleon scattering.

Introduction

The unpolarized quark sea exhibits a flavor asymmetry $\bar{u}(x) \neq \bar{d}(x)$ [1]. Since this asymmetry cannot be generated perturbatively one has to invoke nonperturbative effects in order to explain it. A very appealing explanation provides the pion cloud model [2]. Another, remarkably good quantitative explanation of this phenomenon yields the chiral quark–soliton model (χQSM) [3]. The model is able to reproduce the shape of $\bar{u}(x)-\bar{d}(x)$ in good agreement with data without any free parameter [4]. Moreover it predicts that there is an asymmetry in the polarized quark sea which is even larger than that in the unpolarized quark sea, $\Delta\bar{u}(x)-\Delta\bar{d}(x) > |\bar{u}(x)-\bar{d}(x)|$. Such an asymmetry could not be explained neither perturbatively nor in the pion cloud picture [5].

The chiral quark soliton model

The χQSM is based on two ideas. The one is the spontaneous breaking of chiral symmetry with pions as Goldstone bosons and the other is the limit of large number of colors N_c. A dynamical picture of chiral symmetry breaking is provided by the instanton vacuum. In the large N_c-limit QCD becomes an effective theory of mesons and the nucleon emerges as a soliton of the pion field [6]. Both concepts are realized in the χQSM. Its effective action is given by

$$\exp\left(iS_{\text{eff}}[\pi(x)]\right) = \int \mathcal{D}\psi \int \mathcal{D}\bar\psi \, \exp\left(i\int d^4x\, \bar\psi(i\slashed{\partial} - MU^{\gamma_5})\psi\right)$$

$$\text{with } U^{\gamma_5}(x) = \exp\left(i\gamma_5 \tau^a \pi^a(x)\right). \quad (1)$$

M is the momentum dependent dynamical quark mass. S_{eff} has been derived from the instanton vacuum, see [7] and is valid for momenta $|\mathbf{k}| \ll (\bar\rho)^{-1}$. The inverse of the average size of an instanton $(\bar\rho)^{-1} = \mathcal{O}(600\,\text{MeV})$ defines the scale μ^2 to which the quantities computed in the effective theory refer to, i.e. $\mu^2 \simeq (600\,\text{MeV})^2$. The χQSM allows to compute nucleon matrix elements of non–local bilinear quark operators in an effective $1/N_c$–expansion

$$\langle N|\bar\psi(x)\Gamma\psi(y)|N'\rangle\bigg|_{\mu^2} = \mathcal{O}(N_c^2) + \mathcal{O}(N_c) + \ldots \quad (2)$$

where, however, only LO and NLO can be practically computed. Since the χQSM is essentially an (effective) field theory with explicit quark and antiquark degrees of freedom it gives the opportunity to compute quark and antiquark distribution functions in the large N_c–limit at a low normalization point μ^2.

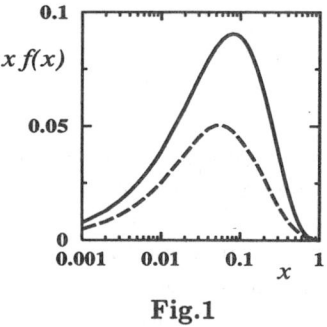
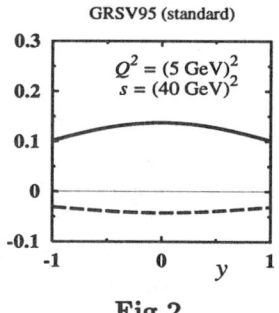

Fig.1 Fig.2

FIGURE 1. The polarized and unpolarized antiquark flavor asymmetries obtained in model calculations in the large–N_c limit (chiral quark–soliton model), evolved (LO) from the low normalization point of $\mu^2 = (600\,\text{MeV})^2$ to a scale of $Q^2 = (5\,\text{GeV})^2$. Dashed line: Unpolarized flavor asymmetry, $x[\bar d(x) - \bar u(x)]$. Solid line: Polarized flavor asymmetry, $x[\Delta\bar u(x) - \Delta\bar d(x)]$.

FIGURE 2. The longitudinal double spin asymmetry in DY pair production through a virtual photon, A_{LL}^γ, in proton–proton collisions, as a function of the rapidity, y. Shown are the results for $s = (40\,\text{GeV})^2, Q^2 = (5\,\text{GeV})^2$ (HERA kinematics). At RHIC, $s = (200\,\text{GeV})^2, Q^2 \sim (20\,\text{GeV})^2$, the situation is very similar. Dashed line: Result obtained using the GRSV95 LO parameterizations [11] as such. Solid line: Result obtained using $(\Delta u + \Delta\bar u)(x)$, $(\Delta d + \Delta\bar d)(x)$, $(\Delta s + \Delta\bar s)(x)$ and $(\Delta\bar u + \Delta\bar d + \Delta\bar s)(x)$ from [11] and $(\Delta\bar u - \Delta\bar d)(x)$ and $(\Delta\bar u + \Delta\bar d - 2\Delta\bar s)(x)$ from χQSM, see [14] for details. The unpolarized distributions are from GRV LO [10] in both cases.

In the model one has to distinguish between large and small distributions

$$\text{large: } \left\{(u+d)(x), (\Delta u - \Delta d)(x)\right\} = N_c^2 f(N_c x)$$
$$\text{small: } \left\{(u-d)(x), (\Delta u + \Delta d)(x)\right\} = N_c\, g(N_c x) \qquad (3)$$

with $f(y)$ and $g(y)$ being stable functions in the limit $N_c \to \infty$. Antiquark distributions exhibit the same large N_c-behavior. It is a non-trivial fact that the large N_c behavior of quantities in the model coincides exactly with the large N_c-counting of QCD. From eq.(3) one immediately concludes that in the large N_c-limit

$$|\Delta \bar{u}(x) - \Delta \bar{d}(x)| = \mathcal{O}(N_c^2) \;>\; |\bar{u}(x) - \bar{d}(x)| = \mathcal{O}(N_c) \,. \qquad (4)$$

The second non-trivial fact to be stressed is that the rather academical conclusion of eq.(4) indeed is supported by practical calculations with finite $N_c = 3$, see fig.(1). This conclusion has been drawn in a similar model [8] and in a statistical model [9].

Measuring polarized antiquark flavor asymmetry

If the flavor asymmetry is really as large as predicted by the χQSM one may wonder why it has not been observed yet. The main source of information on polarized parton distributions is the inclusive polarized deep-inelastic electron-nucleon scattering. From those experiments one can deduce only $(\Delta q + \Delta \bar{q})(x)$. Forced by lack of information it was assumed $\Delta \bar{u}(x) = \Delta \bar{d}(x)$ in parameterizations of polarized parton distributions, e.g. [11].

Only recently spin asymmetries in charged hadron production have been measured by the HERMES Collaboration [12] which in principle – presuming the knowledge of the involved fragmentation functions – allows to disentangle $\Delta \bar{u}(x)$ and $\Delta \bar{d}(x)$. However in the analysis of ref. [12] constraints had to have been imposed on the polarized antiquark distributions in order to improve statistical significance. Two choices have been made i) $\Delta \bar{u}(x) = \Delta \bar{d}(x)$ and ii) $\frac{\Delta \bar{u}(x)}{\bar{u}(x)} = \frac{\Delta \bar{d}(x)}{\bar{d}(x)}$. The result of the analysis with constraint ii) was a polarized antiquark flavor asymmetry consistent with zero. This however is not surprising because the second constraint allows only for a rather small flavor asymmetry in the polarized quark sea due to $\frac{\Delta \bar{u}(x)}{\bar{u}(x)} = \frac{\Delta \bar{d}(x)}{\bar{d}(x)} \Leftrightarrow (\Delta \bar{u} - \Delta \bar{d})(x) = (\bar{u} - \bar{d})(x)\, R(x)$ where the ratio $R(x) \equiv \frac{(\Delta \bar{u} + \Delta \bar{d})(x)}{(\bar{u} + \bar{d})(x)}$ is bound $|R(x)| \le 1$ since $|\Delta \bar{q}(x)| \le \bar{q}(x)$. Thus the second constraint contradicts both, the large N_c-counting and the numerical result found in the χQSM, eq.(4). On the other hand both scenarios – the one based on the assumption $\Delta \bar{u}(x) = \Delta \bar{d}(x)$ and the one based on the model prediction $\Delta \bar{u}(x) \ne \Delta \bar{d}(x)$ – go through the error bars of present data [5]. Whether $\Delta \bar{u}(x) \ne \Delta \bar{d}(x)$ in nature or not could be clarified by future, more exact semi-inclusive DIS data from HERMES.

A process – which is found to be very sensitive to a possible flavor asymmetry in the polarized sea – is the Drell–Yan lepton pair production in polarized nucleon–nucleon scattering. This process will be investigated at RHIC in near future [13]. The observable is the longitudinal double spin asymmetry in DY pair production through a virtual photon, A_{LL}^γ, which is given in the parton model by

$$A_{LL}^\gamma(y,Q^2,s) = \frac{\sum_a e_a^2 \Delta q_a(x_1) \Delta q_{\bar a}(x_2)}{\sum_a e_a^2 q_a(x_1) q_{\bar a}(x_2)} \quad (5)$$

where s is the center of mass energy of the incoming protons, Q^2 is the invariant mass of the produced muon pair, and y is the rapidity which can be deduced from the kinematics of the muons and which allows to reconstruct the parton momenta $x_{1/2} = (Q^2/s)^{1/2} e^{\pm y}$. Fig.(2) demonstrates the effect of "switching on" the non–zero $\Delta \bar u(x) - \Delta \bar d(x)$ as predicted by the model [14]. Apparently the effect is rather large and will be visible in the experiment.

Conclusion

The chiral quark–soliton model predicts a large flavor asymmetry in the polarized quark sea $\Delta \bar u(x) - \Delta \bar d(x)$ at a low normalization point. This asymmetry is larger than the one observed in the unpolarized sea $\bar u(x) - \bar d(x)$, as it is expected from the large N_c–counting in QCD. The large polarized antiquark flavor asymmetry does not contradict to DIS and semi–inclusive DIS data taken so far. A very promising process to measure $\Delta \bar u(x) - \Delta \bar d(x)$ is the Drell–Yan lepton pair production in polarized nucleon–nucleon scattering.

REFERENCES

1. For a review see e.g. : S. Kumano, Phys. Rept. **303** (1998) 183.
2. A. W. Thomas, Phys. Lett. **B126** (1983) 97.
3. D. I. Diakonov et al. Nucl. Phys. **B480** (1996) 341, and Phys. Rev. **D56** (1997) 4069.
4. P. V. Pobylitsa et al. Phys. Rev. **D59** (1999) 034024.
5. B. Dressler, K. Goeke, M. V. Polyakov and C. Weiss, Eur. Phys. J. **C14** (2000) 147.
6. E. Witten, Nucl. Phys. **B223** (1983) 433.
7. D. Diakonov and V. Y. Petrov, Nucl. Phys. **B272** (1986) 457.
8. M. Wakamatsu and T. Watabe, Phys. Rev. **D62** (2000) 017506.
9. R. S. Bhalerao, hep-ph/0003075.
10. M. Glück, E. Reya and A. Vogt, Z. Phys. **C67** (1995) 433.
11. M. Glück, E. Reya, M. Stratmann and W. Vogelsang, Phys. Rev. **D53** (1996) 4775.
12. K. Ackerstaff *et al.* [HERMES Collaboration], Phys. Lett. **B464** (1999) 123.
13. G. Bunce, N. Saito, J. Soffer and W. Vogelsang, hep-ph/0007218.
14. B. Dressler, K. Goeke, M. V. Polyakov, P. Schweitzer, M. Strikman and C. Weiss, hep-ph/9910464.

A Precise Measurement of the g_2 Structure Function of the Proton and Deuteron

Dustin E. McNulty* representing the E155x Collaboration

University of Virginia

Abstract. A precision measurement of the deep inelastic polarized structure function g_2 of the proton and deuteron has been made by the E155x collaboration in the ranges $0.02 < x < 0.9$ and $1(GeV/c)^2 < Q^2 < 25(GeV/c)^2$. The transverse physics asymmetry ($A_\perp$) was measured at SLAC using 29.2 and 32.3 GeV longitudinally polarized electrons incident on transversely polarized target protons ($^{15}NH_3$) and deuterons (^6LiD); the scattered electrons were detected by three fixed angle spectrometers at 2.75°, 5.5°, and 10.5° from the beam line. The g_2 structure function was extracted using the measured $A_\perp$, an E155 fit to g_1, and the most recent fits to world data on F_2 and R. The errors on g_2 for both proton and deuteron are more than three times smaller than those of the previously existing world data set, thus enabling the data to resolve clearly between g_2^{ww} and zero as well as make distinctions between various models.

INTRODUCTION

The deep inelastic polarized nucleon structure functions g_1 and g_2 parameterize the spin dependent part of the hadronic current tensor $W^{\mu\nu}$ and provide information about polarized parton distributions and correlations inside the nucleon. The structure functions also provide a testing ground for QCD, models of nucleon structure, and sum rules. There is already a significant world data set for g_1, and with the results of E155x, there now exists a statistically significant world data set for g_2.

A measurement of g_2 not only probes the nucleon's longitudinal and transverse parton spin distributions, but is also sensitive to higher twist effects such as quark-gluon couplings [1]. The g_2 structure function can be written

$$g_2(x, Q^2) = g_2^{ww}(x, Q^2) + \overline{g_2}(x, Q^2) \tag{1}$$

where

$$g_2^{WW}(x, Q^2) = -g_1(x, Q^2) + \int_x^1 \frac{dy}{y} g_1(y, Q^2) \tag{2}$$

and

$$\overline{g_2} = -\int_x^1 \frac{\partial}{\partial y}\left(\frac{m}{M}h_T(y,Q^2) + \xi(y,Q^2)\right)\frac{dy}{y} \tag{3}$$

g_2^{ww} is the Wandzura Wilczek expression for g_2 and was derived using certain model assumptions that predicted all higher twist terms (beyond twist-2) could be neglected [2]. In general, however, there are additional contributions to g_2 (represented by $\overline{g_2}$). These contributions (shown in Eq. (3)) include the twist-3 term ξ which is related to quark-gluon correlations and another twist-2 term h_T due to the quarks spin distribution transverse to the nucleon momentum. Because the quark transversity h_T is suppressed by the quark mass ratio ($\frac{m}{M}$), any deviation of g_2 from g_2^{ww} will be primarily from the twist-3 quark-gluon term. We evaluate the twist-3 matrix element d_2, defined as the second moment of $\overline{g_2}$, using a world data fit to $\overline{g_2}(x,Q^2) = g_2(x,Q^2) - g_2^{ww}(x,Q^2)$.

EXPERIMENT

The E155x experiment, which took place at SLAC between February and May 1999, was an inclusive measurement of the perpendicular physics asymmetry $A_\perp$. The asymmetry expressed in terms of experimentally measured or calculated quantities is

$$A_\perp = \frac{C_1}{fP_t f_{RC}}\left(\left(\frac{N_L - N_R}{N_L + N_R}\right)\frac{1}{P_b} + C_2 + A_{EW}\right) + A_{RC} \tag{4}$$

where N_L and N_R are the electron rates for left and right beam helicity–corrected for pair-symmetric contributions, pions misidentified as electrons, and electronics dead time, P_b is the beam polarization (≈ 0.83), P_t is the target polarization (≈ 0.75 for protons and ≈ 0.22 for deuterons), C_1 and C_2 are for nuclear corrections and/or target contamination corrections, f is the target dilution factor, A_{EW} is the electroweak asymmetry ($\approx 8 \times 10^{-5}Q^2$), and f_{RC} and A_{RC} are for radiative corrections.

The electrons were accumulated by three independent fixed angle magnetic spectrometers designed to provide wide x and Q^2 kinematic coverage as well as distinguish the high energy electrons from a myriad of background particles. Each spectrometer employed Cerenkov detectors for particle identification, hodoscopes for tracking information and lead-glass shower counters for energy determination and further particle identification.

RESULTS

The preliminary results for $g_2(x)$ for the proton and deuteron are shown in Figure 1. The data points were extracted from the measured asymmetry using an E155 fit to g_1 [3], the most recent NMC fit to the unpolarized structure function $F_2(x, Q^2)$ [4], and the new SLAC fit to $R(x, Q^2)$ [5] defined as the ratio of longitudinal to transverse photo-absorption cross sections (σ_L/σ_T). The data from each spectrometer yielded consistent results in their kinematic overlap regions; the average Q^2 for the 2.75°, 5.5°, and 10.5° spectrometer was approximately 1, 4, and 11 $(GeV/c)^2$ respectively. The results in Figure 1 have been averaged over the spectrometers and two energies used (29 and 32 GeV). For comparison, g_2^p is shown with E143 data and g_2^d is shown with E155 data. Also plotted with the data are g_2^{ww} calculated at the average kinematics of the data and a selection of nuclear structure model predictions for g_2.

The statistical power of the data clearly distinguishes g_2^p from zero and can rule out one of the bag model calculations. The deuteron result has slightly larger errors

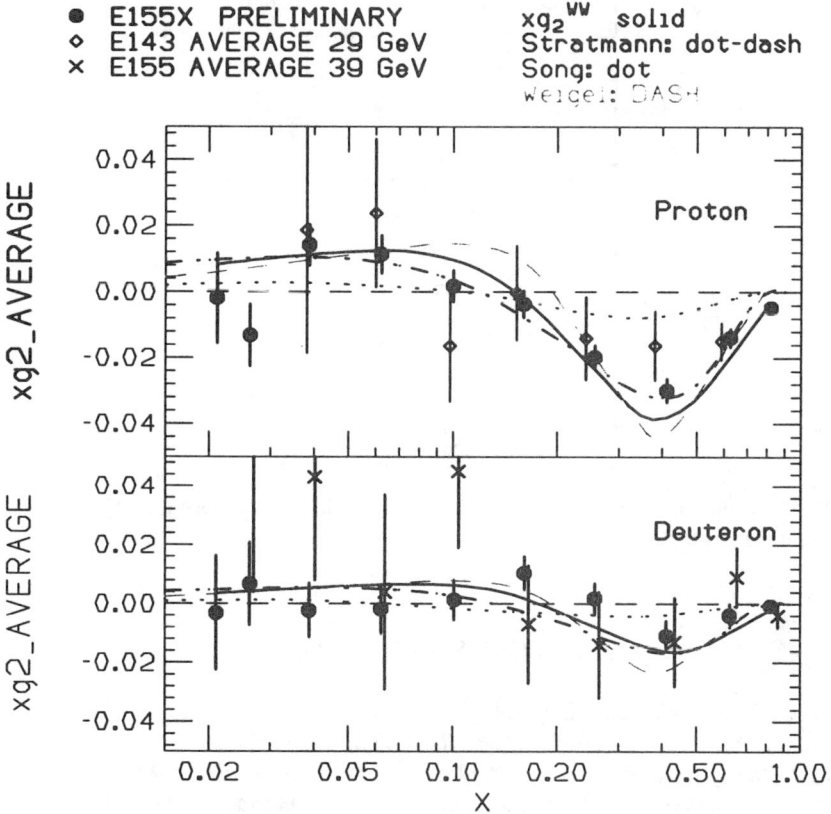

FIGURE 1. The E155x preliminary structure functions xg_2^p (top) and xg_2^d (bottom) averaged over the three spectrometers and two energies shown with E143 and E155 data, g_2^{ww}, the chiral soliton model of Weigel [6], and the bag model calculations of Stratmann [7] and Song [8].

and displays a negative tendency at large x. Both the proton and deuteron results are in qualitative agreement with the chiral soliton calculation of Weigel [6] and the bag model calculation of Stratmann [7]. Also, both data sets exhibit x-dependence similar to g_2^{ww}, although there are small statistical differences at large x possibly indicating that higher twist contributions to g_2 are important.

The preliminary result for the d_2 twist-3 matrix element world data calculation for the proton and neutron are shown in Figure 2 along with the various predictions. The error on this measurement has been improved considerably with the apparent outcome that the twist-3 matrix element is small but non-zero.

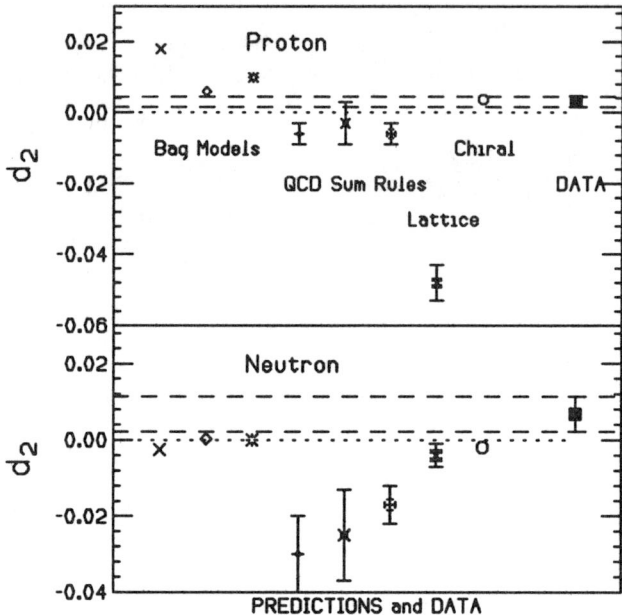

FIGURE 2. Preliminary results for the twist-3 matrix element d_2^p (top) and d_2^n (bottom) shown with theoretical predictions

REFERENCES

1. Shuryak E., and Vainshtein A., *Nucl. Phys.* **B201**, 141 (1982).
2. Wandzura S., and Wilczek F., *Phys. Rev.* **B72**, 195 (1977).
3. E155 collaboration: P. Anthony et al., *Phys. Lett.* **B458**, 529 (1999).
4. NMC collaboration: M. Arneodo et al., *Phys. Lett.* **B364**, 107 (1995).
5. E143 collaboration: K. Abe et al., *Phys. Lett.* **B452**, 194 (1999).
6. Weigel H., Gamberg L., and Reinhart H., *Phys. Rev.* **D55**, 6910 (1997).
7. Stratmann M., *Z. Phys.* **C60**, 763 (1993).
8. Song X., *Phys. Rev.* **D54**, 1955 (1996).

The Q^2–Dependence of the Generalized GDH Integral for the Proton

Björn Seitz* for the HERMES Collaboration

*Department of Physics, University of Alberta, Edmonton, AB, T6G 2J1, Canada
e-mail: bseitz@phys.ualberta.ca

Abstract. The Q^2-dependence of the generalized Gerasimov–Drell–Hearn integral for the proton has been measured in the range $1.2 \text{ GeV}^2 \leq Q^2 \leq 12 \text{ GeV}^2$ and $1 \text{ GeV}^2 \leq W^2 \leq 45 \text{ GeV}^2$ by the HERMES experiment. The contributions of the nucleon-resonance and deep-inelastic regions to this integral have been evaluated separately. At low Q^2 the contributions from the resonance and the deep-inelastic regions are about equal in size, while the latter one dominates for $Q^2 > 3 \text{ GeV}^2$. Since the extrapolation for the unmeasured energy range is small, this measurement provides the first evaluation of the full generalized GDH integral. It shows no significant deviation from a $1/Q^2$ behavior in the measured range and thus no sign of large effects due to either resonance excitations or non-leading twist.

The Gerasimov-Drell-Hearn (GDH) sum rule [1] relates the anomalous contribution κ in the nucleon magnetic moment to an energy-weighted integral of the difference of the nucleon's total spin-dependent absorption cross sections for real photons:

$$\int_0^\infty \frac{\sigma_{1/2}(\nu) - \sigma_{3/2}(\nu)}{\nu} d\nu = -\frac{2\pi^2 \alpha}{M^2} \kappa^2, \qquad (1)$$

where $\sigma_{1/2(3/2)}$ is the photoabsorption cross section for total helicity of the photon-nucleon system equal to $1/2$ ($3/2$), ν is the photon energy and M is the nucleon mass. This sum rule yields a value of -204 μb for the proton.

First experimental data [2] covering the low energy part of Eq. 1 show reasonable agreement with predictions based on recent multipole analysis of single pion photoproduction. These experiments cover the energy region of the lower nucleon resonances. Regge extrapolations predict that sizeable contributions from higher energies and multi-pion photoproduction [3] are needed in order to fulfill the sum rule.

The GDH integral can be generalized to the case of absorption of polarized transverse virtual photons with squared four-momentum $-Q^2$ [4]:

$$I_{GDH}(Q^2) = \int_0^\infty \frac{\sigma_{\frac{1}{2}}(\nu, Q^2) - \sigma_{\frac{3}{2}}(\nu, Q^2)}{\nu} d\nu \qquad (2)$$

$$= 16\pi^2 \alpha \int_0^{x_0} \frac{g_1(x, Q^2) - \gamma^2 g_2(x, Q^2)}{Q^2 \sqrt{1+\gamma^2}} dx$$

$$= \frac{8\pi^2 \alpha}{M} \int_0^{x_0} \frac{A_1(x, Q^2) F_1(x, Q^2)}{K} \frac{dx}{x},$$

where g_1 and g_2 are the polarized structure functions of the nucleon, $\gamma^2 = Q^2/\nu^2 = (2Mx)^2/Q^2$, $x = Q^2/2M\nu$ and $x_0 = Q^2/2M\nu_0$. It should be noted that elastic scattering does not contribute to the integral. The quantity A_1 is the longitudinal asymmetry for virtual photoabsorption, while F_1 is the unpolarized structure function of the nucleon. Note that the Gilman notation [5] for the virtual photon flux factor $K = \nu\sqrt{1+\gamma^2}$ has been used.

The generalization is linked to the first moment of the spin–structure function $g_1(x)$ at large Q^2 [6]. In the deep-inelastic limit $I_{GDH}(Q^2)$ is proportional to the first moment Γ_1 of g_1:

$$I_{GDH}(Q^2, \gamma^2 \to 0) = \frac{16\pi^2 \alpha}{Q^2} \Gamma_1 \equiv \frac{16\pi^2 \alpha}{Q^2} \int_0^1 g_1(x) dx. \qquad (3)$$

This relation yields a small, but positive value for $Q^2 \gg 1$ GeV^2. Therefore a sign change of I_{GDH} at low Q^2 is necessary in order to fulfill both limits. The generalization of the GDH integral to non-zero photon virtuality Q^2 thus provides a way to study the transition from polarized inclusive DIS to polarized real photon photoabsorption, which is dominated by nucleon resonance excitations. Furthermore, this measurement is sensitive to non-leading twist effects at low Q^2.

The data presented are based on a recent analysis of data taken in 1996 and 1997 at the HERMES experiment located at DESY. Details on the spectrometer and the analysis can be found elsewhere [7–9].

The data were divided into six bins in Q^2 in the range 1.2 GeV$^2 \leq Q^2 \leq 12$ GeV^2 and cover an energy range 1 GeV$^2 \leq W^2 \leq 45$ GeV^2, where W^2 is the photon–nucleon invariant mass squared. The data were evaluated in the energy region 1 GeV$^2 \leq W^2 \leq 4.2$ GeV^2 and 4.2 GeV$^2 \leq W^2 \leq 45.0$ GeV^2. These regions define the resonance and DIS regions respectively. The corrections and errors have been evaluated for both region separately. For $W^2 > 45$ GeV^2 an extrapolation following [3] has been used. This extrapolation yields $\approx 3.5 \mu b$ for each Q^2-bin independent of Q^2 and is consistent within 5% with a NLO QCD fit to g_1 [10]. The systematic uncertainties due to detector smearing and resolution were modeled by Monte Carlo simulations. The largest systematic uncertainty stems from the ignorance of A_2 in the resonance region. It was evaluated using three different assumptions ($A_2 = 0$, $A_2 = 0.53 Mx/\sqrt{Q^2}$, $A_2 = 0.06 \pm 16$). The latter one was chosen in the evaluation of the data. The final systematical error for the full integral ranges from 12.4% to 5.8% in the measured Q^2 range and is dominated by the systematical error in the resonance region.

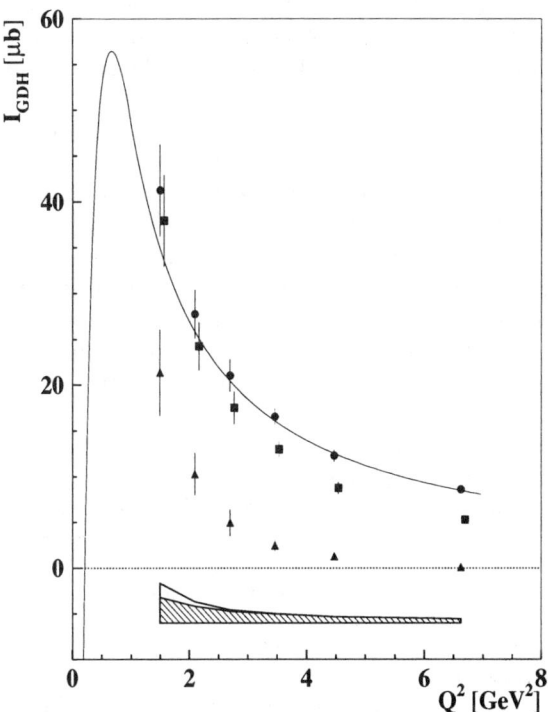

FIGURE 1. I_{GDH} as function of Q^2 for various upper limits of integration: $W^2 \leq 4.2$ GeV^2 (▲), $W^2 \leq 45$ GeV^2 (■), and the full integral I_{GDH} (•). The squares have been slightly shifted to make them more visible. The curve is the Soffer-Teryaev model [11] for the integral I_1. The error bars show the statistical uncertainties, the white and the hatched bands at the bottom represent the systematic uncertainties for the full integral with and without the A_2 uncertainty contribution.

Fig. 1 shows the data taken by HERMES together with a prediction [11] for the integral I_1 based on a Q^2-evolution of the polarized structure functions g_1 and g_2 without considering any explicit nucleon-resonance contribution. The data shown are I_{GDH} for the resonance region ($W^2 \leq 4.2$ GeV^2 (▲)), the whole measured energy range ($W^2 \leq 45$ GeV^2 (■)) and the full integral including the high energy extrapolation (•). The contribution from the resonance region becomes small for $Q^2 > 3$ GeV^2, while the contribution from the DIS region is sizeable throughout the whole measured range.

In the whole energy range, I_{GDH} is consistent within the uncertainties ($\chi^2/N_{df} = 0.4$) with a simple $1/Q^2$ power law. The present results are also in agreement with the deep-inelastic measurements of $\Gamma_1 = 0.120 \pm 0.016$ at $Q^2 = 10$ GeV^2 [13] and $\Gamma_1 = 0.129 \pm 0.010$ at $Q^2 = 5$ GeV^2 [12]. The Q^2-behavior of I_{GDH} suggests that there are no large effects from either resonances or non–leading–twist and indicates that the sign–change of I_{GDH} to meet the real–photon–limit occurs at Q^2 lower than 1.2 GeV^2.

In summary, the Q^2-dependence of the generalized Gerasimov-Drell-Hearn integral for the proton was determined for the first time in both the resonance and the deep-inelastic regions, covering the Q^2-range from 1.2 to 12 GeV^2 and energies up to $W^2 = 45$ GeV^2. The remaining extrapolation to higher energies remains small and constant over the full Q^2–range covered. Even at the lowest measured Q^2 the contribution of the DIS region is of the same order of magnitude as the contribution from the resonance region. The DIS contribution clearly dominates the generalized GDH integral for $Q^2 > 3$ GeV^2. The sizeable contribution from the DIS region at lower Q^2 suggests that this part can not be neglected in the evaluation of the full integral.

The Q^2-dependence of the full GDH integral over the whole measured Q^2 range shows no indication of significant modifications to the $1/Q^2$ evolution of the integral due to resonance excitations or non-leading twist effects.

REFERENCES

1. Gerasimov, S.B. *Sov. J. Nucl. Phys.* **2** 430 (1966); Drell, S.D. and Hearn, A.C. *Phys. Rev. Lett.* **16** 908 (1966).
2. Ahrens, J. et al., *Phys. Rev. Lett.* **84** 5950 (2000) , Thomas, A. *Nucl. Phys.* **B 79** 591c (1999).
3. Bianchi, N. and Thomas, E., *Phys. Lett.* **B 450** 439 (1999).
4. Pantförder, R., *PhD Thesis*, Universität Bonn (1998), BONN-IR-98-06, hep-ph/9805434 and references therein.
5. Gilman, F.J. *Phys. Rev.* **167** 1365 (1968).
6. Anselmino, M., Ioffe, B.L. and Leader, E., *Jad. Fiz.* **49** 214 (1989).
7. HERMES Collaboration, Ackerstaff, K. et al., *Nucl. Instr. and Meth.* **A 417** 230 (1998).
8. HERMES Collaboration, Ackerstaff, K. et al., *Phys. Lett.* **B 444** 531 (1998).
9. HERMES Collaboration, Airapetian, A. et al., DESY 00-096, submitted to *Phys. Lett.* **B**.
10. Glück, M. et al. , *Phys. Rev.* **D 53** 4775 (1996).
11. Soffer, J. and Teryaev, O.V., *Phys. Rev.* **D 51** 25 (1995); Soffer, J. and Teryaev, O.V., *Phys. Rev. Lett.* **70** 3373 (1993).
12. E143 Collaboration, Abe, K. et al., *Phys. Rev.* **D 58** 112003 (1998).
13. SMC Collaboration, Adeva, B. et al., *Phys. Rev.* **D 58** 112001 (1998).

First double polarization measurements towards the test of the Gerasimov-Drell-Hearn sum rule

W. Meyer for the GDH- and A2-Collaborations

Institut für Experimentalphysik der Ruhr-Universität Bochum
Universitätsstr. 150, 44780 Bochum, Germany
E-mail: meyer@ep1.ruhr-uni-bochum.de

Abstract. The helicity dependence of photon induced reactions on the proton has been measured in the energy range from 200 up to 800 MeV for the first time. The experiments have been performed at the Mainz accelerator facility MAMI. A 4π detector system, a circularly polarized tagged photon beam and a newly developed polarized solid target have been used. The data obtained provide new information for multipole analyses of pion photoproduction and give contributions to the Gerasimov-Drell-Hearn (GDH) sum rule and the forward spin polarizability γ_0.

INTRODUCTION

In recent years, the improvements in polarized beam and target techniques have opened new possibilities to investigate the polarization degrees of freedom of the nucleon. In the study of the spin dependent reactions on the nucleon, there is a special interest in the experimental verification of the Gerasimov-Drell-Hearn (GDH) sum rule. This sum rule, derived in the sixties by Gerasimov [1] and independently by Drell and Hearn [2], relates static properties of the nucleon, the anomalous magnetic moment κ and the mass m, to the difference of the total photoabsorption cross sections for circularly polarized photons on longitudinally polarized nucleons. It is written as:

$$\int_{m_\pi}^{\infty} \frac{\sigma_{3/2} - \sigma_{1/2}}{\nu} = \frac{2\pi^2 \alpha}{m^2} \kappa^2 = 204 \mu b \quad (1)$$

where $\sigma_{3/2}$ and $\sigma_{1/2}$ are the photoabsorption cross sections for the total helicity states 3/2 and 1/2 respectively (parallel or antiparallel photon-nucleon spin configuration) and is the photon energy. The left-hand side of eq. 1 is called GDH-integral. In a similar way, the so called forward spin polarizability γ_0 can also be obtained:

$$\gamma_0 = -\frac{1}{4\pi^2} \int_{m_\pi}^{\infty} \frac{\sigma_{3/2} - \sigma_{1/2}}{\nu^3} d\nu \quad (2)$$

The GDH sum rule is based on basic physics principles (Lorentz and gauge invariance, unitarity, causality) applied to the forward Compton scattering amplitude. The only questionable assumption made is that this amplitude becomes spin independent at infinite photon energies. Due to its fundamental character, this prediction provides an excellent test of nucleon spin structure. Due to the technical difficulties associated to this kind of experiments, the GDH sum rule has not been verified previously. Without any direct experimental results, some theoretical evaluations on the GDH integral were made using multipole analyses of (mainly unpolarized) single pion photoproduction data, pion scattering data, and a crude model for double pion photoproduction [3-6]. All models underestimate substantially the sum rule prediction ($204 \mu b$) for the proton.

The measurements have been started at MAMI (Mainz) covering the energy range from the pion threshold up to 800 MeV and will be continued at ELSA (Bonn) up to a photon energy of 3.0 GeV. In this report the experimental set up at Mainz and first preliminary results will be presented.

I EXPERIMENTAL SETUP

The experiment was carried out at the tagged photon facility of MAMI. Circularly polarized photons were produced by bremsstrahlung of longitudinally polarized electrons. The source of polarized electrons, based on the photoeffect on strained GaAs crystals, delivered routinely electrons with a degree of polarization of about 75 % or higher. The degree of polarization was continuously measured during the whole experiment by Møller scattering in a magnetized vacuflux foil. Both electrons were detected in coincidence in the tagging spectrometer. A dedicated trigger ensured that the sum of the electron energies matched the beam energy. Polarized nucleons were available in the frozen spin target [7], that was built and operated by the groups of Bonn, Bochum and Nagoya. The system consists of a horizontal dilution refrigerator and a superconducting polarization magnet, which was used in the polarization phase together with a microwave system for dynamical nuclear polarization (DNP). During the measurement the polarization was maintained in the "frozen spin" mode at temperatures of about 50 mK by an internal superconducting coil ($B \approx 0.4$ T) which is integrated into the dilution refrigerator. The target material was butanol (C_4H_9OH). At 2.5 T maximum polarization values close to 90 % were obtained for the protons with a typical relaxation time in the "frozen spin" mode of about 200 hours. The holding field was homogeneous enough to allow for continuous NMR monitoring of the target polarization during the experiment.

The photon induced reaction products were registered by means of the large acceptance detector DAPHNE [8], built by CEA Saclay and INFN - Sezione di Pavia, which was complemented by forward detectors to increase the solid angle acceptance (see fig. 1). DAPHNE is mainly a charged particle tracking detector with cylindrical symmetry. It consists of 3 coaxial multi-wire proportional cham-

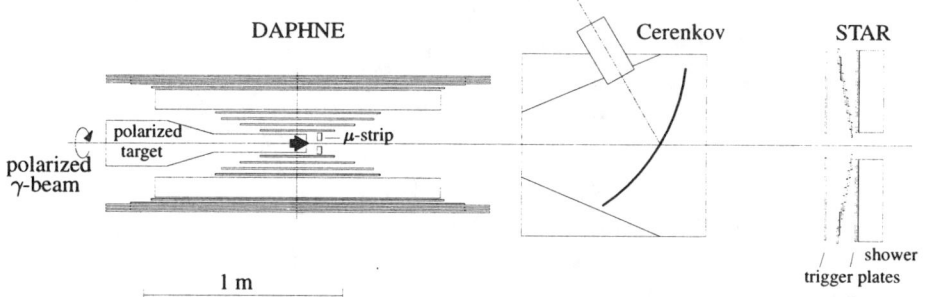

FIGURE 1. Schematic side view of the experimental setup of the GDH experiment in Mainz

bers with cathode readout, surrounded by 16 segments of $\Delta E - E - \Delta E$ plastic scintillator telescopes and by a double scintillator-converter sandwich which allows the detection of neutral pions with a useful efficiency. The silicon microstrip detector MIDAS [9] covers the angular range down to 7 degrees, the aerogel Čerenkov counter serves for online suppression of electrons and positrons, extreme forward angles are covered by the annular ring detector STAR [10] and a lead-scintillator sandwich counter.

II PRELIMINARY RESULTS OF THE TOTAL CROSS SECTION

The identification methods for hadrons have been described in detail previously [11,12]. Using an inclusive analysis method, based on the detection of charged hadrons (p, $\pi^{\pm}$) and neutral pions, the helicity difference of the total photoabsorption cross sections $\sigma_{3/2} - \sigma_{1/2}$ from 200 to 800 MeV has been determined, see fig. 2 (left). This difference starts out negative due to the $E_{0\pi+}^{n+}$ multipole and has, as expected, a strong positive contribution due to the M_{1+} multipole. Above the resonance peak a sizeable contribution of the double pion photoproduction processes can clearly be seen. The curves shown for comparison are from SAID [13] (solution SM99K) and from HDT [14] which both include the single pion production processes only.

Although the data available up to now are still preliminary, we can use them to determine the contribution to the integrals of eqs. 1 and 2. The integration of our data from 200 to 800 MeV yields (228±6) μb and determines the dominant contribution to the GDH integral, see fig. 2 (right). The missing low energy contribution below 200 MeV according to HDT is -27 μb, the high energy contribution above 800 MeV given by single pion photoproduction according to SAID is 25 μb. The integration of our data according to eq. 2 in the energy range from 200 to 800 MeV yields $(-1.68 \pm .7) \cdot 10^{-4} fm^4$. Due to the strong energy weight-

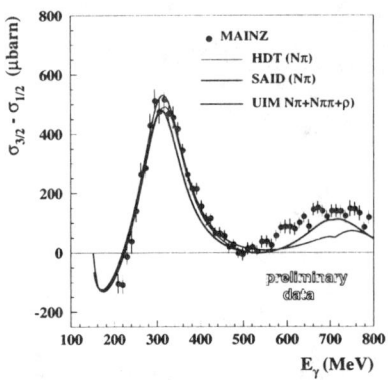 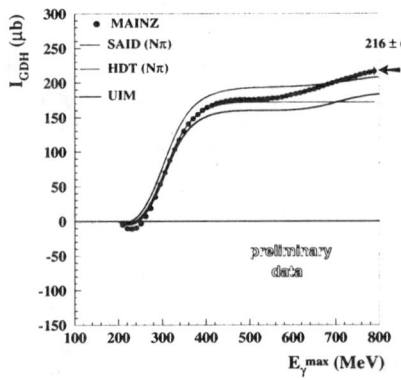

FIGURE 2. Left:Difference of helicity dependent total cross sections $\sigma_{3/2} - \sigma_{1/2}$ on the proton compared to model predictions. Right: GDH integral starting at 200 MeV as a function of the upper integration limit

ing ($\propto \nu^{-3}$) the high energy single pion and double pion processes are expected to contribute very little to the spin polarizability. HDT yields $1.0 \cdot 10^{-4} fm^4$ for the missing contribution below 200 MeV resulting in a preliminary value for γ_0 of about $(-.7 \pm .07) \cdot 10^{-4} fm^4$. This value is significantly smaller than the values extracted from SAID [6] ($-1.34 \cdot 10^{-4} fm^4$) and the result from chiral perturbation theory [15] ($-1.5 \cdot 10^{-4} fm^4$). It is an better agreement with the recent dispersion theoretical result of ref. [16] ($-.80 \cdot 10^{-4} fm^4$).

REFERENCES

1. S.B. Gerasimov, Sov. J. Nucl. Phys.2 (1966) 430
2. S.D. Drell, A.C. Hearn, Phys. Rev. Lett. 16 (1966) 908
3. I. Karliner, Phys. Rev. D 7 (1973) 2717
4. R.L. Workman, R.A. Arndt, Phys. Rev. C 53(1996) 430
5. V. Burkert, Z. Li, Phys. Rev. D 47 (1993) 46
6. A.M. Sandorfi et al., Phys. Rev. D 50 (1994) 11
7. C. Bradtke et al., Nucl. Instr. Meth. A 346 (1999) 430
8. G. Audit et al., Nuc. Instr. Meth A 301 (1991) 473
9. S. Altieri et al., INFN Report TC-98/30 (1998)
10. M. Sauer et al., Nucl. Instr. Meth. A 378 (1996) 143
11. M. MacCormick et al., Phys. Rev. C 55 (1996) 41
12. R. Crawford et al., Nucl. Phys. A 603 (1996) 303
13. R.A. Arndt et al., Phys. Rev. C 53 (1996) 430 (solution SM99K)
14. O.Hanstein et al., Nucl. Phys. A 632 (1999) 561
15. V. Bernard et al., Nucl. Phys. A 388 (1992) 315
16. D. Drechsel et al., Phys. Rev. C 61 (1999) 015204

Transversity Quark Distribution Function in the Large N_c–Limit

P. Schweitzer*, D. Urbano*‡, K. Goeke*, P.V. Pobylitsa*†,
M.V. Polyakov*† and C. Weiss*

*Institut für Theoretische Physik II, Ruhr-Universität Bochum, D-44780 Bochum, Germany
†Petersburg Nuclear Physics Institute, Gatchina, St. Petersburg 188350, Russia
‡Faculdade de Engenharia da Universidade do Porto, 4000 Porto, Portugal

Abstract. The chiral quark–soliton model, which is based on the large N_c–limit, reveals a transversity quark distribution $\delta q(x)$ significantly different from the helicity distribution $\Delta q(x)$ already at a low normalization point, in contrast to the prediction $\delta q(x) \equiv \Delta q(x)$ of the non–relativistic quark model. An estimate for the double spin asymmetry in dilepton production from scattering of transversely polarized protons is given.

Introduction

There are promising prospects of measuring soon the "last unknown" twist two parton distribution in the nucleon – the transversity distribution $\delta q(x)$ – at RHIC [1]. The transversity distribution has not been measured yet. Due to its chiral–odd nature it can be measured only in connection with another chirally odd distribution or fragmentation function. What is known about $\delta q(x)$ are its evolution with increasing Q^2 [2], first moments computed in the sum rule approach [3] or in lattice QCD [4] and the prediction of the non–relativistic quark model that the transversity and the helicity distribution coincide at a low scale. The calculation of $\delta q(x)$ in the large N_c–limit in the framework of the chiral quark–soliton model (χQSM) shows that this is not the case.

Transversity distribution function $\delta q(x)$ in the χQSM

The χQSM allows to compute consistently quark and antiquark distribution functions in the large N_c–limit at a low normalization point $\mu^2 \simeq (600\,\text{MeV})^2$ [5,6]. The effective Lagrangian is valid for low momenta $|\mathbf{k}| \ll 600\,\text{MeV}$ and contains explicit quark and antiquark degrees of freedom and Goldstone meson (pion) degrees of freedom. The nucleon emerges for $N_c \to \infty$ as a soliton of the classical pion–field. There are *no* adjustable parameters in the effective theory. In the framework of the model twist two quark and antiquark distributions can be computed in an effective series in $1/N_c$ which divides distributions into large and small ones. The transversity distribution has the same large N_c-behavior like the helicity distribution

$$\text{large: } \left\{ (\delta u - \delta d)(x), (\Delta u - \Delta d)(x) \right\} = N_c^2 g(N_c x)$$
$$\text{small: } \left\{ (\delta u + \delta d)(x), (\Delta u + \Delta d)(x) \right\} = N_c \tilde{g}(N_c x) \tag{1}$$

where $g(y)$ and $\tilde{g}(y)$ are stable functions for $N_c \to \infty$. The large N_c-behavior eq.(1) corresponds exactly to the large N_c-counting in QCD. In fig.(1) isovector and isoscalar transversity distribution, respectively, are compared to the corresponding helicity distributions. Clearly $\delta q(x, \mu^2) \neq \Delta q(x, \mu^2)$ even at $\mu^2 \simeq (600\,\text{MeV})^2$ [6,7]. Similar calculations have been performed in [8,9] with the same conclusion.

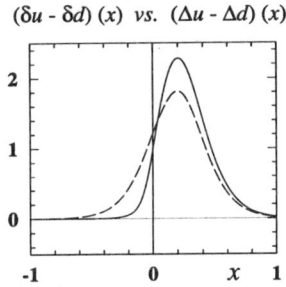 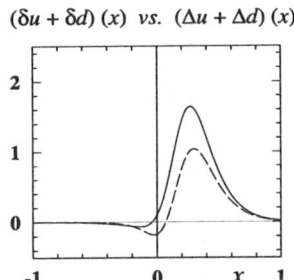

FIGURE 1. Left hand side: The isovector transversity distribution and the isovector helicity distribution at a low normalization point $\mu^2 \simeq (600\,\text{MeV})^2$ in the large N_c-limit, obtained in the framework of the χQSM, versus Bjorken-x. Both distributions are large, i.e. of order $\mathcal{O}(N_c^2)$. Solid line: $(\delta u - \delta d)(x)$ for $x > 0$ and $(-1)(\delta \bar{u} - \delta \bar{d})(-x)$ for $x < 0$. Dashed line: $(\Delta u - \Delta d)(x)$ for $x > 0$ and $(\Delta \bar{u} - \Delta \bar{d})(-x)$ for $x < 0$. **Right hand side:** The same as in the left figure, however for the isoscalar case. The isoscalar distributions are of the order $\mathcal{O}(N_c)$.

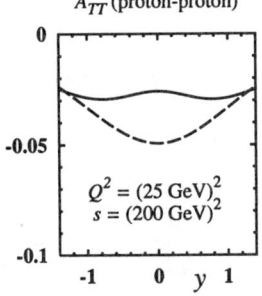

FIGURE 2. The transverse spin asymmetry A_{TT} in the kinematical region $s = (200\,\text{GeV})^2$ and $Q^2 = (25\,\text{GeV})^2$ as it will be available at RHIC. Solid line: A_{TT} computed with $\delta q(x)$ obtained in the large N_c-limit and LO-evolved from the low scale $\mu^2 = (600\,\text{MeV})^2$ to Q^2. Dashed line: The asymmetry computed with $\delta q(x)$ identified with $\Delta q(x)$ (from GRSV LO [12] at the initial scale) and then evolved to Q^2. The numerator is taken in both cases from GRV LO [11].

Measurement of the transversity distribution

Recently azimuthal asymmetries in semi–inclusive hadron production on longitudinally (HERMES) and transversely (SMC) polarized targets have been measured. In those observables essentially $\delta q(x)$ enters in connection with a T-odd and chirally odd fragmentation function. The latter however is only vaguely known, but in principle $\delta q(x)$ could be measured in this way [10]. Alternatively $\delta q(x)$ can be measured in the process of Drell-Yan muon pair production with transversely polarized protons. The corresponding observable is the transverse spin asymmetry A_{TT} [1], which is given in the parton model by

$$A_{TT}(y, Q^2, s) = \frac{\sum_a e_a^2 \delta q_a(x_1) \delta q_{\bar{a}}(x_2)}{\sum_a e_a^2 q_a(x_1) q_{\bar{a}}(x_2)} \tag{2}$$

where s is the center of mass energy of the incoming protons, Q^2 is the invariant dimuon mass. From the kinematics of the muons the rapidity y and from the latter the parton momenta $x_{1/2} = (Q^2/s)^{1/2} e^{\pm y}$ can be reconstructed. Fig.(2) shows the difference between A_{TT} as it has been often estimated assuming $\delta q(x) = \Delta q(x)$ and the A_{TT} as it is predicted by the large N_c–limit.

Conclusion

In the large N_c–limit the transversity quark distribution appears quite different from the helicity distribution alrcady at a low scale. The Drell-Yan process (RHIC) and also semi–inclusive DIS (HERMES) – presuming good knowledge of the involved fragmentation functions – are suitable processes to measure $\delta q(x)$.

REFERENCES

1. G. Bunce, N. Saito, J. Soffer and W. Vogelsang, hep-ph/0007218.
2. X. Artru and M. Mekhfi, Z. Phys. **C45** (1990) 669.
3. B. L. Ioffe and A. Khodjamirian, Phys. Rev. **D51** (1995) 3373.
4. S. Aoki, M. Doui, T. Hatsuda and Y. Kuramashi, Phys. Rev. **D56** (1997) 433.
5. D. I. Diakonov et al. Nucl. Phys. **B480** (1996) 341 and Phys. Rev. **D56** (1997) 4069.
6. P.V. Pobylitsa, M.V. Polyakov, Phys.Lett. B389 (1996) 350-357
7. K. Goeke, P. Pobylitsa, M. Polyakov, P. Schweitzer, D. Urbano, C. Weiss, in prep.
8. L. Gamberg, H. Reinhardt and H. Weigel, Phys. Rev. **D58** (1998) 054014.
9. M. Wakamatsu and T. Kubota, Phys. Rev. **D60** (1999) 034020.
10. A. V. Efremov et al. Phys. Lett. **B478** (2000) 94.
11. M. Glück, E. Reya and A. Vogt, Z. Phys. **C67** (1995) 433.
12. M. Glück, E. Reya, M. Stratmann and W. Vogelsang, Phys. Rev. **D53** (1996) 4775

Spin-Flipping a Stored Polarized Proton Beam with an rf Dipole*

B.B. Blinov[†], Ya.S. Derbenev, T. Kageya, D.Yu. Kantsyrev[‡],
A.D. Krisch, V.S. Morozov[‡], D.W. Sivers[§] and V.K. Wong

*Randall Laboratory of Physics, University of Michigan,
Ann Arbor, Michigan 48109-1120*

V.A. Anferov, P. Schwandt and B. von Przewoski

Indiana University Cyclotron Facility Bloomington, Indiana 47408-0768

Abstract. Frequent polarization reversals, or spin-flips, of a stored polarized high-energy beam may greatly reduce systematic errors of spin asymmetry measurements in a scattering asymmetry experiment. We studied the spin-flipping of a 120 MeV horizontally-polarized proton beam stored in the IUCF Cooler Ring by ramping an rf-dipole magnet's frequency through an rf-induced depolarizing resonance in the presence of a nearly-full Siberian snake. After optimizing the frequency ramp parameters, we used multiple spin-flips to measure a spin-flip efficiency of 86.5±0.5%. The spin-flip efficiency was apparently limited by the rf-dipole's field strength. This result indicates that an efficient spin-flipping a stored polarized beam should be possible in high energy rings such as RHIC and HERA where Siberian snakes are certainly needed and only dipole rf-flipper-magnets are practical.

INTRODUCTION

In a circular accelerator, each proton's spin precesses around the vertical magnetic fields of the ring's bending dipoles, with a frequency called the spin precession frequency f_s that is related to the proton's circulation frequency f_c by:

$$f_s = f_c \nu_s, \qquad (1)$$

where ν_s is the spin tune, which is the number of spin precessions during each turn around the ring.

[*)] Supported by grants from the U.S. Department of Energy and National Science Foundation.
[†)] Corresponding author.
[‡)] also at Moscow State University, Moscow, Russia
[§)] also at Portland Physics Institute, Portland, Oregon 97201.

If there are only vertical magnetic fields, then the vertical beam polarization remains unchanged; however, whenever there is a periodic horizontal magnetic field whose tune is equal to the spin tune a depolarizing resonance occurs, which can destroy the polarization. Since the ν_s is proportional to the proton's energy via:

$$\nu_s = G\gamma, \qquad (2)$$

where G=1.792847 is the proton's anomalous magnetic moment, and γ is the Lorentz energy factor, the protons will encounter many depolarizing resonances as they are accelerated to a high energy.

An elegant method to overcome these depolarizing resonances, proposed by Derbenev and Kondratenko in 1978 [1], involves using a spin rotator called a Siberian snake that makes the spin tune equal to 1/2 independent of the beam energy. This energy-independent ν_s eliminates most of the depolarizing resonances.

Once a polarized proton beam is accelerated to a high energy and stored, it is important to be able to reverse its polarization direction in order to reduce the systematic error in various polarized scattering asymmetry experiments.

Studies at the IUCF Cooler Ring show that these polarization reversals (spin-flips) can be done using an rf magnet [2,3], which creates an rf depolarizing resonance:

$$f_s = f_{rf} + nf_c, \qquad (3)$$

where f_{rf} is the rf magnet's frequency, f_c is the circulation frequency, and n is an integer. After crossing such a resonance, which is varying the rf magnet's frequency from a value below the resonance to a value above the resonance, the final beam polarization P_f is related to the initial beam polarization P_i via the Froissart-Stora formula [4]:

$$P_f = P_i[2e^{-\frac{(\pi \epsilon f_c)^2}{(\Delta f/\Delta t)}} - 1], \qquad (4)$$

where ϵ is the resonance strength, and $\Delta f/\Delta t$ is the resonance crossing rate, while Δf is the frequency's range during its linear variation time Δt. If the resonance is sufficiently strong and/or the crossing rate is sufficiently slow, the final polarization is reversed with respect to the initial polarization, while its absolute value is the same; this is called spin flip.

I EXPERIMENTAL APPARATUS

The apparatus used for this experiment is shown in Fig. 1. The 120 MeV horizontally polarized proton beam in the Cooler Ring was obtained using the new Cooler Injector Polarized IOn Source (CIPIOS) [5] and the new Cooler Injection Synchrotron (CIS) [6]. Also shown are the Siberian snake solenoid with its eight correction quadrupoles, the polarimeter and the rf-dipole magnet, which is one of the Cooler Ring's injection kicker dipoles. In order to increase the strength of the rf-dipole, we connected to it a variable capacitor to form a resonant parallel LC-circuit. The maximum r.m.s. voltage on the rf-dipole that we were able to reach was about 71 V; this corresponded to an $\int B \cdot dl$ of about 0.06 T·mm.

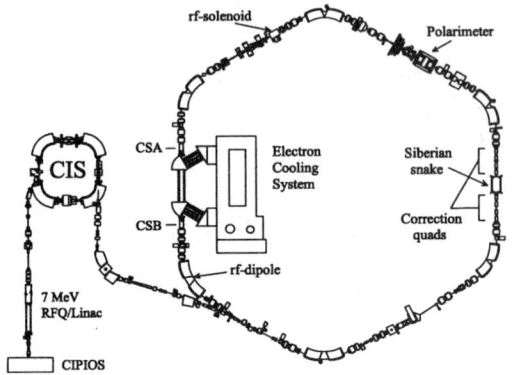

FIGURE 1. IUCF Cooler Ring and its hardware.

II RESULTS AND ANALYSIS

We first determined the rf depolarizing resonance's location by measuring the radial beam polarization at different rf-dipole's frequencies; this measured radial polarization is plotted against the rf-dipole's frequency in Fig. 2 [7]. The curve is a second-order Lorentzian fit to the data with a resonance frequency of 2 384 040±90 Hz and a width 970±20 Hz.

To study spin-flipping, we first set our frequency ramp range Δf to 10 kHz

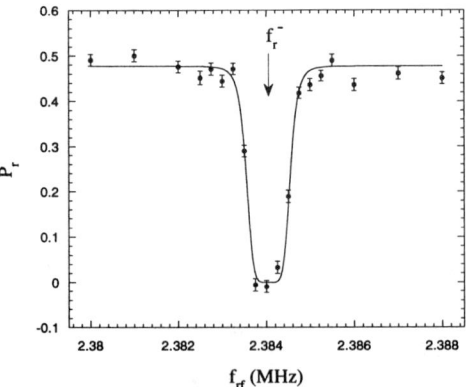

FIGURE 2. RF depolarizing resonance. The arrow shows the resonance's central frequency.

and crossed this rf-induced resonance by linearly ramping the rf-dipole's frequency from $f_r - 5$ to $f_r + 5$ kHz, with various ramp times Δt, while measuring the beam polarization after each crossing. Thus, we determined the optimal ramp time for the most efficient spin-flip. We then tried to further increase the spin-flip efficiency by varying the rf-dipole's frequency range Δf. After setting Δt and Δf to maximize the spin-flip efficiency, we more precisely determined this efficiency by performing many spin-flips. The measured radial beam polarization after several spin-flips is plotted against the number of spin-flips in Fig. 3. We fit this data using:

$$P_n = P_i \cdot \eta^n, \tag{5}$$

where P_n is the measured radial beam polarization after n spin-flips, P_i is the initial polarization, η is the spin-flip efficiency, and n is the number of spin-flips. The best fit gave a spin-flip efficiency of $86.4 \pm 0.5\%$ [7]. The spin-flip efficiency was apparently limited by the strength of the rf-dipole's field.

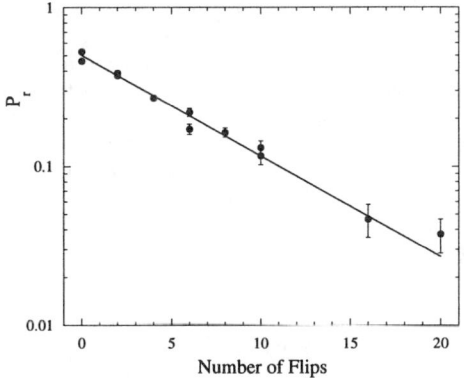

FIGURE 3. Multiple spin-flipping.

In conclusion, we have successfully demonstrated the feasibility of an rf-dipole magnet as a spin-flipper in a storage ring where a full Siberian snake is present. We are currently developing an rf-dipole spin-flipper for the MIT-Bates polarized electron ring.

REFERENCES

1. Ya. S. Derbenev and A. M. Kondratenko, Part. Accel. **8**, 115 (1978).
2. B.B. Blinov et al., Phys. Rev. Lett. **81**, 2906 (1998).
3. V. A. Anferov et al. Phys. Rev. ST-AB **3**, 041001 (2000).
4. M. Froissart and R. Stora, Nucl. Instrum. and Methods **7**, 297 (1960).
5. V.P. Derenchuk and A.S. Belov, in AIP Conf. Proc. **421**, 422 (1998).
6. D.L. Friesel and S.Y. Lee, in Proc. of 1997 Particle Accelerator Conf (PAC-97), Vancouver, 935 (1998).
7. B.B. Blinov et al., Submitted to Phys. Rev. Lett. June 2000.

The HERMES Internal Polarized Deuterium Gas Target

Mark Henoch[†]

on behalf of the HERMES collaboration

[†]*Universität Erlangen-Nürnberg, 91058 Erlangen, Germany*[1]

Abstract. The HERMES experiment (HERa MEasurement of Spin) at DESY uses the HERA polarized lepton beam in combination with the polarized internal target technique in order to determine the spin dependent structure functions of the proton and neutron via deep-inelastic scattering. The HERMES target consists of a storage cell internal to the HERA lepton ring in which polarized deuterium atoms are injected from an atomic beam source (ABS). A sample beam is extracted from the center of the storage cell to analyze the target gas for atomic fraction, hyperfine occupation and atomic polarization using a target gas analyzer (TGA) and a prototype Breit-Rabi-Polarimeter (BRP). The improvement of the target performance with deuterium gas in 2000 will be presented.

INTRODUCTION

The HERMES experiment in the HERA lepton storage ring at DESY is studying the spin structure of the proton and neutron via deep inelastic scattering of polarized leptons off the nucleons in a highly polarized internal gas target. The target was operated with hydrogen gas in 1996 and 1997 and has been modified to use deuterium gas since 1998. A schematic diagram of this target, which uses a storage cell to obtain a high target density, is shown in Figure 1. The Breit-Rabi diagram for deuterium can be seen in its lower left corner.

A beam of deuterium atoms is generated in a surfatron based microwave dissociator [1] which forms part of the atomic beam source (ABS). These atoms are electron polarized by means of Stern-Gerlach spin separation in the sextupole magnet system.

The electron polarization is transfered to the nucleons by means of adiabatic high frequency transitions which interchange the occupations of two different hyperfine states in the atomic beam [2]. The beam of nuclear polarized atoms is injected into the centre of the thin-walled aluminum storage cell via a side tube and the atoms

---

[1] The work is supported by BMBF Germany, 056MU22I(1) and 057ER12P(2).

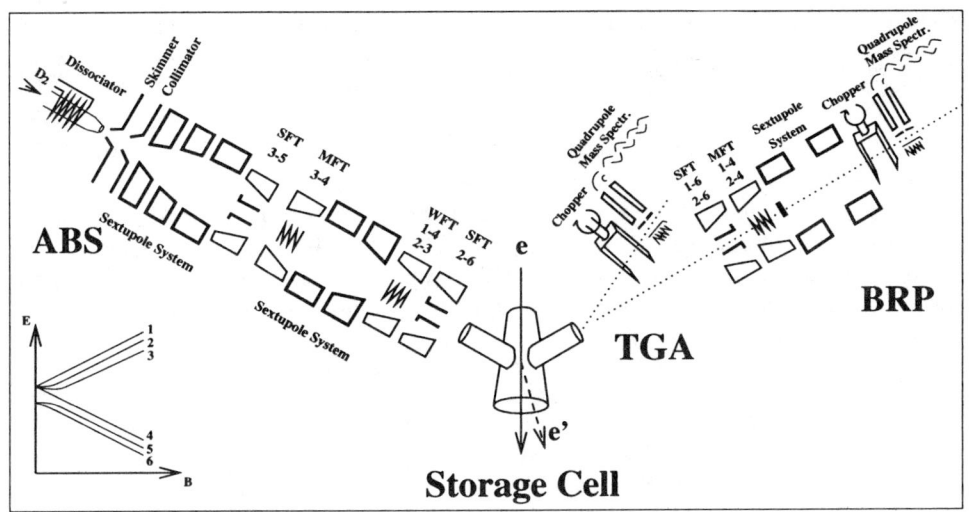

FIGURE 1. Schematic of the HERMES target.

then diffuse to the open ends of the cell where they are removed by a high speed pumping system. In order to minimize wall interaction effects with the polarized gas the cell is coated with Drifilm [3]. A longitudinal magnetic field provides a quantisation axis for the spins and inhibits nuclear spin relaxation by decoupling nucleon and electron spins. The field is produced by 4 coils of a super conducting magnet resulting in a non-uniformity of $\frac{\Delta B}{B} = \pm 1.5\%$ along the cell axis at the working point of $B = 335$ mT.

A second side tube is provided to sample the gas within the target cell. The beam emerging from this tube is analyzed with a Breit-Rabi polarimeter (BRP) to measure its hyperfine occupation and calculate the atomic polarization and a target gas analyzer (TGA) to determine its atomic fraction [4]. During the atom diffusion process relaxation by wall and spin exchange collisions and wall recombination can change the polarization and the atomic fraction of the target gas. The atomic polarization and atomic fraction values measured by the BRP and TGA must be corrected for these effects to obtain the average target polarization. All numbers quoted below refer to the 2000 data taking period with a polarized deuterium target.

THE RUNNING CONDITIONS

Due to improvements to the system the HERMES deuterium target in 2000 was running very stable at doubled atomic density compared to hydrogen in 1996/1997. The cell size has been reduced and its temperature was decreased to 60 K. The RF dissociator used for hydrogen was replaced by a surfatron based microwave dissociator, which can achieve significantly higher throughputs at the same degree of dissociation. The magnetic holding field was operated at $B = 335$ mT which results in a strong decoupling of electron and nucleon spins in the deuterium atom with a decoupling factor of $x = (B/B_c^D) = 28.6$, where $B_c^D = 11.7$ mT is the critical field for deuterium (the decoupling of hydrogen with the same holding field was a factor of ~ 4.33 lower, as the critical field for hydrogen is $B_c^H = 50.7$ mT). The average areal target density was $D_T = 2.6 \times 10^{14}$ nucl. cm^{-2}.

The target polarization was in standard operation randomly flipped between P_z^- (injecting hyperfine states $|3> + |4>$) and P_z^+ ($|1> + |6>$). For certain studies or special measurements any combination of hyperfine states could be injected, e.g. purely tensor polarized deuterium atoms, which provides an experimental access to the still unmeasured structure function b_1.

THE TARGET POLARIZATION

The target polarization P_T is calculated using the following expression:

$$P_T = \alpha_0(\alpha_r + (1-\alpha_r)\beta)P_T^{atom}, \tag{1}$$

where α_0 is the fraction of nucleons entering the storage cell as atoms, α_r is the fraction of atoms surviving recombination and β is the polarization of nucleons in recombined molecules relative to the polarization of atoms P_T^{atom}. In order to determine the target parameters by the corresponding measurement on the sample beam, so-called sampling corrections c_α and c_P have to be introduced, which are defined by

$$\alpha_r = c_\alpha \alpha_r^{TGA} \tag{2a}$$

$$P_T^{atom} = c_P P^{BRP}. \tag{2b}$$

The sampling corrections in equations (2) depend on the sensitivity of the BRP and TGA to the different parts of the cell and are derived by means of Monte-Carlo simulations investigating the history of gas particles travelling through the storage cell and limiting the possible influence of severe surface non-uniformity inside the cell on the average target polarization [5].

The average α_0 was determined to be 0.935 ± 0.020 (*syst.*) where two sources for molecular contribution are taken into account: the ballistic flow of molecules coming from the ABS and the residual gas from the vacuum chamber around the storage cell. Analysing the temperature dependance of the measured polarization and atomic fraction shows, that the storage cell used in 2000 causes no measurable depolarization and recombination on the drifilm surface. The result for the average atomic fraction surviving recombination, α_r, was 0.995 ± 0.005 (*syst.*) for data taking conditions. The range for β is per definition between 0 (the nucleons lose their polarization during a recombination process completely) and 1 (all nucleons keep their polarization). This large uncertainty does not significantly contribute to the target polarization in equation (1) because $\alpha_r \simeq 1$.

A fit of the measured occupation numbers during a scan of the magnetic holding field using the so-called *rate equation model*, a model based on a combination of well established spin relaxation models and a one-dimensional diffusion equation describing the particle transport through the storage cell [6], allows the determination of the injected polarization for deuterium atoms, which was typically 90.5%, and the relaxation parameters within small uncertainties. The average polarization loss by spin exchange collisions was determined to be 1%. Depolarization on the walls is strongly suppressed by the surface coating and was determined to be less than 0.5%. These numbers result in an average atomic polarization in the target of $P_T^{atom} = 0.89 \pm 0.02 \ (syst.)$.

The preliminary result for the average target polarization during the still ongoing data taking period in 2000 can then be determined using equation (1) with the results given above to $P_T = 0.83 \pm 0.02$ at an average areal target density of $D_T = 2.6 \times 10^{14}$ nucl. cm^{-2}.

CONCLUSIONS

The HERMES target is running reliably with stable operation during the 2000 running period. All planned modifications to the system have been implemented and increased the HERMES event rate by a factor of 2. The knowledge gained during the running with hydrogen in 1996/1997 has been successfully transfered to the deuterium target which allows a fast and reliable analysis of the target data.

During the winter shutdown 2000/2001 the target will be modified and upgraded to start running transversally polarized hydrogen in 2001 with improved statistical precision.

REFERENCES

1. Koch, N.; *PhD. Thesis*; Universität Erlangen-Nürnberg 1999.
2. Abragam, A., Winter, J.M.; *Phys. Rev Lett.* 1, 1958, pp. 374.
3. Thomas, G.E. et al.; *Nucl. Instr. Meth.* A 257, 1987, pp. 32.
4. Braun, B.; *AIP Conf. Proc. no. 421: Pol. Beams and Pol. Gas Targets*; Roy, E., Holt, J., Miller, M.A.; Urbana-Champaign 1997, pp. 156.
5. Henoch, M.; *PST 99: International Workshop on Polarized Sources and Targets*; Gute, A., Lorenz, S., Steffens, E.; Universität Erlangen-Nürnberg 1999, pp. 472
6. Baumgarten, C.; *PhD. Thesis*; Universität München 2000.

Analyzing Power in CNI-region at AGS (Experiment E950)

I.G. Alekseev[*], M. Bai[†], B. Bassalleck[‡], G. Bunce[†1],
A. Deshpande[||], J. Doskow[¶], S. Eilerts[‡], D.E. Fields[‡], Y. Goto[$],
K. Imai[**1], M. Ishihara[$1], V.P. Kanavets[*], K. Kurita[$1], H. Huang[†],
V. Hughes[||], K. Kwaitowski[¶], B. Lewis[‡], B. Lozowski[¶], Y. Makdisi[†],
H.O. Meyer[¶], B.V. Morozov[*], M. Nakamura[**], B.V. Przewoski[¶],
T. Rinckel[¶1], T. Roser[†], A. Rusek[†], N. Saito[$1], B. Smith[‡],
D.N. Svirida[*], M. Syphers[†], A. Taketani[$], T.L. Thomas[‡], J. Tojo[**],
D. Underwood[††], D. Wolfe[‡], K. Yamamoto[**], L. Zhu[**]

[*]*Institute for Theoretical and Experimental Physics, B. Cheremushkinskaya 25, Moscow 117259, Russia*
[†]*Brookhaven National Laboratory, PO Box 5000, Upton, NY 11973-5000, USA*
[‡]*University of New Mexico, 800 Yale, Albuquerque, NM 87131, USA*
[||]*Yale University, 217 Prospect, New Haven, CT 06511, USA*
[¶]*Indiana University Cyclotron Facility, 2401 Milo B. Sampson Lane, Bloomington, IN 47408, USA*
[$]*Institute of Chemical and Physical Research RIKEN, 2-1 Hirosawa, Wako-shi, Saitama, 351-0198, Japan*
[**]*Kyoto University, Kitashirakawa-Oiwakecho, Sakyo-ku, Kyoto 606-8502, Japan*
[††]*Argonne National Laboratory, 9700 S. Cass Ave, Argonne, IL 60439, USA*

Abstract. Acceleration of polarized protons is one of the exciting features of the new Relativistic Heavy Ion Collider (RHIC) at Brookhaven National Laboratory. Measurements of beam polarization are required both for experiments and the accelerator tuning. Elastic scattering in the Coulomb nuclear interference (CNI) region of polarized proton beams on a carbon target demonstrates asymmetry which can be used to build a polarimeter. The methods proposed for a RHIC CNI polarimeter were tested with the AGS polarized beam in the E950 experiment. A 21.7 GeV/c polarized proton beam was scattered on an extremely thin carbon ribbon target located in the AGS ring. Two symmetrical arms consisting of silicon strip detectors (SSD) and a micro channel plate (MCP) were used to identify recoil carbon.

Data obtained demonstrates a good identification of the reaction by the apparatus and a significant analyzing power. A RHIC polarimeter setup with 4 SSDs but without MCPs will be used to commission RHIC with polarized protons and for the first spin physics running in 2001.

---

[1] RIKEN BNL Research Center

The E950 experiment was set up using the AGS internal beam. Beam polarization was about 40 % and the sign changed each spill. Bunch full width was ∼25 ns with occupation ∼ $5 \cdot 10^9$ protons and repetition period 2.7 μs. An extremely thin 3.7 μg/cm^2 ribbon carbon target was used to minimize multiple scattering. Two symmetrically placed silicon strip detectors (SSD), measuring left and right scattering, were used to measure both carbon energy and TOF in the angular range near 90° (fig. 1). Each SSD was segmented into 6 strips giving different mean scattering angles. A pair of microchannel plates (MCP) was used to detect electrons kicked by the carbon nuclei out of extra thin foils half way to the SSD, providing another TOF measurement to enhance resolution. (For RHIC polarimetry the bunch length is shorter and MCPs will not be used.) The data acquisition system with 15 μs dead time was based on LeCroy 4300 FERA ADC/TDC with LeCroy 2367 ULM as large FERA memory. Linux-based software read out the data between the AGS spills. More than $2 \cdot 10^7$ events were collected.

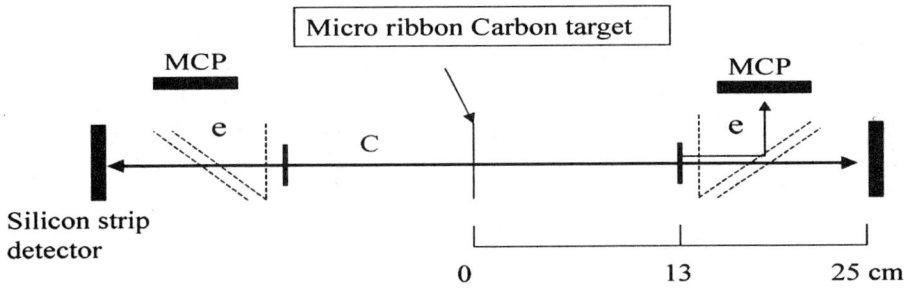

FIGURE 1. E950 layout (view is transverse to beam).

Carbon events were selected by energy/TOF correlation. For this selection it is convenient to use a TOF vs $1/\sqrt{E}$ plot, as in these coordinates the dependence given by the non-relativistic formula $E = M_C V^2/2 = M_C L^2/T^2$ is expected to be linear. Fig. 2a shows that even without MCP time enhancement a good selection can be made by a linear $\pm 2\sigma$ cut. The C-locus width along the T axis corresponds to the bunch length. The main background is due to fast (nearly relativistic) particles that do not stop in the SSD and thus do not reveal an energy/TOF correlation. The momentum transferred spectra for 3 central strips (fig. 2b) are exactly as predicted by the diffraction cone slope for both $-t$ calculated by amplitude and by TOF measurements. The mass spectrum (fig. 2c), is a good test for the self-agreement of the data. Enhanced time resolution, given by SSD-MCP TOF measurements, gives better mass resolution and an α-particle peak is clearly seen. Due to multiple scattering it is difficult to observe energy-angular correlation given by elastic scattering. Nevertheless calculations show that the carbon nucleus from other reactions should go forward and miss the detectors. On the other hand carbon from elastic scattering goes at nearly 90° and hits basically only one strip (see fig. 2d for carbon distributions over strips in the left SSD for different energies).

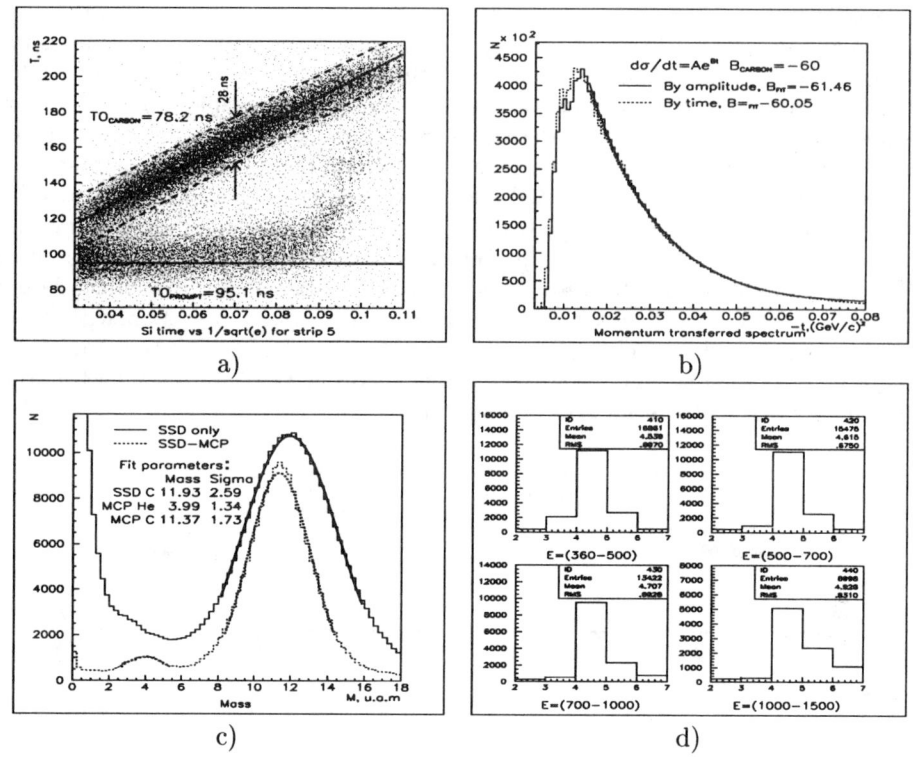

FIGURE 2. TOF vs $1/\sqrt{E}$ correlation (a), momentum transferred spectrum (b), mass spectrum (c) and the distribution over left strips for different carbon energies (d).

The physical asymmetry is given by equation:

$$A = \frac{\sqrt{N_{UL}N_{DR}} - \sqrt{N_{UR}N_{DL}}}{\sqrt{N_{UL}N_{DR}} + \sqrt{N_{UR}N_{DL}}},$$

where N_{ij} are numbers of events collected for beam polarization **Up** and **Down** by the **Left** and **Right** arms correspondingly. the overall raw asymmetry measured was 0.00533 ± 0.00027.

Our preliminary result for analyzing power as function of momentum transferred is shown in fig. 3, where we have used an average beam polarization of $P_{beam} = 40\%$ to convert from asymmetry to analyzing power. This data allows us to extract some information about the parameter r_5, describing the spin flip nuclear amplitude. Asymmetry in the CNI region is described by the Kopeliovich-Lapidus formula [1,2]:

$$\frac{m}{\sqrt{-t}}A = \frac{[2\operatorname{Im} r_5 - (\mu - 1)]\frac{t_c}{t} + 2(\rho \operatorname{Im} r_5 - \operatorname{Re} r_5)}{1 + (\frac{t_c}{t} + \rho + \delta)^2},$$

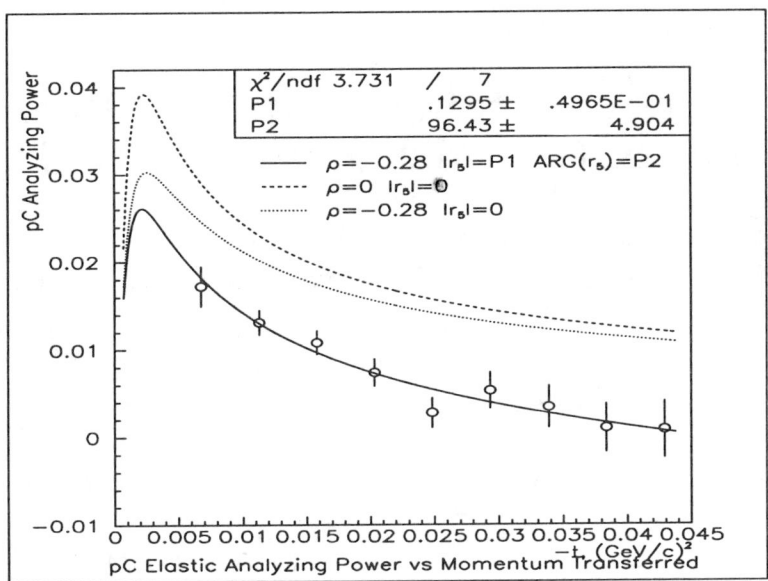

FIGURE 3. Asymmetry of elastic pC-scattering in the CNI region as function of momentum transferred at 21.7 GeV/c. Errors are statistical only. Systematic errors are being evaluated.

where $t_c = \frac{8\pi\alpha Z}{\sigma_{tot}}$, $\rho = \frac{\text{Re } f_{++}(0)}{\text{Im } f_{++}(0)}$, $r_5 = \frac{m}{\sqrt{-t}} \cdot \frac{f_{+-}}{\text{Im } f_{++}}$, μ = proton magnetic moment. The Coulomb phase $\delta = \frac{\alpha Z}{\beta}\left(\ln\frac{1}{B|t|} - 0.57772\right)$, β = proton velocity in c.m., $B = 60\ (\text{GeV}/c)^{-2}$ = diffraction cone slope. The value of $\rho = -0.28$ was taken from experimental data on elastic pd scattering. The result of the fit ($|r_5| = 0.13 \pm 0.05$, $arg(r_5) = 96 \pm 5°$) does not contradict theoretical expectations and is in agreement with experimental data on spin rotation parameters in $\pi^- p$ and pp scattering at 40-45 GeV/c and $|t| \approx 0.2$ [3-5]. The errors given are statistical only and are, therefore, preliminary.

REFERENCES

1. B. Kopeliovich, "High energy polarimetry at RHIC", hep-ph/9801414.
2. N.H. Buttimore, Proc. of RIKEN BNL Research Center Workshop, Vol. 7, 1998, BNL-65615, pp. 365-372.
3. J.Pierrard et al., Phys. Lett. B57, 1975, pp. 393-396.
4. J.Pierrard et al., Phys. Lett. B61, 1976, pp. 107-109.
5. J.Pierrard et al., Czech.J.Phys. 26B, 1976, pp. 13-24.

Polarized Atomic Hydrogen Beam Studies in the Michigan Ultra-cold Jet[1]

R.S. Raymond*, B.B. Blinov*, N.S. Borisov$^\diamond$, J. Cheng*,
A.M. Davidenko$^\triangle$, V.V. Fimushkin$^\diamond$, S.E. Gladycheva*,
V.N. Grishin$^\triangle$, T. Kageya*, D.Yu. Kantsyrev*, D. Kleppner[#],
A.D. Krisch*, V.G. Luppov*, V.S. Morozov*,
J.R. Murray*, J.J. Neumann*, B. Yankama*

* *Randall Laboratory of Physics, University of Michigan, Ann Arbor, MI 48109-1120*
$^\diamond$ *JINR, Dubna, Russia*
$^\triangle$ *IHEP, Protvino, Russia*
[#] *Department of Physics, M.I.T., Cambridge, MA 02139*

Abstract. Studies of an ultra-cold jet of polarized hydrogen atoms are described. This atomic beam is formed by the acceleration of cold (0.3 K) atoms emerging from a region of high magnetic field (12 T). The maximum measured density was about 10^{12} atoms cm^{-3}. The beam's full width half maximum size was less than 4 mm.

We are developing an ultra-cold jet of polarized hydrogen atoms to study spin effects in high energy collisions. The jet, in its test configuration, is shown in Fig. 1. Atoms from a room-temperature dissociator are cooled to 30 K and guided into a 12 T magnetic field by Teflon tubes. In this field, the atoms enter a separation cell in the mixing chamber of a dilution refrigerator and are cooled to 0.3 K by interactions with the superfluid-^{4}He-coated cell walls. Atoms in the two lower-energy hyperfine states are trapped in the high field, recombine, and are pumped away. Hydrogen atoms in the two upper hyperfine states accelerate out of the field to form an atomic beam. The separation cell's exit aperture is at 6 T, which gives the beam an energy corresponding to about 4 K; thus, its internal energy of 0.3 K makes the beam quite monochromatic.

Emerging from the cell, the cold atoms are first focused by a superfluid-^{4}He-coated parabolic mirror; this reflection is mostly specular. The atoms are then further focused by an 11-cm bore sextupole magnet into a compression tube detector for absolute intensity measurements. The compression tube also allows measurements of the azimuthal and radial beam distributions. The maximum flux measured

---

[1] This work is supported by the U.S. Department of Energy.

so far is about 2.8 10^{15} atoms s^{-1} into the compression tube's 12.5 mm x 2.5 mm aperture. This corresponds to a density of about 10^{12} atoms cm^{-3}, within the beam's measured full width half maximum size, which was less than 4 mm.

We are also testing a maser, run in transient mode, to monitor the beam polarization. After injection of a power pulse, we integrate the free induction decay signal to obtain a number proportional to the difference in state populations of the two mixed states. By driving Zeeman transitions between the J=1 hyperfine states before the maser transition, we learn about other state population differences. We have seen the main decay signal from the Jet's beam, but not the signatures of its Zeeman transitions. We are also developing a large-bore, cryogenic rf unit to drive transitions between the mixed hyperfine states, to enhance the final beam polarization. A room-temperature prototype rf unit has a measured transition efficiency of 97% with less than 150 mW of rf power. A new cold prototype rf cavity has a measured Q value of over 16,000 at 5 K. We have also built and tested a 1.2 10^7 liters s^{-1} cryocondensation pump for the jet's use in a storage ring.

Tests to improve the jet's beam intensity and operational characteristics continue.

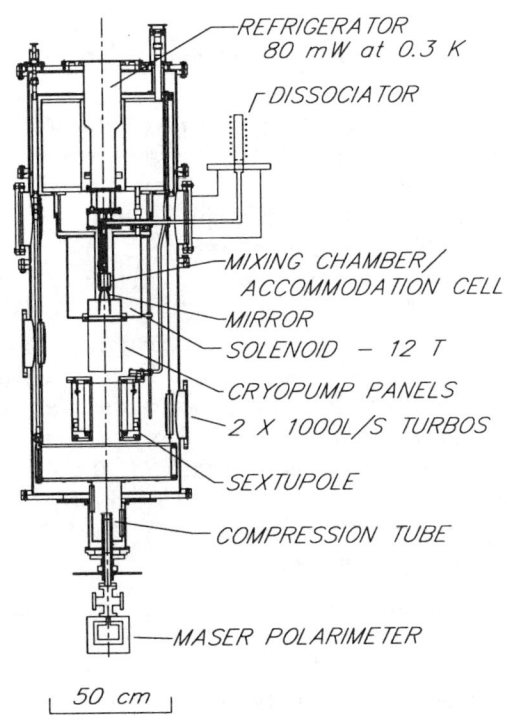

FIGURE 1. The test configuration of the Michigan ultra-cold polarized jet.

Spin Asymmetries in Polarized Hadron Collisions and the Polarized Gluon Density

Werner Vogelsang

RIKEN-BNL Research Center, Brookhaven National Laboratory
Upton, NY 11973, U.S.A.

Abstract. We consider spin asymmetries for prompt photon production and jet production in collisions of longitudinally polarized hadrons. These reactions will be key tools at the BNL-RHIC $\vec{p}\vec{p}$ collider for determining the gluon spin density in a polarized proton. We study the effects of QCD corrections to the corresponding cross sections.

INTRODUCTION

The spin-dependent gluon density in the nucleon, Δg, is currently one of the most interesting quantities in nucleon structure. A large gluon polarization in the nucleon is an exciting possible implication [1] of the measured [2] smallness of the quark contribution to the proton spin. Since Δg is left virtually unconstrained by the inclusive deep-inelastic scattering (DIS) experiments performed so far, several new experiments focus on its measurement. A fixed target DIS experiment, HERMES, measures the process $\vec{e}(\vec{\gamma})\vec{p} \to h^+h^-X$ [3], where $h = \pi, K$, which is in principle sensitive to the gluon polarization. However, the transverse momenta are low, making interpretation in a hard scattering formalism difficult. The DIS experiment COMPASS [4] will measure the same reaction at higher energies, as well as heavy flavor production, to access Δg. Scaling violations and the reaction $\vec{e}(\vec{\gamma})\vec{p} \to \text{jet(s)}X$ will constrain Δg at HERA, if the proton ring is polarized [5]. The best prospects for a measurement of Δg are offered in collisions of longitudinally polarized protons at the BNL 'Relativistic Heavy Ion Collider' (RHIC), where the most promising processes are:

- High-p_T ('prompt') photon production, $\vec{p}\vec{p} \to \gamma X$
- Jet production, $\vec{p}\vec{p} \to \text{jet(s)}X$
- Heavy-flavor production, $\vec{p}\vec{p} \to c\bar{c}X, b\bar{b}X$

In this paper, we will briefly consider the first two of these reactions from a more theoretical point of view.

PROMPT PHOTON PRODUCTION

Prompt-photon production at high transverse momentum, $pp, p\bar{p}, pN \to \gamma X$ [6], has been a classical tool for constraining the unpolarized gluon density. At leading order, a photon can be produced in the reactions $qg \to \gamma q$ and $q\bar{q} \to \gamma g$, giving rise to a supposedly clean electromagnetic probe of QCD hard scattering. Using *polarized* proton beams at RHIC is a very promising method [7–9] to measure Δg.

A recent thorough theoretical next-to-leading order (NLO) QCD study [8] for prompt photon production at RHIC also addressed the unwanted background from photons produced in jet fragmentation, when a parton, resulting from a QCD reaction, fragments into a photon plus a number of hadrons. In particular, the interplay between the fragmentation contribution and the chosen type of 'photon isolation cut', to be imposed in experiment [9] in order to reduce the background from π^0 decay photons, was studied. By comparing to lowest-order results, it was found in [8] that the QCD corrections to the polarized cross section are sizeable and reduce the dependence of the theory predictions on the factorization and renormalization scales; see Fig. 1.

In the unpolarized case, a pattern of disagreement between theoretical predictions and experimental data for prompt photon production has been observed in recent years [11–13], not globally curable by changing factorization and renormalization scales or by 'fine-tuning' the gluon density [14–16]. The main problems reside in

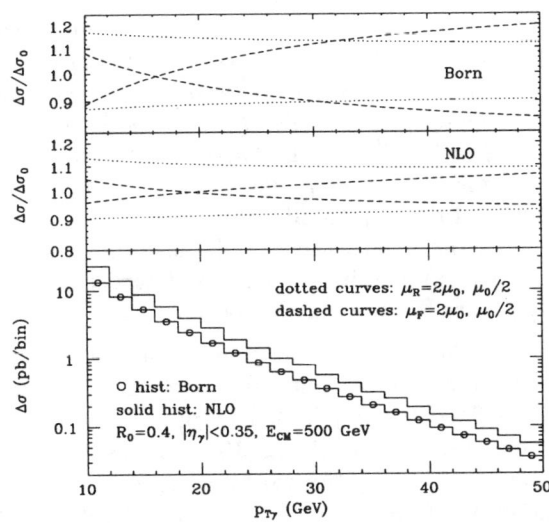

FIGURE 1. Polarized prompt photon cross section in $\bar{p}p$ collisions at RHIC, at Born and NLO levels, as function of photon transverse momentum. The upper half shows the scale dependence. The polarized parton densities were taken from [10]; for further details, see [8].

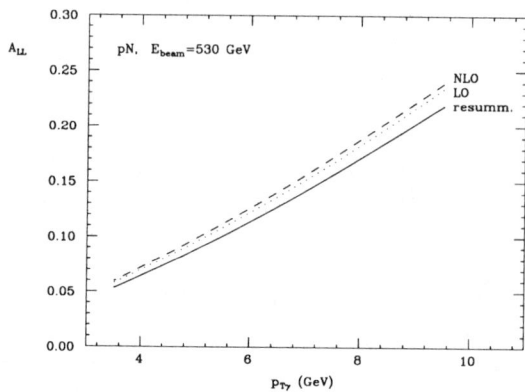

FIGURE 2. Spin asymmetry A_{LL} at LO, NLO, and including NLL threshold resummation.

the fixed-target region, where NLO theory dramatically underpredicts some data sets [12,13]. At collider energies, as relevant to RHIC, there is less reason for concern, but also here the agreement is not satisfactory.

In view of this, various improvements of the theoretical framework have been developed. One of them resorts to applying 'threshold' resummation to the prompt photon cross section [17], which organizes to all orders in α_s large logarithmic corrections to partonic hard scattering, associated with emission of soft gluons. As the partonic center-of-mass energy $\sqrt{\hat{s}}$ approaches its minimum value at $\sqrt{\hat{s}} = 2p_T$, corresponding to 'partonic threshold' when the initial partons have just enough energy to produce the high-p_T photon and the recoiling jet, the phase space available for gluon bremsstrahlung vanishes, resulting in corrections to the partonic cross section $d\hat{\sigma}/dp_T$ as large as $\alpha_s^k \ln^{2k}(1 - 4p_T^2/\hat{s}) \hat{\sigma}^{\text{Born}}$ at k-th order in perturbation theory. Threshold resummation [18,17] allows taking into account this singular, but integrable, behavior of $d\hat{\sigma}/dp_T$ to all orders in α_s. It is carried out in Mellin-N moment space, where the logarithms are of the form $\alpha_s^k \ln^{2k}(N) \hat{\sigma}^{\text{Born}}(N)$. Its application is particularly interesting in the fixed-target regime, since here the highest values of $p_T/\sqrt{s}$ are attained in the data.

In phenomenological applications of threshold resummations to prompt photon production [19,20], one finds a significant, albeit not sufficient, enhancement of the theory prediction in the fixed-target regime at large values of $p_T/\sqrt{s}$, accompanied by a dramatic reduction of scale dependence [19]. Thanks to the universal structure of soft-gluon emission, it is straightforward to apply threshold resummation to the polarized cross section as well. Fig. 2 shows the resulting effects on the *spin asymmetry* A_{LL}, for a 'toy' example that assumes a fictitious polarized set-up of the E706 experiment [12]. Details are as in [19]. Even though Fig. 2 does not directly refer to the case of RHIC, it is good news that resummation effects cancel to a large extent in A_{LL} for our present example.

JET PRODUCTION

At RHIC's highest energy, $\sqrt{s} = 500$ GeV, clearly structured jets will be very copiously produced, and jet observables will show a strong sensitivity to Δg thanks to the dominance of the gg and qg initiated subprocesses in accessible kinematical ranges.

Knowledge of the NLO QCD corrections is expected to be particularly important for the case of jet production, since it is only at NLO that the QCD structure of the jet starts to play a role in the theoretical description, providing for the first time the possibility to realistically match the procedures used in experiment to group final-state particles into jets. The task of calculating the NLO QCD corrections to polarized jet production has been accomplished in [21]. A Monte Carlo code that had been designed in [22], based on [23] and the subtraction method, to calculate any three-parton infrared-safe observable in hadron-hadron unpolarized collisions, was extended to the polarized case in [21]. We emphasize that in the unpolarized case the comparison of NLO theory predictions with jet production data from the TEVATRON is extremely successful [24].

A crucial feature of the NLO corrections to jet production is that they lead to a clear reduction in scale dependence of the cross section. This is shown in Fig. 3. One thereby gains confidence that one is indeed able to calculate reliably the cross section and the spin asymmetry for a given Δg and to confront the result with future experimental data.

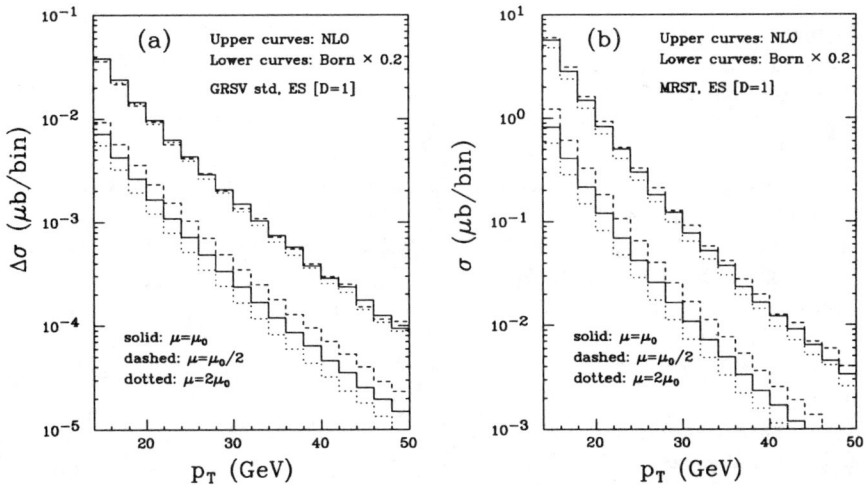

FIGURE 3. Scale dependence of the NLO and Born p_T-distributions for jet production. (a) polarized pp scattering and (b) unpolarized pp scattering, at $\sqrt{s} = 500$ GeV. The range of the pseudo-rapidity is restricted to $|\eta| < 1$. For further details of the calculation, see [21].

ACKNOWLEDGMENTS

I am grateful to D. de Florian, S. Frixione, A. Signer, and G. Sterman for fruitful collaborations. Thanks to RIKEN, Brookhaven National Laboratory and the U.S. Department of Energy (contract number DE-AC02-98CH10886) for providing the facilities essential for the completion of this work.

REFERENCES

1. For recent reviews, see: Anselmino M., Efremov A., and Leader E., *Phys. Rept.* **261**, 1 (1995); E: **281**, 399 (1997); Cheng H.-Y., *Int. J. Mod. Phys.* **A11**, 5109 (1996); hep-ph/0002157; Lampe B., and Reya E., *Phys. Rept.* **332**, 1 (2000); Bass S.D., *Eur. Phys. J.* **A5**, 17 (1999).
2. For a recent review of the data on polarized DIS, see: Hughes E., and Voss R., *Ann. Rev. Nucl. Part. Sci.* **49**, 303 (1999).
3. Airapetian A., et al., HERMES Collab., DESY-99-071, hep-ex/9907020.
4. See, for example, Mallot G.K., hep-ex/9611016.
5. De Roeck A., Deshpande A., Hughes V.W., Lichtenstadt J., and Rädel G., *Eur. Phys. J.* **C6**, 121 (1999).
6. For a compilation of the prompt photon data, see: Vogelsang W., Whalley M., *J. Phys.* **G23**, A1 (1997).
7. Berger E.L., and Qiu J., *Phys. Rev.* **D40**, 778 (1989).
8. Frixione S., and Vogelsang W., *Nucl. Phys.* **B568**, 60 (2000).
9. Bland L.C., hep-ex/9907058; Goto Y., *Nucl. Phys. Proc. Suppl.* **79**, 588 (1999).
10. Glück M., Reya E., Stratmann M., and Vogelsang W., *Phys. Rev.* **D53**, 4775 (1996).
11. Abe F., et al., CDF Collab., *Phys. Rev. Lett.* **73**, 2662 (1994).
12. Apanasevich L., et al., E706 Collab., *Phys. Rev. Lett.* **81**, 2642 (1998).
13. Ballocchi G., et al., UA6 Collab., *Phys. Lett.* **B436**, 222 (1998).
14. Vogelsang W., and Vogt A., *Nucl. Phys.* **B453**, 334 (1995).
15. Huston J., et al., *Phys. Rev.* **D51**, 6139 (1995).
16. Aurenche P., et al., *Eur. J. Phys.* **C9**, 107 (1999).
17. Laenen E., Oderda G., and Sterman G., *Phys. Lett.* **B438**, 173 (1998); Catani S., Mangano M.L., and Nason P., *JHEP* **9807**, 024 (1998).
18. Sterman G., *Nucl. Phys.* **B281**, 310 (1987); Catani S., and Trentadue L., *Nucl. Phys.* **B327**, 323 (1989); **B353**, 183 (1991); Kidonakis N., and Sterman G., *Nucl. Phys.* **B505**, 321 (1997); Bonciani R., Catani S., Mangano M.L., and Nason P., *Nucl. Phys.* **B529**, 424 (1998).
19. Catani S., Mangano M.L., Nason P., Oleari C., and Vogelsang W., *JHEP* **9903**, 025 (1999).
20. Kidonakis N., and Owens J.F., *Phys. Rev.* **D61**, 094004 (2000).
21. de Florian D., Frixione S., Signer A., Vogelsang W., *Nucl. Phys.* **B539**, 455 (1999).
22. Frixione S., *Nucl. Phys.* **B507**, 295 (1997).
23. Frixione S., Kunszt Z., and Signer A., *Nucl. Phys.* **B467**, 399 (1996).
24. see, for example: L. Babukhadia, hep-ex/0005026.

The COMPASS Experiment at CERN

Alessandro Bravar, for the COMPASS Collaboration

Institut für Kernphysik, Universität Mainz, D-55099 Mainz, Germany

Abstract. The COMPASS experiment at CERN is aimed at the study of the nucleon spin structure and hadron spectroscopy. It attempts a measurement of the gluon polarization around $x_g \simeq 0.1$ with a precision better than $\delta(\Delta g/g) < 0.1$. The experiment uses muo-production of open charm and correlated high-p_T hadron pairs to tag the photon-gluon fusion process. In parallel COMPASS will carry out also a rich spin-physics program in polarized DIS.

COMPASS is a fixed-target experiment using a "Common Muon and Proton Apparatus for Structure and Spectroscopy". It has been recently approved at CERN and it is being commissioned this year. First results are expected soon.

One of the most intriguing issues in understanding the nucleon's spin structure is the polarization of gluons, $\Delta g/g$. A sizable value of Δg could explain the smallness of the contribution of the quark spins to the nucleon spin, $\Delta\Sigma$ [1], as well as the correlated appearance of a negatively polarized strange sea. A particularly clean process involving the gluon distribution in leading order is the open charm production via the photon-gluon fusion process, $\gamma g \to c\bar{c}$, shown in Fig. 1. The hard scale is set by the charm quark mass, $4m_c^2 \simeq 10$ GeV^2. An alternative method is the formation of a pair of light quarks with high virtuality detected through a pair of correlated high p_T hadrons, illustrated in Fig. 2. The transverse momenta of the outgoing hadrons give the necessary hard scale of about 10 GeV^2.

Apart from the measurement of $\Delta g/g$ COMPASS offers a rich spin-physics program at high Q^2 with a high luminosity including the measurement of g_1 and g_2, the transversity structure function h_1, spin-flavor decomposition of the structure functions, $\Lambda/\bar{\Lambda}$ polarization in both the current and target fragmentation regions, exclusive vector meson and photon (DVCS) production, etc.

In COMPASS [2] we will access the gluon polarization $\Delta g/g$ by measuring the open charm cross section spin-asymmetry with a polarized muon beam of $100 - 200$ GeV scattering off a fixed polarized target using the full virtual photon spectrum down to quasi-real photons. At these muon energies the contributions of resolved photons, intrinsic charm, and diffractive components to the total charm photoproduction cross section are expected to be very small. The open charm events will be tagged by fully reconstructing a charm hadron in the final state, typically through the $D^0 \to K^-\pi^+ + c.c.$ decay.

The photon-nucleon cross section spin-asymmetry for charm production via PGF

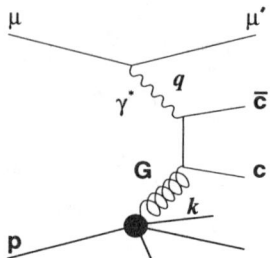

FIGURE 1. The photon gluon fusion process at tree level, leading to the formation of a pair of charm and anti-charm quarks. The fragmentation into charmed hadrons is not shown.

by real (or quasi-real) photons, $A_{\gamma N}^{c\bar{c}}$, is obtained by integrating the cross sections over the kinematically allowed range $x_g^{min} = 4m_c^2/2M\nu \leq x_g \leq 1$

$$A_{\gamma N}^{c\bar{c}}(\nu) = \frac{\Delta\sigma^{\gamma N \to c\bar{c} X}}{\sigma^{\gamma N \to c\bar{c} X}} = \frac{\int_{4m_c^2}^{2M\nu} d\hat{s}\, \Delta\hat{\sigma}(\hat{s})\, \Delta g(x_g, \hat{s})}{\int_{4m_c^2}^{2M\nu} d\hat{s}\, \hat{\sigma}(\hat{s})\, g(x_g, \hat{s})} \sim \langle a_{LL} \rangle \cdot \langle \frac{\Delta g}{g} \rangle. \quad (1)$$

$x_g = \hat{s}/2M\nu$ denotes the nucleon momentum fraction carried by the gluon (note that $x_g \geq x_{Bj}$) and $\hat{s}$ is the invariant mass of the two emerging quarks.

The PGF spin-averaged cross sections, $\hat{\sigma}(\hat{s})$, is known to next-to-leading order since some time [3], while the spin-dependent cross section, $\Delta\hat{\sigma}(\hat{s})$, has become available to next-to-leading order only recently [4]. The PGF scattering spin-asymmetry $a_{LL}(\hat{s}) = \hat{\sigma}/\Delta\hat{\sigma}$ measures the analysis power of this process. At threshold $a_{LL} = 1$ and it crosses zero at about $2 \times \hat{s}_{th}$. Knowing $\hat{\sigma}(\hat{s})$, $\Delta\hat{\sigma}(\hat{s})$ (or $a_{LL}(\hat{s})$), and $g(x_g)$ allows to evaluate $\Delta g(x_g)$ from the measured asymmetry $A_{\gamma N}^{c\bar{c}}$ (Eq. 1).

The COMPASS apparatus is based partially on the existing SMC setup. It exploits the polarized muon beam at CERN with a polarization of ~ -80 %. Figure 3 shows the initial setup of the COMPASS experiment. The wide angular range of the outgoing hadrons requires a two stage magnetic spectrometer with particle identification in both spectrometer stages.

The very high beam intensity of 2×10^8 muons per SPS spill (2×10^8/s) needed for these measurements poses stringent requirements on the tracking detectors from the rate resistance, to readout speed and radiation hardness. Tracking in the beam region will be performed with scintillating fiber detectors. In the inner high rate part of the acceptance close to the beam newly developed gaseous detectors, the *micromegas* and GEMs, will be used. Tracking in the outer region will be performed mainly with straw tubes and more conventional MWPC and drift chambers.

The trigger condition of the experiment will ask for a muon with a minimum energy loss of about 30 % w.r.t. the incident beam energy, and a minimum hadron energy of 5 GeV deposited in the hadronic calorimeters. The principle of the trigger has been already successfully tested and proven in the last year test run.

Two oppositely polarized target cells will be exposed to the same muon beam flux. This will substantially suppress the systematic effects in the extraction of the spin-asymmetries, as has been the case in the past. Two different target materials, lithium deuteride (^{6}LiD) and ammonia (NH$_3$), are foreseen. The materials are

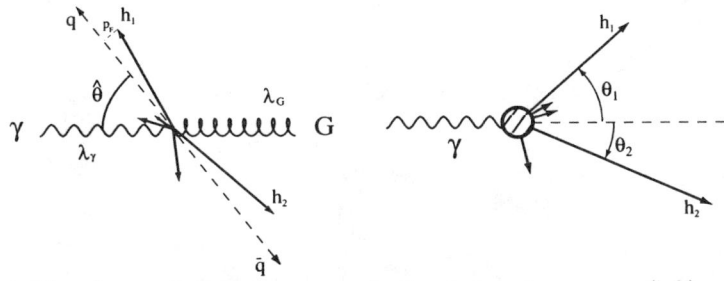

FIGURE 2. The photon-gluon fusion process in the photon gluon c.m.s. (left) and laboratory frame (right) leading to two jets of opposite momentum. At large c.m. angles ϑ^* it leads to the occurance of correlated hadrons of large and opposite p_T.

polarized by the proven dynamic nuclear polarization technology in magnetic field with microwaves at a temperature of a few mK only.

For a running time of 2 years the projected statistical error of the measured asymmetry, based only on the $D^0 \to K^-\pi^+ + c.c.$ detection, is $\delta A_{\gamma N}^{c\bar{c}} = 0.076$. The result can be further improved by tagging the D^0's from $D^{*+} \to D^0 \pi_s^+ \to (K^-\pi^+)\pi_s^+$ decays. The D^{*+} reconstruction leads to a charm signal with very little background. Taking into account the acceptances and possible re-interactions in the target the statistical error of the asymmetry reduces to $\delta A_{\gamma N}^{c\bar{c}} = 0.05$.

As in the inclusive case the muon-nucleon asymmetry is reduced from the virtual-photon asymmetry by the depolarization factor $A_{\mu N}^{c\bar{c}} = D A_{\gamma N}^{c\bar{c}}$ ($D \simeq 0.66$ in the COMPASS kinematics). The correlation between the D meson observables in the final state and the $c\bar{c}$ pair in the c.m.s., which survives the hadronization, can be exploited to enhance the analysis power of the process. A larger sensitivity on the gluon polarization can be obtained by analyzing the dependence of $A_{\gamma N}^{c\bar{c}}$ on the D mesons' p_T, and rejecting D mesons with $p_T > 1$ GeV/c will yield

$$\delta\left(\frac{\Delta g}{g}\right) \simeq 0.10. \qquad (2)$$

The three and four-body decay channels of the D mesons, as well as the semi-leptonic decays may further improve the precision of the measurement, in particular if the D^* tagging can be applied.

The detection of a correlated high p_T hadron pair offers an independent method to study the gluon polarization via PGF [5]. These can be viewed as a pair of degenerated jets where most of the jet momentum is carried by a single hadron (Fig. 2). Selecting hadrons with a high p_T, $p_T > 1.0 - 1.5$ GeV, reduces very effectively the contribution of photon absorption.

The longitudinal spin-asymmetry for this process can be approximated by

$$A_{LL}^{HH} = \frac{\Delta \sigma^{rmHH}}{\sigma_{HH}} \approx \langle a_{LL}^{PGF}\rangle \langle \frac{\Delta g}{g}\rangle \frac{\sigma^{PGF}}{\sigma^{tot}} + \langle a_{LL}^{COM}\rangle \langle \frac{\Delta u}{u}\rangle \frac{\sigma^{COM}}{\sigma^{tot}}. \qquad (3)$$

The parton level asymmetries are about -1 for a_{LL}^{PGF} and about 0.5 for a_{LL}^{COM}.

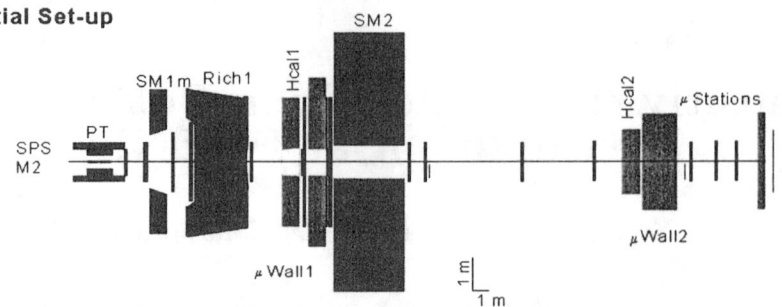

FIGURE 3. COMPASS initial setup. A second RICH detector downstream of SM2, electromagnetic calorimeters, and additional trackers will complete the apparatus.

This channel is not statistic limited, but the precision of the measurement is mostly affected by the *background* coming from gluon radiation. Selecting K^+K^- pairs only, a further reduction of the Compton contribution is possible at the expense of a lower statistics, as strangeness production is suppressed in ordinary fragmentation. In one year COMPASS can achieve a precision of

$$\delta\left(\frac{\Delta g}{g}\right) < 0.05 \qquad (4)$$

for several x_g bins in the region $0.04 < x_g < 0.2$

For a positive Δg the measured spin-asymmetry is expected to be negative, while for the open charm it is expected to be positive, due to the different behavior of the PGF analyzing power a_{LL}^{PGF} for light and heavy quarks, respectively.

COMPASS will also address important and intriguing issues of hadronic structure and dynamics, and hadron spectroscopy in both the perturbative and non-perturbative QCD domains, by studying different reactions with a variety of hadron beams from the polarizabilities of hadrons in Primakoff scattering to centrally produced glueballs and exotic mesons, from the hadro-production of charm to the studies of semi-leptonic and leptonic decays of charm hadrons.

In conclusion, COMPASS offers a broad and rich physics program addressing the nucleon structure and hadron spectroscopy.

REFERENCES

1. B. Adeva *et al.* (SM Collaboration), Phys. Rev. D **58**, 112002 (1998);
   K. Abe *et al.* (E143 Collaboration), Phys. Rev. D **58**, 112003 (1998).
2. The COMPASS Collaboration, COMPASS proposal, CERN/SPSLC 96-14, SPSLC/P297 (Geneva, March 1996); CERN/SPSLC 96-30 (Geneva, May 1996).
3. S. Frixione *et al.*, Nucl. Phys. B **431**, 453 (1994); and references therein.
4. I. Bojak and M. Stratmann, Phys. Lett. B **433**, 411 (1998);
   A.P. Contogouris *et al.*, Phys. Lett. B **482**, 93 (1998).
5. A. Bravar, D. von Harrach, and A. Kotzinian, Phys. Lett. B **421**, 349 (1998).

Sensitivities of the Gluon Polarization Measurement at PHENIX

Yuji Goto, for the PHENIX Collaboration

RIKEN BNL Research Center, Upton, NY 11973, USA

Abstract. We plan to perform direct measurements of the gluon polarization, $\Delta g(x)$, in the proton using polarized proton collisions at RHIC. The measurements are carried out by probing gluon+quark and gluon+gluon reactions with the PHENIX detector system. Sensitivities to the gluon polarization in prompt photon and pion production measurements are discussed.

INTRODUCTION

Polarized lepton–nucleon deep inelastic scattering (DIS) experiments have shown that the fraction of the quark polarization in the nucleon amounts to only about 10–30% of the nucleon spin. The remaining fraction needs to be explained by other carriers. We need to know the contribution of the gluon to the nucleon spin. The DIS experiments are sensitive to the gluon only in the next-to-leading order. They require Q^2 evolution or semi-inclusive measurements to extract the gluon polarization.

Polarized proton collisions which we will perform at RHIC/PHENIX [1] are sensitive to the gluon polarization, $\Delta g(x)$, in the leading order. By measuring longitudinal spin asymmetry, $A_{LL} = (d\sigma_{++} - d\sigma_{+-})/(d\sigma_{++} + d\sigma_{+-})$, where $d\sigma_{++}$ ($d\sigma_{+-}$) represents the cross section with parallel (anti-parallel) beam helicity, $\Delta g(x)$ is extracted.

The PHENIX detector system consists of two Central Arms and two Muon Arms. In the Central Arms which cover $|\eta| < 0.35$ and $\pi/2$ azimuthal angle by each arm, we will measure prompt photon and neutral/charged pion production using fine-segment EM calorimeter (EMCal), tracking chambers and particle-ID detectors. The segmentation of the EMCal is 0.01-radian both in azimuthal and polar directions. The Muon Arms which cover $1.2 < |\eta| < 2.4$ and 2π azimuthal angle consist of muon-tracking chambers and muon-ID detectors. By using both the Central Arms and Muon Arms, we will measure open and bound-state heavy flavor production by detecting single leptons, di-lepton pairs and electron–muon pairs.

PROMPT PHOTON MEASUREMENT

The dominant process for prompt photon production is the gluon Compton process, $gq \to \gamma q$. We can investigate the gluon distribution with a clear theoretical interpretation. In the polarized proton collisions, $\Delta g(x)$ can be obtained through the relation, $A_{LL} = \Delta g(x_g)/g(x_g) \cdot A_1^p(x_q) \cdot a_{LL}^{gq \to \gamma q}(\cos\theta^*)$, using the quark polarization $A_1^p(x_q)$ obtained by the polarized DIS experiments and the calculable parton-level asymmetry $a_{LL}^{gq \to \gamma q}(\cos\theta^*)$.

The measurement is experimentally challenging because there are many backgrounds mainly from the two-photon decay of π^0. The PHENIX detector has good background reduction capability with the fine segmentation of the EMCal. To estimate the sensitivity of the measurement, we have performed PYTHIA simulations [2]. Results using GRV94 leading-order (LO) parton distribution functions (PDF) [3] and GS95 next-to-leading-order (NLO) polarized PDF [4] are shown.

Figure 1 shows the estimated statistical errors of our A_{LL} measurement of the prompt photon in one year of full luminosity, 320pb^{-1}, at $\sqrt{s}$=200GeV. The p_T region from 10 to 30 GeV/c is shown which corresponds to the x_g region from 0.1 to 0.3. The statistics clearly distinguish between three models of GS95 NLO polarized PDF with different integrated gluon polarization values from 1.02 to 1.71 at Q^2=4GeV^2.

The raw ratio of experimental background to the prompt photon yield is 50% to 200% in the p_T region from 10 to 30GeV/c at $\sqrt{s}$=200GeV. Due to the fine-segment EMCal, this ratio can be reduced by mass reconstruction of two photons and an isolation cut down to less than 20% [5].

In order to extract $\Delta g(x)$ from the asymmetry A_{LL} of the measured prompt photon, dilutions of the asymmetry by the experimental backgrounds and processes other than the gluon Compton process need to be evaluated. The annihilation process dilutes the asymmetry by 20% and the higher-order processes like bremsstrahlung dilute it by 10%. The dilution by the experimental backgrounds is

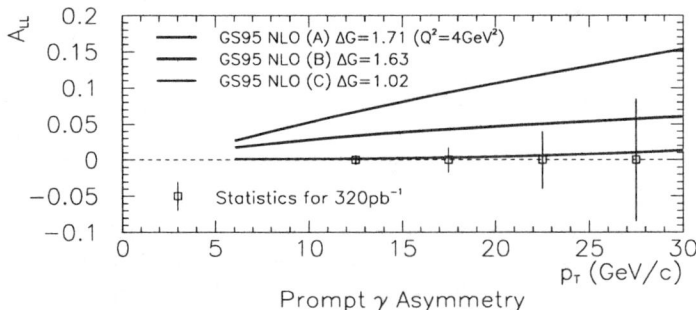

FIGURE 1. Statistical errors of the asymmetry measurement of the prompt photon in one year of full luminosity at $\sqrt{s}$=200GeV. The asymmetry calculations using three models of GS95 NLO polarized PDF are also shown.

negligible after the background reduction show above.

To evaluate other uncertainties, we compare $\Delta g(x)$ derived from the gluon Compton process with input $\Delta g(x)$ in the simulation as shown in Fig.2. There are remaining deviations because of kinematic uncertainties of x_g and $\cos\theta^*$ in the inclusive prompt photon measurement. In our real data analysis, we will fit the asymmetry A_{LL} directly by including these kinematic uncertainties in the global analysis.

PION MEASUREMENT

Pion production measurements are an alternative to a jet measurement in a small acceptance detector. Jet (and pion) production is sensitive to $\Delta g(x)$ through gluon+quark and gluon+gluon reactions and has high statistics. We can start a measurement in RHIC year-2, first year of the polarized proton collision with 10% luminosity. At PHENIX, π^0 is identified clearly by mass reconstruction with the high performance EMCal. We will have excellent statistics to distinguish between three models of GS95 NLO polarized PDF. Because of different fragmentation functions of π^0, π^+ and π^- from each quark, we will observe different asymmetries for each. Figure 3 shows asymmetries of π^0, π^+ and π^- with estimated statistical errors of our A_{LL} measurement of π^0 in one year of 10% luminosity, 32pb^{-1} at $\sqrt{s}$=200GeV. This measurement is similar to the DIS semi-inclusive plus-charged and minus-charged hadron measurement performed in the HERMES and SMC experiments [6], where they measured flavor decomposition of the quark polarization. We will add more information of the flavor decomposition of the quark polarization, in addition to the gluon polarization in this measurement.

SUMMARY

We will perform the direct measurement of $\Delta g(x)$ at PHENIX using many channels and probes. Table 1 summarizes our sensitivities to $\Delta g(x)$ at $\sqrt{s}$=200GeV

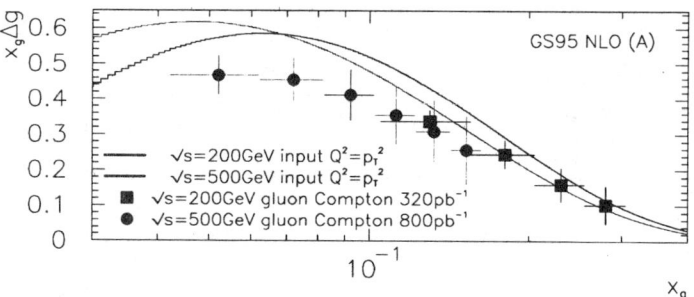

FIGURE 2. Comparison of $\Delta g(x)$ derived from the gluon Compton process and input $\Delta g(x)$ in the simulation.

TABLE 1. Summary of the estimated statistical errors and predicted values of the A_{LL} and the x_g range to be covered by each channel at $\sqrt{s}=200\text{GeV}$. The statistical errors show the absolute values for 10% luminosity 32pb^{-1} for channels indicated by *), and full luminosity 320pb^{-1} for other channels.

Probes	A_{LL} Errors	A_{LL} Predictions			x_g Range
		GS(A)	GS(B)	GS(C)	
prompt γ	0.006	0.05	0.03	0.001	0.1 – 0.3
$\pi^0 / \pi^\pm$	0.001*)	0.02	0.01	0.0001	0.05 – 0.2
$J/\psi \to \mu\mu$	0.006	0.01	0.01	0.002	0.005 – 0.01
$c\bar{c} \to eX$	0.001*)	−0.02	−0.01	−0.0003	0.005 – 0.2
$c\bar{c} \to e\mu X$	0.006	−0.05	−0.02	−0.0001	0.005 – 0.2
$b\bar{b} \to e\mu X$	0.006	0.01	0.01	−0.0004	0.01 – 0.3

including also our heavy flavor production measurement. For pion measurements and single-electron measurements, we expect good statistics in RHIC year–2 with 10% luminosity. These channels cover complementary kinematic ranges of the gluon which in total spread over $0.005 < x_g < 0.3$.

REFERENCES

1. D.P. Morrison, Nucl. Phys. A **638**, 565c (1998), N. Saito, Nucl. Phys. A **638**, 575c (1998).
2. T. Sjöstrand, Comp. Phys. Commun. **82**, 74 (1994).
3. M. Glück, E. Reya, and A. Vogt, Z. Phys. C **67**, 433 (1995).
4. T. Gehrmann and W.J. Stirling, Phys. Rev. D **53**, 6100 (1996).
5. A. Bazilevsky, RIKEN Rev. **28**, 15 (2000).
6. B. Adeva et al., Phys. Lett. B **420**, 180 (1998), K. Ackerstaff et al., Phys. Lett. B **464**, 123 (1999).

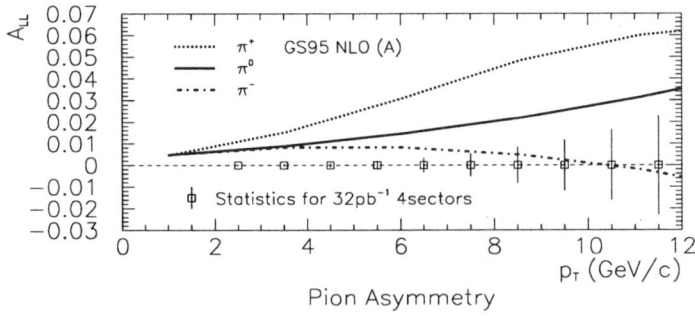

Pion Asymmetry

FIGURE 3. Asymmetries of π^0, π^+ and π^- with estimated errors of our A_{LL} measurement of π^0 in one year of 10% luminosity, 32pb^{-1} at $\sqrt{s}=200\text{GeV}$.

Spin Physics Program at STAR

W. W. Jacobs
and the STAR Collaboration [1]

Indiana University Cyclotron Facility and Department of Physics
Bloomington, Indiana 47408 USA e-mail: jacobs@iucf.indiana.edu

Abstract. The question as to how spin degrees of freedom in nucleons are organized ("spin puzzle") is ongoing. Quantifying the gluon polarization, not measured directly in DIS experiments, has been emphasized as necessary to complete the picture. The role of STAR in providing a definitive measurement of the gluon spin structure function is discussed, and other spin topics of current interest at RHIC energies summarized.

INTRODUCTION

The Relativistic Heavy Ion Collider (RHIC) will accelerate and collide polarized proton beams over the CM energy range, $50 < \sqrt{s} < 500$ GeV. With luminosities up to 10^{32}/cm^2/sec, RHIC will provide the first opportunity to study polarization observables for large p_T processes in a regime where perturbative QCD has successfully described unpolarized phenomena. The large acceptance of baseline STAR [1] and its detector upgrades (barrel EM calorimeter and particularly the endcap EMC), make it the instrument of choice for spin physics studies involving coincident hadron jet detection, such as arise from hard collisions of quarks and gluons [2].

The STAR detector will provide access to critical regions of phase space for colliding polarized beams allowing: 1) definitive determination of the gluon contribution to the proton spin by measurement of the double spin asymmetry A_{LL} in direct photon + jet production; 2) effective separation of u, d, anti-u and anti-d polarization contributions to the proton spin in W$^\pm$ production; 3) a sensitive probe of hyperon spin structure using polarization transfer to hyperon fragments in jets; 4) Standard Model tests via parity-violating helicity asymmetries in hard jet production; 5) calibration of the "$\vec{p}$" in $\vec{p}$QCD; 6) measurements of nucleon transversity and other interesting new spin topics. These measurements will complement and greatly extend the deep inelastic scattering (DIS) studies of the last decade.

---

[1] A recent STAR Collaboration author list is given in Ref. 1

GLUON CONTRIBUTION TO THE PROTON SPIN

Measurement of the gluon contribution to the proton spin is the flagship experiment of STAR spin physics. In this session on gluon polarization, we refer to the summary talk for background and focus on an outline (as presented) of the important points concerning STAR's ability to make a definitive measurement.

1st moment of $\Delta G(x)$: From observed scaling violations in polarized DIS asymmetries, crude constraints on the gluon helicity distribution $\Delta G(Q^2 = 1(\text{GeV}/c^2)) \equiv \int_0^1 \Delta G(x, Q^2) dx \approx +1 \pm 1.5$ are extracted. To do much better requires a probe coupling directly to glue while providing a significant spin-dependent sensitivity. QCD Compton scattering, $q + g \to q + \gamma$, is such a process and the focus of STAR spin measurements with the Endcap EMC [2,3]. A reasonable goal is to measure the integral gluon contribution to the proton spin within error ± 0.5.

$qg \to q\gamma$ is a clean probe of gluons: In leading-order (LO) perturbative QCD, the dominant γ production is from the $q + g \to q + \gamma$ subprocess. A LO physical background ($\sim 10\%$) level comes from $q + \bar{q} \to \gamma + g$ annihilation. Next-to-leading-order (NLO) calculations indicate no qualitative changes from LO expectations; higher-twist corrections should remain negligible at $p_T \gtrsim 10$ GeV/c.

STAR + Endcap: direct extraction of $\Delta G(x)$: The large acceptance of STAR enables the detection of γ-jet *coincidences* (e.g., as opposed to the PHENIX detector at RHIC), allowing event-by-event kinematic reconstruction of the momentum fractions $x_{1,2}$ for the colliding partons. Reconstucted partonic kinematics allows a *direct extraction* of $\Delta G(x)$ at LO from measured spin asymmetries. For polarization measurements, this extraction reduces sensitivity to kinematic (k_T) smearing effects due to transverse components in the initial parton momenta [3].

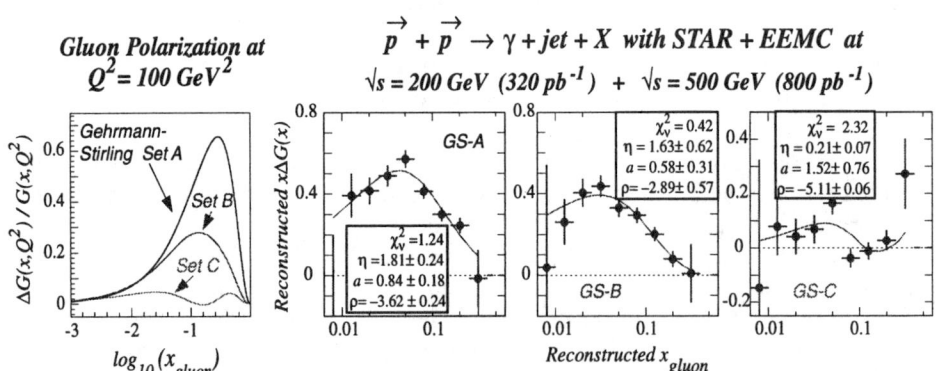

FIGURE 1. $\Delta G(x)$ reconstructed from simulated γ+ jet coincident events, based on three different model inputs for gluon polarization (shown in the left-hand frame), [4] all consistent with the polarised DIS database and evolved to $Q^2 = 100$ $(GeV/c)^2$. Solid curves in the right-hand frames are fits to the reconstructed $\Delta G(x)$ values, where η is the extracted first moment ΔG.

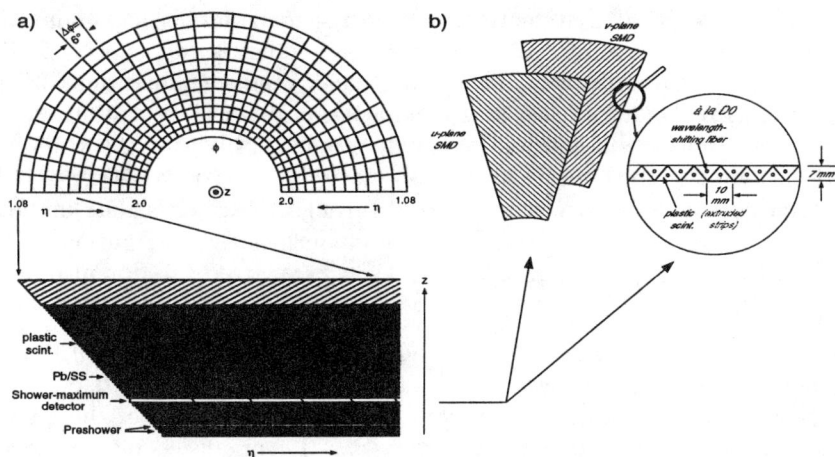

FIGURE 2. *Endcap EMC detector structure. a) Upper view shows detector half subdivision into projective towers in pseudorapidity (η) and azimuthal angle. Lower view gives depth profile slice, with 23 layers of lead/stainless steel absorber and 24 layers of plastic scintillator. b) Schematic of shower maximuam detector (SMD) 30° sector comprising extruded triangular scintillating strips.*

Asymmetric partonic collisions required: At LO, the QCD partonic-level longitudinal spin correlation $\hat{a}_{LL}$ for gluon Compton scattering approaches unity when the γ is detected in the direction of the incident quark (cross section also maximal here). From DIS we know large quark polarizations ($\gtrsim 30\%$) to probe ΔG are available at momentum fractions $x_{quark} \gtrsim 0.2$. Thus, for maximum sensitivity to gluon spin, one needs very asymmetric partonic collisions ($x_q \geq 0.2$ with $x_g \leq 0.1$), in which both products will be boosted forward in the lab frame. Coverage for such asymmetric collisions requires the STAR Endcap EMC [2,3], under construction, allowing polarized quarks (x_{large}) to analyze gluons (x_{small}).

Δ G requires x_{glue} down to 0.01: Measurements of γ-jet coincidences with $p_{T,\gamma} > 10$ GeV/c at STAR with the Endcap EMC at two bombarding energies, $\sqrt{s} = 200$ and 500 GeV, will cover the gluon momentum fraction range $0.01 \lesssim x_g \lesssim 0.3$, required to determine the gluon's contribution $\int \Delta G(x) dx$ to the proton helicity to within $\approx \pm 0.5$. Adequate statistical precision can be obtained with 10-week runs at each energy, at $\vec{p}+\vec{p}$ luminosities $\sim 10^{32}$ cm^{-2}s^{-1}.

Fig. 1 illustrates the sensitivity that STAR will have to the fraction of proton spin carried by the gluon. Three different distributions of $\Delta G(x)$ are indicated at the left. The values of $\Delta G(x)$, reconstructed from simulated A_{LL} γ+ jet coincidences, are shown to the right and correspond to the combined data obtainable from integrated luminosities of 320 (800) pb^{-1} at $\sqrt{s} = 200$ (500) GeV. In the parameterization indicated by the curves, the fitted value of "η" represents $\int \Delta G(x) dx$. It is critical to observe the falloff in $x\Delta G(x)$ at small x to accurately determine η.

Complications? Small systematic errors arising from simplifying assumptions in the direct reconstruction of $\Delta G(x)$ from A_{LL} asymmetries depend on $\Delta G(x, Q^2)$ and can be corrected iteratively. Theoretical backgrounds arising from NLO pQCD diagrams, or from multiple soft gluon radiation prior to hard scattering, are not serious. Experimental backgrounds include abundant high-p_T π^0, $\eta^0 \to \gamma\gamma$ and time projection chamber pileup at high luminosity $\vec{p}+\vec{p}$. The latter requires level 3 event filtering, while the former is attacked through Endcap EMC design features.

STAR ENDCAP EMC AND SPIN TIMELINE

The Endcap is a conventional (~ 21 radiation length) lead and scintillator sheet calorimeter [2]. Its annular shape (outer radius $\sim$215 cm and active depth $\sim$37 cm), is segmented into 720 projective towers (see Fig. 2a). Crucial for the STAR spin program is the shower maximum detector (SMD) – organized into orthogonal U and V planes (see Fig. 2b) made from triangular cross section extruded scintillator strips. The strips help provide the fine position resolution and stability of measured shower profile shape, vital to distinguishing shower profiles characteristic of single photons vs. the close-lying photon pairs from π^0 and η^0 decay. Tests of a prototype detector in-beam at SLAC substantiate the SMD simulated performance and hence its efficacy for the required background reduction. For the EMC towers, the first two scintillator layers are read out separately to serve as a pre-shower detector, aiding (for the highest-energy events) in γ/π^0 discrimination. Endcap construction, bolstered by the release of major NSF funding earlier this year, is underway.

We expect first $\vec{p} + \vec{p}$ collisions at a few percent of design luminosities at RHIC during 2001. Operations should include the Siberian snakes (two per ring), with spin rotators, allowing for collisions with longitudinal polarization, being installed in subsequent years. For STAR, the Barrel calorimetery is projected to grow ~ 30 modules per year, leading to completion by 2004. The first half of the endcap EMC is scheduled to be installed in 2002, with the second half following in 2003.

The STAR spin physics goal in 2001 will be a measurement of A_N for high-p_T inclusive hadron production. Useful constraints on ΔG, from inclusive and di-jet studies, should be forthcoming in 2002, along with a sample of data from γ and $\gamma+$ jet production. The major parts of the photon and $W^{\pm}$ production running at both 200 and 500 GeV (with design luminosities) are anticipated to take place during 2003-4 along with an experiment to calibrate the absolute beam polarization.

REFERENCES

1. K.H. Ackermann et al., Nucl. Physics **A661**, 681c (1999).
2. L.C. Bland, W.W. Jacobs, J. Sowinski, E.J. Stephenson, S.E. Vigdor and S.W. Wissink, *An Endcap EMC for STAR: Conceptual Design*, STAR Note 401 (1999).
3. L.C. Bland, Symp. on "High Energy Spin Physics" (RIKEN, 1999); hep-ex/0002061.
4. T. Gehrmann and W.J. Stirling, Phys. Rev. D**53**, 6100 (1996).

Polarized Fragmentation Functions at RHIC

Daniel de Florian [1]

Institute of Theoretical Physics, ETH CH-8093 Zürich, Switzerland
E-mail: dflorian@itp.phys.ethz.ch

Abstract. In this talk I discuss the prospects for a measurement of Λ polarized fragmentation functions at RHIC.

I INTRODUCTION

The measurement of fragmentation functions provides a complementary look at the internal structure of hadrons. Particularly, in the polarized case, they provide information about how the spin carried by the partons is passed to the hadrons in the fragmentation process and, at variance with DIS, it is possible to measure them even for the case of hadrons which are not available as targets. In ref. [1], 3 different scenarios for the Λ-baryon polarized fragmentation functions, all of them in agreement with available e^+e^- data, were presented in order to cover a rather wide range of plausible models:

Scenario 1 corresponds to the expectations from the non-relativistic naive quark model where only s-quarks can contribute to the fragmentation processes that eventually yield a polarized Λ, even if the Λ is formed via the decay of a heavier hyperon.

Scenario 2 is based on estimates by Burkardt and Jaffe [2,3] for the DIS structure function g_1^Λ of the Λ, predicting sizeable negative contributions from u and d quarks to g_1^Λ by analogy with the breaking of the Ellis-Jaffe sum rule for the proton's g_1^p.

Scenario 3 All the polarized fragmentation functions are assumed to be equal here, contrary to the expectation of the non-relativistic quark model used in scen. 1.

II LONGITUDINAL POLARIZATION

With the advent of RHIC, spin transfer reactions can be studied for the first time also in pp scattering at c.m.s. energies of up to $\sqrt{s} = 500$ GeV. In the following we will demonstrate [4] that measurements with only *one* polarized beam would provide a particularly clean way of discriminating between the various conceivable

---

[1] This work was supported in part by the EU Fourth Framework Programme 'Training and Mobility of Researchers', Network 'Quantum Chromodynamics and the Deep Structure of Elementary Particles', contract FMRX-CT98-0194 (DG 12 - MIHT).

sets of spin-dependent Λ fragmentation functions presented above. The relevant differential polarized cross section can be schematically written as (the subscripts "+","−" below denote helicities)

$$\frac{d\Delta\sigma^{p\vec{p}\to\vec{\Lambda}X}}{d\eta} \equiv \frac{d\sigma^{pp+\to\Lambda+X}}{d\eta} - \frac{d\sigma^{pp-\to\Lambda+X}}{d\eta} \quad (1)$$

$$= \int_{p_T^{min}} dp_T \sum_{ff'\to iX} \int dx_1 dx_2 dz \, f^p(x_1) \times \Delta f'^p(x_2) \times \Delta D_i^\Lambda(z) \times \frac{d\Delta\sigma^{f\vec{f'}\to\vec{i}X}}{d\eta},$$

the sum running over all possible LO subprocesses, and where we have integrated over p_T, with p_T^{min} denoting some suitable lower cut-off.

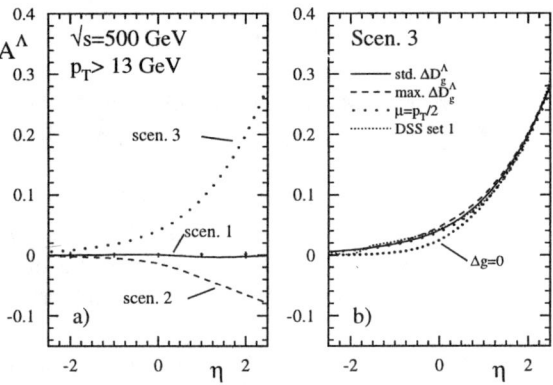

FIGURE 1. (a) The asymmetry A^Λ as a function of rapidity of the Λ at RHIC energies for the various sets of spin-dependent fragmentation functions. (b) same as for scenario 3 in (a), but using the "maximal" ΔD_g^Λ, a hard scale $Q = p_T/2$, $\Delta g = 0$, or the DSS1 distributions

Fig. 1(a) shows our predictions for the spin asymmetry A^Λ (defined as usual by the ratio between polarized and unpolarized cross sections) as a function of rapidity for $\sqrt{s} = 500$ GeV and $p_T^{min} = 13$ GeV. Note that we have counted positive rapidity in the forward region of the *polarized* proton. We have used the three different scenarios for the ΔD_i^Λ discussed above, employing the hard scale $Q = p_T$. The "error bars" should give an impression of the achievable statistical accuracy for such a measurement at RHIC.

The behavior of A^Λ in Fig. 1(a) for the different sets of polarized Λ fragmentation functions can be easily understood from the fact that the process, in this particular kinematical region, is dominated by contributions from u and d quarks, so that the differences between the predictions in Fig. 1(a) are driven by the differences in the corresponding ΔD_u^Λ and ΔD_d^Λ. This immediately implies that the asymmetry has to be close to zero for scenario 1, negative for scenario 2 and positive and larger for scenario 3. The η-dependence is also readily understood: at negative η, the parton densities of the polarized proton are probed at small values of x_2 (i.e., in the "sea region"), where the ratio $\Delta q(x_2)/q(x_2)$ is also small. On the contrary, at large positive η, typical values of x_2 correspond to the valence region where the quarks are polarized much more strongly, resulting in an asymmetry that increases with η.

The results in Fig. 1(a) clearly demonstrate the usefulness of the proposed kind of measurements to determine the polarized Λ fragmentation functions more precisely. The expected statistical errors are much smaller than the differences in A^Λ induced by the various models. Thus an analysis of A^Λ would provide an excellent way of ruling out some of the presently allowed sets of spin-dependent Λ fragmentation functions, *provided* the observed differences in A^Λ are not obscured or washed out by the theoretical uncertainties inherent in this calculation. There are three major sources of uncertainties: the dependence of A^Λ on variations of the hard scale Q, which is of particular importance since we are limited to a LO calculation, our present inaccurate knowledge of the precise x-shape and the flavor decomposition of the polarized densities Δf^p, especially of Δg, and our ignorance of ΔD_g^Λ. Fig. 1(b) gives an example of the scale dependence of A^Λ by changing the scale from $Q = p_T$ to $Q = p_T/2$ for scenario 3. Even though $d\Delta\sigma/d\eta$ and $d\sigma/d\eta$ individually change by as much as a factor 2 at certain values of η, the uncertainty almost cancels in the ratio A^Λ. We also show in the same figure the changes in the predictions resulting from varying the polarized parton distributions, using the recent LO set 1 of Ref. [7], denoted by DSS, instead of the GRSV [6] one. As can be observed, the asymmetry remains practically unchanged. Also, as an extreme way of estimating the impact of the polarized gluon distribution, we have artificially set it to zero ($\Delta g(x,\mu^2) \equiv 0$). We find that changes in our predictions only occur in the region of negative η, but are small in the interesting region $\eta > 0$ where the asymmetries are larger. The expected effect coming from the unknown ΔD_g^Λ distribution is also found to be negligible.

III TRANSVERSE POLARIZATION

In the case of *transverse* polarization no experimental information is available. For instance, the transversity densities, denoted by $\Delta_T q(x, Q^2)$, which are equally fundamental at leading twist as the $\Delta_L q(x, Q^2)$, are completely unknown for the time being. In a similar way, one can define transversity fragmentation functions, denoted by $\Delta_T D_q^h(x, Q^2)$, to describe the fragmentation of a transversely polarized quark into a transversely polarized hadron.

In order to be able to make sensible predictions [5] for the possible spin-transfer asymmetries for this process, we will exploit the positivity constraints derived in [8] to constrain the involved quantities $\Delta_T q(x, Q^2)$ and $\Delta_T D_q^h(x, Q^2)$ in a non-trivial way. Let us first recall that a positivity constraint at the naive parton model level was obtained for the $\Delta_T q(x)$, which reads [8]

$$2|\Delta_T q(x)| \leq q(x) + \Delta_L q(x) \ . \qquad (2)$$

An analogous positivity bound for the fragmentation functions of a quark q into a hadron h holds, namely

$$2|\Delta_T D_q^h(x)| \leq D_q^h(x) + \Delta_L D_q^h(x) \ . \qquad (3)$$

This new result is maintained by the QCD Q^2 evolution at leading order (LO). We will use these non-trivial bounds (2) and (3) to constrain the unmeasured transversity parton densities $\Delta_T q(x, Q^2)$ and fragmentation functions $\Delta_T D_q^h(x, Q^2)$ in our studies of the spin-transfer asymmetry for transversely polarized Λ baryon production at RHIC below.

We will use the approach of *saturating* the positivity inequalities given in Eqs. (2) and (3) at the input resolution scale μ to constrain the unknown transversity parton densities $\Delta_T q(x, Q^2)$ and the Λ fragmentation functions $\Delta_T D_q^\Lambda(x, Q^2)$, respectively.

Fig. 2 shows our predictions for the spin-transfer asymmetry D_{NN}^Λ as a function of rapidity for $\sqrt{s} = 500 \, \text{GeV}$ and $p_T^{min} = 13 \, \text{GeV}$. We have used the three different scenarios for the $\Delta_T D_q^\Lambda$ discussed above, employing the hard scale $Q = p_T$. The possibility to have negative and positive asymmetries of the same size for each scenario reflects the freedom in the choice of the sign for the $\Delta_T D_q^\Lambda$ and the $\Delta_T q$ in Eqs. (3) and (2), respectively. The "error bars" in Fig. 2 should give again an impression of the achievable statistical accuracy for such a measurement at RHIC.

The results shown in Fig. 2 clearly demonstrate the usefulness of studying also the production of transversely polarized Λ hyperons at RHIC. Of course, one should keep in mind that the asymmetries presented in Fig. 2 represent only a rough *upper bound* of what can be expected in an actual measurement. Hence the measured asymmetry will possibly be considerably smaller with respect to our prediction, but even when reduced by a factor of 2 or 4, a measurement of D_{NN}^Λ would still remain feasible since the expected statistical errors are very small.

It is a pleasure to thank J. Soffer, M. Stratmann and W. Vogelsang for enjoyable collaborations.

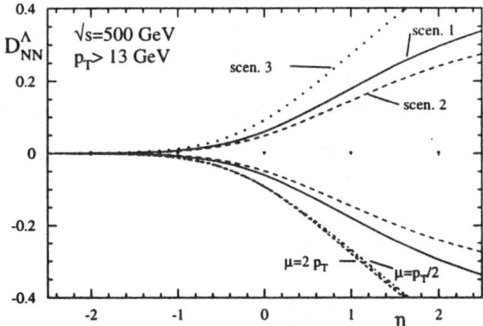

FIGURE 2. Upper bounds for the spin-transfer asymmetry D_{NN}^Λ as functions of the rapidity of the produced Λ at RHIC energies.

REFERENCES

1. D. de Florian, M. Stratmann and W. Vogelsang, *Phys. Rev.* **D57**, 5811 (1998).
2. M. Burkardt and R.L. Jaffe, *Phys. Rev. Lett.* **70**, 2537 (1993).
3. R.L. Jaffe, *Phys. Rev.* **D54**, 6581 (1996).
4. D. de Florian, M. Stratmann, and W. Vogelsang, *Phys. Rev. Lett.* **81**, 530 (1998).
5. D. de Florian, J. Soffer, M. Stratmann, and W. Vogelsang, *Phys. Lett.* **B439**, 176 (1998).
6. M. Glück, E. Reya, M. Stratmann and W. Vogelsang, *Phys. Rev.* **D53**, 4775 (1996).
7. D. de Florian, O. Sampayo and R. Sassot, *Phys. Rev.* **D57**, 5803 (1998).
8. J. Soffer, *Phys. Rev. Lett.* **74**, 1292 (1995).

Measurement of Spin Observables in Exclusive $\bar{p}p \to \bar{\Lambda}\Lambda$ Production

Kent Paschke for the PS185 Collaboration

Carnegie Mellon University
Pittsburgh, PA 15213, USA

Abstract. The PS185 experiment at LEAR has produced a wealth of high precision measurements of cross-sections and final state polarization observables in near-threshold antihyperon-hyperon production from antiproton-proton annihilation. In its most recent run, PS185/3 extended its capabilities by utilizing a transversely polarized frozen spin target to measure exclusive $\bar{\Lambda}\Lambda$ production. This allows access to a broad set of spin observables involving initial state spin. Competing theoretical models for this reaction have differing predictions for some of these newly-accessible spin observables, most notably the depolarization D_{nn}. This data is expected to provide a rigorous test of these models. Current results from the analysis of this data are presented.

INTRODUCTION

The exclusive production of antihyperon-hyperon states from antiproton-proton annihilation ($\bar{p}p \to \bar{Y}Y$) has been extensively studied near threshold by the PS185 collaboration throughout the 13 year lifetime of LEAR at CERN. Results for the total and differential cross-sections and for those spin observables depending on only final state spins have been published for a range of energies and channels. Most of this data has been in the $\bar{\Lambda}\Lambda$ production channel, which has been explored from very near threshold to a center of mass excess energy of about 200 MeV.

Both t-channel meson exchange models and s-channel constituent quark models have been applied to interpret this data, with both approaches providing a reasonable description of the current data. One reason for this is uncertainty in the $\bar{Y}Y$ final state interaction (FSI). The freedom in the FSI parameters, and to a lesser extent in the initial state interaction parameters, obscures the effects of the strangeness production model. This problem is reduced when considering observables that depend on the initial state spin. Specifically, the two models give very different predictions for the depolarization D_{nn} [2]. D_{nn} is the correlation between the initial state proton spin and the Λ spin in the direction normal to the scattering plane. The meson exchange model predicts a strongly negative D_{nn}, while the quark models predicts a positive D_{nn}. These predictions are insensitive to the FSI and ISI, so measurement of this observable provides a sensitive test that can be used to distinguish these models [2,3].

For this reason, the PS185 collaboration modified its apparatus to include a transversely polarized proton target [4]. Data was taken in 1996 at beam momenta of $1525\frac{MeV}{c}$ and $1640\frac{MeV}{c}$. Some preliminary results from the analysis of the $1640\frac{MeV}{c}$ data set are presented here.

EXPERIMENT

A schematic representation of the detector is shown in Fig. 1. A more detailed description of the experimental set-up can be found in Refs. [5]. The PS185 detector recorded the tracks from the charged hyperon decay mode. The boost to the lab frame confines the $\bar{Y}Y$ production reaction to a forward cone, so a detector in the forward region provided 4π C.M.S. coverage. Tracking chambers were used to uniquely identify the events through a fit of the event topology to a kinematic hypothesis. Behind this, three drift planes in a weak magnetic field detected deflection of the tracks in order to distinguish the $\bar{\Lambda}$ from the Λ. For the trigger, scintillators were placed directly upstream and downstream of the target to detect incoming beam particles while vetoing charged particles exiting the target. A hodoscope placed behind the tracking chambers detected the charged final state. These three components were combined to create a simple charge-neutral-charge trigger arrangement. In order to maintain the trigger efficiency, it was crucial to place the veto scintillators as close as possible to the target. A frozen spin butynol target with a cryostat diameter of only 4.2 cm was used, which allowed close placement of the veto counters. The average polarization of the target during the run period was about 62%.

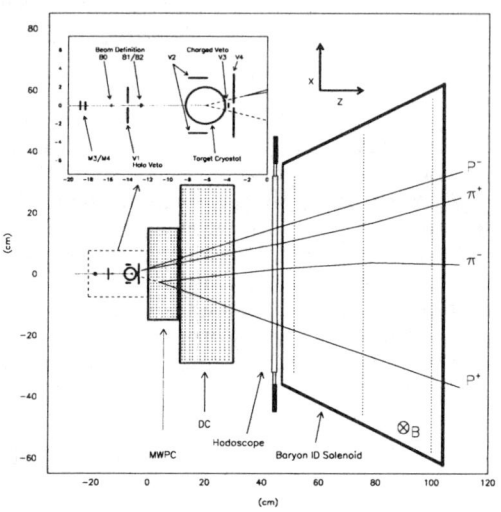

FIGURE 1. The PS185 detector

PRELIMINARY RESULTS

The self-analyzing weak decay of the hyperon correlates the direction of the decay products with the hyperon spin. This allows measurement of the spin observables of the production reaction simply by measuring the angular distribution of the decay products. Without target or beam polarization, the decay distribution for any production angle depends on only 5 spin observables: P_n, C_{mm}, C_{nn}, C_{ll}, and C_{ml}. These observables refer to a frame where $\hat{n}$ is the normal to the scattering plane, $\hat{l}$ is different for each particle and points in the direction of the particle's c.m.s. motion, and $\hat{m} = \hat{n} \times \hat{l}$. $\hat{m}$ and $\hat{l}$ point in opposite directions for the anti-hyperon and hyperon.

In order to discuss the measurement with a polarized target, we introduce a similar frame for the target proton: $\hat{n}$ is the normal to the scattering plane, $\hat{l}$ is the direction of the proton momentum in the CMS frame, and $\hat{m} = \hat{n} \times \hat{l}$. By construction, a transverse target polarization has no component in the $\hat{l}$ (longitudinal) direction, so spin observables involving the $\hat{l}$ direction of the proton spin cannot be directly measured by this experiment.

Many spin observables are restricted to be zero or have trivial relationships with other observables due to parity or C-parity symmetries. Taking these symmetries into account, the angular distribution for the PS185/3 data taken with a transversely polarized target depends on 19 distinct, non-zero spin observables. In principle each of these observables can be extracted. For this report, a reduced angular distribution was generated by averaging over the azimuthal angles of the hyperon decays. This reduced angular distribution depends on only five spin observables (P_n, C_{nn}, A_n, D_{nn}, K_{nn}), two of which (P_n, C_{nn}) have been previously measured. Preliminary results for the three new measurements are presented here.

No predictions have been published for the left-right scattering asymmetry A_n, shown in Fig. 2. It is a rank-1 spin observable, so it can be measured with a high degree of statistical precision. A_n is expected to be sensitive to the initial and final state interactions.

The depolarization D_{nn} and the polarization transfer K_{nn} were the primary focus of this measurement. The polarization transfer, which is a measure of the target proton spin transfer to the $\bar{\Lambda}$, is analogous to the depolarization, which is a measure of the target proton spin transfer to the Λ. The different models of strangeness production produce differing predictions for both K_{nn} and D_{nn}. The preliminary results for K_{nn} and D_{nn} are shown in Figs. 3(a), 3(b), with the predictions [2,3] superimposed. Clearly, the predictions from both models are inconsistent with the results for these observables. This demonstrates that the dynamics of strangeness production are not well understood.

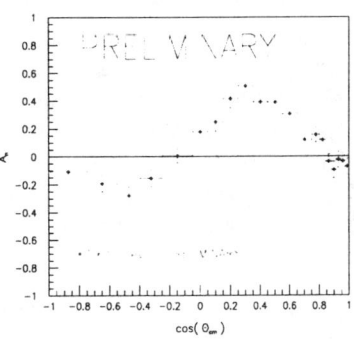

FIGURE 2. The scattering asymmetry A_n. No predictions for A_n are published. Errors shown are statistical only.

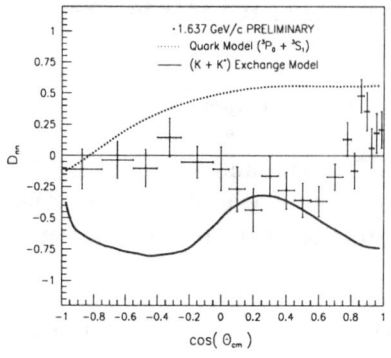 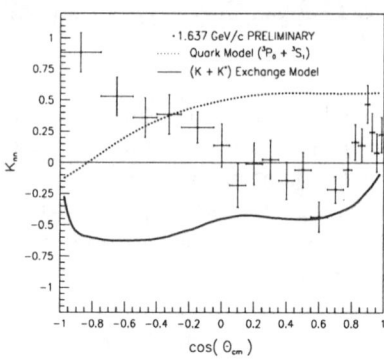

(a) Depolarization D_{nn} (b) Polarization Transfer K_{nn}

FIGURE 3. The Depolarization D_{nn}, and polarization transfer K_{nn} for $\bar{p}p \to \bar{\Lambda}\Lambda$. Meson exchange model predictions are from [2], while quark model predictions are from [3].

FUTURE WORK

The full set of spin observables has yet to be explored for this data. It remains to produce results for all of the available observables. In addition, consideration must be given to potential sources of systematic error such as production off carbon nuclei in the target, but these are expected to have little effect on the presented results. The systematic uncertainty must be estimated before these results can be considered final. Finally, there is a second data set, taken with a beam momentum of $1525 \frac{MeV}{c}$, which is currently being analyzed at the University of New Mexico, Albuquerque. Results from this data set should be available soon.

REFERENCES

1. PS185 Collaboration (P.D. Barnes *et al.*), Phys. Rev. C **54**, 1877 (1996).
2. J. Haidenbauer, K. Holinde, V. Mull and J. Speth, Phys. Lett. **B291** (1992) 223.
   J. Haidenbauer, K. Holinde, V. Mull and J. Speth, Phys. Rev. C **46** (1992) 2158.
3. M.A. Alberg, E.M. Henley, P.D. Kunz, and L. Wilets, Nucl. Phys. **A560** (1993) 365-388.
   M.A. Alberg, E.M. Henley, P.D. Kunz, and L. Wilets, Phys. At. Nucl. **57** (1994) 1608-1613. 113.
4. PS185 Collaboration (B. Bassalleck *et al.*), CERN/SPSLC-95-13.
5. PS185 Collaboration (P.D. Barnes *et al.*), Phys. Lett. B **189**, 249 (1987).
   PS185 Collaboration (P.D. Barnes *et al.*), Phys. Lett. B **229**, 432 (1989).

Transversely polarized Λ production

Daniël Boer

RIKEN-BNL Research Center
Brookhaven National Laboratory, Upton, NY 11973, U.S.A.

Abstract. Transversely polarized Λ production in hard scattering processes is discussed in terms of a leading twist T-odd fragmentation function which describes the fragmentation of an unpolarized quark into a transversely polarized Λ. We focus on the properties of this function and its relevance for the RHIC and HERMES experiments.

INTRODUCTION

Transverse polarization distribution and fragmentation functions parameterize transverse spin effects in hard scattering processes. The question is which of these functions might be relevant for the description of the single transverse spin asymmetry in the process $pp \to \Lambda^\uparrow X$ [1]? The fact that in this asymmetry the transverse spin and the transverse momentum appear to be correlated –they are orthogonal to each other–, indicates that a so-called T-odd function is required.

T-ODD FRAGMENTATION FUNCTIONS

In case there are two large scales ($\sqrt{s}$ and p_T) present in processes like $pp \to \Lambda^\uparrow X$ and $pp^\uparrow \to \pi X$ [2], the description factorizes into a hard subprocess cross section convoluted with two types of soft physics correlation functions. The latter characterize the distribution of quarks inside a hadron (Φ) and quarks fragmenting into

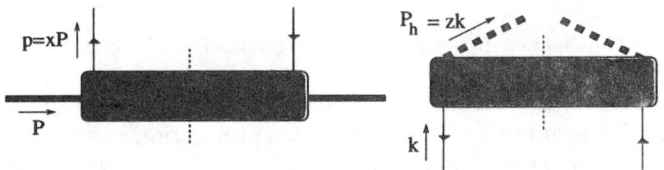

FIGURE 1. The correlation functions $\Phi(x)$ and $\Delta(z)$.

a hadron plus anything (Δ), see Fig. 1. As a function of the lightcone momentum fraction x, Φ is given by (after imposing parity and time reversal)

$$\Phi(x) = \frac{1}{2}[f_1(x)\slashed{P} + g_1(x)\lambda\gamma_5\slashed{P} + h_1(x)\gamma_5\slashed{S}_T\slashed{P}]. \quad (1)$$

The transversity function h_1 [3] is the distribution of transversely polarized quarks inside a transversely polarized hadron. Similarly, the fragmentation correlation function $\Delta(z)$ [4] is parameterized as

$$\Delta(z) = \frac{1}{2}[D_1(z)\slashed{P} + G_1(z)\lambda\gamma_5\slashed{P} + H_1(z)\gamma_5\slashed{S}_T\slashed{P}]. \quad (2)$$

The transversity fragmentation function H_1 is the probability that a transversely polarized quark fragments into a transversely polarized (spin-1/2) hadron plus anything. The transversity functions h_1 and H_1 are not sufficient to describe single spin asymmetries. But if one includes transverse momentum dependence [3], then T-odd functions can lead to unsuppressed single spin asymmetries (for a detailed explanation cf. Ref. [5]). The transverse momentum dependent fragmentation functions[1] are defined through the parameterization of the correlation function $\Delta(z, \mathbf{k}_T)$

$$\Delta(z, \mathbf{k}_T) = \text{T-even part} + \frac{1}{2}\left[D_{1T}^\perp \frac{\epsilon_{\mu\nu\rho\sigma}\gamma^\mu P^\nu k_T^\rho S_T^\sigma}{M} + H_1^\perp \frac{\sigma_{\mu\nu}k_T^\mu P^\nu}{M}\right]. \quad (3)$$

The fragmentation functions $D_{1T}^\perp$ and $H_1^\perp$ are T-odd functions, linking transverse spin –of a hadron and quark, respectively– and transverse momentum with a specific orientation (handedness), cf. Figs. 2 and 3. There are a few experimen-

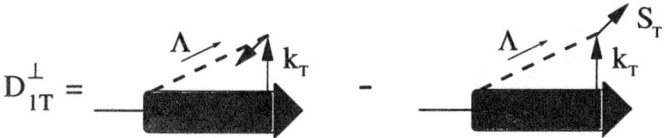

FIGURE 2. The function $D_{1T}^\perp$ signals different probabilities for $q \to \Lambda(\mathbf{k}_T, \pm\mathbf{S}_T) + X$.

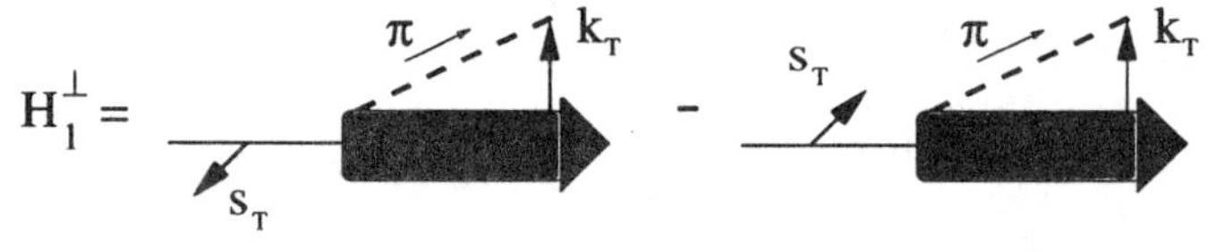

FIGURE 3. The function $H_1^\perp$ signals different probabilities for $q(\pm\mathbf{S}_T) \to \pi(\mathbf{k}_T) + X$.

tal indications that the "Collins effect" function $H_1^\perp$ [6] is indeed nonzero [7–9]; it can account for a number of different pion production asymmetries, including $pp^\uparrow \to \pi X$ [10]. The chiral-even function $D_{1T}^\perp$ [11] is expected to be relevant for transversely polarized Λ production [12], e.g. in $pp \to \Lambda^\uparrow X$.

---

[1] Due to the problematic nature of T-odd *distribution* functions [6], here we will focus only on T-odd *fragmentation* functions, which are expected to arise due to final state interactions, rather than due to time reversal symmetry violation.

TRANSVERSELY POLARIZED Λ PRODUCTION

The single transverse spin asymmetry

$$P_N = \frac{\sigma(pp \to \Lambda^\uparrow X) - \sigma(pp \to \Lambda^\downarrow X)}{\sigma(pp \to \Lambda^\uparrow X) + \sigma(pp \to \Lambda^\downarrow X)} \quad (4)$$

exhibits similar behavior as the pion production single spin asymmetries, namely the magnitude grows as a function of p_T and x_F. The BNL-RHIC collider can reveal whether $pp \to \Lambda^\uparrow X$ persists at larger values of $\sqrt{s}$ and p_T, which would be an important indication that a factorized picture should be applicable. Such a factorized description (requiring $p_T \gtrsim 1$ GeV) in terms of the quark fragmentation function $D_{1T}^\perp$ would imply that the production of a transversely polarized Λ is independent of the details of the initial state and unlike existing models of the above asymmetry [13,14], one could restrict to the modeling of $D_{1T}^\perp$. A detailed study of $pp \to \Lambda^\uparrow X$ within the factorized picture will be presented in Ref. [12]. Here we want to indicate how a fit of $D_{1T}^\perp$ from that data can be used to compare to possible future $\ell p \to \ell' \Lambda^\uparrow X$ data and to $e^+ e^- \to \Lambda^\uparrow \text{jet} X$ data.

1) Recently, an asymmetry in $ep \to \Lambda^\uparrow X$ was reported [15] (preliminary): $P_N = 0.066 \pm 0.011 \pm 0.025$ (arising mainly from $ep \to e'\Lambda^\uparrow X$ events with $Q^2 \simeq 0$). If such an asymmetry would also be found in the semi-inclusive DIS process $ep \to e' \Lambda^\uparrow X$ (for $Q^2 \gtrsim 1 \text{ GeV}^2$) the factorized description [11] would imply[2]

$$\frac{d\Delta\sigma(ep \to e' \Lambda^\uparrow X)}{d\sigma(ep \to e' \Lambda X)} = \sin(\phi_{S_T^\Lambda} - \phi_{P_T^\Lambda})\, P_N, \quad (5)$$

where the analyzing power P_N is a function of f_1, D_1 and $D_{1T}^\perp$. If one makes an Ansatz inspired by a model for the Collins function [6] (η, M are free parameters and Gaussian transverse momentum dependence of the unpolarized fragmentation function D_1 is assumed):

$$D_{1T}^\perp(z, \boldsymbol{k}_T^2) = \eta \frac{M M_\Lambda}{\boldsymbol{k}_T^2 + M^2} D_1(z, \boldsymbol{k}_T^2) \implies P_N \approx \frac{2\eta M Q_T}{Q_T^2 + 4M^2}. \quad (6)$$

This exhibits plausible behavior as a function of the transverse momentum (Q_T) of the Λ and this expression for P_N can be used to fit $D_{1T}^\perp$, allowing for a check of the universality of $D_{1T}^\perp$ if compared to the resulting fit from $pp \to \Lambda^\uparrow X$ data.

2) In the case of $e^+ e^- \to \Lambda^\uparrow \text{jet} X$, one determines the transverse momentum (Q_T) of the Λ compared to the jet (or thrust) axis. In the factorized picture with transverse momentum dependent functions this will yield a contribution [17]

$$\frac{d\sigma(e^+ e^- \to \Lambda^\uparrow \text{jet} X)}{d\Omega dz dQ_T} \propto \sin(\phi_{P_T^\Lambda} - \phi_{S_T^\Lambda}) \frac{Q_T}{M_\Lambda} \sum_{a,\bar{a}} e_a^2\, D_{1T}^{\perp a}(z, z^2 Q_T^2). \quad (7)$$

---

[2] The chiral-even function $D_{1T}^\perp$ can also be probed in charged current exchange processes; for the cross section expressions in semi-inclusive DIS we refer to [16].

Note that the exponential fall-off of the function at larger values of Q_T wins out over the explicit power of Q_T (all under the requirement of $Q_T^2 \ll Q^2$). Also we note that for $e^+e^- \to Z \to \Lambda^\uparrow \text{jet} X$ transverse Λ polarization has been measured to be small (at the percent level) [18]. We expect two effects to contribute to the suppression of the above contribution at large scales such as $Q = M_Z$. The function $D_{1T}^\perp$ might be a decreasing function of Q^2, although this is not yet known. Also, transverse momentum dependent azimuthal asymmetries suffer from effective *power* suppression due to Sudakov factors [19]. The LEP result would then not contradict a possibly large asymmetry in $ep \to e'\Lambda^\uparrow X$ at lower energies (measurable at HERMES).

ACKNOWLEDGEMENTS

I thank M. Anselmino, U. D'Alesio and F. Murgia for collaboration on this topic. Furthermore, I thank RIKEN, Brookhaven National Laboratory and the U.S. Department of Energy (contract number DE-AC02-98CH10886) for providing the facilities essential for the completion of this work.

REFERENCES

1. E.g. R608 Collaboration, A.M. Smith et al., *Phys. Lett.* **B185**, 209 (1987).
2. FNAL E704 Collab., *Phys. Lett.* **B261**, 201 (1991); *Phys. Rev. Lett.* **77**, 2626 (1996).
3. J.P. Ralston and D.E. Soper, *Nucl. Phys.* **B152**, 109 (1979).
4. J.C. Collins and D.E. Soper, *Nucl. Phys.* **B194**, 445 (1982).
5. D. Boer, hep-ph/9912311.
6. J.C. Collins, *Nucl. Phys.* **B396**, 161 (1993).
7. A. Bravar (for the SMC Collaboration), *Nucl. Phys. (Proc. Suppl.)* **B79**, 520 (1999).
8. HERMES Collaboration, A. Airapetian et al., *Phys. Rev. Lett.* **84**, 4047 (2000).
9. A.V. Efremov, O.G. Smirnova and L.G. Tkatchev, hep-ph/9812522.
10. M. Anselmino, M. Boglione, F. Murgia, *Phys. Rev.* **D60**, 054027 (1999); M. Boglione, E. Leader, *Phys. Rev.* **D61**, 114001 (2000).
11. P.J. Mulders and R.D. Tangerman, *Nucl. Phys.* **B461**, 197 (1996).
12. M. Anselmino, D. Boer, U. D'Alesio and F. Murgia, in preparation.
13. J. Félix, *Mod. Phys. Lett.* **A14**, 827 (1999).
14. J. Soffer, hep-ph/9911373.
15. S.L. Belostotski (for HERMES), *Nucl. Phys. (Proc. Suppl.)* **B79**, 526 (1999).
16. D. Boer, R. Jakob and P.J. Mulders, *Nucl. Phys.* **B564**, 471 (2000).
17. D. Boer, R. Jakob and P.J. Mulders, *Nucl. Phys.* **B504**, 345 (1997); *Phys. Lett.* **B424**, 143 (1998).
18. ALEPH Collaboration, *Phys. Lett.* **B374**, 319 (1996); OPAL Collaboration, *Eur. Phys. J.* **C2**, 49 (1998).
19. D. Boer, hep-ph/0004217.

Exploring the Spin of the Gauge Sector with Non-Perturbative Coordinates

John P. Ralston and Roman V. Buniy

Department of Physics and Astronomy, University of Kansas, Lawrence, KS 66045, USA

Abstract. The densitites of Lorentz generators canonically conjugate to the vector potential are not gauge invariant. This has led to confusion of what experiments can measure. Constructing new coordinates for the gauge sector gives gauge invariant densitites of energy, momentum, and spin and orbital angular momentum. The existence of invertible transformations establishes that the same theory can be described in either coordinates.

1 Spin and Canonical Quantities in QCD

There are many fascinating puzzles involving spin and QCD. The proton "spin crisis" has recently focused attention on the spin of the gauge sector. Unfortunately theorists do not yet agree on what is measured. The decomposition of the gluon's angular momentum into spin and orbital terms is not gauge invariant, indicating that theorists may be discussing something unobservable when they address these operators.

Let us think a bit more deeply about this catastrophe. Noether's theorem gives a unique way to calculate canonical quantities. The particular canonical quantities obtained for the energy, angular momentum, spin, and so on are conjugate to the particular variables used in the Noether variations. Since different gauges for the vector potential amount to different choices of canonical variables, it is no wonder that the conjugate momenta depend on the gauge.

The difficulty dates to Poynting's (1884) derivation [?] of $\vec{E} \times \vec{B}$ as the flux of energy in electrodynamics. Textbooks seem to cover up the embarrassment that Poynting's vector is not conjugate to the usual coordinate, the vector potential A^μ. A procedure for forcing expressions that are gauge invariant by adding surface terms is well known. One accomplishes nothing by imitating the same procedure in QCD: in either case, one has changed the currents from what they started out to be. The present situation is one where distinguished theorists have lamented that "what can be interpreted cannot be measured, and what can be measured cannot be interpreted" [1].

If experimentalists cannot measure basic features of the theorist's coordinates, then perhaps theorists should adapt their coordinates to what experimentalists can measure. Indeed the procedure of adding surface terms in canonical physics itself induces canonical transformations: changes of variables. So it is not inconsistent for $\vec{E} \times \vec{B}$ to be the flux of energy in electrodynamics, or for $\vec{r} \times \vec{E} \times \vec{B}$ to represent the angular momentum density, if only one finds those special variables to which these quantities might be conjugate. Yet describing a continuum gauge theory in variables other than the vector potential is challenging. Incorporating some progress on this question [3] we offer an interesting resolution of the paradox.

2 A Canonical Consistency Relation

Let ϕ^k be all the fields of the theory, including the vector potential, with ordinary space-derivative $\phi^k_{,\mu}$ and gauge covariant derivative $i\phi^k_{;\mu} = i\phi^k_{,\mu} - gA_\mu\phi^k$. The gauge group and representations can be left unspecified, the correct color contractions implied. Let the theory be gauge invariant; then by an old theorem of Utiyama [4], every derivative is a gauge-covariant derivative with appropriate charge g. The Lagrangian $L = L(\phi^k, \phi^k_{;\mu})$. The energy momentum tensor from Noether and *using the vector potential as coordinate* is

$$T^{\mu\nu} = \frac{\partial L}{\partial \phi^k_{,\mu}} \phi^k_{,\nu} - g^{\mu\nu} L = \frac{\partial L}{\partial \phi^k_{;\mu}} (\phi^k_{;\nu} - igA_\mu) - g^{\mu\nu} L.$$

The breakdown of gauge invariance is evident in the A_μ term.

Now assume there exists new coordinates $e, \partial e$ so that A is expressed as a function $A(e, \bar{e}; \partial e, \partial \bar{e})$. The bar denotes complex conjugation. Continue the calculation including the variations due to the new coordinates. Can we arrange the calculation so that we find the gauge invariant result?

One can ask for the terms that are not gauge invariant to cancel in a naive way. This gives a consistency relation

$$i\delta^{\alpha\mu} A_\nu = i \sum_\sigma \frac{\partial A^\alpha}{\partial \bar{e}_{\sigma,\mu}} \bar{e}_{\sigma,\nu}. \tag{1}$$

If this consistency relation can be satisfied, then the energy-momentum tensor, and other canonical quantities using derivatives will be gauge invariant.

3 Frames as Coordinates

Our procedure leads directly toward the geometrical meaning of the vector potential as a *connection*. In a geometrical context we identify new coordinates e^a_μ as describing *frames*. Frames need two indices, separated here by Roman and Greek fonts: both indices are internal and refer to the gauge-space. Frames transform homogeneously:

$$e^a_\mu \to U^{ab}(x) e^b_\mu; \quad e^a_\mu \to \Lambda_{\mu\nu}(x) e^a_\nu, \tag{2}$$

where U, Λ are elements of groups. If the frames are a complete set, then their covariant derivatives can be expanded as linear combinations of themselves:

$$\partial e^a_\mu + \Gamma_{\mu\nu} e^a_\nu = -A^{ab} e^b_\mu,$$

where Γ the connection (Christoffel symbol) acting on the components. Lorentz indices have been suppressed. We invert to find $iA = e^{-1}(\partial - \Gamma)e$. The simplest implementation (and making color indices a, b explicit) is

$$A^{ab}(e, \bar{e}) = -i e^a_\mu \partial \bar{e}^b_\mu. \tag{3}$$

A short calculation shows this solves the consistency condition (1). By direct calculation we also find

$$T^{\alpha\beta} = \frac{\partial L}{\partial(D_\alpha \bar{e}^a_\mu)}(D^\beta e^a_\mu) - g^{\alpha\beta} L = F^{\mu\alpha} F^\nu_\alpha - g^{\mu\nu} L$$

The vector potential expressed with the e's must also transform in a consistent way under gauge transformations: from Eq. 3 we find that

$$A \to U(x) A U^{-1}(x) - (i/g) U(x) \partial U^{-1}(x)$$

as a derived property. The local transformations $U(x)$ are the elements of the gauge group.

Thus we have new variables for which the canonical generators are gauge-invariant. However the e's are not entirely new. Many attempts have been made to transform from A to e in the form of "dressing operators" [5] indicated to exist by Gauss' Law. The results are complicated integral expressions. It turns out that our procedure in the reverse direction has a natural geometrical foundation. Now if the e's are Lorentz scalars, which is one possibility, then the covariant angular momentum takes the classic $\vec{r} \times \vec{E} \times \vec{B}$ result of textbooks such as Jackson's: even in the $SU(3)$ non-Abelian theory, $M^{\lambda\mu\nu} = x^\lambda T^{\mu\nu} - x^\mu T^{\nu\lambda}$. Use of this operator is not trivial: when invoked to capture the angular momentum of the gauge sector, it represents a physical hypothesis that the dressing of the quarks and gluons does not depend on their polarization. In all likelihood dressing probably does depend on polarization, which makes the gauge sector very interesting. There is no way to address this question in the perturbative framework because gauge invariance requires all order of approximation.

The old coordinates of the vector potential remain, and seem to be ideally suited for perturbation theory. The spin operator of the vector potential remains as a derived, gauge-dependent and highly theoretical quantity. It is not observable.

Our non-perturbative coordinates may be helpful in describing the non-perturbative character of experiments.

The deepest question is the extent to which our new coordinates for the gauge sector are obliged to describe exactly the same theory as the old one. Lacking any results on quantization rather restricts the claims. Conditions for invertibility are needed to show a one-to-one correspondence of coordinates. Then the theories are exactly the same at the classical level, and need only the same measure to prove the same in quantization. We investigated the conditions of invertibility, which are known as "Frobenius Relations", and find that infinitely many interesting gauge theories can be built upon various kinds of assumptions. For example, if the complementary group Λ is global, then this group must satisfy certain conditions of being larger than the gauge group. Then it is possible to embed gauge theories on a background space that is geometrically flat: the curvatures come entirely from embedding. The $SU(3)$ group offers some interesting challenges in this regard, because it is a rather unusual space. We studied $SU(2)$ and $SO(3)$ theories in detail and found the Frobenius Relations generating A from e and vice-versa explicitly. One can also ask for $\Lambda = \Lambda(x)$ to be a local group, including the general linear groups, producing numerous variations.

Acknowledgments: This work was supported in part under the Department of Energy grant number DE-FG02-98ER41079, by the University of Kansas General Research Fund, and the *Kansas Institute for Theoretical and Computational Science/ K*STAR* program.

REFERENCES

1. R.L. Jaffe, in *Proceedings of the RIKEN BNL Research Center Workshop (RHIC SPIN)*, Volume 25, BNL-52581 Formal Report (2000).
2. J. H. Poynting, Phil. Trans. Roy. Soc. **175** (1884) 343.
3. J. P. Ralston, Found. Phys. (2000, in press); *Proceeding of ISMD 99*, Brown University 1999 (World Scientific, in press); in *Proceedings of the RIKEN BNL Research Center Workshop (RHIC SPIN)*, Volume 25, BNL-52581 Formal Report (2000); R. V. Buniy and J. P. Ralston, *QCD 2000* (Villefranche 2000) (edited by H. M. Fried and B. Mueller, in press).
4. R. Utiyama, *Phys. Rev.* 101, 1957 (1956).
5. K. Haller and L-s. Chen, in *29th International Conference on High-Energy Physics (ICHEP 98)*, (Vancouver, Canada, 1998)(to be published), hep-th/9808044; see also hep-th/9803250. M. Belloni, L-s. Chen, and K. Haller, *Phys.Lett.* **B403**, 316 (1997); *Phys.Lett.* **B373** ,185, (1996); M. Lavelle and D. McMullan, *Physics Reports*, **279** 1, 1997 and references therein; E. Bagan, M. Lavelle, and D. McMullan, *Phys.Rev.* **D57**, 4521(1998) ; *Mod.Phys.Lett* **A12** , 1815, (1997); Erratum-ibid.A12:2317, (1997).

Measurement of g_1^p in the low-x and low-Q^2 Region at HERMES

Ralf Kaiser
on behalf of the HERMES Collaboration

DESY Zeuthen, Platanenallee 6, 15738 Zeuthen, Germany

Abstract. A new analysis has extended the kinematic range of the 1997 HERMES data on the polarized structure function g_1^p to include the region of $0.0021 < x < 0.021$ and $0.1 \text{ GeV}^2 < Q^2 < 0.8 \text{ GeV}^2$. The preliminary results on g_1^p in this region are presented and the observation of scaling violations in polarized deep inelastic lepton-nucleon scattering (DIS) is discussed.

The polarized proton structure function g_1^p has been studied for over a decade, yet the experimental precision is still not satisfactory, in particular at small values of Bjorken-x. New data at low values of x will reduce the extrapolation uncertainty for the integral of g_1^p, which in turn will improve the accuracy in the determination of the quark contribution to the nucleon spin. Low x ($x \leq 0.01$) data in current experiments are typically measured at relatively small values of Q^2 ($Q^2 \approx 1.0 \text{ GeV}^2$). By studying scaling violations it may become possible to draw conclusions on the validity of perturbative QCD (pQCD) at low photon virtualities in polarized DIS.

The HERMES experiment uses the polarized 27.6 GeV positron (or electron) beam of the HERA accelerator at DESY. Spin rotators convert the transverse beam polarization that arises due to the Sokolov-Ternov effect [1] into a longitudinal polarization in the interaction region. The average beam polarization for the 1997 polarized hydrogen data was 0.55 ± 0.02 (syst.). The internal gas target uses a cooled, open-ended storage cell inside the beam vacuum. The storage cell is fed with polarized hydrogen by an atomic beam source. The average target polarization was measured to be 0.88 ± 0.04 (syst.).

The HERMES detector [2] is an open-geometry forward spectrometer. A horizontal septum in the dipole magnet of the spectrometer shields the beam pipe against the magnetic field. As a result, the detector system is split into two identical halves above and below the beam pipe and the minimum vertical acceptance is limited to ± 40 mrad. The maximum acceptances are ± 140 mrad vertically and ± 170 mrad horizontally. The spectrometer has an angular resolution of $\delta\Theta < 0.6$ mrad and a momentum resolution of $\delta p/p < 1.5\,\%$. An excellent hadron-electron separation is provided by the combination of a transition radiation detector, a preshower hodoscope, an electromagnetic calorimeter and a gas threshold Čerenkov counter. The positron identification efficiency is better than 97 % with a hadron contamination of less than 1.0 % for each x-bin of the presented analysis. The DIS trigger consists of a coincidence of signals from three hodoscope planes and the calorimeter with a required energy deposit of at least 1.4 GeV.

The present analysis extends the kinematic range of the HERMES results on $g_1^p(x,Q^2)$ of the previous publication [3] ($0.0212 < x < 0.85$ and $Q^2 > 0.8\,\text{GeV}^2$) down to 0.0021 in x and 0.1 GeV^2 in Q^2. To make this possible, the upper limit on y, the fractional virtual photon energy, was increased from 0.85 to 0.91. This required the detailed understanding of the detector and trigger performance in the low momentum region close to the trigger threshold for the scattered positron. As a result, seven new, preliminary data points for $g_1^p(x,Q^2)$ were obtained in the low-x, low-Q^2 region. The polarized structure function $g_1^p(x,Q^2)$ was determined from the ratio g_1/F_1 by multiplying with $F_1(x,Q^2) = F_2(x,Q^2)(1+\gamma^2)/2x(1+R(x,Q^2))$. The ratio g_1/F_1 has been calculated from the measured longitudinal cross section asymmetry $A_{\parallel}$. F_2 and R were parameterized according to [4] and [5]. More details on the extraction of g_1^p are given in ref. [3].

The experimental uncertainties for the newly analyzed low-x region originate mainly from normalization, background corrections, acceptance cut variations and scattering angle systematics. The systematic uncertainties due to radiative and smearing corrections have been minimized through an iterative procedure. The total systematic uncertainty is about 14 % for the lowest x-value and about 9 % for the other six points.

A comparison of HERMES and SMC data on the x-behaviour of the structure function ratio g_1/F_1 (see ref. [6]) exhibits no Q^2-dependence within the experimental uncertainties. This represents the first comparison of the SMC data at low x to data with improved precision by HERMES. Figure 1 shows the x-dependence of $g_1^p(x,Q^2)$ at its measured average Q^2-values in comparison to recent data from E143 [7] and SMC [8]. The new low-x data points ($0.0021 < x < 0.021$) from HERMES are shown together with the previously published data [3] for $x > 0.0212$. The higher beam energy of SMC compared to HERMES and E143 leads to values of Q^2 that are about one order of magnitude larger in every x-bin. This is shown in the lower panel of figure 1. Hence, the comparison of the HERMES and SMC data allows observations on the Q^2-dependence of g_1^p: Scaling violations are not visible above $x \approx 0.2$, while they are significant for lower x, extending into the region of the new data points. Further low-x data that cover a wide enough Q^2-range are necessary to confirm the latter observation and to test the validity range of pQCD in the spin sector at low photon virtualities.

The author would like to thank all HERMES colleagues that contributed to this analysis, in particular U.Stösslein, H.Böttcher, Y.Gärber and W.-D.Nowak.

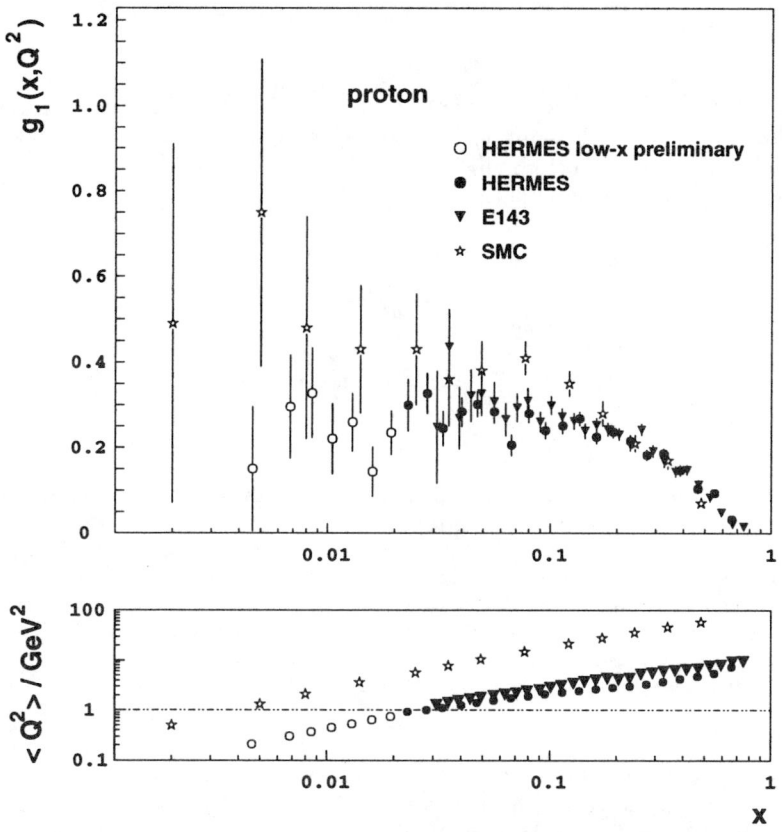

FIGURE 1. *The HERMES polarized proton structure function $g_1^p(x, Q^2)$ (upper panel) and the measured average Q^2-values (lower panel) in comparison to E143 [7] and SMC [8] data. The error bars represent the statistical uncertainties. The four additional SMC points at even lower x from ref. [9] have not been included here.*

REFERENCES

1. A.A. Sokolov, I.M. Ternov, Sov. Phys. Doklady **8**, 1203 (1964)
2. HERMES Coll., K. Ackerstaff *et al.*, Nucl. Instr. and Meth. **A 417**, 230 (1998).
3. HERMES Coll., A. Airapetian *et al.*, Phys. Lett. **B 442**, 484 (1998).
4. H. Abramowicz and A. Levy, hep-ph/9712415.
5. L. W. Withlow *et al.*, Phys. Lett. **B 250**, 193 (1990).
6. H. Böttcher, XIVth Rencontre de Physique de la Vallee d'Aoste, Italy, Feb. 2000
7. E143 Coll., K. Abe *et al.*, Phys. Rev. **D 58**, 112003 (1998).
8. SMC Coll., B. Adeva *et al.*, Phys. Rev. **D 58**, 112001 (1998)
9. SMC Coll., B. Adeva *et al.*, Phys. Rev. **D 60**, 072004 (1999)

Single-Spin/Azimuthal Asymmetries in Semi-Inclusive and Exclusive DIS

C.A. Miller

TRIUMF
Vancouver, BC, Canada V6T 2A3

on behalf of the HERMES Collaboration

Abstract. Recent measurements of pion production in deeply inelastic scattering of positrons have revealed azimuthal distributions of the pions that are proportional to the polarization of only the target nucleon. Such spin-azimuthal asymmetries from semi-inclusive measurements can be interpreted in terms of previously unobserved quark distribution functions related to transversity, which represents the probability of finding a transversely polarized quark in a nucleon polarized transverse to its (infinite) momentum. Much larger asymmetries have now been seen in the exclusive kinematic limit. Their interpretation is more problematic.

We often like to present data as the result of measurements that were carefully planned to investigate the relevant physical phenomena. However, this is the story of how easy it is to find oneself embroiled in the complexities of *transverse* spin physics, even though it may have been the furthest thing from the mind as data were being innocently recorded with a *longitudinally* polarized lepton beam on a *longitudinally* polarized proton target. (I recognize that our colleagues from Yerevan and Frascati who put a lot of effort into developing this field at HERMES may have a different perspective.) This is also the story of fortunate confluence of theory and experiment, since it is only in the last few years that theoretical studies have identified a rich forest of polarized quark distribution functions [1], a few 'trees' of which are central to understanding the new data. These relate to a set of distributions $\delta_q(x)$ known as *transversity* that was first identified and then fully described over the last two decades [2].

Transversity was mentioned in a plenary talk earlier in this conference [3]. However, I will reiterate some key properties. These distributions generate h_1 — the the last completely unmeasured twist-2 structure function among the three (also f_1 and g_1) that survive integration over the intrinsic k_T of the partons. By twist-2, we of course mean that it survives asymptotically as the negative square of the invariant mass of the virtual photon $Q^2 \to \infty$, as its effects are not suppressed by any powers in $1/Q$, and the associated distributions $\delta_q(x)$ can be interpreted as probabilities

— namely that of finding a transversely polarized quark of flavor q in a nucleon polarized transverse to its (infinite) momentum. In non-relativistic quark models of the nucleon, it is equivalent to the familiar longitudinal distribution $\Delta_q(x)$, but in the relativistic world of current quarks, boosts and rotations do not commute. Transverse antiquark polarization contributes to $\delta_q(x)$ with the opposite sign to quark polarization, so $\delta_q(x)$ reflects the polarization of valence quarks. Also, the gluon polarization does not mix with the quark polarization in $\delta_q(x)$ as it does in $\Delta_q(x)$, so its Q^2-evolution is very different. Furthermore, its first moment — the *tensor charge* of the nucleon — is expected to be reliably calculable in lattice QCD. Finally, the reason it is still unmeasured is that it is chiral-odd, and hence is not manifest in inclusive deeply inelastic scattering (DIS), only in processes involving another chiral-odd object. This 'supporting character' can be the same distribution function for another colliding transversely polarized nucleon in *e.g.* the Drell-Yan process, or a chiral-odd fragmentation function in semi-inclusive DIS.

Such a chiral-odd fragmentation function has been identified theoretically [4]. This 'Collins function' is often designated as $H_1^{\perp(1)}(z)$, where z represents the fraction of the virtual photon energy carried by a hadron produced by fragmentation of the struck quark. (See Table 1 for an explanation of the terminology.) The Collins function can be imagined to be a 'polarimeter' for the transverse polarization of the struck quark. For this purpose, the experimenters must distinguish between hadrons produced in directions to the left and right, relative to the spin vector of the struck quark in a frame in which it is pointing 'up'. In other words, the experimental signature for transversity operating in conjunction with the Collins function is an inhomogenous distribution of the hadrons in the azimuthal angle ϕ of the hadron about the direction of the virtual photon's 3-momentum **q**, relative to the plane containing both **q** and the target nucleon polarization axis (see Fig. 1a)). This azimuthal asymmetry is proportional to the target polarization. Such a single-spin asymmetry must involve a time-reversal odd object, which in this case is the Collins function. Its T-odd nature arises through the final-state interactions that it represents, rather than any fundamental symmetry breaking. Up until now, there has been only tentative evidence that the Collins function has a significant magnitude, from both effects on jet structure due to the very small transverse quark polarization produced in Z^0 decay: $e^+ + e^- \to Z^0 \to q\bar{q} \to 2$ jets [5], and from the interpretation of large single-spin asymmetries observed in inclusive pion production from polarized $p + \bar{p}$ collisions [6]. This has been the subject of much interest, since it would offer the opportunity to make the first measurement of transversity.

Complex though this process may appear, it involves only one 'tree' of our forest of quark distributions mentioned above. With a *longitudinally* polarized target, more possibilities appear. First, let us look at the new data from HERMES [7], shown in Fig. 1b). What might this signal have to do with transversity? First there is a relatively trivial effect: the target is polarized along the *lepton beam* axis, and has a (small) component that is transverse to **q**. But this is only a

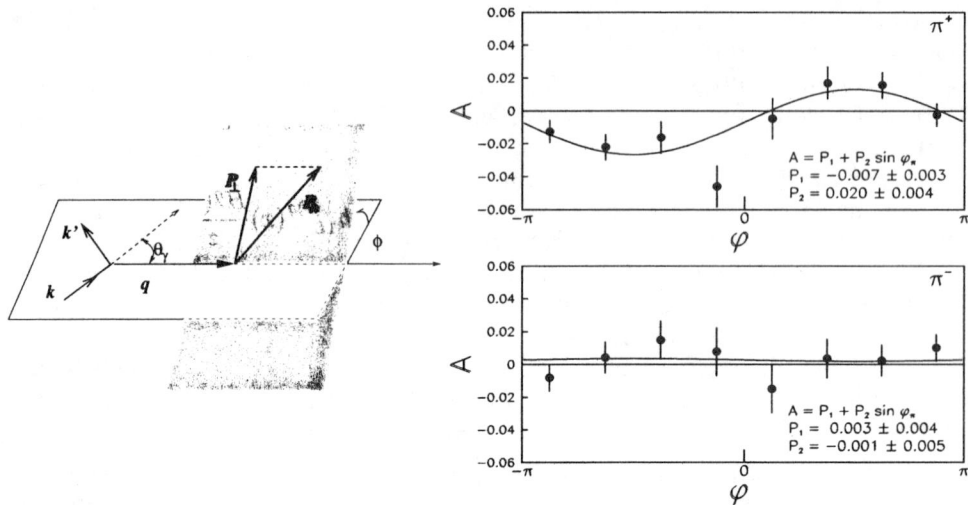

FIGURE 1. a): Geometry for semi-inclusive deeply inelastic scattering b): 'Longitudinal' target single-spin asymmetry for pion lepto-production: $A_{UL}(\phi) = \frac{1}{P} \frac{N^\uparrow(\phi) - N^\downarrow(\phi)}{N^\uparrow(\phi) + N^\downarrow(\phi)}$

small part of the story. The new data prompted the realization [8] that other more esoteric distributions could have larger effects, especially at the relatively small experimental $\langle Q^2 \rangle$ of a few GeV^2, where effects of non-leading twist can appear. Table 1 shows a selection of nucleon-structure and fragmentation functions, classified with respect to twist level and chirality. Of particular interest to us are the chiral-odd structure functions that relate to a target polarized *longitudinal* to q ('L' subscript), as these are candidates for collaborating with the Collins function to produce a spin-azimuthal asymmetry with such a target polarization. There are two candidates, one twist-2 ($h_{1L}^{\perp(1)}$), and one twist-3 (h_L). They are closely related to transversity, as is evident from this relation based on Lorentz covariance [9]:

$$h_L(x) = h_1(x) - \frac{d}{dx} h_{1L}^{\perp(1)}(x) \qquad (1)$$

		T-even DF		T-odd FF	
		chirality		chirality	
		even	odd	even	odd
twist 2	O	f_1		$\mathbf{D_1}$	$H_1^\perp$
	L	g_{1L}	$h_{1L}^{\perp(1)}$		
	T	g_{1T}	$\mathbf{h_1}\ h_{1T}^\perp$	$D_{1T}^\perp$	
twist 3	O	$f^\perp$	e		H
	L	$g_L^\perp$	h_L	$D_L^\perp$	E_L
	T	$g_T\ g_T^\perp$	$h_T\ h_T^\perp$	$\mathbf{D_T}$	E_T

Table 1. A selection of twist-2 and twist-3 distribution and fragmentation functions, sorted according to chirality. A physical observable must involve only chiral-even or two chiral-odd functions. Only the functions in bold-face survive integration over k_T. The subscript '1' indicates leading twist, while the '$\perp$' and '(1)' superscripts indicate respectively an essential role for the intrinsic transverse momentum k_T of the quark in the hadron, and that the quantity represents the k_T^2-moment.

These distribution functions are manifest in specific moments of the spin-azimuthal asymmetry. If $S_{L(T)}$ are the components of the target nucleon spin parallel(orthogonal) to $\mathbf{q}$, then the $\sin\phi$ azimuthal moments appearing with 'U'npolarized beam that are proportional to 'L'ongitudinally target polarization are:

$$\langle\sin\phi\rangle_{UL} \propto S_L \frac{M}{Q} \sum_{q,\bar{q}} e_q^2 x [h_L^q(x) H_1^{\perp(1)q}(z) - x h_{1L}^{\perp(1)q}(x) \frac{\tilde{H}^q(z)}{z}]$$
$$+ S_T \sum_{q,\bar{q}} e_q^2 x h_1^q(x) H_1^{\perp(1)q}(z) \qquad (2)$$

$$\langle\sin 2\phi\rangle_{UL} \propto S_L \sum_{q,\bar{q}} e_q^2 x h_{1L}^{\perp(1)q}(x) H_1^{\perp(1)q}(z) \qquad (3)$$

Note the $1/Q$ factor that multiplies the twist-3 terms appearing in $\langle\sin\phi\rangle_{UL}$. One of these terms involves the interaction-dependent part $\tilde{H}$ of a twist-3 fragmentation function H. However, this need not introduce yet another unknown function, since Lorentz covariance can again be used to relate it to the Collins function:

$$\frac{\tilde{H}(z)}{z} = \frac{d}{dz}[z H_1^{\perp(1)}(z)] \qquad (4)$$

Note also that the definition of the azimuthal angle ϕ must be generalized for the case of target polarization aligned with $\mathbf{q}$. Here the azimuth is measured with respect to the lepton scattering plane. Hence for target spin alignment along the lepton beam direction, the ϕ definitions based on the two spin components coincide. Typically for such measurements, $S_T \ll S_L$.

Can this data be understood quantitatively, with apparently so many unknown functions? The number of unknown distribution functions can be reduced to one, if twist-3 contributions are neglected:

$$h_{1L}^{\perp(1)}(x) \simeq -\frac{x}{2} h_L(x) \simeq -x^2 \int_x^1 dy \frac{h_1(x)}{y^2} \qquad (5)$$

Then these can all be computed from a model for h_1, such as the chiral-soliton model. The Collins function can also be approximated by a model, with the magnitude constrained by the analysis of $e^+ + e^-$ data. Such calculations have been done [10] neglecting all twist-3 contributions including $\tilde{H}$, with results that are consistent with all the HERMES data, as well as less precise data with a tranversely polarized target from SMC [11]. Thus the new data are well enough understood to inspire plans at HERMES to exploit this experimental signal with a transversely-polarized target to measure h_1.

It has been more recently reported by HERMES that these same spin-azimuthal signals become dramatically larger and more difficult to understand as z approaches unity - the exclusive kinematic limit. Fig. 2 shows this data for all 3 pion charge states. The asymmetry for π^+ changes sign before growing in magnitude, that for π^0 grows without changing sign, while for π^-, the asymmetry remains small. Exclusive π^+ production is expected to differ in one respect — it receives a contribution from the 'pole term' representing knockout of cloud pions from the proton. Unfortunately, there are no theoretical predictions of spin-azimuthal asymmetries for exclusive pion production from a *longitudinally* polarized target. However, there exist predictions that asymmetries for production of pions by longitudinal photons from a transversly polarized target are also large [12]. If these could be extracted from future such experimental data, preferably measured at higher Q^2 where the higher-twist production by transverse photons is suppressed, they would constrain the skewed parton distributions that are currently of great interest.

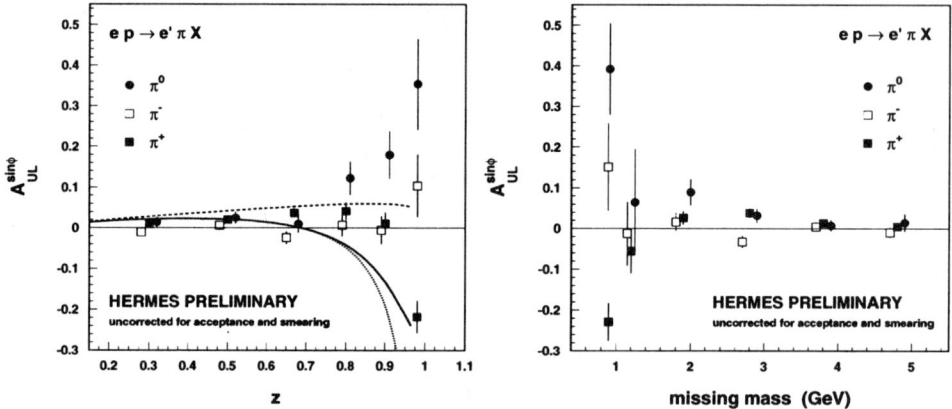

FIGURE 2. The target-related analyzing power for the spin-azimuthal $\sin\phi$ moment for pion production in deeply inelastic scattering of positrons. In the left(right) panel, it is plotted as a function of z(missing mass). The curves in the left panel apply only to π^+, and are described in the text.

The π^0 data in Fig. 2 suggests that the transition to large asymmetries at the exclusive limit might be smooth and continuous in a manner not determined by the experimental resolution of about 300(500) MeV for charged(neutral) pions. It is natural to associate such a trend with the transition from the level of leading twist of 2 for the semi-inclusive region, to 4 for exclusive production, since it is expected that higher-twist contributions progressively appear in the semi-inclusive process as z approaches unity. This observation suggests one way to understand the z-dependence. The dashed curve in Fig. 2 shows the result of a model calculation [13,14,9] for the semi-inclusive production of π^+, similar to the one described above, while the dotted curve shows the effect of including the effect of $\tilde{H}$ according to Eqs. 2 and 4. The change in sign is 'explained', although the effect is too extreme. One more effect can be included. It was first pointed out 20 years ago by Berger [15] that at large z, the struck quark becomes far off shell with invariant mass $k^2 = \frac{P_\perp^2}{z(1-z)}$, with the result that the dominant contribution to the wave function of the produced meson is perturbative one-gluon exchange. It was further shown that the unpolarized fragmention process therefore receives a higher-twist contribution as z approaches unity:

$$D(z \to 1) \to D(z) + \frac{2}{9}\frac{\langle P_\perp^2 \rangle}{Q^2}, \tag{6}$$

where $P_\perp$ is the component of the pion orthogonal to $\mathbf{q}$. When this correction is included, the result is the solid curve in Fig. 2, which better agrees with the π^+ data.[1]

It is clear that single-spin azimuthal asymmetries observed in both semi-inclusive and exclusive leptoproduction of mesons have the potential to become a rich source of information about hadron structure. Both the HERMES and COMPASS experiments will reveal much about this potential, but it will probably require a new high-luminosity polarized accelerator facility such as the proposed EPIC ep collider to fully exploit this field.

REFERENCES

1. Mulders P.J. and Tangerman R.D., *Nucl. Phys.* **B461**, 197 (1996).
2. Ralston J. and Soper D.E., *Nucl. Phys.* **B152**, 109 (1979); Artru X. and Mekhfi M., *Z. Phys* **C 45**, 669 (1990); Jaffe R.L. and Ji X., *Phys. Rev. Lett.* **67**, 552 (1991); Cortes J.L., Pire B. and Ralston J.P., *Z. Phys* **C 55**, 409 (1992).
3. See M.G. Vincter, these proceedings.
4. Collins J., *Nucl. Phys.* **B396**, 161 (1993) and **B420**, 565 (1994).
5. DELPHI: Efremov A.V., Smirnova O.G. and Tkatchev L.G., *Nucl. Phys. (Proc. Suppl.)*, **74**, 49 (1999) and **79**, 554 (1999).
6. Boglione M. and Leader E., *Phys. Rev.* **D 61**, 114001 (2000).
7. HERMES: Airapetian A. et al, *Phys. Rev. Lett.* **84**, 4047 (2000).
8. Oganessyan K.A. et al., hep-ph/9808368.
9. Boglione M. and Mulders P.J., *Phys. Lett.* **B478**, 114 (2000).
10. Efremov A.V. et al., *Phys. Lett.* **B478**, 94 (2000).
11. SMC: Bravar A., *Nucl. Phys. (Proc. Suppl.)*, **B79**, 521 (1999).
12. Frankfurt L.L. et al, *Phys. Rev. Lett.* **84**, 2589 (2000).
13. Avakian, H., DIS2000, Liverpool, April, 2000.
14. Kotzinian A.M. and Mulders P.J., *Phys. Lett.* **B406**, 373 (1997).
15. Berger E.L., *Z. Phys.* **C 4**, 289 (1980).
16. Brandenburg A., Khose V.V. and Müller D., *Phys. Lett.* **B347**, 413 (1995).

---

[1] This kinematic effect has other interesting consequences [16].

Physics Prospects For a Polarized Electron-Ion Collider

J.T. Londergan

Dept. of Physics and Nuclear Theory Center, Indiana University
2401 Sampson Road, Bloomington, IN 47408, USA

Since the advent of Quantum Chromodynamics [QCD], we have made great progress in understanding the properties of the fundamental constituents of matter. Deep inelastic scattering [DIS] reactions, Drell–Yan processes and direct photon processes allow us to map out single-quark and gluon probability densities in the nucleon. Electron accelerators have been powerful tools in extracting this information. Recently, a group of nuclear physicists have considered the essential characteristics of a next-generation electron facility, *EPIC* (*E*lectron *P*olarized-*I*on *C*ollider), which would investigate some outstanding questions regarding the quark-parton structure of matter.

In our opinion, such an electron collider should possess the following characteristics:

- *A polarized electron beam:* An EPIC collider would require a highly polarized beam of electrons with energy in excess of 4 GeV. At present, both a ring configuration and a Linac are under consideration for accelerating the electrons.

- *Polarized nucleon and unpolarized light-ion beams:* Hadrons would be accelerated in a synchrotron, with Siberian Snakes to produce high longitudinal polarization of proton (and deuteron) beams. Beams of heavier ions could be either polarized or unpolarized.

- *Large CM collision energies:* An EPIC facility would have a minimum electron/nucleon total CM energy $\sqrt{s_{eN}} \geq 30$ GeV. This would enable it to surmount charm and bottom quark production thresholds, and would ensure sufficiently high energies and momentum transfers to suppress higher-order transition amplitudes. The asymmetric electron-hadron energies are necessary for the study of target fragmentation.

- *High Luminosity:* High luminosity is required to produce sufficient event rates to measure exclusive and semi–inclusive processes. For $e - p$ collisions, this would require luminosities of the order of $\mathcal{L} \sim 10^{33}$ cm^{-2}s^{-1}. At present, the only known method to achieve such luminosities would involve electron cooling of the hadron beams.

- *Detectors with Excellent particle ID:* The EPIC detectors must be able to detect all particles emerging from hard exclusive reactions. To study target fragmentation, the detectors must be able to resolve both fast and slow-moving particles emerging from the interaction point.

Preliminary studies of possible accelerator design suggest that the desired beam energy, polarization, and luminosities could be achieved without using untested acceleration or beam handling mechanisms.

An EPIC facility would open up a relatively new kinematic region. At present, there are a number of fixed-target machines (e.g., Jefferson Lab, Mainz, Bates) with high luminosities ($\mathcal{L} \geq 10^{37}$) but rather low CM energies ($\sqrt{s} \leq 4$ GeV), and machines such as HERA and COMPASS which have high CM energies ($\sqrt{s} > 20$ GeV) but relatively low luminosity, ($\mathcal{L} \leq 10^{31}$). An EPIC facility would provide optimal control of spin and flavor degrees of freedom over the kinematic region $0.001 \leq x_{Bj} \leq 0.6$, and its detectors would be designed to measure semi–inclusive and exclusive processes over a large interval in rapidity.

The idea of an electron-nucleus collider was considered in some detail at a GSI workshop in 1997 [1]. Roughly a year later a workshop at DESY [2] investigated this idea in more detail. A workshop was held at IUCF in April 1999 [3], and an additional workshop on this topic will be held at MIT in Sept. 2000. Here, we list briefly the important areas, where an EPIC facility would significantly advance our understanding of the partonic structure of matter.

- *The Origin of Spin*: The increasingly precise polarized deep–inelastic scattering data show that quarks carry an anomalously small fraction of the nucleon spin. A critical question is to determine the origin of the nucleon's spin. A leading candidate is polarized glue; although other facilities will measure gluon distributions, an EPIC collider would be capable of exceptionally precise measurements of gluon densities. There may also be significant contributions from the orbital motion of the partonic constituents, and it may eventually be possible to extract this information through knowledge of generalized parton densities [5,6].

    - *Transversity:* In addition to longitudinal spin amplitudes, EPIC could also study transverse spin components. For such studies, azimuthal asymmetries allow a clean separation of effects due to quark transverse momenta [4].

- *Parton Correlations*: pQCD techniques, so successful in describing inclusive processes, have now been extended to semi-inclusive and exclusive final states. Such measurements help to determine the flavor of the quarks which participate in hard collisions: they can also measure the correlations between partons inside a nucleon or nucleus.

    - *Target Fragmentation:* An EPIC facility can investigate the so-called target fragmentation region, and study the behavior of the partonic constituents left behind following a single hard collision. This region is largely unexplored, and has high potential for significantly new discoveries. Initially, target fragmentation experiments would test QCD predictions. Such measurements could ultimately provide access to more complete information about hadronic and nuclear wave functions.

- *Generalized Parton Densities:* There has been much work recently studying the generalized parton densities (GPD) (also termed "skewed" or "off-forward" parton densities), which give information on the correlations between quarks in a hadron. This information can be most easily extracted in exclusive reactions such as deeply virtual Compton scattering [DVCS], and hard meson electroproduction. The information obtained from these measurements is generally in the form of a convolution of the GPD's with the hard scattering amplitudes. However, meson electroproduction can be used to "filter out" certain GPD components. Spin asymmetries can also be used to select specific GPD amplitudes. Certain components could also be extracted by comparing reactions initiated by electrons and positrons [7]. An EPIC collider (which would in principle possess all these features) would be an excellent facility for the study of such processes. This facility could also measure inelastic meson electroproduction, where the nucleon is excited to octet or decuplet baryon final states. Such processes could access the QCD wavefunctions for these excited states, test the validity of the hypothesis of $SU(3)$ flavor symmetry for such states, and study flavor and spin correlations in parton densities.

- *The Structure of the Sea:* A substantial flavor dependence has recently been found in the nucleon sea [8], much larger than predicted by perturbative QCD [pQCD]. An EPIC facility could investigate whether sea quark and antiquark distributions are equal, as is generally assumed. EPIC could also test whether there is flavor dependence in the *polarization* of sea quarks. The EPIC Collider could also examine the role of heavy quark flavors in the nucleon sea. For example, the study of quark fragmentation into hyperons could provide important information regarding the origin of the nucleon's spin. In particular, this facility could isolate target fragments, and study heavy meson production (c and b) at threshold.

- *Nuclear Dynamics:* QCD predicts that in certain hard quasi–elastic processes, only quark components with small transverse size can contribute. The EPIC facility could test whether this phenomenon, termed *color transparency*, is an important feature of nuclear interactions. It would also determine if meson electroproduction is dominated by transitions to the simplest $q-\bar{q}$ components, as has been hypothesized, and whether *hidden color* nuclear configurations become important in processes dominated by small separations. Experiments with light nuclei could also study the modification of the nucleon's quark and gluon distributions when the nucleon is embedded in a nucleus.

Over the years, lepton accelerators have extracted vast amounts of information through inclusive measurements, where only the scattered lepton was observed. More recently, facilities such as HERMES have focused on semi-inclusive measurements, where hadrons (generally mesons) are observed in coincidence with outgoing leptons. Such processes make it easier to determine the flavor of the struck parton. An EPIC facility would concentrate on semi–inclusive and exclusive processes, observing specific final hadronic states in coincidence with the scattered lepton. The asymmetric collider would allow full kinematic coverage, and would especially focus on resolution of events over large intervals in rapidity. Since the spins of both electrons and hadrons can be varied, an EPIC facility would allow optimal control over spin and flavor components for a wide variety of experiments involving either proton or nuclear beams colliding with electrons. With the understanding of current–quark and target fragmentation which we could obtain from such a collider, we may eventually be able to construct a complete "snapshot" of the hadronic light cone wave function.

The workshops held to date demonstrate the promise of an EPIC facility to address these important issues. At present, experimental simulations and theoretical calculations are being undertaken to study the feasibility of measuring various processes. At the Sept. 2000 workshop at MIT, the various accelerator designs will be reviewed, and we will begin detailed studies of various detector configurations. Detailed studies are simultaneously being carried out at Brookhaven, for an $e - A$ collider, "eRHIC". The focus of this effort is on the interesting new physics which could be obtained at low x and high A, using the RHIC facility as a source of high-energy heavy ions. Both of these studies reveal promising new areas where a next-generation electron collider could enrich our understanding of the partonic structure of matter.

The author wishes to thank L. Bland, S. Brodsky, J. Cameron, R. Milner, M. Strikman, A. Szczepaniak and M. Vanderhaeghen for useful discussions. This research was supported in part by the NSF under research contract PHY-0070368.

REFERENCES

1. *Report of the DESY/GSI/NuPECC Workshop on Electron-Nucleus Collisions*, Report GSI 97-04, 1997.
2. *Report of the DESY Workshop on Electron-Nucleus Collider*, Report DESY 98-01, 1998.
3. *Proceedings of workshop on Physics with a High Luminosity Polarized Electron-Ion Collider,* edited by L.C. Bland, J.T. Londergan and A.P. Szczepaniak, World Scientific, Singapore, 2000.
4. Mulders, P., *EPIC Workshop Proceedings*, Ref. [3], 2000, pp.35-49.
5. Ji, X., *Phys. Rev. Lett.* **78**, 610-613 (1997).
6. Radyushkin, A.V., *Phys. Rev.* **D56**, 5524-5557 (1997).
7. Vanderhaeghen, M., eprint hep-ph/0007232, (2000).
8. Peng, J-C. and Garvey, G.T., eprint hep-ph/9912370, (1999).

Nuclear and Particle Astrophysics

Status of the Borexino Solar Neutrino Experiment

J.C. Maneira [1]

INFN – Via Celoria 16 – 20133 Milano – Italy
also FCUL/LIP – Ed. C8, Campo Grande – 1000 Lisboa – Portugal
e-mail: maneira@mi.infn.it

Abstract. Borexino is a new real-time detector for low energy solar neutrinos presently in construction at the Gran Sasso underground laboratory in Italy. The low energy threshold and the low radioactive background rate required for the detection of neutrinos below 1 MeV determines the choice of using liquid scintillator as the detection medium. Several tests on scintillator radiopurity and on various techniques to be used in Borexino were performed in CTF, a prototype detector installed in Gran Sasso.

By performing the first real-time measurement of low energy solar neutrinos, Borexino will supply new information contributing to the solution of the Solar Neutrino Problem. In this paper, the goals, techniques and status of the Borexino experiment are presented, as well as the program for calibrations and monitoring of the detector.

INTRODUCTION

The long-standing solar neutrino problem is confirmed by all the existing experiments and is one strong indication for neutrino mixing and a non-zero neutrino mass. The predictions for solar neutrino fluxes are believed to be very robust and have been confirmed by helioseismologgical observations [1]. The GALLEX and SAGE experiments have checked their results with a neutrino source, finding no discrepancies.

---

[1] on behalf of the Borexino collaboration: G. Alimonti, C. Arpesella, H. Back, M. Balata, G. Bellini, J. Benziger, S. Bonetti, A. Brigatti, B. Caccianiga, L. Cadonati, F. Calaprice, G. Cecchet, M. Chen, M. Deutsch, O. Donghi, F. Elisei, F. von Feilitzsch, R. Fernholtz, R. Ford, C. Galbiati, A. Garagiola, F. Gatti, S. Gazzana, D. Giugni, A. Golubchikov, A. Goretti, C. Hagner, T. Hagner, W. Hampel, J. Handt, F.X. Hartmann, R. von Hentig, G. Heusser, A. Ianni, J. Jochum, H. de Kerret, S. Kidner, J. Kiko, T. Kirsten, G. Korga, G. Korschinek, D. Kryn, V. Lagomarsino, P. Lamarche, M. Laubenstein, F. Loeser, S. Magni, S. Malvezzi, J.C. Maneira, I. Manno, G. Manuzio, F. Masetti, U. Mazzucatto, E. Meroni, P. Musico, H. Neder, M. Neff, S. Nisi, L. Oberauer, M. Obolensky, M. Pallavicini, R. Parsells, L. Perasso, R. Pocar, R. Raghavan, G. Ranucci, W. Rau, A. Razeto, E. Resconi, P. Saggese, R. Scardaoni, C. Salvo, S. Schoenert, K. Schuebeck, T. Shutt, M. Skorokhvatov, A. Sotnikov, O. Smirnov, S. Sukhotine, R. Tartaglia, G. Testera, D. Vignaud, S. Vitale, R.B. Vogelaar, M. Wojcik, O. Zaimidoroga, Y. Zakharov

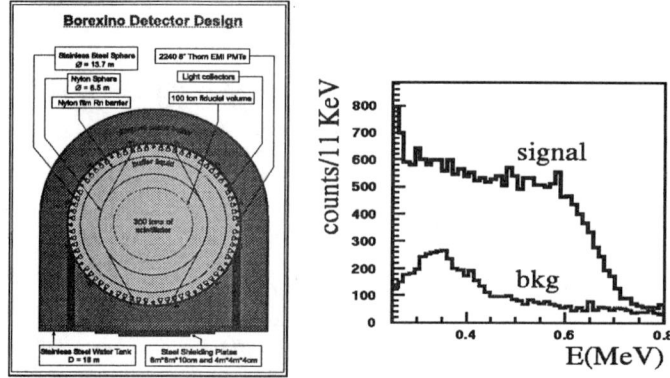

FIGURE 1. a) Borexino design. b) Expected neutrino signal and internal background after one year of data-taking.

However, the difference between measured and expected solar neutrino fluxes has a significance of about 20 σ and the only explanation that can accommodate all the existing data is the existence of neutrino flavor oscillations with small mass-squared differences ($10^{-11}eV^2 < \Delta m^2 < 10^{-5}eV^2$) [2].

New experiments are now aiming to test the oscillation hypothesis and choose between the different solutions allowed by current data. While SuperKamiokande and SNO are sensitive only to ^{8}B neutrinos, Borexino is the only project in an advanced status of construction that will supply data in the low energy region of the solar neutrino spectrum, namely the ^{7}Be 0.861 MeV line, where up to now only integrated measurements were available.

I BOREXINO DESIGN

The main requirements that drive the design of Borexino are the reduction of radioactive backgrounds and the optimization of the light collection efficiency.

The core of the Borexino detector consists of 300 tonnes of liquid scintillator contained in a thin nylon vessel of 8.5 m diameter, observed by 2200 photomultiplier tubes (PMTs) mounted on a 13.7 m diameter stainless steel sphere (SSS), as shown in Figure 1(a). Pseudocumene(PC) – 1,2,4 tri-methyl benzene – with 1.5 g/l of PPO was chosen as the liquid scintillator for Borexino due to its good radiopurity levels, high fluorescence quantum efficiency, low attenuation and fast timing properties. The PMTs (ETL9351) have a high quantum efficiency, low transit time dispersion and are made from low radioactivity Schott glass. They are equipped with highly reflective optical concentrators increasing the total optical coverage to 33 %.

Between the nylon vessel and the SSS there will be a PC buffer, serving both to have the nylon vessel in a neutral buoyancy situation and to shield the scintillator against gamma particles coming from residue radioactivity in the PMT materials.

Additional shielding against external gammas and neutrons is supplied by the water buffer between the SSS and the 18 m diameter external tank. A second nylon vessel will be placed between the inner vessel (IV) and the PMTs, serving as barrier against Radon diffusion.

In order to detect the Čerenkov light produced by muons in the buffer liquid, 17 % of the PMTs do not have concentrators; the muon veto system is completed by 200 PMTs mounted on the outside of the SSS, to collect Cerenkov light produced in the water buffer.

The liquid handling system allows the online purification of the scintillator through different methods: distillation, solid column separation, water extraction and nitrogen sparging. Other strategies that are used to keep the background from radioactive contaminants to a minimum is the accurate selection in terms of radiopurity and Radon emanation (besides chemical compatibility with PC) of all the materials and installation in a clean room environment.

The first phase of the experiment was the construction and operation of the Counting Test Facility (CTF), a small-scale prototype of Borexino dedicated mainly to study the possibility of reaching extremely low levels of radioactive contamination in several tons of liquid scintillator. CTF reached and measured unprecedented low levels of radioactive contamination and gave many insights towards the definitive Borexino design. A detailed description of the apparatus and its results can be found in Refs. [3–5]. CTF is again in operation since June 2000 in order to test samples from batches of Borexino scintillator before filling it.

II EXPECTED PERFORMANCE: SIGNAL AND BACKGROUND

The measured quantities in Borexino are the charge and time distributions of the PMT signals. The total collected charge is proportional to the energy of the recoil electron from a neutrino interaction or from any other electron or alpha particle from radioactive decays in the scintillator. The signature for mono-energetic ^{7}Be neutrinos is the characteristic Compton-like edge in the energy spectrum, shown in Figure 1(b) for one year of data and a fiducial volume of 100 tons. The signal rate is expected to be about 50 events/day in the absence of oscillations and the background at the level of 15 events/day, after the analysis cuts.

The time distributions of the PMT signals are used in three different ways to reject background events: i) Delayed Coincidences – Due to its short lifetime, the decay of ^{214}Po(^{212}Po) in the ^{238}U (^{232}Th) chain can be tagged through an $\alpha/\beta+\gamma$ coincidence; ii) Pulse-shape Discrimination – The different fluorescence decay time allows the rejection of alpha particles; c) Position Reconstruction – The event position is reconstructed from the time-of-flight of detected photons, allowing the definition of a fiducial volume and the rejection of events from external gamma particles, mainly from the PMT glass.

III CALIBRATIONS & MONITORING

The measurement of time variations of the neutrino flux will be important in Borexino, not only as evidence for neutrino oscillations (strong seasonal variations or day-night asymmetry are predicted, respectively, in the vacuum oscillations solution or the LOW oscillations solution to the solar neutrino problem), but also for the detection of the yearly eccentricity modulation, very useful as an overall consistency check. It is therefore essential to monitor the stability of the detector.

Apart from the initial energy calibration with a radioactive source, the program for detector monitoring is focused on non-invasive tools that can be used regularly without contamination risks [6]. We plan to introduce gamma sources at the level of the PMTs, and to use UV laser radiation carried by an optical fiber to excite the scintillator, mimicking the excitation from charged particles. This method has been tested in small scale and we plan to do further tests in CTF2.

A set of optical fibers will be installed in the SSS, pointing laser beams of two different wavelengths in different geometries in order to study the attenuation/scattering length of the buffer liquid and scintillator. A feasibility test of this method was carried in the Two Liquid Test Tank (TLTT), a 7 m^3 Pseudocumene filled tank built in LNGS for a long-term test of the PMTs in realistic conditions. Light was sent in such a way so that two PMTs see the direct light while all the others (46) see reflected or scattered light. Using the time distributions of the PMT signals, it was possible to distinguish between scattered and reflected light and use that to obtain useful parameters. In order to check the sensitivity of the method to PC transparency variations, we compared the parameters with independent spectrophotometric measurements done at Perugia University and verified that while the PC did not change, these oscillated less than 5%, while a change of 13 % occurred when the PC transparency actually degraded.

ACKNOWLEDGMENTS

The author acknowledges the support of the Fundação para a Ciência e a Tecnologia, Portugal and the Istituto Nazionale di Fisica Nucleare, Italy.

REFERENCES

1. Bahcall, J. N. , Basu, S. , Pinsonneault, M. H. , *Phys. Lett. B*, **433**, 1 (1998)
2. Bahcall, J. N. , Krastev, P. I. , Smirnov, A. Y. , *Phys. Rev. D*, **58**, 096016 (1998)
3. Alimonti, G. et al., *Nucl. Inst. and Meth. A*, **406** 411 (1998)
4. Alimonti, G. et al., *Astrop. Phys.*, **8**, 141 (1998)
5. Alimonti, G. et al., *Nucl. Inst. and Meth. A* **440**, 360-371 (1999)
6. Maneira, J., "Sensitivity of Borexino to seasonal variations of the solar neutrino flux", in *New Worlds in Astroparticle Physics – 1998*, edited by A. Mourão et al., World Scientific, Singapore, 1999, pp. 131-135

Charge Conjugation Violation in Supernovae and The Neutron Shortage for R-Process Nucleosynthesis

C. J. Horowitz and Gang Li

Nuclear Theory Center and Dept. of Physics, Indiana University, Bloomington, IN 47405

Abstract. Core collapse supernovae are dominated by energy transport from neutrinos. Therefore, some supernova properties could depend on symetries and features of the standard model weak interactions. The cross section for neutrino capture is larger than that for antineutrino capture by one term of order the neutrino energy over the nucleon mass. This reduces the ratio of neutrons to protons in the ν-driven wind above a protoneutron star by approximately 20 % and may significantly hinder r-process nucleosynthesis.

Core collapse supernovae are perhaps the only present day large systems dominated by the weak interaction. They are so dense that photons and charged particles diffuse very slowly. Therefore energy transport is by neutrinos (and convection).

We beleive it may be useful to try and relate some supernova properties to the symmetries and features of the standard model weak interaction. Parity violation in a strong magnetic field could lead to an asymmetry of the explosion [1]. Indeed, supernovae explode with a dipole asymmetry of order one percent in order to produce the very high 'recoil' velocities observed for neutron stars [2]. However, calculating the expected asymmetry from P violation has proved complicated. Although explicit calculations have yielded somewhat small asymmetries [3–5] it is still possible that more efficient mechanisms will be found.

In this paper we calculate effects of charge conjugation, C, violation in the Standard Model on the difference between neutrino and antineutrino interactions. In Quantum Electrodynamics C symmetry insures the cross section for e^-p is equal to that for e^+p scattering (to lowest order in α). In contrast, the standard model has large parity, P, and C violations (since the product CP is approxamitely conserved). Therefore the $\bar{\nu}$-nucleon cross sections are systematically smaller than ν-nucleon cross sections.

However at the low ν energies in supernovae, time reversal symmetry limits the difference between ν and $\bar{\nu}$ cross sections. Time reversal can relate $\nu - N$ elastic scattering and $\bar{\nu} - N$ where the nucleon scatters from final momentum p_f to initial momentum p_i. If the nucleon does not recoil then the ν and $\bar{\nu}$ cross sections are equal. Thus the difference between ν and $\bar{\nu}$ cross sections are expected to be of recoil order E/M where E is the neutrino energy and M the nucleon mass. This ratio is relatively small in supernovae. However the coefficient multiplying E/M involves the large weak magnetic moment of the nucleon (see below).

The standard model has larger ν cross sections than those for $\bar{\nu}$. For neutral currents, this leads to a longer mean free path for $\bar{\nu}_x$ compared to ν_x (with x=μ or τ). Thus even though ν_x and $\bar{\nu}_x$ are produced in pairs, the antineutrinos escape faster leaving the star neutrino rich. The muon and tau number for the protoneutron star in a supernova could be of order 10^{54} [6]. Supernovae may be the only known systems with large μ and or τ number. For charged currents, the interaction difference can change the equilibrium ratio of neutrons to protons and may have important implications for nucleosynthesis. We discuss this below. To our knowledge, all previous work on nucleosynthesis in supernovae assumed equal ν and $\bar{\nu}$ interactions (aside from the n-p mass difference).

The neutrino driven wind outside of a protoneutron star is an attractive site for r-process nucleosynthesis [7]. Here nuclei rapidly capture neutrons from a low density medium to produce heavy elements [8]. This requires, as a bare minimum, that the initial material have more neutrons than protons. The ratio of neutrons to protons n/p in the wind depends on the rates for the two reactions:

$$\nu_e + n \to p + e^-, \quad (1a)$$

$$\bar{\nu}_e + p \to n + e^+. \quad (1b)$$

The standard model cross sections for Eqs. (1a,1b) to order E/M are,

$$\sigma = \frac{G^2 \cos^2\theta_c}{\pi}(1+3g_a^2)E_e^2[1-\gamma\frac{E}{M} \pm \delta\frac{E}{M}], \quad (2)$$

with G the Fermi constant (and θ_c the Cabbibo angle), $E_e = E \pm \Delta$ the energy of the charged lepton and $\Delta = 1.293$ MeV is the neutron-proton mass difference. The plus sign is for Eq. (1a) and the minus sign for Eq. (1b). We use $g_a \approx 1.26$. Equation (2) neglects small corrections involving the electron mass and coulomb effects while the finite nucleon size only enters at order $(E/M)^2$.

We refer to the γ term as a recoil correction. It is the same for ν and $\bar{\nu}$, $\gamma = (2+10g_a^2)/(1+3g_a^2) \approx 3.10$. Finally the δ term, $\delta = 4g_a(1+2F_2)/(1+3g_a^2) \approx 4.12$, involves the interference of vector $(1+2F_2)$ and axial (g_a) currents. This violates P, which by CP invariance also violates C. This increases the ν and decreases the $\bar{\nu}$ cross section, Note, F_2 is the isovector anomalous moment of the nucleon. (This is the weak magnetism contribution.)

The equilibrium electron fraction per baryon Y_e (which is equal to the proton fraction assuming charge neutrality) is simply related to the rate $\bar\lambda$ for Eq. (1b) divided by the rate λ for Eq. (1a).

$$Y_e = (1+\frac{\bar\lambda}{\lambda})^{-1} \quad (3)$$

The ratio neutrons to protons is, $\frac{n}{p} = \frac{1}{Y_e} - 1$.

In ref. [9] we calculate the reaction rates by averaging Eq. (2) over neutrino spectra to get,

$$Y_e = \left(1 + \frac{L_{\bar{\nu}_e}\bar{\epsilon}}{L_{\nu_e}\epsilon}QC\right)^{-1}. \tag{4}$$

Here ϵ ($\bar{\epsilon}$) is the ν_e ($\bar{\nu}_e$) mean energy, L_{ν_e} ($L_{\bar{\nu}_e}$) is the ν_e ($\bar{\nu}_e$) luminosity, Q is the correction from the reaction Q value,

$$Q = \frac{1 - 2\frac{\Delta}{\bar{\epsilon}} + a_0\frac{\Delta^2}{\bar{\epsilon}^2}}{1 + 2\frac{\Delta}{\epsilon} + a_0\frac{\Delta^2}{\epsilon^2}}, \tag{5}$$

and with C violating one has a factor C,

$$C = \frac{1 - (\delta + \gamma)a_2\frac{\bar{\epsilon}}{M}}{1 + (\delta - \gamma)a_2\frac{\epsilon}{M}}. \tag{6}$$

Simply evaluating Eq. (6) for typical parameters yields $C \approx 0.8$. Thus, *the difference between ν and $\bar{\nu}$ interactions reduces the equilibrium n/p ratio by approximately 20 %.* This is an important result and will be discussed below.

Evaluating Eq. (4) for the neutrino fluxes of a Supernova simulation by Wilson [10] shows that with the C term the neutrino driven wind starts out proton rich and ends up with about equal numbers of neutrons and protons. When C violation is included the wind is never significantlt neutron rich. It is very unlikely that successful r-process nucleosynthesis can take place in the wind of this or similar models.

With the approximately 20 % reduction in n/p from the difference between ν and $\bar{\nu}$ interactions, there appears to be very serious problems with r-process nucleosynthesis in the wind of present supernova models. In addition to the initial lack of neutrons, one has to overcome the effects of neutrino interactions during the assembly of α particles and during the r-process itself [11]. These further limit the available neutrons per seed nucleus. Thus, it is unlikely that present wind models will produce a successful r-process. Of course, the wind in supernovae may not be the r-process site, although this may be unappealing (see for example [8,12]). If the wind is not the site, one must look for alternative environments.

However, the effects of neutrino interactions may be very general. The only requirement is that energy transport from neutrinos plays some role in helping material out of a deep gravitational well. Given this, it is quite likely that the n/p ratio will be determined by the relative rates of Eqs. (1a,1b). Therefore differences in ν and $\bar{\nu}$ interactions may be important for just about any nucleosynthesis site that involves neutrinos.

If the ν-driven wind is the r-process site, it is very likely, present models of the neutrino radiation in supernovae are incomplete. The high values of Y_e make it almost impossible to have a successful r-process by only changing matter properties, such as the entropy. The neutrino fluxes will (almost assuredly) need to be changed.

Changes in the astrophysics used in the simulations or new neutrino physics such as neutrino oscillations [13] could change $\bar{\epsilon}$, ϵ and or the luminosities and lead to a more neutron rich wind. The oscillations of more energetic $\bar{\nu}_x$ with $\bar{\nu}_e$ could increase $\bar{\epsilon}$. However, we have some information on the $\bar{\nu}_e$ spectrum from SN1987a [14]. Thus one can not increase $\bar{\epsilon}$ without limit. Indeed if anything, the Kamiokande data suggest a lower $\bar{\epsilon}$. Any model which tries to solve r-process nucleosynthesis problems by increasing $\bar{\epsilon}$ should first check consistency with SN1987a observations [15]. Alternative modifications could include oscillations of ν_e to a sterile neutrino or a *lowering* of ϵ. (However, we know of no model which lowers ϵ.) Whatever the modification of the neutrino fluxes, one will still need to include the differences between ν and $\bar{\nu}$ interactions in order to accurately calculate n/p.

In conclusion, supernovae are one of the few large systems dominated by energy transport from weakly interacting neutrinos. Therefore, some supernova properties may depend on symmetries and features of the standard model weak interactions. The cross secton for neutrino capture is larger than that for antineutrino capture by a term of order the neutrino energy over the nucleon mass. This difference between neutrino and antineutrino interactions reduces the ratio of neutrons to protons in the ν-driven wind above a protoneutron star by approximately 20 % and may significantly hinder r-process nucleosynthesis.

This work was supported in part by DOE grant: DE-FG02-87ER40365.

REFERENCES

1. A. Vilenkin, 1979, unpublished; Ap J **451** (1995) 700. N. N. Chugai, Pisma Astron. Zh. **10** (1984) 210. [Sov. Astron. Lett. **10** (1984) 87.]
2. A. G. Lyne and D. R. Lorimer, Nature **369** (1994) 127.
3. C. J. Horowitz and J. Piekarewicz, Nuc. Phys. **A640** (1998)281.
4. C.J. Horowitz and Gang Li, Phys. Rev. Lett. **80** (1998) 3694, Erratum-ibid. **81** (1998) 1985.
5. Phill Arras and Dong Lai, astro-ph/9806285.
6. C.J. Horowitz and Gang Li, Phys. Lett. **B443** (1998) 58.
7. S.E. Woosley and R.D. Hoffman, ApJ. **395** (1992) 202. B.S. Meyer, W.M. Howard, G.J. Mathews, S. E. Woosley and R.D. Hoffman, ApJ. **399** (1992) 656.
8. B. S. Meyer, Ann. Rev. Astron. Astrophys. **32** (1994) 153.
9. C. J. Horowitz and Gang Li, Phys.Rev.Lett. **82** (1999) 5198.
10. Y.-Z. Qian and S.E. Woosley, ApJ. **471** (1996) 471.
11. B.S. Meyer, ApJL. **449** (1995) 792. G. M. Fuller and B.S. Meyer, ApJ **453** (1995) 792. B.S. Meyer, G. McLaughlin and G.M. Fuller, astro-ph/9809242.
12. J.J. Cowan, F.-K. Thielemann and J.W. Truran, Phys. Rep. **208**(1991) 267.
13. Yong-Zhong Qian and G.M. Fuller, Phys. Rev. **D52** (1995) 656.
14. K. Hirata et al., Phys. Rev. Lett. **58** (1987) 1490; R. M. Bionta et al., Phys. Rev. Lett. **58** (1987) 1494.
15. A. Yu. Smirnov, D. N. Spergeland and J.N. Bahcall, Phys Rev. **D49** (1994) 1389. B. Jegerlehner, F. Neubig and G. Raffelt, Phys. Rev. **D54** (1996) 1194.

The GZK Bound and Strong Neutrino-Nucleon Interactions above 10^{19} eV: a Progress Report

John P. Ralston [a], Pankaj Jain[b],
Douglas W. McKay[a], and S. Panda[b]

[a] *Department of Physics & Astronomy
University of Kansas, Lawerence, KS 66045, USA*
[b] *Physics Department, I.I.T. Kanpur, India 208016*

Abstract. Cosmic ray events above 10^{19} eV have posed a fundamental problem for more than thirty years. Recent measurements indicate that these events do not show the features predicted by the GZK bound. The events may, in addition, display angular correlations with points sources. If these observations are confirmed for point sources further than 50-100 Mpc, then strong interactions for the ultra-high energy neutrino are indicated. Recent work on extra space-time dimensions provides a context for massive spin-2 exchanges which are capable of generating cross sections in the 1 – 100 mb range indicated by data. Some recent controversies on the applicability of extra-dimension physics are discussed.

I THE GZK PUZZLE

For more than 35 years it has been known extra-galactic proton cosmic rays should not exceed the energy of about 5×10^{19} eV. This is the GZK bound [1]. Year after year, observations have continued to defy this result [2]. Recent observations of showers with reliably determined energies above 10^{20} eV [3] deepen the puzzle.

The GZK bound is as basic as the Big Bang and the Standard Model. The bound is obtained by attenuation of ultra-high energy (UHE) protons on the cosmic microwave background. The threshold of 5×10^{19} eV is kinematic, representing a center of mass energy sufficient for production of electron-positron pairs and nuclear resonances. The cross sections for these reactions and the density of microwave background photons are known. So the attenuation of protons is known, and exponential attenuation will eliminate protons from sources more about 50 Mpc from the Earth. There are not enough sources within 50 Mpc to explain the data, making the origin of the *GZK*-violating events very mystifying. Speculative models such as unstable superheavy relic particles and topological defects can create particles of high enough energy, but also require sources within the same limited sphere. A hint of correlations between observed event directions and cosmologically-distant sources [4], if confirmed, also indicates something very mysterious.

The only elementary particle that could cross the requisite intergalactic distances of about 100 Mpc or more is the neutrino. The only electrically neutral, stable particle that could expain correlations with cosmologically-distant sources is also the neutrino. Flux estimates of UHE neutrinos produced by extra-galactic sources and GZK-attenuated nucleons and nuclei vary widely, but suffice to account for the shower rates observed. The neutrino would be a natural candidate for the events, if its interaction cross section were large enough [5].

The neutrino-nucleon total cross section σ_{tot} is the crucial issue. In the Standard Model σ_{tot} is based on small-x QCD evolution and $W^\pm, Z$ exchange physics. Cross sections of order $10^{-4} - 10^{-5}$ mb are predicted for the region of 10^{20} eV primary energy. This is far too small to explain the data: so perhaps new physical processes may be at work.

A The νN Cross Section with Massive Spin-2 Exchange

Total cross sections at high energies are dominated by characteristics of the t-channel exchanges. The growth of σ_{tot} with energy, in turn, is directly correlated with the *spin* of exchanged particles.

For this reason the GZK-puzzle cannot be explained by considering new spin-1 exchanges. New W- or Z-like massive *vector* bosons would produce σ_{tot} growing at the same rate as the standard one, and normalized below it. Indeed any new neutrino physics must not disturb the agreement of laboratory data with 4-Fermi physics in the low energy region. At the same time the GZK-violating data indicates a cross section growing with energy that must greatly exceed the Standard Model growth by 10^{20} eV. How is this possible?

The paradox is beautifully resolved with *massive spin-2* exchange. The fascinating possibility of massive spin-2 exchange has recently become popular in the context of hidden "extra" dimensions [6]. Kaluza-Klein (KK) excitations of the graviton are believed to act like a tower of massive spin-2 particle exchanges. We calculate the effect using the Feynman rules developed by several groups [7]. We find neutrino-nucleon cross section values at $E_\nu = 10^{20}$ eV in the $1 - 100$ mb range from mass scales in the $1 - 10$ TeV range. Interestingly, this range of masses is just that indicated by the extra-dimensions phenomenology, and this range of cross sections is just that indicated by the GZK-violating data.

What about low energies? With spin-2 exchange, $d\sigma/dt$ grows like s^2 and σ like s^3 in the low-energy, perturbative region (with a cutoff on t, then one power of s in the total cross section is replaced by the cutoff, of course). The new contributions to σ_{tot} are naturally orders of magnitude below the Standard Model component in the entire regime where neutrino cross sections have been measured. Nor is any miracle needed for the neutrino interactions to emerge above ordinary cosmic ray events above 10^{19} eV. That is because the GZK attenuation process must set in,

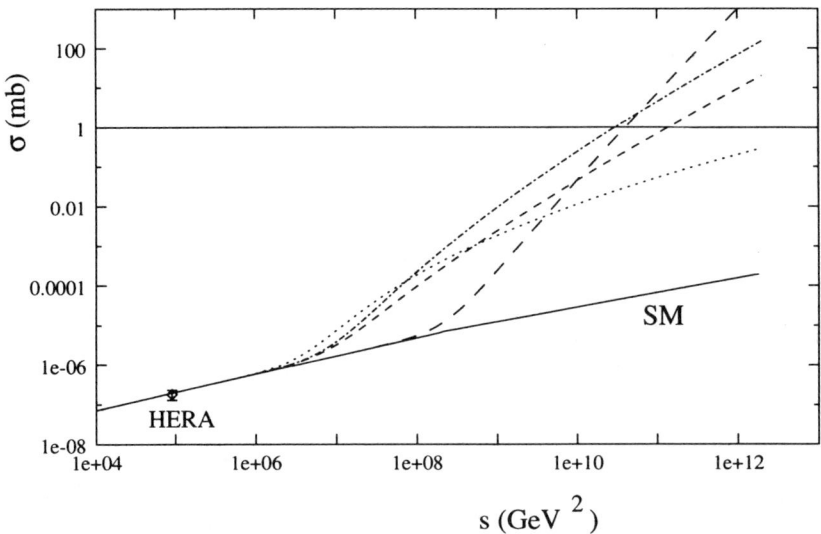

FIGURE 1. The νN cross section in the Standard Model (SM) compared to a theory with large extra dimensions and three different models for the unitarity extrapolation between perturbative to non-perturbative regimes. The dotted line shows the $log(s)$ growth case with $M_S = 1$ TeV and $\xi = 10$. The short dashed and dash-dotted lines show s^1 growth with $M_S = 1$ TeV and $\beta = 1$ and 0.1 respectively. The long dashed line shows s^2 growth with $M_S = 3$ TeV and $\beta = 1$. The contribution from massive graviton exchange is negligible at low energies but rises above the SM contribution when $\sqrt{s} > M_S$, reaching typical hadronic cross sections at incident neutrino energies in the range 5×10^{19} to 5×10^{20} GeV. The HERA data point is shown for comparison. The approximate minimum value required for νN cross-section, $\sigma = 1$ mb, is indicated by the horizontal straight line.

uncovering a spectrum of neutrinos just at the point where protons get attenuated away. Our model of neutrinos and massive spin-2 exchange explain the GZK-violating events consistent with all known experimental limits, as shown in Fig. 1.

B *Updates and Controversies*

This is an exciting area with rapid progress on many applications.

Striking signatures appear in predictions of signals in planned Km^3-scale cosmic neutrino detectors. The 1 TeV to 10 PeV region will be explored, much like the existing AMANDA [8] and RICE [9] detectors. In this regime the predicted [10] ratio of upward to downward neutrino-induced showers differs substantially from the Standard Model, creating a nice diagnostic for new physics.

Horizontal air showers develop over long distances and probe smaller cross sections. Using essentially our enhanced cross section with s^1 behaviour, but with different parameter choices, the authors of Ref. [11] link horizontal shower studies to extra dimension physics. They show how fundamental scales up to 10 TeV can be probed. Detailed studies of shower characteristics generated by our model for various cross sections are also underway [12]. We find that the neutrino-induced and proton-induced showers cannot be distinguished on a shower-by-shower basis. There are exciting prospects for finding signals of new physics in shower angular distributions.

A recent re-calculation within the same Kaluza-Klein KK framework [13] includes the "brane recoil" effects advocated by [14]. After making several technical points, the authors of Ref. [13] present a cross section that is basically equivalent to our s^2 growth model. They conclude that the resulting cross section is too small to explain the UHE cosmic ray showers. However they do not consider all the parameter choices allowed by current experimental limits and hence we find their conclusions unsupported.

Acknowledgments: We thank Tom Weiler, Prashanta Das, Faheem Hussain, and Sreerup Raychaudhuri for useful discussions. This work was supported in part by U.S. DOE Grant number DE-FG03-98ER41079 and the *Kansas Institute for Theoretical and Computational Science*.

REFERENCES

1. K. Greisen, Phys. Rev. Lett. **16**, 748, 1966; G. T. Zatsepin and V. A. Kuzmin, Sov. Phys. JETP Lett. **4**, 78 (1966).
2. J. Linsley, *Phys. Rev. Lett.* **10**, 146 (1963) ; World Data Center for Cosmic Rays, Catalogue for Highest Energy Cosmic Rays, No. 2, Institute of Physical and Chemical Research, Itabashi, Tokyo (1986); Efimov, N.N. et al., Astrophysical Aspects of the Most Energetic Cosmic Rays, M. Nagano and F. Takahara, Eds., (World Scientific, Singapore, 1991) pg. 20.
3. D. Bird et al., *Phys. Rev. Lett.* **71**, 3401 (1993) ; Astrophys. J. **424**, 491 (1994); M. Takeda et al, *Phys. Rev. Lett.* **81**, 1163 (1998) and astro-ph/9902239.
4. G. R. Farrar and P. Biermann, *Phys. Rev. Lett.* **81**, 3579 (1998) .
5. J. Elbert and P. Sommers, Astrophys. J. **441**, 151 (1995).
6. N. Arkani-Hamed, S. Dimopoulos and G. Dvali, *Phys. Lett.* B **429**, 263 (1998) .
7. T. Han, J.D. Lykken and R.-J. Zhang, *Phys. Rev.* D **59**, 105006 (1999) ; G.F. Giudice, R. Rattazzi and J.D. Wells, *Nucl. Phys.* B **544**, 3 (1998) .
8. "From the First Neutrino Telescope, the Antarctic Muon and Neutrino Detector Array AMANDA, to the IceCube Observatory", F. Halzen, for the AMANDA Collaboration, to be published in *Proceedings of International Cosmic Ray Conference 99* (Salt Lake City, 1999); Also see the AMANDA website http://amanda.berkeley.edu/.
9. "Status of the RICE Experiment", G. Frichter, for the AMANDA and RICE Collaborations, to be published in *Proceedings of International Cosmic Ray Conference 99* (Salt Lake City, 1999); Also see the RICE websites http://rice.hep.fsu.edu/ and http://kuhep4.phsx.ukans.edu/ iceman/biblio.html
10. P. Jain, D. McKay, S. Panda and J. Ralston, *Phys. Lett.* B **484**, 267 (2000); hep-ph/0001031.
11. C. Tyler, A. Olinto and G. Sigl, hep-ph/0002257.
12. A. Jain et al, in preparation
13. M. Kachelriess and M. Pluemacher, astro-ph/0005309.
14. M. Bando, T. Kugo, T. Noguchi and K. Yoshioka, Phys. Rev. Lett. 83, 3601 (1999).

Recent Progress in Understanding Nucleosynthesis via Rapid Neutron Capture

Yong-Zhong Qian

School of Physics and Astronomy, University of Minnesota, Minneapolis, MN 55455

Abstract. I discuss the recent progress in our understanding of nucleosynthesis via rapid neutron capture, the r-process, based on meteoritic data for the early solar system and observations of stars at low metallicities. At present, all data require that there be two distinct kinds of r-process events and suggest that supernovae are associated with these events. The diversity of supernova sources for the r-process may depend on whether a neutron star or black hole is formed in an individual supernova. This dependence, if substantiated by future observations discussed here, has important implications for properties of nuclear matter.

INTRODUCTION

The grand scheme for production of various elements was set down more than forty years ago [1,2]. Within this grand scheme, approximately half of the heavy elements with mass numbers $A > 100$ in the solar system were produced via rapid neutron capture, the r-process. A crude picture for the r-process is as follows. One starts with some seed nuclei and lots of neutrons. The seed nuclei then rapidly capture these neutrons to make very neutron-rich unstable progenitor nuclei. After neutron capture stops, the progenitor nuclei successively β-decay towards stability and become the r-process nuclei observed in nature. For a given species of seed nuclei with mass number A_s, the group of nuclei produced by the r-process are determined by the number of neutrons per seed nucleus, or the neutron-to-seed ratio n/s, at the beginning of the r-process. The average mass number of the produced r-process nuclei is $\langle A \rangle = A_s + n/s$.

There are two prominent peaks at $A = 130$ and 195, respectively, in the solar r-process abundance pattern. In order to produce the peak at $A = 130$, we need $n/s \sim 40$ if we start from seed nuclei with $A_s \sim 90$. By comparison, if we start with the same seed nuclei but a higher $n/s \sim 90$, the r-process dominantly produces nuclei with $A > 130$, including the peak at $A = 195$. In general, an astrophysical event would eject, for example, a certain amount of r-process material with $n/s \sim 40$ plus some other amount with $n/s \sim 90$. The ratio of these two amounts then determines the overall r-process abundance pattern produced by this event. A natural question is whether every r-process event produces the same abundance pattern or there should be distinct kinds of events producing very different r-process abundance patterns. In other words, we would like to know whether the solar r-process abundance pattern is produced by every r-process event or just reflects a mixture of different patterns produced by distinct kinds of events. In either case, we also would like to know which astrophysical objects are associated with the r-process.

Here I discuss the progress that we have made recently in answering the above questions. The answers to these questions have important implications for properties of neutrinos, nuclei far from stability, and nuclear matter. I discuss the implications for properties of nuclear matter in particular.

METEORITIC DATA AND DIVERSE SUPERNOVA SOURCES FOR THE R-PROCESS

The meteoritic data on the inventory of radioactive nuclei in the interstellar medium (ISM) at the time of solar system formation play an essential role in our understanding of the r-process. Because a radioactive species decays after it is produced, a finite abundance ratio of a radioactive species to a stable one in the ISM has to be maintained by a series of production events. Consequently, from the measured abundance ratio of a radioactive species to a stable one, we can infer how frequently the radioactive species was replenished in the ISM. We have data on two radioactive nuclei: ^{129}I (e.g., [3]) and ^{182}Hf (e.g., [4]), which are below and above $A = 130$, respectively. If the data indicate that these two species were injected into the ISM at very different frequencies, then the r-process nuclei below and above $A = 130$ must be produced by distinct kinds of events. It was found that over the Galactic history of $\approx 10^{10}$ yr before solar system formation, ^{182}Hf injection occurred at a high frequency $f_\mathcal{H} \sim (10^7 \text{ yr})^{-1}$, while ^{129}I injection occurred at a low frequency $f_\mathcal{L} \sim (10^8 \text{ yr})^{-1}$ [5,6]. Therefore, these are distinct kinds of events.

The frequencies of ^{129}I and ^{182}Hf injection into the ISM inferred from meteoritic data can be naturally explained by associating supernovae with the r-process. A supernova remnant expands to a final radius of ~ 100 pc over a period ($\sim 10^6$ yr) much shorter than the lifetime ($\sim 10^7$ yr) of ^{129}I or ^{182}Hf. Therefore, if we consider a spherical region of ~ 100 pc in radius surrounding an average point in the Galaxy, any supernova within this region can inject fresh radioactive ^{129}I or ^{182}Hf to this point after it is produced by the supernova. For a supernova frequency of $\sim (30 \text{ yr})^{-1}$ over the Galactic volume of $\sim 10^3$ kpc^3, the corresponding frequency in this spherical region is $\sim (10^7 \text{ yr})^{-1}$. Therefore, the meteoritic data on ^{129}I and ^{182}Hf can be explained if we associate the most common supernovae with the r-process events producing ^{182}Hf and a rarer kind with those producing ^{129}I [5–7].

Because ^{129}I is produced together with nuclei at $A \leq 130$ and ^{182}Hf with those at $A > 130$, we conclude that the overall solar r-process abundance pattern is composed of two basic templates characteristic of two distinct kinds of r-process events. These are referred to as the "$\mathcal{H}$" and "$\mathcal{L}$" events hereafter, where "$\mathcal{H}$" stands for the "high" frequency events responsible for "heavy" r-process nuclei with $A > 130$ including ^{182}Hf and "$\mathcal{L}$" for the "low" frequency events responsible for "light" r-process nuclei with $A \leq 130$ including ^{129}I. An average ISM is enriched in r-process elements at a frequency $f_{\mathcal{H}} \sim (10^7 \text{ yr})^{-1}$ by the $\mathcal{H}$ events and at a frequency $f_{\mathcal{L}} \sim (10^8 \text{ yr})^{-1}$ by the $\mathcal{L}$ events.

ABUNDANCES OF R-PROCESS ELEMENTS IN METAL-POOR STARS

The influence of the $\mathcal{H}$ and $\mathcal{L}$ events is best preserved in metal-poor stars formed very early in the Galaxy when only a small number of supernovae had contributed to the r-process and "metal" abundances in these stars. The typical heavy r-process elements observed in these stars are Ba and Eu, and the typical light r-process elements observed are Pd, Ag, and Cd. The observational data are usually given in the spectroscopic notation, e.g., $\log \epsilon(\text{Eu}) \equiv \log(\text{Eu/H}) + 12$ for Eu, where Eu/H is the number abundance ratio of Eu to hydrogen observed in a star. A typical metal is Fe, and the "metallicity" is defined as $[\text{Fe/H}] \equiv \log(\text{Fe/H}) - \log(\text{Fe/H})_\odot$, where $(\text{Fe/H})_\odot$ is the Fe/H ratio in the sun. As the overall abundance of hydrogen has not changed significantly over the history of the universe, hydrogen is a good reference element for considering chemical enrichment of the ISM.

Over the period of $\approx 10^{10}$ yr before solar system formation, an average ISM was enriched in the heavy r-process elements by $\sim 10^3$ $\mathcal{H}$ events and in the light ones by $\sim 10^2$ $\mathcal{L}$ events. The r-process composition of the ISM at the time of solar system formation is reflected by the corresponding solar abundances. Consequently, the r-process abundances resulting from a single $\mathcal{H}$ or $\mathcal{L}$ event (quantities with the subscript "$\mathcal{H}$" or "$\mathcal{L}$") can be predicted directly from the solar system data (quantities with the subscript "$\odot, r$") [6,8]. For example, we have $(\text{Eu/H})_{\odot,r} \sim 10^3 (\text{Eu/H})_{\mathcal{H}}$, and hence $\log \epsilon_{\mathcal{H}}(\text{Eu}) \sim \log \epsilon_{\odot,r}(\text{Eu}) - 3 \approx -2.5$. Likewise, we have $(\text{Ag/H})_{\odot,r} \sim 10^2 (\text{Ag/H})_{\mathcal{L}}$, and hence $\log \epsilon_{\mathcal{L}}(\text{Ag}) \sim \log \epsilon_{\odot,r}(\text{Ag}) - 2 \approx -0.8$.

Figure 1 shows the Eu data for many metal-poor stars [9–12]. The very low metallicities of these stars indicate that they were formed very early in the Galaxy. The band labeled "1 $\mathcal{H}$" corresponds to the Eu abundance resulting from a single $\mathcal{H}$ event predicted by the meteoritic data on ^{129}I and ^{182}Hf as well as the solar r-process abundances of stable nuclei [6,8]. If we pick the centroid of this band, the observed Eu abundances can be explained by contributions from ~ 1–30 $\mathcal{H}$ events, quite consistent with the corresponding low metallicities.

Figure 2 shows the remarkable data on a number of r-process elements for one of the stars shown in Figure 1, CS 22892–052 [13]. The dashed curve labeled "TS" is the solar r-process abundance pattern translated to match the data on the heavy r-process elements. Clearly, this curve cannot match the data on the light r-process elements. Therefore, the conclusion from the meteoritic data that there should be two distinct kinds of r-process events is independently confirmed by stellar observations at low metallicities. The Eu abundance in CS 22892–052 can be explained by $\sim 30\ \mathcal{H}$ events. However, with a frequency ratio of $\sim 10 : 1$

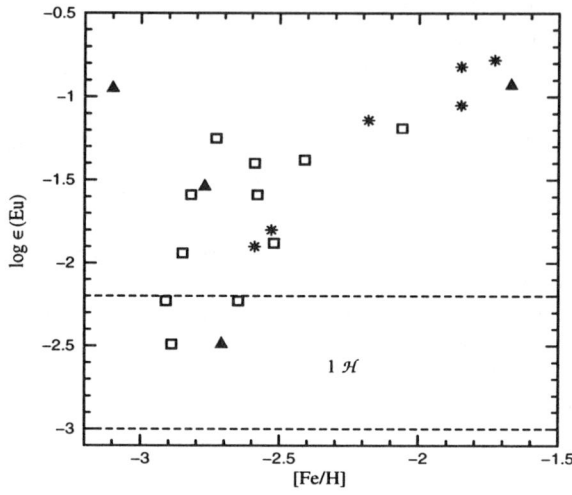

FIGURE 1. Europium data for metal-poor stars (asterisks: [9], squares: [10], triangles: [11,12]) compared with the Eu abundance resulting from a single $\mathcal{H}$ event (the band labeled "1 $\mathcal{H}$") predicted in [6,8].

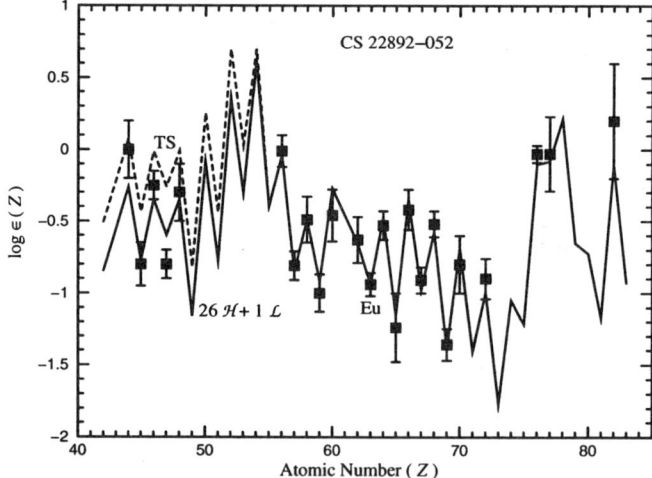

FIGURE 2. Comparison of data for CS 22892–052 with (1) the solar r-process abundance pattern translated to match the Eu data (dashed curve), and (2) the result from a mixture of two distinct kinds of r-process events (solid curve).

between $\mathcal{H}$ and $\mathcal{L}$ events, it is very likely that one $\mathcal{L}$ event also occurred during a period over which ~ 30 $\mathcal{H}$ events took place. Indeed, a mixture of 26 $\mathcal{H}$ events and 1 $\mathcal{L}$ event (solid curve in Fig. 2) can explain all the data rather well [6].

NEUTRON STAR/BLACK HOLE FORMATION AND SUPERNOVA R-PROCESS NUCLEOSYNTHESIS

So far, the questions raised in the introduction have been answered. It is found that there are distinct kinds of r-process events and that they are associated with supernovae. This leads to a new question: what is causing the difference between these supernova r-process events? The answer to the new question may be obtained by further studying r-process enrichment of the Galaxy by supernovae. The present Galactic inventory of either the light r-process nuclei ($100 \lesssim A \leq 130$) or the heavy ones ($A > 130$) is $\sim 4 \times 10^3\,M_\odot$. Assuming a frequency of $\sim (30\,\mathrm{yr})^{-1}$ over the whole Galaxy in its history of $\approx 10^{10}$ yr for the supernovae associated with $\mathcal{H}$ events and a ~ 10 times less frequency for those associated with $\mathcal{L}$ events, we find that in order to account for the present Galactic r-process inventory, each supernova has to eject only $\sim 10^{-5}\,M_\odot$ ($\mathcal{H}$ event) to $\sim 10^{-4}\,M_\odot$ ($\mathcal{L}$ event) of r-process material. By comparison, the total amount of ejecta from a supernova is $\sim 10\,M_\odot$ and the mass of the neutron star produced in a supernova is $\sim 1\,M_\odot$. Despite the striking difference from these comparison mass scales, $\sim (10^{-5}\text{–}10^{-4})\,M_\odot$ of material can be naturally ejected by the neutrinos emitted in a supernova.

A supernova occurs when the core of a massive star at the exhaustion of its nuclear fuel collapses into a compact neutron star. The neutron star has a final radius of ~ 10 km, and a gravitational binding energy of $\sim 10^{53}$ erg is to be released. Due to the high temperatures and densities encountered during the collapse, the most efficient way to release this energy is to emit all three flavors of neutrinos and antineutrinos mainly through electron-positron pair annihilation. In fact, because of the intense scatterings on neutrons and protons common to all neutrino species, even neutrinos have to diffuse out of the neutron star on a timescale of ~ 10 s (as confirmed by the detection of neutrinos from SN 1987A). So the average neutrino luminosity is $\sim 10^{51}$ erg s^{-1} per species.

A few seconds after the core collapse and the subsequent supernova explosion, we have a hot neutron star near the center of the supernova. The neutron star is still cooling by emitting neutrinos. The shock wave which makes the supernova explosion is far away from the neutron star. On its way out to make the explosion, the shock wave has cleared away almost all the material above the neutron star, leaving behind only a thin atmosphere. Close to the neutron star, the temperature is several MeV and the atmosphere is essentially dissociated into neutrons and protons. As the neutrinos emitted by the neutron star free-stream through this atmosphere, some of the ν_e and $\bar{\nu}_e$ are captured by the neutrons and protons and their energy is deposited in the atmosphere. In other words, the atmosphere is heated by the neutrinos. As a result, it expands away from the neutron star and eventually develops into a mass outflow — a neutrino-driven "wind" [14].

Because neutrino heating is driving the mass ejection, the fraction of neutrino luminosity absorbed by the wind material determines the rate at which it is being lifted out of the neutron star gravitational potential. As neutrinos interact weakly, the heating rate is small. On the other hand, the neutron star is a compact object and has a deep gravitational potential. Consequently, we expect that the mass ejection rate is small. Indeed, the typical mass ejection rate in the wind was found to be $\sim 10^{-5}\,M_\odot\,\mathrm{s}^{-1}$ [15]. So provided that r-process nuclei are produced in the neutrino-driven wind, the wind has to last ~ 1 s in an $\mathcal{H}$ event and ~ 10 s in an $\mathcal{L}$ event in order to account for the present Galactic r-process inventory. Note that neutrino emission from a stable neutron star, and hence the corresponding neutrino-driven wind, last ~ 10 s. In order for a wind to last only ~ 1 s, neutrino emission has to be terminated ~ 1 s after the supernova explosion by transformation of the neutron star into a black hole [7]. Therefore, the difference between the $\mathcal{H}$ and $\mathcal{L}$ events may depend on whether a black hole ($\mathcal{H}$ event) or neutron star ($\mathcal{L}$ event) is formed in an individual supernova.

CONCLUSIONS

In summary, meteoritic data on ^{129}I and ^{182}Hf require two distinct kinds of r-process events: the high frequency $\mathcal{H}$ events responsible for heavy r-process nuclei ($A > 130$) and the low frequency $\mathcal{L}$ events responsible for light ones ($100 \lesssim A \leq 130$). The meteoritic data also suggest that supernovae are associated with $\mathcal{H}$ and $\mathcal{L}$ events. These conclusions are either confirmed by or at the very least consistent with observations of r-process abundances in metal-poor stars.

If future studies can show that r-process nuclei are produced in the neutrino-driven wind in a supernova (see e.g., [15–25] for the current unsatisfactory status), then the amount of material required to account for the present Galactic r-process inventory can be adequately provided by the wind. The factor of ~ 10 difference in the amount of r-process ejecta between the $\mathcal{H}$ and $\mathcal{L}$ events calls for a similar difference in the duration of neutrino emission, and hence that of the neutrino-driven wind, between the corresponding supernovae. In turn, the difference in neutrino emission may require transformation of the initial neutron star into a black hole ~ 1 s after the supernova explosion in an $\mathcal{H}$ event and long term stability of the neutron star in an $\mathcal{L}$ event. Consequently, the diversity of supernova sources for the r-process may have profound implications for properties of nuclear matter inside the initial neutron star produced in a supernova.

Two possible observational tests for the association of neutron star/black hole formation with supernova r-process nucleosynthesis have been proposed [25–27]. One test relies on the occurrence of supernovae in binaries consisting of a massive star and a low mass star. Some binaries would survive the supernova explosion of the massive star and become new systems with a neutron star or black hole orbiting around the low mass star. Furthermore, the surface of the low mass star would be contaminated by the r-process ejecta from the supernova. Therefore, we can test black hole and neutron star formation in $\mathcal{H}$ and $\mathcal{L}$ events, respectively, by looking for r-process abundance anomalies on the surface of the binary companion to a neutron star or black hole [25]. This approach is quite promising as large overabundances of O, Mg, Si, and S ejected in supernovae have been observed in the binary companion to a black hole [28].

ACKNOWLEDGMENTS

This work was supported in part by the Department of Energy under grant DE-FG02-87ER40328.

REFERENCES

1. Burbidge, E. M., et al., *Rev. Mod. Phys.* **29**, 547 (1957).
2. Cameron, A. G. W., *Pub. Astron. Soc. Pacific* **69**, 201 (1957).
3. Reynolds, J. H., *Phys. Rev. Lett.* **4**, 8 (1960).
4. Lee, D.-C., and Halliday, A. N., *Nature* **378**, 771 (1995).
5. Wasserburg, G. J., Busso, M., and Gallino, R., *Astrophys. J.* **466**, L109 (1996).
6. Qian, Y.-Z., and Wasserburg, G. J., *Phys. Rep.* **333–334**, 77 (2000).
7. Qian, Y.-Z., Vogel, P., and Wasserburg, G. J., *Astrophys. J.* **494**, 285 (1998).
8. Wasserburg, G. J., and Qian, Y.-Z., *Astrophys. J.* **529**, L21 (2000).
9. Gratton, R. G., and Sneden, C., *Astron. Astrophys.* **287**, 927 (1994).
10. McWilliam, A., et al., *Astron. J.* **109**, 2757 (1995).
11. Sneden, C., et al., *Astrophys. J.* **467**, 819 (1996).
12. Sneden, C., et al., *Astrophys. J.* **496**, 235 (1998).
13. Sneden, C., et al., *Astrophys. J.* **533**, L139 (2000).
14. Duncan, R. C., Shapiro, S. L., and Wasserman, I., *Astrophys. J.* **309**, 141 (1986).
15. Qian, Y.-Z., and Woosley, S. E., *Astrophys. J.* **471**, 331 (1996).
16. Meyer, B. S., et al., *Astrophys. J.* **399**, 656 (1992).
17. Witti, J., Janka, H.-Th., and Takahashi, K., *Astron. Astrophys.* **286**, 841 (1994).
18. Takahashi, K., Witti, J., and Janka, H.-Th., *Astron. Astrophys.* **286**, 857 (1994).
19. Woosley, S. E., et al., *Astrophys. J.* **433**, 229 (1994).
20. Hoffman, R. D., Woosley, S. E., and Qian, Y.-Z., *Astrophys. J.* **482**, 951 (1997).
21. Qian, Y.-Z., et al., *Phys. Rev. C* **55**, 1532 (1997).
22. Haxton, W. C., et al., *Phys. Rev. Lett.* **78**, 2694 (1997).
23. Meyer, B. S., McLaughlin, G. C., and Fuller, G. M., *Phys. Rev. C* **58**, 3696 (1998).
24. Caldwell, D. O., Fuller, G. M., and Qian, Y.-Z., *Phys. Rev. D* **61**, 123005 (2000).
25. Qian, Y.-Z., *Astrophys. J.* **534**, L67 (2000).
26. Qian, Y.-Z., Vogel, P., and Wasserburg, G. J., *Astrophys. J.* **506**, 868 (1998).
27. Qian, Y.-Z., Vogel, P., and Wasserburg, G. J., *Astrophys. J.* **524**, 213 (1999).
28. Israelian, G., et al., *Nature* **401**, 142 (1999).

The rp-Process at X-Ray Burst Conditions

M. Wiescher [a], H. Schatz [b]

[a] Dept. of Physics University of Notre Dame, Notre Dame, IN 46556,
[b] NSCL, Michigan State University, East Lansing, MI 48824

Abstract. X-ray bursts are interpreted as thermonuclear runaways in the atmosphere of an accreting neutron star. The energy release of the burst is driven by the rapid proton capture (rp) process. The characteristics of the rp-process is discussed in this paper. Particular emphasis will be given to the questions of ignition and endpoint of the rp-process.

I X-RAY BURSTS

X-ray bursts are frequently observed phenomena [1]. The observed luminosity suggests a stellar explosion which emits a large amount of energy in the X-ray region on a time scale of tens of seconds. This basic signature has been verified in all of the subsequent observations [1] and a typical example is shown in figure 1 [2]. The interpretation of such X-ray emitting events requires a detailed understanding of the relevant physical processes, in particular the nuclear physics processes which are the main driver of the explosion [3]. The standard models for type-I X-ray bursts are based on accretion processes in a close binary system similar to the nova scenario. In the case of type-I X-ray bursts, accretion takes place from the filled Roche-Lobe of the extended companion star onto the surface of a neutron star. Typical predictions for the accretion rate vary from 10^{-10} to 10^{-9} $M_\odot$ y^{-1} [4,5]. The accreted matter is continuously compressed by the freshly accreted material until it reaches sufficiently high pressure and temperature conditions which allow the thermonuclear ignition. While the basic principle of the scenario seems well accepted, the associated nucleosynthesis and the correlated nuclear energy generation is not fully understood [3]. Presently there exist two fundamental problems. First, what kind of nucleosynthesis mechanism can release the observed amount of energies and can account for the extremely short time scale for the release. Secondly, at what conditions do the nuclear processes freeze out and what are the consequences for the burst and also for the atmosphere conditions of the remaining neutron star.

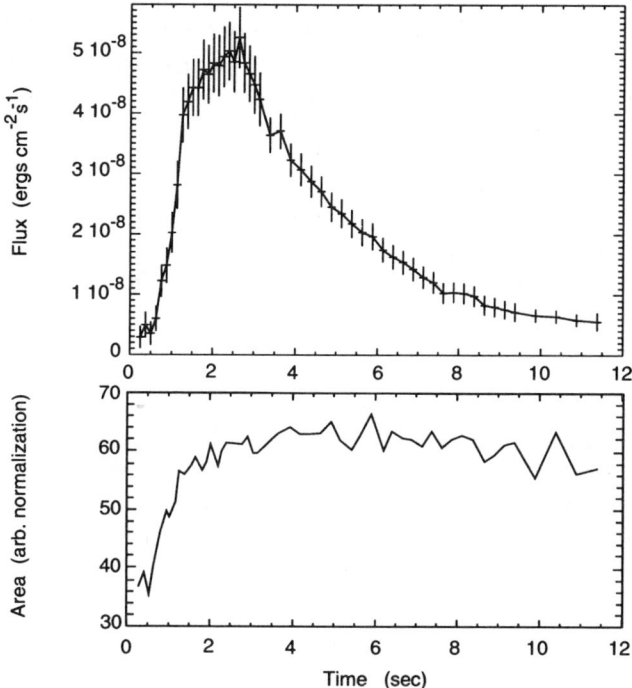

FIGURE 1. A thermonuclear X-Ray Burst from the neutron star in the low mass x-ray binary system 4U 1728-34, as observed with the Rossi X-ray Timing Explorer. The top shows the rapid increase of the x-ray flux, followed by a slower decay. The lower panel indicates the change of the x-ray emitting area calculated for blackbody radiation from a specical system. The initial increase of the area provides strong evidence of the spread of the nuclear burning front over the entire surface of the neutron star. (courtesy: Tod Strohmayer)

II BREAK-OUT REACTIONS AS TRIGGER FOR X-RAY BURSTS

Nuclear burning is ignited at high density conditions via the pp-chains and the hot CNO-cycles in the atmosphere of the accreting neutron star. However, because of their limited energy generation rate neither the pp-chains nor the hot CNO cycles can initiate a thermonuclear runaway [4]. The thermonuclear runaway requires an instability, where the temperature sensitivity of the nuclear energy generation rate $\dot{\epsilon}$ exceeds the temperature sensitivity of the cooling rate $\dot{\epsilon}_{cool}$:

$$\frac{d\dot{\epsilon}}{dT} > \frac{d\dot{\epsilon}_{cool}}{dT}. \qquad (1)$$

Therefore, the explosive nuclear burning is triggered as soon as temperature sensitive reaction sequences like the triple α process or the break-out of the CNO cycles start to contribute significantly to the nuclear energy release. The conditions at which that happens affect the amount of material that can be accreted between bursts and therefore influences burst intervals and burst luminosities. At the typical high densities of $\rho \approx 10^5 - 10^6$ g/cm^3 and temperatures of $T \geq 10^8$ K in the accreted layer, the ^{15}O$(\alpha,\gamma)^{19}$Ne, and the ^{18}Ne$(\alpha,p)^{21}$Na reactions may cause a break-out from the β-limited hot CNO cycles. A comparison of the present reaction rate estimates [6–8] indicates that at rising temperatures the ^{15}O$(\alpha,\gamma)^{21}$Na reaction will be the first break-out reaction to occur. While ^{15}O$(\alpha,\gamma)^{19}$Ne triggers the breakout, ^{18}Ne$(\alpha,p)^{21}$Na dominates the reaction flow initiated by the triple-α-reaction via ^{12}C$(p,\gamma)^{13}$N$(p,\gamma)^{14}$O$(\alpha,p)^{17}$F$(p,\gamma)^{18}$Ne$(\alpha,p)^{21}$Na continuing towards higher masses by the αp-process. This essentially by-passes the ^{15}O$(\alpha,\gamma)^{19}$Ne reaction. The reaction rates of the both processes are determined by single resonance contributions which presently are only based on rather crude estimates. Experimental verification for both reaction rates using radioactive ^{15}O and ^{18}Ne beams is essential to put the calculations on firm experimental ground.

To investigate the significance of the break-out reactions for the structure of the X-ray burst, temperature, energy production and luminosity have been calculated in terms of a one-zone reaction network model [9]. For this model hydrostatic

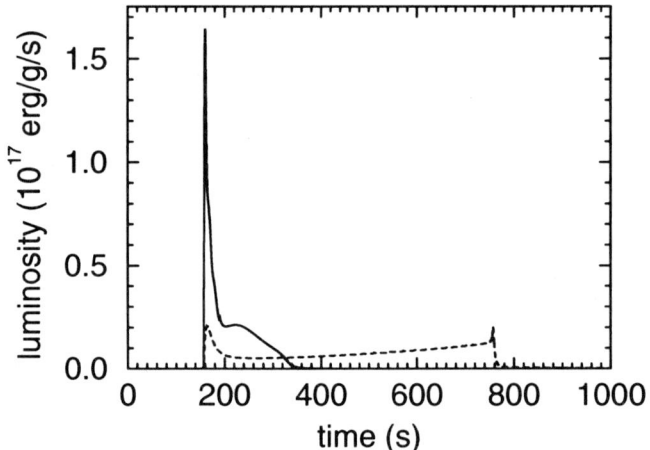

FIGURE 2. The predicted luminosity produced by nuclear reactions in the accreted envelope of a neutron star. The solid line shows the expected luminosity including both break-out reactions ^{18}Ne$(\alpha,p)^{21}$Na, and ^{15}O$(\alpha,\gamma)^{19}$Ne; the dotted line shows the luminosity without break-out reactions.

equilibrium is maintained by keeping the total pressure P in the burning zone constant. The total pressure is determined by the ion pressure, the pressure of the degenerate electron gas, and the radiation pressure. The calculation relies on new opacities which allow improved selfconsistent modeling of the cooling phase compared to the previous approximations [10]. An initial solar abundance was adopted for the accreted material, the network contained 634 nuclei up to Z=50 and included more than 7000 nuclear reactions. Figure 2 shows the result of two independent calculations for the luminosity with and without taking the reaction rates for the break-out processes into account.

Without the break-out reactions, a broad burst is observed which is predominantly fueled by the hot CNO cycles and the feeding triple-alpha process. After an initial rapid increase the burst lasts for $\approx$ 500 s at a very slowly increasing luminosity, before rapidly decaying due to helium depletion. The afterglow is due to the decay of the enriched ^{14}O, ^{15}O, and ^{18}Ne abundances. This luminosity curve differs substantially from the observation shown in figure 1. However, including the break-out reactions shows an entirely different luminosity structure, a rapid increase to an order of magnitude higher luminosity followed by a decline over a 100 s time-scale. This is due to the rapid processing of the CNO material towards ^{56}Ni in the first few seconds of the thermonuclear runaway. It can clearly be seen that this luminosity prediction matches the observed luminosity shown in figure 1 very well. Even though a one zone temperature and density calculation represents a rather simplified X-ray burst model [11], it does give a valuable guide to the importance of break-out reactions.

III RP-PROCESS IN THE COOLING PHASE OF X-RAY BURSTS

The energy generation and luminosity curve is shaped by the time-dependence of the nuclear reactions along the rp-process path. Energy generation comes to a rapid halt with the formation of ^{56}Ni at peak temperatures of T$\approx$2.5 GK. Further processing towards higher masses is prevented by the ^{56}Ni(p,γ)-^{57}Cu(γ,p) equilibrium [3,12]. The temperature drops rapidly due to radiation losses until at T=1 GK the ^{56}Ni$(p,\gamma)^{57}$Cu reaction falls out of thermal equilibrium and a rapid rp-process is initiated beyond ^{56}Ni during the cooling phase of the burst [9]. Recent simulations indicate the the rp-process extends along the N=Z line up to the A=100 mass range [3,9]. Figure 3 shows the reaction path integrated over the entire cooling period. Reliable predictions of the associated nucleosynthesis and energy generation during this cooling phase are handicapped by the lack of reliable experimental knowledge of the associated nuclear structure and nuclear reaction parameters. To verify the present model predictions about the extent and the endpoint of the rp-process, detailed nuclear reaction and structure studies on the neutron deficient side of the line of stability are essential [3]. In particular information about masses, lifetimes, level structures, and proton separation energies are needed. Since the main impedance is associated with the rather long life-times of even-even N=Z isotopes along the reaction path, particular emphasis has to be given to the understanding

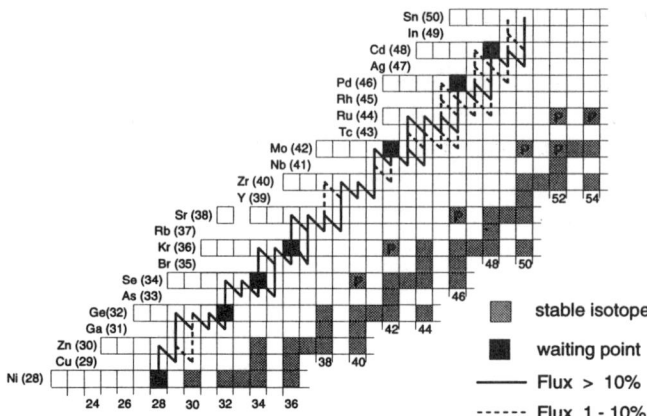

FIGURE 3. Reaction path of the rp-process along the N=Z line during the cooling period of the X-ray burst. The characteristic waiting points are marked.

and interpretation of the 2-proton capture reactions. These processes have been predicted to bridge the proton drip-line and to reduce the effective life-time of these nuclei substantially. Improved mass measurements of the associated isotopes are absolutely necessary to test and confirm the present predictions. [3]. An important question is the one for the endpoint of the rp-process. The endpoint is determined both by the macroscopic time-scale of the burst which depends on the various cooling mechanisms and the microscopic time-scale given by the effective life-times of the waiting point nuclei along the reaction path. For steady state burning scenarios, characterized by long-term high temperature conditions, the endpoint is basically determined by the availability of hydrogen fuel in the accreted layer [10]. All of the predicted reaction and decay rates along the reaction path need to be experimentally tested and verified, yet, the present predictions suggest a reaction flow even beyond ^{100}Sn [3,9,10]. This raises the question for the actual endpoint. Alpha decay studies in the mass range above ^{100}Sn suggest that the neutron deficient isotopes ^{106}Te to ^{108}Te and ^{108}I predominantly decay into the α channel [13]. The isotope ^{109}I has been identified as a short-lived proton emitter [14]. These observations suggest that the actual endpoint of the rp-process might be associated with rapid back-processing of the material via ^{104}Sb$(p,\alpha)^{101}$Sn, ^{105}Sb$(p,\alpha)^{102}$Sn, and possibly also ^{106}Sb$(p,\alpha)^{103}$Sn as shown in figure 4. While some mass models [15,16] predict ^{104}Sb and ^{105}Sb to be proton unbound, the experimental decay-rates are longer than Hauser Feshbach predictions for the (p,α) reaction rates. This multi-cycling in the Sn-Te-I range presents a strong impedance for the rp-process and may represent the actual end-point for the reaction path. To verify this claim, detailed information about the lifetimes and decay branchings of the various Te and I isotopes are necessary. The reaction rates for competing (p,γ) and (p,α) processes need to be calculated to determine the cycle efficiencies. Presently the rates are estimated

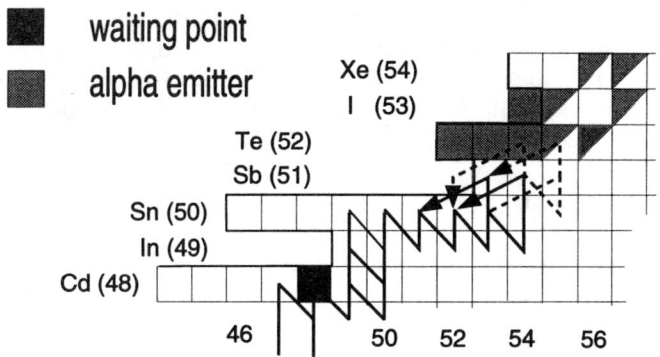

FIGURE 4. Reaction sequences in the Sn-Te mass range. This reaction cycling represents a possible endpoint for the rp-process.

using the Hauser Feshbach code NON-SMOKER [17]. However, the level density in these isotopes near the double magic nucleus ^{100}Sn might be too low to justify a statistical treatment of the reaction rate. The calculations should therefore be complemented by shell-model based level density predictions or by experimental studies of the low energy structure of the Te and I compound isotopes.

REFERENCES

1. W.H.G. Lewin, J. van Paradijs, & R.E. Taam, *Space Sci.Rev.* **62** (1993), 233
2. L. Bildsten L. & T. Strohmayer T., *Physics Today* **52** (1999), 40
3. H. Schatz, A. Aprahamian, J. Görres, M. Wiescher, T. Rauscher, J.F. Rembges, F.-K. Thielemann, B. Pfeiffer, P. Möller, K.-L. Kratz, H. Herndl, B.A. Brown, & H. Rebel, *Phys.Rep.* **294** (1998), 168
4. R.E. Taam, *Ann.Rev.Nucl.Sci.* **35** (1985), 1
5. R.E. Taam, S.E. Woosley, & D.Q. Lamb, *Astrophys.J.* **459** (1996), 271
6. J. Görres, M. Wiescher, & F.-K. Thielemann, *Phys.Rev.C* **51** (1995), 392
7. K.I. Hahn, A. Garcia, E.G. Adelberger, P.V. Magnus, A.D. Bacher, N. Bateman, G.P.A. Berg, J.C. Blackmon, A.E. Champagne, B. Davis, A.J. Howard, J. Liu, B. Lund, Z.Q. Mao, D.M. Markoff, P.D. Parker, M.S. Smith, E.J. Stephenson, K.B. Swartz, S. Utku, R.B. Vogelaar, & K. Yildiz, *Phys.Rev.* C **54** (1996), 1999
8. Z.Q. Mao, H.T. Fortune, & A.G. Lacaze, *Phys.Rev.* C **53** (1996), 1197
9. M. Wiescher, H. Schatz, & A. Champagne, *Phil.Trans.Roy.Soc.* **356** (1998), 1949
10. H. Schatz, L. Bildsten, A. Cummings, & M. Wiescher, *Ap.J.* **524** (1999), 1014
11. F. Rembges, L. Bildsten, M. Liebendörfer, A. Cumming, & F.-K. Thielemann, *Astr.& Astrophys.* (2000), in press
12. K.E. Rehm, F. Borasi, C.L. Jiang, D. Ackermann, I. Ahmad, B.A. Brown, F. Brumwell, C.N. Davids, P. Decrock, S.M. Fischer, J. Görres, J. Greene, G. Hackmann, B. Harss, D. Henderson, W. Henning, R.V.F. Janssens, G. McMichael, V.

Nanal, D. Nisius, J. Nolen, R.C. Pardo, M. Paul, P. Reiter, J.P. Schiffer, D. Seweryniak, R.E. Segel, M. Wiescher, & A.H. Wuosmaa, *Phys.Rev.Lett.* **80** (1998), 676
13. R.D. Page, P.J. Woods, R.A. Cunningham, T. Davinson, N.J. Davis, A.N. James, K. Livingston, P.J. Sellin, & A.C. Shotter, *Phys.Rev. C* **49** (1994), 3312
14. A. Gillitzer, T. Faestermann, K. Hartel, P. Kienle, & E. Nolte, *Z.Phys. A* **326** (1987) 107
15. P. Möller, J.R. Nix, D. Myers, & W.J. Swiatecki, *At.Data.Nucl.Data.Tab.* **59**, 185 (1995)
16. Y. Aboussir, J.M. Pearson, A.K. Dutta, & F. Tondeur, *Nucl.Phys.A* **549**, 155 (1992)
17. T. Rauscher, F.-K. Thielemann, & K.L. Kratz, *Phys. Rev. C* **56** (1997) 1613

Neutrinos

Recent Results from CHORUS and NOMAD Experiments

Stefania Ricciardi

CERN, Geneva, Switzerland and Royal Holloway, University of London, Egham, U.K.

Abstract. CHORUS and NOMAD experiments are searching for ν_τ appearance in the CERN Wide Band Neutrino Beam by using complementary detection techniques. They have both completed the data taking and the first results on the full data sample are reported here. No evidence for oscillation has been found and the new upper limits on the oscillation probability $P(\nu_\mu \to \nu_\tau)$ and $P(\nu_e \to \nu_\tau)$ are more than an order of magnitude lower than the previous best experiment.

INTRODUCTION

CHORUS and NOMAD search for $\nu_\mu \to \nu_\tau$ oscillation through the observation of charged current (CC) interactions $\nu_\tau N \to \tau^- X$, followed by the decay of the τ lepton. The search is sensitive to very small mixing for differences of the squares masses, Δm^2, larger than a few eV^2/c^4, which is the region of cosmological relevance.

The search is performed in the CERN Wide Band Neutrino Beam at about 600 m from the neutrino source, where both detectors are located. The beam contains mainly ν_μ, with a component of $\bar\nu_\mu$ of about 6% and about 1% contamination of electron neutrinos and anti-neutrinos. The estimated ν_τ contamination is below the level of sensitivity of both experiments. The intense ν_μ beam has an average energy of 27 GeV, well above the τ production threshold ($E_\nu \sim 3.5$ GeV), which allows an *appearance* search.

The two experiments adopt complementary techniques.

In CHORUS the signal selection is mainly based on topological criteria. Neutrino interactions occur in a 770 kg target of nuclear emulsions, whose spatial resolution allows one to distinguish the τ production vertex from its decay vertex. The muon decay channel of the τ, as it will be shown later, has a very low background and therefore provides high sensitivity to oscillation.

In NOMAD the τ identification is indirect, through its secondary visible decay products, using kinematic criteria for background rejection. The detector is designed to exploit the low ν_e contamination in the neutrino beam to search for oscillation with higher sensitivity in the electron decay channel of the τ. Drift chambers act both as a low density target with a fiducial mass of 2.7 t and as tracking devices inside a uniform bipolar magnetic field of 0.4 T.

CHORUS

The CHORUS [1] apparatus is composed of an emulsion target, a scintillating fibre tracker system, trigger hodoscopes, a magnetic spectrometer, a lead-scintillator calorimeter and a muon spectrometer.

The search for ν_τ interaction has been performed by looking for two decay modes of the τ lepton, $\mu^-\nu_\tau\bar{\nu}_\mu$ and $h^-(n\pi^0)\nu_\tau$. Both decay modes are characterised by a *kink* topology in the emulsion, that is a change of direction of a track from the vertex of the neutrino interaction after a path of the order of 1mm.

The information of the electronic sub-detectors have been used to define two distinct data set: the 1μ sample containing events with a reconstructed muon of negative charge and the 0μ sample, where no muons are found. In addition to a clear "kink" topology a ν_τ CC candidate has to satisfy the following criteria:

- no other charged leptons at the primary vertex;

- the transverse momentum of the decay daughter with respect to the parent direction is larger than 250 MeV/c (to eliminate K^- decays);

- the distance, L_k, between the decay and interaction vertex should not exceed a maximum value, which varies according to the sample (5 plates for 1μ, a value dependent on the momentum of the daughter for 0μ).

A tighter cut is necessary to suppress the larger background rate for the 0μ sample. In this case negative hadron scattering with no visible recoil or evidence of nuclear break-up generate the so-called *white kink* (WK) background. This background has been measured inside CHORUS for distances between the primary and the secondary vertex outside the signal region and extrapolated to the signal region with the help of a MC simulation. To further reduce this background and optimise the sensitivity in the 0μ sample, a selection was applied in the angle (Φ) between the direction of the τ and the hadronic shower axis in the transverse plane.

The second source of background, common to all decay channels of the τ is due to charm production in CC events where the primary lepton is not identified. Both neutrino and anti-neutrino interactions may give rise to this background. The first, requiring also wrong assignemt of the charge of the charmed particle's daughter, is more important for hadronic decays.

CHORUS has been taking data in the years 1994-1997 and it has collected about 840,000 ν_μ CC interactions in the emulsion target. The first analysis of the complete set of collected data, consisting of 713,000 1μ events and 335,000 0μ events, has been recently completed [1]. A summary of the expected events from different background

---

[1] CHORUS results have been updated in these proceedings to the ones obtained with the full data sample available at the time of writing.

TABLE 1. Summary table of the observed events and expected background for CHORUS. The maximum number of τ events observable is also reported.

	N^τ_{max}	charm	WK	Total exp. background	Observed
1μ	5,014	0.1	-	0.1	0
0μ	2,004	0.3	0.8	1.1	0

sources, the observed number of events and the number of observable τ events (N^{max}_τ) if oscillation probability was equal to one, is given in table 1.

NOMAD

The NOMAD detector is described in Ref. [2]. The active target of drift chambers is inside a 0.4 T magnetic field. It is followed by a transition radiation detector, a preshower and an electromagnetic calorimeter. A hadron calorimeter and two muon stations are located just after the magnet coil. The search for ν_τ CC interactions is performed by identifying the τ decays $e^- \bar{\nu}_e \nu_\tau$, $h^-(n\pi^0)\nu_\tau$ and $h^- h^+ h^- (n\pi^0)\nu_\tau$ for a total branching ratio of 82.8%.

Events containing a primary track identified as a muon by the muon stations are rejected for the oscillation search. However, a sample of these events is used by the so-called Data Simulator to correct the signal and background efficiencies as obtained by the pure Monte Carlo. The procedure is described in Ref. [3].

For each τ decay mode two independent analysis are performed in the DIS and LM samples, which are separated by a cut on the total hadronic momentum at 1.5 GeV/c.

In the electron decay channel, the main source of background is ν_e CC interactions and their rejection is based on the missing momentum and angular relations in the transverse plane. For hadronic decays the main source of background is NC interactions. The isolation between the τ visible decay products and the hadronic jets plays an important role in this case.

In general many discriminating variables are combined into global likelihood functions, for each τ decay mode, describing the probability for an event to be signal or background. Event classification is based on the likelihood ratio between the signal and the background hypothesis.

To avoid bias a "blind analysis" was performed. In this procedure data events inside a pre-defined signal region are not analysed until all selection criteria are defined and the robustness of the background predictions is demonstrated outside this region. The definition of the cuts was made by optimising the sensitivity to oscillation in the frequentist Unified Approach [4].

NOMAD collected about 1,350,000 ν_μ CC interactions in the years 1995-1998. A summary of the observed number of candidate events and the corresponding predicted background in the signal region is shown in table 2. The maximum

TABLE 2. Summary table of the observed events and expected background for NOMAD. The maximum number of observable τ events is also reported.

	Decay channel		N_τ^{max}	Expected background	Observed data
DIS	$\tau \to e$		4110	$5.3^{+0.7}_{-0.5}$	5
	$\tau \to h(n\pi^0)$	ρ	3307	9.5 ± 2.5	7
		π	2022	6.8 ± 2.1	5
		π/ρ	210	$0.0^{+0.74}_{-0.0}$	1
	$\tau \to 3\pi$		1820	9.6 ± 2.4	9
LM	$\tau \to e$		859	5.4 ± 0.9	6
	$\tau \to h(n\pi^0)$	ρ	458	5.2 ± 1.8	7
		π	357	6.7 ± 2.3	5
	$\tau \to 3\pi$		288	3.5 ± 1.2	5

TABLE 3. Summary of the results (90% C.L.) on oscillation probability, Δm^2 for $\sin^2 2\theta = 1$ and $\sin^2 2\theta$ for large Δm^2.

	$P_{\mu\tau}$	Δm^2 (eV^2/c^4)	$\sin^2 2\theta$	$P_{e\tau}$	Δm^2 (eV^2/c^4)	$\sin^2 2\theta$
CHORUS	$3.4 \cdot 10^{-4}$	0.6	$6.8 \cdot 10^{-4}$	$2.6 \cdot 10^{-2}$	7	$5.2 \cdot 10^{-2}$
NOMAD	$2.2 \cdot 10^{-4}$	0.8	$4.4 \cdot 10^{-4}$	$1.1 \cdot 10^{-2}$	6.5	$2.2 \cdot 10^{-2}$

number of ν_τ events is also reported.

RESULTS

CHORUS and NOMAD have recently analysed the full available data sample and do not show any evidence of ν_τ appearance. Limits have been computed on $\nu_\mu \to \nu_\tau$ and $\nu_e \to \nu_\tau$ probabilities of oscillation in the approximated two flavour mixing scheme. For $\nu_e \to \nu_\tau$ oscillation, the assumption was made that any observed ν_τ signal should come from the oscillation of the ν_e component of the beam. The corresponding 90% C.L. limits are shown in table 3.

It has to be emphasized that these limits have been obtained with different statistical methods: CHORUS used the procedure described in [5], while NOMAD relies on [4]. If CHORUS had chosen the same statistical treatment as NOMAD, the upper limit would have been $P_{\mu\tau} \leq 2.0 \cdot 10^{-4}$, which can be compared with the NOMAD result.

The 90% C.L. excluded regions in the ($\sin^2 2\theta$, Δm^2) parameter space are shown in Figure 1 and 2, for $\nu_\mu \to \nu_\tau$ and $\nu_e \to \nu_\tau$ oscillation, respectively.

Both experiments are improving the analysis techniques to achieve the designed sensitivity. CHORUS has started a rescan of all emulsions that, thanks to new and faster automatic scanning techniques and improved offline event reconstruction, will provide better efficiencies in a couple of years.

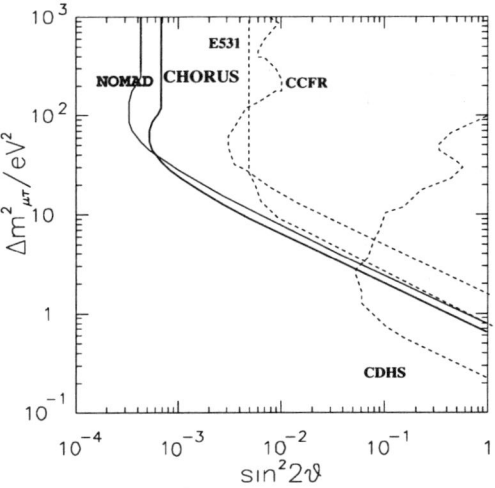

FIGURE 1. Exclusion plots for $\nu_\mu \to \nu_\tau$ oscillation. Previous limits are also shown.

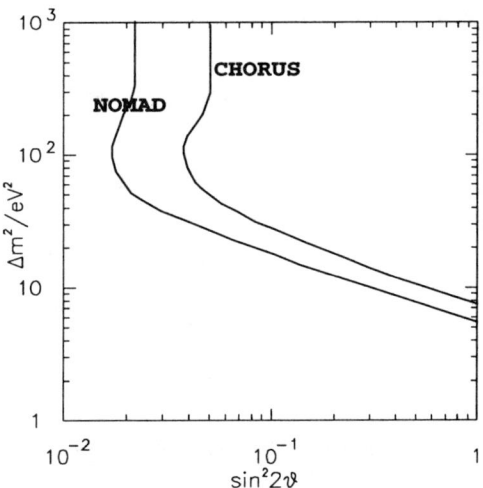

FIGURE 2. Present limits on $\nu_e \to \nu_\tau$ oscillation.

ACKNOWLEDGEMENTS

I would like to thank the organisers of CIPANP 2000 conference for the invitation and financial support to give this talk. I am also very grateful to R. Santacesaria and P. Zucchelli from the CHORUS Collaboration and L. Camilleri from NOMAD for valuable discussions on the latest results.

REFERENCES

1. Eskut E. et al., CHORUS Collaboration, *Nucl. Instr. and Meth.* **A401** (1997) 7.
2. Altegoer, J. et al., NOMAD Collaboration, *Nucl. Instr. and Meth.* **A 404** (1998), 96.
3. Astier P. et al., NOMAD Collaboration, CERN-EP-2000-49, March 17, 2000.
4. Feldman, G. J., and Cousins, R.D., *Phys Rev.* **D 57** (1998), 3873.
5. Junk, T., *Nucl. Instr. and Meth.* **A434** (1999), 435.

Recent results from KARMEN2

Christian Oehler for the KARMEN Collaboration

Forschungszentrum Karlsruhe, Institut für Kernphysik,
und Univeristät Karlsruhe
D-76021 Karlsruhe, Postfach 3640, Germany
e-mail: christian.oehler@bk.fzk.de

Abstract. The results of the search for neutrino oscillations by the KARMEN experiment since Feb.1997 until Feb.1999 are presented. We find 8 events conditioning all cuts of the $\bar{\nu}_\mu \to \bar{\nu}_e$ analysis in good agreement with the background expectation of 7.8 ± 0.6 events. From an event based maximum likelihood method an upper limit for the mixing angle is deduced to $\sin^2(2\Theta) < 21 \cdot 10^{-4}$ (90% C.I.) for large $\Delta m^2 (=100 \text{eV}^2)$.

I INTRODUCTION

The physics program of KARMEN (KArlsruhe Rutherford Medium Energy Neutrino Experiment) includes the investigation of neutrino-nucleus interactions [1], the search for neutrino oscillations in the appearance modes ($\bar{\nu}_\mu \to \bar{\nu}_e$, $\nu_\mu \to \nu_e$ [2]) as well as the search for lepton violating decays of pions and muons and tests of the V-A structure of μ^+ decay [4]. Here we concentrate on the KARMEN search for $\bar{\nu}_\mu \to \bar{\nu}_e$ oscillations. The KARMEN experiment is sensitive to neutrino mass differences in the range 0.1-100 eV^2 and therefore has the potential to check the LSND evidence region [8].

II NEUTRINO SOURCE AND DETECTOR

The KARMEN experiment is located at the neutron spallation facility ISIS of the Rutherford Appleton Laboratory (Chilton,UK). The neutrinos are produced by spallation processes if 800 MeV protons are shot on a Ta–D$_2$O–target: ν_μ, ν_e and $\bar{\nu}_\mu$ emerge with equal intensities from the decay chain $\pi^+ \to \mu^+ + \nu_\mu$ and $\mu^+ \to e^+ + \nu_e + \bar{\nu}_\mu$. The ν_μ from π^+–decay at rest are monoenergetic ($E_\nu = 30$ MeV) whereas the continuous energy distributions of ν_e and $\bar{\nu}_\mu$ up to 52.8 MeV can be calculated using the V–A theory. ISIS produces two parabolic proton pulses of 100 ns base width separated by 325 ns. The repetition frequency of the proton pulses is 50 Hz. In the 20 ms beam pause background measurements can be undertaken with high accuracy. The different lifetimes of π^+ ($\tau = 26$ ns) and μ^+

($\tau = 2.2\,\mu$s) allows a clear separation in time of the ν_μ-burst from the following ν_e and $\bar{\nu}_\mu$ and provide duty cycles of 1×10^{-5} and 5×10^{-4} respectively.

Neutrinos are detected in a 56 ton liquid scintillation calorimeter with an active volume of 96% and an excellent energy resolution of $\sigma_E = 11.5\%/\sqrt{E[\text{MeV}]}$. The detector is segmented by double acrylic walls. Gd_2O_3 coated paper within the module walls provide efficient detection of thermal neutrons via Gd(n,γ) capture. A massive 7000 ton iron blockhouse in combination with two layers of active anti counters provides shielding against beam correlated spallation neutron background and suppression of hadronic cosmic radiation as well as reduction of the flux of cosmic muons. In 1996 an additional third VETO counter system with total area of 300 m^2 was installed within the 3 m thick roof and the 2-3 m thick walls of the iron shielding [7]. Muons interacting elastically or deep inelastically with the iron nuclei of the steel blockhouse up to a distance of 1 m from the main detector can now be recognized and successive events as the main background for the oscillation search be eliminated. The background reduction compared to the level of the KARMEN1 experiment before this upgrade is a factor of 40.

III $\bar{\nu}_\mu \rightarrow \bar{\nu}_e$ SIGNATURE

If a $\bar{\nu}_e$ appears due to oscillation it can be detected by a unique signature of a spatially and in time correlated coincidence between a high energetic positron from p($\bar{\nu}_e, e^+$)n und γ emission from the subsequent neutron capture. The time distribution of the positrons must exactly reflect the 2.2 μs lifetime structure of the μ^+. Their energies are expected up to 51 MeV. The shape of the measured energy spectrum strongly depends on Δm^2 via the modulation through the oscillation formular. In this analysis we demand the positron to have energy from 16-50 MeV and to show up in the detector in the time range of 0.6-10.6 μs. Neutrons are either captured on protons or on Gd, giving rise to the emission of monoenergetic 2.2 MeV photons, or up to 3 γ quanta with a continuous energy distribution up to 8 MeV respectively. The time difference between e$^+$ and γ's from neutron capture is characterized by the thermalisation time of neutrons in the szintillator parameterized with an exponential slope of 120 μs. We demand a spatial correlation within a coincidence volume of 1.2 m^3.

IV DATA ANALYSIS

Within well defined cuts, optimized to get highest efficieny for the oscillation search, we find 8 sequences fulfilling all conditions. With a total background expectation of 7.8 ± 0.5 sequences the data sample shows no oscillation signal, neither in the number of events nor in the spectral distribution. The indivdual background sources are 1.9 ± 0.1 sequences induced by cosmic μ, 2.6 ± 0.3 sequences induced by charged current reactions of ν_e on ^{12}C, 2.3 ± 0.3 ν–induced accidental coincidences and 1.1 ± 0.1 sequences from the intrinsic ISIS $\bar{\nu}_e$ contamination. Besides the intrinsic contamination all background contributions can be measured in their strength and spectral distributions with high precission outside the beam window. To take the excellent detector resolution and the full information of each event into account we use an event based maximum likelihood method (ML) to determine a possible $\bar{\nu}_\mu \to \bar{\nu}_e$ signal. The different background contributions are fixed and only the ratio between oscillation signal and total background is varied.

As the statistical method to derive a 90% confidence interval for the (Δm^2,$\sin^2(2\Theta)$) parameter space from the Likelihoodfunction (LF) we use the Unified Approach [6].

V RESULTS

The global maximum of the likelihood function is at the mixing angle $\sin^2 2\Theta = 0$. The best description of the KARMEN data is therefore the *no oscillation hypothesis*. Under the assumption of maximum mixing ($\sin^2(2\Theta)=1$) we had expected 1604 oscillation events, thus setting an upper limit on $\sin^2(2\Theta)$:

$$\sin^2 2\Theta < 2.1 \times 10^{-3} \qquad \Delta m^2 = 100 eV^2 \quad (90\% C.I) \qquad (1)$$

This experimental result is in very good agreement with the expected sensitivity of $\sin^2 2\Theta < 2.3 \times 10^{-3} (\Delta m^2 = 100 eV^2)$. The sensitivity is defined as the average upper limit that would be obtained by an ensemble of experiments with the expected background and no signal. The 90% C.I. exclusion curve for the entire Δm^2 range is plotted as black solid line in figure 1 in comparison to the LSND [10] result. The LSND areas correspond to 90 % and 95% C.I., which also have been derived according reference [6]. As can be seen, KARMEN excludes the entire LSND regions above 2 eV^2 not only in the 90% C.I. but even in the 95% C.I. (dashed line). For lower Δm^2 we set the most stringent upper limits on $\bar{\nu}_\mu \to \bar{\nu}_e$ mixing and exclude most of the LSND favored parameter space.

Due to the extremely small background level of KARMEN2 after the VETO upgrade, the collected amount of data and the powerfull analyzing technique based on a maximum likelihood method KARMEN's sensitivity has nearly reached its final (in spring 2001) expected sensitivity of $\sin^2(2\Theta) = 1.4 \cdot 10^{-3}$ for large Δm^2 and is now probing for the first time the small Δm^2 favored regions of LSND leaving only very little statistical space for speculations that LSND is due to anBetween the proton pulses there is a nearly 20 ms beam pause where background measurements can be undertaken with high accuracy. oscillation signal.

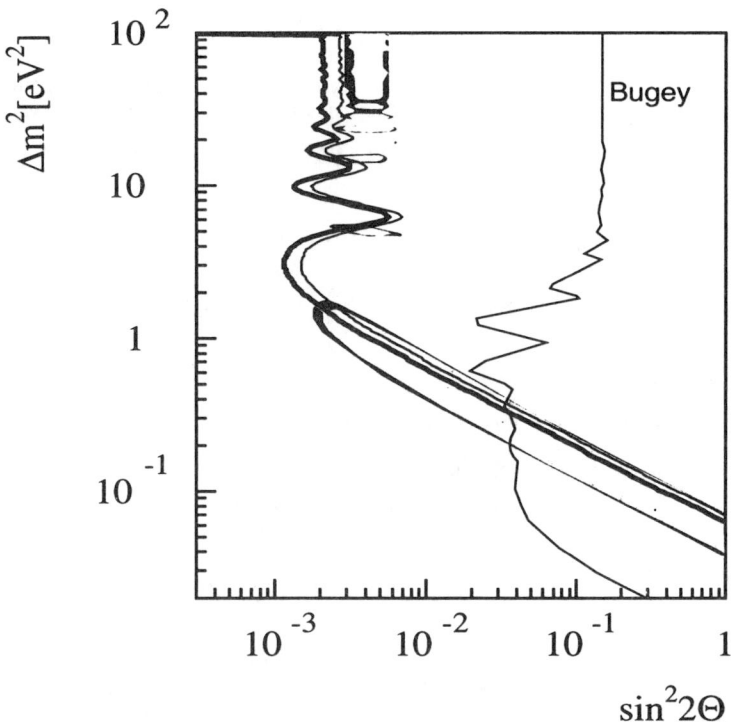

FIGURE 1. This plot shows the upper limit of the KARMEN 2 90 % confidence intervall (solid curve) and 95 % C.I. (dashed line). The 90% and 95% confidence intervalls of the LSND experiment are shown as grey areas. In addition the 90% C.L. exclusion curve by Bugey is plotted, excluding mixing angles $\sin^2 2\Theta > 0.04$.

REFERENCES

1. R. Maschuw et al., Prog. Part. Nucl. Physics **40**, 183 (1998).
2. B. Zeitnitz et al., Prog. Part. Nucl. Physics **40**, 169 (1998).
3. B. Armbruster et al., Phys. Rev. C **57**, 3414 (1998).
4. B. Armbruster et al., Phys. Rev. Lett. **81**, 520 (1998).
5. R.L. Burman et al., Nucl. Instr. Meth. A **368**, 416 (1996).
6. G.J. Feldmann and R.D. Cousins, Phys. Rev. D **57**, 3873 (1998).
7. G. Drexlin et al., Prog. Part. Nucl. Physics **40**, 193 (1998).
8. C. Athanassopoulos et al., Phys. Rev. C **54**, 2685 (1996).
9. B. Achkar et al., Nucl. Phys. B **434**, 503 (1995).
10. R. Tayloe et al., Proc. of the Lake Louise Winter Institute (1999)

Recent Results Addressing the KARMEN Timing Anomaly

E. D. Zimmerman

Columbia University, New York, New York

Abstract. Recent resuls from experiments at Fermilab and the Paul Scherrer Institute have constrained the parameter space available for a hypothetical particle Q^0 produced in the decay $\pi^+ \to \mu^+ Q^0$. This decay has been invoked to explain a peculiar feature of an event arrival time distribution observed in the KARMEN neutrino experiment.

THE KARMEN SIGNAL

In 1995, the KARMEN collaboration at the Rutherford Appleton Laboratory's ISIS spallation source published evidence [1] for an anomaly in the arrival-time distribution of neutrinos from pion and muon decay in a pulsed beam-stop source. The anomaly consisted of an excess of events with a small amount of electromagnetic energy ($<\sim$35 MeV) delayed 3.6 μs with respect to the beam arrival.

In KARMEN's first data set (1990-95) the anomaly comprised 83 ± 28 events; data through 1998 [2] increased the excess to 103 ± 34 events.

The particle interpretation

An explanation of the anomaly is that an exotic particle, referred to in this paper as Q^0, is produced by the rare decay $\pi^+ \to \mu^+ Q^0$ near the kinematic threshold for that process. From a π^+ decay at rest, the Q^0 would travel 17.5 meters to the KARMEN detector in 3.6 μs (corresponding to a velocity $v \approx 0.016c$). The anomalous signal in the detector would be due to the decay of the Q^0 to an electromagnetic final state. Based on the measured time of flight, the mass of the Q^0 must be 33.91 MeV. The visible energy of the anomaly favors a three-body decay to an electron (or possibly photon) pair and an invisible final-state particle, likely a neutrino. A likelihood analysis [2] has shown evidence for a correlation between position and arrival time for events in the anomaly, consistent with the slow-moving particle interpretation.

If interpreted as such a particle, the KARMEN signal corresponds to a curve in Q^0 lifetime versus $BR(\pi^+ \to \mu^+ + Q^0) \cdot BR(Q^0 \to \text{visible})$. This branching ratio is minimized at $\sim 10^{-16}$ for a lifetime $\tau_{Q^0} = 3.6$ μs. For any larger branching ratio, two solutions exist, one at a longer and one at a shorter lifetime. A previous experiment at the Paul Scherrer Institute (PSI) has ruled out at 90% C.L any exotic π^+ decays to muons above a branching ratio of 2.1×10^{-8} [3]. This constrains the lifetime of the Q^0 to be between $\sim 10^{-7}$ and $\sim 10^3$ s (see Figure 1).

Theoretical explanations

Most theoretical speculation on the anomaly has been along one of two lines: neutral heavy leptons [4,5] and light neutralinos [8]. Both are discussed briefly below:

A "standard" neutral heavy lepton ("heavy" or "sterile neutrino"), which would be produced and decayed solely through mixing with ν_μ, is consistent with the KARMEN data alone. However, the KARMEN result would require the particle to have a relatively large mixing with ν_μ [6,7]; such a large mixing is not consistent with the PSI branching ratio limit. A neutral heavy lepton explanation for the Q^0 is still allowed if its production is dominated by a small mixing with ν_μ and its decay is dominated by a much larger mixing with ν_τ.

Another scenario consistent with the KARMEN data is an R-parity violating neutralino decay. This scenario is allowed in an unconstrained supersymmetric model, but chargino mass limits [9] exclude a such light neutralino in models such as SUGRA, which introduce a chargino-neutralino mass relation.

FERMILAB E815 (NUTEV)

Fermilab experiment 815 (NuTeV) has performed a direct search for Q^0 decay using a beam created by the decays of high-energy pions and kaons. The experiment took data during Fermilab's 1996-97 Tevatron fixed target run, accumulating 2.54×10^{18} 800 GeV protons on a BeO target with the detector configured to search for exotic particle decays. Sign-selected secondary pions and kaons in the 100−400 GeV range were focused down a 440 m decay pipe, where they could decay before hitting a steel beam dump. A total of $(1.4 \pm 0.1) \times 10^{15}$ pion decays occurred in the pipe. Neutral weakly-interacting decay products (neutrinos and possibly Q^0's) traveled through approximately 900 m of earth berm shielding before arriving at the detector.

The E815 detector consisted of an instrumented decay channel followed by an iron-target neutrino detector. The decay channel was composed of an upstream charged particle veto followed by a series of helium bags totaling 34 m in length, interspersed with 3 m × 3 m multiwire argon-ethane drift chambers. The chambers were used to track charged particles from decays in the helium fiducial volume. The neutrino detector, which consisted of iron plates interspersed with drift chambers and liquid scintillator counters, provided calorimetry, particle identification, and triggering. A toroid spectrometer at the downstream end of the detector measured the momentum of muons too energetic to range out in the calorimeter.

In this experiment, the experimental signature of Q^0 decay was the spontaneous appearance of a low-mass, low-transverse momentum electron-positron pair in the helium decay channel. The analysis requirements were that two well-reconstructed tracks form a vertex in the decay channel fiducial volume, and that the tracks be identified as electrons based on the shape of their shower in the calorimeter. Because of the low mass of the Q^0, signal events typically had a very small opening angle and thus formed a single merged electron-like cluster in the calorimeter.

Backgrounds to the Q^0 search were primarily due to interactions between the high-flux neutrino beam and material in the upstream berm, veto, and decay channel drift chambers. Interactions in the berm and veto wall could produce photons or neutral kaons which could then convert or decay to charged particles in the decay channel. Interactions in the drift chambers themselves could produce charged tracks directly in the fiducial volume. These backgrounds were reduced by removing events with activity in the upstream veto wall, by making tight cuts on electron identification, and by a series of kinematic cuts designed to discriminate against the high-Q^2 events with large invariant mass final states typical of neutrino deep inelastic scattering.

With the final requirements, the expected background level from all sources was (0.06±0.05) events with a signal acceptance of 16% for decays in the fiducial region. After a blind analysis, the signal region contained no events. The result excluded the short-lifetime solution to the KARMEN anomaly above a branching ratio of $\approx 5 \times 10^{-12}$ at 90% C.L. (see Figure 1 and Ref. [10]).

THE SEARCH AT PSI

A new indirect search for Q^0 production was recently conducted at PSI by M. Daum et al. [11]. The experiment exploited the near-zero Q-value in the $\pi^+ \to \mu^+ Q^0$ decay by searching for muons from π^+ decay emerging parallel to and with the same velocity as the parent π^+.

The experiment began with a proton beam aimed at a production target; a narrow-band beam of pions with momentum 150 MeV/c was selected and focused toward a decay region. Downstream of the decay region a beamline selected charged particles of a particular momentum traveling in the forward direction. A series of scintillation counters in the downstream beamline measured the velocity of particles in the beamline, allowing identification of decay products.

The signature of Q^0 production is an excess of forward muons with a momentum of 113.5 MeV. The search technique was to scan the analysis beamline to select different momenta around 113.5 MeV/c, in 0.5 MeV/c steps. The number of forward muons was counted in each momentum configuration, and the dependence of the muon rate was fit to a Q^0 plus background distribution.

The final fit indicated a branching ratio $BR(\pi^+ \to \mu^+ Q^0) = (1.3 \pm 2.3) \times 10^{-10}$,

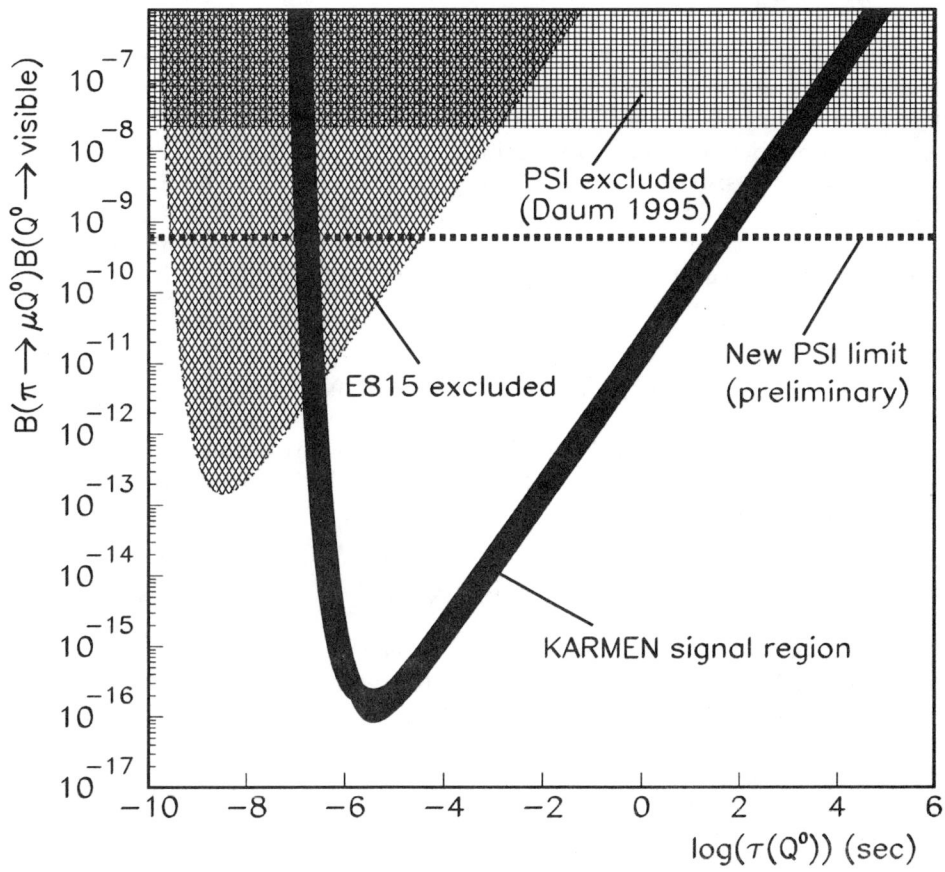

FIGURE 1. The KARMEN signal region and limits from other experiments.

reported as a limit $BR(\pi^+ \to \mu^+ Q^0) < 6.0 \times 10^{-10}$ at 95% C.L. Because the Q^0 decay is not detected, this limit applies to both the short and long lifetime solutions of the KARMEN anomaly. Unlike the E815 limit, it is independent of $BR(Q^0 \to \text{visible})$.

STATUS OF THE ANOMALY

At present, the timing anomaly remains an apparently significant feature of the KARMEN data set. New data from KARMEN over the next year will be interesting, but will not be a major addition to the existing data. Neither the PSI nor the Fermilab groups expect to take more data. Their current results are already approaching inherent limitations in their technique: the unavailability of arbitrarily high beam intensities in the Fermilab case, and the intrinsic $\pi^+ \to \mu^+ \nu \gamma$ background at PSI. Some upcoming experiments, including BooNE (Fermilab E898) will have some sensitivity to the Q^0, particularly in the short-lifetime region. A dedicated search for Q^0 decay has been proposed at ISIS, and it is possible that the muon source at a muon storage ring may be used to search for Q^0 production. No experiment currently approved is likely to have sufficient sensitivity to confirm or rule out the existence of the Q^0.

ACKNOWLEDGMENTS

The author thanks the E815 (NuTeV) collaboration, M. Daum and J. Koglin of PSI, and C. Oehler of the University of Karlsruhe. This work was supported by the National Science Foundation.

REFERENCES

1. B. Armbruster et al., *Phys. Lett.* **B348**, 19 (1995).
2. C. Oehler, *Nucl. Phys. B, Proc. Suppl.* **85**, 101 (2000).
3. M. Daum et al., *Phys. Lett.* **B361** 179 (1995).
4. J. Govaerts, J. Deutsch, and P.M. Van Hove, *Phys. Lett.* **B389** 700 (1996).
5. V. Barger, R.J.N. Phillips, and S. Sarkar, *Phys. Lett.* **B352** 365 (1995) and erratum: *Phys. Lett.* **B356** 617 (1995).
6. M.Gronau, C.N. Leung, and J.L. Rosner, *Phys. Rev.* **D29** 2539 (1984).
7. R. E. Shrock, *Phys. Rev.* **D24** 1232 (1981).
8. D. Choudhury et al., *Phys. Rev.* **D61** 095009 (2000).
9. D. E. Groom et al., *Eur. Phys. J.* **C15** 1 (2000).
10. J. Formaggio et al., *Phys. Rev. Lett.* **84** 4043 (2000).
11. J. Koglin, PhD Dissertation, The University of Virginia (2000).

Atmospheric Neutrino Results from Super-Kamiokande Experiment

Yoshihisa Obayashi for the Super-Kamiokande collaboration

Kamioka Observatory, Institute for Cosmic Ray Research, University of Tokyo, Japan.
ooba@icrr.u-tokyo.ac.jp

Abstract. Super-Kamiokande has been observing atmospheric neutrinos since April 1, 1996. The observation livetime is now 990 days, corresponding to an exposure of 61 kt-yr. Observed flavor, energy and direction distributions of are inconsistent with atmospheric neutrino flux assuming no neutrino oscillation; they are, however, consistent with ν_μ to ν_τ oscillation. The allowed ν_μ to ν_τ oscillation parameter space is $2 \times 10^{-3} < \Delta m^2 < 5 \times 10^{-3} eV^2$ and $\sin^2 2\theta > 0.88$, at 90% confidence level. This result agrees well with other recent underground atmospheric neutrino experiments. The data have also been analyzed for ν_μ to $\nu_{sterile}$ oscillation and $\nu_e - \nu_\mu - \nu_\tau$ 3-flavor oscillation. Both analyses favor ν_μ to ν_τ oscillation.

INTRODUCTION

Atmospheric neutrinos are produced through the interaction of primary cosmic ray with nuclei in the atmosphere and the decay chain like $\pi^\pm \to \mu^\pm + \nu_\mu(\bar{\nu}_\mu), \mu^\pm \to e^\pm + \nu_e(\bar{\nu}_e) + \bar{\nu}_\mu(\nu_\mu)$.

Neutrino oscillation occurs if a finite mass difference and mixing angle between exists between two neutrino flavors; the probability for this oscillation given by $P(\nu_\alpha \to \nu_\beta) = \sin^2 2\theta \cdot \sin^2(1.27 \frac{L(km)}{E_\nu(GeV)} \Delta m^2 (eV^2))$, where θ is the mixing angle between two neutrinos, L is the flight length of the neutrino, E_ν is the energy of the neutrino, Δm^2 is the mass squared difference ($\equiv |m_i^2 - m_j^2|$). Typical values of E_ν and L for atmospheric neutrinos are 100MeV - 100GeV and 10km - 13000km, respectively. Flux of atmospheric neutrinos are calculated by several parties [1] [2].

The flavor ratio ($N(\nu_\mu/\nu_e)$) and zenith angle distribution of these neutrinos depend very much on whether or not neutrino oscillation occurs. Several recent underground experiments have observed the neutrino flavor ratio and zenith angle distribution; all [3] [4] [5] [6] disagree with the no oscillation hypothesis, but agree well with hypotheses incorporating neutrino oscillation. This paper reports on recent results of the Super-Kamiokande atmospheric neutrino oscillation analysis.

TABLE 1. Event summary for 990day contained sample

Sub-GeV ($E_{vis} < 1.33 GeV$)		
	Data	MC(Honda flux)
1ring e-like	2185	2081.8
1ring μ-like	2178	3137.4
Multi ring	1637	2011.5
Total	6000	7230.7

Multi-GeV FC($E_{vis} > 1.33 GeV$)		
	Data	MC(Honda flux)
1ring e-like	492	481.3
1ring μ-like	421	640.0
Multi ring	1027	1273.8
Total	1940	2395.1

Partially Contained		
Total	563	818.9

EVENT SAMPLE

Super-Kamiokande is a 50,000 ton water Cherenkov detector located under Mt. Ikenoyama, giving it a rock over-burden of 2,700 m water-equivalent. The fiducial mass of the detector for atmospheric neutrino interactions is 22,500 tons.

Neutrino events observed in the Super-Kamiokande are categorized as fully-contained (FC) or partially-contained (PC), according to the amount of anti-counter activity. FC events are subdivided into Sub-GeV and Multi-GeV samples according to the amount of visible energy. Events with only one reconstructed ring are subdivided into e-like and μ-like based on likelihood analysis of the detected Cherenkov light. The number of events in each category for 990 days of livetime is shown in TABLE 1 and their zenith angle distribution is shown in FIGURE 1.

The flavor ratio is evaluated by the ratio between data and expectation without oscillation: $R \equiv \frac{(\mu-like/e-like)_{data}}{(\mu-like/e-like)_{MC}}$. The observed values of R for 990days are $0.661 \pm 0.020 \pm 0.052$ for sub-GeV and $0.660^{+0.038}_{-0.035} \pm 0.078$ for Multi-GeV samples.

Muons from the high-energy interaction of $\nu_\mu(\bar{\nu}_\mu)$ with rock below the detector are also used in the oscillation analysis; these events are called "up through-going" and "up stopping" muons, depending on whether the muon penetrates the detector or stops in it.

The up/down ratio for the combined Multi-GeV μ-like + PC event sample after 990 days is $U/D = 0.56 \pm 0.04 \pm 0.01$ where U (D) is number of events with $\cos \Theta < -0.2 (> 0.2)$. The observed up/down ratio differs from the no-oscillation hypothesis by $\sim 8\sigma$.

$\nu_\mu \rightarrow \nu_\tau$ OSCILLATION ANALYSIS

The most likely value of the neutrino oscillation parameters $\sin^2 2\theta$ and Δm^2 for $\nu_\mu \rightarrow \nu_\tau$ oscillation are obtained using 1ring-FC, PC, and up-going muon samples. The fitting method is described elsewhere [4]. The minimum χ^2 is found to be 61.1 with 82 degrees of freedom (d.o.f.) at $\Delta m^2 = 2.8 \times 10^{-3} eV^2$, $\sin^2 2\theta = 1.0$. Allowed

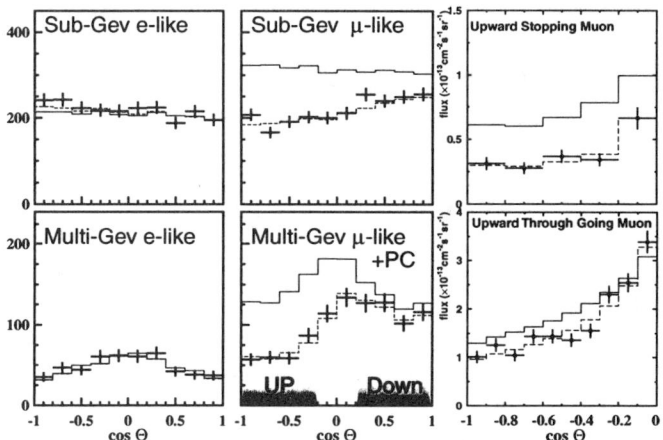

FIGURE 1. Zenith angle distribution of Super-Kamiokande 990day FC, PC and UPMU samples. Cross, solid line and dashed line correspond to Data, no oscillation MC and MC with best fit oscillation parameters, respectively. up/down ratio from the Multi-GeV μ-like + PC sample is calculated as $0.56 \pm 0.04 \pm 0.01$.

oscillation parameter region is shown in FIGURE 2. χ^2 for no oscillation was found to be 232.5 for 84 d.o.f.

THREE FLAVOR OSCILLATION ANALYSIS

The results described in the last section agree well with the 2-flavor, $\nu_\mu \to \nu_\tau$ oscillation hypothesis. The data, however, does not preclude the possibility of 3-flavor oscillation. If the mass-degeneracy condition, $\Delta m_{23}^2 \gg \Delta m_{12}^2$ is assumed (not unreasonable for $\Delta m_{23}^2 = \Delta m_{atm}^2 > 10^{-3}$ and $\Delta m_{12}^2 = \Delta m_{sol}^2 < 10^{-4}$), the 3-flavor neutrino oscillation probability is given to a good approximation as: $P(\nu_\mu \leftrightarrow \nu_e) = \sin^2 2\theta_{13} \cdot \sin^2 \theta_{23} \cdot \Psi$, $P(\nu_\mu \leftrightarrow \nu_\tau) = \cos^4 \theta_{13} \cdot \sin^2 2\theta_{23} \cdot \Psi$, $P(\nu_e \leftrightarrow \nu_\tau) = \sin^2 2\theta_{13} \cdot \cos^2 \theta_{23} \cdot \Psi$, $\Psi = \sin^2(1.27 \cdot \Delta m^2 \cdot L/E)$. In these equations, $\Delta m^2 = \Delta m_{23}^2$, and Δm_{12}^2 is set to zero. Note that there are only three oscillation parameters $(\sin^2 \theta_{13}, \sin^2 \theta_{23}, \Delta m^2)$. A 3-flavor oscillation analysis was performed on the Super-Kamiokande 990days, FC + PC data. The binning in this analysis is the same as the 2-flavor case, except that the up-going muon samples were not used. The allowed parameter region for $\sin^2 \theta_{13}$ and $\sin^2 \theta_{23}$ is shown in FIGURE 3. The limit on θ_{13} from the CHOOZ [7] is shown as well. The best-fit parameter values are $(\sin^2 \theta_{13}, \sin^2 \theta_{23}, \Delta m^2) = (0.03, 0.63, 3 \times 10^{-3} eV^2)$; this is consistent with the CHOOZ result. Finally, while CHOOZ allows for $\sin^2 \theta_{13} > 0.97$, this study excludes such a large $\sin^2 \theta_{13}$ value which correspond to completely no mixing even between ν_μ and ν_τ.

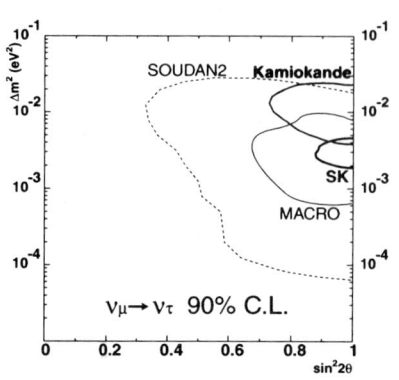 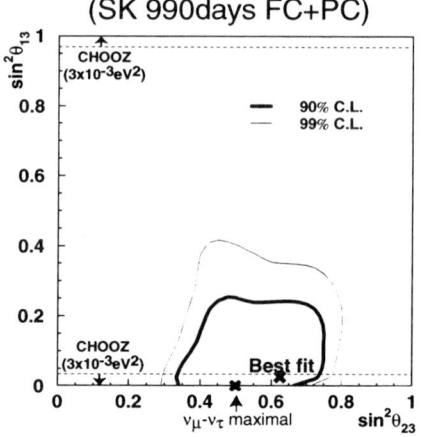

FIGURE 2. 90% confidence level allowed regions for $\nu_\mu \to \nu_\tau$ oscillation obtained by Super-Kamiokande 990day result, KAMIOKANDE [3], MACRO [5], and SOUDAN2[6].

FIGURE 3. 90% and 99% confidence level allowed region for three flavor oscillation obtained by Super-Kamiokande 990day FC and PC sample.

$\nu_\mu \to \nu_\tau$ OR $\nu_\mu \to \nu_{STERILE}$ ANALYSIS

Some models predict that ν_μ oscillates into "sterile" neutrino (ν_s) that does not interact even via neutral current (N.C.). If the observed deficit in ν_μ is due to $\nu_\mu \to \nu_s$ oscillation, then the number of events produced via N.C. interaction for up-going neutrino should also be reduced. Also, in the case of $\nu_\mu \to \nu_s$, matter effect will suppress oscillation in the high energy ($E_\nu > 15 GeV$) region. We used the following data sample to observe these effects: (a) N.C. enriched sample; (b) the high-energy (visible energy > 5GeV) part of the PC sample; and (c) up-through-going muons. The sterile-neutrino hypothesis is tested using the UP/DOWN ratio in samples (a) and (b) and the UP/HORIZONTAL ratio in sample (c). Zenith angle distributions for each sample is shown in FIGURE 4. As shown in FIGURE 5, $\nu_\mu \to \nu_s$ oscillation is excluded at almost 99% C.L. while $\nu_\mu \to \nu_\tau$ oscillation is well favored.

SUMMARY

Super-Kamiokande results from 990days of contained events and up-going muons events give 90% C.L. allowed parameter regions $\sin^2 2\theta > 0.88$ and $2 \times 10^{-3} eV^2 < \Delta m^2 < 5 \times 10^{-3} eV^2$. The 3-flavor oscillation analysis limits $\sin^2 \theta_{13} < 0.25$ at 90% C.L.. Finally, the hypothesis test between $\nu_\mu \to \nu_\tau$ versus $\nu_\mu \to \nu_s$ disfavors latter at 99% C.L..

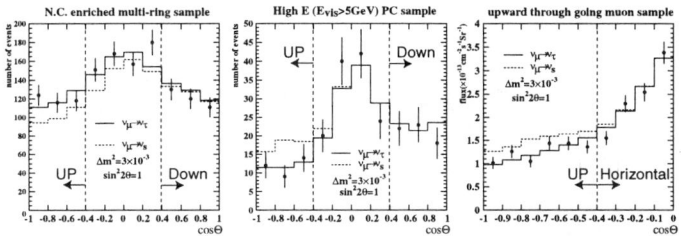

FIGURE 4. Zenith angle distributions of left: NC enriched sample, center: PC high-E_ν sample, right: up-through-going muon sample.

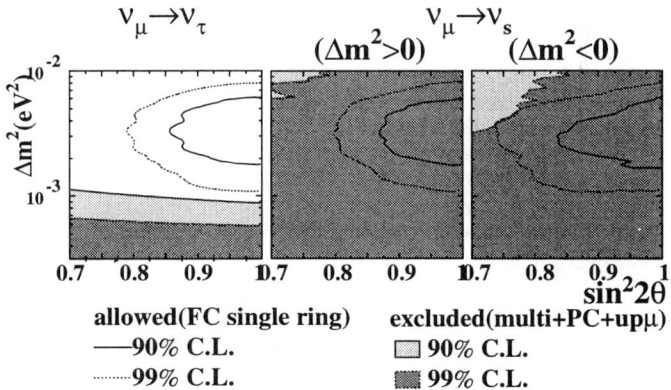

FIGURE 5. Allowed regions obtained by FC sample analysis and excluded regions obtained by hypothesis tests on NC-enriched sample and High-energy sample.

REFERENCES

1. M. Honda et al., Phys. Lett. **B248**, 193 (1990); M. Honda et al., Phys. Rev. **D52**, 4985 (1995)
2. G. Barr et al., Phys. Rev. **D39** 3532 (1989); V. Agrawal et al., Phys. Rev. **D53** 1313 (1996); T. K. Gaisser and T. Stanev, Proc. 24th Int. Cosmic Ray Conf. (Rome) Vol.1 694 (1995)
3. K. S. Hirata et al., Phys. Lett. **B205**, 416 (1988); Y. Fukuda et al. Phys. Lett. **B335**, 237 (1994).
4. Y. Fukuda et al., Phys. Rev. Lett. **81**, 1562 (1998)
5. M. Ambrosio et al., Phys. Lett. **B434**, 451 (1998)
6. T. Kafka, Proceeding of the 6th International Workshop on Topics in Astroparticle and Underground Physics (TAUP 99), Paris, France, 6-10 Sep 1999 (hep-ex/9912060)
7. M. Apollonio et al., Phys. Lett. **B466**, 415 (1999)

The First Results of K2K long-baseline Neutrino Oscillation Experiment

Taku Ishida*, representing K2K collaboration

*Institute for Particle and Nuclear Studies(IPNS)
High Energy Accelerator Research Organization(KEK)
1-1 Oho, Tsukuba-shi, Ibaraki 305-0801, Japan

Abstract. The first results of the K2K(KEK to Kamioka) long-baseline neutrino oscillation experiment are presented in this talk. In 1999 7.2×10^{18} protons on target were delivered to the experiment. During this period of running there were 3 events fully contained in the Super-Kamiokande inner detector fiducial area which occurred during the beam spill timing window. In the case of no oscillations the expected number of events during this period was $12.3^{+1.7}_{-1.9}$. The near detectors located at KEK also have begun detailed measurements of neutrino interactions in water at around 1 GeV.

INTRODUCTION

The atmospheric neutrino anomaly observed by Super-Kamiokande(SK) and other recent underground experiments strongly suggests $\nu_\mu \leftrightarrow \nu_\tau$ neutrino oscillation. The allowed region of the oscillation parameters obtained by SK are in the range of $\Delta m^2 = 2 \sim 5 \times 10^{-3}$ eV^2 and $sin^2(2\theta) > 0.88$ with 90% confidence level [1], where Δm^2 is the mass difference squared between two neutrino mass eigenstates and θ is the mixing angle between two neutrinos.

The principal goal of the K2K experiment is to confirm neutrino oscillation with a man-made neutrino beam and to measure the oscillation parameters. Fig. 1 is a schematic of the setup. We use the 12 GeV KEK-PS as a neutrino source and SK as the far detector. The distance between KEK and SK is $250 km$, and the neutrino beam has an average energy of 1.4 GeV. The neutrinos are produced by charged pion decays and expected to be 99% pure ν_μ with an angular deviation $\leq 3\ mrad$. In order to measure the effects of oscillation we compare the ν_μ spectrum observed by SK to the one measured in the front detectors at the point of production.

Fig. 2(a) shows the Monte Carlo simulation of reconstructed neutrino spectra at SK for 1×10^{20} protons on target(pot). This corresponds to approximately 5 years of measurement. Oscillated spectra for four specific oscillation parameter sets are shown. The effects of oscillation are seen in the figures as a significant divergence from the spectra of the null oscillation case, which is given by open histograms.

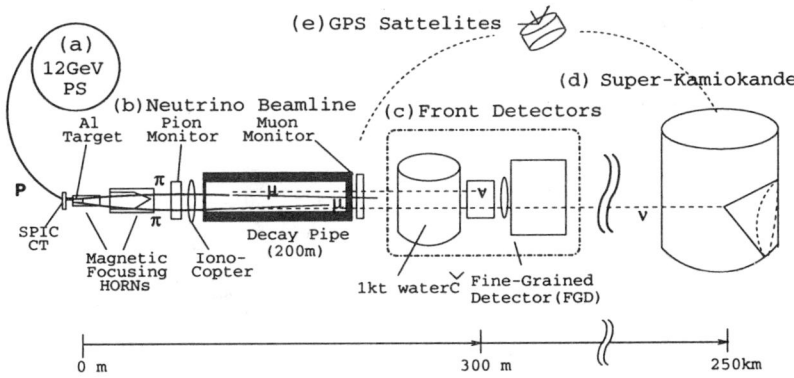

FIGURE 1. Schematic overview of the K2K experiment. (a) 12 GeV KEK-PS. A spill has 9 bunches in 1.1 μsec, 7×10^{12} *ppp* for every 2.2 sec. (b) Neutrino beam line. Protons, bent to the direction of Kamioka in the "arc section" of the beam line, are injected into Aluminum target of 3cmϕ × 66cm, embedded in the 1st horn. The proton beam profile and strength before the target are measured by 2 SPICs and 2 CTs, respectively. Two horn magnets for π focusing, a gas Cherenkov counter for π momentum distribution measurement (π monitor), and ϕ-symmetry monitor (iono-copter) are in a target station. πs decay into μ and ν_μ in 200m of decay pipe filled with helium gas. The μ profile is measured by ionization chambers and a silicon pad detector located behind the beam dump (μ monitor). (c) Front detector system to measure neutrino beam properties at the production. It is composed of a SK-like water Cherenkov detector (1kt detector) and a fine-grained detector(FGD) for detailed study of neutrino interaction with water. (d) Super-Kamiokande(SK) as a far detector, 50 kt water Cherenkov detector under stable operation since 1996, 250 km downstream of the target. (e) The GPS system is used to look for events at SK during the KEK PS beam pulse. The precision of $\Delta(T_{\rm KEK} - T_{\rm SK})$ is within 300 $nsec$, calibrated by an atomic clock.

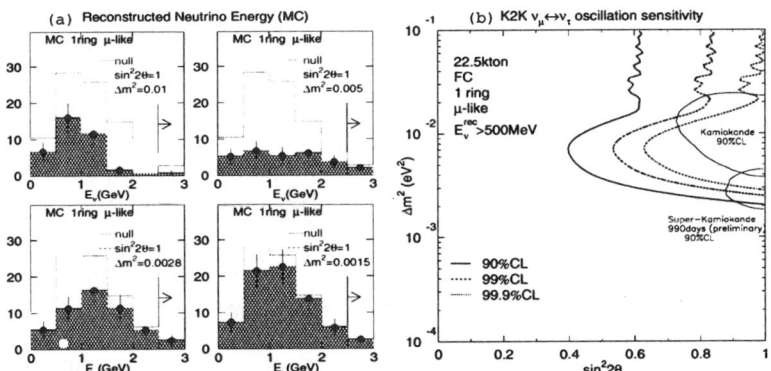

FIGURE 2. (a) Monte Carlo simulated reconstructed neutrino spectrum at SK, with four oscillation parameter sets. 10^{20} protons on target correspond to 190 neutrino interactions in the 22.5 kt SK inner fiducial volume in case of null oscillation. Plots are for 90 fully contained single-ring μ-like events. (b) Sensitivity plots of the K2K experiment for $\nu_\mu \leftrightarrow \nu_\tau$ oscillation.

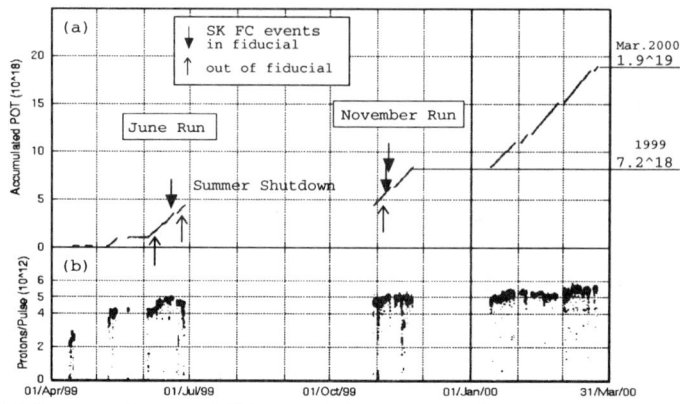

FIGURE 3. (a) Integrated *pot* and (b) *ppp* at the target, as functions of date from April 1999 to March 2000. Dates of SK fully-contained events are marked with arrows for 1999.

K2K's 90% confidence level sensitivity covers almost all of the 90% allowed regions obtained by Kamiokande and SK, as shown in Fig. 2(b).

Fig. 3 shows the record of protons on target. After the completion of the front detector construction in Feb. 1999, the fast extraction of protons for the experiment started on Feb. 3. On Mar. 4, the neutrino beam DAQ started with a horn current of 175 kA and with proton intensity of $3 \times 10^{12} ppp$. After engineering runs to study neutrino beam operations in April through May, stable data taking began in June with an aluminum target with $2cm\phi$, a horn current of 200 kA, and a proton intensity of $4.5 \times 10^{12} ppp$. After the summer shutdown, continuous data taking began again in November, this time with an aluminum target of $3cm\phi$ and a horn current of 250 kA with $5 \times 10^{12} ppp$. In 1999 we accumulated $7.2 \times 10^{18} pot$ in total.

OBSERVATION BY NEAR DETECTORS

The near detector system consists of a 1kt water Cherenkov detector(1kt) and a fine-grained detector(FGD). The latter consists of a scintillating fiber tracker(SFT) [2], plastic scintillator veto counters, an electromagnetic calorimeter of 600 lead glass blocks, and a muon ranger of 12 iron plates ($10cm \times 4 + 20cm \times 8$) with drift chambers(MUC). The SFT is composed of 19 layers of $6cm$-thick water containers sandwiched with $20 \times (yy\text{-}xx)$ layers of $700\mu m\phi$ scintillating fibers.

1kt is used for normalizing predicted beam flux at SK. The fact that the two detectors are so similar cancels out systematic errors that would otherwise be present.

MUC contained events are used to check the stability of the neutrino beam. This is because the large fiducial volume mass of the MUC results in a very large event rate. The beam center is observed to be stable to within $1mrad$.. In addition, spill by spill beam centering, monitored by the profile of the muons produced by pion-decay that penetrate the beam dump, was also found to be within $0.5mrad$.

Fig. 4(a) is an example of quasi-elastic(QE) neutrino event candidates with vertex

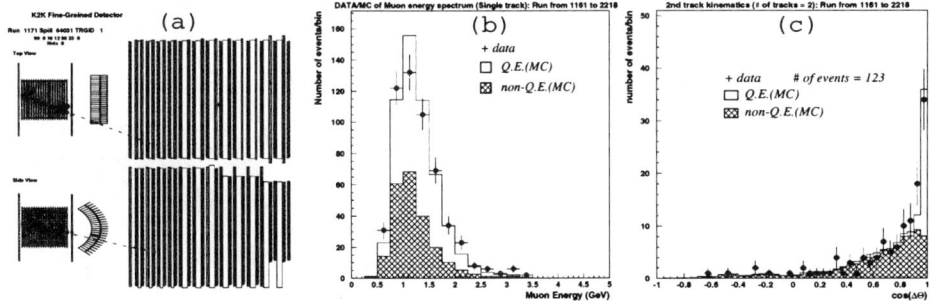

FIGURE 4. (a) A typical ν_μ event observed with the FGD. (b) Muon energy distribution for single track samples. (c) $cos(\Delta\theta_P)$ distribution for 2 track samples. In (b)/(c), histograms show MC, where open area is for quasi-elastic events, and hatched area is for non-QE (charged current inelastic + neutral current) events.

in the SFT. The primary goal of the FGD is to reconstruct the neutrino spectrum by using QE samples. (b) shows the reconstructed muon energy distribution for single track events, and (c) is $cos(\Delta\theta_P)$ distribution for 2 track samples, where $\Delta\theta_P$ is angular difference between reconstructed proton track and that calculated from muon momentum. The single track samples and 2 track samples with $cos(\Delta\theta_P) \geq$ 0.95 contain a large fraction of QE events, ~60% and ~80% respectively.

We also study the ratio of inelastic events to QE events, in order to reduce the uncertainty of the calculated neutrino interaction cross section to less than 10%. This information is important not only for K2K, but also for the SK atmospheric neutrino analyses. Table 1 gives summary of the front detector results. Event numbers normalized by MC predictions agree very well to each other.

NEUTRINO EVENTS IN SUPER-KAMIOKANDE

On June 19, 1999, 18:42(JST), K2K observed its first neutrino event due to the KEK neutrino beam in Super-Kamiokande. This was the first time that an artificially produced particle was detected after traveling such a large distance. Fig. 5(c) shows ΔT distributions at each reduction step, where ΔT is the time difference between SK event and KEK beam pulse, taking time of flight between

TABLE 1. Summary of the near detector observation.

	Mass(t)		$pot(10^{18})$	Events	Data/MC$\pm st.\pm sys.$
1kt	50.3	June	2.03	8,157	$0.83\pm.01\pm.07$
		November	2.62	11,337	$0.84\pm.01\pm.11$
MUC contained	445.	June	3.02	27,985	$0.83\pm.01\pm.11$
		November	2.75	28,077	$0.86\pm.01\pm.11$
SFT+MUC	4.9	June	2.28	315	$0.83\pm.05\pm^{.08}_{.09}$
		November	1.94	347	$0.86\pm.05\pm^{.08}_{.09}$

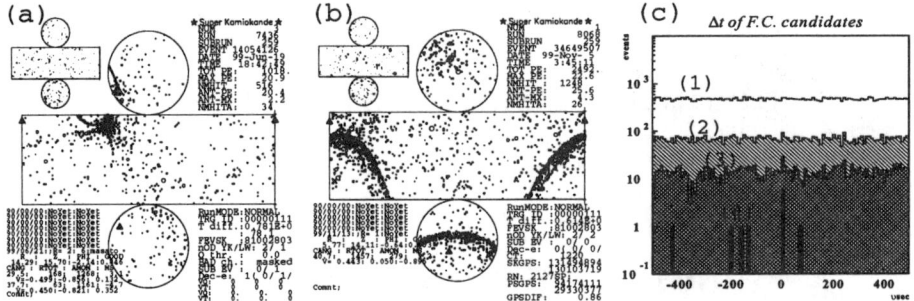

FIGURE 5. SK fully contained event examples (a) the 1st event in June, and (b) in November. (c) ΔT ($\equiv T_{SK} - T_{KEK} - T_{T.O.F.}$) distributions at each reduction stage: $\pm 500\mu sec$ time cut, (1) μ-decay electron cut (2) high energy trigger condition (3) inner counter total p.e. cut $200 < Q_{TOT} < 50,000 p.e.$, and (4) outer detector cut.

KEK and SK into account. It is obtained by GPS within the precision of 300 $nsec$. During the period of running we have 6 events within $1.3\mu sec$ time window, 3 of which are within $22.5 kt$ fiducial volume (vertex distance from wall $\geq 2m$).

We can estimate the number (N_{SK}^{pred}) using the observed number of events at 1kt(N_{KEK}): $N_{SK}^{pred} = (N_{KEK}/\epsilon_{KEK}) \cdot R \cdot \epsilon_{SK}$, where R is a factor of extrapolation from KEK to SK, ϵ_{KEK} and ϵ_{SK} are the detection efficiency of 1kt and SK, respectively. Since the target material (water) is common and the systematic uncertainty due to the cross section cancels, the value of R depends mostly on the flux ratio between SK and KEK, whose uncertainty is $^{+8\%}_{-10\%}$. N_{SK}^{pred} was estimated to be $12.3^{+1.7}_{-1.9}$ in total. Table 2 summarizes these numbers, with expectations for the case of three typical Δm^2 with $sin^2(2\theta) = 1$. We will accumulate $> 2 \times 10^{19} pot$ in this summer, leading to an increase in the statistical power of these results.

REFERENCES

1. Obayashi, Y., *this conference*.
2. Suzuki, A. *et al*, to be published in *Nucl. Instrum. Methods* A, hep-ex/0004024.

TABLE 2. Summary of SK events. Fully-Contained(FC) are events with the significant light detected in the inner detector only. Outer Detector(OD) events, with light detected in the outer detector is also tabulated for reference (systematic uncertainty~ 40%).

	Data	No oscillation	$\Delta m^2 = 3 \times 10^{-3}$	5×10^{-3}	7×10^{-3} $(eV)^2$
FC in fiducial	3	$12.3^{+1.7}_{-1.9}$	8.0	5.4	4.6
out of fiducial	3	$5.5^{+1.1}_{-1.2}$	3.5	2.4	2.1
OD contained	4	8.7 ± 3.3	5.5	3.5	2.9
(inner crossing)	2	4.2 ± 1.6	3.2	2.0	1.3

Upcoming Neutrino Oscillation Experiments at Fermilab

C. James

Fermi National Accelerator Laboratory
Batavia, Illinois 60510

Abstract. Two experiments to investigate neutrino oscillations are now under construction at Fermilab. An overview of the experiments is given.

INTRODUCTION

Recent results from atmospheric and reactor-based neutrino oscillation experiments have spurred the development of several accelerator based experiments, which offer more control over the energy spectrum and production rate of the particles. Two neutrino oscillation experiments are now under construction at Fermilab, the Booster Neutrino Experiment, or BooNE, and the Main Injector Neutrino Oscillation Search, or MINOS. While both experiments have the similar goal of observing neutrino oscillations and measuring the oscillation parameters Δm^2 and $sin^2 2\theta$, they each have a different emphasis.

BOONE

BooNE is primarily motivated by the reactor-based LSND and KARMEN experiments [1,2], whose results are interpreted as $\bar{\nu}_\mu \to \bar{\nu}_e$. Both experiments however were limited by low statistics. The BooNE experiment proposes to construct a beamline and detector facility designed explicitly to model the LSND and KARMEN setups, but produce neutrinos at far higher rates. If the LSND signal is due to $\nu_\mu \to \nu_e$ oscillations, then BooNE expects to collect about 1000 such events per year.

The neutrino beam uses 8 GeV protons produced from the FNAL Booster, thus giving the experiment its name. The Booster is a highly reliable machine and delivers protons to all the other sections of the FNAL accelerator complex; projections show that the Booster is capable of accommodating the BooNE, MINOS, and collider programs simultaneously. The proton rate delivered to BooNE is proposed to be 5×10^{12} protons per pulse at 5 Hz, which allows the production of sufficient quantities of low energy neutrinos, $0.10 < E_\nu < 2.0$ GeV. The beamline design is typical for accelerator production of neutrinos, see Figure 1.

The BooNE detector consists primarily of a 12 m diameter spherical tank filled with 769 tons of mineral oil, divided into an inner fiducial volume and an outer veto volume. The fiducial volume uses 1200 phototubes which line the tank, pointing inwards; the phototubes, electronics, and data acquisition are the same as used by the LSND experiment. The events of interest are quasi-elastic neutrino interactions with the nuclei making up the oil, $\nu_e C \to e^- X$ and $\nu_\mu C \to \mu^- X$, detecting the cerenkov light from the interaction products. Since the beamline produces almost entirely ν_μ, to first order the experiment reconstructs events and identifies and tallies ν_μ and ν_e interactions. The main backgrounds come from some intrinsic ν_e contamination in the beam, mis-identified ν_μ quasi-elastic scattering, and mis-identified neutral current π^0 production. Extensive modeling of both the beam production and the detector have shown that these backgrounds can in fact be measured and their contribution to the event sample subtracted. Excess events are evidence for neutrino oscillations; the energy spectrum of the excess events gives a means of extracting oscillation parameters.

MINOS

MINOS is motivated by results observed from various atmosperic neutrino experiments, particularly the recent results from Super-K [3]. The oscillation parameter space explored by MINOS is different from, but somewhat overlapping, the parameter space BooNE is designed for, with MINOS being designed to reach down to smaller values of Δm^2. This is done by creating a neutrino beam with a higher and variable energy spectrum, and a much longer distance between the neutrino source and detector. The neutrino beam for MINOS is created at FNAL, and aimed toward the Soudan Mine in northern Minnesota, 730 km away.

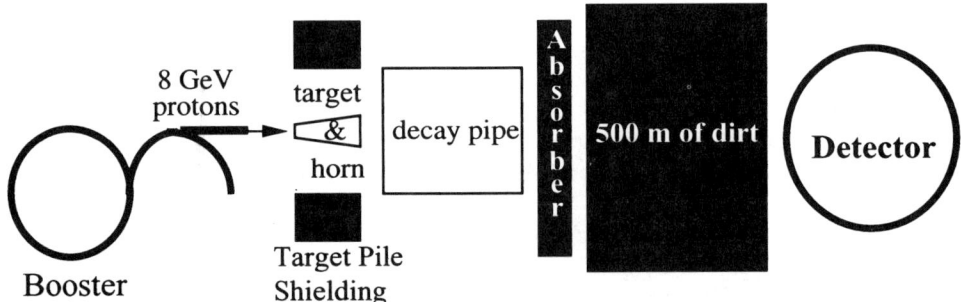

FIGURE 1. Schematic of the BooNE beamline.

The beamline to produce neutrinos for MINOS is called NuMI, for Neutrinos at the Main Injector. As the name implies, it relies on 120 GeV protons from Main Injector accelerator; plans call for up to 4×10^{13} protons every 2 seconds, delivered in a 10μsec pulse. Schematically, the beamline appears much like the schematic for the BooNE beam with some details changed: the focusing system consists of two horns rather than one, and the decay length is 725m. The energy spectrum of the ν_μs can be "tuned" by adjusting the relative positions of the target and the two horns, giving the experiment considerable flexibility in Δm^2 range.

MINOS uses a two detector philosophy: one detector is located at the source of the neutrinos at FNAL, and the other 730 km away in the Soudan Mine. Both detectors have the same basic design. The near detector at FNAL measures the content and energy spectrum of the un-oscillated neutrino beam; the far detector at Soudan also measures the content and energy spectrum of the beam, which will be different from what is seen in the near detector if the neutrinos oscillate between FNAL and Soudan. This method reduces systematic uncertainties as much as possible.

The two detectors have the same composition, both being iron-scintillator sandwiches, with toroidal magnetic fields in the 2.5-cm thick steel plates. The scintillator has sufficiently fine transverse granularity (4-cm wide strips) so that it provides both calorimetry and tracking information. The far detector is a total of 5400 metric tons, split into two modules. The cross-section shape is an 8-m wide octagon, as shown in Figure 2. The near detector is smaller in dimension and weight, being 980 metric tons, and the shape is a slightly "squashed" octagon, so that the region

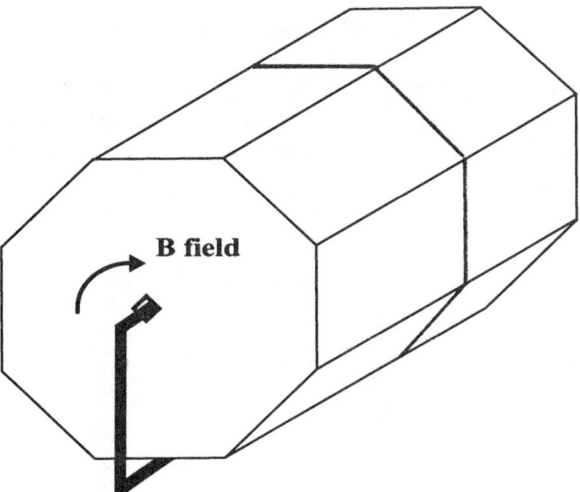

FIGURE 2. Schematic of the MINOS far detector, showing the 8-m diameter octagon shape. There are 486 layers of steel and scintallator in two supermodules; each supermodule is 15-m long. The coil generates about a 1T field over most of the ocatgon area.

where the beam intersects the detector is away from the region occupied by the magnetic coil.

The detectors are used to observe both charged current (long event topology) and neutral current (short event topology) interactions. Any mode of ν_μ oscillations will result in the depletion of ν_μ CC events seen in the far detector as compared to the near detector; the CC energy spectrum can be used to measure the oscillation parameters. The ratio of NC/CC events can likewise be used to study oscillations. Additionally, the CC and NC energy spectra of events can be used to distinguish oscillations to ν_τ, ν_e, and $\nu_{sterile}$ CC, providing a means of identifying the flavor of neutrino which the ν_μ oscillated into. The number of ν_μ events expected at the far detector (no oscillations) depends upon the selected energy spectrum of the beam, ranging from 3000 CC events/kt/yr for the high energy beam to several hundred for the low energy beam.

SUMMARY

Construction is now underway at FNAL on both BooNE and the NuMI beamline. At Soudan, construction is proceeding on the new underground cavern which will house the MINOS far detector. BooNE expects to be taking data by late 2001, and MINOS by late 2003. Many more details on both projects, and updates on their construction, is available on-line : http://www.hep.anl.gov/ndk/hypertext/numi.html for MINOS, and http://www-boone.fnal.gov/ for BooNE.

REFERENCES

1. LSND Homepage - http://www.neutrino.lanl.gov/LSND/
2. KARMEN Homepage - http://www-ik1.fzk.de/www/karmen/karmen_e.html
3. SuperK Homepage - http://www-sk.icrr.u-tokyo.ac.jp/doc/sk/index.html

Neutrino Factories– Physics Potentials

Zohreh Parsa

Brookhaven National Laboratory *
Physics Department 510 A,
Upton, NY 11973-5000, USA

Abstract. The recent results from Super-Kamiokande atmospheric and solar neutrino observations opens a new era in neutrino physics and has sparked a considerable interest in the physics possibilities with a Neutrino Factory based on the muon storage ring. We present physics opportunities at a Neutrino Factory, and prospects of Neutrino oscillation experiments. Using the precisely known flavor composition of the beam, one could envision an extensive program to measure the neutrino oscillation mixing matrix, including possible CP violating effects. These and Neutrino Interaction Rates for examples of a Neutrino Factory at BNL (and FNAL) with detectors at Gran Sasso, SLAC and Sudan are also presented.

INTRODUCTION

A muon storage ring based Neutrino Source (Neutrino Factory) beside providing a first phase of a muon collider facility, it would generate more intense and well collimated neutrino beams than currently available. The BNL- AGS or some other proton driver would provide an intense proton beam that hits a target, produces pions that decay into muons. The muons must be cooled, accelerated and injected into a storage ring with a long straight section where they decay. The decays occurring in the straight sections of the ring would generate neutrino beams that could be directed to detectors located thousands of kilometers away, allowing studies of neutrino oscillations with precisions not currently accessible [4].

The composition and spectra of an intense neutrino beam from a muon storage ring depends on momentum, polarization and charge of the stored muons, through the decays $\mu^- \to e^- \nu_\mu \bar{\nu}_e$ or $\mu^+ \to e^+ \bar{\nu}_\mu \nu_e$.

The neutrino fluxes from the proposed muon-based beams would be higher than ever previously achieved with a much better-understood flavor composition. In ad-

---

*) Supported by US Department of Energy contract DE-AC02-98CH10886.
†) E-mail: parsa@bnl.gov

dition, since the neutrino beams from these sources would be secondary beams from high energy muon decays, they would be extremely well collimated. Distances between production and detection could, therefore span the globe. Using the precisely known flavor composition of the beam, one could envision an extensive program to measure the neutrino oscillation mixing matrix, including possible CP violating effects.

NEUTRINO OSCILLATION

With only two massive neutrinos, with mass difference $\Delta m^2 = m_2^2 - m_1^2$, mass eigenstates ν_1 and ν_2 with mixing angle θ, the flavor eigenstates become:

$$\begin{pmatrix} \nu_a \\ \nu_b \end{pmatrix} = \begin{pmatrix} \cos\theta & \sin\theta \\ -\sin\theta & \cos\theta \end{pmatrix} \begin{pmatrix} \nu_1 \\ \nu_2 \end{pmatrix}. \quad (1)$$

The probability that a neutrino of flavor ν_a and energy E appears as flavor ν_b after traversing distance L in vacuum is

$$P(\nu_a \to \nu_b) = \sin^2\left(1.27 \Delta m^2 [\mathrm{eV}^2] \frac{L[\mathrm{km}]}{E[\mathrm{GeV}]}\right) \sin^2 2\theta. \quad (2)$$

Since the atmospheric neutrino data involves GeV muon neutrinos with distance scales of the Earth's diameter, this suggests Δm^2 of order 10^{-3} (eV)^2 for $\sin^2 2\theta \approx 1$. The solar neutrino data involves MeV electron neutrinos and distance scales of the radius of the Earth's orbit, suggesting Δm^2 of order 10^{-10} (eV)^2 with $\sin^2 2\theta \approx 1$ for vacuum oscillations [15]. The LSND result involves 30-MeV muon antineutrino and a distance scale of 30 m, suggesting Δm^2 of order 1 (eV)^2; large mixing angles are excluded by reactor data [16], thus, $\sin^2 2\theta$ can only be of order 10^{-2} in this case. Obviously, four different massive neutrinos are required to accommodate all three results, given their disparate scales of Δm^2. The Standard Model presently includes only three neutrinos with standard electroweak couplings and $m_\nu < m_Z/2$, so a "sterile" neutrino is required if all the data are correct citesterile. Even discarding the LSND result, three massive neutrinos are required with a corresponding 3×3 mixing matrix, e.g.

$$\begin{pmatrix} \nu_e \\ \nu_\mu \\ \nu_\tau \end{pmatrix} = \begin{pmatrix} c_{12}c_{13} & s_{12}c_{13} & s_{13}e^{-i\delta} \\ -s_{12}c_{23} - c_{12}s_{13}s_{23}e^{i\delta} & c_{12}c_{23} - s_{12}s_{13}s_{23}e^{i\delta} & c_{13}s_{23} \\ s_{12}s_{23} - c_{12}s_{13}c_{23}e^{i\delta} & -c_{12}s_{23} - s_{12}s_{13}c_{23}e^{i\delta} & c_{13}c_{23} \end{pmatrix} \begin{pmatrix} \nu_1 \\ \nu_2 \\ \nu_3 \end{pmatrix}. \quad (3)$$

a MNS matrix [18], where $c_{12} = \cos\theta_{12}$, etc.., In the three massive neutrino model, the neutrino oscillation probabilities of interest depends on six measurable parameters: three mixing angles ($\theta_{12}, \theta_{13}, \theta_{23}$); a phase δ related to CP violation as indicated in eq. (3); and two differences of the squares of the neutrino masses (Δm_{12}^2 and Δm_{23}^2 for instance). The interpretation of the solar and atmospheric neutrino data in terms of the three-neutrino oscillation hypothesis suggests $|\Delta m_{12}^2| \ll |\Delta m_{23}^2|$, with Δm_{12}^2 and Δm_{23}^2 being responsible for the transitions and/or oscillations of the solar and atmospheric neutrinos, respectively.

The description of the atmospheric neutrino data requires $\Delta m_{23}^2 \approx (2-6) \times 10^{-3}$ eV^2 and large mixing angle θ_{23}: $\sin^2 2\theta_{23} \approx (0.9-1.0)$. For $|\Delta m_{12}^2| \ll |\Delta m_{23}^2|$ and with Δm_{23}^2 in the above range, the non-observation of oscillations of the reactor electron antineutrinos in the CHOOZ experiment [19] implies a limit on the angle θ_{13}: $\sin^2 \theta_{13} < 0.11$. Given these constraints, the transitions/oscillations of the solar neutrinos in the three-neutrino mixing scheme under discussion depend largely on the remaining two parameters: Δm_{12}^2 and $\sin^2 2\theta_{12}$.

Further, the presence of matter may modify the oscillations of electron neutrinos because of their charged-current interaction (MSW effect [20]). In particular, the oscillations can be resonantly enhanced by the matter effects even when the oscillation probabilities are small in vacuum. This leads to additional interpretations of the solar neutrino data in which Δm_{12}^2 can be of order 10^{-5} (eV)^2 [21]. In effect at the present time, there are four viable interpretations of the solar neutrino data:

1) Vacuum oscillation (VO) solution with $\Delta m_{12}^2 \approx (0.5 - 5.0) \times 10^{-10}$ eV^2 and $\sin^2 2\theta_{12} \approx (0.7 - 1.0)$,

2) Low MSW solution corresponding to $\Delta m_{12}^2 \approx (0.5 - 2.0) \times 10^{-7}$ eV^2 and $\sin^2 2\theta_{12} \approx (0.9 - 1.0)$,

3) Small mixing angle (SMA) MSW solution with $\Delta m_{12}^2 \approx (4.0 - 9.0) \times 10^{-6}$ eV^2 and $\sin^2 2\theta_{12} \approx (0.001 - 0.01)$,

4) Large mixing angle (LMA) MSW solution, $\Delta m_{12}^2 \approx (0.2 - 2.0) \times 10^{-4}$ eV^2 and $\sin^2 \theta_{12} \approx (0.65 - 0.96)$.

With four interpretations of the solar neutrino data, and the two interpretations of the LSND data as either right or wrong, there are a total of eight scenarios for explanations of the data. The experimental challenge is to reduce these to a single scenario, and to make accurate measurements of the parameters of that scenario.

Thus, with the available experimental guidelines as to the parameters of neutrino masses and mixings, one can begin to plan for more extensive studies namely, with neutrino beams derived from the decay of muons in a storage ring. Both μ^- and μ^+ can be stored in the ring, but only one sign would be used at a time. For example if μ^- are stored, their decay

$$\mu^- \to e^- \nu_\mu \bar{\nu}_e, \qquad (4)$$

leads to beams with nearly equal numbers of ν_μ and $\bar{\nu}_e$ with spectra that are well known.

At the detectors, the neutrino and the antineutrino may or may not have changed their flavor, leading to the appearance of a different flavor or the disappearance of the initial flavor, respectively. When detected by a charged-current interaction, there are 6 classes of signatures in a three-neutrino model:

1) $\nu_\mu \to \nu_e \to e^-$ (appearance);

2) $\nu_\mu \to \nu_\mu \to \mu^-$ (disappearance);

3) $\nu_\mu \to \nu_\tau \to \tau^-$ (appearance);

4) $\bar{\nu}_e \to \bar{\nu}_e \to e^+$ (disappearance);

5) $\bar{\nu}_e \to \bar{\nu}_\mu \to \mu^+$ (appearance);

6) $\bar{\nu}_e \to \bar{\nu}_\tau \to \tau^+$ (appearance).

For operation with positive muons, a similar list of processes may be written. The 5th process where a muon of different sign from the parent muon appears, has a very unique possibilities at a neutrino factory based on muon storage rings. Since they are the only sources of intense high energy electron (anti)neutrino beams. The τ appearance (cases 3 and 6) are practical only for neutrino beams with 10's of GeV energy.

Experiments carried out at a neutrino factory within the next decade can add compelling new information to our understanding of neutrino oscillations, if the number of useful muon decays exceeds 10^{19} per year, and energy is $\gtrsim 20\ GeV$.

It is anticipated that by the time a muon storage ring would be built the two angles (θ_{23} and θ_{12}), and the magnitudes of two mass squared differences (Δm_{23}^2 and Δm_{12}^2) would be known, from the solar and atmospheric neutrino measurements (which would have been verified by long baseline and reactor experiments), for example, MINOS and KamLAND. The remaining pieces of the puzzle would be θ_{13}, the CP-violating phase δ and the signs of the Δm_{ij}^2. Moreover, the indicated long-baseline experiments will not be sensitive to the matter effects in neutrino oscillations because the distances between the sources and detectors are not sufficiently large. Verifying the existence of matter effects in neutrino oscillations by observing directly the modification of the neutrino oscillation probabilities by these effects, would also be fundamental and interesting.

The third mixing angle θ_{13} can be measured in several channels at a neutrino factory The detector must be far to avoid background but not too far (< 1000 km) so that the effects of Δm_{12}^2 remain negligible and thus δ can formally be set to zero. Fig. 1 shows the achievable sensitivity to the yet-unknown value of θ_{13}, and illustrates sensitivity reach in the ($\sin^2 \theta_{13}, \Delta m_{23}^2$) plane for a 10 kton detector and a neutrino beam from 2×10^{20} decays of 20 GeV muons in a storage ring at distance 732 km. The appearance process $\bar{\nu}_e \to \bar{\nu}_\mu \to \mu^+$, shown by the lines on the left, has much greater sensitivity than the disappearance process $\nu_\mu \to \nu_\mu \to \mu^-$, shown by the lines on the right. The interior of the box is the approximate region allowed by Super-Kamiokande data [15].

CP VIOLATION

The three-neutrino scenario [23] can lead to CP violation in for example

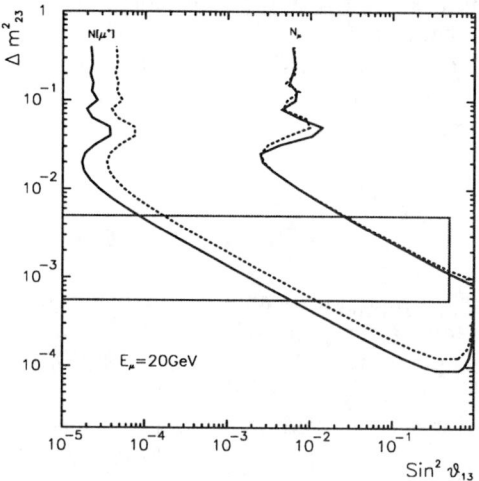

FIGURE 1. Sensitivity reach in the $(\sin^2 \theta_{13}, \Delta m_{23}^2)$ plane.

$$A_{\rm CP} = \frac{P(\nu_e \to \nu_\mu) - P(\bar{\nu}_e \to \bar{\nu}_\mu)}{P(\nu_e \to \nu_\mu) + P(\bar{\nu}_e \to \bar{\nu}_\mu)}, \quad (5)$$

or time-reversal violation

$$A_{\rm T} = \frac{P(\nu_e \to \nu_\mu) - P(\nu_\mu \to \nu_e)}{P(\nu_e \to \nu_\mu) + P(\nu_\mu \to \nu_e)}. \quad (6)$$

The asymmetry (5) can be measured using wrong-sign muons and the two charges of the muon beam. However, the genuine CP violating contribution to (5) due to a non-vanishing phase δ competes with terms related to matter effects, i.e., to the different rates of evolution for ν_e and $\bar{\nu}_e$ between source and detector. The relative strength of the matter-induced asymmetry increases quadratically with distance, and dilutes the signal of CP violation in a far detector.

If the solution to solar neutrino problem involves, large mixing angles and matter enhancement (LMA MSW, $\sin^2 2\theta_{12} \approx \sin^2 2\theta_{23} \approx 1$), then there is a possibility of measuring the CP violating asymmetry (5), with expression

$$|A_{\rm CP}| \approx \left| \frac{2 \sin \delta}{\sin 2\theta_{13}} \sin\left(\frac{1.27 \Delta m_{12}^2 L}{E}\right) \right|, \quad (7)$$

provided the detector is located sufficiently far and high statistics ($> 10^{21}$ muons per year) are available. For all the other solar neutrino solutions $A_{\rm CP}$ is extremely small, being suppressed by a factor of either $\sin^2 2\theta_{12}$ or Δm_{12}^2.

The asymmetry (6) is not sensitive to matter effects, but relies on distinguishing the process $\nu_\mu \to \nu_e \to e^-$ from $\bar{\nu}_e \to \bar{\nu}_e \to e^+$. In the detector, it will be very difficult to distinguish electrons from positrons but the relative ν_μ and $\bar{\nu}_e$ fluxes can be varied by varying the polarization of the muons in the storage ring [24].

If future experiments confirm the interpretation of the LSND data that there exist more than three light neutrinos, then use of the neutrino factory flavor-rich beams would be even more crucial, because the parameter space for CP/T violating effects would be considerably enlarged and could be explored in experiments with such beams [25].

PRECISION PHYSICS

Muon storage ring based neutrino beams would bring about new neutrino oscillation measurements, and a new era for high-precision neutrino scattering experiments [26]. For example, with a detector located 30 m from a 150 m straight section of a 50-GeV, 10^{21}-μ/yr muon storage ring, the event rate is 40 million events per kilogram per year over a 10 cm radius. Oscillation-related measurements may be interpreted precision measurements of the total neutrino and antineutrino cross sections, as well as of the beam divergence. As precision probes of nuclear and nucleon structure, the neutrinos may be used to provide additional information to that obtained with charged lepton beams, in related studies. It is known that, neutrino scattering allows a clean separation of the valence and sea quark distributions, and use of a polarized target permits characterization of the spin dependence of these distributions. Thus, near detectors are the natural successor to nucleon structure measurements presently underway at HERA, HERMES, Jefferson Lab, RHIC and elsewhere. For example, scattering of the four neutrino types ν_μ, $\bar{\nu}_\mu$, ν_e, and $\bar{\nu}_e$ off electrons could lead to measurements of the Weinberg angle ten times better than known at present.

Note that, a high-flux multi-GeV neutrino beam is a charm factory, in which a ν_μ beam leads to c quarks that are tagged by a final-state μ^- ($\nu_\mu d \to \mu^- c$), while $\bar{\nu}_\mu$ beam leads only to tagged $\bar{c}$ quarks. For example, for the above described beam parameters, there would be 10^7 leptonic tagged charm decays in only 40 kg-years (not kton-years!), permitting measurements of V_{cd} to fraction of a percent, and perhaps even direct observation of $D^0 - \overline{D}^0$ mixing.

EVENT RATES – POTENTIALS

The number of neutrino interactions per unit mass of a detector at distance L from a muon storage ring operating at energy E_μ scales as

$$N_{\text{events}} \propto N_\mu \, E_\mu^3 \, L^{-2} \tag{8}$$

for the example of a proton source with 1.5 MW power, in one year (10^7 s) of operation, there would be about 4×10^{20} muons per year decaying in the storage ring. Assuming the fraction of the ring pointing to a given detector to be about

TABLE 1. Neutrino Interaction Rates at a Neutrino Factory.

Case		Source at Detector at L (km)	Mode	BNL G. Sasso 6528	BNL SLAC 4139	BNL Soudan 1712	FNAL G. Sasso 7332	FNAL SLAC 2899	FNAL Soudan 732
1)	μ^+		$\nu_e \to \nu_\mu$	90	160	190	63	180	200
			$\nu_e \to \nu_e$	1400	3600	16000	1100	8000	1.2×10^5
				(2.4σ)	(2.7σ)	(1.5σ)	(1.9σ)	(2.0σ)	(0.6σ)
			$\bar\nu_\mu \to \bar\nu_\mu$	890	2200	9300	700	4800	7.0×10^4
2)	μ^+		$\nu_e \to \nu_\mu$	5×10^{-2}	0.86	1.5	3×10^{-5}	1.3	1.6
			$\nu_e \to \nu_e$	1500	3800	16000	1200	8200	1.2×10^5
			$\bar\nu_\mu \to \bar\nu_\mu$	890	2200	9400	700	4800	7.0×10^4
3)	μ^+		$\nu_e \to \nu_\tau$	31	60	70	20	67	73
			$\nu_e \to \nu_e$	1400	3700	1.6×10^4	1100	8000	1.2×10^5
				(2.4σ)	(2.7σ)	(1.5σ)	(1.9σ)	(2.0σ)	(0.6σ)
			$\bar\nu_\mu \to \bar\nu_\mu$	890	2200	9400	700	4800	7.0×10^4
4)	μ^-		$\nu_\mu \to n\nu_\tau$	450	570	650	410	620	680
			$\nu_\mu \to \nu_\mu$	760	3100	1.7×10^4	490	8000	1.4×10^5
				(35σ)	(23σ)	(12σ)	(40σ)	(16σ)	(4.6σ)
			$\bar\nu_e \to \bar\nu_e$	770	1900	8100	600	4100	6.1×10^4

0.25 then the number of decays pointing to the given detector will be approximately 10^{20}. It may be noted that the number of events with the 1.5 MW neutrino factory, in a detector at the same 730 km, is approximately 100 times that in the proposed CERN - Gran Sasso experiment (NGS) [9], and about 40 times the maximum event rate that MINOS [10] can expect. Upgrading the proton driver to 4 MW, the factors become about 300 and 100 for Gran Sasso and Soudan, respectively.

Table 1, gives charged current neutrino interaction rates (per kiloton-year) as a function of baseline length L for an $E_\mu = 50$ GeV muon storage ring in which there are 1×10^{20} unpolarized muon decays per year within a neutrino beam-forming straight section [14]. The rates are listed for oscillations:

1) $\nu_e \to \nu_\mu$: $\Delta m^2 = 3.5 \times 10^{-3}$ eV$^2/c^4$ & $\sin^2 2\theta = 0.1$,

2) $\nu_e \to \nu_\mu$: $\Delta m^2 = 1 \times 10^{-4}$ eV$^2/c^4$ & $\sin^2 2\theta = 1$,

3) $\nu_e \to \nu_\tau$: $\Delta m^2 = 3.5 \times 10^{-3}$ eV$^2/c^4$ & $\sin^2 2\theta = 0.1$,

4) $\nu_\mu \to \nu_\tau$: $\Delta m^2 = 3.5 \times 10^{-3}$ eV$^2/c^4$ & $\sin^2 2\theta = 1$.

Table 1, also gives the rates for the unoscillated neutrino interactions, the corresponding statistical significance of the disappearance signal (numbers in parenthesis), and the rates for the antineutrino interactions.

If future experiments confirm the interpretation of the LSND data that there exist more than three light neutrinos, then use of the neutrino factory flavor-rich beams would be even more crucial, because the parameter space for CP/T violating effects would be considerably enlarged and could be explored in experiments with such beams [25].

SUMMARY

A Muon storage ring based neutrino factory has a strong physics case, but require additional feasibility studies. The recent more ambitious ideas for utilizing high intensity muon sources are being explored. Indeed, if very high intensities, $\sim 10^{21} \frac{\nu}{year}$, are attained and nature has been kind in her neutrino mass and mixing parameters, one could envision a complete exploration of the 3 × 3 neutrino mixing matrix and even the detection of CP violation in the oscillation phenomena. [2] - [29].

REFERENCES

1. A.C. Melissinos unpublished note (1960).
2. The Neutrino Factory Collaboration http://www.cap.bnl.gov/mumu/
3. http://www.awa.tohoku.ac.jp/html/KamLAND/
4. See e.g., [5] - [29]; and references therein.
5. Z. Parsa, *Muon Sources*, this Proceedings (CIPANP2000) and references therein
6. Z. Parsa, *Muon Storage Rings - Neutrino Factories*, in *Next Generation Nucleon decay and Neutrino detector (NNN99)* AIP CP 533, p. 181–195 (2000).
7. K.T. McDonald, ed. for the Neutrino Factory and Muon Collider Collaboration, physics/9911009, 6 Nov, 1999, and references therein
8. C.M. Ankenbrandt et al., *Status of muon collider research development and future plans*, Phys. Rev. ST Accel. Beams **2**, 081001 (1999); refrences therein.
9. http://www.cern.ch/NGS/ngs99.pdf
10. MINOS TDR, http://www.hep.anl.gov/NDK/Hypertext/minos_tdr.html
11. The MiniBooNE project: http://www.neutrino.lanl.gov/BooNE
12. The Oak Ridge Large ν Detector http://www.orau.org/orland/
13. http://chorus01.cern.ch/~pzucchel/loi/
14. S. Geer, MUC0051 (1999)
15. E.g.: S. Geer, Phys. Rev. D **57**, 6989 (1998). A. Buena, M. Campanelli, and A. Rubbia, beam from muon decay, hep-ph /9808485; and A. Buena, M. Campanelli, and A. Rubbia, hep-ph/9809252; A. De Rujula, M.B. Gavela, and P. Hernandez, Nucl. Phys. **B547**, 21 (1999), hep-ph/9811390; M. Campanelli, A. Buena, and A. Rubbia, hep-ph/9905420; V. Barger, S. Geer, and K. Whisnant, hep-ph/9906487; O. Yasuda, hep-ph/9910428;
    I. Mocioiu and R. Shrock, hep-ph/9910554.
16. G.S. Vidaykin et al., JETP Lett. **59**, 390 (1994);
    B. Achkar et al., Nucl. Phys. **B434**, 503 (1995).
17. E.g.: A. Zee, Phys. Lett. **B93**, 389 (1980); V. Barger, P. Langacker, J. Leveille, and S. Pakvasa, Phys. Rev. Lett. **45**, 692 (1980); V. Barger, S. Pakvasa,

T.J. Weiler, and K. Whisnant, Phys. Rev. D **58**, 093016 (1998); S.M. Bilenky, G. Giunti, hep-ph/9905246.
18. Z. Maki, M. Nakagawa, and S. Sakata, Prog. Theor. Phys. **28**, 970 (1962). M. Nakagawa, hep-ph/9811358.
19. http://www.hep.anl.gov/ndk/hypertext/chooz.html
20. L. Wolfenstein, Phys. Rev. D **17**, 2369 (1978); S.P. Mikheyev and A.Y. Smirnov, Sov. J. Nuc. Phys. **42**, 913 (1986).
21. N. Hata and P. Langacker, Phys. Rev. D **56**, 6107 (1997), J.N Bahcall, P.I. Krastev, and A.Y. Smirnov, Phys. Rev. D **58**, 096016 (1998), J.N. Bahcall, P. Langacker, J. Bahcall and P. Krastev, Phys. Lett. **B436**, 243 (1998), hep-ph/9807525.
22. S. Geer, Phys. Rev. D **57**, 6989 (1998), hep-ph/9712290; A. Buena, M. Campanelli, and A. Rubbia, hep-ph /9808485; and hep-ph/9811390; http://fnalpubs.fnal.gov/archive/1999 hep-ph/9905420; V. Barger, S. Geer, and K. Whisnant, hep-ph/9906487; O. Yasuda, hep-ph/9910428; I. Mocioiu and R. Shrock, hep-ph/9910554.
23. M. Tanimoto, Prog. Theor. Phys. **97**, 9091 (1997),J. Arafune, M. Koike, and J. Sato, Phys. Rev. D **56**, 3093 (1997),S.M. Bilenky, C. Giunti, and W. Grimus, hep-ph/9705300;

 H. Minakata and H. Nunokawa, Phys. Lett. **B413**, 369 (1997),H. Minakata and H. Nunokawa, *CP Violating vs. Matter Effect in Long-Baseline Neutrino Oscillation Experiments*, Phys. Rev. D **57**, 4403 (1998),S.M. Bilenky, C. Giunti, and W. Grimus, Phys. Rev. D **58**, 033001 (1998), M. Tanimoto, hep-ph/9906375; K.R. Schubert, hep-ph/9902215;

 K. Dick, M. Freund, M. Lindner, and A. Romanino, hep-ph/9903308; J. Bernabeu, hep-ph/9904474;

 M. Tanimoto, hep-ph/9906516; A. Donini, M.B. Gavela, P. Hernandez, and S. Rigolin, hep-ph/9909254;H. Fritzsch and Z.-Z. Xiang, hep-ph/9909304;A. Romanino, hep-ph/9909425; M. Koike and J. Sato, hep-ph/9909469; J. Sato, hep-ph/9910442. hep-ph/9910442.
24. A. Blondel, http://alephwww.cern.ch/~bdl/muon/nufacpol.ps
25. V. Barger, Y.-B. Dai, K. Whisnant and B.-L. Young, Phys. Rev. D **59**, 113010 (1999), hep-ph/9901388;A. Kalliomaki, J. Mallampi, and M. Tanimoto, hep-ph/9909301;A. Donini, M.B. Gavela, P. Hernandez, and S. Rigolin, hep-ph/9910516.
26. See, for example, B.J. King, AIP Conf. Proc. **435**, 334 (1998). D.A. Harris and K.S. McFarland, *ibid.* p. 505;
27. The $\mu^+\mu^-$ Collider Collaboration, $\mu^+\mu^-$ *Collider Feasibility Study*, BNL-52503, FERMILAB-Conf-96/092, LBNL-38946 (July 1996).
28. Z. Parsa, New High Intensity Muon sources and Flavor Changing Neutral Currents, World scientific Publishing, pp 147-153 (1998).
29. *Neutrino Factory Feasibility Studies at Fermilab*, S. Geer and N. Holtkamp et al., http://www.fnal.gov/projects/muon_collider/nu_factory/.

Matter Effects on Neutrino Oscillations

Irina Mocioiu[a] and Robert Shrock[b]

(a) C.N. Yang Institute for Theoretical Physics
State University of New York, Stony Brook, NY 11794
(b) Physics Department, Brookhaven National Laboratory
Upton, NY 11973

Abstract. We calculate matter effects on neutrino oscillations in long baseline experiments, using actual density profiles in the Earth. We study the dependence of the signal on $E/\Delta m^2_{atm}$, the angles in the leptonic mixing matrix and the influence of Δm^2_{sol} and CP phase on the oscillations. The results show quantitatively how matter effects can cause significant changes in the oscillation probabilities. These effects can be useful in amplifying certain neutrino oscillation signals and helping one to obtain measurements of mixing parameters and the magnitude and sign of Δm^2_{atm}.

An important effect that must be taken into account in long baseline neutrino oscillation experiments concerns the matter-induced oscillations which neutrinos undergo along their flight path through the Earth from the source to the detector. Given the typical density of the Earth, matter effects are important for the neutrino energy range $E \sim O(10)$ GeV and $\Delta m^2_{atm} \sim 3 \times 10^{-3}$ eV^2, values relevant for the long baseline experiments. Many studies of these effects have been done over the years: [1]- [3]. Here we shall present calculations from [3] that show matter effects on oscillation probabilities for parameters relevant to possible long baseline neutrino experiments envisioned for the muon storage ring [4]. These are calculated using the actual density profiles in the Earth [5] rather than assumptions of constant density. These calculations yield some important conclusions concerning how one can take advantage of these matter effects to achieve good sensitivity to the leptonic mixing matrix element U_{13}, or equivalently, to the associated rotation angle θ_{13}.

In a hypothetical world in which there were only two neutrinos, ν_μ and ν_τ, the $\nu_\mu \to \nu_\tau$ oscillations in matter would be the same as in vacuum, since both have the same forward scattering amplitude, via Z exchange, with matter. However, in the realistic case of three generations, because of the indirect involvement of ν_e due to a nonzero U_{13}, and because of the fact that ν_e has a different forward scattering amplitude off of electrons, involving both Z and W exchange, there will be a matter-induced oscillation effect on $\nu_\mu \to \nu_\tau$ (as well as other channels). We consider here the three flavors of active neutrinos, with no light sterile (= electroweak-singlet) neutrinos.

The evolution of the flavor eigenstates of neutrinos is given by

$$i\frac{d}{dx}\nu = \left(\frac{1}{2E}UM^2U^\dagger + V\right)\nu \tag{1}$$

where $\nu = U\nu_m$, $\nu_m = \begin{pmatrix}\nu_1\\ \nu_2\\ \nu_3\end{pmatrix}$, $M^2 = \begin{pmatrix}m_1^2 & 0 & 0\\ 0 & m_2^2 & 0\\ 0 & 0 & m_3^2\end{pmatrix}$, $V = \begin{pmatrix}\sqrt{2}G_F N_e & 0 & 0\\ 0 & 0 & 0\\ 0 & 0 & 0\end{pmatrix}$

Here N_e is the electron number density and we have $\sqrt{2}G_F N_e$ [eV]$= 7.6 \times 10^{-14} Y_e \rho$ [g/cm^3]. For $\bar{\nu}$, V is replaced by $(-V)$.

The atmospheric neutrino data suggests almost maximal mixing in the 2 − 3 sector. However, a small but non-zero U_{13} is still allowed, and this produces the matter effect in the traversal of neutrinos through the Earth. We use here the bound $\sin^2(2\theta_{13}) \leq 0.1$, consistent with both Chooz [6] and the atmospheric neutrino data [7]. We have taken here $\theta_{13} \in [0, \pi/4]$. One could also consider $\theta_{13} \in [\pi/4, \pi/2]$; this is essentially equivalent to reversing the sign of Δm_{atm}^2.

There are several motivations for very long baseline experiments, since, with sufficiently high-intensity sources, these can be sensitive to quite small values of Δm^2 and since the matter effects, being larger, can amplify certain oscillations and can, in principle, be used to get information on the sign of Δm_{atm}^2. Hence we concentrate here on these very long baseline experiments; for these, the neutrino flight path goes through several layers of the Earth with different densities. We show results for the Fermilab to SLAC path length $L \simeq 2900$ km and for $L \simeq 7330$ km, the distance from Fermilab to Gran Sasso.

If one assumes the large mixing angle (LMA) solution to the solar neutrino problem, then for the $L = 2900$ km baseline, both Δm_{atm}^2 and Δm_{sol}^2 have to be considered. In Fig.1 we show $P(\nu_\mu \to \nu_e)$ as function of energy for $\Delta m_{atm}^2 = 3.5 \times 10^{-3}$ eV^2 and $\sin^2 2\theta_{23} = 1$, as suggested by the atmospheric data, and $\sin^2 2\theta_{13} = 0.1$, the maximum value allowed, and two different choices of Δm_{sol}^2 and $\sin^2 2\theta_{12}$. One choice corresponds to the LMA solution, with $\Delta m_{sol}^2 = 5 \times 10^{-5}$ eV^2 and $\sin^2 2\theta_{12} = 0.8$. For this case, the choice of the CP violating phase δ is relevant; here we take $\delta = 0$ and compare with nonzero δ below. The other choice is for the vacuum oscillation (VO) solution, with $\Delta m_{sol}^2 = 10^{-10}$ eV^2 and $\sin^2 2\theta_{12} = 1$. The small mixing angle (SMA) solution gives the same results as VO. One sees that the terms involving Δm_{sol}^2 can have non-negligible effects on the $\nu_\mu \to \nu_e$ oscillation probability for this path-length, especially at lower energies. When we take into account Δm_{sol}^2 (LMA), we also have to consider CP-violating effects. We present in Fig.1 a comparison showing the results for the probability $P(\nu_\mu \to \nu_e)$ for $\delta = 0$ and $\delta = \pi/2$. We can see that the effects of the CP-violating phase are small. Note however that for non-zero CP-violation, $P(\nu_\mu \to \nu_e) \neq P(\nu_e \to \nu_\mu)$. For no CP-violation, even with matter effects, there is no difference between these two probabilities. Since with a muon storage ring, by switching between μ^- and μ^+ beams, one could obtain both ν_μ and ν_e beams, there is the possibility of searching for the CP (actually T) violating difference $P(\nu_\mu \to \nu_e) - P(\nu_e \to \nu_\mu)$. In practice,

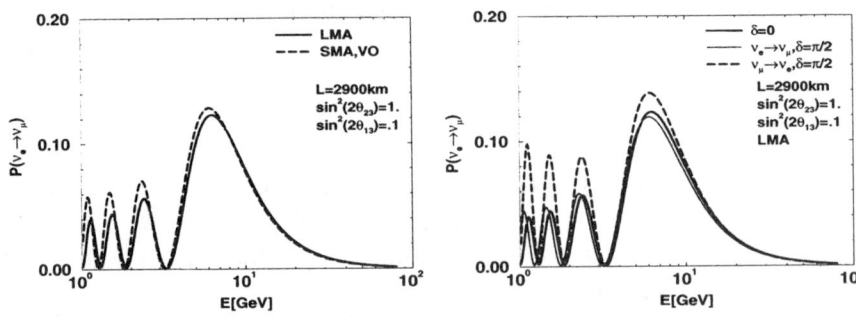

FIGURE 1.

however, it would be difficult to identify the e^- from ν_e, given that the μ^- stored beam that yields the initial ν_μ also yields $\bar\nu_e$, which produce e^+ in the detector, and given that it would be quite difficult to measure the sign of the $e^\pm$ in planned detectors. An alternate method, to measure the asymmetry $D = \frac{P(\nu_e \to \nu_\mu) - P(\bar\nu_e \to \bar\nu_\mu)}{P(\nu_e \to \nu_\mu) + P(\bar\nu_e \to \bar\nu_\mu)}$ is, in principle, possible, although it is complicated by the fact that D is rendered nonzero by matter effects even in the absence of CP violation (see also [2]). If the solar neutrino problem is solved by the SMA or VO, CP-violation effects are not observable in the experiments of interest here. Indeed, even for the LMA solution, the CP violation would be very hard to detect for path lengths larger than ~ 3000 km because of matter effects.

For the Fermilab to Gran Sasso distance $L \simeq 7330$ km, the Δm_{sol}^2 corrections are negligible, so we can analyze the problem using fewer relevant parameters: Δm_{atm}^2, θ_{13} and θ_{23}. We calculate the oscillation probabilities as a function of $E/\Delta m^2$, rather than using a particular value for Δm^2 or the energy. This way of presenting the results is useful since, for a given L value, it shows the matter effect for a wide range of E and Δm^2 and hence can serve as an input in the choice of optimal beam energy (along with other considerations such as the cross section dependence $\sigma \sim E$, beam divergence $\sim (LE)^{-2}$, effects of cuts, consequent event rates, and cost estimates). We calculate the oscillation probabilities for different values of the mixing angles θ_{13} and θ_{23} allowed by the atmospheric neutrino data and the CHOOZ experiment. In order to study the effects at different distances, we show the same type of graphs for both $L = 7330$ km and $L = 2900$ km. For $L = 2900$ km, the probabilities can be expressed as functions of $E/\Delta m_{atm}^2$ only for the SMA and VO solutions to the solar neutrino problem. For LMA, small Δm_{sol}^2 and CP violation corrections are added, as shown in Fig.1.

The relative effects of matter can be especially dramatic in the oscillation probability $P(\nu_e \to \nu_\mu)$, since these directly involve ν_e, so we concentrate here on this channel. Since the ν_e beam would arise from a stored μ^+ beam, and the $\bar\nu_\mu$'s from the decays of the μ^+'s would produce μ^+'s in the detector, the signature for the $\nu_e \to \nu_\mu$ oscillation would be wrong-sign muons. Planned detectors would be capable of searching for such wrong-sign muons. Since this is a sub-dominant channel, the oscillation effect is small. If the beam went through the vacuum, $P(\nu_e \to \nu_\mu)$ would be very small. Because of the matter effect however, this probability can be strongly enhanced (Fig.2 and Fig.3). For $L = 7330$ km, the enhancement is largest for $E/\Delta m^2 \simeq 2.5 \times 10^3$ GeV/eV^2. This is close to the ratio that one would get for a neutrino energy of $E \sim 10$ GeV, given the indication from the data that $\Delta m^2_{atm} = 3.5 \times 10^{-3}$ eV^2. For $L = 2900$ km, the largest enhancement is obtained for $E/\Delta m^2$ a factor of 3 lower. This probability is quite sensitive to the value of

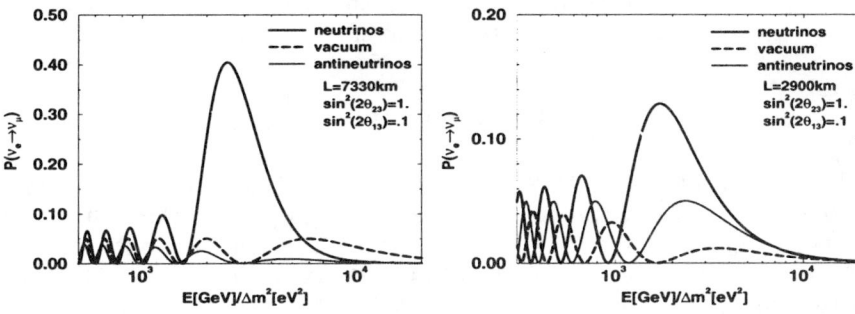

FIGURE 2.

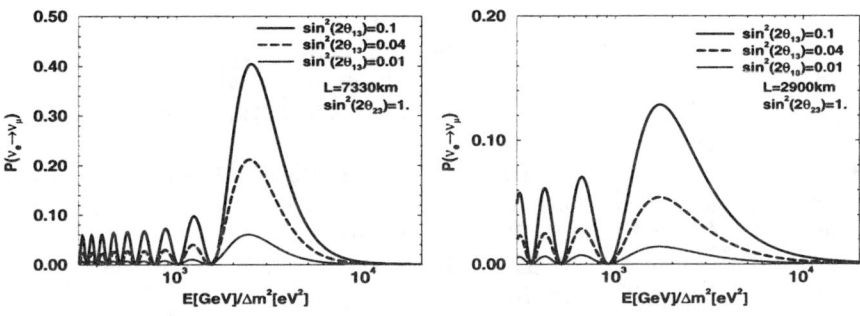

FIGURE 3.

θ_{13} (Fig.3), so one should be able to use it for a good determination of this angle.

We consider both neutrinos and antineutrinos. The matter effects reverse sign in these two cases. This implies that if Δm^2 is positive (as considered here), one can get a resonant enhancement of the oscillations for neutrinos, while for antineutrinos the matter effects would suppress the oscillations (Fig.2). The situation would be reversed if Δm^2 were negative. In consequence, an independent measurement of the two channels ($\nu_e \to \nu_\mu$ and $\bar\nu_e \to \bar\nu_\mu$) would be very valuable, giving a quantitative measure of the matter effects and allowing to determine the sign of Δm^2.

To summarize, in planning for very long baseline neutrino oscillation experiments, it is important to take into account matter effects. These are significant for the range of neutrino energies E of order 10's of GeV that are planned for these experiments, given the density of the Earth and the value of Δm^2_{atm} indicated by current atmospheric neutrino data. Matter effects can be useful in amplifying neutrino oscillation signals and helping one to obtain measurements of mixing parameters and the magnitude and sign of Δm^2_{atm}. We have only showed here results for $P(\nu_e \to \nu_\mu)$, since matter effects are most important in this channel. We want to emphasize however that the allowed parameter space can be strongly constrained, leading to precise measurements of all mixings and Δm^2, only by combining results from different types of measurements, in different channels of oscillations.

REFERENCES

1. L. Wolfenstein, Phys. Rev. **D17**, 2369 (1978). S. P. Mikheyev and A. Smirnov, Yad. Fiz. **42**, 1441 (1985) [Sov.J. Nucl. Phys. **42**, 913 (1986)], Nuovo Cim., **C9**, 17 (1986). V. Barger, K. Whisnant, S. Pakvasa, and R. J. N. Phillips, Phys. Rev. **D22**, 2718 (1980). P. Krastev, Nuovo Cimento **103A**, 361 (1990). R. H. Bernstein and S. J. Parke, Phys. Rev. **D44**, 2069 (1991). P. Lipari, hep-ph/9903481. M. Campanelli, A. Bueno, A. Rubbia, hep-ph/9905240, hep-ph/0005007 S. Dutta, R. Gandhi, and B. Mukhopadhyaya, hep-ph/9905475. V. Barger, S. Geer, K. Whisnant, Phys.Rev. **D61**, 053004 (2000). D. Dooling, C. Giunti, K. Kang, C. W. Kim, hep-ph/9908513. V. Barger, S. Geer, R. Raja, K. Whisnant, Phys.Rev.**D62**, 013004 (2000), hep-ph/0003184, hep-ph/0004208. M. Freund, T. Ohlsson, hep-ph/9909501; T. Ohlsson, H. Snellman, hep-ph/9910546, hep-ph/9912295. M. Freund, M. Lindner, S.T. Petcov, A. Romanino, hep-ph/9912457. A. Cervera, A. Donini, M.B. Gavela, J.J. Gomez Cadenas, P.Hernandez, O. Mena, S. Rigolin, Nucl.Phys.**B579**,17 (2000). M. Freund, P. Huber, M. Lindner, hep-ph/0004085
2. S.M.Bilenky, C.Giunti, W.Grimus, Phys.Rev.**D58**, 033001 (1998); K.Dick, M.Freund, M.Lindner, A.Romanino, Nucl. Phys. **B562**, 29 (1999); M.Tanimoto, Phys. Lett. **B462**, 115 (1999); A.Donini, M.B.Gavela, P.Hernandez, S.Rigolin, hep-ph/9909254; M.Koike, J.Sato, hep-ph/9909469; P.F.Harrison, W.G.Scott, hep-ph/9912435.
3. I.Mocioiu, R. Shrock, Phys.Rev.D,in press (hep-ph/0002149), Proc.Conf.on Next Nucleon Decay and Neutrino Detectors 1999, A.I.P. Proc.,!in press (hep-ph/9910554).
4. S. Geer, Phys. Rev. **D57**, 6989 (1998). De Rujula, M.B. Gavela, P. Hernandez, Nucl. Phys. **B547**, 21 (1999).
   http://www.fnal.gov/projects/muon_collider/nu/study/study.html
   http://www.cap.bnl.gov/mumu. http://www.cern.ch/ autin/nufact99/whitepap.ps.
   physics/9911009, http://puhep1.princeton.edu/mumu/NSFLetter/nsfmain.ps.
5. A.Dziewonski, Earth Structure, in: "The Encyclopedia of Solid Earth Geophysics", D.E.James (Ed.), (Van Nostrand Reinhold, New York, 1989), p.331.
6. M. Apollonio et al., Phys. Lett. **B420**, 397 (1998); Phys. Lett. **B466**, 415 (1999).
7. Super-Kamiokande Collab., Y. Fukuda et al., Phys. Lett. **B433**, 9 (1998); Phys. Rev. Lett. **81**,1562 (1998); *ibid.*, **82**, 2644 (1999); Phys. Lett. **B467**, 185 (1999).

RESULTS FROM THE PALO VERDE NEUTRINO EXPERIMENT

J. Wolf (Palo Verde Collaboration)

*Department of Physics and Astronomy,
University of Alabama, Tuscaloosa AL 35487*

Abstract. I report on the initial results from a measurement of the anti-neutrino flux and spectrum from the three reactors of the Palo Verde Nuclear Generating Station using a segmented gadolinium-loaded scintillation detector at a distance of about 800 m. We find that the anti-neutrino flux agrees with that predicted in the absence of oscillations, excluding at 90% CL $\bar{\nu}_e - \bar{\nu}_x$ oscillations with $\Delta m^2 > 1.12 \times 10^{-3}$ eV^2 for maximal mixing and $\sin^2 2\theta > 0.21$ for large Δm^2. Our results support the conclusion that the atmospheric neutrino oscillations observed by Super-Kamiokande do not involve ν_e. I will also give a short overview of the present status of the next generation long baseline reactor neutrino experiment, KamLAND.

I INTRODUCTION

Nuclear reactors have been used as intense sources of $\bar{\nu}_e$ in experiments searching for neutrino oscillations [1]. These experiments usually detect $\bar{\nu}_e$ by the process $\bar{\nu}_e + p \to n + e^+$, where the cross-section-weighted energy spectrum of $\bar{\nu}_e$, peaking at about 4 MeV, can be deduced from the measured e$^+$ spectrum. Any $\bar{\nu}_e$ flux deficit or distortions of the $\bar{\nu}_e$ energy spectrum would indicate oscillations. The low energy of reactor $\bar{\nu}_e$ allows these experiments to reach very small mass parameters, albeit with modest mixing-angle sensitivity. Past experiments [2] with detectors at 50-100 m from a reactor have explored the mass-parameter range down to 10^{-2} eV^2. The work described here, and a similar experiment at the Chooz reactor in France [3], are the first long baseline ($\sim$1 km) searches, designed to explore the parameter range down to 10^{-3} eV^2 as suggested by the early Kamiokande atmospheric neutrino anomaly [4]. Although later results from Super-Kamiokande [5], which appeared while this work was in progress, seem to disfavor the $\nu_\mu - \nu_e$ channel, a direct experimental exploration amply motivated this work.

The Palo Verde neutrino oscillation experiment [6] is located at the Palo Verde Nuclear Generating Station near Phoenix, Arizona. The total thermal power from three identical pressurized water reactors is 11.6 GW. Two of the reactors are 890 m from the detector, while the third is at 750 m. Our detector is placed in a shallow underground site (32 meter-water-equivalent overburden), thus eliminating the hadronic component of cosmic radiation and reducing the muon flux by a factor of ~ 5. The fiducial mass, segmented to reject the remaining background, consists of 11.3 tons of 0.1% Gd-loaded liquid scintillator contained in a 6×11 array of 9 m-long acrylic cells, as shown in Fig. 1. Each cell is viewed by two 5-inch photomultiplier tubes, one at each end. A $\bar{\nu}_e$ is identified by space- and time-correlated e^+ and n signals. Positrons deposit their energies in the scintillator and annihilate, yielding two 511 keV γ's, giving a triple coincidence. Neutrons thermalize and are captured in Gd, giving a γ-ray shower of 8 MeV total energy.

The Gd loading of the scintillator has two advantages: it reduces the neutron capture time from 170 μs (on protons) to 30 μs and provides a high energy γ shower to tag the neutron capture, resulting in a substantial background reduction. Both the positron and the neutron are triggered by a triple coincidence requiring at least one cell above a "high" threshold set at about 600 keV (positron ionization or neutron shower core), and two cells above a "low" threshold set at about 40 keV (Compton scattering from annihilation photons or neutron shower tails). The triple coincidences are required to be within a 3×5 matrix anywhere in the detector.

The central detector is surrounded by a 1 m water shield to moderate background neutrons produced by muons outside the detector and to absorb γ's from the laboratory walls. Outside the water tanks are 32 large liquid scintillator counters and two end-caps to veto cosmic muons. The rate of cosmic muons is approximately 2 kHz. In order to reduce natural radioactivity, all building materials for the detector were carefully selected, including the aggregate (marble) used in the concrete of the underground laboratory.

II EFFICIENCY, CALIBRATION, NEUTRINO FLUX

Since the ultimate sensitivity of the experiment relies on a disappearance measurement, precise knowledge of the detector efficiency and of the expected $\bar{\nu}_e$ flux from the reactors is essential.

The efficiency calculation is based upon a primary measurement performed a few times per year with a calibrated ^{22}Na β^+ source and an Am-Be neutron source, placed at various positions inside the detector. The ^{22}Na source is inserted trough calibration pipes inside the detector and mimics the effects of the positron from the $\bar{\nu}_e$ interaction by providing annihilation radiation and a 1.275 MeV photon which simulates the e^+ ionization in the scintillator. The neutron detection efficiency is measured by scanning the detector with the Am-Be source where the 4.4 MeV γ associated with the neutron emission is tagged with a miniaturized NaI(Tl) counter.

Other radioactive sources are used to measure the energy response of the detector. A ^{228}Th source placed at 7 positions along each cell is used more frequently to track the scintillator transparency. Weekly runs of fiber-optic and LED flasher systems are used to monitor the gain and linearity of photomultipliers and the timing/position relationship along the cells.

The $\bar{\nu}_e$ flux and spectrum from a fission reactor and the $\bar{\nu}_e + p \to n + e^+$ cross section are well known [1,2,7] and are calculated by tracking the ^{235}U, ^{238}U, ^{239}Pu, and ^{241}Pu fission rates in the three plant reactors, taking into account both power level and fuel age. The uncertainty in the $\bar{\nu}_e$ reaction rate is less than 3%.

III NEUTRINO SELECTION

The data presented here were collected in periods of 67.3 days in 1998 and 134.4 days in 1999. During the 98 (99) data taking one of the far (near) reactors was off for 31.3 (23.4) days. Here we outline the principles of the analysis and the results. A detailed description can be found elsewhere [6].

Neutrino candidates were selected by requiring an appropriate pattern of energy to be present in the detector for the positron- and the neutron-like parts of the events. In addition the two sub-events are required to occur within 1 m from each other.

At our depth the background to $\bar{\nu}_e$ events consists of two types of events: uncorrelated hits from cosmic-rays and natural radioactivity and correlated ones from cosmic-muon-induced neutrons. The first type can be measured by studying the time difference between positron-like and neutron-like parts of an event. By requiring that the time lapse between the two sub-events t_{en} be 5 μs $< t_{en} <$ 200 μs, the uncorrelated background is reduced to 3.4$\pm$0.2 events d^{-1} (4.8$\pm$0.2 events d^{-1}) for 1998 (1999), as measured from a fit to an appropriate combination of exponential functions.

From the distribution of time intervals between a cosmic-ray μ and a $\bar{\nu}_e$-like event we infer that the majority of correlated background is produced by pairs of neutrons, where the capture of the first neutron in each pair mimics the positron signature. The requirement that no cosmic-ray hits be present in a window of 150 μs preceding the $\bar{\nu}_e$ candidate completes the event selection.

The resulting rates N of $\bar{\nu}_e$ candidates per day in different periods are given in Tab. 1. Along with the $\bar{\nu}_e$ events this final data set contains both random and correlated background, as mentioned above. Two independent techniques were used to estimate and subtract the background. The most straightforward method ("Method 1") relies on the changes of the $\bar{\nu}_e$ signal when different reactors are turned off. A χ^2-analysis is performed, comparing the expected rate to the efficiency-corrected measured rate, assuming a constant background. The systematic error for "Method 1" is 10%. We find good agreement with the hypothesis of no oscillation, as shown in Fig 1 (curve a).

In "Method 2", that is described in detail elsewhere [8], we make use of the intrinsic symmetry of the dominant background components under interchange of the energy cuts for positron and neutron sub-events. The rate of $\bar{\nu}_e$ candidates after all cuts can be written as $N = B_{unc} + B_{nn} + B_{pn} + S_\nu$, where the contribution of the uncorrelated B_{unc}, two-neutron B_{nn} and other correlated backgrounds B_{pn} are explicitly represented, along with the $\bar{\nu}_e$ signal S_ν. The event selection with

TABLE 1. Summary of results from the Palo Verde experiment. Method 2: $B = B_{unc} + B_{nn} + B_{np}$. R_{Obs} and R_{Calc} are the observed and calculated $\bar{\nu}_e$ rates corrected by the efficiencies η for the case of no-oscillations. Uncertainties are statistical only. (one reactor off at 890 m* or 750 m†)

Period	98 "on"	98 "off"*	99 "on"	99 "off"†
Duration (d)	36.0	31.3	111.0	23.4
Efficiency η	0.0746	0.0772	0.112	0.111
Event rate N (d^{-1})	38.2 ± 1.0	32.2 ± 1.0	52.9 ± 0.7	43.9 ± 1.4
$\bar{\nu}_e$ rate S_ν (d^{-1})	16.5 ± 1.4	13.4 ± 1.4	25.2 ± 0.9	15.1 ± 1.9
Background B (d^{-1})	21.7 ± 1.0	18.8 ± 1.0	27.7 ± 0.6	28.8 ± 1.3
R_{Obs} (d^{-1})	221 ± 18	174 ± 17	225 ± 8	136 ± 17
R_{Calc} (d^{-1})	218	155	218	130

swapped cuts results in a rate $N' = B_{unc} + B_{nn} + \epsilon_1 B_{pn} + \epsilon_2 S_\nu$ where ϵ_1 and ϵ_2 account for the different efficiency for selecting asymmetric events after the swap.

The dominant background B_{nn} and B_{unc} is symmetric under exchange of sub-events, so that it cancels when we subtract $N - N' = (1 - \epsilon_1)B_{pn} + (1 - \epsilon_2)S_\nu$. The efficiency correction for the neutrino signal is $\epsilon_2 \simeq 0.2$. Monte Carlo simulations and data above a 10 MeV energy cut are used to estimate $(1 - \epsilon_1)B_{pn}$. We find that the processes of μ-spallation in the laboratory walls and capture of μ's passing the veto counter untagged contribute to $(1 - \epsilon_1)B_{pn}$, while other backgrounds are negligible. We obtain $(1 - \epsilon_1)B_{pn} = -0.3 \pm 0.7$ d^{-1} (-0.4 ± 0.8 d^{-1}) in 1998 (1999). This represents only a small correction to $N - N'$. This technique makes the best possible use of the statistical power of all data collected. The systematic error for "Method 2" is 8%. The results are shown in the second part of Tab. 1 for different running periods. Clearly Method 2 is also in agreement with the no-oscillation hypothesis (Fig. 1 curve b).

In conclusion, the data from the first period of running from the Palo Verde detector show no evidence for $\bar{\nu}_e - \bar{\nu}_x$ oscillations. This result, together with the data already reported by Super-Kamiokande [5] and a more stringent limit by Chooz [3], exclude the channel $\nu_\mu - \nu_e$ as being responsible for the atmospheric neutrino anomaly reported by Kamiokande [4]. Data-taking at Palo Verde[1] is scheduled to continue until the July 29, 2000.

IV THE STATUS OF KAMLAND

KamLAND [10] is the next generation very long baseline reactor neutrino experiment being built by a Japanese-US collaboration in the Kamioka mine in Japan.

---

[1] We would like to thank the Arizona Public Service Company for the generous hospitality at the Palo Verde plant. This project was supported in part by the US DoE. It also received support from the Hungarian OTKA fund and from the ARCS foundation.

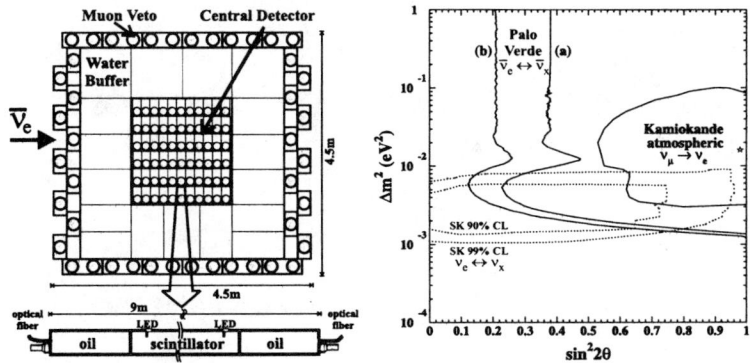

FIGURE 1. Left: Schematic view of the Palo Verde neutrino detector Right: Limits on Δm^2 and $sin^2 2\theta$ from the present work (90% CL). Curves (a) and (b) are based on Method 1 and 2 for background subtraction, respectively as described in the text. Also shown: Kamiokande allowed region and Super-Kamiokande [9] bounds on $sin^2 2\theta_{13}$, using a simplified 3-flavor analysis, assuming $\Delta m_{13}^2 = \Delta m_{23}^2 = \Delta m^2$ and $\Delta m_{12}^2 = 0$.

Construction of the 1000 ton scintillation detector is proceeding as planned. The detector vessel has been completed. In a successful test, conducted in May 2000, the containment balloon for the scintillator has been deployed and the detector tank was filled with water. The experiment is scheduled to start data-taking on April 1, 2001. It will measure neutrinos from 16 nuclear power plants at distances between 81 km and 824 km. These long baselines will extend the sensitivity to $\Delta m^2 < 10^{-5}$, allowing the first direct check of the MSW large mixing angle solution of the solar neutrino problem.

REFERENCES

1. See e.g. F. Boehm and P. Vogel, "Physics of Massive Neutrinos" Cambridge University Press 1992 (2nd ed.).
2. G. Zacek et al., Phys. Rev. D34 (1986) 2621 ;
   B. Achkar et al., Nucl. Phys. B434 (1995) 503.
3. M. Apollonio et al., Phys. Lett. B466 (1999) 415.
4. Y. Fukuda et al., Phys. Lett. B335 (1994) 237.
5. Y. Fukuda et al., Phys. Rev. Lett. 81 (1998) 1562.
6. F. Boehm et al., hep-ex/0003022, accepted for publ. in Phys. Rev. D.
7. P. Vogel and J.F. Beacom, Phys. Rev. D60 (1999) 053003.
8. Y.F. Wang et al., hep-ex/0002050, accepted for publ. in Phys. Rev. D.
9. K. Nakamura, Super-Kamiokande Collaboration, Talk at CIPANP-2000
10. P. Alivisatos et al., KamLAND proposal (US), Stanford-HEP-98-03

Active-Sterile Neutrino Transformation and r-Process Nucleosynthesis

G. C. McLaughlin

TRIUMF, 4004 Wesbrook Mall, Vancouver, BC, V6T2A3

Abstract. The type II supernova is considered as a candidate site for the production of heavy elements. Since the supernova produces an intense neutrino flux, neutrino scattering processes will impact element formation. We examine active-sterile neutrino conversion in this environment and find that it may help to produce the requisite neutron-to-seed ratio for synthesis of the r-process elements.

The r-process of nucleosynthesis accounts for the most neutron rich of the heavy elements. The most likely environment for this type of synthesis is the late time ($t > 10$ s post-core bounce) supernova environment. Many studies have explored this 'neutrino driven wind' as a candidate environment and found it to be potentially viable [1,2]. However, to date, no model correctly reproduces the observed abundance pattern.

In the neutrino driven wind, material in the form of free nucleons is 'lifted' off of the surface of the neutron star by energy deposited by neutrino interactions. Analytic and semianalytic parameterizations of the thermodynamic and hydrodynamic conditions in the wind can be obtained [3,4]. Models of this type may be used to explore the range of conditions within the context of the wind which will produce the solar system distribution of r-process elements. The key determinant of whether a given scenario will produce the r-process is the neutron to seed nucleus ratio at the onset of the neutron capture phase. This ratio must be quite high ($R > 100$) in order to produce the very neutron-rich r-process elements. The factors which determine the neutron-to-seed ratio are the entropy of the material, the hydrodynamic outflow timescale and the electron fraction, $Y_e = 1/(1 + n/p)$ where n/p is the neutron-to-proton ratio. A study of many possible model parameters shows that one must decrease the electron fraction, and/or increase the entropy and/or decrease the hydrodynamic outflow timescale, relative to the conditions found in typical wind models, in order to produce the neutron-to-seed ratio necessary for the r-process [5,6].

Including the effects of neutrino interactions in general tends to make the requisite conditions for r-process element production more extreme [7,8]. In particular, neutrino capture on free nucleons during alpha particle formation increases the electron fraction [8]. This is the "alpha effect". Other neutrino process are discussed in [2,9].

There are three possible solutions to this problem. The first is that the supernova is the site of r-process synthesis, but it does not occur in the neutrino driven wind as it is currently modeled. The second is that the r-process elements are made at some other site such as neutron star-neutron star mergers. However, timescale arguments combined with isotopic abundance measurements and observations of old halo star metallicity show that this site is unlikely to account for the entire r-process distribution [10,11].

The third solution is the one that is investigated here: active-sterile ($\nu_e \leftrightarrow \nu_s$, $\bar{\nu}_e \leftrightarrow \bar{\nu}_s$) neutrino transformation. The ν_s in our study is defined as a particle which mixes with the ν_e (and possibly also with ν_μ, and/or ν_τ) but does not contribute to the width of the Z boson.

If we neglect ν-ν forward scattering contributions to the weak potentials, then the equation which governs the evolution of the neutrinos as they pass though the material in the wind can be written as:

$$i\hbar \frac{\partial}{\partial r} \begin{bmatrix} \Psi_e(r) \\ \Psi_s(r) \end{bmatrix} = \begin{bmatrix} \varphi_e(r) & (\delta m^2/4E)\sin 2\theta_v \\ (\delta m^2/4E)\sin 2\theta_v & -\varphi_e(r) \end{bmatrix} \begin{bmatrix} \Psi_e(r) \\ \Psi_s(r) \end{bmatrix}, \quad (1)$$

where

$$\varphi_e(r) = \frac{1}{4E}\left(\pm 2\sqrt{2}\, G_F \left[N_e^-(r) - N_e^+(r) - \frac{N_n(r)}{2}\right] E - \delta m^2 \cos 2\theta_v \right) \quad (2)$$

The upper sign is relevant for neutrino transformations; the lower one is for antineutrinos. In these equations $\delta m^2 \equiv m_2^2 - m_1^2$ is the vacuum mass-squared splitting, θ_v is the vacuum mixing angle, G_F is the Fermi constant, and $N_e^-(r)$, $N_e^+(r)$, and $N_n(r)$ are the total proper number densities of electrons, positrons, and neutrons respectively in the medium. Resonances can occur when the on-diagonal terms in the wave equation are zero. The quantity in the brackets in Eq. 2 is proportional to $Y_e - 1/3$. Since the electron fraction can take on values between zero and one, the bracketed quantity in Eq. 2 can be either positive or negative. Therefore, for a given choice of δm^2, resonances may occur for neutrinos or antineutrinos depending on the value of the electron fraction.

In order to determine the survival probabilities of the neutrinos and antineutrinos, we must know the electron fraction. In the neutrino driven wind, neutrino and antineutrino capture are the most important reactions in determining the electron fraction. However, near the surface of the protoneutron star there is also a contribution from electrons and positrons:

$$\nu_e + n \rightleftharpoons p + e^-; \quad \bar{\nu}_e + p \rightleftharpoons n + e^+. \quad (3)$$

Therefore, the problem involves a feedback mechanism. The neutrino capture rates determine the electron fraction and the electron fraction determines the potential which is used in the neutrino transformation equations. These equations then determine survival probabilities of the neutrinos, and therefore their capture rates.

We note that we can not assume weak equilibrium, (very fast capture rates compared with outflow rate) or weak freeze-out (very slow rates compared with the outflow rate). The rates must be tracked numerically. A previous study took the limit of weak equilibrium and found different behavior [12] than presented here.

We perform calculations by tracking a mass element in the neutrino driven wind. We use the results of analytic models [4], where $r \propto \exp(-t/\tau)$ and $\rho \propto r^{-3}$ where τ is the outflow timescale. Close to the surface, before the wind begins to operate we use the density profile of Wilson and Mayle [13]. At each time step we calculate all thermodynamic quantities, calculate the weak rates and evolve the neutrino transformation equations forward. We assume, since the outflow timescale is short, $t \sim \tau \lesssim 0.5\,\mathrm{s}$, that each mass element will have the same evolution as the previous one. Since we use a nuclear statistical equilibrium calculation, we cut our calculations off at the time when heavy nuclei begin to form. More detail is contained in [14].

A calculation for one mass element is shown in Fig. 1. The upper curve shows the evolution for the case of no neutrino oscillations. The wind parameters were $\tau = 0.3\,\mathrm{s}$, $s = 100$. The solid line shows the actual electron fraction, while the

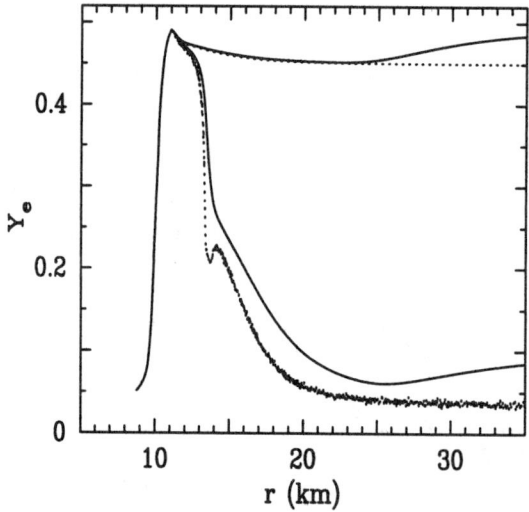

FIGURE 1. The electron fraction is plotted against distance from the center of the neutron star. The upper line shows the evolution with no transformation. The lower line shows the evolution of active-sterile mixing parameters of $\sin^2\theta = 0.01$ and $\delta m^2 = 20\,\mathrm{eV}^2$. The dotted line shows the value that the electron fraction would take on if weak equilibrium obtained.

dotted line shows what the electron fraction would be if weak equilbrium obtained. The initial rise is due to Pauli unblocking of the electrons at the surface of the proto-neutron star. The lower curve shows the evolution of the electron fraction for mixing parameters of $\delta m^2 = 20\,\mathrm{eV}^2$, $\sin^2 2\theta_v = 0.01$. In the latter case there is a rapid drop in the electron fraction when the neutrinos begin to transform. In fact, there are three neutrino transformations. Initially, the electron neutrinos transform to steriles. Later antineutrinos transform, and finally, when the density falls far enough the antineutrinos transform back. The first transformation of the antineutrinos is seen in the small bump in the equilibrium electron fraction. A small "alpha effect" can be see as the slight rise in both solid lines at large distance.

The active-sterile transformation scenario successfully reproduces a low electron fraction which is beneficial to the r-process. We now consider a range of δm^2, $\sin^2 \theta$ parameters. A contour plot of the electron fraction is shown in Fig. 2. Inside the dashed contour shows the region where conditions are neutron-rich enough to be favorable for r-process nucleosynthesis. In the bottom left corner of the plot, the solution is approaching the case without neutrino transformation.

Although not shown here, we have studied a range of timescales for the neutrino driven wind models, and seen that the qualitative features of this effect are reproduced [14].

We have used several approximations in our calculations, which we are continuing to study. These include the importance of the neutrino-neutrino scattering

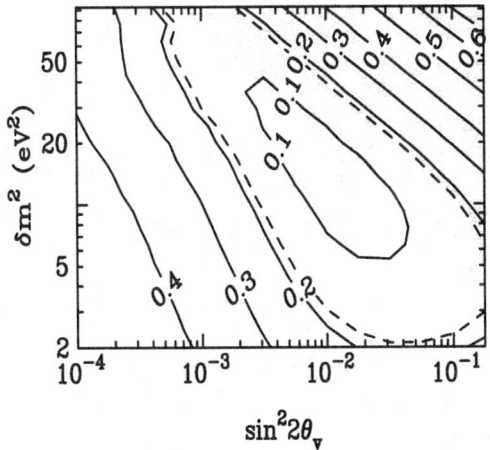

FIGURE 2. Contour plot of electron fraction as measured at the point where heavy nuclei begin to form. Neutrino driven wind parameters employed here are $s/k = 100$, $\tau = 0.3\,\mathrm{s}$.

background in the oscillation equations, the importance of nonradial paths and feedback in the dense region near the proto-neutron star. The region where the latter two are important is the shaded region in Fig. 2. These problems are being studied in [15].

Conclusions: Meteoritic and observational evidence points to supernovae as the source of the r-process elements, although a self-consistent model of the neutrino driven wind which will produce the r-process elements is still elusive. In the next few years significant advances are expected in supernova modeling. If and when potential hydrodynamic solutions are exhausted and the caveats above have been explored, then the r-process may provide a signature for new neutrino physics.

ACKNOWLEDGEMENTS

This work was done in collaboration with J. M. Fetter, A. B. Balantekin and G. M. Fuller.

REFERENCES

1. Woosley, S. E., Wilson, J. R., Mathews, G. J., Hoffman, R. D. and Meyer, B. S. *Astrophys. J.* **433**, 229-246 (1994), Takahashi, K., Witti, J., and Janka, H.-Th *Astron. Astrophys.* **286**, 857-869 (1994).
2. Meyer, B. S., Howard, W. M. Mathews, G. J., Woosley, S. E., and Hoffman, R. D. *Astrophys. J.* **399**, 656-664 (1992).
3. Duncan, R. C., Shapiro, S. L., & Wasserman, I. *Astrophys. J.*, 309, 141-160 (1986).
4. Qian, Y.-Z. and Woosley, S. E. *Astrophys. J.*, **471**, 331-351 (1996).
5. Meyer, B. S. and Brown, J. S. *Astrophys. J. Suppl.* **112**, 199-220, (1997).
6. Hoffman, R. D., Woosley, S. E. and Qian, Y.-Z.*Astrophys. J.* **482**, 951-962 (1997).
7. Meyer, B. S. *Astrophys J.*, **449**, L55-L58 (1995)
8. Fuller, G. M., and Meyer, B. S., it Astrophys J., **453**, 792-809, (1995), McLaughlin, G. C., Fuller, G. M. and Wilson, J. R., it Astrophys J., **472**, 440-451, (1996), Meyer, B. S., McLaughlin, G. C. and Fuller, G. M., *Phys. Rev. C*, **59** 2873-2887 (1998).
9. Haxton, et. al., *Phys. Rev.Lett.* 78, 2694-2697 (1997)
10. Qian, Y.-Z. and Wasserburg, G. J., *Phys. Rept.* **333-334**, 77-108 (2000)
11. Sneden C. et al., *Astrophys. J.*, **467**, 819-840, (1996), Sneden, C. et al. *Astrophys. J.*, **496**, 235-245, (1998).
12. Nunokawa, H., Rossi, A., Semikoz, V. and Valle, J. W. F. *Phys. Rev.* **D56** 1704-1713 (1997).
13. Mayle, R. W. and Wilson, J. R., unpublished (1993).
14. McLaughlin, G. C., Fetter, J. M., Balantekin, A. B. and Fuller, G. M., *Phys. Rev.* **C59**, 2873-2887. (1999)
15. Fetter, J. M., Balantekin, A. B. and Fuller, G. M., and McLaughlin, G. C. in preparation (2000).

The Sudbury Neutrino Observatory

Fraser Duncan *for the SNO Collaboration*

Queen's University, Kingston Ontario

Abstract. Located 2,000 meters below the surface of the earth in the Creighton Nickel Mine near Sudbury, Ontario, Canada, is the Sudbury Neutrino Observatory (SNO). Operational for almost a year now, SNO is a 1000 tonne heavy water Cerenkov detector designed to observe solar neutrinos. The use of heavy water allows SNO to detect neutrinos with an interaction sensitive only to electron neutrinos and with another interaction that is sensitive to all neutrino flavours. SNO's unique ability to separately measure the total solar neutrino flux and electron neutrino fluxes allows the experiment to make a search for flavour oscillations in solar neutrinos in a model independent fashion. The status of the experiment will be described.

SNO uses deuterium in the form of heavy water (D_2O) to detect neutrinos through several interactions. Of particular interest is a Charged Current (CC) interaction,

$$\nu_e + d \to p + p + e^-, \qquad (1)$$

sensitive only to electron neutrinos and a Neutral Current (NC) interaction,

$$\nu_x + d \to n + p + \nu_x, \qquad (2)$$

by which any active neutrino (ν_e, ν_μ, ν_τ) interacts with a deuteron. As well, the neutrinos can undergo an Elastic Scattering (ES) interaction with the electrons in the water,

$$\nu_x + e^- \to e^- + \nu_x. \qquad (3)$$

which is primarily sensitive to electron neutrinos but has a smaller sensitivity (approximately 1/6) to other neutrino flavours.

The CC interaction has a Q value of 1.4 MeV and produces an energetic electron whose energy is strongly correlated to the incident neutrino energy. This electron is detected by the Cerenkov light it emits while slowing down in the water. The design energy threshold for the recoil electrons is 5 MeV giving the CC interaction a neutrino energy threshold of 6.4 MeV. The NC interaction (threshold 2.2 MeV) produces a free neutron which can be detected by several different methods. The energy thresholds are such that SNO is only sensitive to ^{8}B and ^{3}He + p solar neutrinos.

The means by which the neutron from the neutral current interaction is detected define the three phases of the experiment:

1. In the *Pure D_2O Phase* the neutron is detected when it captures on deuterium in the heavy water. This produces a 6.2 MeV γ which undergoes Compton scattering in the water, resulting in an energetic electron which produces Cerenkov light. The average capture efficiency for neutrons in pure D_2O is 27%.

2. The *Salt Phase* of the experiment will include either NaCl or $MgCl_2$ in the heavy water. The neutron then captures on Cl producing a γ cascade of total energy 8.6 MeV. It will increase the capture efficiency of the neutrons to 80%

3. The *Neutral Current Detector Phase* will use approximately 100 dedicated Neutral Current Detectors (NCDs) in the form of proportional counters in the heavy water to detect the neutron capture on ^{3}He. The NC detectors have a neutron capture efficiency of order 50%.

The intent is for the pure D_2O and salt phases to last approximately one year each followed by an extended running period with the neutral current detectors.

The primary physics goal of SNO is to measure both the CC and NC fluxes of ^{8}B neutrinos from the sun. A comparison of these fluxes will show in a model independent way if there are flavour oscillations from the electron neutrinos produced in the solar interior to other active flavours. In addition, the good energy correlation between the recoil electron and the incident neutrino in the CC interaction will allow a precise measurement of the ^{8}B neutrino spectral shape which will constrain the allowed mixing parameters if oscillations are present. SNO will measure day/night variation in the neutrino signal and look for variations in the neutrino flux correlated with the earth's distance from the sun. SNO will also make measurements of atmospheric neutrinos and is sensitive to neutrino bursts from supernovae.

THE DETECTOR

The SNO detector is located 2 km underground in a laboratory constructed at the 6800 ft level of INCO's Creighton Mine near Sudbury Ontario. The detector (figure 1) consists of 1000 tonnes of heavy water contained in a 5 cm thick 12 m diameter spherically shaped acrylic vessel. The Cerenkov light from interactions in the heavy water are observed by 9438 Hamimatsu R1408 20 cm photomultiplier tubes mounted in a 18 m diameter geodesic sphere surrounding the acrylic vessel. Reflectors around the tubes increase the light collection solid angle to approximately 65% of 4π. An additional 91 photomultiplier tubes look outward from the geodesic to act as a cosmic ray veto. Both acrylic vessel and photomultiplier

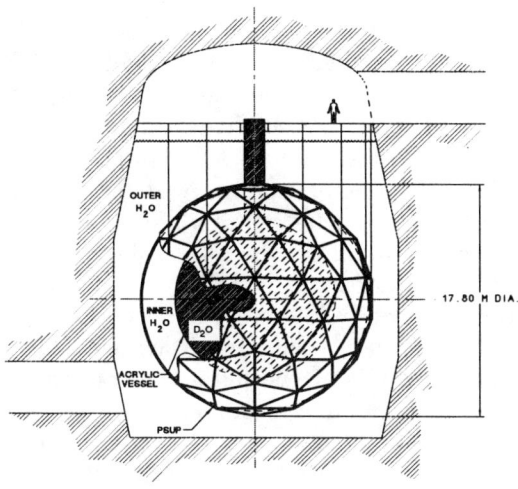

FIGURE 1. Schematic Diagram of the SNO Detector

support structure are suspended in 7000 tonnes of ultrapure light water contained in a cavity 22m in diameter and 34 m high. The light water shields the D_2O volume from gamma-rays and neutrons from the cavity wall and from the photomultiplier support structure. An extensive description of the detector can be found in [1].

Because of the depth of the detector, cosmic ray muons are infrequent (of order 70 per day) and the dominant source of background is radioactivity. The design goal of the experiment is to have less than 10^{-14} gram of either ^{238}U or ^{232}Th (or their daughter products) per gram of water in the detector. Elaborate measures were taken to achieve this goal. The detector materials (acrylic, phototube glass, support structures etc) where carefully chosen for low activity. The norite rock that the laboratory is hewn out of contains of order 10^{-6} gm/gm U and Th so precautions are taken to prevent mine dust from entering the detector. The laboratory is operated as a Class 2000 clean room. The detector cavity is lined with an 8-mm thick polyurethane liner to prevent radon in the rock from diffusing into the detector. Due to emanation from the rock walls, the laboratory air contains ^{222}Rn at the level of 3 pCi per litre of air. A cover gas system is in place to prevent this radon from entering the detector. In addition to the efforts to choose clean materials and to prevent the ingress of contaminants, there is an extensive water purification system to clean both the heavy water and light water on a continual basis.

STATUS

The water fill of the detector was completed in April 1999 and the pure D_2O data taking began in November 1999. Presently 98.5% of all channels are fully operational with an overall photomultiplier tube failure rate of 0.5% per year. The average channel threshold is 0.25 photoelectrons. The mean phototube noise rate is 500 Hz resulting in approximately 2 noise tubes per event. The total trigger rate for the detector is 15 Hz and is comprised of several different triggers. The primary trigger for neutrino analysis is a coincidence of 18 or more photomultipliers in a 100 ns window. Since the beginning of the pure D_2O production data taking, the detector live time has been approximately 85%.

Since the completion of water fill, there have been extensive assays of the radioactive contamination in both the light and heavy water. The preliminary measurements indicate that the levels are within a factor of three of the design goals for the experiment and are continuing to improve.

A large part of the SNO collaboration's effort since the completion of water fill of the detector has been directed towards calibration. The calibration plan includes measurement of the optical properties of the detector with a pulsed dye laser system, energy calibrations with a variety of gamma-ray sources and neutral current efficiency measurements with a neutron source. The calibration sources can be place through out the D_2O volume (and in selected locations within the light water) with a manipulator system that allows deployment of sources without breaking either the light or radon seal on the detector. To date the calibrations that have been done are: optical measurements at several wavelengths between 337 and 500 nm; energy calibrations with 6.13 MeV gamma-rays from a ^{16}N source; and neutron calibrations with a ^{252}Cf source. Analysis of the calibration data is ongoing.

SUMMARY

The Sudbury Neutrino Observatory has now been operational for approximately a year. The initial commissioning of the experiment is complete and the detector is performing well. Measurements indicate that the target levels for radioactive backgrounds will be reached. The project is now approximately half way through the pure D_2O phase and the intent is to go to the salt phase of the experiment near the end of this year.

REFERENCES

1. J. Boger et. al. Nucl.Instrum.Meth. A449 (2000) 172-207

Electroweak Physics at NuTeV

G.P. Zeller

for the NuTeV Collaboration

Northwestern University, Evanston IL 60208

Abstract. The NuTeV experiment at Fermilab presents a determination of the electroweak mixing angle. High purity, large statistics samples of $\nu_\mu N$ and $\bar\nu_\mu N$ events allow the use of the Paschos-Wolfenstein relation, a technique which considerably reduces systematic errors associated with charm production and other sources. Within the Standard Model, this measurement of $\sin^2\theta_W$ indirectly determines the W boson mass to a precision comparable to direct measurements from high energy e^+e^- and $p\bar p$ colliders. NuTeV measures $\sin^2\theta_W^{(\text{on-shell})} = 0.2253 \pm 0.0019(\text{stat.}) \pm 0.0010(\text{syst.})$, which implies $M_W = 80.26 \pm 0.11$ GeV. Outside the Standard Model, this result can be used to explore the possibility of new physics; in particular, we present limits on both neutrino oscillations and the presence of extra neutral vector gauge bosons.

In deep inelastic neutrino-nucleon scattering, the weak mixing angle can be extracted from the ratio of neutral current (NC) to charged current (CC) total cross sections. A method for determining $\sin^2\theta_W$ that is much less dependent on sources of model uncertainty and the the details of charm production (the largest source of uncertainty in the previous neutrino measurement [1]) employs the Paschos-Wolfenstein relation [2]:

$$R^- = \frac{\sigma^\nu_{NC} - \sigma^{\bar\nu}_{NC}}{\sigma^\nu_{CC} - \sigma^{\bar\nu}_{CC}} = \frac{R^\nu - rR^{\bar\nu}}{1-r} = \frac{1}{2} - \sin^2\theta_W \qquad (1)$$

where $R^{\nu,\bar\nu} = \sigma^{\nu,\bar\nu}_{NC}/\sigma^{\nu,\bar\nu}_{CC}$, and $r = \sigma^{\bar\nu}_{CC}/\sigma^\nu_{CC}$. Unfortunately, the substantially reduced uncertainties come at a price: R^- is a more difficult quantity to measure experimentally because neutral current neutrino and antineutrino events have identical observed final states. The two samples can only be separated by knowing the incoming neutrino beam type.

High-purity neutrino and antineutrino beams were provided by the Sign Selected Quadrupole Train (SSQT) at the Fermilab Tevatron during the 1996-1997 fixed target run. Neutrinos are produced from the decay of pions and kaons resulting from interactions of 800 GeV protons in a BeO target. Dipole magnets immediately downstream of the proton target bend pions and kaons of specified charge in the direction of the NuTeV detector, while wrong-sign and neutral mesons are stopped in beam dumps. The resulting beam is almost purely neutrino or antineutrino depending on the selected sign of the parent mesons (opposite particle contamination is ~0.1%). In addition, the beam is almost purely muon neutrinos with a small (~1%) contamination of electron neutrinos.

Neutrino interactions are then observed in the NuTeV detector [3], which is located approximately 1.5 km downstream of the proton target. The detector consists of an 18m long, 690 ton steel-scintillator target followed by an instrumented iron-toroid spectrometer. The target calorimeter is composed of 168 3m x 3m x 5.1cm steel plates interspersed with liquid scintillation counters and drift chambers. The scintillation counters provide triggering information as well as a determination of the longitudinal event vertex, event length and visible energy deposition. The mean position of hits in the drift chambers help establish the transverse event vertex. The toroidal spectrometer, which determines muon sign and momentum, is not directly used for this analysis. In addition, because the detector was continuously calibrated through exposure to a wide energy range of test beam hadrons, electrons and muons, many systematics related to detector effects were substantially reduced.

WITHIN THE STANDARD MODEL

In order to measure $\sin^2 \theta_W$, observed neutrino events must be separated into charged current (CC) and neutral current (NC) categories. Both CC and NC neutrino interactions initiate a cascade of hadrons in the target that is registered in both the scintillation counters and drift chambers. However, muon neutrino CC events are distinguished by the presence of a final state muon, which typically penetrates well beyond the hadronic shower and deposits energy in a large number of consecutive scintillation counters. These differing event topologies enable the statistical separation of CC and NC interactions based solely on event length (i.e., on the presence or absence of a muon in an event). Events with a long length (spanning more than 20 counters) are identified as CC candidates; those with a short length (spanning less than 20 counters) as NC candidates. The experimental quantity measured in both neutrino and antineutrino modes is the ratio:

$$R_{meas} = \frac{\# \text{ SHORT events}}{\# \text{ LONG events}} = \frac{\# \text{ NC candidates}}{\# \text{ CC candidates}} \qquad (2)$$

The ratios of short to long events (R_{meas}) measured in the NuTeV data are 0.4198 ± 0.0008 in the neutrino beam and 0.4215 ± 0.0017 in the antineutrino beam. A Standard Model value of $\sin^2 \theta_W$ can be directly extracted from these measured ratios by using a detailed Monte Carlo simulation of the experiment. The Monte Carlo must include the integrated neutrino fluxes, the neutrino cross section, and a detailed description of the NuTeV detector. More detailed information on the specific components of the Monte Carlo simulation can be found elsewhere [4].

Using our separate high-purity neutrino and antineutrino data sets, NuTeV measures the following linear combination of R^ν and $R^{\bar\nu}$:

$$R^- = R^\nu - xR^{\bar\nu} \tag{3}$$

The value for x is selected using the Monte Carlo in order to minimize uncertainties related to charm quark production. The single remaining free parameter in the Monte Carlo, $\sin^2\theta_W$, is then varied until the model calculation of R^- agrees with what is measured in the data. The preliminary result from the NuTeV data sample for $M_{\text{top}}=175$ GeV and $M_{\text{Higgs}}=150$ GeV is:

$$\sin^2\theta_W^{(\text{on-shell})} = 0.2253 \pm 0.0019(\text{stat.}) \pm 0.0010(\text{syst.}) \tag{4}$$

Having chosen the convention $\sin^2\theta_W^{(\text{on-shell})} \equiv 1 - \frac{M_W^2}{M_Z^2}$, and given the well-determined Z mass from LEP, our result implies:

$$M_W = 80.26 \pm 0.10(\text{stat.}) \pm 0.05(\text{syst.}) = 80.26 \pm 0.11 \text{GeV} \tag{5}$$

Our measurement is in good agreement with Standard Model expectations, and is consistent with current measurements from W and Z production as well as from other neutrino experiments (Figures 1, 2). The data tend to collectively favor a light Higgs mass. The central value from recent global fits to all precision data is $M_{\text{Higgs}} = 98^{+57}_{-38}$ GeV with an upper bound of $M_{\text{Higgs}} \leq 235$ GeV at 95% CL [5].

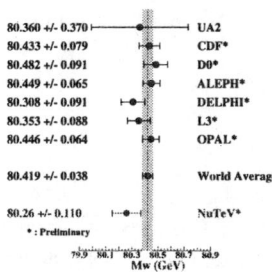

FIGURE 1. Direct W boson mass measurements compared with this result.

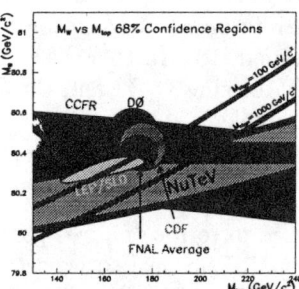

FIGURE 2. Experimental constraints presented on the M_W-M_{top} plane. The two narrow bands indicate the Standard Model predictions for M_{Higgs}=100 and 1000 GeV.

BEYOND THE STANDARD MODEL

Outside the Standard Model, deviations between electroweak measurements in νN scattering and those from other processes are sensitive to new physics. In these proceedings we discuss two such possibilities: neutrino oscillations and extra neutral vector gauge bosons.

The presence of neutrino oscillations will directly shift our measured ratios, R_{meas}, from their Standard Model predictions. Since ν_μ's oscillating to either ν_e's or ν_τ's would be less likely to produce a final state muon, we would expect to observe an excess of short events. Since no such excess is observed, single mode (ν or $\bar{\nu}$) limits can be set for both $\nu_\mu \to \nu_e$ and $\nu_\mu \to \nu_\tau$ oscillations. One advantage to this type of search is that the Paschos-Wolfenstein quantity, R^-, is particularly sensitive to CP-violating oscillations because it is formed from a difference in neutrino and antineutrino rates. As a result, NuTeV is presently the only experiment with direct limits on $\bar{\nu}_\mu \to \bar{\nu}_\tau$. More details on this analysis can be found elsewhere [6].

Extra Z bosons (Z') are of interest not only because they are predicted by many Grand Unified Theories and superstring models, but also in light of recent experimental developments. Erler and Langacker have shown in a recent global fit that the precision electroweak data are better described if an extra TeV-scale Z boson is included [7]. Of course, a large portion of this improvement arises from the fact that the 2.5 σ deviation of the new atomic parity violation (APV) measurement [8] from the Standard Model prediction can be explained by including an additional Z boson [9].

In our case, extra Z bosons would manifest themselves as shifts in the neutrino-quark couplings away from their Standard Model values. These shifts can arise from both pure-Z' exchange as well as from Z-Z' mixing contributions. If we consider constraints on extra Z bosons in E_6 models, then these coupling shifts are well-determined [10]. In this case, the lightest extra Z boson is a linear combination of the SO(10) singlet Z_χ and the SU(5) singlet Z_ψ:

$$Z' = Z_\chi \cos\beta + Z_\psi \sin\beta \quad (6)$$

expressed in terms of a free parameter β. Our data tend to disfavor the inclusion of additional Z bosons, so we set a 95% CL lower limit on the mass of such an extra Z' plotted as a function of β (Figure 3). Limits are displayed for two cases: the Z' has no mixing with the Standard Model Z, and the more realistic case that allows for some level of mixing. For the latter, we input the Z pole data constraint that the level of Z-Z' mixing is small (10^{-3}) [7]. Note that the inclusion of Z-Z' mixing, even at this small level, weakens the limits. At 95% CL and assuming a 10^{-3} level of Z-Z' mixing, we set lower mass limits of 675 and 380 GeV for the Z_χ and Z_η respectively. Direct searches have already excluded masses below ~ 600 GeV [11].

As can be seen from Figure 3, our maximum sensitivity is to the Z_χ, and our exclusion peaks in the most viable region suggested by the APV data. Figure 4 shows a direct comparison of the region NuTeV excludes in Z' mass-β space to the favored central value from the APV data analyses [9].

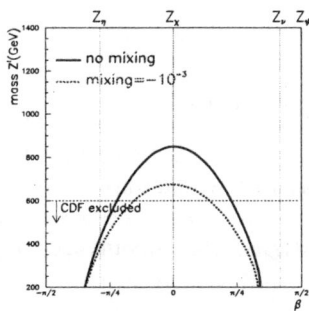
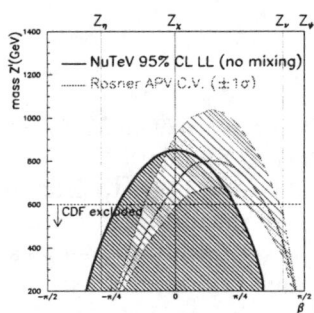

FIGURE 3. NuTeV 95% CL lower limits on the mass of the Z' (in GeV) as a function of the Z_χ, Z_ψ mixing angle β. Limits are shown for both no-mixing and allowed-mixing cases.

FIGURE 4. Central value from Rosner's APV data analysis [9]. NuTeV excludes the darker shaded region at 95% CL. For purposes of comparison, no Z-Z' mixing has been allowed.

CONCLUSIONS

NuTeV has successfully completed its data taking and has extracted a value of $\sin^2\theta_W$. The precision of this result represents a factor of two improvement over previous measurements in νN scattering, because of reduced uncertainties associated with measuring the Paschos-Wolfenstein ratio, R^-. Interpreted within the framework of the Standard Model, this result is equivalent to a determination of the W mass and is consistent with direct measurements of M_W. Outside the Standard Model, this measurement can be used to set limits on neutrino oscillations as well as extra neutral vector gauge bosons.

REFERENCES

1. K.S. McFarland, *et al.*, Eur. Phys. Jour. **C31**, 509 (1998).
2. E.A. Paschos and L. Wolfenstein, Phys. Rev. **D7**, 91 (1973).
3. D.A. Harris, J. Yu, *et al.*, Nucl. Instr. Meth. **A447**, 373 (2000).
4. K.S. McFarland, *et al.*, hep-ex/9806013; G. Zeller, *et al.*, hep-ex/9906037.
5. D.E. Groom, *et al.*, Eur. Phy. J. **C15**, 1 (2000).
6. D.A. Harris, *et al.*, submitted to the proceedings of the International Europhysics Conference on High-Energy Physics, Tampere Finland (1999).
7. J. Erler and P. Langacker, Phys. Rev. Lett. **84**, 212 (2000).
8. S.C. Bennett and C.E. Wieman, Phys. Rev. Lett. **82**, 2484 (1999).
9. J. Rosner, hep-ph/9907524; R. Casalbuoni, *et al.*, hep-ph/0001215.
10. P. Langacker, *et al.*, Rev. Mod. Phys. **64**, 87 (1991)
    G.-C. Cho, *et al.*, Nucl. Phys. **B531**, 65 (1998).
    D. Zeppenfeld and K. Cheung, hep-ph/9810277.
11. F. Abe, *et al.*, Phys. Rev. Lett. **79**, 2192 (1997).

The MUNU Experiment

A. Tadsen for the MUNU Collaboration

Instituto Nazionale di Fisica Nucleare-Sezione di Padova, via Marzolo 8, I-35131 Padova, Italy

Abstract. The MUNU experiment has been built to measure the neutrino magnetic moment at a reactor. Its central part, at the same time target and detector, is a 1 m^3 TPC filled with 3 bar of CF4. This TPC is surrounded by active and passive vetos. Data have been taken for more than a year, giving first preliminary results.

THE NEUTRINO MAGNETIC MOMENT

Although the standard model predicts the neutrino magnetic moment to be zero, there exist other models leading to relatively large values [1-6]. With a non vanishing magnetic moment neutrinos will have electromagnetic interactions scattering them into a sterile state in the magnetic field of the sun, thus making them invisible for experiments. This may be an explanation for the observed low neutrino flux from the sun in the Homestake [7], GALLEX [8], SAGE [9] and KAMIOKANDE experiments [10] if neutrinos have a magnetic moment in the order of $10^{-10} - 10^{-12} \mu_B$.

To measure the neutrino magnetic moment directly in an experiment we are using $\bar{\nu}_e e^-$ scattering. The cross section for this interaction increases for a non vanishing magnetic moment. It is given by:

$$\frac{d\sigma}{dT} = \frac{G_F^2 m_e}{2\pi} \left[(g_A^2 - (g_V - x)^2) \frac{m_e T}{E_\nu^2} \right.$$
$$+ (g_V - x + g_A)^2 + (g_V - x - g_A)^2 \left(1 - \frac{T}{E_\nu}\right)^2 \right] \quad (1)$$
$$+ \frac{\pi \alpha^2 \mu_\nu^2}{m_e^2} \frac{1 - T/E_\nu}{T}$$

where E_ν is the neutrino energy and T the electron recoil energy. The first two lines give the contribution from the standard model due to the weak interaction with

$$x = \frac{2M_W^2}{3} \langle r^2 \rangle \sin^2 \theta_W \quad (2)$$

where $\langle r^2 \rangle$ is the square charge radius of the neutrino. The last line of eq.1 is the additional contribution due to the magnetic moment μ_ν.

PRESENT EXPERIMENTAL SITUATION

The first experiment measuring $\bar{\nu}_e e^-$ scattering has been performed by Reines at the Savannah River Plant in the mid seventies [12] giving a limit of $\mu_\nu < 14 \times 10^{-10} \mu_B$. In the meantime two other reactor experiments have given limits of $\mu_\nu < 2.4 \times 10^{-10} \mu_B$ at Kurchakov [14] and $\mu_\nu < 1.8 \times 10^{-10} \mu_B$ at Rovno [15] An experiment has been done at the LAMPF beam-dump giving a limit of $\mu_\nu < 1.08 \times 10^{-9} \mu_B$ [11].

Astrophysical observations lead to much lower limits but with larger theoretical errors. The neutrino burst from SN1987A limits the neutrino magnetic moment to a value of $\mu_\nu < 1 - 20 \times 10^{-13} \mu_B$ assuming neutrinos to be Dirac particles [16–18]. Limits from stellar cooling, $\mu_\nu < 10^{-12} - 10^{-11} \mu_B$, apply to Dirac and Majorana neutrinos but are less stringent [19,20].

THE MUNU EXPERIMENT

To measure the neutrino magnetic moment with high precision we have built a new detector with a low energy threshold [21]. Its main part is a TPC filled with CF4 gas which serves as neutrino target and detector. Due to its tracking capabilities background events can be identified and rejected. To complete the detector the TPC is surrounded by a liquid scintillator veto and a passive shielding. Figure 1 shows the detector setup.

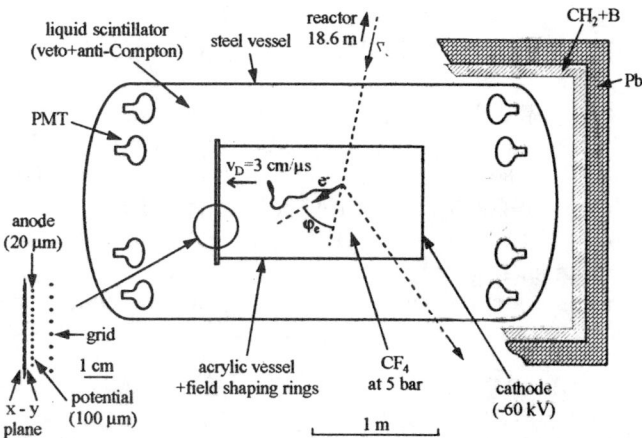

FIGURE 1. The MUNU detector, at the center the TPC which is surrounded by the anti-Compton and the passive shielding

The experiment is situated at a nuclear power plant in Bugey, France, inside the reactor building at 18 m from the core. At the experimental site the neutrino flux is $5 \times 10^{20} \bar{\nu}_e/sec$. The shielding due to the concrete of the building reduces the muon flux to $32/m^2s$ and the soft cosmic ray component to zero. The detector is housed in a shielding of 15 cm of lead and 8 cm of polyethylene to protect it against gammas and neutrons from the laboratory. It consists of a 10 m^3 stainless steel tank with the TPC mounted inside at the center. The tank is filled with a mineral oil based liquid scintillator. It is viewed by 48 photomultiplier tubes of 20 cm diameter directly immersed inside and used to veto cosmic muons and as an anti-Compton detector.

The 1 m^3 TPC is filled with 3 bar of CF4. This gas has a very high density of 3.68 g/l at 1 bar. It has a relatively low Z which reduces multiple scattering and thus improves angular resolution, and no free hydrogen to avoid the $\bar{\nu}_e + p \to e^+ + n$ background reaction. CF4 is also a suitable gas for a TPC having a high drift speed of 4 cm/μs and a low diffusion of 2.5 mm after 1 m, both at 120V/bar. The attenuation length has been measured to be 22^{+14}_{-6} m.

The TPC itself is made from a cylindrical acrylic vessel with an inner diameter of 90 cm and a length of 162 cm. Inside the TPC vessel a copper plate is screwed at one end to form the cathode. At the other end a wire grid, anode wires and a xy plane from perpendicular strips are fixed. Field shaping rings outside the TPC define the drift field. The anode wires have a spacing of 4.95 cm and are read out all together. The strips have a 3.5 mm pitch. Each of the 512 strips is read out with a 8 bit FADCs with a 80 ns sampling and a depth of 1024 words. The anode information delivers the deposited energy and the strip information together with the time the xyz coordinates of the events. In this way tracks are seen in the TPC, allowing direction and fiducial volume determination and particle identification.

DETECTOR PERFORMANCE

Due to the tracking capability of the TPC, single electrons have a clear signature. They can be easily separated from alphas, muons or coincidences from radioactive chains. This reduces the background. A fiducial volume smaller than the TPC has been defined without problems to identify background electrons coming from the walls or entering the TPC.

For the first time in such an experiment electron energy and direction are measured and this is a large advantage. From these values the neutrino energy can be calculated:

$$E_\nu = Tm_e/(p_e \cos(\Theta) - T) \qquad (3)$$

and the neutrino spectrum reconstructed, reducing the systematic error. Electrons not coming from the neutrino direction give negative values for the neutrino energy, this is a powerful cut against background. For electrons above 700 keV 85 % of the events can be rejected. In addition in this way the background is measured together with the signal with high statistics instead of measuring it in the usually very short reactor off periods.

The liquid scintillator surrounding the TPC is used to veto muons, the rate in the TPC is 65 Hz because it is not underground. In addition it is used to veto Compton events, reducing the rate by at least an order of magnitude. The measured attenuation length of the scintillator is 8 m at 430 nm. We trigger at the one photon level and the veto efficiency is 97 % for $E_\gamma > 100 keV$.

To limit the background from natural activities the whole detector has been made from radiochemical pure materials. The liquid scintillator and the acrylic have been tested with neutron activation. Their contamination is lower than 10^{-12} g/g for ^{232}Th and ^{238}U and 3×10^{-9} g/g (scintillator) and 2×10^{-7} g/g (acrylic) for ^{40}K. All other materials have been carefully selected using γ activity measurements with Ge detectors. With these contaminations the expected background is 6 events per day above 500 keV from the neutrino direction.

STATUS

The MUNU experiment is running and taking data since 1.5 y. The anti-Compton works as foreseen and no unexpected contamination has been seen. In the TPC at startup there was a Radon contamination from an Oxisorb gas filter which subsequently was changed against a getter, solving the problem. From this Radon exposition the cathode go contaminated with ^{210}Pb leading to a rate higher than expected. Now the cathode has been changed and the rate of electrons from the cathode side decreased by an order of magnitude. The date from the period before were analyzed with a higher electron energy threshold, of 800 keV instead of 300 keV and with severe cuts. Figure 2 shows the difference in the electron energy spectrum

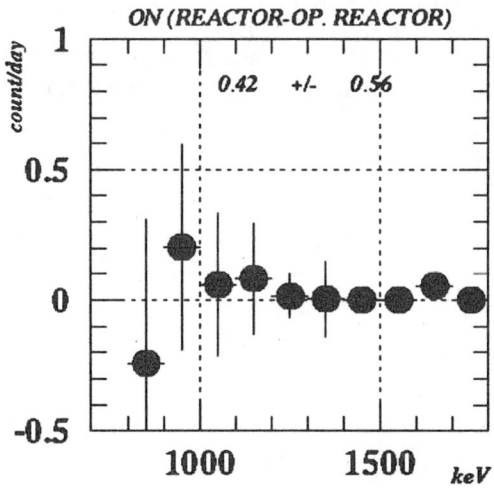

FIGURE 2. Electron energy spectrum from reactor direction - opposite direction

between the reactor direction (signal) and the opposite direction (background). No signal has been seen. For a neutrino magnetic moment of $10^{-10}\mu_B$ 0.9 events per day are expected where 0.18+/-0.78 events are seen. This leads to a preliminary limit of $\mu_\nu < 3.3 \times 10^{-10}\mu_B$ at 90 %C.L.

After the exchange of the cathode data taking resumed and a better limit will be reachable after some months of data taking.

REFERENCES

1. Voloshin M.B., *Sov. J. Nucl. Phys.* **48**, 804 (1988).
2. Babu K.S., Mohapatra R.N., *Phys. Rev. Lett.* **63** 228 (1989)
3. Leurer M. et al., *Phys. Lett. B* **237** 81 (1990)
4. Barbieri R. et al., *Phys. Lett. B* **252** 251 (1990)
5. Barr S.M., Friere E.M., Zee A., *Phys. Rev. Lett.* **65** 2626 (1990)
6. Pal B.K., *Phys. Rev. D* **44** 2261 (1991)
7. B.T. Cleveland et al., *Nucl. Phys. B (proc. Suppl.)* **38** 47 (1995)
8. Vignaud D., *Nucl. Phys. B (proc. Suppl.)* **60** 20 (1998)
9. Abdurashitov J.N. et al., *Nucl. Phys. B (proc. Suppl.)* **38** 60 (1995)
10. Swoboda R., *Nucl. Phys. B (proc. Suppl.)* **66** 271 (1998)
11. Krakauer D.A. et al., *Phys. Lett. B* **252** 117 (1990)
12. Reines F., Gurr H.S., Sobel H.W., *Phys. Rev. Lett.* **37** 315 (1976)
13. Vogel P., Engels J., *Phys. Rev. D* **39** 3378 (1998)
14. Vidyakin G.S. et al., *JETP Lett.* **49** 740 (1989)
15. Derbin A.I. et al., *JETP Lett.* **57** 768 (1993)
16. Lattimer J.M., Cooperstein J., *Phys. Rev. Lett.* **61** 23 (1988)
17. Barbierei R., Mohapatra R.N., *Phys. Rev. Lett.* **61** 27 (1988)
18. Notzold D., *Phys. Rev. D* **38** 1658 (1988)
19. Raffelt G., *Phys. Rev. Lett.* **64** 2856 (1990)
20. Castellani V., Deg'Innocenti S., *The Astr. Phys. J.* **402** 574 (1993)
21. Amsler C. et al., *Nucl. Instr. and Meth. A* **396** 115 (1997)

Status of the NEMO 3 experiment for the study of neutrinoless double-β decay

Corinne Augier* for the NEMO collaboration

*Laboratoire de l'Accélérateur Linéaire,
CNRS-IN2P3 et Université de Paris-Sud, BP 34 F-91898 Orsay Cedex, FRANCE.

Abstract. The NEMO collaboration is presently mounting the NEMO3 detector in the Fréjus Underground Laboratory (Laboratoire Souterrain de Modane or LSM). This detector is devoted to search of neutrinoless double beta decay of various isotopes and especially Molybdenium 100. It will be able to reach a sensitivity to neutrino effective mass in the order of 0.1 eV. The NEMO3 detector is composed of 20 sectors. Three of them take data in Modane since the end of April, without shielding or magnetic field. The full detector is planned to be in operation by the end of the year.

INTRODUCTION

The neutrinoless double beta decay $\beta\beta(0\nu)$ is the only way to prove the existence of Majorana neutrinos. From the limit on the $\beta\beta(0\nu)$ half-life, an upper limit on $\langle m_\nu \rangle$ can be inferred with the relation $(T_{1/2}^{0\nu})^{-1} = (\langle m_\nu \rangle / m_e)^2 \times |M_{0\nu}|^2 \times F_{0\nu}$, where $M_{0\nu}$ is the nuclear matrix element of the relevant isotope of which calculations have large theoretical uncertainties and $F_{0\nu}$ is the phase-space factor analytically calculable and proportional to $Q_{\beta\beta}^5$.

The aim of the NEMO 3 detector, which will operate in the LSM is to search for $\beta\beta(0\nu)$ with various isotopes with large $Q_{\beta\beta}$ values.

THE NEMO 3 DETECTOR

The NEMO 3 experiment is based on the direct detection of the two electrons by a tracking device and on the measurement of their energies by a calorimeter [1]. The NEMO 3 detector is similar in function to the earlier prototype NEMO 2 [2] and is able to accomodate at least 10 kg of double beta decay isotopes.

The detector is cylindrical in design and divided into 20 equals sectors. Thin source foils ($\sim 50\mu m$) are fixed vertically between two concentric cylindrical tracking volumes composed of 6180 open octagonal drift cells, 270 cm long, operating in Geiger mode and providing three-dimensional tracking. The tracking volume is covered with calorimeters made of 1940 large blocks of plastic scintillators coupled to very low radioactivity 3" and 5" PMTs. The energy resolution is $\sigma(E)/E = 5.6\%$ at 1 MeV.

A solenoid surrounding the detector produces a magnetic field of 30 Gauss in order to recognize (e^+e^-) pair production events in the source foils. An external shield, in the form of 20 cm thick low radioactivity iron, covers the detector to reduce γ-rays and thermal neutron fluxes. Outside of this shield, an additional shield is added to thermalize fast neutrons. With those characteristics the NEMO 3 detector is able to identify e^-, e^+, γ, n and α particles.

CURRENT STATUS OF CONSTRUCTION

The construction of the 20 sectors of the NEMO 3 detector has been completed. Currently, 12 sectors are in the LSM and 6 of them are equipped with their source foils and mounted on the detector frame.

The double-β decay isotopes which are being mounted in the detector are the following : 7 kg of ^{100}Mo (corresponding to 12 sectors), 1 kg of ^{82}Se (2.3 sectors), 0.6 kg of ^{116}Cd (1 sector), 0.7 kg of ^{130}Te (1.8 sectors), and small quantities of ^{150}Nd, ^{96}Zr and ^{48}Ca. Also, 2.7 sectors are devoted to external background measurements : one is equipped with an ultra-pure copper foil and 1.7 sectors with 0.9 kg of ^{nat}TeO$_2$.

Three sectors installed on the detector frame have been succesfully running since the end of April 2000 (without a magnetic field and an external shield). The wire chamber and the PMTs coupled to the scintillators are running well. Geiger β tracks obtained with the finalized NEMO 3 trigger and acquisition system are shown in Figure 1.

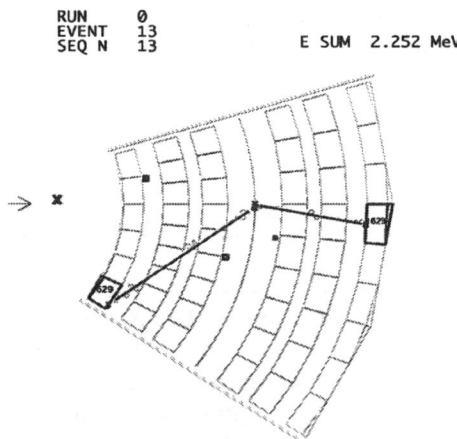

FIGURE 1. Transverse view of two Geiger tracks with associated PM measured with the first 3 sectors of NEMO 3 running in the LSM.

EXPECTED BACKGROUND

There are three origins of expected background which can occur in this search for a $\beta\beta 0\nu$ signal around 3 MeV. The first background comes from the beta decays of ^{214}Bi ($Q_\beta = 3.2$ MeV) and ^{208}Tl ($Q_\beta = 5.0$ MeV) which are present in the source, from the Uranium and Thorium decay chains. They can mimic $\beta\beta$ events by β emission followed by Möller effect or by a $\beta - \gamma$ cascade followed by a Compton interaction. Thus, the experiment requires ultra-pure enriched $\beta\beta$ isotopes as explained below.

A second origin of $\beta\beta 0\nu$ background is due to high energy gamma rays (> 2.6 MeV) interacting with the source foil. Their origin is from neutron captures occuring inside the detector. The interactions of these gammas in the foil can lead to 2 electrons by e^+e^- pair creation, double Compton scattering or Compton followed by Möller scattering. However it was demonstrated with the neutron simulations for NEMO 3 that an appropriate neutron shield (like paraffin) and a 30 Gauss magnetic shield will make the neutron background negligible [4].

Finally, given the energy resolution, the ultimate background is the tail of the $\beta\beta 2\nu$ decay distribution. It defines the half-life limits to which the $\beta\beta 0\nu$ can be studied.

Additionally, the components of the detector have to be ultra-pure in ^{214}Bi, ^{208}Tl and ^{40}K to have a low background in the $\beta\beta 2\nu$ energy spectrum. As expected, the radioactive contamination in the detector is dominated by the low radioactivity glass in the PMTs, of which activities are three orders of magnitude below standard PMT levels.

Radiopurity of the ^{100}Mo sources in ^{214}Bi and ^{208}Tl

Maximum levels of ^{214}Bi and ^{208}Tl contamination in the molybdenum source have been calculated to insure that $\beta\beta 2\nu$ is the limiting background. These limits are ^{214}Bi < 0.3 mBq/kg and ^{208}Tl < 0.02 mBq/kg. These activities in ^{214}Bi and ^{208}Tl correspond to a level of $2 \ 10^{-11}$ g/g in ^{238}U and 10^{-11} g/g in ^{232}Th respectively when we assume the natural radioactive families of ^{238}U and ^{232}Th are in equilibrium.

To reach these specifications, two methods have been developed to purify the enriched Molybdenum isotope. The first one developed by ITEP (Moscow, Russia), is a purification by local melting of solid Mo with an electron beam and drawing a monocrystal from the liquid portion. One gets an ultra-pure ^{100}Mo monocrystal.

The second purification method is a chemical process done at INEEL (Idaho, USA) which leaves the Mo in a powder form that is then used to produce foils with a binding paste and mylar strips which have been etched with an ion beam and a chemical process. To date 3 kg of ^{100}Mo have been purified. No activity has been observed in the purified ^{100}Mo after 1 month of HP-Ge measurements in the LSM and the most stringent limits obtained for radiopurities are ^{214}Bi < 0.2 mBq/kg and ^{208}Tl < 0.05 mBq/kg. However, the chemical extraction factors defined as the ratio of contamination before and after purification were measured with a ^{nat}Mo sample. Applying the ^{208}Tl extraction factor to the ^{208}Tl activity measured in the ^{100}Mo before purification, one obtains after purification an expected level in ^{208}Tl of 0.01 mBq/kg which is again lower than the design specifications.

CONCLUSION

The sensitivity that the NEMO-3 detector will reached after 5 years of data collection, has been calculated with 7 kg of ^{100}Mo and 1 kg of ^{82}Se. After 5 years, in the energy window 2.8 to 3.2 MeV, a total of 6 background events are expected with 7 kg of ^{100}Mo and no background events are expected with 1 kg of ^{82}Se. The $\beta\beta 0\nu$ detection efficiency in the same energy window, 2.8 to 3.2 MeV, is $\epsilon(\beta\beta 0\nu) = 14\%$. The expected numbers of background events and sensitivities are summarized in Table 1.

The NEMO 3 detector will be completed by the end of the year 2000. To date, ^{82}Se, ^{116}Cd, ^{nat}TeO$_2$ and the copper foils are mounted. We are now starting to mount the ^{100}Mo sources.

TABLE 1. Expected number of background events per year and per kg and expected sensitivity (90% C.L.) for NEMO 3 after 5 years in the energy window 2.8 to 3.2 MeV around the $\beta\beta 0\nu$ signal peak.

	7 kg ^{100}Mo	1 kg ^{82}Se
^{214}Bi (in events/yr/kg)	< 0.03	negl.
^{208}Tl (in events/yr/kg)	< 0.04	negl.
$\beta\beta 2\nu$ (in events/yr/kg)	0.11	0.01
External neutrons (in events/yr/kg)	< 0.01	< 0.01
TOTAL	< 0.18	0.01
Number of events in the energy window 2.8 to 3.2 MeV	6 bkg evts expected 6 events observed 5 $\beta\beta 0\nu$ excluded	0 bkg evts expected 0 events observed 2.5 $\beta\beta 0\nu$ excluded
$T_{1/2}^{0\nu}$	> 4. 10^{24} yr	> 1.5 10^{24} yr
$\langle m_\nu \rangle$	< 0.25 - 0.7 eV	< 0.6 - 1.2 eV

REFERENCES

1. Ch. Marquet, "Double beta decay with the NEMO experiment : status of the NEMO 3 detector," in *TAUP 99*, edited by J. Dumarchez, M. Froissart, D. Vignaud, *Nucl. Phys.* B (Proc. Suppl.) 87, (2000) 298-300
2. R. Arnold et al., *Nucl. Inst. Meth.*, **A 354**, (1995) 338-351
3. R. Arnold et al., *Nucl. Phys.*, **A 636**, (1998) 209-223
4. Ch. Marquet et al., accepted for publication in *Nucl. Inst. Meth.*

Observation of high energy atmospheric neutrinos with AMANDA

A. Karle[11], E. Andres[11], P. Askebjer[13], X. Bai[1], G. Barouch[11], S.W. Barwick[8], R.C. Bay[7], K.-H. Becker[2], L. Bergström[13], D. Bertrand[3], D. Bierenbaum[8], A. Biron[4], J. Booth[8], O. Botner[12], A. Bouchta[4], M.M. Boyce[11], S. Carius[5], A. Chen[11], D. Chirkin[7,2], J. Conrad[12], J. Cooley[11], C.G.S. Costa[3], D.F. Cowen[10], J. Dailing[8], E. Dalberg[13], T. DeYoung[11], P. Desiati[4], J.-P. Dewulf[3], P. Doksus[11], J. Edsjö[13], P. Ekström[13], B. Erlandsson[13], T. Feser[9], M. Gaug[4], A. Goldschmidt[6], A. Goobar[13], L. Gray[11], H. Haase[4], A. Hallgren[12], F. Halzen[11], K. Hanson[10], R. Hardtke[11], Y. D. He[7], M. Hellwig[9], H. Heukenkamp[4], G.C. Hill[11], P.O. Hulth[13], S. Hundertmark[8], J. Jacobsen[6], V. Kandhadai[11], J. Kim[8], B. Koci[11], L. Köpke[9], M. Kowalski[4], H. Leich[4], M. Leuthold[4], P. Lindahl[5], I. Liubarsky[11], P. Loaiza[12], D.M. Lowder[7], J. Ludvig[6], J. Madsen[11], P. Marciniewski[12], H.S. Matis[6], A. Mihalyi[10], T. Mikolajski[4], T.C. Miller[1], Y. Minaeva[13], P. Miocinovic[7], P.C. Mock[8], R. Morse[11], T. Neunhöffer[9], F.M. Newcomer[10], P. Niessen[4], D.R. Nygren[6], H. Ogelman[11], C. Pérez de los Heros[12], R. Porrata[8], P.B. Price[7], K. Rawlins[11], C. Reed[8], W. Rhode[2], A. Richards[7], S. Richter[4], J. Rodriguez Martino[13], P. Romenesko[11], D. Ross[8], H. Rubinstein[13], H.-G. Sander[9], T. Scheider[9], T. Schmidt[4], D. Schneider[11], E. Schneider[8], R. Schwarz[11], A. Silvestri[2,4], M. Solarz[7], G. Spiczak[1], C. Spiering[4], N. Starinsky[11], D. Steele[11], P. Steffen[4], R. G. Stokstad[6], O. Streicher[4], Q. Sun[13], I. Taboada[10], L. Thollander[13], T. Thon[4], S. Tilav[11], N. Usechak[8], M. Vander Donckt[3], C. Walck[13], C. Weinheimer[9], C.H. Wiebusch[4], R. Wischnewski[4], K. Woschnagg[7], W. Wu[8], G. Yodh[8], S. Young[8]

(1) Bartol Research Institute, University of Delaware, Newark, DE, USA
(2) Fachbereich Physik, BUGH Wuppertal, Germany
(3) Brussels Free University, Brussels, Belgium
(4) DESY-Zeuthen, Zeuthen, Germany
(5) Dept. of Technology, Kalmar University, Kalmar, Sweden
(6) Lawrence Berkeley National Laboratory, Berkeley, CA, USA
(7) Dept. of Physics, University of California, Berkeley, CA, USA
(8) Dept. of Physics and Astronomy, University of California, Irvine, CA, USA
(9) Institute of Physics, University of Mainz, Mainz, Germany
(10) Dept. of Physics and Astronomy, University of Pennsylvania, Philadelphia, PA, USA
(11) Dept. of Physics, University of Wisconsin, Madison, WI, USA
(12) Dept. of Radiation Sciences, Uppsala University, Uppsala, Sweden
(13) Fysikum, Stockholm University, Stockholm, Sweden

Abstract. In 1997 the Antarctic Muon and Neutrino Detector Array (AMANDA) started operating with 10 strings. In an analysis of data taken during the first year of operation 188 atmospheric neutrino candidates were found. Their zenith angle distribution agrees with expectations based on MonteCarlo simulations. A preliminary upper limit is given on a diffuse flux of high energy neutrinos of astrophysical origin.

I STATUS OF AMANDA

In the Austral summer of 1996-97 the construction of the first generation Antarctic Muon and Neutrino Detector Array (AMANDA) was completed. This detector, referred to as AMANDA-B10, consists of 302 optical sensors on 10 strings located at depths of 1500 to 2000 m in the deep Antarctic ice. The array forms a cylinder of 400 m height and 120 m diameter. Figure 1 shows an optical module, together with a schematic view of the array and a neutrino event observed in 1997.

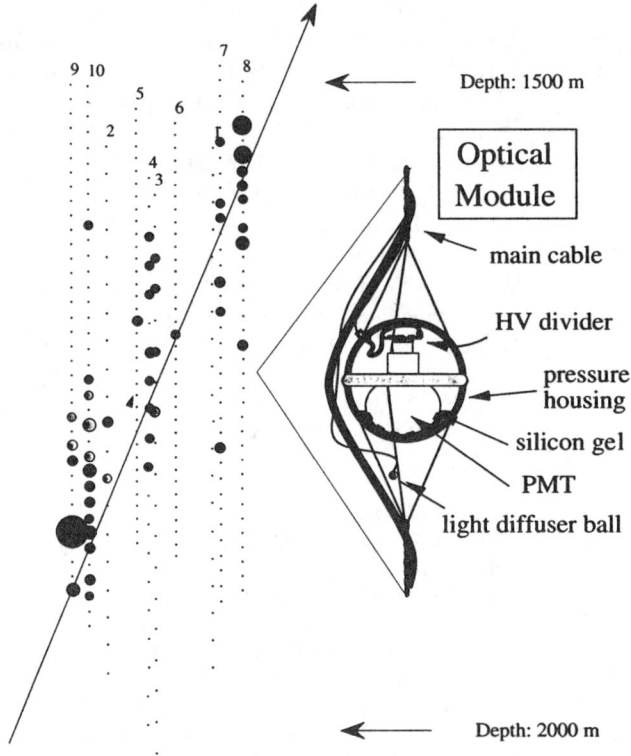

FIGURE 1. A schematic view of the AMANDA-B10 array with an event display of an upgoing muon track. Each dot represents an optical module, a schematic of which is shown. The circles show pulses from the photomultipliers; the size of the circles indicates the amplitude of the pulse and the shading shows the timing.

An optical module consists of an 8-inch photomultiplier tube (Hamamatsu R5912-02) and its voltage divider, housed in a glass pressure vessel. A cable provides the high voltage and transmits the anode current signals of the photomultiplier to the data acquisition electronics at the surface.

In January 2000 the detector was upgraded to an array of 19 strings consisting of 677 optical sensors. The additional strings, located on an outer ring of 200 m diameter, use fiber-optic cables for calibration and for analog signal transmission of the photomultiplier pulses. A proposal exists to construct the IceCube detector which would consist of 4800 photomultipliers to be deployed on 80 strings. It will allow us to reach a 1 km^2 effective telescope area. It's energy threshold is 100 GeV.

II ATMOSPHERIC NEUTRINOS

The results presented in this report were obtained from data collected by AMANDA-B10 during 1997. The effective livetime has been determined to $1.20 \cdot 10^7$ seconds for the selected data. The method of calibration and the characteristics of the optical sensors are very similar to the 4 string prototype array described in ref. [2]. Figure 1 shows a neutrino-generated muon with a recorded track length of about 400 m inside the array. Simulations predict a rate of 24 events per day from atmospheric neutrinos above a threshold of 30-50 GeV, compared to $6 \cdot 10^6$ events from cosmic ray muons. We use a maximum likelihood method, incorporating a detailed description of the scattering and absorption of photons in the ice, to reconstruct muon tracks from the measured photon arrival times.

A small fraction of the downgoing muons ($5 \cdot 10^{-5}$) are reconstructed as upward and form a background to the neutrino-induced events. This background is removed by applying a total of six quality criteria to the time profiles of the observed photons as well as to their spatial distribution in the array. The application of the quality criteria reduces the background by a factor of approximately 10^8, while retaining about 5% of the neutrino signal. When the quality cuts are applied 188 events remain, a rate of approximately 1.4 events per day of livetime. The following table gives the total number of events at trigger level and after the background has been rejected, both, for the experimental data and the prediction of atmospheric neutrino induced events.

	Experimental Data	MC: Atmospheric Neutrinos
Triggered	$1.2 \cdot 10^9$	4574
Upward going	188	235

Based on this analysis and on background studies under way, the backgrounds are estimated to contribute less than 15% of the signal. A complete simulation of the background generated by cosmic ray muons is under way.

The zenith angle distribution for the 188 events is shown in Figure 2, and compared to that for the signal simulation. In the figure the MonteCarlo events (235) was normalized to the observed events. The achieved agreement in the absolute flux of atmospheric neutrinos is consistent with the systematic uncertainties of the absolute sensitivity and the flux of high energy atmospheric neutrinos. The shape of the distribution, which agrees well with the prediction reflects the angular acceptance of the narrow but tall detector. The observation of atmospheric neutrinos at a rate consistent with Monte-Carlo prediction establishes AMANDA-B10 as a neutrino telescope.

III SEARCH FOR DIFFUSE FLUX OF NEUTRINOS OF ASTROPHYSICAL ORIGIN

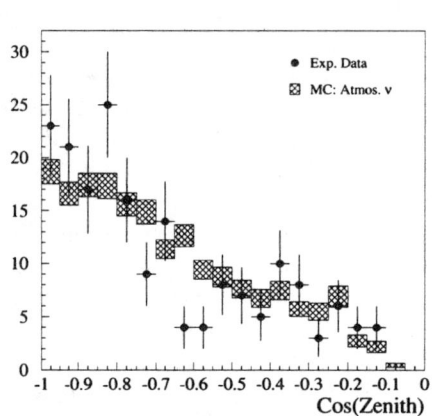

FIGURE 2. Reconstructed zenith angle distribution. The points mark the data and the shaded boxes a simulation of atmospheric neutrino events, the widths of the boxes indicating the error bars.

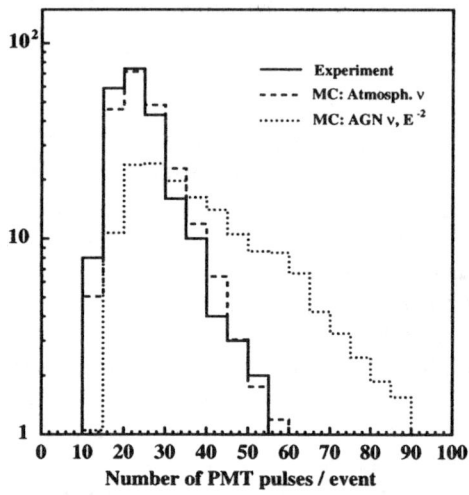

FIGURE 3. Number of observed photomultiplier pulses for events that pass the neutrino search criteria for a) experimental data b) MC simulation for atmospheric neutrinos, and c) MC of an E-2 type diffuse neutrino flux.

Searches for neutrinos from point sources and gamma ray bursts, for magnetic monopoles, and for a cold dark matter signal from the center of the Earth are in progress [4] and yield limits comparable or better than those obtained from smaller underground neutrino detectors run for many years.

The search for a diffuse neutrino flux of astronomical origin follows naturally from the observation of diffuse flux of neutrinos generated in the atmosphere. Neutrinos from generic astrophysical sources are expected to extend to higher energies while the energy spectrum of atmospheric neutrinos falls off steeply with increasing energy. A very simple and robust measure of the energy of the observed muons is the number of optical sensors that observed at least one photoelectron in a given event. Figure 3 shows the distribution of the number of photomultiplier signals as predicted for a) atmospheric neutrinos, b) an assumed energy spectrum for astrophysical neutrinos following a power law of $dN/dE_\nu = 10^{-5} E_\nu^{-2} \, \text{cm}^{-2} \, \text{s}^{-1} \, \text{sr}^{-1} \, \text{GeV}^{-1}$

and c) experimental data. The experimental data agree well with the atmospheric neutrino spectrum. The assumed astronomical neutrino flux would generate a significant excess at high multiplicities of fired photomultipliers. No such excess is observed. From the non-observation of an excess of high energy events, we derive an upper limit on an assumed diffuse E^{-2} spectrum of astrophysical neutrinos. Our preliminary 90% confidence limit is

$$dN/dE_\nu \leq 1.6 \cdot 10^{-6} E_\nu^{-2} \, \text{cm}^{-2} \, \text{s}^{-1} \, \text{sr}^{-1} \, \text{GeV}^{-1}.$$

This limit on the diffuse neutrino flux is below previously stated upper limits by experiments such as BAIKAL [7], SPS-DUMAND [5], AMANDA-A [6], and FREJUS [8]. It is comparable to the AGN prediction by Salamon and Stecker [9] and approaches the prediction of Protheroe [10]. The analysis of the data taken in 1998 and 1999 is under way and will improve the present results significantly.

ACKNOWLEDGMENTS

This research was supported by the following agencies: 1. U.S. National Science Foundation, Office of Polar Programs; 2. U.S. National Science Foundation, Physics Division; 3. University of Wisconsin Alumni Research Foundation; 4. U.S. Department of Energy; 5. Swedish Natural Science Research Council; 6. Swedish Polar Research Secretariat; 7. Knut and Alice Wallenberg Foundation, Sweden; 8. German Ministry for Education and Research; 9. U.S. National Energy Research Scientific Computing Center (supp. by the Office of Energy Research of the U.S. Department of Energy); 10. UC-Irvine AENEAS Supercomputer Facility. 11. Deutsche Forschungsgemeinschaft (DFG) 12. D.F. Cowen greatfully acknowledges the support of the NSF CAREER program.

REFERENCES

1. D.M. Lowder, et al., Nature **353**, 331 (1991)
2. E.Andres et al., *Astroparticle Physics* **13**, 1 (2000).
3. P.Askjeber et al., *Science* **267**, 1147 (1995).
4. Preliminary results in Proc. 26th Int. Cosmic Ray Conf., Salt Lake City (1999), Vol.2, pp.108, 196, 200, 221, 229, 344, 348, 432.
5. Bolesta J.W. et al., Proc. 25-th ICRC, Durban South Africa, 7:29 (1997)
6. Porrata et al., Proc. 25-th ICRC, Durban South Africa, 7:9 (1997)
7. Balkanov et al., TAUP99, September 1999, Paris, France, astro-ph/0001145 (2000)
8. Rhode W. et al. (Frejus-Coll.), AstroPart. Phys. 4:217 (1996)
9. Stecker F.W., Salamon M.H., astro-ph/9501064 (1995)
10. Protheroe RJ, astro-ph/9809144 (1998)

Accelerators, Facilities and Detectors

Muon Sources

Zohreh Parsa

Brookhaven National Laboratory [1]
Physics Department 510 A,
Upton, NY 11973-5000, USA

Abstract. A full high energy muon collider may take considerable time to realize. However, intermediate steps in its direction are possible and could help facilitate the process. Employing an intense muon source to carry out forefront low energy research, such as the search for muon - number non - conservation, represents one interesting possibility. For example, the MECO proposal at BNL aims for 2×10^{-17} sensitivity in their search for coherent muon - electron conversion in the field of a nucleus. To reach that goal requires the production, capture and stopping of muons at an unprecedented $10^{11} \frac{\mu}{sec}$. If successful, such an effort would significantly advance the state of muon technology. More ambitious ideas for utilizing high intensity muon sources are also being explored. Building a muon storage ring for the purpose of providing intense high energy neutrino beams is particularly exciting. We present an overview of muon sources and example of a muon storage ring based Neutrino Factory at BNL with various detector location possibilities.

INTRODUCTION

High intensity muon sources are needed in exploring neutrino factories, lepton flavor violating muon processes, and lower energy experiments [1] as the stepping phase towards building higher energy $\mu^+\mu^-$ colliders.

Atmospheric-neutrino results suggest that the long-baseline accelerator experiments such as MINOS [4], K2K [3], and NGS [5] should also find neutrino oscillations. Further, the LSND experiment that was conducted at a short-baseline accelerator facility, can be confirmed by future accelerator experiments such as MiniBooNE [6], ORLanD [7], and CERN P311 [8]. Moreover, physics associated with some interpretations of the solar-neutrino deficit may be accessible to studies in accelerator-based experiments, if neutrino-beam fluxes can be improved by 1-2 orders of magnitude.

---

[1] Supported by US Department of Energy contract DE-AC02-98CH10886.
[†] E-mail: parsa@bnl.gov

To obtain a factor of 100 improvement in neutrino flux, the best prospect appears to be neutrino-beams derived from a muon-storage-ring, rather than from direct pion decays. However, such an approach requires considerable development before it can be realized in the laboratory.

A schematic concept of a Neutrino Factory Facility based on a muon storage ring, its components and a possible upgrade to a full muon collider is discussed in the following sections. The examples described are based on some of the scenarios being explored by our Neutrino Factory and Muon Collider Collaboration, [10].

NEUTRINO FACTORY - FACILITY

A neutrino factory based on a muon storage ring is a challenging extension of present accelerator technology. Conventionally, neutrino beams employ a proton beam on a target to generate pions, which are focused and allowed to decay into neutrinos and, muons [4]. The muons are stopped in the shielding, while the muon-neutrinos are directed toward the detector. In a neutrino factory, pions are made the same way and allowed to decay, but it is the decay muons that are captured and used. The initial neutrinos from pion decay are discarded, or used in a parasitic low-energy neutrino experiment. But the muons are accelerated and allowed to decay in a storage ring with long straight sections. It is the neutrinos from the decaying muons (both muon-neutrinos and anti-electron-neutrinos) that are directed to a detector.

In a Neutrino Factory, a proton driver of moderate energy (< 50 GeV) and high average power,(e.g., 1-4 MW), similar to that required for a muon collider, but with a less stringent requirements on the charge per bunch and power is needed. This is followed by a target and a pion-muons capture system. A longitudinal phase rotation is performed to reduce the muon energy spread at the expense of spreading it out over a longer time interval. The phase rotation system may be designed to correlate the muon polarization with time, allowing control of the relative intensity of muon and anti-electron neutrinos. Some cooling may be needed, to reduce phase space, about a factor of 50 in six dimensions. This is much smaller than the factor of 10^6 needed for a muon collider. Production is followed by fast muon acceleration to 50 GeV (for example), in a system of linac and two recirculating linear accelerators (RLA's), which may be identical to that for a first stage of muon collider such as a Higgs Factory. A muon-storage ring with long straight sections could point to one or more distant neutrino detectors for oscillation studies, and to one or more near detectors for high intensity scattering studies.

A planar bowtie - shaped ring (illustrated in Figure 1) can be designed and oriented to send neutrino beams to any two detector sites. Since, there is no net bending, the polarization may be preserved. (A disadvantage of the Bowtie - shaped ring is that it may need extra bending. Since there is geometry constrains on the ratio of short to long straight sections, the ring circumference may increase.) With the ring in a tilted plane, both long straight sections would point down into

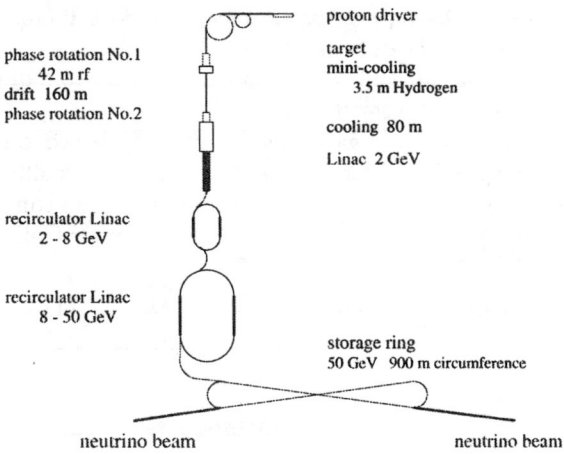

FIGURE 1. A schematic concept of a Neutrino Factory Facility based on a Bowtie Muon Storage lattice.

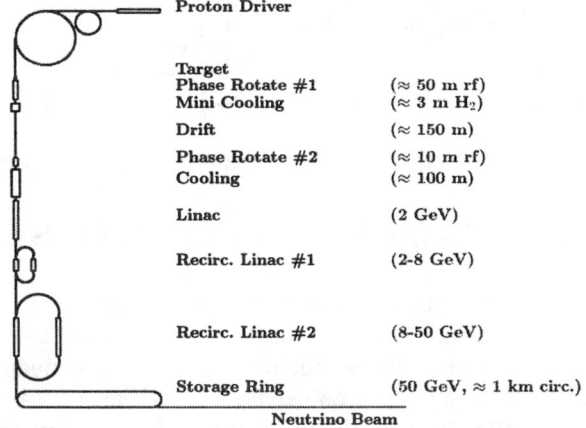

FIGURE 2. Overview of a Neutrino Factory Concept, with a Racetrack Muon - Storage Ring

the earth, such that neutrinos can be directed into two very distant detectors. Triangular-shaped storage rings also have this advantage.

Figure 2, illustrates components of a Neutrino Factory based on a Racetrack - shaped Muon Storage lattice Figure 1 and Figures 2 show examples of the scenarios being explored by our Collaboration, [10].

In the following sections, a description and simulation of target through cooling-channel and a bowtie-shaped muon storage lattice will be discussed.

TABLE 1. Example of parameters for various Proton driver scenarios at BNL and FNAL.

	BNL_1	BNL_2	$FNAL_1$	$FNAL_2$
Energy [GeV]	24	24	16	16
Power [MW]	1	4	1	4
Rep. Rate [Hz]	2.5	5	15	15
p's/fill	10^{14}	$2\ 10^{14}$	$2.5\ 10^{13}$	10^{14}
Bunches	6	6	4	4
Circumference [m]	807	807	474	474
Bunch spacing [m]	135	135	118	118
σ_t [nsec]	1	1	1	1

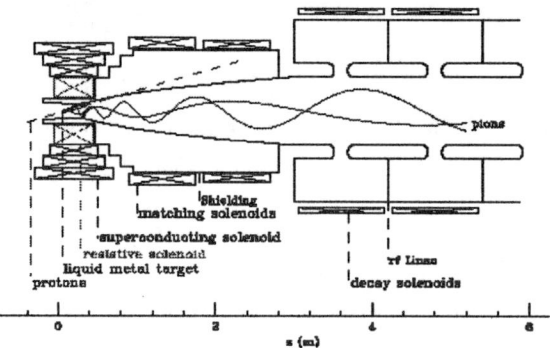

FIGURE 3. A Schematic of Targetry, Pion Capture, and beginning of Phase Rotation.

NEUTRINO FACTORY FRONT-END SYSTEM

The number of pions per proton produced with an optimized system varies linearly with the proton energy, Thus, the number of pions, and the number of muons into which they decay, is essentially proportional to the proton beam power. Table 1 presents possible parameters for proton drivers at BNL and FNAL. The target requirements are very similar to those for the muon collider, except the instantaneous shock heating is somewhat less because protons are distributed in a larger number of bunches. In the scheme presented here, it is assumed that the liquid mercury jet solution is used. The capture solenoid is likely to be the same as described in the muon collider status report [13]. Figure 3, shows the pion production target, solenoidal capture, decay channel and beginning of phase rotation. At the end of this first phase rotation stage, the bunch length increases by about a factor of 6 and the energy spread decreases by the same amount. Whether this first stage of phase rotation can be eliminated is being investigated. Figure 4 illustrates schematics of a Muon source front-end compenents, and Fig. 5 shows the muon emittance variation in the target-to-linac channel.

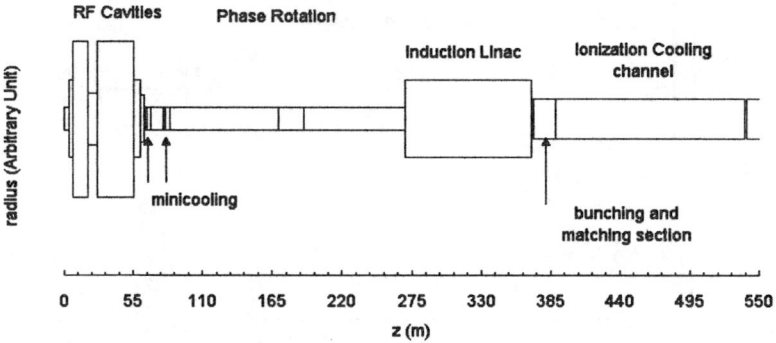

FIGURE 4. Schematics of the Muon Source from Target to Linac.

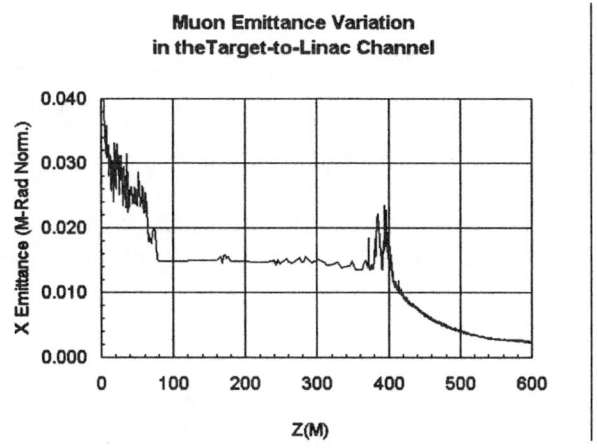

FIGURE 5. Muon emittance variation in Target to Linac channel.

COOLING, ACCELERATION, STORAGE

The challenges of further acceleration and storage of the muon beam will be substantially easier if we reduce the transverse phase area of the beam by an additional factor of 10. This may not be accomplished in a single step of ionization cooling, but involves alternating ionization cooling and rf acceleration, all in a magnetic channel. The acceleration from $\sim$ 100 MeV to e.g., $\sim$ 50 GeV may be accomplished in recirculating linacs with superconducting rf cavities, after which muons are injected into a muon storage ring. The desire for multiply directed neutrino beams with very small angular divergence may require a more novel design for the storage ring, with a plane that is far from horizontal. The R&D needs for a muon collider are very similar, but with additional challenges in cooling and storage ring design. At least four orders of magnitude more cooling (including continual exchange between transverse and longitudinal emittance) are required for a muon collider than a neu-

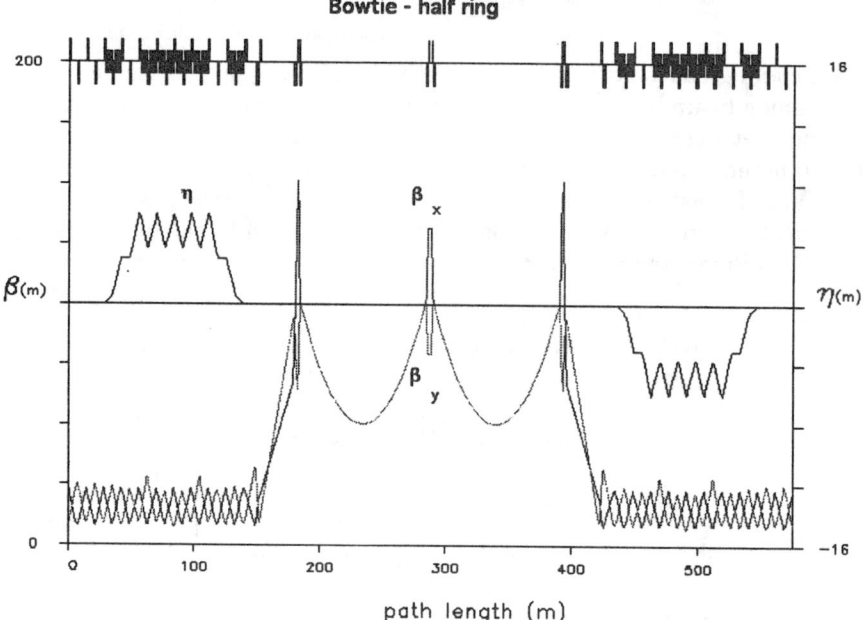

FIGURE 6. Example of Lattice Functions for Bowtie-shaped Half Ring.

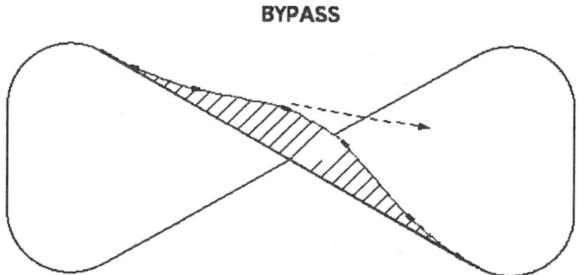

FIGURE 7. Lattice functions for Bowtie-shaped Ring with Bypass. The arrow illustrates direction of a neutrino beam to additional detector site(s) via the Bypass.

trino factory. Also, a different ring is needed to maximize collider luminosity than simply to hold the muons while they decay.

Ionization cooling that has been proposed involves passing the beam through an absorber in which the muons lose transverse- and longitudinal-momentum by ionization loss (dE/dx). The longitudinal momentum is then restored by coherent re-acceleration, leaving a net loss of transverse momentum (transverse cooling). The process is repeated many times to achieve a large cooling factor. The beam energy spread can also be reduced using ionization cooling by introducing a transverse variation in the absorber density or thickness (e.g. a wedge) at a location where there is dispersion (the transverse position is energy dependent). Theoretical studies have shown that, assuming realistic parameters for the cooling hardware, ionization cooling can be expected to reduce the phase-space volume occupied by the initial muon beam by a factor of $10^5 - 10^6$. Ionization cooling is a new technique that has not yet been demonstrated. Special hardware needs to be developed to perform transverse and longitudinal cooling. It is recognized that understanding the feasibility of constructing an ionization cooling channel that can cool the initial muon beams by factors of $10^5 - 10^6$ is on the critical path to the overall feasibility of the muon collider concept.

MUON STORAGE RING

A racetrack–shaped muon storage - ring, with two long straight sections (illustrated in Figure 2), can be configured to deliver one neutrino beam to an arbitrary detector site.

A planar bowtie-shaped ring (illustrated in Figure 1), lattice has two long-straight sections, two short-straight sections and two arcs. Bowtie - shaped and triangle shaped rings can be configured to deliver neutrino beams to two arbitrarily selected detector sites. This can be done by appropriate choice of, 1) the ring plane, 2) the orientation of the ring in that plane and 3) the angle at the crossing point between the two long straight sections. By inclusion of bypasses, additional detector sites may be accessible from a single muon storage-ring source.

A bypass would lie in a plane that includes the original long straight section (but differs from that of the ring), and begin and end on one of the long straight sections. Its magnets would be powered when one desires to send the muons along the deformed bypass path rather than along the normal straight path. In such a bypass, dipoles would produce a roughly triangular path in the bypass plane, one of whose sides would point to the desired detector. The two necessary degrees of freedom are provided by the angle between the bypass and ring planes and by the magnitude of the deflection given by the bypass dipoles. To suppress the dispersion pairs of dipoles should be placed 180 deg apart, in FODO cells.

MUON SOURCE AT BNL

As known, the BNL-AGS proton beam parameters are very suited for use as a source for muon storage ring based neutrino factory and muon collider. Table 2 illustrates basic BNL-AGS proton beam properties. With a muon storage ring - neutrino source at BNL (Figure 8), detectors at Fermilab or Soudan, Minnesota (1715 km), become very interesting possibilities. The feasibility of constructing and operating such a muon-storage-ring based Neutrino-Factory, including geotechnical questions related to building non-planar storage rings (e.g. for BNL-fermilab; at 8° angle for BNL-Soudan, and 31° angle for BNL-Gran Sasso) along with the design of the muon capture, cooling, acceleration, and storage ring for such a facility is being

TABLE 2. BNL- AGS Proton Beam Properties

Parameters	BNL-AGS	Muon Collider
Proton Energy [GeV]	24	16 - 24
Proton/Bunch	1.6×10^{13}	5×10^{13}
Bunch No.	6	2
Proton/cycle	1.0×10^{14}	1.0×10^{14}
Bunch Length [μs]	2.2	1
Bunch spacing [ns]	440	1000

explored by our Neutrino Factory and Muon Collider Collaboration, and requires additional studies for a BNL site specific example.

Figure 8 shows schematics of space angles [20] and baselines for example of a muon storage neutrino source at BNL, with detectors (placed at Fermilab; Soudan; Minnesota (1715 km); or Gran Sasso, Italy (6527 km)) at various global locations.

DISCUSSION

Employing an intense muon source to carry out forefront low energy research, such as the search for muon - number non - conservation, represents one interesting possibility, which requires the production, capture and stopping of muons at an unprecedented $10^{11} \frac{\mu}{sec}$. If successful, such an effort would significantly advance the state of muon technology.

If a neutrino factory is successfully accomplished, it would provide a major advancement. Its ambitious goals would test essentially all aspects of the muon collider concept, muon production, collection, cooling and acceleration. Furthermore, if properly coordinated, the neutrino factory complex might be suitably expanded into the First Muon Collider, perhaps a Higgs factory with center of mass energy $\sim$ 100 GeV.

A 20 GeV muon storage ring intense muon (neutrino) source at BNL is very interesting but expensive? An alternative source of intense muons are the conventional Horn Beams which may be not only competitive with the lower energy muon storage rings but also at a lower cost. For example, with the same number of proton (p) on target and same size (kTon) detector the BNL – AGS 1 $GeV \nu_\mu^{peak}$ Horn $\simeq$ 10GeV Muon Storage Ring (statistically if L/E is fixed). Upgraded Horn facility is potentially powerful. Further R&D on $6 \times 10^{14} p/sec$ driver and target at BNL are important for both the muon storage ring and Horn. [2]- [25].

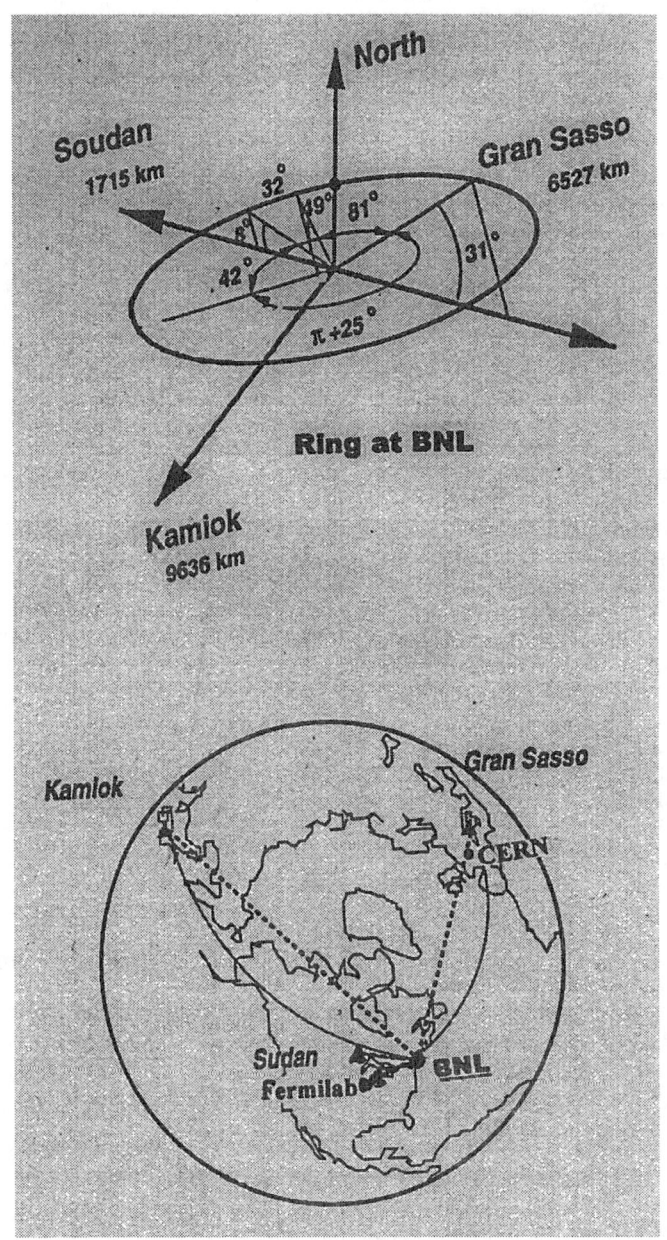

FIGURE 8. Shows space angles and baselines for a Muon - Storage Ring at BNL and possible detector sites (at Fermilab, Sudan, CERN, Kamioka and Gran Saso).

REFERENCES

1. See e.g., MECO Presentation in this Proceedings, [25].
2. The Neutrino Factory and Muon Collider Collaboration http://www.cap.bnl.gov/mumu/
3. http://www.awa.tohoku.ac.jp/html/KamLAND/
4. http://www.hep.anl.gov/NDK/Hypertext/minos_tdr.html
5. http://www.cern.ch/NGS/ngs99.pdf
6. The MiniBooNE project: http://www.neutrino.lanl.gov/BooNE
7. The Oak Ridge Large ν Detector http://www.orau.org/orland/
8. Search for $\nu_\mu \to \nu_e$ Oscillations at CERN PS,http://chorus01.cern.ch/~pzucchel/loi/
9. See e.g., Z. Parsa,*Neutrino Factory – Physics Potentials* in *this (CIPANP2000) Proceedings* and references therein.
10. See e.g., [11] - [15]; and references therein
11. Z. Parsa, *Muon Storage Rings - Neutrino Factories*, in *Next Generation Nucleon decay and Neutrino detector (99)*, ed. M. Diwan and C. Jung, AIP CP 533, pp. 181-195, N.Y. (2000).
12. Z. Parsa, *Intense Muon Beams and Neutrino Factories* in *MUMU99*, ed. D. Cline (2000).
13. C.M. Ankenbrandt et al., *Status of muon collider research and development and future plans*, Phys. Rev. ST Accel. Beams **2**, 081001 (1999), and refrences therein; (1993) (unpublished).
14. K.T. McDonald, ed. for the Neutrino Factory and Muon Collider Collaboration, physics/9911009, 6 Nov, 1999, and references therein; ibid, Private comm.
15. http://lyoninfo.in2p3.fr/nufact99/ R.B. Palmer, C. Johnson, E Keil, BNL-66971,
16. R.B. Palmer, Draft Paramets of a Neutrino Factory, MUC0046.
17. C. Kim, simulation of the target to linac, (99).
18. A. Garren, Private Comm.: Bowtie-lattice with code SYNCH.
19. S. Geer, *Neutrino beams from muon storage rings*, Phys. Rev. D **57**, 6989 (1998)
20. Drawing from Y. Fukui was modified for Fig. 1. See MUMu99 Procd., ed. D. Cline.
21. D. B. Cline (ed.), *Physics Potential and Development of $\mu^+\mu^-$ Colliders* AIP CP **352** (1996).
22. The $\mu^+\mu^-$ Collider Collaboration, *$\mu^+\mu^-$ Collider Feasibility Study*, BNL-52503, FERMILAB-Conf-96/092, LBNL-38946 (July 1996);
23. Z. Parsa, ed., Future High Energy Colliders, AIP CP **397**, AIP-Press, Woodbury, NY (1997).
24. Z. Parsa, Ionization cooling and Muon Dynamics, AIP CP 441, 289-294 (1997).
25. Z. Parsa, New High Intensity Muon sources and Flavor Changing Neutral Currents, World scientific Publishing, pp 147-153 (1998).
26. Z. Parsa, Lasers and Future High Energy Collliders, STS-Press, pp 823-830 (1997).
27. Kamal, B., Marciano, W., Parsa, Z., Resonant Higgs enhancement at the first muon collider, AIP CP 441, pp174- (1997); ibid, AIP 435 pp567-662 (1997).
28. Z. Parsa, *New Ideas for Particle Accelerators* Program Santa Barbara CA (1996); Z. Parsa, W. Marciano, W. Molzon, Y. Kuno, Y. Ogata, R. Djilkibaev et al, *High Intensity Muon Source* worksop, Santa Barbara CA (July 22- Aug 23, 1996).
29. Z. Parsa, *Polarization and Luminosity requirements for the First Muon Collider*, in AIP Conf. Proc. 472, pp. 251-259, (1998).
30. Z. Parsa, Muon Dynamics and Ionization Cooling at Muon Colliders, Procd. of EPAC98, Stockholm, Sweden, Vol 2, pp.1055- (1998).
31. C.N. Ankenbrandt et al., *Ionization Cooling Research and Development Program for a High Luminosity Muon Collider*, FNAL-P904 (April 15, 1998),
32. *Neutrino Factory Feasibility Studies at Fermilab:*
 http://www.fnal.gov/projects/muon_collider/nu_factory/
33. N. Mokhov, π/μ *Yield and Power Dissipation for Carbon and Mercury Targets in 20-T Solenoid with Matching*, MUC0061; ibid, private comm.

HERA-B Status and First Results

Lluís Garrido

Universitat de Barcelona[1]
Diagonal 647, 08028 Barcelona (Spain)
E-mail: garrido@ecm.ub.es

for the HERA-B Collaboration

Abstract. HERA-B is an experiment designed primarily to study CP violation looking at decays of B mesons into the "gold plated" decay $B^0 \to J/\psi K_S^0$. These B mesons are produced in interactions of 920 GeV protons of the HERA proton beam with an internal wire target in the beam halo. In addition to the CP violation measurement, the experiment addresses a wide range of topics in heavy quark physics, complementing the physics accessible at e^+e^- colliders, which are already in operation. A brief overview of the detector and trigger, its status, and some of commissioning results achived as of late spring 2000 are presented.

INTRODUCTION

One of the most interesting observations in particle physics is the violation of the CP symmetry in decays of neutral kaons [1]. In neutral B mesons the CP violation effects are expected to be larger than in kaons, but the decay channels which can exhibit CP asymmetries are extremely rare, typically are suppressed by four to five orders of magnitude. This requires a machine acting as a B factory with a detector designed to select such rare channels.

As opposed to e^+e^- colliders, where the B signature is very clean, experiments at hadron colliders are more challenging, due to the small B production cross section compared to the total one. In HERA-B the background of normal inelastic interactions dominates B production by six orders of magnitude. Assuming $\sigma_{b\bar{b}} \approx 12$ nb at 920 GeV, we arrive at the conclusion that $4 \cdot 10^{14}$ interactions are needed to get the 1500 reconstructed $B^0 \to J/\psi K_S^0$ and achieve an error of 0.13 in sin 2β. To achieve this in 10^7 seconds of running time with a bunch crossing frequency of 10 MHz, 4 interactions per bunch crossing are needed. This imposes serious requirements on the detector and trigger design.

---

[1] The HERA-B protect at the University of Barcelona is supported in part by a CICYT contract AEN99-0483

DETECTOR AND TRIGGER SYSTEM

The HERA-B detector [2] is shown in Fig 1. Its main tracking system consists of 13 stations. The ones located outside the dipole magnet serve for triggering and pattern recognition, while the stations inside the dipole magnet help to propagate tracks trough the magnetic field and match them to the vertex detector. The granularity and technology vary with distance from the beam in order to limit the occupancy of each detector cell and minimize the total number of channels. Complementary to it, a silicon Vertex Detector System (VDS) measures tracks over the first 2 m and is used to find the primary and secondary vertices.

The electron identification is done by the Electromagnetic Calorimeter (ECAL), that uses a shashlik design with scintillating fibers, and a Transition Radiation Detector (TRD) using fiber and straw detector cells in the very forward region. The muon identification is done by a conventional muon system (MUON) with four chamber layers at different depths in the absorber.

A Ring Imaging CHerenkov (RICH) counter is used to identify a tagging kaon with momentum between a few GeV up to about 50 GeV. Finally, a set of pixel chambers installed inside the magnet is used to trigger on high-p_T hadrons.

The charge particle fluxes passing currently through HERA-B tracking devices, are comparable with ones that are expected by LHC experiments in year 2005. Due to this severe radiation environment, HERA-B has suffered a significant delay compared to its original schedule. However, working solutions have been found, and now HERA-B is completely installed and commissioned. A good progress in alignment have been achieved, and tracks reconstructed in the tracking system are easily matched with clusters in ECAL, muon candidates in the muon system, tracks on the VDS and centers of rings in the RICH. The average number of these reconstructed tracks is pretty similar to the expected one.

The event rate of 10 MHz should be reduced by five orders of magnitude before data can be stored. Three different detectors can supply fast pretriggers: clusters in ECAL and coincidences in the muon or high-p_t systems. Any of such pretrigger

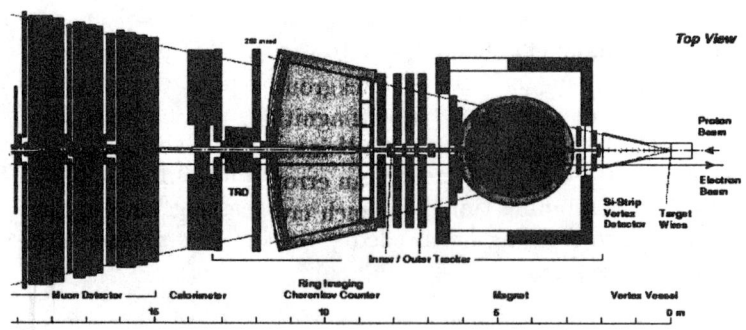

FIGURE 1. The HERA-B detector

defines a geometrical seed Region on Interest (ROI) for the first level trigger (FLT). The FTL attempts to follow the track trough tracking stations downstream of the target.

At present the muon and ECAL pretriggers are completed, and the commissioning of the high-p_t has started with several pretrigger boards. The complete chain of the DAQ has been running stably for several months, and the FLT hardware has been installed and its commissioning is ongoing.

FIRST RESULTS

Since year 1999, a trigger setup consisting of the ECAL pretrigger (requiring one E_t cluster) and tracking search algorithms implemented in the second level trigger (SLT) hardware was operational. Similar trigger has been setup recently for the muon system and, in both cases, a clear $J/\psi \to l^+l^-$ signal has been found (see Fig 2).

Continuing the commisioing of the FLT, a test using the ECAL pretrigger and the full FLT hardware has been performed. Analysis offline show that real tracks have been reconstructed by the FLT. Next steps are related to compute its efficiency. We expect to collect a reasonable sample of J/ψ, with the use of the FLT, before the next shutdown in September 2000.

REFERENCES

1. J.H. Christenson *et al*, *Phys. Rev. Lett.* **97**, 1387 (1955).
2. E. Hartouni *et al*, *HERA-B Design Report* DESY-PRC 95/01 (1995).
3. Maxim Titov, HERA-B Status and First Results, In La Thuile 2000, Results and perspectives in particle physics.

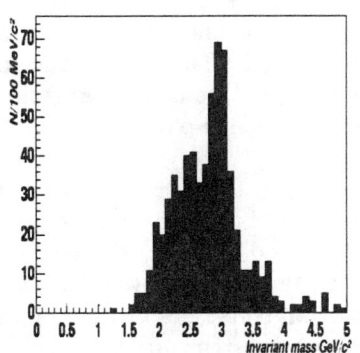 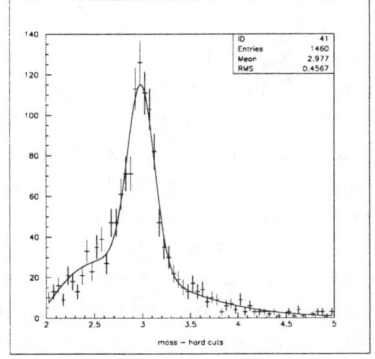

FIGURE 2. Online $J/\psi \to l^+l^-$ signal. Electrons at left and muons at right.

Performance of the BABAR Detector

Urs Langenegger

Stanford Linear Accelerator Center
P.O. Box 4349, MS 95
Stanford, CA 94309, USA
(for the BABAR Collaboration)

Abstract. The performance of the BABAR experiment at SLAC as of May 2000 is presented. The asymmetric e^+e^- accelerator PEP-II achieved record luminosities in very short time and has delivered more than $9\,\text{fb}^{-1}$ so far, half of which has been analyzed. This data set is used for a detailed study of all the detector subsystems and the preparation of a multitude of analyses.

INTRODUCTION

The primary goal of the BABAR experiment [1] is a comprehensive study of CP violation in the decays of neutral B mesons [2]. This involves a measurement of time-dependent asymmetries for the decay of a neutral B meson into a CP eigenstate f_{CP} of the form

$$A(B \to f_{CP}) = \frac{\Gamma(B^0(t) \to f_{CP}) - \Gamma(\overline{B}^0(t) \to f_{CP})}{\Gamma(B^0(t) \to f_{CP}) + \Gamma(\overline{B}^0(t) \to f_{CP})}. \quad (1)$$

Three basic techniques facilitate this measurement:
- A final state with a clean experimental signature and small theoretical uncertainties is the decay $B \to J/\psi K_S^0$. The low branching ratio of the entire decay chain necessitates a high-luminosity accelerator and a large-acceptance detector.
- The flavor of the B decaying into the CP eigenstate is inferred from the flavor of the other B. This *tagging* requires high-performance particle identification over the entire kinematically accessible phase space.
- The time dependent signal for CP violation in this decay vanishes when integrating over all decay time differences (in the case of coherent production of the two B mesons). It is thus mandatory to measure the difference of decay times, which is not possible with current state-of-the-art vertex detectors without an additional boost. This boost is either acquired through large energies of the B mesons or by boosting the $\Upsilon(4S)$ such that its center-of-mass system no longer coincides with the laboratory frame.

Both BABAR and PEP-II have been built with these requirements in mind.

PEP-II B-FACTORY

PEP-II is a storage ring colliding 9.0 GeV/c electrons onto 3.1 GeV/c positrons at a single interaction point within the BABAR detector. Special attention has been paid to the efficiency and stability of the machine (the "factory" aspect). The peak luminosity achieved until now is 2.17×10^{33} cm^{-2}s^{-1} which is about 72% of the design luminosity. Since the start of collisions on May 26, 1999, PEP-II has delivered more than 9 fb^{-1} of which BABAR has recorded 8 fb^{-1}. The total luminosity delivered within 24 hours has already exceeded the design value of 135 pb^{-1}/day.

THE BABAR DETECTOR

Tracking: Silicon Vertex Tracker and Drift Chamber

The silicon vertex tracker consists of five concentric cylindrical layers with AC-coupled double-sided silicon detectors read out by a low-noise radiation-hard custom IC. The radiation absorbed by the detector is monitored by an array of 12 PIN diodes and is well below the budget (based on a lifetime of 10 years). The angular acceptance is limited by machine components to $-0.87 < \cos\theta_{lab} < 0.96$. As the silicon vertex tracker is mounted on PEP-II dipole permanent magnets and not fixed with respect to the drift chamber, it can move up to 70 μm per day. This is corrected with an automatic run-by-run global alignment. The silicon vertex tracker is performing at design values: A single hit reconstruction efficiency of more than 98% is achieved on both views. The single hit resolution amounts to 12 μm at normal incidence both in data and Monte Carlo simulation.

The drift chamber is a 280 cm long cylinder with 7104 hexagonal cells arranged between an inner radius of 23.5 cm and an outer radius of 79 cm. The angular coverage is $-0.92 < \cos\theta_{lab} < 0.96$. A gas mixture of 20% : 80% isobutane:helium is used to minimize multiple scattering. The average single cell hit resolution

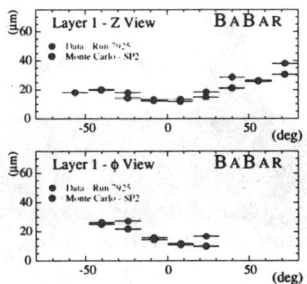
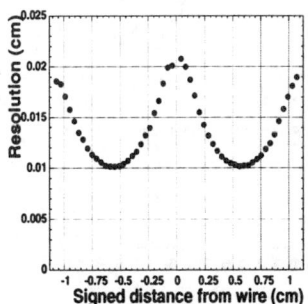

FIGURE 1. Hit resolution as a function of the incident track angle in the vertex detector (left). Drift chamber single cell resolution (right).

exceeds the design specifications of 140 μm. The dE/dx resolution determined with Bhabha events is 7.5% (design: 7%), providing a pion-kaon separation of two standard deviations up to a momentum of 700 MeV/c.

The combined tracking performance can be demonstrated by the mass resolutions obtained: In the decay mode $D^0 \to K^- \pi^+$, the invariant mass of the kaon and pion is at $m_{K\pi} = 1863$ MeV/c^2 with a width of $\sigma_m = 8.8$ MeV/c^2 for the weighted mean of a narrow and wide Gaussian. A mass difference of $m_{K\pi\pi} - m_{K\pi} = 252$ keV/c^2 is achieved in the reconstruction of D^* decays with beam-spot constrained refits of the decay particles.

Detection of Internally Reflected Cerenkov light (DIRC)

The novel feature of the *BABAR* detector is the charged hadron identification device, which is based on the detection of internally reflected Cerenkov light. This detector is constructed as a thin cylinder made of 144 quartz bars, arranged in 12 modules. Charged particles above Cerenkov threshold radiate photons in the quartz bars, which are internally reflected and transmitted to the backward end of the detector where they are measured by an array of 10752 photomultiplier tubes. The average Cerenkov angle resolution per track is 2.8 mrad, giving a pion-kaon separation of three standard deviations at 3 GeV/c. Within the angular acceptance of $-0.84 < \cos\theta_{lab} < 0.9$, the DIRC provides a background suppression by a factor of five with a kaon identification efficiency of roughly 80%.

Electromagnetic Calorimeter

The electromagnetic calorimeter is made of 6580 CsI(Tl) crystals, arranged as a barrel outside the DIRC and a forward endcap. The energy resolution measured with Bhabhas and π^0 is within 10% of the expectation from Monte Carlo simulation.

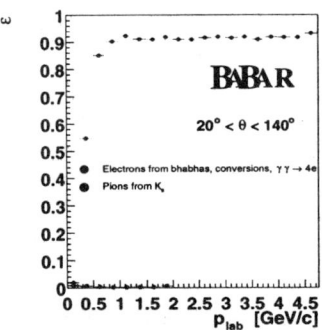
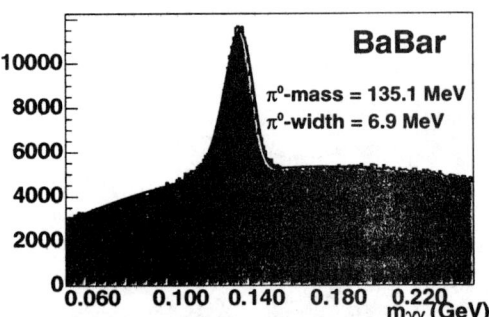

FIGURE 2. Electron identification efficiency and pion misidentification probability as a function of momentum (left). Mass of $\gamma\gamma$ pairs in hadronic events for $E_{\gamma\gamma} > 300$ MeV and $E_\gamma > 30$ MeV (right).

The calorimeter plays an important role in the identification of electrons, mainly with the measurement of E/p, the ratio of the deposited energy and the momentum. The lateral shower development and its expansion into azimuthal moments are used to refine the electron identification. Mean efficiencies of more than 90% are achieved with a pion misidentification of less than 0.2%.

Instrumented Flux Return

The iron flux return for the superconducting main solenoidal magnet is instrumented with nearly 900 resistive plate chambers and is used in the identification of muons and K_L. The muon identification as determined from dimuon events is on the level of 75% with a pion misidentification of a few percent.

PRELIMINARY LOOK AT ANALYSES

The main analysis at BABAR is the measurement of CP violation in the neutral B system, where we expect a yield of roughly 14 "gold-plated" $B \to J/\psi K_S^0$ decays per fb^{-1}. Another interesting analysis measures the time dependence of $B^0 \overline{B}^0$ oscillations using dilepton events. This will allow a very competitive determination of the mixing parameter Δm_B. The good performance of the electromagnetic calorimeter results in an excellent detection of photons, providing a clean measurement of, e.g., $B^0 \to K^{*0} \gamma$.

CONCLUSIONS AND OUTLOOK

PEP-II has been performing exceedingly well so far, and we expect to collect more than $10\,fb^{-1}$ of data before the summer. Many subdetectors are performing at their design specifications already. The data analyses are advancing rapidly, with the first preliminary results expected for the summer conferences 2000.

Acknowledgments: The work presented in this report was supported by the Department of Energy contract DE-AC03-76SF00515.

REFERENCES

1. BABAR Techical Design Report, BABAR Collaboration (D. Boutigny et al.), SLAC-R-457, March 1995.
2. The BABAR Physics Book: Physics at an Asymmetric B Factory, BABAR Collaboration (P.F. Harrison and H. Quinn, eds.), SLAC-R-504, October 1998.

Status of the Belle Experiment

Eric Prebys* for the Belle Collaboration

*Princeton University, Princeton, NJ, USA

Abstract. In this paper, we present the status and first data from the Belle experiment. A brief introduction will summarize the physics motivations for the detector and the current state of the physics analysis will be presented. Because Belle is currently taking and analyzing data, the reader is referred to the experiment's web page [1] for the most up-to-date information on the experiment.

INTRODUCTION

Motivation

The physics motivations for the Belle experiment are discussed in great detail elsewhere [2]. From the experimental standpoint, the primary goal of the experiment is to reconstruct B mesons, tag their flavor, and analyze in detail the time dependence of a specific set of rare decay channels. This goal has guided the specifications for the Belle detector. The requirements and technology choices for the various subsystems are summarized in table 1. The detector is located at the KEKB accelerator at the Japanese High Energy Accelerator Research Lab (KEK), which is an asymmetric e^+e^- collider with a center-of-mass energy corresponding to the $\Upsilon(4S)$ $b\bar{b}$ resonance.

A History and Current Status

The feasibility of a Japanese B-Factory was being seriously studied in Japan in the early 1990's. The Belle collaboration formed in January of 1994 and currently consists of roughly 300 scientists from 50 institutions in 10 countries. Construction began in 1995 and the detector was completed on December 18, 1998. On May 1, 1999, the detector rolled into place and beam commissioning began. The first hadronic event was seen on June 1, 1999, and data collection has continued since then. At the time of this conference, approximately 3.5 pb^{-1} had been collected. As of that time, the peak instantaneous luminosity which had been achieved was 1.8×10^{33} cm^{-2}s^{-1}.

Measurement	Motivation	Technology
Vertex Measurement	About 50 μm resolution to measure time dependent B-meson decay.	Double-sided Silicon microstrip detector.
Charged Tracking	$\Delta P/P \approx .5-1\%$ to kinematically distinguish $B \to \pi\pi$ from $B \to \pi K$ and $B \to KK$.	Single volume drift chamber.
Particle ID ($p <$ 1.5 GeV)	Tag flavor of B-meson decay.	Dedicated time-of-flight counters and dE/dX from drift chamber.
Particle ID ($p >$ 1.5 GeV)	Distinguish $B \to \pi\pi$ from $B \to \pi K$ and $B \to KK$.	Threshold aerogel detectors.
EM Calorimetry	Efficiency down to 20 MeV to detect asymmetric π^0's.	CsI Calorimeter.
Hadronic Calorimetry	Muon identification for flavor tagging and K_L identification for $B^0 \to J/\psi K_L$.	Resistive plate chambers (RPC's) in the magnet return steel.

TABLE 1. Motivation for detector parameters.

I SUBDETECTOR PERFORMANCE

All subdetectors were in place and functioning for the entire running period. In no case is the performance of any subdetector substantially worse than design, and all are considered adequate to achieve our physics goals.

II CURRENT STATUS OF CP SIGNAL

The decays of most initial interest in the study of CP violation are the decays $B^0 \to J/\psi K_S$ and $B^0 \to J/\psi K_L$.

Figure 1a shows the current sample of J/Ψ's which decayed to e^+e^- or $\mu^+\mu^-$. The K_S signal is shown in figure 1b for those events which contain a J/ψ.

We combine J/ψ's and K_S's to form B^0 candidates. Because of the kinematic constraints, both the energy and momentum are determined in the center-of-mass frame if a correct combination is made. Figure 2a shows the reconstructed mass of the $J/\psi K_S$ as well as a plot of momentum versus energy in the center of mass frame. A small but clear signal is seen.

The search for $B^0 \to J/\psi K_L$ is a bit more complicated because we can only measure the direction of the K_L in our detector. If we assume that the K_L combines with a J/ψ to form a particle with a mass of a B^0, this constrains the momentum of the K_L in the lab frame. We then look at the total momentum of the B^0 in the center-of-mass frame, remembering that real B^0's have a fixed momentum of .45 GeV/c when produced at the $\upsilon(4S)$ resonance. Figure 2b shows the center-of-mass momentum spectrum for $B^0 \to J/\psi K_L$ candidates. Again, a clear signal is seen.

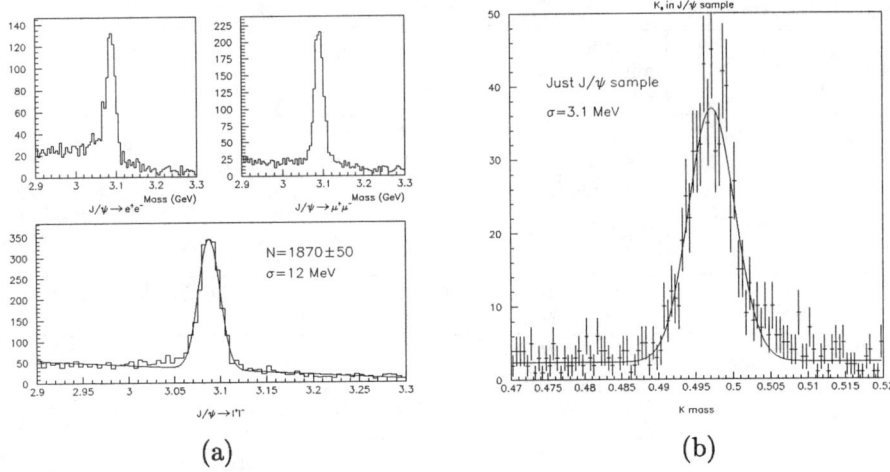

FIGURE 1. Figure (a) shows the current J/ψ sample as observed in l^+l^- decays. Figure (b) shows the K_S signal from events containing a J/ψ.

III SUMMARY AND OUTLOOK

The KEKB accelerator and the Belle detector are functioning well. We have demonstrated our ability to analyze the data to look for the physics of interest. We look forward to the rich results which await us when we have accumulated more data.

IV ACKNOWLEDGEMENTS

We gratefully acknowledge the efforts of the KEKB group in providing us with excellent luminosity and running conditions and the help with our computing and network systems provided by members of the KEK computing research center. We thank the staffs of KEK and collaborating institutions for their contributions to this work, and acknowledge support from the Ministry of Education, Science, Sports and Culture of Japan and the Japan Society for the Promotion of Science; the Australian Research Council and the Australian Department of Industry, Science and Resources; the Department of Science and Technology of India; the BK21 program of the Ministry of Education of Korea and the Basic Science program of the Korea Science and Engineering Foundation; the Polish State Committee for Scientific Research; the Ministry of Science and Technology of Russian Federation; the National Science Council and the Ministry of Education of Taiwan; the Japan-Taiwan

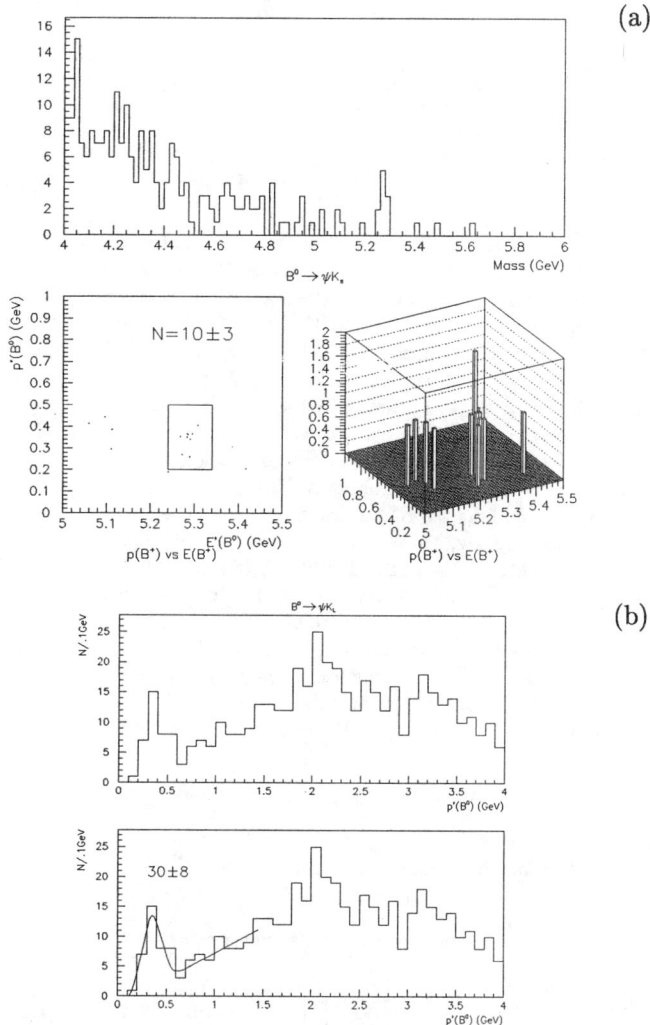

FIGURE 2. Figure (a) shows the $B^0 \to J/\psi K_S$ signal and (b) shows the $B^0 \to J/\psi K_L$ signal.

Cooperative Program of the Interchange Association; and the U.S. Department of Energy.

REFERENCES

1. http://bsunsrv1.kek.jp/
2. M.T. Cheng et al., (BELLE), "Letter of Intent for a Study of CP Violation in B Meson Decays", KEK Report 94-2, April 1994.

KLOE at DAΦNE

the KLOE Collaboration

M. Adinolfi, A. Aloisio, F. Ambrosino, A. Andryakov, A. Antonelli, M. Antonelli, F. Anulli,
C. Bacci, A. Bankamp, G. Barbiellini, G. Bencivenni, S. Bertolucci, C. Bini, C. Bloise, V. Bocci,
F. Bossi, P. Branchini, S.A. Bulychjov, G. Cabibbo, A. Calcaterra, R. Caloi, P. Campana,
G. Capon, G. Carboni, A. Cardini, M. Casarsa, G. Cataldi, F. Ceradini, F. Cervelli, F. Cevenini,
G. Chiefari, P. Ciambrone, S. Conetti, S. Conticelli, E. De Lucia, G. De Robertis, R. De Sangro,
P. De Simone, G. De Zorzi, S. Dell'Agnello, A. Denig, A. Di Domenico, S. Di Falco, A. Doria,
E. Drago, V. Elia, O. Erriquez, A. Farilla, G. Felici, A. Ferrari, M. L. Ferrer, G. Finocchiaro,
C. Forti, A. Franceschi, P. Franzini, M. L. Gao, C. Gatti, P. Gauzzi, S. Giovannella,
V. Golovatyuk, E. Gorini, F. Grancagnolo, W. Grandegger, E. Graziani, P. Guarnaccia,
U.v. Hagel, H.G. Han, S.W. Han, X. Huang, M. Incagli, L. Ingrosso, Y. Y. Jiang, W. Kim,
W. Kluge, V. Kulikov, F. Lacava, G. Lanfranchi, J. Lee-Franzini, T. Lomtadze, C. Luisi,
C. S. Mao, M. Martemianov, M. Matsyuk, W. Mei, L. Merola, R. Messi, S. Miscetti,
A. Moalem, S. Moccia, M. Moulson, S. Mueller, F. Murtas, M. Napolitano, A. Nedosekin,
M. Panareo, L. Pacciani, P. Pagès, M. Palutan, L. Paoluzi, E. Pasqualucci, L. Passalacqua,
M. Passaseo, A. Passeri, V. Patera, E. Petrolo, G. Petrucci, D. Picca, G. Pirozzi, C. Pistillo,
M. Pollack, L. Pontecorvo, M. Primavera, F. Ruggieri, P. Santangelo, E. Santovetti,
G. Saracino, R. D. Schamberger, C. Schwick, B. Sciascia, A. Sciubba, F. Scuri, I. Sfiligoi,
J. Shan, T. Spadaro, S. Spagnolo, E. Spiriti, C. Stanescu, G.L. Tong, L. Tortora, E. Valente,
P. Valente, B. Valeriani, G. Venanzoni, S. Veneziano, Y. Wu, Y.G. Xie, P.P. Zhao, Y. Zhou.

presented by Patrizia de Simone

Laboratori Nazionali di Frascati dell'INFN, Frascati, Italy

INTRODUCTION

At the Frascati ϕ-factory DAΦNE [1] the kaon pairs are produced almost [1] back-to-back with a momentum of $\simeq$ 110 MeV/c, and with well defined quantum numbers, those of the photon. Because of these unique features of the ϕ-factory, the kaons can be efficiently tagged [2], and kaon quantum interferometry experiments can be performed.

The main physics motivation of the KLOE experiment is the measurement of the CP and CPT violation parameters from the classical method of the double ratio and from interferometry [2]. At full luminosity, $\mathcal{L}_o = 5 \times 10^{32}$ cm^{-2} s^{-1}, given that

---

[1] Due the small crossing angle of the e^+e^- beams, 12.5 mrad, the ϕ mesons are produced with a momentum of $\simeq$ 13 MeV.
[2] The observation of one kaon guarantees the existence of the other with determined direction and identity.

the cross section for $e^+e^- \to \phi$ at the ϕ resonance peak is $\simeq 3.2\mu$b, about 1600 ϕ's will be produced per second. This will allow the measurement of $\Re(\varepsilon'/\varepsilon)$ from the double ratio with a statistical error of $\simeq 10^{-4}$ in two years (assuming 1 year of 10^7 s) of data taking, while keeping the systematic error down to the same level.

THE KLOE DETECTOR

Concerning the CP studies in K° decays, the main tasks of the KLOE detector are the control of the efficiencies for the decays of interest, and the rejection of the background from the copious K_L decays to states other than $\pi\pi$. The main components of the KLOE detector [3] [4] are the electromagnetic calorimeter (EmC) and the drift chamber (DC), both embedded in a magnetic field of 0.6 Tesla, generated by a superconducting solenoid.

The beam pipe at the interaction region is a 10 cm ($\simeq 16\lambda_S$) beryllium sphere 0.5 mm thick. This allows to define a fiducial volume for K_S without complications from regeneration and to minimize the multiple scattering and energy loss for charged particles.

Two low-beta quadrupole triplets are placed inside the inner cylinder of the DC, close to the interaction region. They are instrumented with lead-scintillator tile calorimeters to improve the photon acceptance. This will help the rejection of $K_L \to 3\pi^o$.

THE ELECTROMAGNETIC CALORIMETER

The main tasks of the KLOE EmC are the measurement of the kaon neutral vertices with a resolution of $\simeq 1$ cm, and the rejection of the background due to the $3\pi^o$ mode. To reach these purposes the EmC has to provide good energy resolution, excellent timing performance, good resolution for the γ apices measurements, high efficiency for the detection of low energy γ (down to 20 MeV), and hermetic coverage. The KLOE EmC is a very fine sampling lead/scintillating fiber calorimeter. The composite (lead:fiber:glue) of the modules has a density of $\simeq 5$ gr/cm^3 resulting to a sampling fraction of $\simeq 13\%$ for a minimum ionizing particle. The barrel is a cylinder of 4 m inner diameter, made of 24 modules with fibers running parallel to the beam line. The end-caps are made of 32 C shaped modules of variable length with fibers running perpendicular to the beam line. The total acceptance is 98% of the full solid angle. The modules are read-out at both ends by photomultipliers. The read-out granularity is 4.4 × 4.4 cm^2 varying slowly in size across the modules. The spatial resolution in reconstructing the e.m. shower apices is ~ 1.2 cm, and ~ 2.5 cm along the direction of the fibers as obtained by time difference. The energy calibration is done equalizing the response of the photomultipliers with MIPs and deducing the energy scale with Bhabha and $\gamma\gamma$ events.

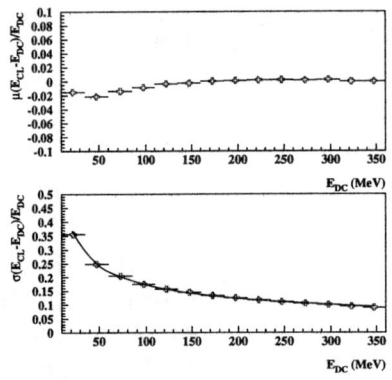
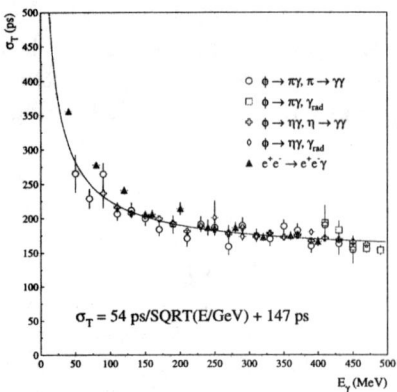

FIGURE 1. The non-linearity in the energy response (up). The measured EmC energy resolution ($\sigma_E/E \simeq 5.7\%/\sqrt{E(GeV)}$) (down).

FIGURE 2. The measured EmC time resolution ($\sigma_t \simeq 54 \text{ ps}/\sqrt{E(GeV)} + 147 \text{ ps}$).

The energy resolution curve has been obtained using $\phi \to \pi^+\pi^-\pi^o$ events and it is in fine agreement with the design request ($\sigma_E/E \simeq 5.\%/\sqrt{E(GeV)}$). The non-linearity of the energy response is within $\sim 1\%$ over the entire energy range of interest (fig. 1). After time equalization with MIPs, the time resolution as a function of the energy is deduced from the ϕ radiative decays and Bhabha radiative decays (fig. 2). Such a good timing performance allows to measure the flight path of the $K_L \to$ neutrals, thanks to the method illustrated in figure 3.

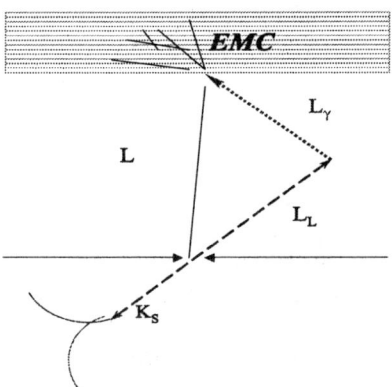

FIGURE 3. Neutral vertex position measurement: L is deduced from the γ conversion point, L_K is deduced from the direction of K_S that decays in $\pi^+\pi^-$. In addition the γ arrival time is measured and β_{K_L} is known.

THE DRIFT CHAMBER

Because of the decay length of the K_L ($\lambda_L \simeq 340$ cm), to maximize the fiducial volume for the charged decay detection, the chosen radius of the DC is 2 m. The K_L decay vertices are uniformly distributed in the tracking volume, and to obtain high and uniform track and vertex reconstruction efficiency the wires are in all-stereo configuration, in order to have an uniform filling of the chamber with almost square cells. The stereo angles varies with radius from 50 to 120 mrad going outward, and the dimension of the cells are $\simeq 2 \times 2$ cm^2 for the first 12 layers and $\simeq 3 \times 3$ cm^2 for the outer 46 layers. Because of the low momenta of the particles (between 50 and 300 MeV/c) a helium based gas mixture has been chosen to minimize the multiple scattering: $90\% He - 10\% iC_4H_{10}$ for a total radiation length of ~ 900 m, including $\simeq 52000$ wires. The KLOE DC needs all its walls to be very thin in order for them to be almost transparent to γ's from π^o decays to be detected in the calorimeter. The whole mechanical structure is made of carbon-fiber/epoxy $\simeq 8$ mm thick, for a total material ≤ 0.1 X_o.

The time response of the KLOE DC is fully calibrated using cosmic ray events. The spatial resolution in the r, ϕ plane is $\leq 150 \mu$m over a large part of the cell. The hit efficiency is $\geq 99\%$. The efficiency for associating an hit to a track is $\simeq 97\%$. The measured momentum resolution, obtained using Bhabha events, agrees with the expected value $\sigma_{p_\perp}/p_\perp \simeq 0.3\%$ (fig. 4). The good DC performance is also shown by the invariant mass resolution for $K_S \to \pi^+\pi^-$, $\sigma \simeq 1$ MeV/c^2 (see fig. 5).

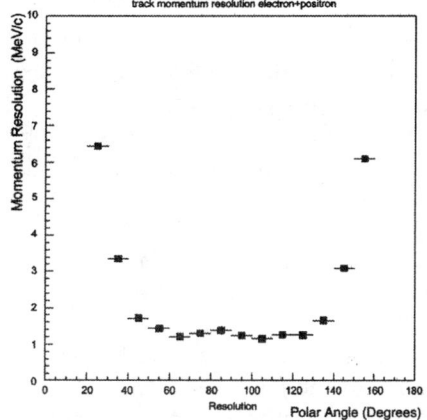

FIGURE 4. $\sigma_{p_\perp}/p_\perp$ against the polar angle.

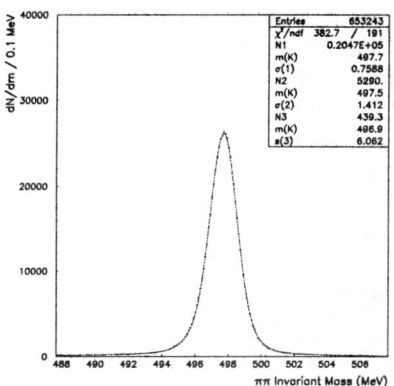

FIGURE 5. Invariant mass for $K_S \to \pi^+\pi^-$.

TRIGGER AND DAQ

The KLOE trigger [5] is designed to have an inefficiency $few \times 10^{-3}$ on CP violation decays and to reject/down-scale to few KHz the Bhabha events (~ 20 KHz), cosmic rays and machine background ($\sim$ 3 KHz). The bunch crossing occurs at DAΦNE every 2.7 ns and thus the KLOE trigger must operate in continuous mode. The KLOE trigger is a two level trigger. The *fast trigger* T1 (about 200 ns after the ϕ) is synchronized with the DAΦNE clock and provides the start to the digitizing electronics of the EmC. The *validation trigger* T2 (1.5 μs after T1) provides the stop to the TDCs of the DC (allowing for drift time) and the start to the Data Acquisition system. T1 and T2 triggers are based on the topology and energy deposits in the EmC and/or on the number and distribution of the DC hits. The final trigger configuration is fully operational and the study of its performance is in progress.

The Data Acquisition (DAQ) system [6] has to handle $\simeq$ 23000 FEE channels on a $\sim$ 5 KHz data rate, half due to the ϕ decays and half due to properly scaled down Bhabha and cosmic rays events. The signal digitization occurs in 2μs, and the bandwidth is 50 MByte/s (at 5 KByte/event). The system has been operated successfully up to more than 24 hours continuous running.

FIRST DATA TAKING OF KLOE AT DAΦNE

The grand total of the events has been collected mainly in November and December 99 ($\mathcal{L} \simeq 3.5 \times 10^{30}$ cm^{-2} s^{-1}) and correspond to an integrated luminosity of 2.4 pb^{-1}, about 8 millions of ϕ decays. The events collected has been used to fully calibrate and to study the main performances of KLOE in detecting the K mesons decays. Here are reported few results related with the K_L and K_S tagging procedures.

The K_L can be tagged identifying the $K_S \to \pi^+\pi^-$ events, looking for charged vertecies with 2 tracks of opposite sign reconstructed inside a cylinder ($r \leq 6$ cm and $|z| \leq 8$ cm) around the interaction region, and invariant mass between 400 and 600 MeV/c^2. The K_L tag efficiency looking for $K_S \to \pi^+\pi^-$ events is $\simeq 50\%$. The background contamination is less than 1%, and is mainly due to $\phi \to K^+K^-$ events. The $K_S \to \pi^+\pi^-$ sample has been used to evaluate the K_S lifetime: $\lambda_S = 5.70 \pm 0.05 \pm 0.03$ mm.

The excellent time resolution is used to tag the K_S looking for K_L's interacting in the calorimeter (K_L crash) by measuring their speed. The selection looks for one neutral cluster with speed $0.195 \leq \beta \leq 0.245$, and energy $E_{cluster} \geq 100$ MeV. The K_S tag efficiency looking for K_L crash is $\simeq 30\%$, and the background contamination, mainly due to machine background and cosmic μ's, is $\sim 15\%$.

The figure 6 shows the plot of the square of the missing invariant mass in the $\pi\pi$ hypothesis, against the missing momentum of the charged track pairs coming from the K_L vertecies [3]. Thanks to the good momentum resolution of the KLOE DC, the three main K_L charged decay modes are nicely separated. Applying a very loose cut, $P_{miss} \leq 10$ MeV/c and $|M|^2_{miss} \leq (70$ MeV/c$^2)^2$, the first CP violating events distributed on a narrow peak (494.9 ± 2.5 MeV/c^2) has been identified (fig. 7).

---

[3] The K_L momentum and direction are given by the reconstruction of the $K_S \to \pi^+\pi^-$ events.

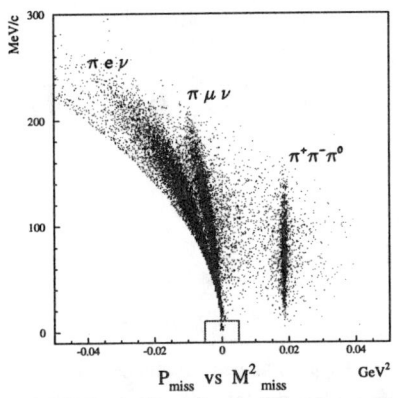

FIGURE 6. K_L charged decay modes in the kinematic plane. The small box defines the CP violation region of interest.

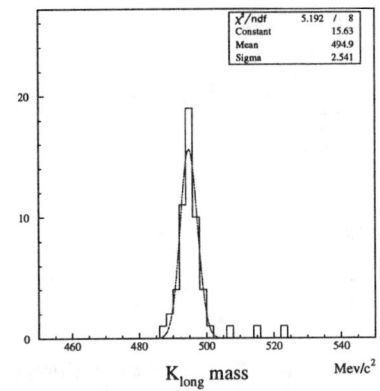

FIGURE 7. Mass distribution of the selected $K_L \to \pi^+\pi^-$ events.

CONCLUSION

The KLOE experiment at the Frascati ϕ-factory DAΦNE has collected $\simeq 2.4$ pb^{-1} of integrated luminosity. These data has been used to study the detector performances in reconstructing the kaon decays, as introduced here and more in detail in [7], and also to produce the first physical results concerning the radiative decays of the ϕ's [8].

The main physics item of the KLOE experiment is the measurement of the CP and CPT violation parameters in the kaon system. However an initial luminosity of $\mathcal{L} \simeq 10^{31}$ cm^{-2}s^{-1} will permit to collect in 1 year of running, enough data to study the structure functions of the weak and e.m. kaon decays (still unknown or contradictory like in $K_{\mu3}$), and to perform tests of the Chiral Perturbation Theory.

REFERENCES

1. C.Milardi, proc. *Physics and Detector for DAΦNE*, ed. S.Bianco et al., (Frascati 1999).
2. J.Lee-Franzini, *The second DAΦNE Physics Handbook*, ed. L.Maiani et al., (Frascati 1995).
3. The KLOE Collaboration, *A general purpose detector for DAΦNE*, LNF-92/019.
4. The KLOE Collaboration, *The KLOE detector, Technical Proposal*, LNF-93/002.
5. The KLOE Collaboration, *The KLOE trigger system*, LNF-96/043.
6. The KLOE Collaboration, *The KLOE data acquisition system*, LNF-95/014.
7. A.Passeri, for the KLOE Collaboration, proc. *XXXth International Conference on High Energy Physics*, July 27 - August 2, 2000, Osaka, Japan.
8. P.Gauzzi, for the KLOE Collaboration, proc. *XXXth International Conference on High Energy Physics*, July 27 - August 2, 2000, Osaka, Japan.

Future Kaon Programs at BNL, FNAL

S.H. Kettell

Brookhaven National Laboratory

Abstract. Future kaon decay programs at BNL and FNAL are discussed. The primary focus of these programs is the measurement of the golden modes, $K_L^\circ \to \pi^\circ \nu \bar{\nu}$ and $K^+ \to \pi^+ \nu \bar{\nu}$. The observation of $K^+ \to \pi^+ \nu \bar{\nu}$ by E787 at BNL is the first step in a series of measurements which will completely determine the unitarity triangle within the kaon system.

The next step after E787 in the measurement of $B(K^+ \to \pi^+ \nu \bar{\nu})$ will be the E949 experiment at BNL that is currently under construction. This experiment, building on the experience of E787 and making use of the intense AGS proton beam, is scheduled to run in FY01–03 and to observe $\mathcal{O}(10)$ SM events with a small and well-understood background. The proposed CKM experiment at FNAL would take the next step, using a decay-in-flight technique and a 22 GeV/c RF-separated kaon beam from the Main Injector, to observe $\mathcal{O}(100)$ SM events.

At the same time, two concepts for the measurement of $B(K_L^\circ \to \pi^\circ \nu \bar{\nu})$ have been developed. One of these, building on the experience of KTeV with a 'pencil' K_L beam, has been proposed at FNAL as KAMI. The other, with a measurement of the kaon momentum in a large angle K_L beam derived from a bunched proton beam, has been proposed at BNL as KOPIO.

INTRODUCTION

The decays $K_L^\circ \to \pi^\circ \nu \bar{\nu}$ and $K^+ \to \pi^+ \nu \bar{\nu}$ are two of the 'golden modes' for measuring CKM parameters. Measurement of the branching ratio $B(K^+ \to \pi^+ \nu \bar{\nu})$ provides a clean and unambiguous determination of the CKM matrix element $|V_{td}|$, in particular of the quantity $|\lambda_t| \equiv |V_{ts}^* V_{td}|$. Measurement of the direct-CP-violating decay $K_L^\circ \to \pi^\circ \nu \bar{\nu}$ will cleanly determine the imaginary part of λ_t, $Im(\lambda_t)$.

The theoretical uncertainty in $K^+ \to \pi^+ \nu \bar{\nu}$ is quite small ($\sim 7\%$) and even smaller in $K_L^\circ \to \pi^\circ \nu \bar{\nu}$ ($\sim 2\%$), as the hadronic matrix element can be extracted from the $K \to \pi e \nu_e$ branching ratio. The $K \to \pi \nu \bar{\nu}$ branching ratios have been calculated to next-to-leading-log approximation [1], complete with isospin violation corrections [2] and two-loop-electroweak effects [3]. Fits based on the best current data for the CKM matrix elements give branching ratios of [4]

$$B(K^+ \to \pi^+ \nu \bar{\nu}) = (8.2 \pm 3.2) \times 10^{-11} \qquad (1)$$
$$B(K_L^\circ \to \pi^\circ \nu \bar{\nu}) = (3.1 \pm 1.3) \times 10^{-11}.$$

These branching ratios are very small and, with two neutrinos in the final state, both of these experiments are challenging.

MEASUREMENT OF $|V_{td}|$ FROM $K^+ \to \pi^+ \nu \bar{\nu}$

The E787 experiment at BNL was designed to search for $K^+ \to \pi^+ \nu \bar{\nu}$ and reported the first observation of $K^+ \to \pi^+ \nu \bar{\nu}$ from analysis of the 1995 data set [5]. A new analysis of the 1995 data combined with the 1996 and 1997 data sets, has reduced the background levels by about a factor of three. A plot from the 1995-97 data set [6] of the range vs. energy of events passing all other $K^+ \to \pi^+ \nu \bar{\nu}$ criteria is shown in Figure 1. One event is observed in the signal

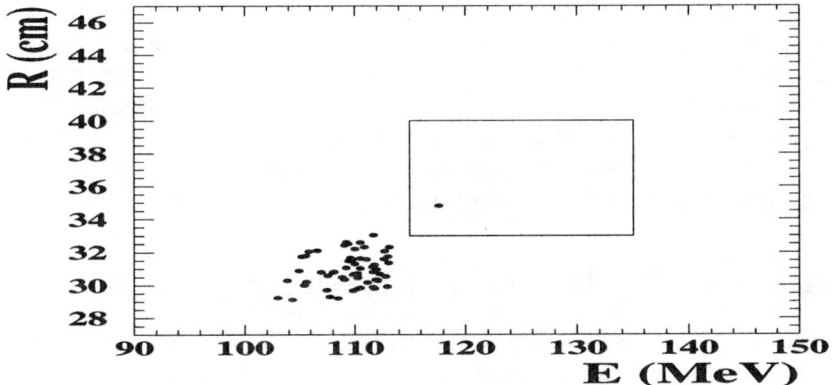

FIGURE 1. Range vs. Kinetic energy plot of the final sample. The events near $E = 108$ MeV are $K_{\pi 2}$ background. The box indicates the accepted region for $K^+ \to \pi^+ \nu \bar{\nu}$ events.

region with a measured background of 0.08 ± 0.02 events. The branching ratio is $B(K^+ \to \pi^+ \nu \bar{\nu}) = (1.5^{+3.4}_{-1.2}) \times 10^{-10}$. From this measurement, a limit on $|V_{td}|$ of $0.002 < |V_{td}| < 0.04$ can be derived, as well as the following limits on $\lambda_t \equiv V_{ts}^* V_{td}$: $|Im(\lambda_t)| < 1.22 \times 10^{-3}$, $-1.10 \times 10^{-3} < Re(\lambda_t) < 1.39 \times 10^{-3}$, and $1.07 \times 10^{-4} < |\lambda_t| < 1.39 \times 10^{-3}$. The E787 experiment has finished running and the final sensitivity, based on the complete 1995-98 data set, should reach the SM expectation for $K^+ \to \pi^+ \nu \bar{\nu}$.

E949 at BNL

A new experiment under construction, E949, is expected to run in 2001-03, symbiotically with RHIC. E949 will use the AGS proton beam, between fills of RHIC, for approximately 20 hours/day. The E787 experiment has already demonstrated sufficient background rejection ($\sim 10\%$ of the SM signal) for a very precise measurement of $B(K^+ \to \pi^+ \nu \bar{\nu})$. Taking advantage of the very large AGS proton flux and the experience gained with the E787 detector, E949 with modest upgrades should observe $\mathcal{O}(10)$ SM events in a two year run. The background is small and well-understood.

CKM at FNAL

The CKM experiment was proposed in 1998 and has been pursuing R&D towards a full technical proposal in 2001 as E905 at FNAL. It would run simultaneously with the Tevatron collider using protons from the Main Injector that are not needed for the collider and extract them over a long spill (~1 sec). CKM plans to collect $\mathcal{O}(100)$ SM events with a background of $\mathcal{O}(10)$ events, starting sometime after 2005. This experiment will use an intense RF-separated 22 GeV/c kaon beam derived from the Main Injector. This novel K^+ decay-in-flight technique will obtain redundant kinematic measurements from independent momentum and velocity spectrometers. The kaon momentum will be measured in a Si spectrometer and the pion momentum in straw-tube drift chambers in the vacuum decay region. The velocities of the kaon and pion will be measured in RICH counters. Two large Pb-scintillator photon veto systems reduce backgrounds from $K^+ \to \pi^+\pi^\circ$ decays and a muon veto system reduces background from $K^+ \to \mu^+\nu_\mu$ decays.

MEASUREMENT OF $Im(V_{td})$ FROM $K_L^\circ \to \pi^\circ \nu \bar{\nu}$

Presently, the best limit on $K_L^\circ \to \pi^\circ \nu \bar{\nu}$ is derived in a model-independent way [7] from the E787 measurement of $K^+ \to \pi^+ \nu \bar{\nu}$:

$$B(K_L^\circ \to \pi^\circ \nu \bar{\nu}) < 4.4 \times B(K^+ \to \pi^+ \nu \bar{\nu}) \qquad (2)$$
$$< 2.6 \times 10^{-9} \text{ (90\% CL)}.$$

The best direct limits come from the KTeV experiment at FNAL. KTeV used a narrow 'pencil' beam to define the transverse vertex position of $\pi^\circ \to \gamma\gamma$ decays in a one-day test run and observed one background event, probably from a neutron interaction. From this special run, a 90%-CL limit [8] of $B(K_L^\circ \to \pi^\circ \nu \bar{\nu}) < 1.6 \times 10^{-6}$ was established. A better limit is obtained using the $\pi^\circ \to e^+e^-\gamma$ decay, which is inherently a factor of 80 less sensitive but has the significant advantage of a precise vertex location. Since the vertex location is known a larger, more intense kaon beam can be used; and the background levels are lower as the transverse momentum is known with better precision. In the full 1997 KTeV data set no events were seen, and at the 90% confidence level, $B(K_L^\circ \to \pi^\circ \nu \bar{\nu}) < 5.9 \times 10^{-7}$ [9]. The P_T distribution of $\pi^\circ \to e^+e^-\gamma$ events passing all other cuts can be seen in Figure 2. The expected background was $0.12^{+0.05}_{-0.04}$, mainly from $\Lambda \to n\pi^\circ$ and $\Xi^\circ \to \Lambda\pi^\circ$.

The next generation of $K_L^\circ \to \pi^\circ \nu \bar{\nu}$ experiments, all using the $\pi^\circ \to \gamma\gamma$ decay mode, will start with E391a at KEK, which hopes to reach a sensitivity of $\sim 10^{-10}$, using a technique similar to KTeV. Although the reach of E391a is not sufficient

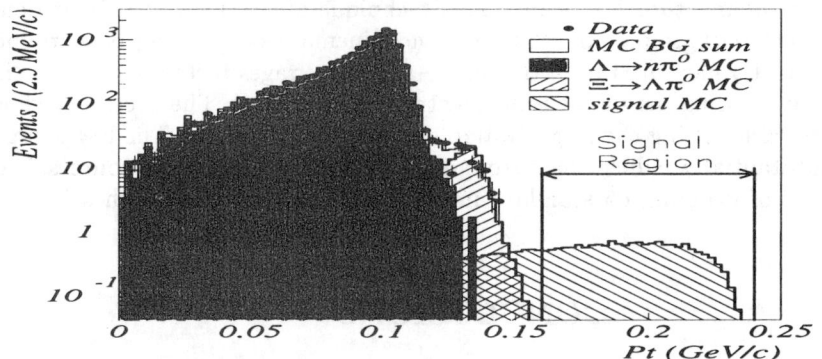

FIGURE 2. Final KTeV $K_L^\circ \to \pi^\circ \nu \bar{\nu}$ ($\pi^\circ \to e^+ e^- \gamma$) data sample collected during 1996–1997 after all cuts. No $K_L^\circ \to \pi^\circ \nu \bar{\nu}$ events are seen above $P_T = 160$ MeV/c.

to observe a signal at the standard model level, the experiment will be able to rule out large enhancements from new physics and learn more about how to do this difficult experiment. It is designed around a pencil K_L beam, a high-resolution crystal calorimeter, and very efficient photon veto systems. This experiment would eventually move to the JHF and aim for a sensitivity of $\mathcal{O}(10^{-14})$. Two other major efforts to observe and measure $K_L^\circ \to \pi^\circ \nu \bar{\nu}$ are KAMI and KOPIO.

KAMI at FNAL

The KAMI collaboration submitted an expression of interest at FNAL for an experiement to measure B($K_L^\circ \to \pi^\circ \nu \bar{\nu}$). KAMI, like CKM, will make use of a slow extracted spill from the Main Injector, simultaneous with the operation of the Tevatron. KAMI plans to reuse the excellent KTeV CsI calorimeter, with a single high intensity pencil K_L beam directed through a hole in the middle. The decay volume upstream of the calorimeter will be surrounded by a hermetic, highly efficient array of photon veto detectors. An additional photon detector will catch photons escaping along the beam. The current design of the KAMI detector includes a fiber tracker to expand the number of secondary modes to be studied. KAMI expects to detect $\mathcal{O}(100)$ SM events in a couple of years of running with a background of ~40% of the SM signal.

KOPIO at BNL

The KOPIO experiment at BNL has been given scientific approval and is currently undergoing funding review. It plans to run at the AGS after the completion of E949 and in the same mode, with ~20 hours per day available between RHIC fills. KOPIO will reconstruct the kaon center of mass using a bunched proton beam and a very low momentum K_L beam. This technique allows for two independent criteria to reject background, photon veto and kinematics—allowing background levels to be directly measured from the data—and encourages further confidence in the signal by measuring the momentum spectrum of the decay. The necessary kaon flux will be obtained using the large available AGS proton current. The low-energy beam also substantially reduces backgrounds from neutrons and other sources. After three years of running, 65 standard-model events are expected with a S/B $\geq$ 2:1.

CONCLUSIONS

The unprecedented sensitivities of rare kaon decay experiments and the recent discovery of $K^+ \to \pi^+ \nu \bar{\nu}$ have opened doors to the measurement of the unitarity triangle completely within the kaon system. Significant progress in the determination of the fundamental CKM parameters will come from the generation of experiments that is now starting. These measurements can provide critical, unambiguous determination of the standard-model CP violation parameters. Comparison with the B-system will then over-constrain the triangle and test the SM explanation of CP violation:

- Comparison of the angle 2β from the ratio B($K_L^\circ \to \pi^\circ \nu \bar{\nu}$)/B($K^+ \to \pi^+ \nu \bar{\nu}$) and the CP asymmetry in the decay $B_d^\circ \to \psi K_S^\circ$ will provide one of the most important tests [7,10].

- Comparison of the magnitude $|V_{td}|$ from $K^+ \to \pi^+ \nu \bar{\nu}$ and the ratio of the mixing frequencies of B_s to B_d mesons will also provide an important test with small theoretical uncertainty [4].

ACKNOWLEDGMENTS

I wish to thank many people for discussions regarding this talk: particularly, Greg Bock, Laurie Littenberg, Ron Ray, Tony Barker, Bob Tschirhart, Robin Appel and Peter Cooper. This work was supported under U.S. Department of Energy contract #DE-AC02-98CH10886.

REFERENCES

1. Bucahalla G. and Buras A., *Nucl. Phys.* **B412**, 106 (1994).
2. Marciano W.J. and Parsa Z., *Phys. Rev.* **D53**, R1 (1996).
3. Bucahalla G. and Buras A., *Phys. Rev.* **D57**, 216 (1998).
4. Bucahalla G. and Buras A., *Nucl. Phys.* **B548**, 309 (1999).
5. Adler S., *et al.*, *Phys. Rev. Lett.* **79**, 2204 (1997).

6. Adler S., et al., *Phys. Rev. Lett.* **84**, 3768 (2000).
7. Grossman Y., and Nir Y., *Phys. Lett.* **B398**, 163 (1997).
8. Adams J, et al., *Phys. Lett.* **B447**, 240 (1999).
9. Alavi-Harati A, et al., *Phys. Rev.* **D61**, 072006 (2000).
10. Bucahalla G. and Buras A., *Phys. Lett.* **B333**, 221 (1994); Bucahalla G. and Buras A., *Phys. Rev.* **D54**, 6782 (1996); Nir Y. and Worah M.P., *Phys. Lett.* **B423**, 319 (1998); Bergmann S. and Perez G., hep-ph/0007170.

A Polarized Electron-Nucleon Scattering Experiment at TESLA

Ralf Kaiser
for the TESLA-N Study Group[1]

DESY Zeuthen, Platanenallee 6, 15738 Zeuthen, Germany

Abstract. Longitudinally polarized electrons, accelerated as a small fraction of the total current in the e$^+$ arm of the TESLA collider, can be directed onto a solid state target that may be either longitudinally or transversely polarized. In this way, polarized electron-nucleon scattering measurements can be realized with projected luminosities that are about two orders of magnitude higher than those of comparable experiments. This will allow the measurement of a large variety of polarized parton distribution and fragmentation functions with unprecedented accuracy, many of them for the first time. A main result will be the precise measurement of the x- and Q^2-dependence of the as yet experimentally totally unknown transverse quark spin distributions.

Within the nuclear and particle physics communities there exists an increasing conviction in the necessity of a new facility to study polarized lepton-nucleon/nucleus scattering with very high luminosity and a high enough center-of-mass energy to cover a sufficient kinematic domain. A fixed-target experiment at TESLA [1] offers a highly competitive and cost-effective option.

The electrons for the fixed-target experiment TESLA-N [2] will be accelerated together with the positrons in the North arm of the TESLA collider. This makes it possible to use a static magnet system for the separation of the beams (cf. figure 1). To maximize the luminosity, it is foreseen to fill every bucket in the bunch train (one every 0.77 ns) with about 20,000 electrons each. The resulting 20 nA represent a negligible fraction of the total TESLA beam. In combination with polarized solid state targets of NH_3 and ^6LiD with an areal density of about 1 g/cm^2 luminosities of up to $7.5 \cdot 10^{34}$ cm^{-2} s^{-1} can be reached. Even with a conservatively estimated overall efficiency of 0.25 and a 'safety factor' of 6 (accounting for possible technical difficulties), this corresponds to a very high integrated luminosity of 100 fb^{-1} per effective year. At a beam energy of 250 GeV, TESLA-N combines this very high

---

[1] A list of contributors to the Study Group can be found under http://www.ifh.de/hermes/future/.

luminosity with a high center-of-mass energy of 22 GeV. In addition, data taking at higher (400-500 GeV) and lower (45 GeV) beam energies may be possible as well.

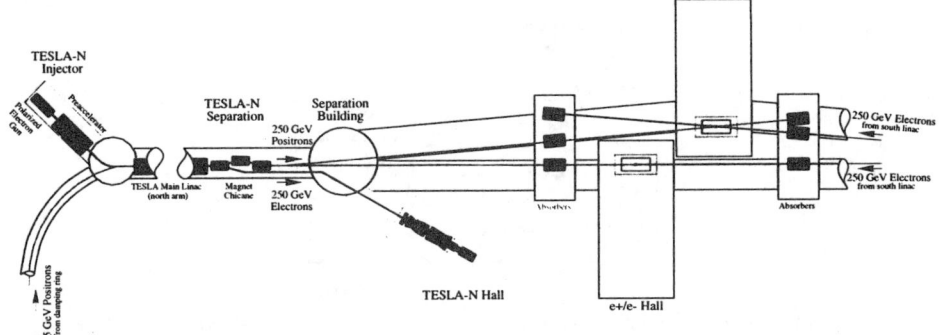

FIGURE 1. Schematic top view of the machine-related elements, showing the polarized source with pre-accelerator, the beam separation and the TESLA-N hall.

The physics program of a polarized fixed-target experiment at TESLA will have several parts. The most obvious possibility is the extension of existing measurements with the much higher luminosity and kinematic range of the experiment. This includes a precision mapping of the x- and Q^2-dependence of the polarized structure function g_1, which in turn allows the indirect extraction of the gluon polarization ΔG from a QCD fit (see figure 2). A direct measurement of $\Delta G(x)$ will be possible using pairs of high-p_T hadrons [3] with results that are competitive with the projections for RHIC [4]. The key part of the physics program is the measurement of the still completely unknown transverse spin structure of the nucleon. The main difference between longitudinal (Δq) and transverse quark spin distributions (δq) is their different QCD evolution. The tensor charge $\delta\Sigma(Q^2) = \sum_q \int_0^1 dx (\delta q(x, Q^2) - \delta \bar{q}(x, Q^2))$ is an all-valence object and is expected to be of similar size as the prediction of the simple, relativistic quark model [5]. While first results on transverse quark spin distributions can be expected from HERMES [6,7] and COMPASS [8] within 3-5 years from now, a complete high precision mapping of their Q^2- and x-dependence requires high statistics measurements that are beyond their scope. Transverse quark spin distributions are not accessible in an inclusive measurement. The necessity of semi-inclusive measurements increases the requirements on the particle identification of the experiment. Measurements of skewed parton distribution functions (SPDs) may have the potential to determine the total angular momentum of quarks and gluons. This would in turn allow the determination of the orbital angular momentum contributions from quarks and gluons to the spin of the nucleon [9]. However, this topic is very challenging for both experimentalists and theorists and may well require a decade of intense research on both sides. Finally, the additional possibilities of using unpolarized targets and of experiments with a real photon beam turn TESLA-N into a versatile next-generation facility at the intersection of particle and nuclear physics.

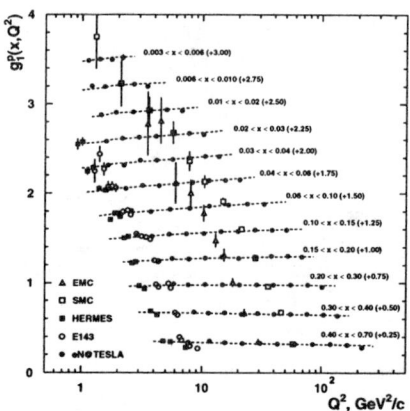

FIGURE 2. Projected statistical accuracy for a measurement of $g_1^p(x, Q^2)$ at TESLA-N, based on a luminosity of 100 fb^{-1} and a minimum detector acceptance of 5 mrad.

In a fixed-target electron-nucleon scattering experiment at 250 GeV, acceptable resolutions in particle momentum and scattering angle may only be achieved by using a multi-stage spectrometer. It is anticipated that the spectrometer will have three stages, each equipped with a dipole magnet. Each stage will need adequate particle identification capabilities, which will be realized with RICH detectors, TRDs and electromagnetic calorimeters. The conditions of the experiment make it impossible for a single tracking device to have both the required very fast response and the necessary position resolution. Therefore it is planned to adopt the COMPASS approach to combine fast (e.g. scintillating fibres), with precise tracking detectors (e.g. drift chambers) [8]. About 20% of the incoming electrons produce bremsstrahlung in the target. Due to the deflection in the spectrometer magnets the resulting lower momentum electrons form a 'sheet of flame' on their way down the spectrometer. This will be taken into account in the design of the spectrometer by a horizontal orientation of the dipole fields that deflect the electrons towards a beam dump in the ground, and by providing a vacuum chamber that contains also the sheet-of-flame electrons and the radiated photons. As a result, the spectrometer will be split into two halves, left and right of the beam.

The author would like to thank all collaborators of the TESLA-N study group.

REFERENCES

1. R. Brinkmann *et al.*, DESY 1997-048, ECFA 1997-182
2. TESLA-N Internal Report 00-01
3. A. Bravar, D. v.Harrach, A. Kotzinian, *Phys.Lett.* **B421**, 349-359 (1998)
4. L. Bland, hep-ex/9907058
5. S. Aoki, M. Doui, T. Hatsuda and Y. Kuramashi, *Phys. Rev.* **D56**, 433 (1997)
6. HERMES Collaboration, DESY-PRC 99/08 (1999)
7. V.A. Korotkov, W.-D. Nowak, K.A. Oganessyan, DESY 99-176, hep-ph/0002268
8. COMPASS Collaboration, CERN/SPSLC 96-14 (1996).
9. X. Ji, Proc. of SPIN-96, p.68-74, (QCD161:S921:1996)

Development of the ATHENA Vertex Detector

P. Riedler

University of Zürich
Zürich CH-8057, Switzerland[1]

Abstract.
The ATHENA experiment [1] is a new experiment at the anti-proton decelerator (AD) at CERN. Its goal is the production, detection and spectroscopic investigation of anti-hydrogen in a magneto-static trap. The experimental technique used will allow to compare the atomic structure of matter and anti-matter to a level of 10^{-15} and thus provides a test for CPT conservation. In order to unambiguously detect the annihilation of anti-hydrogen atoms a detector system consisting of double sided silicon strip detectors and CsI-crystals was developed. The complete vertex detector has to be operated at 77 K. An overview of the experiment is given and the development of the vertex detector is presented.

I INTRODUCTION

The most precise tests of CPT symmetry are up to now provided by comparison of the magnetic moments of electrons and positrons [2] and the comparison of the neutral kaon masses [3]. An experiment with hydrogen atoms and their anti-atoms is the first direct comparison of matter and antimatter. The ATHENA experiment will approach this test of matter-antimatter symmetry in several steps.

In the first step $\approx 10^7$ (100 MeV/c) anti-protons from the AD are extracted into the experiment and pass through a degrading system consisting of a silicon beamcounter and several degrading foils before entering the $\bar{p}$ catching-trap. This trap consists of 10 cylindrical electrodes inside a 3 T superconducting magnet. The low-energetic tail of the $\bar{p}$-cloud is trapped by switching a 15 kV electrode and further cooled to cryogenic temperatures using electron cooling. In parallel positrons from a 8 mCi Na^{22}-source are accumulated and cooled using a solid Neon layer and N_2-buffer-gas in a positron accumulator. As soon as the two plasmas ($\bar{p}$, e^+) are sufficiently cold both clouds are loaded into the recombination trap. All traps are operated at less than 4.2 K to minimise the kinetic energy of the

---

[1] email: petra.riedler@cern.ch

particles. First tests have successfully shown the capture of several thousand $\bar{p}$ from one extracted spill in the ATHENA $\bar{p}$ catching trap.

Recombination to anti-hydrogen can occur e.g. via spontaneous recombination or three-body-recombination [4]. In both cases the density of the plasmas and the kinetic energy of the particles are crucial for the recombination rate. These effects can be studied in detail in the ATHENA apparatus. The final step for the ATHENA experiment is the comparison of the 1S-2S transition of the $\bar{H}$-atom and H-atoms in a reference cell via 2-photon spectroscopy. The spectroscopic measurements are planned for phase two of the experiment, starting in 2002. The first recombination of $\bar{H}$ is foreseen for autumn 2000.

II THE VERTEX DETECTOR

$\bar{H}$-atoms are neutral and once created are not confined in the recombination trap and will hence annihilate on the walls of the trap. The annihilation produces several charged and neutral pions from the $\bar{p}$-annihilation and two 511 keV γ's emitted back-to-back from the e^+-annihilation. In order to unambiguously identify a $\bar{H}$ annihilation, the vertex detector must measure the charged annihilation vertex and check the collinearity of the two γ's with respect to this vertex. The signals also have to be registered in coincidence within a time window of less than 1 μs.

A double layer silicon vertex detector (SVD) is used to register the charged minimum ionising pions from the $\bar{p}$-annihilation. Outside the SVD a crystal calorimeter (CC) consisting of pure CsI crystals read out by silicon photo-diodes is placed to register the 511 keV γ's. The vertex detector consisting of the SVD and the CC is situated around the recombination trap. The detector can also be placed around the $\bar{p}$-catching-trap in order to study the interaction of very low energetic $\bar{p}$ (less than 15 keV) with different gases or rest-gas-atoms. The configuration of the traps and the 3T magnet allow only a very limited space for the vertex detector. The radial space is 6.5 cm and the overall length is 24.1 cm. A sketch of the cross-section of the vertex detector with its 16-fold symmetry can be seen in Figure 1.

The operating conditions are 77 K and 10^{-7} mbar. In total 8960 electronic channels are read out using VA2_TA ASIC chips with a signal shaping time of $\approx 2\mu$s.

The CC consists of 192 CsI crystals. This high granularity is necessary to unambiguously identify 511 keV γ's from $\bar{H}$-annihilations. A 511 keV γ-background results from π^0 from the $\bar{p}$-annihilation decaying into 2 high energetic γ which produce e^+e^--pairs in the superconducting solenoid. 511 keV γ's from the e^+-annihilation are then registered in the CsI crystals, indicated in Figure 1. The high granularity of the CC thus allows to select real back-to-back events.

Thickness	380±15 μm
Area	81.6 mm × 19 mm (+0.05/-0 mm)
p^+-side	128 readout strips with two floating strips
n^+-side	64 readout pads
Coupling	AC (p^+-side), DC (n^+-side)
Biasing	FOXFET (p^+-side)
Guard-ring	6 guard-rings

TABLE 1. Parameters of the double sided silicon strip detectors.

A The Silicon Vertex Detector

The SVD consists of two layers of double sided silicon strip detector modules (16 modules per layer). Each module consists of two double sided silicon strip detectors, a ceramic hybrid with two front-end readout chips (VA2_TA) and a silicon support-structure for the detectors (bus). The 64 silicon detectors were produced by SINTEF, Norway. The parameters of the silicon detectors are summarised in Table 1.

The operating conditions of the SVD (77 K, 10^{-7} mbar) and the limited space required a careful study of the mechanical and electric properties of the detector.

To ensure quality and long-term stability of each detector a series of measurements were carried out on each detector before constructing the modules. The measurements included current-voltage measurements, current-stability measurements at 80 V over several hours to days, pin-hole tests and measurements of the depletion voltage of each detector. Figure 2 shows current-voltage characteristics measured on ATHENA detectors. The high uniformity of the total leakage current of all detectors shown in the histogram in the bottom right corner reflects the high quality of the detectors.

On the modules the readout strips are connected to the inputs of VA2_TA chips. First measurements of a full-sized module using cosmics gave a signal to noise ratio of the $\approx$16 cm long p^+ strips of 46:1. This high S/N ratio ensures excellent spatial resolution of passing tracks. The spatial resolution in r-ϕ direction is expected to be better than 20 μm and in z-direction to be in the order of 300-400 μm. This greatly exceeds the required resolution which is experimentally limited to about 1 mm.

B The Crystal Calorimeter

Twelve crystals are placed in each of the 16 rows to provide sufficient granularity to verify the collinearity of the 2×511 keV γ's. A photosensitive diode is connected to each crystal. At 77 K the light-output of a CsI crystal has a wavelength of $\approx$350 nm. To allow the blue light to enter into the active volume of the silicon diode, special diodes with a n^+-layer of only 150 nm were produced by SINTEF.

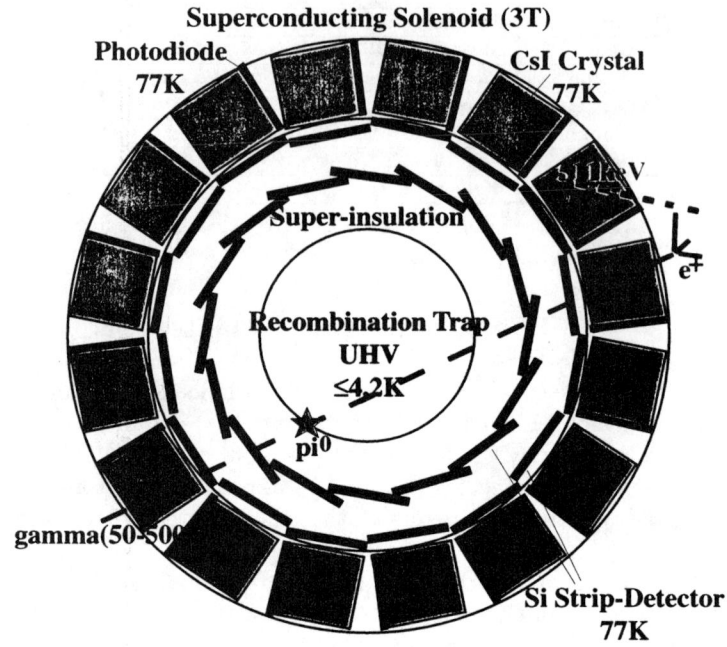

FIGURE 1. Cross-section of the ATHENA vertex detector.

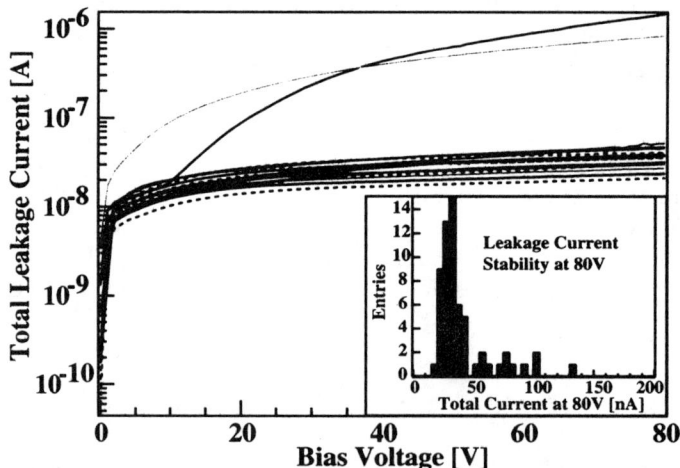

FIGURE 2. Total leakage current vs. voltage characteristic of ATHENA detectors. Data are normalised to 20°C. Histogram of the total leakage current at 80V after the stability measurement for all ATHENA detectors in the bottom right corner of this figure.

The photo-diodes are separated into four pads which are connected to inputs of VA2_TA chips by ultrasonic-wire-bonding. The smaller area per pad thus reduces the noise on each channel caused by load capacitance.

III SUMMARY

The ATHENA experiment is a new experiment at CERN with the goal to produce and investigate $\bar{\text{H}}$-atoms. To unambiguously detect $\bar{\text{H}}$-annihilations a complete vertex detector was developed with a two layer high resolution silicon vertex detector and a CsI-crystal calorimeter. Due to the low temperature required for the traps and the superconducting magnet the operating temperature of the vertex detector is set to 77 K. To our knowledge this is the first complete vertex detector operated under these conditions and thus required a detailed study of the mechanical and electric properties. The detector will be installed in the experiment in autumn 2000 and first physics results are expected later this year.

REFERENCES

1. ATHENA Collaboration, *CERN SPSC Report* **013**, (2000).
2. Van Dyck R.S., and Schwinberg P.B., and Dehmelt H.G., *Phys. Rev. Letters* **59**, 26 (1987).
3. Adler R., et al. *Phys. Lett. B* **363**, 237 (1995).
4. Holzscheiter M.H., and Charlton M., *Rep. Prog. Phys.* **62**, 1-60 (1999).

The ATLAS Liquid Argon Electromagnetic Calorimeter

Pascal Pralavorio (for the ATLAS Liquid Argon group)

CPPM,CNRS/IN2P3 - Univ. Méditerrannée, Marseille, France

Abstract. The ATLAS detector will start operation on the LHC in 2005. The collaboration has chosen a Liquid Argon electromagnetic calorimeter with accordion shape. Modules 0 of the barrel and the endcap were tested under electron beam at CERN during summer 99. The results of these tests are presented as well as the status of the modules' production.

INTRODUCTION

ATLAS is a general purpose detector under construction that will operate on a proton-proton collider, the Large Hadron Collider (LHC) [1]. This new machine will be installed in the LEP tunnel at CERN and provide 14 TeV center of mass energy collisions at a luminosity of 10^{34} cm^{-2}s^{-1}. The bunch crossing frequency will be 40 MHz and first collisions are foreseen for July 2005.

The ATLAS detector [2] is a typical high energy physics detector with a very large discovery potential for new physics such as Higgs bosons and supersymmetric particles [3]. In most of these physics channels, the electromagnetic (EM) calorimeter will play a key role in energy reconstruction and position measurement of electrons and photons. This is the case for example in the decay channels of Higgs boson: H$\to \gamma\gamma$ and H$\to eeee$.

The EM calorimeter [4] is divided in two parts, see Fig.1: i) the *barrel*, located in the pseudorapidity range $|\eta| < 1.4$, with an internal radius of ~1.5 m and a total length of ~6 m; ii) the *endcap*, located in the region $1.4 < |\eta| < 3.2$, made of two wheels with an external radius of ~2 m.

A presampler detector in front of the calorimeter for $|\eta| < 1.8$, and is used to correct for the energy lost in the material upstream.

TEST BEAM RESULTS

This EM calorimeter is a sampling calorimeter where absorbers are made of lead and liquid argon is the ionising medium – thus the detector will be placed

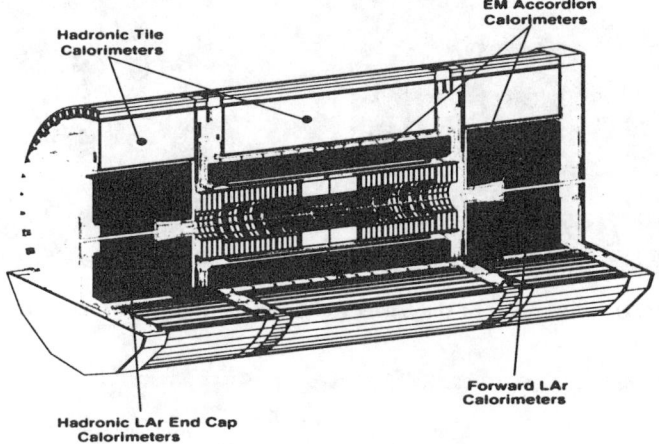

FIGURE 1. Schematic view of the ATLAS calorimeter system.

in a cryostat. Electrodes, separated from the absorbers by honeycomb spacers, collect the signal and set the high voltages. Both electrodes and absorbers have an 'accordion' geometry [3]. The calorimeter is segmented in 3 longitudinal samplings: i) *S1* made of narrow strips to perform γ-π^0 separation and position measurement. It has a depth of 6 radiation lengths (X_0); ii) *S2* made of square towers (4×4 cm^2 at $\eta = 0$) with a depth of 18 X_0 where most of the energy of e/γ shower is collected; iii) *S3* with 2 to 12 X_0 is used for high energies.

The readout electronics is made of current preamplifiers and shapers, placed outside the cryostat. The signal is sampled, digitised every 25 ns and sent to the control room if accepted by the trigger. The electronics chain has been calibrated continually for each readout channels during the test beam period.

Barrel and endcap modules 0, equipped with ATLAS-like electronics, were intensively tested during summer 1999 at CERN with 5 to 300 GeV electron and muon beams. The Fig. 2 shows these modules before insertion in the cryostat.

For data analysis, a cluster (typically 3×3) is formed in S2 around the most energetic cell and then add up with an adequate number of strips and S3 cells to compute the total energy, E. Corrections are applied to take into account the finite size of the cluster and the accordion shape of the absorbers. After these corrections, typical resolutions for both modules are shown in Fig.3, where a represents the stochastic term and c the local constant term. The electronics noise is estimated to be $\sim$300 MeV per cluster and has been subtracted from the energy resolution. For muons a signal-to-noise ratio of 4 was obtained.

FIGURE 2. View of the barrel (left) and endcap (right) modules 0.

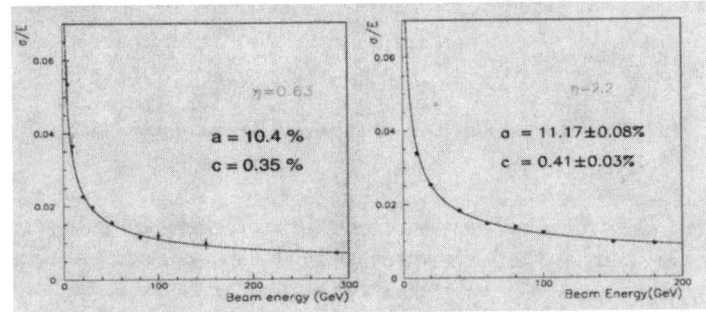

FIGURE 3. Energy resolution of barrel (left) and endcap (right) modules 0.

CONCLUSIONS

Barrel and endcap modules 0 have undergone 3 weeks of intensive beam test at CERN, last year. An energy resolution of $\frac{\sigma_E}{E} \sim \frac{10\%}{\sqrt{E}} \oplus 0.4\% \oplus \frac{300 MeV}{E}$ consistent with simulation has been found for both modules. Overall constant term is still to be obtained.

The module production (32 modules for the barrel and 16 for the endcap) has just started and will last 3 years.

REFERENCES

1. The Largon Hadron Collider, CERN/AC/95-05 (1995)
2. The ATLAS Technical Proposal, CERN/LHCC/94-43 (1994)
3. The ATLAS Detector and Physics Performance, Technical Design Report, CERN/LHCC/99-14 and 99-15 (1999)
4. The ATLAS Liquid Argon Calorimeter, Technical Design Report, CERN/LHCC/96-41 (1996)

Status and Physics Prospects of the Joint Project of KEK and JAERI in Japan

Tomofumi Nagae

*High Energy Accelerator Research Organization (KEK),
1-1 Oho, Tsukuba, Ibaraki 305-0801, Japan*

Abstract. The Japan Hadron Facility (JHF) was proposed by KEK with a high-intensity 50-GeV proton synchrotron as a main accelerator. The project is now proposed as the Joint Project between KEK and Japan Atomic Energy Institute (JAERI). The status of the Joint Project is described.

FROM JHF TO THE JOINT PROJECT

The Japan Hadron Facility (JHF) was proposed by KEK in 1997 [1,2], after merging three institutes, the former KEK, the Institute for Nuclear Study (INS), and the Meson Science Laboratory at University of Tokyo into one research organization (the present KEK). The budget request was officially submitted to the Ministry of Education, Culture, Sports, and Science (Monbu-sho) by KEK in the summer of 1997, for the first time. However, the construction budget has not been approved, whereas construction of an injector part of the proton linac up to ∼60 MeV was funded as an R&D project.

In 1998, when Monbu-sho and KEK sought a way of how to start the JHF project, a new idea came up to start this project under the sponsorship of both Monbu-sho and Science and Technology Agency (STA). These two agencies will be merged into one Ministry at the beginning of 2001 on the occasion of a major reorganization of the Japanese Government. Through intense discussions, KEK and JAERI identified each other as a counter partner.

JAERI had the Neutron Science Project (NSP) as their future plan in the Tokai site. The project aimed to construct a 5-MW pulsed spallation neutron source for neutron science and a 3-MW continuous proton beam for nuclear transmutation research. The accelerator composed of a ∼1.5-GeV superconducting proton linac and a proton storage ring. It was recognized that further R&D efforts would be needed before the official construction of high-intensity superconducting proton linac. In the JHF project, neutron science at a 3-GeV proton synchrotron with an 0.6-MW pulsed spallation neutron source was one of the main research fields. Here, rather moderate design parameters were selected for the neutron source. The accelerator complex was based on the currently established technologies in order to start an immediate construction. Therefore, it seemed reasonable to create a joint project between KEK's JHF and JAERI's NSP with high-intensity proton accelerators by making a MW-class neutron source as a key link between the two.

In the fall of 1998, significant efforts were made to start the Joint Project by using the second Supplemental Budget for JFY98. Unfortunately, this trial did not work. Nevertheless, JAERI and KEK agreed to continue to pursue the Joint Project. Under the MoU between JAERI and KEK signed in March, 1999, the Joint Project Team was initiated, and the proposal of the Joint Project was written [3]. The proposal was reviewed in April, 1999, by the international review committee chaired by Y. Cho of ANL. The Joint Project received a very positive support. Subsequently, an R&D budget for the Joint Project was requested from both Monbu-sho and STA in September, 1999, to the Ministry of Finance.

The major review of the Joint Project at the government level started from December, 1999. It reviewed various aspects of the project not only on the scientific significance but also on the sociological and economical impacts of the project. The final report will be published in August, 2000.

WHAT IS THE JOINT PROJECT ?

The accelerator complex of the Joint Project consists of

- 400-MeV normal-conducting linac,
- 600-MeV linac(superconducting) to increase the energy from 400 to 600 MeV,
- 3-GeV synchrotron ring, which provides proton beams at 333 μA (1MW), and
- 50-GeV synchrotron ring, which provides proton beams at 15 μA (0.75 MW).

In addition, an upgrade in the proton beam power to 5-MW in a few-GeV energy region is proposed as a "Phase 2" project of the present proposal.

At the initial stage, the normal conducting 400-MeV linac will be used as an injector to the 3-GeV ring. At the stage when the superconducting 600-MeV linac becomes stable, this 600-MeV linac will be switched as the injector to the 3-GeV ring to enhance the beam power for the 3 GeV.

At the 50-GeV Proton Synchrotron (PS), nuclear/particle physics experiments using kaon beams, antiproton beams, hyperon beams, neutrino beams and primary proton beams are planned. The 3-GeV ring will be used as a booster synchrotron for the 50-GeV main ring. In addition, it is designed to provide the beam power of 1 MW. Extensive physics programs which cover condensed matter physics, material sciences and structural biology will be carried out by using neutrons and muons. The high-current 600-MeV linac will be used for R&D of the accelerator-driven nuclear transmutation.

The site of the proposed accelerators is chosen to the JAERI Tokai site, which is located at about 60 km north from KEK. Since the ultimate goal of the project is to

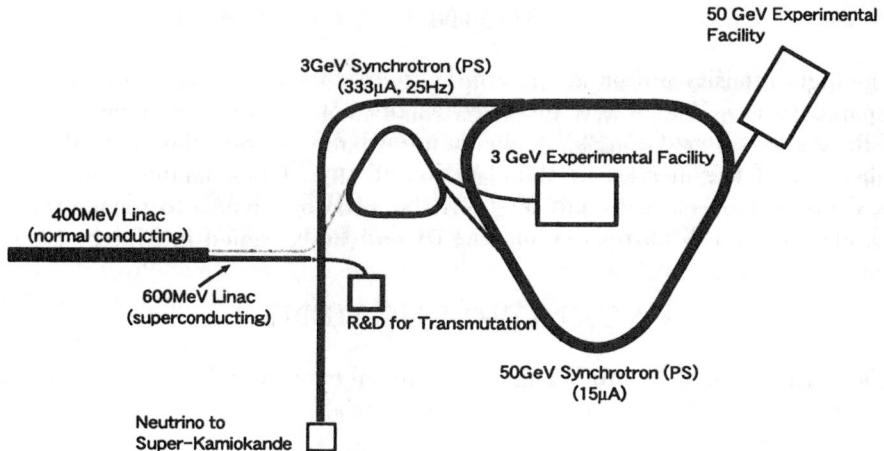

FIGURE 1. Schematic layout of the accelerator complex and the experimental facilities of the Joint Project in the Tokai site.

seek a 5-MW beam power including R&D experiments for accelerator-driven transmutation, JAERI-Tokai has advantages in the expertise handling of the radioactive waste and the use of nuclear fuel materials.

In Figure 2, a schematic layout of the experimental area for the slow extraction beams from the 50-GeV PS is shown. We have another experimental area for neutrino beams with fast extraction. As shown in the figure, there are two primary beam lines (A and B) in this hall. There, we could have several secondary beam lines such as a low momentum K^- beam(K1.1), a medium-energy K^- beam(K1.8), a neutral kaon beam(KL), etc. The experimental hall is designed to allow an extension to a larger room in the future.

One of major research subjects here is strangeness nuclear physics [4]; precise hypernuclear gamma-ray spectroscopy, spectroscopy of Ξ-hypernuclei, high-statistics hyperon-nucleon scattering experiments, etc. As for particle physics, precision tests of the standard model are carried out through the rare kaon decays such as $K^+ \rightarrow \pi^+ \nu \bar{\nu}$ and $K_L \rightarrow \pi^0 \nu \bar{\nu}$. The high-intensity neutrino beam available here is also very powerful to determine the possible mixing parameters among three generations of neutrino with a very high precision.

The upgrade of the 50-GeV PS is also discussed intensively. An upgrade to gradually increase the primary proton beam intensity from 15.6 μA to 30, 43, and 74 μA is possible. New facilities such as an antiproton accumulator ring, a muon storage ring for the neutrino factory, etc. are also proposed for the future extensions.

SUMMARY

The high-intensity proton accelerator complex with an expanded capabilities as compared to the JHF is now proposed as the Joint Project between KEK and JAERI. As for the 50-GeV PS, the beam intensity is increased from 10 μA to 15.6 μA because of the increase of the injection energy from the proton linac to the 3-GeV PS, which is already not far from the TRIUMF KAON project. We hope the budget request be approved for construction from JFY2001.

ACKNOWLEDGMENTS

The Author would like to thank all the members of the Joint Project team in KEK and JAERI: Prof. Shoji Nagamiya, in special.

REFERENCES

1. JHF Project Office, *Proposal for Japan Hadron Facility*, KEK Report 97-3, May 1997.
2. JHF Project Office, *JHF Accelerator Design Study Report*, KEK Report 97-16, March 1998.
3. The Joint Project Team of JAERI and KEK, *The Joint Project for High-Intensity Proton Accelerators*, KEK Report 99-4/JAERI-Tech 99-056, July 1999.
4. Nagae T., *Nucl. Phys.* **A639**, 551c (1998).

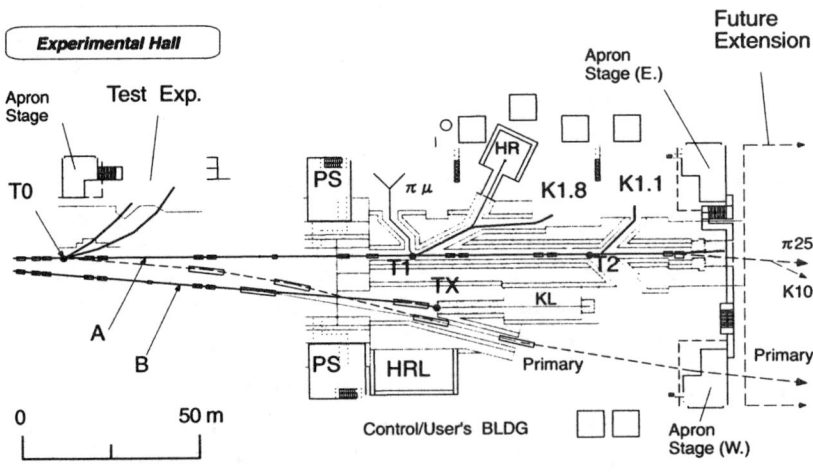

FIGURE 2. Schematic layout of the experimental area at the 50-GeV PS.

The Hall D Detector at Jefferson Lab

Curtis A. Meyer*
for the Hall D Collaboration

Carnegie Mellon*, Catholic University, Christopher Newport, University of Connecticut, Florida International, Florida State, Indiana University, IHEP Protvino, Jefferson Lab, Los Alamos, Moscow State, Norfolk State, Budker Inst, Ohio University, Old Dominion, University of Pittsburgh, University of Regina, Rensselaer, Saclay

Abstract. The Hall D experiment at Jefferson Lab is part of the proposed CEBAF upgrade to 12 GeV beam energy. The Experiment will study gluonic excitations of mesons in the 1.5 to 2.5 GeV/c^2 mass region using an 8 to 9 GeV beam of linearly polarized photons.

The Hall D detector [1] is a major new initiative to study light quark mesons and to search for gluonic excitations using a beam of linear polarized photons. The experiment is a key piece of the proposed 12 GeV upgrade to the Jefferson Lab accelerator, (see Figures 1). Current experimental studies have started to show evidence for exotic mesons, both mesons with manifestly exotic quantum numbers, and for the scalar glueball. Understanding the spectrum of these states, and how they are related and mixed with the normal mesons is a critical element in understanding QCD in the non-perturbative regime. This will also provide stringent milestones for theory, in particular the lattice, to be able to explain both the spectrum and decay properties of such states.

Recent progress in the field of meson spectroscopy has been spear-headed by efforts in $p\bar{p}$ annihilation at LEAR [2], central production experiments at CERN [3], and pion scattering experiments at BNL and at VES [4]. These efforts have produced several large, clean data sets that have shown that meson states beyond the naive quark model exist, however, the exact interpretation of these states is not clear. In the exotic meson sector, there appear to be two isovector states with $J^{PC} = 1^{-+}$. However, a hybrid interpretation expects exactly one such state. In addition, both states are lighter than what is expected by current calculations.

While one could in principal try to continue exploiting current reactions, it is useful to step back and look at all production processes. One quickly notices that for historical reasons of either low rates, duty factor, or poor beam definition, there is virtually no data on the photoproduction of mesons. In addition, photoproduction is fairly unique in these processes in that the incident particle carries one unit of intrinsic angular momentum into the reaction. Typically, photoproduction is viewed through the eyes of Vector Meson Dominance, where before a t–exchange process with the nuclear target, the photon is transformed into a vector meson such as a ρ, ω or a ϕ. This is also unique in that these mesons have the $q\bar{q}$ in a Spin 1 configuration, (compared to a spin 0 configuration for π's and K's. This has an added advantage that in many models of hybrid mesons, these gluonic excitations are built from a q–$\bar{q}$ system that is in a spin 1 system. Photoproduction is quite likely to be exactly the place to look to observe many of these reactions [5].

The last piece of this is the need for linearly polarized photons to carry out the analysis of the the final states. Naively, the linear polarization picks out a new direction in addition to the the photon momentum, thus reducing the complications in the analysis by a factor of two. In fact, it can also be used to select the naturality of the exchange particle, and give us additional information on how the mesons are observed. The goal of the Hall D experiment is to use 8 to 9 GeV linearly polarized photons on a liquid hydrogen target to study the meson spectrum from roughly $1\,\text{GeV}/c^2$ up to about $2.5\,\text{GeV}/c^2$. This will be done using a new hermetic detector with excellent charged particle and photon capabilities and very good particle identification.

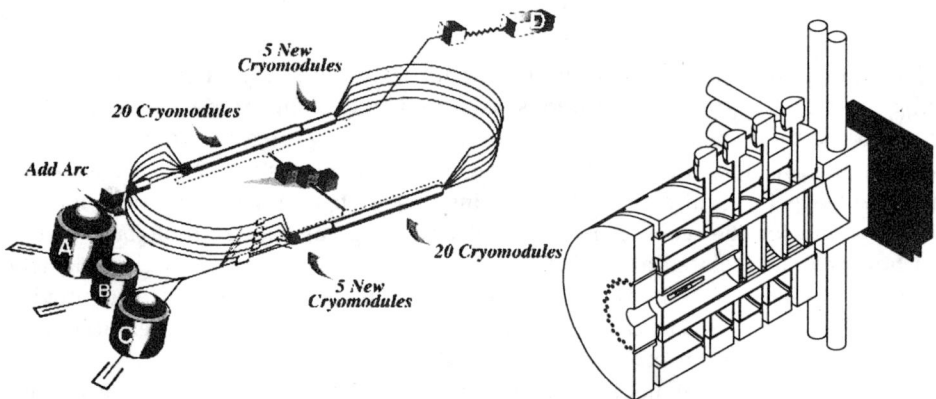

FIGURE 1. The left–hand figure is the CEBAF accelerator after the 12 GeV upgrade. The upgrade is accomplished by adding 10 new cryomodules, and a fifth arc to one end of the accelerator. Hall D extracts electrons at the highest beam energy and delivers them to a tagger complex. Photons are produced via coherent Bremsstrahlung off a thin diamond target, and the deflected electrons are then detected in a tagger. Only photons can be delivered to the Hall D building which houses the Detector (right–hand figure). The detector has nearly 4π coverage for both charged particles and photons and sits in a 2.2 T solenoidal field. It uses a combination of time of flight and threshold cherenkov detectors to do K–π separation.

The is new detector would be situated in a new Hall, (Hall D), at the Jefferson Lab CEBAF accelerator. The linearly polarized photons will be produced using a tagged coherent Bremsstrahlung process, and in order to achieve both good rates and sufficient polarization a 12GeV primary electron will be needed. The new Hall will be located at the opposite end of the accelerator as the current Halls, and with the addition of a 5'th arc of magnets, the electrons delivered to the Hall D photoproduction target would have $5\frac{1}{2}$ passes through the CEBAF linacs, rather than the 5 as seen by the other halls. This leads to 12 GeV beam in Hall D, and 11 GeV beam in Halls A, B and C.

The Hall D physics program would start with $10^7 \gamma/s$ tagged linearly polarized photons in the 8 to 9 GeV energy region. The ultimate goal will be to push this to $10^8 \gamma/s$ as the experiment is understood and additional computer power is added to the trigger system. Photoproduction cross sections of interest range from about $1\,\mu$b down to the nb level. Even at the initial rates, in one year of running the statistics will exceed the best in π scattering experiments by at least a factor of 100, and lead to on the order of 10^7 events from a $1\,\mu$b cross section. The detector is optimized to do excellent reconstruction for almost all final states, so when coupled with the high statistics, this will allow us not only to find these exotic mesons, but also to map out their decays in detail. It is the decay patterns which are the challenge for QCD to explain, in particular in cases where different states can mix.

Understanding the bound states of QCD in the light-quark sector is an important goal of both theorists and experimentalists trying to understand the strong limit of QCD. Recent lattice calculations [6], [7] have started to address these, and improvements in theory and modeling over the next few years will rely heavily on understanding the role that glue plays in QCD, and information on gluonic excitations will be a crucial test in our understanding. The Hall D detector will fill an important role in this regard, and with high quality, high statistics data available in a previously very poorly studied channel, is very likely to make crucial discoveries in this field.

REFERENCES

1. See http://dustbunny.physics.indiana.edu/HallD/ for additional information. In particular, **The Hall D Design Report, Version 2**, http://dustbunny.physics.indiana.edu/HallD/DR.html.
2. Claude Amsler, Rev. Mod. Phys. **70**, 1293, (1998).
3. D. Barberis, *et al.*, (The WA102 Collaboration), hep-ex/0003033, March 2000.
4. See J. Napolitano in these proceedings and Stephen Godfrey and Jim Napolitano, Rev. Mod. Phys. **71**, 1411, (1999).
5. Nathan Isgur, Phys. Rev. D**60**, 114016, (1999).
6. C. Bernard, *et al.* (MILC Collaboration), Phys. Rev D**56**, 7039, (1997).
7. K, Juge, J. Kuti and C. J. Morningstar, Nucl. Phys. Proc. Suppl., **63**, 326, (1997).

Physics Prospects with an Intense Neutrino Experiment

Nickolas Solomey

Enrico Fermi Institute, The University of Chicago, Illinois 60637

Abstract. With new forthcoming intense neutrino beams, for the study of neutrino oscillations, it is possible to consider other physics experiments that can be done with these extreme neutrino fluxes available close to the source.

INTRODUCTION

The objectives of my talk are to bring to the attention of both the particle and nuclear physics communities the unique physics potential that an intense neutrino beam experiment might be able to perform, other than that of neutrino oscillation which is covered elsewhere in these proceedings [1]. The list of non-oscillation physics is large and will not be completely covered in detail in such a short writeup, but the hope is that it might interest physicists, both theoretical and experimental, to contribute their ideas to help formulate a new experiment that can take full advantage of the intense neutrino beams planned in the NuMI project at Fermilab [2] and eventually the muon collider prototype storage ring [3] which might be located at Fermilab along the NuMI beamline direction.

PHYSICS PROSPECTS

Possible physics from an intense neutrino beam are:

Scattering: The first such experiments were performed by Rutherford with alpha scattering which showed the nuclei exists as a concentrated center of positive charge, while direct evidence of the quarks themselves came from experiments using the highest possible energy electrons to probe the nucleon structure. Muon scattering experiments have also contributed similar results, as well as offered a more massive probe. Furthermore, greatly improved statistics of electron scattering reactions are now being done at CEBAF. The physics of these experiments aim to measure structure functions with high precision. This is done by measuring the electron scattering cross-section:

$$\frac{d^2\sigma}{d\Omega dE'} = \frac{4\alpha^2(E')^2}{Q^4}\cos^2(\theta/2) \times [\frac{F_2(x,Q^2)}{(E-E')} + \frac{2F_1(x,Q^2)}{M}\tan^2(\theta/2)] \quad (1)$$

Where $\alpha = 1/137$, M is the proton rest mass, E is the electron energy before and E' after the scattering, θ is the electron scattering angle, and the Bjorken scaling variable is $x \equiv \frac{Q^2}{2M(E-E')}$. With neutrino scattering the expression changes a little: $\frac{4\alpha^2(E')^2}{Q^4}$ becomes $\frac{G^2}{2\pi}$ where G is the Fermi constant of weak coupling, and another structure function F_3 is needed because of parity violation. Neutrino scattering experiments are capable of measuring in detail the internal structure of the nucleon threw such experiments, with the advantage that the neutrino interactions are purely weak processes. They too can look towards further improvements with higher statistics. Eventually with a sufficiently large data sample that can permit the data to be separated into ν and $\bar{\nu}$ this can be used to study the similarities between quarks and antiquarks.

Strangeness Production: The most interesting new measurement would be strange particle production by neutrino interactions. This is suppressed by a factor of $tan^2\theta_c$ (where θ_c is the Cabibbo mixing angle) and can only be done once an intense neutrino source is available with an experiment capable of identifying strange particles. With a small dedicated experiment of good resolution, events can be seen of the charged current (CC) and neutral current (NC) type:

$$\bar{\nu}_l + p^+ \to l^+ + \Lambda^0$$
$$\to l^+ + \Sigma^0$$
$$\to l^+ + \pi^0 + \Lambda^0$$
$$\to l^+ + \pi^0 + \Sigma^0$$
$$\to l^+ + K^- + p^+$$
$$\vdots$$
$$\bar{\nu}_l + n^0 \to l^+ + \Sigma^-$$
$$\to l^+ + \pi^- + \Lambda^0$$
$$\to l^+ + \pi^0 + \Sigma^-$$
$$\to l^+ + K^- + \Lambda^0$$
$$\vdots$$

$$\bar{\nu} + p^+ \to \bar{\nu} + \Sigma^+ \quad SCNC$$
$$\to \bar{\nu} + \pi^0 + \Sigma^+ \quad SCNC$$
$$\to \bar{\nu} + K^0 + \Sigma^+$$
$$\to \bar{\nu} + K^0 + p^+ \quad SCNC$$
$$\to \bar{\nu} + K^+ + n^0 \quad SCNC$$
$$\vdots$$
$$\bar{\nu} + n^0 \to \bar{\nu} + \Lambda^0 \quad SCNC$$
$$\to \bar{\nu} + \Sigma^0 \quad SCNC$$
$$\to \bar{\nu} + K^0 + \Lambda^0$$
$$\to \bar{\nu} + K^0 + \Sigma^0$$
$$\vdots$$

To date only a handful of experiments have measured a few of these reactions, all with bubble chambers where the particle interaction and secondary particles produced could be explicitly identified. The best results and only cross sections measured is from the CERN-PS Gargamelle bubble chamber experiment [4] with 15 events of Λ^0 CC production at 2×10^{-40} cm^2/nucleon; while 7 events exist from the ZGS bubble chamber which includes one neutral current strange particle production event [5]. More recent experiments of the 80s and 90s have not been able to measure such interactions because they use large dense detectors to increase their neutrino interaction rate. With the advent of intense neutrino beams this can be over come by using a single thin target with a high precession experiment downstream to identify all of the secondary particles. Not only would such an experiment be able to measure strange particle production cross sections for charged and neutral current, but the comparison of the two could give an independent measurement of the weak mixing angle for a few channels, and is also possible to improve the main interactions by explicitly removing these strange particle production reactions from the sample. Table 1 lists the total neutrino and antineutrino interaction rate, as well as the expected rate of Λ^0 production in the NuMI medium energy neutrino beam. Some of the NC interactions are forbidden and if seen at a low level would be an indication of strangeness changing neutral currents (SCNC) which is a worthy physics objective. Also the production of hyperons can permit a study of hyperon polarization when produced by neutrinos, the polarization of hyperons in fixed target experiments by protons beams was originally a surprise and is still theoretically unexplained [6].

TABLE 1. Neutrino interactions expected per year in a solid target 2.5 cm thick and 1x1 m^2, or for liquid target with a volume of 10 m^3. The Lambda yield is for the reaction $\bar{\nu}_\mu + p \to \mu^+ + \Lambda^0$.

target	ν interactions/year	$\bar{\nu}$ interactions/year	Lambda yield/year
Fe	1.0 M	0.25 M	40 k
C	0.3 M	75 k	12 k
W	2.5 M	0.63 M	110 k
H$_2$	2.2 M	0.55 M	90 k
D$_2$	5.0 M	1.25 M	250 k

Hyperon beta decays $A \to B\, e^-\, \bar{\nu}_e$ are important to study for their weak decay form-factors which give an understanding of their underlying structure [7]. In the V-A formulation the transition amplitude is:

$$M = \frac{G}{\sqrt{2}} <B|J^\lambda|A> \bar{u}_e \gamma_\lambda (1+\gamma_5) u_\nu \qquad (2)$$

The V-A hadronic current can be written as:

$$<B|J^\lambda|A> = \mathcal{C}\, i\, \bar{u}(B)\, [\, f_1(q^2)\gamma^\lambda + f_2(q^2)\frac{\sigma^{\lambda\nu}\gamma_\nu}{M_A} + f_3(q^2)\frac{q^\lambda}{M_A} + $$
$$[\, g_1(q^2)\gamma^\lambda + g_2(q^2)\frac{\sigma^{\lambda\nu}\gamma_\nu}{M_A} + g_3(q^2)\frac{q^\lambda}{M_A}]\gamma_5\,]u(A) \quad (3)$$

where $\mathcal{C}$ is the CKM matrix element, and q is the momentum transfer. There are 3 vector form factors: f_1 (vector), f_2 (weak magnetism) and f_3 (an induced scaler); plus 3 axial-vector form factors: g_1 (axial vector), g_2 (weak electricity) and g_3 (an induced pseudo-scaler). This is also possible to study with neutrino interactions that produce hyperons, because the hyperon's decay itself has a self analyzing power its polarization. To obtain a large and clean data sample such measurements can be unambiguous avoiding the common problem of the missing neutrino which introduces multiple solutions and missing momentum.

Comparison: The production of strangeness from nucleon scattering using electrons or photons started in the 50s, but there is yet no comprehensive theory [8]. This data is modeled by many theorists, but the models only work for the explicit interaction it was developed for. The final goal is to have a description of all of the underlying reaction mechanisms for strangeness production, and this is far from attained, especially since the strangeness production by neutrino scattering are a grand sum of 26 events: 15 from one experiment, 7 from another and a few other experiments with one event each. However, an important message from the electron and photon experiments doing similar studies is the importance to investigate simultaneously all production channels [9]:

$$\gamma + p^+ \to K^+ + \Lambda^0$$
$$\to K^+ + \Sigma^0$$
$$\to K^0 + \Sigma^+$$
$$\vdots$$

$$e^- + p^+ \to e^- + K^+ + \Lambda^0$$
$$\to e^- + K^+ + \Sigma^0$$
$$\to e^- + K^0 + \Sigma^+$$
$$\vdots$$

This is not a comprehensive list, other strange particle production experiments have also been done with charged mesons: $\pi^\pm$ and $K^\pm$. Plus these reactions can be expanded to more complicated ones, but it is the simple processes that are most interesting. By extending these studies to include neutrino scattering experiments of strange particle production, it would give the ability to expand our understanding of the mechanism, aid in improving the underlying nucleon, hyperon and strange meson structure, and the best improvement will be the ability to compare what is found here with that of electromagnetic interactions. This is promising to expand our knowledge of the quark sea in the nucleon.

Target Dependency: The most elementary particle structure information will come from the simplest targets which is the lightest elements: hydrogen. While neutrino scattering off of the atomic shell electron with high statistics would be useful with both CC and NC interactions. However, slightly more complicated reactions with deuterium, tritium, ^{3}He and ^{4}He are also useful. For example the deuterium data could be combined with the hydrogen results to yield the neutron scattering information. While the tritium and ^{3}He data provides clues to the rule of the simpler nucleon structure of an extra proton or neutron in the nucleus. By using heavier targets, which give higher rates, it is possible to study the A dependency (number of nucleons, protons and neutrons, in the target nuclei), e^- and $\mu^\pm$ scattering experiments have found some intriguing results that show the quarks in heavy nuclei are not simply confined to their protons or neutrons as some models have suggested. Testing this with high statistics neutrino scattering experiments is important. This data could also be used to compare with nuclear models that exist for N-N, π-N, e-N, γ-N and μ-N interactions; and eventually to formulate models of

ν-N and $\bar{\nu}$-N interactions. With the ultimate goal of formulating a comprehensive theory that works for all interactions simultaneously.

τ *Neutrino:* With the high energy neutrino beam option of NuMI there is the possibility to do ν_τ detection from a point source which would further require a small emulsion active target, but due to particle rates it would have to be changed often. The physics prospects are higher statistics ν_τ detection, and short baseline neutrino oscillation. In this case the detector would be used to trigger on those events to search for in the emulsion. Due to the intense neutrino beams these emulsion targets should be highly segmented and easily changed. The multiple drawer system envisioned for the MINOS far detector is a possible choice [10]. If CP violation studies are ever to be done in the lepton sector, then the oscillation rate of $\bar{\nu}_\mu \to \bar{\nu}_\tau$ from disappearance will have to be compared to $\bar{\nu}_\tau \to \bar{\nu}_\mu$ from appearance, but it will be necessary to have a small emulsion experiment at the near site to determine the $\bar{\nu}_\tau$ initial intensity.

DETECTOR REQUIREMENTS

An experiment to look for strange particle production by neutrino interaction does not have to be very large, but does have several things it must perform well. A simple basic sketch of the conceptional detector design is shown in figure 1. There is the need for a veto array in front of the target and around the sides to

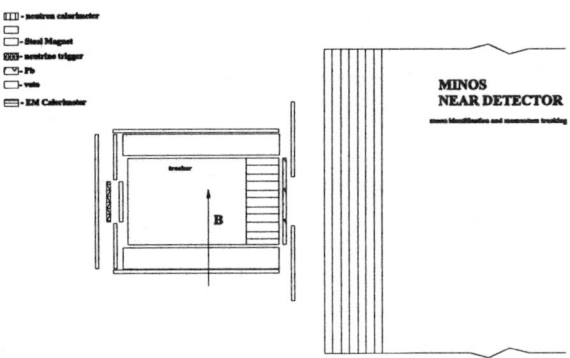

FIGURE 1. Conceptual design of a neutrino strange-particle production experiment.

catch any escaping particles, a target which could possible be: liquid, different materials or maybe even active. A tracking chamber in a magnetic field to be able to track and measure the momentum of all the charged particles, with the ability to reconstruct the decay vertex of the neutral strange particles (Λ^0, Σ^0, K^0 ...). A good electromagnetic calorimeter, for electrons and especially photon pairs to be able to see $\pi^0 \to \gamma\gamma$. There is the need for the ability to distinguish protons, $\pi^\pm$, and $K^\pm$. Since the neutrino beam is not extremely high energy these particles will be relatively low in momentum. Permitting their identification by dE/dx in the tracking chamber or TOF. By having the experiment sit in front of the MINOS near detector, information from it can be used to identify muons and measure their charge and momentum. Also, the first 3-5 interaction lengths in the MINOS near detector can make a nice hadron calorimeter for identifying neutrons.

There are many advantages for both experiments to be located together. The cost effectiveness of producing the maximal physics from the NuMI neutrino beam is the most obvious. Each experiment can use information from the other to help calibrate. This experiment could measure more precisely the neutrino beam content and its radial distribution. Also protons and π^- from Λ^0 decay can be identified and then the response of the MINOS near detector studied so that hadron/muon identification in MINOS can be better understood as a function of momentum.

CONCLUSIONS

It is hoped that such an experiment can be organized in the next year with detailed detector designs, physics goals achievable and a sufficiently funded collaboration formed. Such an experiment can be of diverse interest to both the high energy and nuclear physics communities and represents an opportunity of cooperation. Furthermore, it will give another reason for the NuMI project to operate intense neutrino beams, which increase the physics yield from the neutrino flux that will be provided for the other experiment.

I would like to acknowledge discussion with Paul Schoessow, Amol Dighe, Philippe Mine, Malcolm Derrick and Tom Fields.

REFERENCES

1. James talk at **these proceedings**
2. K. Anderson *et al.*, The NuMI Facility Technical Design Report, October 1998.
3. S. Greer, Phys. Rev. D **57** (1998) 6989.
4. K. Myklebost, Quasielastic Production Of Lambda Hyperons In Anti-Neutrino Interactions At The CERN PS, Proceedings, Neutrinos 78 (1978) c37-c41.
5. S.J. Barish et al., Phys. Rev. Lett. **33** (1975) 1446.
6. L. Pondrom, Phys. Rep. **122** (1985) 57.
7. Solomey other talk at **these proceedings**
8. J.C. David *et al.*, Phys. Rev. C 53 (1996) 2613.
9. B. Saghai, Nuclear Physics **A639** (1998) 217c-226c.
10. P. Adamson *et al.*, The Hybrid Emulsion Detector for MINOS, Fermilab Report NuMI-L-473, April 1999.

An Ultracold Neutron Facility at PSI

Manfred Daum[1] for the UCN collaboration[1,2,3,4,5,6]

[1] PSI, Paul-Scherrer-Institut, CH 5232 Villigen, Switzerland [2] PNPI, St. Petersburg Nuclear Physics Institute, Gatchina, RU
[3] ETHZ, Eidgenössische Technische Hochschule, Zürich, CH
[4] Jagellonian University, Cracow, PL
[5] University of Virginia, Charlottesville, USA
[6] ILL, Institute von Laue Langevin, Grenoble, FR

Abstract. We build a new type of ultracold neutron source, based on the spallation process. The essential elements of the source are a pulsed proton beam with a high intensity and a very low duty cycle, a heavy element spallation target and a large moderator consisting of solid deuterium kept at a temperature of about 8 K. Recent experimental studies of the production of ultracold neutrons in solid deuterium open prospects for densities of about 1000 ultracold neutrons per cm^3. As a first experiment at the new facility, we intend to measure the electric dipole moment of the neutron with a sensitivity of about 10^{-27} ecm, an improvement by more than one order of magnitude.

INTRODUCTION

During the recent years most attention in physics with ultracold neutrons was focused on two experiments: high sensitivity search for time reversal violation via measurement of the electric dipole moment of the neutron (EDM), and the decay asymmetry and lifetime of the neutron. These experiments rely heavily on the long observation time of an ensemble of stored neutrons, and were the driving force behind the development of techniques for the production and storage of *ultracold neutrons* (UCN), i.e. neutrons with an extremely low speed (≤ 6 m/s). Ultracold neutrons have brought significant progress in the reduction of systematic effects in these experiments, however the achieved statistical precision is still below expectations.

I THE NEW METHOD

These important experiments are hindered by a low density of the ultracold neutrons available for the experiments. Recently, an attractive solution of these problems has been proposed [1]. It is based on a pulsed spallation source, customized to the needs of UCN production. The proposed scheme solves a contradiction between the high neutron thermal flux flowing into the solid deuterium moderator and an average heat load in such a moderator.

We propose to produce neutrons in a spallation process using a high intensity, proton beam in a pulsed mode. This new scheme ("Macro-Pulsing"), with its characteristic very low duty cycle, brings important advantages. The beam is directed for a time period of the order of few seconds onto the lead spallation target to generate a high density neutron pulse in a moderator assembly *dedicated* to the production of ultracold neutrons. Then, the beam is deflected back to the main beamline and can be used without further disturbance during approx. 10 min. eg. for continuous production of mesons or cold neutrons. The target station, incorporating a water cooling system, a compact premoderator at liquid nitrogen temperature, neutron reflectors and a shield against gamma rays, is surrounded by a block of solid deuterium or other efficient moderator cooled with liquid helium. The emerging neutron gas, with a density reaching saturation level in the solid may be directed to a dedicated experiment located as close to the production target as only the secondary radiation field allows. Alternatively, UCN are accumulated in the intermediate storage vessel and distributed over a period of few tens of seconds to experimental stations located around the source (UCN-factory version [1]). During the filling time the intense prompt background dies away and also a large portion of the delayed radiation is reduced. The ultracold neutrons are then observed and/or counted over a period of about the neutron lifetime in the actually running experiments. After the density of the UCN in the experimental stations has dropped significantly, a new proton pulse generates the next ultracold neutron packet, and the whole cycle is repeated.

Estimations [2] and Monte Carlo studies [3] show that a UCN density in the order of $10^3 - 10^4$ UCN/cm^3 can be delivered to the experiments using the proposed method. This is $2-3$ orders of magnitude more than in experiments at the reactors in Institute Laue Langevin, Grenoble and in St. Petersburg Nuclear Physics Institute, Gatchina, which are at present the world leading centers in ultracold neutron research. The Paul Scherrer Institute with its superior proton beam is an ideal place to realize such a new UCN source, which is a prerequisite for a next leap in the EDM and lifetime experiments.

REFERENCES

1. A.P. Serebrov, V.A. Mityukhlaev, A.A. Zakharov, T. Bowles, G. Greene and J. Sromicki, JETP Lett. 66 (1997) 803.
2. J. Sromicki, A.P. Serebrov, International Symposium on Weak and Electromagnetic Interactions in Nuclei, WEIN 98, Santa Fe, NM, June 1998, World Scientific, in press.
3. I. Potapov, V. Kuzminov, contributions to the "First UCN Factory Workshop",Pushkin, RU, Jan. 19-22, 1998.

Recent Highlights at the MIT-Bates Linear Accelerator Center[1]

Kenneth D. Jacobs

MIT Bates Linear Accelerator Center
P.O. Box 846 21 Manning Rd. Middleton, MA 01949 USA

Abstract. The MIT-Bates Linear Accelerator Center is an electron scattering facility for medium energy nuclear physics. Recent and ongoing developments are extending the range of physics which can be studied there. These include several projects associated with the Bates South Hall Ring (SHR), as well as upgrades to the linac. Details of these developments are presented.

INTRODUCTION

The MIT-Bates Linear Accelerator Center comprises an S-band electron linac, an energy doubling recirculation system, a storage/pulse stretcher ring, and two main experimental halls. The facility layout is shown in Figure 1. Much of the recent experimental activity has been focused in the North Hall, on the SAMPLE experiment [1,2]. In addition, we are nearing completion of commissioning the South Hall Ring (SHR), which will provide both high current stored beam for internal target experiments, and high duty factor beam for external target coincidence experiments. Polarized electrons are available in all operating modes.

ACCELERATOR

Recent upgrades to the accelerator radio frequency (RF) system have increased the energy reach of Bates. Early in 2000, a beam of 1.0 GeV energy, at a peak current of 14 mA and 1.3 μs pulse length, was demonstrated at a beam repetition rate 600 Hz. This is the nominal repetition rate, and would provide an average current of 10 μA extracted from the SHR. At low peak current the beam energy would be close to 1.1 GeV, which could be stacked in the SHR for stored beam.

---

[1] Work supported by the U.S. Department of Energy cooperative agreement DE-FC02-94ER40818.A000.

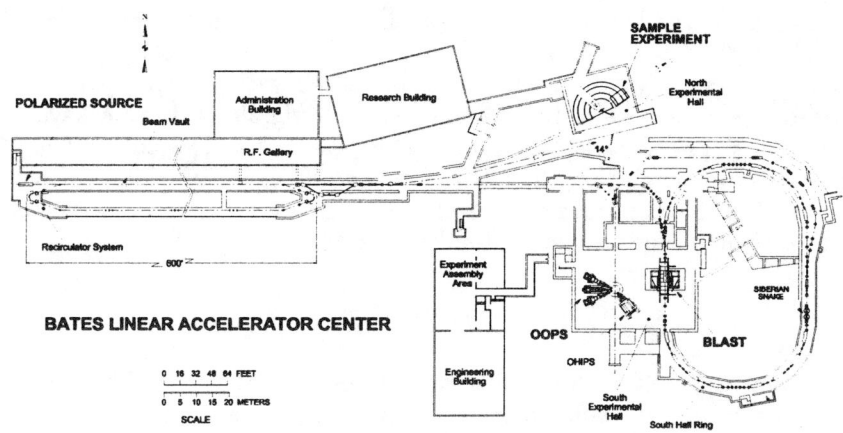

FIGURE 1. MIT-Bates facility layout.

SOUTH HALL RING

The SHR serves two functions. One is to accumulate and store beam from the accelerator, for experiments using thin targets internal to the ring. The second is to take the pulsed beam from the accelerator, and use "pulse stretching" techniques to convert it to high duty factor beam for use by fixed, external target experiments.

Internal target experiments will utilize the Bates Large Acceptance Spectrometer Toroid (BLAST), now under construction [3,4]. Stored beam capabilities are already sufficient for BLAST experiments. We have stored up to 200 mA, with $1/e$ lifetimes of tens of minutes, with an internal target cell in place. BLAST is scheduled to be complete by the end of FY'01.

Experiments are presently underway utilizing beam resonantly extracted from the SHR. This process takes the low duty factor beam from the accelerator, and converts it to quasi-CW beam. The OOPS experimental program at Bates [5] will use this beam for an extensive study of the five response functions accessible through the $(\vec{e}, e'p)$ reaction. To date, $4\,\mu$A of beam at 60 % duty factor, has been used by experiments. Further increases in both average current and duty factor are expected in the near future.

BEAM POLARIZATION

Polarized electrons are obtained from the Bates polarized electron source [6]. In order to maintain longitudinal polarization of the beam in the SHR at arbitrary energies, a Siberian Snake has been installed. The location of the snake will maintain the polarization for both stored and extracted beam. First tests of the snake were done in early 2000 using unpolarized beam in storage mode, and demonstrated the expected behavior: Ring tunes and optics were changed by the predicted amounts.

Tests using polarized beam, and also tests in extraction mode, are planned in the latter half of 2000.

Beam polarization in the ring will be measured with a Compton backscatter polarimeter. Recent tests of the polarimeter with unpolarized beam have demonstrated good signal to background, and false asymmetries are understood. Backgrounds are low enough that the polarimeter can be operated simultaneously with BLAST and an internal target. We plan to test the polarimeter with polarized beam at the same time we test the snake.

FUTURE

A program is in place to install a spin-flipper in the SHR. The design is to reverse the spin of the beam stored in the ring on a time scale short compared to the lifetime of the beam. This will facilitate the reduction of systematic errors in the measurement of spin observables. Tests to date have demonstrated the practicality of such a device [7].

With the RF power presently installed in the SHR, and the present SHR optical configuration, beam lifetimes adequate for BLAST experiments cannot be sustained beyond roughly 1 GeV. We plan to extend this limit by reconfiguring the SHR lattice (which will extend the energy limit to 1.3 GeV), and installing additional RF power in the ring (to reach 1.5 GeV). Because this is higher than the maximum energy from the linac, it will be necessary to ramp the ring. Maintaining beam polarization while ramping remains to be investigated.

SUMMARY

Recent accomplishments at Bates have extended the range of physics which can be studied there. Future experiments will utilize polarized beam and the South Hall Ring, in both storage and pulse-stretching modes.

REFERENCES

1. Spayde, D., et al., *Phys. Rev. Lett.* **84**, 1106–1109 (2000).
2. Ito, T., "Parity Violating Electron Scattering and Strange Magnetic Moment", *these proceedings*.
3. Alarcon, R., *Nucl. Phys. A* **663 & 664**, 1111c-1114c (2000).
4. Alarcon, R., "Status of the BLAST Project at Bates", *these proceedings*.
5. Dolfini, S. M., et al., *Nucl. Instr. Methods Phys. Res. A* **344**, 571 (1994).
6. Farkhondeh, M., et al., "Polarized Electrons at MIT-Bates", in *Proceedings of International Workshop on Polarized Gas Targets and Polarized Beams*, edited by R. J. Holts and M. A. Miller, AIP Conf. Proc. 421, 1998, p. 240.
7. Blinov, B. B., et al., *Phys. Rev. Lett.* **81**, 2906–2909 (1998).

Status of the BLAST Project

Ricardo Alarcon* and the BLAST Collaboration

*Arizona State University, Department of Physics and Astronomy, Tempe, AZ 85287-1504
and
MIT-Bates Linear Accelerator Center, Middleton, MA 01949-2846

Abstract. The BLAST detector is presently under construction by an international collaboration and it is scheduled to start commissioning in the summer of 2001. The initial scientific program calls for comprehensive measurements of the spin dependent electromagnetic response of the few body systems. A status report of the construction is presented as well as highlights of some of the planned measurements.

INTRODUCTION

BLAST is a detector designed to study in a comprehensive and precise way the spin-dependent electromagnetic response of few-body nuclei at momentum transfers up to 1 $(GeV/c)^2$ at the MIT-Bates South Hall Ring. It will be used to measure spin-dependent scattering from elastic kinematics to the nucleon resonance region from the proton, deuteron and ^{3}He nucleus using the longitudinally polarized electron beam of the South Hall Ring at beam energies up to 1 GeV. For these measurements we will use polarized internal gas targets of hydrogen, deuterium and ^{3}He which have been developed by our groups and which are at present operating at Bates. Polarized electrons are available at the MIT-Bates accelerator and have been injected into the ring. The longitudinal polarization at the internal target location will be maintained with a Siberian snake. The ability with BLAST to carry out multiparticle detection over a large solid angle using polarized internal targets will provide an unprecedented and unique opportunity to study simultaneously the spin structure of the few-body nuclear ground state, the reaction mechanism and nucleon form factors.

CONSTRUCTION PROJECT

The BLAST detector consists of an eight sector copper coil array producing a toroidal magnetic field, instrumented with two opposing wedge-shaped sectors of wire chambers, scintillation detectors, Čerenkov counters and neutron detectors. The open geometry maximizes acceptance while allowing good momentum and angular resolution and a luminosity capability matched to the densities of the polarized internal targets. The design emphasizes proven technology, commercial electronics, and existing data acquisition system software to achieve low cost and a short implementation time. Clear upgrade possibilities exist so that the detector can evolve to match developing physics priorities.

Construction of BLAST is well underway. The coils have been constructed and, with the appropriate mechanical supports, are scheduled for installation at Bates during the fall of 2000. A polarized proton and deuteron target has been installed and it will be tested in the BLAST configuration as soon as the coils are in operation. A Compton back-scattering polarimeter has been built and tested with an unpolarized beam. The beam polarimeter commissioning will be carry out during the 2000 running cycle. Full production of the detectors is underway following construction and testing of full-size prototypes. Commissioning of the full detector with polarized beam and target is scheduled to begin during the summer of 2001.

SCIENTIFIC PROGRAM

BLAST will carry out a precise measurement of the spin-dependent momentum distribution in few-body nuclei in order to understand the spin structure of few-body systems in terms of the successful theoretical framework which has been developed primarily for unpolarized scattering. Effects such as final-state interactions, meson exchange currents, and the off-shell nature of the bound nucleon can be studied over the broad kinematics range provided by the BLAST detector.

BLAST will provide precise information on nucleon form factors for momentum transfers up to about 1 $(GeV/c)^2$. In particular, BLAST will provide data on the neutron magnetic and electric form factors with both deuteron and ^{3}He targets. These are fundamental quantities which are essential to any description of electromagnetic scattering from nuclei. Because of the significantly larger binding energy of the neutron in ^{3}He compared to that in deuterium, it is possible that the neutron charge distribution in ^{3}He may be modified from its free value. BLAST has the necessary precision to probe for such an effect.

With a tensor polarized deuterium target BLAST will provide precise data from elastic e-d scattering (T_{20}), particularly in the region of the first minimum of the charge form factor of the deuteron. In addition, data will be obtained in the same experiment for the exclusive scattering channels.

BLAST will carry out measurements of spin-dependent electron-proton elastic scattering at low momentum transfers in order to extract the charge radius of the proton. This is a fundamental property of the nucleon and a precise measurement has implications for understanding proton structure as well as being a crucial input for modern tests of QED.

BLAST will measure the $^{16}O(e,e'\alpha)^{12}C$ reaction in order to extract the astrophysical S-factor at stellar energies. Internal targets offer the possibility of detecting the recoil products at very low energy.

Tests of Fundamental Symmetries

Recent Nucleon Decay Results from Super–Kamiokande

Matthew Earl
For the Super–Kamiokande Collaboration

Boston University Department of Physics
Boston Massachusetts 02215

Abstract. The results of searches for various nucleon decay modes using the 61 kton·year (991 days) exposure of the Super-Kamiokande water Cherenkov detector will be presented. In particular, nucleon decay into modes favored by supersymmetric (SUSY) grand unified theories (GUTs) will be stressed.

INTRODUCTION

Grand Unified Theories (GUTs) are an attempt to unify the fundamental forces observed in nature into a single force at some large energy scale. This is seen in the apparent merging of the coupling constants of the strong and electroweak forces at this scale. A consequence of GUTs is the instability of protons and bound neutrons. The prototypical GUT, minimal $SU(5)$ [1], predicts nucleon decay via the exchange of an ultra-heavy gauge boson whose mass is on the order of the energy scale at which the coupling constants merge. The favored mode is $p \to e^+\pi^0$ with a very long lifetime predicted to be around 10^{32} years. Despite this long lifetime, IMB excluded this possibility with a limit of $\tau(p \to e^+\pi^0) > 5.5 \times 10^{32}$ years (90% CL) [2]. Super–Kamiokande has set an even more stringent limit of 1.6×10^{33} years (90% CL) [3].

Although minimal $SU(5)$ has been excluded, GUTs with supersymmetry (SUSY) incorporated provide some hope of observing nucleon decay [4]. In SUSY, the existence of an entirely new spectrum of particles extends the scale at which the coupling constants merge. Raising the mass of the new vector boson. This results in a suppression of $p \to e^+\pi^0$ by about 4 orders of magnitude. Despite this suppression in SUSY, there exist different mechanisms by which nucleons can decay [5–7]. A feature of these mechanisms is that decays into kaons is favored. Models based on $SO(10)$ incorporate neutrino mass and predict that proton decay via $p \to \bar{\nu}K^+$ should be right around the corner [8,9]. In fact, the decay rate into this mode predicted by Pati, Babu, and Wilczek is predicted to be no more than

5.0×10^{33} years, in observable range of Super–Kamiokande [9]. In addition, their model predicts comparable rates for $p \to \mu^+ K^0$. This talk will therefore stress the search for the modes $p \to \nu K^+$ and $p \to \mu^+ K^0$.

I EVENT SAMPLE

Super–Kamiokande is a 50 kiloton (22.5 kiloton fiducial) water Cherenkov detector located at a depth of 1000 meters (2700 meters water equivalent) in the Mozumi mine of the Kamioka Mining and Smelting Company in Gifu Prefecture, Japan. The detector is divided into two regions, the inner and outer detectors (ID and OD). For high energy events, the OD is used to either veto incoming cosmic ray muons or to tag outgoing particles which originated in the ID.

In the search for rare nucleon decays, νs induced by cosmic ray interactions in the upper atmosphere are the limiting background. We detect about 8 fully-contained (FC) atmospheric ν events per day in the fiducial volume (7940 total for the 61 kton·year exposure). FC events are events with the vertex reconstructing inside the ID and all energy deposited in the ID After FC event selection, parameters such as vertex position, number and momenta of particle tracks, particle ID of non-showering (μ-like) or showering (e-like), and the number of decay electrons are determined (see [10] for details).

II NUCLEON DECAY SEARCH

In general, the strategy for nucleon decay searches is straightforward. Monte Carlo samples of the decay mode being studied and atmospheric νs are generated to provide a picture of how signal and background will look inside the detector. In this talk, a 40-year equivalent sample of atmospheric ν MC was used to estimate the background. This sample was normalized to the 61 kton·year livetime of SK and to the observed deficit of ν_μs. More details of the MC generation can be found in [10].

To search for $p \to \bar{\nu} K^+$ three methods were used, two using the primary decay mode of the K^+, $K^+ \to \mu^+ \nu_\mu$ (63.5% B.R.) and one using the mode $K^+ \to \pi^+ \pi^0$ (21.2% B.R.). The final results are shown in table 1 and figure 1.

The first method utilizing the decay $K^+ \to \mu^+ \nu_\mu$ exploits the fact that if a proton not in the outermost shell of O^{16} decays, the remaining N^{15} nucleus is left in an excited state which immediately decays sometimes emitting γ-rays. A proton of O^{16} in the $p_{3/2}$ state decaying leads to the emission of a 6.3-MeV γ-ray from N^{15} and is most prominent (41% BR). Due to the K^+ lifetime of 12 ns, the signal from its decay products will be separated from the signal from the de-excitation γs. The selection criteria were decided to be 1) one μ-like ring 2) one decay 3) 215 MeV/c $< p_\mu <$ 260 MeV/c 4) $N_{hit}^{12ns} > 7$. Criterion 3 represents the fact that most K^+s stop before decaying and therefore in two-body decays the daughter particles will have monochromatic momenta. In criterion 4, N_{hit}^{12ns} is the maximum number of PMTs hit within a 12-ns width sliding timing window starting 100 ns before the signal from the μ^+. The second method takes events which passed criteria 1 and 2 but failed criterion 4 and bins them according to momentum. The spectrum is then searched for an excess around $p_\mu = 236$ MeV/c. The χ^2 method was used to search for this excess by fitting normalization parameters for the proton decay and atmospheric ν MC samples to the data. More details can be found in our previous paper [12]. The final method to search for $p \to \nu K^+$ uses the decay mode $K^+ \to \pi^+ \pi^0$. The selection criteria were decided to be: 1) two e-like rings 2) one decay electron 3) 85 MeV/c^2 < $M_{\gamma\gamma}$ < 185 MeV/c^2 4) 175 MeV/c < $p_{\gamma\gamma}$ < 250 MeV/c 5) 40 PE < Q_{back} < 100 PE. Criteria 1,3, and 4 search for the π^0 with the monochromatic momentum expected. The parameter Q_{back} in criterion 5 is defined as the number of PE detected by the PMTs lying within a cone of 40° opening angle whose axis is opposite the reconstructed direction of the π^0. This parameter is a search for the low momentum π^+ whose direction is opposite that of the π^0. The results are shown in table 1 and figure 1.

Unlike the decay $p \to \nu K^+$ which has an invisible ν, the decay $p \to \mu^+ K^0$ is completely visible. To search for this mode, three methods were used, one utilizing $K_S^0 \to \pi^0 \pi^0$ and two utilizing $K_S^0 \to \pi^+ \pi^-$. The selection criteria for all three methods exploit the kinematics of such events. The results for the 61 kiloton·year data for the search for $p \to \mu^+ K^0$; $K_S^0 \to \pi^0 \pi^0$ are shown in figure 2 and the results for all three methods are shown in table 1

The results presented in this talk are beginning to put a stringent limit on the model by Pati, Babu, and Wilczek [9]. If their model is correct, Super–Kamiokande should begin to see hints of nucleon decay within the next couple of years. Since we are beginning to be limited by background in these modes, our sensitivity is starting to scale as $\sqrt{exposure}$. Because of this, we need to think of new ways to reduce the background.

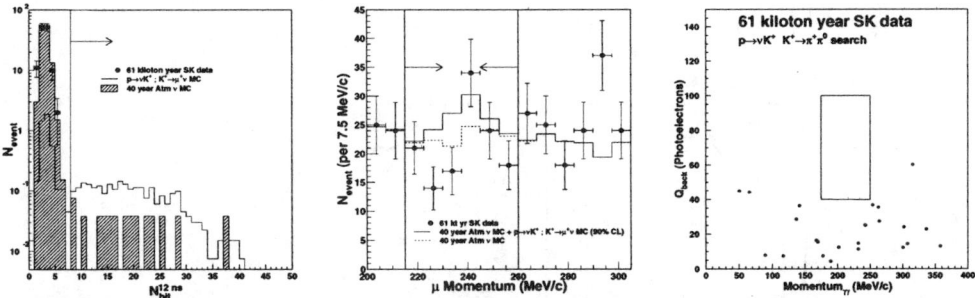

FIGURE 1. Search for $p \to \nu K^+$. First figure shows N_{hit}^{12ns} for proton decay MC (solid), atmν MC (dashed), and 61 kton·year data (points) for the first method. No events pass. The second figure shows the momentum distribution for single ring μ-like events for atmν MC (dotted), atmν + proton decay (solid), and data (points). No excess can be seen. The final figure shows Q_{back} vs. momentum 61 kton·yr. data using the third method.

Mode	Method	efficiency	N_{BG}^{exp}	N_{obs}	τ/B (90%CL)
$p \to \nu K^+$	$K^+ \to \pi^+ \pi^0$	6.8%	1.7	0	6.0×10^{32} yr
	$K^+ \to \mu^+ \nu_\mu$ prompt γ	9.3%	1.0	0	8.2×10^{32} yr
	$K^+ \to \mu^+ \nu_\mu$ spectrum fit	33%	137	128	4.3×10^{32} yr
	combined	49%	-	-	1.7×10^{33} yr
$n \to \nu K^0$	$K_S^0 \to \pi^0 \pi^0$	6.1%	11	5	2.8×10^{32} yr
	$K_S^0 \to \pi^+ \pi^-$	2.0%	2.0	4	0.5×10^{32} yr
	combined	8.1%	-	-	2.5×10^{32} yr
$p \to \mu^+ K^0$	$K_S^0 \to \pi^0 \pi^0$	6.1%	1.0	0	5.4×10^{32} yr
	$K_S^0 \to \pi^+ \pi^-$ (3-ring)	2.8%	0.2	0	2.5×10^{32} yr
	$K_S^0 \to \pi^+ \pi^-$ (2 ring)	5.3%	1.3	0	4.7×10^{32} yr
	combined	14%	-	-	1.2×10^{33} yr

TABLE 1. Summary of results for $p \to \nu K^+$, $p \to \mu^+ K^0$, and $n \to \nu K^0$.

REFERENCES

1. H. Georgi and S. Glashow, *Phys. Rev. Lett* **32**, 438 (1974).
2. C. McGrew et al, *Phys. Rev.* **D59**,052004 (1999).
3. M. Shiozawa et al, *Phys. Rev. Lett.* **81**,3319 (1998).
4. S. Dimopoulos and S. Raby, *Nuc. Phys.* **192**,353 (1981).
5. S. Weinberg, *Phys. Rev.* **D26**,287 (1982).
6. N. Sakai and T. Yanagida, *Nuc. Phys.* **197**,83 (1982).
7. S. Dimopoulos, S. Raby, and F. Wilczek, *Phys. Lett* **112B**,133 (1982).
8. V. Lucas and S. Raby, *Phys. Rev.* **D55**,6986 (1997).
9. K.S. Babu, J.C. Pati, and F. Wilczek, *Nuc. Phys.* **B1566**,33 (2000).
10. Y. Fukuda et al, *Phys. Lett.* **B433**, 9 (1998).
11. H. Ejiri, *Phys. Rev.* **C48**,1442 (1993).
12. Y. Hayato et al, *Phys. Rev. Lett* 83, 1529 (1999).

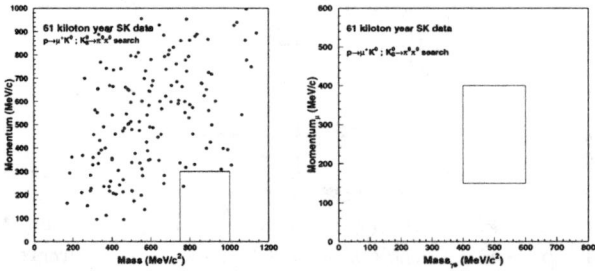

FIGURE 2. Search for $p \to \mu^+ K^0$; $K_S^0 \to \pi^0 \pi^0$. First figure shows the total momentum vs. total mass for events with one μ-like ring and two to four e-like rings with zero or one decays. The second figure shows p_μ vs. e-like mass. e-like mass should be the mass of the kaon.

On the Particle Oscillations in a Medium

V.I.Nazaruk

Institute for Nuclear Research of RAS, 60th October
Anniversary Prospect 7a, 117312 Moscow, Russia.
E-mail: nazaruk@al20.inr.troitsk.ru

Abstract

The new approach to the particle oscillations which allows one to work out problems with variable number of particles is considered. It predicts an enhancement of oscillations in the medium.

1 Introduction

The theory of ab oscillations [1] are based on single-particle model. The interaction of particles a and b with the matter is described by potentials $U_{a,b}$ (potential model). ImU_b is responsible for loss of b-particle intensity. The wave functions $\Psi_{a,b}$ are given by equations of motion.

In some instances there is a need to consider the ab conversion in the matter followed by reaction

$$(a - \text{medium}) \to (b - \text{medium}) \to b + c \to f. \tag{1}$$

Here c is the particle of medium; $b + c \to f$ represents the reaction. The whole process (ab transition, b-medium interaction) takes place in the same layer of matter. An example is the $n\bar{n}$ transitions in the medium followed by annihilation

$$(n - \text{medium}) \to (\bar{n} - \text{medium}) \to f, \tag{2}$$

where f are the annihilation products which should be detected. What this means is b-particle absorption is essential and ImU_b cannot be ignored. The problems of this type are solved indirectly, by means of $\Psi_{a,b}$ and unitarity condition. As is shown below this way is incorrect, because U_b is non-Hermitian. (One can neglect by ReU_b.) This obvious flow should be remedied in any case, that implies a fundamental modification to the model. The direct calculation, i.e. the calculation with variable number of particles is needed. Approach of this kind (field approach with finite time interval (FTA)) is considered below. It reproduces the all results in neutrino oscillations in which $Imf_b(0)$ is ignored as well as standard calculation corresponding to the process (2). We will also show that this standard calculation is wrong.

2 Approach with finite time interval

For definiteness, we deal with the process (2). In this case the annihilation plays a crucial role and distinctions between calculations are extreme. The neutron is described by wave function $n(x) = V^{-1/2} \exp(-i\epsilon_n t + i\mathbf{p}_n \mathbf{x})$, $\epsilon_n = \mathbf{p}_n^2/2m + U_n$, where m and $U_n = const$ are the neutron mass and potential, respectively. The interaction Hamiltonian involves two terms:

$$H_{n\bar{n}}(t) = \epsilon \int d^3 x (\bar{\Psi}_{\bar{n}} \Psi_n + H.c.),$$
$$H(t) = \text{(all } \bar{n} - \text{medium interactions)} - U_n, \qquad (3)$$

$H_I = H_{n\bar{n}} + H$, $\epsilon = (m_2 - m_1)/2 = 1/\tau_{n\bar{n}}$. Here $H_{n\bar{n}}(t)$ is the oscillation Hamiltonian, Ψ_n and $\Psi_{\bar{n}}$ are the fields of n and $\bar{n}$, $m_{\bar{n}} = m$, $m_{1,2}$ are the masses of the stationary states, $\tau_{n\bar{n}}$ is a free-space $n\bar{n}$ oscillation time; ϵ is a small parameter. This model coincides with potential one (see [2,3] for future references) exepting the expression for H. In the potential model $H = \delta U = U_{\bar{n}} - U_n$, $U_{\bar{n}} = ReU_{\bar{n}} - i\Gamma/2 = const$, where $U_{\bar{n}}$ and Γ are the optical potential and annihilation width of $\bar{n}$. We deal with H in the general form (3) as well as in the symplest form $H = \delta U$.

First of all we consider the $n\bar{n}$ transitions with $\bar{n}$ in the final states ($\bar{n}$ are detected). The similar problem takes place in neutrino oscillations. Due to the zero momentum transfer in the vertex corresponding to $H_{n\bar{n}}(t)$ the process amplitude is singular [2]. Let us introduce the evolution operator $U(t) = I + iT(t)$. The interval $(t, 0)$ is determined by the problem conditions. In the lowest order in ϵ we have

$$<\bar{n}0|\, U(t) - I\, |0n> = iT_{\bar{n}i}(t) = (-i) <\bar{n}_p 0|\int_0^t dt_\beta H_{n\bar{n}}(t_\beta) + T^{\bar{n}}(t-0) \int_0^{t_k} dt_\beta H_{n\bar{n}}(t_\beta)\, |0n_p>$$

$$T^{\bar{n}}(t - t_\beta) = \sum_{k=1}^{\infty} (-i)^k \int_{t_\beta}^t dt_1 ... \int_{t_\beta}^{t_{k-1}} dt_k H(t_1)...H(t_k),$$

where $|0n_p>$ and $|0\bar{n}_p>$ are the states of the medium containing the neutron and antineutron with 4-momenta $p = (\epsilon_n, \mathbf{p}_n)$. One obtains

$$T_{\bar{n}i}(t) = -\epsilon t - \epsilon \int_0^t dt_\beta i T_{ii}^{\bar{n}}(t - t_\beta), \qquad (4)$$
$$iT_{ii}^{\bar{n}}(\tau) = <\bar{n}_p 0|\, T^{\bar{n}}(\tau)\, |0\bar{n}_p>,$$

where $\tau = t - t_\beta$. For verification of FTA we calculate the amplitude $T_{ii}^{\bar{n}}$ in the framework of the potential model: $H = \delta U$. The process probability $W_{\bar{n}}(t)$ is

$$W_{\bar{n}}(t) = |\, T_{\bar{n}i}(t)\,|^2 = \frac{\epsilon^2}{|\delta U|^2} [1 - 2\cos(Re\delta U t)e^{-\Gamma t/2} + e^{-\Gamma t}]. \qquad (5)$$

On the other hand potential model gives [3]

$$W_{pot}(t) = 1 - |U_{ii}(t)|^2 = 2Im i(\epsilon/\delta U)^2 [1 - i\delta U t - \exp(-i\delta U t)]. \qquad (6)$$

When $\Gamma = 0$, $W_{pot}(t) = W_{\bar{n}}(t)$. (When $ImU_{\bar{n}} = 0$, $W_{pot}(t)$ is beyond question. If $\Gamma \neq 0$, $W_{pot}(t)$ is wrong.) So the FTA verified by the example of exactly solvable potential model. This fact means that *FTA should reproduce the all results in neutrino oscillations* in which $Imf_b(0)$ is ignored.

Let us return to the process (2) wherein annihilation products are detected, namely, the $n\bar{n}$ transitions in the nuclear matter. We consider the imaginary process $n \to \bar{n} + \Phi$ on the interval $(t/2, -t/2)$. (Such scheme allows to verify and study the FTA.) For decay to be permissible in vacuum put $m_{\bar{n}} = m - 2m_\Phi$. Instead of Eqs.(3) we have $H_I = H'_{n\bar{n}} + H$,

$$H'_{n\bar{n}}(t) = \epsilon' \int d^3x (\bar{\Psi}_{\bar{n}} \Phi^* \Psi_n + H.c.). \tag{7}$$

Introducing the multipliere $\exp(-\alpha |t_\beta|)$, $\alpha > 0$ for realization of adiabatic hypothesis we have

$$T_b(t) = i\epsilon' \frac{1}{\Delta q - i\alpha} <f | T^{\bar{n}}(t/2, -t/2) | 0\bar{n}_{p-q}> NF,$$

$$F = e^{-\alpha|t_k|} - e^{-\alpha t/2} e^{-i\Delta q(t_k + t/2)},$$

$$\Delta q = q_0 - \epsilon_n + (\mathbf{p}_n - \mathbf{q})^2/2m + U_n, \tag{8}$$

Here $<f|$ represents the annihilation products with (n) mesons, $|0\bar{n}_{p-q}>$ is the state of the medium containing the $\bar{n}$ with 4-momenta $p-q$. The normalizing factors of the wave functions of $\bar{n}$ and annihilation mesons included in $<f | T^{\bar{n}}(t) |0\bar{n}_{p-q}>$ and the other those in the multiplier N.

It is easy to verify that: (a) When $q \neq 0$ (q is 4-momenta of particle escaped in the $n\bar{n}$ transition vertex) and $t \to \infty$, the result coincides with S-matrix one. (b) When $q \to 0$ the result converts to one corresponding to process (2). The probability to find the annihilation products is $W_{ann}(t) \approx \epsilon^2 t^2$ [3]. The value $\epsilon^2 t^2 = t^2/\tau_{n\bar{n}}^2 = W_f$ is the free-space $n\bar{n}$ transition probability. Due to the annihilation channel $n\bar{n}$ conversion is practically unaffected by the medium. So $\tau_{n\bar{n}} \sim T_{n\bar{n}}$, where $T_{n\bar{n}}$ is the oscillation time of neutron bound in a nucleus. (c) The functional structure of W_{pot} is wrong.

3 On an anhancement of oscillations

In the region $|Im\delta Ut| \gg 1$, $W_{ann}(t)/W_{pot}(t) \sim \Gamma t \gg 1$ that means an anhancement of oscillationst. In the standard approach $ImU_{\bar{n}}$ leads to suppression of oscillations, in our one this is not the case.

For large part of actual problems the both approaches give an identical result. For two-step processes of the type (1), when b-particle absorption is essential, the potential model is inapplicable ($U_{ii}(t)$ is in error). This has the effect of anhancement of oscillations in our calculation.

The presence of open channels of $\bar{n}$-medium interaction (which are described by $ImU_{\bar{n}}$ in the potential model) entails an anhancement of oscillations, that is also true for any ab transitions.

References

[1] M.L.Good, Phys.Rev. **106** (1957) 591; Phys.Rev. **110** (1958) 550; L.Wolfenstein, Phys. Rev. **D17** (1978) 2369; E.D.Commins and P.H.Bucksbaum, *Weak Interactions of Leptons and Quarks* (Cambridge University Press, 1983); F.Boehm and P.Vogel, *Physics of Massive Neutrinos* (Cambridge University Press, 1987).

[2] V.I.Nazaruk, Phys. Lett. **B337** (1994) 328.

[3] V.I.Nazaruk, Phys. Rev. **C58** (1998) R1884.

Tests of CPT with CPLEAR

CPLEAR Collaboration[1]

University of Athens - University of Basle - Boston University - CERN - LIP and University of Coimbra - Delft University of Technology - University of Fribourg - University of Ioannina - University of Liverpool - J. Stefan Inst. and Phys. Dep., University of Ljubljana - CPPM, IN2P3-CNRS et Université d'Aix-Marseille II - CSNSM, IN2P3-CNRS, Orsay - Paul Scherrer Institut(PSI) - CEA, DSM/DAPNIA, CE-Saclay - KTH-Stockholm - University of Thessaloniki - ETH-IPP Zürich

Presented by Marko Mikuž
University of Ljubljana and Jožef Stefan Institute, Ljubljana, Slovenia

Abstract. Measurements of $K^0/\bar{K}^0$ time-dependent asymmetries in pionic and semileptonic decays with the CPLEAR detector enabled a complete determination of CPT violation parameters in the neutral kaon system. A global fit using the Bell-Steinberger relation yielded values for the CPT violation parameter in K^0 mixing $\text{Im}(\delta) = (2.4 \pm 5.0) \times 10^{-5}$ and $\text{Re}(\delta) = (2.4 \pm 2.8) \times 10^{-4}$, the determination of the latter relying essentially on CPLEAR measurements. In addition, CPT violation parameters in semileptonic decays $\text{Re}(y)$ and $\text{Re}(x_-)$ could be measured for the first time. From these results, the equality of K^0 and $\bar{K}^0$ masses and lifetimes is tested at the 10^{-18} GeV level. Assuming CPT conservation in decays, the 90 % C.L. limit on $|m_{K^0} - m_{\bar{K}^0}|$ can be pushed down to 5×10^{-19} GeV. CPT violation amplitudes in $I = 0$ and $I = 2$ decays into two pions were excluded at the level of 10^{-4} of the corresponding CPT conserving decay amplitudes.

Introduction

The CPT theorem [2] is regarded as an axiom of physics. But nevertheless, it should be subject to stringent experimental verification which can be carried out with highest precision in the neutral kaon system. In the absence of a CPT violating theory, the interpretation of results in terms of phenomenological parameters should be made with as little prejudice as possible.

---

[1] See [1] for the complete list of authors.

Phenomenology of the neutral kaon system

Here only a brief description of the parameters used is given. For a thorough discussion, the reader is directed to numerous reviews, as for example [3].

The time evolution of $|\psi> = a(\tau)|K^0> + \bar{a}(\tau)|\bar{K}^0>$ is given by

$$i\frac{d}{d\tau}\left[\begin{array}{c}a(\tau)\\ \bar{a}(\tau)\end{array}\right] = \hat{H}\left[\begin{array}{c}a(\tau)\\ \bar{a}(\tau)\end{array}\right], \quad (1)$$

with $\hat{H} = \hat{M} - i/2 \cdot \hat{\Gamma}$. Here $\hat{M}$ is the mass and $\hat{\Gamma}$ the decay matrix with eigenvalues $\Lambda_{S,L} = m_{S,L} - i/2 \cdot \Gamma_{S,L}$. CPT and T violation in $K^0 \leftrightarrow \bar{K}^0$ mixing result from the parameters

$$\epsilon_T = \frac{2i}{\Delta\Gamma}\frac{|H_{12}|^2 - |H_{21}|^2}{2(\Lambda_L - \Lambda_S)} \qquad \delta_{CPT} = \frac{H_{22} - H_{11}}{2(\Lambda_L - \Lambda_S)} \quad (2)$$

where $\Delta\Gamma = \Gamma_S - \Gamma_L$. Both T and CPT violation result in CP violation. The phase of ϵ_T is fixed by CP invariant quantities to $\phi_{SW} = \arctan 2\Delta m/\Delta\Gamma \approx \pi/4$. The component of δ_{CPT}, resulting from the mass difference $m_{K^0} - m_{\bar{K}^0} = M_{11} - M_{22}$, denoted $\delta_\perp$, is orthogonal to ϕ_{SW} and the one resulting from $\Gamma_{K^0} - \Gamma_{\bar{K}^0}$ parallel to ϕ_{SW} ($\delta_\parallel$).

Amplitudes for two-pion decays to isospin eigenstates $I_{\pi\pi} = 0, 2$ can be parametrized as

$$<\pi\pi_{I=0,2}|T|K^0> = (A_{0,2} + B_{0,2})e^{i\delta_{0,2}} \quad <\pi\pi_{I=0,2}|T|\bar{K}^0> = (A^*_{0,2} - B^*_{0,2})e^{i\delta_{0,2}}, \quad (3)$$

where $\mathcal{R}eA$ conserves CP, T and CPT, B violates CPT and imaginary parts of A and B violate T. Large $\mathcal{I}mB$ violate unitarity and were not retained in the analysis.

Parameterization of semileptonic decays

$$\begin{array}{ll}<\pi^-\ell^+\nu|T|K^0> = a(1-y) & <\pi^+\ell^-\bar{\nu}|T|\bar{K}^0> = a^*(1+y^*)\\ <\pi^+\ell^-\bar{\nu}|T|K^0> = a^*(x_+ - x_-)^* & <\pi^-\ell^+\nu|T|\bar{K}^0> = a(x_+ + x_-)\end{array} \quad (4)$$

introduces $\mathcal{R}e(a)$ as the dominant amplitude, conserving CP, T and CPT. y induces CPT violation and all the imaginary parts violate T. Violation of the $\Delta Q = \Delta S$ rule in CPT conserving decays is described by x_+ for CPT conserving and by x_- for CPT violating decays.

CPLEAR Experiment

The reader is directed to [4] for a detailed description of the apparatus. The feature of CPLEAR is strangeness tagging at production (see [1] for a brief) enabling measurements of time evolution of an initially pure K^0 or $\bar{K}^0$ state. Parameters were extracted from decay asymmetries, measured for various final states f

$$A_f(\tau) = \frac{R(\bar{K}^0 \to f, \tau) - R(K^0 \to f, \tau)}{R(\bar{K}^0 \to f, \tau) + R(K^0 \to f, \tau)}. \qquad (5)$$

In an asymmetry acceptances cancel to first order and any asymmetry is a manifestation of CP violation.

Direct test of CPT in mixing

In semileptonic decays, CPLEAR could tag the strangeness of the kaon at decay time via the charge of the decay electron using the $\Delta Q = \Delta S$ rule. Four decay rates: $R^+, R^-, \bar{R}^+$ and $\bar{R}^-$, labelled by the initial neutral kaon strangeness and the decay electron charge, could be measured. A direct test of CPT in mixing in principle consists of a comparison of transition probabilities of $K^0(\tau = 0) \to \bar{K}^0(\tau)$ with $\bar{K}^0(\tau = 0) \to K^0(\tau)$. CPLEAR needs to correct for relative detection efficiencies of the primary $K\pi$ pair $\xi = \epsilon(K^+\pi^-)/\epsilon(K^-\pi^+)$, used for strangeness tagging at production, and the secondary $e\pi$ pair $\eta = \epsilon(\pi^+e^-)/\epsilon(\pi^-e^+)$, labelling strangeness at decay. It was shown in [5] that the combined asymmetry

$$A_\delta = \frac{\eta \bar{N}^+ - \alpha_{2\pi} N^-}{\eta \bar{N}^+ + \alpha_{2\pi} N^-} + \frac{\bar{N}^- - \eta \alpha_{2\pi} N^+}{\bar{N}^- + \eta \alpha_{2\pi} N^+} \xrightarrow{\tau \gg \tau_S} 8\mathcal{R}e(\delta_{CPT}) \qquad (6)$$

allows a direct measurement of CPT violation in mixing regardless of CPT in decays or of the validity of the $\Delta Q = \Delta S$ rule. The parameter $\alpha_{2\pi}$, extracted from $\pi^+\pi^-$ decays, is directly related to ξ, for a complete discussion refer to [5]. The published result [5] from the complete sample of 1300000 semileptonic decays reads

$$\mathcal{R}e(\delta_{CPT}) = (3.0 \pm 3.3_{stat} \pm 0.6_{syst}) \times 10^{-4},$$

an improvement of almost two orders of magnitude over the previous value in [6].

Global fit with the Bell-Steinberger equation

To perform a global assessment of the improved measurements in the neutral kaon system, achieved by CPLEAR, an analysis involving unitarity through the Bell-Steinberger equation has been performed [7]. In the Bell-Steinberger equation

$$\mathcal{R}e(\epsilon_T) - i\mathcal{I}m(\delta_{CPT}) = \frac{1}{2(i\Delta m + \frac{\Gamma_S + \Gamma_L}{2})} \sum A_{fS}^* A_{fL} \qquad (7)$$

it was shown that in the amplitude product it is sufficient to consider the dominant two pion as well as the three pion and semileptonic decays

$$\sum A_{fS}^* A_{fL} = \sum BR_{\pi\pi}^S \Gamma_S \eta_{\pi\pi} + \sum BR_{\pi\pi\pi}^L \Gamma_L \eta_{\pi\pi\pi}^* \\ + 2BR_{\pi\ell\nu}^L \Gamma_L [\mathcal{R}e(\epsilon_T - y) - i\mathcal{I}m(\delta_{CPT} + x_+)]. \qquad (8)$$

Here BR are the branching ratios and $\eta_{\pi\pi(\pi)}$ is the CP violation parameter in two(three) pion decays. The sums run over neutral and charged decay channels.

CPLEAR semileptonic data were combined into two asymmetries

$$A'_T = \frac{\eta \bar{N}^+ - \alpha_{2\pi} N^-}{\eta \bar{N}^+ + \alpha_{2\pi} N^-} \quad A_{CPT} = \frac{\bar{N}^- - \eta \alpha_{2\pi} N^+}{\bar{N}^- + \eta \alpha_{2\pi} N^+}, \quad (9)$$

which were fitted with the appropriate phenomenological expressions using the Bell-Steinberger equation and the semileptonic charge asymmetry

$$\delta_\ell = 2\mathcal{R}e(\epsilon_T - \delta_{CPT}) - 2\mathcal{R}e(y + x_-) = (3.27 \pm 0.12) \times 10^{-3} \quad (10)$$

as constraints. Parameter values were taken from [6], enhanced with the latest published CPLEAR results and suitably combined. Care was taken not to include data assuming CPT conservation. The fit yielded

$$\begin{aligned}
\mathcal{R}e(\epsilon_T) &= (164.9 \pm 2.5) \times 10^{-5} & \mathcal{I}m(\delta_{CPT}) &= (2.4 \pm 5.0) \times 10^{-5} \\
\mathcal{I}m(x_+) &= (-2.0 \pm 2.7) \times 10^{-3} & \mathcal{R}e(y) &= (0.3 \pm 3.1) \times 10^{-3} \\
\mathcal{R}e(\delta_{CPT}) &= (2.4 \pm 2.8) \times 10^{-4} & \mathcal{R}e(x_-) &= (-0.5 \pm 3.0) \times 10^{-3}.
\end{aligned}$$

This represents the first evaluation of the CPT violation parameters y and x_- in semileptonic decays and substantial improvements on others except for $\mathcal{R}e(\delta_{CPT})$ which essentially retained its precision given by the direct CPLEAR measurement described above. The limiting quantity in the evaluation of $\mathcal{I}m(\delta_{CPT})$ is the precision of the CP violation parameter in $\pi^0\pi^0\pi^0$ decays. Assuming $\mathcal{I}m(\eta_{000}) = \mathcal{I}m(\eta_{+-0})$ improves it to $\mathcal{I}m(\delta_{CPT}) = (0.5 \pm 2.0) \times 10^{-5}$.

K^0 versus $\bar{K}^0$: mass and lifetime differences

Having measured both $\mathcal{R}e(\delta_{CPT})$ and $\mathcal{I}m(\delta_{CPT})$, CPLEAR is in the unique position to determine simultaneously the mass and decay width differences between the neutral kaon and its anti-particle

$$\delta_\| = \frac{\sin\phi_{SW}}{4\Delta m}(\Gamma_{K^0} - \Gamma_{\bar{K}^0}) = \mathcal{R}e(\delta_{CPT})\cos\phi_{SW} + \mathcal{I}m(\delta_{CPT})\sin\phi_{SW} \quad (11)$$

$$\delta_\perp = \frac{\sin\phi_{SW}}{2\Delta m}(m_{K^0} - m_{\bar{K}^0}) = -\mathcal{R}e(\delta_{CPT})\sin\phi_{SW} + \mathcal{I}m(\delta_{CPT})\cos\phi_{SW} \quad (12)$$

The result of the evaluation, published in [8], is depicted in Fig. 1 and reads

$$m_{K^0} - m_{\bar{K}^0} = (-1.5 \pm 2.0) \times 10^{-18} \text{ GeV} \quad \Gamma_{K^0} - \Gamma_{\bar{K}^0} = (3.9 \pm 4.2) \times 10^{-18} \text{ GeV}$$

with a correlation of $\rho = -0.94$. Using the common assumption of CPT invariance in decay and setting $\mathcal{I}m(\eta_{000}) = \mathcal{I}m(\eta_{+-0})$ leads to the limit on the mass difference

$$|m_{K^0} - m_{\bar{K}^0}| < 5. \times 10^{-19} \text{ GeV} \quad (90\% \text{ C.L.}).$$

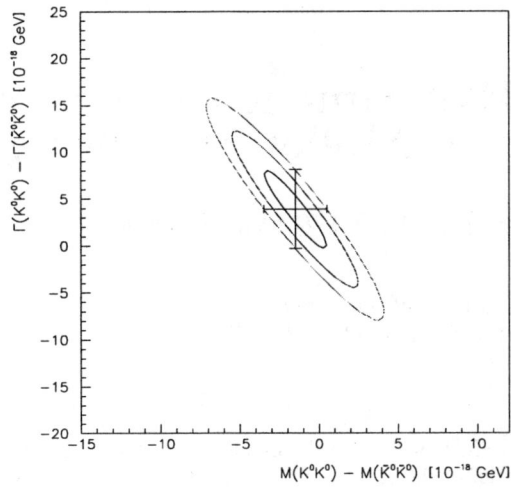

FIGURE 1. K^0 - $\bar{K}^0$ decay-width versus mass difference. The 1 σ, 2 σ and 3 σ ellipses are also shown.

For completeness let us quote, in addition to results on CPT violation parameters in semileptonic decays y and x_-, results on CPT violation in two pion decays [8]

$$\frac{\mathcal{R}eB_0}{\mathcal{R}eA_0} = (2.6 \pm 2.9) \times 10^{-4} \qquad \frac{\mathcal{R}eB_2}{\mathcal{R}eA_2} = (1.3 \pm 4.5) \times 10^{-4}.$$

REFERENCES

1. CPLEAR Collab.: "First direct observation of time-reversal violation," *these proceedings*.
2. J.S. Bell, *Proc. Royal Soc.* **A 231** (1955) 479;
   G.L. Lüders, *Ann. Phys.* **2** (1957) 1;
   R. Jost, *Helv. Phys. Acta* **30** (1957) 409.
3. C. D. Buchanan, *Phys.Rev.* **D45** (1992) 4088.
4. CPLEAR Collab.: R. Adler et al.,*Nucl.Instr.Meth.* **A379** (1996) 76.
5. CPLEAR Collab.: A. Angelopoulos et al., *Phys.Lett.* **B444** (1998) 52.
6. Particle Data Group, *Eur. Phys. J.* **C3** (1998) 1.
7. CPLEAR Collab.: A. Apostolakis et al., *Phys.Lett.* **B456** (1999) 297.
8. CPLEAR Collab.: A. Apostolakis et al., *Phys.Lett.* **B471** (1999) 332.

SLAC E158: An Experiment to Measure Parity Violation in Moller Scattering

M. Woods* (Representing the E158 Collaboration)

Stanford Linear Accelerator Center
Stanford University, Stanford, CA 94309, USA

Abstract. The E158 experiment at SLAC will make the first measurement of parity violation in Moller scattering. The left-right cross-section asymmetry (A_{LR}) in the elastic scattering of a 45-GeV polarized electron beam with unpolarized electrons in a liquid hydrogen target will be measured. This will give a precise measurement of the weak mixing angle, with $\delta(\sin^2 \theta_W^{eff}) \approx 0.0008$ (at $Q^2 = 0.03 GeV^2$).

INTRODUCTION

The weak mixing angle has been precisely measured at the Z-pole by experiments at CERN's LEP and SLAC's SLC machines. But precise measurements away from the Z-pole are needed to probe for certain classes of new physics, and to test the Standard Model predictions for the running of $\sin^2 \theta_W^{eff}$ with Q^2. [1] E158 will measure A_{LR} for small angle Moller scattering with very high statistics. [2] The expected asymmetry at tree level is approximately $3 \cdot 10^{-7}$. Radiative corrections reduce this asymmetry by about 40%. E158 will measure A_{LR} with a relative accuracy of 7%. This will give the best measurement of $\sin^2 \theta_W^{eff}$ away from the Z-pole, and will be sensitive (at 95% confidence level) to additional Z' bosons in the range 600-900 GeV and to compositeness scales to ≈ 10 TeV.

POLARIZED ELECTRON BEAM

The electron beam is produced by photoemission, using a circularly polarized laser beam and a strained GaAs photocathode. The expected electron polarization is 75%. The beam is accelerated to high energy in the two-mile SLAC Linac and then transported through a 24.5° bend angle in the A-line to a liquid hydrogen (LH$_2$)target in End Station A (ESA). The electron beam spin is longitudinal at the source and remains longitudinal in the Linac. In the A-line bend magnets, the spin

---

*) Work supported by the Department of Energy, Contract DE-AC03-76SF00515.

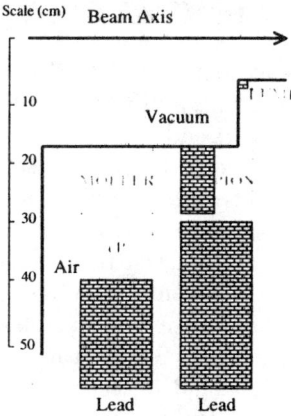

FIGURE 1. Detector Layout for E158 in one azymuthal section.

precesses and becomes longitudinal at the target for a beam energy of 45 GeV. The beam is pulsed at 120 Hz with an intensity of $4 \cdot 10^{11}$ electrons in a 300ns pulse.

The E158 Data Acquisition (DAQ) will control three Pockels cells and a piezo mirror in the polarized laser source. Two of these Pockels cells are used to polarize the laser beam (similar to SLAC's Compton polarimeter laser system). [3] A third Pockels cell is used in a feedback to achieve equal intensities for the left and right beams, and a piezo mirror is used in a feedback to achieve equal steering for the left and right beams. Careful attention is also paid to the laser's residual linear polarization, since photoemission from strained GaAs can have a significant dependence on this. [4]

TARGET, SPECTROMETER AND DETECTORS

The LH_2 target is 1.5 meters long (0.18 radiation lengths), with a volume of 47 liters and a flow rate of 10 meters/s. Eight wire-mesh annular disks in the target cell region introduce turbulence at the 2mm scale (comparable to the beam size) and also induce a transverse velocity component to ensure mixing of the liquid between beam pulses.

A spectrometer consisting of a 3-dipole chicane and 4 quadrupoles is used to spatially separate the Moller-scattered electrons from the Mott (electron-proton scattering) background at the detector plane 60 meters downstream of the LH_2 target. The detector layout is shown in Fig. 1. There are 4 principle detectors: MOLLER to look at the Moller signal, MOTT to look at the Mott background, PION to look at the pion background and LUMI to look at the very forward Mott and Moller electrons with small asymmetry. The MOLLER and MOTT detectors will be quartz fiber calorimeters; the PION detector will be a quartz bar radiator; and LUMI will be a threshold gas Cherenkov detector.

EXPERIMENTAL FEATURES

To perform an accurate measurement of the small Moller physics asymmetry, left-right beam asymmetries must be minimized and accurately measured. First, the beam helicity is chosen in a pseudo-random sequence of pulse quadruplets, $R_1 R_2 \overline{R_1 R_2}, R_3 R_4 \overline{R_3 R_4}...$ The helicity of the first two pulses are chosen pseudo-randomly, while the next two states are complements of the first two. Then two more pseudo-random helicity states are chosen and so on. The experimental analysis will compute the asymmetry from each pulse quadruplet and then average all the quadruplet asymmetry results. Second, the physics asymmetry can be reversed either by inserting a halfwave plate in the polarized light source or by changing the beam energy 3.2 GeV. Third, the LUMI detector should measure an asymmetry a factor 10 smaller than the MOLLER detector. Lastly, we are planning to implement laser and beam optics to allow reversing of some false asymmetries, while leaving the physics asymmetries unchanged. This can be achieved by alternately running with "+I" and "-I" optics configurations. An "I" transformation maps $x \to +x$ and $x' \to +x'$, while an "-I" transformation maps $x \to -x$ and $x' \to -x'$.

RUN SCHEDULE

E158 will perform a series of beam tests in preparation for a first physics run in 2001. The polarized source will be commissioned in a week-long test with full current beam to a diagnostic station at 1.2 GeV, with 3 cavity BPMs and 2 toroids read out by the E158 DAQ. In January 2001, compatible running of ESA beam with PEP-II beam will be studied, and beam will be brought to ESA for spectrometer checkout and background measurements. Beam dithering will be commissioned for mapping detector and BPM sensitivities to energy, position and angle effects. The pulse-to-pulse fluctuations of beam and detector signals will be analyzed. It will be challenging to achieve the goal of $< 2 \cdot 10^{-4}$ ($< 1 \cdot 10^{-4}$) fluctuations in the normalized MOLLER (LUMI) signals.

E158 plans to have a 2-month physics run in Spring 2001. Goals for this run are to make the first observation of parity violation in Moller scattering and a modest measurement of the weak mixing angle, with an uncertainty $\delta(sin^2\theta_W)$ of 0.0025. The main physics run is expected to occur one year later in Spring 2002. In this run, E158 hopes to make the best measurement of the weak mixing angle away from the Z-pole with an uncertainty of $\delta(sin^2\theta_W) = 0.0008$.

REFERENCES

1. W. Marciano, hep-ph/0003049 (2000).
2. SLAC-PROPOSAL-E-158 (1997).
3. M. Woods, hep-ex/9611005 (1996).
4. R.A. Mair et al., Phys. Lett. **A212**, 231 (1996).

Parity Violating Electron Scattering on the Proton and Deuteron at Backward Angles

Takeyasu M. Ito
for the SAMPLE collaboration

W.K.Kellogg Radiation Laboratory
California Institute of Technology
Pasadena, CA 91125

Abstract. The parity violating asymmetry in quasielastic electron scattering from the deuteron at backward scattering angles has been recently measured for the first time. Combined with the previously performed similar measurement on the proton, this measurement provides a determination of both the proton's strange magnetic form factor G_M^s and the axial vector e-N form factor G_A^e. A preliminary analysis indicates that G_M^s is slightly positive but consistent with zero and that $G_A^e(T=1)$ is in substantial disagreement with the theoretical estimate.

INTRODUCTION

The measurement of the neutral weak magnetic form factor of the proton provides an important clue to the quark flavor structure of the nucleon: combined with the known (electromagnetic) magnetic form factors of the proton and neutron, it allows a separation of the proton's magnetic form factor into the three contributing flavors of quarks (up, down and strange) [1]. To the lowest order, the neutral weak magnetic form factor of the proton G_M^Z can be related to the known electromagnetic form factors and a contribution from strange quarks as follows:

$$G_M^Z = (G_M^p - G_M^n) - 4\sin^2\theta_W G_M^p - G_M^s, \qquad (1)$$

where G_M^p and G_M^n are the electromagnetic magnetic form factors of the proton and neutron, θ_W is the weak mixing angle, and G_M^s is the contribution from strange quarks. Thus, the measurement of G_M^Z provides unique window to study the role of the strange quark-antiquark "sea" in the electromagnetic structure of the nucleon at low energies.

It is well established that parity violating electron scattering is sensitive to the neutral weak current [2]. Not only is it sensitive to the neutral weak vector current, but it is also sensitive to the axial current. Unlike the case of ν-N scattering, in e-N scattering, the axial form factor G_A^e receives an additional contribution from the anapole form factor and can be written as

$$G_A^e = G_A^Z + \eta F_A + R^e, \qquad (2)$$

where G_A^Z is the contribution from Z-exchange, η is a constant ($\eta = \frac{8\pi\sqrt{2}\alpha}{1-4\sin^2\theta_W} = 3.45$), F_A is the nucleon anapole form factor [3], and R^e is a radiative correction. The anapole form factor is the parity violating coupling of the photon to the nucleon and is generated at the fundamental level from the weak interaction between quarks in the nucleon. Thus, parity violating electron scattering also provides iteresting and unique information on the axial vector structure of the nucleon.

At the backward angles, the parity-violating asymmetry for quasielastic scattering on the deuteron for the incident electron energy of 200 MeV can be written as

$$A_d = \left[\frac{0.049}{\sigma_d}\right]\left[\frac{-G_F Q^2}{\pi\alpha\sqrt{2}}\right]\left[1 - 0.22 G_A^e(T=1) - 0.10 G_M^s\right]. \qquad (3)$$

The similar expression for elastic electron scattering on the proton at 200 MeV is

$$A_p = \left[\frac{0.026}{\sigma_p}\right]\left[\frac{-G_F Q^2}{\pi\alpha\sqrt{2}}\right]\left[1 - 0.24 G_A^e(T=1) - 0.61 G_M^s\right]. \qquad (4)$$

G_F is the Fermi coupling constant and α is the fine structure constant. σ_d and σ_p are defined from $\sigma_{(p,n)} = \epsilon(G_E^{(p,n)})^2 + \tau(G_M^{(p,n)})^2$ and $\sigma_d = \sigma_p + \sigma_n$, where ϵ and τ are kinematic factors. The asymmetries are given in units of ppm and the form factors in units of nuclear magnetons. Thus, measurement of both A_p and A_d provides a determination of both the strange magnetic form factor G_M^s and the isovector axial form factor $G_A^e(T=1)$. The contribution from the isoscaler piece of G_A^e, although also uncertain, is small.

The measurement of A_p has been already published [4]. Below, we present our new measurement on A_d and the results from preliminary combined analysis of A_p and A_d.

EXPERIMENT AND RESULTS

The experiment was performed using the SAMPLE apparatus at the MIT/Bates Linear Accelerator Center. The apparatus was essentially the same as in Ref. [4]; the hydrogen target was replaced with liquid deuterium and borated polyethylene shielding was installed between the photomultiplier tubes and the target. This additional shielding was neccessary to reduce the background from neutrons produced in the target from $d(\gamma,n)p$ reactions.

A beam of longitudenally polarized electrons was generated from circularlly polarized laser light incident on a bulk GaAs crystal, accelarated to 200 MeV, and

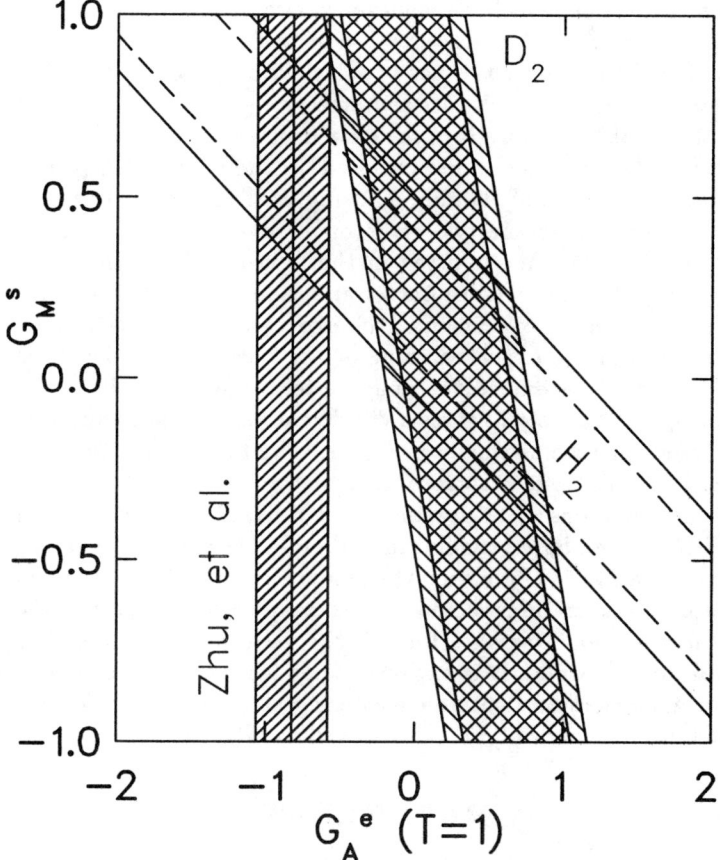

FIGURE 1. A result of a combined analysis of the data from the two SAMPLE measurements. The two error bands from the hydrgeon experiment [4] and the preliminary deuterium experiment are indicated. The inner hatched region includes the statistical error and the outer represents the systematic uncertainty added in quadrature. Also plotted is the estimate of the isovector axial e-N form factor $G_A^e(T=1)$ obtained by using the anapole form factor and radiative corrections of Ref. [5].

then introduced into the deuterium target. The beam was pulsed at 600 Hz and each pulse had a duration of 25 μs. The helicity of the beam was randomly chosen for each of ten consecutive pulses and the complement helicities were used for the next ten pulses. Electrons scattered at backward angles were detected by ten large solid angle air Čerenkov detectors. The asymmetry in the detector signal yield, normalized to incident beam charge, was computed for pulse pairs separated by 1/60 s to minimize systematic errors due to 60 Hz line noise. The measured asymmetry was corrected for the beam polarization ($\sim 36\%$) and the background dilution factor to obtain the physics asymmetry.

A result of a combined analysis of the data from the two SAMPLE measurements is shown in Fig. 1. The constraints imposed on the values of G_M^s and $G_A^e(T=1)$ from the measured values of A_d and A_p using Eqs. (3) and (4) are shown as error bands. The region where the two error bands overlap provides a determination of G_M^s and $G_A^e(T=1)$. Also plotted in the figure is the estimate of $G_A^e(T=1)$ obtained by using the anapole form factor and radiative corrections of Zhu et al. [5].

Prior to running these experiments, the expected value of G_M^s was in the range of -0.5 to 0 [6], and the expected value of G_A^e was $\sim -0.71 \pm 0.20$ as a result of substantial modification due to the anapole term and the radiative correction [7] (their recent update in Ref. [5] gives a consistent value). The experiments indicate a rather different picture as shown in Fig. 1. The best value of G_M^s appears to be slightly positive, consistent with zero, and the best value of G_A^e indicates that the substantial modifications of G_A^e predicted in Refs. [7,5] are not only present, but probably with an even larger magnitude. From a theoretical standpoint, the most uncertain contribution to G_A^e is from the anapole term and the experimental results can be interpreted as an unexpectedly large anapole form factor of the nucleon.

Clearly the situation warrants further theoretical study as well as additional experimental investigation. It is expected that a new measurement of parity violating quasielastic electron-deuteron scattering at lower energy [8] will provide an improved determination of $G_A^e(T=1)$ as well as G_M^s.

REFERENCES

1. Kaplan, D., and Manohar, A., *Nucl. Phys.* **B310**, 527 (1988).
2. McKeown, R. D., *Phys. Lett.* **B219**, 140 (1989), Beck, D. H., *Phys. Rev. D* **39**, 3248 (1989).
3. Zel'dovich, I., *JETP Lett.* **33**, 1531 (1957).
4. Mueller, B. et al., *Phys. Rev. Lett.* **78**, 3824 (1997), Spayde, D. T. et al., *Phys. Rev. Lett.* **84**, 1106 (2000).
5. Zhu, S.-L. et al., hep-ph/0002252, to appear in Phys. Rev. D.
6. McKeown, R.D., in *Parity Violation in Atoms and Polarized Electron Scattering*, Frois, B. and Bouchiat, M. A., Eds. World Scientific, 1999, p.423.
7. Musolf, M. J. and Holstein, B. R., *Phys. Lett.* **B242**, 461 (1990).
8. MIT-Bates experiment 00-04, SAMPLE Collaboration (Ito, T. M., spokesperson).

Recent Results on the Muon Anomalous Magnetic Moment from BNL E821

C.J. Gerco Onderwater[1], H.N. Brown[3], G. Bunce[3], R.M. Carey[2], P. Cushman[8],
G.T. Danby[3], P.T. Debevec[1], H. Deng[12], W. Deninger[1], S.K. Dhawan[12],
V.P. Druzhinin[10], L. Duong[8], W. Earle[2], E. Efstathiadis[2], F.J.M. Farley[12],
G.V. Fedotovich[10], S. Giron[8], F.E. Gray[1], M. Grosse-Perdekamp[12], A. Grossmann[6],
U. Haeberlen[7], E.S. Hazen[2], D.W. Hertzog[1], V.W. Hughes[12], K. Jungmann[6],
D. Kawall[12], B.I. Khazin[10], J. Kindem[8], F. Krienen[2], I. Kronkvist[8], R. Larsen[3],
Y.Y. Lee[3], I. Logashenko[2], R. McNabb[8], W. Meng[3], J. Mi[3], J.P. Miller[2], W.M. Morse[3],
Y. Orlov[4], C. Özben[3], C. Pai[3], J. Paley[2], C.C. Polly[1], J. Pretz[12], R. Prigl[3],
G. zu Putlitz[6], S.I. Redin[12], O. Rind[2], B.L. Roberts[2], N.M. Ryskulov[10], R. Sanders[3],
S. Sedykh[1], Y. Semertzidis[3], Yu.M. Shatunov[10], E. Solodov[10], A. Steinmetz[12],
L.R. Sulak[2], C. Timmermans[8], A. Trofimov[2], D. Urner[1], D. Warburton[3], D. Winn[5],
A. Yamamoto[9], D. Zimmerman[8],

[1] *Illinois,* [2] *Boston,* [3] *Brookhaven.* [4] *Cornell,* [5] *Fairfield,* [6] *Heidelberg,* [7] *Max Planck Inst. für Med. Forschung, Heidelberg* [8] *Minnesota,* [9] *KEK,* [10] *Budker Inst. of Nucl. Phys., Novosibirsk,* [11] *Tokyo Inst. of Tech.,* [12] *Yale.*

Abstract. The new muon $(g - 2)$ experiment at BNL aims at a final precision of 0.35 ppm on the muon anomaly. The experiment has completed four runs beginning in 1997 and has sub-ppm data on tape. Analyzed data include a published 13 ppm result from our initial commissioning run and a recently completed 5 ppm result from our 1998 initial muon injection run. We are actively working on the remainder of the data analysis and on systematic studies.

INTRODUCTION

The quest for phenomena that cannot be explained by the Standard Model (SM) of fundamental interactions is the focal point of present day particle physics. Although the SM is consistent with all available experimental data, there is increasingly strong theoretical reason and also circumstantial evidence that points to the existence of new physics below the TeV scale [1] (*e.g.* the indication of neutrino oscillations at Super-Kamiokande, the solar neutrino problem, the apparent

TABLE 1. Standard Model predictions for a_μ [5].

Variable	Value ($\times 10^{11}$)	relative contribution (ppm)
a_μ(SM)	116 591 628(77)	± 0.66
a_μ(QED)	116 584 705.7(1.8)	± 0.015
a_μ(had)	6771(77)	57.79 ± 0.66
a_μ(weak)	151(4)	1.30 ± 0.03

non-unitarity of the Cabibbo-Kobayashi-Maskawa matrix and the consequences of atomic parity non-conservation). Precision experiments at low energy can probe for deviations from the SM predictions and provide complementary constraints to those obtained in high-energy collision experiments. The uncertainty principle mandates that even heavy particles make small contributions to low-energy observables, such as the magnetic moment of the muon.

THEORY

The magnetic moment of a particle is related to its spin, $\vec{\mu} = g\left(\frac{e}{2m}\right)\vec{s}$. According to Dirac's theory, the proportionality constant g, or gyromagnetic ratio, is equal to 2 for pointlike, *i.e.* structureless, particles. The discovery that $g_e \neq 2$ for electrons [2] and the calculation by Schwinger [3] predicting that (to first order in α) the radiative correction to g_e is α/π, were the first steps towards the very complex present day calculations.

The muon magnetic moment expressed as [4]

$$\mu_\mu = (1 + a_\mu)\frac{e\hbar}{2m_\mu} \quad \text{where} \quad a_\mu = \frac{(g-2)}{2}. \tag{1}$$

The SM prediction for the anomaly a_μ is traditionally split in three parts, $a_\mu(\text{SM}) = a_\mu(\text{QED}) + a_\mu(\text{hadronic}) + a_\mu(\text{weak})$. Any contribution from "new" physics will be reflected in a measured value which does not agree with this prediction. The presently most accurate predictions for each of these contributions are given in table 1. The uncertainty is dominated by that in a_μ(hadronic), which cannot be calculated directly, but must be determined from data.

EXPERIMENT

For polarized muons moving in a uniform magnetic field $\vec{B}$ which is perpendicular to the muon spin direction and to the plane of the orbit, and with an electric quadrupole field $\vec{E}$ for vertical focusing, the difference angular frequency, ω_a, between the spin precession frequency ω_s and the cyclotron frequency ω_c, is given by [6]

$$\vec{\omega}_a = -\frac{e}{m}\left[a_\mu \vec{B} - \left(a_\mu - \frac{1}{\gamma^2-1}\right)\vec{\beta}\times\vec{E}\right]. \qquad (2)$$

The dependence of ω_a on the electric field is eliminated by storing muons with the "magic" γ_μ=29.3, which corresponds to a muon momentum $p_\mu = 3.09$ GeV/c. Hence measurement of B and of ω_a determines a_μ.

The magnetic field strength is measured using proton NMR, whereas ω_a is extracted from the modulation of the positron rate of in-flight decay $\mu^+ \to e^+\nu_e\bar{\nu}_\mu$. For an energy threshold E, this rate is [6,7]

$$N(t) = N(E)e^{-t/\gamma\tau}\left[1 + A(E)\cos\left(\omega_a t + \phi(E)\right)\right], \qquad (3)$$

with $N(E)$ a normalization constant and $A(E)$ the parity violating asymmetry parameter. Maximum statistical power is obtained for $E = 1.8$ GeV.

RESULTS

Our experiment measures the frequency ratio $R = \omega_a/\omega_p$, where ω_p is the free proton NMR frequency in our magnetic field. Including the pitch and electric field corrections, we obtain $R = 3.707\ 201(19) \times 10^{-3}$, where the 5 ppm error includes a 1 ppm systematic error estimate. We obtain a_{μ^+} from $a_{\mu^+} = R/(\lambda - R)$ = 116 591 91(59) $\times 10^{-10}$ in which $\lambda = \mu_\mu/\mu_p = 3.183\ 345\ 39(10)$. [4,8] This new result is in good agreement with the mean of the CERN measurements for a_{μ^+} and a_{μ^-}, [4] and our previous measurement of a_{μ^+}, [9].

Assuming CPT symmetry, the weighted mean of the four measurements gives a new world average of $a_\mu = 116\ 592\ 10(46) \times 10^{-10}$ (± 3.9 ppm), which agrees with the standard model to within one standard deviation.

REFERENCES

1. P. Herczeg, in: *Precision Tests of the Standard Electroweak Model*, Advanced Series on Directions in High-Energy Physics, Vol. 14, edited by P. Langacker, Singapore: World Scientific, 1995, pp. 786.
2. J.E. Nafe, et al., Phys. Rev. **71**, 914 (1947), D.E. Nagl, et al., Phys. Rev. **72**, 971 (1947) P. Kusch and H.M. Foley, Phys. Rev. **74**, 250 (1948).
3. J. Schwinger, Phys. Rev. **73**, 416 (1948) and Phys. Rev. **75**, 898 (1949).
4. Particle Data Group, Eur. Phys. J. C **3**, 1 (1998).
5. V.W. Hughes and T. Kinoshita, Rev. Mod. Phys. **71**, S133 (1999).
6. J. Bailey, et al., Nucl. Phys. B **150**, 1 (1979).
7. F.J.M. Farley and E. Picasso in *Quantum Electrodynamics*, ed. T. Kinoshita, (World Scientific, Singapore, 1990), p. 479.
8. W. Liu, et al., Phys. Rev. Lett. **82**, 711 (1999).
9. R.M. Carey et al., Phys. Rev. Lett. **82**, 1632 (1999).

A Precision Measurement of the Michel Parameter ξ'' in Polarized Muon Decay

R. Prieels*, P. Van Hove*, J. Deutsch*, J. Govaerts*, P. Knowles [1],
R. Medve*, A. Ninane*, J. Egger†, F. Foroughi†, X. Morelle†,
L. Simons†, N. Danneberg+, W. Fetscher+, M. Hadri+, C. Hilbes+,
K. Kirch [2], J. Lang+, O. Naviliat [3], J. Sromicki+

*Université catholique de Louvain, B-1348 Louvain-la-Neuve, Belgium
† Paul Scherrer Institut, CH-5232 Villigen, PSI, Switzerland
+Institut für Mittelenergiephysik, Eidgenossische Technische Hochschule, Zurich, Switzerland

Abstract.
Muon decay allows for the most precise experiments in testing the Standard Model of electroweak interactions. An experiment aimed to measure one of the Michel parameters with an improvement of 70 is described. A production run is foreseen for summer 2000.

I INTRODUCTION

The Michel parameters are phenomenological quantities which describe the various observables in muon-decay and they can be related to the leptonic coupling constants, only one of them (g_{LL}^V) being non-zero in the Standard Model (SM). Most of the Michel parameters are known to have values close to the SM ones with precisions better than a few percent [1]. One notable exception is the parameter ξ'', or the combination $(\xi''/\xi\xi' - 1)$ which is zero in the SM. Its present experimental value is $(\xi''/\xi\xi' - 1) = -0.35 \pm 0.39$ [2]. This combination governs the angular and energy dependence of the positron longitudinal polarization in polarized muon decay:

$$P_L(x,z) = \xi' + \frac{P_\mu z \xi \xi'(2x-1)}{(3-2x) + P_\mu z \xi (2x-1)} \left(\frac{\xi''}{\xi\xi'} - 1\right), \qquad (1)$$

where $0 \leq x \leq 1$ is the normalized energy and $z = \cos\theta$, θ being the angle between the muon spin and the positron momentum; the parameters ξ, ξ', ξ'' are all equal

---
[1] On leave at Uni. Fribourg, Switzerland
[2] Presently at LANL, USA
[3] Presently at LPC Caen, France

to 1 in the SM. As can be seen in the previous formula, values of x close to 1 and z close to -1, strongly enhance the impact of a non vanishing $(\xi''/\xi\xi' - 1)$ on P_L for highly polarized muons. We will undertake relative P_L measurements. The longitudinal polarization will be measured versus the momentum of the positrons near the end point, and it will also be compared to its value when the muon is unpolarized, yielding the relative positron longitudinal polarization $R(P_\mu) = P_L(x, z, P\mu \neq 0)/P_L(x, z, P\mu = 0)$. This procedure will lead to the evaluation of the combination $(\xi''/\xi\xi' - 1)$. Our aim is to reach this quantity with a precision of $5.\,10^{-3}$ i.e. two orders of magnitude better than the precision obtained in [2].

II THE SETUP

The apparatus needed is show in figure 1. A surface muons beam selected by the $\pi E3$ beam line at the Paul Sherrer Institute (PSI) cross a Wien filter in order to divert the positrons it contains. The filtered muons, backward polarized at 95%, are then stopped in targets that either maintain (Al) or destroy (S) the muon polarization. The muon polarization and rate are monitored using three telescopes (Ti). A spectrometer consisting in three solenoidal magnets collects the high energy decay positrons, provides a region of uniform magnetic field for energy measurement with a resolution of about 1 MeV FWHM, and finally realizes a good focusing on the polarimeter. The positron polarimeter, based on the spin dependence of Bhabha scattering (BB) and annihilation in flight (AIF) of positrons on polarized electrons, consists of 2 Vacoflux foils with opposite magnetization and interleaved wire chambers allowing to determine the nature of the interaction and the foil in which it took place. Follows a hodoscope in coincidence with a wall of 127 BGO crystals which identifies e^+e^- or $\gamma\gamma$ clusters and measures their energies. The full polarimeter is magnetically well shielded with a Big Iron Box. The BB and AIF event rate asymmetry from the two foils, with opposite magnetization, are proportional to the longitudinal polarization of the positrons.

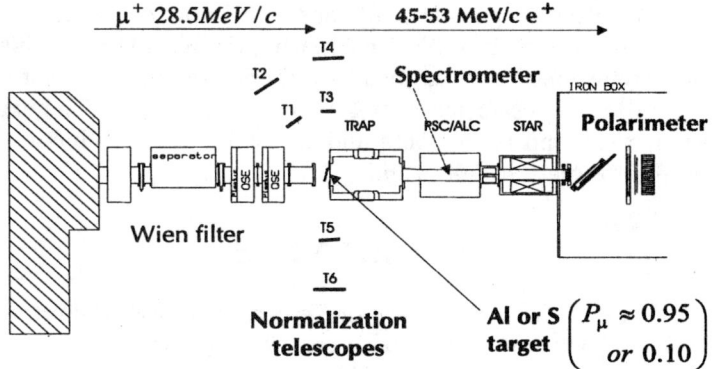

FIGURE 1. *Full experimental setup.*

TABLE 1. *Measured analyzing powers for weak and strong software requirements applied to a fraction of the data. The mean result for the strong requirements is compared to the asymmetry expected from the Monte Carlo simulation. All numbers are in 10^{-2} unit. Errors are in parenthesis.*

	weak	strong	Expected
ANI foil1	-1.96(0.69)	-3.75(0.97)	-3.2
ANI foil2	0.94(0.82)	1.56(1.12)	2.6
Bhabha foil1	1.13(0.64)	1.75(0.76)	1.2
Bhabha foil2	-0.83(0.36)	-1.25(0.41)	-1.1

III THE TEST RESULTS

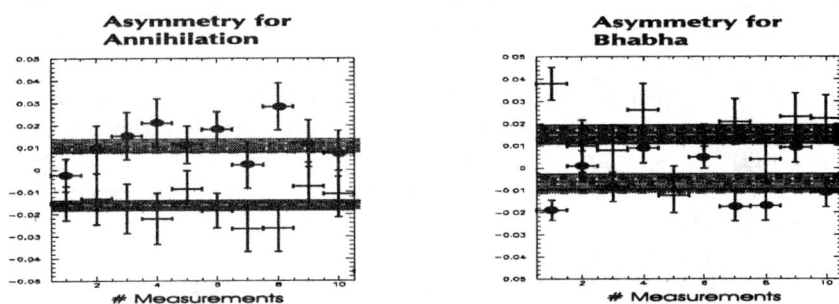

FIGURE 2. *ANI and Bhabha experimental asymmetry. Each point corresponds to events occurring in one foil and is computed from two data files of opposite magnetization. The big dots correspond to processes occurring in the 2nd foil. As is clearly seen, the analyzing power changes sign between Bhabha and ANI events*

Results of 20 hours of data taking, obtained from a very crude analysis with no trajectory retracing are shown in figure 2. In this case we expect a global reduction of the analyzing powers due to the uncorrelated background and to misinterpretations of real events. Severe (geometrical) constraints applied to the data eliminate this problem. This is shown in table 1 comparing the asymmetry values resulting from weak and strong constraints applied to a fraction of the data. Not shown here, the positron momentum resolution is of 1 MeV/c as expected. The experiment is now well tested, the setup is complete and ready for data production. The goal can be achieved in 6 weeks of full beam.

REFERENCES

1. W. Fetscher and H.-J. Gerber, in *Precision Tests of the Standard Electroweak Model*, ed. P. Langacker (World Scientific, Singapore, 1995), pp. 657-705.
2. H. Burkard et al, *Phys. Lett.* **B150** (1985) 242.

New Results on Strange Form Factors of the Proton

R. Holmes* for the HAPPEX Collaboration

*Syracuse University
Syracuse, New York 13244*

Abstract. At the Thomas Jefferson National Accelerator Facility, we have studied the elastic scattering of polarized electrons from hydrogen. The resulting parity-violating electroweak asymmetry is sensitive to the contributions of strange quarks to the nucleon form factors at a level that is of theoretical interest. Using events at a laboratory scattering angle of 12.3° and $\langle Q^2 \rangle = 0.477$ GeV/c, we measure the linear combination of strange form factors $(G_E^s + 0.39 G_M^s)/(G_M^{p\gamma}/\mu_p) = 0.091 \pm 0.054 \pm 0.039$, where the first error is the quadratic sum of our systematic and statistical errors and the second error is due to uncertainty in nucleon form factors.

The nucleon consists of three valence quarks embedded in a "sea" of gluons, quarks, and antiquarks. Over a decade ago the EMC collaboration published data on spin-dependent structure functions [1] which indicated the valence quark contribution to the nucleon spin is small. It then is reasonable to ask whether sea quarks make significant contributions to these and other nucleon observables. In particular, what effects do strange quarks have on nucleon properties?

Kaplan and Manohar [2] have suggested strange quarks might contribute to the charge radius and magnetic moment of the nucleon. These strange matrix elements cannot be isolated in unpolarized scattering experiments. However, measuring the parity-violating electroweak asymmetry in elastic electron-proton scattering allows us to disentangle the strange quark contributions to these quantities. Several experiments have reported significantly nonzero measurements of this quantity [3-6]. I report here on the most precise measurement of this asymmetry to date and discuss its implications.

Because the electron coupling to the Z boson is helicity dependent, unlike its coupling to the photon, the parity violating asymmetry

$$A^{PV} = \frac{\sigma_R - \sigma_L}{\sigma_R + \sigma_L} \quad (1)$$

probes the weak part of the electron-proton scattering amplitude. Here $\sigma_{R(L)}$ is the cross section for scattering right (left) polarized electrons.

The electromagnetic and weak form factors of the proton $F_{1,2}^{\gamma(Z)p}$ may be written in terms of quark form factors $F_{1,2}^{u(d)(s)}$. Assuming isospin symmetry, the neutron electromagnetic form factors $F_{1,2}^{\gamma n}$ can also be written in terms of the same quark form factors. Then A^{PV} may be expressed in terms of $F_{1,2}^{\gamma p}$, $F_{1,2}^{\gamma n}$, and $F_{1,2}^{s}$. In practice one uses the Sachs form factors $G_E = F_1 - \tau F_2$ and $G_M = F_1 + F_2$, where $\tau = Q^2/4M_p$. Then A^{PV} is given by

$$A^{PV} = \left[\frac{-G_F M_p^2 \tau}{\pi\alpha\sqrt{2}}\right] \times$$
$$\left\{(1 - 4\sin^2\theta_W) - \frac{[\varepsilon G_E^{p\gamma}(G_E^{n\gamma} + G_E^s) + \tau G_M^{p\gamma}(G_M^{n\gamma} + G_M^s)]}{\varepsilon(G_E^{p\gamma})^2 + \tau(G_M^{p\gamma})^2}\right\} + A_A \quad . \quad (2)$$

Here $\varepsilon = [1 + 2(1 + \tau)\tan^2(\theta/2)]^{-1}$, where θ is the electron scattering angle in the laboratory, and A_A is an axial term which is small at forward angles.

In the $\tau \to 0$ limit, G_E^s and G_M^s are related to the strange charge radius parameter ρ_s and strange magnetic moment μ_s by the limits $G_E^s \to \tau\rho_s$ and $G_M^s \to \mu_s$.

Data were taken in 1999 using a 3.3 GeV, 67-76% polarized electron beam of about 35 μA incident on a 15 cm liquid hydrogen target in Jefferson Laboratory's Hall A. The polarized electron beam was produced using circularly polarized laser light incident on a strained GaAs cathode. To reduce false asymmetries, a half-wave plate was inserted in the laser beam or removed every 1-2 days during the run; a total of 10 data sets with half-wave plate in, alternating with 10 data sets with half-wave plate out, were taken. Beam polarization was measured using a Møller polarimeter and a new Compton polarimeter.

Both of Hall A's High Resolution Spectrometers were configured to focus elastically scattered electrons at $\theta = 12.3°$ onto lead-Lucite sandwich Čerenkov counters installed close to the focal planes. The beam polarization was changed pseudo-randomly at 15 Hz, with a 30 ms "window" of the randomly chosen polarization followed by another "window" with the opposite polarization. The signals from the two Čerenkov counter phototubes, a beam current monitor, five beam position monitors (two x and y measurements near the target and one x measurement at a high dispersion point to monitor energy fluctuations), and several other diagnostic devices were integrated over each window in custom-built integrating 16-bit ADC modules.

Helicity-correlated changes in the beam intensity, position, angle, or energy could give rise to false asymmetries in the integrated rates. We instead used rates normalized by the measured beam intensity, and reduced our sensitivity to higher order effects by using a feedback system to minimize the beam intensity asymmetry. By monitoring the beam position and calibrating our sensitivity using steering coils and an energy vernier, we were able to compute corrections due to position, angle, and energy differences.

Figure 1 shows the measured asymmetries in the normalized detector rates as a function of data set number; the asymmetry changes sign corresponding to the

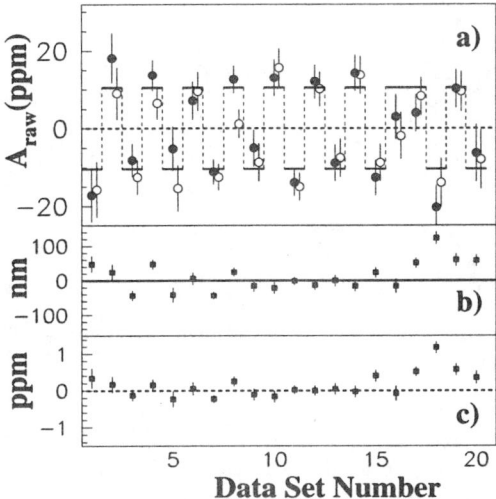

FIGURE 1. a) Raw asymmetries in the two arms versus data set. The $\chi^2 = 33.7$ for 39 degrees of freedom. b) Helicity-correlated horizontal position difference measured near the target. c) Correction to asymmetry for one spectrometer due to all of the beam parameter differences.

presence or absence of the half-wave plate. Also shown in Figure 1 are the most significant beam position differences and the total correction for each data set. Corrections averaged over all data sets were 0.02 ± 0.02 ppm.

We measured backgrounds giving a correction of $1.2 \pm 0.6\%$, and uncertainty in Q^2 contributed 3.8% to our error. Polarimetry contributed 3.2% to our systematic error. The final measured physics asymmetry for the 1999 data was $A = -14.62 \pm 1.07(\text{stat}) \pm 0.54(\text{syst})$ ppm at $\langle Q^2 \rangle = 0.477$ $(\text{GeV}/c)^2$. Combining with the 1998 data [4] yields a final result of $A = -14.60 \pm 0.94(\text{stat}) \pm 0.54(\text{syst})$ ppm.

Using published values for the nucleon form factors [7–14], we obtain $(G_E^s + 0.39 G_M^s)/(G_M^{p\gamma}/\mu_p) = 0.091 \pm 0.054 \pm 0.039$, where the first error is the quadratic sum of our statistical and systematic errors and the second is due to the nucleon form factors. (Normalizing to $G_M^{p\gamma}/\mu_p$ reduces the variation with Q^2 and reduces our sensitivity to uncertainties in the nucleon form factors.) If we assume the $\tau \to 0$ limit is valid at our Q^2, the result is $\rho_s + 2.9\mu_s = 0.67 \pm 0.41 \pm 0.30$. This band is shown in Figure 2, together with several model predictions. We note that the nucleon form factors are a major source of uncertainty in the interpretation of our results; furthermore, there are data for G_M^n [15] which are inconsistent with the value we chose and would lead to $(G_E^s + 0.39 G_M^s)/(G_M^{p\gamma}/\mu_p) = 0.143 \pm 0.054 \pm 0.047$. Experiments now in progress should soon reduce these uncertainties.

This work was supported by DOE contract DE-AC05-84ER40150 under which

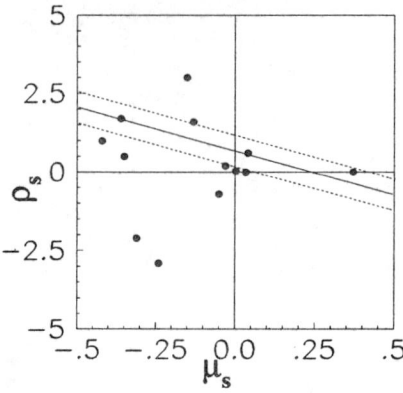

FIGURE 2. Band: allowed region from our results with assumptions listed in text. Points: various estimates from models.

the Southeastern Universities Research Association (SURA) operates the Thomas Jefferson National Accelerator Facility, and by the Department of Energy, the National Science Foundation, the Korean Science and Engineering Foundation (Korea), the INFN (Italy), the Natural Sciences and Engineering Research Council of Canada, the Commissariat à l'Énergie Atomique (France), and the Centre National de Research Scientifique (France).

REFERENCES

1. Ashman, J., *et al.*, *Phys. Lett.* **B206**, 364 (1988); *Nucl. Phys.* **B328**, 1 (1989).
2. Kaplan, D. B., and Manohar, A., *Nucl. Phys.* **B310**, 527 (1988).
3. Mueller, B. *et al.*, *Phys. Rev. Lett.* **78**, 3824 (1997).
4. Aniol, K. *et al.*, *Phys. Rev. Lett.* **82**, 1096 (1999).
5. Spayde, D. T. *et al.*, *Phys. Rev. Lett.* **84**, 1106 (2000).
6. Aniol, K. *et al.*, submitted to *Phys. Rev. Lett.* (2000).
7. R. C. Walker *et al.*, *Phys. Rev. D* **49**, 5671 (1994).
8. M. K. Jones *et al.*, *Phys. Rev. Lett.* **84**, 1398 (2000).
9. H. Anklin *et al.*, *Phys. Lett.* **B428**, 248 (1998).
10. I. Passchier *et al.*, *Phys. Rev. Lett.* **82**, 4988 (1999).
11. C. Herberg *et al.*, *Eur. Phys. Jour. A* **5**, 131 (1999).
12. M. Ostrick *et al.*, *Phys. Rev. Lett.* **83**, 276 (1999).
13. J. Becker *et al.*, *Eur. Phys. Jour. A* **6**, 329 (1999).
14. D. Rohe *et al.*, *Phys. Rev. Lett.* **83**, 4257 (1999).
15. E. E. W. Bruins *et al.*, *Phys. Rev. Lett.* **75**, 21 (1995). We have corrected these data down by 25% of the difference of these data and those of Ref. [9], based on a private communication from B. Schoch.

Parity Violating Measurements of Neutron Densities

C. J. Horowitz

Dept. of Physics and Nuclear Theory Center, Indiana University, Bloomington, IN 47405 USA
E-mail: charlie@iucf.indiana.edu

Abstract. Parity violating electron nucleus scattering is a clean and powerful tool for measuring the spatial distributions of neutrons in nuclei with unprecedented accuracy. Parity violation arises from the interference of electromagnetic and weak neutral amplitudes, and the Z^0 of the Standard Model couples primarily to neutrons at low Q^2. Experiments are now feasible at existing facilities. We show that theoretical corrections are either small or well understood, which makes the interpretation clean. A neutron density measurement may have many implications for nuclear structure, atomic parity nonconservation experiments, and the structure of neutron stars.

INTRODUCTION

The size of a heavy nucleus is one of its most basic properties. However, because of a neutron skin of uncertain thickness, the size does not follow from measured charge radii and is relatively poorly known. For example, the root mean square neutron radius in ^{208}Pb, R_n is thought to be about 0.25 fm larger then the proton radius $R_p \approx 5.45$ fm. An accurate measurement of R_n would provide the first clean observation of the neutron skin in a stable heavy nucleus. This is thought to be an important feature of all heavy nuclei.

Ground state charge densities have been determined from elastic electron scattering, see for example ref. [1]. Because the densities are both accurate and model independent they have had a great and lasting impact on nuclear physics. They are, quite literally, our modern picture of the nucleus.

In this paper we discuss future parity violating measurements of neutron densities. These purely electro-weak experiments follow in the same tradition and can be both accurate and model independent. Neutron density measurements have implications for nuclear structure, atomic parity nonconservation (PNC) experiments, isovector interactions, the structure of neutron rich radioactive beams, and neutron rich matter in astrophysics. It is remarkable that a single measurement has so many applications in atomic, nuclear and astrophysics.

Donnelly, Dubach and Sick [2] suggested that parity violating electron scattering can measure neutron densities. This is because the $Z-$boson couples primarily to the neutron at low Q^2. Therefore one can deduce the weak-charge density and the closely related neutron density from measurements of the parity-violating asymmetry in polarized elastic scattering.

Of course the parity violating asymmetry is very small, of order a part per million. Therefore measurements were very difficult. However, a great deal of experimental progress has been made since the Donnelly *et. al.* suggestion, and since the early SLAC experiment [3]. This includes the Bates ^{12}C experiment [4], Mainz ^{9}Be experiment [5], SAMPLE [6] and HAPPEX [7]. The relative speed of the HAPPEX result and the very good helicity correlated beam properties of CEBAF show that very accurate parity violation measurements are possible. Parity violation is now an established and powerful tool.

It is important to test the Standard Model at low energies with atomic parity nonconservation (PNC), see for example the Colorado measurement in Cs [8,9]. These experiments can be sensitive to new parity violating interactions such as additional heavy $Z-$bosons. Furthermore, by comparing atomic PNC to higher Q^2 measurements, for example at the Z pole, one can study the momentum dependence of Standard model radiative corrections. However, as the accuracy of atomic PNC experiments improves they will require increasingly precise information on neutron densities [10,11]. This is because the parity violating interaction is proportional to the overlap between electrons and neutrons. In the future the most precise low energy Standard Model test may involve the combination of an atomic PNC measurement and parity violating electron scattering to constrain the neutron density.

There have been many measurements of neutron densities with strongly interacting probes such as pion or proton elastic scattering, see for example ref. [12]. Unfortunately, all such measurements suffer from potentially serious theoretical systematic errors. As a result no hadronic measurement of neutron densities has been generally accepted by the field. Because of the uncertain systematic errors, modern mean field interactions are typically fit without using any neutron density information.

Finally, there is an important complementarity between neutron radius measurements in a finite nucleus and measurements of the neutron radius of a neutron star. Both provide information on the equation of state of dense matter. In a nucleus, R_n is sensitive to the density dependence of the symmetry energy. Likewise the neutron star radius depends also on the density dependence of the symmetry energy at normal and somewhat higher densities. In the future, we expect a number of improving radius measurements for nearby isolated neutron stars such as Geminga [13] and RX J185635-3754 [14].

We now present general considerations for neutron density measurements, discusses possible theoretical corrections and then conclude.

GENERAL CONSIDERATIONS

In this section we illustrate how parity violating electron scattering measures the neutron density and discuss the effects of Coulomb distortions and other corrections. These corrections are either small or well known so the interpretation of a measurement is clean.

Born Approximation Assymetry

The effect of the parity-violating part of the weak interaction may be isolated by measuring the parity-violating asymmetry in the cross section for the scattering of left(right) handed electrons. In Born approximation the parity-violating asymmetry is,

$$A_{LR} = \frac{G_F Q^2}{4\pi\alpha\sqrt{2}} \left[4\sin^2\theta_W - 1 + \frac{F_n(Q^2)}{F_p(Q^2)} \right], \tag{1}$$

with G_F the Fermi constant and θ_W the weak mixing angle. The Fourier transform of the proton distribution is $F_p(Q^2)$ while that of the neutron distribution is $F_n(Q^2)$ and Q^2 is the momentum transfer squared. The asymmetry is proportional to $G_F Q^2/\alpha$ which is just the ratio of Z^0 to photon propagators. Since 1-4$\sin^2\theta_W$ is small and $F_p(Q^2)$ is known we see that A_{LR} directly measures $F_n(Q^2)$. Therefore, A_{LR} provides a practical method to cleanly measure the neutron form factor and hence R_n.

Coulomb distortions

By far the largest known correction to the asymmetry comes from coulomb distortions. By coulomb distortions we mean repeated electromagnetic interactions with the nucleus remaining in its ground state. All of the Z protons in a nucleus can contribute coherently so distortion corrections are expected to be of order $Z\alpha/\pi$. This is 20 % for ^{208}Pb.

Distortion corrections have been accurately calculated in ref. [16]. Here the Dirac equation was numerically solved for an electron moving in a coulomb and axial-vector weak potentials. From the phase shifts, all of the elastic scattering observables including the asymmetry can be calculated.

Other theoretical corrections from meson exchange currents, parity admixtures in the ground state, dispersion corrections, the neutron electric form factor, strange quarks, the dependence of the extracted radius on the surface shape, etc. are discussed in reference [15]. These are all small. Therefore the interpretation of a parity violating measurement is very clean.

CONCLUSION

With the advent of high quality electron beam facilities such as CEBAF, experiments for accurately measuring the weak density in nuclei through parity violating electron scattering (PVES) are feasible. The measurements are cleanly interpretable, analogous to electromagnetic scattering for measuring the charge distributions in elastic scattering. From parity violating asymmetry measurements in elastic scattering, one can extract the weak charge density in nuclei and from this the neutron density.

By a direct comparison to theory, these measurements test mean field theories and other models of the size and shape of nuclei. They therefore can have a fundamental and lasting impact on nuclear physics. Furthermore, PVES measurements have important implications for atomic parity nonconservation (PNC) experiments. In the future it may be possible to combine atomic PNC experiments and PVES to provide a precise test of the Standard Model at low energies.

ACKNOWLEDGMENTS

This work was done in collaboration with Robert Michaels, Steven Pollock and Paul Souder. It was supported in part by DOE grant: DE-FG02-87ER40365.

REFERENCES

1. B. Frois et. al., Phys. Rev. Lett. **38**, 152 (1977).
2. T.W. Donnelly, J. Dubach and Ingo Sick, Nuc. Phys. **A 503** (1989) 589.
3. C. Y. Prescott *et al.*, Phys. Lett. **84B**, 524 (1979).
4. P. A. Souder *et al.*, Phys. Rev. Lett. **65**, 694 (1990).
5. W. Heil *et al.*, Nucl. Phys. **B327**, 1 (1989).
6. B. Mueller *et al.*, Phys. Rev. Lett. **78**, 3824 (1997).
7. K. A. Aniol *et al.*, Phys. Rev. Lett. **82**, 1096 (1999).
8. C. S. Wood et al, Science **275**, 1759 (1997).
9. S. C. Bennett and C. E. Wieman, Phys. Rev. Lett.**82**, 2484 (1999).
10. S. J. Pollock, E. N. Fortson, and L. Wilets, Phys. Rev. C **46**, 2587 (1992), S.J. Pollock and M.C. Welliver, Phys. Lett. **B** 464, 177 (1999).
11. P. Q. Chen and P. Vogel, Phys. Rev. **C 48** 1392 (1993).
12. L.Ray and G.W.Hoffmann, Phys. Rev. C **31**, 538 (1985).
13. Patrizia A. Caraveo et. al., ApJ **461**, L91 (1996); A. Golden and A. Shearer, astro-pn/9812207.
14. F. Walter, S. Wolk and R. Neuhauser, Nature **379**, 233 (1996); Bennett Link, Richard I. Epstein and James M. Lattimer, PRL **83**, 3362 (1999).
15. C.J. Horowitz, S. Pollock, P.A. Souder and R. Michaels, Nucl-th/9912038 submitted to Phys Rev C.
16. C.J. Horowitz, Phys. Rev. **C57** (1998) 3430.

Search for right-handed currents in the β^+-decay of ^{118}Sb

B. Vereecke[a], D. Beck[a], M. Beck[a], B. Delauré[a], T. Phalet[a],
P. Schuurmans[a], N. Severijns[a], S. Versyck[a], J. Deutsch[b], R. Prieels[b],
the NICOLE and ISOLDE collaborations

[a] I.K.S., K.U.Leuven, B-3001 Leuven, Belgium
[b] I.P.N., U.C. de Louvain, B-1348 Louvain-la-Neuve, Belgium

Abstract. We report on a recent experiment searching for right-handed currents in the ^{118}Sb nuclear β-decay. The correlation between the spin polarization and the β-emission asymmetry of the positrons from the ^{118}Sb decay is sensitive to the helicity structure of the weak interaction. A precision measurement of this correlation improves the limit on right-handed currents in β-decay.

INTRODUCTION

The standard model for the weak interaction is based on the $SU(2)_L \otimes U(1)$ gauge group for the electroweak sector, violating left-right symmetry in weak decays. Left-right symmetric (LRS) extensions based on $SU(2)_L \otimes SU(2)_R \otimes U(1)$ have been proposed to restore this symmetry. Adding the right-handed sector introduces extra parameters in the model. The relevant parameters for nuclear β-decay in a minimal extension of the standard model, the so called manifest left-right symmetric model [1] (MLRS), are a second gauge boson W' and a mixing angle ζ between the mass eigenstates and the gauge eigenstates of the W bosons.

$$W_L = W \cos\zeta + W' \sin\zeta \qquad W_R = -W \sin\zeta + W' \cos\zeta \qquad (1)$$

More general extensions [2] also allow for different coupling constants and different Cabibbo-Kobayashi-Maskawa (CKM) mixing matrices for the left- and right-handed sectors. It can be shown that in these general LRS models β-decay experiments are complementary [3] to direct searches for heavy W bosons.

METHOD AND FORMALISM

In the standard model, the ratio $R = \frac{P^-}{P^0}$ of the longitudinal spin polarization of positrons emitted from oriented (P^-) and unoriented (P^0) ^{118}Sb nuclei is given by

$$R_{SM} = \frac{\beta^2 - \overline{\beta} \cdot \overline{J} A_{exp} \Psi_3}{\beta^2(1 - \overline{\beta} \cdot \overline{J} A_{exp})} \qquad (2)$$

where $\beta = \frac{v}{c}$ the average velocity of the positrons and the solid angle factor Ψ_3 depend on the geometry of the set-up. The experimental β-asymmetry $\overline{\beta} \cdot \overline{J} A_{exp}$ can be calculated from count rates in the detectors. The ratio R can also be determined directly by measuring the longitudinal polarizations of the positrons for an oriented and an unoriented ensemble of ^{118}Sb nuclei. Any deviation between both measurable quantities R and R_{SM} indicates physics beyond the standard model:

$$R = R_{SM}(1 + k\Delta) \qquad (3)$$

where k is an enhancement factor given by

$$k = \frac{4\overline{\beta} \cdot \overline{J} A_{exp}}{\beta^2 - \overline{\beta} \cdot \overline{J} A_{exp} \Psi_3} \qquad (4)$$

and Δ is the contribution of non standard model interactions. Interpreted in general LRS models

$$\Delta = \frac{g_R^4 |V_{ud}^R|^2 m_{W_L}^4}{g_L^4 |V_{ud}^L|^2 m_{W_R}^4} \qquad (5)$$

where we assume $\zeta = 0$ and all neutrinos sufficiently light, such that their production is not kinematically suppressed. Assuming equal coupling constants g and CKM matrices (V_{ud}) in left- and right-handed sectors the expression reduces to $\Delta = \frac{m_{W_L}^4}{m_{W_R}^4}$ from which mass limits for the second gauge boson W_R can be deduced.

EXPERIMENT AND RESULTS

At the ISOLDE separator, ^{118}Xe was implanted in a high purity iron foil. After several β-decays a ^{118}Te ($t_{1/2} = 6.0d$) source of $\approx 10^7 at/s$ was obtained. This activity decays via electron capture (EC) into ^{118}Sb ($t_{1/2} = 3.5m$) which finally decays by EC and β^+-decay into stable ^{118}Sn. The only positrons coming from the Te source are thus those of Sb. The main contribution ($\approx 97\%$) in this β^+ spectrum is the $1^+ \to 0^+$ ground state to ground state transition with $log ft = 4.5$.

The Te source was mounted in the NICOLE dilution refrigerator coupled to a positron spectro-polarimeter. The spectrometer focuses the positrons into the detector set-up and assures an energy selection. The polarimeter is based on the method of time-resolved positronium hyperfine spectroscopy [4] to determine the spin of the positrons along the magnetic field axis. By cooling the Sb to temperatures around $8mK$ a nuclear polarization of about 70% was reached, yielding a

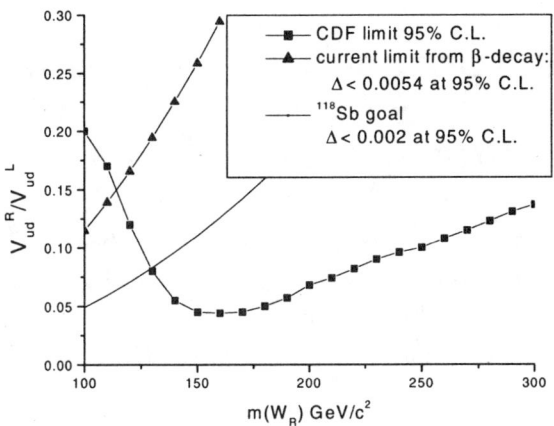

FIGURE 1. Exclusion region for this experiment compared to previous β-decay experiments and the the CDF result, the region above the curves is excluded.

β-asymmetry $\overline{\beta} \cdot \overline{J} A_{exp} \approx 0.55$. An unpolarized source was obtained by raising the temperature to about $3K$.

Results from previous polarization-asymmetry correlation experiments on ^{12}N [5] and ^{107}In [3] yielded $\Delta = 0.0004(26)$ or a lower mass limit of 306 GeV/c^2 (90% C.L. $\zeta = 0$) for W_R in the MLRS model. Preliminary analysis of the ^{118}Sb data shows that a sensitivity of about 400 GeV/c^2 (90% C.L.) is within reach.

This limit is still below the 720 GeV/c^2 from $p\overline{p}$ collisions at CDF [6] and D0 [7], but since in β-decays other combinations of coupling constants, masses and CKM matrix elements are probed, the results are complementary in more general LRS models as can be seen in figure 1. For small values of m_{W_R} and V_{ud}^R collider experiments are less sensitive then β-decay experiments.

REFERENCES

1. Beg M.A.B et al.,*Phys. Rev. Lett.* **38** 1252 (1977)
2. Langacker P. and Sankar U.,*Phys. Rev.* **D40** 1569 (1989)
3. Severijns N.et al., *Nucl. Phys.* **A629**, 423 (1998).
4. Prieels R. "Test of left right symmetric models through beta polarisation measurements" in *Proceedings of the 24th rencontre de Moriond, January 1989* edited by O. Fackler and J. Trân Thanh Van, Eds. Frontières, Gif-sur-Yvette, 1989, p287
5. Allet M. et al., *Phys. Lett.* **B383**, 139 (1996).
6. Abe F. et al., *Phys. Rev. Lett.* **74**, 2900 (1995).
7. Abachi S. et al., *Phys. Rev. Lett.* **76**, 3271 (1996).

A Novel Approach For Measuring the Beta-Neutrino Angular Correlation in Nuclear Beta Decay

M.Beck[1], F.Ames[2], D.Beck[1], B.Delauré[1], J.Deutsch[3], G.Bollen[4], O.Forstner[5], T.Phalet[1], W.Quint[6], P.Schmidt[2], P.Schuurmans[1], N.Severijns[1], B.Vereecke[1], S.Versyck[1] and the Eurotraps collaboration

[1] *Instituut voor Kern- en Stralingsfysica, K.U.Leuven, Celestijnenlaan 200D, B-3001 Leuven, Belgium*
[2] *Institut für Physik, Universität Mainz, D-55099 Mainz, Germany*
[3] *Insitut de physique nuclaire, UCL, Chemin du Cyclotron 2, B-1348 Louvain-la-Neuve, Belgium*
[4] *LMU München, Sektion für Physik, D-85748 Garching, Germany*
[5] *CERN, CH-1211 Geneva 23, Switzerland*
[6] *GSI-Darmstadt, p.o. box 110552, D-64220 Darmstadt, Germany*
E-mail: Marcus.Beck@fys.kuleuven.ac.be

Abstract. The experiment described here will search for deviations from the V-A structure of the standard electroweak model. It is based on measuring the recoil energy spectrum in nuclear beta decay which is determined by the electron-neutrino angular correlation. For pure Fermi decays this is exactly known in the standard model and any deviation will point to additional scalar interaction. The experiment consists of a Penning trap coupled to a retardation spectrometer to measure the energy of the recoiling daughter nuclei. The current status will be presented.

I BACKGROUND

The standard model of the electroweak interaction contains just vector (V) and axialvector (A) interaction. Scalar (S), tensor (T) and pseudoscalar (PS) interaction are absent. Parity violation is maximal. However, the absence of S and T interactions has experimentally only been proven down to the level of about 10% (90%CL) in nuclear beta decay [1,2]. Assuming standard gauge couplings, this corresponds to masses of possible new gauge bosons of 250GeV [2]. In the case of scalar couplings model independent searches for charged scalar bosons in medium and high energy particle physics yield limits in the range of 300GeV [3,4].

These limits leave significant room for scalar or tensor contributions to the charged current of the electroweak interaction. Nuclear beta decay experiments are complementary to high energy physics experiments in the sense that former probe the structure of the interaction independent of any particle model whereas latter search for specific particles.

The different possible forms of weak interactions result in different beta-neutrino angular correlations in nuclear beta decay. The $\beta - \nu$ angular correlation is given by [5]:

$$\omega_{(\vartheta_{\beta\nu})} \sim 1 + a \cdot v/c \cdot cos(\vartheta_{\beta\nu}) \cdot (1 - \Gamma \cdot m/E \cdot b)$$

with $\vartheta_{\beta\nu}$ the angle between the electron and the neutrino, v/c the velocity, m the restmass and E the total energy of the electron and $\Gamma = \sqrt{(1 - (\alpha \cdot Z)^2)}$. The Fierz interference term b has experimentally been shown to be small [6].

For pure Fermi decays ($0^+ - 0^+$ transitions) there is only V interaction, i.e. $a = 1$ in the standard model. Any admixture of S interaction will yield $a < 1$. For $a = 1$ the electron and the neutrino are emitted predominantly parallel, giving a high energy and momentum transfer to the recoiling daughter nucleus. $a < 1$ will decrease the number of decays with electron and neutrino emitted parallel and increase the number where they are emitted antiparallel, shifting the recoil energy spectrum to lower energies (fig.1). Thus, by measuring the shape of the recoil energy spectrum of the daughter nuclei for Fermi decays, the $\beta - \nu$ angular correlation and therefore a possible admixture of S to V interaction can be determined.

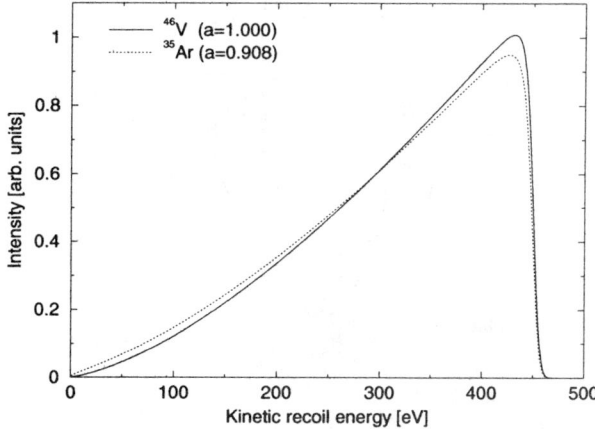

FIGURE 1. Spectral shape of the differential recoil energy spectrum for two decays of interest. ^{46}V is a Fermi emitter, ^{35}Ar a mirror nucleus that decays in a mixed Fermi/Gamow-Teller transition. The energy-axis for ^{46}V has been scaled to the endpoint energy of ^{35}Ar.

II EXPERIMENT

The ions are produced by ISOLDE [7] and mass-selectively cooled by REXTRAP [8]. From here they are transmitted in bunches to the Penning trap of the retardation spectrometer setup (fig.2) where they are captured in flight and cooled a second time. They are stored for typically one half-life in the trap. After β-decay in the trap the recoiling daughter ions generally leave it and spiral along the magnetic field lines from a strong magnetic field of $\approx 9T$ in the Penning trap into a weak magnetic field of $\approx 0.1T$, thereby converting nearly all of their radial energy into axial energy due to the conservation of the magnetic moment of the ion orbit. The ions with axial energies above the retardation potential between the strong and the weak field region are then reaccelerated to get off the field lines. They are finally focused electrostatically onto the detector where they are counted. By changing the retardation voltage the integral recoil energy spectrum is measured. This is described in more detail in [9] and the references therein.

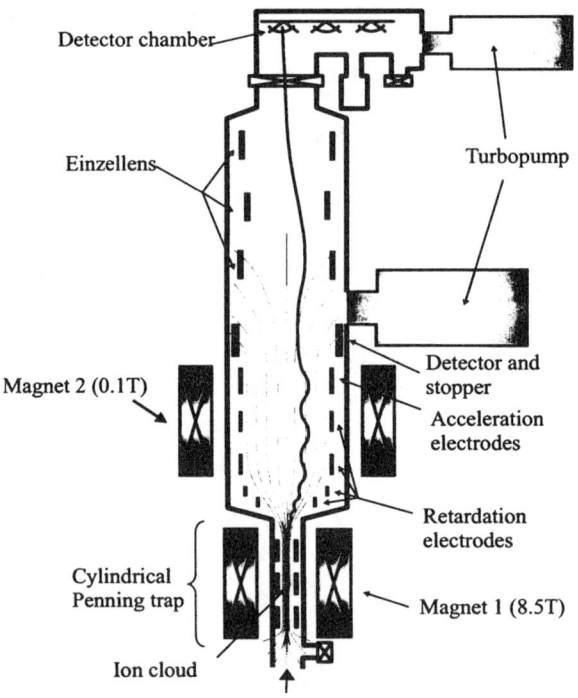

FIGURE 2. Schematic setup of the retardation spectrometer

The advantage of using an ion trap is that the recoiling daughter ions do not loose any energy in the surrounding medium compared with the usual technique of implanting the activity in a thin catcher. Therefore the recoils can be measured for a wide variety of beta emitters whereas previous $\beta - \nu$ angular correlation studies in nuclear beta decay were limited to just a few special cases [1,2].

The experiment is approved by the ISOLDE Scientific Committee as test experiment P111 and will be located close to REXTRAP. The construction of the set-up has started.

Possible candidates to be investigated are ^{26}Al, ^{35}Ar and ^{46}V. Simulations show that e.g. with ^{35}Ar about 10^7 to 10^8 events in the differential recoil energy spectrum will be needed to reach the current limits on S interaction. With reasonable assumptions this will be achieved in a few days of beamtime.

The first measurement is expected to take place by the end of 2001.

III CONCLUSION

The experiment that was described aims at improving the sensitivity for $\beta - \nu$ correlation measurements through a precise determination of the recoil energy spectrum. In a first step it will be used to search for a possible scalar weak interaction in Fermi beta decays or, if this is not found, to improve the limits. Other experiments will follow. Among them will be the measurement of Fermi-to-Gamow Teller ratios, EC branching ratios, the distribution of ionization states after beta decay and the determination of Q_β-values.

REFERENCES

1. C.H. Johnson et al., *Phys. Rev.* **132**, 1149 (1963)
2. E. G. Adlerberger et al., *PRL* **83**, 1299 (1999), F. Glück, *Nucl. Phys. A* **628**, 493 (1998)
3. see e.g. The H1 Collaboration, C. Adloff et al., *PLB* **479**, 358 (2000)
4. see e.g. The OPAL Collaboration, K. Ackerstaff et al., *Eur. Phys. J. C* **2**, 441 (1998), The OPAL Collaboration, G. Abbiendi et al., *Eur. Phys. J. C* **13**, 553 (2000)
5. J.D. Jackson, S.B.Treiman and H.W. Wyld, *Nucl. Phys.* **4**, 206 (1957)
6. J.C. Hardy et al., *Nucl. Phys. A* **509**, 429 (1990), A.S. Carnoy et al., *Nucl. Phys. A* **568**, 265 (1994), I.S. Towner and J.C. Hardy, Proc. of the Fifth International WEIN Symposium, Santa Fe, NM, USA, June 1998, eds. P. Herczeg, C.M. Hoffman, H.V. Klapdor-Kleingrothaus (World Scientific, Singapore 1999) 338
7. E. Kugler et al., *NIM B* **70**, 41 (1992)
8. F. Ames et al., Proc. Exotic Nuclei and Atomic Masses (Bellaire, USA, 1998), A.I.P. Conf. Proc. 445 (1998) 927
9. D. Beck et al., Proposal to the ISOLDE scientific committee, CERN/ISC 99-13, ISC/P111, April 1999

Muon-electron conversion in nuclei

Andrzej Czarnecki and William J. Marciano

Brookhaven National Laboratory
Upton, New York 11973

Kirill Melnikov

Stanford Linear Accelerator Center
Stanford University, Stanford, CA 94309

Abstract. Transition rates for coherent muon–electron conversion in the field of a nucleus (muonic atoms), $\mu N \to eN$, are computed for various possible muon–number violating amplitudes. Attention is paid to relativistic atomic effects, Coulomb wave-function distortions, finite nuclear size, and nucleon distributions.

A very sensitive test of the muon number conservation can be performed by considering the $\mu \to e$ conversion in a muonic atom [1] (for a review of other muon number conservation tests see e.g. [2]). Coherent conversion is the process $\mu^- N \to e^- N$ in which the nucleus N remains in its initial state (up to recoil effects). The rate of the coherent conversion is enhanced with respect to processes with nuclear excitation by a factor of the order of the number of nucleons.

Since the pioneering papers on the conversion theory [1,3,4], there have been a number of theoretical and experimental studies in connection with the appropriate description and observation of coherent muon conversion. Two types of theoretical issues arise: short distance effects which are responsible for the muon number violation, caused by some as yet unknown "new physics" and the rate of the transition which depends on the long distance atomic physics of the muonic atom.

The first issue has been studied in many extensions of the standard model. In particular, in ref. [5] the possibility of coherent conversion $\mu^- N \to e^- N$ was examined in a variety of gauge models (for a review see [6]).

The atomic physics aspects were studied in [7–9]. A relatively thorough theoretical description of the atomic effects in $\mu^- N \to e^- N$ was given in [7], where relativistic wave functions, Coulomb distortion, and finite nuclear size effects were taken into account.

Recently, we have re-examined the atomic physics aspect of $\mu^- N \to e^- N$ conversion [10,11]. Our primary goal was to extend the analysis of [7] and to re-analyze possible sources of theoretical uncertainty. Here, we briefly summarize our results.

We begin with an effective Lagrangian for muon–number violating processes and derive general formulas for the coherent conversion rate. We also illustrate the dependence of the conversion branching ratio on the nuclear target with a particular example of a model which predicts lepton flavor violation. Finally we re-derive bounds on muon number violating couplings using experimental results from the SINDRUM collaboration [12].

The muon–electron transition can occur via various mechanisms (see e.g. [6,13,14]). To keep the description as general as possible it is convenient to write down a low energy effective Lagrangian for the $\mu \to e$ transition:

$$\mathcal{L} = \bar{e}\hat{O}\mu + h.c., \tag{1}$$

$$\hat{O} = -\sqrt{4\pi\alpha}\left[\gamma_\alpha(f_{E0} - f_{M0}\gamma_5)\frac{q^2}{m_\mu^2} + i\sigma_{\alpha\beta}\frac{q^\beta}{m_\mu}(f_{M1} + f_{E1}\gamma_5)\right]A^\alpha(q) \tag{2}$$

$$+ \frac{G_F}{\sqrt{2}}\gamma_\alpha(a - b\gamma_5)J^\alpha + \frac{2G_F}{\sqrt{2}}\frac{\sqrt{m_e m_\mu}\eta_{He\mu}}{m_H}\frac{\eta_{Hpp}m_N}{m_H}(1 - \kappa\gamma_5)(p\bar{p} + n\bar{n}),$$

$$J^\alpha = \bar{u}\gamma^\alpha u + c_d\bar{d}\gamma^\alpha d. \tag{3}$$

In this equation $a, b, f_{E0,1}, f_{M0,1}, \eta_{He\mu}$ are model-dependent dimensionless coupling constants and m_N is the nucleon mass. The coupling constant η_{Hpp} describes the interaction of the Higgs boson with proton or neutron and is estimated to be $0.2 - 0.5$.

In their pioneering work on muon–electron conversion Weinberg and Feinberg [1] performed an approximate calculation of the coherent conversion rate for some of the above interaction operators. However, their approximations fail for sufficiently heavy nuclei and a more precise treatment requires solving the Dirac equation for both muon and electron wave functions in the field of the nucleus [10,11,15].

It is convenient to express the conversion rate in terms of several integrals, which contain the information about the atomic physics of the muonic atom. The necessary integrals which describe muon and electron wave-function overlaps with the nucleons are

$$I_1^p = -\frac{4\pi Z\alpha}{m_\mu^2}\int dr r^2 \rho_p(r) g_\mu^- g_e^-, \qquad I_2^p = -\frac{4\pi Z\alpha}{m_\mu^2}\int dr r^2 \rho_p(r) f_\mu^- f_e^-$$

$$I_1^n = -\frac{4\pi Z\alpha}{m_\mu^2}\int dr r^2 \rho_n(r) g_\mu^- g_e^-, \qquad I_2^n = -\frac{4\pi Z\alpha}{m_\mu^2}\int dr r^2 \rho_n(r) f_\mu^- f_e^-$$

$$I_3 = \frac{1}{m_\mu}\int dr r^2 \frac{dV(r)}{dr} g_\mu^- f_e^-, \qquad I_4 = \frac{1}{m_\mu}\int dr r^2 \frac{dV(r)}{dr} f_\mu^- g_e^-. \tag{4}$$

We have tabulated the values of these integrals for various nuclei in Table 2 of ref. [11]. For practical applications, we illustrate how these values should be used.

We write the conversion rate as:

$$\omega_{\text{conv}} = 3 \cdot 10^{15} (\omega_{\text{conv}}^{(1)} + \omega_{\text{conv}}^{(2)}) \cdot 10^8 \text{ sec}^{-1} \tag{5}$$

$$\omega_{\text{conv}}^{(1)} = \left| f_{E0} I_p - \frac{G_F}{\sqrt{2}} \frac{m_\mu^2}{4\pi Z\alpha} a \Big(Z(2+c_d) I_p + N(1+2c_d) I_n \Big) + f_{M1} I_{34} \right|^2 ,$$

$$\omega_{\text{conv}}^{(2)} = \left| f_{M0} I_p - \frac{G_F}{\sqrt{2}} \frac{m_\mu^2}{4\pi Z\alpha} b \Big(Z(2+c_d) I_p + N(1+2c_d) I_n \Big) + f_{E1} I_{34} \right|^2 . \tag{6}$$

More general formulas, including scalar interactions, will be presented in [15].

The following notations were introduced here: $I_p = -(I_1^p + I_2^p)$, $I_n = -(I_1^n + I_2^n)$, and $I_{34} = I_3 + I_4$. All dimensional parameters in the rate formula should be used in Fermi units. For example, the muon mass is $m_\mu = 0.5354$ fm^{-1} and the Fermi coupling constant is $G_F = 0.4542 \cdot 10^{-6}$ fm^{-2}.

Examples of supersymmetric GUT models which predict lepton flavor violation were considered in Ref. [14]. For those particular models, the couplings f_{M1} and f_{E1} (cf. Eq. (2)) are significantly enhanced in a large region of the parameter space in comparison with other couplings in the effective Lagrangian. Therefore, we neglect the other couplings and analyze the dependence of $\omega_{\text{conv}}/\omega_{\text{capt}}$ on the choice of target.

With these approximations the ratio of the conversion rate to the capture rate becomes

$$\frac{\omega_{\text{conv}}}{\omega_{\text{capt}}} = 3 \cdot 10^{12} \Big(|f_{E1}|^2 + |f_{M1}|^2 \Big) B_{SG}(A, Z), \tag{7}$$

where B_{SG} is

$$B_{SG}(A, Z) = 10^3 \frac{I_{34}^2}{(\omega_{\text{capt}}/\text{sec}^{-1}) \, 10^{-8}}.$$

The target dependence of the conversion rate can now be analyzed using our results for the integrals I_3 and I_4 and the experimental data on the capture rate. We find: for ^{27}Al, $B_{SG} = 1.0 - 1.2$, taking into account uncertainties in the input parameters; for ^{48}Ti, $B_{SG} = 1.8$; for ^{142}Ce, $B_{SG} = 1.9$; and for ^{208}Pb, $B_{SG} = 1.1 - 1.4$. We conclude that the ratio of the coherent conversion rate to the capture rate does not change significantly with changing the target.

Using updated results for the atomic part of the theoretical description of the coherent muon electron conversion, we reconsider the analysis performed in Ref. [12]. In that paper the upper limit for the ratio $\omega_{\text{conv}}/\omega_{\text{capt}}$ has been reported based on the measurement on ^{208}Pb. These results were combined with previous measurement on ^{48}Ti target to obtain the bounds on muon–number violating coupling constants. In that analysis it was assumed that the effective Lagrangian does not contain a piece which corresponds to the photon mediated conversion. We therefore put $f_{E0,1}$ and $f_{M0,1}$ equal to zero in our conversion rate formula.

Using notations of [12], the new bounds for the coupling constants g_V^0, g_V^1 are $|g_V^0| < (8 \pm 2) \cdot 10^{-7}$ and $|g_V^1| < (40 \pm 13) \cdot 10^{-6}$. The variation in the boundaries shown above is the estimate of the uncertainty of the result induced by uncertaintes in the input nuclear parameters. These new bounds should be compared with the values $|g_V^0| < 3.9 \cdot 10^{-7}$ and $|g_V^1| < 9.7 \cdot 10^{-6}$ which were given in Ref. [12]. There are two reasons for the difference. First, in Ref. [12] the results of Ref. [7] have been used to interpret experimental data. We note in this respect that Table I of Ref. [7] does not provide correct results – the numbers for $\omega_{\text{conv}}/\omega_{\text{capt}}$ quoted there are too large by about a factor of 2. Second, in the case of a Pb target the difference in proton and nucleon distributions becomes quite noticeable. This fact was also not properly discussed in [7], in particular the numbers in Table I of [7] were obtained using identical proton and neutron distributions. These two reasons conspire to give quite a large discrepancy.

In summary, we have reanalyzed the atomic physics aspects of a search for $\mu \to e$ coherent conversion in muonic atoms. We have tabulated the matrix elements that are required to compute the conversion rate and have discussed the dependence of a signal on the charge and atomic number of the nuclei for some models of lepton flavor violation [10,11,15].

Acknowledgments: This work was supported in part by DOE under grants number DE-AC02-98CH10886 and DE-AC03-76SF00515.

REFERENCES

1. S. Weinberg and G. Feinberg, Phys. Rev. Lett. **3**, 111 (1959), erratum: ibid., p. 244.
2. A. van der Schaff, Prog. Part. Nucl. Phys. **31**, 1 (1993).
3. N. Cabibbo and R. Gatto, Phys. Rev. **116**, 1334 (1959).
4. S. P. Rosen, Nuovo Cim. **15**, 7 (1960).
5. W. J. Marciano and A. I. Sanda, Phys. Rev. Lett. **38**, 1512 (1977).
6. W. J. Marciano, in A. Perlmutter and L. F. Scott, eds., *New frontiers in High Energy Physics* (Plenum, New York, 1978), proc. of Orbis Scientiae, Coral Gables, 1978.
7. O. Shanker, Phys. Rev. **D20**, 1608 (1979).
8. H. C. Chiang, E. Oset, T. S. Kosmas, A. Faessler, and J. D. Vergados, Nucl. Phys. **A559**, 526 (1993).
9. T. S. Kosmas, A. Faessler, F. Simkovic, and J. D. Vergados, Phys. Rev. **C56**, 526 (1997), nucl-th/9704021.
10. A. Czarnecki, W. J. Marciano, and K. Melnikov, in S. Geer and R. Raja, eds., *Physics at the first muon collider* (AIP, Woodbury, 1998), pp. 409–418, hep-ph/9801218.
11. A. Czarnecki, W. J. Marciano, and K. Melnikov, *Coherent muon–electron conversion in the field of a nucleus*, to appear in the Proc. of the KEK International Workshop on High Intensity Muon Sources (HIMUS99).
12. W. Honecker *et al.* (SINDRUM II), Phys. Rev. Lett. **76**, 200 (1996).
13. J. Bernabéu, E. Nardi, and D. Tommasini, Nucl. Phys. **B409**, 69 (1993).
14. R. Barbieri, L. Hall, and A. Strumia, Nucl. Phys. **B445**, 219 (1995), hep-ph/9501334.
15. A. Czarnecki, W. J. Marciano, and K. Melnikov, work in progress.

Muon Conversion Experiments - Current and Future

Allen I. Mincer [1] for the MECO Collaboration [2]

New York University
New York, NY 10003

Abstract.
Many extensions of the standard model predict muon to electron conversion in the process $\mu + Nucleus \to e + Nucleus$ at ratios, R, of 10^{-14} to 10^{-17} muon captures. A long history of experiments which have improved upper limits on R, to the current value of $< 6.1 \times 10^{-13}$ set by SINDRUM II, have been limited by muon statistics rather than background.

A new experiment, MECO, has been proposed to search for muon to electron conversion at levels below 5×10^{-17}. MECO takes advantage of a pulsed, high-flux muon beam produced when a proton beam at the Brookhaven AGS is incident on a target in a graded solenoidal field. The detector has 800 KeV momentum resolution and is designed to minimize the background from the beam and maximize efficiency for signal electrons. With 10^{11} stopped muons per second, R= 10^{-16} would give 5 detected events in a year of running with less than 0.5 background events expected.

INTRODUCTION

Many extensions of the standard model predict violation of individual lepton generation number. Predicted rates [1] for the muon conversion process, $\mu + Nucleus \to e + Nucleus$, can be large or small compared to $\mu \to e\gamma$. The very clean experimental signature of a single electron with energy equal to the muon mass less the muon nucleus binding energy and the small nuclear recoil energy does not suffer from the accidental coincidences which contribute to $e\gamma$ background. Attainable experimental results therefore may be competitive with those from $e\gamma$ even for some models for which predicted rates are lower [2].

---

[1] Research supported by NSF Contract PHY-9901496.
[2] M. Brennan, K. Brown, W. Morse, Y. Semertzidis, P. Yamin (Brookhaven National Laboratory); J. Miller, B.L. Roberts, O. Rind (Boston University); T. Liu, W. Molzon, V. Tumakov (University of California, Irvine); E. Hungerford, K.A. Lan, W. Mayes, L. Pinsky, J. Wilson (University of Houston); V. Lobashev, A.N.Toropin (Institute for Nuclear Research, Moscow); R. Djilkibaev (NYU/INR), A.I. Mincer, P. Nemethy, J. Sculli, J. Popp (New York University); W. Wales (University of Pennsylvania); D. Koltick (Purdue University); M. Eckhause, J. Kane, and R. Welsh (College of William and Mary).

TABLE 1. History of improvements in muon conversion limits.

R	Ref	Number of Events[a]	Stopped muons	Energy Cut (MeV)	Experimental Improvements
$< 5 \times 10^{-4}$	[3]	0	8K [b]	~ 53	First accelerator experiment
$4^{+3}_{-2} \times 10^{-6}$	[4]	3	1.3×10^9	~ 90	measure momentum veto prompts
$< 5.9 \times 10^{-6}$	[5]	0	6.25×10^8	90	cosmic ray veto
$< 2.2 \times 10^{-7}$	[6]	0	2.4×10^9	90	NaI energy measurement cerenkov veto beam e^-
$< 2.2 \times 10^{-7}$	[7]	0		80	
$< 2.6 \times 10^{-8}$ [c]	[8]	1	6.22×10^{10}	~ 80	
$< 7 \times 10^{-11}$	[9]	0	8.9×10^{11}	93	Multiple [d]
$< 4.6 \times 10^{-12}$	[10]	0	$\sim 1.2 \times 10^{13}$	96.5 to 106	TPC
$< 6.1 \times 10^{-13}$	[11]	0	3×10^{13}	99	Drift Chamber

[a] Number of events in signal region
[b] Flux times detector solid angle
[c] $\mu^- \to e^+$ Search
[d] Thin stopping target, pulsed beam (allows better prompt veto), target in center of detector (increasing solid angle).

THE PAST AND PRESENT

The dominant background in conversion experiments is muon decay in orbit (MDIO) which has an electron energy endpoint equal to the energy of conversion electrons. This is higher than in free muon decay due to the nucleus which can balance momentum opposite the emitted electron. However, the MDIO spectrum falls rapidly as $(E_{max} - E_{electron})^5$, so that background is essentially determined by momentum resolution. Although other background sources exist, they are suppressed by careful design of the beam and detector. Since the muon lifetime is typically much longer than the travel time of energetic beam related background, vetoing on prompt particles eliminates the bulk of the background.

Past experiments which successively improved upper limits on R, the ratio of muon conversions to muon captures ($\mu^- p \to n\nu_\mu$), employed increased muon intensity, which in turn made necessary better energy resolution to avoid MDIO background (see Table 1). In the end this was sufficient: each of the experiments had at most a few events in the signal region. This is also true of the Sindrum II experiment which has set the current limit.

Using a drift chamber and hodoscopes in a 1.2 Tesla field, the Sindrum II resolution (FWHM) is 2.3 MeV. The 1993 run used the PSI 88 MeV/c beam which provided 1.2×10^7 muons/sec, 28% of which stopped on a titanium target. A 1999 titanium run [12] in the new 58 MeV/c beam line is currently being analyzed, while a recently begun run with a gold target is expected to set limits of about 10^{-13}.

The MELC proposal [13] suggested using a pulsed beam incident on a target in a graded magnetic field to achieve an intensity of 10^{11} stopped muons/sec. Background supression in an S-shaped transport region and a graded field in the detection region would allow the possibility of measurements of R as low as 10^{-16}.

THE FUTURE

The currently proposed MECO experiment incorporates several of the MELC ideas to obtain a sensitivity to R as low as 5×10^{-17}. The AGS at Brookhaven National Laboratory is used to provide a 7 to 8 GeV/c beam in two bunches, each less than 100 ns wide and containing 2×10^{13} protons spaced 1.35 µs apart. The signal window is the last 700 ns. of the bunch cycle. The beam energy is chosen below beam transition (for stability), to maximize π production, and to minimize $\bar{p}$ production. Extinction of off bunch particles plus kickers, if necessary, will reduce the off bunch proton ratio to below 10^{-9}.

The beam is incident, pointing away from the transport solenoid, on a tungsten target ($^{184}_{74}W$) designed to maximize pion yield, maximize radiative cooling, and minimize secondary scatter. Pion yield has been estimated using tantalum ($^{181}_{73}Ta$) data [14]. The graded production region field reflects low momentum, quickly decaying pions and muons back toward an S-shaped solenoid which transports particles to the conversion target. Collimation in this solenoid uses drift in the third dimension to eliminate particles with transverse momentum $P_t > 50$ MeV/c, total momentum $P \gtrsim 100$ MeV/c, and particles of the wrong sign. A small grade in the transport field prevents particles from accumulating large delay times.

Muons entering the detection region are captured on an aluminum target yielding 0.025 stopped muons/proton. The 880 ns. muon lifetime and particle transport delay gives 49% of electrons arriving during the signal gate. A graded field insures that 100 MeV/c particles originating upstream of the target have P_t below signal threshold at the detector.

The straw tube detector consists of an octagonal barrel with vanes extending radially outward from the the vertices. Each barrel or vane plane consists of 3 layers of 5 mm straw tubes read out on the wires and capacitively coupled strips. A single helix turn gives 3 to 4 clusters (multi-hit layers), with two full turns in the detector for 39% (100%) of electrons produced with $P_t > 91$ MeV/c for a 2.4 m (2.9 m) detector. Particles with $P_t < 60$ MeV/c miss the detector and go through the barrel center. Using various fitting strategies and the resolution of $\sim 200\,\mu m$ in ϕ and 1.5 mm in z provides ~ 800 KeV FWHM momentum measurement. Cosmic ray shielding is both passive and active (two scintillation layers, each 99% efficient). A scintillator barrel (or crystal vane) electron trigger calorimeter is 63% (87%) efficienct for electrons with energy > 60 MeV (80 MeV). Overall, about 20% of conversion electrons are detected.

A one year (10^7 sec) run with 4×10^{13} protons/sec would give 5 detected signal events for $R = 10^{-16}$. Table 2 lists background sources, reduction methods, and the expected background in the signal region ($E > 103.5$ MeV) for such a run.

TABLE 2. Potential Background Sources and Rate estimates

Background Source	Rejection [a]	Expected Number
Muon Decay in Orbit	σ	0.25
MDIO plus noise hits	σ	< 0.006
Radiative Muon Capture	σ	< 0.005
Electrons from production or Transport	B, C, P	~ 0.04
Muon decay in flight, no electron scatter	B, C, P	< 0.03
Pion decay in flight, no electron scatter	B, C, P	< 0.001
Muon decay in flight, electron scatter	C, P, L	< 0.04
Pion decay in flight, electron scatter	C, P, L	< 0.001
Prompt radiative pion capture	P, σ	~ 0.07
Delayed radiative pion capture	σ	~ 0.001
Anti proton induced	Be Window	~ 0.007
Other late arriving backgrounds	σ, B, C	negligible
Cosmic Rays	Cosmic veto	~ 0.004
Total		0.45

[a] Experimental background reduction via: P = Prompt vetoe, σ = Momentum resolution, B = Graded detector field and P_t cuts, C = Transport solenoid collimation L = Low detector region total material

REFERENCES

1. See for example A. Czarnecki, W.J. Marciano, and K. Melnikov, in *Workshop on the Front End of a Muon Collider*, edited by S. Geer and R. Raja, New York: American Institue of Physics, 1998, p.409.
2. See for example, J. Hisano et. al., *Phys. Lett.* **B391**, 341 (1997) for predictions in some supersymmetric models.
3. J. Steinberger and H.B. Wolfe *Phys. Rev.* **100**, 1490 (1955).
4. R.D. Sard, K.M. Crowe, and H. Kruger, *Phys. Rev.* **121**, 619 (1961).
5. M. Conversi et. al., *Phys. Rev.* **122**, 687 (1961).
6. G. Conforto et. al., *Nuovo Cim.* **26**, 261 (1962).
7. J.H. Bartley et. al., *Phys. Lett.* **13**, 258 (1964).
8. D.A. Bryman et. al., *Phys. Rev. Lett.* **28**, 1469 (1972).
9. A. Badertscher et. al., *Phys. Rev. Lett.* **39**, 1385 (1977); *Nucl. Phys.* **A377**, 406 (1982).
10. S. Ahmad et. al., *Phys. Rev.* **D38**, 2102 (1988).
11. C. Dohman et. al., *Phys. Lett.* **B317**, 631 (1993). F. Riepenhausen, in *Proceedings of the Sixth Conference on the Intersections of Particle and Nuclear Physics*, edited by T. Donnelly, New York: American Institue of Physics, 1997, p.34.
12. Peter Wintz, private communication.
13. R. Djilkibaev and V.M. Lobashev, *Sov. J. Nucl. Phys.* **49(2)**, 384 (1989); V.S. Abadjev et. al., *MELC Experiment to Search for the $\mu^- A \to e^- A$ Process*, INR Preprint 786/92, 1992.
14. D. Armutliski et. al., *JINR* P1-91-191 (1991).

Searches for lepton flavor violation in τ and B^0 decays

Ilya Narsky[†]

[†]*Physics Department, Southern Methodist University, Dallas, Texas 75275-0175, USA*

Abstract. Theoretical predictions and recent experimental searches for lepton flavor violating decays of τ and B^0 are reviewed. So far, no evidence for such processes has been found. Up-to-date experimental upper limits on these branching fractions constrain the parameter space of some models.

THEORY

Lepton number and lepton flavor are almost exact symmetries in the Standard Model. However, there is no fundamental motivation for these symmetries. In the Standard Model lepton flavor violating (LFV) decays can proceed via tunneling transitions between topologically distinct vacua due to instantons [1] or via neutrino mixing [2]. The amplitudes of these processes are well beyond the sensitivity of the present experimental technology. More optimistic predictions can be obtained in the framework of the Standard Model extended by one heavy right-handed neutrino. In this case, however, amplitudes of LFV processes are inversely proportional to the mass of the heavy neutrino and are still negligibly small to allow experimental observation.

On the other hand, LFV is expected in many extensions of the Standard Model through mechanisms involving new couplings and/or new particles. An incomplete list of such models includes: SUSY, Grand Unified Theories, left-right symmetric models, supergravity, superstrings, technicolor, and composite models.

One of the most promising channels favored by several models is the LFV decay $\tau \to \mu\gamma$. Predictions based on the Minimal SuperSymmetric Model with heavy right-handed neutrinos [3] are shown in Fig. 1. Here, dominant contributions to the decay $\tau \to \mu\gamma$ are due to Higgsino and wino exchanges at $\tan\beta \geq 1$. Other models that predict high branching fractions for this decay are SU(5) SUSY GUT [3] and a string inspired version of the MSSM without a simple unified gauge group [4]. Large branching fractions of neutrinoless τ decays into a lepton and two pseudoscalars have been calculated in Ref. [5], and τ decays into three charged leptons have been studied in Ref. [6]. All these predictions are quite close to the present experimental bounds. For example, the parameter space of models [3,4] is constrained by an upper limit on the branching fraction $\mathcal{B}(\tau \to \mu\gamma)$ recently set by the CLEO Collaboration [7].

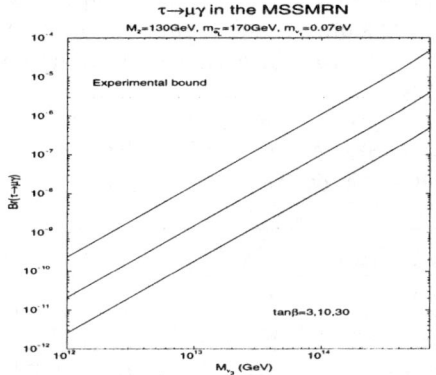

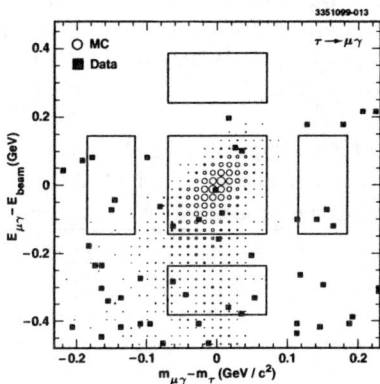

FIGURE 1. Predicted branching fraction $\mathcal{B}(\tau \to \mu\gamma)$ versus third-generation right-handed neutrino mass in the MSSM. The horizontal line is an old CLEO bound [8].

FIGURE 2. Signal and sideband region definitions for the $\tau \to \mu\gamma$ search [7] by CLEO.

LFV in B^0 decays can occur, e.g., in the leptoquark model [9]. The LFV aspects of this model are discussed in more detail in Ref. [10].

EXPERIMENT

Most stringent experimental bounds on neutrinoless decays of leptons have been obtained in searches [8] for muon decays and μ to e conversion on nucleus. However, predicted branching fractions for LFV decays of τ are typically several orders of magnitude larger than those for μ decays. Due to this enhancement, constraints on new physics obtained in the studies of τ and μ decays are competitive for some models.

Limits for LFV τ and B^0 decays are dominated by CLEO [11]. This is due to the fact that until recently CLEO possessed the largest recorded samples of $\tau^+\tau^-$ and $B\bar{B}$ events.

Search for neutrinoless τ and B^0 decays at a $\Upsilon(4S)$ symmetric e^+e^- machine is relatively easy. Standard selection cuts are used to minimize backgrounds from radiative Bhabha and muon pairs, and from two-photon interactions. A multiplicity cut is imposed to separate hadronic events from non-hadronic. Due to the τ boost, each τ event candidate is topologically divided into signal and tag hemispheres and each hemisphere is analyzed separately. Since there are no neutrinos, signal decay products have to satisfy two kinematic constraints: their invariant mass has to be equal to the mass of the τ (or B^0), and their total energy has to be equal to the energy of the beam. A typical analysis establishes a signal region defined by the energy and mass resolutions of the detector and estimates background by extrapolating from sidebands in the energy-vs-mass plane. The signal region and sidebands used in the recent CLEO's search for $\tau \to \mu\gamma$ [7] are shown in Fig. 2.

Experimental bounds on LFV decays of τ are summarized in Table 1. All present limits [8] on LFV decays of B^0 have been set by CLEO, with a recently updated limit $\mathcal{B}(B^0 \to e\mu) < 5.9 \times 10^{-6}$ at 90% CL [12].

TABLE 1. Upper limits on branching fractions of LFV decays of τ^- at 90% confidence level. The only numbers obtained not by CLEO are the bounds on the decays $\tau^- \to e^- K^0$ and $\tau^- \to \mu^- K^0$ and the angular momentum violating decay $\tau^- \to \pi^-\pi^0$.

Channel	UL, 10^{-6}	Channel	UL, 10^{-6}	Channel	UL, 10^{-6}	Channel	UL, 10^{-6}
$e^-\gamma$	2.7	$e^-e^+e^-$	2.9	$e^-e^+\mu^-$	1.7	$e^-\mu^-e^-$	1.5
$e^-\pi^0$	3.7	$e^-\eta$	8.2	$e^-\rho^0$	2.0	e^-K^{*0}	5.1
$e^-\bar{K}^{*0}$	7.4	$e^-\phi$	6.9	$e^-\pi^+\pi^-$	2.2	$e^-\pi^+K^-$	6.4
$e^-K^+\pi^-$	3.8	$e^-K^+K^-$	6.0	$e^+\pi^-\pi^-$	1.9	$e^+\pi^-K^-$	2.1
$e^+K^-K^-$	3.8	$e^-\pi^0\pi^0$	6.5	$e^-\pi^0\eta$	24	$e^-\eta\eta$	35
$\mu^-\gamma$	1.1	$\mu^-\mu^+\mu^-$	1.9	$\mu^-\mu^+e^-$	1.8	$\mu^-e^+\mu^-$	1.5
$\mu^-\pi^0$	4.0	$\mu^-\eta$	9.6	$\mu^-\rho^0$	6.3	μ^-K^{*0}	7.5
$\mu^-\bar{K}^{*0}$	7.5	$\mu^-\phi$	7.0	$\mu^-\pi^+\pi^-$	8.2	$\mu^-\pi^+K^-$	7.5
$\mu^-K^+\pi^-$	7.4	$\mu^-K^+K^-$	15	$\mu^+\pi^-\pi^-$	3.4	$\mu^+\pi^-K^-$	7.0
$\mu^+K^-K^-$	6.0	$\mu^-\pi^0\pi^0$	14	$\mu^-\pi^0\eta$	22	$\mu^-\eta\eta$	60
$\bar{p}\gamma$	3.5	$\bar{p}2\pi^0$	33	$\bar{p}\pi$	15	$\bar{p}\pi^0\eta$	27
$\bar{p}\eta$	8.9	$\pi^-\pi^0$	370	e^-K^0	1300	μ^-K^0	1000

In the future, searches for LFV decays of τ and B^0 are likely to remain the monopoly of low energy e^+e^- colliders. For a decisive test of some extensions of the Standard Model the present experimental sensitivity has to be improved by a few orders of magnitude.

REFERENCES

1. G. t'Hooft, Phys. Rev. Lett. **37**, 8 (1976).
2. S.M. Bilenkii and B.M. Pontecorvo, Phys. Lett. **61 B**, 248 (1976).
3. J. Hisano and D. Nomura, Phys. Rev. **D 59**, 116005, (1999).
4. S.F. King and M. Oliveira, Phys. Rev. **D 60**, 035003 (1999).
5. A. Ilakovac, Phys. Rev. **D54**, 5653 (1996).
6. M. Bisset et al., Phys. Rev. **D 62**, 035001 (2000).
7. CLEO Collaboration, S. Ahmed et al., Phys. Rev. **D 61**, 071101R (2000).
8. Particle Data Group, C. Caso et al., The Eur. Phys. J. **C 3**, 1-4, (1998).
9. J. Pati and A. Salam, Phys. Rev. **D 10**, 275 (1974).
10. G. Valencia, S. Willenbrock, Phys.Rev. **D 50**, 6843 (1994).
11. Y. Kubota et al., Nucl. Instr. and Meth. **A 320**, 66 (1992); T.S. Hill, Nucl. Instr. and Meth. **A 418**, 32 (1998).
12. CLEO Collaboration, T. Bergfeld et al., CLNS 00/1679 (2000).

Measurement of Direct CP Violation in K meson Decays at KTeV

A. Glazov

Abstract. The first measurement of direct CP violation parameter $Re(\epsilon'/\epsilon)$ from the KTeV experiment is presented in this paper. Using a subset of data collected in 1996 and in 1997 we find that $Re(\epsilon'/\epsilon) = [28.0 \pm 3.0(stat) \pm 2.8(syst)] \times 10^{-4}$ [1].

INTRODUCTION

The neutral K meson system is an extremely precise natural laboratory to study various small effects which have fundamental importance for physics. This is due to mixing of two strangeness states $(K^0, \overline{K}^0)$ to produce the observable long- and short-lived K mesons (K_S, K_L) with small mass splitting. One of the most striking discoveries was observation of decays $K_L \to \pi\pi$ in 1964 [2] revealing the CP symmetry violation.

The main features of this effect is an asymmetry in the $K^0 - \overline{K}^0$ mixing, parameterized by ϵ and called *indirect* CP violation. The other possibility, CP violation in decay process itself, called *direct* CP violation has been of the great interest. It can be measured comparing the decay rate into different two π final states:

$$Re(\epsilon'/\epsilon) = \frac{1}{6}\left[\frac{\Gamma(K_L \to \pi^+\pi^-)/\Gamma(K_S \to \pi^+\pi^-)}{\Gamma(K_L \to \pi^0\pi^0)/\Gamma(K_S \to \pi^0\pi^0)}\right]. \quad (1)$$

The Standard Model describes CP violation in a natural way with a complex phase in the CKM matrix. However, the calculations of $Re(\epsilon'/\epsilon)$ are very involved and rather uncertain. Most of recent estimates give values near or below 10^{-3}, for example, $(4.6 \pm 3.0) \times 10^{-4}$ [3] and $(8.5 \pm 5.9) \times 10^{-4}$ [4]. There are also estimates with somewhat larger range of values, $(17^{+14}_{-10}) \times 10^{-4}$ [5]. Alternatively, the indirect CP violation can be also described by additional "superweak" interaction. In this case $Re(\epsilon'/\epsilon) = 0$. Therefore, a nonzero value of $Re(\epsilon'/\epsilon)$ rules out the possibility that a superweak interaction is the sole source of CP violation [6].

The two most precise past measurements of $Re(\epsilon'/\epsilon)$ performed at Femilab (E731) and at CERN (NA31) reported $Re(\epsilon'/\epsilon) = (7.4 \pm 5.9) \times 10^{-4}$ [7] and $Re(\epsilon'/\epsilon) = (23 \pm 6.5) \times 10^{-4}$ [8], respectively. Because of importance to definitively establish direct CP violation new experiments have been undertaken at Fermilab, CERN and Frascati to measure $Re(\epsilon'/\epsilon)$ with precision of about 1×10^{-4}. This paper describes the measurement performed using 23% of the data taken during 1996-1997 run by the KTeV collaboration at Fermilab.

THE EXPERIMENT AND DATA ANALYSIS

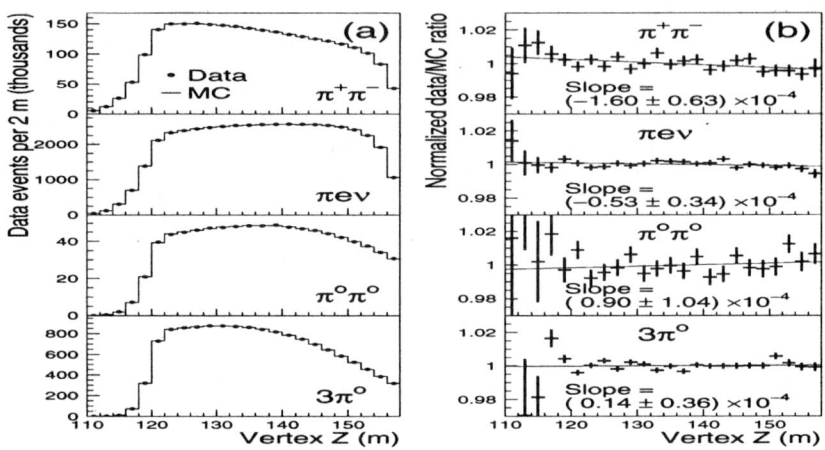

FIGURE 1. (a) Monte Carlo compared to Data of vacuum beam Z distributions for $\pi^+\pi^-$, $\pi e\nu$, $\pi^0\pi^0$ and $\pi^0\pi^0\pi^0$ decays. (b) Linear fits to ratio of data/MC distributions for these samples.

Descriptions of the KTeV detector can be found elsewhere [9]. The detector design is similar to one of E731 [10] but with many improvements of various components. Two parallel K_L beams enter the fiducial decay region 110 meters from the production target. One of the beams passes through scintillator regenerator to produce K_S. We will call this beam as regenerator or K_S beam compared to vacuum or K_L beam. The regenerator position alternates once per minute between the two input beams in order to minimize systematic differences between K_S and K_L events. The regenerator parameters are optimized to increase relative fraction of coherent regeneration.

The decay volume, held in vacuum, extends up to 159 meters from the target. The $\pi^+\pi^-$ decays are measured in drift chamber spectrometer consisting of two detectors before and two detectors after an analysis magnet with horizontal P_t kick of 0.41 GeV/c. The $\pi^0\pi^0$ decays are detected by a 3100 channel CsI electromagnetic calorimeter located at 186 meters from the target. The decay and detector volumes are surrounded by a series of photon vetoes.

In addition to signal $\pi^+\pi^-$ and $\pi^0\pi^0$ events high statistics K_L decay modes, $K \to \pi\nu e$, $K \to \pi^0\pi^0\pi^0$ and $K \to \pi^0\pi^+\pi^-$ are collected for detector calibration and study of systematic effects.

The $\pi^0\pi^0$ samples used for this analysis are from data collected in 1996 while the $\pi^+\pi^-$ samples are from beginning of KTeV data taking in 1997. This is due to 22%

online software inefficiency in 1996 for $\pi^+\pi^-$ events because of too strict reconstruction criteria for the tracks in the drift chambers. Using $\pi^+\pi^-$ and $\pi^0\pi^0$ data from different time periods practically does not increase systematic uncertainty in $Re(\epsilon'/\epsilon)$ since the two modes are reconstructed in essentially independent detector subsystems, detector inefficiencies and dead time effects cancel in single ratio of K_L to K_S yields for either mode.

The essence of KTeV detector design is to minimize differences in detection of K_S and K_L events. But some of them are fundamentally irreducible, the most important one being the difference in the life time. For KTeV energies the mean decay length of K_S is on the level of few meters compared to few kilometers for K_L. This leads to very different distributions of the decay position along the beam (Z-coordinate). Clearly, understanding of detector acceptance in Z is very important for $Re(\epsilon'/\epsilon)$ measurement. We use Monte Carlo simulation to determine it.

Figure 1 shows the comparison of vacuum beam Z distributions for the signal and calibration decay modes. One can see that the overall agreement for all modes is excellent. This is especially true in case of $\pi^0\pi^0$ and $\pi^0\pi^0\pi^0$ events, where the ratios of data to Monte Carlo Z distributions show no slope within statistical error. On the other hand, there is about 2.5σ slope in case of $\pi^+\pi^-$ events. Since the mean Z positions for K_L and K_S decays differ by about 6 m, this slope corresponds to 1.6×10^{-4} bias in $Re(\epsilon'/\epsilon)$ – the largest source systematic error in this analysis.

We have finished analyzing of $\pi^+\pi^-$ decay mode for the rest of the 1997 data sample. For this data, we find no slope in data to Monte Carlo simulation Z overlay. This reduces systematic uncertainty in $Re(\epsilon'/\epsilon)$ coming from the charged mode by about 2/3.

Other sources of systematic uncertainties are discussed in [1]. The total systematic uncertainty in $Re(\epsilon'/\epsilon)$ is calculated as a quadratic sum of all individual sources. It is found to be equal to 2.8×10^{-4} compared to statistical uncertainty of 3.0×10^{-4}. Many sources of systematic uncertainty, including the dominant ones, have essentially statistical nature. They can be reduced using the total event sample collected at KTeV.

RESULTS

$Re(\epsilon'/\epsilon)$ and other K meson system parameters are extracted from the background subtracted data using a fitting program which calculates decay vertex distributions, properly treating regeneration and $K_S - K_L$ interference. K_L and K_S mass difference, ΔM; K_S life time, τ_S; decay phase for $\pi^0\pi^0$, ϕ_{00} and for $\pi^+\pi^-$ events, ϕ_{+-} are calculated for both $\pi^+\pi^-$ and $\pi^0\pi^0$ samples. Consistent results are found leading to the following preliminary KTeV result:

$$\begin{aligned} \Delta m &= (0.5200 \pm 0.0013) \times 10^{-10} \hbar s^{-1} \\ \tau_S &= (0.8967 \pm 0.0007) \times 10^{-10} s \\ \Delta\phi &= \phi_{+-} - \phi_{00} = 0.09^o \pm 0.46^o \end{aligned} \quad (2)$$

Fitting of $Re(\epsilon'/\epsilon)$ was done "blind", by hiding the value of $Re(\epsilon'/\epsilon)$ with an unknown offset, until after the systematic error evaluation were finalized. The result is:

$$Re(\epsilon'/\epsilon) = [28.0 \pm 3.0(stat) \pm 2.8(syst)] \times 10^{-4} = (28.0 \pm 4.1) \times 10^{-4} \quad (3)$$

where the final error is obtained by adding the statistical and systematic uncertainty in quadrature.

This result definitively establishes the existence of direct CP violation and shows that a superweak interaction cannot be the sole source of CP violation in the K meson system. Recently NA48 collaboration at CERN published a new result $Re(\epsilon'/\epsilon) = (18.5 \pm 7.3) \times 10^{-4}$ [11] based on 1997 data and announced a new preliminary result $Re(\epsilon'/\epsilon) = (12.2 \pm 4.9) \times 10^{-4}$ based on 1998 data (see this proceedings). The average of all five recent measurements, $(19.1 \pm 2.4) \times 10^{-4}$, supports the notion of a nonzero phase in the CKM matrix.

New results on $Re(\epsilon'/\epsilon)$ from KTeV collaboration are expected soon. They will be based on data sample taken in 1997 and 1999 which is about eight times bigger than reported here allowing to reduce statistical uncertainty in $Re(\epsilon'/\epsilon)$ to 1×10^{-4} level. The analysis of the 1997 $\pi^+\pi^-$ sample has been finalized. We find result on the single ratio consistent with the presented here with systematic uncertainty reduced by about 2/3. The final goal of the $Re(\epsilon'/\epsilon)$ analysis is to reduce total systematic error below 1×10^{-4} leading to combined uncertainty in $Re(\epsilon'/\epsilon)$ of about 1.5×10^{-4}.

REFERENCES

1. A. Alavi-Harati et al. *Phys. Rev. Lett.* **83**, 22 (1999).
2. J.H Chistenson, J.W. Cronin, V.L. Fitch, and R. Turlay, *Phys. Rev. Lett.* **13**, 138 (1964).
3. M. Ciuchini, *Nucl. Phys.* (Proc. Suppl.) **59**, 149 (1997).
4. A.J. Buras, in *Probing the Standard Model of Particle Interactions*, edited by R. Gupta et al. (Elsevier, New York, 1999); hep-ph/9806471.
5. S. Bertolini et al., *Nucl. Phys.* **B514**, 93 (1998).
6. L. Wolfenstein, *Phys. Rev. Lett.* **13**, 562 (1964); *Comments Nucl. Part. Phys.* **21**, 275 (1994).
7. L.K. Gibbons et al., *Phys. Rev. Lett.* **70**, 1203 (1993).
8. G.D. Barr et al., *Phys. Lett.* B **317**, 233 (1993).
9. R. Kessler, *Proc. VII of Heavy Quarck at Fixed Target (HQ96)* (ed. L. Kopke, Rhinefels 1996) 321 (1997).
10. L.K. Gibbons et al. *Phys. Rev. D* **55**, 6625 (1997).
11. V. Fanti et al., *Phys. Lett.* B **465**, 335 (1999).

New Measurement of Direct CP Violation Parameter Re(ε'/ε) by the NA48 Experiment at CERN

Isabelle Wingerter-Seez

on behalf of the NA48 collaboration[1]

Laboratoire de l'Accélérateur Linéaire, 91898 Orsay, France

Abstract. The NA48 experiment at CERN detects concurrently the four decay modes of the long and short lived neutral kaons. During 1998 data taking, statistics in each mode was increased by at least a factor two, with respect to the 1997 published data. A new measurement of the double ratio of the four modes has been obtained from these data. The new preliminary result on the direct CP violation parameter Re(ε'/ε) is $(12.2 \pm 4.9) \times 10^{-4}$ which, combined with the NA48 published result gives Re(ε'/ε) = $(14.0 \pm 4.3) \times 10^{-4}$.

DIRECT CP VIOLATION.

The CP-violating decay of the long-lived neutral kaon, K_L, into two pions is interpreted to be mainly due to the CP-conserving decay of the K_1 ($CP = +1$) component (CP violation in the K^0-$\bar{K}^0$ mixing) and is parametrized by the ε parameter ($|\varepsilon| = (2.28 \pm 0.02) \times 10^{-3}$).

The interference of the amplitudes for final states with different isospin could also induce a direct CP violation in the decay of the $K_2(CP = -1)$ component; this contribution to CP violation is quantified by parameter ε'. Previous mesurments can be found in [1–4].

In the Standard Model, a value of ε' different from zero arises naturally from the complex phase of the CKM matrix. Theoretical predictions for the ratio Re(ε'/ε) are typically in the range $(0 - 20) \times 10^{-4}$. The difficulty of the theoretical calculations come from QCD penguin diagrams and interactions in the final state [5].

---

[1] Cagliari, Cambridge, CERN, Dubna, Edinburgh, Ferrara, Firenze, Mainz, Orsay, Perugia, Pisa, Saclay, Siegen, Torino, Vienna, Warsaw

NA48 METHOD AND DETECTOR.

The aim of the NA48 experiment is to evaluate $\mathrm{Re}(\varepsilon'/\varepsilon)$, the direct CP violation parameter, with an accuracy $\simeq 2 \times 10^{-4}$, by measuring the double ratio

$$R = \frac{\Gamma(K_L \to \pi^0\pi^0)}{\Gamma(K_S \to \pi^0\pi^0)} \bigg/ \frac{\Gamma(K_L \to \pi^+\pi^-)}{\Gamma(K_S \to \pi^+\pi^-)} \simeq 1 - 6 \times \mathrm{Re}(\varepsilon'/\varepsilon)$$

As the K_L CP-violating decays are suppressed $\simeq 200$ times compared to 3-body decays, **high rate** is required.

To minimize corrections to the ratio, the four modes are measured:

- **simultanously** so that effects due to fluxes, inefficiencies, dead time, accidental activities cancel.
- from the **same fiducial volume**: lifetime $< 3.5 \times \tau_s$. To equalize longitudinal distribution between K_S and K_L, a weight, function of lifetime, is applied so that the acceptance correction is minimum.
- with **high resolution**: magnetic spectrometer for $\pi^+\pi^-$ with uniform magnetic field and homogenous liquid Krypton calorimeter for $\pi^0\pi^0$ to reject backgrounds.
- in **bins of Kaon energy** to account for differences of K_S and K_L energy spectra (20 bins between 70 and 170 GeV).

The NA48 beam line was designed to fulfil these requirements; a schematic view is presented on figure 1.

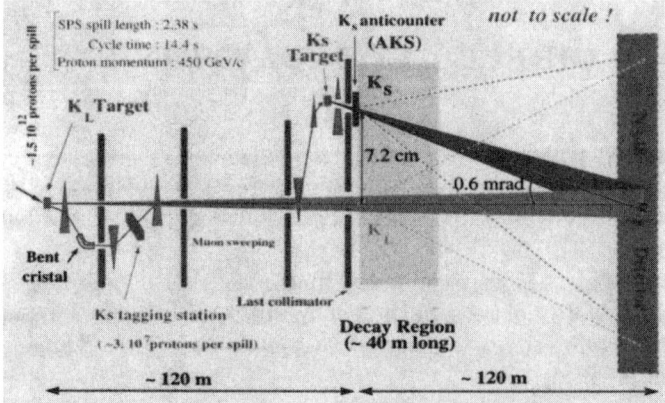

FIGURE 1. Schematic layout of the simultaneous K_S and K_L beams used by NA48

K_S and K_L decays are distinguished by time of flight: the fraction of protons directed to the K_S target cross the tagging station which measures the proton time with a resolution better than 200 ps. Events with a coincidence inside $\pm 2ns$ are labelled K_S. In charged decays, the vertex position allows to identify vertices coming from the K_S and K_L targets and therefore to measure:

- The **tagging inefficiency**, the probability that a K_S decay is not in coincidence with a proton: $(1.95 \pm 0.05) \times 10^{-4}$.

- The **accidental tagging probability**, the probability that a K_L decay is in coincidence with a proton: $(11.05 \pm 0.01\%)$.

The beginning of the fiducial region is precisely defined by an anticounter in the K_S line.

Charged decays are measured by a magnetic spectrometer made of four large drift chambers; the invariant mass resolution is 2.5 MeV/c^2. The time measured with a plastic scintillator hodoscope has a resolution better than 200 ps.

Neutral decays are reconstructed by a quasi-homogeneous liquid Krypton calorimeter; the π^0 invariant mass resolution is $\simeq 1$ MeV/c^2 and the time resolution around 270 ps.

1998 DATA ANALYSIS

Analysis consists in counting events in each mode to evaluate the double ratio R. Table 1 summarizes the statistics accumulated in 1998 (and 1997 for comparison). Corrections are then applied (table 2 summarizes the corrections applied on R_{raw}):

- **Trigger inefficiencies**: if K_S-K_L asymetric they can induce a bias on R. The neutral trigger is a fully pipelined system with a negligible dead time and an efficiency of $(99.93 \pm 0.02)\%$. No correction is applied. The charged trigger inefficiencies induce a correction on R of $(-1 \pm 11) \times 10^{-4}$.

- **Backgrounds** come from K_L three-body decays. To reject $K_L \to \pi^\pm \mu^\mp \nu_\mu$, $K_L \to \pi^\pm e^\mp \nu_e$ and K_L scattered on collimators, mass and transverse momentum cuts are applied; the charged background level after these cuts is $(19 \pm 3) \times 10^{-4}$. Scattered events are not rejected in neutral decays; the correction for this effect is $(10 \pm 3) \times 10^{-4}$. To reject $K_L \to 3\pi^0$ with two lost γ, a cut on a χ^2, computed comparing the $\gamma\gamma$ masses and the π^0 mass ($Relli$), is applied; the remaining background is at a level of $(7 \pm 2) \times 10^{-4}$.

- The **acceptance** correction is reduced by K_L weighting technique to a level (measured from Monte-Carlo) of $(31 \pm 6(stat) \pm 6(syst)) \times 10^{-4}$.

- The **neutral energy scale** is constrained by the reconstruction of the begining of the decay region to an accuracy of $\pm 10 \times 10^{-4}$.

- The **accidental activity** can induce asymetric losses and gains in K_S and K_L; we measure an effect on R of $(2 \pm 6) \times 10^{-4}$, by superimposing events triggered proportionally to the beam intensity, to good $\pi^+\pi^-$ and $\pi^0\pi^0$ events. The uncertainty on accidental effects is estimated by studying differential activities in K_S and K_L events, to a level of $(\pm 10) \times 10^{-4}$.

Statistics (millions)	1997	1998
$K_L \to \pi^0\pi^0$	0.49	1.14
$K_S \to \pi^0\pi^0$	0.98	1.80
$K_L \to \pi^+\pi^-$	1.07	4.87
$K_S \to \pi^+\pi^-$	2.09	7.46

Table 1: 1998 statistics.

Source		
Charged trigger	-1	± 11
Accidental tagging	+1	±8
Tagging efficiency	–	± 3
Neutral rec. systematics	–	±10
Charged vertex	+2	±2
Acceptance	+31	±9
Neutral BKG	-7	±2
Charged BKG	+19	±3
Beam scattering	-10	±3
Accid. activity	+2	±12
Total	+37	±24

Table 2: Corrections to R (10^{-4}).

RESULTS

The preliminary measured value of Re(ε'/ε) based on data presented here is

$$\mathrm{Re}(\varepsilon'/\varepsilon) = (12.2 \pm 2.9(\mathrm{stat}) \pm 4.0(\mathrm{syst})) \times 10^{-4}$$

where the first error is statistical, the second systematic. Figure 2 shows R as a funtion of Kaon energy. Combining this preliminary result with the final NA48 result based on data collected during 1997 [4] gives:

$$\mathrm{Re}(\varepsilon'/\varepsilon) = (14.0 \pm 4.3) \times 10^{-4}$$

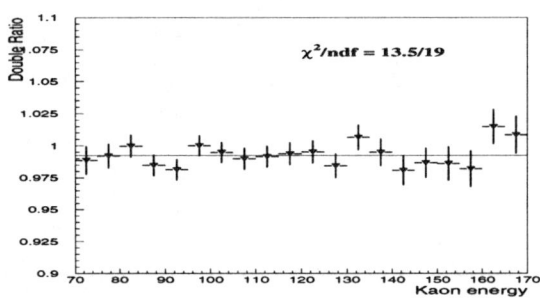

FIGURE 2. Double Ratio R as a function of Kaon Energy (GeV)

REFERENCES

1. G. Barr et al, *Phys. Lett.* B **317**, 233 (1993).
2. L.K. Gibbons et al, *Phys. Rev. Lett.* **70**, 1203 (1993).
3. A. Alavi-Harati et al, *Phys. Rev. Lett.* **83**, 22 (1999).
4. V. Fanti et al, *Phys. Lett.* B **465**, 335 (1999).
5. S. Bosch et al, hep-ph/9904408. M. Ciuchini, G. Martinelli, hep-ph/0006056. See also presentation by C. Wolfe at this conference.

First direct observation of time-reversal violation

CPLEAR Collaboration

A. Angelopoulos[a], A. Apostolakis[a], E. Aslanides[k], G. Backenstoss[b], P. Bargassa[m], O. Behnke[r],
A. Benelli[b], V. Bertin[k], F. Blanc[g,m], P. Bloch[d], P. Carlson[o], M. Carroll[i], E. Cawley[i],
M.B. Chertok[c], M. Danielsson[o], M. Dejardin[n], J. Derre[n], A. Ealet[k], C. Eleftheriadis[p],
W. Fetscher[r], M. Fidecaro[d], A. Filipčič[j], D. Francis[c], J. Fry[i], E. Gabathuler[i], R. Gamet[i],
H.-J. Gerber[r], A. Go[d], A. Haselden[i], P.J. Hayman[i], F. Henry-Couannier[k], R.W. Hollander[f],
K. Jon-And[o], P.-R. Kettle[m], P. Kokkas[b], R. Kreuger[f], R. Le Gac[k], F. Leimgruber[b], I. Mandić[j],
N. Manthos[h], G. Marel[n], M. Mikuž[j], J. Miller[c], F. Montanet[k], A. Muller[n], T. Nakada[m],
B. Pagels[r], I. Papadopoulos[p], P. Pavlopoulos[b], G. Polivka[b], R. Rickenbach[b], B.L. Roberts[c],
T. Ruf[d], M. Schäfer[r], L.A. Schaller[g], T. Schietinger[b], A. Schopper[d], L. Tauscher[b], C. Thibault[l],
F. Touchard[k], C. Touramanis[i], C.W.E. Van Eijk[f], S. Vlachos[b], P. Weber[r], O. Wigger[m],
M. Wolter[r], D. Zavrtanik[j], and D. Zimmerman[c]

[a]University of Athens, Greece, [b]University of Basle, Switzerland [c]Boston University, USA
[d]CERN, Geneva, Switzerland [e]LIP and University of Coimbra, Portugal, [f]Delft University of Technology, Netherlands [g]University of Fribourg, Switzerland, [h]University of Ioannina, Greece,
[i]University of Liverpool, UK, [j]J. Stefan Inst. and Phys. Dep., University of Ljubljana, Slovenia,
[k]CPPM, IN2P3-CNRS et Université d'Aix-Marseille II, France [l]CSNSM, IN2P3-CNRS, Orsay, France, [m]Paul Scherrer Institut (PSI), Villigen, Switzerland, [n]CEA, DSM/DAPNIA, CE-Saclay, France, [o]Royal Institute of Technology, Stockholm, Sweden, [p]University of Thessaloniki, Greece, [r]ETH-IPP Zürich, Switzerland.

Presented by Marko Mikuž
University of Ljubljana and Jožef Stefan Institute, Ljubljana, Slovenia

Abstract. Using its unique capability of strangeness tagging at K^0 production in $p\bar{p} \to K^{\mp}\pi^{\pm}K^0(\bar{K}^0)$ and at decay with the lepton charge in semileptonic decays, CPLEAR measured the semileptonic decay-rate asymmetry

$$\frac{R(\bar{K}^0 \to e^+\pi^-\nu) - R(K^0 \to e^-\pi^+\bar{\nu})}{R(\bar{K}^0 \to e^+\pi^-\nu) + R(K^0 \to e^-\pi^+\bar{\nu})}.$$

The asymmetry, fitted over the eigentime interval 1-20 τ_S, yielded a non-zero result of $(6.6 \pm 1.3_{\text{stat}} \pm 1.1_{\text{syst}}) \times 10^{-3}$. A thorough phenomenological analysis identifies T violation in K^0 mixing and/or CPT violation in semileptonic decays as possible interpretations. A confrontation with world data on neutral kaon decays, however, excludes the latter with sufficient precision to establish the result as the first direct observation of time reversal non-invariance.

Introduction

Since the discovery of CP violation in the neutral kaon system back in 1964 [1] it was realized that validity of the CPT theorem [2] would imply time reversal violation in the same system. Time reversal violation manifests itself as a difference in rates of a process and its time-reversed counterpart. $\bar{K}^0 \leftrightarrow K^0$ mixing represents such a pair of processes and in 1970 a direct test of time reversal was proposed [3] in the asymmetry

$$A_T(\tau) = \frac{R(\bar{K}^0(\tau=0) \to K^0(\tau)) - R(K^0(\tau=0) \to \bar{K}^0(\tau))}{R(\bar{K}^0(\tau=0) \to K^0(\tau)) + R(K^0(\tau=0) \to \bar{K}^0(\tau))}. \quad (1)$$

With CPT conserved, the asymmetry is expected to amount to $A_T = 2\delta_\ell = 6.6 \times 10^{-3}$ where δ_ℓ is the semileptonic charge asymmetry. Experimental realization of the proposed measurement was held up by the unavailability of pure $\bar{K}^0$ and K^0 beams until the advent of LEAR.

CPLEAR Experiment

CPLEAR (see [4] for a description of the apparatus) made use of strangeness conservation in $\bar{p}p$ annihilations to tag the strangeness of the neutral kaon at production time with the charge of the accompanying kaon in the reactions

$$\bar{p}p \to K^0 K^- \pi^+ \quad \bar{p}p \to \bar{K}^0 K^+ \pi^-. \quad (2)$$

In semileptonic decays, only positive leptons resulting from K^0 have been observed (and negative from $\bar{K}^0$), establishing the $\Delta Q = \Delta S$ rule. By measuring the charge of the lepton, CPLEAR can thus determine the strangeness of the neutral kaon at decay time and has thus all the experimental tools needed to observe time reversal violation from eq. (1) for the first time.

The Measured Asymmetry

CPLEAR measured four semileptonic decay rates, labelled by the initial neutral kaon strangeness and the decay electron (muons could not be resolved) charge:

$$R^+ = R(K^0(\tau) \to e^+ \pi^- \nu) \quad R^- = R(K^0(\tau) \to e^- \pi^+ \bar{\nu})$$
$$\bar{R}^- = R(\bar{K}^0(\tau) \to e^- \pi^+ \bar{\nu}) \quad \bar{R}^+ = R(\bar{K}^0(\tau) \to e^+ \pi^- \nu).$$

Obviously, R^- signifies a $K^0 \to \bar{K}^0$ transition and $\bar{R}^+$ its time reversed process $\bar{K}^0 \to K^0$. Before building the asymmetry, however, the rates have to be corrected for the relative detection efficiencies of the strangeness tag at production, parametrized by the efficiency ratio $\xi = \epsilon(K^+\pi^-)/\epsilon(K^-\pi^+)$ as well as for the efficiency ratio $\eta = \epsilon(\pi^+e^-)/\epsilon(\pi^-e^+)$ for the strangeness tag at decay. With the measured number of events the time-reversal asymmetry thus reads

$$A_T^{\exp} = \frac{\eta \bar{N}^+ - \xi N^-}{\eta \bar{N}^+ + \xi N^-}. \tag{3}$$

Calibration data were used to obtain η in bins of (p_e, p_π). Its average value was $<\eta>= 1.014 \pm 0.002$. The parameter ξ was extracted from $\pi^+\pi^-$ decay weights $\alpha_{2\pi} = [1 + 4\mathcal{R}e(\epsilon_T - \delta_{CPT})] \times \xi$, as described in [5]. For that transition, external information on $\mathcal{R}e(\epsilon_T - \delta_{CPT})$ was taken from the semileptonic charge asymmetry [6]

$$\delta_\ell = 2\mathcal{R}e(\epsilon_T - \delta_{CPT}) - 2\mathcal{R}e(x_- + y) = (3.27 \pm 0.12) \times 10^{-3},$$

which is equal to $2\mathcal{R}e(\epsilon_T - \delta_{CPT})$ in the limit of CPT symmetry in semileptonic decays. Using that, an event-by-event correction was applied according to primary pair kinematics, leading to an average value of $<\xi>= 1.12023 \pm 0.00043$. The

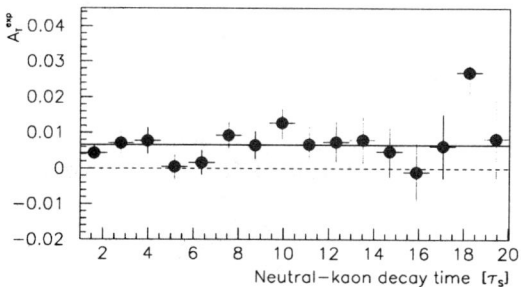

FIGURE 1. Time-violation asymmetry $A_T^{\exp}$. Full line represents a constant fit from 1-20 τ_S.

experimental asymmetry from the complete data-set with $\approx 640\,000$ events [7] is shown in Fig.1. There is a clear constant offset from the zero value, representing a surplus of $\bar{K}^0 \to K^0$ transitions in the whole lifetime range. Its average value

$$<A_T^{\exp}>_{(1-20)\tau_S} = (6.6 \pm 1.3_{stat} \pm 1.0_{syst}) \times 10^{-3}$$

represents the first direct measurement of time-reversal violation. The significance of this measurement exceeds four standard deviations.

Interpretation of the Result

A measurement of such fundamental importance is bound to raise doubts on the validity of assumptions, leading to its success. The key question arising from a comparison of eqs. (1) and (3) is: what are the implications of a possible $\Delta Q = \Delta S$ and/or CPT violation on the interpretation of A_T^{exp} ? To answer that, full blown phenomenology without any prior assumptions has to be involved, like in [8]. Expanding A_T^{exp} to first order in small T, CPT and/or $\Delta Q = \Delta S$ violation parameters yields

$$A_T^{\text{exp}}(\tau) \approx \frac{\bar{R}^+(\tau) - R^-(\tau)}{\bar{R}^+(\tau) + R^-(\tau)} - 2\mathcal{R}e(x_- + y) = 4\mathcal{R}e(\epsilon_T) - 4\mathcal{R}e(x_- + y)$$
$$+ 2\frac{\mathcal{R}e(x_-)(e^{-\frac{1}{2}\Delta\Gamma\tau} - \cos(\Delta m\tau)) + \mathcal{I}m(x_+)\sin(\Delta m\tau)}{\cosh(\frac{1}{2}\Delta\Gamma\tau) - \cos(\Delta m\tau)}. \quad (4)$$

Here, ϵ_T is the T violation parameter in mixing and $\Delta\Gamma = \Gamma_S - \Gamma_L$. In semileptonic decays y parameterizes CPT violation and x_+ and x_- $\Delta Q = \Delta S$ violation in CPT conserving and CPT violating decays, respectively. The transition $\alpha_{2\pi} \to \xi$ using δ_ℓ resulted in the term $-2\mathcal{R}e(x_- + y)$ added to A_T^{exp}.

From the expression for A_T^{exp} it is clear, that the measured offset could in principle result either from T violation in mixing and/or CPT violation in decay. The latter possibility is, however, not consistent with existing experimental data on the neutral kaon system. A global analysis performed by CPLEAR [9] invoking unitarity through the Bell-Steinberger equation places a strict limit on the possible amount of CPT violation in semileptonic decays

$$\mathcal{R}e(x_- + y) = (-0.2 \pm 0.3) \times 10^{-3}.$$

With this, the CPLEAR measurement of A_T^{exp} is thus unambiguously established as the first direct experimental observation of the arrow of time in the world of elementary particles.

REFERENCES

1. J.H. Christenson, J.W. Cronin, V.L. Fitch, and R. Turlay,
   Phys.Rev.Lett. **13** (1964) 138.
2. J.S. Bell, *Proc. Royal Soc.* **A 231** (1955) 479;
   G.L. Lüders, *Ann. Phys.* **2** (1957) 1;
   R. Jost, *Helv. Phys. Acta* **30** (1957) 409.
3. P.K. Kabir, *Phys.Rev.* **D2** (1970) 540;
   A. Aharony, *Lett. Nuovo Cimento* **3** (1970) 791.
4. CPLEAR Collab.: R. Adler et al., *Nucl.Instr.Meth.* **A379** (1996) 76.
5. CPLEAR Collab.: A. Angelopoulos et al., *Phys.Lett.* **B444** (1998) 52.
6. Particle Data Group, *Eur. Phys. J.* **C3** (1998) 1.
7. CPLEAR Collab.: A. Angelopoulos et al., *Phys. Lett.* **B444** (1998) 43.
8. C.D. Buchanan, *Phys.Rev.* **D45** (1992) 4088.
9. CPLEAR Collab.: A. Apostolakis et al., *Phys.Lett.* **B456** (1999) 297.

Measurement of B($K^+ \to \pi^+ \nu \bar{\nu}$)

S.H. Kettell[1]

Brookhaven National Laboratory

Abstract. The experimental measurement of $K^+ \to \pi^+ \nu \bar{\nu}$ is reviewed. New results from experiment E787 at BNL are presented: with data from 1995-97 the branching ratio has been measured to be B($K^+ \to \pi^+ \nu \bar{\nu}$) = $(1.5^{+3.4}_{-1.2}) \times 10^{-10}$. The future prospects for additional data in this mode are examined.

INTRODUCTION

The unprecedented sensitivities of rare kaon decay experiments and the recent observation of $K^+ \to \pi^+ \nu \bar{\nu}$ have opened doors to the measurement of the unitarity triangle completely within the kaon system. The decay $K^+ \to \pi^+ \nu \bar{\nu}$ is one of the 'golden modes' for measuring CKM parameters. Measurement of the branching ratio B($K^+ \to \pi^+ \nu \bar{\nu}$) provides a clean and unambiguous determination of the CKM matrix element $|V_{td}|$, in particular of the quantity $|\lambda_t| \equiv |V_{ts}^* V_{td}|$.

The theoretical uncertainty in $K^+ \to \pi^+ \nu \bar{\nu}$ is quite small (~7%) as the hadronic matrix element is extracted from the $K \to \pi e \nu_e$ branching ratio B(K_{e3}). The $K^+ \to \pi^+ \nu \bar{\nu}$ branching ratio has been calculated to next-to-leading-log approximation [1], with isospin violation corrections [2] and two-loop-electroweak effects [3]. The branching ratio can be expressed as [4]

$$B(K^+ \to \pi^+ \nu \bar{\nu}) = \frac{\kappa_+ \alpha^2 B(K_{e3})}{2\pi^2 \sin^4\theta_W |V_{us}|^2} \sum_l |X_t \lambda_t + X_c \lambda_c|^2 \quad (1)$$
$$= 8.88 \times 10^{-11} A^4 [(\bar{\rho}_0 - \bar{\rho})^2 + (\sigma \bar{\eta})^2]$$
$$= 3.6 \times 10^{-4} ([Re(\lambda_t) - 1.4 \times 10^{-4}]^2 + [Im(\lambda_t)]^2)$$
$$= (8.2 \pm 3.2) \times 10^{-11},$$

with the error determined by the present uncertainty of λ_t. The current limit on B_s–$\bar{B}_s$ mixing, combined with the measured frequency of B_d–$\bar{B}_d$ mixing, implies a limit on $K^+ \to \pi^+ \nu \bar{\nu}$ that has very small theoretical uncertainty [4]: B($K^+ \to \pi^+ \nu \bar{\nu}$) < 1.67×10^{-10}.

---

[1] for the E787 collaboration

EXPERIMENT E787

The E787 experiment at BNL [5] was designed to search for $K^+ \to \pi^+ \nu \bar{\nu}$ and collected data in 1989–91 [6] and then again in 1995–98 after an upgrade to the detector and beamline [7].

The first observation of $K^+ \to \pi^+ \nu \bar{\nu}$ was reported in the analysis of the 1995 data set [8]. A new and improved analysis of the 1995–97 data set has reduced the background levels by almost a factor of three. A plot of range vs. energy for events passing all other $K^+ \to \pi^+ \nu \bar{\nu}$ criteria is shown in Figure 1(a). One event

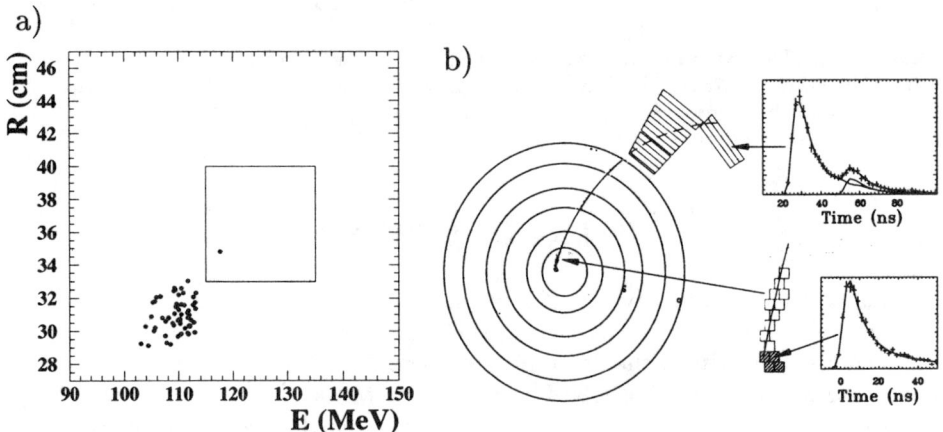

FIGURE 1. (a) Range vs. Kinetic energy plot of the final sample. The events near $E = 108$ MeV are $K_{\pi 2}$ background. The box indicates the accepted region for $K^+ \to \pi^+ \nu \bar{\nu}$ events. (b) Event display of the $K^+ \to \pi^+ \nu \bar{\nu}$ signal event.

is observed in the signal region (the same one as observed in the 1995 data: see Ref. 8). A reconstruction of the event is shown in Figure 1(b). The background is measured to be 0.08 ± 0.02 events. Based on this one event, the branching ratio is $B(K^+ \to \pi^+ \nu \bar{\nu}) = (1.5^{+3.4}_{-1.2}) \times 10^{-10}$ [9]. From this measurement, a limit of

$$0.002 < |V_{td}| < 0.04 \qquad (2)$$

is determined; in addition, the following limits on $\lambda_t \equiv V_{ts}^* V_{td}$ can be set:

$$|Im(\lambda_t)| < 1.22 \times 10^{-3} \qquad (3)$$
$$-1.10 \times 10^{-3} < Re(\lambda_t) < 1.39 \times 10^{-3}$$
$$1.07 \times 10^{-4} < |\lambda_t| < 1.39 \times 10^{-3}.$$

The final sensitivity of the E787 experiment, based on data from 1995–98, should reach a factor of two further to the standard-model (SM) expectation for $K^+ \to \pi^+ \nu \bar{\nu}$.

The 90% CL upper limit on $K^+ \to \pi^+ \nu \bar{\nu}$ is $B(K^+ \to \pi^+ \nu \bar{\nu}) < 5.8 \times 10^{-10}$ and a model-independent limit [10] on the neutral mode, $K_L^\circ \to \pi^\circ \nu \bar{\nu}$, can be derived from this result:

$$B(K_L^\circ \to \pi^\circ \nu \bar{\nu}) < 4.4 \times B(K^+ \to \pi^+ \nu \bar{\nu}) \quad (4)$$
$$< 2.6 \times 10^{-9} \; (90\% \, \text{CL})$$

The $K \to \pi \pi \nu \bar{\nu}$ decays, such as $K^+ \to \pi^+ \pi^\circ \nu \bar{\nu}$, can also, in principle, provide a clean determination of CKM matrix parameters; however, due to the small SM branching ratios, e.g. $B(K^+ \to \pi^+ \pi^\circ \nu \bar{\nu}) = (1-2) \times 10^{-14}$, their usefulness is limited. E787 has recently set the first limit on any of these modes [11], with $B(K^+ \to \pi^+ \pi^\circ \nu \bar{\nu}) < 4.3 \times 10^{-5}$. Additional new results from E787 include a measurement of the direct emission (DE) radiation in $K^+ \to \pi^+ \pi^\circ \gamma$ decay ($K_{\pi 2 \gamma}$) and the first observation of structure-dependent radiation (SD) in $K^+ \to \mu^+ \nu_\mu \gamma$ decays ($K_{\mu 2 \gamma}$). With eight times higher statistical sensitivity than previous experiments and better kinematic constraints E787 has measured a branching ratio for direct emission $B(K_{\pi 2 \gamma}:DE)$ that is four times smaller than previous results [12], $B(K_{\pi 2 \gamma}:DE, 55 \, \text{MeV} < T_+ < 90 \, \text{MeV}) = (4.7 \pm 0.8 \pm 0.3) \times 10^{-6}$ (T_+ is the kinetic energy of the π^+). E787 has also measured the branching ratio for structure-dependent $K_{\mu 2 \gamma}$ [13] to be $B(K_{\mu 2 \gamma} : SD^+) = (1.33 \pm 0.12 \pm 0.18) \times 10^{-5}$. The vector and axial-vector form factors are $|F_V + F_A| = 0.165 \pm 0.007 \pm 0.011$ and $-0.04 < F_V - F_A < 0.24$ at 90% CL.

FUTURE PROSPECTS

Significant progress in the determination of the fundamental CKM parameters from the $K \to \pi \nu \bar{\nu}$ system will be made in the generation of experiments that is now starting: E949 and KOPIO at BNL, CKM and KAMI at FNAL and E391 at KEK. These measurements can unambiguously determine SM CP violation parameters. Comparison with the B-system will then over-constrain the triangle and test the SM explanation of CP violation. The two most important tests are expected to be:

- Comparison of the angle 2β from the ratio $B(K^+ \to \pi^+ \nu \bar{\nu})/B(K_L^\circ \to \pi^\circ \nu \bar{\nu})$ and the CP asymmetry in the decay $B_d^\circ \to \psi K_S^\circ$ [10,14].

- Comparison of the magnitude $|V_{td}|$ from $K^+ \to \pi^+ \nu \bar{\nu}$ and the ratio of the mixing frequencies of B_s and B_d mesons [4].

Improvement in the charged mode $K^+ \to \pi^+ \nu \bar{\nu}$ will come in two steps: E949 and CKM.

The E787 experiment has already demonstrated sufficient background rejection for a very precise measurement of $B(K^+ \to \pi^+ \nu \bar{\nu})$. A new experiment under construction, E949, is expected to run in 2001–03. Taking advantage of the very large AGS proton flux and the experience gained with the E787 detector, E949 with modest upgrades should observe $\mathcal{O}(10)$ SM events in a two year run. The background is well-understood and is $\sim 10\%$ of the SM signal.

A proposal for a further factor of 10 improvement has been initiated at FNAL. The CKM experiment (E905) plans to collect $\mathcal{O}(100)$ SM events, with $\mathcal{O}(10)$ background events, in a two year run starting after 2005. This experiment will use a new technique, with K$^+$ decay-in-flight and independent momentum (Si and straw-tube trackers) and velocity (kaon and pion RICH) spectrometers. CKM is situated in a high flux 22 GeV/c RF-separated kaon beam, derived from the Main Injector at FNAL.

Improvements in the sensitivities of $K^+ \to \pi^+ \nu \bar{\nu}$ experiments over time is shown in Figure 2. Published data are shown as solid points and future projections to

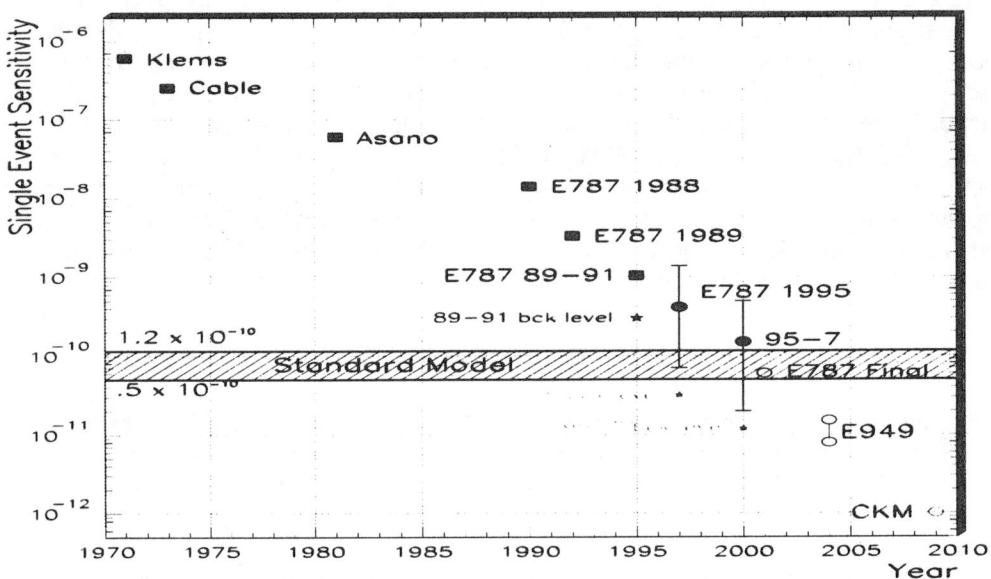

FIGURE 2. Sensitivities of experiments searching for $K^+ \to \pi^+ \nu \bar{\nu}$. The sensitivities of experiments setting limits are shown as red squares. The projected sensitivities are shown as open circles. Background measurements for several of the recent E787 searches are shown as blue stars.

E949 and CKM are shown as open points.

ACKNOWLEDGMENTS

This work was supported under U.S. Department of Energy contract #DE-AC02-98CH10886.

REFERENCES

1. Bucahalla G. and Buras A., *Nucl. Phys.* **B412**, 106 (1994).
2. Marciano W.J. and Parsa Z., *Phys. Rev.* **D53**, R1 (1996).
3. Bucahalla G. and Buras A., *Phys. Rev.* **D57**, 216 (1998).
4. Bucahalla G. and Buras A., *Nucl. Phys.* **B548**, 309 (1999).
5. Atiya M., et al., *Nucl. Instrum. Methods* **A321**, 129 (1992); Atiya M., et al., *Nucl. Instrum. Methods* **A279**, 180 (1989).
6. Adler S., et al., *Phys. Rev. Lett.* **76**, 1421 (1996).
7. Chiang I.H., et al., *IEEE Trans. Nucl. Sci.* **NS-42**, 394 (1995); Bryman D.A., et al., *Nucl. Instrum. Methods* **A396**, 394 (1997); Blackmore E.W., et al., *Nucl. Instrum. Methods* **A404**, 295 (1998); Komatsubara T.K., et al., *Nucl. Instrum. Methods* **A404**, 315 (1998); Doornbos J. et al., *Nucl. Instrum. Methods* **A444**, 546 (2000).
8. Adler S., et al., *Phys. Rev. Lett.* **79**, 2204 (1997).
9. Adler S., et al., *Phys. Rev. Lett.* **84**, 3768 (2000).
10. Grossman Y., and Nir Y., *Phys. Lett.* **B398**, 163 (1997).
11. Ng C. *The Search for the Rare Decay of $K^+ \to \pi^+\pi^0\nu\bar{\nu}$*. PhD thesis. SUNY Stonybrook (2000); also submitted to *Phys. Rev. D*.
12. Adler S., et al., *Phys. Rev. Lett.* in press; also hep-ex/0007021.
13. Adler S., et al., *Phys. Rev. Lett.* in press; also hep-ex/0003019..
14. Bucahalla G. and Buras A., *Phys. Lett.* **B333**, 221 (1994); Bucahalla G. and Buras A., *Phys. Rev.* **D54**, 6782 (1996); Nir Y. and Worah M.P., *Phys. Lett.* **B423**, 319 (1998); Bergmann S. and Perez G., hep-ph/0007170.

MEASUREMENT OF THE TRANSVERSE POLARIZATION OF POSITRONS FROM THE DECAY OF POLARIZED MUONS

Norbert Danneberg

ETH Zurich-Institute for Particle Physics, Hoenggerberg, CH-8093 Zurich, Switzerland

Abstract. In the standard model (SM) of electroweak interactions the positron from the decay of polarized positive muons is mainly longitudinal polarized. However the model also predicts a small transverse polarization component P_{T_1}, which lies in the plane spanned by muon-spin and positron momentum. Interference with additional scalar couplings, as they appear in supersymmetric extensions of the SM, would result in substantial values for P_{T_1} as well as in a non zero value of the transverse component P_{T_2} which is perpendicular to the above mentioned plane. A non zero component P_{T_2}, proportional to the imaginary part of a possible scalar coupling, would be the first observation of time reversal violation in a purely leptonic decay. Measuring P_{T_1}, which is proportional to the real part, amounts to a model independent determination of the Fermi coupling constant. The μ_{P_T} experiment at the Paul Scherrer Institute will improve the current experimental limits $P_{T_1} = (16 \pm 22) \cdot 10^{-3}$, $P_{T_2} = (7 \pm 23) \cdot 10^{-3}$ by almost one order of magnitude. Besides a description of the experiment detailed results from the last run in August 1999 will be given.

Introduction

Measurements of muon decay are low energy tests of the standard model. In fact, only a few years ago it has been shown that $V - A$, as one of the basic assumptions of the standard model, follows from the results of a selected set of muon decay experiments (including inverses muon decay) [1]. The experimental limits obtained up to now, however still allow for substantial contributions from non-standard couplings which differ in their spin structure from the $V - A$ interaction. The limits on these couplings can be efficiently reduced by performing experiments with polarized muons and positrons.

This experiment aims at improving present limits [2] on the transverse positron polarization from muon decay by one order of magnitude and to obtain better limits on non-standard couplings and time reversal invariance in weak interactions.

The transverse polarization of positrons from muon decay can be split into two orthogonal components P_{T_1} and P_{T_2}, where P_{T_1} is in the plane of muon polarization

P_μ and positron momentum k_{e^+}, and P_{T_2} is perpendicular to that plane. Both components can be expressed in terms of the Michel parameters $\frac{\beta}{A}$ and $\frac{\beta'}{A}$ assuming that the michel parameter α is small [2]

$$P_{T_1} = \frac{(1-x)x_0}{2x^2 - 3x + x_0^2} + \frac{2\eta}{2x-3} \quad P_{T_2} = \frac{-\sqrt{x^2 - x_0^2}\frac{\beta'}{A}\sqrt{1-x_0^2}}{2x^2 - 3x + x_0^2 + 6\eta(x-1)x_0}. \quad (1)$$

Here x is the reduced energy and x_0 the mimimal positron energy. Using the fact that $\eta = -\beta/A$ and assuming only one additional scalar coupling beyond the standard model it can be shown that $\beta \propto \Re(g_{RR}^S)$ and $\beta' \propto \Im(g_{RR}^S)$ The first component P_{T_1} is therefore proportional to the real part of a possible new scalar coupling and can be used to look for new gauge bosons beyond the standard model. The second component P_{T_2} is proportional to the imaginary part of such a coupling and therefore would indicate time reversal violation in a purely leptonic decay.

Setup

Polarized μ^+ with momentum k_μ arrive in bunches every 20 ns and are stopped in a beryllium target yielding a stop rate of $2 \cdot 10^7 \text{s}^{-1}$. They precess with a frequency of 50 MHz in a magnetic field of 0.37 T. Due to the fact that the accelerator frequency is equal to the precession frequency (50 MHz) the high polarization P_μ of the muons ($\approx 91\%$) is preserved. The decay positrons (momentum k_{e^+}) passing a start counter are tracked with drift chambers, and annihilate in a magnetized VacofFlux foil which serves as a polarimeter. Two additional small chambers and veto counters ensure that the annihilation took place in the magnetized foil. The two annihilation photons are detected in the BGO calorimeter consisting of 127 hexagonal modules.

Longitudinal Polarization

In order to verify that the apparatus is sensitive to polarization effects, the longi-

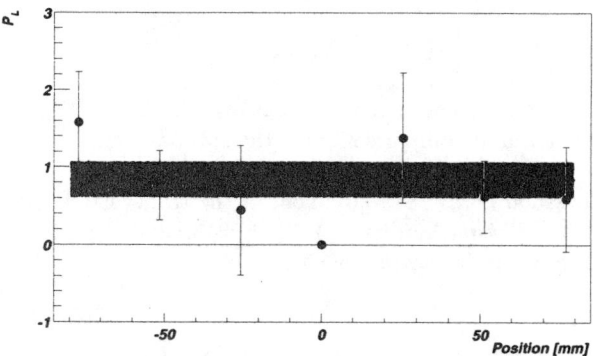

FIGURE 1. Measured longitudinal polarization of the positrons from muon decay. The straight line is the average of the data, the gray area indicates the average error

tudinal polarization of the positrons can be measured via the spin dependece of the annihilation cross section. Figure 1 shows the resulting values for P_L in the different sectors of the foil. The average longitudinal polarization is $P_L = 0.92 \pm 0.11$ which is consistent with previous measurements [3] as well as with the standard model. Furthermore it prooves the sensitivity of the apparatus to polarization measurements.

Transverse Polarization

Due to the μ-spin rotation, the effect of a transverse polarization component will manifest in a rotation of the photon intensity distribution on the BGO calorimeter. Figure 2 shows the result of a log-likelihood to 16 Mio. annihilation events

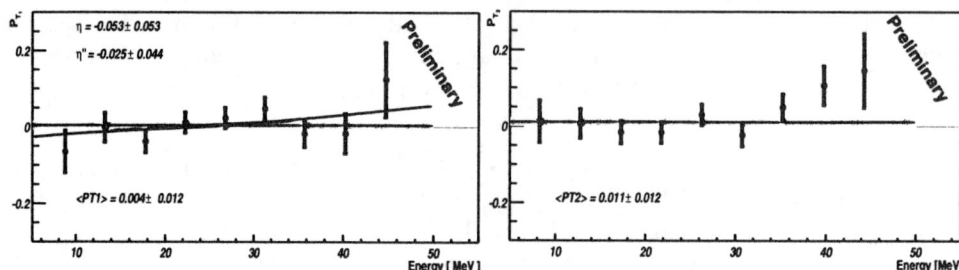

FIGURE 2. Transversal positron polarization components P_{T1} and P_{T2} as a function of the positron energy at the time of annihilation. The gray bar indicates the one sigma average, the line the best fit for η.

taken in August 99 . Proper averaging of P_{T1} and P_{T2} over the energy dependence gives a preliminary result of $< P_{T1} > = 0.033 \pm 0.029$, $< P_{T2} > = 0.011 \pm 0.029$ $\eta = -0.053 \pm 0.053$ and $\eta'' = -0.025 \pm 0.044$. Using this subset of the available data, both transversal polarization components are compatible with the standard model.

This Project is supported by the Suisse National Science Foundation.

REFERENCES

1. W. Fetscher, H.J. Gerber, and K.F. Johnson. Muon decay: Complete determination of the interaction and comparison with the standard model. *Physics Letters B*, 173(1):102, 1986.
2. H. Burkard et al. Muon decay: Measurement of the transverse positron polarization and general analysis. *Physics Letters B*, 160(4,5):343, 1985.
3. H. Burkard et al. *Physics Letters B*, 150:242, 1985.

A Test of Time Reversal Invariance in Stopped K^+ Decay

Michael D. Hasinoff
for the KEK-E246 Collaboration[1]

Dept. of Physics and Astronomy, Univ. of British Columbia, Vancouver, Canada V6T 1Z1

Abstract. An experiment to search for the T–violating transverse muon polarization in $K^+ \to \mu^+\pi^o\nu_\mu$ decay ($K^+_{\mu 3}$) is now underway at the KEK 12 GeV proton synchrotron in Japan. The experiment uses a stopped K^+ beam and a large Superconducting Toroidal Spectrometer. The expected limit on $P_\mu^\perp$ will set constraints on non-standard models of CP–violation such as the multi–Higgs doublet model. The experiment should finish data–taking by the end of 2000.

INTRODUCTION

For more than 20 years the holy grail in subatomic physics has been to perform high precision tests of the Standard Model (SM) and to search for new physics beyond the SM. The transverse polarization ($P_\mu^\perp$) of the muon in K^+ decay is one such observable which is predicted to be negligibly small ($\sim 10^{-6}$) in the SM but there are many extensions of the SM such as multi–Higgs doublets [1], leptoquarks [2], R–parity violating or Squark family mixing SUSY [3] which make predictions for ($P_\mu^\perp$) as high as 10^{-3}, just below the current experimental upper limit. Consequently this triple–vector correlation, $P_\mu^\perp = \hat{s}_\mu \cdot (\hat{p}_\mu \times \hat{p}_\pi)$ in $K_{\mu 3}$ decay has been studied using both neutral and charged kaons [4–6] since it was first suggested by Sakurai [7] as a test of time reversal invariance (TRI). The K^+ decay mode is especially sensitive since the final state interaction, which can mimic a T–violation effect, has been calculated [8] to be $\leq 10^{-6}$.

In V–A theory, the decay matrix element can be written [9] in terms of two form factors, $f_\pm(q^2)$. Since these form factors arise from the hadronic current there is no relative phase between them. Hence, if the ratio $\xi = f_-/f_+$ has any non-zero imaginary part there will be a violation of TRI and this will signify the presence of

---

[1] The E246 collaboration includes IPNS/KEK, Univ. of Tsukuba, Tokyo Institute of Technology, Inst. for Nuclear Research (Moscow), Univ. of British Columbia, Montréal, Saskatchewan, TRIUMF, Korea Univ., Princeton Univ., Yonsei Univ., Virginia Poly. Inst. & State Univ., Nat. Taiwan Univ.

a new scalar(S) or pseudoscalar(P) interaction. $P_\mu^\perp$ is proportional to Im(ξ) with a proportionality factor given by kinematics (integrated over the experimentally available phase space). For the present experiment, $P_\mu^\perp \sim 0.3\times$ Im(ξ).

The previous measurement [6] of $P_\mu^\perp$ in $K_{\mu 3}^+$ decay used in-flight decays and obtained $P_\mu^\perp = (-3.1 \pm 5.4) \times 10^{-3}$ with a corresponding value of the T–violating parameter, Im(ξ) = -0.016 ± 0.025, for the maximal case when Re(ξ) = 0.

KEK–E246 EXPERIMENT

The present KEK experiment uses a separated K^+ beam from the low momentum (660 MeV/c) beam channel (K5) along with the high acceptance Superconducting Toroidal Spectrometer(STS). A Fitch-type Cherenkov counter is used to select K's which are then degraded in BeO and stopped in an active target (consisting of 256–5 mm square fibres) located at the centre of the magnet (see Fig.1). The charged decay products can also be tracked inside the target before being momentum analyzed (by C1-C4) in any one of the 12 identical magnet gaps. Upon exiting the magnet gap the μ's are degraded by a wedge-shaped Cu degrader and then stopped in a segmented high-purity Al target. The muon polarization is measured by detecting an asymmetry in the e$^+$ angular distribution. A small transverse magnetic field ($\sim$ 130 G) is applied at the position of the muon stoppers so that P_T can be preserved while, at the same time, the in-plane muon polarization can be precessed. This serves to reduce any spurious effect due to the large in-plane μ^+ polarization. The π^0's from $K_{\mu 3}^+$ decay are detected in a highly segmented (768 crystal) CsI(Tl) barrel detector which completely surrounds the target except for 12 openings at the entrance to the STS magnet gaps. The rotationally symmetric STS spectrometer ensures a powerful cancellation of many systematic effects. For a $K_{\mu 3}^+$ event having the π^0 moving along the detector axis (forward or backward), the decay plane is almost radial to the target and parallel to the beam axis. $P_\mu^\perp$ is then directed azimuthally in a screw sense around the detector axis and it will show up as a difference in the e$^+$ counting rate between the clockwise(cw) and counter-clockwise(ccw) side counters surrounding that muon stopper in the polarimeter. Summing all the cw– and ccw–counters for both forward or backward going π^0's we can then form a double ratio

$$\frac{[\sum_{i=1}^{12} N_i(cw)/\sum_{i=1}^{12} N_i(ccw)]_{fwd}}{[\sum_{i=1}^{12} N_i(cw)/\sum_{i=1}^{12} N_i(ccw)]_{bwd}} \cong 1 + 4 <\cos\theta_T> \alpha\, P_\mu^\perp, \qquad (1)$$

which provides a factor of 4 increased sensitivity to $P_\mu^\perp$ as well as a substantial reduction in nearly all potential systematic errors. The attenuation factor $<\cos\theta_T>$ is due to the finite angular acceptance ($\cos\theta_{\pi^0,\gamma} \geq 0.342$) of π^0's detected either as 2γ's or 1γ (if E$_\gamma$ > 70 MeV); this is obtained from Monte Carlo. The analyzing power, α (the ratio of $A_{e^+}/P_\mu^\perp$) = 0.197 $\pm$ 0.005, is obtained by analyzing $K_{\mu 3}^+$ events where the normal polarization, P_N, lies in the decay plane.

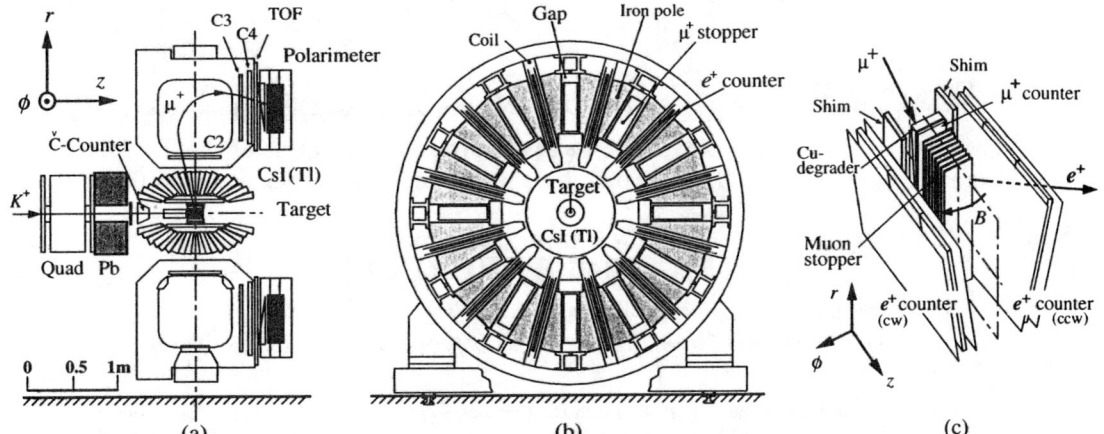

FIGURE 1. (a) Side view and (b) beam view of the E246 setup. (c) Detailed view of a muon polarimeter for each or the 12 magnet sectors.

PRESENT RESULTS AND FUTURE PLANS

The '96 and '97 data contained $\sim 3.9 \times 10^6$ good $K_{\mu 3}^+$ events. The various sources of systematic errors were carefully evaluated (with real data wherever possible). The Monte Carlo studies were checked against real data and excellent agreement was obtained. These first results [10] give

$$P_\mu^\perp = -0.0042 \pm 0.0049(stat) \pm 0.0009(syst) \quad \text{and}$$

$$\text{Im}(\xi) = -0.0130 \pm 0.0160(stat) \pm 0.0030(syst)$$

which are consistent with no violation of TRI. The small systematic errors imply that further data taking is justified. The results from our '98–99 run will be available shortly and we have been approved to collect even more data during 2000. We expect our final statistical error on $\text{Im}(\xi)$ should be ≤ 0.007.

REFERENCES

1. Garisto, R., and Kane, G., *Phys. Rev.* **D44**, 2038 (1991).
2. Bélanger, G. and Geng, C.Q., *Phys. Rev.* **D44**, 2789 (1991).
3. Wu, G.-H., and Ng, J. N., *Phys. Lett.* **B392**, 93 (1997).
4. Schmidt, M. *et al.*, *Phys. Rev. Lett.* **43**, 556 (1979).
5. Morse, W.M. *et al.*, *Phys. Rev.* **D21**, 1750 (1980).
6. Blatt, S.R., *et al.*, *Phys. Rev.* **D27**, 1056 (1983).
7. Sakurai, J.J., *Phys. Rev.* **109**, 980 (1958).
8. Zhitnitskii, A.R., *Sov. J. Nucl. Phys.* **31**, 52 (1980).
9. Cabibbo, N., and Maksymowicz, A., *Phys. Lett.* **9**, 352 (1964).
10. Abe, M. *et al.*, *Phys. Rev. Lett.* **83**, 4253 (1999).

Search For T-violation In Reactions With Slow Neutrons

E. Korobkina[*], G. Danilyan[¶]

[*]Johannes Gutenberg Universitaet-Mainz, Institue fuer Physik, 55099 Mainz, Germany
[¶]SSC-Institute for Theoretical and Experimental Physics, 117259 Moscow, Russia

Abstract. The present state of art in search for T-violation in neutron induced reactions is briefly observed. The interest is focused on one interesting and still not well understood result, which is a recent observation of a non-vanished T-odd triple correlation in ternary fission $D(^{233}U) = (-1.8 \pm 0.1) \times 10^{-3}$ and $D(^{235}U) = (+0.52 \pm 0.06) \times 10^{-3}$. The puzzle is a value quite large even for the mimic final state effect. Since the theoretical estimation of the latter is rather complicated due to the fission complexity, we discuss an experimental approach to reveal the origin of this correlation.

INTRODUCTION

In general, two means to search for Time Reversal Invariance (TRI) violation are available. One is to check the basic TRI consequence like a comparison of cross sections of direct and reverse reactions or comparison of the polarization and asymmetry in cinematically identical scattering experiments. The disadvantage is a need to perform two independent measurements and to find proper reversible reactions. In physics with slow neutrons usually the other means is employed, i.e. to measure the only parameter, namely a correlation coefficient of the T-odd product of momenta and/or angular momenta of in/outgoing particles. Two simplest T-odd correlations are triple ones D ($\mathbf{J} \cdot [\mathbf{p_1} \times \mathbf{p_2}]$) and R ($\mathbf{p} \cdot [\mathbf{J}_1 \times \mathbf{J}_2]$), where $\mathbf{J}$ stands for spins and $\mathbf{p}$ for momenta, D and R are correlation coefficients to be measured. Since the first one is P-even whereas the second is P-odd, they provide null test for two different type of T-violation interaction beyond Standard Model.

To search for T-violation by means of correlation measurement looks similar to investigations of Parity Non Conservation (PNC) effects. However, unlike those the restriction on appearance of the non-vanishing value of T-odd correlation's is only an approximation in the first oder of the basic interaction. That is to say, if investigated reaction is not reversible in time, the mimic effect may arise roughly proportional to the interaction constant square. Therefore, only the value exceeding the mimic effect can be an evidence of T-violation. The technical obstacles and the need to account the mimic effects constrains new experimental results since the first T-violation search in neutron induced reactions were performed about 20 years ego. Just over last two years, we are seeing a renovation of activity in this field owing to improvement of experimental technique, neutron facilities and some new ideas.

Search for T-violation in β-decay

Quite accesible for theoretical estimations, the β-decay of neutron is does attractive for T-violation search as like as the neutron EDM.

The present averaged experimental result of D coefficient has been obtained in 1974-1978 and is equal to $(-0.5\pm1.4) \times 10^{-3}$ [1]. Being mostly of weak magnetizm origin, the mimic effect of the D-coefficient was estimated to be $D_f \le 10^{-5}$ [2,3]. Therefore, the range down to 10^{-5} is still to be achieved. Two collaborations, TRINE based in Munich and emiT based in NIST, are working in this field claiming to lower the limit down to $D \le 10^{-4}$ [4,5]. Their first runs seem to be successful and the outcome will be published soon.

The first experiment to measure the R-coefficient in neutron β-decay is under way at Paul- Scherrer Institute [6], Switzerland, where a new cold neutron facility SINQ came into operation last year. In the case of the R measurement, the experimental technique becomes much more complicate since one have to analyze the electron polarization. The PSI collaboration is developing a Mott polarimeter with the track reconstruction technique. The detector prototype has been already constructed and tested.

Search for T-violation in Nuclear Reactions with Neutrons

The interest to search for TRI violation in neutron interaction with heavy nuclei has been trigged by the quite impressive enhancement of PNC effects. The experiment proved a compound state to be responsible for this gain. Futher theoretical estimations draw attention to the possible enhancement of T-violating effects by a factor of 10^3 in the vicinity of weak resonances [7].

Until now there is no experimental result in the case of triple R-correlation. The inavoidable obstacle is the mimic effect arising due to the need in neutron and nuclei polarization, i. e. the pseudomagnetic spin-spin interaction [8]. Nevertheless, the main point of existing proposals is to find such experimental conditions, when the real T-violation is gained, whereas the mimic effect is not affected. For example, the ^{139}La target with a most significant resonance enhancement looks promising [9].

Tracing the first observation of PNC effect, the T-odd correlation has been measured firstly in the radiative capture of polarized neutrons. Only the upper limit of the D-coefficient at the level $D < 10^3$ in correlation $\mathbf{J} \cdot [\mathbf{K_1} \times \mathbf{K_2}])$ has been found. Here $\mathbf{J}$ is the nuclear polarization after polarized neutron capture, $\mathbf{K_1}$ and $\mathbf{K_2}$ are wave vectors of the first and the second gamma quantum in a mixed (M1, E2) cascade transition of ^{35}Cl [10].

The upper limit on the ratio of the T-odd to T-even admixture in the effective nucleon-nucleon interaction was measured to be $< 5 \cdot 10^{-3}$ by means of 5-fold correlation in polarized neutron scattering on aligned nuclei of ^{165}Ho [11]. The new experiments are under way in Dubna [12] and Jülich [13].

The use of 5-fold correlation allows overcoming the lack of suitable reactions to measure the D – correlation. Indeed, in the range of the slow neutron energy one needs at least three outgoing particles to have two independent momentum vectors. In fact, apart the neutron β-decay the only accessible reactions are neutron radiative capture and ternary fission, where third light particles accompanies two heavy fragments. Although the first reaction was measured 30 years ago, the interest to the second one appears only recently, after the observation of an unexpected large non-vanishing D – correlation in ternary fission of ^{233}U and ^{235}U [14]. The raw values are $D(^{233}U) = (-1.8 \pm 0.1) \times 10^{-3}$ and $D(^{235}U) = (+0.52 \pm 0.06) \times 10^{-3}$. This result seems to be a new manifestation of the coherent formation of angular correlations, in which fission fragments are involved. The first hint came from the fission of aligned nuclei. Then it was the observation of P-odd asymmetry and P-even s-p waves interference in the fragment angular distribution, whereas the same correlations but with the ternary α-particle momentum showed only the uper limit at the level 10^{-4}.

The origin of T-odd correlation is not clear yet. If there is any T-violating contribution, it is worthwhile to asume its strong dependence on the resonanse neutron energy, E_n, like it was observed in P-odd and P-even effects mentioned above. The proposal to measure $D(E_n)$ in ^{239}Pu ternary fission has been accepted and a test run has been scheduled in August, 2000 at the High Flux Reactor of the Institute Laue-Langevin, France. In addition to the technical test of set-up, we are going to check an asumption about the unknown D-coefficient value of ^{239}Pu that is expected to be $D(^{239}Pu) = +2 \times 10^{-3}$. It is based on following observations: (i) signs of both $D(^{233}U)$ and $D(^{235}U)$ coefficients coincident with ones of compound state polarizations $P(^{233}U) = -0.32$ and $P(^{235}U) = +0.13$; (ii) the values of ratio $D(^{233}U)$ to $D(^{235}U)$ and $P(^{233}U)$ to $P(^{235}U)$ are the same in the limit of experimental uncertainties. If this is do the case, it would be a good evidence the origin of D-coefficient really lies only in the final state interaction, for example, like it is proposed in [15].

ACKNOWLEDGMENTS

The experiment would not be possible without the kind support of E. Lelièvre-Berna. We are also very grateful to W. Heil and O. Zimmer. The work is supporting by RFBR, grant N 00-02-16011.

REFERENCES

1. Barnet, R. M., et al., *Phys. Rev. D* **54**, 1-721 (1996).
2. Curtis G. Callan, Jr., and Treimen, S.B., *Phys. Rev.* **162**, 1494-1497 (1967).
3. Herczeg, P., and Khriplovich, I.V., *Phys. Rev. D* **56**, 80-89 (1997).
4. Soldner, T., et al., *NIM A* **440**, 643-647 (2000).
5. Lising, L. J., et al., "New Limit on the D Coefficient in Polarized Neutron Decay", *nucl-ex/0006001-2000*, submitted to Phys. Rev. C, 2000, pp. 1-11.
6. Sromicki, J., et al., *NIM A* **440**, 609-617 (2000).
7. Bunakov, V.E., *Phys. Rev. Lett.* **60**, 2250-2253 (1988).
8. Larmoreaux, S. K., and Golub, B., *Phys. Rev. D* **50**, 5632-5638 (1994).
9. Hautle, P., and Iinuma, M., *NIM A* **440**, 638-641 (2000).
10. Bulgakov, M.I., et al., *Phys. Lett. B* **42**, 375-379 (1972).
11. Koster, J.E., et al., *Phys. Lett. B* **267**, 23-28 (1991).
12. Atsarkin, V.A., Barabanov, A.L., Beda, A.G., and Novitsky, V.V., *NIM A* **440**, 626-631 (2000).
13. Bisplinghoff, J., et al., "Test of Time Reversal Invariance in Polarized Proton-Deuteron Scattering" *nucl-ex/9810003*, contribution to the fifth international WEIN symposium, 1998, pp.1-6.

14. Jesinger, P., et al., "Ternary fission as a testing ground for conservation laws", in *Annual Report 1999* editing by H. Büttner and C. Vettier, Institute Laue-Langevin, Grenoble, 2000, pp. 59-61.
15. Bunakov, V.E., and Petrov, G.A., "A possible explanation of the triple correlation origin in ternary fission", will be publiched in *Proceedings of VIII International Seminar on Interaction of Neutrons with Nuclei*, ISINN-8, Dubna, 2000.

Determination of γ in $B^- \to D^0 K^-$ and Related Processes

David Atwood*, Isard Dunietz[†] and Amarjit Soni[‡]

*Dept. of Physics, Iowa State University, Ames, IA 50011 (presenting author).
[†]Fermilab, Batavia IL 60500.
[‡]Theory Group, Brookhaven National Laboratory, Upton, NY 11973.

Abstract.
Direct CP violation in decays such as $B^- \to D^0 K^-$ are sensitive to the CKM angle γ since these decays allow the interference of b-quark to c-quark and b-quark to u-quark transitions. To determine γ in this way, one must simultaneously extract the strong phases in the system. In this talk, I will discuss the measuring γ through the use of various 2 and 3 body decay modes of the D^0 as well as the possible impact of D^0 mixing. Such methods should find natural applications at B-factories and hadronic B facilities.

The advent of colliders which are capable of producing a large number of B mesons provides a opportunity to expand the repertory of CP violating signals beyond the K^0 system. It is expected that CP violation in B decays will provide an important test of the Standard Model (SM), by fixing the angles α, β and γ of the Unitarity Triangle. In this talk I will discuss strategies considered in [1,2] for determining the angle γ from interference of the quark level processes $b \to c\bar{u}s$ and $b \to \bar{c}us$ which leads to direct CP violation and is therefore applicable to a large number of facilities such as BTeV, CDF, D0 and LHCB or a high luminosity Z factory.

Since the final state quark content is different, in the absence of $D\overline{D}$ mixing interference only possible if a common final state is observed. This may be accomplished through the interference of decay channels $B^- \to K^- D^0$ with $B^- \to K^- \overline{D}^0$ where both D^0 and $\overline{D}^0$ decay to a common final state X (e.g. $X = K^+ K^-$ or $K^+ \pi^-$), together with the charge conjugate process.

Thus, to determine γ one can measure the rates $d(X) = Br(B^- \to K^-[X])$ and $\bar{d}(\overline{X}) = Br(B^+ \to K^+[\overline{X}])$ (here $[X]$ means a decay to X via D^0 or $\overline{D}^0$ channel) provided one also knows the branching ratios $a = Br(B^- \to K^- D^0)$; $b = Br(B^- \to K^- \overline{D}^0)$; $c(X) = Br(D^0 \to X)$ and $\bar{c}(X) = Br(\overline{D}^0 \to X)$. With this information, one can solve (up to an eight fold ambiguity) for the weak phase γ as well as the total strong phase difference ξ.

In practice, however, b is not easy to determined directly because it is difficult to find a prominent tag for $\overline{D}^0$. For instance, a leptonic tag has a large background from leptonic decay of B^- while if one tries to tag it through decays such as $\overline{D}^0 \to K^+\pi^-$ the signal is subject to interference effects from $D^0 \to K^+\pi^-$. In fact it is the existence of such interference effects we wish to exploit as a source of CP violation that make the direct determination of b via a hadronic tag impossible.

Theoretical models might give estimates for b but it is desirable to extract γ from experiment alone. This may be done if one uses two separate modes, X_1 and X_2. In this case, one has four experimental inputs: $d(X_{1,2})$ and $\overline{d}(X_{1,2})$ while one needs to solve for two total strong phase differences ($\xi_{1,2}$) as well as γ and b.

With exactly two mode, there is a ≤ 16 fold ambiguity in γ. However if 3 or more modes are taken together, $Q \equiv \sin^2\gamma$ may be determined uniquely giving four fold ambiguity in γ. This may be further reduced to two fold if one uses phase information about D decays obtained from other experiments [1–3].

Although one cannot precisely determine γ from a single mode, if there is large CP violation in a given mode, X, one can derive a lower bound on Q: $Q \geq Q_{min} = (1+z)(1-\sqrt{1-|A|})/2$ where $1+z = (d(X) + \overline{d}(\overline{X}))/(2ac(X))$ and $A = (d(X) - \overline{d}(\overline{X}))/(d(X) + \overline{d}(\overline{X}))$. Likewise one can obtain a bound on b: $1 - \sqrt{(1+z)(1+|A|)} \leq \sqrt{b\overline{c}(X)/ac(X)} \leq 1 + \sqrt{(1+z)(1-|A|)}$.

Two classes of D decay modes are particularly useful: Doubly Cabibbo suppressed (DCS) [1,4] and CP eigenstates (CPES) [5]. There is the potential for large CP asymmetries with DCS modes since the ratio $\overline{c}(X)/c(X) \sim 10^{-2}$ is roughly offset by $a/b \sim 10^2$. The amplitude ratio is thus $(ac(X)/b\overline{c}(X))^{1/2} \sim 1$ giving $O(1)$ interference. Some modes of this type are: $K^+\pi^-$, $K_S\pi^0$, $K^+\rho^-$, $K^+a_1^-$, $K_S\rho^0$, $K^{*+}\pi^-$. CP asymmetries for CPES modes is likely only $O(10\%)$ but the statistics can be enhanced by combining different modes which include: $3K_s$, $K_s\eta$, $K_s\rho^0$, $K_s\omega$, $K_s\eta'$, K_sf_0, $K_s\phi$, $\pi^+\pi^-$, K^+K^-. In [2] it is estimated that if the number of $B's$ times acceptance and efficiency is 10^8 then the accuracy in determining γ could be about $10°$ at 90% c.l. depending on the strong phases.

Since it is useful [2,6] to include as many modes of each type as possible in any analysis one should also consider analogous decays with K and D resonances such as $B^- \to K^{*-}D^0$ and $B^- \to K^-D^{*0}$ [7]. Another approach to gather more data is to consider three body decays [2] of the D^0 such as $D^0 \to K^+\pi^-\pi^0$. In such decays each point of the Dalitz plot may be regarded as a separate mode. For instance, the largest lower bound on Q for each point provides a global lower bound. To see how well this approach might work in practice, let us consider the phenomenological model used to fit the experimental data in [8] where the decay $D^0 \to K^-\pi^+\pi^0$ is modeled by a K^{-*} channel, ρ^+ channel, K^{0*} channel and a continuum while the (as yet unobserved) decay $D^0 \to K^+\pi^-\pi^0$ is related by $SU(3)$. In this model, for all combinations of the overall strong phase and γ, the maximum value of Q_{min} is equal to Q. In fact about 20% of the Dalitz plot has $Q_{min} > 0.9Q$. More generally, γ may be extracted from 3 body decays by fitting the Dalitz plot from B^+ and B^- decays to a resonance model which would have enough information to over determine γ, b and the strong phase. Using these method one is taking advantage of CP asymmetries in the total decay rate and its distribution.

If a DCS mode is used for the D^0 decay, $D\overline{D}$ mixing on the order of 1% could lead to contamination from the evolution of a D^0 to $\overline{D}^0$ followed by CBA decay. Given a mixing rate expected [9,10] to be $\lesssim 1\%$ the decay time dependence takes the form [2,11]: $\Gamma(\tau) \propto (U + V\tau)exp(-\tau)$ where $\tau = \Gamma_D t$, and t is the time between the B decay and the D decay. If we fit the time dependence of such decays to this form, then the parameter U represents prompt decay rate, free of mixing effects. It can be shown [2] that using this strategy produces the same statistical errors as the case were mixing disregarded given about twice as many B's.

In conclusion, the interference of $b \to c\bar{u}s$ with $b \to \bar{c}us$ via the processes $B^- \to K^- D^0$ and $B^- \to K^- \overline{D}^0$ can give CP violation provided a final state X common to D^0 and $\overline{D}^0$ decay is detected. If $D^0 \to X$ is a DCS decay, then large asymmetries are possible since the DCS offsets the color suppression of $B^- \to K^- \overline{D}^0$. Even if the rate for $B^- \to K^- \overline{D}^0$ is not known, a lower bound on $\sin^2 \gamma$ may be established with a large asymmetry in such a mode. With additional modes, one can fit for γ and b simultaneously. This will also apply to B^- decays to higher K^* and D^* resonances and one can generalize these methods to 3- (or n-) body decays of the D^0. The effects of $D\overline{D}$ mixing near the current bound may be eliminated by using information about the time between the B^- and D^0 decay.

REFERENCES

1. D. Atwood, I. Dunietz and A. Soni, Phys. Rev. Lett. **78**, 3257 (1997).
2. D. Atwood, I. Dunietz and A. Soni, hep-ph/0008xxx.
3. A. Soffer, hep-ex/9801018.
4. I. Dunietz, Phys. Lett. **B270**, 75 (1991). I. Dunietz, Z. Phys. **C56**, 129 (1992). I. Dunietz, "CP Violation with Additional B Decays", published in B Decays, S. Stone ed. (World Scientific, Singapore, 1992).
5. M. Gronau and D. Wyler, Phys. Lett. **B265** (1991); M. Gronau and D. London., Phys. Lett. **B253**, 483 (1991).
6. A. Soffer, Phys. Rev. **D60**, 054032 (1999).
7. Which are strictly analogous if the decay is controlled by one amplitude.
8. P. L. Frabetti et al., Phys. Lett. **B331**, 217 (1994).
9. E. M. Aitala et al [E791 Collab.], Phys. Rev. Lett. 83, 32 (1999); R. Godang et al. [CLEO Collab.], hep-ex/0001060; J. M. Link et al. [FOCUS Collab.], hep-ex/0004034.
10. Z. Xing, Phys. Rev. **D55**, 196 (1997); T. Liu, *Tau Charm Factory Workshop*, Argonne IL (1995); J. F. Donoghue, E. Golowich, B. R. Holstein and J. Trampetic, Phys. Rev. **D33**, 179 (1986); M. Gronau, Phys. Rev. Lett. 83, 4005 (1999); H. N. Nelson, hep-ex/9908021.
11. J. P. Silva and A. Soffer, Phys. Rev. **D61**, 112001 (2000).

Electric dipole moments of neutron and heavy atoms and constraints on CP SUSY phases

Maxim Pospelov

Theoretical Physics Institute, School of Physics and Astronomy
University of Minnesota, 116 Church St., Minneapolis, MN 55455, USA

Abstract. Electric dipole moments provide a relatively unexpensive yet powerful tool of searching for new physics beyond the Standard Model. I discuss current problems in calculations of EDMs, mainly related to the strong interactions effects at the transition from the quark-gluon to hadronic description. These problems become especially important in the case of supersymmetric models whith several CP-violating phases, so that all existing experimental limits need to be used in order to extract reliable constraints on these phases.

I EFFECTIVE CP-ODD LAGRANGIAN AT 1 GEV

The null experimental results for the electric dipole moments (EDMs) of the electron, neutron, heavy atoms and diatomic molecules can in general place very strong constraints on the CP-violating sector of a new theory and probe energy scales which are inaccessible for direct observations at colliders [1,2]. In general, the relevant contribution to the dipole moments at scales of ~1 GeV can be parameterized in terms of effective operators of different dimensions suppressed by corresponding powers of a high scale M where these operators were generated:

$$\mathcal{L}_{eff} = \sum_{n \geq 4} \frac{c_{ni}}{M^{n-4}} \mathcal{O}_i^{(n)}, \tag{1}$$

Here $\mathcal{O}_i^{(n)}$ are operators of dimension n, with its field content, Lorentz structures, etc., denoted by i. The fields relevant for the low-energy dynamics of interest are gluons, the three light quarks, the electron, and the electromagnetic field. This general form is independent of the particular construction of the new theory, and the details of a given model enter only through the values of the coefficients c_{ni}/M^{n-4}.

In the MSSM, the number of operators which can generate an EDM is considerably smaller than in the generic case. The relevant part of the effective Lagrangian at the scale of 1 GeV contains the theta term, the three-gluon Weinberg operator, the EDMs of quarks and electron and the color EDMs (CEDMs) of quarks,

$$\mathcal{L}_{eff} = \theta \frac{g_s^2}{32\pi^2} G_{\mu\nu}^a \tilde{G}_{\mu\nu}^a + w \frac{g_s^3}{6} f^{abc} G_{\mu\nu}^a \tilde{G}_{\nu\alpha}^b G_{\alpha\mu}^c \quad (2)$$

$$+ i \sum_{i=u,d,s} \frac{d_i}{2} \bar{q}_i F_{\mu\nu} \sigma_{\mu\nu} \gamma_5 q_i + i \sum_{i=u,d,s} \frac{\tilde{d}_i}{2} \bar{q}_i g_s t^a G_{\mu\nu}^a \sigma_{\mu\nu} \gamma_5 q_i + i \frac{\tilde{d}_e}{2} \bar{e} F_{\mu\nu} \sigma_{\mu\nu} \gamma_5 e.$$

The coefficients in front of the operators in Eq. (2) can be calculated for any given model of CP-violation and then evolved down to the low-energy scale, using standard renormalization group techniques. In the MSSM, in particular, one can compute effective Lagrangian (2) for any given point in the supersymmetric parameter space. Then, to get the final predictions for EDMs, one has to take various matrix elements for these operators over hadronic, nuclear and atomic states. In most cases this is a source of major uncertainty, especially when hadronic physics is involved.

The EDMs of paramagnetic atoms [3], mostly sensitive to d_e, are free from hadronic uncertainties and give the most reliable limits on CP-violating phases in the MSSM. It is clear, however, that the electron EDM limit alone cannot exclude the possibility of large CP-violating phases. This is because d_e, as any other coefficient in Eq. (2), is in general a function of *several* CP-violating phases, and mutual cancellations are possible.

Unfortunately, at the present stage of knowledge about the strong dynamics at the hadronic scale, there are substatial difficulties in calculation of the EDM of the neutron as the function of CP CUSY phases. The major obstacle is that almost all the operators from the effective Lagrangian (2) are important and the coefficients with which they contribute to d_N are known only within an order of magnitude. Since d_i, $\tilde{d}_i$ and w are in general *different* functions of CP SUSY phases, the exact dependence of d_N on SUSY phases is not known.

II EDMS OF HEAVY ATOMS

The EDM of mercury atom, strongly limited by Seattle experiment [4], provides the second best quality limits on the CP SUSY phases. This is due to the fact that the leading contribution to d_{Hg} comes from the triplet combination of the up and down color EDM operators, as it was shown in [1]. This combination induces the CP-odd pion-nucleon constant and thus generates the T-odd nucleon-nucleon interaction. This interaction induces the so called Schiff moment of Hg nucleus [2]. The Schiff moment leads to the $s-p$ mixing of atomic orbitals, and induces the EDM of a diamagnetic atom (mercury, in particular). Even though the atomic and nuclear calculations of these effects are bound to have uncertainties, which can be as large as 100%, these are *overall* uncertainties, and the dependence of d_{Hg} on CP SUSY phases is given by $\tilde{d}_d - \tilde{d}_u$. Skipping the details of the calculation [1], I give the final result for d_{Hg}:

$$d_{Hg} = -(\tilde{d}_d - \tilde{d}_u - 0.012\tilde{d}_s) \times 3.2 \cdot 10^{-2} e, \qquad (3)$$

Using the experimental limit [4], $d_{Hg} < 9 \cdot 10^{-28} e \cdot$cm, the very strong constraint on the combinations of the color EDMs of quarks can be deduced [1]:

$$|\tilde{d}_d - \tilde{d}_u - 0.012\tilde{d}_s| < 3.0 \cdot 10^{-26} \text{cm}. \qquad (4)$$

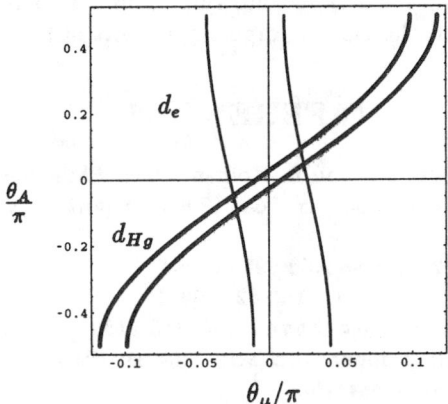

FIGURE 1. Constraints on the SUSY CP phases, θ_A and θ_μ, provided by electron and mercury EDMs. The masses of superpartners are set to 500 GeV

This constraint can be used very effectively in combination with those, provided by d_e, and reduces considerably the MSSM parameter space, where the cancellation of the large CP phases in d_e may occur [1]. Figure 1 illustrates this assertion, and gives the combined constraint in case of degenerate soft-breaking masses equal to 500 GeV and two CP-violating phases, θ_A and θ_μ. The allowed values of the SUSY phases should fit inside the rectangular intersection of the two bands, provided by d_e and d_{Hg}.

III DISCUSSION

I have shown that the EDM of the mercury atom provides indispensable constraints on the CP-violation in SUSY models, complementary to those, coming from the EDM of the electrons. Unlike the EDM of the neutron which recieves a variety of comparable contributions, the EDM of mercury is mostly sensitive to the triplet combination of the color electric dipole moments of up and down quarks. This operator induces the T-odd pion nucleon coupling constant, leading to the Schiff moment of the nucleus, and ultimately to the EDM of the whole atom.

Near future progress in attaining new accuracy levels ($\sim$ factor of 10 improvement) in the measurements of mercury EDM and new limits on the EDM of Xe atom are plausible [6]. Together with expected progress in d_e limits, it will probe the CP violation in supersymmetry for the supersymmetric masses as heavy as few TeV.

The main theoretical problem in calculations of EDMs of neutrons and diamagnetic atoms is the transition from quark-gluon to hadronic description. The quality of constraints provided by the neutron EDM limit can be improved *if* the contributions of different effective operators in Eq. (2) to d_N are calculated/estimated within the same method to at least a 100% accuracy level. The most promising analytical tool in this case is the QCD sum rule approach, which has been recently applied for the calculation the electric dipole moments of hadrons [5].

REFERENCES

1. T. Falk, K. Olive, M. Pospelov and R. Roiban, *Nucl. Phys.* B560 3 (1999).
2. I.B. Khriplovich and S.K. Lamoreaux, *"CP Violation Without Strangeness"*, Springer, 1997.
3. E.D. Commins *et al.*, *Phys. Rev.* A50 2960 (1994).
4. J.P. Jacobs *et al.*, *Phys. Rev. Lett.*71 3782 (1993).
5. M. Pospelov and A. Ritz, *Nucl. Phys.* B558 243 (1999); *Phys. Rev. Lett.* 83 2526 (1999); hep-ph/9908508, to appear in *Nucl. Phys.* B; *Phys. Lett.* B471 388 (2000).
6. Mike Romalis, *private communication.*

Physics with Polarized Cold Neutrons at the Spallation Source SINQ

K. Bodek[ae], P. Böni[b], N. Danneberg[a], W. Fetscher[a], W. Haeberli[f],
C. Hilbes[a], St. Kistryn[e], J. Lang[a], M. Lüthy[c], M. Markiewicz[a],
A. Pusenkov[d], A. Schebetov[d], A. Serebrov[d] and J. Sromicki[a]

[a] *Institute of Particle Physics, ETH, Zurich, Switzerland*
[b] *Laboratory for Neutron Scattering, ETH and PSI, Switzerland*
[c] *Paul Scherrer Institute, Villigen, Switzerland*
[d] *St. Petersburg Nuclear Physics Institute, Gatchina, Russia*
[e] *Institute of Physics, Jagellonian University, Cracow, Poland*
[f] *University of Wisconsin, Madison, USA*

Abstract. A new facility for particle physics has been installed at the spallation source SINQ. An experimental area has been constructed for a series of experiments fed by a high intensity, polarized cold neutron beam. The physics program for this new facility will focus on free neutron decay studies addressing mainly the questions of fundamental symmetries (time reversal and parity) in the weak interactions.

The neutron decay process has always posed questions at the forefront of particle physics. Nowadays it attracts great attention as a tool for investigating subtle effects in the interaction between quarks and leptons [1]. Extensions of the Standard Model postulate right handed intermediate vector bosons, leptoquarks, other spin-zero gauge bosons and supersymmetric particles. Their existence may be manifested by the scalar or tensor weak interaction (with or without time reversal violating terms). These questions are addressed in well designed and dedicated experiments in the field of neutron and nuclear beta decay. A range of masses approaching 1 TeV/c^2 and an interesting range of coupling constants for the exotic particles may be probed in these studies. In the recent decade great progress has been achieved in the determination of the neutron lifetime and decay asymmetry parameter A. However, the more complicated observables involving the spin of the neutron (B, D, R, G) are not yet known with desired precision; R and G have not yet been measured at all. These experiments bear first order sensitivity to the so called "Physics Beyond Standard Model".

Neutron decay experiments are limited by counting statistics. With present day technology, the highest spatial density of free neutrons is obtained in a beam of *cold neutrons*. Recently, a new facility for fundamental physics with polarized cold neu-

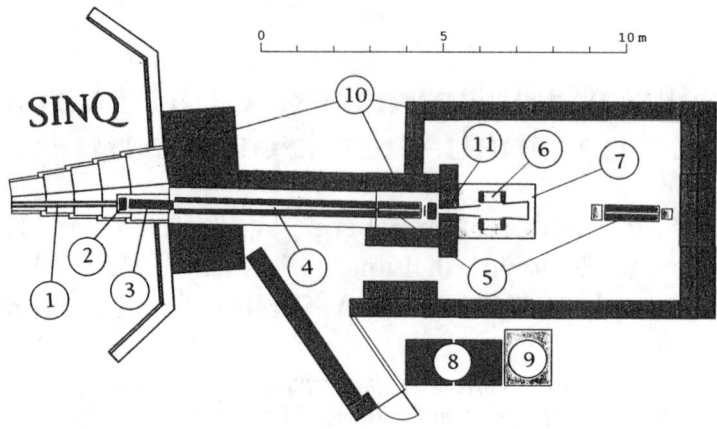

FIGURE 1. Layout of the Polarized Cold Neutron Facility at PSI. (1) internal beam guide, (2) beam shutter, (3) polarizer, (4) focusing guide, (5) spin flipper and beam polarimeter, (6) detection system, (7) helium box, (8) electronics, (9) gas system, (10) shielding, (11) collimator.

trons has been added to the spallation complex SINQ at the Paul Scherrer Institute, Villigen, Switzerland. At present SINQ operates with 1.2 mA of 0.6 GeV proton beam. We have equipped one of the neutron channels, facing the liquid deuterium cold source, with a large aperture and large momentum acceptance beam guide (Fig. 1). The guide is covered, right from the exit of the cold source tubing inside the D_2O moderator tank, by supermirrors consisting of 450 layers of Ni/Ti, with a high index of angular reflection ($m \approx 3$). The cold neutron flux density measured at the border of the SINQ shielding is around 10^9 n/(cm^2·s·mA); the total number of unpolarized neutrons with the characteristic "cold" spectrum is $\sim 10^{11}$ n/(s·mA). The main neutron beam shutter is integrated into the SINQ shielding block. The external beam tract, which guides the neutrons to the experimental station, consists of a multislit, supermirror polarizer and bender, a cold neutron beam stop, a focusing beam guide (tapered from 8×15 cm^2 down to 4×15 cm^2), radiofrequency spin flippers, a chopper device for time-of-flight measurements, and two additional cold neutron beam dumps placed downstream. The beam tract is immersed in magnetic field for guiding the neutron spin. The front of this beam channel, with polarizer and bender, is integrated into the bunker of SINQ, to absorb the radiation associated with the neutron capture reactions. A sandwich radiation shielding (brass/polyethylene/iron/lead) slows down and absorbs fast neutrons penetrating the walls of the guide as well as gamma rays. The residual radiation is contained in a massive, biological shield arranged around the neutron guide. The commissioning data measured at the place of experiments are: 2×10^8 n/(cm^2·s·mA) with a polarization in the central part of the beam cross section exceeding 97% (Fig. 2). Detailed measurements of the beam properties, i.e. divergence, flux and polarization distributions, are in progress.

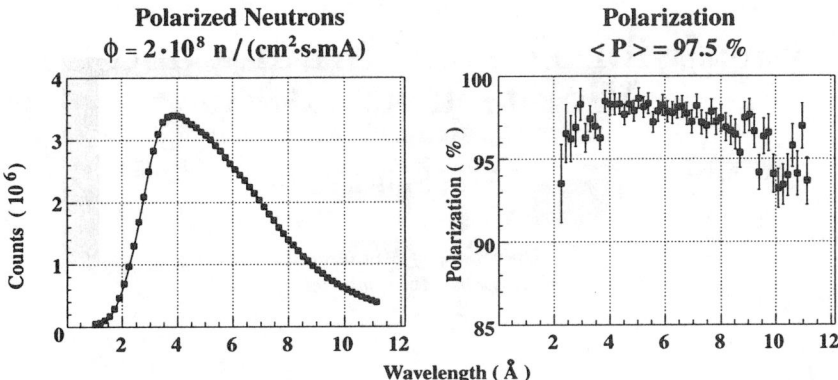

FIGURE 2. Left panel: Neutron spectrum measured using the TOF technique and a thin ^{3}He detector. Right panel: Wavelength dependence of the beam polarization.

The physics program for this new facility will focus on free neutron decay studies. The first experiment is a search for time reversal violation [2]. The transverse polarization of electrons emitted by the neutron decaying in flight will be determined with a precision of 0.005. This observable, known as the "R-correlation", is sensitive to a combination of the tensor and the scalar couplings [3]:

$$R = \frac{Im\left[(C_V^* + 2C_A^*)(C_T + C_T') + C_A^*(C_S + C_S')\right]}{|C_V|^2 + 3|C_A|^2}, \quad (1)$$

where C_V, C_A, C_S, C_T are the vector, axial vector, scalar and tensor coupling constants as defined in [4].

The challenging issue in this experiment is the polarimetry of low energy electrons (200-700 keV) in the presence of background induced by stray neutrons captured around the apparatus. Low mass, low Z detectors will be used for tracking the electrons scattered from a lead (or gold) analyzer foil in the Mott process. Helium based prototype multiwire proportional chambers passed successfully laboratory tests and are presently investigated in neutron beam conditions.

REFERENCES

1. For details refer to the Int. Workshop on Particle Physics with Slow Neutrons, ILL, Grenoble, 1998; Nucl. Instr. Meth. **A440** (2000).
2. Barnett I.C., Bodek K., Sromicki J. et al., *Search for Time Reversal Violation in the Decay of Free, Polarized Neutrons*, PSI Proposal, June 1997.
3. Sromicki J., Nucl. Instr. Meth. **A440**, 609 (2000).
4. Jackson J.D., Trieman S.B. and Wyld H.W., Phys. Rev., **106**, 517 (1957).

Search for $D^0 - \overline{D^0}$ Mixing and CP Violation at CLEO

Alex Smith*

*University of Minnesota[1]
Minneapolis Minnesota 55455

Abstract. We present recent results of searches for $D^0 - \overline{D^0}$ mixing, doubly-Cabibbo-suppressed decays, and CP violation using 9 fb^{-1} of e^+e^- collisions collected with the CLEO II.V detector on and just below the $\Upsilon(4S)$ resonance. We have observed the first signal for the "wrong-sign" decay $D^0 \to K^+\pi^-\pi^0$. Using the $D^0 \to K^+\pi^-$ final state, we limit the mixing parameters x' and y' to $(1/2)x'^2 < 0.041\%$ and $-5.8\% < y' < 1.0\%$, respectively, and the CP asymmetry in $D^0 \to K^+\pi^-$ to $-0.36 < A_D < 0.30$ at the 95% confidence level. We also measured the CP asymmetry in the decays of D^0 to the CP-even K^+K^- and $\pi^+\pi^-$ final states to be $(0.04 \pm 2.18 \text{ (stat)} \pm 0.84 \text{ (sys)})\%$ and $(1.94 \pm 3.22 \text{ (stat)} \pm 0.84 \text{ (sys)})\%$, respectively.

$D^0 - \overline{D^0}$ mixing and CP violation provides an excellent laboratory in which to search for physics beyond the Standard Model. The amplitudes for long- and short-range contributions to $D^0 - \overline{D^0}$ mixing are given by parameters $-iy$ and x, respectively, in units of $\Gamma_{D^0}/2$ [1]. Non-Standard Model particles which couple to the quark sector, could manifest themselves as an enhancement to x at a level which is within reach of present experiments. Since the Standard Model prediction for x is beyond the sensitivity of present experiments, if this type of mixing is observed, it is due to new physics. Furthermore, observation of CP violation in the D^0 system would be evidence for physics beyond the Standard Model.

Great improvements to the limits on $D^0 - \overline{D^0}$ mixing and CP violation from CLEO [2] using the decay channel $D^0 \to K^+\pi^-$ have renewed interest in these searches. D^0's from the decay $D^{*+} \to D^0\pi^+$ were used in this analysis. The flavor of the D^0 at production is tagged by the charge of the slow pion from the D^{*+} decay. Events were categorized as "wrong-sign" ("right-sign") if the slow pion and kaon had the same (opposite) charge. Right-sign (RS) events come from Cabibbo-

---

[1] We gratefully acknowledge the effort of the CESR staff in providing us with excellent luminosity and running conditions. This work was supported by the National Science Foundation, the U.S. Department of Energy, the Research Corporation, the Natural Sciences and Engineering Research Council of Canada, the A.P. Sloan Foundation, the Swiss National Science Foundation, the Texas Advanced Research Program, and the Alexander von Humboldt Stiftung.

favored decay (CFD). In decays to hadronic final states, wrong-sign (WS) events may come either from doubly-Cabibbo-suppressed decays (DCSD) or from the D^0 evolving to $\overline{D^0}$ through x or y mixing.

The time-dependent WS rate is given by

$$r(t) \equiv [R_D + \sqrt{R_D}y't + \frac{1}{4}(x'^2 + y'^2)t^2]e^{-t},$$

where R_D ($R_M \equiv \frac{1}{2}(x'^2 + y'^2) = \frac{1}{2}(x^2 + y^2)$) is the DCSD (mixing) rate normalized to the CFD rate [3] [4]. The mixing amplitudes x' and y' are related to x and y by $x' \equiv x\cos\delta_s + y\sin\delta_s$ and $y' \equiv y\cos\delta_s - x\sin\delta_s$, where δ_s allows for a difference in strong phase between the two processes. By studying the time evolution of the WS rate, it is possible to disentangle the three terms R_D, $\sqrt{R_D}y'$, and $R_M/2$.

Figure 1a shows the two-dimensional distribution in $m(K\pi)$ and Q for the WS $D^0 \to K^+\pi^-$ data, where the variable Q is the energy released in the $D^{*+} \to D^0\pi^+$ decay. Large Monte Carlo samples were used to model the shapes of the backgrounds, which come from $\overline{D^0} \to K^+\pi^-$ decays combined with an uncorrelated slow pion, non-signal $e^+e^- \to c\bar{c}$, and $e^+e^- \to u\bar{u}$, $d\bar{d}$, or $s\bar{s}$. A maximum likelihood fit to the $m(K\pi)$-Q distribution was used to extract the signal and background contributions shown in Fig. 1a. From this fit we measured the time-integrated rate of the WS decay $D^0 \to K^+\pi^-$ normalized to the RS decay $\overline{D^0} \to K^+\pi^-$, R_{WS}, to be $(0.332^{+0.063}_{-0.065}$ (stat) ± 0.040 (sys))%.

Using the results of the $m(K\pi) - Q$ fit to constrain the background and total WS signal normalizations, the composition of the WS signal was resolved by fitting the D^0 proper time distribution, as shown in Fig. 1b. The CP asymmetries A_M, A_D, and ϕ, which parameterize CP violation in mixing, direct decay, and the interference between these two processes, respectively, were free parameters in the fit. The best fit values were found to be $R_D = (0.48 \pm 0.12$ (stat) ± 0.04 (sys))%, $y' = (-2.5^{+1.4}_{-1.6}$ (stat) ± 0.3 (sys))%, $x' = (0.0 \pm 1.5$ (stat) ± 0.2 (sys))% leading to 95% confidence level limits of $(1/2)x'^2 < 0.041\%$ and $-5.8\% < y' < 1.0\%$. We measured the CP asymmetry to be consistent with zero, and set the first constraint on A_D of $-0.36 < A_D < 0.30$ at the 95% confidence level.

A very similar method was used to search for WS events from the decay channel $D^0 \to K^+\pi^-\pi^0$. We have made the first observation of a signal of 35 ± 9 events in the WS data, shown in Fig. 2, compared with a RS yield of ~ 8750 events. The statistical significance of this WS signal was found to be 4.6σ.

In the two-body decay $D^0 \to K^+\pi^-$, R_{WS} is simply the ratio N_{WS}/N_{RS}, since the RS and WS samples have identical efficiencies. In the case of the three-body $D^0 \to K^+\pi^-\pi^0$ decay, possible differences in resonant substructure of the RS and WS decays may lead to kinematic differences in the final state particles. In calculating R_{WS}, the total efficiencies of these two samples do not necessarily cancel:

$$R_{WS} = \frac{\varepsilon_{RS}}{\varepsilon_{WS}} \cdot \frac{N_{WS}}{N_{RS}}.$$

Thus, an analysis of the WS Dalitz plot must be performed in order to measure this ratio of efficiencies. In a recent CLEO analysis, the RS decay $D^0 \to K^-\pi^+\pi^0$ was found to have a rich resonant substructure, with contributions from ρ^+, K^{*-}, K^{*0}, $\rho^+(1700)$, $K^{*-}(1430)$, $K^{*0}(1430)$, and $K^{*0}(1680)$, in addition to the non-resonant component [5]. The presence of many interfering contributions greatly complicates analysis of this final state.

Work is nearing completion on a fit to the WS Dalitz plot, which will be used to determine the ratio $\varepsilon_{RS}/\varepsilon_{WS}$, leading to a measurement of R_{WS}. In addition, first measurements of the dominant resonant substructure will be made, albeit with large statistical errors. As in the case of $D^0 \to K^+\pi^-$, a fit to the D^0 proper time distribution will be used to resolve the contributions to the WS signal.

Decays of the D^0 meson to CP-even (K^+K^-, $\pi^+\pi^-$) and CP-odd ($K_s^0\phi$, $K_s^0\rho^0$, $K_s^0\omega$) final states can be used to directly measure y, which causes differences in the lifetime between the heavy and light eigenstates. In practice, one measures the lifetime of the CP-even (CP-odd) eigenstates and compares them to the well-measured $D^0 \to K^-\pi^+$ ($D^0 \to K^{*-}\pi^+$) lifetimes.

Measurements of y using these CP eigenstates are in progress. In the CP-even eigenstates K^+K^- and $\pi^+\pi^-$, we observe signals of ~ 2200 and ~ 900 events on small backgrounds. In the CP-odd final states, we observe ~ 1850, ~ 1900, and ~ 4300 events in the $K_s^0\phi$, $K_s^0\rho^0$, and $K_s^0\omega$ final states, respectively. We expect a combined sensitivity of 0.02 or better in y using these CP eigenstates.

CP violation would result in differences between the rates of $D^0 \to f$ and $\overline{D^0} \to f$, where f is an eigenstate of CP. Experimentally, we measure the asymmetry, A_{CP}, in the decays of D^0 to K^+K^- and $\pi^+\pi^-$ to be $(0.04\pm2.18\,(\text{stat})\pm0.84\,(\text{sys}))\%$ and $(1.94\pm3.22\,(\text{stat})\pm0.84\,(\text{sys}))\%$, respectively.

We have measured some of the best limits on $D^0 - \overline{D^0}$ mixing and CP violation using $D^0 \to K^+\pi^-$ and CP-even final states in an environment that is quite complementary to fixed target photo- and hadro-production experiments. Due to our low-background e^+e^- environment and superb resolution for π^0's, we are able to observe and study the WS decay $D^0 \to K^+\pi^-\pi^0$ for the first time. We have access to many decay channels in which to search for $D^0 - \overline{D^0}$ mixing at CLEO, including two-, three-, and four-body hadronic, CP even, CP odd, and semileptonic final states. We expect many of these analyses to bear results in the near future.

REFERENCES

1. T. Lee, R. Oehme, and C. Yang *Phys. Rev.* **106**, 340 (1957); A. Pais and S. Treiman *Phys. Rev.* D **12**, 2744 (1975).
2. CLEO Collaboration, *Phys. Rev. Lett.* **84**, 22 (2000).
3. S. Treiman and R. Sachs *Phys. Rev.* **103**, 1545 (1956).
4. G. Blaylock, A. Seiden, and Y. Nir *Phys. Lett.* B **355**, 555 (1995).
5. T. Bergfeld, "D^0 Decays From e^+e^- Collisions: Precision Rare and Substructure

Results", University of Illinois at Urbana-Champaign, Ph.D thesis (2000), to be submitted to *Phys. Rev. D*.

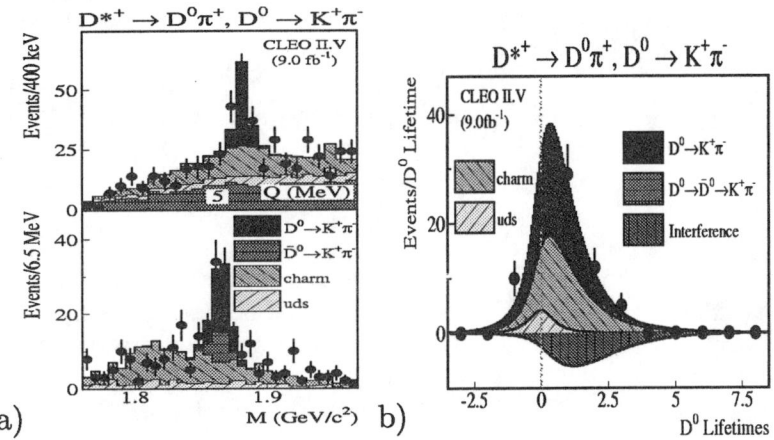

FIGURE 1. a) Distribution of the WS $D^0 \to K^+\pi^-$ data in Q and D^0 candidate mass. b) Results of a fit to the $D^0 \to K^+\pi^-$ proper time distribution.

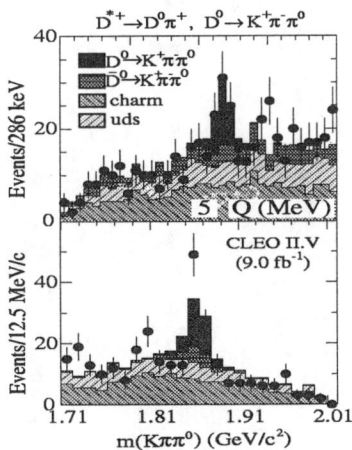

FIGURE 2. Distribution of the WS $D^0 \to K^+\pi^-\pi^0$ data in Q and D^0 candidate mass.

CIPANP2000 Conference Schedule

Monday, May 22, 2000

5:00-9:00pm   Registration (Panorama Room)
7:00-9:00pm   Welcoming Reception (Panorama Room)

Tuesday, May 23, 2000   Porte du Palais   Chairman: E. Berger, ANL

7:30am -	Registration (Lauzon Room)	
8:30	Opening Remarks	S. Kowalski (MIT)
8:40	Precision Electroweak Physics	H. Przysiezniak (CERN)
9:25	CP Violation: Beyond the Standard Model	R. Mohapatra (UMd)
10:10	Coffee Break	
10:30	Heavy Quarks: Strong Sector	A. B. Wicklund (ANL)
11:15	Heavy Quarks: Weak Sector	J. Richman (UC SB)
12:00pm	Lunch (Panorama Room)	
1:30	Parallel Sessions	
3:10	Coffee Break	
3:40	Parallel Sessions	

Wednesday, May 24, 2000   Porte du Palais   Chairman: T. W. Donnelly, MIT

8:30am	Structure Functions High Q^2- Deep Inelastic Scattering at HERA and the Tevatron	A. Doyle (Glascow)
9:15	Polarized Structure Functions	M. Vincter (UAlberta)
10:00	Coffee Break	
10:30	Nucleon and Deuteron Electromagnetic Form Factors	G. Petratos (KSU)
11:15	Exotic Hadrons	J. Napolitano (RPI)
12:00pm	Lunch (Panorama Room)	
1:30	Parallel Sessions	
3:10	Coffee Break	
3:40	Parallel Sessions	
6:00-7:30	Reception (Panorama Room)	

Thursday, May 25, 2000   Porte du Palais   Chairman: D. Skopik (JLab)

8:30am	Exploring the QCD Phase Diagram	K. Rajagopal (MIT)
9:15	Relativistic Heavy Ion Collisions: The Past Through the Future	W. Zjac (Columbia)
10:00	Coffee Break	
10:30	A Consumer's Guide to Lattice QCD Results	A. el-Khadra (UIll)
11:15	Effective Field Theory	M. Savage (UWash)
12:00pm	Lunch (not provided)	
1:30	Parallel Sessions	
3:10	Coffee Break	
3:40	Parallel Sessions	

Friday, May 26, 2000   Porte du Palais   Chairman: R. Alarcon, (ASU)

8:30am	Parity Violation and Nucleon Strangeness	P. Souder (Syracuse)
9:15	Tests of Discrete Symmetries	D. Bryman (TRIUMF)
10:00	Session ends	

Saturday, May 27, 2000   Porte du Palais   Chairman: H. C. Walter (PSI)

8:30am	Neutrino Oscillation Experiments	K. Nakamura (KEK)
9:15	Rare Isotope Physics	W. Henning (GSI)
10:00	Coffee Break	
10:30	Detection of Dark Matter	M. Isaac (UCBerkeley)
11:15	Accelerating Universe/Cosmological Constant	P. Garnavich (Notre Dame)
12:00pm	Lunch (not provided)	
1:30	Parallel Sessions	
3:10	Coffee Break	
3:40	Parallel Sessions	
6:30	Conference Banquet (Ballroom)	

Sunday, May 28, 2000   Porte du Palais   Chairman: W. Marciano, BNL

8:30am	New Particle Searches at LEP 200++	R. McPherson (CERN)
9:00	First Results from the Alpha Magnetic Spectrometer	P. Fisher (MIT)
9:30	Coffee Break	
10:00	What's the Matter?: Physics Opportunities and Future Facilities	M. Shaevitz (FNAL)
10:45	Intersections 2000: What's Happening in Hadron Physics	J. Bjorken (SLAC)
11:30	Conference ends	

List of Participants

Last Name	First Name	Initial	Affiliation
Alarcon	Ricardo	O.	Arizona State University
Alekseev	Igor	G.	Institute for Theoretical and Experimental Physics
Appel	Robin		Hofstra University
Arakelyan	Gevorg	H.	Yerevan Physics Institute
Arkhipov	Andrei		Institute for High Energy Physics
Arnold	Roger		Institut de Recherches Subatomiques
Ashery	Daniel		Tel Aviv University
Atwood	David	M.	Iowa State University
Augier	Corinne		Laboratoire de l'Accélérateur Linéaire
Balantekin	Akif Baha		University of Wisconsin - Madison
Balashov	Vsevolod	V.	Institute of Nuclear Physics, Moscow State University
Bar-Touv	Jacob		Ben-Gurion University
Barker	Paul	H.	University of Auckland
Bauer	Christian	W.	University of Toronto
Beck	Marcus		K.U. Leuven
Beckers	Jules		University of Liege
Beckmann	Marc	O.	Universitat Freiburg
Bellwied	Rene	K.	Wayne State University
Benslama	Kamal		University of Regina
Berger	Edmond	L.	Argonne National Laboratory
Bernstein	Aron	M.	M.I.T.
Besson	Dave	Z.	University of Kansas
Betts	R. Russell		University of Illinois at Chicago
Bjorken	James	D.	Stanford Linear Accelerator Center
Black	Deirdre	M.	Syracuse University
Bland	Leslie	C.	Indiana University Cyclotron Facility
Blinov	Boris	B.	University of Michigan
Blokland	Ian	R.	University of Alberta
Bock	Greg		Fermilab
Bodek	Kazimierz		Swiss Federal Institute of Technology Zurich
Bodek	Arie		University of Rochester
Boer	Daniel		Riken - BNL Research Center
Bordalo	Paula		LIP/CERN
Boridy	Elie		Université du Québec a Montréal
Bowles	Jeanne	M.	Los Alamos National Laboratory
Bravar	Alessandro		University of Mainz
Bruell	Antje		MIT
Bryman	Douglas		University of British Columbia
Bullard	Ginny		MIT - Bates Lab
Caliandro	Rocco		Istituto Nazionale Di Fisica Nucleare
Carroll	Alan	S.	Brookhaven National Laboratory
Cason	Neal	M.	University of Notre Dame

Chen	Zhang		Manhattanville College
Chiu	Mickey	G.	Columbia University
Choi	Seonho		Temple University
Clark	Jessica	HD	College of William and Mary
Cole	Brian	A.	Columbia University
Conrad	Janet	M.	Columbia University
Cooper	Martin	D.	Los Alamos National Laboratory
Cormack	Christopher	M.	Rutherford Appleton Laboratory
Cotanch	Stephen	R.	North Carolina State University
Crannell	Hall		The Catholic University of America
Czarnecki	Andrzej		Brookhaven National Lab
Danneberg	Norbert	C.G.	ETH Zurich Institute of Particle Physics
Dasgupta	Sudebsankar		Burdwan University
Daum	Manfred		PSI, Paul-Scherrer-Institut
Davies	David	A.	State University of New York - OCC
de Florian	Daniel	E.	Theoretische Physik - ETH Zurich
De Schepper	Dirk		Argonne National Lab
de Simone	Patrizia		Laboratori Nazionali di Frascati deff' INFN
Deustua	Susana		Lawrence Berkeley National Laboratory
Dodson	George	W.	Spallation Neutron Source / ORNL
Donnelly	T.	W.	M.I.T.
Doyle	Anthony	T.	University of Glasgow
Dumitru	Adrian		Columbia University
Duncan	Fraser	A.	Queen's University
Dutta	Dipangkar		Massachusetts Institute of Technology
Dytman	Steven	A.	University of Pittsburgh
Earl	Matthew	A.	Boston University
Efremenko	Yuri	V.	University of Tennessee
El-Khadra	Aida	X.	University of Illinois
Ershov	Alexey	V.	Harvard University
Eugenio	Paul	M.	Carnegie Mellon University
Fisher	Peter		M.I.T.
Fox	Brendan	D.	University of Colorado
Franklin	Gregg	B.	Carnegie Mellon University
Frolov	Valery	V.	University of Minnesota
Funsten	Herbert	O.	College of William and Mary
Gadiyak	Valeriya	G.	University of Maryland
Gale	Charles		McGill University
Garcia	Alejandro		University of Notre Dame
Garnavich	Peter	M.	University of Notre Dame
Garrido	Lluis		Universitat de Barcelona
Glazov	Alexandre		
Golwala	Sunil	R.	Center for Particle Astrophysics

Goto	Yuji		RIKEN BNL Research Center
Griest	Kim		University of California, San Diego
Gulkarov	Ilia	S.	Paradise Valley Community College
Halkiadakis	Eva		Rutgers University
Hasell	Douglas	K.	MIT
Hasinoff	Michael	D.	University of British Columbia
Hatakeyama	Kenichi		The Rockefeller University
Heintz	Ulrich		Boston University
Heinz	Tammy	A.	Riken BNL Research Center
Henning	Walter	F.	Wissenschaftlicher Geschaftsfuhrer
Henoch	Mark		University of Erlangen-Nurnberg, Germany
Heppelmann	Steven	F.	Penn State University
Hiller	Gudrun		SLAC
Hiller	John	R.	University of Minnesota - Duluth
Hirosky	Robert	J.	University of Illinois, Chicago
Holmes	Richard	S.	Syracuse
Horowitz	Charles	J.	Indiana University - Nuclear Theory Center
Isaac	Maria	C.P.	Center for Particle Astro Physics
Ishida	Taku		Institute for Particle and Nuclear Studies (IPNS)
Ito	Takeyasu		California Institute of Technology
Jacobs	Ken		MIT-Bates Linear Accelerator Center
Jacobs	William	W.	Indiana University Cyclotron Facility
James	Catherine	C.	Fermi National Accelerator Lab
Jeon	Sangyong		Lawrence Berkeley National Lab
Jones	Christopher	R.	University of Cambridge
Jones	Richard	T.	University of Connecticut
Joo	Kyungseon		University of Virginia
Junghans	Harald		Forschungszentrum Julich
Kaiser	Ralf		DESY Zeuthen
Karle	Albrecht		University of Wisconsin-Madison
Karn	Santosh	K.	University of Delhi, Delhi 110007
Katsouleas	Tom		University of Southern California
Keppel	Cynthia	E.	Hampton University/Jefferson Lab
Kettell	Steve		Brookhaven National Laboratory
Kharzeev	Dmitri	E.	Brookhaven National Laboratory
Kloet	Willem		Rutgers University
Koetke	Donald	D.	Valparaiso University
Konobeevski	Eugene	S.	Inst. for Nuclear Res., Russian Academy of Sciences
Kornicer	Mihajlo		University of Connecticut
Korobkina	Ekaterina		Johannes Gutenberg Universitat-Mainz
Kowalski	Stanley		MIT
Krishnaswami	Govind	S.	University of Rochester
Kyle	Gary	S.	New Mexico State University

Labelle	Patrick		McGill University & Champlain College
Laget	Jean-Marc		CEA / Saclay
Lajoie	John	G.	Iowa State University
Landsman	Hagar	Yael	Technion
Langenegger	Urs		SLAC
Lawrence	David	W.	University of Massachusetts
Ledovskoy	Alexander	A.	University of Virginia
Leksanov	Aleksey	V.	The Pennsylvania State University
Levonian	Serguei		DESY
Lisa	Michael	A.	The Ohio State University
Londergan	Timothy		Indiana University
Lopatin	Igor	V.	Petersburg Nuclear Physics Institute
Lorenzon	Wolfgang	B.	University of Michigan
MacInnis	Anne	B.	MIT - Bates
Maggiora	Angelo		INFN-Torino
Mahlon	Gregory	D.	McGill University
Maksimovic	Petar		Harvard
Mal'tsev	Anatoliy		Joint Institute for Nuclear Research
Mal'tsev	Mikhail	A.	Joint Institute for Nuclear Research
Maltman	Kim	R.	York University
Malvezzi	Sandra		I.N.F.N. Milano
Maneira	Jose	C.	INFN / FCUL/LIP
Marciano	William	J.	Brookhaven National Laboratory
Mayer	Benjamin		CEA - Saclay
McLaughlin	Gail	C.	TRIUMF
McLean	Kenneth	W.	CLEO Collaboration, Vanderbilt Physics
McNulty	Dustin	E.	University of Virginia
McPherson	Robert	A.	University of Victoria & IPP
Menary	Scott	R.	York University
Meyer	Werner	P.	Ruhr-Univ. Bochum
Meyer	Curtis	A.	Carnegie Mellon University
Mikuz	Marko		University of Ljubljana
Miller	C. Andrew		TRIUMF
Milner	Richard	G.	MIT
Mincer	Allen	I.	New York University
Mintz	Stephan	L.	Florida International University
Mischke	Richard	E.	Los Alamos National Laboratory
Mocioiu	Irina		C.N. Yang Institute for Theoretical Physics
Mohapatra	Rabindra	N.	University of Maryland
Molzon	William	R.	University of California, Irvine
Mori	Toshinori		University of Tokyo
Mueller	James	A.	University of Pittsburgh
Musser	James	R.	Texas A&M University

Nagae	Tomofumi		KEK
Nagle	James	L.	Columbia University
Nakamura	Kenzo		KEK
Napier	Austin		Tufts University
Napolitano	James	J.	Rensselaer Polytechnic Institute
Narsky	Ilya	V.	Southern Methodist Univ., Dallas, TX, USA
Nazaruk	Valery	I.	Institute for Nuclear Research of RAS
Nie	Shuquan		College of William and Mary
Obayashi	Yoshihisa		Kamioka Observatory, ICRR, University of Tokyo
Oehler	Christian		
Olsson	Martin	G.	University of Wisconsin, Madison
Onderwater	Cornelis	J.G.	University of Illinois at Urbana-Champaign
Ortiz	Chris	E.	University of Notre Dame
Page	Philip	R.	Los Alamos National Laboratory
Parsa	Zohreh		Brookhaven National Laboratory
Paryev	Eduard	Y	Inst. for Nuclear Res., Russian Academy of Sciences
Paschke	Kent	D.	Carnegie Mellon University
Paul	Thomas	C.	Northeastern University
Peng	Jen-Chieh		Los Alamos National Laboratory
Petratos	Gerassimos	G.	Kent State University
Phinney	Nan		SLAC
Piasetzky	Eliazer		Tel Aviv University
Pospelov	Maxim		Theoretical Physics Institute, UMN
Pralavorio	Pascal		CPPM
Prebys	Eric	J.	Princeton University
Preedom	Barry	M.	University of South Carolina
Prieels	René		Université Catholique de Louvain
Przysiezniak	Helenka		CERN
Qian	Yongzhong		University of Minnesota
Quinn	Brian	P.	Carnegie Mellon University
Rajagopal	Krishna		MIT
Ralston	John	P.	University of Kansas
Ramos	Sergio		LIP/CERN
Ramsey-Musolf	Michael	J.	University of Connecticut
Raymond	Richard	S.	University of Michigan
Ricciardi	Stefania		RHBNC, University of London
Richman	Jeffrey	D.	University of California, Santa Barbara
Riedler	Petra		University of Zurich
Roberts	B. Lee		Boston University
Roethel	Wilhelm	J.	Northwestern University
Rosati	Marzia		Iowa State University
Roser	Thomas		Brookhaven National Lab
Rothstein	Ira	Z.	Carnegie-Mellon University

Sammarruca	Francesca		University of Idaho
Sarwar	Shahzad		Instituto Nazionale Di Fisica Nucleare L.N.F.
Savage	Martin	J.	University of Washington
Schumacher	Reinhard	A.	Carnegie Mellon University
Schwarthoff	Hubert		Ohio State University
Schweitzer	Peter		Petersburg NPI and Bochum University
Schwindt	Oliver	F.	UMIST
Seitz	Bjorn		University of Alberta
Shaevitz	Michael	H.	Fermilab
Sharma	Lalit Kumar		University of Botswana
Sheldon	Paul	D.	Vanderbilt University
Shigaki	Kenta		KEK
Simes	Fern	M.	Brookhaven National Laboratory
Skabelin	Alexander	V.	Massachusetts Institute of Technology
Skopik	Dennis	M.	Thomas Jefferson National Accelerator Facility
Smith	Alex	B.	University of Minnesota
Sobol	Andrei		LAPP
Sokol	Garry	A.	Lebedev Physical Inst., Russian Academy of Sciences
Solomey	Nickolas		University of Chicago, EFI
Souder	Paul	A.	Syracuse University
Stanislaus	Shirvel		Valparaiso University
Stephenson	Edward	J.	Indiana University Cyclotron Facility
Stewart	Iain	W.	University of California, San Diego
Stibunov	Victor	N.	Tomsk Politechnical University
Strauch	Steffen		Rutgers, The State University of New Jersey
Swanson	Eric	S.	University of Pittsburgh
Szczepaniak	Adam	P.	Indiana University
Tadsen	Almut		INFN Padova
Takizawa	Makoto		Showa Pharmaceutical University
Thomas	Anthony	W.	CSSM, University of Adelaide
Thomson	Gordon		Rutgers University
Trueman	T. Laurence		Brookhaven National Laboratory
Truran	James	W.	University of Chicago
Tung	Wu-Ki		Michigan State University
Tuominen	Kimmo	I.	University of Jyvaskyla
Uretsky	Jack	L.	Argonne
Urheim	Jon	E.	University of Minnesota
Vaandering	Eric	W.	University of Colorado, Boulder
Valenti	Giovanni		CERN
Vance	Stephen	E.	Brookhaven National Laboratory
Vasiliev	Maxim	A.	Texas A & M University
Vereecke	Bart	W.	IKS - Ku Leuven
Vincter	Manuella	G.	University of Alberta

Vogelsang	Werner		Riken - BNL Research Center
Voloshin	Mikhail	B.	University of Minnesota
Walter	Hans-Christian		TRIUMF
Watson	John		Kent State University
Webb	Jason	C.	New Mexico State University
Weigel	Herbert		Massachusetts Institute of Technology
Weygand	Dennis		Jefferson Lab
Wicklund	Arthur	B.	Argonne National Laboratory
Wiescher	Michael		University of Notre Dame
Wingerter-Seez	Isabelle		Laboratoire de l'Accelerateur Linéare - ORSAY
Wolf	Joachim	W.	University of Alabama, Tuscaloosa
Wolfe	Carl	E.	Indiana University Nuclear Theory Center
Woods	Michael	B.	Stanford Linear Accelerator Center
Wuerthwein	Frank		MIT
Yang	Un-ki		University of Cnicago
Yelton	John	M.	University of Florida
Yudichev	Valeri		Joint Institute for Nuclear Research
Zajc	William	A.	Columbia University
Zeier	Markus	F.	University of Virginia
Zeller	Geralyn	P.	Northwestern University
Zielinski	Marek		University of Rochester
Zimmerman	Eric	D.	Columbia University Nevis Labs

Author Index

A

Abreu, M. C., 341
Adams, T., 509
Adinolfi, M., 852
Aibergenov, T. A., 411
Alarcon, R., 893
Alekseev, I. G., 670
Alessandro, B., 341
Alexa, C., 341
Allen, T. J., 283
Aloisio, A., 852
Alster, J., 306, 310, 447, 451
Alton, A., 509
Ambrosino, F., 852
Ames, F., 934
Andres, E., 823
Andryakov, A., 852
Anferov, V. A., 662
Angelopoulos, A., 957
Antinori, F., 363
Antonelli, A., 852
Antonelli, M., 852
Anulli, F., 852
Apostolakis, A., 957
Armstrong, D. S., 403, 407
Arnaldi, R., 341
Arroyo, C. G., 509
Arsyan, G., 451
Ashery, D., 295
Askebjer, P., 823
Aslanides, E., 957
Asryan, G., 306, 310, 447
Atayan, M., 341
Atwood, D., 976
Augier, C., 819
Averichev, Y., 306, 310, 447, 451
Avvakumov, S., 509
Awes, T. C., 532

B

Bacci, C., 852
Backenstoss, G., 957
Baglin, C., 341
Bai, M., 670
Bai, X., 823
Bakke, H., 363
Balashov, V. V., 475
Baldit, A., 341
Bankamp, A., 852
Barbiellini, G., 852
Bargassa, P., 957
Barouch, G., 823
Barsov, S., 421
Barton, D., 306, 310, 447, 451
Barwick, S. W., 823
Bass, S. A., 359
Bassalleck, B., 670
Baturin, V., 306, 310, 447, 451
Bay, R. C., 823
Bazarko, A. O., 509
Bechstedt, U., 421
Beck, D., 931, 934
Beck, M., 931, 934
Becker, K.-H., 823
Beckmann, M., 639
Beddo, M. E., 346, 532
Bedjidian, M., 341
Behnke, O., 957
Bellwied, R., 374
Bencivenni, G., 852
Benelli, A., 957
Beolè, S., 341
Bergström, L., 823
Bernstein, A. M., 442
Bernstein, R. H., 509
Bertin, V., 957
Bertolucci, S., 852
Bertrand, D., 823
Beusch, W., 363
Bibikov, A. V., 475
Bierenbaum, D., 823
Bini, C., 852
Biron, A., 823
Bjorken, J. D., 211
Black, D., 241
Blanc, F., 957
Blinov, B. B., 662, 674
Bloch, P., 957
Bloise, C., 852
Blokland, I. R., 483
Bloodworth, I. J., 363

Bocci, V., 852
Bodek, A., 509
Bodek, K., 983
Boer, D., 701
Boldea, V., 341
Bollen, G., 934
Bolton, T., 509
Böni, P., 983
Booth, J., 823
Borchert, G., 421
Bordalo, P., 341
Borgs, W., 421
Borisov, N. S., 674
Boros, C., 479
Bossi, F., 852
Botner, O., 823
Bouchta, A., 823
Boyce, M. M., 823
Branchini, P., 852
Brau, J., 509
Bravar, A., 681
Brooks, M. L., 532
Brown, C. N., 346, 532
Brown, H. N., 917
Bryman, D., 150
Buchholz, D., 509
Budd, H., 509
Bugel, L., 509
Bukhtoyarova, N., 306, 310, 447, 451
Bulychjov, S. A., 852
Bunce, G., 670, 917
Buniy, R. V., 302, 705
Burakovsky, L., 249
Burkert, V., 259
Büscher, M., 421
Bush, J. D., 532
Bussière, A., 341

C

Cabibbo, G., 852
Calcaterra, A., 852
Caliandro, R., 363
Caloi, R., 852
Campana, P., 852
Capelli, L., 341
Capon, G., 852
Carboni, G., 852
Cardini, A., 852

Carey, R. M., 917
Carey, T. A., 346, 532
Carius, S., 823
Carlson, P., 957
Carrer, N., 363
Carroll, A., 306, 310, 447, 451
Carroll, M., 957
Casagrande, L., 341
Casarsa, M., 852
Cason, N. M., 318
Castor, J., 341
Cataldi, G., 852
Cawley, E., 957
Ceradini, F., 852
Cervelli, F., 852
Cevenini, F., 852
Chambon, T., 341
Chang, T. H., 346, 532
Chaurand, B., 341
Chen, A., 823
Chen, Z., 514
Cheng, J., 674
Chertok, M. B., 957
Chevrot, I., 341
Cheynis, B., 341
Chiavassa, E., 341
Chiefari, G., 852
Chirkin, D., 823
Choi, S., 520
Ciambrone, P., 852
Cicalò, C., 341
Clark, J. H. D., 403, 407
Claudino, T., 341
Coleman, A., 259
Comets, M. P., 341
Conetti, S., 852
Conrad, J., 509, 823
Constans, N., 341
Constantinescu, S., 341
Conticelli, S., 852
Cooley, J., 823
Cooper, W. E., 346, 532
Cormack, C. M., 501
Costa, C. G. S., 823
Cotanch, S. R., 275
Cowen, D. F., 823
Cushman, P., 917
Czarnecki, A., 938

D

Dailing, J., 823
Dalberg, E., 823
Danby, G. T., 917
Danielsson, M., 957
Danilyan, G., 972
Danneberg, N., 920, 966, 983
Daum, M., 888
Davidenko, A. M., 674
de Barbaro, L., 509
de Barbaro, P., 509
Debevec, P. T., 917
Debowski, M., 421
De Falco, A., 341
de Florian, D., 693
Dejardin, M., 957
Delauré, B., 931, 934
Dellacasa, G., 341
Dell'Agnello, S., 852
De Lucia, E., 852
De Marco, N., 341
Deng, H., 917
Denig, A., 852
Deninger, W., 917
Derbenev, Y. S., 662
De Robertis, G., 852
Derre, J., 957
De Sangro, R., 852
De Schepper, D., 547
Deshpande, A., 670
Desiati, P., 823
De Simone, P., 852
Deutsch, J., 920, 931, 934
Devaux, A., 341
Dewulf, J.-P., 823
DeYoung, T., 823
De Zorzi, G., 852
Dhawan, S. K., 917
Di Bari, D., 363
Di Domenico, A., 852
Di Falco, S., 852
Di Liberto, S., 363
Dita, S., 341
Doksus, P., 823
Dolinov, V. K., 475
Doria, A., 852
Doskow, J., 670
Doyle, A. T., 49
Drago, E., 852

Drapier, O., 341
Dressler, B., 643
Drucker, R. B., 509
Druzhinin, V. P., 917
Ducroux, L., 341
Dumitru, A., 359
Duncan, F., 805
Dunietz, I., 976
Duong, L., 917
Dutta, D., 562
Dutt-Mazumder, A. K., 369

E

Ealet, A., 957
Earle, W., 917
Earl, M., 897
Edsjö, J., 823
Efstathiadis, E., 917
Egger, J., 920
Eilerts, S., 670
Ekström, P., 823
Eleftheriadis, C., 957
Elia, D., 363
Elia, V., 852
Erlandsson, B., 823
Erriquez, O., 852
Ershov, A., 628
Erven, W., 421
Eßer, R., 421
Espagnon, B., 341
Eugenio, P., 233
Evans, D., 363

F

Fanebust, K., 363
Fargeix, J., 341
Fariborz, A. H., 241
Farilla, A., 852
Farley, F. J. M., 917
Fedorets, P., 421
Fedotovich, G. V., 917
Felici, G., 852
Ferrari, A., 852
Ferrer, M. L., 852
Feser, T., 823
Fetscher, W., 920, 957, 983

Fiase, J. O., 291
Fidecaro, M., 957
Fields, D. E., 670
Filipčič, A., 957
Fil'kov, L. V., 267
Fimushkin, V. V., 674
Fini, R. A., 363
Finocchiaro, G., 852
Fleming, B. T., 509
Force, P., 341
Formaggio, J. A., 509
Foroughi, F., 920
Forstner, O., 934
Forti, C., 852
Franceschi, A., 852
Francis, D., 957
Franzini, P., 852
Frey, R., 509
Fry, J., 957
Ftáčnik, J., 363
Funsten, H., 259

G

Gabathuler, E., 957
Gadiyak, V., 554
Gagliardi, C. A., 346, 532
Gale, C., 369
Gallio, M., 341
Gamberg, L., 467
Gamet, R., 957
Gao, M. L., 852
Garrido, L., 841
Garvey, G. T., 346, 532
Gatti, C., 852
Gaug, M., 823
Gauzzi, P., 852
Gavrilov, Y. K., 341
Geesaman, D. F., 346, 532
Gerber, H.-J., 957
Gerschel, C., 341
Ghidini, B., 363
Ginocchio, J. N., 471
Giovannella, S., 852
Giron, S., 917
Giubellino, P., 341
Gladycheva, S. E., 674
Glazov, A., 949
Go, A., 957

Goeke, K., 643, 659
Goldman, J., 509
Goldman, T., 471
Goldschmidt, A., 823
Golovatyuk, V., 852
Golubeva, M. B., 341
Goncharov, M., 509
Gonin, M., 341
Goobar, A., 823
Gorini, E., 852
Gorringe, T. P., 403, 407
Goto, Y., 670, 685
Gotta, D., 421
Govaerts, J., 920
Grancagnolo, F., 852
Grandegger, W., 852
Gray, F. E., 917
Gray, L., 823
Graziani, E., 852
Grella, G., 363
Grigorian, A. A., 341
Grishin, V. N., 674
Grosse-Perdekamp, M., 917
Grossiord, J. Y., 341
Grossmann, A., 917
Guarnaccia, P., 852
Guber, F. F., 341
Guichard, A., 341
Gulkanyan, H., 341
Gulkarov, I. S., 495

H

Haase, H., 823
Hadri, M., 920
Haeberlen, U., 917
Haeberli, W., 983
Hagel, U. v., 852
Hakobyan, R., 341
Halkiadakis, E., 624
Hallgren, A., 823
Halzen, F., 823
Han, H. G., 852
Han, S. W., 852
Hanson, K., 823
Hardtke, R., 823
Haroutunian, R., 341
Harris, D. A., 509
Hartmann, M., 421

Haselden, A., 957
Hasinoff, M. D., 403, 407, 969
Hatakeyama, K., 542
Hawker, E. A., 346, 532
Hayman, P. J., 957
Hazen, E. S., 917
He, X. C., 346, 532
He, Y. D., 823
Heintz, U., 593
Hellwig, M., 823
Helstrup, H., 363
Henoch, M., 666
Henry-Couannier, F., 957
Heppelmann, S., 306, 310, 447, 451
Hertzog, D. W., 917
Heukenkamp, H., 823
Hilbes, C., 920, 983
Hill, G. C., 823
Hiller, G., 614
Hiller, J. R., 399
Hollander, R. W., 957
Holme, A. K., 363
Holmes, R., 923
Horowitz, C. J., 491, 729, 927
Huang, H., 670
Huang, X., 852
Hughes, V., 670
Hughes, V. W., 917
Hulth, P. O., 823
Hundertmark, S., 823
Huss, D., 363

I

Idzik, M., 341
Imai, K., 670
Incagli, M., 852
Ingrosso, L., 852
Isenhower, L. D., 346, 532
Ishida, T., 772
Ishihara, M., 670
Issac, M. C. P., 177
Ito, T. M., 913

J

Jacholkowski, A., 363
Jacobs, K. D., 890

Jacobs, W. W., 689
Jacobsen, J., 823
Jain, P., 302, 455, 733
James, C., 777
Jeon, S., 337
Ji, X., 554
Jiang, Y. Y., 852
John, V., 524
Johnson, R. A., 509
Jon-And, K., 957
Jones, G. T., 363
Jones, R. T., 237
Joo, K., 434
Jouan, D., 341
Jung, C., 554
Junghans, H., 421
Jungmann, K., 917

K

Kacharava, A., 421
Kageya, T., 662, 674
Kaiser, R., 709, 864
Kambor, J., 575
Kamys, B., 421
Kanavets, V. P., 670
Kandhadai, V., 823
Kantsyrev, D. Y., 662, 674
Kaplan, D. M., 346, 532
Karavitcheva, T. L., 341
Karle, A., 823
Kashevarov, V. L., 267
Kaskulov, M. M., 475
Kaufman, S. B., 346, 532
Kawabata, T., 306, 310, 447, 451
Kawall, D., 917
Kettell, S. H., 858, 961
Kettle, P.-R., 957
Khazin, B. I., 917
Kim, J., 823
Kim, J. H., 509
Kim, W., 852
Kindem, J., 917
King, B. J., 509
King, P. M., 407
Kinnel, T., 509
Kinson, J. B., 363
Kirch, K., 920
Kirk, P. N., 532

Kistryn, S., 983
Klehr, F., 421
Klein, F., 259
Kleppner, D., 674
Kluberg, L., 341
Kluge, W., 852
Knowles, P., 920
Knudson, K., 363
Koch, H. R., 421
Koci, B., 823
Koetke, D. D., 346, 532
Kokkas, P., 957
Koltsov, A. V., 411
Komarov, V. I., 421
Konobeevski, E. S., 267
Köpke, L., 823
Koptev, V. P., 421
Korobkina, E., 972
Koutsoliotas, S., 509
Kovash, M., 403
Kowalski, M., 823
Králik, I., 363
Kravtsov, A. V., 411
Kreuger, R., 957
Krienen, F., 917
Krisch, A. D., 662, 674
Krishnaswami, G. S., 524
Kronkvist, I., 917
Krutov, Y. I., 411
Kulessa, P., 421
Kulikov, A., 421
Kulikov, V., 852
Kundu, B., 455
Kurbatov, V., 421
Kurepin, A. B., 341
Kurita, K., 670
Kwaitowski, K., 670
Kyle, G., 532

L

Lacava, F., 852
Lajoie, J. G., 350
Lamm, M. J., 509
Landsman, H., 609
Lanfranchi, G., 852
Lang, J., 920, 983
Langenegger, U., 844
Larabee, A., 259

Larsen, R., 917
Le Bornec, Y., 341
Lee, D. M., 532
Lee, W. M., 346, 532
Lee, Y. Y., 917
Lee-Franzini, J., 852
Le Gac, R., 957
Leich, H., 823
Leimgruber, F., 957
Leitch, M. J., 346, 532
Leksanov, A., 306, 310, 447, 451
Lenti, V., 363
Leuthold, M., 823
Lewis, B., 670
Li, G., 729
Lietava, R., 363
Lindahl, P., 823
Lisa, M. A., 355
Liubarsky, I., 823
Llanes-Estrada, F. J., 275
Loaiza, P., 823
Loconsole, R. A., 363
Logashenko, I., 917
Lomtadze, T., 852
Londergan, J. T., 479, 718
Lourenço, C., 341
Løvhøiden, G., 363
Lowder, D. M., 823
Lozowski, B., 670
Ludvig, J., 823
Luisi, C., 852
Luppov, V. G., 674
Lüthy, M., 983
L'vov, A. I., 411

M

Macciotta, P., 341
Mac Cormick, M., 341
Macharashvili, G., 421
Madsen, J., 823
Mahlon, G., 388
Maier, R., 421
Makdisi, Y., 306, 310, 447, 451, 670
Makins, N., 532
Malki, A., 306, 310, 447, 451
Maltman, K., 245, 287, 575
Malvezzi, S., 569
Mandić, I., 957

Maneira, J. C., 725
Manthos, N., 957
Manzari, V., 363
Mao, C. S., 852
Marciano, W. J., 938
Marciniewski, P., 823
Marel, G., 957
Markiewicz, M., 983
Marsh, W., 509
Martemianov, M., 852
Marzari-Chiesa, A., 341
Masera, M., 341
Mason, D., 509
Masoni, A., 341
Matis, H. S., 823
Matsyuk, M., 852
Mazzoni, M. A., 363
McFarland, K. S., 509
McGaughey, P. L., 346, 532
McKay, D. W., 733
McLaughlin, G. C., 800
McNabb, R., 917
McNicoll, G., 233
McNulty, C., 509
McNulty, D. E., 647
McPherson, R. A., 189
Mecking, B., 259
Meddi, F., 363
Medve, R., 920
Mehrabyan, S., 341
Mei, W., 852
Melnikov, K., 938
Melnitchouk, W., 463
Meng, W., 917
Merola, L., 852
Merzliakov, S., 421
Messi, R., 852
Meyer, H. O., 670
Meyer, C. A., 326, 879
Meyer, W., 655
Mi, J., 917
Michalon, A., 363
Michalon-Mentzer, M. E., 363
Mihalyi, A., 823
Mikirtychiants, S., 421
Mikolajski, T., 823
Mikuž, M., 905, 957
Miller, C. A., 712
Miller, J. P., 917, 957
Miller, T. C., 823

Minaeva, Y., 823
Mincer, A. I., 942
Minina, E., 306, 310, 447, 451
Mintz, S. L., 558
Miocinovic, P., 823
Miscetti, S., 852
Mishra, S. R., 509
Moalem, A., 852
Moccia, S., 852
Mocioiu, I., 790
Mock, P. C., 823
Mohapatra, R. N., 16
Montanet, F., 957
Monteno, M., 341
Morando, M., 363
Mordovskoy, M. V., 267
Morelle, X., 920
Morozov, B. V., 670
Morozov, V. S., 662, 674
Morse, R., 823
Morse, W. M., 917
Moss, J. M., 346, 532
Moulson, M., 852
Mueller, B. A., 346, 532
Mueller, J. A., 255
Mueller, S., 852
Muller, A., 957
Müller, H., 421
Murray, J. R., 674
Murtas, F., 852
Mussgiller, A., 421
Musso, A., 341

N

Nagae, T., 875
Nagle, J. L., 549
Naito, K., 417
Nakada, T., 957
Nakamura, K., 164
Nakamura, M., 670
Naples, D., 509
Napolitano, J., 86
Napolitano, M., 852
Narsky, I., 946
Naviliat, O., 920
Navon, I., 306, 310, 447, 451
Nazaruk, V. I., 901
Nedosekin, A., 852

Nemoto, Y., 417
Neumann, J. J., 674
Neunhöffer, T., 823
Newcomer, F. M., 823
Nicholson, H., 306, 310, 447, 451
Nienaber, P., 509
Niessen, P., 823
Nigam, B. P., 495
Ninane, A., 920
Nioradze, M., 421
Nord, P. M., 532
Norman, P. I., 363
Nygren, D. R., 823

O

Obayashi, Y., 767
Oehler, C., 758
Ogawa, A., 306, 310, 447, 451
Ogelman, H., 823
Ohm, H., 421
Oka, M., 417
Olsson, M. G., 283
Onderwater, C. J. G., 917
Orlov, Y., 917
Özben, C., 917

P

Pacciani, L., 852
Page, P. R., 249, 471
Pagels, B., 957
Pagès, P., 852
Pai, C., 917
Paley, J., 917
Palutan, M., 852
Panareo, M., 852
Panda, S., 733
Panebratsev, Y., 306, 310, 447, 451
Paoluzi, L., 852
Papadopoulos, I., 957
Papavassiliou, V., 346, 532
Park, B. K., 532
Parlyuchenko, V. P., 411
Parsa, Z., 781, 831
Paryev, E. Y., 425
Paschke, K., 697
Pasqualucci, E., 852

Passalacqua, L., 852
Passaseo, M., 852
Passeri, A., 852
Pastirčák, B., 363
Patera, V., 852
Paul, T., 579
Pavlopoulos, P., 957
Pavlyuchenko, L. N., 411
Peng, J. C., 346, 532
Pérez de los Heros, C., 823
Petiau, P., 341
Petitt, G., 346, 532
Petratos, G. G., 74, 438
Petrolo, E., 852
Petrucci, G., 852
Petrus, A., 421
Phalet, T., 931, 934
Piasetzky, E., 306, 310, 447, 451
Picca, D., 852
Piccotti, A., 341
Piekarewicz, J., 491
Pirozzi, G., 852
Pistillo, C., 852
Pizzi, J. R., 341
Pobylitsa, P. V., 659
Polivka, G., 957
Pollack, M., 852
Polly, C. C., 917
Polyakov, M. V., 643, 659
Pontecorvo, L., 852
Porrata, R., 823
Pospelov, M., 979
Potashev, S. I., 267
Pralavorio, P., 872
Prasuhn, D., 421
Prebys, E., 848
Pretz, J., 917
Price, P. B., 823
Prieels, R., 920, 931
Prigl, R., 917
Primavera, M., 852
Prino, F., 341
Przysiezniak, H., 1
Puddu, G., 341
Pusenkov, A., 983
Pysz, K., 421

Q

Qian, Y.-Z., 737
Quercigh, E., 363
Quint, W., 934
Quintans, C., 341

R

Rajagopal, K., 95
Rajeev, S. G., 524
Ralston, J. P., 302, 455, 705, 733
Ramello, L., 341
Ramos, S., 341
Rathmann, F., 421
Rato Mendes, P., 341
Rawlins, K., 823
Raymond, R. S., 674
Redin, S. I., 917
Reed, C., 823
Reimer, P. E., 346, 532
Rhode, W., 823
Riccati, L., 341
Ricciardi, S., 753
Richards, A., 823
Richman, J. D., 29
Richter, S., 823
Rickenbach, R., 957
Riedler, P., 867
Rimarzig, B., 421
Rinckel, T., 670
Rind, O., 917
Roberts, B. L., 917, 957
Rodriguez Martino, J., 823
Roethel, W., 322
Romana, A., 341
Romano, G., 363
Romenesko, P., 823
Romosan, A., 509
Ropotar, I., 341
Roser, T., 670
Ross, D., 823
Rothstein, I. Z., 604
Rozen, Y., 609
Rubinstein, H., 823
Rudy, Z., 421
Ruf, T., 957
Ruggieri, F., 852
Rusek, A., 670
Ryskulov, N. M., 917

S

Sadler, M. E., 346, 532
Šafařík, K., 363
Saito, N., 670
Sakumoto, W. K., 509
Sammarruca, F., 459
Samuelsson, J., 455
Sander, H.-G., 823
Sanders, R., 917
Šándor, L., 363
Santangelo, P., 852
Santovetti, E., 852
Saracino, G., 852
Saturnini, P., 341
Savage, M. J., 120
Schäfer, M., 957
Schaller, L. A., 957
Schamberger, R. D., 852
Schatz, H., 744
Schebetov, A., 983
Schechter, J., 241
Scheider, T., 823
Schellman, H., 509
Schetkovsky, A., 306, 310, 447, 451
Schietinger, T., 957
Schleichert, R., 421
Schmidt, P., 934
Schmidt, T., 823
Schneider, C., 421
Schneider, D., 823
Schneider, E., 823
Schneider, H., 421
Schopper, A., 957
Schult, O. W. B., 421
Schuurmans, P., 931, 934
Schwandt, P., 662
Schwarthoff, H., 619
Schwarz, R., 823
Schweitzer, P., 643, 659
Schwick, C., 852
Schwindt, O., 487
Sciascia, B., 852
Sciubba, A., 852
Sciulli, F. J., 509

Scomparin, E., 341
Scuri, F., 852
Sedykh, S., 917
Segato, G., 363
Seitz, B., 651
Selden, J., 532
Seligman, W. G., 509
Semertzidis, Y., 917
Serci, S., 341
Serebrov, A., 983
Severijns, N., 931, 934
Seyfarth, H., 421
Sfiligoi, I., 852
Shaevitz, M. H., 201, 509
Shahoyan, R., 341
Shan, J., 852
Sharma, L. K., 291
Shatunov, Y. M., 917
Sherif, H. S., 483
Shigaki, K., 378
Shimanskiy, S., 306, 310, 447, 451
Shrock, R., 790
Sidorin, S. S., 411
Silva, S., 341
Silvestri, A., 823
Simonov, V. A., 267
Simons, L., 920
Sistemich, K., 421
Sitta, M., 341
Sivers, D. W., 662
Skorkin, V. M., 267
Smith, A., 986
Smith, B., 670
Smith, W. H., 509
Soave, C., 341
Sobol, A., 314
Sokol, G. A., 411
Solarz, M., 823
Solodov, E., 917
Solomey, N., 263, 882
Sonderegger, P., 341
Sondheim, W. E., 346, 532
Soni, A., 976
Souder, P. A., 138
Spadaro, T., 852
Spagnolo, S., 852
Spentzouris, P., 509
Spiczak, G., 823
Spiering, C., 823
Spiriti, E., 852

Sromicki, J., 920, 983
Stanescu, C., 852
Stankus, P. W., 346, 532
Starinsky, N., 823
Staroba, P., 363
Steele, D., 823
Steffen, P., 823
Steffens, F. M., 463
Stein, H. J., 421
Steinmetz, A., 917
Stephenson, E. J., 459
Stern, E. G., 509
Stewart, I. W., 598
Stocki, T. J., 407
Stokstad, R. G., 823
Strauch, S., 430
Streicher, O., 823
Ströher, H., 421
Sulak, L. R., 917
Sun, Q., 823
Svirida, D. N., 670
Swanson, E. S., 330
Syphers, M., 670
Szczepaniak, A. P., 330

T

Taboada, I., 823
Tadsen, A., 814
Taketani, A., 670
Takizawa, M., 417
Tang, A., 306, 310, 447, 451
Tarrago, X., 341
Tauscher, L., 957
Teodorescu, O., 369
Thibault, C., 957
Thollander, L., 823
Thomas, A. W., 463, 479
Thomas, T. L., 670
Thompson, M., 363
Thompson, T. N., 532
Thon, T., 823
Thorsteinsen, T. F., 363
Tilav, S., 823
Timmermans, C., 917
Tojo, J., 670
Tong, G. L., 852
Topilskaya, N. S., 341
Torrieri, G. D., 363

Tortora, L., 852
Touchard, F., 957
Touramanis, C., 957
Towell, R. S., 346, 532
Tribble, R. E., 346, 532
Tripathi, S., 403, 407
Trofimov, A., 917
Tung, W.-K., 536
Tuominen, K., 384
Tveter, T. S., 363

U

Underwood, D., 670
Urbán, J., 363
Urbano, D., 659
Urner, D., 917
Usai, G. L., 341
Usechak, N., 823

V

Vaandering, E. W., 583
Vaitaitis, A., 509
Vakili, M., 509
Valente, E., 852
Valente, P., 852
Valeriani, B., 852
Vander Donckt, M., 823
Van Eijk, C. W. E., 957
Van Hove, P., 920
Vasiliev, M. A., 346, 532
Veneziano, G., 852
Vercellin, E., 341
Vereecke, B., 931, 934
Versyck, S., 931, 934
Veseli, S., 283
Villalobos Baillie, O., 363
Villatte, L., 341
Vincter, M. G., 62
Virgili, T., 363
Vlachos, S., 957
Vogelsang, W., 676
Volkov, M. K., 279
Voloshin, M. B., 632
von Przewoski, B., 662, 670
Votruba, M. F., 363

W

Walck, C., 823
Walet, N. R., 487
Wang, Y. C., 532
Wang, Z. F., 532
Warburton, D., 917
Watson, J. W., 306, 310, 447, 451
Webb, J. C., 346, 532
Weber, P., 957
Weigel, H., 271, 467
Weinheimer, C., 823
Weiss, C., 643, 659
Wiebusch, C. H., 823
Wiescher, M., 744
Wigger, O., 957
Willis, J. L., 346, 532
Willis, N., 341
Wingerter-Seez, I., 953
Winn, D., 917
Wischnewski, R., 823
Wise, D. K., 532
Wolf, J., 795
Wolfe, C. E., 287
Wolfe, D., 670
Wolter, M., 957
Wong, V. K., 662
Woods, M., 910
Woschnagg, K., 823
Wright, D. H., 403, 407
Wu, V., 509
Wu, W., 823
Wu, Y., 852

X

Xie, Y. G., 852

Y

Yamamoto, A., 917
Yamamoto, K., 670
Yang, U. K., 509
Yankama, B., 674
Yelton, J., 588
Yodh, G., 823
Yoshida, H., 306, 310, 447, 451
Young, G. R., 346, 532

Young, S., 823
Yu, J., 509
Yudichev, V. L., 279

Z

Závada, P., 363
Zavrtanik, D., 957
Zeier, M., 395
Zeller, G. P., 509, 809

Zhalov, D., 306, 310, 447, 451
Zhao, P. P., 852
Zhou, Y., 852
Zhu, L., 670
Zieliński, M., 527
Zimmerman, D., 917, 957
Zimmerman, E. D., 509, 762
Zołnierczuk, P. A., 403, 407
Zuev, S. V., 267
zu Putlitz, G., 917
Zychor, I., 421

Previous Proceedings in the Series of Conferences on Intersections Between Particle and Nuclear Physics

	Year	Held in	Publisher	ISBN
6th	1997	Big Sky, Montana	AIP Conf. Proceedings vol. 412	1-56396-712-X
5th	1994	St. Petersburg, Florida	AIP Conf. Proceedings vol. 338	1-56396-335-3
4th	1991	Tucson, Arizona	AIP Conf. Proceedings vol. 243	0-88318-950-X
3rd	1989	Rockport, Maine	AIP Conf. Proceedings vol. 176	0-88318-376-5
2nd	1986	Lake Louise, Canada	AIP Conf. Proceedings vol. 150	0-88318-349-8
1st	1984	Steamboat Springs, CO	AIP Conf. Proceedings vol. 123	0-88318-322-6

Other Related Titles from AIP Conference Proceedings

541   Theoretical High Energy Physics: MRST 2000
Edited by C. R. Hagen, November 2000, 1-56396-966-1

539   Symmetries in Subatomic Physics: 3rd International Symposium
Edited by X.-H. Guo, A. W. Thomas, and A. G. Williams, October 2000, 1-56396-964-5

508   Hadron Physics: Effective Theories of Low Energy QCD
Edited by A. H. Blin, B. Hiller, M. C. Ruivo, C. A. Sousa, and E. van Beveren, March 2000, 1-56396-927-0

496   Workshop on Instabilities of High Intensity Hadron Beams in Rings
Edited by T. Roser and S. Y. Zhang, December 1999, 1-56396-910-6

494   New Directions in Quantum Chromodynamics
Edited by Chueng-Ryong Ji and Dong-Pil Min, November 1999, 1-56396-908-4

482   RHIC Physics and Beyond: Kay Kay Gee Day
Edited by Berndt Müller and Robert Pisarski, July 1999, 1-56396-878-9

459   Heavy Quarks at Fixed Target
Edited by Harry W. K. Cheung and Joel N. Butler, February 1999, 1-56396-864-9

432   Hadron Spectroscopy: Seventh International Conference
Edited by Suh-Urk Chung and Hans J. Willutzki, June 1998, 1-56396-765-0

To learn more about these titles, or the AIP Conference Proceedings Series, please visit the webpage http://www.aip.org/catalog/aboutconf.html